ANNUAL REVIEW OF ECOLOGY AND SYSTEMATICS

ANNUAL REVIEW OF ECOLOGY AND SYSTEMATICS

VOLUME 33, 2002

DOUGLAS J. FUTUYMA, *Editor*
State University of New York at Stony Brook

H. BRADLEY SHAFFER, *Associate Editor*
University of California, Davis

DANIEL SIMBERLOFF, *Associate Editor*
University of Tennessee

www.annualreviews.org   science@annualreviews.org   650-493-4400

ANNUAL REVIEWS
4139 El Camino Way • P.O. Box 10139 • Palo Alto, California 94303-0139

Ⓐ
R
ANNUAL REVIEWS
Palo Alto, California, USA

International Standard Serial Number: 0066-4162
International Standard Book Number: 0-8243-1433-6
Library of Congress Catalog Card Number: 71-135616

TYPESET BY TECHBOOKS, FAIRFAX, VA
PRINTED AND BOUND BY MALLOY INCORPORATED, ANN ARBOR, MI

Ⓡ *Annual Review of Ecology and Systematics*
Volume 33, 2002

Contents

ERRATA
An online log of corrections to *Annual Review of Ecology and Systematics*
chapters may be found at http://ecolsys.annualreviews.org/errata.shtml

RELATED ARTICLES

Microbial Communities and Their Interactions in Soil and Rhizosphere Ecosystems, Angela D. Kent and Eric W. Triplett

What Are Bacterial Species? Frederick M. Cohan

From the *Annual Review of Phytopathology*, Volume 40 (2002)

Evolutionary Ecology of Plant Diseases in Natural Ecosystems, Gregory S. Gilbert

Viral Sequences Integrated into Plant Genomes, Glyn Harper, Roger Hull, Ben Lockhart, and Neil Olszewski

Comparative Genomic Analysis of Plant-Associated Bacteria, M. A. Van Sluys, C. B. Monteiro-Vitorello, L. E. A. Camargo, C. F. M. Menck, A. C. R. da Silva, J. A. Ferro, M. C. Oliveira, J. C. Setubal, J. P. Kitajima, and A. J. Simpson

Pathogen Population Genetics, Evolutionary Potential, and Durable Resistance, Bruce A. McDonald and Celeste Linde

ANNUAL REVIEWS is a nonprofit scientific publisher established to promote the advancement of the sciences. Beginning in 1932 with the *Annual Review of Biochemistry*, the Company has pursued as its principal function the publication of high-quality, reasonably priced *Annual Review* volumes. The volumes are organized by Editors and Editorial Committees who invite qualified authors to contribute critical articles reviewing significant developments within each major discipline. The Editor-in-Chief invites those interested in serving as future Editorial Committee members to communicate directly with him. Annual Reviews is administered by a Board of Directors, whose members serve without compensation.

Annu. Rev. Ecol. Syst. 2002. 33:1–23
doi: 10.1146/annurev.ecolsys.33.010802.150507
Copyright © 2002 by Annual Reviews. All rights reserved
First published online as a Review in Advance on August 6, 2002

SAPROXYLIC INSECT ECOLOGY AND THE SUSTAINABLE MANAGEMENT OF FORESTS

Simon J. Grove

*Division of Forest Research and Development, Forestry Tasmania, GPO Box 207, Hobart,
Tasmania 7001, Australia; email: simon.grove@forestrytas.com.au*

Key Words   dead wood, insect conservation, logging, beetles, forestry

■ **Abstract**   Saproxylic insects comprise a diverse, species-rich and dominant functional group that share a dependence on dead wood and the old trees that generate it (mature timber habitat). Recent research has highlighted their sensitivity to forest management, with managed or secondary forests generally supporting fewer individuals, fewer species, and different assemblages compared to old-growth or primary forests. This sensitivity is a product of their association with a habitat that tends to diminish in managed forests. Many species also have low powers of dispersal relative to human-induced fragmentation, making breaks in habitat continuity particularly harmful. In western Europe, many species are now regionally extinct. Information is largely lacking elsewhere, but similar ecological and management principles should apply. Measures taken to protect the habitat of hollow-dependent vertebrates may ensure the survival of some saproxylic insects, but unless their needs are expressly considered, there remains the risk that many others may be lost as forest areas shrink and management of remaining areas intensifies.

INTRODUCTION

Scope of Review

In this paper I review the literature on saproxylic insect ecology and conservation with particular reference to their relationships with forest management. I dwell particularly on the western European experience and also consider my own research findings from the tropical rainforests of northeastern Queensland, Australia. In so doing I highlight some of the common ecological traits shared by many saproxylic insects (especially the beetles, the subject of my own research) and relate these to the forest management practices that have so often led to their demise. Assuming that saproxylic insects are likely to respond similarly to these practices wherever they occur, I offer some thoughts on the extent to which they have a secure long-term future in managed forests worldwide, together with some suggestions for how their needs could better be addressed.

0066-4162/02/1215-0001$14.00   **1**

What are Saproxylic Insects and Why do They Have Their Own Word?

Saproxylic insects are defined as those that are "dependent, during some part of their life cycle, upon the dead or dying wood of moribund or dead trees (standing or fallen), or upon wood-inhabiting fungi, or upon the presence of other saproxylics" (Speight 1989). The saproxylic habit includes representatives from all major insect orders (especially beetles and flies), and accounts for a large proportion of the insect fauna in any natural forest (Table 1). For example, 56% of all forest beetle species in a region of Germany were considered saproxylic (Köhler 2000). Beetles alone may represent almost 40% of all arthropod species (Grove & Stork 2000), and there are at least twice as many species of saproxylic beetles as there are terrestrial verte-brates (Parker 1982)—probably many more. Saproxylic insects, then, are a species-rich and functionally important component of forest ecosystems. Other invertebrate groups, fungi, and other microorganisms could equally be described as saproxylic.

The word saproxylic was coined in France (Dajoz 1966) and has been widely adopted in Europe since the publications of Harding & Rose (1986) and Speight (1989). Publications from Canada (Hammond 1997) and Australia (Michaels & Bornemissza 1999, Grove & Stork 1999, Yee et al. 2001, Grove 2002a) have recently followed suit. Frequently used equivalent terms include deadwood (Elton 1966) and wood-living (SG Nilsson 1997). Saproxylic is a particularly useful word, because included in the definition (besides wood-feeders) are bark-feeders, feeders on wood-decomposing fungi, associated predators, parasitoids, detritivores feeding on their waste products, and other commensals. Hence, it refers to an entire functional group that has obligate associations with an array of dead-wood habitats. Mature timber habitat (Grove 2001a) is perhaps the only term in frequent use that can be used to sum up all the key habitat features on which saproxylic insects depend. Examples include standing and fallen dead wood of various diameters and in various states and stages of decay, wood-rotting and other dependent fungi (hyphae and sporocarps), fissures and crevices in bark, water- or humus-filled rot-holes and other tree cavities, sap-runs, and the tunnels and frass of wood-borers. In fire-adapted forests, charred wood is another example (Wikars 2001), as is waterlogged wood for aquatic saproxylic insects (Braccia & Batzer 2001).

"Deadwoodology": The Basics

In the context of forest management, specific concern for saproxylic insects is acute in Europe but remains at a much lower level elsewhere. Because the study of mature timber habitat and saproxylic insects may therefore be unfamiliar to many readers, I offer the following six guiding principles:

THE RICHNESS OF LIFE IN THE "ARBOREAL MEGALOPOLIS"   Larger-diameter, over-mature, senescent, moribund, decadent, or veteran trees (Read 1991) form the centerpiece of the mature timber habitat concept. They may be commercially overmature but are in their ecological prime of life and have been likened to an

TABLE 1 Examples demonstrating the richness of the saproxylic insect fauna

Country	Reference	Observations
Australia	Hammond et al. (1996)	21% of all beetle species in canopy samples from SE Queensland were xylophagous
Australia	Yee et al. (2001)	104 beetle species hand-collected from *Eucalyptus* logs in a single forest region in Tasmania
Australia	Grove (2002a)	339 saproxylic beetle species recorded from flight intercept traps in rainforest in NE Queensland
French Guyana	Tavakilian et al. (1997)	500 species of cerambycid beetle were reared from dead wood from leguminous forest trees
Indonesia	Hammond (1990)	20% of the 3488 beetle species collected in north Sulawesi were xylophagous or xylomycophagous
United Kingdom	Elton (1966)	456 species of invertebrates were recorded from dead wood habitats in a single woodland
Finland	Hanski & Hammond (1995)	287 species of saproxylic beetles were recorded in a single forest
Finland	Siitonen (2001)	20–25% of all forest-dwelling species (not just insects) were saproxylic, including about 800 beetle species
Finland	Martikainen et al. (2000)	42% of 553 beetle species sampled in old spruce forest were saproxylic
Finland	Martikainen (2001)	42% of 780 beetle species collected from dead aspen trees were saproxylic
Sweden	Palm (1959)	342 species of beetle associated with dead aspen in southern and central Sweden
Sweden	Ehnström (2001)	405 saproxylic beetle species associated with dead birch, 389 with dead Scotch pine and 354 with dead Norway spruce
Norway	Økland et al. (1996)	Nearly 700 obligate saproxylic beetle species nationally and 200 facultatively saproxylic species
Germany	Derksen (1941)	217 species of saproxylic insects in dead beech in a single forest
Germany	Köhler (2000)	56% of all forest-dwelling beetle species in north Rhineland considered saproxylic
Germany	Blab et al. (1994)	25% of German beetle species considered saproxylic
Canada	Hammond (1997)	257 species of saproxylic beetle recorded from dead aspens in one forest type
United States	Deyrup (1976)	More than 300 saproxylic insect species recorded from Douglas-fir in Washington
United States	Howden & Vogt (1951)	188 species of beetles recorded from dead pines in a single Maryland forest
United States	Savely (1939)	162 species of beetles recorded from pine and oak logs in a single North Carolina forest

arboreal megalopolis (Speight 1989) or termed megatrees (Nilsson & Baranowski 1994). They probably assume greatest relative importance in forests dominated by broadleaves, whether temperate (Speight 1989) or tropical (Grove 2002b) (Figure 1). In boreal and other coniferous or sclerophyllous forests, the standing dead trees (snags) and dead wood on the forest floor derived from megatrees may be more important (Berg et al. 1994). However, despite the importance of megatrees, we know relatively little about what lives in them, compared with what lives in snags or in dead wood on the forest floor. Some saproxylic insects are found in all three types, but many have distinct preferences (Palm 1959). In Swedish broadleaved forests megatrees support more substrate-specific species than either snags or dead wood on the forest floor and include most of the rarest and most threatened saproxylic insects (Nilsson & Baranowski 1997, Jonsell et al. 1998).

THERE'S DIVERSITY IN DECAY   A key source of saproxylic insect diversity is the range of decay stages and types that occur within dead wood, each providing habitats for different assemblages. As wood decomposes it is colonized by a succession of saproxylic insect species (Blackman & Stage 1924, Kletecka 1996, Hammond et al. 2001), as are the fruiting-bodies of wood-rotting ascomycetes and basidiomycetes (Jonsell et al. 1999). Insects that colonize living or freshly dead wood often show narrow host specificity (Hamilton 1978); the same applies for species on wood-decaying fungi (Kaila et al. 1994). Host specificity is rare at the tree species level but common at higher plant taxonomic levels (e.g., Tavakilian et al. 1997). In German forests Köhler (2000) found that 13% of the saproxylic beetle species were tree genus specific. As decomposition proceeds host specificity drops off, though there remains a big difference between the fauna of hardwoods and that of softwoods (Savely 1939, Ås 1993). Within either, decay type is important (Harmon et al. 1986, Jonsell et al. 1998). For instance, there is a big difference in the fauna depending on whether the wood was decayed by white-rot or brown-rot fungi (Araya 1993, Wood et al. 1996, Yee et al. 2001). Further differences are attributable to wood moisture content and exposure to sun (Martikainen 2001).

ALL DEAD WOOD IS GOOD, BUT BIGGER IS BETTER   Size is important for saproxylic insects. Some are able to make use of their chosen substrate whatever its dimensions (Palm 1959), but most are more particular (Elton 1966), which results in different assemblages in substrates of different sizes. Most studies suggest a positive relationship between tree or dead-wood diameter and species richness, incidence, or abundance (Table 2), with larger-diameter material being especially important for rare and threatened species (e.g., Warren & Key 1991). Various explanations exist for this phenomenon. First, larger-diameter trees and dead wood are highly heterogeneous habitats, allowing many specialist species to occupy them at the same time (Kolström & Lumatjärvi 2000). Second, larger-diameter

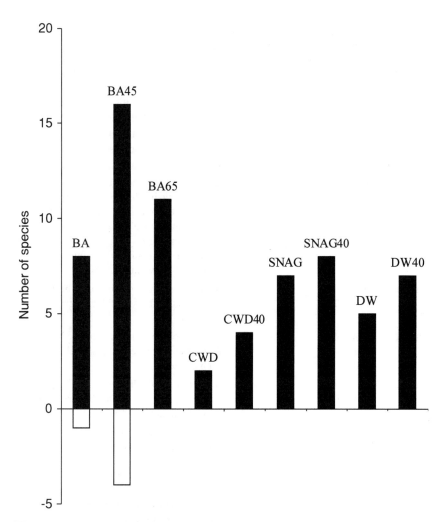

Figure 1 Numbers of saproxylic beetle species showing significant correlations ($p < 0.05$; Pearson two-tailed) in abundance with mature timber habitat attributes across the author's 81 study plots in the Daintree lowlands of northeastern Queensland, Australia. Based on all species represented by five or more individuals (n = 118). Correlations above the horizontal axis (*black bars*) are positive; correlations below the horizontal axis (*white bars*) are negative. Note that some species may contribute to more than one column. BA, basal area; BA45, basal area of trees >45 cm dbh (diameter at 1.5 m); BA65, basal area of trees >65 cm dbh; CWD, volume of coarse woody debris on the forest floor; CWD40, volume of CWD >40 cm dia (mid-point diameter); SNAG, volume of snags; SNAG40, volume of snags >40 cm dia; DW, total volume of dead wood; DW40, total volume of dead wood >40 cm dia.

TABLE 2   Examples demonstrating relationships between saproxylic insects and mature
timber habitat diameter

Country	Reference	Relationship[a]	Observation
Australia	Grove (2002b)	+	Larger-diameter living trees a better predictor of saproxylic beetle species richness and assemblage composition than total basal area in lowland tropical rainforest
Australia	Yee et al. (2001)	+	Saproxylic beetle fauna of rotting logs varied by rot type, with some types primarily found in larger-diameter logs
Finland	Kolström & Lumatjärvi (2000)	+	Model of forest-stand structure linked to database of saproxylic beetle habitat requirements Predicted that retaining aspen trees in managed conifer stands would result in a progressive increase in saproxylic species as tree diameter increased. Thinning aspen along with conifers would not allow this increase
Finland	Väisänen et al. (1993)	+	Number of species of subcortical beetles in trees in a managed forest positively correlated with diameter
Finland	Siitonen & Saaristo (2000)	+	Incidence of threatened saproxylic beetle *Pytho kolwensis* in spruce logs positively correlated with diameter
Finland	Martikainen et al. (2000)	+	Number of saproxylic beetle species per plot positively correlated with abundance of larger-diameter living and dead trees
Sweden	Jonsell et al. (1998)	+	Models based on known habitat associations of threatened saproxylic invertebrates predicted that most would occur in dead wood in the largest diameter class, including 178 not found in smaller-diameter classes. Only 94 species would occur in the smallest, with only 13 of these not found in larger classes
Sweden	Ranius & Jansson (2000)	+	Species richness of saproxylic beetles in hollow oaks positively related to tree diameter, especially amongst threatened species and those associated with sporocarps of wood-rotting fungi
Sweden	T Nilsson (1997)	+	Probability of occurrence of saproxylic beetle *Bolitophagus reticulatus* increased with tree diameter
Norway	Thunes et al. (2000)	+	Sporocarps of wood-rotting fungus *Fomitopsis pinicola* harbored more saproxylic beetle species on larger trees

(Continued)

TABLE 2 (*Continued*)

Country	Reference	Relationship[a]	Observation
Germany	Kleinevoss et al. (1996)	+	Higher saproxylic beetle species richness on larger-diameter dead wood
United Kingdom	Elton (1966)	−/+	Assemblage composition of smaller-diameter dead wood intermediate between that of larger-diameter dead wood and leaf litter
United States	Edmonds & Eglitis (1989)	+	Higher incidence of the cerambycid beetle *Monochamus scutellatus* on larger-diameter Douglas-fir logs
United States	Torgersen & Bull (1995)	+	*Camponotus* (carpenter) ants prefer larger logs to smaller ones
United States	Hespenheide (1976)	−/+	Some species of buprestid beetle have particular branch-diameter preferences, whereas others make use of a wide range of diameters
Japan	Araya (1993)	−	The lucanid beetle *Prismognathus acuticollis* occurred more predictably in smaller-diameter logs

[a]+, positive relationship; −, negative relationship

pieces take longer to decompose, and a more stable microclimate is maintained within, favoring many species (Palm 1959). Third, larger-diameter pieces support more species of fungi, including species specific to larger-diameter wood (Kruys & Jonsson 1999, Nordén & Paltto 2001), on which many saproxylic insects depend.

DEAD WOOD IS DYNAMIC AND SO ARE ITS INHABITANTS   Dead-wood abundance partly depends on its rates of input and output (wood growth and decay) and whether these are in equilibrium. Equilibrium is most likely in those old-growth forests in which the normal disturbance regime is one of small-scale gap replacement. If disturbance dynamics are more sporadic (e.g., occasional wildfires, windstorms, or earthquakes), then dead wood abundance may vary greatly depending on the time since the last disturbance event (Siitonen 2001, Spies et al. 1988).

A region's saproxylic insect fauna will presumably be pre-adapted to its normal disturbance regime (McPeek & Holt 1992). Species whose habitat is naturally long-lived and/or abundant are usually poor dispersers (e.g., many specialists of tree hollows) (Nilsson & Baranowski 1997; Ranius & Hedin 2001) compared with those whose habitat is scarce and/or ephemeral (e.g., many scolytine beetles on wind- or fire-damaged trees) (Nilssen 1984). Alteration of disturbance dynamics of a forest through management may have knock-on effects on saproxylic insects

(Schiegg 2000). The most likely alteration is a reduction in mature forest stands, potentially threatening many species (Ranius 2002) while turning a few others into pests (Heliövaara & Väisänen 1984, Safranyik & Linton 1999). Alternatively, fire suppression can lead to a reduction in early successional stages and an increase in stand density (Johnson et al. 2001), threatening many fire-adapted, thermophilic or shade-intolerant saproxylic insect species, which can comprise a large proportion of the regional fauna (Økland et al. 1996, Martikainen 2001).

Ecologists are familiar with metapopulation dynamics operating over the scale of kilometers among habitat patches measured in hectares (Hanski 1999, Thomas 2000), but for some saproxylic insect species, metapopulation dynamics may operate among habitat patches measured in cubic meters (Table 3). Individual trees, logs, snags, or fungal sporocarps can host several generations of a particular species, but eventually the entire population will become extinct once the resource decomposes, while new substrate patches appear for colonization elsewhere. If forest management alters patch abundance and distribution, then the population dynamics of species dependent on them may break down, leading to landscape-level extinction. For example, a common harvesting practice in forests pre-adapted to wildfires is clearfelling. Although often justified on the grounds of mimicking natural disturbance (Attiwill 1994), the resulting dead wood dynamics may be very different (Harmon et al. 1986, Price et al. 1998, Grove et al. 2002). Not only has much of the wood been harvested, but the remainder is fragmented and positioned entirely on the ground, where it may decay so rapidly that little remains within a few decades. Saproxylic insects may then face a critical habitat continuity gap, during which metapopulation dynamics might well break down, even though dead wood levels might subsequently increase later in the silvicultural cycle.

CONNECTIONS IN SPACE AND TIME: CONTINUITY IS THE KEY   One of the key management issues for saproxylic insects is the maintenance of ecological continuity in space (connectivity) and in time. In much of western Europe ecological continuity (or lack of it) is thought to be a major limiting factor for many of the less vagile saproxylic insect species (Speight 1989). Following millennia of intensive use (Thirgood 1989), there are probably few remaining patches of forest that have experienced sufficient ecological continuity to support anything close to the natural complement of saproxylic insect species (see "Saproxylic Insects and the Unsustainable Management of Forests," below). This has led to European conservationists paying particular attention to old-forest species, which are confined to primary forest, despite there being many mature secondary forests in the region (Speight 1989, Martikainen et al. 1999, Siitonen & Saaristo 2000). In the United Kingdom a suite of such species can be used to discriminate between mature woodland dating from the eighteenth century or earlier and the other 78% of the forest estate, which consists of more recently established plantations or regrowth (Fowles 1997). However, where forests are less fragmented and/or

TABLE 3 Examples demonstrating relationships between saproxylic insects and availability of mature timber habitat

Country	Reference	Observation
Australia	Grove (2002b)	Basal area of larger-diameter trees the best predictor of saproxylic beetle abundance and assemblage composition; coarse woody debris volume a better predictor of species richness in lowland tropical rainforest
Finland	Martikainen et al. (2000)	Total volume of dead wood provided the best predictor of beetle species richness
Finland	Siitonen (1994a)	Specialist beetle species were more abundant in a forest containing abundant dead wood than in one in which dead wood was scarce
Finland	Sippola & Kallio (1995)	Positive relationship between dead-wood volume and beetle species richness at the stand (1 ha) scale
Finland	Komonen et al. (2000)	Number of saproxylic trophic levels based on the bracket fungus *Fomitopsis rosea* decreased from three in intact old-growth forest to one in fragmented old-growth forest
Finland/ Russia	Siitonen (1994b)	On the Russian side of the border dead aspen trees were common, but on the Finnish side they were rare. On the Russian side they supported a much richer beetle fauna than on the Finnish side, including 2 species extinct in Finland and 18 species regarded as nationally rare in Finland, compared with 5 rare species on the Finnish side
Sweden	Jonsell et al. (1999)	Fungus beetles showed differing abilities to colonize sporocarps, dependent on distance and the nature of the intervening land. Most of the species studied could persist in the managed forest landscape if suitable breeding substrate were created or maintained at the landscape (1 km²) scale. To some extent larger habitat patches compensated for isolation
Sweden	Nilsson & Baranowski (1997)	Beetle species specializing on tree-hollows in living trees (a rare but long-lived resource) less vagile than species living in dead wood on the forest floor (a commoner but less persistent resource). Logged forest supported all species living in dead wood on the forest floor but not several tree-hollow species, even though the trees there were now large enough to support suitable hollows
Norway	Thunes et al. (2000)	The major factor influencing the number of species of beetles and the number of threatened species on dead sporocarps of the bracket fungus *Fomitopsis pinicola* was the amount of dead wood in and around the sampling site. Some of the species colonizing dead sporocarps were also able to live in dead wood and only occurred on sporocarps when dead wood was abundant

(Continued)

TABLE 3   *(Continued)*

Country	Reference	Observation
Norway	Rukke (2000)	Incidence of beetle species associated with the bracket fungus *Fomes fomentarius* related to degree of isolation of habitat trees, as well as to habitat size at tree and landscape scale
Norway	Sverdrup-Thygeson & Midtgaard (1998)	Probability of occurrence of the tenebrionid fungus-beetle *Bolitophagus reticulatus* increased with tree diameter and with the number of dead sporocarps on the tree and decreased as the distance to surrounding inhabited trees increased
Norway	Økland et al. (1996)	Positive relationship between dead-wood volume and beetle species richness at the landscape (32 ha) but not the stand (1 ha) scale. At all scales diversity of dead tree parts, number of large-diameter dead trees and number of polypore fungi species all correlated with beetle species richness and with abundance of many species; several species absent below a certain density of dead wood
Switzerland	Schiegg (2000)	Different species of beetle and fly found in forests with differing degrees of connectivity of dead wood; plots with higher dead-wood connectivity associated with higher species richness
Canada	Kehler & Bondrup-Nielsen (1999)	Probability of occurrence of the tenebrionid fungus-beetle *Bolitotherus cornutus* decreased with distance at the scale of sporocarps, logs, and forest blocks in an agricultural matrix. At any scale, distance proved the most consistent indicator of isolation
USA	Chandler (1991)	Greater abundance of many beetle species in old-growth compared with regrowth forest attributed to the greater prevalence of suitable dead-wood habitat in old growth

where occasional catastrophic disturbance is the norm, the present availability of the right dead-wood habitat may have greater significance for saproxylic species than does its continuity over time. This is held to be the case in the boreal forests of Fennoscandia (Kouki et al. 2001) and might also be true elsewhere.

DEAD WOOD IS NOT WASTE WOOD   For classically trained foresters and for many members of the public, wood allowed to rot or burn is wood that has gone to waste ("zero waste tolerance") (Lofroth 1998). Yet in nature waste recycling is a critical ecological process. Decomposing dead wood is the means by which a large proportion of the nutrients and energy accumulated by the living tree is returned to the soil (Laiho & Prescott 1999). Dead wood also acts as a medium-term store of carbon (Mackensen & Bauhus 1999). As decomposition progresses dead wood becomes incorporated into soil, helping to maintain levels of organic matter and carbon (Tate et al. 1993). Decomposition is brought about primarily by the activity of fungi and other microorganisms but is often mediated by saproxylic invertebrates (Swift 1977, Edmonds & Eglitis 1989, Schowalter et al. 1992, Hanula 1996). Quite apart from their own rights to existence, saproxylic insects are a principal source of food for other forest-dwelling organisms such as woodpeckers (Hanula & Franzreb 1998, Torgersen & Bull 1995).

CONSERVATION AND MANAGEMENT ISSUES

Death by a Thousand Cuts: How Humans and Saproxylic Insects Compete for Trees

Over the past few thousand years humans have increasingly ended up competing with saproxylic insects for timber (Hagan & Grove 1999). A small minority of saproxylic species such as some bark beetles (Peltonen 1999, Weslien & Schroeder 1999) can compete successfully with humans and are therefore termed pests. The vast majority are more likely to end up on the losing side, resulting in managed forests supporting altered saproxylic insect species assemblages, often coupled with lower numbers and fewer species overall (Table 4). By managing forests so that we can remove wood before it is lost to decay, we progressively eliminate mature timber habitat.

Elimination of mature timber habitat may be most rapid and thorough through excessive forest hygiene (Schmitt 1992), salvage logging (Maser 1996), or fuel-wood or biomass harvesting (Tritton et al. 1987, Wall 1999, Grove et al. 2002). In other situations its loss may be more insidious and may start with a temporary increase in visible dead wood through wasteful logging practices (Harmon 2001) or fire suppression and pest outbreaks brought on by management (Edmonds & Marra 1999). However, with successive silvicultural cycles an incremental loss is likely: As existing old trees die, the dead wood they produce decomposes, and the younger, more healthy trees of the managed forest do not live long enough to generate sufficient replacement habitat. Mature timber habitat may thus be reduced from abundance to rarity. Goodburn & Lorimer (1998) reported that levels of dead wood on the forest floor of selectively managed hardwood forests in Wisconsin and Michigan were only about 60% those of old-growth forests, with differences in larger-diameter material being even more pronounced. Fennoscandian production forests provide a more extreme example, with coarse woody debris levels falling by 90–98% over the past century (Siitonen 2001). In other parts of Europe it is not even possible to say how much dead wood an old-growth forest would have had, because there are none left. Writing from a British perspective, Elton (1966) stated that "if fallen timber and slightly decayed trees are removed [from a natural forest] the whole system is gravely impoverished of perhaps more than a fifth of its fauna." If anything, this is likely an underestimate, as the studies in Table 1 suggest.

TABLE 4   Examples of relationships between saproxylic insects and forest management[a]

Country	Reference	N[b]	S[c]	C[d]	Observations
Australia	Grove 2002a	+	+	+	Analyses of data concerning 339 saproxylic beetle species in old-growth, selectively logged and regrowth tropical rainforest. Differences subtle but consistent
Australia	Michaels & Bornemissza 1999	+	+	NR	Lucanid beetle species richness and abundance lower in regrowth forest than in recent clear-fells owing to continued presence of old-growth logs in clear-felled areas
Finland	Martikainen et al. 2000	+	+	+	78% of 232 saproxylic beetle species more abundant in old-growth than in managed mature forest; almost no overlap in assemblage composition
Finland	Martikainen et al. 1999	+	+	NR	Scolytine beetles in old-growth, overmature, and managed mature forest
Finland	Väisänen et al. 1993	NR	NR	+	Among saproxylic beetles living subcortically in dead trees, proportion of scolytines much higher in managed forest; proportion of rare species higher in old growth
Poland	Gutowski 1986	+	0	+	Cerambycid beetles in old-growth versus managed forest. More species present only in old-growth than only in managed forest
Germany	Schmitt 1992	NR	+	+	Recorded more species of saproxylic beetle in logs and more threatened species in unmanaged than in adjacent managed forest. Managed forest assemblages more dominated by scolytines
Canada	Spence et al. 1996	+	+	+	Several saproxylic beetle species found in old-growth that were rare or absent in mature managed forest
United States	Chandler & Peck 1992	+	+	NR	More species and individuals of leiodid beetles in old-growth versus "regrowth" (former clear-cut) forest
United States	Chandler 1991	+	NR	NR	More individuals of saproxylic beetles in old-growth versus regrowth forest

[a] Relationships are expressed with respect to "old-growthness"; +, positive relationship; 0, no relationship; NR, not reported
[b] N, abundance of individuals
[c] S, species richness
[d] C, assemblage composition

Referring to North American forests, Huston (1996) stated, "No other manageable property of the forest environment has a greater impact on biodiversity than coarse woody debris. Even [timber] harvesting . . . probably has a greater effect on total forest biodiversity through the alteration and removal of coarse woody debris than through its effect as a disturbance that 'resets' forest succession."

The situation may be equally critical wherever native forests are heavily exploited for fuelwood [e.g., China (Nalepa et al. 2001), South Africa (DuPlessis 1995), Australia (Driscoll et al. 2000)]. It may also be comparable to parts of the eastern United States, where the extensive mature forests are the result of "old-field" succession and are less than 200 years old. Within my study area in northeastern Queensland, the volume of dead wood on the forest floor in selectively logged tropical rainforest averaged about 80% that of nearby old growth, with much greater deficiencies among larger-diameter material (Grove 2001b). Differences were more marked among living trees, and in these forests it appeared to be the loss of larger-diameter trees from logged forest that was the main cause of the differences in the saproxylic beetle fauna (Grove 2002b) (Figure 2). This is likely also the case in forests managed by selection silviculture elsewhere.

Saproxylic Insects and the Unsustainable Management of Forests: A 5000-Year European Experiment

The history of forest use and abuse in western Europe is not unique but is well documented. Forests had scarcely reached their maximum post-glacial extent when farmers started clearing them. Over the following millennia forest cover was drastically reduced, and the structure and composition of remaining fragments greatly altered (Thirgood 1989). By 1000 A.D., there was probably no truly natural forest left in Europe outside Fennoscandia (Greig 1982).

It was not just wolves, bears, and lynx that retreated. As forests were cleared and the remaining fragments were managed for firewood, poles, and timber, mature timber habitat and saproxylic insects also began to vanish (Speight 1989). Their prehistoric extinction in the United Kingdom is well documented. Buckland & Dinnin (1993) list 17 saproxylic beetle species known in the United Kingdom only from subfossils, mostly from peat deposits dating from about 2900 B.C., and speculate that the list will grow with further excavations. None of these species is yet globally extinct, but most now survive only in tiny refugia elsewhere in Europe. Many belong to genera containing species that are currently common in other parts of the world (e.g., *Rhysodes, Prostomis, Pycnomerus, Dromaeolus, Platycerus, Cerambyx,* and *Eremotes*) and provide a clear reminder of what any region stands to lose if the processes that operated in western Europe are repeated elsewhere.

The 5000-year-long trend in forest loss in western Europe has now largely been reversed, with the expansion of secondary woodlands and plantations. However, the extinction of saproxylic insects continues apace. For instance, Hammond (1974) reported the loss from the United Kingdom of a further 20 beetle species over

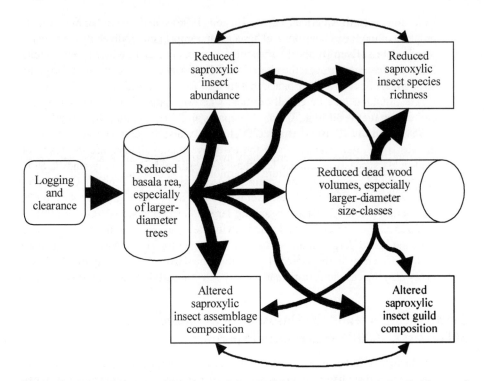

Figure 2 Likely impacts of logging and forest clearance on mature timber habitat, and consequences for the saproxylic insect fauna. Based on the main findings of the author's research on saproxylic beetles in northeastern Queensland. The thickness of each arrow is approximately proportional to the strength of the relationship as gauged by correlations, analyses of variance, or multivariate analyses as discussed in Grove (2002b). Relationships concerning guild composition are derived from the author's unpublished data.

the previous two centuries, and a fifth of the remaining UK cerambycid beetle fauna is now considered threatened (Twinn & Harding 1999). Today extinction is mostly driven by increasingly intensive forest management (Hanski & Hammond 1995, Siitonen 2001) coupled with the delayed effects of past fragmentation. In Fennoscandia, fire suppression and no-burn silviculture is one of the main threats to saproxylic species (Martikainen 2001, Johnson et al. 2001). Saproxylic insects now comprise a disproportionately large percentage of nationally rare and threatened species in Europe (Geiser 1983, Shirt 1987). For instance, in Sweden, of the 739 threatened forest invertebrate species, old living trees are considered a key habitat for 33%, logs 28%, and snags 35% (Berg et al. 1994). The removal of mature timber habitat is considered the main threat for 65% of the United Kingdom's 150 threatened woodland insect species (Hambler & Speight 1995).

Many saproxylic species now survive in Europe only as relictual populations, "hanging on by the tips of their tarsi" (Buckland & Dinnin 1993) in small patches of forest or pasture woodland—or even single trees—which for historical reasons did not experience the same loss of mature timber habitat. In the absence of positive management, the ultimate extinction of some such species (truly the "living dead") is almost inevitable through stochastic events (Thomas 2000).

Silviculture, Morticulture, and the Living Dead: Catering for Saproxylic Insects in Managed Forests

The forestry profession is increasingly aware of its responsibilities towards dead wood management ["morticulture" (Harmon 2001)], especially in North America (Franklin 1989) and northern Europe (Siitonen 2001). Done well, morticulture should cater for saproxylic insects. Whereas it is unrealistic to expect even sustainably managed commercial forests to retain a full complement of saproxylic insect species at the stand level, they ought to be able to at the landscape level. Some brief suggestions for how to do so follow.

At the landscape level the bottom line is that management should retain sufficient elements of natural forest dynamics to maintain the ecological processes in which mature timber habitat is involved—especially the chance for trees to live to ecological maturity and decay without intervention. Sufficient trees or patches should be retained to allow for some landscape-level temporal and spatial habitat continuity (Martikainen 2001, Grove et al. 2002). Management should also allow for the occasional occurrence of pest and disease outbreaks, windthrow, and fire if these are normal disturbance events. At the stand level there are many measures that can be taken (e.g., Grove 2001a, Ehnström 2001). For all forest types these should include using reduced-impact logging techniques, avoiding incidental damage to logs and snags, and avoiding harvesting of woody debris. Additional measures for selectively logged forests should include retaining habitat trees and legacy trees and avoiding silvicultural refinement. Additional measures for forests subjected to clearfelling should include extending rotations long enough for mature timber habitat to accumulate between felling cycles (at least over part of the production forest estate), a preference for small logging coupes and/or aggregated retention, and separating consecutive logging coupes to ensure that there is always some mature forest in the production forest matrix.

In regions where species are regionally extinct through a history of forest fragmentation and intensive use (as in western Europe), reintroductions are the only option. If they still survive as "living dead" in refugia, then morticultural management of refugia and the surrounding matrix may yet save them from extinction. Managing refugia is perhaps best developed in the United Kingdom, where most of the remaining key sites are cultural landscapes (such as royal hunting parks), now owned by nature conservation organizations (Alexander 1995), allowing a level of intensive care that would be uneconomical for commercial forestry. An example is the reinstatement of pollarding (Read 1991). Originally carried out to provide a supply of pole-wood above the reach of browsing animals, it incidentally helps prolong tree life, providing the necessary mature timber habitat and habitat continuity for a wide range of saproxylic insects and other organisms. In North America other suggestions for enhancing mature timber habitat in production forests have included the use of explosives (Bull & Partridge 1986) to produce snags from living trees, attracting bark beetles to living trees with pheromones to produce snags for "wildlife" (Ross & Niwa 1997), and the felling of "cull" trees to supplement coarse woody debris (Franklin 1989). Killing trees with herbicides has also been promoted, but the resultant snags provide much poorer conditions for saproxylic insects than do those resulting from natural tree death (Aulen 1991).

Management Indicators: The Way Forward for Sustaining Saproxylic Insects?

In Europe enough is now known about saproxylic insects for their potential use in identifying key sites for nature conservation or for monitoring the sustainability of forest management. Many taxonomic groups have been promoted for this purpose, especially beetles (Harding & Rose 1986, Harding & Alexander 1994, Nilsson et al. 1995). However, their use still remains the domain of entomologists and academics, rather than forestry practitioners. No such level of knowledge—or indeed interest—appears to exist in any other part of the world, so their chances of being widely adopted as indicators are slim.

As an alternative, mature timber habitat could well provide useful structural surrogates for saproxylic insects (e.g., Thunes et al. 2000), and incorporating such surrogates into guidelines, criteria, and indicators is perhaps our best chance of addressing the needs of saproxylic insects worldwide. One advantage is that habitat is more readily measurable than are the saproxylic insects themselves (Hodge & Peterken 1998). Measuring the basal area or standing volume of larger-diameter living trees is a simple procedure and merely an extension of a standard forestry activity. Snags and dead wood are slightly more difficult to measure accurately owing to their natural patchiness (Siitonen 1994a, Sippola et al. 1998, Grove 2001b). Currently, most initiatives assume that snags and dead wood on the forest floor are the key features. This may be so in forests where occasional large-scale disturbances are the norm, but as this review has demonstrated, it is not the whole picture, especially in forests where smaller-scale disturbances operate.

European guidelines for certification and sustainable management increasingly promote the retention and maintenance of dead wood and old trees (UKFC 1998, PEFC 1999). The level of management concern for coarse woody debris, snags, and dependent vertebrates in North America (Fenger 1996, Hagan & Grove 1999, Harmon 2001) and Australia (Gibbons & Lindenmayer 2001, Woldendorp et al. 2001) provides some hope that the needs of some saproxylic insects will be addressed by default. In the tropics nearly all available sustainable forest management guidelines are too general to refer specifically to the mature timber habitat, with one main exception (CIFOR 1999). Much work remains to be done worldwide to develop more appropriate guidelines and to get them implemented (Grove 2001a).

CONCLUSION

From European studies it is clear that saproxylic insects are peculiarly susceptible to forest management, in no small part owing to their dependence on a suite of habitats that is naturally abundant in unmanaged forest but often rare in long-managed forests. Many species may have low powers of dispersal, with populations governed by metapopulation dynamics. These characteristics make evolutionary sense, but do not favor survival in managed and fragmented forest settings, where spatial and temporal breaks in habitat continuity can lead to population declines and extinctions.

I hope this review has demonstrated that saproxylic insects deserve much greater consideration amongst forest managers and their ecologist advisers. Sustaining saproxylic insects may not be any more challenging than sustaining other forest biota, but unless their needs are specifically addressed there remains a serious risk that we will lose a large proportion of the forest fauna from our managed forests before we even know it. One only has to look at western Europe to see what might happen if we fail to act.

ACKNOWLEDGMENTS

I thank the Cooperative Research Centre for Tropical Rainforest Ecology and Management, James Cook University, and the Australian Tropical Research Foundation for funding my doctoral research in northeastern Queensland, which formed the basis of this review. I also thank Christine Nalepa, Donna Scheungrab, Tim Schowalter, Roger Sandquist, and Christine Niwa for following up on my requests for advice on the status of research on saproxylic insects in North America. I thank Petri Martikainen, Anne Sverdrup-Thygeson, Jim Hammond, and Marie Yee for their helpful comments on a previous version of this manuscript.

The *Annual Review of Ecology and Systematics* is online at
http://ecolsys.annualreviews.org

LITERATURE CITED

Alexander KNA. 1995. Historic parks and pasture-woodlands: the National Trust resource and its conservation. *Biol. J. Linn. Soc. London* 56 (Suppl.):155–75

Araya TE. 1993. Relationship between decay types of dead wood and occurrence of lucanid beetles (Coleoptera: Lucanidae). *Appl. Entomol. Zool.* 28:27–33

Ås S. 1993. Are habitat islands islands? Woodliving beetles (Coleoptera) in deciduous forest fragments in boreal forest. *Ecography* 16:219–28

Attiwill PM. 1994. Ecological disturbance and the conservation management of eucalypt forests in Australia. *For. Ecol. Manage.* 63:301–46

Aulen G. 1991. Increasing insect abundance by killing deciduous trees—a method of improving the food situation for endangered woodpeckers. *Holarct. Ecol.* 14:68–80

Berg A, Ehnström B, Gustafsson L, Hallingbäck T, Jonsell M, Weslien J. 1993. Threatened plant, animal and fungus species in Swedish forests: distribution and habitat associations. *Conserv. Biol.* 8:718–31

Blab J, Nowak E, Trautmann W, Sukopp H. 1994. *Rote Liste der gefährdeten Tiere und Pflanzen in der Bundesrepublik Deutschland.* Greven, Ger: Kilda

Blackman MW, Stage HH. 1924. On the succession of insects living in the bark and wood of dying, dead and decaying hickory. *Tech. Bull. NY State Coll. For.* 17:3–268

Braccia A, Batzer DP. 2001. Invertebrates associated with woody debris in a southeastern US forested floodplain wetland. *Wetlands* 21:18–31

Buckland PC, Dinnin MH. 1993. Holocene woodlands, the fossil insect evidence. In *Dead Wood Matters: The Ecology and Conservation of Saproxylic Invertebrates in Britain*, ed. KJ Kirby, CM Drake, pp. 6–20. Peterborough, UK: English Nat.

Bull EL, Partridge AD. 1986. Methods of killing trees for use by cavity nesters. *Wildl. Soc. Bull.* 14:142–46

Chandler DS. 1991. Comparison of some slime-mold and fungus feeding beetles (Coleoptera) in an old-growth and 40-year-old forest in New Hampshire. *Coleopt. Bull.* 45:239–56

Chandler DS, Peck SB. 1992. Diversity and seasonality of leiodid beetles (Coleoptera: Leiodidae) in an old-growth and a 40-year-old forest in New Hampshire. *Environ. Entomol.* 21:1283–91

CIFOR. 1999. *C&I Toolbox Series No. 2: The CIFOR Criteria and Indicators Generic Template.* Bogor, Indonesia: Cent. Int. For. Res. 53 pp.

Dajoz R. 1966. Ecologie et biologie des coléoptères xylophages de la hêtraie. [Ecology and biology of xylophagous beetles of the beechwood] *Vie Milieu* 17:525–636 In French with English summ.

Derksen W. 1941. Die Succession der pterygoten Insekten im abgestorbenen Buchenholz [The succession of pterygote insects in fallen beech wood]. *Z. Morphol. Ökol. Tier.* 37:683–734. In German

Deyrup MA. 1976. *The insect community of dead and dying Douglas-fir: Diptera, Coleoptera, and Neuroptera.* PhD thesis. Univ. Seattle

Driscoll D, Mikovits G, Freudenberger D. 2000. *Impact and Use of Firewood in Australia.* Canberra, ACT: CSIRO Sustainable Ecosystems

DuPlessis MA. 1995. The effects of fuelwood removal on the diversity of some cavity-using birds and mammals in South Africa. *Biol. Conserv.* 74:77–82

Edmonds RL, Eglitis A. 1989. The role of the Douglas-fir beetle and wood borers in the decomposition of and nutrient release from Douglas-fir logs. *Can. J. For. Res.* 19:853–59

Edmonds RL, Marra JL. 1999. Decomposition of woody material: nutrient dynamics, invertebrate/fungi relationships and management in Northwest forests. In *Proc. Pac. Northwest For. Rangeland Soil Organism Symp.*, ed. RT Meurisse, WG Ypsilantis, C Seybold, pp. 68–79. Corvallis, OR: USDA For. Ser. Gen. Tech. Rep. PNW-GTR-461

Ehnström B. 2001. Leaving dead wood for insects in boreal forests—suggestions for the future. *Scand. J. For. Res. Suppl.* 3:91–98

Elton C. 1966. Dying and dead wood. In *The Pattern of Animal Communities*, ed. C Elton, pp. 279–305. London: Methuen

Fenger M. 1996. Implementing biodiversity conservation through the British Columbia Forest Practices Code. *For. Ecol. Manage.* 85:67–77

Fowles AP. 1997. The Saproxylic Quality Index: an evaluation of dead wood habitats based on rarity scores, with examples from Wales. *Coleopterist* 6:61–66

Franklin J. 1989. Toward a new forestry. *Am. For.* 95:37–44

Geiser R. 1983. Rote Liste ausgewählter Familien xylobionter Käfer (Coleoptera) in Österreich [Red list of threatened xylobiont beetle families (Coleoptera) in Austria] (In German). In *Rote Listen Gefährdeter Tiere*

[*Red Lists of Threatened Species*], ed. J Gepp, pp. 131–37. Wien, Austria: Österr. Bundesminist. Gesundh. Umweltschutz

Gibbons P, Lindenmayer D. 2001. *Tree Hollows and Wildlife Conservation in Australia.* Canberra: CSIRO. 240 pp.

Goodburn JM, Lorimer CG. 1998. Cavity trees and coarse woody debris in old growth and managed northern hardwood forests in Wisconsin and Michigan. *Can. J. For. Res.* 28:427–38

Greig J. 1982. Past and present limewoods in Europe. In *Archaeological Aspects of Woodland Ecology*, ed. M Bell, S Limbrey, pp. 89–99. Oxford: Br. Archaeol. Rep.

Grove SJ. 2001a. Developing appropriate mechanisms for sustaining mature timber habitat in managed natural forest stands. *Int. For. Rev.* 3:272–84

Grove SJ. 2001b. Extent and composition of dead wood in Australian lowland tropical rainforest with different management histories. *For. Ecol. Manage.* 154:35–53

Grove SJ. 2002a. The influence of forest management history on the integrity of the saproxylic beetle fauna in an Australian lowland tropical rainforest. *Biol. Conserv.* 104:149–71

Grove SJ. 2002b. Tree basal area and dead wood as surrogate indicators of saproxylic insect faunal integrity: a case study from the Australian lowland tropics. *Ecol. Indicators* 1:171–88

Grove SJ, Meggs J, Goodwin A. 2002. *A review of biodiversity conservation issues relating to coarse woody debris management in the wet eucalypt production forests of Tasmania.* Hobart, Aust.: For. Tasmania Tech. Rep. 22:1–72

Grove SJ, Stork NE. 1999. The conservation of saproxylic insects in tropical forests: a research agenda. *J. Insect Conserv.* 3:67–74

Grove SJ, Stork NE. 2000. An inordinate fondness for beetles. *Invertebr. Taxon.* 14:733–39

Gutowski JM. 1986. Species composition and structure of the communities of longhorn beetles (Col., Cerambycidae) in virgin and managed stands of Tilio-Carpinetum sta-

chetosum association in Bialowieza Forest (NE Poland). *J. Appl. Entomol.* 102:380–91

Hagan JM, Grove SL. 1999. Coarse woody debris: humans and nature competing for trees. *J. For.* 97:6–11

Hambler C, Speight MR. 1995. Biodiversity conservation in Britain—science replacing tradition. *Br. Wildl.* 6:137–47

Hamilton WD. 1978. Evolution and diversity under bark. In *Diversity of Insect Faunas*, ed. LA Mound, N Waloff, pp. 154–75. London: R. Entomol. Soc. London

Hammond HEJ. 1997. Arthropod biodiversity from *Populus* coarse woody material in north-central Alberta—a review of taxa and collection methods. *Can. Entomol.* 129:1009–33

Hammond HEJ, Langor DW, Spence JRS. 2001. Early colonization of *Populus* wood by saproxylic beetles. *Can. J. For. Res.* 31:1175–83

Hammond PM. 1974. Changes in the British coleopterous fauna. In *The Changing Flora and Fauna of Britain*, ed. DL Hawksworth, pp. 323–69. London: Academic

Hammond PM. 1990. Insect abundance and diversity in the Dumoga-Bone National Park, N. Sulawesi, with special reference to the beetle fauna of lowland rainforest in the Toraut region. In *Insects and the Rain Forests of South East Asia (Wallacea)*, ed. WJ Knight, JD Holloway, pp. 197–254. London: R. Entomol. Soc. London

Hammond PM, Kitching RL, Stork NE. 1996. The composition and richness of the tree-crown Coleoptera assemblage in an Australian subtropical forest. *Ecotropica* 2:99–108

Hanski I. 1999. *Metapopulation Ecology.* Oxford: Oxford Univ. Press. 313 pp.

Hanski I, Hammond PM. 1995. Biodiversity in boreal forests. *Trends Ecol. Evol.* 10:5–6

Hanula JL. 1996. *Relationship of wood-feeding insects and coarse woody debris.* See McMinn & Crossley 1996, pp. 55–81

Hanula JL, Franzreb K. 1998. Source, distribution and abundance of macroarthropods on the bark of longleaf pine: potential prey of the red-cockaded woodpecker. *For. Ecol. Manage.* 102:89–102

Harding PT, Alexander KNA. 1994. The use of saproxylic invertebrates in the selection and conservation of relic forest in pasture-woodland. *Br. J. Entomol. Nat. Hist.* 7:21–26

Harding PT, Rose F. 1986. *Pasture-Woodlands in Lowland Britain.* Huntingdon, UK: Inst. Terr. Ecol. 89 pp.

Harmon ME. 2001. Moving towards a new paradigm for woody detritus management. *Ecol. Bull.* 49:269–78

Harmon ME, Franklin JF, Swanson FJ, Sollins P, Gregory SV, et al. 1986. Ecology of coarse woody debris in temperate ecosystems. *Adv. Ecol. Res.* 15:133–302

Heliövaara K, Väisänen R. 1984. Effects of modern forestry on northwestern European forest invertebrates: a synthesis. *Acta For. Fenn.* 189:5–30

Hespenheide HA. 1976. Patterns in the use of single plant hosts by wood-boring beetles. *Oikos* 27:61–64

Hodge SJ, Peterken GF. 1998. Deadwood in British forests: priorities and a strategy. *Forestry* 71:99–112

Howden HF, Vogt GB. 1951. Insect communities of standing dead pine (*Pinus virginiana* Mill.). *Ann. Entomol. Soc. Am.* 44:581–95

Huston MA. 1996. *Modeling and management implications of coarse woody debris impacts on biodiversity.* See McMinn & Crossley, 1996, pp. 55–81

Johnson EA, Miyanishi K, Bridge SRJ. 2001. Wildfire regime in the boreal forest and the idea of suppression and fuel buildup. *Conserv. Biol.* 15:1554–57

Jonsell M, Nordlander G, Jonsson M. 1999. Colonization patterns of insects breeding in wood-decaying fungi. *J. Insect Conserv.* 3:145–61

Jonsell M, Weslien J, Ehnström B. 1998. Substrate requirements of red-listed saproxylic invertebrates in Sweden. *Biodivers. Conserv.* 7:749–64

Kaila L, Martikainen P, Punttila P, Yakovlev E. 1994. Saproxylic beetles (Coleoptera) on dead birch trunks decayed by different polypore species. *Ann. Zool. Fenn.* 31:97–107

Kehler D, Bondrup-Nielsen S. 1999. Effects of isolation on the occurrence of a fungivorous forest beetle, *Bolitotherus cornutus*, at different spatial scales in fragmented and continuous forests. *Oikos* 84:35–43

Kleinevoss K, Topp W, Bohac J. 1996. Buchen-Totholz im Wirtschaftswald als Lebensraum für xylobionte Insekten [Dead beech wood in the commercial forest as habitat for xylobiont insects]. *Z. Ökol. Nat.schutz* 5:85–95. In German

Kletecka Z. 1996. The xylophagous beetles (Insecta, Coleoptera) community and its succession on Scotch elm (*Ulmus glabra*) branches. *Biologia (Bratislava)* 51:143–52

Köhler F. 2000. *Totholzkäfer in Naturwaldzellen des nördlichen Rheinlandes. Vergleichende Studien zur Totholzkäferfauna Deutschlands und deutschen Naturwaldforschung [Saproxylic Beetles in Nature Forests of the Northern Rhineland. Comparative Studies on the Saproxylic Beetles of Germany and Contributions to German Nature Forest Research].* Recklinghausen: Landesamt Agrarordnung NRW. 351 pp. In German

Kolström M, Lumatjärvi J. 2000. Saproxylic beetles on aspen in commercial forests: a simulation approach to species richness. *For. Ecol. Manage.* 126:113–20

Komonen A, Penttilä R, Lindgren M, Hanski I. 2000. Forest fragmentation truncates a food chain based on an old-growth forest bracket fungus. *Oikos* 90:119–26

Kouki J, Lofman S, Martikainen P, Rouvinen S, Uotila A. 2001. Forest fragmentation in Fennoscandia: linking habitat requirements of wood-associated threatened species to landscape and habitat changes. *Scand. J. For. Res. Suppl.* 3:27–37

Kruys N, Jonsson BG. 1999. Fine woody debris is important for species richness on logs in managed boreal spruce forests of northern Sweden. *Can. J. For. Res.* 29:1295–99

Laiho R, Prescott CE. 1999. The contribution of coarse woody debris to carbon, nitrogen, and phosphorus cycles in three Rocky Mountain coniferous forests. *Can. J. For. Res.* 29:1592–1603

Lofroth E. 1998. The dead wood cycle. In *Conservation Biology Principles for Forested Landscapes*, ed. J Voller, S Harrison, pp. 185–214. Vancouver, Can.: UBC Press

Mackensen J, Bauhus J. 1999. *The Decay of Coarse Woody Debris*. Canberra, Aust.: Aust. Greenhouse Off. 41 pp.

Martikainen P. 2001. Conservation of threatened saproxylic beetles: significance of retained aspen *Populus tremula* on clearcut areas. *Ecol. Bull.* 49:205–18

Martikainen P, Siitonen J, Kaila L, Punttila P, Rauh J. 1999. Bark beetles (Coleoptera, Scolytidae) and associated beetle species in mature managed and old-growth boreal forests in southern Finland. *For. Ecol. Manage.* 116:233–45

Martikainen P, Siitonen J, Punttila P, Kaila L, Rauh J. 2000. Species richness of Coleoptera in mature managed and old-growth boreal forests in southern Finland. *Biol. Conserv.* 94:199–209

Maser C. 1996. Salvage logging—the loss of ecological reason and moral restraint. *Int. J. Ecofor.* 12:176–78

McMinn J, Crossley DA, eds. 1996. *Biodiversity and Coarse Woody Debris in Southern Forests*. Athens GA: USDA For. Ser. Gen. Tech. Rep. SE-94

McPeek MA, Holt RD. 1992. The evolution of dispersal in spatially and temporally varying environments. *Am. Nat.* 140:1010–27

Michaels K, Bornemissza G. 1999. Impact of clearfell harvesting on lucanid beetles (Coleoptera: Lucanidae) in wet and dry sclerophyll forests in Tasmania. *J. Insect Conserv.* 3:85–95

Nalepa CA, Li LI, Wen-Hua LU, Lazell J. 2001. Rediscovery of the wood-eating cockroach *Cryptocercus primarius* (Dictyoptera: Cryptocercidae) in China, with notes on ecology and distribution. *Acta Zootaxon. Sin.* 26:184–90

Nilssen AC. 1984. Long-range aerial dispersal of bark beetles and bark weevils (Coleoptera, Scolytidae and Curculionidae) in northern Finland. *Ann. Ent. Fenn.* 50:37–42

Nilsson SG. 1997. *Grynocharis oblonga* L. (Coleoptera: Trogossitidae): a specialized wood beetle with a relict distribution. *Entomol. Tidskr.* 118:1–9. In Swedish with English summ.

Nilsson SG, Arup U, Baranowski R, Ekman S. 1995. Tree-dependent lichens and beetles as indicators in conservation forests. *Conserv. Biol.* 9:1208–15

Nilsson SG, Baranowski R. 1994. Indikatorer på jätteträdskontinuitet—svenska förekomster av knäppare som är beroende av grova, levande träd [Indicators of megatree continuity—Swedish distribution of click beetles (Coleoptera, Elateridae) dependent on hollow trees]. *Entomol. Tidskr.* 115:81–97. In Swedish with English summ.

Nilsson SG, Baranowski R. 1997. Habitat predictability and the occurrence of wood beetles in old-growth beech forests. *Ecography* 20:491–98

Nilsson T. 1997. *Spatial population dynamics of the black tinder fungus beetle*, Bolitophagus reticulatus (*Coleoptera: Tenebrionidae*). PhD thesis. Univ. Uppsala, Sweden. 44 pp.

Nordén B, Paltto H. 2001. Wood-decay fungi in hazel wood: species richness correlated to stand age and dead wood features. *Biol. Conserv.* 101:1–8

Økland B, Bakke A, Hågvar S, Kvamme T. 1996. What factors influence the diversity of saproxylic beetles? A multiscaled study from a spruce forest in southern Norway. *Biodivers. Conserv.* 5:75–100

Palm T. 1959. Die Holz- und Rindenkäfer der süd- und mittelschwedischen Laubbäume [The wood- and bark-beetles of south and mid-Swedish broadleaves]. *Opusc. Entomol.* 16 (Suppl.):1–374. In German

Parker SP. 1982. *Synopsis and Classification of Living Organisms*. New York: McGraw-Hill. 1232 pp.

PEFC. 1999. Pan Eur. For. Certif. Website: http://www.pefc.org/

Peltonen M. 1999. Windthrows and dead-standing trees as bark-beetle breeding material at forest-clearcut edge. *Scand. J. For. Res.* 14:505–11

Price K, Pojar J, Roburn A, Brewer L, Poirier N. 1998. Windthrown or clearcut—what's the difference? *Northwest Sci.* 72:30–32

Ranius T. 2002. Influence of stand size and quality of tree hollows on saproxylic beetles in Sweden. *Biol. Conserv.* 103:85–91

Ranius T, Hedin J. 2001. The dispersal rate of a beetle, *Osmoderma eremita*, living in tree hollows. *Oecologia* 126:363–70

Ranius T, Jansson N. 2000. The influence of forest regrowth, original canopy cover and tree size on saproxylic beetles associated with old oaks. *Biol. Conserv.* 95:85–94

Read HJ. 1991. *Pollard and Veteran Tree Management.* London: Corp. London. 60 pp.

Ross DW, Niwa CG. 1997. Using aggregation and antiaggregation pheromones of the Douglas-fir beetle to produce snags for wildlife habitat. *West. J. Appl. For.* 12:52–54

Rukke BA. 2000. Effects of habitat fragmentation: increased isolation and reduced habitat size reduces the incidence of dead wood fungi beetles in a fragmented forest landscape. *Ecography* 23:492–502

Safranyik L, Linton DA. 1999. Spruce beetle (Coleoptera: Scolytidae) survival in stumps and windfall. *Can. Entomol.* 131:107–13

Savely HE. 1939. Ecological relations of certain animals in dead pine and oak logs. *Ecol. Monogr.* 9:321–85

Schiegg K. 2000. Effects of dead wood volume and connectivity on saproxylic insect species diversity. *Ecoscience* 7:290–98

Schmitt M. 1992. Buchen-Totholz als Lebensraum für xylobionte Käfer [Dead beech wood as habitat for xylobiont beetles]. *Waldhygiene* 19:97–191. In German

Schowalter TD, Caldwell BA, Carpenter SE, Griffiths RP, Harmon ME, et al. 1992. Decomposition of fallen trees: effects of initial conditions and heterotroph colonization rates. In *Tropical Ecosystems: Ecology and Management*, ed. KP Singh, JS Singh, pp. 373–83. New Delhi, India: Wiley Eastern

Shirt DB. 1987. *British Red Data Books: 2.*

Insects. Peterborough, UK: Nat. Conserv. Counc.

Siitonen J. 1994a. Decaying wood and saproxylic Coleoptera in two old spruce forests: a comparison based on two sampling methods. *Ann. Zool. Fenn.* 31:89–95

Siitonen J. 1994b. Occurrence of rare and threatened insects living on decaying *Populus tremula*: a comparison between Finnish and Russian Karelia. *Scand. J. For. Res.* 9: 185–91

Siitonen J. 2001. Forest management, coarse woody debris and saproxylic organisms: Fennoscandian boreal forests as an example. *Ecol. Bull.* 49:11–42

Siitonen J, Saaristo L. 2000. Habitat requirements and conservation of *Pytho kolwensis*, a beetle species of old-growth boreal forest. *Biol. Conserv.* 94:211–20

Sippola A, Kallio R. 1995. Species composition of beetles (Coleoptera) in different habitats within old-growth and managed forests in Finnish Lapland. In *Northern Wilderness Areas: Ecology, Sustainability, Management*, ed. A Sippola, P Alaraudanjoki, B Forbes, V Hallikainen, pp. 59–77. Rovaniemi, Finland: Arctic Cent.

Sippola A, Siitonen J, Kallio R. 1998. Amount and quality of coarse woody debris in natural and managed coniferous forests near the timberline in Finnish Lapland. *Scand. J. For. Res.* 13:204–14

Speight MCD. 1989. *Saproxylic Invertebrates and their Conservation.* Strasbourg, Fr: Counc. Eur. 79 pp.

Spence JR, Langor DW, Niemelä J, Carcamo H, Currie CR. 1996. Northern forestry and carabids: the case for concern about old-growth species. *Ann. Zool. Fenn.* 33:173–84

Spies TA, Franklin JF, Thomas TB. 1988. Coarse woody debris in Douglas-fir forests of western Oregon and Washington. *Ecology* 69:1689–702

Sverdrup-Thygeson A, Midtgaard F. 1998. Fungus-infected trees as islands in boreal forest: spatial distribution of the fungivorous beetle *Bolitophagus reticulatus* (Coleoptera, Tenebrionidae). *Ecoscience* 5:486–93

Swift MJ. 1977. The ecology of wood decomposition. *Sci. Prog.* 64:175–99

Tate KR, Ross DJ, O'Brien BJ, Kelliher FM. 1993. Carbon storage and turnover, and respiratory activity, in the litter and soil of an old-growth southern beech (*Nothofagus*) forest. *Soil Biol. Biochem.* 25:1601–12

Tavakilian G, Berkov A, Meurer-Grimes B, Mori S. 1997. Neotropical tree species and their faunas of xylophagous longicorns (Coleoptera: Cerambycidae) in French Guyana. *Bot. Rev.* 63:303–55

Thirgood JV. 1989. Man's impact on the forests of Europe. *J. World For. Resour. Manage.* 4:127–67

Thomas CD. 2000. Dispersal and extinction in fragmented landscapes. *Proc. R. Soc. London Ser. B* 267:139–45

Thunes KH, Midtgaard F, Gjerde I. 2000. Diversity of Coleoptera of the bracket fungus *Fomitopsis pinicola* in a Norwegian spruce forest. *Biodivers. Conserv.* 9:833–52

Torgersen TR, Bull EL. 1995. Down logs as habitat for forest-dwelling ants—the primary prey of pileated woodpeckers in northeastern Oregon. *Northwest Sci.* 69:294–303

Tritton LM, Martin CW, Hornbeck JW, Pierce RS. 1987. Biomass and nutrient removals from commercial thinning and whole-tree clear-cutting of central hardwoods. *Environ. Manage.* 11:659–66

Twinn PFG, Harding PT. 1999. *Provisional Atlas of the Longhorn Beetles (Coleoptera, Cerambycidae) of Britain.* Abbots Ripton, UK: Nat. Environ. Res. Counc. 96 pp.

UK For. Comm. 1998. *The UK Forestry Standard: The UK Government's Approach to Sustainable Forestry.* Edinburgh: For. Comm.

Väisänen R, Biström O, Heliövaara K. 1993. Sub-cortical Coleoptera in dead pines and spruces: Is primeval species composition maintained in managed forests? *Biodivers. Conserv.* 2:95–113

Wall J. 1999. Fuelwood in Australia: impacts and opportunities. In *Temperate Eucalypt Woodlands in Australia: Biology, Conservation, Management and Restoration*, ed. RJ Hobbs, CJ Yates, pp. 372–81. Chipping Norton, Aust.: Surrey Beatty & Sons

Warren MS, Key RS. 1991. Woodlands: past, present and potential for insects. In *The Conservation of Insects and their Habitats*, ed. NM Collins, JA Thomas, pp. 155–203. London: Academic

Weslien J, Schroeder LM. 1999. Population levels of bark beetles and associated insects in managed and unmanaged spruce stands. *For. Ecol. Manage.* 115:267–75

Wikars L-O. 2001. The wood-decaying fungus *Daldinia loculata* (Xylariaceae) as an indicator of fire-dependent insects. *Ecol. Bull.* 49:263–68

Woldendorp G, Keenan RJ, Ryan MF. 2001. *Coarse Woody Debris in Australian Forest Ecosystems.* Canberra, Aust.: AFFA Bur. Rural Sci.

Wood GA, Hasenpusch J, Storey RI. 1996. The life history of *Phalacrognathus muelleri* (Macleay) (Coleoptera, Lucanidae). *Aust. Entomol. Mag.* 23:37–48

Yee M, Yuan Z, Mohammed C. 2001. Not just waste wood: decaying logs as key habitats in Tasmania's wet sclerophyll *Eucalyptus obliqua* production forests: the ecology of large and small logs compared. *Tasforests.* 13:119–28

Annu. Rev. Ecol. Syst. 2002. 33:25–47
doi: 10.1146/annurev.ecolsys.33.010802.150424

CONUS VENOM PEPTIDES: Reflections from the Biology of Clades and Species

Baldomero M. Olivera

Department of Biology, University of Utah, Salt Lake City, Utah 84112;
email: olivera@biology.utah.edu

Key Words conotoxins, speciation, hypermutation

■ **Abstract** The 500 cone snail species (*Conus*) use complex venoms to capture prey, defend against predators and deter competitors. Most biologically active venom components are small, highly structured peptides, each encoded by a separate gene. Every *Conus* species has its own distinct repertoire of 100–200 venom peptides, with each peptide presumably having a physiologically relevant target in prey or potential predators/competitors. There is a remarkable interspecific divergence observed in venom peptide genes, which can be rationalized because of biotic interactions that are species specific. The peptide families/subfamilies characteristic of clades of related *Conus* species are potentially useful clade markers and can be used to indicate common biological mechanisms characterizing that clade. By knowing both the distribution and the physiological function of venom peptides, a type of reverse ecology becomes possible; the peptides in a *Conus* venom are a molecular readout of the biotic interactions of a species or clade.

INTRODUCTION

The marine gastropods known as cone snails (*Conus*) constitute an unusually species-rich group of venomous predators, one of the largest single genera (>500 species) of living marine invertebrates. Species of *Conus* inject a complex venom into their prey. Some cone snails have evolved what *a priori*, would seem to be exceedingly unpromising evolutionary directions for relatively slow snails that are unable to swim: A significant complement of *Conus* species (>50) catch fish as either their major or exclusive prey. The sting of at least one of these, *Conus geographus*, the geography cone, has resulted in a high frequency of human fatality. Although information regarding the detailed biology of most species is sparse, a group of snails that has evolved both the ability to hunt fish and venom that can kill people has clearly taken some intriguing evolutionary directions. The comprehensive investigation of the venoms of *Conus* carried out by our laboratory and by others provides an emerging framework for understanding the history and evolutionary biology of this species-rich group.

The biological activity of a *Conus* venom is due to a large complement (100–200) of unusually small, highly structured peptides, which are the major focus of this review. Each *Conus* species has its own distinct repertoire of venom peptides (Olivera 1997, Olivera & Cruz 2001).

The great majority of *Conus* species are highly specialized predators; in some cases, only a single species of prey is envenomated. However, collectively, cone snails have a remarkably broad spectrum of prey that fall into at least four phyla (Röckel et al. 1995). Despite the many attempts to split the genus, taxonomists generally use *Conus* as the generic taxon for the entire group because no alternative scheme has gained wide acceptance. Quite apart from the intrinsic interest in any large group of animals, the unusual biology of *Conus* has provided unexpected opportunities for insights that have broader biological significance. To introduce the reader to these aspects of *Conus* biology, we first provide a general overview of the genus.

TAXONOMIC STATUS AND EVOLUTIONARY HISTORY OF *CONUS*

In most taxonomic treatments of gastropods, cone snails (family Conidae) are regarded as advanced and generally placed within the order Neogastropoda. Among the neogastropods, several families are regarded as more closely related to the cone snails in the traditional taxonomy—in particular, other venomous marine gastropods such as tower snails (Terebridae) and turrid snails (traditionally in the family Turridae, but in recent taxonomic proposals split into multiple families). These venomous marine snails are generally referred to as the toxoglossate molluscs (Toxoglossa) or conoideans (Conoidea, or alternatively superfamily Conacea—see Taylor et al. 1993). In total, there are probably well over 3000 species of venomous marine snails, making the conoideans a significant component of gastropod diversity.

However, molecular data from neogastropods has failed to support the hypothesis of a close relationship among the major conoidean genera (Espiritu et al. 2002, Harasewych et al. 1997). In one study the Conidae was found to be as closely related to other neogastropod families such as the Mitridae, Costellaridae, and Olividae as to the Terebridae and Turridae; the data were consistent with the theory that all six neogastropod families examined are essentially phylogenetically equidistant from one another, leading to the suggestion of a star phylogeny (Espiritu et al. 2002) rather than the branching tree phylogeny implied by the conventional taxonomy. These preliminary conclusions need to be confirmed.

The fossil evidence suggests that the first radiation of cone snails took place in the Eocene period; no reliable fossils predate the K-T boundary. Although a number of Cretaceous *Conus* fossils have been proposed in the literature, Kohn (1990) has examined these and concluded that no reliable *Conus* can be dated before the early Eocene. A second larger radiation occurred starting in the Miocene, and except for a set-back in the lower Pliocene when a significant extinction occurred, the expansion of the genus has basically continued to the present.

OVERVIEW OF THE BIOLOGY OF *CONUS*

The living cone snails are found in all tropical marine environments and are particularly prominent around coral reefs and other shallow-water tropical marine habitats (Kohn 1959, 1960, 1985; Kohn & Nybakken 1975). A single coral reef in the tropical Indo-Pacific may have over 30 different species of *Conus* (Kohn et al. 2001). A few species have adapted to cooler waters, but species diversity falls sharply outside the tropics. The most familiar *Conus* species live in relatively shallow waters, but there are many forms, probably including a large complement that remain undescribed, that live in deeper marine habitats (>150 m).

All *Conus* species use venom to capture prey. The genus has evolved a highly specialized venom apparatus (see, e.g., Kohn et al. 1960). The biologically active components of *Conus* venoms are synthesized in epithelial cells lining the long, tubular venom duct, and the contents of the duct are expelled into the proboscis by a muscular bulb. All cone snails have specialized, hollow radular teeth that are individually moved into the proboscis, which serve as both harpoon and hypodermic needle. The radular teeth have been used for *Conus* taxonomy because different groups of cone snails have anatomically distinct ones.

The venom apparatus of a fish-hunting *Conus* species, *C. striatus*, is illustrated in Figure 1. Also illustrated in Figure 1 are two teeth, each from a different *Conus* species. Most shallow-water tropical cone snail species are highly specialized with regard to their prey. *Conus* species have traditionally been divided into three groups, depending on their major prey: Members of the largest group specialize on various types of worms (vermivorous *Conus*), mostly polychaetes; those of a second group prey on other gastropod molluscs (molluscivorous *Conus*); and those of the final group, the major focus of this review, capture fish (piscivorous *Conus*). In addition, some *Conus* are also known to feed on hemichordates, echiuroids, and bivalve molluscs. A few *Conus* species, such as *C. californicus*, which has colonized a cooler-water habitat and has no congeners to compete with, are generalists (Kohn 1966). Remarkably little is known about the biology of the vast majority of *Conus* species, with most published literature coming from the work of Alan Kohn and his colleagues.

Some of the fish-hunting species, notably *Conus geographus*, are clearly dangerous to humans: Approximately three dozen fatalities have been recorded in the medical literature, with a high frequency of fatality in untreated human envenomation cases. Although it has been recorded for three centuries in the scientific literature that cone snails are venomous and capable of killing people (Rumphius 1705), it was not until Kohn started investigating the ecology of cone snails that it was discovered that some species were specialists in envenomating and capturing fish (Kohn 1956).

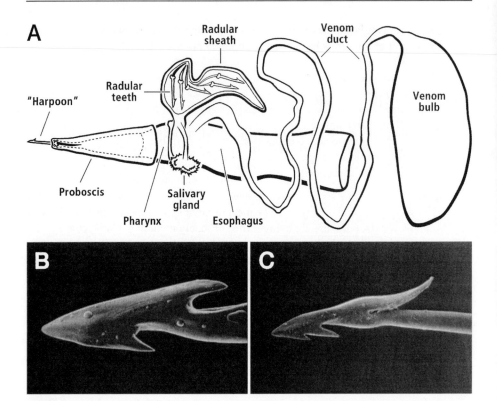

Figure 1 (*Top*) The venom apparatus of *Conus*. The venom bulb (*extreme right*), the venom duct (*long thin tube*) and the radular sac in which harpoon-like teeth are stored. A single tooth is moved into the proboscis (shown in place at *left*). (*Bottom*) Scanning electron micrographs of the anterior end of harpoon-like radular teeth from two *Conus* species, *C. purpurascens* (left) and *C. obscurus* (right).

CONUS VENOM PHARMACOLOGY AND TOXIN BIOCHEMISTRY

A significant amount of information has been collected for several gene families expressed in the venom ducts of cone snails, but most of the data are for a small subset of *Conus* species and were originally acquired for reasons unrelated to either the taxonomy or ecology of these snails. The lethality of *C. geographus* envenomation to humans attracted the attention of the toxinology community, and a physiological and pharmacological characterization of the venoms of a few cone snails was carried out. A first comprehensive study of the effects of different *Conus* venoms (Kohn et al. 1960) demonstrated that there are striking differences in potency; in particular, venoms of fish-hunting *Conus* are much more lethal than those of the other groups when tested in vertebrates. Early work by Endean and coworkers suggested that there might be components with unusual pharmacological properties not previously observed in other venoms (Endean et al. 1974, 1976, 1977, 1979).

The first biochemical attempt at characterizing the biologically active components of cone snail venoms was reported by Spence et al. (1977), who purified a toxin from *C. geographus* that paralyzed muscle and inhibited muscle action potentials, without detectable effects on the electrical activity of nerves; these workers determined the amino acid composition but did not fully characterize the toxin. In retrospect, there is little doubt that they were the first to purify a peptide in a class of *Conus* venom components now known as {μ}-conotoxins, novel polypeptides that block sodium channels by binding the same site as tetrodotoxin (from the pufferfish, fugu) and saxitoxin (from dinoflagellates that cause "red tides"), but with much higher selectivity.

A systematic biochemical analysis of *Conus* venoms was initiated in the 1970s and originally focused on a few fish-hunting species, in particular *C. geographus*, *C. striatus* and *C. magus*. This early biochemical work, which was reviewed in 1985 (Olivera et al. 1985), showed definitively that the major biologically active components of *Conus* venoms are unusually small, highly structured peptides and that many of these potently alter the function of ion channels.

MOLECULAR STUDIES OF VENOM COMPONENTS

Woodward et al. (1990) reported the first molecular analysis of cone snail cDNA clones from the messenger RNA of venom ducts. This study revealed that peptides expressed in the venom duct of *C. textile*, were initially translated as prepropeptides, with a specific proteolytic cleavage required to generate the mature toxins. This study determined the sequences of several related *C. textile* toxin precursors. Although some elements of these precursors, notably the signal sequence at the N-terminal end, were conserved, the mature peptide toxins, encoded at the C-terminus of the precursors, had diverged remarkably from each other in amino acid sequence. Thus, this work established the striking juxtaposition of conserved and highly divergent regions in a *Conus* precursor; this "focal hypermutation" is generally observed in the C-terminal, mature-toxin–encoding region whenever sequences of two *Conus* peptide precursor genes in the same family are compared.

The discovery that some *Conus* venom peptides have unique pharmacological properties accelerated the characterization of venom components; by 1990 (Olivera et al. 1990) it had become clear that the venom of all cone snails is remarkably complex and that an individual species could express 100–200 different venom peptides. Furthermore, interspecific focal hypermutation resulted in a different complement of peptides in the venom of each cone species. As had been established for intraspecific divergence, the interspecific hypermutation occurs only in the C-terminal mature toxin region and not in the N-terminal section of the precursor, nor in the 3' untranslated region of the mRNA.

Thus, a consensus picture emerged that the biologically active components of *Conus* venom are mostly small peptides (referred to as conopeptides or conotoxins, terms used interchangeably in this review). An enormous diversity of peptides are expressed in the venoms of living *Conus* (a conservative estimate is over ~50,000 peptides altogether), each apparently encoded by a separate gene. However, as the molecular genetic analysis continued, it also became clear that this massive peptide diversity has been generated by diversification of only a few gene superfamilies (~20–30), with members of a given gene superfamily sharing highly conserved sequence features such as the signal sequence that become a molecular signature for all members of that superfamily (Olivera 1997).

The conserved sequence elements in a *Conus* peptide gene superfamily have made it feasible to use PCR to uncover new conopeptides sequences that are now easily accessed, resulting in the availability of an ever-expanding venom peptide database from many *Conus* species. The hypermutation of the C-terminal, mature toxin region as speciation occurs leads to different complements of venom peptides in different species—the growing list of available toxin sequences provides an unparalleled opportunity to analyze several gene families that have undergone accelerated evolution over a large number of congeneric species. It is arguably the largest such molecular database available at the present time for interspecific divergence within a single genus of animals. Thus, *Conus* peptides provide a window into molecular events that accompany speciation.

CONOPEPTIDES ARE A SPECIALIZED ADAPTATION OF CONE SNAILS

Each conopeptide is the final functional gene product of an individual *Conus* gene; like conventional proteins, conopeptides have specific conformations essential for their high affinity and high-specificity interactions with the target protein. Collectively, the ~50,000 conopeptides probably comprise the smallest highly structured polypeptidic gene products known. Although there are certainly smaller peptides (for example, neuropeptides in the central nervous system), these usually function as signal molecules, secreted by one cell to bind with a specific receptor on another cell, and cause a physiological change. Most endogenous intercellular signals are designed to have a relatively short half-life. The flexibility and lack of a specific three-dimensional structure in smaller peptides is the feature that gives peptide signal molecules a relatively short half-life; such peptides generally equilibrate rapidly between many conformations, making them susceptible to rapid degradation by proteases in the extracellular environment.

The biochemical innovation in cone snail peptides arises from the combination of being unusually small and highly structured. This biochemical niche is one that cone snails have become expert at filling; their venoms are loaded with extensively disulfide-cross-linked peptides, a large fraction of which are in a size range of 10–30 amino acids.

In this size range, the disulfide cross-links provide an essential scaffold for stabilizing structures of what would otherwise be flexible peptides. Thus, the pattern of disulfide cross-links is a decisive determinant of peptide structure. Each *Conus* venom peptide gene superfamily has one or two characteristic patterns of disulfide cross-links. In most cases, the disulfide connectivity can be deduced from the arrangement of Cys residues in the primary sequence of the toxin—thus, members of a *Conus* peptide superfamily can usually be defined by two sequence elements: A characteristic consensus signal sequence (that differs between gene superfamilies, but is conserved within a superfamily) and a characteristic arrangement of Cys residues in the primary amino acid sequence of the mature toxin region, resulting in a conserved pattern of disulfide cross-links (see Table 1). Most of the *Conus* venom peptides that have been characterized have two or three disulfide

TABLE 1   Conotoxin superfamilies and peptide scaffolds

Gene super-family	Scaffold	(Previous nomenclature)	Pattern
A	A-1	I/II	CC———C———C
	A-2	IV	CC——C——C——C——C
M	M-1	III	CC———C———C———CC
	M-2	XII	CC———C———C———CC
O	O-1	VI/VII	C———C———CC———C———C
P	P-1	IX	C———C———C———C———C———C
T	T-1	V	CC———CC
	T-2	X	CC———C—C
I	I-1*	XI	C———C———CC——CC———C———C
S	S-1*	VII	C—C—C—C—C—C—C———C—C—C
C	C-1*		γγ——γ/κ——γ——γ

*The disulfide connectivity for these scaffolds has not been established.

cross-links, and are 12–30 AA in length (although the I- and S-superfamilies have four or five disulfide cross-links, respectively, and are larger (30–50 AA)).

EVOLUTIONARY SUCCESS THROUGH NEUROPHARMACOLOGY

A useful, if somewhat idiosyncratic conceptual framework for viewing the venom peptides of cone snails is to assume a pharmacological perspective. From this point of view, cone snails are specialists in neuropharmacology; venoms are, in essence, mixtures of potent drugs that have evolved through natural selection for their powerful pharmacological effects. One biological end point is envenomation of prey to allow a cone snail to capture and devour it. Alternatively, in defensive situations (almost certainly the case when humans get stung), venom components provide an effective deterrent to potential predators.

In many respects the evolution of *Conus* venom components parallels the drugs produced by a pharmaceutical company, particularly with regard to some recent trends in drug development. The use of combinatorial library strategies to identify a lead drug candidate, coupled with sophisticated medicinal chemistry to optimize each lead for drug development, are avant-garde technical developments in the pharmaceutical industry that parallel strategies evolved in cone snails by at least the early Eocene, 55 million years ago. The hypermutation observed in the mature toxin region as *Conus* speciation occurs is the equivalent of a combinatorial library search in a pharmaceutical company for new drug leads. Each female cone snail lays tens or hundreds of thousands of fertilized eggs at one time (in some cases, over a million). Thus, any genetic mechanism that introduces a high rate of mutation in venom peptides results in more offspring with variant peptides; these are immediately subject to natural selection. The fact that the life history of cone snails involves an enormous number of offspring means that mechanisms for the hypermutation of venom peptides could generate a substantial combinatorial library in a few generations, and the subsequent natural selection is akin to the "biopanning" of synthetic combinatorial libraries that has proven to be so useful to the drug industry in recent years.

Another parallel is the use of medicinal chemistry by the pharmaceutical industry to optimize an initial lead compound. In a large drug company, optimization of the lead compound is the responsibility of medicinal chemists, who experiment with additional functional groups in order to improve potency and efficacy. A similar development occurs in emerging *Conus* peptide genes; to sharpen specificity and potency, a lead conopeptide undergoes a separate focus for selection: The introduction of functional groups not found in the standard 20 amino acids. This is accomplished through a variety of posttranslational modifications. Posttranslational modification enzymes are recruited to act on potential precursor substrates (Hooper et al. 2000). Appendix I summarizes the presently understood posttranslational modifications in *Conus* peptides and what is known about the enzymes that catalyze these modifications. Thus, one view of *Conus* is that this is a taxon that has specialized in neuropharmacology—by rapidly evolving highly effective pharmacological agents in their venoms, and having the capacity to efficiently deliver these potent "drugs" (through their harpoons, that are equivalent to disposable hypodermic needles), cone snails have diversified into an exceptionally species-rich genus of marine predators.

The evolutionary imperatives that have led this group of animals to specialize in producing a large and extremely diverse set of pharmacologically active small peptides can be rationalized in general terms. Cone snails are slow-moving predators that lack strong mechanical aids to capture prey or defend against predators; they likely routinely encounter much more agile potential prey or predators in their environment. This provides an ecological scenario for the evolution of venoms that act very quickly. The complex venoms comprising unusually small, highly structured peptides that potently affect receptor targets primarily in the nervous system of the envenomated animals are presumably a response to this strong selective pressure. In order to evolve this biochemical specialization, a number of barriers probably had to be overcome (such as properly folding small peptides), barriers that apparently have kept other taxa from also generating a large diversity of structured peptides in such a small size range. Thus, over the last 50 million years, the cone snails have used specialized biochemical and genetic adaptations that in effect comprise a highly successful drug-development strategy.

OVERVIEW OF *CONUS* PEPTIDES

The small size of conopeptides makes chemical synthesis possible in amounts sufficient to investigate the biochemistry, pharmacology, and physiology of each gene product. Thus, a relatively large set of venom peptides from cone snails has been investigated to an extent that the physiological mechanisms underlying their biological activity are understood; for a small subset of these, molecular interactions with their target biomolecule have been defined. The database that has accumulated is sufficient to predict, on the basis of sequence homologies alone, the probable general physiological mechanisms that underlie the biological activity of many additional *Conus* peptides. Thus, the extensive molecular analysis of the genes encoding venom peptides of *Conus* has provided a database of over 1000 genes from close to 100 different *Conus* species, a large enough sequence dataset to provide a window into general patterns of interspecific divergence of the gene families represented.

Recently, molecular data have provided a preliminary picture of the phylogenetic relationships among the major *Conus* species discussed in this review. This work reveals several distinct clades of cone snails, with reasonable agreement between datasets from different labs as to which species belong to a clade (Duda et al. 2001, Duda & Palumbi 1999, Espiritu et al. 2001, Monje et al. 1999). In this review I refer to putative clades using the nomenclature of Espiritu et al. (2001).

Conus Species Clades

Alan Kohn concluded on the basis of the fossil record that there were two radiations of *Conus*: An initial radiation in the Eocene, followed by a decrease in diversity at the end of the Eocene and in the Oligocene, and a larger radiation starting at the beginning of the Miocene. The molecular data indicate that the great majority of tropical, shallow-water *Conus* species are derived from a single lineage that survived an apparent end-of-the-Eocene/Oligocene extinction. A small minority of the *Conus* species analyzed belong to other early lineages, which presumably diverged in the first Eocene radiation from the main lineage that has generated most living, shallow-water *Conus* species. Of the 77 species included in one study (Espiritu et al. 2001), over 70 are presumably descended from the major lineage that had the explosive Miocene radiation; the other "early-diverging lineages" are very poorly represented in the *Conus* species analyzed so far.

The molecular data define 17 clades of species that originated from the second, post-Eocene radiation of *Conus* (Espiritu et al. 2001). Four of these clades (I to IV) comprise fish-hunting species (see Figure 2, color insert) and two (V, VI) comprise snail-hunting species. The most species-rich fish-hunting (Clade I) and snail-hunting (Clade V) clades are illustrated in Figures 3 and 4 (see color insert). There is a larger number of worm-hunting (vermivorous) clades (VII–XVII; see Figure 5, color insert). Some of the worm-hunting clades are highly specialized, such as clade XVII, in which all species apparently devour amphinomids ("fireworms"), a distinctive group of errant polychaete annelids defended by sharp spicules that would seem to make them singularly unattractive prey (see Kohn et al. 2001). Examples of species in a given clade and in different clades are shown in Figures 2 to 5 (see color insert), and summarized in Table 2.

The discussion that follows focuses on fish-hunting *Conus* that belong to Clade I (such as *C. magus* and *C. striatus*), Clade II (such as *C. geographus* and *C. obscurus*) and Clade III (such as *C. purpurascens* and *C. ermineus*). Clade III is geographically isolated from the others and diverged from them in the Miocene or earlier (Espiritu et al. 2001). Extensive research has been done on venom peptides in species belonging to each clade.

CONUS PEPTIDES, BEHAVIOR AND ECOLOGY

Increasingly, the conopeptides in a venom can be correlated with ecological/behavioral parameters. In some cases observed biological differences between *Conus* species can be used to rationalize differences between conopeptides expressed in their venom ducts. Functional divergence between venoms was anticipated by earlier work (Endean et al. 1974, Kohn et al. 1960) that demonstrated that the effects of injecting crude venom could be correlated with the major prey of that particular *Conus* species. Not surprisingly, the venom from fish-hunting cone snails was found to be more potent than the venom from worm-hunting species when tested on fish or on other vertebrates. The molecular information that has been subsequently collected generally corroborates the correlation with prey type. Presumably, individual conotoxins are under selective pressure to evolve high affinity for the physiologically-relevant target, and affinity falls significantly when apparently homologous targets in other animal phyla are tested.

TABLE 2   *Conus* clades

Clades	*Conus* species example	Prey
	Group A Clades*	
Fish-hunting		
I	*striatus*	Fish
II	*geographus*	Fish
III	*purpurascens*	Fish
IV	*radiatus*	Fish
Mollusc-hunting		
V	*textile*	Gastropods
VI	*marmoreus*	Gastropods
Worm-hunting		
VII	*lividus*	Hemichordates; polychaetes
VIII	*glans*	Errant polychaetes (Eunicidae)
IX	*planorbis*	Errant polychaetes (Eunicidae)
X	*betulinus*	Sedentary polychaetes (Capitellidae (?))
XI	*ebraeus*	Errant polychaetes (Eunicidae)
XII	*vexillum*	Errant polychaetes (Eunicidae)
XIII	*virgo*	Sedentary polychaetes (Terebellidae)
XIV	*arenatus*	Sedentary polychaetes (Capitellidae)
XV	*sponsalis*	Errant polychaetes (Nereidae)
XVI	*tessulatus*	Errant polychaetes
XVII	*imperialis*	Errant polychaetes (Amphinomidae)

*Early-diverging groups (see Figure 5 and text):
   A Group—(see clades above).
   B Group—*C. distans*.
   C Group—*C. memiae*.
   D Group—*C. californicus*.
Clade classification from Espiritu et al. (2001); prey assignments from Duda Jr. et al. (2001). Examples of different species in Clades I, V, and XII are shown in Figures 3, 4, and 5 respectively.

A molecular correlation with ecological/behavioral factors is best documented at present for certain fish-hunting cone snail species. A particularly illuminating example are the differences uncovered between *C. geographus* (a Clade II species, see Figure 2, color insert) and *C. striatus* (a Clade I species, see Figures 2 and 3, color insert) (Olivera 1997). These two Indo-Pacific cones are large, widely distributed piscivorous species, but with divergent behavior and ecology.

C. striatus buries itself in the sand and ambushes its fish prey by extending its proboscis which acts like a harpoon line; the prey is harpooned as the barbed radular tooth (see Figure 1) is thrust out of the proboscis. Typically, *C. striatus* will tightly grasp the harpoon with circular musculature at the end of the proboscis and jerk the proboscis back, thereby tethering the fish through the barbed radular tooth. As the fish is harpooned and tethered, it is injected with venom through the hollow radular tooth, which acts as a hypodermic needle. Venom injection results in an almost instantaneous tetanic paralysis; in a good strike, the fish jerks violently, after which it is immobilized with fins extended, in a characteristic tetanic state. *C. striatus* has been routinely observed in an aquarium to capture fish while remaining largely buried, with only its proboscis sticking above the sandy substrate. Thus, a characteristic feature of *C. striatus* behavior is that in the presence of a fish, it will immediately extend its proboscis, which can survey a wide region around the snail (see Figure 3, color insert, and Figure 6). However, this species seeks out fish aggressively at night, when fish are relatively inactive. In the field, *C. striatus* has been observed to approach fish with its long proboscis fully extended, which

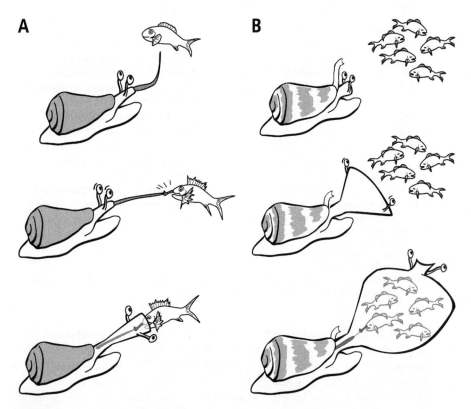

A   **B**

Figure 6 A cartoon representing two different strategies for catching fish: *C. striatus*, a Clade I species (*A*) and *C. geographus*, a Clade II species (*B*). *C. striatus* extends its proboscis, harpoons the fish and through a "lightning-strike cabal" of toxins causes an almost instantaneous immobilization of prey. *C. geographus* probably goes after schools of fish hiding in reefs at night using a net strategy; once it has engulfed the school with its highly distensible rostrum, it uses a "nirvana cabal of toxins" to make the fish quiescent for stinging them and causing an irreversible neuromuscular paralysis. These species are shown live in Figure 2 (color insert). Panel A could also represent the fish-hunting strategy of Clade III species such as *Conus purpurascens* (see Terlau et al. 1996).

can be over three times the length of the shell, ready to harpoon its prey (M. Silva, personal communication).

The contrast between *C. geographus* and *C. striatus* behavior is very striking. *C. geographus* engulfs fish with its highly distensible rostrum (sometimes called the false mouth), which looks like a large net. It will characteristically engulf the fish before any stinging occurs (the fact that the fish are stung within the rostrum of the snail can be verified because a radular tooth is always found when the snail regurgitates the scales and bones of the fish). Furthermore, *C. geographus*, when presented with many fish will try to capture more than one. It is likely, given how active this snail is in aquaria, that *C. geographus* stalks schools of smaller fish and attempts to engulf several individuals of the school; the snail can then pick them off one by one with its proboscis. *C. geographus* never buries itself when in an aquarium, but generally hides in crevices of rocks or coral rubble if these are provided. The contrast between *C. striatus* and *C. geographus* is shown in the cartoon in Figure 6.

These differences in the biology of *C. striatus* and *C. geographus* are reflected in differences in venom peptides. We refer to groups of toxins that appear to act together to cause a particular physiological response as toxin "cabal" (Olivera & Cruz 2001); there are very different toxin cabals in these two species. In *C. striatus*, a cabal of conotoxins causes a massive depolarization of axons in the neighborhood of the injection site, eliciting the almost-immediate tetanic paralysis that can cause even a large fish to be totally immobilized in less than a second. We call the groups of toxins that cause this the "lightning-strike cabal." The molecular mechanisms that underlie the activity of individual toxins have also been elucidated; one major component is the δ-conotoxins that keep sodium channels open, while other components block various K channels (Bulaj et al. 2001, Walker et al. 1999; J. Imperial and H. Terlau, unpublished). The overall effect of the lightning-strike cabal is analogous to electrocuting the prey (as an electric eel does in the Amazon).

In contrast, the major components that comprise the lightning-strike cabal appear absent in *C. geographus*. Instead, many conopeptides in this venom elicit hypoactivity in sensory neuronal circuitry. When a fish is placed into the rostrum of a *C. geographus*, it seems sedated, and we have referred to the group of peptides responsible for this state as the "nirvana cabal." These do not depolarize axons to cause increased electrical activity, but instead appear to suppress electrical activity in the targeted circuitry. The nirvana cabal is a complex mixture of peptides, each of which targets a specific receptor, most of which are located in the peripheral neuronal circuitry. Thus, the different physiological targets of the venom peptides in *C. striatus* and *C. geographus* can be rationalized on the basis of their different behavioral strategies for catching potential fish prey.

CONOPEPTIDE FAMILIES: CLADE PATTERNS

When conopeptide families in different *Conus* clades are compared, several contrasting patterns are observed. In some cases, a conopeptide family does not exhibit significant divergence between clades. However, the more common pattern is that some clade-specific differences can be discerned; in the most extreme cases, a *Conus* peptide family may be clade-specific, i.e., found in all species of a given clade, but in no other *Conus* species.

Some differences between species clades in patterns of *Conus* peptide families expressed can be rationalized by differences in behavior and other biological parameters of members between different species clades, as in the example given in the preceding section (*C. striatus* vs. *C. geographus*). However, even when such behavioral/biological differences are not apparent, divergence in the venom peptides of different clades is observed; this may be the consequence of the evolutionary history of each clade.

The specific example we will discuss here are conopeptide families in two groups of fish-hunting *Conus* clades: The *C. striatus* group (Clade I), the major clade of piscivorous Indo-Pacific species and the *C. purpurascens* clade (Clade III) found in the Atlantic and Eastern Pacific. The two clades probably diverged in the Miocene or earlier (Espiritu et al. 2001). In both cases, the snails characteristically extend their proboscis in the presence of a fish, and cause a rapid tetanic immobilization through a lightning strike cabal of conopeptides, followed by an irreversible neuromuscular block, using a "motor cabal" of conopeptides that act on various ion channels essential for signaling between motor axons and skeletal muscle (Terlau et al. 1996, Olivera 1997).

In both the lightning-strike cabal and the motor cabal, there are conopeptide families that are essentially conserved between Clade I and Clade III *Conus* species. Examples are the μ-conotoxins, which block voltage-gated sodium channels and are an important component of the motor cabal. A second conserved family are the δ-conotoxins, which block inactivation of voltage-gated sodium channels, and are a key component of the lightning-strike cabal of conopeptides (see Table 3). Both of these conopeptide families are found in all species of Clade I and Clade III *Conus* that have been examined. The mature conotoxin sequences are very similar in both clades (see (Bulaj et al. 2001)).

A striking contrast is seen with one component of the lightning-strike cabal in Clade I species: The κA-conotoxins (Craig et al. 1998). These peptides, which have been found in every Clade I species examined (M. Marsh, J. Garrett, G. Bulaj, B. Olivera, unpublished results) appear to be completely absent from Clade III species. Indeed, so far there has been a one-to-one correlation between assignment of a species to Clade I and the presence of κA-conotoxins (no other *Conus* species except those that belong to Clade I have κA-conotoxins in the large set that has been examined so far). Thus, this family of conopeptides appears to be a highly specialized adaptation of Clade I *Conus* species.

The third pattern observed is divergence between clades in conopeptide families, leading to characteristic sequence features found in the conopeptide family in the venom of species in that clade. This is best typified by the α-conotoxin family of venom peptides, which are widespread throughout the genus. All α-conotoxins are nicotinic antagonists (for a review, see McIntosh et al. 1999) which belong to the A-conopeptide gene superfamily. In most non-fish-hunting *Conus* species, including most of the mollusc-hunting and worm-hunting species examined to

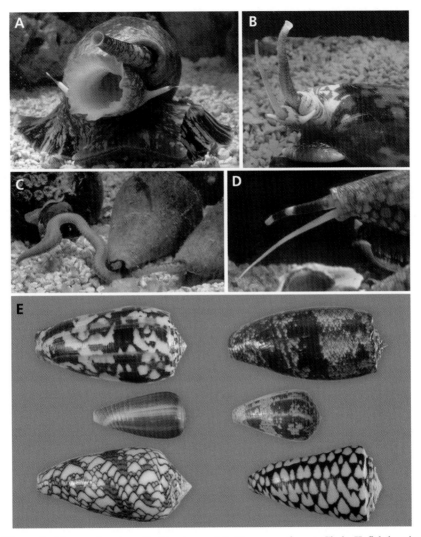

Figure 2 *Conus* of various feeding types. (*A*) *C. geographus*, a Clade II fish-hunting species, foraging. The rostrum the snail uses as a net is partially open; the eye stalks on both sides of the rostrum have the eyes attached. The siphon, which the snail uses to detect prey is above and to the right of the partially open rostrum. (*B*) *C. striatus*, a Clade I fish-hunting species, with its yellow proboscis extended, ready to strike. (*C*) *C. quercinus*, a Clade VII vermivorous *Conus*, attacking a polychaete worm. (*D*) *C. marmoreus*, a Clade VI molluscivorous species, with its striped siphon and white translucent proboscis extended, about to sting a prey snail. (*E*) Shells of representative species of fish-hunting and mollusc-hunting clades. Top two rows, fish-hunting species—from top left clockwise: *C. striatus* (Clade I); *C. geographus* (Clade II); *C. purpurascens* (Clade III); *C. radiatus* (Clade IV). Bottom row, mollusc-hunting species—left, *C. textile* (Clade V) and right, *C. marmoreus* (Clade VI).

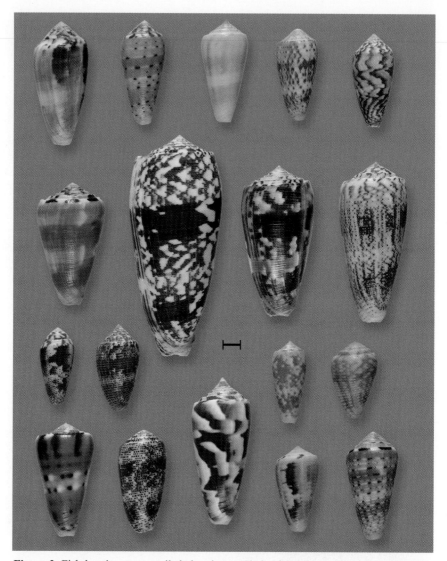

Figure 3 Fish-hunting cone snails belonging to Clade I (the *C. striatus* clade). The plate shows species related to the large, fish-hunting *Conus* species, *C. striatus*, the striated cone (two specimens in the center). Top row, left to right: *C. fulmen*, Japan; *C. circumcisus*, Cebu, Philippines; *C. consors*, Manila Bay, Philippines; *C. floccatus*, Marshall Islands; *C. monachus*, Marinduque Island, Philippines. Second row, left to right: *C. gauguini*, Nuku Hiva, Marquesas; *C. striatus*, Bohol Island, Philippines; *C. striatus*, Reunion Islands; *C. magus*, Cuyo Island, Philippines. Third row, left to right: *C. striolatus*, Marinduque Island, Philippines; *C. achatinus*, Marinduque Island, Philippines; *C. timorensis*, Indonesia; *C. catus*, Sulu Sea, Philippines. Bottom row, left to right: *C. barthelemyi*, Mauritius; *C. stercusmuscarum*, Marinduque, Philippines; *C. gubernator*, Zanzibar, East Africa; *C. magus*, Palawan Island, Philippines; *C. aurisiacus*, Manado, Indonesia.

Figure 4 Shell of *Conus* species belonging to Clade V, a mollusc-hunting clade. Top row, left to right: *C. episcopatus*, Philippines; *C. natalis*, South Africa; *C. omaria*, Philippines; *C. pennaceus*, Madagascar; *C. aureus*; Philippines. Second row, left to right: *C. dalli*, Galapagos Islands; *C. textile* (two specimens), Philippines; *C. gloriamaris*, Philippines. Third row, left to right: *C. auricomus*, Philippines; *C. victoriae*, Australia; *C. victoriae* (albino), Australia; *C. victoriae* (dark variety), Australia. Bottom row, left to right: *C. bengalensis*, Andaman Sea; *C. ammiralis*, Philippines; *C. telatus*, Philippines; *C. victoriae* (light variety), Australia; *Conus retifer*, Philippines, *C. aulicus*, Philippines.

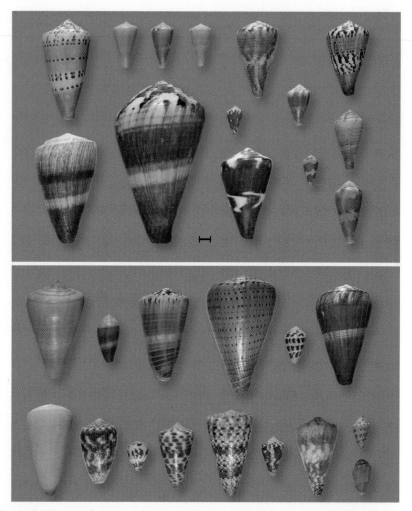

Figure 5 (*Top panel*) A worm-hunting clade (the *C. vexillum* clade, Clade XII). Top row, left to right: *C. mustelinus*, Marinduque Island, Philippines; next three specimens are *C. suzanka* (various color forms), Panglao Island, Philippines; *C. namocanus*, Red Sea; *Conus rattus*, East Africa (slightly lowered). *C. capitaneus*, Cebu, Philippines. Bottom row, left to right: *C. miles*, Sulu Sea, Philippines; *C. vexillum*, Sulu Sea, Philippines; *C. capitanellus*, Panglao, Bohol, Philippines (small specimen); *C. vexillum*, Sulu Sea, Philippines; three specimens on the lower right are *C. pertusus* (pinkish-red specimens), Olango, Philippines. (*Bottom panel*) Representatives of worm-hunting clades and early-diverging groups. Top row, from left: *C. quercinus* (Clade VII); *C. glans* (Clade VIII); *C. planorbis* (Clade IX); *C. betulinus* (Clade X); *C. ebraeus* (Clade XI); *C. vexillum* (Clade XII). Bottom row from left: *C. virgo* (Clade XIII); *C. arenatus* (Clade XIV); *C. sponsalis* (Clade XV); *C. tessulatus* (Clade XVI); *C. imperialis* (Clade XVII); *C. distans* (early-diverging group B); top right, *C. memiae* (early-diverging group C); bottom right, *C. californicus* (early-diverging group D). All specimens were collected from the Philippines except *C. californicus* which is from California, U.S.A. See Table 2 for worm prey of each clade.

date, the characteristic $\alpha-$conotoxin belongs to the so-called $\alpha4/7$ subfamily, structurally represented as ($-CCX_4CX_7C-$). In all Clade I species, the major α-conotoxin expressed in the venom (which targets the muscular nicotinic receptor, and is therefore part of the motor cabal) does not belong to the $\alpha4/7$ subfamily, but to the $\alpha3/5$ subfamily, i.e., ($-CCX_3CX_5C-$).

In Clade III *Conus* species there also has been a divergence from the $\alpha4/7$ subfamily in conopeptides targeting the neuromuscular junction. The major venom peptides that target muscle nicotinic receptors are the αA-conotoxins—a distinctive branch of the A-superfamily—which as seen in Table 3, have a structural topology distinct from either the $\alpha4/7$ or $\alpha3/5$ subfamilies. Thus, both clades of fish-hunting *Conus* species evolved specialized A-superfamily peptides to target the muscle nicotinic receptor subtype, but the $\alpha3/5$ conotoxins and the αA-conotoxins are as divergent from each other as from A-superfamily conopeptides in other *Conus* clades.

Thus, the types of results shown in Table 3 illustrate that some peptide families such as the αA-conotoxins will be useful taxonomic markers. αA-conotoxins have only been found in Clade III *Conus* species and should be a useful marker for this clade. In contrast, the mature δ-conotoxins sequences are not likely to be very useful for resolving any *Conus* clades that diverged from each other during the Miocene or later.

As the sequence database expands, it becomes clearer which conotoxins are diagnostic for a particular *Conus* clade. Thus, the venoms of Clade III species, in

TABLE 3   A comparison of functionally analogous conopeptides from two clades of piscivorous *Conus*

A. Major peptides targeted to nicotinic receptors
Clade I fish-hunting species: $\alpha3/5$ conotoxin subfamily

Conus striatus	α-SI	ICCNPACGPKYSC*
	α-SIA	YCCHPACGKNFDC*
Conus magus	α-MI	GRCCHPACGKNYSC*

Clade III fish-hunting species: $\{\alpha\}$A-conotoxins

Conus purpurascens	αA-PIVA	GCCGSYONAACHOCSCKDROSYCGQ*
Conus ermineus	αA-EIVA	GCCGPYONAACHOCGCKVGROOYCDROSGG*

Mollusc-hunting species: $\alpha4/7$ conotoxin subfamily

Conus pennaceus	α-PnIA	GCCSLPPCAANNPDYC*

B. Major δ-conotoxins that delay Na channel inactivation
Clade I fish-hunting species

Conus striatus	δ-SVIE	DGCSSGGTFCGIHOGLCCSEFCFLWCITFID
Conus magus	δ-MVIA	DGCYNAGTFCGIROGLCCSEFCFLWCITFVDS*

Clade III fish-hunting species

Conus purpurascens	δ-PVIA	EACYAOGTFCGIKOGLCCSEFCLPGVCFG*
Conus ermineus	δ-EVIA	EACYPOGTFCGIKOGLCCSELCLPAVCVG*

Mollusc-hunting species

Conus textile	δ-TxVIA	WCKQSGEMCNLLDQNCCDGYCIVLVCT

*, C-terminal amidation; O, 4-hydroxyproline.

addition to having αA-conotoxins that are competitive nicotinic antagonists, are characterized by the ψ-conotoxins, which are noncompetitive nicotinic receptor antagonists, and the κ-conotoxins, which are K channel inhibitors. None of these conopeptides have been found in *Conus* species belonging to any other clade. For Clade I, the κA-conotoxins (Craig et al. 1998), which are believed to inhibit K channels (although the precise molecular targets are not yet defined), are also diagnostic for species in this clade. Some families such as the δ-conotoxins and the ω-conotoxins are found in several different clades. This may be useful in assessing relationships between clades in which the rooting is ambiguous; thus, preliminary data suggest that Clades I, II, and IV comprise fish-hunting species with ω-conotoxins that target N-type calcium channels, whereas Clade III venoms lack this group.

The three parts of a toxin gene (N-terminal signal sequence, middle pro region and C-terminal mature toxin region) can be used to assess both interspecific and intraspecific relationships between genes to provide a historical perspective. Signal sequences change very slowly, propeptide regions change at an intermediate rate, and the mature toxin region is the focal target of hypermutation. The δ-conotoxins in Clade I and Clade III *Conus* species were recently analyzed (see Bulaj et al. 2001, Espiritu et al. 2001). The most detailed historical reconstruction possible was that for the four genes present in one species, *Conus magus*.

A comparison of four δ-conotoxin precursor sequences in *C. magus* revealed an ancient duplication, probably even before the species itself had emerged, as well as much more recent duplications. A recombination event was identified between two of the genes, with a cross-over point in the mature toxin region; this generated a third peptide of the δ-conotoxin family in this species. Finally, there was the type of hypermutation that occurs when species diverge that was also observed for the fourth δ-conotoxin found within the species. Thus, the analysis of δ-conotoxin precursors could provide historical information regarding the origins of these genes within a single species.

As more genes are sequenced from different *Conus* species, the history of a class of peptides within a clade of *Conus*, and possibly even between groups of related clades, can be similarly reconstructed. A parallel analysis of the evolution of a gene family within a species, as well as orthologous genes that have segregated into different species, can be carried out. These types of data should begin to allow us to evaluate origins of new genes, as well as the interspecific and intraspecific factors that affect divergence after a novel gene has emerged.

THE COMPLEXITY OF *CONUS* VENOM PEPTIDES: A RATIONALE

One surprising feature of *Conus* venoms is their complexity (Olivera et al. 1990): each species has a large repertoire of peptides that can be expressed. The complex behavioral repertoire of different *Conus* species, coupled with the complexity of their environment may be the primary factors that drive the complexity of cone snail venoms.

One strategy for setting up hypotheses regarding the function of diverse venom components is to summarize the likely general utility of various venom components. Most obvious of all is the use of venom to capture prey. There is also excellent (if somewhat indirect) evidence that venom is used by many, if not all, species of *Conus* to defend themselves against potential predators. Finally, there are preliminary indications that at least certain species use their venom for competitive interactions. Although detailed data are not available for the latter two cases, these venom uses could well contribute most of the complexity observed in cone snail venoms.

Human stinging cases are almost certainly defensive in nature. It is easy to demonstrate the defensive behavior of *C. geographus*; when this species is feeding, it opens it rostrum wide to capture an individual fish or a school of fish. However, when the species is disturbed, it sticks out its proboscis (and there is considerable variation in how readily a particular individual will react to irritating stimuli by sticking out its proboscis; some individuals seem oblivious to the presence of other *Conus* in the aquarium, and others when approached by a member of another *Conus* species will stick out the proboscis as if in a warning display). Finally, if individuals of *C. geographus* are starved and placed in a large, deep aquarium with fish that do not hide in rock crevices but are constantly swimming, the fish are consumed one at a time. In one instance, we observed an individual *C. geographus* approaching such a fish, but the fish had a very peculiar behavior: Although it was clearly not paralyzed, it seemed unable to swim away. The snail, which had approached the fish from a tube in the aquarium, had its mouth extended, shaped like the barrel of a gun pointing towards the fish, a pose that it kept up continuously. The fish remained in an almost hypnotic state until the snail was able to maneuver close enough to engulf it with its rostrum.

The behavioral poses of *C. geographus* are illustrated in Figure 2 (color insert) and Figure 7, with the "sieving" net pose with its rostrum fully open, the "barrel-of-a-gun" pose with its rostrum in a very cylindrical shape, and finally the defensive pose with its proboscis out and its rostrum withdrawn. All of these behaviors have implications for corresponding venom components. As suggested above, when *C. geographus* is net hunting there is a need for both nirvana cabal and motor cabal toxins. When the snail is aiming its rostrum like a barrel of a gun, are there *Conus* venom components that induce the "hypnotic" effects? When the snail stings various predators/competitors defensively or competitively, what *Conus* venom components would be most efficacious?

Although we have posed these questions rhetorically, the general answer is that the various individual *Conus* venom peptides must be linked to these behaviors. Thus, there may be specialized defensive peptides against common predators of *C. geographus* and peptides that permit the snail to capture fish at a distance when it assumes the barrel-of-a-gun pose. Although we cannot specifically identify the *Conus* venom peptides used in conjunction with each of these behaviors, the

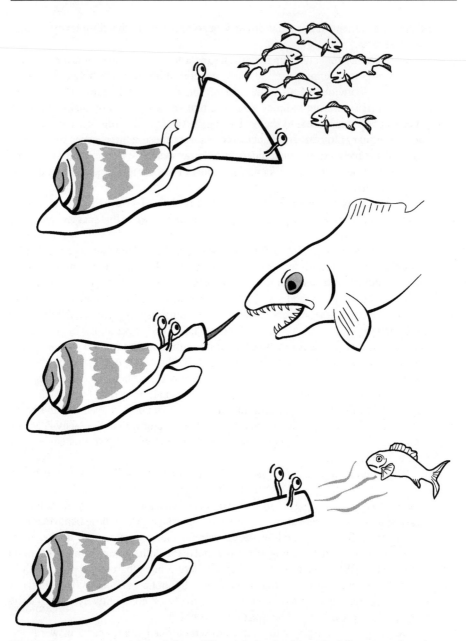

Figure 7   A cartoon version of three different behavioral poses of *C. geographus*. At top, the snail is attempting to net a school of fish. In the middle panel, the snail has its proboscis extended, ready to strike a potential predator. At the bottom, the snail has its rostrum in the "barrel-of-a-gun" mode, during which it apparently aims venom components that affect fish behavior, such that potential prey, though not paralyzed, does not swim away. Different venom peptides are presumably associated with each behavioral mode.

different behaviors imply separate venom components. The general perception that emerges is that cone snail peptides mediate complex biotic interactions of cone snails with their prey, with their predators, and with their competitors.

Thus, the underlying general reason for accelerated evolution of *Conus* venom peptides is that *Conus* peptide genes are a molecular readout of the biotic interactions of each *Conus* species. Because different species have divergent biotic interactions, hypermutation resulting in the divergence of *Conus* venom peptides is a molecular mirror of the underlying biology; these are the genes exquisitely tuned to the interactions of an individual *Conus* species with the other organisms in its environment, and these are the genes that must change when environmental changes, and the attendant changes in biotic interactions, occur to such an extent that new species evolve to exploit the new ecological opportunities.

APPENDIX

Posttranslational Modification

One defining characteristic of *Conus* venom peptides is the high frequency of posttranslational modification found (see Table 4).

Sufficient data has been collected recently to suggest a paradigm for how cone snails effectively apply the principles of medicinal chemistry to their venom peptides. One intensively studied posttranslational modification is the conversion of glutamate to $\{\gamma\}$-carboxyglutamate (Bandyopadhyay et al. 1998, 2002). It appears that the snails have hijacked an already existing enzyme, presumably used for some other purpose, $\{\gamma\}$-glutamyl carboxylase. Closely related homologs of the enzyme are present in vertebrates and in animals in other phyla such as insects and tunicates, suggesting an ancient lineage.

Studies of this enzyme indicate that binding to a potential substrate is mediated through a recognition signal sequence in the propeptide region—a region that gets clipped off as the mature toxin is generated. Thus, cone snail peptide genes have the potential to evolve signals in the pro region of the precursor that bind posttranslational modification enzymes; these can then act on a particular standard amino acid in the mature toxin region to convert it into an unusual amino acid with nonstandard functional groups (see Hooper et al. 2000). The presence of docking signals for proteins and enzymes important for *Conus* peptide precursor maturation in the intervening pro region between the signal sequence and the mature toxin region at the C-terminus may be a general theme, for which only very preliminary evidence has been obtained.

The most unexpected discovery made in the characterization of the *Conus* $\{\gamma\}$-carboxylase (Bandyopadhyay et al. 2002) was that most of the introns in this gene were clearly present before the Cambrian explosion; most introns are found in precisely the same loci in the cone snail and human γ-glutamyl carboxylase genes. In contrast, when the *Drosophila* enzyme was similarly analyzed, most introns in the *Drosophila* $\{\gamma\}$-carboxylase gene were absent, but the three that remain are in

TABLE 4 Modifications present in conopeptides

Modification	Example peptide	Sequence	Enzyme	Ref.
Disulfide bridge formation	GI	ECCNPACGRHYSC*	Disulfide isomerase	Gray et al. 1981
Hydroxylation of proline	GIIIA	RDCCTOOKKCKDRQCKOQRCCA*	Proline hydroxylase	Cruz et al. 1985
Amidation of C-terminus	MI	GRCCHPACGKNYSC*	Protein amidating monooxygenase	McIntosh et al. 1982
Carboxylation of glutamic acid	Conantokin-G	GE{γγ}LQ{γ}NQ{γ}LIR{γ}KSN*	{γ}-Glutamate carboxylase	McIntosh et al. 1984
Bromination of tryptophan	Bromocontryphan	GCOW̲EPW⁺C*	Bromoperoxidase	Jimenez et al. 1997
Isomerization of tryptophan	Contryphan	GCOW̲EPWC*	Tryptophan epimerase	Jimenez et al. 1996
Cyclization of N-terminal Gln	Bromoheptapeptide	ZCGQAW⁺C*	Glutaminyl cyclase	Craig et al. 1997
Sulfation of tyrosine	EpI	GCCSDPRCNMNNPY‡C*	Tyrosyl sulfotransferase	Loughnan et al. 1998
O-Glycosylation	SIVA	ZKSLVPS§VITTCCGYDOGTMCOOCRCTNSC*	Polypeptide HexNAc transferase	Craig et al. 1998

{γ}, {γ}-Carboxyglutamate (Gla); W⁺, bromotryptophan; W̲, D-tryptophan; S§, O-glycosylated serine; Y‡, sulfated tyrosine; Z, pyroglutamate; O, hydroxyproline; *, amidation.

the same loci as are their corresponding *Conus* and human counterparts. Thus, the picture that emerges is that the posttranslational modification enzyme to convert Glu to Gla is not only very ancient but is also highly conserved across animal phyla. Even more surprisingly, what was thought to be junk DNA in introns is retained exactly where the introns were in the very early stages of evolution.

The size distributions of the introns in the three genes do not overlap: The *Conus* introns analyzed are the largest (>1 kb), whereas the corresponding introns in human DNA are significantly smaller (100–400 nucleotides). In *Drosophila* the remaining introns are extremely small (<100 amino acids), and at most positions no introns are found at all. The picture that emerges is that an ancient ancestral enzyme was already present and had presumably important functions more than 500 million years ago, and that this gene had >14 different introns. The precise conservation of intron loci when mammalian and *Conus* genes are compared definitively establishes their common ancestry. The loss of most of these introns in *Drosophila* provides some taxonomic opportunities, because once an intron is lost it is unlikely to ever be regained at the same locus. Thus, the pattern of loss of introns may be used as a taxonomic tool, providing data that are in effect a set of binary markers that serve as a readout of evolutionary history of broad taxonomic groups.

The *Annual Review of Ecology and Systematics* is online at
http://ecolsys.annualreviews.org

LITERATURE CITED

Bandyopadhyay PK, Colledge CJ, Walker CS, Zhou L-M, Hillyard DR, Olivera BM. 1998. Conantokin-G precursor and its role in γ-carboxylation by a vitamin K-dependent carboxylase from a *Conus* snail. *J. Biol. Chem.* 273:5447–50

Bandyopadhyay PK, Garrett JE, Shetty RP, Keate T, Walker CS, Olivera BM. 2002. γ-Glutamyl carboxylation: an extracellular post-translational modification that antedates the divergence of molluscs, arthropods and chordates. *Proc. Natl. Acad. Sci. USA.* 99:1264–69

Bulaj G, DeLaCruz R, Azimi-Zonooz A, West P, Watkins M, et al. 2001. δ-Conotoxin structure/function through a cladistic analysis. *Biochemistry* 40:13201–208

Craig AG, Jimenez EC, Dykert J, Nielsen DB, Gulyas J, et al. 1997. A novel posttranslational modification involving bromination of tryptophan: identification of the residue, L-6-bromotryptophan, in peptides from *Conus imperialis* and *Conus radiatus* venom. *J. Biol. Chem.* 272:4689–98

Craig AG, Zafaralla G, Cruz LJ, Santos AD, Hillyard DR, et al. 1998. An O-glycosylated neuroexcitatory *Conus* peptide. *Biochemistry* 37:16019–25

Cruz LJ, Gray WR, Olivera BM, Zeikus RD, Kerr L, et al. 1985. *Conus geographus* toxins that discriminate between neuronal and muscle sodium channels. *J. Biol. Chem.* 260: 9280–88

Duda TF Jr, Kohn AJ, Palumbi SR. 2001. Origins of diverse feeding ecologies within *Conus*, a genus of venomous marine gastropods. *Biol. J. Linn. Soc.* 73:391–409

Duda TF Jr, Palumbi SR. 1999. Molecular genetics of ecological diversification: duplication and rapid evolution of toxin genes of the venomous gastropod *Conus*. *Proc. Natl. Acad. Sci. USA* 96:6820–23

Endean R, Gyr P, Surridge J. 1977. The pharmacological actions on gunea-pig ileum of crude venoms from the marine gastropods *Conus striatus* and *Conus magus*. *Toxicon* 15:327–37

Endean R, Gyr P, Surridge J. 1979. The effects of crude venom of *Conus magus* and *Conus striatus* on the contractile response and electrical activity of guinea-pig cardiac musculature. *Toxicon* 17:381–95

Endean R, Parrish G, Gyr P. 1974. Pharmacology of the venom of *Conus geographus*. *Toxicon* 12:131–38

Endean R, Williams H, Gyr P, Surridge J. 1976. Some effects on muscle and nerve of crude venom from the gastropod *Conus striatus*. *Toxicon* 14:267–74

Espiritu DJD, Watkins M, Dia-Monje V, Cartier GE, Cruz LJ, Olivera BM. 2001. Venomous cone snails: molecular phylogeny and the generation of toxin diversity. *Toxicon* 39:1899–916

Espiritu JD, Cruz LJ, Cartier GE, Olivera BM. 2002. Venomous gastropods: *Conus*, conoideans and other neogastropod families. *Boll. Malacol.* In press

Gray WR, Luque A, Olivera BM, Barrett J, Cruz LJ. 1981. Peptide toxins from *Conus geographus* venom. *J. Biol. Chem.* 256:4734–40

Harasewych MG, Adamkewicz SL, Blake JA, Saudek D, Spriggs T, Bult CJ. 1997. Neogastropod phylogeny: a molecular perspective. *J. Molluscan Stud.* 63:327–51

Hooper D, Lirazan MB, Schoenfeld R, Cook B, Cruz LJ, et al. 2000. Post-translational modification: a two-dimensional strategy for molecular diversity of *Conus* peptides. In *Peptides for the New Millennium: Proceedings of the Sixteenth American Peptide Symposium*, ed. GB Fields, JP Tam, G Barany, pp. 727–29. Dordrecht, The Netherlands: Kluwer Acad.

Jimenez EC, Craig AG, Watkins M, Hillyard DR, Gray WR, et al. 1997. Bromocontryphan: post-translational bromination of tryptophan. *Biochemistry* 36:989–94

Jimenez EC, Olivera BM, Gray WR, Cruz LJ. 1996. Contryphan is a D-tryptophan-containing *Conus* peptide. *J. Biol. Chem.* 281:28002–5

Kohn AJ. 1956. Piscivorous gastropods of the genus *Conus*. *Proc. Natl. Acad. Sci. USA* 42:168–71

Kohn AJ. 1959. The Ecology of Conus in Hawaii. *Ecol. Monogr.* 29:47–90

Kohn AJ. 1960. Ecological notes on *Conus* (Mollusca, Gastropoda) in the Trincomalee region of Ceylon. *Ann. Mag. Nat. Hist.* 13:309–20

Kohn AJ. 1966. Food specialization in *Conus* in Hawaii and California. *Ecology* 47:1041–43

Kohn AJ. 1985. Evolutionary ecology of *Conus* on Indo-Pacific coral reefs. *Proc. 5th Int. Coral Reef Congr.* 4:139–44

Kohn AJ. 1990. Tempo and mode of evolution in Conidae. *Malacologia* 32:55–67

Kohn AJ, Nybakken JW. 1975. Ecology of *Conus* on eastern Indian Ocean fringing reefs: diversity of species and resource utilization. *Mar. Biol.* 29:211–34

Kohn AJ, Omori M, Yamakawa H, Koike Y. 2001. Maximal species richness in *Conus*: diversity, diet and habitat on reefs of northeast Papua New Guinea. *Coral Reefs* 20:25–38

Kohn AJ, Saunders PR, Wiener S. 1960. Preliminary studies on the venom of the marine snail *Conus*. *Ann. NY Acad. Sci.* 90:706–25

Loughnan M, Bond T, Atkins A, Cuevas J, Adams DJ, et al. 1998. {α}-Conotoxin EpI, a novel sulfated peptide from *Conus episcopatus* that selectively targets neuronal nicotinic acetylcholine receptors. *J. Biol. Chem.* 273:15667–74

McIntosh JM, Cruz LJ, Hunkapiller MW, Gray WR, Olivera BM. 1982. Isolation and structure of a peptide toxin from the marine snail *Conus magus*. *Arch. Biochem. Biophys.* 218:329–34

McIntosh JM, Olivera BM, Cruz LJ, Gray WR. 1984. γ-Carboxyglutamate in a neuroactive toxin. *J. Biol. Chem.* 259:14343–46

McIntosh JM, Santos AD, Olivera BM. 1999. *Conus* peptides targeted to specific nicotinic acetylcholine receptor subtypes. *Annu. Rev. Biochem.* 68:59–88

Monje VD, Ward R, Olivera BM, Cruz LJ. 1999. 16S mitochondrial ribosomal RNA gene sequences: a comparison of seven *Conus* species. *Philos. J. Sci.* 128:225–37

Olivera BM. 1997. EE. Just Lecture. 1996. *Conus* venom peptides, receptor and ion channel targets and drug design: 50 million years of neuropharmacology. *Mol. Biol. Cell* 8:2101–109

Olivera BM, Cruz LJ. 2001. Conotoxins, in retrospect. *Toxicon* 39:7–14

Olivera BM, Gray WR, Zeikus R, McIntosh JM, Varga J, et al. 1985. Peptide neurotoxins from fish-hunting cone snails. *Science* 230:1338–43

Olivera BM, Rivier J, Clark C, Ramilo CA, Corpuz GP, et al. 1990. Diversity of *Conus* neuropeptides. *Science* 249:257–63

Röckel D, Korn W, Kohn AJ. 1995. *Manual of the Living Conidae*. Wiesbaden, Ger.: Christa Hemmen. 517 pp.

Rumphius GE. 1705. *D'Amboinsche Rariteikamer*. Amsterdam: Fr. Halma

Spence I, Gillessen D, Gregson RP, Quinn RJ. 1977. Characterization of the neurotoxic constituents of *Conus geographus* (L) venom. *Life Sci.* 21:1759–70

Taylor JD, Kanton YI, Sysoev AV. 1993. Foregut anatomy, feeding mechanisms, relationships and classification of the Conoidea (= Toxoglossa) (Gastropoda). *Bull. Nat. Hist. Mus. London (Zool.)* 59:125–70

Terlau H, Shon K, Grilley M, Stocker M, Stühmer W, Olivera BM. 1996. Strategy for rapid immobilization of prey by a fish-hunting cone snail. *Nature* 381:148–51

Walker C, Steel D, Jacobsen RB, Lirazen MB, Cruz LJ, et al. 1999. The T-superfamily of conotoxins. *J. Biol. Chem.* 274:30664–71

Woodward SR, Cruz LJ, Olivera BM, Hillyard DR. 1990. Constant and hypervariable regions in conotoxin propeptides. *EMBO J.* 1: 1015–20

Annu. Rev. Ecol. Syst. 2002. 33:49–72
doi: 10.1146/annurev.ecolsys.33.010802.150509
First published online as a Review in Advance on August 6, 2002

TROUBLESHOOTING MOLECULAR PHYLOGENETIC ANALYSES

Michael J. Sanderson and H. Bradley Shaffer

*Section of Evolution and Ecology, University of California, Davis, California 95616;
email: mjsanderson@ucdavis.edu, hbshaffer@ucdavis.edu*

Key Words   algorithm, model, data partition, rooting, long-branch attraction

■ **Abstract**   The number, size, and scope of phylogenetic analyses using molecular data has increased steadily in recent years. This has simultaneously led to a dramatic improvement in our understanding of phylogenetic relationships and a better appreciation for an array of methodological problems that continue to hinder progress in phylogenetic studies of certain data sets and/or particular parts of the tree of life. This review focuses on several persistent problems, including rooting, conflict among data sets, weak support in trees, strong but evidently incorrect support, and the computational issues arising when methods are applied to the large data sets that are becoming increasingly commonplace. We frame each of these issues as a specific problem to be overcome, review the relevant theoretical and empirical literature, and suggest solutions, or at least strategies, for further investigation of the issues involved.

INTRODUCTION

The explosive growth in the number of molecular phylogenetic studies can be attributed to two advances: the development of rigorous tree-building methods, and the advent of rapid and relatively inexpensive technology for obtaining comparative molecular data. However, the former predates the latter by some time. Parsimony, likelihood, and distance methods for building trees were all introduced and applied to real data in the 1960s (Edwards & Cavalli-Sforza 1967), whereas the explosive growth of GenBank did not begin until the late 1980s. Consequently, the literature on phylogenetic methods is now extensive, with important work appearing not just in journals devoted to systematics and molecular evolution, but also genetics, mathematics, and computer science. In the past decade the mathematical complexity of phylogenetic methodology has taken a quantum leap forward owing to a determined effort to place much of phylogenetic inference in a statistical framework (e.g., Huelsenbeck & Crandall 1997 for review; Felsenstein 2001). A less well advertised but equally significant infusion of ideas from computer science (e.g., Nakleh et al. 2001) has also impacted theoretical phylogenetics. Together these methodological advances have presented investigators who are now finding it easy to obtain molecular sequence data with an unprecedented and often bewildering set of choices on methodological issues.

Acquiring comparative sequence data is now as simple as accessing GenBank or other databases via the internet, and the number of scientists interested in reconstructing phylogenies as a secondary research interest is growing rapidly. Whereas the inclusion of molecular, developmental, and evolutionary biologists into the world of practical phylogenetics is clearly beneficial for all, a growing knowledge gap between the leading edge of phylogenetic theory and this growing pool of relative novices has developed. Within our own university, and in our interactions with a wide range of colleagues not formally trained in phylogenetic methods, we find ourselves constantly asked how to analyze data sets, what strategies are sufficient, and how to deal with different sorts of comparative information. This review focuses on several issues that repeatedly confront molecular phylogenetic studies, and our view of the recent literature on solutions to these problems. We aim at the relative newcomer to the field who desires advice on troubleshooting rather than an in-depth understanding of all the technical details. The risk of this strategy is that it may give a false impression: Hard problems do not always have simple solutions. Still, the resolution of many problems seems to be taking shape, and solutions to others can be narrowed in scope to a handful of analytical strategies.

Below we list a series of commonly cited problems, and discuss potential strategies for overcoming them. Among the most important of these have been questions about the performance and accuracy of tree-building methods, whether and when to combine disparate sources of data, how to assess levels of support for trees, and how to analyze large data sets with algorithms that often require embarrassingly long running times.

PARTS OF THE TREE SEEM TO BE "WRONG"

Sometimes a phylogenetic analysis will imply relationships that appear dead wrong in the context of previously published results or long-held views. A celebrated example in vertebrates is the placement of bony fish within tetrapods in an analysis of complete mitochondrial genomes (Naylor & Brown 1998). An equally pathological case in plants is the placement of various aquatic or parasitic angiosperms as the solitary sister group to all other angiosperms (*Ceratophyllum*: Chase et al. 1993, *Rafflesia*: Lipscomb et al. 1998). Of course, the possibility always exists that the widely held view is in fact wrong, and that a new phylogenetic paradigm has been uncovered, as appears to be the case for turtles within the amniotes (Rieppel & Reisz 1999), but weight of evidence suggests such cases are probably the exception. Because weak data and sampling error can always lead to apparently incorrect results, the onus is on the investigator to document that a result is both strongly supported and apparently incorrect. Troubleshooting such cases then involves demonstrating that a relationship is questionable and attempting to overcome the cause of the mistaken inferences. This has been an area of active interchange between theoreticians and empirical phylogeneticists and some recommendations have emerged.

Check for Pre-Processing Errors: Alignment, Orthology/Paralogy

The assumption of all phylogenetic inference methods is that the input data consist of a set of homologous characters. For sequence data this means that the sequences are homologs and that their sites have been correctly aligned. As an analytical problem, multiple sequence alignment is of the same order of complexity as phylogenetic inference (Gusfield 1997), yet it rarely receives the same level of attention from phylogeneticists (Lee 2001). Problems can range from fairly trivial to quite serious. Manual sequence alignments, which are commonplace in studies of noncoding DNA data, for example, sometimes can have a misplaced or mistakenly inferred gap that "frameshifts" a large part of a sequence or set of sequences. This can introduce a large number of substitutions into inferred trees, producing false clades that appear strongly supported. Routinely checking alignments with alignment programs (e.g., CLUSTALW; Thompson et al. 1994) can help identify such cases.

Other problems are stickier. Certain genes are notoriously difficult to align either manually or with the assistance of algorithms (e.g., small subunit rDNA), and alignment choices clearly impact phylogenetic inference strongly (Morrison & Ellis 1997). Little consensus is available regarding strategies for alignment in problematic cases (see review in Lee 2001). Three possibilities are (*a*) removing problematic sites, (*b*) concatenating sequences derived from multiple alignment replicates with different parameters (Elision method: Wheeler et al. 1995), and (*c*) extracting common phylogenetic information from trees constructed under different alignments (multiple analysis method: Lee 2001). Of these, removing problematic regions of the alignment is the most conservative but may lead to a loss of resolution.

Phylogenetic error can also arise if paralogs (genes derived via gene duplication) are mistakenly interpreted to be orthologs (genes derived via speciation) in the construction of species trees from sequence data. Paralogs and orthologs are all homologous, but species relationships can only be directly inferred from orthologous sequences (though indirect inferences are possible; see Page & Charleston 1997). Orthology is difficult to discern from sequence data alone, but introns and flanking regions can leave telltale clues. For example, the number and position of introns is a useful diagnostic for distinguishing paralogs from orthologs in the *Adh* gene family of cotton (Small & Wendel 2000). Formal algorithms for inferring species trees from gene family trees (Page & Cotton 2002), are available but are very sensitive to the completeness of sampling of the gene family, a problem with many data sets. The complexities of working with gene families partly explain the popularity of organellar single copy genes in phylogenetics. However, as the data from relatively small organellar genomes is exploited fully, and additional complete nuclear genomes become available, the impetus to identify and use single-copy nuclear genes free of paralogy problems will undoubtedly increase.

Sample Additional Genes with Different Substitution Rates

Yang (1998) and Bininda-Emonds et al. (2001) showed via simulation studies that accuracy—the probability of inferring the correct or nearly correct tree—is maximized at an intermediate nucleotide substitution rate; it drops quickly at lower rates, and more slowly at higher rates. Optimal rates are dependent on the details of the substitution process of the gene sampled and therefore must themselves be estimated for any given data set. However, a useful strategy might be to compare a data set's bootstrap support to that obtained with a faster gene and a slower gene. Given that "too fast" is better than "too slow," a rule of thumb might be to always add a gene with higher rate first. Because rates vary between lineages in most data sets, however, it is necessary to estimate an average rate to permit comparisons between genes. Conventionally, such comparisons are made by reference to a single comparison between two terminal taxa. A more comprehensive estimate of mean rate can be obtained (in PAUP*, for example, Swofford 1999) by rooting a tree, enforcing a molecular clock, obtaining branch lengths, and summing these along one path from root to tip. This value can then be compared to those obtained for the same taxa for other genes. If the age of the root is known it can be put in units of substitutions per site per unit time, but this is not necessary for comparisons between genes. An important caveat is that rates that are too high can lead to long-branch attraction (see next).

Check if Rate Heterogeneity is Sufficient to Cause Long-Branch Attraction

The most widely cited reason for mistaken phylogenetic inferences is long-branch attraction (LBA). Strictly speaking, LBA is an "asymptotic" property: a failure of a method to reconstruct the right tree as the number of characters goes to infinity. This is also known as statistical inconsistency. First noticed for maximum parsimony (MP; Felsenstein 1978), it is now understood to be a problem for parametric methods as well, including maximum likelihood (ML) and neighbor-joining (NJ) using parametric corrections, when model assumptions are violated (Lockhart et al. 1994, Chang 1996). Just as a model of evolution can be constructed in which ML is consistent and MP is not, models containing mixtures of simple substitution models can be constructed in which MP is consistent and ML is not (J. Kim, personal communication). The term long-branch attraction refers to a set of model conditions known to make MP fail. High rates along two branches separated by a branch with low rate will fool MP into joining the long branches together. The LBA problem shows up in many guises, including tree rooting and base compositional bias (see below); it is one of the more pervasive problems in phylogenetics.

However, it is also difficult to document. Huelsenbeck (1997) argued that it is not enough to point to "long" branches in the vicinity of a suspect relationship and suggested two strategies for implicating LBA rigorously. One is to test whether branches are long enough and heterogeneous enough to cause mistaken inferences in a data set given the number of characters observed. The procedure is to choose a model tree, infer reasonable model parameters (including branch lengths) using ML, and then conduct replicate simulations with those parameter values to determine whether or not the tree would be reconstructed with high probability (Huelsenbeck 1997, Maddison et al. 1999, Sanderson et al. 2000, Omilian & Taylor 2001). The Seq-Gen program (Rambaut & Grassly 1997) is an excellent tool for conducting these and other tree-based simulations.

A second strategy is to try different inference methods. In theory, if the model is close to being correct, ML and NJ should be less sensitive to long-branch attraction than MP. However, care should be taken in cases in which MP provides different answers than ML/NJ to make sure the differences are strongly supported—i.e., that the explanation is not just sampling error. The issue of model selection is also critical (Posada & Crandall 2001). The number of substitution models for DNA sequence data is quite large, and minimally one should explore automated model selection methods such as those available in MODELTEST (Posada & Crandall 1998), which balance model complexity with improved likelihood. A nonparametric alternative is provided by Willson's Higher Order Parsimony method (Willson 1999: software available on author's website), which uses characters conventionally viewed as uninformative in standard MP analysis (autapomorphies and constant characters), just as ML does. It is limited to four-taxon trees but may be useful as part of more general quartet methods in larger trees. Simulations suggest it outperforms MP under classical LBA conditions and ML when the model used in ML is incorrect.

Assess the Extent of Base Composition Bias and Heterogeneity

Some genes and genomes are biased in favor of AT or GC base compositions. For example, many animal mitochondrial genomes are AT rich, and this has been proposed as a partial explanation for the occasional reconstruction of incorrect relationships (Steel et al. 1993). Although Conant & Lewis (2001) conducted simulations suggesting that bias must be quite strong before standard methods would be misled, compositional bias has been blamed for many aberrant results, such as the pairing of honeybee and nematode complete mitochondrial protein sequences to the exclusion of other insects (Foster & Hickey 1999), and the removal of *Drosophila willistoni* from its "true" subgenus because its low GC content is similar to more distant outgroups (Tarrio et al. 2001). Inspection of simple base composition statistics provided by many programs (e.g., TREE-PUZZLE, Strimmer & Von Haeseler 1996; PAUP*, Swofford 1999), including χ^2 tests of homogeneity across taxa, should indicate whether compositional bias is a potential issue. More sensitive likelihood ratio tests can then be undertaken to confirm that differences in substitution pattern among extant taxa indicate changes in base composition deeper in the tree (Tarrio et al. 2001). This compares a standard stationary homogeneous model [e.g., the general time-reversible model (GTR); Swofford et al. 1996], to a nonstationary, nonhomogeneous model (Galtier & Gouy 1998).

Possible solutions to the problems induced by composition bias include transforming the data to transversions only (Woese et al. 1991) or translating to amino acid sequences (Hasegawa & Hashimoto 1993). However, Foster & Hickey (1999) showed that compositional bias can propagate to the protein level, producing similar artifacts. Some reconstruction methods are explicitly designed to be robust to composition bias, including distance methods using LogDet distances (Lockhart et al. 1994), modifications of the Tamura (1992) distance that take composition variation into account (Galtier & Gouy 1995), and ML inference with a nonstationary model (Galtier & Gouy 1998). Results from these methods have been mixed, however, with some workers reporting that they helped (Tarrio et al. 2001) and others reporting no improvement (e.g., Foster & Hickey 1999). One problem is that multiple biases may be present and interact, and these problems may be confounded with LBA issues (Whitfield & Cameron 1998, Foster & Hickey 1999). A particularly vexing factor is site-to-site rate variation. Tree reconstruction is exceptionally problematic in the face of both changing base composition through time and different rates between sites (Baake 1998). Until further progress is made with reconstruction methods, simple composition variation can be identified and dealt with, but complex bias remains a difficult obstacle to overcome.

Test for a Covariotide/Covarion Effect

Evidence is mounting that rates of molecular evolution vary not only between sites and between lineages but between both simultaneously, an idea first put forth by Fitch & Markowitz (1970) as the covarion hypothesis (for proteins, covariotides for DNA). An extreme manifestation occurs when some sites are apparently unable to vary in certain taxa but are free to vary in others and vice versa. Several recent studies of deep phylogenies have uncovered such patterns in rDNA (Philippe & Germot 2000) and protein coding genes (Lockhart et al. 1998, 2000; Lopez et al. 1999). Even maximum likelihood inference with complex models can be misled by sequences evolving according to this model (Lockhart et al. 1996). Lockhart and colleagues (Lockhart et al. 1998) provide a relatively simple inequality test for covariotide patterns based on the expected distribution of invariant sites in two sets of sequences. Omilian & Taylor (2001) used this test on large subunit rDNA in the relatively shallow phylogeny of *Daphnia* and found evidence of a covariotide pattern, but they were unable to overcome its effects by removing misleading character classes.

BOOTSTRAP VALUES ARE LOW

Molecular phylogenies sometimes contain clades with low statistical support, even with large amounts of sequence data. Support can be assessed in a variety of ways, but bootstrap resampling (Felsenstein 1985, Efron et al. 1996) is the most commonly used method for molecular data, and its assumptions and interpretation have been scrutinized more intensely than any other (Hillis & Bull 1993, Felsenstein & Kishino 1993, Sanderson 1995, Efron et al. 1996). The general strategy is to resample, with replacement, a data matrix of aligned sequences, where each site (character, column) is a sampling unit. Pseudoreplicate data sets of the same size as the original are constructed and analyzed, and the resulting trees are summarized as a consensus of generally >100 replicates. The bootstrap proportion (BP) has been interpreted in at least two ways: (*a*) as a measure of accuracy, which is the probability that repeated sampling with the same sample size of real data from the universe of characters would recover the true clade; and (*b*) as an estimate of $1 - \alpha$, where α is the conventional type I error (the probability of mistakenly concluding the group is a clade when it is not). Other interpretations of BPs are also possible, including Bayesian ones (see Efron et al. 1996).

Simulation and theory indicate that BP tends to underestimate accuracy when the clade is correct (Zharkikh & Li 1992a,b; Hillis & Bull 1993; Felsenstein & Kishino 1993; Newton 1996; Efron et al. 1996), but that to a first order approximation it does estimate the $1 - \alpha$ significance level (Efron et al. 1996). However, a number of factors can make the latter approximation poor, necessitating second-order corrections. Low BPs may also arise for reasons other than statistical biases. Felsenstein (1985) showed that in the absence of homoplasy BPs are positively correlated with the number of characters changing along the branch subtending a clade, and meta-analyses of real data sets indicate this is true in the presence of homoplasy as well (Sanderson & Donoghue 1996, Bremer et al. 1999). Thus, gathering additional sequence data for the same taxa is one potential cure for low BPs. Slightly less obvious strategies are highlighted below.

Use Bootstrap Correction Factors

Several modified bootstrap procedures have been suggested to obtain BPs closer to true accuracy, including iterated bootstrapping (Rodrigo et al. 1993), the complete-and-partial bootstrap (Zharkikh & Li 1992a) and the approximately unbiased (AU) method (Shimodaira & Hasegawa 2001). Efron et al. (1996) also describe a kind of iterated bootstrap method that explicitly estimates type I error by resampling from data sets constructed to lie on the boundary between regions of tree space that have and do not have the clade of interest. Felsenstein's simple bootstrap method requires some number, K, of iterations, where K is usually >100. These more complex methods often require on the order of K^2 iterations, and thus are computationally very expensive. This is unfortunate given that the bias in BP appears to be exacerbated in intensively sampled clades of large trees that require heavy computation in every replicate (Sanderson & Wojciechowski 2000). Some software is available (e.g., Shimodaira & Hasegawa 2001), but none of these methods are currently implemented in widely used phylogenetics packages.

Check for Rogue Taxa

Because a bootstrap proportion refers to a precise hypothesis about a potentially long list of clade members, it is very sensitive to one or a few "rogue" taxa whose position is unstable. They may be unstable because of missing data, an elevated substitution rate causing homoplasy, or extremely low rates inside and outside the clade, all of which can cause low BPs. Part of the problem is that exact monophyly of a group is a very specific hypothesis, which is all too easily rejected in cases of even a single unstable taxon. Sanderson (1989) suggested relaxing the notion of strict monophyly by putting confidence bounds on r, the minimum number of species needed to make a paraphyletic group monophyletic. Wilkinson (1994, 1996) suggested putting (bootstrap) confidence levels on n-taxon statements rather than clades. An n-taxon statement is a statement of relationship about a subset of taxa on a tree. For example, on the tree $(a, b, c, d, (e, f, g, h))$, an example of a 4-taxon statement is $(a, b, (e, f))$. The support for this 4-taxon statement might well be much higher than it is for the monophyly of (e, f, g, h), especially if taxon g, for example, is problematic. Thorley & Wilkinson (1999) describe a test based on triplets (3-taxon statements) specifically designed to assess the stability of individual taxa (implemented in RadCon: Thorley & Page 2000).

A more general solution is to find maximum agreement subtrees (Finden & Gordon 1985, Steel & Penny 2000) in the collection of bootstrap trees. These are sets of smaller trees constructed by pruning a minimum number of rogue taxa such that the relationships among the remaining taxa agree among all trees. Any clades on the output trees are then supported at the 100% level, although the original tree might not have had any clades supported at this level. Better still would be a fuzzier version of the maximum agreement subtree that returned a majority rule consensus with values higher than that on the original large tree. These approaches extract signal about some relationships at the expense of remaining agnostic about others.

Experiments with taxon sampling from the data available to the investigator can also provide hints about the robustness of BPs. Many studies have found BPs are inversely correlated with the number of taxa (Bremer et al. 1999) or that adding taxa to a study decreases BPs (Sanderson & Wojciechowski 2000, Mitchell et al. 2000), although Omland et al. (1999) found that including more sequences from within each species improved BPs.

Avoid Using "Fast" Bootstrap Methods in Large Trees (if Possible)

In larger data sets, it is common to use heuristic shortcuts in search strategies, such as in PAUP's "fast bootstrap" option, which perform less rigorous searches in the interest of saving computer time. Debry & Olmstead (2000) performed an exhaustive series of simulations, and showed that these strategies lower BPs except in extremely well-supported clades (BP > 90%), and that the deterioration worsened as the number of taxa increased. Mort et al. (2000) found similar results in two real data sets, although those in data clades with BPs of 80% or more were relatively immune. To avoid this deterioration, increasing the stringency of heuristic searches to include at least some minimal branch swapping may be necessary.

When Bootstrapping in Maximum Likelihood Analyses, Experiment with Model Complexity

Bootstrapping in conjunction with maximum likelihood inference is computationally expensive but is a viable option if the number of taxa is not too large (perhaps less than about 25 sequences using commonly available hardware). It is generally assumed that increasing model complexity decreases bias but at the expense of increasing variance (Berry & Gascuel 1996, Burnham & Anderson 1998). Thus, as the number of parameters increases BPs might be expected to decline (Berry & Gascuel 1996). However, Buckley et al. (2001) found little evidence of a consistent pattern in which parameter-rich models had lower BPs than parameter-poor models, although they did find large differences in BPs between models. Computational limitations when bootstrapping ML runs make exploration of the bias-variance tradeoff difficult, and much more work is needed to determine in what direction model choice should be aimed.

SIGNALS FROM DIFFERENT DATA SETS OR PARTITIONS CONFLICT

Since molecular data (including allozymes and DNA sequences) became an important element of phylogenetics, the question of how to best utilize the phylogenetic information in different data partitions has loomed large on the horizon. The problem of conflicts among characters in phylogenetics goes back at least to Hennig (1966), who felt that character conflicts must represent investigator error in making homology statements. However, the current methodological attention to the resolution of conflicts among data partitions dates to Kluge (1989). In that key paper, Kluge made at least two important observations and recommendations. First, to increase the likelihood of character independence, different classes, or kinds of characters should be sampled for a given phylogeny. Second, these different data "partitions" should be combined into a single analysis (often referred to as a "total evidence" approach), rather than testing each partition separately for the presence of different signals. In response to this view, a decade of research has led to new tools for determining whether different partitions have divergent phylogenetic signals, and (to a lesser extent) methods for treating data incongruence. For the practicing phylogeneticist, at least three problems regularly arise. First, how does one determine if data partitions are in conflict? Second, if there is a conflict, how should one obtain a single, best tree from the conflicting data? Third, are there empirical or theoretical guidelines for when one should worry about single-partition trees, given that most practitioners still have data from only a single gene? That is, under some conditions one may feel quite confident that single-gene trees represent true "species trees," whereas in other cases one may feel very uncomfortable with only a single data partition.

Do Different Data Partitions Agree or Disagree in Their Phylogenetic Information?

Over the past two decades, a number of strategies have been proposed to examine the level of disagreement among data partitions. Swofford (1991) provides an excellent review of the first decade. Of the several methods that have stood the test of time, the incongruence length difference test (ILD; also known as the partition homogeneity test), which was originally outlined some 20 years ago (Mickevich & Farris 1981; see also Swofford 1991) is by far the most widely used. The ILD is based on the difference in tree length (that is, the total number of inferred changes under parsimony) between a single tree in which two data partitions are combined, compared to the sum of the tree lengths of each partition on its own maximum parsimony tree. Thus,

$$D_{XY} = L_{(X+Y)} - (L_X + L_Y),$$

where D_{XY} is the length difference, $L_{(X+Y)}$ is the total length of the tree where data partitions X and Y are combined, and L_X, L_Y are the lengths of the most parsimonious trees for each partition separately. Significance is judged by creating random data partitions from the original data set of the same size as X and Y, and asking how frequently a value of D_{XY} as great or greater occurs from this simulated distribution (Farris et al. 1995). When significant incongruence is found, it may be due to differences in the phylogenetic signal of the two data partitions, differences in the level of noise (that is, the randomness of a data set caused by saturation), or a combination of the two (Dolphin et al. 2000). Although the interpretation of the ILD as a test of combinability has received increased scrutiny (Yoder et al. 2001), it remains the test of choice for the moment. It is implemented in PAUP* (Swofford 1999) and other programs.

Several other tests of partition homogeneity have been suggested recently and have considerable appeal. They include an alternative randomization test to the ILD (Rodrigo et al. 1993), the use of congruence among trees rather than data sets (Miyamoto & Fitch 1995), Partitioned Bremer Support (which quantifies the positive or negative contribution of each data partition to the character support for a particular node in a combined data analysis; Baker & Desalle 1997), and a likelihood ratio test that compares differences in likelihood with and without the constraint that the same phylogeny underlies all data partitions (Huelsenbeck & Bull 1996). While each of these recent alternatives has merit, none are as easily implemented as the ILD, and we presume that they will enjoy less widespread use than the ILD test in the near future.

Resolving Conflicts Among Data Partition

As molecular data sets grow to include more genes, conflicts among some or all data partitions appear to be something of a rule. Early comparisons of biochemical and morphological data often found this to be the case (Kluge 1989, reanalyzed in Swofford 1991; Shaffer et al. 1991). However, a small sampling of recent DNA-based analyses with or without morphology on a wide range of plant (Olmstead & Sweere 1994, Nickrent et al. 2000, Sanderson et al. 2000) and metazoan (Baker et al. 1998, Cao et al. 1998, Shaffer et al. 1997, Wiegmann et al. 2000, O'Grady et al. 1998) taxa emphasize that virtually any outcome is possible, with some data sets showing strong conflicts among partitions, and others not. A clear result from these and many other studies is that partitions often conflict in their phylogenetic signal, leaving it to the practitioner to decide how to handle such conflicts.

Fundamentally, there are two strategies for resolving conflicts among data partitions. The first is that advocated by Kluge (1989)—based on first principles: All data should always be included in a combined analysis for any phylogenetic problem. Even when a problematic data partition is successfully identified, some argue that the clearest and most reasonable strategy is to combine (Baker & Desalle 1997). Data conflicts that lead to two well-resolved, but different topologies may result in an unresolved bush when combined (e.g., Shaffer et al. 1991), but even so, advocates feel that this correctly summarizes the current state of knowledge. The alternative view is most clearly laid out by Bull et al. (1993), who advocate a "conditional combinability" strategy. They propose that one should test for homogeneity, and either combine (if the null of homogeneity cannot be rejected), remove the cause of heterogeneity (if the null is rejected and the cause can be identified), or give up (if the null is rejected but the cause cannot be identified). The last option comes down to the "No Obvious Resolution" in Bull et al.'s flow chart. This conditional combinability strategy has been embraced by most systematists (but notable exceptions exist, e.g., Miyamoto & Fitch 1995), and later methodological and simulation work has largely focused on what to do when conflict exists.

Two reasonable solutions appear to be emerging. First, using the ILD (Cunningham 1997), Partitioned Bremer Support (Baker et al. 1998), or visual examination of data conflicts, attempt to identify one gene or data partition within a gene that accounts for much of the conflict (Cao et al. 1998). If a single problematic partition exists, eliminate that partition (Huelsenbeck & Bull 1996). In a similar vein, Johnson (2001) argues for using the ILD to identify problematic taxa and remove them from the analysis. However, as noted by Baker & Desalle (1997) as more gene partitions become available, the likelihood of complex patterns of overlapping conflict becomes ever greater, making it difficult to determine which, if any, partitions should be eliminated (Krzywinski et al. 2001). An emerging theme from several recent analyses is that saturated data (that is, very noisy data suffering from many multiple substitutions per site) is a major source of data conflict (Baker & Desalle 1997, Dolphin et al. 2000). Perhaps because of the problem of long-branch attraction (see above), saturated data partitions are a real source of conflict, and should be eliminated. Unfortunately, these same partitions may be informative in the more recent parts of a tree, but misleading deeper in the tree, making their complete elimination a less desirable choice (Källersjö et al. 1999). In this case, restricting the use of such genes to recently diverged sets of taxa is an attractive analytical solution.

Theoretical Guidelines on Potential Conflicts in New Data Sets

Over the past decade, several guidelines have been proposed regarding the predicted relationship between monophyly in different genome partitions, and how those individual gene partitions reflect the underlying species phylogeny. These guidelines are spread across two divergent literatures, that of phylogenetics/

phylogeography (Ball et al. 1990, Maddison 1997, Moore 1995, Palumbi et al. 2001) and of theoretical population genetics/coalescent modeling (Wu 1991, Hudson 1992). The fundamental issue is the extent to which intra- or interspecific monophyly for a single gene partition predicts monophyly in other partitions, or for the species as a whole. Because single gene mitochondrial (mtDNA) sequences have been used so extensively in metazoan phylogenetics, Moore (1995) used a neutral coalescent framework and the four-fold difference in effective population size of mitochondrial and nuclear genes to calculate the difference in "time to monophyly" (the age of the most recent common ancestor of the sample of sequences) for these two classes of gene partition. Moore concluded that it would take, on average, 16 times as much nuclear DNA sequence information as mitochondrial to achieve equivalent phylogenetic resolution. He thus recommended using mtDNA trees, although some fraction of the time the species trees based on mtDNA will be incorrect. Using very similar logic, Palumbi et al. (2001) formulated their "three times rule" for the predicted concordance of nuclear and mtDNA gene trees. The rule states that, on average, nuclear coalescence (that is, monophyly) is predicted when mtDNA shows a pattern of relatively great divergence between clades compared to the within-clade diversity. In particular, when clades for mtDNA are monophyletic and the ratio of the length of the branch subtending a clade to the depth of the within-clade mtDNA gene tree is greater than about three, nuclear genes may often be monophyletic. Additional theoretical work (R.R. Hudson, M. Turelli, personal communication) has questioned the assumptions leading to the "three times" part of the rule, and emphasizes that guidelines may be lineage-specific with high variances, and therefore have little predictive value. However, the basic premise that low levels of within-clade divergence compared to among-clade divergence in one gene should be associated with similar patterns across loci remains a potentially important result (Hudson 1992, Wu 1991).

The determination of an "x-times rule," including what x may be, is clearly an empirical question, as emphasized by most authors. However, the general framework offers the opportunity to determine, a priori, when a given level of intraclade sampling and a resultant gene tree may predict concordance in other gene partitions. If general rules emerge from empirical results, this could allow molecular systematists to predict when a single gene should suffice, and when many, potentially conflicting, partitions need to be examined to determine organismal phylogenies.

ROOTING THE TREE IS PROBLEMATIC

Rooting phylogenetic trees is a relatively neglected component of many phylogenetic studies. Often considered straightforward, rooting has been identified as "frequently the most precarious step in any phylogenetic analysis" (Swofford et al. 1996, p. 478). Virtually all users of phylogenetic trees want rooted (rather than unrooted) trees, because directionality of evolutionary change and the concept of monophyly itself only have meaning on rooted trees (Smith 1994). By default, software for tree reconstruction typically generates unrooted trees (although they may be displayed as rooted). Rooting, the process of selecting a branch of the unrooted tree into which a root node is inserted, is accomplished afterwards. Although several strategies for rooting trees exist, the one most generally embraced by the phylogenetics community is outgroup analysis, in which some taxa are assigned to a monophyletic ingroup, and one or more taxa to the outgroup. After the analysis is complete, the (unrooted) ingroup tree is rooted at the branch where the outgroup(s) joins the ingroup tree. In many ways, outgroup analysis is nothing more than adding a few additional taxa to an analysis and building a tree, although the initial assumption of monophyly of the ingroup is a strong one. The repercussions of outgroup choice are enormous, however, for they can determine which character states are interpreted as shared derived features and which are ancestral for the entire ingroup (Nixon & Carpenter 1994). In addition, outgroup choice can affect ingroup topology, even for nodes far removed from the presumed root placement (Milinkovitch & Lyons-Weiler 1998, Tarrio et al. 2000).

We discuss two of the more difficult problems with rooting a tree. First, what can be done if reasonable choices exist for potential outgroups, but they are distantly related to the ingroup? Second, what strategies are useful when no obvious potential outgroup(s) can be identified?

Dealing with Distantly Related Outgroups

For previously studied groups, one often knows or can make an educated guess on a set of taxa that are outside of, but closely related to, the ingroup. One or more of these candidates will generally comprise the outgroup (Smith 1994, Nixon & Carpenter 1994). Unfortunately, outgroup sequences are sometimes very divergent from those in the ingroup (Johnson 2001). In the extreme, outgroups may represent little more than randomized, fully-saturated sequences with respect to the ingroup (Wheeler 1990, Maddison et al. 1992). In such cases, outgroup placement reduces to a long-branch attraction problem, and the root will often fall on the longest internal branch of the ingroup tree. As Smith (1994) notes, when a molecular clock is in effect (unusual for most molecular data), this will generally lead to the correct placement of the root. However, the root placement will still result from two long branches (the internal branch and the outgroup branch) attracting, not because of correct phylogenetic signal.

A number of strategies have been suggested that improve the chances of an outgroup correctly rooting a tree. First, attempt to identify and include the sister group to the ingroup for rooting purposes. Generally, this taxon's sequence will be closer to the ingroups than more distantly related outgroups, and will therefore more reliably root the ingroup tree. However, if the sister group has experienced a severe rate speedup, more distantly related—but less divergent—outgroups will provide more reliable evidence on ingroup rooting than the sister group (Lyons-Weiler et al. 1998), although this may be an uncommon scenario.

Second, add additional outgroup taxa (Maddison et al. 1984). Ideally, these additional outgroups should be chosen in such a way that they will break up the branch between the first outgroup and the ingroup, and thus reduce any long-branch attraction problem associated with root placement. The alternative, adding additional outgroups that fall outside of the ingroup-plus-first-outgroup, may simply add long branches to the base of the tree, contributing little to the rooting solution. Unfortunately, choosing multiple outgroups correctly requires prior knowledge of their relationships to each other and to the ingroup, and this knowledge is often not available. Although most recent studies strongly advocate using multiple outgroups, this is not universally accepted (Lyons-Weiler et al. 1998). Sometimes extinct fossil lineages are more relevant outgroups than extant ones, in which case morphological data alone or in combination with molecular data can be used to root the tree (Shaffer et al. 1997).

Finally, if the sister group is only a single species, or a set of very closely related species (Johnson 2001, Maddison et al. 1992), one should test whether the rooting is relatively robust. Strategies include: (*a*) adding different, divergent outgroups to determine if root placement changes, in which case it is suspect (Tarrio et al. 2000, Hutcheon et al. 1998, Dalevi et al. 2001), and (*b*) identifying and removing hypervariable sites from the data matrix, because they probably contribute to any long-branch attraction (Hendy & Penny 1989, Smith 1994). In this latter case, the hypervariable regions may well be informative within the ingroup, in which case the ingroup tree can be built with all sites. This topology can then be fixed, the hypervariable sites removed, and the outgroup(s) added to the analysis. A related problem recently emphasized by Hutcheon et al. (1998) and Tarrio et al. (2000) occurs when codon usage in an outgroup is very different from the ingroup. This is a special case of an outgroup that is so divergent from the ingroup that it has little value in identifying the ingroup root. Midpoint rooting of the ingroup tree (see below) may be preferable to using such a divergent outgroup (Tarrio et al. 2000).

What to Do When No Obvious Outgroup is Known

When studying the phylogeny of very poorly known groups, or groups at the base of the tree of life (Dalevi et al. 2001), there may be no particularly close outgroup with which to root a tree. In this case, exploratory experiments with different strategies may be useful. One possibility is to use midpoint rooting, which assigns the root to the midpoint of the longest path between two terminal taxa in the tree. Under the assumption that the two most divergent lineages in the ingroup tree have evolved at an equal rate, this method correctly roots a tree without reference to an outgroup (Swofford et al. 1996). More generally, reconstructing a tree under the assumption of a molecular clock will also infer the root, although this is a slightly stronger assumption. Alternatively, a variety of outgroups can be tested and compared for their effect on both the ingroup topology and placement of the root. If all give the same result (and even better if it matches the midpoint root), then this lends some support to the root placement being correct. This type of experiment has led to a variety of results ranging from constancy across outgroups (Hutcheon et al. 1998,

Dalevi et al. 2001) to considerable heterogeneity (Milinkovitch & Lyons-Weiler 1998). Tarrio et al. (2000) found that, at least in the case of the *willistoni* and *saltans* groups of *Drosophila*, using a simple model of DNA sequence evolution (as opposed to a more complex model) seemed to eliminate outgroup instability, although the generality of this result remains untested.

COMPUTATION TIME IS TOO LONG

Phylogeneticists are spending more time waiting for analyses to finish than ever before. Molecular data sets containing hundreds to thousands of sequences (e.g., Källersjö et al. 1998, Lipscomb et al. 1998, Savolainen et al. 2000, Tehler et al. 2000, Johnson 2001) are increasingly common, pushing the computational ability of phylogenetic algorithms to their limits and beyond. Parametric inference methods such as maximum likelihood are often invoked with complex models of molecular evolution, which also demand time-consuming computation. Assessment of confidence limits in a tree with bootstrapping adds several orders of magnitude to processing time over and above inferring the tree. Besides taking increasingly long coffee breaks, several strategies may help.

Use Fast(er) Algorithms

The dominant factor in determining the running time of phylogenetic algorithms is the number of sequences, N. No known algorithms are guaranteed to find solutions for optimization-based tree-building procedures, such as MP or ML, in running times that scale better than exponential (e^N). Commonly used heuristics, such as those implemented in PAUP* (Swofford 1999) and other programs, have running times that scale polynomially, as N^k, where k is some constant, and this represents a huge improvement in run times. However, these provide no guarantees about how close they will come to finding the optimal tree. Simulation results suggest that the ML heuristics using sequential addition and branch-swapping scale worse than MP, by a factor of about N in standard implementations (Sanderson & Kim 2000). Other heuristics have been proposed more recently (Goloboff 1999, Nixon 1999, Quicke et al. 2001), which may find better scoring trees faster than more widely used heuristics, but these ultimately have the same worst-case running times and likewise lack guarantees on their solutions. Parsimony heuristics using sequential addition of sequences without rearrangement steps scale as N^2, so this may be the method of choice in desperate circumstances. Note that even non-optimization methods such as NJ (Saitou & Nei 1987), which are often regarded as being fast, scale no better than N^3.

A variety of "divide and conquer" methods have been proposed, which break data sets into subsets (often quartets), build the corresponding trees, and then reassemble them. Quartet Puzzling (Strimmer & Von Haeseler 1996) and a recent variant, Weighted Optimization (Ranwez & Gascuel 2001) can be implemented in running times that scale as N^4. They are thus mainly useful for ML searches; they can be faster than standard heuristic search strategies because finding many small trees is much faster than one larger tree. Ota & Li (2000, 2001) took this strategy in a rather different direction with a hybrid between ML and NJ, which identifies well-supported subtrees via NJ and then performs constrained searches across the less-well supported nodes via ML.

One of the main problems in all optimization strategies is the existence of multiple local optima, which are well known in MP (Maddison 1991) and ML (Salter 2001). Stochastic optimization methods such as simulated annealing (Barker 1997, Salter & Pearl 2001) or genetic/evolutionary algorithms (Matsuda 1996; Lewis 1998; Goloboff 1999; Moilanen 1999, 2001; Katoh et al. 2001) strive to avoid being trapped at local solutions by randomly moving away from local optima according to various schemes. Some of these can be significantly faster than traditional heuristic searches in large data sets (Salter & Pearl 2001), but have not been road-tested much on real data.

Thus, a speed-up in search time can generally be obtained by switching algorithms or computer programs. As always, the difficult question concerns whether accuracy is sacrificed. The literature on accuracy of tree-building algorithms is exceptionally large and difficult to distill, mainly because it has relied on simulations done under different conditions of substitution models, parameter values, data set sizes, and measures of accuracy. A new framework for understanding performance of these methods is provided by the notion of fast convergence (Huson et al. 1999, Nakleh et al. 2001). A method is said to be fast converging if the number of characters needed to guarantee a certain level of accuracy grows at worst polynomially with the number of taxa. Unfortunately, the only theoretical results regarding this concept relate to nonstandard methods such as Disk-Covering (Huson et al. 1999), which is fast converging. It is not yet known whether any standard tree-building method is fast converging, but simulation studies hint that NJ and MP are, at least within the class of phylogenies generated by simple branching processes (Bininda-Emonds et al. 2001, Nakleh et al. 2001).

Reduce the Complexity of Parametric Models

Workers using ML methods have developed many shortcuts that decrease running times when using complex models with many parameters, by using approximations for many of the parameter estimates. A typical strategy is to estimate parameters of the substitution model once on a single "seed" tree, such as one obtained by fast NJ or MP analyses, and then fix parameters at the estimates obtained on that one tree during heuristic searches across many trees (e.g., Moncalvo et al. 2000). Various features in PAUP* facilitate these shortcuts. In moderate-sized data sets (<100 sequences), running times can be improved by an order of magnitude or more.

Sequence More Genes for the Same Taxa

A long-term solution for improving running times for data sets may be to acquire more sequence data for the same taxa. Addition of data can dramatically decrease running times of heuristic searches in data sets with many taxa (Hillis 1996, Soltis et al. 1998, Savolainen et al. 2000), presumably by enhancing the signal to noise ratio (Hillis 1996). Thus, although one might assume that running times would be positively correlated with number of characters, because tree-score calculations must be done for all characters independently, this increase is apparently offset by decreases in the volume of tree space that must be examined during heuristic searches. One caveat is that the addition of genes with different evolutionary histories (whether due to recombination, lineage sorting of ancestral polymorphisms, or other processes) may have the opposite effect, because algorithms are forced to try to sort out significantly conflicting partitions in the data. Tests for homogeneity in the phylogenetic signal (see above) between genes should therefore be performed prior to combining.

Estimate the "Consensus Support" Tree Rather Than the Optimal Tree

A fundamentally different strategy is to abandon attempts to obtain the optimal (i.e., best) tree and estimate instead only the best-supported elements of that tree (Goloboff & Farris 2001). Implementations generally involve randomization of some type. The parsimony jackknife (PJ) runs repeated heuristic searches using data matrices randomly subsampled from the original, while discarding clades that appear in fewer than some cutoff frequency of replicates. Bayesian Markov Chain Monte Carlo (MCMC) methods (Mau et al. 1999, Larget & Simon 1999) estimate the marginal posterior probabilities of clades. Both methods are much quicker than optimization per se, but not surprisingly, probably do not provide good estimates of the optimal tree. This is particularly true for MCMC in the case of large data sets (Salter & Pearl 2001), because the Markov chains rarely visit any specific tree, rendering the estimate of its posterior probability subject to large sampling error. However, estimating well-supported groups may often be sufficient or even preferable, because they emphasize areas where the data are informative. Consensus trees obtained by these methods do have limitations if they are to be used in subsequent evolutionary studies. However, reconstructions of character evolution or applications of the comparative method, which generally require resolved trees (Maddison & Maddison 1992), can be weighted by posterior probabilities sampled from an MCMC chain (Huelsenbeck et al. 2001).

Use Faster Hardware

Although computer hardware improvements are unlikely to permit solution of large phylogenetic optimization problems exactly, they can certainly speed up heuristic searches. Processor speed continues to double every 1.5 years (Moore's law), but high performance computing now almost always entails some form of parallel processing. Practical and inexpensive commodity hardware in the form of Beowulf clusters (Sterling et al. 1999) have generated much interest among phylogeneticists, although software to take advantage of these architectures is not yet widely used. Some phylogenetic problems such as bootstrapping are trivially parallelizable, whereas others, such as sequential addition and branch swapping are not (Janies & Wheeler 2001, Stewart et al. 2001). Many stochastic methods requiring simple replication, such as bootstrapping, PJ, genetic algorithms, or MCMC, can be distributed at the operating system level without writing true parallel code. However, truly parallelized optimization software is now available for MP (Gladstein & Wheeler 1996), ML (Ceron et al. 1998, Stewart et al. 2001), Quartet Puzzling (Schmidt et al. 2002) and genetic algorithms (Katoh et al. 2001), but development of additional tools is badly needed.

CONCLUSIONS

The diversity and complexity of methods available for phylogeny reconstruction presents both a steep learning curve for newcomers and a considerable investment of time and energy in exploring and exploiting the information content of a data set. Just as statisticians in the twentieth century were increasingly called upon to provide advice and guidance to biologists, phylogeneticists play a similar role in the twenty-first century as diverse biologists come to terms with the abundance of comparative sequence data. Particularly as data sets become larger and more diverse, creative and user-friendly strategies, computer programs, and trouble-shooting guidelines are necessary for empiricists to continue building the tree of life. Solutions to some of these emerging problems already exist, but many are unresolved challenges worthy of attention. As these problems attract a progressively broader assemblage of workers interested in studying them, we look forward to new solutions to some of the more recalcitrant problems described here.

The *Annual Review of Ecology and Systematics* is online at
http://ecolsys.annualreviews.org

LITERATURE CITED

Baake H. 1998. What can and cannot be inferred from pairwise sequence comparisons? *Math. Biosci.* 154:1–21

Baker RH, Desalle R. 1997. Multiple sources of character information and the phylogeny of Hawaiian drosophilids. *Syst. Biol.* 46:654–73

Baker RH, Yu XB, DeSalle R. 1998. Assessing the relative contribution of molecular and morphological characters in simultaneous analysis trees. *Mol. Phylogenet. Evol.* 9:427–36

Ball RM Jr, Neigel JE, Avise JC. 1990. Gene genealogies within the organismal pedigrees of random-mating populations. *Evolution* 44:360–70

Barker D. 1997. *LVB 1.0: Reconstructing Evolution with Parsimony and Simulated Annealing.* Edinburgh: Univ. Edinburgh

Berry V, Gascuel O. 1996. On the interpretation of bootstrap trees: appropriate threshold of clade selection and induced gain. *Mol. Biol. Evol.* 13:999–1011

Bininda-Emonds ORP, Brady SG, Kim J, Sanderson MJ. 2001. Scaling of accuracy in extremely large phylogenetic trees.

Presented *Pac. Symp. Biocomput.* 6:547–58

Bremer B, Jansen RK, Oxelman B, Backlund M, Lantz H, Kim K-J. 1999. More characters or more taxa for a robust phylogeny: case study from the coffee family (Rubiaceae). *Syst. Biol.* 48:413–35

Buckley TR, Simon C, Chambers GK. 2001. Exploring among-site rate variation models in a maximum likelihood framework using empirical data: effects of model assumptions on estimates of topology, branch lengths, and bootstrap support. *Syst. Biol.* 50:67–86

Bull JJ, Huelsenbeck JP, Cunningham CW, Swofford DL, Waddell PJ. 1993. Partitioning and combining data in phylogenetic analysis. *Syst. Biol.* 42:384–97

Burnham KP, Anderson DR. 1998. *Model Selection and Inference.* New York: Springer

Cao Y, Janke A, Waddell PJ, Westerman M, Takenaka O, et al. 1998. Conflict among individual mitochondrial proteins in resolving the phylogeny of eutherian orders. *J. Mol. Evol.* 47:307–22

Ceron C, Dopazo J, Zapata EL, Carazo JM, Trelles O. 1998. Parallel implementation of DNAml program on message-passing architectures. *Parallel Comput.* 24:701–16

Chang JT. 1996. Inconsistency of evolutionary tree topology reconstruction methods when substitution rates vary across characters. *Math. Biosci.* 134:189–215

Chase MW, Soltis DE, Olmstead RG, Morgan D, Les DH, et al. 1993. Phylogenetics of seed plants: an analysis of nucleotide sequences from the plastid gene rbcL. *Ann. Mo. Bot. Gard.* 80:528–80

Conant GC, Lewis PO. 2001. Effects of nucleotide composition bias on the success of the parsimony criterion in phylogenetic inference. *Mol. Biol. Evol.* 18:1024–33

Cunningham CW. 1997. Can three incongruence tests predict when data should be combined? *Mol. Biol. Evol.* 14:733–40

Dalevi D, Hugenholtz P, Blackall LL. 2001. A multiple-outgroup approach to resolving division-level phylogenetic relationships using 16S rDNA data. *Int. J. Syst. Evol. Microbiol.* 51:385–91

Debry RW, Olmstead RG. 2000. A simulation study of reduced tree-search effort in bootstrap resampling analysis. *Syst. Biol.* 49:171–79

Dolphin K, Belshaw R, Orme CDL, Quicke DLJ. 2000. Noise and incongruence: Interpreting results of the incongruence length difference test. *Mol. Phylogenet. Evol.* 17:401–6

Edwards AWF, Cavalli-Sforza LL. 1967. Phylogenetic analysis: models and estimation procedures. *Evolution* 21:550–70

Efron B, Halloran E, Holmes S. 1996. Bootstrap confidence levels for phylogenetic trees. *Proc. Natl. Acad. Sci. USA* 93:13429–34. Erratum. 1996. *Proc. Natl. Acad. Sci. USA* 93(14):7085–96

Farris JS, Källersjö M, Kluge AG, Bult C. 1995. Constructing a significance test for incongruence. *Syst. Biol.* 44:570–72

Felsenstein J. 1978. A likelihood approach to character weighting and what it tells us about parsimony and compatibility. *Biol. J. Linn. Soc.* 16:183–96

Felsenstein J. 1985. Confidence limits on phylogenies: an approach using the bootstrap. *Evolution* 39:783–91

Felsenstein J. 2001. The troubled growth of statistical phylogenetics. *Syst. Biol.* 50:465–67

Felsenstein J, Kishino H. 1993. Is there something wrong with the bootstrap on phylogenies? A reply to Hillis and Bull. *Syst. Biol.* 42:182–92

Finden CR, Gordon AD. 1985. Obtaining common pruned trees. *J. Classif.* 2:255–76

Fitch WM, Markowitz E. 1970. An improved method for determining codon variability in a gene and its application to the rate of fixation of mutations in evolution. *Biochem. Genet.* 4:579–93

Foster PG, Hickey DA. 1999. Compositional bias may affect both DNA-based and protein-based phylogenetic reconstructions. *J. Mol. Evol.* 48:284–90

Galtier N, Gouy M. 1995. Inferring phylogenies from DNA sequences of unequal base compositions. *Proc. Natl. Acad. Sci. USA* 92:11317–21

Galtier N, Gouy M. 1998. Inferring pattern and process: maximum-likelihood implementation of a nonhomogeneous model of DNA sequence evolution for phylogenetic analysis. *Mol. Biol. Evol.* 15:871–79

Gladstein D, Wheeler W. 1996. POY. Software for direct optimization of DNA and other data. New York: Am. Mus. Nat. Hist.

Goloboff PA. 1999. Analyzing large data sets in reasonable times: solutions for composite optima. *Cladistics* 15:415–28

Goloboff PA, Farris JS. 2001. Methods for quick consensus estimation. *Cladistics* 17: S26–34

Gusfield D. 1997. *Algorithms on Strings, Trees and Sequences.* New York: Cambridge Univ. Press

Hasegawa M, Hashimoto T. 1993. Ribosomal RNA trees misleading? *Nature* 361:23

Hendy MD, Penny D. 1989. A framework for the quantitative study of evolutionary trees. *Syst. Zool.* 38:297–309

Hennig W. 1966. *Phylogenetic Systematics.* Chicago: Univ. Chic. Press. 263 pp.

Hillis DM, Bull JJ. 1993. An empirical test of bootstrapping as a method for assessing confidence in phylogenetic analysis. *Syst. Biol.* 42:182–92

Hudson RR. 1992. Gene trees, species trees and the segregation of ancestral alleles. *Genetics* 131:509–12

Huelsenbeck JP. 1997. Is the Felsenstein zone a fly trap? *Syst. Biol.* 46:69–74

Huelsenbeck JP, Bull JJ. 1996. A likelihood ratio test to detect conflicting phylogenetic signal. *Syst. Biol.* 45:92–98

Huelsenbeck JP, Crandall KA. 1997. Phylogeny estimation and hypothesis testing using maximum likelihood. *Annu. Rev. Ecol. Syst.* 28: 437–66

Huelsenbeck JP, Ronquist F, Nielsen R, Bollback JP. 2001. Bayesian inference of phylogeny and its impact on evolutionary biology. *Science* 294:2310–14

Huson DH, Nettles SM, Warnow TJ. 1999. Disk-covering, a fast-converging method for phylogenetic tree reconstruction. *J. Comp. Biol.* 6:369–86

Hutcheon JM, Kirsch JAW, Pettigrew JD. 1998. Base-compositional biases and the bat problem. III. The question of microchiropteran monophyly. *Philos. Trans. R. Soc. London Ser. B* 353:607–17

Janies DA, Wheeler WC. 2001. Efficiency of parallel direct optimization. *Cladistics* 17: S71–82

Johnson KP. 2001. Taxon sampling and the phylogenetic position of Passeriformes: evidence from 916 avian cytochrome b sequences. *Syst. Biol.* 50:128–36

Källersjö M, Albert VA, Farris JS. 1999. Homoplasy increases phylogenetic structure. *Cladistics* 15:91–93

Källersjö M, Farris JS, Chase MW, Bremer B, Fay MF, et al. 1998. Simultaneous parsimony jackknife analysis of 2538 rbcL DNA sequences reveals support for major clades of green plants, land plants, seed plants and flowering plants. *Plant. Syst. Evol.* 213:259–87

Katoh K, Kuma K-i, Miyata T. 2001. Genetic algorithm-based maximum-likelihood analysis for molecular phylogeny. *J. Mol. Evol.* 53:477–84

Kluge AG. 1989. A concern for evidence and a phylogenetic hypothesis of relationships among Epicrates (Boidae, Serpentes). *Syst. Zool.* 38:7–25

Krzywinski J, Wilkerson RC, Besansky NJ. 2001. Toward understanding Anophelinae (Diptera, Culicidae) phylogeny: insights from nuclear single-copy genes and the weight of evidence. *Syst. Biol.* 50:540–56

Larget B, Simon DL. 1999. Markov chain Monte Carlo algorithms for the Bayesian analysis of phylogenetic trees. *Mol. Biol. Evol.* 16:750–59

Lee MSY. 2001. Unalignable sequences and molecular evolution. *Trends Ecol. Evol.* 16: 681–85

Lewis PO. 1998. A genetic algorithm for maximum-likelihood phylogeny inference using nucleotide sequence data. *Mol. Biol. Evol.* 15:277–83

Lipscomb DL, Farris JS, Källersjö M, Tehler A. 1998. Support, ribosomal sequences and the phylogeny of the eukaryotes. *Cladistics* 14:303–38

Lockhart PJ, Huson D, Maier U, Fraunholz MJ, Van de Peer Y, et al. 2000. How molecules evolve in eubacteria. *Mol. Biol. Evol.* 17:835–38

Lockhart PJ, Larkum AWD, Steel MA, Waddell PJ, Penny D. 1996. Evolution of chlorophyll and bacteriochlorophyll: the problem of invariant sites in sequence analysis. *Proc. Natl. Acad. Sci. USA* 93:1930–34

Lockhart PJ, Steel MA, Barbrook AC, Huson DH, Charleston MA, Howe CJ. 1998. A covariotide model explains apparent phylogenetic structure of oxygenic photosynthetic lineages. *Mol. Biol. Evol.* 15:1183–88

Lockhart PJ, Steel MA, Hendy MD, Penny D. 1994. Recovering evolutionary trees under a more realistic model of sequence evolution. *Mol. Biol. Evol.* 11:605–12

Lopez P, Forterre P, Philippe H. 1999. The root of the tree of life in the light of the covarion model. *J. Mol. Evol.* 49:496–508

Lyons-Weiler J, Hoelzer GA, Tausch RJ. 1998. Optimal outgroup analysis. *Biol. J. Linn. Soc.* 64:493–511

Maddison DR. 1991. The discovery and importance of multiple islands of most-parsimonious trees. *Syst. Zool.* 40:315–28

Maddison DR, Baker MD, Ober KA. 1999. Phylogeny of carabid beetles as inferred from 18S ribosomal DNA (Coleoptera: Carabidae). *Syst. Entomol.* 24:103–38

Maddison DR, Ruvolo M, Swofford DL. 1992. Geographic origins of human mitochondrial DNA phylogenetic evidence from control region sequences. *Syst. Biol.* 41:111–24

Maddison WP. 1997. Gene trees in species trees. *Syst. Biol.* 46:523–36

Maddison WP, Donoghue MJ, Maddison DR. 1984. Outgroup analysis and parsimony. *Syst. Zool.* 33:83–103

Maddison WP, Maddison DR. 1992. *MacClade: Analysis of Phylogeny and Character Evolution*. Sunderland, MA: Sinauer

Matsuda H. 1996. Protein phylogenetic inference using maximum likelihood with a genetic algorithm. In *Pacific Symposium on Biocomputing*, ed. L Hunter, TE Klein, pp. 512–13. Singapore: World Sci.

Mau B, Newton MA, Larget B. 1999. Bayesian phylogenetic inference via Markov chain Monte Carlo methods. *Biometrics* 55:1–12

Mickevich MF, Farris JS. 1981. The implications of congruence in *Menidia. Syst. Zool.* 30:351–70

Milinkovitch MC, Lyons-Weiler J. 1998. Finding optimal ingroup topologies and convexities when the choice of outgroups is not obvious. *Mol. Phylogenet. Evol.* 9:348–57

Mitchell A, Mitter C, Regier JC. 2000. More taxa or more characters revisited: combining data from nuclear protein-encoding genes for phylogenetic analyses of Noctuoidea (Insecta: Lepidoptera). *Syst. Biol.* 49:202–24

Miyamoto MM, Fitch WM. 1995. Testing species phylogenies and phylogenetic methods with congruence. *Syst. Biol.* 44:64–76

Moilanen A. 1999. Searching for most parsimonious trees with simulated evolutionary optimization. *Cladistics* 15:39–50

Moilanen A. 2001. Simulated evolutionary optimization and local search: introduction and application to tree search. *Cladistics* 17:S12–25

Moncalvo J-M, Lutzoni FM, Rehner SA, Johnson J, Vilgalys R. 2000. Phylogenetic relationships of agaric fungi based on nuclear large subunit ribosomal DNA sequences. *Syst. Biol.* 49:278–305

Moore WS. 1995. Inferring phylogenies from mtDNA variation: mitochondrial-gene trees versus nuclear-gene trees. *Evolution* 49:718–26

Morrison DA, Ellis JT. 1997. Effects of nucleotide sequence alignment on phylogeny estimation: a case study of 18S rDNAs of apicomplexa. *Mol. Biol. Evol.* 14:428–41

Mort ME, Soltis PS, Soltis DE, Mabry ML. 2000. Comparison of three methods for estimating internal support on phylogenetic trees. *Syst. Biol.* 49:160–71

Nakleh L, Roshan U, St. John K, Sun J, Warnow

TJ. 2001. Designing fast converging phylogenetic methods. *Bioinformatics* 1:1–9

Naylor GJP, Brown WM. 1998. Amphioxus mitochondrial DNA, chordate phylogeny, and the limits of inference based on comparisons of sequences. *Syst. Biol.* 47:61–76

Newton MA. 1996. Bootstrapping phylogenies: large deviations and dispersion effects. *Biometrika* 83:315–28

Nickrent DL, Parkinson CL, Palmer JD, Duff RJ. 2000. Multigene phylogeny of land plants with special reference to bryophytes and the earliest land plants. *Mol. Biol. Evol.* 17: 1885–95

Nixon KC. 1999. The Parsimony Ratchet, a new method for rapid parsimony analysis. *Cladistics* 15:407–14

Nixon KC, Carpenter JM. 1994. On outgroups. *Cladistics* 9:413–26

O'Grady PM, Clark JB, Kidwell MG. 1998. Phylogeny of the *Drosophila saltans* species group based on combined analysis of nuclear and mitochondrial DNA sequences. *Mol. Biol. Evol.* 15:656–64

Olmstead RG, Sweere JA. 1994. Combining data in phylogenetic systematics: an empirical approach using three molecular data sets in the Solanaceae. *Syst. Biol.* 43:467–81

Omilian AR, Taylor DJ. 2001. Rate acceleration and long-branch attraction in a conserved gene of cryptic daphniid (Crustacea) species. *Mol. Biol. Evol.* 18:2201–12

Omland KE, Lanyon SM, Fritz SJ. 1999. A molecular phylogeny of the New World orioles (Icterus): the importance of dense taxon sampling. *Mol. Phylogenet. Evol.* 12:224–39

Ota S, Li W-H. 2000. NJML: a hybrid algorithm for the neighbor-joining and maximum-likelihood methods. *Mol. Biol. Evol.* 17:1401–9

Ota S, Li W-H. 2001. NJML+: an extension of the NJML method to handle protein sequence data and computer software implementation. *Mol. Biol. Evol.* 18:1983–92

Page RDM, Charleston MA. 1997. From gene to organismal phylogeny: reconciled trees and the gene tree/species tree problem. *Mol. Phylogenet. Evol.* 7:231–40

Page RDM, Cotton JA. 2002. Vertebrate phylogenomics: reconciled trees and gene duplications. *Pac. Symp. Biocomput.* 7:536–47

Palumbi SR, Cipriano F, Hare MP. 2001. Predicting nuclear gene coalescence from mitochondrial data: the three-times rule. *Evolution* 55:859–68

Philippe H, Germot A. 2000. Phylogeny of eukaryotes based on ribosomal RNA: long-branch attraction and models of sequence evolution. *Mol. Biol. Evol.* 17:830–34

Posada D, Crandall KA. 1998. MODELTEST: testing the model of DNA substitution. *Bioinformatics* 14:817–18

Posada D, Crandall KA. 2001. Selecting the best-fit model of nucleotide substitution. *Syst. Biol.* 50:580–601

Quicke DLJ, Taylor J, Purvis A. 2001. Changing the landscape: a new strategy for estimating large phylogenies. *Syst. Biol.* 50:60–66

Rambaut A, Grassly NC. 1997. Seq-Gen: an application for the Monte Carlo simulation of DNA sequence evolution along phylogenetic trees. *Cabios* 13:235–38

Ranwez V, Gascuel O. 2001. Quartet-based phylogenetic inference: improvements and limits. *Mol. Biol. Evol.* 18:1103–16

Rieppel O, Reisz RR. 1999. The origin and early evolution of turtles. *Annu. Rev. Ecol. Syst.* 30:1–22

Rodrigo AG, Kelly-Borges M, Bergquist PR, Bergquist PL. 1993. A randomisation test of the null hypothesis that two cladograms are sample estimates of a parametric phylogenetic tree. *NZ J. Bot.* 31:257–68

Saitou N, Nei M. 1987. The neighbor-joining method: a new method for reconstructing phylogenetic trees. *Mol. Biol. Evol.* 4:406–25

Salter LA. 2001. Complexity of the likelihood surface for a large DNA dataset. *Syst. Biol.* 50:970–78

Salter LA, Pearl DK. 2001. Stochastic search strategy for estimation of maximum likelihood phylogenetic trees. *Syst. Biol.* 50:7–17

Sanderson MJ. 1989. Confidence limits on phylogenies the bootstrap revisited. *Cladistics* 5: 113–30

Sanderson MJ. 1995. Objections to bootstrapping phylogenies: a critique. *Syst. Biol.* 44: 299–320

Sanderson MJ, Donoghue MJ. 1996. The relationship between homoplasy and confidence in phylogenetic trees. In *Homoplasy: The Recurrence of Similarity in Evolution*, ed. MJ Sanderson, L Hufford, pp. 67–89. New York: Academic

Sanderson MJ, Kim J. 2000. Parametric phylogenetics? *Syst. Biol.* 49:817–29

Sanderson MJ, Wojciechowski MF. 2000. Improved bootstrap confidence limits in large-scale phylogenies, with an example from Neo-Astragalus (Leguminosae). *Syst. Biol.* 49:671–85

Sanderson MJ, Wojciechowski MF, Hu JM, Khan TS, Brady SG. 2000. Error, bias, and long-branch attraction in data for two chloroplast photosystem genes in seed plants. *Mol. Biol. Evol.* 17:782–97

Savolainen V, Chase MW, Hoot SB, Morton CM, Soltis DE, et al. 2000. Phylogenetics of flowering plants based on combined analysis of plastid atpB and rbcL gene sequences. *Syst. Biol.* 49:306–62

Schmidt HA, Strimmer K, Vingron M, von Haeseler A. 2002. TREE-PUZZLE: maximum likelihood phylogenetic analysis using quartets and parallel computing. *Bioinformatics* 18:502–4

Shaffer HB, Clark JM, Kraus F. 1991. When molecules and morphology clash: a phylogenetic analysis of the North American ambystomatid salamanders (Caudata: Ambystomatidae). *Syst. Zool.* 40:284–303

Shaffer HB, Meylan P, McKnight ML. 1997. Tests of turtle phylogeny: molecular, morphological, and paleontological approaches. *Syst. Biol.* 46:235–68

Shimodaira H, Hasegawa M. 2001. CONSEL: for assessing the confidence of phylogenetic tree selection. *Bioinformatics* 17:1246–47

Small RL, Wendel JF. 2000. Copy number lability and evolutionary dynamics of the Adh gene family in diploid and tetraploid cotton (Gossypium). *Genetics* 155:1913–26

Smith AB. 1994. Rooting molecular trees: problems and strategies. *Biol. J. Linn. Soc.* 51:279–92

Soltis DE, Soltis PS, Mort ME, Chase MW, Savolainen V, et al. 1998. Inferring complex phylogenies using parsimony: an empirical approach using three large DNA data sets for angiosperms. *Syst. Biol.* 47:32–42

Steel MA, Lockhart PJ, Penny D. 1993. Confidence in evolutionary trees from biological sequence data. *Nature* 364:440–42

Steel MA, Penny D. 2000. Parsimony, likelihood, and the role of models in molecular phylogenetics. *Mol. Biol. Evol.* 17:839–50

Sterling T, Salmon J, Becker D, Savarese D. 1999. *How To Build A Beowulf: A Guide to the Implementation and Application of PC Clusters.* Cambridge, MA: MIT Press

Stewart CA, Hart D, Berry DK, Olsen GJ, Wernert EA, Fischer W. 2001. *Parallel implementation and performance of fastDNAml-a program for maximum likelihood phylogenetic inference.* Presented at SC2001, Denver

Strimmer K, Von Haeseler A. 1996. Quartet puzzling: a quartet maximum-likelihood method for reconstructing tree topologies. *Mol. Biol. Evol.* 13:964–69

Swofford DL. 1991. When are phylogeny estimates from molecular and morphological data incongruent? In *Phylogenetic Analysis of DNA Sequences*, ed. MM Miyamoto, J Cracraft, pp. 295–333. New York: Oxford Univ. Press

Swofford DL. 1999. PAUP * 4.0. *Phylogenetic Analysis Using Parsimony and Other Methods.* Sunderland, MA: Sinauer

Swofford DL, Olsen GJ, Waddell PJ, Hillis DM. 1996. Phylogenetic inference. In *Molecular Systematics*, ed. DM Hillis, C Moritz, BK Mable, pp. 407–514. Sunderland, MA: Sinauer

Tamura K. 1992. Estimation of the number of nucleotide substitutions when there are strong transition-transversion and G-plus-C content biases. *Mol. Biol. Evol.* 9:678–87

Tarrio R, Rodriguez-Trelles F, Ayala FJ. 2000. Tree rooting with outgroups when they differ in their nucleotide composition from the ingroup: the *Drosophila saltans* and *willistoni* groups, a case study. *Mol. Phylogenet. Evol.* 16:344–49

Tarrio R, Rodriguez-Trelles F, Ayala FJ. 2001. Shared nucleotide composition biases among species and their impact on phylogenetic reconstructions of the Drosophilidae. *Mol. Biol. Evol.* 18:1464–73

Tehler A, Farris JS, Lipscomb DL, Källersjö M. 2000. Phylogenetic analyses of the fungi based on large rDNA data sets. *Mycologia* 92:459–74

Thompson JD, Higgins DG, Gibson TJ. 1994. CLUSTAL W: improving the sensitivity of progressive multiple sequence alignment through sequence weighting, position-specific gap penalties and weight matrix choice. *Nucleic Acids Res.* 22:4673–80

Thorley JL, Page RDM. 2000. RadCon: phylogenetic tree comparison and consensus. *Bioinformatics* 16:486–87

Thorley JL, Wilkinson M. 1999. Testing the phylogenetic stability of early tetrapods. *J. Theor. Biol.* 200:343–44

Wheeler WC. 1990. Nucleic acid sequence phylogeny and random outgroups. *Cladistics* 6:363–68

Wheeler WC, Gatesy J, Desalle R. 1995. Elision: a method for accommodating multiple molecular sequence alignments with alignment-ambiguous sites. *Mol. Phylogenet. Evol.* 4:1–9

Whitfield JB, Cameron SA. 1998. Hierarchical analysis of variation in the mitochondrial 16S rRNA gene among hymenoptera. *Mol. Biol. Evol.* 15:1728–43

Wiegmann BM, Mitter C, Regier JC, Friedlander TP, Wagner DM, Nielsen ES. 2000. Nuclear genes resolve Mesozoic-aged divergences in the insect order Lepidoptera. *Mol. Phylogenet. Evol.* 15:242–59

Wilkinson M. 1994. Common cladistic information and its consensus representation: reduced Adams and reduced cladistic consensus trees and profiles. *Syst. Biol.* 43:343–68

Wilkinson M. 1996. Majority-rule reduced consensus trees and their use in bootstrapping. *Mol. Biol. Evol.* 13:437–44

Willson SJ. 1999. A higher order parsimony method to reduce long-branch attraction. *Mol. Biol. Evol.* 16:694–705

Woese CR, Achenbach L, Rouviere P, Mandelco L. 1991. Archaeal phylogeny: reexamination of the phylogenetic position of *Archaeoglobus fulgidus* in light of certain composition-induced artifacts. *Syst. Appl. Microbiol.* 14:364–71

Wu CI. 1991. Inferences of species phylogeny in relation to segregation of ancient polymorphisms. *Genetics* 127:429–36

Yang Z. 1998. On the best evolutionary rate for phylogenetic analysis. *Syst. Biol.* 47:125–33

Yoder AD, Irwin JA, Payseur BA. 2001. Failure of the ILD to determine data combinability for slow loris phylogeny. *Syst. Biol.* 50:408–24

Zharkikh A, Li WH. 1992a. Statistical properties of bootstrap estimation of phylogenetic variability from nucleotide sequences. I. Four taxa with a molecular clock. *Mol. Biol. Evol.* 9:1119–47

Zharkikh A, Li WH. 1992b. Statistical properties of bootstrap estimation of phylogenetic variability from nucleotide sequences. II. Four taxa without a molecular clock. *J. Mol. Evol.* 35:356–66

Annu. Rev. Ecol. Syst. 2002. 33:73–90
doi: 10.1146/annurev.ecolsys.33.020602.095426

THE EARLY RADIATIONS OF CETACEA (MAMMALIA): Evolutionary Pattern and Developmental Correlations

J. G. M. Thewissen and E. M. Williams

Department of Anatomy, Northeastern Ohio Universities College of Medicine, Rootstown, Ohio 44272; email: Thewisse@neoucom.edu, emw@neoucom.edu

Key Words marine mammal, fossil record, evolution, development

■ **Abstract** The origin and early evolution of Cetacea (whales, dolphins, and porpoises) is one of the best examples of macroevolution as documented by fossils. Early whales are divided into six families that differ greatly in their habitats, which varied from land to freshwater, coastal waters, and fully marine. Early cetaceans lived in the Eocene (55–37 million years ago), and they show an enormous morphological diversity. Toward the end of the Eocene the modern cetacean body plan originated, and this body plan remained more or less the same in the subsequent evolution. It is possible that some aspects of this body plan are rooted in constraints that are dictated by cetacean embryologic development and controlled by genes that affect many organ systems at once. It may be possible to use a study of patterns of correlations among morphological traits to test hypotheses of developmental links among organ systems.

INTRODUCTION

Cetaceans (whales, dolphins, and porpoises) are obligate aquatic mammals and invariably are described in superlatives. Members of the order range over four orders of magnitude in size with the largest, the Blue Whale, averaging 24 m in length and weighing between 80,000 and 150,000 kg (maximum length of approximately 33 m and weight of 190,000 kg). The smallest member, the Vaquita, reaches a length of about 1.4 m and weighs less than 40 kg. Cetaceans also enjoy the greatest geographic range of any mammal; single species, such as sperm whales or killer whales, are distributed over more surface area of the earth (all oceans) than any other mammalian species. With the possible exception of humans, they occupy a greater latitudinal range (tropical through polar) as well. In addition, deep diving species cover a greater elevational range than any other mammal (sperm whales can dive to a depth of 1 km below the sea surface, a record shared with some seals). The animal with the largest brain is also a cetacean (Blue Whale, 7 kg), and until the late Pliocene when big-brained humans evolved (Marino 1998, Marino et al. 2000), cetaceans also held the record for the largest relative brain size.

0066-4162/02/1215-0073$14.00

In spite of these records, there are only 83 species of modern cetaceans, grouped into 14 families (Rice 1998). This places them near the middle of a ranking of diversity of mammalian orders, in which they are similar in species richness to such families as the American possums (Didelphidae, 70 species). Whereas their body plan is remarkably different from many other orders of mammals (with the exception of sirenians), cetaceans are notably average in their similarity of body shape among members of the order. All modern cetaceans have a streamlined body and lack a neck, and their forelimbs do not show individual fingers but form a paddle, the flipper, used mainly for steering. External hind limbs are absent, and the tail is modified into a flat hydrofoil, or fluke, that is moved through the water in a vertical plane and is the main propulsive organ.

Morphological diversity of other organs is also limited. There are only two basic feeding types, employed respectively by the two modern suborders. The members of the suborder Odontoceti (toothed whales, which includes dolphins and porpoises) hunt single large prey animals and use echolocation to find them. The members of the suborder Mysticeti (baleen whales) collect small prey in bulk using a series of baleen plates hanging down from the left and right upper jaw.

The reasons for the low taxonomic and morphological diversity may be the strict demands of life in the water, which greatly constrains body form, and the lack of extremes (in temperature for instance) in the watery environment. However, it could also be that cetacean development put limits on morphological diversification and partly determined the homogeneity of morphospace among the modern forms.

The modern cetacean body plan, including morphologies indicating modern modes of locomotion and food procurement, originated around the Eocene-Oligocene boundary (approximately 34 million years ago) (Fordyce & de Muizon 2001). This stable body plan was preceded by approximately 15 million years of experimentation in which all organ systems, including those relating to locomotion and feeding, were adapted to life in the sea. These experimental Eocene whales are collectively known as archaeocetes, and they document the transition from land to water—one of the most remarkable examples of macroevolutionary change documented by the fossil record (Thewissen & Bajpai 2001). These fossil cetaceans therefore present an opportunity to examine the evolutionary, and perhaps developmental, patterns and constraints that existed as these animals took to the water.

In this article we describe the Eocene whales and discuss the differences in their biology as they changed from land mammals to obligate swimmers. Then we discuss some developmental mechanisms that may underlie the unusual features and homogeneity of the cetacean body plan.

THE EOCENE RADIATION OF CETACEANS

There are six families of Eocene cetaceans: pakicetids, ambulocetids, remingtonocetids, protocetids, dorudontids, and basilosaurids (Figure 1, color insert). Taken together they describe the transition from a terrestrial quadruped to a fully aquatic marine mammal. The oldest known cetaceans are the pakicetids, which lived in the floodplains of present-day South Asia 50 million years ago. Pakicetids are followed by the semiaquatic ambulocetids, which inhabited the bays and estuaries of the Tethys Ocean in northern Pakistan. The slender-snouted remingtonocetids are also found in shallow marine deposits of Pakistan and India; however, they appear to have been more aquatic than the older ambulocetids. Contemporaries of the remingtonocetids, the protocetids are more aquatic, although they retain hind limbs, and are distributed globally with fossil remains uncovered in South Asia, Africa, and North America. By the middle to late Eocene, dorudontids and basilosaurids were swimming in an open marine environment and had nearly lost all vestiges of life on land.

Although the cast of the early cetaceans has been filled in the past 20 years with new discoveries (Thewissen & Bajpai 2001), the cetacean sister group has remained a hotly contested issue until recently. Given that cetaceans are mammals, the precursor of aquatic forms must certainly be terrestrial, and there were a number of viable candidates, including the extinct ungulates, the mesonychians, or one of the many artiodactyls. Whereas historically morphologists fell in the camp of a mesonychian sister relationship, molecular evidence suggests overwhelmingly that hippopotamids are the modern sister group of cetaceans (Nikaido et al. 1999, Gatesy & O'Leary 2001). Recent morphological studies (Gingerich et al. 2001) agree with the hippopotamid hypothesis, whereas another (Thewissen et al. 2001b) suggests sister group relations between Cetacea and Artiodactyla as a whole. These theories converge if the hippopotamid lineage is ancient and includes forms that are distinctly morphologically unlike modern hippos. Both cetaceans and known artiodactyls could then be derived from a form that resembled the most archaic artiodactyls known: an animal that skeletally resembled a small, stocky mousedeer (Rose 1982).

As to the systematic relationship within Cetacea, there is little doubt that the modern suborders of cetaceans, odontocetes, and mysticetes form a monophyletic clade (Fordyce 2001) and that their sister group is one of the archaeocetes, most likely one of the dorudontids (Uhen & Gingerich 2000). The phylogenetic relations of the archaeocetes are a bit more complex, but it is generally accepted that pakicetids are the most basal, followed by ambulocetids, remingtonocetids, paraphyletic protocetids and dorudontids, and basilosaurids. The divergence between the modern suborders of cetaceans occurred in the late Eocene (around 37 million years ago).

The following is a more in-depth description of each of the earliest families of cetaceans, including some history of their description, discussion of important elements of their skeletons, and their presumed habitats. To see what these animals might have looked like, refer to Figure 2. Table 1 provides details of the diagnostic anatomy for each of the families.

Pakicetidae

The Pakicetidae are the earliest cetaceans known and they occupy the basal branch on the cetacean cladogram (Figure 1, color insert). They include three genera,

Ichthyolestes, *Pakicetus*, and *Nalacetus*. The first pakicetid to be described was *Ichthyolestes* (Dehm & Oettingen-Spielberg 1958), and although these authors considered the taxon to be a fish-eater (the name means fish-robber), they did not recognize it as a cetacean. West (1980) was the first to describe pakicetid material as cetacean. His work was quickly followed by the discovery of a braincase and the naming of the genus *Pakicetus* (Gingerich & Russell 1981). Additional pakicetid cranial material was described by Gingerich & Russell (1981), Gingerich et al. (1983), Thewissen & Hussain (1993, 1998), Luo & Gingerich (1999), and Thewissen et al. (2001b). Thewissen et al. described postcranial material for this family, and the last revision of the group was made by Thewissen & Hussain (1998). Virtually all elements of the pakicetid skeleton are known, but no skeletons of a single individual are available. Bajpai & Gingerich (1998) described a new and presumed older genus of pakicetid *Himalayacetus*; however, morphologically this specimen appears to be an ambulocetid and its age is contested (Thewissen et al. 2001a).

Pakicetids are found in late-early Eocene fluvial deposits in northern Pakistan and northwestern India. Aslan & Thewissen (1996) suggested that it was an arid environment with ephemeral streams and moderately developed floodplains. The pakicetid fossils are always recovered from localities of river channel deposits. The most well known of these, Howard University–Geological Survey of Pakistan (H-GSP) Locality 62 (Figure 3), contains an abundance of pakcetids where they are buried with freshwater fish (catfish, channids) and land mammals (didelphids, rodents, raoellid artiodactyls, and anthracobunid proboscideans). At this locality pakicetids are by far the most abundant mammals, constituting approximately 60% of the recovered mammals (Thewissen et al. 2001).

Pakicetids have never been found associated with marine fauna or marine deposits. They were clearly terrestrial or freshwater animals. Their limb morphology of long thin legs and relatively short hands and feet suggests that they were poor swimmers, and the sedimentology of the H-GSP Locality 62 indicates that it was a very shallow river, probably too shallow for a pakicetid to swim. However, the great abundance of pakicetids in riverine deposits does suggest an affinity for water. In addition, some pakicetid bones are osteosclerotic, with the shafts of some long bones consisting completely of compact bone. It could be interpreted that their heavy bones were used for ballast, and it is clear that pakicetids were not fast on land, in spite of cursorial body proportions (Thewissen et al. 2001b). It appears most likely that pakicetids lived in and near freshwater bodies and that their diet

←

Figure 2 Reconstructions of representative Eocene cetaceans. Clockwise from top: a beached *Dorudon* (Dorudontidae), *Ambulocetus* (Ambulocetidae), *Pakicetus* (Pakicetidae), *Kutchicetus* (Remingtonocetidae), and *Rodhocetus* (Protocetidae). These cetaceans are shown together for comparison, but they were not contemporaries and lived in different environments. Artwork by Carl Buell.

TABLE 1   Some diagnostic features for Eocene cetacean families*

	Pakicet.	Ambuloce.	Remington.	Protocet.	Dorudont.	Basilosaur.
Position of nasal opening	Over I	?	Over I	Over C or P^1	Over P^1–P^2	Over P^1–P^2
Length of rostrum	Short	Short	Long	Short	Short	Short
Size of orbit	Large	Large	Small	Large	Large	Large
Position of orbits	Close to the midline	Close to the midline	Moderately spaced	Widely spaced	Widely spaced	Widely spaced
Shape of supraorbital region	Concave	Slightly concave	Convex	Supraorbital shield	Supraorbital shield	Supraorbital shield
M^3	Present	Present	Present	Present, small	Absent	Absent
Protocone on molars	Present	Present	Absent	Present/ Absent	Absent	Absent
Trigonid/ talonid distinct	Yes	Yes	No	Yes	No	No
Mandibular foramen	Small	Medium	Large	Large	Large	Large
Extent of mandibular symphysis	P_2	P_1	P_2–M_1	P_2	P_2	P_2
Shape of ascending ramus	Abrubtly ascending	Abruptly ascending	Gradually ascending	Gradually ascending	Gradually ascending	Gradually ascending
Sacrum	4 verts.	4 verts.	4 verts.	1–4 verts.	No fusion	No fusion
Lumbar vertebrae	Short	Short	Short	Short	Short	Long
Hind limbs	Long	Intermediate	Intermediate	Intermediate	Small	Small

*I–incisor, P–premolar, M–molar

included land animals approaching water for drinking or some freshwater aquatic organisms. The long narrow snout suggests a specialized feeding method.

Ambulocetidae

Ambulocetids are the most basal amphibious marine cetaceans. They are known only from a limited number of specimens, all of which were recovered in northern Pakistan and northwestern India in middle-Eocene deposits. The most complete of these is a skeleton of a single individual, *Ambulocetus natans* (Thewissen et al. 1994, 1996; Madar et al. 2002), for which the only unknown elements include

Figure 3 Composite photograph of fossiliferous layer at Howard University–Geological Survey of Pakistan Locality 62 in the Kala Chitta Hills of northern Pakistan (~50 million years ago). Bones recovered from this locality are mapped on the photo; ~60% of fossils are pakicetid cetaceans (Thewissen et al. 2001). There are five fragmentary pakicetid skulls in this image (H-GSP 96231, 96254, 96343, 96386, and 96565).

the humerus, some tail vertebrae, and the rostrum. Other genera are *Gandakasia* and *Himalayacetus*, each known from one or a few dental specimens. *Gandakasia* was the first ambulocetid to be described (Dehm & Oettingen-Spielberg 1958), but like the pakicetid *Ichthyolestes*, it was not recognized as a cetacean. *Himalayacetus* was initially described as a marine pakicetid; however, its morphology and the lithologic setting in which it was recovered are most consistent with ambulocetids.

Ambulocetids are always found in near-shore shallow marine deposits associated with abundant marine plant fossils and littoral molluscs. It is likely that they lived in estuaries or bays (Williams 1998). In spite of these shallow marine habitats, stable oxygen isotopes indicate that ambulocetids were still tied to freshwater habitats and probably were partly dependent on freshwater at some stages of their life (Roe et al. 1998).

Ambulocetids were amphibious mammals, well able to locomote on land and in water. They were large, similar in size to a male sea lion. They probably swam by paddling with their large feet (Thewissen & Fish 1997). This mode of swimming is relatively inefficient, and ambulocetids probably did not chase prey. Unlike most cetaceans, ambulocetids had a narrow head with eyes facing laterally, a morphology common in theropod dinosaurs. Their lower jaw shows some of the characteristics relating to sound transmission that occur in later whales but are absent in pakicetids. The suggestion that ambulocetids were hairless (Gatesy & O'Leary 2001) was based on cladistic arguments and lacks physiological support. It is unlikely that a marine mammal lacking blubber could sustain the high metabolic costs of hairlessness.

Remingtonocetidae

Remingtonocetids are a diverse family of middle-Eocene archaeocetes found in north and central Pakistan and western India. The most ancient remingtonocetid is *Attockicetus*, a contemporary of *Ambulocetus*. Other remingtonocetid genera are *Remingtonocetus, Dalanistes, Andrewsiphius*, and *Kutchicetus*. Cranial fossils are common for remingtonocetids (Sahni & Mishra 1972, 1975; Kumar & Sahni 1986; Gingerich et al. 1993, 1995a; Bajpai & Thewissen 1998), but dental material is relatively rare (Bajpai & Thewissen 1998, Thewissen & Bajpai 2001). Postcranial material is associated with some remingtonocetids (Gingerich et al. 1995b, Bajpai & Thewissen 2001), although some of it was initially attributed to the Protocetidae (Gingerich et al. 1997). One of the earliest records of remingtonocetids is by Sahni & Mishra (1975), who proposed that remingtonocetids were closely related to odontocetes. This idea is now discounted, and most authors consider them an early radiation with distinct specializations (but see Thewissen & Hussain 2000).

Remingtonocetids were amphibious whales well adapted to swimming. They had a long narrow snout, widely set, small eyes, and a long sinuous body. Cranially, *Remingtonocetus* is the best-known remingtonocetid (Kumar & Sahni 1986, Gingerich et al. 1995a, Bajpai & Thewissen 1998). Our understanding of remingtonocetid postcrania is primarily based upon a single specimen of the genus *Kutchicetus*. *Kutchicetus* is a small form with a long and muscular back and tail. No hands or feet are known for *Kutchicetus* (or any other remingtonocetid), but the long bones are short and stocky. Swimming in *Kutchicetus* may have been similar to *Pteronura*, the South American giant freshwater otter (Bajpai & Thewissen 2000). *Pteronura* has a long, flat tail with which it swims, but lacks a fluke.

Remingtonocetid fossils have been recovered from a variety of coastal marine environments, including near-shore and lagoonal deposits (Gingerich et al. 1995b, Misra 1992). They are often found in association with marine catfish and crocodilians, as well as protocetid whales and sirenians. Stable oxygen isotopes suggest that most remingtonocetids were independent of freshwater (Roe et al. 1998).

Protocetidae

Protocetids were marine cetaceans possibly similar in life style to modern seals. This family contains many genera, most of which are poorly known. It appears that there is great variation among them and that they represent a range of aquatic adaptations. Some were probably able to support their weight on land, whereas others could not. Protocetids are the earliest and most basal cetaceans to leave the South Asian subcontinent and disperse across the globe. Indo-Pakistani protocetids are the oldest in the family, they include *Indocetus*, *Rodhocetus* (a possible synonym of *Indocetus*, as implied by Gingerich et al. 1995b), *Babiacetus*, *Takracetus*, and *Artiocetus*. African protocetids are *Protocetus*, *Eocetus*, and *Pappocetus*, and described North American forms are *Georgiacetus*, *Eocetus*, and *Nachitochia* (for an overview see Williams 1998).

Until the 1980s the most well-known protocetid was *Protocetus*, for which a skull and axial skeletal elements were described (Fraas 1904, Stromer 1908). Therefore, for much of the past century, it provided the main source of information about basal Cetacea (Kellogg 1928, 1936). At present the best-known protocetid is *Rodhocetus*, which is primarily known from two partial skeletons (Gingerich et al. 1994, 2001). Together, however, these specimens provide a complete image of the animal. *Rodhocetus* had short limbs but long hands and feet that were probably webbed (Gingerich et al. 2001). *Rodhocetus* was initially thought to have a tail fluke (Gingerich et al. 1994), but new analyses (Buchholtz 1998) and new finds (Gingerich et al. 2001) contradict this notion. The sacrum of *Rodhocetus* (Gingerich et al. 1994) consists of four vertebrae that have fused transverse processes (or arches) but unfused centra. Functionally this implies that the sacrum was immobile, and it is likely that the centra would fuse in older individuals, as they do in ambulocetids and remingtonocetids. Other protocetids, such as *Georgiacetus*, lacked a fused sacrum (Hulbert 1998). All protocetids had a supraorbital shield (a large flat part of the frontal over the eyes, causing the eyes to face laterally, although this shield is relatively narrow in *Artiocetus* (Gingerich et al. 2001). Unlike remingtonocetids, protocetids had large orbits. Bajpai et al. (1996) described an endocast for *Indocetus*.

Protocetids almost certainly represent a paraphyletic family (Hulbert et al. 1998, Uhen 1998, Geisler 2001) and encompass great morphological variety. Indo-Pakistani protocetids are mainly known from coastal and lagoonal facies, but African and North American forms also include open marine forms that were able to colonize continents outside of Indo-Pakistan (Williams 1998).

Dorudontidae

Dorudontids were dolphin-like in body shape and appearance. They are the dominant cetaceans of the late Eocene and are best known from northern Africa and North America, although some have been reported from Indo-Pakistan, Europe, and New Zealand. Included genera are *Dorudon*, *Pontogeneus*, *Zygorhiza*, *Saghacetus*, *Ancalecetus*, and *Chrysocetus*. The familial attribution of the genus *Gaviacetus* is disputed; although originally described as a protocetid (Gingerich et al. 1995a), it is now considered a dorudontid by some (Thewissen 1998b), on the basis of the absence the M3 (Bajpai &Thewissen 1998) and the single sacral vertebra (Gingerich et al. 1995b). If this is correct, it would represent the oldest dorudontid (middle Eocene). Uhen (1998, 2001) presented an excellent summary of dorudontids and basilosaurids. Dorudontids and basilosaurids were discovered before any other Eocene cetaceans, in the first half of the nineteenth century. Kellogg (1936) and Uhen (1998) discussed the history of discovery of these families.

The complete skeleton is known for dorudontids (Kellogg 1936, Uhen 1998). They had a powerful vertebral column and short, probably flipper-shaped, forelimbs. The external hind limbs were tiny and are certainly not involved in locomotion (Buchholtz 1998, Uhen 1998), and the pelvis was not attached to the axial skeleton. The tail had a vertebra that was rounded and had aberrant proportions, the ball vertebra (Uhen 1998). It indicates that dorudontids had a fluke, and they probably swam similarly to modern cetaceans. Dorudontids are usually found in deposits indicative of fully marine environments, lacking any freshwater influx.

Basilosauridae

Basilosaurids are late Eocene cetaceans, commonly found in association with dorudontids. Because the skull morphology, environment, and age of dorudontids and basilosaurids are similar, they are often included in a single family (Basilosauridae). In body form, however, these families are very different. While dorudontids are proportionally like modern small cetaceans, basilosaurids are large and have long (up to 16 m) (Uhen 1998), snake-like bodies. This is a result of the extreme elongation of their lumbar vertebrae. The family includes two genera: *Basilosaurus* for which several skeletons are known (Kellogg 1936, Gingerich et al. 1990), and *Basiloterus*, which is based on a single caudal vertebral centrum (Gingerich et al. 1997). Uhen (1998) questioned the validity of the latter. Basilosaurids are known from North America, the Middle East, and Pakistan. Dorudontids and basilosaurids were probably distributed throughout the tropical and subtropical seas of the world.

Basilosaurus was the best known Eocene whale in the time of Kellogg's 1936 revision of archaeocetes, and its large size speaks to the imagination. Its name *Basilosaurus* means king lizard and was based on an early misidentification (discussed by Kellogg 1936 and Uhen 1998). Its tiny pelvis and femur were found over a century ago (Lucas 1900), and the complete foot has also been discovered (Gingerich et al. 1990). It is likely that *Basilosaurus* had a tail fluke, but its body proportions suggest that it swam by means of undulating its back and that the fluke was not the propulsive organ (Buchholtz 1998). Stomach contents indicate that *Basilosaurus* ate fish.

EVOLUTIONARY MECHANISMS

Variation in the morphology of the cetacean body is almost certainly narrowly constrained by selection. The locomotor organs of all modern cetaceans are relatively similar (Fish 1998), because the locomotor modes are similar. Hind-limb locomotion has been eliminated in favor of locomotion by movements of the vertebral column. Feeding variation is also limited within each of the suborders. Modern odontocetes have a large (polydonty), single (monophyodonty) series of repeated similar (homodonty) elements: simple peg-like teeth. There are no teeth in mysticetes, but food procurement in these forms also occurs by a large series of repeated, similar elements: the baleen plates.

Eocene cetaceans were very different; they had a wide variety of swimming modes and morphologies, from quadrupedal paddling in pakicetids to caudal oscillation in dorudontids. Similarly, feeding morphologies are different; all Eocene cetaceans were heterodont and diphyodont and had a stable number of teeth (10 or 11, depending on species, per jaw half). Whereas several authors (Lipps & Mitchell 1976, Gingerich et al. 1983, O'Leary & Uhen 1999) have suggested that diet was one of the main driving forces of early cetacean evolution, it is not clear what the early cetaceans consumed.

Although selection obviously played a major role in determining the course of locomotor and food processing evolution, it is likely that development provided a mechanism that generated variation and constrained the breadth of possible Eocene morphological experiments in cetaceans. A study of the patterns of morphological diversity can therefore be used to explore the possible causal correlation between the evolutionary pattern and genic controls on development. Figure 4 summarizes morphological data possibly indicative of developmental constraints.

Locomotion: Function and Evolution

All modern cetaceans are propelled by dorso-ventral movements of the tail fluke that are concentrated in a relatively small region of the vertebral column (oscillation, Fish 1993; undulation, Slijper 1960, Buchholtz 2001). Morphological correlates of fluke-based locomotion (Watson & Fordyce 1993, Fish & Hui 1991, Fish 1998, Uhen 1998) were already present in dorudontids and basilosaurids (Kellogg 1936, Buchholtz 1998, Uhen 1998), suggesting that fluke-based locomotion in cetaceans dates to the late Eocene. There is, however, no evidence for fluke-based locomotion in early and middle Eocene cetacean families (Thewissen & Fish 1997, Buchholtz 1998, Hulbert 1998, Bajpai & Thewissen 2001, Gingerich et al. 2001, Thewissen et al. 2001).

Theories about the evolution of cetacean locomotion predate the discovery of fossils of Eocene cetaceans. Fish (1993, 1996, 2001; Fish & Hui 1991) developed an explicit theory for cetacean locomotor evolution based on hydrodynamic and kinematic data on modern swimming mammals (Figure 5, color insert). Thewissen & Fish (1997) used modern mammalian swimmers to identify morphological correlates of each swimming mode and tested Fish's model by using these predicted morphologies to study the skeleton of *Ambulocetus*. *Ambulocetus*' morphology was consistent with Fish's model. Bajpai & Thewissen (2001), following methods developed by Buchholtz (e.g., 1998), studied the vertebral column of the remingtonocetid *Kutchicetus* and concluded that its inferred locomotor behavior was also consistent with Fish's hypothesis.

Locomotion: Development

Fish's hypothesis suggests that selection for enhanced swimming efficiency drove the modification of the postcranial skeleton. However, development may have constrained the course of evolution and therefore may be the mechanism that

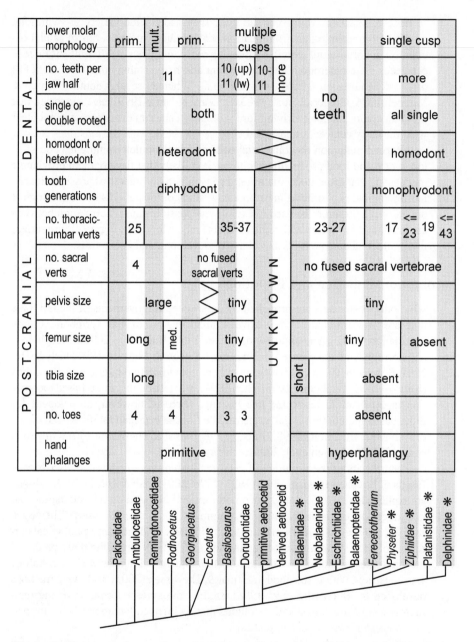

Figure 4   Summary of some features of the vertebral column, limbs, and dentition in fossil and modern cetaceans. Such a table can be used to study patterns of correlation among features and could be used to identify potential developmental constraints in cetacean evolution. Families with modern representatives are indicated with an asterisk. Sawtooth pattern indicates documented transitional morphologies.

underlies the observed evolutionary pattern. Cohn & Tickle (1999) studied the development of pythons and found that progressive limb loss coincides with loss of regionalization and addition of elements to the vertebral column. They also determined which genes controlled these processes. In embryonic pythons (Gasc 1966, Cohn & Tickle 1999), as well as in cetaceans (Ogawa 1953, Deimer 1977, Omura 1980, Sedmera et al. 1997), the hind-limb bud degenerates soon after its formation, and in adults only some remnants of pelvic girdle and hind-limb bones remain (Struthers 1881, 1893; Amasaki et al. 1989). This demonstrates that the ontogenetic pattern in cetaceans and snakes is very similar.

In addition, there is a suggestive similarity between the evolutionary pattern of cetaceans and the developmental pattern of snakes. Vertebral regionalization (differentiation between classes of vertebrae, such as thoracic and lumbar) and hind limb patterning decrease as fore limb size decreases in the development of snakes, and these processes are controlled by the same genes (Cohn & Tickle 1999). A similar pattern occurs in an evolutionary series of archaic cetaceans, in which hind limb patterning (such as digit count) and vertebral patterning (differentiation between classes of vertebrae) are both minimized (Figure 4). The number of toes decreases in archaic cetaceans [*Ambulocetus*, four toes (Thewissen et al. 1996)] to late Eocene forms [*Basilosaurus*, three toes (Gingerich et al. 1990)] to modern cetaceans (no distal hind limb, only internal bones) (Struthers 1893, Amasaki et al. 1989).

Vertebral regionalization is also mostly lost between early Eocene and modern times. Pakicetid lumbar vertebrae have large, recurved articular processes, and their thoracic vertebrae have smaller, unrecurved processes (Thewissen et al. 2001). Ambulocetid lumbar vertebrae have large processes but lack recurvature (Madar et al. 2002) and are thus more similar to thoracic vertebrae. Dorudontids have even smaller processes (Kellogg 1936, Uhen 1998). In most modern cetaceans articular processes of lumbar vertebrae are tiny to absent and do not differ from those of the thoracic, sacral, or caudal vertebrae (Slijper 1936). Regionalization of the vertebral column decreases as the hind limb becomes smaller.

Similarity between the developmental pattern of snakes and the evolutionary pattern of cetaceans raises the question of whether the same genes control development of these regions in both animals. It also begs the question whether two distinct features of great functional importance (vertebral column morphology and hind limb morphology) are developmentally constrained to co-evolve at some level. If this is the case, then selection for one could drive evolution, whereas the other would passively be pushed toward modern morphologies, without it being actively selected.

Feeding: Dental Evolution and Development

Loss of regionalization and increase in repeated elements (Figure 4) is also a theme in the evolution of the dentition (and its eventual loss in mysticetes). Odontocete dentitions are monophyodont, polydont, and homodont and have simple tooth crowns (consisting of a single cusp). Developing odontocete teeth appear to all be part of one tooth generation (Misek et al. 1996). Mysticete embryonic teeth are also more or less homodont (Dissel-Scherft & Vervoort 1954), and fossil mysticetes are homodont and polydont to some degree (Emlong 1966, Barnes et al. 1994). Homodonty and polydonty are gradually acquired, and early odontocetes and mysticetes are heterodont to some degree (Fordyce 1983) and have a limited number of excess teeth.

Eocene cetaceans show intermediate stages in the evolution of dentition. Molars of the earliest forms are primitive (lower molars have a trigonid and talonid), whereas late Eocene forms have simpler teeth more closely resembling those of early mysticetes and odontocetes (Kellogg 1936; Uhen 1998, 2001; Thewissen & Bajpai 2001). No Eocene cetacean is polydont, and all have two tooth generations (Uhen 2000).

Our current understanding of the genes that control development of dentition is limited (e.g., Jernvall & Jung 2000, Sharpe 2000, Teixeira 1999), and nothing is known about the differences in developmental control of the dental apparatus in cetaceans. However, it is clear that the main changes, loss of regionalization and repetition of similar elements, must have a developmental basis.

CONCLUSIONS

It has been pointed out that there is a surprising similarity in factors controlling limb and tooth development (Jernvall & Thesleff 2000, Sharpe 2000). There are no specific studies that show that such developmental correlations actually changed or constrained the evolutionary history of an organism. We contend that cetaceans may be such a case.

In cetaceans there is a historical correlation between hind limb reduction and homogenization of the vertebral column. Both may be underlain by the same genetic changes, as they are in snakes. Furthermore, the similarities in genetic cassettes that are used in both limb and tooth development make it tempting to speculate that the similarities in historical pattern (loss of regionalization, repetition of elements) in locomotor system and dentition are underlain by the similar developmental patterns. Patterns of correlation in evolutionary history can be used to identify correlations, and the cetacean fossil record is beginning to show intriguing patterns. Figure 4 is an attempt at listing characters that can form the basis for a study of such correlations.

It is clear that selection played a pivotal role in identifying morphologies useful for the survival of Eocene cetaceans in their new, watery, environment, but development is at the root of producing the morphologies available for sorting. Studying the interaction of development and evolution is one of the most exciting challenges in evolutionary biology and one in which historical patterns in cetaceans may play a pivotal role.

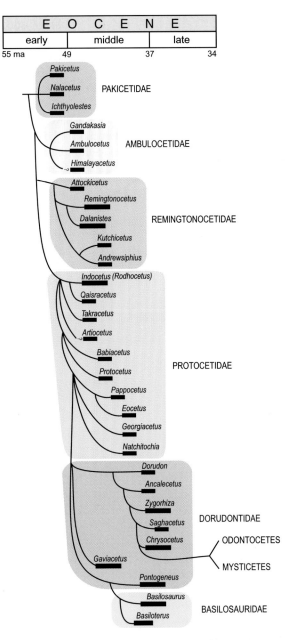

Figure 1 Phylogenetic hypothesis for the relations among all Eocene genera of cetaceans. This phylogeny combines results from several explicit cladistic analyses (Uhen 1998, 1999; Geisler 2001; Gatesy & O'Leary 2001; Thewissen & Hussain 2000; Thewissen et al. 2001a). No phylogenetic analysis of all these taxa is available, and several families are probably paraphyletic.

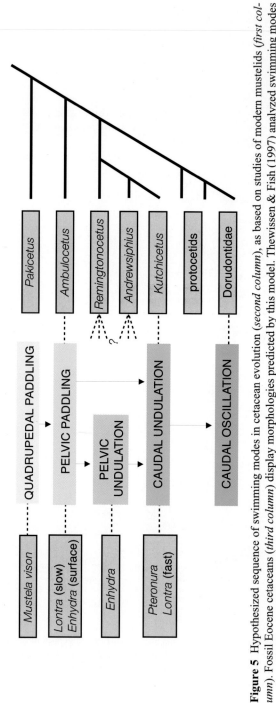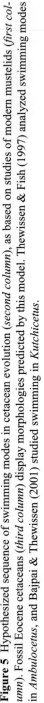

Figure 5 Hypothesized sequence of swimming modes in cetacean evolution (*second column*), as based on studies of modern mustelids (*first column*). Fossil Eocene cetaceans (*third column*) display morphologies predicted by this model. Thewissen & Fish (1997) analyzed swimming modes in *Ambulocetus*, and Bajpai & Thewissen (2001) studied swimming in *Kutchicetus*.

The *Annual Review of Ecology and Systematics* is online at
http://ecolsys.annualreviews.org

LITERATURE CITED

Amasaki H, Ishikawa H, Daigo M. 1989. Developmental changes of the fore- and hind-limb in the fetuses of the Southern Minke whale. *Anat. Anz.* 169:14–148

Aslan A, Thewissen JGM. 1996. Preliminary evaluation of paleosols and implications for interpreting vertebrate fossil assemblages, Kuldana Formation, northern Pakistan. *Palaeovertebrata* 25(2–4):261–77

Bajpai S, Gingerich PD. 1998. A New Eocene archaeocete (Mammalia, Cetacea) from India and the time of origin of whales. *Proc. Natl. Acad. Sci. USA* 95:15464–68

Bajpai S, Thewissen JGM. 1998. Middle Eocene cetaceans from the Harudi and Subathu Formations of India. See Thewissen 1998a, pp. 213–33

Bajpai S, Thewissen JGM. 2001. A new, diminutive Eocene whale from Kachchh (Gujarat, India) and its implications for locomotor evolution of cetaceans. *Curr. Sci.* 79(10, 25):1478–82

Bajpai S, Thewissen JGM, Sahni A. 1996. *Indocetus* (Cetacea, Mammalia) endocasts from Kachchh (India). *J. Vertebr. Paleontol.* 16:582–84

Barnes LG, Kimura M, Furusawa H, Sawamura H. 1994. Classification and distribution of Oligocene Aetiocetidae (Mammalia; Cetacea; Mysticeti) from western North America and Japan. *Island Arc.* 3:392–431

Buchholtz EA. 1998. Implications of vertebral morphology for locomotor evolution in early Cetacea. See Thewissen 1998a, pp. 325–51

Buchholtz EA. 2001. Modeling whales: functional interpretation of delphinoid postcrania. *J. Vertebr. Paleontol.* 21:36A

Cohn MJ, Tickle C. 1999. Developmental basis of limblessness and axial regionalization in snakes. *Nature* 399:474–79

Dehm R, Oettingen-Spielberg T. 1958. Paläontologische und geologische Untersuchungen im Tertiär von Pakistan. 2. Die mitteleocanen Saügetiere vom Ganda Kas bei Basal in Nordwest Pakistan. *Abh. Bayer. Akad. Wiss. Math. Naturwiss. Kl. NF* 91:1–54

Deimer VP. 1977. Der rudimentäre hintere Extremitätengurtel des Pottwals (*Physeter macrocephalus* Linnaeus, 1758), seine Varabilität und Wachstumallometrie. *Z. Säugetierk.* 42:88–101

Dissel-Scherft MCV, Vervoort W. 1954. Development of the teeth in fetal *Balaenoptera physalus* (L.) (Cetacea, Mysticeti). *Proc. Ned. Akad. Wet. B* 197–210

Emlong D. 1966. A new archaic cetacean from the Oligocene of Northwest Oregon. *Bull. Mus. Nat. Hist. Univ. Ore.* 3:1–51

Fish FE. 1993. Influence of hydrodynamic design and propulsive mode on mammalian swimming energetics. *Aust. J. Zool.* 42:79–101

Fish FE. 1996. Transitions from drag-based to lift-based propulsion in mammalian swimming. *Am. Zool.* 36:628–41

Fish FE. 1998. Biomechanical perspective on the origin of cetacean flukes. See Thewissen 1998a, pp. 303–24

Fish FE. 2001. A mechanism for evolutionary transition in swimming mode by mammals. See Mazin & de Buffrénil 2001, pp. 261–87

Fish FE, Hui CA. 1991. Dolphin swimming—a review. *Mamm. Rev.* 21:181–95

Fordyce RE. 1983. Dental anomaly in a fossil squalodont dolphin from New Zealand and the evolution of polydonty in whales. *NZ J. Zool.* 9:419–26

Fordyce RE. 2001. Neoceti. See Perrin et al. 2002, pp. 787–91

Fordyce RE, de Muizon C. 2001. Evolutionary history of the cetaceans: a review. See Mazin & de Buffrénil 2001, pp. 261–87

Fraas E. 1904. Neue Zeuglodonten aus dem

unteren Mitteleocän vom Mokattam bei Cairo. *Geol. Paläontol. Abh. N. F.* 6:199–220

Gasc JP. 1966. Les rapports anatomiques du member pelvien vestigial chez les squamates serpentiformes. *Bull. Mus. Nat. Hist. Ser. 2* 38:99–110

Gatesy J, O'Leary MA. 2001. Deciphering whale origins with molecules and fossils. *Trends Ecol. Evol.* 16:562–70

Geisler JH. 2001. New morphological evidence for the phylogeny of Artiodactyla, Cetacea, and Mesonychidae. *Am. Mus. Novit.* 3344:1–53

Gingerich PD, Arif M, Bhatti MA, Anwar M, Sanders WJ. 1995a. *Protosiren* and *Babiacetus* (Mammalia, Sirenia and Cetacea) from the middle Eocene Drazinda Formation, Sulaiman Range, Punjab (Pakistan). *Contrib. Mus. Paleontol. Univ. Mich.* 29:331–57

Gingerich PD, Arif M, Bhatti MA, Anwar M, Sanders WJ. 1997. *Basilosaurus drazindai* and *Basiloterus hussaini*, new Archaeoceti (Mammalia, Cetacea) from the middle Eocene Drazinda Formation, with revised interpretation of ages of whale-bearing strata in the Kirthar Group of the Sulaiman Range, Punjab (Pakistan). *Contrib. Mus. Paleontol. Univ. Mich.* 30:55–81

Gingerich PD, Arif M, Clyde WC. 1995b. New archaeocetes (Mammalia, Cetacea) from the middle Eocene Domanda Formation of the Sulaiman Range, Punjab (Pakistan). *Contrib. Mus. Paleontol. Univ. Mich.* 29(11):291–330

Gingerich PD, Haq M, Zalmout IS, Khan IH, Malkani MS. 2001. Origin of whales from early artiodactyls: hands and feet of Eocene Protocetidae from Pakistan. *Science* 293:2239–42

Gingerich PD, Raza SM, Arif M, Anwar M, Zhou X. 1993. Partial skeletons of *Indocetus ramani* (Mammalia, Cetacea) from the lower Middle Eocene Domanda Shale in the Sulaiman Range of Punjab (Pakistan). *Contrib. Mus. Paleontol. Univ. Mich.* 28:393–416

Gingerich PD, Raza SM, Arif M, Anwar M, Zhou X. 1994. New whale from the Eocene of Pakistan and the origin of cetacean swimming. *Nature* 368:844–47

Gingerich PD, Russell DE. 1981. *Pakicetus inachus*, a new archaeocete (Mammalia, Cetacea). *Contrib. Mus. Paleontol. Univ. Mich.* 25:235–46

Gingerich PD, Smith BH, Simons AL. 1990. Hind limbs of Eocene *Basilosaurus*: evidence of feet in whales. *Science* 249:154–57

Gingerich PD, Wells NA, Russell DE, Shah SMI. 1983. Origin of whales in epicontinental remnant seas: new evidence from the early Eocene of Pakistan. *Science* 220:403–6

Hulbert RC. 1998. Postcranial osteology of the North American middle Eocene protocetid *Georgiacetus*. See Thewissen 1998a, pp. 235–67

Hulbert RC Jr, Petkewich RM, Bishop GA, Bukry D, Aleshire DP. 1998. A new middle Eocene protocetid whale (Mammalia: Cetacea: Archaeoceti) and associated biota from Georgia. *J. Paleontol.* 72(5):907–26

Jernvall J, Jung HS. 2000. Genotype, phenotype and developmental biology of molar tooth characters. *Am. J. Phys. Anthropol.* 43S:171–90

Jernvall J, Thesleff I. 2000. Return of lost structure in the developmental control of tooth shape. In *Development, Function, and Evolution of Teeth*, ed. MF Teaford, MM Smith, MWJ Ferguson. New York: Cambridge Univ. Press. 314 pp.

Kellogg R. 1928. The history of whales—their adaptation to life in the water. *Q. Rev. Biol.* 3:29–76

Kellogg AR. 1936. A review of the Archaeoceti. *Carnegie Inst. Wash. Publ.* 482:1–366

Kumar K, Sahni A. 1986. *Remingtonocetus harudiensis*, new combination, a middle Eocene archaeocete (Mammalia, Cetacea) from western Kutch, India. *J. Vertebr. Paleontol.* 6:326–49

Lipps JH, Mitchell E. 1976. Trophic model for the adaptive radiations and extinctions of pelagic marine mammals. *Paleobiology* 2:147–55

Lucas FA. 1900. The pelvic girdle of zeuglodon *Basilosaurus cetoides* (Owen), with notes on other portions of the skeleton. *Proc. US Natl. Mus.* 23:327–31

Luo Z, Gingerich PD. 1999. Terrestrial Mesonychia to aquatic Cetacea: transformation of the basicranium and evolution of hearing in whales. *Univ. Mich. Pap. Paleontol.* 31:1–98

Madar SI, Thewissen JGM, Hussain ST. 2002. Additional holotype remains of *Ambulocetus natans* (Cetacea, Ambulocetidae), and their implications for locomotion in early whales. *J. Vertebr. Paleontol.* 22:405–22

Marino L, Uhen MD, Frohlich B, Aldag JM, Blane C, et al. 2000. Endocranial volume of mid-late Eocene Archaeocetes (Order: Cetacea) revealed by computed tomography: implications for Cetacean brain evolution. *J. Mamm. Evol.* 7:81–93

Mazin JM, de Buffrénil V, eds. 2001. *Secondary Adaptation of Tetrapods to Life in Water: Proc. Int. Meet. Poitiers, 1996.* München: Pfeil. 367 pp.

Mísek I, Witter K, Peterka M, Lesot H, Sterba O, et al. 1996. Initial period of tooth development in dolphins (*Stenella attenuata*, Cetacea)—a pilot study. *Acta Vet. Brno* 65:277–84

Misra BK. 1992. Genesis of Indian Tertiary coals and lignites: a biopetrological and palaeobotanical view point. *Palaeobotanist* 40:490–513

Nikaido M, Pooney AP, Okada N. 1999. Phylogenetic relationships among cetartiocdactyls based on insertions of short and long interspersed elements: Hippopotamuses are the closest extant relatives of whales. *Proc. Natl. Acad. Sci. USA* 96:10261–66

Ogawa T. 1953. On the presence and disappearance of the hind limb in the cetacean embryos. *Sci. Rep. Whale Res. Inst.* 8:127–32

Omura H. 1980. Morphological study of pelvic bones of the Minke Whale from the Antarctic. *Sci. Rep. Whale Res. Inst.* 32:25–37

O'Leary MA, Uhen MD. 1999. The time of origin of whales and the role of behavioral changes in the terrestrial-aquatic transition. *Paleobiology* 25:534–56

Perrin WF, Würsig B, Thewissen JGM, eds. 2002. *Encyclopedia of Marine Mammals.* San Diego: Academic. 1414 pp.

Rice DW. 1998. *Marine Mammals of the World: Systematics and Distribution.* Soc. Marine Mammalogy. Lawrence, KS 231 pp.

Roe LJ, Thewissen JGM, Quade J, O'Neil JR, Bajpai S, et al. 1998. Isotopic approaches to understanding the terrestrial-to-marine transition of the earliest Cetaceans. See Thewissen 1998a, pp. 399–419

Rose KD. 1982. Skeleton of *Diacodexis*, oldest known artiodactyl. *Science* 216:621–23

Sahni A, Mishra VP. 1972. A new species of *Protocetus* (Cetacea) from the middle Eocene of Kutch, western India. *Palaeontology* 15:490–95

Sahni A, Mishra VP. 1975. Lower Tertiary vertebrates from western India. *Monogr. Palaeontol. Soc. India* 3:1–48

Sedmera D, Mísek I, Klima M. 1997. On the development of cetacean extremities: I. Hind limb rudimentation in the spotted dolphin (*Stenella attenuata*). *Eur. J. Morphology* 35:25–30

Sharpe PT. 2000. Homeobox genes in initiation and shape of teeth during development in mammalian embryos. In *Development, Function, and Evolution of Teeth*, ed. MF Teaford, MM Smith, MWJ Ferguson. New York: Cambridge Univ. Press. 314 pp.

Slijper EJ. 1936. Die Cetaceen: vergleichend-anatomisch und systematisch. *Capita. Zool.* VI/VII:1–590

Slijper EJ. 1960. Locomotion and locomotory organs in whales and dolphins (Cetacea). *Symp. Zool. Soc. London* 5:77–94

Stromer E. 1908. Die Archaeoceti des Ägyptischen Eozäns. *Beitr. Paläontol. Geol. Österreich-Ungarns Orients* 21:106–77

Struthers J. 1881. On the bones, articulations, and muscles of the rudimentary hind-limb of the Greenland Right whale (*Balaena mysticetus*). *J. Anat. Physiol.* 15:141–32

Struthers MD. 1893. On the rudimentary hind-limb of a great fin-whale (*Balaenoptera musculus*) in comparison with those of the Humpback whale and the Greenland Right-whale. *J. Anat. Physiol.* 27:291–335

Thewissen JGM, ed. 1998a. *The Emergence of Whales: Evolutionary Patterns in the Origin of Cetacea*. New York: Plenum. 477 pp.

Thewissen JGM. 1998b. Cetacean origins: evolutionary turmoil during the invasion of the oceans. See Thewissen 1998a, pp. 451–64

Thewissen JGM, Bajpai S. 2001. Whale origins as a poster child for macroevolution. *Bioscience* 51:1017–29

Thewissen JGM, Fish FE. 1997. Locomotor evolution in the earliest cetaceans: functional model, modern analogues, and paleontological evidence. *Paleobiology* 23:482–90

Thewissen JGM, Hussain ST. 1993. Origin of underwater hearing in whales. *Nature* 361:444–45

Thewissen JGM, Hussain ST. 1998. Systematic review of the Pakicetidae, early and middle Eocene Cetacea (Mammalia) from Pakistan and India. *Bull. Carnegie Mus.* 34:220–38

Thewissen JGM, Hussain ST. 2000. *Attockicetus praecursor*, a new remingtonocetid cetacean from marine Eocene sediments of Pakistan. *J. Mamm. Evol.* 7:133–46

Thewissen JGM, Hussain ST, Arif M. 1994. Fossil evidence for the origin of aquatic locomotion in archaeocete whales. *Science* 263:210–12

Thewissen JGM, Madar SI, Hussain ST. 1996. *Ambulocetus natans*, an Eocene cetacean (Mammalia) from Pakistan. *Cour. Forsch.-Inst. Senckenberg* 190:1–86

Thewissen JGM, Williams EM, Hussain ST. 2001a. Eocene mammal faunas from northern Indo-Pakistan. *J. Vertebr. Paleontol.* 21: 347–66

Thewissen JGM, Williams EM, Roe LJ, Hussain ST. 2001b. Skeletons of terrestrial cetaceans and the relationship of whales to artiodactyls. *Nature* 413:277–81

Teixeira CC. 1999. New horizons in understanding early tooth development. *Clin. Orthod. Res.* 2:171–74

Uhen MD. 1998. Middle to Late Eocene Basilosaurines and Dorudontines. See Thewissen 1998a, pp. 29–61

Uhen MD. 1999. New species of protocetid archaeocete whales, *Eocetus wardii* (Mammalia: Cetacea) from the middle Eocene of North Carolina. *J. Paleontol.* 73:512–28

Uhen M. 2000. Replacement of deciduous first premolars and dental eruption in archaeocete whales. *J. Mamm.* 81:123–33

Uhen MD. 2001. Basilosaurids. See Perrin et al. 2002, pp.78–81

Uhen MD, Gingerich PD. 2000. New genus of dorudontine archaeocete (Cetacea) from the middle-to-late Eocene of South Carolina. *Marine Mamm. Sci.* 17:1–34

Watson AG, Fordyce RE. 1993. Skeletons of two Minke whales, *Balaenoptera acutorostrata*, stranded on the south-east coast of New Zealand. *NZ Nat. Sci.* 20:1–14

West RM. 1980. Middle Eocene large mammal assemblage with Tethyan affinities, Ganda Kas Region, Pakistan. *J. Paleontol.* 54:508–33

Williams EM. 1998. Synopsis of the earliest cetaceans: Pakicetidae, Ambulocetidae, Remingtonocetidae, and Protocetidae. See Thewissen 1998a, pp. 1–28

Annu. Rev. Ecol. Sys. 2002. 33:91–124
doi: 10.1146/annurev.ecolsys.33.010802.150517
Copyright © 2002 by Annual Reviews. All rights reserved
First published online as a Review in Advance on August 6, 2002

THE MESOZOIC RADIATION OF BIRDS

Luis M. Chiappe[1] and Gareth J. Dyke[2]

[1]*Department of Vertebrate Paleontology, Natural History Museum of Los Angeles County,
900 Exposition Boulevard, Los Angeles, California 90007; email: lchiappe@nhm.org*
[2]*Division of Vertebrate Zoology (Ornithology), American Museum of Natural History,
Central Park West at 79th Street, New York, New York 10024; email: gdyke@amnh.org*

Key Words birds, Mesozoic, origin, evolution, extinction, aves, flight, feathers, phylogeny

■ **Abstract** Until recently, most knowledge of the early history of birds and the evolution of their unique specializations was based on just a handful of diverse Mesozoic taxa widely separated in time and restricted to marine environments. Although *Archaeopteryx* is still the oldest and only Jurassic bird, a wealth of recent discoveries combined with new phylogenetic analyses have documented the divergence of a number of lineages by the beginning of the Cretaceous. These and younger Cretaceous fossils have filled much of the morphological chasm that existed between *Archaeopteryx* and its living counterparts, providing insights into the evolutionary development of feathers and other important features of the avian flight system. Dramatic new perceptions of the life history, growth and development of early birds have also been made possible by the latest data. Although no primitive birds are known to have survived beyond the end of the Cretaceous, the present fossil record provides no evidence for a sudden disappearance. Likewise, a Mesozoic origin for extant birds remains controversial.

INTRODUCTION

Birds are the most speciose group of land vertebrates. Today's 10,000 species are the extant members of an ancient radiation that can be traced back 150 million years, to the famous *Archaeopteryx lithographica* from the Late Jurassic Solnhofen limestones of Germany. The taxonomic diversity and genealogical relationships of early birds, the origin and refinement of flight, the timing of divergence of extant lineages, and the origin of avian functional and physiological specializations are just some of the evolutionary issues that have captured the interest of decades of paleoornithological research. For most of this time, evidence for investigating these issues was limited to a small number of fossils greatly separated both temporarily and morphologically, and largely restricted to near-shore and marine environments. This situation has continued to change over the past two decades, as increasing discoveries of Cretaceous birds have begun to reveal an unexpected diversity of lineages (Figure 1). The number of new species of Mesozoic birds discovered and described over the past 10 years more than triples those known for much of the past

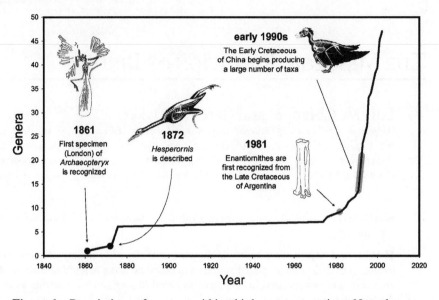

Figure 1 Descriptions of non-neornithine bird genera over time. Note the steep increment of new discoveries during the past two decades.

two centuries (Table 1). This burst of fossil discoveries has been summarized in a number of recent publications (Chiappe 1995, Chatterjee 1997, Padian & Chiappe 1998, Feduccia 1999). Although essentially examining the same fossil record, these studies reveal a variety of interpretations resulting from different methodological approaches to reconstructing phylogeny. Our review is one guided by the principles of phylogenetic systematics.

ORIGINS

The origin of birds—Aves, the clade including the last common ancestor of *Archaeopteryx* and living birds—has been the subject of debate almost since the advent of evolutionary thought. Historical proposals for the ancestry of birds have included turtles, lizards, crocodylomorphs, and pterosaurs, as well as both ornithischian and theropod dinosaurs (Witmer 1991, Padian & Chiappe 1998). Today, in spite of a few remaining and poorly substantiated hypotheses identifying crocodiles (Martin 1983, Martin et al. 1980, Martin & Stewart 1999) or a variety of basal archosauromorphs (Feduccia & Wild 1993, Welman 1995) as birds' closest relatives, most researchers agree that birds are theropod dinosaurs (Chatterjee 1997, Chiappe 2001, Gauthier & Gall 2001). Although this idea had its roots in the nineteenth century, modern hypotheses stem from the detailed work of J. Ostrom (1969, 1973, 1976). Since then, a wealth of osteological evidence has accumulated in support of Ostrom's hypothesis that birds originated within small and predominantly terrestrial coelurosaurian theropods (Ostrom 1976;

TABLE 1 Described genera of Mesozoic birds (those of questionable validity are not listed). Environment and year refer to the depositional environment and year of original description, respectively; taxonomic position of *Rahonavis ostromi* follows Forster et al. 1998a

Genera	Clade	Stratigraphic age	Material	Environment	Distribution	Year
Archaeopteryx	Archaeopterygidae	L. Jurassic	Several specimens	Near shore	Germany	1861
Rahonavis	Archaeopterygidae	L. Cretaceous	Single specimen	Inland	Madagascar	1998
Confuciusornis	Confuciusornithidae	E. Cretaceous	Many specimens	Inland	China	1995
Changchengornis	Confuciusornithidae	E. Cretaceous	Single specimen	Inland	China	1999
Nanantius	Enantiornithes	E. Cretaceous	Isolated bones	Near shore	Australia	1986
Noguerornis	Enantiornithes	E. Cretaceous	Single specimen	Inland	Spain	1989
Iberomesornis	Enantiornithes	E. Cretaceous	Single specimen	Inland	Spain	1992
Sinornis	Enantiornithes	E. Cretaceous	Several specimens	Inland	China	1992
Otogornis	Enantiornithes	E. Cretaceous	Single specimen	Inland	China	1994
Boluochia	Enantiornithes	E. Cretaceous	Single specimen	Inland	China	1995
Concornis	Enantiornithes	E. Cretaceous	Single specimen	Inland	Spain	1995
Eoalulavis	Enantiornithes	E. Cretaceous	Single specimen	Inland	Spain	1996
Eoenantiornis	Enantiornithes	E. Cretaceous	Single specimen	Inland	China	1999
Protopteryx	Enantiornithes	E. Cretaceous	Single specimen	Inland	China	2000
Gobipteryx	Enantiornithes	L. Cretaceous	Several specimens	Inland	Mongolia	1974
Alexornis	Enantiornithes	L. Cretaceous	Isolated bones	Inland	Mexico	1976
Enantiornis	Enantiornithes	L. Cretaceous	Isolated bones	Inland	Argentina	1981
Avisaurus	Enantiornithes	L. Cretaceous	Isolated bones	Inland	USA	1985
Soroavisaurus	Enantiornithes	L. Cretaceous	Isolated bones	Inland	Argentina	1993
Yungavolucris	Enantiornithes	L. Cretaceous	Isolated bones	Inland	Argentina	1993

(Continued)

TABLE 1 (*Continued*)

Genera	Clade	Stratigraphic age	Material	Environment	Distribution	Year
Lectavis	Enantiornithes	L. Cretaceous	Isolated bones	Inland	Argentina	1993
Neuquenornis	Enantiornithes	L. Cretaceous	Single specimen	Inland	Argentina	1994
Halimornis	Enantiornithes	L. Cretaceous	Single specimen	Offshore	USA	2002
Enaliornis	Ornithuromorpha	E. Cretaceous	Isolated bones	Offshore	UK	1876
Ambiortus	Ornithuromorpha	E. Cretaceous	Single specimen	Inland	Mongolia	1982
Gansus	Ornithuromorpha	E. Cretaceous	Single specimen	Inland	China	1984
Longipteryx	Ornithuromorpha	E. Cretaceous	Single specimen	Inland	China	2001
Yanornis	Ornithuromorpha	E. Cretaceous	Single specimen	Inland	China	2001
Yixianornis	Ornithuromorpha	E. Cretaceous	Single specimen	Inland	China	2001
Hesperornis	Ornithuromorpha	L. Cretaceous	Many specimens	Offshore	USA/Canada	1872
Ichthyornis	Ornithuromorpha	L. Cretaceous	Many specimens	Offshore	USA/Canada	1872
Apatornis	Ornithuromorpha	L. Cretaceous	Single specimen	Offshore	USA	1876
Baptornis	Ornithuromorpha	L. Cretaceous	Several specimens	Offshore	USA/Canada	1877
Parahesperornis	Ornithuromorpha	L. Cretaceous	Single specimen	Offshore	USA	1984
Patagopteryx	Ornithuromorpha	L. Cretaceous	Several specimens	Inland	Argentina	1992
Vorona	Ornithuromorpha	L. Cretaceous	Single specimen	Inland	Madagascar	1996
Limenavis	Ornithuromorpha	L. Cretaceous	Single specimen	Inland	Argentina	2001
Apsaravis	Ornithuromorpha	L. Cretaceous	Single specimen	Inland	Mongolia	2001
Chaoyangia	Controversial status	E. Cretaceous	Single specimen	Inland	China	1995
Liaoningornis	Controversial status	E. Cretaceous	Single specimen	Inland	China	1996
Sapeornis	Controversial status	E. Cretaceous	Single specimen	Inland	China	2002

Abbreviations: E., Early; L., Late.

Figure 2   Reconstructions of the dromaeosaurid *Velociraptor mongoliensis* (*left*; after Paul 1988) and the oviraptorosaur *Caudipteryx zoui* (*right*; after Currie 2000) scaled to a rock pigeon (*Columba livia*).

Gauthier 1986; Holtz 1998, 2001; Sereno 1999; Norell et al. 2001) (Figure 2). Alternative hypotheses, however, compete regarding the exact sister-taxon of Aves among coelurosaurians, with dromaeosaurids (e.g., *Deinonychus*, *Velociraptor*, *Sinornithosaurus*), troodontids (e.g., *Troodon*, *Byronosaurus*), oviraptorids (e.g., *Oviraptor*, *Khan*), and alvarezsaurids (e.g., *Mononykus*, *Shuvuuia*), being commonly cited (Ostrom 1976; Gauthier 1986; Perle et al. 1993a, 1994; Holtz 1998, 2001; Sereno 1999; Chiappe et al. 1996, 1998; Elzanowski 1999; Xu et al. 1999b, 2000; Norell et al. 2001).

For decades, interpretation of birds as living dinosaurs was based on osteological comparisons, but a series of recent discoveries have provided additional evidence in support of this hypothesis (Chiappe 2001). Discoveries of embryonic remains of coelurosaurians inside their eggs have provided evidence of the egg morphology of theropod dinosaurs (Norell et al. 1994, Varricchio et al. 1997) and have shown that features of the shell microstructure are uniquely shared between birds and these non-avian theropod lineages (Grellet-Tinner & Chiappe, 2002). Specimens associating adult coelurosaurians with their clutches of eggs have allowed for inferences of nesting behavior and suggest the presence of avian brooding behavior among these dinosaurs (Norell et al. 1995, Dong & Currie 1996, Varricchio et al. 1997, Clark et al. 1999). Finally, a number of exquisitely preserved coelurosaurians from the Early Cretaceous of China (Chen et al. 1998; Ji et al. 1998; Xu et al. 1999b, 2000, 2001; Zhou & Wang 2000; Zhou et al. 2000; Norell et al. 2002) have compellingly shown that feathers had their origin within theropod dinosaurs. This wealth of data accumulated over decades of osteological research and recently corroborated by studies of oology, behavior, and integument, suggests beyond any reasonable doubt that birds evolved from coelurosaurian dinosaurs sometime before the Late Jurassic. However, the temporal context for this divergence hinges upon identifying the oldest records for the group.

For much of the history of paleoornithology, the Late Jurassic *Archaeopteryx* stood unchallenged as the oldest known bird. In recent years, however, several alleged fossil birds have been claimed to be older than this taxon, although none

Figure 3   Reconstruction of the controversial *Protoavis texensis* from the Late Triassic of Texas (after Chatterjee 1999) scaled to a rock pigeon (*Columba livia*). In spite of comprehensive reconstructions such as this, the known material of this taxon is extremely fragmentary.

has yet been proved convincingly (Molnar 1985, Chiappe 1995, Padian & Chiappe 1998). The most publicized and contentious is *Protoavis texensis* from the Late Triassic Dockum Group of West Texas (Chatterjee 1991, 1997, 1999) (Figure 3). Most of the available material of this taxon comprises two disarticulated skeletons collected in 1983 from a bone-bed (Post Quarry) that has produced a diverse array of tetrapods (Chatterjee 1997). A few more isolated bones were collected years later from another site (Kirkpatrick Quarry) some 50 km away and 60 m lower than, Post Quarry (Chatterjee 1999). Because *Protoavis* is more than 75 million years older than *Archaeopteryx* and supposedly a member of Ornithothoraces (Figure 4) (Chatterjee 1991, 1999), acceptance of this taxon as a bird would mean that the origin of the group would have to have occurred deep in the Triassic, if not earlier. Yet the available material of *Protoavis* is problematic. The poor preservation of several elements precludes their osteological identification and a number of features have been misinterpreted (Chiappe 1998, Witmer 2001). Some of the additional material from Kirkpatrick Quarry—in particular a large keeled bone interpreted as a sternum (Figure 3)—does not overlap with either the type or referred specimen from Post Quarry, thus rendering impossible their referral to *Protoavis*. Even the placement of these specimens within a single species has been questioned (Ostrom 1991, Chiappe 1998, Renesto 2000, Sereno 2000, Witmer 2001), which some believe to be a composite formed by disparate taxa (Sereno 2000, Witmer 2001). Although it is true that the issue to be addressed is not whether the skeletons of *Protoavis* are made of several disparate taxa but rather whether any of those bones are avian (Witmer 1997, 2001), such a conclusion should

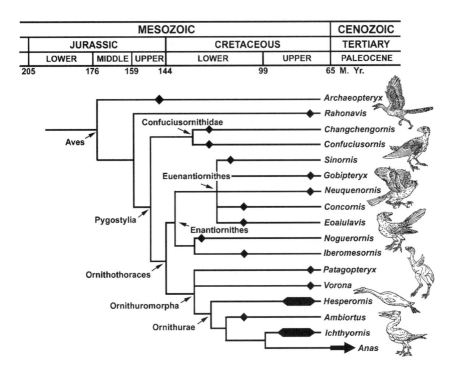

MESOZOIC					CENOZOIC
JURASSIC			CRETACEOUS		TERTIARY
LOWER	MIDDLE	UPPER	LOWER	UPPER	PALEOCENE
205	176	159 144		99	65 M. Yr.

Figure 4 Phylogenetic relationships of principal lineages of Mesozoic birds (after Chiappe 2001).

be supported by a cladistic analysis, difficult to perform with incomplete single bones (Clarke & Chiappe 2001). Interestingly, one of the most compelling avian features of *Protoavis*, the near-heterocoelic (saddle-shaped) cervical vertebrae, has recently been reported for *Megalancosaurus preonensis*, a chameleon-bodied diapsid reptile of uncertain relationships (Renesto 2000), but also from the Late Triassic.

No reliable evidence has yet come to light in support of the existence of fossil birds prior to the Late Jurassic; *Archaeopteryx lithographica* remains the oldest known and most primitive bird. Eight skeletal specimens and a feather (Mäuser 1997, Elzanowski 2002) presumed to belong to this taxon are in existence but in spite of their spectacular preservation, only limited anatomical information is available for certain areas of the skeleton (e.g., braincase, palate, orbit, sternum, feet). Even the placement of these specimens in one or more closely related species remains controversial. Although new names have occasionally been applied to some of them (e.g., Howgate 1984, Wellnhofer 1993, Elzanowski 2001), support for the existence of another bird as well as *Archaeopteryx lithographica* in the Solnhofen limestones remains unconvincing. The most debated aspect of this bird, however, concerns its mode of life. Was it predominantly arboreal or terrestrial? Did it climb trees using its forelimbs? Could it fly, and if so, how well? Was it a "glider" or a "flapper"? Although numerous and diverse hypotheses have been proposed for these and other ecological and functional questions (Padian & Chiappe 1998), they remain conjectural and are unlikely ever to be tested. At this point, we concur with functional and aerodynamic considerations that interpret *Archaeopteryx* as a predominantly terrestrial bird (Ostrom 1974, Chiappe 1995), able to take off from the ground (Burgers & Chiappe 1999) and to fly by flapping its asymmetrically feathered wings (Rayner 2001).

BASAL LINEAGES

For almost a century, knowledge of the Cretaceous diversity of birds was limited to a series of fossils from the marine deposits of the Late Cretaceous Western Interior Seaway of North America. Although toothed, these ichthyornithiforms and hesperornithiforms (Figure 5) were distinctly modern in many aspects of their skeletons (Marsh 1880), a fact noticed by early studies that placed them either close to recent birds (Heilmann 1926) or as basal forms of different extant lineages (Marsh 1880, Brodkorb 1971, Simpson 1980). The morphology of *Hesperornis* and *Ichthyornis* testified to an enormous gap in the early history of the group when

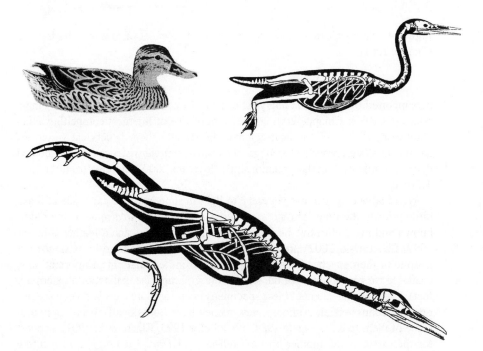

Figure 5 Reconstructions of the marine ornithurine birds *Hesperornis regalis* (*bottom*) and *Baptornis advenus* (*top right*) scaled to a mallard (*Anas platyrhynchos*).

compared to the much older *Archaeopteryx*. Many of the discoveries of recent years have served to fill in this morphological and temporal gap and now more than 30 valid avian taxa, representing different intermediate lineages, have been recognized from the Mesozoic (Table 1). Conversely the fossil record of birds close to the divergence of the extant lineages (Neornithes) has remained largely unchanged (Norell & Clarke 2001).

Cretaceous Diversity

Much recent controversy surrounded the discovery of two Late Cretaceous lineages claimed to be only slightly more advanced than *Archaeopteryx*: alvarezsaurids (Figure 6) and *Rahonavis ostromi* (Figure 7). Alvarezsaurids were first recognized in 1991 on the basis of an incomplete specimen from Patagonia, named *Alvarezsaurus calvoi* and interpreted as an enigmatic non-avian theropod (Bonaparte 1991). The subsequent discovery of *Mononykus olecranus* (Perle et al. 1993a,b) from the Gobi Desert (Figure 6), a long-legged cursor with short and stout forelimbs, and a multi-toothed and highly kinetic skull (Perle et al. 1994, Chiappe et al. 1998), shed much light both on the bird-like osteology and phylogenetic position of the group. Nonetheless, the close relationship between *Alvarezsaurus* and *Mononykus* was not recognized until the discovery of the Patagonian *Patagonykus puertai* (Novas 1996, 1997). Since then other fossil discoveries as well as new taxa have documented that alvarezsaurids were a diverse group of agile cursors inhabiting South and North America as well as Central Asia (Chiappe et al. 2002a). In spite of the superficially non-avian morphology of the abbreviated forelimb of *Mononykus*, initial cladistic analyses (Perle et al. 1993a, Chiappe et al. 1996, Novas 1996) interpreted it to be phylogenetically closer to extant birds than *Archaeopteryx*. This "avian hypothesis" immediately resulted in a great deal of opposition (Martin & Rinaldi 1994, Ostrom 1994, Wellnhofer 1994, Zhou 1995a, Feduccia & Martin 1996), although dissent was not framed in a cladistic context (Chiappe et al. 2002a). Studies of additional fossils and cladistic analyses following that of Perle et al. (1993a) consolidated the avian relationship of the group (Chiappe et al. 1996, 1998; Novas 1996, 1997; Forster et al. 1998a; Holtz 1998, 2001). The "avian hypothesis" received further support when structures found surrounding a skeleton of the Asian alvarezsaurid *Shuvuuia deserti* were immunologically shown to be composed of only β-keratin (Schweitzer et al. 1999), just like the feathers of birds (Brush 2001). However, given the subsequent discovery of feathers in non-avian theropods, the presence of these integumentary structures in alvarezsaurids would not alone be sufficient to support their avian relationship (Schweitzer et al. 1999).

Over the past few years, however, more serious questions have been raised contrary to the "avian hypothesis." Several cladistic analyses (Sereno 1999, 2000, 2001; Chiappe 2002a; Norell et al. 2001; Clark et al. 2002; Novas & Pol 2002) have argued for a non-avian relationship of alvarezsaurids, resulting in different placements of the group within coelurosaurian theropods. Whereas Sereno (1999,

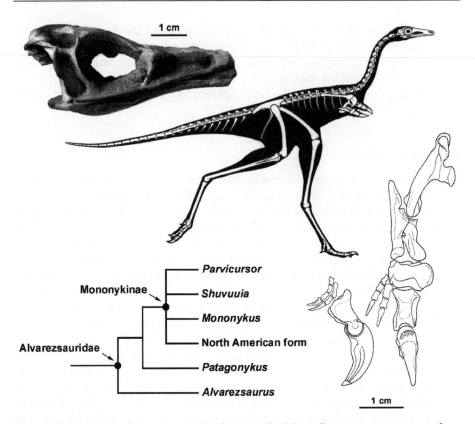

Figure 6   Alvarezsaurids are known by five described Late Cretaceous genera, two from southern South America (*Alvarezsaurus* and *Patagonykus*) and three from central Asia (*Mononykus*, *Shuvuuia*, and *Parvicursor*, although the last two could be synonyms), as well as a fragmentary, unnamed specimen from North America. The Asian and North American forms comprise a monophyletic group, mononykines (centered skeletal reconstruction), with the South American taxa forming successive outgroups (Novas 1996, Chiappe et al. 1998) (cladogram on *lower left corner*). The skull of *Shuvuuia* (*top left corner*) shows the apomorphic nature of the alvarezsaurid cranium, with large orbits, an elongated snout, and a prokinetic type of skull kinesis (Chiappe et al. 1998). A recently described specimen of *Shuvuuia* (Suzuki et al. 2002) has a nearly complete hand preserved in articulation that provides evidence that mononykines, and presumably all alvarezsaurids, had two manual digits (fingers 2 and 3) in addition to the stout digit 1 characteristic of the group (*lower right corner*).

2000), Norell et al. (2001), Clark et al. (2002), and Novas & Pol (2002) have proposed a more basal placement within theropods, Chiappe (2002a) placed alvarezsaurids as the immediate outgroup of Aves. Among the diverse non-avian hypotheses that have been put forward, Sereno's (1999, 2001) claim for a close relationship between alvarezsaurids and ornithimimid coelurosaurs (i.e., "ostrich-like" theropods) is the most radical, since under this hypothesis, alvarezsaurids

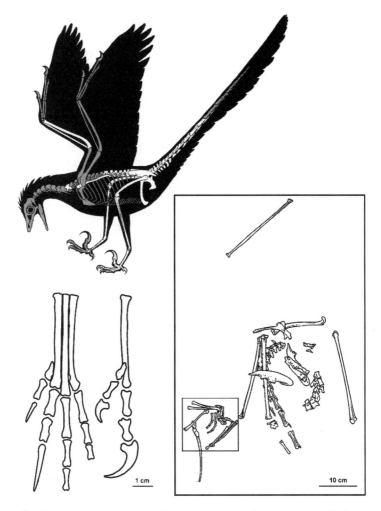

Figure 7   The Late Cretaceous *Rahonavis ostromi* of Madagascar. Skeletal reconstruction (*top left*) of the singly known specimen of this taxon. Note the proximity
in which the skeletal elements of this specimen were found (*outer box*). *Rahonavis'*
sickle-clawed foot is shown in dorsal and medial views (*bottom left*). (After Forster
et al. 1998a).

are placed outside maniraptoriforms within a new clade, the Ornithomimoidea
(alvarezsaurids + ornithomimids). However, critical examination of the character
evidence for this hypothesis has shown that support is, at best, weak (Suzuki et al.
2002). Although at this point the precise phylogenetic placement of alvarezsaurids
within coelurosaurians remains in question, it is evident that the group has relevance for understanding the evolutionary transformations that led to the origin of
extant birds.

The Malagasy *Rahonavis ostromi* (Forster et al. 1998a,b) (Figure 7) is another recently discovered basal taxon whose avian status has subsequently been challenged. On the basis of a cladistic analysis including several non-avian theropods and basal birds, Forster et al. (1998a) placed this sickle-clawed bird as the sister-taxon of *Archaeopteryx*, although cautioning about the possibility of it being closer to neornithine (extant) birds (an alternative concordant with its placement by Chiappe 2002a; see Figure 4). Combining a suite of avian features such as a reversed first toe (Figure 7) and quill knobs on the forearm with the sickle-clawed pedal specializations of dromaeosaurid theropods, *Rahonavis* provided further evidence in support of a coelurosaurian origin of birds (Forster et al. 1998a). Because of this apparently mosaic combination of characters, critics of the theropod hypothesis of avian ancestry claimed the holotype to be a composite (Feduccia 1999, Geist & Feduccia 2000): the forelimb and shoulder girdle of a bird mixed up with the hindlimb, pelvis, and tail of a non-avian theropod. Nevertheless, the taphonomic context of *Rahonavis* suggests otherwise. Although mostly disarticulated, the single specimen has no duplicated elements and was found on a surface of less than 0.14 m^2 (Figure 7). This alone strongly suggests that all described bones belong to a single individual (Forster et al. 1998a). Further, Forster et al. (1998a) were careful to address this issue by conducting separate cladistic analyses, one scoring the whole specimen and another excluding the forelimb and shoulder from the data set. Both analyses produced the same result, placing *Rahonavis* in a basal position within birds. Although subsequent studies also supported this basal placement for *Rahonavis* (Holtz 1998, Chiappe 2002a), more recent work by Holtz (2001) and Clark et al. (2002) have hypothesized that this taxon is the most immediate outgroup of Aves.

Although the avian relationship of alvarezsaurids and *Rahonavis* has been hotly debated, the paramount issue within basal avian systematics remains whether or not there is a deep dichotomy separating two main evolutionary radiations, which acquired in parallel, a suite of increasingly derived (i.e., modern) characters. A number of workers (e.g., Martin 1983, 1995; Kurochkin 1995, 1996, 2000, 2001; Hou et al. 1995, 1996; Feduccia 1999) have argued that all basal lineages of birds can be classified within two major subdivisions. On the one hand, "Sauriurae" encompassing *Archaeopteryx*, Confuciusornithidae (toothless birds from the Early Cretaceous of China; Chiappe et al. 1999) (Figure 8), and Enantiornithes (a flighted, cosmopolitan group recorded throughout the Cretaceous; Chiappe & Walker 2002) (Figure 9), and on the other hand, the Ornithurae, including several lineages of Cretaceous birds (Hesperornithiformes and Ichthyornithiformes

---

→

Figure 8   The Early Cretaceous *Confuciusornis sanctus* of China. Skeletal reconstruction of this crow-sized bird (*top*). Specimen displaying a pair of long tail feathers (*bottom left*) and a close up of one of these feathers (*bottom right*). Whereas specimens preserving these long tail feathers are not uncommon, whether their presence documents the existence of sexual dimorphs or not remains unclear (Chiappe et al. 1999) (after Chiappe et al. 1999).

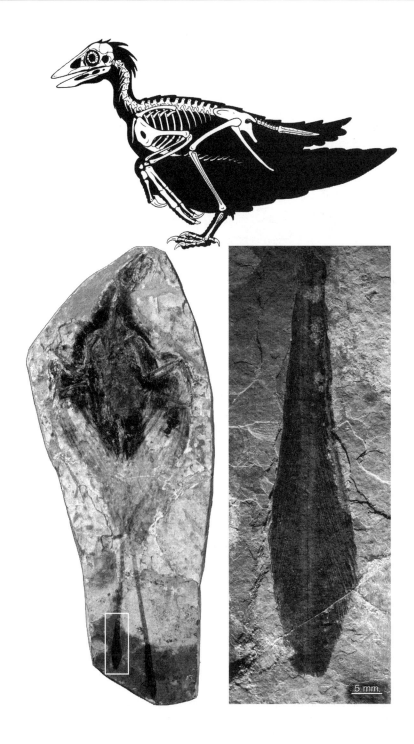

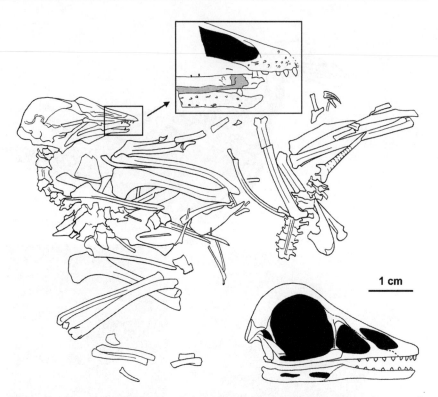

Figure 9   One of several known specimens of the Early Cretaceous *Sinornis santensis* of China (*center*) (after Hou 1997), with a close up of the tip of its rostrum (*top*). Skull reconstruction of an enantiornithine hatchling from the Early Cretaceous of Spain (*bottom right*) (after Sanz et al. 1997). Inset not to scale.

among others) (Figure 10), plus all extant taxa. Whereas most supporters of "Sauriurae" would agree on the monophyletic status of Aves, some workers have carried this view of a basal dichotomy to an extreme (Kurochkin 2001), arguing for two completely separate origins for these groups, and thus a diphyletic Aves. However, such a notion is entirely based on primitive characters not evaluated in a cladistic framework. Several cladistic analyses have strongly supported the notion that confuciusornithids and enantiornithines are successively more closely related to extant birds (i.e., Neornithes; Figure 4) and that neither of these two basal lineages shares a common ancestor with *Archaeopteryx* that is not the common ancestor of all birds (Cracraft 1986; Chiappe 1991, 1995, 2002a; Forster et al. 1998a; Sereno 1999). Acceptance of a monophyletic "Sauriurae" requires explaining the numerous morphological and functional similarities seen between the "sauriurine" lineages (in particular Enantiornithes) and extant birds as evolutionary convergences. Parsimony analyses show, however, that these shared features are most simply explained as a stepwise series of synapomorphies diagnosing nodes within

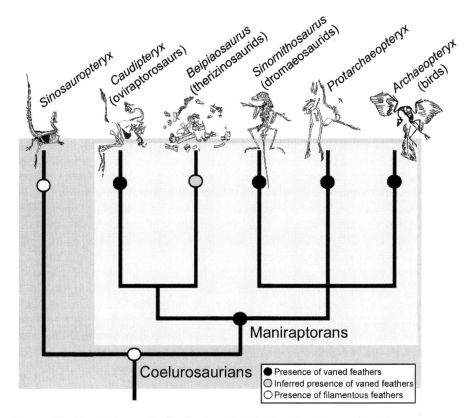

Figure 10 The phylogenetic distribution of the feathered coelurosaurians *Sinosauropteryx prima, Caudipteryx zoui, Beipiaosaurus inexpectus, Sinornithosaurus milleni,* and *Protarchaeopteryx robusta* (phylogenetic relationships simplified from Chen et al. 1998, Ji et al. 1998, and Norell et al. 2001). We agree with a number of previous studies (e.g., Ji et al. 1998, Sereno 1999, Norell et al. 2001, Clark et al. 2002) in regarding the undeniably feathered *Caudipteryx* as a non-avian coelurosaurian, specifically a member of Oviraptorosauria. Filamentous and vaned feathers are interpreted as coelurosaurian and maniraptoran synapomorphies, respectively.

a pectinate cladogram (Figure 4). The conceptual impasse underlying this phylogenetic disagreement likely resides in how workers have chosen to formulate and test homology (Chiappe 1999).

By far the most diverse group of Cretaceous birds was Enantiornithes, with nearly 20 valid species reported to date (Table 1; Figure 9) (Chiappe & Walker 2002). Basal members of this lineage are well represented in Early Cretaceous lake deposits of Spain (Sanz et al. 1995, 1996, 1997) and China (Zhou 1995b, Zhang & Zhou 2000, Zhang et al. 2001). These early enantiornithines are toothed and of small size (from that of a sparrow to that of a thrush). Their flight apparatus approaches that of their living counterparts in proportions and shares with them several derived characters (e.g., alula, mobile scapulocoracoid articulation, narrow interclavicular). In spite of the fact that definitive evidence for perching is lacking in more basal birds, these capabilities were clearly present among the earliest enantiornithines as evidenced by their pedal morphology (Sereno & Rao 1992, Chiappe & Calvo 1994, Chiappe 1995, Martin 1995, Sanz et al. 1995, Zhou 1995b). There is a distinct size increment between these taxa and later enantiornithines such as the Late Cretaceous *Enantiornis* (with a wing-span of about 1 m; Walker 1981, Chiappe 1996), *Avisaurus* (Chiappe 1993), and *Gobipteryx* (Chiappe et al. 2001), the only known toothless member of the group. Although mostly recorded from inland deposits, enantiornithines also occupied littoral and marine environments, and even extended into polar regions (Chiappe 1996).

A number of other taxa, including one of the earliest secondarily flightless birds, the Late Cretaceous *Patagopteryx deferrariisi* from southern Argentina (Alvarenga & Bonaparte 1992, Chiappe 2002b), fall between the enantiornithine radiation and the divergence of another major group, Ornithurae, which contains the most immediate outgroups of extant birds (Figure 4). The record of these close relatives of Neornithes is mostly limited to the familiar Hesperornithiformes and Ichthyornithiformes. Hesperornithiforms comprise several species of toothed and highly specialized divers with extremely abbreviated forelimbs (Figure 5). Although primarily known from marine environments, some fossils have been recovered from estuarine and near shore deposits. Some of the most derived members (e.g., *Hesperornis regalis*) would have been about the size of an Emperor Penguin. The flighted ichthyornithiforms are represented by a handful of incompletely known taxa from essentially the same deposits as the hesperornithiforms. A revision of ichthyornithiforms has begun to cast doubts on the monophyly of this group as well as the validity of some species of *Ichthyornis* (Clarke 1999, 2002). In spite of being toothed, *Ichthyornis* and allies are morphologically very similar to their extant relatives, and although among the earliest discovered Mesozoic birds, many aspects of their anatomy, taxonomy, and inter-relationships remain unclear. One significant recent addition to the Mesozoic record of ornithurines is the exquisitely preserved *Apsaravis ukhaana* (Norell & Clarke 2001) from the Mongolian Late Cretaceous that adds critical data to understanding evolutionary transformations leading to the modern avian wing. Other recently described ornithurines include the Early Cretaceous *Yanornis martini* and *Yixianornis grabaui* from China (Zhou & Zhang 2001), which although only preliminarily studied appear to be the earliest informative members of this lineage. Finally, despite several reports of Cretaceous fossils of extant avian lineages, the existence of neornithine birds in pre-Tertiary times remains dubious (Dyke 2001).

Do Feathers Make a Bird?

Under the accepted phylogenetic definition of Aves (i.e., common ancestor of *Archaeopteryx*, extant birds and all descendants), the existence of a suite of differentiated and structurally modern feather types (e.g., primary and secondary remiges, retrices, and covertors) in *Archaeopteryx* strongly suggests that the origin of feathers predated the origin of birds (Sereno 1999, Rayner 2001). Recently, this prediction has been confirmed by a number of discoveries from the Early Cretaceous of China including at least six taxa of non-avian theropod dinosaurs preserving integumentary structures interpreted as feather homologes (Chen et al. 1998; Ji et al. 1998; Xu et al. 1999a,b, 2000, 2001; Norell et al. 2002). These new taxa, of a wide range of morphologies and sizes, represent lineages phylogenetically spread throughout the non-avian coelurosaurian portion of the theropod tree (Figure 10). Often preserved associated with the integumentary covering of the skull, axial and appendicular skeleton, these structures range from simple and filament-like (e.g., *Sinosauropteryx*; Chen et al. 1998, Prum 1999) to tufts joined at their bases or serially arranged along a central filament (e.g., *Sinornithosaurus*; Xu et al. 2001) to more complex structures with vanes and distinct shafts (*Caudipteryx*, *Protarchaeopteryx*; Ji et al. 1998, Zhou et al. 2000). Support for the homologous interpretation of these structures as feathers comes from their complex branched arrangement, characteristic of avian feathers (Prum 1999, Xu et al. 2001, Norell et al. 2002). Dissenters to the theropod hypothesis of bird origins have countered by claiming that in some instances these structures are not feathers but frayed internal composite fibers of the structural protein collagen (Feduccia 1999, Geist & Feduccia 2000). In other cases they have argued that the implicated taxa (e.g., *Caudipteryx* and *Protoarchaeopteryx*) are avian (Feduccia 1999, Geist & Feduccia 2000, Jones et al. 2000, Martin & Czerkas 2000, Ruben & Jones 2000), and thus irrelevant for understanding the origin of feathers. The morphology and length of the filaments, however, are inconsistent with their interpretation as frayed internal composite fibers (Prum 1999, Xu et al. 2001) and there is no doubt that these structures are external. Equally well supported is the non-avian identification of taxa for which integumentary structures have been unquestionably identified as vaned feathers, such as *Caudipteryx* and *Protarchaeopteryx* (Ji et al. 1998). Jones et al. (2000) compared the hindlimb proportions and position of the gravity center of *Caudipteryx* to those of non-avian theropods and living flightless birds, concluding that this taxon is more similar to extant flightless birds than it is to non-avian theropods. This claim remains unsubstantiated, however, since the majority of the specimens used in this study are too incompletely preserved to allow the measurements cited to be replicated and in some instances (e.g., *Carnotaurus*) the given measurements are for elements that simply do not exist. Most significantly, qualitative osteological data prevents placing both *Caudipteryx* and *Protarchaeopteryx* among birds, as shown by several recent cladistic analyses (Ji et al. 1998, Holtz 1998, Sereno 1999, Xu et al. 2000, Norell et al. 2001, Clark et al. 2002). Whereas the phylogenetic placement within the non-avian coelurosaurian tree of the incomplete *Protarchaeopteryx* remains controversial (Ji et al. 1998), the much better represented *Caudipteryx* has been consistently nested within oviraptorosaurs (Holtz 1998, Sereno 1999, Clark et al. 2001, Norell et al. 2001) (Figure 10). Morphological arguments in favor of an avian placement of *Caudipteryx* have relied on characters that are either incorrect, circular, or also found among other non-avian coelurosaurians (Table 2). A number of other characters are ambiguous because

TABLE 2   Our interpretation of the characters used to argue for an avian rather than an oviraptorid relationship for the feathered *Caudipteryx*

Character/interpretation

---

1. **Shortened, incipiently fused tail ("protopygostyle")** (Geist & Feduccia 2000, Martin & Czerkas 2000).
   Irrelevant. This character is absent in *Caudipteryx*. The distalmost tail vertebrae are fused in other oviraptorids (Barsbold et al. 2000).

2. **Ventrally oriented foramen magnum** (Geist & Feduccia 2000).
   Ambiguous. The known skulls of *Caudipteryx* are not well enough preserved to identify the orientation of the foramen magnum (Ji et al. 1998).

3. **Vaned feathers** (Geist & Feduccia 2000, Martin & Czerkas 2000).
   Circular. This character cannot be used in favor of the avian relationship of *Caudipteryx* since the hypothesis to be tested is that vaned feathers occur in non-avian theropods.

4. **Ligamental quadratojugal-quadrate articulation** (Geist & Feduccia 2000, Ruben & Jones 2000).
   Ambiguous. The preservation of the known specimens, in which pertinent bones are disarticulated, makes this character problematic.

5. **Lack of contact between quadratojugal and squamosal** (Geist & Feduccia 2000, Ruben & Jones 2000).
   Ambiguous. The preservation of the known specimens, in which pertinent bones are disarticulated, makes this character problematic.

6. **Absence of obturator process on ischium** (Ruben & Jones 2000).
   Incorrect. A distinct and large obturator process is present at least in the holotype specimen of *Caudipteryx* (Ji et al. 1998).

7. **Expanded roots on premaxillary teeth** (Martin & Czerkas 2000).
   Ambiguous. The teeth of *Caudipteryx* are highly apomorphic, greatly differing from both avian and non-avian theropod teeth.

8. **Carpus containing at least four bones** (Martin & Czerkas 2000).
   Irrelevant. Three carpals are preserved in the holotype of *Caudipteryx*, these correspond to the radiale, semilunate, and element X (Hinchliffe 1985). The 'absence' of an ulnare (i.e., fourth carpal), otherwise present in all tetrapods, is clearly due to a preservation bias. Four or more carpals are known for other nonavian theropods (Xu et al. 1999b).

9. **Absence of pubic foot** (Martin & Czerkas 2000).
   Ambiguous. The distal ends of the pubes of the known specimens of *Caudipteryx* (Ji et al. 1998; Zhou et al. 2000) are not preserved.

10. **Opposable first toe** (Martin & Czerkas 2000).
    Incorrect. This character cannot be confirmed in any of the known specimens of *Caudipteryx* (Z. Zhou 2001, personal communication).

11. **Loss of teeth in maxilla and mandible** (Martin & Czerkas 2000).
    Irrelevant. Teeth are absent in the maxilla and mandible of all oviraptorids (Barsbold et al. 1990, Clark et al. 2001) and present in most non-neornithine lineages (Chiappe et al. 1999).

12. **Mandibular foramen** (Martin & Czerkas 2000).
    Irrelevant. A mandibular foramen is present in most archosaurs including theropod dinosaurs.

(Continued)

TABLE 2   *(Continued)*

Character/interpretation

13. **Enlargement of premaxilla** (Martin & Czerkas 2000).
    Irrelevant. An expanded premaxilla is typical of oviraptorid theropods (Barsbold et al. 1990, Clark et al. 2001).

14. **Reduction of maxilla** (Martin & Czerkas 2000).
    Irrelevant. An expanded premaxilla is typical of oviraptorid theropods (Barsbold et al. 1990, Clark et al. 2001).

15. **Reduction of hyperpubic spoon** (Martin & Czerkas 2000).
    Ambiguous. The distal ends of the pubes are not preserved in the known specimens of *Caudipteryx* (Ji et al. 1998, Zhou et al. 2000).

16. **Ball-shaped head of femur** (Martin & Czerkas 2000).
    Irrelevant. A ball-shaped femoral head is present in all dinosaurs including oviraptorids and birds.

17. **Reduction of fibula** (Martin & Czerkas 2000).
    Incorrect. A distinct socket on the calcaneum of *Caudipteryx* indicates that although the fibulae are incomplete these bones reached the proximal tarsals (Ji et al. 1998) and thus, they were not reduced.

18. **Enlargement of astragalus at the expense of calcaneum** (Martin & Czerkas 2000).
    Irrelevant. An much larger astragalus than calcaneum is primitive for coelurosaurian theropods including oviraptorids and birds.

---

their presence cannot be confirmed in the ten known specimens of *Caudipteryx* (Z. Zhou 2001, personal communication). Maryanska et al.'s (2002) recent interpretation of oviraptorosaurs as birds still needs to be critically evaluated, but obvious problems of taxonomic sampling (e.g., no neornithine lineage included and *Archaeopteryx* and *Confuciusornis* were the only birds of the analysis) makes this phylogenetic inference dubious.

Overlooked by the supporters of an avian relationship for *Caudipteryx* is the fact that a great deal of homoplasy would have to be explained to nest this taxon within birds. For example, if Feduccia's (1999) interpretation of *Caudipteryx* as a flightless enantiornithine were correct, it would require the re-elongation of a bony tail, separation of the vertebral elements co-ossified into a pygostyle, development of new phalanges and finger re-elongation, substantial transformations in the sternum (from the single large and keeled element of Enantiornithes to the two small and separate plates of *Caudipteryx*), and loss of fusion of several compound bones (e.g., carpometacarpus, tibiotarsus, tarsometatarsus). The most parsimonious explanation, then, is that *Caudipteryx* is a non-avian theropod dinosaur with vaned feathers. The hypothesized homology between these feathers and those of extant birds has been explained by the developmental model of Prum (1999) that provides a framework for homologizing the various feather morphologies of non-avian theropods and for understanding the evolution of feather complexity. Further support for the presence of vaned feathers among non-avian theropods has recently been provided by Norell et al. (2002), who described this type of feathers in an Early Cretaceous dromaeosaurid from China.

Flight

Birds are characterized by their ability to move by active flapping flight, the functional and ecological contexts of which have been at the center of a long and heated debate (Hecht et al. 1985, Chatterjee 1997, Padian & Chiappe 1998, Feduccia 1999). Studies have demonstrated that aerodynamic features (e.g., elongated forelimb with flexible wrists, vaned feathers arranged in remiges and retrices, lateral orientation of shoulder socket) evolved prior to the origin of birds (Gauthier & Padian, 1985, Novas & Puerta 1997, Padian & Chiappe 1998, Sereno 1999). Over the course of bird evolution, the refinement of flight entailed the appearance of a complex suite of morphological specializations, which are understood within the context of phylogenetic relationship. Despite the fact that numerous other transformations (physiological, muscular, behavioral, neurological) must have been required to develop such a sophisticated locomotor system, each major lineage of Mesozoic birds provides direct evidence of the changes in the skeleton and plumage that led to the development of this system. *Archaeopteryx* had a wing without an alula (a feathered tuft attached to the first finger) that retained an ancestral configuration (humerus and hand longer than ulna-radius, first finger longer than metacarpals, complete set of manual phalanges) as well as a long bony tail supporting a frond-like array of feathers, a small and unkeeled sternum, a stout boomerang-shaped furcula (wishbone), and no evidence of postcranial pneumatization beyond the vertebral column. Although unquestionably interpreted as a flying bird (Hecht et al. 1985, Padian & Chiappe 1998, Burgers & Chiappe 1999, Feduccia 1999), the significantly smaller lift produced by its frond-like tail (Gatesy & Dial 1996) (Figure 11), the less developed flight musculature (Rayner 1991, 2001) inferred from the size and shape of its sternum, the apparent absence of a uniquely sophisticated and efficient air-sac respiratory system pneumatizing much of the skeleton, combined with the absence of a number of aerodynamic structures [e.g., alula (Sanz et al. 1996), spring-like U-shaped furcula (Jenkins et al. 1988)], suggests that *Archaeopteryx* was probably weaker and less maneuverable in flight compared to most of its extant relatives. The primitive wing of *Archaeopteryx* was essentially retained in the more derived confuciusornithids, but the long bony tail was shortened into a few free vertebrae and a stump, or pygostyle, formed by the fusion of the last vertebrae (Figure 8). The confuciusornithid wing had a propatagium, the skin fold joining the shoulder and wrist, a feature apparently unknown in *Archaeopteryx* and non-avian coelurosaurians (Clarke et al. 2001) that would have increased the area of the lift-generating airfoil. Estimates of important aerodynamic parameters such as wing loading also hint at refinements in the flight capabilities of these birds. Sanz et al. (2000) calculated the wing loading of *Confuciusornis* to be lower than that of *Archaeopteryx* and less than half that

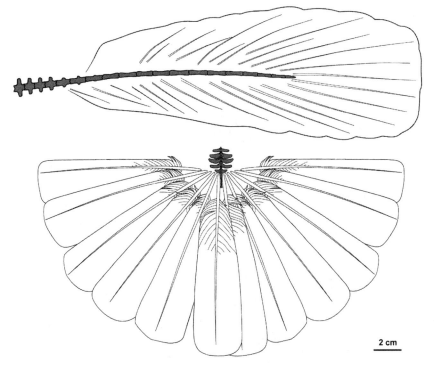

2 cm

Figure 11   A scaled comparison between the frond-like tail of *Archaeopteryx* (*top*) and the fan-like tail of a pigeon (*bottom*) to illustrate the much larger airfoil attained by the latter design (after Gatesy & Dial 1996). Tail vertebrae in gray.

estimated for *Caudipteryx*. Although some confuciusornithid specimens preserve a pair of long tail feathers (Chiappe et al. 1999) (Figure 8), the aerodynamic function of their otherwise short-feathered tail would have been limited, in contrast to the tail fans typical of extant birds (Rayner 2001). The existence of a pair of long tail feathers in specimens of confuciusornithids (Chiappe et al. 1999) and enantiornithines (Zhang & Zhou 2000), however, provides evidence for elaborate feather differentiation as early as the Early Cretaceous, although it remains unclear whether this documents the evolution of marked sexual dimorphisms among these primitive birds (Chiappe et al. 1999). Many more features associated with the aerodynamic capabilities of living birds are recorded for the first time in Enantiornithes (Chiappe 1991, Sanz et al. 1996, Rayner 2001). These birds had modern wing configurations with the ulna-radius the longest segment, much shorter fingers (yet still clawed), and an alula (Figure 12). The alula plays an important aerodynamic role in controlling airflow over the dorsal surface of the wing during slow flight, thus enhancing maneuverability as well as performance during take-off and landing (Sanz et al. 1996). Although enantiornithines had large pygostyles that

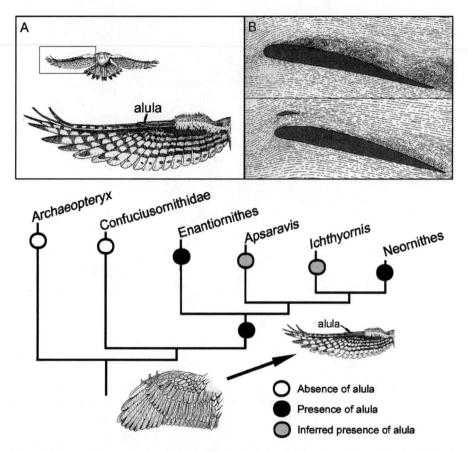

Figure 12 Aerodynamic function and phylogenetic distribution of the alula in birds. A small tuft of feathers attached to manual digit I (*A*), the alula functions by creating a slot along the leading edge of the wing. This slot (*B, lower cross-section*) decreases the turbulence originating on the dorsal wing surface at high angles of attack, thus enhancing control and maneuverability at low speeds, especially during take off and landing. The presence of an alula is considered a synapomorphy of Ornithothoraces (see Figure 4).

could have supported a feathered tail fan (Gatesy & Dial 1996, Rayner 2001), the available fossil material does not confirm this.

If an overall trend in size reduction is visible during the transition from non-avian coelurosaurians (e.g., dromaeosaurids, troodontids, oviraptorids) to basalmost birds (e.g., *Archaeopteryx*, *Rahonavis*), this pattern is further evidenced by the earliest enantiornithines, which were of sizes comparable to modern sparrows (e.g., *Iberomesornis*, *Sinornis*) and thrushes (e.g., *Concornis*, *Eoenantiornis*). General flight performance is often correlated with size reduction. Wing loading (for a given airfoil) decreases and power-to-weight ratio increases as body size decreases.

Furthermore, smaller birds are more maneuverable and flight is for them energetically less expensive. The advanced skeletal features of enantiornithines coupled with the presence of an alula and their small size suggest that even the earliest of these birds had aerodynamic abilities approaching those seen in extant forms.

A further step toward the modern flight condition is seen in the ornithurine *Apsaravis ukhaana*—that had a well-developed extensor process on the first metacarpal for the insertion of muscles involved in the automatic extension of the wing during upstroke-downstroke transition (Norell & Clarke 2001). This taxon along with more advanced flighted members of this clade [e.g., *Ichthyornis* (Marsh 1880, Clarke 2001), *Ambiortus* (Kurochkin 1985, 1999)] show the majority of the skeletal features associated with advanced flapping flight, suggesting that they had comparable abilities to those of extant birds.

Growth and Development

Extant birds grow rapidly, and although their rates of growth vary depending on mode of development (altricial forms growing much faster than precocial ones), they normally reach adult size within a year. Although the bones of living birds are known to be fast-deposited and typically uninterrupted woven tissues (fibrolamellar bone), Mesozoic avians exhibited a great range of variation. The limb bones of *Rahonavis* (Chinsamy & Elzanowski 2001) and enantiornithines (Chinsamy et al. 1995) had thick cortices of slowly formed lamellar tissue (e.g., in *Rahonavis* 44% of the femoral wall was lamellar) that have multiple lines of arrested growth or LAGs, skeletal chronological indicators typically interpreted as representing annual pauses of postnatal bone deposition (Chinsamy et al. 1995, Ericson et al. 2001, Padian et al. 2001). Conversely, *Patagopteryx* (Chinsamy et al. 1995), *Hesperornis* (Houde 1987, Chinsamy et al. 1998), and *Ichthyornis* (Chinsamy et al. 1998) had bone walls formed primarily by fast growing fibrolamellar bone, with only *Patagopteryx* retaining a single LAG. The presence of LAGs in most non-neornithines suggests that in contrast to their living relatives, and perhaps all ornithurines, these basal birds may have required more than a year to reach adult size (Chinsamy et al. 1995, 1998; Chinsamy & Elzanowski 2001, Padian et al. 2001). This infererence is consistent with known growth series where individuals lacking neonate features (fully fledged) are very differently sized, [e.g., in *Archaeopteryx lithographica* (Houck et al. 1990) and *Confuciusornis sanctus* (Chiappe et al. 1999), the smallest specimens are 50% and 60%, respectively, the size of the largest]. Histological evidence suggests that growth rates of basal birds were distinctly lower than those of their extant counterparts and that these elevated rates developed early in ornithurine history (Chinsamy et al. 1995, Padian et al. 2001, Chinsamy 2002).

Relatively rapid growth rates have been inferred in non-avian dinosaurs, which approached the condition seen in extant mammals and birds more than that of living non-avian reptiles (de Ricqlès 1980, Chinsamy 1990, 1993; Erickson et al.

2001; Padian et al. 2001). Histological data collected for basal birds indicates that their rates of growth were slower not only than those of extant birds but also those of their non-avian dinosaur relatives (Chinsamy & Elzanowski 2001, Padian et al. 2001). Causes underlying this evolutionary shift are not clear, but recent studies have proposed that this reduction in growth rates from non-avian dinosaurs to basal birds could have been achieved by shortening the duration of the rapid growth phase (i.e., deposition of fibrolamellar bone) that is characteristic of early ontogenetic stages in postnatal development (Padian et al. 2001, Chinsamy 2002). This shift could have led to the distinct decrease in size observed during the transition from non-avian theropods to enantiornithines (Padian et al. 2001) and has been used (Chinsamy & Elzanowski 2001) to support previous ideas (Elzanowski 1981) advocating superprecociality as the developmental strategy of enantiornithines and other basal birds. On the basis of the high degree of ossification seen in enantiornithine embryos, Elzanowski (1981) proposed a superprecocial developmental mode (independent young able to fly soon after hatching) for this group, because the skeletons of extant precocial hatchlings are significantly more ossified than those of altricial forms. A precocial developmental strategy for basal birds is supported by the distribution of this strategy among extant avian lineages (Chiappe 1995), where all the most basal taxa (e.g., paleognaths, galliforms, anseriforms) are characterized by having hatchlings that fall on the precocial side of the precocial-altricial spectrum. However, correlations between the extent of bone formation in fossil embryos and specific developmental strategies are weakly supported by embryogenetic studies of extant birds, which show that differences in the degree of ossification between superprecocial, precocial, and semiprecocial hatchlings are minor (Starck & Ricklefs 1998). Thus, although the shortening in the duration of the rapid growth phase could have been correlated to superprecociality and the development of precocial flight in basal, Mesozoic birds (Chinsamy & Elzanowski 2001), identifying this specific mode of development by recording degrees of ossification in fossil embryos is problematic. Interestingly, support for this proposal may come from the presence of fledged wings in some newly discovered enantiornithine juveniles (L. Chiappe, personal observations), suggesting the early onset of aerodynamic capabilities.

DIVERSIFICATION AND EXTINCTION

Basal Lineage Dynamics

Because of the incompleteness of the fossil record, lineage dynamics across time cannot be accurately observed simply from the temporal distribution of fossils. The oldest fossil representatives of a lineage provide evidence only for the minimal age of that particular taxon, but a more precise picture of the temporal pattern of lineage origination could emerge through calibration with a phylogenetic hypothesis. Definitive records of Enantiornithes (e.g., *Eoenantiornis*) and Confuciusornithidae are known from the 125–million-year-old Yixian Formation of China (see Swisher et al. 2002 for radioisotopic dates). Enantiornithines (e.g., *Noguerornis* and an unnamed hatchling; Sanz et al. 1997, Chiappe & Lacasa-Ruiz 2002) may even be known from the apparently older lithographic limestones of Montsec (~130 mya) in Spain (Martín-Closas & López-Morón 1995). Even though the oldest ornithuromorphs (*Yixianornis* and *Yanornis*) come from deposits slightly younger than the Yixian Formation, calibration of our phylogeny (Figure 4) on the basis of the oldest records of confuciusornithids and enantiornithines implies that all these lineages must have diverged at least 130 million years ago. Indeed, it would not be surprising if early members of these lineages were to be found in Jurassic sediments in the future. The paucity of Jurassic deposits (e.g., less than 20% compared to the Cretaceous in the United States) (Clark et al. 2002), in particular those of continental origin (Benton 1994), may explain the fact that no bird of this age has yet been found with the exception of *Archaeopteryx*. The timing of divergences within Ornithuromorpha, however, is more controversial given the fragmentary nature of many specimens, in particular those closely related to extant birds (see below).

Although a wealth of new discoveries has shown that the Cretaceous was a time of active diversification for birds, this period has also proven to be one of widespread extinction. Indeed, no lineage of non-neornithine birds has ever been recorded in post-Mesozoic deposits (Figure 4). Feduccia (1995, 1999) has argued in favor of a dramatic extinction of all primitive Mesozoic lineages at the Cretaceous-Tertiary (K-T) boundary. Indeed, the last occurrence of enantiornithines (Chiappe 1993, Stidham & Hutchison 2001, Chiappe & Walker 2002) and other non-neornithine birds in strata very close to the K-T boundary (e.g., Lance and Hell Creek Formations of North America and Maastricht Formation of Europe) suggests that some primitive lineages may have become extinct along with larger dinosaurs in the terminal Cretaceous mass extinction (Stidham & Hutchison 2001). Nonetheless, the record of birds for the latest Maastrichtian (~67–65 mya) remains incomplete and the precise stratigraphic provenance of most specimens is not well documented. Further, most avian Late Cretaceous lineages are known by single occurrences (e.g., *Patagopteryx*, *Vorona*, *Apsaravis*) and of those known by ranges, some (e.g., ichthyornithiforms; Clarke 2002) have last occurrences in pre-Maastrichtian deposits, more than 10 million years before the K-T boundary. Thus, generalizations about a dramatic bottleneck in birds at the K-T boundary (Feduccia 1995, 1999), and the sudden eradication of all non-neornithine lineages, are not presently supported by the fossil record.

Temporal Origin of Extant Lineages

A heated debate in current evolutionary biology involves the question of the timing of the origination of the extant lineages of birds. Estimates for the divergence of these neornithine birds based on the "clock-like" modeling of molecular sequence data predict that lineages (i.e., the extant traditional orders and families) originated deep in the Cretaceous (in some cases up to 90–100 million years ago; Hedges et al. 1996, Cooper & Penny 1997, Kumar & Hedges 1998). However, the actual record of Cretaceous specimens that have been referred to extant lineages is sparse: Only a few of these are represented by more than an isolated element (e.g., *Ambiortus, Gansus*) (Table 3). This handful of fossils is either controversial, have not been studied in any detail, or their placement within Neornithes has been rejected. The incompleteness of most of the specimens referred as Cretaceous neornithines

TABLE 3   Alleged records of Neornithes (modern birds) from the Mesozoic

Alleged taxonomy	Material	Formation and age
NEORNITHES *incertae sedis*		
Ceramornis major	Coracoid	Lance Fm. (M)
Gallornis straeleni	Femur	Hateg Basin (M)
PALAEOGNATHAE		
Ambiortus dementjevi	Partial skeleton	Andaikhudag Fm. (Ha-Ba)
NEOGNATHAE		
Galliformes		
Palintropus retusus	Coracoids	Lance Fm. (M)
Anseriformes		
Apatornis celer	Synsacrum	Niobrara Fm. (Ca)
sp. Indet	Partial skeleton	López de Bertodano Fm. (M)
sp. Indet.	Tarsometatarsus	Barun-Goyot Fm. (Ca)
Anatalavis rex	Humeri	Hornerstown Fm. (M*)
Charadriiformes		
sp. indet.	Coracoid	Lance Fm. (M)
Cimoloperyx rara	Coracoids	Lance Fm. (M)
Cimoloperyx rara	Carpometacarpi	Lance Fm. (M)
Cimoloperyx maxima	Coracoids	Lance Fm. (M)
Graculavus velox	Humerus	Hornerstown Fm. (M*)
Graculavus velox?	Carpometacarpus	Hornerstown Fm. (M*)
Graculavus augustus	Humerus	Lance Fm. (M)
Telmatornis priscus	Humerus	Hornerstown Fm. (M*)
	Carpometacarpus	Hornerstown Fm. (M*)
	Ulna	Hornerstown Fm. (M*)
Laornis edvardsianus	Tibiotarsus	Hornerstown Fm. (M*)
Palaeotringa littoralis	Tibiotarsus	Hornerstown Fm. (M*)
Palaeotringa littoralis?	Humerus	Hornerstown Fm. (M*)
Palaeotringa vagans	Tibiotarsus	Hornerstown Fm. (M*)
Procellariiformes		
Lonchodystes estesi	Tibiotarsus	Lance Fm. (M)
Tyttostonyx glauciniticus	Humerus	Hornerstown Fm. (M*)
sp. indet.	Furcula	Nemegt Fm. (Ca)

(*Continued*)

TABLE 3   *(Continued)*

Alleged taxonomy	Material	Formation and age
Gaviiformes		
"Polarornis gregorii" [†]	Partial skeleton	López de Bertodano Fm. (M)
Neogaeornis wetzeli	Tarsometatarsus	Quiriquina Fm. (M)
Pelecaniformes		
sp. indet.	Scapula	Nemegt Fm. (Ca)
Elopteryx nopscai	Femora	Hateg Basin Fm. (M)
Psittaciformes		
sp. indet.	Dentary	Lance Fm. (M)
Gruiformes		
Horezmavis eocretacea	Tibiotarsus	Khodzhakul Fm. (Al)

[*]A Late Cretaceous age for the Hornerstown Formation of New Jersey is questionable; Olson (1994) discusses palynological evidence for the Palaeocene age of this deposit.

[†]*'Polarornis gregorii'* was only informally named and its Cretaceous age has been questioned (Clarke & Chiappe 2001).

Abbreviations: Al, Albian (112–97 mya); Ca, Campanian (83–74 mya); Ha-Ba, Hauterivian-Barremian (135–124.5 mya); M, Maastrichtian (74–65 mya). See Dyke (2001) for citations.

renders few characters useful for phylogenetic analyses (Clarke 1999, Dyke & Mayr 1999, Clarke & Chiappe 2001, Dyke 2001), and the poor understanding of the higher-level phylogenetic relationships of extant lineages further complicates their systematic consideration (Livezey & Zusi 2001). This scanty fossil record has, however, been used to hypothesize the existence of a number of extant lineages prior to the end of the Mesozoic (i.e., Pelecaniformes, Charadriiformes, Anseriformes, Gaviiformes, Galliformes, Psittaciformes) (Table 3), either by taking it at face value (Feduccia 1999) or by using it for the temporal calibration of morphological (Chiappe 1995) or molecular (Hedges et al. 1996, Cooper & Penny 1997) phylogenies. Nonetheless, it is imperative that existing Cretaceous reports of neornithine birds are treated with extreme caution (Clarke 1999, Dyke & Mayr 1999, Clarke & Chiappe 2001). Indeed, the earliest neornithine birds that are complete enough to be informative for cladistic analyses, and hence potentially informative for estimating the temporal divergence of extant lineages, come from rocks that are roughly 55 million years old (i.e., the Early Eocene Green River and London Clay formations of the United States and England, respectively; Dyke 2001), deposited some 10 million years after the end of the Cretaceous. Although a few specimens consisting of more than single bones do fill this temporal gap, they have not yet been considered within cladistic analyses. Certainly, it would not be surprising if future studies of these or even older specimens support their original placement within extant lineages, but such a work has yet to be undertaken. Although the presence of several immediate neornithine outgroups in the Late Cretaceous (e.g., *Ichthyornis, Limenavis, Apatornis*) does imply that the lineage leading to extant birds must have differentiated prior to the end of the Mesozoic (Clarke & Chiappe 2001), this inference does not provide any information about the temporal divergence among extant lineages. To date, all Cretaceous specimens that have been submitted to rigorous cladistic analyses lay outside Neornithes (Clarke & Chiappe 2001). More well-preserved fossils and a better understanding of the relationships of modern clades are necessary to unravel the temporal divergence of extant avian lineages in pre-Tertiary times.

ACKNOWLEDGMENTS

We are grateful to W. Evans, N. Frankfurt, and M. Schwengle for rendering the illustrations and to S. Chatterjee, A. Chinsamy, C. Forster, S. Gatesy, M. Norell, and G. Paul for granting us permission to use their art. We also thank J. Clarke, R. Prum, J. Rayner, and Z. Zhou for providing information. This research was funded by grants from the National Science Foundation (DEB-9873705), the Frank M. Chapman Fund of the American Museum of Natural History, and the Natural History Museum of Los Angeles County.

The *Annual Review of Ecology and Systematics* is online at
http://ecolsys.annualreviews.org

LITERATURE CITED

Alvarenga HMF, Bonaparte JF. 1992. *A new flightless landbird from the Cretaceous of Patagonia. Sci. Ser.* 36:51–64. Nat. Hist. Mus. LA County

Barsbold R, Maryanska T, Osmólska H. 1990. Oviraptorosauria. In *The Dinosauria*, ed. DB Weishampel, P Dodson, H Osmólska, pp. 249–58. Berkeley: Univ. Calif. Press

Barsbold R, Osmólska H, Watabe M, Currie PJ, Tsogtbataar K. 2000. A new oviraptorosaur (Dinosauria: Theropoda) from Mongolia: the first dinosaur with a pygostyle. *Acta Paleontol. Pol.* 45:97–106

Benton MJ. 1994. Late Triassic to Middle Jurassic extinctions among continental tetrapods: testing the pattern. In *In the Shadow of the Dinosaurs*, ed. N Frasier, H-D Sues, pp. 366–97. New York: Cambridge Univ. Press

Bonaparte JF. 1991. Los vertebrados fósiles de la Formación Río Colorado de Neuquén y cercanias, Cretácico Superior, Argentina. *Rev. Mus. Argent. Cienc. Nat. Paleontol.* 4:17–123

Brodkorb P. 1971. Origin and evolution of birds. In *Avian Biology*, ed. DS Farner, JR King, pp. 19–55. New York: Academic

Brush AH. 2001. The beginnings of feathers. See Gauthier & Gall 2001, pp. 171–79

Burgers P, Chiappe LM. 1999. The wing of *Archaeopteryx* as a primary thrust generator. *Nature* 399:60–62

Chatterjee S. 1991. Cranial anatomy and relationships of a new Triassic bird from Texas. *Philos. Trans. R. Soc. London Ser. B* 332: 277–346

Chatterjee S. 1997. *The Rise of Birds*. Baltimore: Johns Hopkins Press

Chatterjee S. 1999. *Protoavis* and the early evolution of birds. *Palaeontogr. Am.* 254:1–100

Chen PJ, Dong ZM, Zhen SN. 1998. An exceptionally well-preserved theropod dinosaur from the Yixian Formation of China. *Nature* 391:147–52

Chiappe LM. 1991. Cretaceous avian remains from Patagonia shed new light on the early radiation of birds. *Alcheringa* 15:333–38

Chiappe LM. 1993. Enantiornithine (Aves) tarsometatarsi from the Cretaceous Lecho Formation of northwestern Argentina. *Am. Mus. Novit.* 3083:1–27

Chiappe LM. 1995. The first 85 million years of avian evolution. *Nature* 378:349–55

Chiappe LM. 1996. Late Cretaceous birds of southern South America: anatomy and systematics of Enantiornithes and *Patagopteryx deferrariisi*. *Münch. Geowissen. Abh.* 30: 203–44

Chiappe LM. 1998. Review [of Chatterjee 1997], The rise of birds. *Am. Zool.* 38:797–98

Chiappe LM. 1999. Early avian evolution: roundtable report. *Smithson. Contr. Paleobiol.* 88:335–40

Chiappe LM. 2001. The rise of birds. In *Palaeobiology II: A Synthesis*, ed. DEG Briggs, PR Crowther, pp. 102–6. Cambridge, UK: Cambridge Univ. Press

Chiappe LM. 2002a. Basal bird phylogeny: problems and solutions. See Chiappe & Witmer 2002, pp. 448–72

Chiappe LM. 2002b. Osteology of the flightless *Patagopteryx deferrariisi* from the Late Cretaceous of Patagonia. See Chiappe & Witmer 2002, pp. 281–316

Chiappe LM, Calvo JO. 1994. *Neuquenornis volans*, a new Upper Cretaceous bird (Enantiornithes: Avisauridae) from Patagonia, Argentina. *J. Vert. Paleontol.* 14(2):230–46

Chiappe LM, Ji S, Ji Q, Norell MA. 1999. Anatomy and systematics of the Confuciosornithidae (Aves) from the Late Mesozoic of northeastern China. *Bull. Am. Mus. Nat. Hist.* 242:1–89

Chiappe LM, Lacasa-Ruiz A. 2002. *Noguerornis gonzalezi* (Aves) from the Early Cretaceous of Spain. See Chiappe & Witmer 2002, pp. 230–39

Chiappe LM, Norell MA, Clark JM. 1996. Phylogenetic position of *Mononykus* from the Upper Cretaceous of the Gobi Desert. *Mem. Queensl. Mus.* 39:557–82

Chiappe LM, Norell MA, Clark JM. 1998. The skull of a new relative of the stem-group bird *Mononykus*. *Nature* 392:272–78

Chiappe LM, Norell MA, Clark JM. 2001. A new skull of *Gobipteryx minuta* (Aves: Enantiornithines) from the Cretaceous of the Gobi Desert. *Am. Mus. Novit.* 3346:1–17

Chiappe LM, Norell MA, Clark JM. 2002. The Cretaceous, short-armed Alvarezsauridae: *Mononykus* and its kin. See Chiappe & Witmer 2002, pp. 87–120

Chiappe LM, Walker CA. 2002. Skeletal morphology and systematics of the Cretaceous Euenantiornithes (Ornithothoraces: Enantiornithes). See Chiappe & Witmer 2002, pp. 240–67

Chiappe LM, Witmer LM, eds. 2002. *Mesozoic Birds: Above the Heads of Dinosaurs*. Berkeley, Univ. Calif. Press

Chinsamy A. 1990. Physiological implications of the bone histology of *Syntarsus rhodesiensis* (Saurischia: Theropoda). *Palaeontol. Afr.* 27:77–82

Chinsamy A. 1993. Bone histology and growth trajectory of the prosauropod dinosaur *Massospondylus*. *Mod. Geol.* 18:319–29

Chinsamy A. 2002. Bone microstructure of early birds. See Chiappe & Witmer 2002, pp. 421–31

Chinsamy A, Chiappe LM, Dodson P. 1995. Mesozoic avian bone microstructure: physiological implications. *Paleobiology* 21:561–74

Chinsamy A, Elzanowski A. 2001. Evolution of growth pattern in birds. *Nature* 412:402–3

Chinsamy A, Martin LD, Dodson P. 1998. Bone microstructure of the diving *Hesperornis* and the volant *Ichthyornis* from the Niobrara Chalk of western Kansas. *Cret. Res. Soc. Q.* 19:225–35

Clark JM, Norell MA, Barsbold R. 2001. Two new oviraptorids (Theropoda: Oviraptorosauria), Upper Cretaceous Djadokhta Formation, Ukhaa Tolgod, Mongolia. *J. Vert. Paleontol.* 21:209–13

Clark JM, Norell MA, Chiappe LM. 1999. An oviraptorid skeleton from the Late Cretaceous of Ukhaa Tolgod, Mongolia, preserved in an avian-like brooding position over an oviraptorid nest. *Am. Mus. Novit.* 3265:1–36

Clark JM, Norell MA, Makovicky PJ. 2002. Cladistic approaches to the relationship of birds to other theropod dinosaurs. See Chiappe & Witmer 2002, pp. 31–64. Berkeley: Univ. Calif. Press

Clarke JA. 1999. New information on the type material of *Ichthyornis*: of chimeras, characters and current limits of phylogenetic inference. *J. Vert. Paleontol.* 19:A38

Clarke JA. 2002. *The morphology and systematic position of* Ichthyornis *marsh and the phylogenetic relationships of basal Ornithurae*. PhD thesis. Yale Univ., New Haven, CT. 532 pp.

Clarke JA, Chiappe LM. 2001. A new carinate bird from the Late Cretaceous of Patagonia. *Am. Mus. Novit.* 3323:1–23

Cooper A, Penny D. 1997. Mass survival of birds across the Cretaceous-Tertiary boundary: molecular evidence. *Science* 275:1109–13

Cracraft J. 1986. The origin and early diversification of birds. *Paleobiology* 12:383–99

Currie PJ. 2000. Feathered dinosaurs. In *The Scientific American Book of Dinosaurs*, ed. GS Paul, pp. 183–189. New York: St. Martin's

de Ricqlès AJ. 1980. Tissue structure of dinosaur bone. Functional significance and possible relation to dinosaur physiology. In *A Cold Look at the Warm Blooded Dinosaurs*, ed. RDK Thomas, EC Olson, pp. 103–39. Boulder, CO: Westview

Dong Z-M, Currie PJ. 1996. On the discovery of an oviraptorid skeleton on a nest of eggs at Bayan Mandahu, Inner Mongolia, People's Republic of China. *Can. J. Earth. Sci.* 33:631–36

Dyke GJ. 2001. The evolution of birds in the Early Tertiary: systematics and patterns of diversification. *Geol. J.* 36:305–15

Dyke GJ, Mayr G. 1999. Did parrots exist in the Cretaceous period? *Nature* 399:317–18

Elzanowski A. 1981. Embryonic bird skeletons from the Late Cretaceous of Mongolia. *Acta Paleontol. Pol.* 42:147–76

Elzanowski A. 1999. A comparison of the jaw skeleton in theropods and birds, with a description of the palate in Oviraptoridae. *Smithson. Contrib. Paleobiol.* 311–23

Elzanowski A. 2001. A new genus and species for the largest specimen of *Archaeopteryx*. *Acta Paleontol. Pol.* 46:519–32

Elzanowski A. 2002. Archaeopterygidae. See Chiappe & Witmer 2002, pp. 129–59

Erickson GM, Curry Rogers KC, Yerby SA. 2001. Dinosaurian growth patterns and rapid avian growth rates. *Nature* 412:429–33

Feduccia A. 1999. *The Origin and Evolution of Birds*. New Haven, CT: Yale Univ. Press. 2ⁿᵈ ed.

Feduccia A, Martin LM. 1996. Jurassic *Urvogels* and the myth of the feathered dinosaurs. In *The Continental Jurassic*, ed. M Morales, pp. 185–91. Flagstaff: Bull. Mus. Ariz.

Feduccia A, Wild R. 1993. Bird-like characters in the Triassic archosaur *Megalancosaurus*. *Naturwissenschaften* 80:564–66

Forster CA, Sampson SD, Chiappe LM, Krause DW. 1998a. The theropodan ancestry of birds: new evidence from the Late Cretaceous of Madagascar. *Science* 279:1915–19

Forster CA, Sampson SD, Chiappe LM, Krause DW. 1998b. Genus correction. *Science* 280:185

Gatesy SM, Dial KP. 1996. From frond to fan: *Archaeopteryx* and the evolution of short tailed birds. *Evolution* 50:2037–48

Gauthier JA. 1986. Saurischian monophyly and the origin of birds. In *The Origin of Birds and the Evolution of Flight*, ed. K Padian, pp. 1–55. Berkeley: Calif. Acad. Sci.

Gauthier JA, Gall LF, eds. 2001. *New Perspectives on the Origin and Early Evolution of Birds: Proc. Int. Symp. Honor of John H. Ostrom*. New Haven: Peabody Mus. Nat. Hist., Yale Univ.

Gauthier JA, Padian K. 1985. Phylogenetic, functional, and aerodynamic analyses of the origin of birds and their flight. See Hecht et al. 1985, pp. 185–97

Geist NR, Feduccia A. 2000. Gravity defying behaviors: identifying models for protoaves. *Am. Zool.* 40:664–75

Grellet-Tinner G, Chiappe LM. 2002. Dinosaur eggs and nesting: implications for understanding the origin of birds. In *Symposium on the Origin of Birds*, ed. RT Bakker, PJ Currie. Bloomington: Indiana Univ. Press. In press

Hecht MK, Ostrom JH, Viohl G, Wellnhofer P, eds. 1985. *The Beginnings of Birds: Proc. Int. Archaeopteryx Conf.* Eichstätt: Jura Mus.

Hedges SB, Parker PH, Sibley CG, Kumar S. 1996. Continental breakup and the ordinal diversification of birds and mammals. *Nature* 381:226–29

Heilmann G. 1926. *Origin of Birds*. London: Witherby

Hinchliffe JR. 1985. 'One, two, three' or 'two, three, four': an embryologist's view of the homologies of the digits and carpus of modern birds. See Hecht et al. 1985, pp. 141–47

Holtz TR Jr. 1998. A new phylogeny of the carnivorous dinosaurs. *Gaia* 15:5–61

Holtz TR Jr. 2001. Arctometatarsalia revisited: the problem of homoplasy in reconstructing theropod phylogeny. See Gauthier & Gall 2001, pp. 99–122

Hou L. 1997. *Mesozoic Birds of China*. Nan-Tou: Taiwan Prov. Feng Huang Ku Bird Park

Hou L, Martin LD, Zhou Z, Feduccia A. 1996. Early adaptive radiation of birds: evidence from fossils from northeastern China. *Science* 274:1164–67

Hou L, Zhou Z, Martin LD, Feduccia A. 1995. A beaked bird from the Jurassic of China. *Nature* 377:616–18

Houck MA, Gauthier JA, Strauss RE. 1990. Allometric scaling in the earliest fossil bird, *Archaeopteryx lithographica*. *Science* 247:195–98

Houde P. 1987. Histological evidence for the systematic position of *Hesperornis* (Odontornithes: Hesperornithiformes). *Auk* 104:125–29

Howgate ME. 1984. The teeth of *Archaeopteryx* and a reinterpretation of the Eichstätt specimen. *Zool. J. Linn. Soc.* 82:159–75

Jenkins FA Jr, Dial KP, Goslow GE Jr. 1988. A cineradiographic analysis of bird flight: the wishbone in starlings is a spring. *Science* 241:1495–98

Ji Q, Currie P, Norell MA, Ji S-A. 1998. Two feathered dinosaurs from northeastern China. *Nature* 393:753–61

Jones TD, Farlow JO, Ruben JA, Henderson DM, Hillenius WJ. 2000. Cursoriality in bipedal archosaurs. *Nature* 406:716–18

Kumar S, Hedges SB. 1998. A molecular timescale for vertebrate evolution. *Nature* 392:917–20

Kurochkin EN. 1985. A true carinate bird from Lower Cretaceous deposits in Mongolia and other evidence of Early Cretaceous birds in Asia. *Creat. Res. Soc. Q.* 6:271–78

Kurochkin EN. 1995. Synopsis of Mesozoic birds and early evolution of class Aves. *Archaeopteryx* 13:47–66

Kurochkin EN. 1996. A new enantiornithid of the Mongolian Late Cretaceous, and a general appraisal of the Infraclass Enantiornithes (Aves). *Russ. Acad. Sci. Palaeont. Inst. Spec. Issue*. 50 pp.

Kurochkin EN. 1999. The relationships of the Early Cretaceous *Ambiortus* and *Otogornis* (Aves: Ambiortiformes). *Smithson. Contrib. Paleobiol.* 89:275–84

Kurochkin EN. 2000. Mesozoic birds of Mongolia and the former USSR. In *The Age of Dinosaurs in Russia and Mongolia*, ed. MJ Benton, MA Shiskin, DM Unwin, EN Kurochkin, pp. 533–59. Cambridge, UK: Cambridge Univ. Press

Kurochkin EN. 2001. New ideas about the origin and early evolution of birds. In *Achievements and Problems of Ornithology of Northern Eurasia on a Boundary of Centuries*, ed. EN Kurochkin, I Rakchimov, pp. 68–96. Kazan: Magarif. In Russian with English summary

Livezey BC, Zusi RL. 2001. Higher-order phylogenetics of modern Aves based on comparative anatomy. *Neth. J. Zool.* 51:179–205

Marsh OC. 1880. *Odontornithes: A Monograph on the Extinct Toothed Birds of North America. U.S. Geol. Explor., 40th Parallel*. Washington, DC: GPO

Martin LD. 1983. The origin and early radiation of birds. In *Perspectives in Ornithology*, ed.

AH Brush, GA Clark Jr, pp. 291–338. New York: Cambridge Univ. Press

Martin LD. 1995. The Enantiornithes: terrestial birds of the Cretaceous. *Cour. Forsch. Senckenb.* 181:23–26

Martin LD, Czerkas SA. 2000. The fossil record of feather evolution in the Mesozoic. *Am. Zool.* 40:687–94

Martin LD, Rinaldi C. 1994. How to tell a bird from a dinosaur. *Maps Digest* 17:190–96

Martin LD, Stewart JD. 1999. Implantation and replacement of bird teeth. *Smithson. Contrib. Paleobiol.* 89:295–300

Martin LD, Stewart JD, Whetstone KN. 1980. The origin of birds: structure of the tarsus and teeth. *Auk* 97:86–93

Martin-Closas C, López-Morón N. 1995. The charophyte flora. In *Montsec and Mont-Ral-Alcover, Two Konservat-Lagerstätten. Field Trip Guide Book.* Int. Symp. Lithogr. Limestones, pp. 29–31. Catalonia, Spain.

Maryanska T, Osmólska H, Wolsan M. 2002. Avialan status for oviraptorosauria. *Acta Palaeontol. Pol.* 47(1):97–116

Mäuser M. 1997. Der achte *Archaeopteryx. Fossilien* 3:156–57

Molnar RE. 1985. Alternatives to *Archaeopteryx*: a survey of proposed early or ancestral birds. See Hecht et al. 1985, pp. 209–17

Norell MA, Clarke JA. 2001. Fossil that fills a critical gap in avian evolution. *Nature* 409:181–84

Norell MA, Clark JM, Chiappe LM, Dashzeveg D. 1995. A nesting dinosaur. *Nature* 378:774–76

Norell MA, Clark JM, Dashzeveg D, Barsbold R, Chiappe LM. 1994. A Theropod dinosaur embryo, and the affinities of the Flaming Cliffs dinosaur eggs. *Science* 266:779–82

Norell MA, Clark JM, Makovicky PJ. 2001. Phylogenetic relationships among coelurosaurian theropods. See Gauthier & Gall 2001, pp. 49–68

Norell MA, Ji Q, Gao K, Yuan C, Zhao Y, Wang L. 2002. "Modern" feathers on a non-avian dinosaur. *Nature* 416:36–37

Novas FE. 1996. Alvarezsauridae, Cretaceous maniraptorans from Patagonia and Mongolia. *Mem. Queensl. Mus.* 3939:675–702

Novas FE. 1997. Anatomy of *Patagonykus puertai* (Theropoda, Maniraptora, Alvarezsauridae) from the Late Cretaceous of Patagonia. *J. Vert. Paleontol.* 17:137–66

Novas FE, Pol D. 2002. Alvarezsaurid relationships reconsidered. See Chiappe & Witmer 2002, pp. 121–28

Novas FE, Puerta P. 1997. New evidence concerning avian origins from the Late Cretaceous of Patagonia. *Nature* 387:390–92

Olson SL. 1994. A giant *Presbyornis* (Aves: Anseriformes) and other birds from the Paleocene Aquia Formation of Maryland and Virginia. *Proc. Biol. Soc. Wash.* 107:429–35

Ostrom JH. 1969. Osteology of *Deinonychus antirrhopus*, an unusual theropod from the lower Cretaceous of Montana. *Bull. Peabody Mus. Nat. Hist.* 30:1–165

Ostrom JH. 1973. The ancestry of birds. *Nature* 242:136

Ostrom JH. 1974. *Archaeopteryx* and the origin of flight. *Q. Rev. Biol.* 49:27–47

Ostrom JH. 1976. *Archaeopteryx* and the origin of birds. *Biol. J. Linn. Soc.* 8:91–182

Ostrom JH. 1991. The bird in the bush. *Nature* 353:212

Ostrom JH. 1994. On the origin of birds and of avian flight. In *Major Features of Vertebrate Evolution*, ed. DR Prothero, RM Schoch, pp. 160–77. Austin, TX: Paleontol. Soc.

Padian K, Chiappe LM. 1998. The origin and early evolution of birds. *Biol. Rev. Camb. Philos. Soc.* 73:1–42

Padian K, de Ricqlès AJ, Horner JR. 2001. Dinosaurian growth rates and bird origins. *Nature* 412:405–8

Paul GS. 1988. *Predatory Dinosaurs of the World.* New York: Simon & Schuster

Perle A, Chiappe LM, Barsbold R, Clark JM, Norell MA. 1994. Skeletal morphology of *Mononykus olecranus* (Theropoda: Avialae) from the Late Cretaceous of Mongolia. *Am. Mus. Novit.* 3105:1–29

Perle A, Norell MA, Chiappe LM, Clark JM. 1993a. Flightless bird from the Cretaceous of Mongolia. *Nature* 362:623–26

Perle A, Norell MA, Chiappe LM, Clark JM. 1993b. Correction to flightless bird from the Cretaceous of Mongolia. *Nature* 363:628

Prum RO. 1999. Development and evolutionary origin of feathers. *J. Exp. Zool.* 285:291–306

Rayner JMV. 1991. Avian flight evolution and the problem of *Archaeopteryx*. In *Biomechanics in Evolution*, ed. JMV Rayner, RJ Wooton, pp. 183–212. Cambridge, UK: Cambridge Univ. Press

Rayner JMV. 2001. On the origin and evolution of flapping flight aerodynamics in birds. See Gauthier & Gall 2001, pp. 363–85

Renesto S. 2000. Bird-like head on a chameleon body: new specimens of the enigmatic diapsid reptile *Megalancosaurus* from the Late Triassic of Northern Italy. *Rev. Ital. Paleontol. Strat.* 106:157–80

Ruben JA, Jones TD. 2000. Selective factors associated with the origin of fur and feathers. *Am. Zool.* 40:585–96

Sanz JL, Álvarez JC, Meseguer J, Soriano C, Hernández-Carrasquilla F, Pérez-Moreno BP. 2000. Wing loading in primitive birds. *Vert. Palaeontol. Asiat.* 38:27

Sanz JL, Chiappe LM, Buscalioni A. 1995. The osteology of *Concornis lacustris* (Aves: Enantiornithes) from the Lower Cretaceous of Spain and a re-examination of its phylogenetic relationships. *Am. Mus. Novit.* 3133:1–23

Sanz JL, Chiappe LM, Pérez-Moreno BP, Buscalioni AD, Moratalla J. 1996. A new Lower Cretaceous bird from Spain: implications for the evolution of flight. *Nature* 382:442–45

Sanz JL, Chiappe LM, Pérez-Moreno BP, Moratalla J, Hernández-Carrasquilla F, et al. 1997. A nestling bird from the Early Cretaceous of Spain: implications for avian skull and neck evolution. *Science* 276:1543–46

Schweitzer MH, Watt JA, Avci R, Knapp L, Chiappe LM, et al. 1999. Beta-keratin specific immunological reactivity in feather-like structures of the Cretaceous alvarezsaurid, *Shuvuuia deserti*. *J. Exp. Zool.* 285:146–57

Sereno PC. 1999. The evolution of dinosaurs. *Science* 284:2137–47

Sereno PC. 2000. Origin and early evolution of Aves: data and hypotheses. *Vert. Palaeontol. Asiat.* 38:27–28

Sereno PC. 2001. Alvarezsaurids: birds or ornithomimosaurs? See Gauthier & Gall 2001, pp. 69–98

Sereno PC, Rao C. 1992. Early evolution of avian flight and perching: new evidence from Lower Cretaceous of China. *Science* 255:845–48

Simpson GG. 1980. Fossil birds and evolution. In *Papers in Avian Paleontology Honoring Hildegarde Howard*, ed. KE Campbell, pp. 3–8. Contrib. Sci. 330 Nat. Hist. Mus. LA County

Starck JM, Ricklefs RE. 1998. Patterns of development: the altricial-precocial spectrum. In *Avian Growth and Development*, ed. JM Starck, RE Ricklefs, pp. 3–26. Oxford, UK: Oxford Univ. Press

Stidham TA, Hutchison JH. 2001. The North American avisaurids (Aves: Enantiornithes): new data on biostratigraphy and biogeography. *Assoc. Paleontol. Argentina, Pub. Especial* 7:175–77

Suzuki S, Chiappe LM, Dyke GJ, Watabe M, Barsbold R, Tsogtbaatar K. 2002. A new specimen of *Shuvuuia deserti* from the Mongolian Late Cretaceous with a discussion of the relationships of alvarezsaurids to other theropod dinosaurs. *Sci. Ser. Nat. Hist. Mus. LA County*

Swisher CC III, Wang X, Zhou Z, Wang Y, Jin F, et al. 2002. Further support for a Cretaceous age for the feathered-dinosaur beds of Liaoning, China: new ^{40}Ar/^{39}Ar dating of the Yixian and Tuchengzi Formations. *Chin. Sci. Bull.* 47:135–38

Varricchio DV, Jackson F, Borkowski JJ, Horner J. 1997. Nest and egg clutches of the dinosaur *Troodon formosus* and the evolution of avian reproductive traits. *Nature* 385:247–50

Walker CA. 1981. New subclass of birds from the Cretaceous of South America. *Nature* 292:51–53

Wellnhofer P. 1993. Das siebte exemplar von *Archaeopteryx* aus den Solnhofener Schichten. *Archaeopteryx* 11:1–48

Wellnhofer P. 1994. New data on the origin and evolution of birds. *R. Acad. Sci. Paris* 319:299–308

Welman J. 1995. *Euparkeria* and the origin of birds. *S. Afr. J. Sci.* 91:533–37

Witmer LM. 1991. Perspectives on avian origins. In *Origins of the Higher Groups of Tetrapods*, ed. H-P Schultze, L Trueb, pp. 427–66. Ithaca, NY: Comstock

Witmer LM. 1997. Forward. See Chatterjee 1997, pp. vii–xii

Witmer LM. 2001. The role of *Protoavis* in the debate on avian origins. See Gauthier & Gall 2001, pp. 537–48

Xu X, Tang Z, Wang X. 1999a. A therizinosaurid dinosaur with integumentary structures from China. *Nature* 399:350–54

Xu X, Wang X-L, Wu X-C. 1999b. A dromaeosaur dinosaur with filamentous integument from the Yixian Formation of China. *Nature* 401:262–66

Xu X, Zhou Z, Prum RO. 2001. Branched integumentary structures in *Sinornithosaurus* and the origin of feathers. *Nature* 410:200–4

Xu X, Zhou Z, Wang X. 2000. The smallest known non-avian theropod dinosaur. *Nature* 408:705–8

Zhang F, Zhou Z. 2000. A primitive enantiornithine bird and the origin of feathers. *Science* 290:1955–59

Zhang F, Zhou Z, Hou L, Gu G. 2001. Early diversification of birds: evidence from a new opposite bird. *Chin. Sci. Bull.* 46:945–49

Zhou Z. 1995a. Is *Mononykus* a bird? *Auk* 112:958–63

Zhou Z. 1995b. Discovery of a new enantiornithine bird from the Early Cretaceous of Liaoning, China. *Vert. Palaeontol. Asiat.* 33:99–113

Zhou ZH, Wang XL. 2000. A new species of *Caudipteryx* from the Yixian Formation of Liaoning, northeast China. *Vert. Palaeontol. Asiat.* 38:111–27

Zhou ZH, Wang XL, Zhang Z, Xu X. 2000. Important features of *Caudipteryx*: evidence from two nearly complete specimens. *Vert. Palaeontol. Asiat.* 38:241–54

Zhou ZH, Zhang F. 2001. Two new ornithurine birds from the Early Cretaceous of western Liaoning, China. *Chin. Sci. Bull.* 46:1258–64

Annu. Rev. Ecol. Syst. 2002. 33:125–59
doi: 10.1146/annurev.ecolsys.33.010802.150452
Copyright © 2002 by Annual Reviews. All rights reserved
First published online as a Review in Advance on August 6, 2002

PLANT ECOLOGICAL STRATEGIES: Some Leading Dimensions of Variation Between Species

Mark Westoby, Daniel S. Falster, Angela T. Moles,
Peter A. Vesk, and Ian J. Wright
*Department of Biological Sciences, Macquarie University, Sydney, New South Wales
2109, Australia; email: mwestoby@rna.bio.mq.edu.au*

Key Words   seed mass, leaf mass per area, foliage height, leaf size, leaf lifespan, twig size

■ **Abstract**   An important aim of plant ecology is to identify leading dimensions of ecological variation among species and to understand the basis for them. Dimensions that can readily be measured would be especially useful, because they might offer a path towards improved worldwide synthesis across the thousands of field experiments and ecophysiological studies that use just a few species each. Four dimensions are reviewed here. The leaf mass per area–leaf lifespan (LMA-LL) dimension expresses slow turnover of plant parts (at high LMA and long LL), long nutrient residence times, and slow response to favorable growth conditions. The seed mass–seed output (SM-SO) dimension is an important predictor of dispersal to establishment opportunities (seed output) and of establishment success in the face of hazards (seed mass). The LMA-LL and SM-SO dimensions are each underpinned by a single, comprehensible tradeoff, and their consequences are fairly well understood. The leaf size–twig size (LS-TS) spectrum has obvious consequences for the texture of canopies, but the costs and benefits of large versus small leaf and twig size are poorly understood. The height dimension has universally been seen as ecologically important and included in ecological strategy schemes. Nevertheless, height includes several tradeoffs and adaptive elements, which ideally should be treated separately. Each of these four dimensions varies at the scales of climate zones and of site types within landscapes. This variation can be interpreted as adaptation to the physical environment. Each dimension also varies widely among coexisting species. Most likely this within-site variation arises because the ecological opportunities for each species depend strongly on which other species are present, in other words, because the set of species at a site is a stable mixture of strategies.

INTRODUCTION

Plant species all use the same major resources of light, water, CO_2, and mineral nutrients. Ecological differences among vascular land plant species arise from different ways of acquiring the same resource rather than from use of alternative foodstuffs. Leaves, stems, roots, and seeds vary between species in construction, in lifespan, and in relative allocation. This review discusses four major dimensions of variation across vascular land plant species. The dimensions affect ecological strategy, that is, the manner in which species secure carbon profit during vegetative growth and ensure gene transmission into the future. Every plant ecologist will have his or her own list of traits that are informative about a species, and his or her own ranking among those traits. Nevertheless, three of the traits discussed—leaf mass per area, seed mass, and height—rank near the top of most plant ecologists' lists (Vendramini et al. 2002, Weiher et al. 1999, Westoby 1998, Wilson et al. 1999).

ABOUT ECOLOGICAL STRATEGIES

A variety of ecological strategy schemes have been proposed (see review in Westoby 1998). One type expresses response or distribution in relation to single environmental factors. Examples include increasers and decreasers in relation to livestock grazing (Dyksterhuis 1949), the requirement for canopy gaps during establishment (Denslow 1980), and reestablishment potential in relation to time since fire or other disturbance (Noble & Slatyer 1980). Raunkiaer's (1934) life-form system involves the location of buds from which regrowth occurs following winter or dry season.

Currently, the International Geosphere-Biosphere Program aims to model vegetation dynamics under future global change. To this end a broad-purpose scheme of plant functional types is seen as essential, and various committees and workshops have been discussing this matter (McIntyre et al. 1999). Among the schemes that have included more than one dimension (Begon et al. 1996, Loehle 2000, Smith & Huston 1989), the best developed is Grime's CSR triangle (Grime 1974, 1977, 1979; Grime et al. 1988). The R (ruderal) dimension expresses response to disturbance, and the S-C (stress tolerator–competitor) dimension expresses capacity to take advantage of favorable growth conditions. The merits of the CSR scheme have been vigorously debated. Nevertheless, the underlying idea is simply that coping with disturbance and adapting to fast versus slow growth opportunities are two major dimensions of ecological variation. This much would have been accepted by most plant ecologists since the 1800s.

Strategy schemes can have different aims. Some have region-specific applications in range or forest management. Some express concepts about the most important factors and opportunities shaping the ecology of plants. Here we wish to focus on a particular role for ecological strategy schemes: their potential for drawing together and organizing the knowledge gained from hundreds of experiments worldwide, each covering one or a few species. Southwood (1977) likened ecology to what chemistry must have been like before the periodic table of the elements was invented. As he put it, "each fact had to be discovered by itself, and each fact remembered in isolation."

To benefit from synthesizing experimental results across different continents and environments, potential indicators of the ecology of species need to be measured easily and consistently worldwide. The conceptual strategy schemes such as

CSR have not met this need. Rather, species are related by comparing performance or distribution in a landscape where they occur together. For this reason attempted syntheses have been forced back to growth-form, life-form, or habitat categorizations in the attempt to make sense of the accumulated experimental literature (e.g., Connell 1983, Crawley 1983, Goldberg 1996, Goldberg & Barton 1992, Gurevitch et al. 1992, Schoener 1983, Vesk & Westoby 2001, Wilson & Agnew 1992). With this in mind, Westoby (1998) previously suggested a "leaf-height-seed" scheme, with the three dimensions readily quantifiable.

Here we put this idea in a different way. Rather than setting up a named three-dimensional strategy scheme, we present a shortlist of dimensions that might be helpful for literature synthesis. At least one of the traits associated with each dimension can be readily quantified. The list need not stop at three, and there need be no requirement for absolute consensus about rankings. Still, if reasonably wide agreement can be achieved about a few traits worth measuring consistently, then we may hope for considerable benefits from using these traits as predictors of ecological behavior.

ABOUT DIMENSIONS OF VARIATION

Criteria for Ranking a Dimension of Variation

ECOLOGICAL SIGNIFICANCE   The position of a species along the dimension should be known to have an important influence with regard to how the species makes a living or where it does best. Preferably, there should be solid experimental evidence about this. Often there may be cross-species correlations among traits, such that information about one measurable trait carries with it broader knowledge about the ecology of species.

SPREAD AND CONSISTENCY   Species should be spread widely along the dimension (breadth of variation is discussed in "Spread of Species Along the Dimensions," below). Further, rankings of species along the dimension should be consistent (at least approximately) in the face of within-species variation due to plasticity, acclimation, or ecotypic variation. Traits need not be constant within species. Indeed, it would be surprising if natural selection had not endowed species with some capacity to adjust traits of ecological importance depending on the situation. The consistent-ranking criterion means that plasticity and other variation within species should not be a conceptual problem, though they may cause complications in measurement.

PRACTICALITY FOR LITERATURE SYNTHESIS   It should be practical to quantify the dimension in a manner that does not depend on the local context of physical environment or co-occurring species.

Cross-species correlations among traits can arise in different ways. Most straightforward and desirable (for the purpose at hand) is a physically enforced tradeoff. An example is that seed output per gram of seed produced cannot be increased without a decrease in mass of individual seed. This is a matter of logic. Another example is that longer leaf lifespan seems nearly always to demand a more robust structure and hence greater leaf mass per area.

Trait correlation can also arise because available niches favor it. For example, lifestyles or habitats involving tall stems might tend also to select for larger seed mass. These correlations might be expected to be looser than correlations enforced by a physical tradeoff.

Trait correlations across species arise also by correlated evolutionary divergence of traits at a phylogenetic branch-point deep in the past, with the trait combinations persisting within each of the descendant lineages (Felsenstein 1985, Lord et al. 1995, Prinzing et al. 2001, Wright et al. 2000). These old divergences should not be regarded as a causation that is distinctly separate from present-day ecological selection or from a physically enforced tradeoff, because the evolution of species into present-day opportunities often has a large element of phylogenetic niche conservatism (Harvey & Rambaut 2000, Price 1997, Westoby 1999, Westoby et al. 1995). Only if the traits in question were incapable of responding to selection through millions or tens of millions of years would it be useful to regard the present-day correlation as owing to the old divergence rather than to continuing selection. It makes sense to regard the most recent causative process as the effective cause of a present-day pattern.

All three criteria for importance are met by leaf mass per area trading off with leaf lifespan (LMA-LL) and by mass of individual seed trading off with seed output per gram of reproductive effort (SM-SO). Height of a species at maturity and the spectrum from small to large leaf size and twig size meet the criteria of broad spread and practical measurement. They are also known to be ecologically significant. However, height is a complex trait with several components, and the costs and benefits of leaf size are poorly understood.

LEAF MASS PER AREA AND LEAF LIFESPAN

Species with higher leaf mass per leaf area (LMA) have thicker laminas, veins that protrude more, higher tissue density, or combinations of these (Niinemets 1999, Pyankov et al. 1999, Shipley 1995, Wilson et al. 1999, Witkowski & Lamont 1991). High-LMA species tend to achieve longer average leaf lifespan in a variety of habitats (Figure 1) (Diemer 1998a,b; Reich et al. 1997; Ryser & Urbas 2000; Williams-Linera 2000; Wright et al. 2002), suggesting that longer leaf lifespans require extra structural strength (Coley 1988, Reich et al. 1991, Wright & Cannon 2001). In short, the LMA-LL spectrum is a trade-off between potential rate of return per leaf mass and duration of return.

Higher LMA protects against wear and tear and also deters herbivory. Under any of several concepts about allocation to defense (Bryant et al. 1983, Coley et al. 1985, Herms & Mattson 1992), species with slower leaf turnover should spend more to discourage herbivores. Thicker, tougher leaves are themselves the

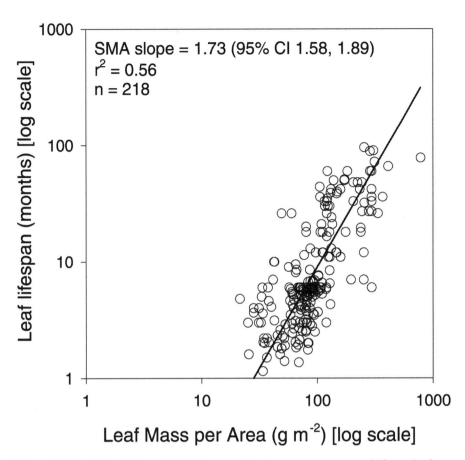

Figure 1   Correlation between leaf lifespan and leaf mass per area across 218 species from several habitats and continents. Regraphed from Reich et al. (1997); data kindly provided by the authors. SMA = Standard Major Axis; CI = confidence internal.

most common and general-purpose form of defense (Coley 1983, Cunningham et al. 1999), but long leaf lifespan may also be correlated with greater relative allocation to tannins, phenols, or other defensive compounds (Coley 1988). Slow leaf turnover should favor strength against wear and tear for the same reason it favors defense against herbivores. For both these reasons, defense against herbivory can be regarded as part of the LMA-LL spectrum.

Leaf Economics and Theory for Leaf Lifespan

A leaf represents an investment on the part of a plant. Kikuzawa's theory for leaf lifespan (Figure 2) can be understood through the curve of cumulative return from the investment. This return is expressed as net dry-mass gain per unit leaf

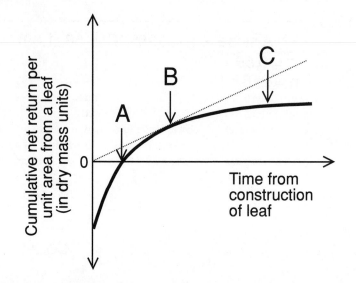

Figure 2 Essentials of existing theory for leaf lifespan (Kikuzawa 1995). Curve shows cumulative dry-mass return from a unit leaf area, net of costs of leaf respiration and of root and stem activity to support the leaf's photosynthesis. Curve is initially negative owing to construction costs (leaf mass per area), then increases through a leaf's lifetime. Payback time for the investment is at A. Net dry-mass return per time per leaf area is the slope of a line from the origin to the curve. It is maximized at the lifespan B. This optimum at B, and also payback time A, shift to longer lifespan if the cumulative dry-mass gain curve is shallower (slow-revenue environments) or if the initial investment is greater (higher leaf mass per area). At C, approximately, the leaf is no longer returning net dry-mass revenue.

area. Construction costs per unit leaf mass vary relatively little between species (Chapin 1989, Poorter & De Jong 1999, Villar & Merino 2001). Cumulative return is initially negative by the amount of LMA, which is dry-mass investment per leaf area, then rises through the lifespan. The slope becomes shallower over time owing to deterioration of the leaf's position within the canopy or of its physiological capacity. Eventually a leaf returns no further net dry-mass gain (at C in Figure 2) when photosynthesis no longer exceeds costs of leaf respiration and of root and stem activity to support the leaf's photosynthesis.

Early verbal formulations (Chabot & Hicks 1982, and more recently Niinemets 2001, Poorter 1994, Williams et al. 1989) were to the effect that leaf lifespan needed to be long enough to pay back the initial investment costs (point A in Figure 2). This is true, but plainly lifespans need to be longer than the payback time if plants are to grow. In Kikuzawa's model (Kikuzawa 1991, 1995; Kikuzawa & Ackerly 1999), revenue per unit time per unit leaf area is maximized. Replacing an old leaf with a new leaf becomes justified when the current return on an old leaf (marginal return) falls below the expected rate of return averaged over the life of a new leaf (point B in Figure 2). Optimizing per unit leaf area assumes (in effect) that new leaves can be produced only by closing down old leaves. At point B in Figure 2 leaves are closed down while still producing net dry-mass revenue. However, plants do generate new leaves without closing down the same area of old leaves. Longer lifespans (point C rather than B in Figure 2) would be favored if leaves were only discarded when their net revenue had deteriorated to zero. A shoot carbon budget model (Ackerly 1999) favored keeping leaves this long to maximize extension growth. However, building new leaves demands nitrogen as well as photosynthate, and some of this nitrogen is reallocated from old leaves prior to leaf fall (Aerts & Chapin 2000). This should favor closing down old leaves sooner, somewhere between points B and C in Figure 2.

In environments where revenue accrues more slowly, the curve in Figure 2 is shallower, and points A and B are both shifted to longer lifespan. In species with higher LMA the curve starts more negative, and similarly points A and B are both shifted to longer lifespan. Thus, although point A (payback) does not predict the correct lifespan, it predicts the same direction of lifespan response to alternative curves as does point B (maximum return per leaf area). However, point C (no further return) does not predict longer lifespan in response to slower revenue accrual or higher LMA. The pace at which the cumulative net return curve flattens is what decides lifespan if this depends on the point of no further return.

Correlates, Costs, and Benefits

Species with low LMA tend also to have higher photosynthetic capacity per unit leaf mass, A_{mass} (Field & Mooney 1986, Niinemets 1999, Reich et al. 1997, Wright et al. 2001). As well as having more light-capture area deployed per mass, low-LMA species generally have higher leaf N concentrations (Diemer 1998a,b; Field & Mooney 1986; Reich et al. 1997) and shorter diffusion paths from stomata to chloroplasts (Parkhurst 1994). Leaf N reflects the concentration of Rubisco and other photosynthetic proteins (Lambers et al. 1998). Leaf N is more similar across species per unit leaf area than per unit leaf mass (Reich et al. 1997). Probably the lower N_{mass} in leaves of high-LMA species is partly a forced tradeoff (because a greater concentration of fibers, cell walls, etc. leaves less room for N-rich mesophyll) and partly reflects evolutionary coordination between N allocation and LMA and leaf lifespan.

Variation in LMA and leaf lifespan among coexisting species is 3- to 50-fold, strikingly greater than the 2- to 3-fold shifts between habitat types (Table 1). What are the relative advantages or disadvantages to species at different positions along the LMA–LL spectrum, and why is there not a clear advantage at some particular LMA-LL compromise in a given habitat?

Low LMA, high photosynthetic capacity, and generally faster turnover of plant parts permit flexible response to the spatial patchiness of light and soil resources (Grime 1994), giving short-term advantages over high-LMA species. However, high-LMA–long-LL species have longer-term advantages. Longer mean residence time of nutrients (Aerts & Chapin 2000, Eckstein et al. 1999, Escudero et al.

TABLE 1 Summary of three dimensions of ecological variation between species discussed here. (The height dimension is not included in this table.)[a]

	LMA-LL	SM-SO	LS-TS
Basis of relationship	Presumed that greater engineering strength of leaf is required for longer LL. There are several parallel LMA-LL relationships, for example ~40% shorter LL for a given LMA was achieved at low rainfall (Wright et al. 2002).	SO = (mass allocated to reproduction)/(SM plus accessory mass per live seed). Because SM varies much more widely than mass allocated to reproduction per unit canopy area or leaf area and because accessory mass is closely correlated with SM, SM is the dominant influence on SO.	Presumed owing to requirements of mechanical support, hydraulic conductance, and leaf spacing. Possibly also large leaves require large meristems, which cannot be accommodated in very thin twigs.
Slope of relationship (standardized major axis on log-log scales)	LL increases faster than LMA. Across all available data log-log slope 1.73, 95% confidence intervals 1.6–1.9 (Reich et al. 1997); within-site slope 1.3 common to 4 sites, 95% confidence intervals 1.1–1.6 (Wright et al. 2002).	Seed output versus seed mass log-log slope approx. −1 (Henery & Westoby 2001).	Leaf area versus twig cross-section log-log slopes 1.01 (deciduous angiosperms), 1.14 (evergreen angiosperms), 1.44 (gymnosperms) (Brouat et al. 1998), 1.69–2.05 at 3 sites (Westoby & Wright in press 2002).
Other correlated traits	Mass-based leaf N (N-mass), leaf diffusive conductance G(s), and net photosynthetic capacity A(max) all negatively correlated with LMA and LL (Reich et al. 1999) LL correlated with tissue density (Ryser 1996, Ryser & Urbas 2000, Schlapfer & Ryser 1996).	Correlates of SM include dispersal mode ($r^2 \approx 0.29$) (Leishman et al. 1995), height or growth form ($r^2 \approx 0.20$) (Leishman et al. 1995, Levin 1974, Mazer 1989, Metcalfe & Grubb 1995), leaf size (Cornelissen 1999), relative growth rate, LMA (Reich et al. 1998), seedling size (Grime & Jeffrey 1965). Smaller, rounded seeds are likely to have persistent soil seed banks in England (Hodkinson et al. 1998, Thompson et al. 1993), Argentina (Funes et al. 1999), and New Zealand (Moles et al. 2000) but not in Australia (Leishman & Westoby 1998).	LS with height $r^2 = 0.29$ (gymnosperms included), with seed mass 0.26, with infructescence size 0.79 (gymnosperms excluded) (Cornelissen 1999). Relationship between LS and LMA seems complex (see text).

(Continued)

TABLE 1 *(Continued)*

	LMA-LL	SM-SO	LS-TS
Variation across sites, between species within sites, and within species.	~50-fold variation in LMA (12–560 g m^{-2}) (Fonseca et al. 2000, Niinemets 1999), >100-fold variation in LL (Eckstein et al. 1999, Reich et al. 1997). Between-species variation in LL is much larger than within-species (factor of >200 compared with 2, respectively) (Eckstein et al. 1999). LMA variation among coexisting species is greater than between habitats, e.g., 3- to 50-fold versus 2- to 3-fold (Reich et al. 1999).	Within temperate zone differences between communities account for only ~4% of variation in seed mass between species (Leishman et al. 1995). Differences between the tropics and the temperate are somewhat larger (Lord et al. 1997). Species establishing in deep shade tend to have larger seeds. Range within sites 4–5 orders of magnitude (log$_{10}$ units). Central 66% of species (± 1 SD) spans ~2 log$_{10}$ units (mean of 7 floras) (Leishman et al. 2000, Lord et al. 1995). Within species ± 1 SD (66% of seeds) spans ~4-fold (median of 39 species) (Michaels et al. 1988).	6 classes defined by Raunkiaer (1934) from leptophyll (<25 mm^2) to megaphyll (>164,000 m^2), spanning 5 orders of magnitude. Range among coexisting species 2.5–5 orders of magnitude (Ackerly & Reich 1999, Brouat et al. 1998, Cornelissen 1999, Fonseca et al. 2000, Niinemets & Kull 1994, Westoby & Wright 2002, White 1983a,b).

[a]LMA, leaf mass per area (g m^{-2}) = 1/SLA, specific leaf area; LL, leaf lifespan or leaf longevity; tissue density = dry mass/volume; SM, seed mass; SO, seed output (numbers of viable seeds), per area occupied or per leaf area or per plant mass; LS, leaf size, either leaf area or leaf width (refers to green surfaces, hence leaflets within compound leaves) TS, twig size (diameter or cross-sectional area) at base of current year's growth.

1992) permits a progressively greater share of nitrogen pools in a habitat to be sequestered (Aerts & van der Peijl 1993). Further, slow decomposition of high-LMA litter may restrict opportunities for potentially fast-growing competitors (Berendse 1994, Cornelissen et al. 1999). Second, over time high-LMA–long-LL species accumulate greater total leaf mass than low-LMA species (Bond 1989, Midgley & Bond 1991). Despite the offsetting effect of less leaf area per unit leaf mass, high-LMA species tend to generate a larger total leaf area as well (Gower et al. 1993, Haggar & Ewel 1995, Reich et al. 1992). Combining their lower photosynthetic capacity per leaf area (Reich et al. 1999) with this greater accumulation of leaf area may result in above-ground net primary production similar to or higher than that of low-LMA species (Bond 1989, Gower et al. 1993, Haggar & Ewel 1995, Matyssek 1986, Midgley & Bond 1991, Reich et al. 1992).

The slope of LL-LMA relationships among coexisting species has been significantly steeper than 1.0 in several datasets (evaluated using Standard Major Axis "model 2" slope fitting from Diemer 1998b, Reich et al. 1999, Wright et al. 2002).

That is, species with twice the leaf mass per area typically had more than twice the leaf lifespan. Would this not lead to runaway selection for ever-increasing LMA and leaf lifespan? Not necessarily. The revenue stream generated by a leaf unavoidably diminishes in value as time passes ("time-discounting") (Westoby et al. 2000) for a combination of reasons. Leaves suffer damage from herbivores and pathogens (Coley & Barone 1996, Landsberg & Gillieson 1995, Lowman & Box 1983, Showalter et al. 1986) and are colonized by epiphylls (Clark et al. 1992, Coley et al. 1993). Their light-interception position deteriorates owing to over-shading by leaves produced subsequently, by competitors and by the plant itself (Ackerly & Bazzaz 1995, Hikosaka 1996, Kitajima et al. 1997, Koike 1988). A given export rate of photosynthate from the leaf becomes less valuable, because if obtained earlier it could have been reinvested sooner (Harper 1989). Taking all these factors together, there may be no clear-cut advantage to either long leaf life-span or low-LMA strategies among coexisting species, in terms of fitness value of carbon gain over the lifetime of a unit of leaf mass (Westoby et al. 2000).

Leaf Mass per Area: Leaf Lifespan Tradeoff in Different Environments

Greater leaf mass per area represents greater cost to the plant. If it were possible to achieve the same leaf lifespan for lower LMA, plants would be selected to do so. In our view, the underlying reason why leaf lifespan and LMA are correlated is that long leaf lifespan nearly always requires leaves to be strong in an engineering sense. Depth and material strength are the two main influences on the strength of a horizontal beam (Vogel 1988), and both of these are reflected in LMA.

Within the overall LMA-LL correlation shown in Figure 1, there are seemingly a number of parallel relationships. Shifts in the LMA required to achieve a given leaf lifespan could arise from two causes. First, the wear and tear on a leaf might be more severe in some environments than others. Second, leaf tissue might be softer in some environments than others, such that a greater lamina depth is required to achieve a given overall structural strength. It has recently been shown how differences between rainfall environments can be traced to this second cause (Wright & Westoby 2002a, Wright et al. 2002). Average LMA is well known to be higher at low rainfall, owing to thicker leaves, denser tissue, or both (Cunningham et al. 1999, Fonseca et al. 2000, Mooney et al. 1978, Niinemets 2001, Schulze et al. 1998, Specht & Specht 1989). It has now been shown that this does not achieve longer leaf lifespan [two rainfall comparisons in Australia (Wright et al. 2002a) and one in the United States (Reich et al. 1999)]. That is, a shift to higher LMA is required at low rainfall to achieve a given leaf lifespan. Surprisingly, the higher-LMA leaves at low rainfall did not show greater structural strength. Rather, low-rainfall species tended to be built from softer tissue (Wright & Westoby 2002a). Low-rainfall species had higher leaf N per mass and per area (Wright et al. 2001). This was associated with stronger drawdown of internal CO_2 concentrations, leading to economy of transpiration, but was associated also with softer tissue, requiring higher LMA for a given overall leaf strength and leaf lifespan (Wright & Westoby 2002a).

Species characteristic of shaded understorey usually have longer leaf lifespan in association with lower LMA than species from well-lit habitats (Bongers & Popma 1990, Hladik & Miquel 1990, King 1994, Lusk & Contreras 1999, Suehiro & Kameyama 1992, Valladares et al. 2000, Xu et al. 1990; but see Williams et al. 1989). Similarly, within species, individuals or leaves growing in shade often have longer leaf lifespan and lower LMA (Miyaji et al. 1997, Reich et al. 2002, Steinke 1988). The low wind, high humidity, and reduced risk of wilting that characterize dense-shade environments may make it possible to achieve longer leaf lifespan without physical reinforcement expressed as a cost in increased LMA (see also Bongers & Popma 1990). Slow revenue has been proposed to explain the increased leaf lifespan in shade (Figure 2) but does not explain decreasing LMA at the same time as increasing leaf lifespan. The most plausible explanation for increased leaf lifespan in conjunction with lower LMA is that there are a number of separate but parallel LMA-LL relationships (as shown for rainfall in Figure 3). The shift to humid, low-wind, low-radiation conditions of shaded understorey is a shift toward the upper left in Figures 1 and 3, but within shaded

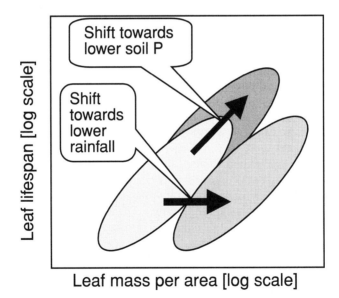

Figure 3 Schematic of leaf lifespan: leaf mass per area (LMA) relationships observed by Wright et al. (2002). Each oval cloud represents the scatter of species in a given habitat. Species occurring at lower soil P tend to have higher LMA, and leaf lifespan is also higher, corresponding to the same LMA-LL relationship observed across species within habitat. Species occurring at lower rainfall also tend to higher LMA but have shifted to a parallel relationship achieving shorter leaf lifespan for a given LMA.

understorey there would still be a positive relationship between LMA and leaf lifespan.

Species occurring on nutrient-poor soils shift toward higher average LMA and longer leaf lifespan than species on more fertile soils (Chapin 1980, Monk 1966, Reich et al. 1992). It has recently been shown (Figure 3) that the shift toward higher leaf lifespan and LMA on low-nutrient soils occurs along the same LMA-LL relationship as on higher nutrients, rather than shifting to a different parallel LMA-LL relationship, as at low rainfall and deep shade. Most likely there is stronger selection in nutrient-poor habitats to extend nutrient retention (Aerts & van der Peijl 1993), favoring extended leaf lifespan, which is in turn made possible by higher LMA.

LMA-Related Leaf Traits that Have Been Suggested as Alternative Strategy Indices

Historically, LMA first attracted attention as a significant descriptor of plant strategies because of its relationship to potential relative growth rate, potRGR. PotRGR is the exponential growth rate (dry mass gain per mass per time) measured on fast-growing seedlings given plentiful water and nutrients. PotRGR has been seen as a bioassay of responsiveness to favorable conditions (Grime & Hunt 1975, Lambers & Poorter 1992). PotRGR is made up of net assimilation rate $\times$ leaf mass fraction $\times$ specific leaf area (SLA). Hence, LMA $(= 1/SLA)$ necessarily influences potRGR. Indeed, in most comparative studies SLA $(= 1/LMA)$ has been the largest of the three sources of variation between species in potRGR (Lambers & Poorter 1992, Wright & Westoby 1999, and references therein). It has now become apparent that high LMA, long leaf lifespan, slow turnover of plant parts, and long nutrient residence times are associated with adaptation to slow-growth situations in a more fundamental way than is slow seedling potRGR (Aerts & van der Peijl 1993, Chapin 1980, Cunningham et al. 1999, Poorter & Garnier 1999).

LMA is made up of lamina depth multiplied by tissue density (Witkowski & Lamont 1991). Both components, or measures closely related to them, have been advocated as better indices of plant strategies than LMA. Leaf volume is made up of solid (cell walls), liquid (cell contents), and gas (intercellular space). Roderick et al. (1999a,b, 2000) argued that liquid volume of leaves should be considered fundamental, because the metabolically active components are in liquid phase. Further, because light capture is area-based while gas exchange is volume-based (Charles-Edwards 1978), the surface area-to-volume ratio of leaves should be considered a fundamental descriptor of leaf structure and function. In effect, this argues that leaf thickness is more informative than LMA or SLA. Dry mass/fresh mass (dry matter content) approximates tissue density for leaves with little intercellular space and has been used in several studies (e.g., Ryser 1996, Wright & Westoby 1999). Wilson et al. (1999) found dry matter content more tightly correlated than LMA with a "primary axis of specialization" that Grime et al. (1997) identified by ordination of 67 traits among 43 British herbaceous species, corresponding to the C-S axis of the CSR scheme, but Vendramini et al. (2002) found the reverse in Argentina.

In our view, LMA remains the most useful single indicator of leaf strategy, although thickness, dry matter content, and volume components should also be measured where possible (Garnier et al. 1997, Niinemets 1999). First, LMA is the construction cost of a unit leaf area, a fundamental quantity in leaf economics. Second, a leaf's physical strength depends on both its thickness and its tissue density. Third, LMA generally appears to be correlated as least as strongly with traits such as leaf lifespan, residence time of nutrients, and photosynthetic capacity as are alternative indicators (Niinemets 1999, Ryser 1996, Ryser & Urbas 2000, Schlapfer & Ryser 1996, Vendramini et al. 2002, Wright et al. 2002). Fourth, internal volumes are much harder to measure than mass and area. Most workers would be unwilling to shift to expressing traits on a volume rather than a mass basis, except for species without internal gas volume, where the two are interchangeable.

SEED MASS AND SEED OUTPUT

The mass of an individual seed ranges 5–6 orders of magnitude across species, even within communities (Figure 4). This is much larger than within-species variation (Table 1). Within-species variation occurs mainly within individuals rather than among plants or populations (Obeso 1998, Vaughton & Ramsey 1998), indicating environmental effects during development more than genetic differences between mothers.

Vegetation can be thought of as species competing to occupy patches. Under this "sessile dynamics" theoretical tradition (Fagerström & Westoby 1997), the ground available is imagined as a set of patches. The abundance of each species is represented by the proportion of patches it occupies. For species to persist in these patch-occupancy models, they need to colonize vacant patches at least as rapidly as they vacate patches by the death of individuals. A crucial parameter in the criteria for invasion and coexistence is the chance that a vacated patch will be reached by one or more seeds from a patch already occupied by the species. Seed output per occupied patch (or per unit area) is a central quantity for understanding the differences in potential colonization ability between species within communities of sessile, patch-occupying species.

Several studies have found seed output negatively correlated with seed mass, sometimes after adjusting for plant size (Greene & Johnson 1994 across 17 woody species, Jakobsson & Eriksson 2000 across 72 seminatural grassland species, Shipley & Dion 1992 across 57 herbaceous angiosperms, Turnbull et al. 1999 across 7 annuals in limestone grassland, Werner & Platt 1976 across 6 *Solidago* species). In a study that expressed seed output per square meter occupied, log seed mass could predict about three fourths of the variation between species in log seed output across 47 woody evergreen species (Henery & Westoby 2001), and slope was -1, i.e., directly inversely proportional. By definition, seed output per square meter is equal to mass devoted to reproduction per square meter

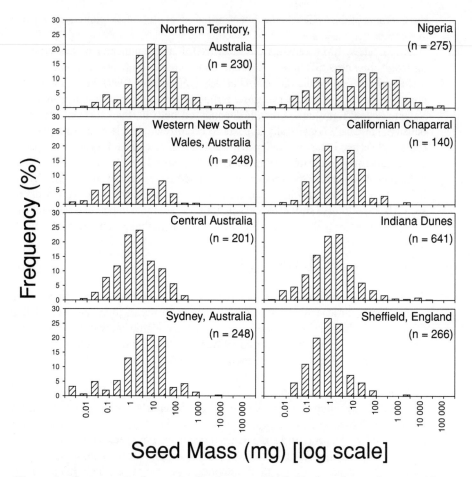

Figure 4 Cross-species frequency distributions of individual seed mass for several locations (Leishman et al. 2000). Two bars per order of magnitude of seed mass.

divided by (seed mass + accessory costs per seed), where accessory costs include fruit structures, dispersal structures, and early aborted seeds. The influence of these different components on seed output must depend on how widely each varies and on any cross-correlations between them. In Henery & Westoby's (2001) dataset, seed mass varied across three orders of magnitude, but reproductive production varied across only one (even allowing for limited sampling during a single season); hence, seed mass accounted for three fourths of the variation in output. Accessory costs varied in proportion to seed mass and so did not change the −1 slope.

The large output advantage of small-seeded over large-seeded species must be counterbalanced at some stage of the life cycle. The most obvious influence of seed mass is on the initial size of the seedling and on the provisions available during early seedling life. Therefore, it is logical to expect the countervailing advantage to be principally at seedling establishment phase. Strong evidence supports this.

Under sufficiently favorable conditions, 100% of viable seeds make seedlings, no matter how small. Hence, it must be under different kinds of hazards that the advantage of larger seeds becomes apparent. A substantial body of manipulative experiments (Table 2) shows clearly that larger-seeded species usually perform better under hazards during seedling establishment. Further, in four of the six experiments where larger-seeded species did not perform significantly better, the reason is understood. Three in deep shade are discussed below. In a field drought experiment (Leishman & Westoby 1994) conditions were particularly harsh, so that survival was low even in watered treatments.

The benefit of larger seed size applies across many different hazards (Table 2). That is, larger seed size might initially evolve in a lineage owing to one type

TABLE 2   Between-species experiments that have tested the proposition that seedlings from species having larger seeds perform better than seedlings from species having smaller seeds, under various hazards*

Hazard	Larger-seeded species performed better	Larger-seeded species did not perform significantly better
Competition from established vegetation	Gross (1984), Bakker (1989), Reader (1993), Ryser (1993), Burke & Grime (1996), Eriksson & Eriksson (1997), George & Bazzaz (1999; first year)	Thompson & Baster (1992), George & Bazzaz (1999; second year)
Competition from other seedlings	Leishman (1999), Turnbull et al. (1999)	
Deep shade	Grime & Jeffrey (1965), Leishman & Westoby (1994a), Osunkoya et al. (1994), Saverimuttu & Westoby (1996b; cotyledon-phase), Walters & Reich (2000; early phase)	Augspurger (1984), Saverimuttu & Westoby (1996b; first-leaf phase); Walters & Reich (2000; later phase)
Defoliation	Armstrong & Westoby (1993), Harms & Dalling (1997)	
Mineral nutrient shortage	Jurado & Westoby (1992), Milberg et al. (1998)	
Depth under soil or litter	Gulmon (1992), Jurado & Westoby (1992), Jurik et al. (1994), Seiwa & Kikuzawa (1996)	
Soil drought	Leishman & Westoby (1994b; glasshouse)	Leishman & Westoby (1994b; field)

*Updated from Westoby et al. (1996). Studies were included only if they covered at least five species, with seed mass spanning at least one order of magnitude

of advantage but having evolved, would be functional also in relation to other hazards.

Many studies (reviewed in Leishman et al. 2000) have shown that initial seedling size is positively related to seed size across species as well as within species. Seedlings from larger-seeded species also tend to have greater reserves relative to the autotrophic functioning parts of the seedling, and therefore can support respiration longer under carbon deficit. This has been called the "larger-seed-later-deployment" interpretation (Ganade & Westoby 1999, Kidson & Westoby 2000, Leishman et al. 2000) or "cotyledon functional morphology hypothesis" (Garwood 1995, Hladik & Miquel 1990, Kitajima 1996a,b). For competition from established vegetation, depth under soil or litter, and soil drought, better seedling survival might be attributable to larger initial size of the seedling. In most deep shade experiments, for defoliation and for mineral nutrients there is no resource gradient away from the soil surface and consequently no built-in advantage to a larger initial seedling that reaches further away from the surface. Advantages of larger seed mass owing to greater reserves (as distinct from owing to larger initial seedling size) apply during cotyledon phase but not later during seedling life. This has been directly demonstrated under deep shade (Saverimuttu & Westoby 1996, Walters & Reich 1999) and can account for all three cases in which larger seed mass was not associated with better survival under deep shade (Table 2).

There are some consistent shifts in mean log seed size between different environments, though not so many as might be expected from the experimental results. Many studies have shown that species establishing in closed or shaded environments tend to have larger seeds than those in open environments [e.g., Baker 1972, Foster 1986, Foster & Janson 1985, Grubb & Metcalfe 1996 (genera within families but not species within genera), Hewitt 1998, Hodkinson et al. 1998 (angiosperms but not gymnosperms)], Mazer 1989, Metcalfe & Grubb 1995, Salisbury 1942, Thompson & Hodkinson 1998]. Evidence about nutrient-poor versus nutrient-rich soils is contradictory. Westoby et al. (1990) found little effect comparing sclerophyll woodland on low-nutrient soils with temperate rainforest on higher-nutrient soils in Australia. Parolin (2000) found larger seeds in nutrient-poor environments among 58 Central American floodplain species. It has been suggested that species in drought-prone environments are likely to have larger seeds than those in moist environments (Baker 1972, Rockwood 1985, Salisbury 1942). However, the correlation Baker found was mainly caused by a tendency for herbaceous species in flood-prone habitats to have small seeds (Westoby et al. 1992), and Mazer (1989) failed to find a relationship between aridity and seed mass. Seed mass tends to be greater toward low latitudes (Baker 1972, Levin 1974, Lord et al. 1997), and this is only in part a correlate of shifts in growth form, dispersal mode, and shading. No relationship with altitude has been found (e.g., Rockwood 1985).

Despite some shifts in the mean between habitats, especially shaded versus open, it is striking that the spread of seed mass between species within a habitat is very wide (>95% of all variation is within rather than between temperate zone floras) (Table 1 and Figure 4).

LEAF-SIZE–TWIG-SIZE SPECTRUM

Corner (1949) pointed out two cross-species patterns. The thicker the plant axis (stem), the larger the individual appendages (leaves, inflorescences, fruits). The more closely spaced the ramification (branching), the thinner the ultimate axes and the smaller the appendages. Halle et al. (1978) called these patterns Corner's Rules (White 1983a,b), and recent work has confirmed and quantified them (Ackerly & Donoghue 1998, Brouat et al. 1998, Cornelissen 1999).

Twig cross-sectional area is expected to be coordinated with the total leaf area supported on the twig, for both hydraulic and mechanical reasons. However, a given total leaf area might logically be made up of many small or of few large leaves. Hence, coordination of twig cross-sectional area with transpiration demand from the leaves or with requirement for mechanical support does not, in itself, force a correlation between the size of twigs and the size of individual leaves. Presumably, when twigs are closer together, then leaves should be smaller because otherwise they would overlap wastefully. Perhaps also they might damage each other by thrashing together in the wind.

Although the leaf-size–twig-size spectrum is apparent all around us, and quantifications of it are accumulating, its adaptive significance is far from clear. Theory about leaf size has been developed by Parkhurst & Loucks (1972), Givnish & Vermeij (1976), and Givnish (1978, 1979, 1984). Larger leaves have thicker boundary layers of still air. Consequently their convective heat loss is slower, and they tend to be heated above air temperature by a wider margin. This may be a disadvantage, increasing respiration rates more than photosynthesis rates. This effect will be more important for leaves under stronger radiation and where water is in short supply (because transpiration also sheds heat). Givnish (1978, 1979, 1984, 1987) expressed the costs of slow convective heat loss in terms of the carbon expenditure in the root system that would be needed to supply the leaf with sufficient water for cooling by transpiration. These costs were balanced against photosynthetic carbon gain in assessing optimal leaf size. As rainfall decreased, increasing water acquisition costs were expected to favor smaller leaves. Under low soil nutrient, Givnish invoked a flatter response to temperature of mass-based photosynthesis, dropping optimal temperature for net carbon gain and favoring smaller leaves.

Empirically, leaf size tends to decrease toward dry, sunny, or nutrient-poor habitats (Givnish 1984, 1986, 1987; Hall & Swaine 1981; Raunkiaer 1934; Schimper 1903; Shields 1950; Walter 1973; Webb 1968), as expected under Givnish's treatment. It is widely held that closely ramified branching is better suited to strong light environments (Cornelissen 1993, Givnish 1984, Horn 1971, Kempf & Pickett 1981). A substantial paleobotanical literature uses fossil leaf sizes to estimate past precipitation (Gregory-Wodzicki 2000, Jacobs 1999, Wiemann et al. 1998, Wilf et al. 1998, Wolfe 1995).

Other potential costs and benefits of a position high or low along the leaf-size–twig-size spectrum remain little investigated. Leaf-size–twig-size is well correlated with the size of infructescences and weakly with mass of individual seeds

(Cornelissen 1999). These relationships also occur ontogenetically and between genders in dioecious species (Bond & Midgley 1988, Midgley & Bond 1989), suggesting that under some circumstances selection for larger fruits or flowers could drive selection on leaf size. Smaller-leaved species may suffer less herbivory during leaf expansion, because the duration of expansion is shorter (Moles & Westoby 2000). Ritchie & Olff (1999) modeled coexistence of species that forage at different scales. Species that could orient their foraging accurately into small patches required higher resource concentrations to support their populations. Therefore, fine-scale foraging species were superior in local resource concentrations but unable to occupy most of the space available, and a series of coarser-scale species could coexist. One of their datasets used leaf width as an indicator of foraging scale. Given that leaf size is correlated with twig size and degree of ramification, it could be argued that leaf size acts as a surrogate for something to do with the scale of twig or branch systems, which conceivably reflect the natural scale of foraging for light (Cornelissen 1993, Givnish 1984, Horn 1971, Kempf & Pickett 1981). However, Bragg & Westoby (2002) did not find support for this idea. Although smaller-leaved species were positioned in higher light relative to their height, the light patches in question were several-fold larger than whole plants.

Surprisingly, the dry-mass economics of deploying leaf area as few large versus many small leaves seem to have been little investigated. Dry-mass gain per shoot mass can be partitioned into dry-mass gain/leaf area × leaf area/shoot mass. Given that leaf area ratio (leaf area per plant mass) has proved to be the largest source of variation in seedling growth between species (Poorter & van der Werf 1998), it would not be surprising if leaf area per shoot mass were important in the economy of branch systems. We have found a substantial tendency for larger-leaved species to have higher leaf mass fraction beyond 10 mm^2 twig cross-section, but this was counterbalanced by larger-leaved species having lower leaf area per leaf mass (higher LMA) (M. Pickup, A. Basden, M. Westoby, unpublished data). It is already clear that this second pattern is not a universal rule. Grubb (1998) and Shipley (1995) found that among species with similar ecology those with larger leaves tend to have higher LMA. They suggested this was a requirement for mechanical support. Niinemets (1996, 1998; Niinemets & Kull 1994), however, found lower LMA in larger-leaved species; this pattern may occur within a vegetation type when species are differentiated mainly along a shade-tolerance spectrum. Using data from several habitats, Ackerly & Reich (1999) found LMA and leaf size negatively correlated, but this consisted almost entirely of the difference between angiosperms and conifers; there was little correlation within either clade. Across sites, SLA and leaf size both tend to decline towards lower rainfall and lower soil nutrient (Ackerly et al. 2002, Fonseca et al. 2000). In datasets spanning a range of habitats, within-habitat and between-habitat patterns are liable to cancel each other out, leaving little or no overall relationship between leaf size and LMA (Ackerly et al. 2002, Fonseca et al. 2000, Wilson et al. 1999). In summary, if larger-leaved species turn out consistently to have higher leaf mass fractions in their shoots, this will not always be counterbalanced by higher LMA. Higher leaf mass fraction may deliver a substantial leaf area per shoot mass advantage to large leaf size and large twig size in some comparisons.

POTENTIAL CANOPY HEIGHT

Potential or maximum canopy height can range from ~1 cm to >50 m, four orders of magnitude. Sometimes this full range can be found at a single site. Here we consider self-supporting species only. The height-parasites, climbers and epiphytes, are important in some vegetation types. Their strategies testify to the advantage of height but do not really have characteristic heights of their own.

The benefit of supporting oneself at a height cannot be understood by considering a single strategy in isolation (Iwasa et al. 1985). At any moment in the successional process, being taller than neighbors confers competitive advantage through prior access to light. However, height incurs costs from past investment in stems and support structures, from continuing maintenance costs for the stems and vasculature, and from disadvantages in the transport of water to height. In the absence of competition, a single strategy could maximize productivity per leaf area by minimizing the costs associated with height and growing close to the ground. However, plants using this strategy would be open to invasion by plants using a strategy that diverted some resources to nonproductive tissues such as stem and grew slightly taller. Plants using that strategy in turn would be open to invasion by plants using a third strategy that grew slightly taller still, and so forth. Height is the classic case of a strategy that needs to be understood in the context of game theory (Kawecki 1993).

Height is the one quantitative trait that has been adopted by virtually everyone doing comparative plant ecology (e.g., Bugmann 1996, Chapin et al. 1996, Grime et al. 1988, Hubbell & Foster 1986, Keddy 1989, Weiher et al. 1999, Westoby 1998, Wilson et al. 1999). Nevertheless, unlike the LMA-LL and seed mass–seed output spectra, ideas about canopy height strategies include several trade-offs (Givnish 1995). The upper limit on height, the pace at which species grow upward (which may decide relative height advantage at different times after disturbance), and the duration over which stems persist at their upper height, have costs and benefits that are at least partly separate. If the costs and benefits could be elucidated, and if easily measurable key traits could be identified, these components would ideally be separated out.

Some studies have considered the biomass cost for supporting a unit of leaf area at a given height. In herbaceous vegetation taller species have lower leaf area ratios (leaf area per aboveground biomass) owing to lower leaf mass fraction or higher SLA or both. Although shorter species intercept less light per unit leaf area, they are not necessarily inferior with regard to light interception per aboveground biomass (Anten & Hirose 1999, Hirose & Werger 1995).

A weakness of considering light interception per unit biomass is that biomass of stems and branches accumulates over time. Accumulated biomass is not in the same units as current photosynthesis for assessing the budget at a particular stage of growth. The argument can be improved in two possible directions. One is to express stem and branch biomass as current costs. The other is to consider the time-dynamics of competition for height and light.

Along the current-budget direction, Givnish (1988, 1995) used allocation equations from Whittaker & Woodwell (1968) to estimate current growth of non-leaf tissues. Allocation to non-leaf increased with the height of the tree. Taken together with leaf respiration and with leaf construction costs amortized over the season, these costs can be balanced against current photosynthesis. Through this reasoning a maximum height can be estimated as the height at which costs fully consume all current photosynthesis.

This respiration hypothesis for the upper limit to plant height has been contrasted with a hydraulic-limitation hypothesis (Gower et al. 1996, Ryan & Yoder 1997), leading to a recent outburst of debate (Becker et al. 2000a,b; Bond & Ryan 2000; Mencuccini & Magnani 2000). In reality, arguments about single limiting factors for plant height are unlikely to be profitable, at least for thinking about differences between species. First, species are not expected to evolve to grow as tall as is physiologically possible. It is the density and height of shading by competitors that determines how much benefit can accrue from height growth. Second, natural selection shapes strategies in such a way that no single capability of the plant is limiting. Rather, strategies are characterized by coordination of different capabilities and quantities. Sapwood cross-sectional area is coordinated with total leaf area, the ratio of the two being adjusted to evaporative climate (Mencuccini & Grace 1995, Schafer et al. 2000). Stomatal conductance (and hence assimilation and transpiration) are coordinated with vascular conductance (Hubbard et al. 2001, Magnani et al. 2000, Nardini & Salleo 2000, Sperry 2000) and with the cavitation risk of the vasculature (Whitehead 1998). Constrictions in vessels to lower branches can ensure that sufficient water flow is directed to upper branches (Hacke & Sperry 2001). Vessel tapering can be coordinated with height to buffer the effect that height would otherwise have on conductance of the vascular pathway (Becker et al. 2000a, West et al. 1999). Similarly, vessel length (frequency of cross-walls) is modulated according to height within a tree to balance conductance against cavitation risk (Comstock & Sperry 2000). Pore diameters between vessels control air-seeding of embolisms from one vessel to another and the linkage between xylem pressure and xylem conductance (Sperry 1995, Tyree 1999). If one of these traits were consistently limiting relative to another, then alternative genotypes or species strategies would be favored that reallocated their efforts, because this could relieve the limitation without equivalent disadvantage elsewhere. This is the principle of equalization of marginal returns on alternative expenditures, familiar to economists and evolutionary ecologists (e.g., Venable 1996) but yet to be fully assimilated into physiological thinking. A species' strategies are expected to evolve to the point at which many factors are limiting simultaneously, or no single factor is limiting, depending how one likes to think about it.

Whereas some aspects of height strategy may be understood via the costs and benefits of height at a point in time, other aspects need to be considered over cycles of disturbance and growth. Disturbances open the canopy, daylight becomes available near the ground, and a race upwards for the light ensues. Races are restarted when a new disturbance destroys the accumulated stem height. Winning strategies are not only those that eventually result in the tallest plants. Any species that is near the lead at some stage during the race may make sufficient carbon profit to ensure that it runs also in subsequent races (which is the criterion for a viable strategy). Entry in subsequent races may occur via vegetative regeneration, via a stored seed bank, or via dispersal to other locations, but the prerequisite for any of these is sufficient carbon accumulation at some stage during the race for light. Within a race series having some typical race duration, one finds successful height strategies that have been designed by natural selection to be among the leaders early in a race, and other successful strategies that join the leaders at various later stages. Among tree species, those with faster growth usually have lower wood densities, associated with less strength and decay resistance (Loehle 1988, 1996). Toward the later stages of height races, tree species with very persistent stems reach the lead because faster-growing species are disintegrating above them, rather than by overtaking the faster-growing species.

SPREAD OF SPECIES ALONG THE DIMENSIONS

For each of the dimensions discussed, species traits tend to shift along climatic gradients and between sites in a landscape, as briefly outlined above. However, a striking feature is the wide spread of traits among species growing interspersed within a single vegetation type.

The fact that assemblage averages shift in response to physical conditions of the site indicates that the species sifted into a site are drawn selectively from the regional flora ["filtered" (Diaz et al. 1998)] with regard to the value of the trait. However, if one compares two sites with different assemblage means, the site with lower mean will include some species with a higher trait value than the assemblage mean of the other site. Further, the spread of trait values seems at least as wide in harsh as in favorable habitats. This contradicts the idea that physical properties of the two sites determine in a simple way what trait values are permitted. Something spreads out the frequency distribution across species within a site into a broad mixture, at the same time that physical properties of the site somehow position the whole frequency distribution. In principle, three sources might contribute to the wide spread within sites: (a) a broad frequency distribution of physical conditions across microsites within each site, (b) continuing immigration from sites with different physical conditions, and (c) some game-theoretic or frequency-dependent process. Data are not available to partition the contributions of these three forces to spreading out the within-site frequency distributions, but we think it likely that game-theoretic processes are the most important. The field experiment to test for a frequency-dependent process is simple in principle. One needs to selectively remove species from a particular band of, for instance, seed mass, then show that recolonization is drawn selectively from that same band of species more than from among the seed masses left behind. The difficulty is that the processes to be measured are only expected to operate consistently over several generations (which for woody vegetation would mean tens to hundreds of years), over several cycles of disturbance (because using different times since disturbance is one of the main forms of game-theoretical interaction), and over a large enough area of space for realistic population dynamics within all of the species involved (which would mean hectares up to square kilometers for each replicate). This experiment, which would be the natural next step in terms of research logic, is unfortunately nearly unachievable in practice.

The qualitative conditions for coexistence of a broad mixture of strategies for a trait are reasonably well understood. Evolutionarily stable strategy-mixture theory developed by Geritz for seed mass (Fagerström & Westoby 1997; Geritz 1995, 1998; Geritz et al. 1999; Rees & Westoby 1997) illustrates the principles. Species at one end of the spectrum of the trait need to be competitively superior, but at the same time unable to sequester all the space available. In Geritz's models large-seed strategies defeat smaller-seed strategies in seedling competition for space. Hence, strategy mixtures can always be invaded by a larger seed mass, up to the size at which seed output is so low that the population can no longer have a positive rate of increase, even though the strategy is never outcompeted. At the same time, plants using strategies with smaller seeds and higher seed output from each occupied patch are able to persist, because their seeds reach some establishment opportunities that are not reached by any larger seed. In other words, the broad spread of strategies is made possible by a colonization-competition tradeoff (Chesson 2000, Pacala 1997). The specific assumptions of this model are probably not met in most vegetation types (Leishman 2001), but the point is that game-theoretic processes with this general structure are the most likely forces generating broad strategy mixtures within assemblages.

For the LMA-LL and leaf size-twig size dimensions, ESS models have yet to be developed, but appropriate spectra have been suggested. For LMA-LL, a model by Aerts & van der Peijl (1993) is underpinned by a spectrum of time-since-disturbance. Low LMA species grow faster during early succession, but high LMA species have longer nutrient retention times, and over time they take over an increasing proportion of the nutrient pool. For leaf size-twig size, the model by Ritchie & Olff (1999) discussed above is underpinned by an assumption that high-value resources occur in small patches, and species exploiting these small patches are therefore unable to occupy the whole environment.

There have been ESS treatments of height (Givnish 1982, King 1991, Makela 1985, Sakai 1995), of the shading relationship between strategies as affected by SLA or leaf angle (Hikosaka & Hirose 1997, Schieving & Poorter 1999), and of the pace of height growth as affected by allocation between roots and shoots and between stem and foliage (Givnish 1995, Sakai 1991, Vincent & Vincent 1996). All of these treatments predict a single winning strategy. This illustrates that the observed wide spread within sites on these dimensions is not inevitable.

The ESS treatment of Iwasa et al. (1985) accounts for a mixture of height strategies. This shows that provided the costs of maintaining leaves at a given height rise with height in an accelerating manner, leaves should be spread through a continuous range of heights. It can easily be imagined that this strategy mixture could be made up of many species of different heights, though Iwasa et al. left open the possibility of a single species with a deep canopy.

CONCLUSION AND SOME FUTURE DIRECTIONS

In comparing ecology to chemistry before the periodic table, Southwood (1977) pointed both to the disorganized condition of ecological knowledge and to the hope for integration via an agreed-upon ecological strategy scheme. Actually, ecological strategy schemes are unlikely to look like a periodic table (Steffen 1996). A closer analogy might be personality schemes as used in psychology (Atkinson et al. 1990). Personality schemes have spectra such as introvert to extrovert, analytical to emotional. Key issues in personality scheme research have been about the number of spectra that convey different, useful information (consensus is approximately 5) and about the meaning of particular spectra and their predictive power for what people will do in different situations.

Similarly in ecology, the need is to identify leading dimensions of variation that seem useful. The list can be open-ended: There is no need to decide in advance on a scheme with a fixed number of dimensions. The present situation is that a degree of consensus is emerging about LMA-LL, seed mass–seed output, and height. These three dimensions of variation capture important generalities about how plant species make a living. They can be measured relatively easily, which is important if they are to serve for coordinating information worldwide.

What are the likeliest directions for future progress? First, these dimensions have not been tried out much as potential explanatory variables during literature synthesis. Plainly they do have some explanatory power in relation to some kinds of phenomena, for example LMA-LL in relation to ecophysiological traits such as leaf nitrogen and potential photosynthetic rates. At this stage we cannot be sure the dimensions will be much help in pulling together the large literature on a few species at a time in ecophysiology, succession, or experimental ecology, but it is worth trying. Otherwise we default to the discouraging conclusion that each species is idiosyncratic and a separate problem.

A second direction for further progress is through adding dimensions to a list of those that are most informative. An obvious deficiency is the lack of something indicating temperature preferences, whether as optima for growth or as lower or upper tolerances. This is crucial for the position of a species on global climate maps and for its future under climate change. Unfortunately, there is no obvious candidate for a simple measurement that would capture temperature preference.

A third direction for research progress would be to clarify and separate the different elements of height: the pace of height growth, the asymptotic height, and the capacity to persist at a height. A fourth direction would be to clarify costs and benefits along the leaf-size-twig-size dimension, which is conspicuous but poorly understood. A fifth direction would be to develop convincing models that predict both the spread of species traits observed and shifts in the upper and lower bounds of the spread from one environment to another. It is understood what is needed in principle to support a broad mixture of coexisting strategies, but this has yet to be modeled in detail and with convincing experimental evidence for any particular strategy dimension.

There is much to be done. There is also a real hope that we may be getting somewhere.

ACKNOWLEDGMENTS

We thank David Ackerly, Hans Cornelissen, Sandra Diaz, and Jessica Gurevitch for exceptionally helpful comments on the manuscript. Research was funded by the Australian Research Council.

The *Annual Review of Ecology and Systematics* is online at
http://ecolsys.annualreviews.org

LITERATURE CITED

Ackerly DD. 1999. Self-shading, carbon gain and leaf dynamics: a test of alternative optimality models. *Oecologia* 119:300–10

Ackerly DD, Bazzaz FA. 1995. Seedling crown orientation and interception of diffuse radiation in tropical forest gaps. *Ecology* 76:1134–46

Ackerly DD, Donoghue MJ. 1998. Leaf size, sapling allometry, and Corner's rules: phylogeny and correlated evolution in maples (*Acer*). *Am. Nat.* 152:767–91

Ackerly DD, Knight CA, Weiss SB, Barton K, Starmer KP. 2002. Leaf size, specific leaf area and microhabitat distribution of chaparral woody plants: contrasting patterns in species level and community level analyses. *Oecologia* 130:449–57

Ackerly DD, Reich PB. 1999. Convergence and correlations among leaf size and function in seed plants: a comparative test using independent contrasts. *Am. J. Bot.* 86:1272–81

Aerts R, Chapin FS III. 2000. The mineral nutrition of wild plants revisited: a re-evaluation of processes and patterns. *Adv. Ecol. Res.* 30:1–67

Aerts R, van der Peijl MJ. 1993. A simple model to explain the dominance of low-productive perennials in nutrient-poor habitats. *Oikos* 66:144–47

Anten NPR, Hirose T. 1999. Interspecific differences in above-ground growth patterns result in spatial and temporal partitioning of light among species in a tall-grass meadow. *J. Ecol.* 87:583–97

Armstrong DP, Westoby M. 1993. Seedlings from large seeds tolerate defoliation better: a test using phylogenetically independent contrasts. *Ecology* 74:1092–100

Atkinson RL, Atkinson RC, Smith E, Bem D. 1990. *Introduction to Psychology*. San Diego: Harcourt Brace Jovanovich. 788 pp.

Augspurger CK. 1984. Seedling survival of tropical tree species: interactions of dispersal distance, light-gaps, and pathogens. *Ecology* 65:1705–12

Baker HG. 1972. Seed weight in relation to environmental conditions in California. *Ecology* 53:997–1010

Bakker JP. 1989. *Nature Management by Grazing and Cutting: On the Ecological*

Significance of Grazing and Cutting Regimes Applied to Restore Former Species-Rich Grassland Communities in the Netherlands. London: Kluwer

Becker P, Gribben RJ, Lim CM. 2000a. Tapered conduits can buffer hydraulic conductance from path-length effects. *Tree Physiol.* 20:965–67

Becker P, Meinzer FC, Wullschleger SD. 2000b. Hydraulic limitation of tree height: a critique. *Funct. Ecol.* 14:4–11

Begon M, Mortimer M, Thomson DJ. 1996. *Population Ecology: A Unified Study of Plants and Animals.* Oxford: Blackwell Sci. 247 pp. 3rd ed.

Berendse F. 1994. Litter decomposability—a neglected component of plant fitness. *J. Ecol.* 82:187–90

Bond BJ, Ryan MG. 2000. Comment on 'Hydraulic limitation of tree height: a critique' by Becker, Meinzer & Wullschleger. *Funct. Ecol.* 14:137–40

Bond WJ. 1989. The tortoise and the hare: ecology of angiosperm dominance and gymnosperm persistence. *Biol. J. Linn. Soc.* 36:227–49

Bond WJ, Midgley JJ. 1988. Allometry and sexual differences in leaf size. *Am. Nat.* 131:901–10

Bongers F, Popma J. 1990. Leaf characteristics of the tropical rain forest flora of Los Tuxtlas, Mexico. *Bot. Gaz.* 151:354–65

Bragg JG, Westoby M. 2002. The leaf size spectrum in relation to light environment and dark respiration in a sclerophyll woodland. *Funct. Ecol.* In press

Brouat C, Gibernau M, Amsellem L, McKey D. 1998. Corner's rules revisited: ontogenetic and interspecific patterns in leaf-stem allometry. *New Phytol.* 139:459–70

Bryant JP, Chapin FS III, Klein DR. 1983. Carbon/nutrient balance of boreal plants in relation to vertebrate herbivory. *Oikos* 40:357–68

Bugmann H. 1996. Functional types of trees in temperate and boreal forests: classification and testing. *J. Veg. Sci.* 7:359–70

Burke MJW, Grime JP. 1996. An experimental study of plant community invasibility. *Ecology* 77:776–90

Chabot BF, Hicks DJ. 1982. The ecology of leaf lifespans. *Annu. Rev. Ecol. Syst.* 13:229–59

Chapin FS III. 1980. The mineral nutrition of wild plants. *Annu. Rev. Ecol. Syst.* 11:233–60

Chapin FS III. 1989. The cost of tundra plant structures: evaluation of concepts and currencies. *Am. Nat.* 133:1–19

Chapin FS III, Bret-Harte M, Syndonia M, Hobbie SE, Zhong H. 1996. Plant functional types as predictors of transient responses of arctic vegetation to global change. *J. Veg. Sci.* 7:347–58

Charles-Edwards DA. 1978. Photosynthesis and crop growth. In *Photosynthesis and Plant Development*, ed. R Marcelle, H Clijsters, M Van Poucke, pp. 111–24. The Hague: Junk

Chesson P. 2000. Mechanisms of maintenance of species diversity. *Annu. Rev. Ecol. Syst.* 31:343–66

Clark DB, Clark DA, Grayum MH. 1992. Leaf demography of a neotropical rain forest cycad, *Zamia skinneri* (Zamiaceae). *Am. J. Bot.* 79:28–33

Coley PD. 1983. Herbivory and defensive characteristics of tree species in a lowland tropical forest. *Ecol. Monogr.* 53:209–29

Coley PD. 1988. Effects of plant growth rate and leaf lifetime on the amount and type of antiherbivore defense. *Oecologia* 74:531–36

Coley PD, Barone JA. 1996. Herbivory and plant defenses in tropical forests. *Annu. Rev. Ecol. Syst.* 27:305–35

Coley PD, Bryant JP, Chapin FS III. 1985. Resource availability and plant antiherbivore defense. *Science* 230:895–99

Coley PD, Kursar T, Machado J-L. 1993. Colonization of tropical rain forest leaves by epiphylls: effects of site and host plant leaf lifetime. *Ecology* 74:619–23

Comstock JP, Sperry JS. 2000. Theoretical considerations of optimal conduit length for water transport in vascular plants. *New Phytol.* 148:195–218

Connell JH. 1983. On the prevalence and relative importance of interspecific competition:

evidence from field experiments. *Am. Nat.* 122:661–96

Cornelissen JHC. 1993. Aboveground morphology of shade-tolerant *Castanopsis fargesii* saplings in response to light environment. *Int. J. Plant Sci.* 154:481–95

Cornelissen JHC. 1999. A triangular relationship between leaf size and seed size among woody species: allometry, ontogeny, ecology and taxonomy. *Oecologia* 118:248–55

Cornelissen JHC, Perez-Harguindeguy N, Diaz S, Grime JP, Marzano B, et al. 1999. Leaf structure and defence control litter decomposition rate across species and life forms in regional floras on two continents. *New Phytol.* 143:191–200

Corner EJH. 1949. The durian theory, or the origin of the modern tree. *Ann. Bot.* 13:368–414

Crawley MJ. 1983. *Herbivory.* Oxford: Blackwell Sci. 437 pp.

Cunningham SA, Summerhayes B, Westoby M. 1999. Evolutionary divergences in leaf structure and chemistry, comparing rainfall and soil nutrient gradients. *Ecol. Monogr.* 69:569–88

Denslow JS. 1980. Gap partitioning among tropical rainforest trees. *Biotropica* 12:47–55

Diaz S, Cabido M, Casanoves F. 1998. Plant functional traits and environmental filters at a regional scale. *J. Veg. Sci.* 9:113–22

Diemer M. 1998a. Leaf lifespans of high-elevation, aseasonal Andean shrub species in relation to leaf traits and leaf habit. *Global Ecol. Biogeogr. Lett.* 7:457–65

Diemer M. 1998b. Lifespan and dynamics of leaves of herbaceous perennials in high-elevation environments—news from the elephant's leg. *Funct. Ecol.* 12:413–25

Dyksterhuis EJ. 1949. Condition and management of range land based on quantitative ecology. *J. Range Manage.* 2:104–15

Eckstein RL, Karlsson PS, Weih M. 1999. Leaf lifespan and nutrient resorption as determinants of plant nutrient conservation in temperate-arctic regions. *New Phytol.* 143: 177–89

Eriksson A, Eriksson O. 1997. Seedling recruitment in semi-natural pastures: the effects of disturbance, seed size, phenology and seed bank. *Nord. J. Bot.* 17:469–82

Escudero A, del Arco JM, Sanz IC, Ayala J. 1992. Effects of leaf longevity and retranslocation efficiency on the retention time of nutrients. *Oecologia* 90:80–87

Fagerström T, Westoby M. 1997. Population dynamics in sessile organisms: some general results from three seemingly different theory-lineages. *Oikos* 80:588–94

Felsenstein J. 1985. Phylogenies and the comparative method. *Am. Nat.* 125:1–15

Field CB, Mooney HA. 1986. The photosynthesis-nitrogen relationship in wild plants. See Givnish 1986a, pp. 25–55

Fonseca CR, Overton JM, Collins B, Westoby M. 2000. Shifts in trait-combinations along rainfall and phosphorus gradients. *J. Ecol.* 88:964–77

Foster SA. 1986. On the adaptive value of large seeds for tropical moist forest trees: a review and synthesis. *Bot. Rev.* 52:260–99

Foster SA, Janson CH. 1985. The relationship between seed size and establishment conditions in tropical woody plants. *Ecology* 66:773–80

Funes G, Basconcelo S, Diaz S, Cabido M. 1999. Seed size and shape are good predictors of seed persistence in soil in temperate mountain grasslands of Argentina. *Seed Sci. Res.* 9:341–45

Ganade G, Westoby M. 1999. Seed mass and the evolution of early seedling etiolation. *Am. Nat.* 154:469–80

Garnier E, Cordonnier P, Guillerm J-L, Soni L. 1997. Specific leaf area and leaf nitrogen concentration in annual and perennial grass species growing in Mediterranean old-fields. *Oecologia* 111:490–98

Gartner BL, ed. 1995. *Plant Stems: Physiology and Functional Morphology.* San Diego, CA: Academic

Garwood NC. 1996. Functional morphology of tropical tree seedlings. In *The Ecology of Tropical Forest Tree Seedlings*, ed. MD Swaine, pp. 59–129. New York: Parthenon

George LO, Bazzaz FA. 1999. The fern understory as an ecological filter: emergence and establishment of canopy-tree seedlings. *Ecology* 80:833–45

Geritz SAH. 1995. Evolutionarily stable seed polymorphism and small-scale spatial variation in seedling density. *Am. Nat.* 146:685–707

Geritz SAH. 1998. *The evolutionary significance of variation in seed size*. PhD thesis. Rijksuniversiteit, Leiden, The Netherlands. 151 pp.

Geritz SAH, van der Meijden E, Metz JAJ. 1999. Evolutionary dynamics of seed size and seedling competitive ability. *Theor. Popul. Biol.* 55:324–43

Givnish TJ. 1978. Ecological aspects of plant morphology: leaf form in relation to environment. *Acta Biotheor.* 27:83–142

Givnish TJ. 1979. On the adaptive significance of leaf form. In *Topics in Plant Population Biology*, ed. OT Solbrig, S Jain, GB Johnson, PH Raven, pp. 375–407. London: Macmillan

Givnish TJ. 1982. On the adaptive significance of leaf height in forest herbs. *Am. Nat.* 120:353–81

Givnish TJ. 1984. Leaf and canopy adaptations in tropical forests. In *Physiological Ecology of Plants of the Wet Tropics*, ed. E Medina, HA Mooney, C Vazquez-Yanez, pp. 51–84. The Hague: Junk

Givnish TJ. 1986. Optimal stomatal conductance, allocation of energy between leaves and roots, and the marginal cost of transpiration. See Givnish 1986a, pp. 171–213

Givnish TJ, ed. 1986a. *On the Economy of Plant Form and Function*. Cambridge: Cambridge Univ. Press

Givnish TJ. 1987. Comparative studies of leaf form: assessing the relative roles of selective pressures and phylogenetic constraints. *New Phytol.* 106 (Suppl.):131–60

Givnish TJ. 1988. Adaptation to sun and shade: a whole-plant perspective. *Aust. J. Plant. Physiol.* 15:63–92

Givnish TJ. 1995. Plant stems: biomechanical adaptation for energy capture and influence on species distributions. See Gartner 1995, pp. 3–49

Givnish TJ, Vermeij GJ. 1976. Sizes and shapes of liane leaves. *Am. Nat.* 110:743–78

Goldberg DE. 1996. Competitive ability: definitions, contingency and correlated traits. *Philos. Trans. R. Soc. London Ser. B* 351:1377–85

Goldberg DE, Barton AM. 1992. Patterns and consequences of interspecific competition in natural communities: a review of field experiments with plants. *Am. Nat.* 139:771–801

Gower ST, McMurtrie RE, Murty D. 1996. Aboveground net primary production decline with stand age: potential causes. *Trends Ecol. Evol.* 11:378–82

Gower ST, Reich PB, Son Y. 1993. Canopy dynamics and aboveground production of five tree species with different leaf longevities. *Tree Physiol.* 12:327–45

Greene DF, Johnson EA. 1994. Estimating the mean annual seed production of trees. *Ecology* 75:642–47

Gregory-Wodzicki KM. 2000. Relationships between leaf morphology and climate, Bolivia: implications for estimating paleoclimate from fossil floras. *Paleobiology* 26:668–88

Grime JP. 1974. Vegetation classification by reference to strategies. *Nature* 250:26–31

Grime JP. 1977. Evidence for the existence of three primary strategies in plants and its relevance to ecological and evolutionary theory. *Am. Nat.* 111:1169–94

Grime JP. 1979. *Plant Strategies and Vegetation Processes*. Chichester, UK: Wiley. 222 pp.

Grime JP. 1994. The role of plasticity in exploiting environmental heterogeneity. In *Exploitation of Environmental Heterogeneity by Plants: Ecophysiological Processes Above- and Below-Ground*, ed. MM Caldwell, RW Pearcy, pp. 1–19. New York: Academic

Grime JP, Hodgson JG, Hunt R. 1988. *Comparative Plant Ecology*. London: Unwin-Hyman. 742 pp.

Grime JP, Hunt R. 1975. Relative growth-rate: its range and adaptive significance in a local flora. *J. Ecol.* 63:393–422

Grime JP, Jeffrey DW. 1965. Seedling establishment in vertical gradients of sunlight. *J. Ecol.* 53:621–42

Grime JP, Thompson K, Hunt R, Hodgson JG, Cornelissen JHC, et al. 1997. Integrated screening validates primary axes of specialisation in plants. *Oikos* 79:259–81

Gross KL. 1984. Effects of seed size and growth form on seedling establishment of six monocarpic perennial plants. *J. Ecol.* 72:369–87

Grubb PJ. 1998. A reassessment of the strategies of plants which cope with shortages of resources. *Perspect. Plant Ecol. Evol. Syst.* 1:3–31

Grubb PJ, Metcalfe DJ. 1996. Adaptation and inertia in the Australian tropical lowland rain-forest flora: contradictory trends in intergeneric and intrageneric comparisons of seed size in relation to light demand. *Funct. Ecol.* 10:512–20

Gulmon SL. 1992. Patterns of seed germination in Californian serpentine grassland species. *Oecologia* 89:27–31

Gurevitch J, Morrow L, Wallace A, Walsh J. 1992. A meta-analysis of field experiments on competition. *Am. Nat.* 140:539–72

Hacke UG, Sperry JS. 2001. Functional and ecological xylem anatomy. In *Perspect. Plant Ecol. Evol. Syst.* 4:97–115

Haggar JP, Ewel JJ. 1995. Establishment, resource acquisition, and early productivity as determined by biomass allocation patterns of three tropical tree species. *For. Sci.* 41:689–708

Hall JB, Swaine MD. 1981. *Distribution and Ecology of Vascular Plants in a Tropical Rain Forest*. The Hague: Junk. 383 pp.

Halle F, Oldeman RAA, Tomlinson PB. 1978. *Tropical Trees and Forests: An Architectural Analysis*. Berlin: Springer. 441 pp.

Harms KE, Dalling JW. 1997. Damage and herbivory tolerance through resprouting as an advantage of large seed size in tropical trees and lianas. *J. Trop. Ecol.* 13:617–21

Harper JL. 1989. The value of a leaf. *Oecologia* 80:53–58

Harvey PH, Rambaut A. 2000. Comparative analyses for adaptive radiations. *Philos. Trans. R. Soc. London Ser. B* 355:1599–605

Henery M, Westoby M. 2001. Seed mass and seed nutrient content as predictors of seed output variation between species. *Oikos* 92:479–90

Herms DA, Mattson WJ. 1992. The dilemma of plants: to grow or defend. *Q. Rev. Biol.* 67:283–335

Hewitt N. 1998. Seed size and shade-tolerance: a comparative analysis of North American temperate trees. *Oecologia* 114:432–40

Hikosaka K. 1996. Effects of leaf age, nitrogen nutrition and photon flux density on the organization of the photosynthetic apparatus in leaves of a vine (*Ipomoea tricolor* Cav) grown horizontally to avoid mutual shading of leaves. *Planta* 198:144–50

Hikosaka K, Hirose T. 1997. Leaf angle as a strategy for light competition: optimal and evolutionarily stable light-extinction coefficient within a leaf canopy. *Ecoscience* 4:501–7

Hirose T, Werger M. 1995. Canopy structure and photon flux partitioning among species in a herbaceous plant community. *Ecology* 76:466–74

Hladik A, Miquel S. 1990. Seedling types and plant establishment in an African rain forest. In *Reproductive Ecology of Tropical Plants*, ed. KS Bawa, M Hadley, pp. 261–82. Paris/Carnforth, UK: UNESCO/Parthenon

Hodkinson DJ, Askew AP, Thompson K, Hodgson JG, Bakker JP, Bekker RM. 1998. Ecological correlates of seed size in the British flora. *Funct. Ecol.* 12:762–66

Horn HS. 1971. *The Adaptive Geometry of Trees*. Princeton, NJ: Princeton Univ. Press. 144 pp.

Hubbard RM, Ryan MG, Stiller V, Sperry JS. 2001. Stomatal conductance and photosynthesis vary linearly with plant hydraulic conductance in ponderosa pine. *Plant Cell Environ.* 24:113–21

Hubbell SP, Foster RB. 1986. Commonness and rarity in a tropical forest: implications for tropical tree diversity. In *Conservation*

Biology, ed. ME Soule, pp. 205–31. Sunderland, MS: Sinauer

Iwasa Y, Cohen D, Leon J. 1985. Tree height and crown shape, as results of competitive games. *J. Theor. Biol.* 112:279–97

Jacobs BF. 1999. Estimation of rainfall variables from leaf characters in tropical Africa. *Palaeogeogr. Palaeoclimatol. Palaeoecol.* 145:231–50

Jakobsson A, Eriksson O. 2000. A comparative study of seed number, seed size, seedling size and recruitment in grassland plants. *Oikos* 88:494–502

Jurado E, Westoby M. 1992. Seedling growth in relation to seed size among species of arid Australia. *J. Ecol.* 80:407–16

Jurik TW, Wang SC, Vandervalk AG. 1994. Effects of sediment load on seedling emergence from wetland seed banks. *Wetlands* 14:159–65

Kawecki TJ. 1993. Age and size at maturity in a patchy environment: fitness maximization versus evolutionary stability. *Oikos* 66:309–17

Keddy PA. 1989. *Competition.* London: Chapman & Hall. 202 pp.

Kempf J, Pickett STA. 1981. The role of branch length and branching angle in branching pattern of forest shrubs along a successional gradient. *New Phytol.* 88:111–16

Kidson R, Westoby M. 2000. Seed mass and seedling dimensions in relation to seedling establishment. *Oecologia* 125:11–17

Kikuzawa K. 1991. A cost-benefit analysis of leaf habit and leaf longevity of trees and their geographical pattern. *Am. Nat.* 138:1250–63

Kikuzawa K. 1995. The basis for variation in leaf longevity of plants. *Vegetation* 121:89–100

Kikuzawa K, Ackerly DD. 1999. Significance of leaf longevity in plants. *Plant Species Biol.* 14:39–45

King DA. 1991. The allometry of trees in temperate and tropical forests. *Res. Explor.* 7:342–51

King DA. 1994. Influence of light level on the growth and morphology of saplings in a Panamanian forest. *Am. J. Bot.* 81:948–57

Kitajima K. 1996a. Cotyledon functional morphology, patterns of seed reserve utilization and regeneration niches of tropical tree seedlings. In *The Ecology of Tropical Forest Tree Seedlings*, ed. MD Swaine, pp. 193–210. New York: Parthenon

Kitajima K. 1996b. Ecophysiology of tropical tree seedlings. In *Tropical Forest Plant Ecophysiology*, ed. SS Mulkey, RL Chazdon, AP Smith, pp. 559–96. New York: Chapman & Hall

Kitajima K, Mulkey SS, Wright SJ. 1997. Decline of photosynthetic capacity with leaf age in relation to leaf longevities for five tropical canopy tree species. *Am. J. Bot.* 84:702–8

Koike T. 1988. Leaf structure and photosynthetic performance as related to the forest succession of deciduous broad-leaved trees. *Plant Species Biol.* 3:77–87

Lambers H, Chapin FS III, Pons TL. 1998. *Plant Physiological Ecology.* New York: Springer-Verlag. 540 pp.

Lambers H, Poorter H. 1992. Inherent variation in growth rate between higher plants: a search for ecological causes and consequences. *Adv. Ecol. Res.* 23:187–261

Landsberg J, Gillieson DS. 1995. Regional and local variation in insect herbivory, vegetation and soils of eucalypt associations in contrasted landscape positions along a climatic gradient. *Aust. J. Ecol.* 20:299–315

Leishman MR. 1999. How well do plant traits correlate with establishment ability? Evidence from a study of 16 calcareous grassland species. *New Phytol.* 141:1–10

Leishman MR. 2001. Does the seed size/number trade-off model determine plant community structure? An assessment of the model mechanisms and their generality. *Oikos* 93:294–302

Leishman MR, Westoby M. 1994a. The role of large seed size in shaded conditions: experimental evidence. *Funct. Ecol.* 8:205–14

Leishman MR, Westoby M. 1994b. The role of seed size in seedling establishment in dry soil conditions: experimental evidence from semi-arid species. *J. Ecol.* 82:249–58

Leishman MR, Westoby M. 1998. Seed size and

shape are not related to persistence in soil in Australia in the same way as in Britain. *Funct. Ecol.* :480–85

Leishman MR, Westoby M, Jurado E. 1995. Correlates of seed size variation: a comparison among five temperate floras. *J. Ecol.* 83:517–29

Leishman MR, Wright IJ, Moles AT, Westoby M. 2000. The evolutionary ecology of seed size. In *Seeds: The Ecology of Regeneration in Plant Communities*, ed. M Fenner, pp. 31–57. Wallingford, UK: CAB Int.

Levin DA. 1974. The oil content of seeds: an ecological perspective. *Am. Nat.* 108:193–206

Loehle C. 1988. Tree life history strategies: the role of defenses. *Can. J. For. Res.* 18:209–22

Loehle C. 1996. Optimal defensive investments in plants. *Oikos* 75:299–302

Loehle C. 2000. Strategy space and the disturbance spectrum: a life-history model for tree species coexistence. *Am. Nat.* 156:14–33

Lord J, Egan J, Clifford T, Jurado E, Leishman M, et al. 1997. Larger seeds in tropical floras: consistent patterns independent of growth form and dispersal mode. *J. Biogeogr.* 24:205–11

Lord J, Westoby M, Leishman M. 1995. Seed size and phylogeny in six temperate floras: constraints, niche conservatism, and adaptation. *Am. Nat.* 146:349–64

Lowman MD, Box JD. 1983. Variation in leaf toughness and phenolic content among 5 species of Australian rain forest trees. *Aust. J. Ecol.* 8:17–26

Lusk CH, Contreras O. 1999. Foliage area and crown nitrogen turnover in temperate rain forest juvenile trees of differing shade tolerance. *J. Ecol.* 87:973–83

Magnani F, Mencuccini M, Grace J. 2000. Age-related decline in stand productivity: the role of structural acclimation under hydraulic constraints. *Plant Cell Environ.* 23:251–63

Makela A. 1985. Differential games in evolutionary theory: height growth strategies of trees. *Theor. Popul. Biol.* 27:239–67

Matyssek R. 1986. Carbon, water and nitrogen relations in evergreen and deciduous conifers. *Tree Physiol.* 2:177–87

Mazer SJ. 1989. Ecological, taxonomic, and life history correlates of seed mass among Indiana Dune angiosperms. *Ecol. Monogr.* 59:153–75

McIntyre S, Lavorel S, Landsberg J, Forbes TDA. 1999. Disturbance response in vegetation—towards a global perspective on functional traits. *J. Veg. Sci.* 10:621–30

Mencuccini M, Grace J. 1995. Climate influences the leaf area/sapwood area ratio in Scots pine. *Tree Physiol.* 15:1–10

Mencuccini M, Magnani F. 2000. Comment on 'Hydraulic limitation of tree height: a critique' by Becker, Meinzer and Wullschleger. *Funct. Ecol.* 14:135–36

Metcalfe DJ, Grubb PJ. 1995. Seed mass and light requirement for regeneration in South-East Asian rain forest. *Can. J. Bot.* 73:817–26

Michaels HJ, Benner B, Hartgerink AP, Lee TD, Rice S, et al. 1988. Seed size variation: magnitude, distribution, and ecological correlates. *Evol. Ecol.* 2:157–66

Midgley JJ, Bond WJ. 1989. Leaf size and inflorescence size may be allometrically related traits. *Oecologia* 78:427–29

Midgley JJ, Bond WJ. 1991. Ecological aspects of the rise of the angiosperms: a challenge to the reproductive superiority hypotheses. *Biol. J. Linn. Soc.* 44:81–92

Milberg P, Perez-Fernandez MA, Lamont BB. 1998. Seedling growth response to added nutrient depends on seed size in three woody genera. *J. Ecol.* 86:624–32

Miyaji KI, Dasilva WS, Alvim PD. 1997. Longevity of leaves of a tropical tree, *Theobroma cacao*, grown under shading, in relation to position within the canopy and time of emergence. *New Phytol.* 135:445–54

Moles AT, Hodson DW, Webb CJ. 2000. Do seed size and shape predict persistence in soil in New Zealand? *Oikos* 89:541–45

Moles AT, Westoby M. 2000. Do small leaves expand faster than large leaves, and do shorter expansion times reduce herbivore damage? *Oikos* 90:517–26

Monk CD. 1966. An ecological significance of evergreenness. *Ecology* 47:504–5

Mooney HA, Ferrar PJ, Slatyer RO. 1978. Photosynthetic capacity and carbon allocation patterns in diverse growth forms of *Eucalyptus*. *Oecologia* 36:103–11

Nardini A, Salleo S. 2000. Limitation of stomatal conductance by hydraulic traits: sensing or preventing xylem cavitation? *Trees* 15:14–24

Niinemets U. 1996. Plant growth-form alters the relationship between foliar morphology and species shade-tolerance ranking in temperate woody taxa. *Vegetatio* 124:145–53

Niinemets U. 1998. Are compound-leaved woody species inherently shade-intolerant: an analysis of species ecological requirements and foliar support costs. *Plant Ecol.* 134:1–11

Niinemets U. 1999. Components of leaf dry mass per area—thickness and density—alter leaf photosynthetic capacity in reverse directions in woody plants. *New Phytol.* 144:35–47

Niinemets U. 2001. Global-scale climatic controls of leaf dry mass per area, density, and thickness in trees and shrubs. *Ecology* 82:453–69

Niinemets U, Kull K. 1994. Leaf weight per area and leaf size of 85 Estonian woody species in relation to shade tolerance and light availability. *For. Ecol. Manage.* 70:1–10

Noble IR, Slatyer RO. 1980. The use of vital attributes to predict successional changes in plant communities subject to recurrent disturbances. *Vegetatio* 43:5–21

Obeso JR. 1998. Patterns of variation in *Ilex aquifolium* fruit traits related to fruit consumption by birds and seed predation by rodents. *Ecoscience* 5:463–69

Osunkoya OO, Ash JE, Hopkins MS, Graham AW. 1994. Influence of seed size and seedling ecological attributes on shade-tolerance of rainforest tree species in northern Queensland. *J. Ecol.* 82:149–63

Pacala SW. 1997. Dynamics of plant communities. In *Plant Ecology*, ed. MJ Crawley, pp. 532–55. Oxford: Blackwell Sci.

Parkhurst DF. 1994. Diffusion of CO_2 and other gases inside leaves. *New Phytol.* 126:449–79

Parkhurst DF, Loucks OL. 1972. Optimal leaf size in relation to environment. *J. Ecol.* 60:505–37

Parolin P. 2000. Seed mass in Amazonian floodplain forests with contrasting nutrient supplies. *J. Trop. Ecol.* 16:417–28

Poorter H. 1994. Construction costs and payback time of biomass: a whole plant perspective. In *A Whole Plant Perspective on Carbon-Nitrogen Interactions*, ed. E Roy, E Garnier, pp. 111–27. The Hague, The Netherlands: SPB Academic

Poorter H, De Jong R. 1999. A comparison of specific leaf area, chemical composition and leaf construction costs of field plants from 15 habitats differing in productivity. *New Phytol.* 143:163–76

Poorter H, Garnier E. 1999. Ecological significance of inherent variation in relative growth rate and its components. See Pugnaire & Valladares 1999, pp. 81–120

Poorter H, van der Werf A. 1998. Is inherent variation in RGR determined by LAR at low irradiance and by NAR at high irradiance? A review of herbaceous species. In *Inherent Variation in Plant Growth. Physiological Mechanisms and Ecological Consequences*, ed. H Lambers, H Poorter, MMI Van Vuuren, pp. 309–36. Leiden, The Netherlands: Backhuys

Price T. 1997. Correlated evolution and independent contrasts. *Philos. Trans. R. Soc. London Ser. B* 352:519–29

Prinzing A, Durka W, Klotz S, Brandl R. 2001. The niche of higher plants: evidence for phylogenetic conservatism. *Proc. R. Soc. Lond. Ser. B* 268:2383–89

Pugnaire FI, Valladares F, eds. 1999. *Handbook of Functional Plant Ecology*. New York: Dekker

Pyankov VI, Kondratchuk AV, Shipley B. 1999. Leaf structure and specific leaf mass: the alpine desert plants of the Eastern Pamirs, Tadjikistan. *New Phytol.* 143:131–42

Raunkiaer C. 1934. *The Life Forms of Plants*

and Statistical Plant Geography. London: Clarendon. 632 pp.

Reader RJ. 1993. Control of seedling emergence by ground cover and seed predation in relation to seed size for some old field species. *J. Ecol.* 81:169–75

Rees M, Westoby M. 1997. Game theoretical evolution of seed mass in multi-species ecological models. *Oikos* 78:116–26

Reich PB, Ellsworth DS, Walters MB, Vose JM, Gresham C, et al. 1999. Generality of leaf trait relationships: a test across six biomes. *Ecology* 80:1955–69

Reich PB, Tjoelker MG, Walters MB, Vanderklein DW, Buschena C. 1998. Close association of RGR, leaf and root morphology, seed mass and shade tolerance in seedlings of nine boreal tree species grown in high and low light. *Funct. Ecol.* 12:327–38

Reich PB, Uhl C, Walters MB, Ellsworth DS. 1991. Leaf lifespan as a determinant of leaf structure and function among 23 Amazonian tree species. *Oecologia* 86:16–24

Reich PB, Walters MB, Ellsworth DS. 1992. Leaf life-span in relation to leaf, plant, and stand characteristics among diverse ecosystems. *Ecol. Monogr.* 62:365–92

Reich PB, Walters MB, Ellsworth DS. 1997. From tropics to tundra: global convergence in plant functioning. *Proc. Natl. Acad. Sci. USA* 94:13730–34

Ritchie ME, Olff H. 1999. Spatial scaling laws yield a synthetic theory of biodiversity. *Nature* 400:557–60

Rockwood LL. 1985. Seed weight as a function of life form, elevation and life zone in neotropical forests. *Biotropica* 17:32–39

Roderick ML, Berry SL, Noble IR. 2000. A framework for understanding the relationship between environment and vegetation based on the surface area to volume ratio of leaves. *Funct. Ecol.* 14:423–37

Roderick ML, Berry SL, Noble IR, Farquhar GD. 1999a. A theoretical approach to linking the composition and morphology with the function of leaves. *Funct. Ecol.* 13:683–95

Roderick ML, Berry SL, Saunders AR, Noble IR. 1999b. On the relationship between the composition, morphology and function of leaves. *Funct. Ecol.* 13:696–710

Ryan MG, Yoder BJ. 1997. Hydraulic limits to tree height and tree growth. *Bioscience* 47:235–42

Ryser P. 1993. Influences of neighbouring plants on seedling establishment in limestone grassland. *J. Veg. Sci.* 4:195–202

Ryser P. 1996. The importance of tissue density for growth and lifespan of leaves and roots: a comparison of five ecologically contrasting grasses. *Funct. Ecol.* 10:717–23

Ryser P, Urbas P. 2000. Ecological significance of leaf lifespan among Central European grass species. *Oikos* 91:41–50

Sakai S. 1991. A model analysis for the adaptive architecture of herbaceous plants. *J. Theor. Biol.* 148:535–44

Sakai S. 1995. Evolutionarily stable growth of a sapling which waits for future gap formation under closed canopy. *Evol. Ecol.* 9:444–52

Salisbury EJ. 1942. *The Reproductive Capacity of Plants*. London: Bell & Sons. 244 pp.

Saverimuttu T, Westoby M. 1996. Seedling longevity under deep shade in relation to seed size. *J. Ecol.* 84:681–89

Schafer KVR, Oren R, Tenhunen JD. 2000. The effect of tree height on crown level stomatal conductance. *Plant Cell Environ.* 23:365–75

Schieving F, Poorter H. 1999. Carbon gain in a multispecies canopy: the role of specific leaf area and photosynthetic nitrogen-use efficiency in the tragedy of the commons. *New Phytol.* 143:201–11

Schimper AFW. 1903. *Plant-Geography Upon a Physiological Basis*. Trans. WR Fisher. Oxford: Clarendon. 839 pp.

Schlapfer B, Ryser P. 1996. Leaf and root turnover of three ecologically contrasting grass species in relation to their performance along a productivity gradient. *Oikos* 75:398–406

Schoener TW. 1983. Field experiments on interspecific competition. *Am. Nat.* 122:240–85

Schulze ED, Williams RJ, Farquhar GD, Schulze W, Langridge J, et al. 1998. Carbon and nitrogen isotope discrimination and nitrogen nutrition of trees along a rainfall gradient in northern Australia. *Aust. J. Plant. Physiol.* 25:413–25

Seiwa K, Kikuzawa K. 1996. Importance of seed size for the establishment of seedlings of five deciduous broad-leaved tree species. *Vegetatio* 123:51–64

Shields LM. 1950. Leaf xeromorphy as related to physiological and structural influences. *Bot. Rev.* 16:399–447

Shipley B. 1995. Structured interspecific determinants of specific leaf area in 34 species of herbaceous angiosperms. *Funct. Ecol.* 9: 312–19

Shipley B, Dion J. 1992. The allometry of seed production in herbaceous angiosperms. *Am. Nat.* 139:467–83

Showalter TD, Hargrove WW, Crossley DA. 1986. Herbivory in forested ecosystems. *Annu. Rev. Entomol.* 31:177–96

Smith T, Huston M. 1989. A theory of the spatial and temporal dynamics of plant communities. *Vegetatio* 83:49–69

Southwood TRE. 1977. Habitat, the templet for ecological strategies. *J. Anim. Ecol.* 46:337–65

Specht RL, Specht A. 1989. Canopy structure in *Eucalyptus*-dominated communities in Australia along climatic gradients. *Acta Oecol.* 10:191–213

Sperry JS. 1995. Limitations on stem water transport and their consequences. See Gartner 1995, pp. 105–24

Sperry JS. 2000. Hydraulic constraints on plant gas exchange. *Agric. For. Meteorol.* 104:13–23

Steffen WL. 1996. A periodic table for ecology? A chemist's view of plant functional types. *J. Veg. Sci.* 7:425–30

Steinke TD. 1988. Vegetative and floral phenology of three mangroves in Mgeni Estuary. *S. Afr. J. Bot* 54:97–102

Suehiro K, Kameyama K. 1992. Leaf age composition of evergreen broadleaved trees. *Jpn. J. Ecol.* 42:137–47

Thompson K, Band SR, Hodgson JG. 1993. Seed size and shape predict persistence in the soil. *Funct. Ecol.* 7:236–41

Thompson K, Baster K. 1992. Establishment from seed of selected Umbelliferae in unmanaged grassland. *Funct. Ecol.* 6:346–52

Thompson K, Hodkinson DJ. 1998. Seed mass, habitat and life history: a re-analysis of Salisbury (1942, 1974). *New Phytol.* 138:163–67

Turnbull LA, Rees M, Crawley MJ. 1999. Seed mass and the competition/colonization trade-off: a sowing experiment. *J. Ecol.* 87:899–912

Tyree MT. 1999. Water relations and hydraulic architecture. See Pugnaire & Valladares 1999, pp. 221–68

Valladares F, Wright SJ, Lasso E, Kitajima K, Pearcy RW. 2000. Plastic phenotypic response to light of 16 congeneric shrubs from a Panamanian rainforest. *Ecology* 81:1925–36

Vaughton G, Ramsey M. 1998. Sources and consequences of seed mass variation in *Banksia marginata* (Proteaceae). *J. Ecol.* 86: 563–73

Venable DL. 1996. Packaging and provisioning in plant reproduction. *Philos. Trans. R. Soc. London Ser. B* 351:1319–29

Vendramini F, Diaz S, Gurvich DE, Wilson PJ, Thompson K, Hodgson JG. 2002. Leaf traits as indicators of resource-use strategy in floras with succulent species. *New Phytol.* 154:147–57

Vesk PA, Westoby M. 2001. Predicting plant species responses to grazing from published studies. *J. Appl. Ecol.* 38:897–909

Villar R, Merino J. 2001. Comparison of leaf construction costs in woody species with differing leaf life-spans in contrasting ecosystems. *New Phytol.* 151:213–26

Vincent TLS, Vincent TL. 1996. Using the ESS maximum principle to explore root-shoot allocation, competition and coexistence. *J. Theor. Biol.* 180:111–20

Vogel S. 1988. *Life's Devices: The Physical World of Animals and Plants*. Princeton, NJ: Princeton Univ. Press. 367 pp.

Walter H. 1973. *Vegetation of the Earth in Relation to Climate and the Eco-Physiological Conditions*. New York: Springer-Verlag. 237 pp.

Walters MB, Reich PB. 1999. Low-light carbon balance and shade tolerance in the seedlings of woody plants: Do winter deciduous and broad-leaved evergreen species differ? *New Phytol.* 143:143–54

Walters MB, Reich PB. 2000. Seed size, nitrogen supply, and growth rate affect tree seedling survival in deep shade. *Ecology* 81:1887–901

Webb LJ. 1968. Environmental relationships of the structural types of Australian rainforest vegetation. *Ecology* 49:296–311

Weiher E, van der Werf A, Thompson K, Roderick ML, Garnier E, Eriksson O. 1999. Challenging Theophrastus: a common core list of plant traits for functional ecology. *J. Veg. Sci.* 10:609–20

Werner P, Platt WJ. 1976. Ecological relationships of co-occurring goldenrods (*Solidago*: Compositae). *Am. Nat.* 110:959–71

West GB, Brown JH, Enquist BJ. 1999. A general model for the structure and allometry of plant vascular systems. *Nature* 400:664–67

Westoby M. 1998. A leaf-height-seed (LHS) plant ecology strategy scheme. *Plant Soil* 199:213–27

Westoby M. 1999. Generalization in functional plant ecology: the species sampling problem, plant ecology strategies, schemes, and phylogeny. See Pugnaire & Valladares 1999, pp. 847–72

Westoby M, Jurado E, Leishman M. 1992. Comparative evolutionary ecology of seed size. *Trends Ecol. Evol.* 7:368–72

Westoby M, Leishman M, Lord J. 1996. Comparative ecology of seed size and dispersal. *Philos. Trans. R. Soc. London Ser. B* 351: 1309–18

Westoby M, Leishman MR, Lord JM. 1995. On misinterpreting the 'phylogenetic correction'. *J. Ecol.* 83:531–34

Westoby M, Rice B, Howell J. 1990. Seed size and plant stature as factors in dispersal spectra. *Ecology* 71:1307–15

Westoby M, Warton D, Reich PB. 2000. The time value of leaf area. *Am. Nat.* 155:649–56

Westoby M, Wright IJ. 2002. The spectrum of twig size and related traits. *Oecologia.* In press

White PS. 1983a. Corner's Rules in eastern deciduous trees: allometry and its implications for the adaptive architecture of trees. *Bull. Torrey Bot. Club* 110:203–12

White PS. 1983b. Evidence that temperate east north American evergreen woody plants follow Corner's Rules. *New Phytol.* 95:139–45

Whitehead D. 1998. Regulation of stomatal conductance and transpiration in forest canopies. *Tree Physiol.* 18:633–44

Whittaker RH, Woodwell GM. 1968. Dimension and production relations of trees and shrubs in the Brookhaven Forest, New York. *Ecology* 56:1–25

Wiemann MC, Manchester SR, Dilcher DL, Hinojosa LF, Wheeler EA. 1998. Estimation of temperature and precipitation from morphological characters of dicotyledonous leaves. *Am. J. Bot.* 85:1796–802

Wilf P, Wing SL, Greenwood DR, Greenwood CL. 1998. Using fossil leaves as paleoprecipitation indicators: an Eocene example. *Geology* 26:203–6

Williams K, Field CB, Mooney HA. 1989. Relationships among leaf construction cost, leaf longevity, and light environment in rainforest plants of the genus *Piper. Am. Nat.* 133:198–211

Williams-Linera G. 2000. Leaf demography and leaf traits of temperate-deciduous and tropical evergreen-broadleaved trees in a Mexican montane cloud forest. *Plant Ecology* 149:233–44

Wilson JB, Agnew ADQ. 1992. Positive-feedback switches in plant communities. *Adv. Ecol. Res.* 23:263–336

Wilson PJ, Thompson K, Hodgson JG. 1999. Specific leaf area and leaf dry matter content as alternative predictors of plant strategies. *New Phytol.* 143:155–62

Witkowski ETF, Lamont BB. 1991. Leaf specific mass confounds leaf density and thickness. *Oecologia* 88:486–93

Wolfe JA. 1995. Palaeoclimatic estimates from Tertiary leaf assemblages. *Annu. Rev. Earth Planet. Sci.* 23:119–42

Wright IJ, Cannon K. 2001. Relationships between leaf lifespan and structural defences in a low nutrient, sclerophyll flora. *Funct. Ecol.* 15:351–59

Wright IJ, Clifford HT, Kidson R, Reed ML, Rice BL, Westoby M. 2000. A survey of seed and seedling characters in 1744 Australian dicotyledon species: cross-species trait correlations and correlated trait-shifts within evolutionary lineages. *Biol. J. Linn. Soc.* 69:521–47

Wright IJ, Reich PB, Westoby M. 2001. Strategy-shifts in leaf physiology, structure and nutrient content between species of high and low rainfall, and high and low nutrient habitats. *Funct. Ecol.* 15:423–34

Wright IJ, Westoby M. 1999. Differences in seedling growth behaviour among species: trait correlations across species, and trait shifts along nutrient compared to rain gradients. *J. Ecol.* 87:85–97

Wright IJ, Westoby M. 2002. Leaves at low versus high rainfall: coordination of structure, lifespan and physiology. *New Phytol.* In press

Wright IJ, Westoby M, Reich PB. 2002. Convergence towards higher leaf mass per area in dry and nutrient-poor habitats has different consequences for leaf lifespan. *J. Ecol.* 90:534–43

Xu G, Ninomiya I, Ogino K. 1990. The change of leaf longevity and morphology of several tree species grown under different light conditions. *Bull. Ehime Univ. For.* 28:35–44

Annu. Rev. Ecol. Syst. 2002. 33:161–79
doi: 10.1146/annurev.ecolsys.33.010802.150439
First published online as a Review in Advance on August 6, 2002

REPRODUCTIVE PROTEIN EVOLUTION

Willie J. Swanson[1,2] and Victor D. Vacquier[2]

[1]*Department of Biology, University of California, Riverside, California 92521;
email: willies@citrus.ucr.edu*
[2]*Department of Genome Sciences, University of Washington, Box 357730, Seattle,
Washington 98195-7730*
[3]*Center for Marine Biotechnology and Biomedicine, Scripps Institution of
Oceanography, University of California, San Diego, La Jolla, California 92093;
email: vvacquier@ucsd.edu*

Key Words   positive Darwinian selection, fertilization, speciation, sexual conflict, lysin

■ **Abstract**   The evolution of proteins involved in reproduction is only now beginning to be studied. A reoccurring observation is the rapid evolution of the molecules mediating reproductive events following the release of gametes. We review the examples where rapid evolution of reproductive proteins has been documented, covering taxa ranging from diatoms to humans. The selective pressures causing this divergence remain unknown, but several hypotheses are presented. The functional consequences of this rapid divergence could be involved in speciation.

INTRODUCTION

An emerging trend in the study of reproductive proteins is the observation of rapid, adaptive evolution in many genes that mediate post-copulatory gamete usage/storage, signal transduction and fertilization (Singh & Kulathinal 2000, Swanson & Vacquier 2002). For example, mammalian egg coat proteins are among the 10% most rapidly evolving proteins in comparison of orthologs between humans and rodents (Makalowski & Boguski 1998). Likewise, the most rapidly evolving proteins in the genome *Drosophila* are involved in reproduction (Schmid & Tautz 1997, Ting et al. 1998, Tsaur & Wu 1997). In this review, we demonstrate that this is a general phenomenon that occurs in diverse taxa. We present several hypotheses about the selective forces that may be driving this rapid evolution, but stress that the selective pressure remains unknown and may differ between taxa. Finally, we point out potential functional consequences of this rapid evolution, including the possibility of speciation.

We define rapidly evolving genes as those that have a higher than average percentage of amino acid substitutions—typically in the 10% most rapidly evolving class of genes within a genome. We recognize that this is a crude measure, because many rapidly evolving genes may show extreme divergence only in a portion of their sequence, such as a binding site. Additionally, defining rapid evolution solely by percent divergence does not provide information about the potential causes of rapid evolution. For example, rapid evolution might be due to a lack of functional constraint, e.g., a pseudogene accumulating mutations. Alternatively, rapid evolution might be due to positive Darwinian selection, which occurs when natural selection promotes amino acid divergence resulting in protein adaptation. A clear signal of positive Darwinian selection is an excess in the number of nonsynonymous substitutions per nonsynonymous sites (d_N; amino acid altering) compared with the number of synonymous substitutions per synonymous sites (d_S; silent changes). Because d_N and d_S are normalized to the number of sites, in cases of no selection, $d_N/d_S = 1$ (for example in a pseudogene). For most proteins, d_N/d_S is much less than one [average d_N/d_S ratio of $\sim$0.2 is found between humans and mice (Li 1997)] indicating purifying selection. Averaging the d_N/d_S ratio across all sites and lineages is a conservative test for positive Darwinian selection (Yang & Bielawski 2000). The classic example is the class one major histocompatibility locus, where the average d_N/d_S ratio across all sites is $\sim$0.5, but those sites directly involved in binding foreign antigens have d_N/d_S ratios >1 (Hughes & Nei 1988, Yang & Swanson 2002).

If sequences are available from multiple species, new methods are available to use maximum likelihood predictions to detect selection acting upon a subset of codons (Nielsen & Yang 1998, Yang et al. 2000a). These methods can find a subset of sites showing extreme divergence, if there are sufficient data (Anisimova et al. 2001). Importantly, these new methods do not require a priori knowledge of the sites under selection. They can also be used to predict functionally important sites of a gene subject to positive Darwinian selection (Swanson et al. 2001c). There are two steps in using these methods. First is the identification of the presence, or absence, of a subset of sites being subjected to positive selection. This is accomplished by a likelihood ratio test comparing the likelihood of a neutral model with that of a selection model. The neutral models range from those having two classes of d_N/d_S ratios set at 0 and 1 (Nielsen & Yang 1998), to modeling a beta distribution of d_N/d_S ratios between the values of 0 and 1 (Yang et al. 2000a). The selection models add an additional class of sites with a d_N/d_S ratio that is estimated from the data and can be greater or less than one. This appears to be a robust and powerful method for detecting positive selection, as indicated by analysis of simulated (Anisimova et al. 2001) and empirical data (Yang & Swanson 2002). The second step occurs if a subset of sites is predicted to be under positive selection. Then, an empirical Bayesian approach is used to assign the posterior probabilities of sites falling within the selected class of codons (Nielsen & Yang 1998, Yang et al. 2000a). A signal of positive Darwinian selection indicates that there is an adaptive advantage to changing the amino acid sequence, and this signal can be used to identify functionally important gene regions, such as binding sites (Swanson et al. 2001c, Yang & Swanson 2002).

Rapidly Evolving Reproductive Proteins

Many eukaryotes have reproductive proteins, the sequences of which have extensively diverged between closely related species (Table 1). Many of these genes are rapidly evolving but do not show a clear signal of positive Darwinian selection. However, future analyses, perhaps with improved statistical methods or additional

TABLE 1 Rapidly evolving genes involved in reproduction

Gene (locus)	Organism	Evidence for + selection	Reference
Pheromones	*Euplotes* (ciliate protozoa)	None	(Luporini et al. 1995)
mid1	*Chlamydomonas* (green alga)	None	(Ferris et al. 1997)
fus1	*Chlamydomonas* (green alga)	None	(Ferris et al. 1996)
Sig1	*Thalassiosira* (Diatoms)	None	(Armbrust & Galindo 2001)
Pheromones	*Basidiomycete* (fungi)	None	(Brown & Casselton 2001)
SCR	Brassicaceae	None	(Schopfer et al. 1999)
S-locus	Solanaceae	None	(Richman & Kohn 2000)
Pollen coat proteins	Arabidopsis	None	(Mayfield et al. 2001)
Lysin	*Tegula & Haliotis* (turban snails and abalone)	Overall $d_N/d_S > 1$	(Hellberg & Vacquier 1999, Lee et al. 1995)
sp18	*Haliotis* (abalone)	Overall $d_N/d_S > 1$	(Swanson & Vacquier 1995a)
TMAP	Tegula (*turban snails*)	Overall $d_N/d_S > 1$	(Hellberg et al. 2000)
Bindin	Sea urchins	Region with $d_N/d_S > 1$	(Metz & Palumbi 1996)
Acp26Aa	Drosophila	Lineage with $d_N/d_S > 1$	(Tsaur & Wu 1997)
Acp36DE	Drosophila	Polymorphism survey	(Begun et al. 2000)
ZP3	Mammals	Class of sites with $d_N/d_S > 1$	(Swanson et al. 2001c)
ZP2	Mammals	Class of sites with $d_N/d_S > 1$	(Swanson et al. 2001c)
OGP	Mammals	Class of sites with $d_N/d_S > 1$	(Swanson et al. 2001c)
Zonadhesin	Mammals	None	(Gao & Garbers 1998)
TCTE1	Mammals	None	(Juneja et al. 1998)
Protamines	Mammals	$d_N/d_S > 1$	(Wyckoff et al. 2000)

data, may reveal the signal of positive Darwinian selection. Nevertheless, they are of interest because their rapid evolution may result in functional differences affecting reproduction.

Marine ciliates of the genus *Euplotes* secrete protein pheromones of 40–43 amino acids that mediate sexual conjugation and vegetative growth. An alignment of these pheromone sequences from different mating types of one species shows that the mature protein has diverged extensively (Luporini et al. 1995). However, the signal sequences of these genes remain highly conserved. Additionally, a region that is cleaved off to make the mature pheromone also shows high conservation, suggesting different selective pressures acting upon the different regions of this molecule (Figure 1). These pheromones are involved in both cell mitotic proliferation and mating pair formation in *E. raikovi*. Interestingly, the receptor for the pheromones has been identified as an alternatively spliced, membrane bound form of the pheromone, with the external region structurally identical to the secreted form. It has been hypothesized that binding of the same pheromone induces mitotic division, while binding of an alternative pheromone induces mating pair formation (Ortenzi et al. 2000).

Two genes that control mating in the unicellular green alga, *Chlamydomonas reinhardtii*, show extensive divergence between species. The product of the *Chlamydomonas mid* gene determines if a cell will be of mating type plus or minus, whereas *fus1* encodes a protein needed for fusion of plus and minus cells. No homologues of *C. reinhardtii fus1*, and only one homologue of *mid*, were found in

```
Euplotes mating pheromone

1  MKAIFIILAILMVTQAFKMTSKVNTKLQSQIQSKFQSKNKLASTFQTSSQLKYY--CWEEPYTSSITGCSTS-LACYEAS-
2  ...............................................--.....V.............-.....-
3  ...............................M.NLKT.DAPD.YSQT.---L...N.NPDKCW.NSN


1  DCSVTGNDQDKCNNVGQNMIDK-------FFELWG-VCINDYETCLQYVDRAWIHYSDSEFCGCTNPE-QESAFRDAMDCLQF-
2  ......D.T....D..Y..YY.-------.NS....N.-................NE.G......QA-L..Q..ELF..W..S
3  G.GS.DGSLGE.YTNDPENPGVGDNIAQVI.DR.MLG.YE.VSN.VVDAGAMYAIF.SQYL.N.GYEFGNVND.TYGF-.GVYP


SCR gene

1  MKSVLYALLCFIFIVSSHAQDVEANLMNR--CTRELPFPGKCGSSEDGGCIKLYSSEKKLHPSRCECEPRYKARF-CRCKI
2  ...AV...........G.I.E......MP--.G-SFM.-.N.RNIGARE.E..N.PG.R-K..H.K.TDTQMGTYS.D..L
3  ...AI...........V.E.....RKT--.VHR.NSG.S..K.GQHD.EAF.TNKTNQKAFY.N.TSPFR---------
4  -.ATI......L.....RG.E....V.KN--.PIQFNLG.Q..N.GGDA.VEE.NRK..KKKIF.S.GG-VRVGQ-.-.--
5  ...AI...........L....G.E.....KKN--.VGKTRL..P..D.GASS.RD..NQTE.TM.VS.R.V.TGR----.F.SL
6  ...AI.........IL.RS.ELTEVGADKQQ.KKNF--..H.ET..R--.ENT.KRLN.-KVFD.H.Q.FGRR.L-.T..-
```

Figure 1 Alignments of two reproductive proteins. The *Euplotes* mating pheromones (Luporini et al. 1995) and the plant SCR proteins (Schopfer et al. 1999). In both cases, the functional secreted proteins are extensively divergent while the portions cleaved off to make the secreted proteins are well conserved. Signal sequences of both proteins are shown in bold. The pre-region that is cleaved off of the *Euplotes* functional secreted pheromone is underlined. Dots denote identity to the first sequence and dashes are inserted for alignment.

12 other *Chlamydomonas* species (Ferris et al. 1996, 1997). These genes are located in the mating type locus (*MT*), a ~1 Mb domain under recombinational suppression. This region is extremely dynamic, having large indels, translocated genes, gene duplications, and gene inactivation events. Although this region does have a generally high level of mutatgenic change, there are also several conserved housekeeping genes in the region (Ferris et al. 2002).

An extracellular matrix protein encoded by the *Sig1* gene of the diatom *Thalassiosira* is upregulated during mating, and is thought to function in the mating process. *Sig1* is highly divergent both within and between species and there are well-documented differences that distinguish between *Sig1* from the Atlantic and the Pacific Oceans (Armbrust & Galindo 2001). *Sig1* is present as a multicopy gene, with some individuals displaying as many as 19 gene variants of *Sig1*, suggesting a minimum of 10 loci. *Sig1* shows extreme intraspecific, interspecific, and inter-individual variation. Although the exact function of the Sig1 protein remains unknown, its extreme divergence suggests that it might be a barrier to reproduction between different diatom strains.

Mating compatibility in *Basidiomycete* fungi requires secretion of protein pheromones that bind to cell surface receptors to mediate signal transduction to induce expression of mating genes (Brown & Casselton 2001). The pheromones and their receptors show extreme sequence variation (Casselton & Olesnicky 1998), which could underlie species-specific gamete interaction.

The reproductive proteins from *Euplotes*, *Chlamydomonas*, *Thalassiosira*, and *Basidiomycete* show extreme divergence between species; however, there is no evidence that this divergence is promoted by positive Darwinian selection. There are currently not enough sequences available, or those that are available are too divergent to reliably align (Figure 1), making it impossible to perform likelihood ratio tests to determine if any sites are subjected to positive selection.

Many species of flowering plants cannot self-fertilize because their pollen (male) is incompatible with stylar (female) tissue, and this self-incompatibility prevents inbreeding depression. In sporophytic self-incompatibility in the genus *Brassica*, the pollen component is encoded by the highly variable S-locus cysteine rich gene *SCR* (Schopfer et al. 1999). The stylar recognition S-locus receptor kinase, *SRK*, is also highly variable and encodes a membrane-spanning protein kinase (Schopfer & Nasrallah 2000) that has been shown to bind SCR in an allele-specific manner (Kachroo et al. 2001). SCR is similar to the *Euplotes* pheromones in that, although the signal sequences of SCR proteins are relatively conserved, the mature SCR proteins have diverged extensively (Figure 1) (Nasrallah 2000). In gametophytic self-incompatibility in the *Solanaceae*, the pollen component has yet to be identified, but the stylar product of the self-incompatibility gene encodes an extracellular RNase encoded by the *S* locus. However, the single S-allele RNase is sufficient to determine self-incompatibility (Richman & Kohn 2000), indicating it may function as a receptor and ligand in a manner analogous to the *Euplotes* mating pheromone. *S* alleles can differ by 50% in amino acid identity within the same species and can show a clear signature of positive Darwinian selection by

TABLE 2   Example reproductive proteins with overall d_N/d_S ratios >1[a]

Gene	d_N	d_S	d_N/d_S
Protamine			
Hsp-OWM	13.1	4.6	2.89
Lysin			
Hr-Hs	5.8	1.8	3.63
sp18			
Hr-Hs	8.1	1.8	4.50
TMAP			
Tb-Tr	17.6	5.0	3.51
S-alleles			
within Pc	40.2	16.1	2.49
Acp26Aa			
Dm-Dy	47.7	30.2	1.57

[a]Data from: protamine (Wyckoff et al. 2000); lysin and sp18 (Vacquier et al. 1997); TMAP (Hellberg et al. 2000); S-alleles (Richman et al. 1996); and Acp26Aa (Tsaur & Wu 1997). Species abbreviations are: *Homo sapiens* (Hsp), Old world monkey (OWM), *Haliotis sorenseni* (Hs), *H. rufescens* (Hr), *Tegula brunnea* (Tb), *T. regina* (Tr), *Physalis crassifolia* (Pc), *Drosophila melanogaster* (Dm), and *D. yakuba* (Dy).

having d_N/d_S ratios >1 (Table 2) (Richman & Kohn 2000). This indicates that there is an advantage for sequence diversity at this locus.

Components of the pollen coat from *Arabidopsis thaliana* also show extensive variability (Mayfield et al. 2001). A recent study identified all abundant pollen coat proteins greater than 10 kDa from *Arabidopsis* by protein sequencing and comparison with the completed genome. A total of 10 proteins were isolated, several of which were duplicated genes arranged in clusters. For example, five oleosin genes were found that clustered together in both *A. thaliana* and *Brassica oleracea*. Like the diatom gene *Sig1*, there was extensive intraspecific, interspecific, and inter-individual sequence variation (Mayfield et al. 2001).

Immediately before fertilization, sperm of marine gastropods of the genus *Haliotis* (abalone) and *Tegula* (turban snail) release a soluble protein called lysin onto the surface of the egg vitelline envelope (VE). In a species-specific, nonenzymatic process, lysin dissolves a hole in the VE through which the sperm passes to reach the egg cell membrane. The amino acid sequences of lysins from different species are extremely divergent, which has resulted from adaptive evolution (Hellberg & Vacquier 1999, Metz et al. 1998b, Yang et al. 2000b). Pairwise comparisons show d_N/d_S ratios significantly greater than one (Table 2). Additionally, site-specific analyses using maximum likelihood ratio tests (Yang et al. 2000a) show that a subset of sites on the surface of lysin have been the target of natural selection (Yang & Swanson 2002, Yang et al. 2000b). Some of these sites, such as the hypervariabe N-terminal, have been implicated in the species-specific function of lysin-mediated VE dissolution by site directed mutagenesis (Lyon & Vacquier 1999). Sites predicted to be under positive selection from five California abalone species are located primarily on the N- and C-terminal regions (Figure 2). Although the driving force behind this rapid evolution is not yet clear, it has been suggested that the rapid diversification of sperm lysin is driven by the need for its adaptation to a constantly changing egg receptor that is evolving neutrally (Swanson et al. 2001a, Swanson & Vacquier 1998). The VE receptor for lysin, VERL, is known from several closely related species of the abalone. Having been released from the acrosome, lysin binds to VERL molecules of the egg vitelline envelope in a species-specific manner (Kresge et al. 2001a, Swanson & Vacquier 1997). The fibrous VERL molecules lose their cohesion and splay apart, creating a hole through which the sperm swims (Kresge et al. 2001a, Swanson & Vacquier 1998). At ~1 million Daltons, VERL is a large glycoprotein that contains 22 tandem repeats of a ~153–amino acid sequence. In contrast to lysin, VERL shows no evidence of positive Darwinian selection; instead, it appears to be evolving neutrally. VERL repeat d_N/d_S ratios are less than one, and there is no evidence for a class of sites subjected to positive Darwinian selection using likelihood ratio tests (Swanson et al. 2001a). Additionally, several tests of neutrality based on polymorphism surveys also failed to reject equilibrium neutral expectations (Swanson et al. 2001a). The tandem VERL repeats show high levels of DNA sequence identity (>95%) within a molecule, indicating the repeats are subject to concerted evolution (Swanson & Vacquier 1998). Concerted evolution is the process in which unequal crossing over and gene conversion randomly homogenize a sequence of tandem repeats within the gene and within a population, a mechanism exemplified by ribosomal genes (Elder & Turner 1995, McAllister & Werren 1999). The end result is that the repeats within a molecule from one species are more similar to each other than they are to the homologous repeats from other species.

Abalone sperm also release a protein (sp18) that is thought to mediate the fusion of the sperm and egg (Swanson & Vacquier 1995b). In five Californian abalone species, sp18 proteins are up to 73% different at the amino acid sequence level (Swanson & Vacquier 1995a) and there is evidence that this protein might evolve up to 50 times faster than the fastest evolving mammalian proteins (Metz et al. 1998b). Like lysin, estimates of d_N/d_S ratios show evidence for positive Darwinian selection with d_N/d_S ratios as high as 4.7 averaged across the entire molecule (Table 2). Likelihood ratio tests also indicate a subset of sites subjected to positive selection, which when placed on the 3-D structure of sp18, fall near the top and bottom of the molecule (Figure 1). Comparison of sp18 to the database using BlastP or Psi-Blast reveals no similarity to other proteins. However, more detailed analyses detected similarity to abalone lysin (Swanson & Vacquier 1995a), which was later confirmed by intron mapping (Metz et al. 1998a) and determination of the three-dimensional structure of both molecules (Kresge et al. 2001b). Figure 2 shows the 3D structure of both molecules. The similarity of the alpha-helical arrangement

sp18 front

sp18 back

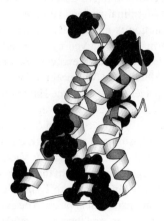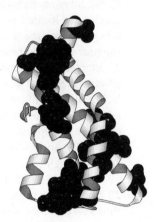

lysin front

lysin back

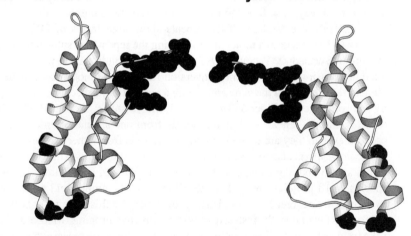

Figure 2   Three-dimensional structures of abalone sperm lysin and sp18. The lysin structure is from the red abalone (Kresge et al. 2000a), and the structure of sp18 is from the green abalone (Kresge et al. 2000b). The black residues shown in spacefill (displaying the size of the residue side-chains) are the sites predicted to be subjected to positive Darwinian selection using maximum likelihood methods. Only sites with a posterior probability greater than 0.95 are shown. The backbone alpha-helices of the two molecules are strikingly similar, indicating the two proteins arose by a gene duplication event. However, the angles of the helices are quite different, perhaps indicating selection for different functions.

is apparent. However, the angles of the helices are significantly different resulting in a root mean square deviation of 6.45 angstroms for the comparison of the alpha-carbon traces of the two molecules. The structural high similarity between the two molecules indicates that the two abalone fertilization proteins arose by gene duplication. In addition to lysin and sp18, *Tegula* sperm also release a major acrosomal protein of unknown function that is highly divergent and subject to adaptive evolution (Table 2) (Hellberg et al. 2000).

Rapid, extensive evolution of reproductive proteins has also been seen in sea urchins, the sperm of which use a protein called bindin to attach to the egg surface and possibly to fuse with the egg cell membrane (Vacquier et al. 1995). *Echinometra mathaei* and *E. oblonga* are two sympatric sea urchin species that live in the Pacific, and based on the mitochondrial DNA sequence comparisons, they are the most closely related sea urchin species from all known urchin species. Because adhesion of bindin to eggs has evolved to be species-specific (Palumbi & Metz 1991) few interspecies hybrids are formed. Bindin sequences show remarkable divergence both within and between *Echinometra* species (Metz & Palumbi 1996), as well as between species of another sea urchin genus, *Strongylocentrotus* (Biermann 1998). In both *Echinometra* and *Strongylocentrotus* bindin (Biermann 1998), a region with elevated d_N/d_S ratio has been identified as a target of positive selection. The exact function of this region remains unknown, but it might be involved in the species-specific adhesion of sperm to eggs.

Nonmarine invertebrates also show rapid adaptive evolution of reproductive proteins, and the accessory gland proteins of *Drosophila* are the best characterized example (Partridge & Hurst 1998, Wolfner 1997). During copulation, an estimated 83 protein products of the *Drosophila* male accessory glands (Swanson et al. 2001b) are transferred along with sperm to the female reproductive tract (Wolfner 1997). These proteins have a variety of effects on the physiology of the mated female (Wolfner 1997). It has been shown that the accessory gland proteins are twice as diverse between species as are the nonreproductive proteins (Civetta & Singh 1995, Singh & Kulathinal 2000). Although DNA analysis confirms this two-fold increase in the rate of amino acid replacement between species (Swanson et al. 2001b), the molecular evolution of only a few accessory gland proteins has been studied in detail. In particular, the accessory gland protein *Acp26Aa* is one of the fastest evolving genes in the *Drosophila* genome, with a d_N/d_S ratio of 1.6 between *D. melanogaster* and *D. yakuba* (Table 2). This indicates that its evolution is driven by positive Darwinian selection (Tsaur et al. 1998, Tsaur & Wu 1997). Other accessory gland proteins that show signs of positive selection include *Acp36DE* (Begun et al. 2000) and *Acp29AB* (Aguade 1999). The divergence of accessory gland proteins has been shown to be partly responsible for species-specific usage of gametes in some *Drosophila* species, where injections of conspecific accessory gland extracts can rescue an infertile hybrid cross (Fuyama 1983).

The rapid, adaptive evolution of reproductive proteins and species-specific fertilization is not limited to invertebrates, as similar phenomena have also been described in mammals (Swanson et al. 2001c, Wyckoff et al. 2000). This was first demonstrated in mammalian protamines (Rooney et al. 2000, Wyckoff et al.

2000). An excess of amino acid replacements was observed in primates, which was also accompanied by low amino acid polymorphism within species (Wyckoff et al. 2000). Comparison between humans and Old World monkeys showed d_N/d_S ratios >1 (Table 2). Adaptive evolution has also been observed in mammalian zona pellucida egg coat proteins ZP2 and ZP3 by using maximum likelihood ratio tests to identify a subset of sites with a d_N/d_S ratios greater than one (Swanson et al. 2001c). The zona pellucida is an elevated glycoproteinaceous envelope that protects the mammalian egg (Wassarman et al. 2001). The zona pellucida glyco-proteins bind sperm and induce the acrosome reaction, one of the first steps in mammalian fertilization. A region of ZP3 that was previously implicated in the species-specific induction of the acrosome reaction (Chen et al. 1998) was identi-fied to be the target of positive selection by predicting the sites subjected to positive selection using an empirical Bayes approach (Swanson et al. 2001c).

NOT ALL REPRODUCTIVE PROTEINS EVOLVE RAPIDLY

We have stressed the generality of rapidly evolving reproductive proteins above. However, it is clear that not all reproductive proteins show rapid evolution. Addi-tionally, reproductive proteins may evolve rapidly in some lineages, but be con-served in other lineages. This was exemplified by a study of the sea urchin protein bindin (Metz et al. 1998a). As detailed above, bindin was shown to be subjected to positive Darwinian selection from an analysis of species in the genera *Echinometra* and *Strongylocentrotus*. However, a study of bindin in the genus *Arbacia* showed no signs of positive Darwinian selection. In fact, bindin sequences are quite con-served among species of *Arbacia*. Consistent with this observation, two different species, living in different oceans, are interfertile. Why is bindin so conserved in the genus *Arbacia*? One possibility is that the four species of *Arbacia* examined are all allopatric, so there is no chance for hybridization. Species of the genera *Echinometra* and *Strongylocentrotus* have overlapping habitats with congeneric species, which may increase the chance of hybridization. Alternatively, bindin from *Arbacia* species may be under increased functional constraint as indicated by a lack of indels and acquisition of a hydrophobic domain not found in bindin from other genera (Metz et al. 1998a). Further comparisons of nonrapidly evolving versus rapidly evolving reproductive proteins, may shed light on the forces affect-ing their evolution. For example, are there some characteristics (either ecological or molecular) that are associated with rapidly evolving reproductive proteins?

WHAT DRIVES THE EVOLUTION OF REPRODUCTIVE PROTEINS?

The selective pressure driving the evolution of reproductive proteins remains un-known. However, several hypotheses have been proposed. Below, we list some of these hypotheses and discuss how they may apply to the systems described above. We stress that no single selective pressure may account for all the examples of rapidly evolving reproductive proteins. Additionally, several of these selective pressures may be acting simultaneously. Finally, we note that many of the hypotheses have overlapping predictions and could be considered a subset of each other. For example, sperm competition may be intimately coupled with sexual conflict in some situations (Rice & Holland 1997).

Relaxed Constraint

It is possible that the repetitive nature of some reproductive proteins leads to relaxed constraint and neutral evolution. This would produce a continually changing sequence. For example, the abalone VERL protein may change as a result of a mutation that occurs in one of the 22 VERL repeats. This change might result in a lower affinity of the mutant repeat for lysin, but the mutant repeat is tolerated and fertilization occurs because there are still 21 unchanged VERL repeats in each VERL molecule. Thus, the redundant nature of VERL leads to relaxed selection on each repeat unit, such that mutations do not have any fitness consequences, being neither beneficial, nor harmful. Such tolerance has been suggested for gamete recognition in sea urchins (Metz & Palumbi 1996). In successive generations, concerted evolution randomly spreads the mutant repeat within the VERL gene by unequal crossing over and gene conversion (Elder & Turner 1995, McAllister & Werren 1999, Swanson & Vacquier 1998). Sperm would be selected for that interact most efficiently with the new forms of VERL, because there would be enormous competition between sperm to be the first to fertilize the egg. Additionally, because specificity of sperm-egg interaction is not absolute, the eggs will most likely be fertilized if there is sperm around—which may further reduce the selective pressure on eggs. This creates a continuous selective pressure on lysin to adapt to the ever-changing VERL, and provides an explanation for the adaptive evolution of lysin (Metz et al. 1998b, Yang et al. 2000b). In other organisms there are also repeated, redundant, reproductive genes within the genome—such as in the diatom *Sig1* (Armbrust & Galindo 2001) gene and the *Arabidopsis* pollen coat proteins (Mayfield et al. 2001). It may be that positive Darwinian selection will be found to enhance diversity at these loci, which would refute this hypothesis in certain cases.

Reinforcement

If two allopatric populations come back in contact, hybrids may be formed if the reproductive recognition system has not diverged. If the hybrid mating results in less fit offspring owing to genomic incompatibilities, there would be selection to favor differentiation of the reproductive recognition system (Dobzhansky 1940, Howard 1993). This process has been referred to as reinforcement, and could lead to the rapid evolution of proteins mediating reproduction. It should be possible to test this hypothesis by comparisons of reproductive proteins from allopatric and sympatric population/species. The finding that bindin from sea urchins of the genus

Arbacia are not rapidly evolving may be consistent with reinforcement, because this represents a clear case of allopatric speciation. However, other scenarios could also be consistent with this observation (Metz et al. 1998a).

Gene Duplication

One hypothesis for the rapid, adaptive evolution of these molecules could be selection for new function. This may have occurred following the gene duplication event that gave rise to abalone sp18 and lysin. Presumably, there was an ancestral mollusk that had one sperm acrosomal protein. This protein may have performed two tasks—dissolution of the egg VE and fusion between the sperm and egg. Following a gene duplication event, the two molecules could have specialized—sp18 for fusion, and lysin for VE dissolution. Currently, lysin dissolves the VE, but sp18 has no effect on the integrity of the VE. Both molecules are capable of fusing artificial liposomes, but sp18 is a much more potent fusagen and localizes to the acrosomal process—the region of the abalone sperm that fuses with the egg (Swanson & Vacquier 1995b). However, we feel this would not lead to the continual evolution that has been observed. Rather it should lead to a burst of positive selection followed by purifying selection to maintain function. It is of interest that several of the reproductive proteins described above are members of duplicated gene families. It will be of interest to determine if reproductive proteins arise from gene duplication events more frequently than nonreproductive proteins.

Sperm Competition

Sperm competition (Clark et al. 1999) occurs because every sperm competes with all the other sperm to be the first to fuse with the egg. This competition can be fierce; for example, in the male sea urchin there are 200 billion sperm cells per five milliliters of semen (a typical amount of semen released in a spawning event). Sperm competition can exert a selective pressure at many steps in the fertilization cascade. Individual sperm could be selected for being the best or the fastest at initiating and maintaining swimming, responding to chemoattractants that diffuse from the egg, binding to the egg's surface, binding to the egg components that induce the acrosome reaction, penetrating the egg envelope, and fusing with the egg. Sperm competition can also occur in internal fertilizers when the female mates multiple times. In many cases, the majority of offspring are sired by one of the males with whom the female mated. Sperm precedence is a similar but distinct mechanism that occurs when females preferentially utilize sperm from homo-specific males (Howard 1999). One question regarding sperm competition and sperm precedence is the involvement of the female. Genetic studies have shown that female genotype can play an important role in sperm competition (Clark & Begun 1998, Clark et al. 1999, Price 1997). What provides the selective force for the female reproductive proteins to change? Possibilities include neutrally drifting targets as mentioned above for abalone VERL, or perhaps sexual conflict as discussed below.

Sexual Conflict

Sexual conflict could come into play when sperm cells are too abundant (Gavrilets 2000, Rice & Holland 1997). Sperm competition presents some problems for the egg. For example, fusion with multiple sperm (polyspermy) will result in arrest of the zygote's development. How do eggs prevent polyspermy? Eggs can prevent multiple fusions by a variety of methods. For example, some eggs have a fast block to polyspermy that is accomplished by reversing the electrical potential of the egg membrane that prevents fusion with additional sperm (Gould-Somero & Jaffe 1984). Additionally, some eggs undergo a slow block to polyspermy by altering the egg receptors to prevent further sperm-egg interaction. It is intriguing that in organisms like mammals with only a slow block to polyspermy (Gould-Somero & Jaffe 1984, Jaffe et al. 1983), we see adaptive evolution in egg coat proteins. However, in organisms that have a fast block to polyspermy such as abalone (Gould-Somero & Jaffe 1984) we see relaxed selection on the egg coat proteins. It may be that adaptive evolution is observed in mammals to slow down fertilization, allowing the egg to more efficiently regulate sperm entry to prevent polyspermy. However, in organisms with fast blocks to polyspermy this selective pressure is absent resulting in neutral evolution of egg coat proteins.

Sexual Selection

Sexual selection at the cellular level is known as cryptic female choice (Eberhard 1996). Cryptic female choice might come into play when one egg preferentially binds with a sperm that carries a particular allele of a sperm surface protein, as opposed to the alleles carried by other sperm. One example of assortative female choice is the preference of *Echinometra* eggs to be fertilized by sperm that carry the same bindin allele as they do (Palumbi 1999). However, the question remains why females might change their preference. This question is being explored at the organismal level, using models of Fisherian and/or good genes sexual selection (Iwasa & Pomiankowski 1995) and empirical evidence of preference change in response to environmental factors. Could some of these hypotheses be appropriate at the cellular level?

Microbial Attack

It has been proposed that eggs may be subjected to invasion by pathogens, particularly in marine invertebrates where the eggs settle into microrganism-rich marine sediments to undergo development (Vacquier et al. 1997). In such cases, there may be selective pressure to change egg coats to prevent invasion by pathogens. This would lead to a continual selective pressure on sperm to evolve to keep up with the ever-changing egg surface. This could also occur in internal fertilizers, perhaps mediated by sexually transmitted microbial diseases. The potential for parasites to impact mating preferences has been long discussed in a variety of systems (Hamilton & Zuk 1989).

Self-/Non-Self-Recognition

Molecules involved in self-recognition often show extreme variability within a species. It is possible that molecules involved in reproduction could also have been involved in self-recognition. For example, the *Euplotes* mating pheromone is involved in both self and non-self-recognition (Luporini et al. 1995), and the differing conditions result in either induction of cell growth or sexual mating. In plants, the selective pressure to prevent self-fertilization is thought to be one mechanism by which great diversity arises at the *SCR* (Schopfer et al. 1999) and *S*-allele loci (Richman & Kohn 2000).

Within-Population Variation

The rapid evolution between species discussed above suggests that there may be interesting evolutionary dynamics occurring at these loci within a species. Several population surveys have been performed on reproductive proteins, and the results can be quite different. For example, the abalone sperm proteins lysin and sp18 show signs of selective sweeps with little or no variation within a species (Metz et al. 1998b). In contrast, the sea urchin sperm protein bindin shows extensive polymorphism within species (Metz & Palumbi 1996). High divergence accompanied by high polymorphism is also observed in the *Drosophila* accessory gland proteins (Begun et al. 2000). An intriguing question is whether the extensive polymorphism and divergence of some reproductive proteins could be involved in speciation.

Population surveys of abalone fertilization are starting to hint at the possibility of assortative mating based upon reproductive loci. The last abalone egg VERL repeat in the repeat array of 22 repeats was identified and sequenced from 11 sympatric pink abalone (*H. corrugata*) and two different variants of the last VERL repeat sequences were found (Figure 3). These types are different by several amino acids as well as a large indel. Five individuals were homozygous for each of the two variants, and only one was heterozygous for both variants (Swanson et al. 2001a). The small number of heterozygotes indicates that assortative mating may take place in the same population of pink abalone, or that there is postmating selection against heterozygotes. Theoretically, this molecular differentiation could eventually lead to a sympatric speciation event—the splitting of the current pink abalone population into two new species. Larger samples of abalone need to be analyzed before we can say whether this prediction will be borne out.

A similar assortative mating phenomenon has also been found in the *Echinometra* sea urchins. Individual *E. mathaei* have two different alleles of bindin, and homozygotes for each variant can be distinguished by PCR and restriction mapping. Eggs of *E. mathaei* are preferentially fertilized by sperm that carry the same bindin allele (Palumbi 1999). This result indicates that the genes that encode bindin and its egg surface receptor might be linked and inherited as one unit, as occurs in reproductive gene pairs in fungi (Casselton & Olesnicky 1998) and plants (Schopfer & Nasrallah 2000, Schopfer et al. 1999). Quantitative fertilization specificity has also been documented in the sea urchin *Strongylocentrotus pallidus*

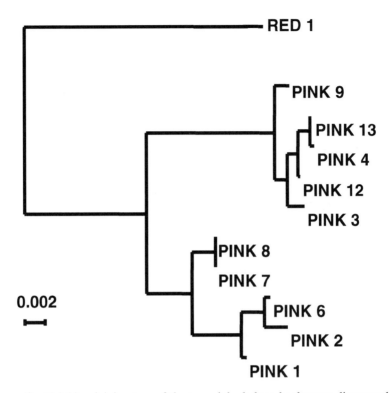

Figure 3 Neighbor-joining tree of the two pink abalone haplotypes discovered by sequencing the last VERL repeat from a population of pink abalone. The haplotypes are rooted with the VERL repeat from the red abalone. The separation of the two pink clades is supported by 100% bootstrap values. The scale bar shows total nucleotide divergence.

(Biermann 2000) and other *Echinometra* species (Rahman & Uehara 2000), indicating that the differentiation of the gamete recognition system might have a crucial role in reproductive isolation in many sea urchin genera.

CONCLUSIONS

In this review, we have demonstrated that the rapid evolution of reproductive proteins occurs in several taxonomic groups. In some instances, the rapid evolution is driven by positive Darwinian selection, but in other cases the patterns do not depart from neutrality (Table 1). This variation in selective pressure occurs at all taxonomic levels as well as varying temporally. The selective pressure driving the evolution of reproductive loci remains unknown. We have listed a few, nonexclusive hypotheses that several researchers are in the process of testing. Several of these hypotheses have overlapping predictions, making it difficult to distinguish between them. Undoubtedly, there are other selective pressures not discussed here that could drive the rapid evolution of reproductive proteins. Further studies on the evolution and function of reproductive proteins could provide answers to the nature of the selective pressure driving the evolution of reproductive proteins. We may also learn what the functional consequence of the rapid evolution could mean.

ACKNOWLEDGMENTS

We thank the National Institutes of Health and National Science Foundation for grants supporting this work: NIH Grant HD12896 to V.D.V. and NSF grant DEB-0111613 to W.J.S. Drs. C.F. Aquadro, M.F. Wolfner, J.D. Calkins, and J.P. Vacquier are thanked for their criticisms of the manuscript.

The *Annual Review of Ecology and Systematics* is online at
http://ecolsys.annualreviews.org

LITERATURE CITED

Aguade M. 1999. Positive selection drives the evolution of the Acp29AB accessory gland protein in Drosophila. *Genetics* 152:543–51

Anisimova M, Bielawski JP, Yang Z. 2001. Accuracy and power of the likelihood ratio test in detecting adaptive molecular evolution. *Mol. Biol. Evol.* 18:1585–92

Armbrust EV, Galindo HM. 2001. Rapid evolution of a sexual reproduction gene in centric diatoms of the genus *Thalassiosira. Appl. Environ. Microbiol.* 67:3501–13

Begun DJ, Whitley P, Todd BL, Waldrip-Dail HM, Clark AG. 2000. Molecular population genetics of male accessory gland proteins in *Drosophila. Genetics* 156:1879–88

Biermann CH. 1998. The molecular evolution of sperm bindin in six species of sea urchins (*Echinodia: Strongylocentrotidae*). *Mol. Biol. Evol.* 15:1761–71

Biermann CH. 2000. Geographic divergence of gamete recognition systems in two species in the sea urchin genus *Strongylocentrotus. Zygote* 8:S86–87

Brown AJ, Casselton LA. 2001. Mating in mushrooms: increasing the chances but prolonging the affair. *Trends Genet.* 17:393–400

Casselton LA, Olesnicky NS. 1998. Molecular genetics of mating recognition in basidiomycete fungi. *Microbiol. Mol. Biol. Rev.* 62:55–70

Chen J, Litscher ES, Wassarman PM. 1998. Inactivation of the mouse sperm receptor, mZP3, by site-directed mutagenesis of individual serine residues located at the combining site for sperm. *Proc. Natl. Acad. Sci. USA* 95:6193–97

Civetta A, Singh RS. 1995. High divergence of reproductive tract proteins and their association with postzygotic reproductive isolation in Drosophila melanogaster and Drosophila virilis group species. *J. Mol. Evol.* 41:1085–95

Clark AG, Begun DJ. 1998. Female genotypes affect sperm displacement in Drosophila. *Genetics* 149:1487–93

Clark AG, Begun DJ, Prout T. 1999. Female × male interactions in Drosophila sperm competition. *Science* 283:217–20

Dobzhansky T. 1940. Speciation as a stage in evolutionary divergence. *Am. Nat.* 312–21

Eberhard WG. 1996. *Female Control: Sexual Selection by Cryptic Female Choice.* Princeton, NJ: Princeton Univ. Press

Elder JF Jr, Turner BJ. 1995. Concerted evolution of repetitive DNA sequences in eukaryotes. *Q. Rev. Biol.* 70:297–320

Ferris PJ, Armbrust EV, Goodenough UW. 2002. Genetic structure of the mating-type locus of Chlamydomonas reinhardtii. *Genetics* 160:181–200

Ferris PJ, Pavlovic C, Fabry S, Goodenough UW. 1997. Rapid evolution of sex-related genes in Chlamydomonas. *Proc. Natl. Acad. Sci. USA* 94:8634–39

Ferris PJ, Woessner JP, Goodenough UW. 1996. A sex recognition glycoprotein is encoded by the plus mating-type gene fus1 of Chlamydomonas reinhardtii. *Mol. Biol. Cell* 7:1235–48

Fuyama Y. 1983. Species-specificity of paragonial substances as an isolating mechanism in Drosophila. *Experientia* 39:190–92

Gao Z, Garbers DL. 1998. Species diversity in the structure of zonadhesin, a sperm-specific membrane protein containing multiple cell adhesion molecule-like domains. *J. Biol. Chem.* 273:3415–21

Gavrilets S. 2000. Rapid evolution of reproductive barriers driven by sexual conflict. *Nature* 403:886–89

Gould-Somero M, Jaffe LA. 1984. Control of cell fusion at fertilization by membrane potential. In *Cell Fusion and Transformation*, ed. RF Beers, EG Bassett, pp. 27–38. New York: Raven

Hamilton WD, Zuk M. 1989. Parasites and sexual selection. *Nature* 341:289–90

Hellberg ME, Moy GW, Vacquier VD. 2000. Positive selection and propeptide repeats promote rapid interspecific divergence of a gastropod sperm protein. *Mol. Biol. Evol.* 17:458–66

Hellberg ME, Vacquier VD. 1999. Rapid evolution of fertilization selectivity and lysin cDNA sequences in teguline gastropods. *Mol. Biol. Evol.* 16:839–48

Howard DJ. 1993. *Reinforcement: Origin, Dynamics, and Fate of an Evolutionary Hypothesis*, pp. 46–69. Oxford, England: Oxford Univ. Press

Howard DJ. 1999. Conspecific sperm and pollen precedence and speciation. *Annu. Rev. Ecol. Syst.* 30:109–32

Hughes AL, Nei M. 1988. Pattern of nucleotide substitution at major histocompatibility complex class I loci reveals overdominant selection. *Nature* 335:167–70

Iwasa Y, Pomiankowski A. 1995. Continual changes in mate preferences. *Nature* 377:420–22

Jaffe LA, Sharp AP, Wolf DP. 1983. Absence of an electrical polyspermy block in the mouse. *Dev. Biol.* 96:317–23

Juneja R, Agulnik SI, Silver LM. 1998. Sequence divergence within the sperm-specific polypeptide TCTE1 is correlated with species-specific differences in sperm binding to zona-intact eggs. *J. Androl.* 19:183–88

Kachroo A, Schopfer CR, Nasrallah ME, Nasrallah JB. 2001. Allele-specific receptor-ligand interactions in *Brassica* self-incompatibility. *Science* 293:1824–26

Kresge N, Vacquier VD, Stout CD. 2000a. 1.35 and 2.07 A resolution structures of the red abalone sperm lysin monomer and dimer reveal features involved in receptor binding. *Acta Crystallogr. D Biol. Crystallogr.* 56:34–41

Kresge N, Vacquier VD, Stout CD. 2000b. The high resolution crystal structure of green abalone sperm lysin: implications for species-specific binding of the egg receptor. *J. Mol. Biol.* 296:1225–34

Kresge N, Vacquier VD, Stout CD. 2001a. Abalone lysin: the dissolving and evolving sperm protein. *Bioessays* 23:95–103

Kresge N, Vacquier VD, Stout CD. 2001b. The crystal structure of a fusagenic sperm protein reveals extreme surface properties. *Biochemistry* 40:5407–13

Lee YH, Ota T, Vacquier VD. 1995. Positive selection is a general phenomenon in the evolution of abalone sperm lysin. *Mol. Biol. Evol.* 12:231–38

Li W-H. 1997. *Molecular Evolution*. Sunderland, MA: Sinauer

Luporini P, Vallesi A, Miceli C, Bradshaw RA. 1995. Chemical signaling in ciliates. *J. Eukaryot. Microbiol.* 42:208–12

Lyon JD, Vacquier VD. 1999. Interspecies chimeric sperm lysins identify regions mediating species-specific recognition of the abalone egg vitelline envelope. *Dev. Biol.* 214:151–59

Makalowski W, Boguski MS. 1998. Evolutionary parameters of the transcribed mammalian genome: an analysis of 2,820 orthologous rodent and human sequences. *Proc. Natl. Acad. Sci. USA* 95:9407–12

Mayfield JA, Fiebig A, Johnstone SE, Preuss D. 2001. Gene families from the Arabidopsis thaliana pollen coat proteome. *Science* 292:2482–85

McAllister BF, Werren JH. 1999. Evolution of tandemly repeated sequences: What happens at the end of an array? *J. Mol. Evol.* 48:469–81

Metz EC, Gomez-Gutierrez G, Vacquier VD. 1998a. Mitochondrial DNA and bindin gene sequence evolution among allopatric species of the sea urchin genus Arbacia. *Mol. Biol. Evol.* 15:185–95

Metz EC, Palumbi SR. 1996. Positive selection and sequence rearrangements generate extensive polymorphism in the gamete recognition protein bindin. *Mol. Biol. Evol.* 13:397–406

Metz EC, Robles-Sikisaka R, Vacquier VD. 1998b. Nonsynonymous substitution in abalone sperm fertilization genes exceeds substitution in introns and mitochondrial DNA. *Proc. Natl. Acad. Sci. USA* 95:10676–81

Nasrallah JB. 2000. Cell-cell signaling in the self-incompatibility response. *Curr. Opin. Plant Biol.* 3:368–73

Nielsen R, Yang Z. 1998. Likelihood models for detecting positively selected amino acid sites and applications to the HIV-1 envelope gene. *Genetics* 148:929–36

Ortenzi C, Alimenti C, Vallesi A, Di PB, La TA, Luporini P. 2000. The autocrine mitogenic loop of the ciliate Euplotes raikovi: The pheromone membrane-bound forms are the cell binding sites and potential signaling receptors of soluble pheromones. *Mol. Biol. Cell* 11:1445–55

Palumbi SR. 1999. All males are not created equal: Fertility differences depend on gamete recognition polymorphisms in sea urchins. *Proc. Natl. Acad. Sci. USA* 96:12632–37

Palumbi SR, Metz EC. 1991. Strong reproductive isolation between closely related tropical sea urchins (Genus *Echinometra*). *Mol. Biol. Evol.* 8:227–39

Partridge L, Hurst LD. 1998. Sex and conflict. *Science* 281:2003–8

Price CS. 1997. Conspecific sperm precedence in Drosophila. *Nature* 388:663–66

Rahman MA, Uehara T. 2000. Experimental hybridization between two tropical species of sea urchins (genus *Echinometra*) in Okinawa. *Zygote* 8:S90

Rice WR, Holland B. 1997. The enemies within: intergenomic conflict, interlocus contest evolution (ICE), and the intraspecific Red Queen. *Behav. Ecol. Sociobiol.* 41:1–10

Richman AD, Kohn JR. 2000. Evolutionary genetics of self-incompatibility in the Solanaceae. *Plant Mol. Biol.* 42:169–79

Richman AD, Uyenoyama MK, Kohn JR. 1996. Allelic diversity and gene genealogy at the self-incompatibility locus in the Solanaceae. *Science* 273:1212–16

Rooney AP, Zhang J, Nei M. 2000. An unusual form of purifying selection in a sperm protein. *Mol. Biol. Evol.* 17:278–83

Schmid KJ, Tautz D. 1997. A screen for fast evolving genes from Drosophila. *Proc. Natl. Acad. Sci. USA* 94:9746–50

Schopfer CR, Nasrallah JB. 2000. Self-incompatibility. Prospects for a novel putative peptide-signaling molecule. *Plant Physiol.* 124:935–40

Schopfer CR, Nasrallah ME, Nasrallah JB. 1999. The male determinant of self-incompatibility in Brassica. *Science* 286:1697–700

Singh RS, Kulathinal RJ. 2000. Sex gene pool evolution and speciation: a new paradigm. *Genes Genet. Syst.* 75:119–30

Swanson WJ, Aquadro CF, Vacquier VD. 2001a. Polymorphism in abalone fertilization proteins is consistent with the neutral evolution of the egg's receptor for lysin (VERL) and positive Darwinian selection of sperm lysin. *Mol. Biol. Evol.* 18:376–83

Swanson WJ, Clark AG, Waldrip-Dail HM, Wolfner MF, Aquadro CF. 2001b. Evolutionary EST analysis identifies rapidly evolving male reproductive proteins in Drosophila. *Proc. Natl. Acad. Sci. USA* 98:7375–79

Swanson WJ, Vacquier VD. 1995a. Extraordinary divergence and positive Darwinian selection in a fusagenic protein coating the acrosomal process of abalone spermatozoa. *Proc. Natl. Acad. Sci. USA* 92:4957–61

Swanson WJ, Vacquier VD. 1995b. Liposome fusion induced by a M(r) 18,000 protein localized to the acrosomal region of acrosome-reacted abalone spermatozoa. *Biochemistry* 34:14202–8

Swanson WJ, Vacquier VD. 1997. The abalone egg vitelline envelope receptor for sperm lysin is a giant multivalent molecule. *Proc. Natl. Acad. Sci. USA* 94:6724–29

Swanson WJ, Vacquier VD. 1998. Concerted evolution in an egg receptor for a rapidly evolving abalone sperm protein. *Science* 281:710–12

Swanson WJ, Vacquier VD. 2002. Rapid evolution of reproductive proteins. *Nat. Rev. Genetics* 3:137–44

Swanson WJ, Yang Z, Wolfner MF, Aquadro CF. 2001c. Positive Darwinian selection drives the evolution of several female reproductive proteins in mammals. *Proc. Natl. Acad. Sci. USA* 98:2509–14

Ting C-T, Tsaur S-C, Wu M-L, Wu CI. 1998. A rapidly evolving homeobox at the site of a hybrid sterility gene. *Science* 282:1501–4

Tsaur SC, Ting CT, Wu CI. 1998. Positive selection driving the evolution of a gene of male reproduction, Acp26Aa, of Drosophila: II. Divergence versus polymorphism. *Mol. Biol. Evol.* 15:1040–46

Tsaur SC, Wu CI. 1997. Positive selection and the molecular evolution of a gene of male reproduction, Acp26Aa of Drosophila. *Mol. Biol. Evol.* 14:544–49

Vacquier VD, Swanson WJ, Hellberg ME. 1995. What have we learned about sea urchin sperm bindin? *Dev. Growth Differ.* 37:1–10

Vacquier VD, Swanson WJ, Lee YH. 1997. Positive Darwinian selection on two homologous fertilization proteins: What is the selective pressure driving their divergence? *J. Mol. Evol.* 44:S15–22

Wassarman PM, Jovine L, Litscher ES. 2001. A profile of fertilization in mammals. *Nat. Cell Biol.* 3:59–64

Wolfner MF. 1997. Tokens of love: functions and regulation of Drosophila male accessory gland products. *Insect Biochem. Mol. Biol.* 27:179–92

Wyckoff GJ, Wang W, Wu CI. 2000. Rapid evolution of male reproductive genes in the descent of man. *Nature* 403:304–9

Yang Z, Bielawski JP. 2000. Statistical methods for detecting molecular adaptation. *Trends Ecol. Evol.* 15:496–503

Yang Z, Nielsen R, Goldman N, Pedersen AM. 2000a. Codon-substitution models for heterogeneous selection pressure at amino acid sites. *Genetics* 155:431–49

Yang Z, Swanson WJ. 2002. Codon-substitution models to detect adaptive evolution that account for heterogeneous selective pressures among site classes. *Mol. Biol. Evol.* 19:49–57

Yang Z, Swanson WJ, Vacquier VD. 2000b. Maximum-likelihood analysis of molecular adaptation in abalone sperm lysin reveals variable selective pressures among lineages and sites. *Mol. Biol. Evol.* 17:1446–55

Annu. Rev. Ecol. Syst. 2002. 33:181–233
doi: 10.1146/annurev.ecolsys.33.010802.150444

THE CAUSES AND CONSEQUENCES OF ANT INVASIONS

David A. Holway,[1] Lori Lach,[2] Andrew V. Suarez,[3] Neil D. Tsutsui,[4] and Ted J. Case[1]

[1]Section of Ecology, Behavior, and Evolution, Division of Biological Sciences,
University of California, San Diego, La Jolla, California 92093;
email: dholway@ucsd.edu, case@biomail.ucsd.edu
[2]Department of Ecology and Evolutionary Biology, Cornell University, Ithaca,
New York 14853; email: ljl13@cornell.edu
[3]Department of Environmental Science, Policy and Management & Division of
Insect Biology, University of California, Berkeley, California 94720-3112;
email: asuarez@nature.berkeley.edu
[4]Center for Population Biology & Section of Evolution and Ecology, University
of California, Davis, One Shields Ave., Davis, California 95616;
email: ndtsutsui@ucdavis.edu

Key Words ants, biological invasion, indirect effects, interspecific competition

■ **Abstract** Invasions by non-native ants are an ecologically destructive pheno-menon affecting both continental and island ecosystems throughout the world. Invasive ants often become highly abundant in their introduced range and can outnumber native ants. These numerical disparities underlie the competitive asymmetry between inva-sive ants and native ants and result from a complex interplay of behavioral, ecological, and genetic factors. Reductions in the diversity and abundance of native ants resulting from ant invasions give rise to a variety of direct and indirect effects on non-ant taxa. Invasive ants compete with and prey upon a diversity of other organisms, including some vertebrates, and may enter into or disrupt mutualistic interactions with numerous plants and other insects. Experimental studies and research focused on the native range ecology of invasive ants will be especially valuable contributions to this field of study.

INTRODUCTION

Ants play a diversity of roles in terrestrial ecosystems. Ants act as predators, scav-engers, herbivores, detritivores, and granivores (Hölldobler & Wilson 1990) and participate in an astonishing array of associations with plants and other insects (Beattie 1985, Hölldobler & Wilson 1990, Huxley & Cutler 1991, Jolivet 1996). Ants, in turn, are preyed upon by a variety of specialist predators, including reptiles (Pianka & Parker 1975), mammals (Redford 1987), spiders (Porter & Eastmond 1982), and insects (Gotelli 1996) and are host to both dipteran (Feener & Brown

1997) and hymenopteran parasitoids (Heraty 1994). Ants also serve as important agents of soil turnover, nutrient redistribution, and small-scale disturbance (Hölldobler & Wilson 1990, Folgarait 1998, MacMahon et al. 2000). For these reasons, and because they can be sampled and identified with relative ease, ants figure prominently in ecological studies and have become a key indicator group in studies of diversity and ecosystem function (Agosti et al. 2000). The widespread success of ants stems in large part from their elaborate social behavior, which is itself a tremendously rich source of information for studies of kin selection, reproductive skew, levels of selection, foraging behavior, and self-organization (Wilson 1971, Hölldobler & Wilson 1990, Keller 1993, Bourke & Franks 1995, Crozier & Pamilo 1996).

Invasive ants form a small and somewhat distinct subset of the at least 150 species of ants introduced into new environments by humans (McGlynn 1999a). A majority of introduced ants remain confined to human-modified habitats and some of these species are often referred to as tramp ants because of their reliance on human-mediated dispersal and close association with humans generally (Hölldobler & Wilson 1990, Passera 1994). Although also largely dependent on humans to reach new environments, invasive ants differ from most other introduced ants in several key respects. Invasive ants penetrate natural ecosystems where they often reduce native ant diversity and affect other organisms both directly and indirectly. A minor caveat regarding this definition is that species invading oceanic islands with few or no native ants may exhibit patterns of invasion different from those observed in regions with indigenous ants. In Hawaii, for example, species such as *Hypoponera opaciceps* and *Solenopsis papuana* have spread into undisturbed forest (Reimer 1994), and their occurrence in natural environments on this archipelago may be due in part to the lack of native ants there (Zimmerman 1970, Cole et al. 1992, Reimer 1994).

In this review, we focus on (*a*) the causes underlying the ecological success of invasive ants (especially their interactions with native ants), and (*b*) the direct and indirect effects that occur following invasion. This perspective differs from previous reviews on introduced ants, most of which are limited to *Solenopsis invicta* and have primarily addressed urban and agricultural impacts (Vinson 1986, 1997, Vander Meer et al. 1990a, Williams 1994, Taber 2000). Here, in contrast, we highlight studies conducted in more natural ecosystems. Given their broad and steadily increasing geographical range, high local abundance, and potential to disrupt ecosystems, invasive ants are an important conservation concern. This problem is particularly worrisome given that, once established, invasive ants have proven difficult to control and virtually impossible to eradicate. Obtaining a better understanding of the causes and consequences of ant invasions remains crucial to achieving the ultimate goal of reducing problems associated with these invaders and preventing the introduction of other species that possess similar characteristics. We also submit, as have others (Ross & Keller 1995, Tschinkel 1998, Chapman & Bourke 2001), that these introductions present unique opportunities for research in ecology, behavior, and evolution.

INVASIVE ANTS AND THEIR GENERAL CHARACTERISTICS

Table 1 lists characteristics of six of the most widespread, abundant, and damaging invasive ants. A great disparity exists with respect to how much is known about each of these species. For example, the red imported fire ant (*S. invicta*) ranks as one of the most well studied social insects (Ross & Keller 1995, Tschinkel 1998), whereas *Anoplolepis gracilipes* remains poorly studied by comparison, so much so that its native range is not even known. As an inevitable result of this difference, we devote more attention to the relatively well-known *S. invicta* and the Argentine ant (*Linepithema humile*) than to other species but caution against drawing the conclusion that other invasive ants resemble *L. humile* and *S. invicta* or pose less

TABLE 1   Characteristics of the six most widespread, abundant, and damaging invasive ants

Species	Subfamily[a]	Geographical Range[b] Native	Geographical Range[b] Introduced	Poly- morphic workers	Colony Structure[c] Native Range	Colony Structure[c] Introduced Range	Diet[d]
Anoplolepis gracilipes Long-legged ant, crazy ant	F	AF? (1), AS? (2)	AF, AS, AU, CAR, IO, PO (3)	N	?	U (4,5)	OM (4,5)
Linepithema humile Argentine ant	D	SA (6,7)	AF, AO, AS, AU, ME, NA, PO, SA (6)	N	M-V (8)	U (8)	OM (9–11)
Pheidole megacephala Big-headed ant	M	AF (1)	AF, AU, NA, CAR, IO, ME, PO, SA (3)	Y	?	U (5,12)	OM, GR (5,12)
Solenopsis invicta Red imported fire ant	M	SA (13)	CAR, NA (14), AU (15), NZ (16)	Y	V (17)	V (17,18)	OM (19)
Solenopsis geminata Tropical fire ant	M	CA, NA, SA (20)[e]	AF, AS, AU, PO, CAR (?), IO (21)	Y	V (22)	V (?)	OM, GR (19,23)
Wasmannia auropunctata Little fire ant	M	CA, SA (3)	AF, CAR, PO, SA, NA (3)	N	?	U (24)	OM (24)

[a]D = Dolichoderinae, F = Formicinae, M = Myrmicinae.

[b]AF = Africa (subsaharan), AO = Atlantic Ocean (islands), AS = Asia, AU = Australia, CA = Central America, CAR = Caribbean, IO = Indian Ocean (islands), ME = Mediterranean, NA = North America, PO = Pacific Ocean (islands), SA = South America.

[c]M = Multicolonial, U = Unicolonial, V = Variable (see text).

[d]OM = omnivorous, GR = granivorous.

[e]The native range of *S. geminata* is disputed, in part because the species is continuously distributed from the SE United States to northern South America. Some of these populations (including those in the Caribbean) may be the result of human introductions. See Ross et al. 1987 for more information.

1. Wilson & Taylor 1967; 2. Wheeler 1910; 3. McGlynn 1999a; 4. Haines & Haines 1978a; 5. Greenslade 1972; 6. Suarez et al. 2001; 7. Tsutsui et al. 2001; 8. Tsutsui et al. 2000; 9. Newell & Barber 1913; 10. Markin 1970a; 11. Human et al. 1998; 12. Hoffmann 1998; 13. Ross & Trager 1990; 14. Williams et al. 2001; 15. http://www.dpi.qld.gov.au/fireants/; 16. http://www.maf.govt.nz/biosecurity/pests-diseases/animals/fire-ants/; 17. Ross & Keller 1995; 18. Tschinkel 1998; 19. Tennant & Porter 1991; 20. Ross et al. 1987; 21. Taber 2000; 22. MacKay et al. 1990; 23. Torres 1984; 24. Clark et al. 1982.

serious threats. Our focus on the species listed in Table 1 is not meant to dismiss the potential importance of other, less well known or more locally distributed ants; some of these may emerge as problematic invaders in the future. In this section, we provide an overview of the general characteristics of invasive ants; we return to many of these same topics later in our review.

Invasive ants originate in both the New and Old Worlds but now occur in many areas outside their original ranges (Table 1). For all species, the climate of the introduced range approximately matches that of the native range. *L. humile*, for example, is native to sub-tropical and mild-temperate portions of Argentina and surrounding regions and is not known to invade tropical and cold-temperate regions but is widespread in areas with mild-temperate climates (e.g., Mediterranean-type ecosystems) (Suarez et al. 2001). *S. invicta*, also from central South America, likewise prevails as an invader primarily in areas of the southeastern United States with mild-temperate or subtropical climates (Tschinkel 1983, Korzukhin et al. 2001). The remaining invasive ants are from the tropics or subtropics and have primarily invaded regions with similar climates. In Hawaii, where several invasive ants now occur, tropical species (e.g., *Pheidole megacephala* and *A. gracilipes*) occur at low to mid elevations, while *L. humile* occurs at intermediate to high elevations (Fluker & Beardsley 1970, Reimer 1994). Cold-temperate climates appear unsuitable for the invasive ants listed in Table 1; in areas with such climates these ants occur only around human habitation (Ulloa-Chacon & Cherix 1990, Taber 2000, Suarez et al. 2001).

Invasive ants exhibit both phylogenetic and morphological diversity (Table 1). Although both the Ponerinae and Pseudomyrmicinae contain widespread tramps (McGlynn 1999a), the invasive ants listed in Table 1 derive from the three most species-rich subfamilies of ants: Dolichoderinae, Formicinae, and Myrmicinae. Although *Solenopsis*, *Wasmannia*, and *Pheidole* are all Myrmicines, these genera are placed in different tribes (Brown 2001). The fire ants, however, include three invasive species, *S. invicta*, *S. richteri*, and the widely distributed *S. geminata*. Compared to ants as a whole, invasive ants are small to medium-sized; workers range in length from 1–2 mm (*Wasmannia auropunctata*) to >5 mm (*A. gracilipes*). Across species, invasive ants also vary in the extent of physical polymorphism among workers, ranging from monomorphism (*L. humile*, *W. auropunctata*, *A. gracilipes*) to pronounced dimorphism (*P. megacephala*) and polymorphism (*Solenopsis*) (Table 1). Although McGlynn (1999b) argued that introduced ants are smaller than congeners that tend not to be introduced, he excluded *A. gracilipes* (monomorphic but with polymorphic congeners), *P. megacephala*, *S. invicta*, and *S. geminata* from his analysis because of caste polymorphism. It is unclear whether these ants are smaller than their non-invasive congeners.

A striking feature shared by many invasive ants is the tendency for introduced populations to be unicolonial, that is, to form expansive and polygynous (multiple-queened) supercolonies that lack distinct behavioral boundaries among physically separate nests. Unicoloniality appears to be over-represented among invasive ants (Table 1) given that only a tiny minority of ant species exhibits this form of colony structure. Hölldobler & Wilson (1977) suggested that unicoloniality allows species such as *L. humile*, *W. auropunctata*, and *P. megacephala* to attain high local abundances and consequently to dominate entire habitats. Like these species, introduced populations of *A. gracilipes* maintain populous supercolonies within which intraspecific aggression is largely absent (Haines & Haines 1978a). The situation for fire ants, however, is more complex. In North America, *S. invicta* colonies now occur in both monogyne and polygyne forms (Ross & Keller 1995, Tschinkel 1998), although it should be noted that the monogyne form appears responsible for the initial and rapid invasion of most of the southeastern United States. Monogyne colonies defend territories against neighboring conspecifics (Tschinkel et al. 1995, Adams 1998), whereas polygyne colonies exhibit reduced intraspecific aggression (Morel et al. 1990, Vander Meer et al. 1990b) and maintain high densities of interconnected nests (Bhatkar & Vinson 1987, Porter et al. 1988, Porter & Savignano 1990, Greenberg et al. 1992, Macom & Porter 1996). *Solenopsis geminata* also occurs in both monogynous and polygynous forms (Banks et al. 1973, Adams et al. 1976, MacKay et al. 1990, Williams & Whelan 1991). Although little is published concerning the extent to which polygynous *S. geminata* defends territories intraspecifically, like polygyne *S. invicta*, polygynous *S. geminata* can reach high densities (MacKay et al. 1990, Way et al. 1998).

Another characteristic shared by invasive ants is omnivory. Like many aboveground foraging ants (Hölldobler & Wilson 1990), invasive ants opportunistically scavenge dead animals, prey upon small invertebrates, and harvest carbohydrate-rich plant and insect exudates. Differences exist, of course, in the proportional representation of different food items in the diets of each species. *S. geminata* (Tennant & Porter 1991) and perhaps *P. megacephala* (Hoffmann 1998), for example, commonly include a high proportion of seeds in their diets. *S. geminata*, *S. invicta*, and *W. auropunctata* all possess a venomous sting that may give these species a greater ability to subdue vertebrate and large invertebrate prey. In Argentina, *L. humile* is commonly referred to as the sugar ant, a fitting name given its strong preference for sweet substances (Newell & Barber 1913). Because invasive ants feed extensively on liquid food (Markin 1970a, Tennant & Porter 1991, Human et al. 1998), it is often difficult to obtain an understanding of the composition and seasonal variability of the diets of invasive ants and how these diets might differ from those of native ants.

Other introduced species not listed in Table 1 also possess attributes of invasive ants, but are either poorly studied or currently exhibit localized distributions. Some of these species may become troublesome invaders. For example, Boomsma et al. (1990) and Van Loon et al. (1990) describe the species *Lasius neglectus* from urban areas of central Europe (see also Seifert 2000). *L. neglectus* resembles other invasive ants in that it forms populous, unicolonial supercolonies and appears to outcompete other ants, at least in urban environments. The black imported fire ant (*Solenopsis richteri*) was introduced into the southeastern United States from Argentina several decades prior to the introduction of its more notorious congener, *S. invicta*, but is now confined to only a small portion of northern Alabama and

Mississippi, where it hybridizes with *S. invicta* (Shoemaker et al. 1996). The impact of *Solenopsis richteri* on ants and other organisms native to the southeastern United States is poorly known. Native to Brazil, *Paratrechina fulva* has invaded the Magdalena Valley of Colombia where it reduces native ant diversity (Zenner-Polania 1994); this species is also reported from Cuba (Fontenla Rizo 1995). Another potentially invasive species is *Anoplolepis custodiens*. Native to southern Africa, this species has invaded Zanzibar where it reaches extremely high densities and displaces the native weaver ant *Oecophylla longinoda* (Way 1953). *Technomyrmex albipes* is likewise highly invasive in edge habitats in Mauritius and Madagascar and is associated with substantially reduced ant species richness in these areas (P.S. Ward, unpublished observation). The African *Paratrechina longicornis* is a widely distributed tramp that is sometimes considered invasive (McGlynn 1999a, Wetterer et al. 1999), but few studies have documented its effects on native ants, and in no case is it known to be a competitive dominant (Levins et al. 1973, Torres 1984, Banks & Williams 1989, Morrison 1996). Successful identification of future invaders will be greatly facilitated by careful regional studies that document new introductions and incipient invasions (e.g., Deyrup et al. 2000).

DISPERSAL AND COLONIZATION

All the ants listed in Table 1 have been introduced—most of them worldwide—as a result of human commerce. Species differ, however, in the importance of human-mediated dispersal versus natural dispersal in determining large-scale patterns of spread. In introduced populations of some invasive ants, winged dispersal of female reproductive forms is rare or absent and colonies often reproduce by budding. Colony reproduction by budding alone greatly limits rates of spread: 37–402 m/yr for *A. gracilipes* in the Seychelles (Haines & Haines 1978a), 10–40 m/yr for the polygyne-form of *S. invicta* in central Texas (Porter et al. 1988), 15–270 m/yr for *L. humile* in northern California (Suarez et al. 2001), and approximately 15 m/yr for *P. megacephala* in northern Australia (Hoffmann et al. 1999). In the absence of human-mediated dispersal, introduced populations of *Wasmannia auropunctata* are also believed to spread predominantly by budding (Clark et al. 1982). Because rates of spread by budding are so low [i.e., on the low end for terrestrial organisms (Grosholz 1996)], species that undergo colony reproduction solely by budding depend largely on human-mediated dispersal to colonize new and distant locations (Suarez et al. 2001). Moreover, budding limits the initial spread of invasive ants to areas adjacent to points of introduction or to source habitats. In such cases, invasive ants occur as localized edge effects (Suarez et al. 1998, Human et al. 1998, Bolger et al. 2000, Holway et al. 2002).

Important exceptions to the above pattern include *S. geminata* and the monogyne form of *S. invicta*, two species for which winged dispersal of female reproductive forms is common (see De Heer et al. 1999 and Goodisman et al. 2000 for discussions of dispersal in the polygyne form of *S. invicta*). In such cases, regional-scale patterns of spread may be driven both by human-assisted transport and by the winged dispersal of female alates, which can travel kilometers from natal nests during mating flights (Wilson & Brown 1958, Vinson & Greenberg 1986, Porter et al. 1988). In contrast to colony reproduction by budding, winged dispersal of female reproductive forms enables new beachheads to be established in areas distant from the colony of origin.

Nesting behavior also influences the importance of human-assisted transport. Invasive ants exhibit general and somewhat flexible nesting habits, allowing them to associate closely with humans. Incipient colonies occupy an especially wide range of nesting substrates, including nursery stock and other products of commerce. Some invasive ants readily relocate nests in response to physical disturbance or to exploit favorable but ephemeral sites (Hölldobler & Wilson 1977, Passera 1994). *L. humile*, for example, which often uses ephemeral nest sites, engages in nest relocation in response to changes in the physical environment (Newell & Barber 1913, Markin 1970b, Passera 1994, Gordon et al. 2001) and the distribution of food resources (Newell & Barber 1913, Holway & Case 2000, Silverman & Nsimba 2000). For species with budding, such opportunistic nesting behavior must play a key role in distributing nests to locations where they are likely to be transported by humans. Lastly, the ability of colonies of both *S. invicta* (Morrill 1974, Tschinkel 1983) and *L. humile* (Barber 1916) to raft in response to flooding may pre-adapt them to live in disturbed or well-watered urban environments.

Following dispersal to a new environment, propagules face a host of obstacles that can impede successful establishment. As for other introduced insects (Lawton & Brown 1986, Simberloff 1989), the factors influencing the probability of successful colonization by invasive ants remain poorly known. Below, we discuss attributes that may affect colonization success but acknowledge that the colonization process itself remains inadequately studied. This discussion focuses on ants that reproduce by budding; independent colony founding by the monogyne form of *S. invicta* has been studied in detail (Markin et al. 1972, Tschinkel & Howard 1983) and is reviewed elsewhere (Tschinkel 1993, 1998; Bernasconi & Strassmann 1999).

Although little information exists on the sizes of propagules transported by human commerce, it seems likely that the probability of successful establishment increases steeply with propagule size, at least for small colonies. Propagules with no workers must commonly fail because queens in species with dependent colony founding often lack sufficient metabolic reserves to found colonies on their own (Chang 1985, Keller & Passera 1989, Ross & Keller 1995, Hee et al. 2000). For incipient colonies containing workers and at least one queen, the number of workers present is likely an important determinant of colony-level survival. Disparities in the size of incipient colonies are important since, relative to smaller propagules, larger ones may be able to better withstand stressful physical environments (Markin et al. 1973), starvation (Kaspari & Vargo 1995), and competition from established, neighboring colonies. Although much remains to be learned about the factors governing colonization success, important insights could be obtained through manipulative experiments involving founding queens or small colonies.

Such experimental approaches have been employed with great success as a means to examine the dynamics of independent colony founding in *S. invicta* (Tschinkel & Howard 1983; Adams & Tschinkel 1995a,b; Bernasconi & Strassman 1999).

For at least some species of invasive ants, propagules need not even contain queens to establish successfully (Aron 2001). Although workers of all invasive ants are sterile, in species such as *L. humile*, workers can rear eggs and early instar larvae into sexuals in the absence of queens (Passera et al. 1988, Vargo & Passera 1991). Moreover, in *L. humile*, production of sexuals does not require overwintering, queens produce haploid eggs throughout the year, and mating occurs in the nest (Aron 2001). In combination, these attributes make it possible for dequeened propagules that contain eggs and larvae to produce both male and female sexuals. There is little evidence of inbreeding in *L. humile* colonies despite intranidal mating (Krieger & Keller 2000) (probably because males disperse by flight among nests), and therefore dequeened propagules could in theory develop into reproductively viable colonies (Aron 2001). It will be of interest to determine if other invasive ants share this remarkable capability. [See Tschinkel & Howard (1978) for a discussion of queen replacement in orphaned colonies of *S. invicta*.]

CHARACTERISTICS OF INVADED HABITATS

Natural ecosystems vary greatly in the extent to which they are affected by invasive ants. An ecosystem's physical environment (Tremper 1976, Ward 1987, Holway 1998b, Holway et al. 2002) and its history of anthropogenic disturbance (Tschinkel 1988) and fragmentation (Suarez et al. 1998) all influence susceptibility to invasion. Introduced populations of both *S. invicta* and *S. geminata*, for example, often favor open and disturbed habitats (Tschinkel 1988, Morrison 1996, De Heer et al. 1999). The extent to which biotic resistance from native ants influences the vulnerability of communities to invasion appears to vary regionally. In riparian woodlands in northern California, native ant richness and rate of spread of Argentine ants were uncorrelated over a 4-year period (Holway 1998b). Majer (1994), Andersen (1997), and Hoffmann et al. (1999), however, suggest that the *Iridomyrmex*-rich fauna of Australia may be resistant to invasion owing to the strong competitive ability of ants in this genus (Andersen 1992, Andersen & Patel 1994). Moreover, the absence of native ants on Hawaii and other islands in the Pacific Ocean undoubtedly makes these areas vulnerable to invasion (Cole et al. 1992, Reimer 1994).

EFFECTS OF INVASIVE ANTS ON NATIVE ANTS

The competitive displacement of native ants by invasive ants is the most dramatic and widely reported effect of ant invasions (Hölldobler & Wilson 1990, Williams 1994). In invaded areas, the abundance of native ants can be reduced by over 90% [Porter & Savignano 1990 (but see Morrison 2002), Cammell et al. 1996, Human & Gordon 1997, Holway 1998a, Hoffmann et al. 1999]. Evidence of this phenomenon comes from a variety of sources: (*a*) historical accounts (Newell & Barber 1913, Van Der Goot 1916, Haskins & Haskins 1965, Brandao & Paiva 1994), (*b*) longitudinal studies documenting the advance of invasive ants at the expense of natives (Greenslade 1971; Erickson 1971; Tremper 1976; Porter et al. 1988; Holway 1995, 1998b; Human & Gordon 1996; Hoffmann et al. 1999; Sanders et al. 2001), (*c*) studies documenting mutually exclusive distributions between native ants and invasive ants (Tremper 1976; Clark et al. 1982; Ward 1987; Porter & Savignano 1990; Morris & Steigman 1993; Cammell et al. 1996; Human & Gordon 1997; Heterick 1997; Way et al. 1997; Holway 1998a,b; Suarez et al. 1998; Kennedy 1998; Hoffmann et al. 1999; Vanderwoude et al. 2000), (*d*) large-scale studies illustrating the effects of invasive ants on latitudinal gradients in ant diversity (Gotelli & Arnett 2000), and (*e*) lab and field studies (only some experimental) demonstrating that invasive ants differ from native ants with respect to competitive ability, periods of activity, behavioral characteristics, or resource use (Clark et al. 1982; Jones & Phillips 1987, 1990; Porter & Savignano 1990; Morrison 1996; Human & Gordon 1996, 1999; Holway 1999; Morrison 1999, 2000; Holway & Case 2001).

Invasive ants may have the greatest effect on ecologically similar native ants. In the southeastern United States, for example, the fire ants *Solenopsis xyloni* and *S. geminata* appear highly sensitive to displacement by *S. invicta* (Wilson & Brown 1958; Porter et al. 1988; Morrison 2000, 2002). Likewise, in California, Argentine ants and ecologically similar native Dolichoderines (e.g., *Tapinoma sessile*, *Liometopum occidentale*) rarely, if ever, coexist (Ward 1987, Holway 1998a). Native ants that resemble invasive ants in their ecologies are by no means the only ants displaced. In California, for example, a variety of harvester ant species succumb to Argentine ants (Erickson 1971, Human & Gordon 1996, Suarez et al. 1998) despite little apparent overlap in food resources. *S. invicta* has also been reported to eradicate colonies of the harvester ant *Pogonomyrmex barbatus* in central Texas (Hook & Porter 1990).

Although invasive ants displace many species of native ants, some often persist. Hypogeic ants, for example, may persevere in areas occupied by invasive ants (Ward 1987) or persist longer than other taxa (Hoffmann et al. 1999). Tschinkel (1988) speculated that predation by the hypogeic *Solenopsis molesta* upon *S. invicta* brood may restrict monogyne colonies of *S. invicta* to disturbed habitats lacking *S. molesta*. Species resistant to displacement also include those adapted to physical conditions not tolerated by invaders. In California, the cold-tolerant *Prenolepis imparis* appears relatively immune to displacement by the Argentine ant (Tremper 1976, Ward 1987, Holway 1998a, Suarez et al. 1998), whereas the heat-tolerant *Dorymyrmex insanus* and *Forelius mccooki* coexist locally with Argentine ants along the periphery of hot, exposed areas not colonized by *L. humile* (Holway et al. 2002). Both *D. insanus* and *F. mccooki* co-occur with *S. invicta* as well (Summerlin et al. 1977, Camilo & Philips 1990, Morrison 2002). Heat tolerant native ants also appear to coexist with Argentine ants in South Africa (Witt & Giliomee 1999).

Other native ants may resist displacement by invasive ants primarily because of their strong competitive ability. In North America, for example, many authors have reported that species in the genus *Monomorium* can, to some extent, withstand interspecific competition by invasive ants. Several authors report *Monomorium minimum* co-occurring with red imported fire ants in the southeastern United States (Baroni-Urbani & Kannowski 1974, Howard & Oliver 1979, Stein & Thorvilosen 1989, Porter & Savignano 1990), and *M. ergatogyna* was the only native ant (of seven species tested) to resist displacement by Argentine ants at baits in northern California (Holway 1999). *Monomorium* species may persist both through their use of potent chemical defensive compounds (Adams & Traniello 1981, Andersen et al. 1991) and their tolerance of warm temperatures (Adams & Traniello 1981).

Mechanisms

Although the ability of invasive ants to displace native ants is well known, the mechanisms involved have, until recently, received relatively little attention. Even with a number of recent and informative studies, much remains to be learned. For example, native ants succumb to the combined effects of interference and exploitative competition, but an understanding of the relative importance of these two forms of competition is often unclear and undoubtedly varies depending on the invading species, characteristics of the native ant community, and other environmental variables. Furthermore, invasive ants are unusual in that introduced populations of most species typically maintain populous and expansive supercolonies. The abundance of invasive ants can exceed that of all native ant species combined in comparable uninvaded areas (Porter & Savignano 1990, Holway 1998a, Hoffmann et al. 1999). Although disparities in colony size are an important determinant of the competitive asymmetry between native and invasive ants, there exists only a rudimentary understanding of why invasive ants differ from native species in this important respect.

As with ants generally (Hölldobler & Wilson 1990), the interference repertoire of invasive ants includes both worker-level behaviors such as physical aggression and the use of chemical defensive compounds as well as colony-level behaviors such as recruitment of nestmates, interspecific territoriality, and nest raiding (Table 2). Despite the pervasiveness of physical aggression among competing ants as a whole, invasive ants are commonly described as exhibiting pronounced interspecific aggression (Hölldobler & Wilson 1977, 1990; Passera 1994; Human & Gordon 1999). Even though invasive ants may be more aggressive than the native ants they displace, differential aggression provides only a partial explanation for their interference prowess. In ants generally, interference competition, especially for valuable food finds or nest sites, is typically a colony-level activity. The outcome of such inter-colony contests depends primarily on asymmetries in recruitment response or local density (often reflective of differences in colony size) (Hölldobler & Lumsden 1980, Hölldobler & Wilson 1990, Adams 1990). Numerical advantages enjoyed by invasive ants contribute greatly to their interference ability (Greenslade 1971, Tremper 1976, Holway 1999, Morrison 2000, Holway & Case 2001).

TABLE 2   Competitive mechanisms and behaviors reported to be important in interactions between invasive ants and native ants

Species	Competitive mechanism or behavior	Study
Anoplolepis gracilipes	Use of chemical defensive compounds	Fluker & Beardsley 1970
	Use of physical aggression by workers	Haines & Haines 1978a
	Active both day and night	Haines & Haines 1978a
Linepithema humile	Use of chemical defensive compounds	Lieberburg et al. 1975, Holway 1999, Holway & Case 2001
	Use of physical aggression by workers	Newell 1908, Erickson 1971, De Kock 1990, Human & Gordon 1996, 1999, Holway 1999
	Initiates one-on-one interactions more often than do native ants	Human & Gordon 1999, Holway & Case 2001
	Raids nests of other species	Fluker & Beardsley 1970, De Kock 1990, D.A. Holway, unpublished observation, P.S. Ward, unpublished observation
	Workers prey upon winged queens of native ant species	Human & Gordon 1996
	Remains at baits longer than do native ants	Human & Gordon 1996
	Proficient at displacing native ants from baits	Human & Gordon 1996, Holway 1999
	Active both day and night	Human & Gordon 1996
	Active throughout the year	Holway 1998a
	Recruits to baits in higher numbers than do native ants	Human & Gordon 1996, Holway 1998b
	Recruits to more baits than do native ants	Holway 1998b
	Discovers and recruits to baits more quickly than do native ants	Holway 1998a, Holway 1999
	Adjusts foraging behavior to local worker density	Gordon 1995
Pheidole megacephala	Use of physical aggression by workers	Fluker & Beardsley 1970, Lieberburg et al. 1975
	Use of physical aggression by soldiers	Fluker & Beardsley 1970
	Recruitment of many workers	Lieberburg et al. 1975
	Active both day and night	Hoffmann 1998
Solenopsis invicta	Use of chemical defensive compounds (gaster flagging and stinging)	Bhatkar et al. 1972, Obin & Vander Meer 1985, King & Phillips 1992, Morrison 2000
	Use of physical aggression by workers	Bhatkar et al. 1972, Jones & Phillips 1987, Bhatkar 1988, King & Phillips 1992, Morrison 1999, Morrison 2000
	Raids nests of other species	Bhatkar et al. 1972, Hook & Porter 1990
	Retrieves baits in the lab more rapidly than do 2 species of native ants	Jones & Phillips 1990
	Discovers and recruits to baits more quickly than do native ants	Porter & Savignano 1990
	Recruits to baits in higher numbers than do native ants	Porter & Savignano 1990
	Recruits to baits in higher numbers than does native *S. geminata*	Morrison 1999

(Continued)

TABLE 2   *(Continued)*

Species	Competitive mechanism or behavior	Study
Solenopsis invicta (continued)	Retrieves more food than does native *S. geminata*	Morrison 1999
	Colony-level interference ability in the lab superior to that of native fire ants (*S. geminata* and *S. geminata* × *S. xyloni* hybrids)	Morrison 2000
	Active both day and night	Morrison 2000
Solenopsis geminata	Use of physical aggression by workers	Morrison 1996
	Recruits to baits in higher numbers than do native ants	Morrison 1996
	Reduces the access of *Monomorium monomorium* to baits	Morrison 1996
	Most dominant interference competitor (out of 19 species) at baits	Torres 1984
Wasmannia auropunctata	Remains at baits longer than do other ants	Clark et al. 1982
	Proficient at displacing other ants from baits	Clark et al. 1982
	Recruits to baits in higher numbers than do other ants	Clark et al. 1982; Tennant 1994
	Active both day and night	Clark et al. 1982
	Pirates food from other ants	Brandao & Paiva 1994

A poorly studied but potentially important component of the competitive displacement of native ants by invasive ants involves colony-level battles and nest raiding. Colonies of both *S. invicta* and *L. humile* engage in aggressive, episodic raids on nests of other species (Table 2). Although effects of colony-level battles remain difficult to quantify because individual events may occur only infrequently, such raids may eradicate native ant colonies. Hook & Porter (1990), for example, estimated that a colony of *P. barbatus* in central Texas lost over 1200 workers to *S. invicta* over a period of six to seven weeks; this colony was believed to die out eventually as a direct result of these incursions. Although the distinction between interference competition and intra-guild predation in such cases is not clear-cut, at present, there is little evidence that invasive ants consume native ants during these events. The role of nest raiding in this context should be investigated in greater detail.

Invasive ants also compete with native ants indirectly via exploitative competition. As with interference ability, large colony size enhances exploitative ability because large colonies can simultaneously maintain large forces of scouts (i.e., workers actively searching for food) and recruits (i.e., workers in the nest available to help exploit rich food discoveries) (Johnson et al. 1987). Compared to the native ants they displace, invasive ants commonly excel at behaviors correlated with exploitative ability: rapid discovery of food, rapid recruitment, recruitment of large numbers of workers, extended duration of recruitment, and 24-hour activity (Table 2). Somewhat surprisingly, few field studies measure exploitative ability directly (e.g., food retrieval rates) (Morrison 1999). Despite putative differences in exploitative ability between native and invasive ants, the precise role of exploitative competition in the demise of native ants remains unclear. Exploitative competition probably plays a minor role in situations where both invasive ants and native ants maintain separate and well-delineated territories (Morrison 2000) and where large or immobile food items (e.g., carrion, aggregations of Homoptera) need to be usurped from or defended against other colonies. In contrast, exploitative competition may be more important in situations where territories are actively being carved out (e.g., at the leading edge of an invasion front) and where small, particulate food items can be retrieved without recruitment of nestmates (e.g., small, dead insects).

Compared to the native species they supplant, invasive ants may excel at the two forms of competition simultaneously, allowing them both to exploit and to monopolize a majority of food resources. Coexistence among competing species of ants within a community can result from species-specific differences in competitive ability: An inverse relationship often exists between the ability of a species to discover food and its ability to dominate resources behaviorally or numerically (Wilson 1971, Fellers 1987, Banks & Williams 1989, Perfecto 1994, Morrison 1996; see Johnson 1981 and Nagamitsu & Inoue 1997 for a similar relationship among Meliponine bees). Relative to native ants, however, invasive ants may excel at both resource discovery and resource dominance, effectively breaking the trade-off (Davidson 1998, Holway 1999, Feener 2000). For example, compared to a northern California native ant community that was subject to the discovery-dominance trade-off, *L. humile* discovered food in less time than did all native species and also displaced a majority (six out of seven species) from baits (Holway 1999) suggesting that *L. humile* secures a majority of food resources in areas where it meets native ants. Although discovery-dominance trade-offs offer a simplistic view of community structure in that they ignore mechanisms known to affect coexistence (e.g., species-specific tolerances to the physical environment), they provide a conceptual framework for testing alternative hypotheses concerning the success of ant invasions (Davidson 1998, Adler 1999, Feener 2000). Moreover, given that competitive trade-offs are frequently invoked to explain species coexistence, their usefulness as a tool to explain cases where species fail to coexist (as in invasions) may well be general [see Tilman (1999) for a similar argument pertaining to plant invasions]. Numerical advantages probably allow invasive ants to achieve simultaneous proficiency at both resource dominance and resource discovery (Holway 1999, Morrison 2000), providing a proximate mechanism for why invasive ants break the dominance-discovery trade-off. A key unresolved question, however, is why invasive ants become so much more abundant than the native ants they displace.

Hypotheses to Explain the Abundance of Invasive Ants

Several hypotheses have been advanced to account for the disproportionate abundance of invasive ants. First, as with other introduced species, invasive ants have escaped competitors and natural enemies and may achieve larger colony sizes and increased colony densities as a consequence (Buren 1983, Porter et al. 1997). Second, because unicolonial ants do not defend territorial boundaries against conspecifics, they can allocate workers to tasks other than colony defense and shunt resources to worker production that would otherwise be expended on fighting neighboring colonies of the same species (Hölldobler & Wilson 1977, Macom & Porter 1996, Holway et al. 1998). The loss of intraspecific territoriality and the formation of supercolonies may allow such species to monopolize resources and to displace competitors; this would lead to further increases in local density (Macom & Porter 1996). Third, invasive ants may consume resources, such as plant and homopteran exudates that native ants either fail to exploit or do so less efficiently compared to invasive ants. In this sense, invasive ants could function partly as herbivores, as has been suggested for tropical arboreal ants that maintain densities greatly in excess of what scavenging and predation alone could support (Tobin 1994, Davidson 1997). Access to carbohydrate-rich food resources such as plant and homopteran exudates may allow invasive ants to fuel workers at a high tempo, making possible the maintenance of high dynamic densities (ants/area/time), the defense of absolute territories, and further monopolization of resources (Davidson 1997, 1998). Because these processes in all likelihood operate simultaneously (Porter & Savignano 1990, Holway 1999, Morrison 2000), an outstanding challenge will be to uncover their relative importance. As a start, we review current evidence bearing on this issue for the two best-known invasive ants, *L. humile* and *S. invicta*.

LINEPITHEMA HUMILE   Early reports from the Argentine ant's introduced range described workers, queens, and brood moving freely among spatially separate nests (Newell & Barber 1913), and more recent studies have shown that a single large supercolony appears to occupy nearly the entire introduced range of *L. humile* in California (Tsutsui et al. 2000, Tsutsui & Case 2001). Argentine ants also appear unicolonial in every other part of their introduced range where this behavior has been studied (Passera 1994, Way et al. 1997, Tsutsui et al. 2000, Krieger & Keller 2000, Giraud et al. 2002). In contrast, although native populations of *L. humile* are also polygynous and maintain multiple nests, high levels of intraspecific aggression are commonly observed between nests over short (<100 m) spatial scales (Tsutsui et al. 2000, Tsutsui & Case 2001).

Despite gross differences in colony structure between the native and introduced ranges, workers in both areas tend to display intraspecific aggression toward workers from genetically different colonies (Tsutsui et al. 2000). This observation is of interest for at least two reasons. First, it suggests that nestmate recognition in Argentine ants has an underlying genetic basis, as in other ants (e.g., Stuart 1987; Carlin & Hölldobler 1986, 1987; Beye et al. 1998). Second, it provides an explanation for the widespread absence of intraspecific aggression in introduced populations of this species. Because introduced populations experienced a loss of genetic diversity during their introduction and establishment (Tsutsui et al. 2000, 2001), they do not appear to have sufficient genetic variation to elicit fighting between workers from different nests. Workers in the introduced range rarely encounter genetically different individuals; as a result, intraspecific aggression seldom occurs, and almost all nests function in an apparently cooperative, uni-colonial fashion.

The shift in colony organization between native and introduced populations of Argentine ants helps explain their success as invaders. In lab experiments with pairs of colonies that either did or did not exhibit intraspecific aggression, aggressive pairs experienced higher mortality, lower foraging activity, and lower rates of colony growth after 70 days (Holway et al. 1998). Moreover, in 60% of nonaggressive nest pairs, colonies fused, even though some of these pairings were composed of colonies that were originally collected from locations up to 100 km apart (Holway et al. 1998). Although caution seems warranted in extrapolating the results of this lab study to the population level, these findings show in principle how the loss of intraspecific aggression could lead to increases in colony size and the formation of supercolonies—both of which would enhance interspecific competitive ability.

Other factors also contribute to the abundance of Argentine ants in their introduced range. In its native Argentina, *L. humile* coexists with many other species of ants (Suarez et al. 1999), including competitive dominants such as *S. invicta* and *S. richteri*. It seems likely that the absence of strong competitors such as *S. invicta* and *S. richteri* throughout most of the Argentine ant's introduced range contributes to its high abundance. *S. invicta*, for example, appears to have displaced *L. humile* where introduced populations of both species overlapped in the southeastern United States (Wilson 1951, Glancey et al. 1976). At present, there is no evidence concerning the role of escape from natural enemies in the success of introduced populations of Argentine ants. *L. humile* appears to lack phorid fly parasitoids (Orr et al. 2001) despite earlier reports to the contrary (Orr & Seike 1998, Feener 2000). Recent work suggests that previous studies of phorid fly parasitism involved different species of *Linepithema* (Orr et al. 2001, Tsutsui et al. 2001).

An additional factor that may promote high worker densities in introduced populations is the ability of *L. humile* to exploit resources that native ants either fail to consume or consume less efficiently than invasive ants. For example, Argentine ants have a strong predilection for homopteran honeydew. When Argentine ants tend Homoptera in agricultural systems (especially citrus orchards), protected Homoptera can attain locally high densities, possibly further increasing *L. humile* populations (Newell & Barber 1913, Way 1963). However, it remains unclear whether Argentine ants differ from native ants in their preference for honeydew. In riparian woodlands in northern California, for example, native ants such as *Liometopum occidentale* commonly tend Homoptera (Ward 1987).

SOLENOPSIS INVICTA   The monogyne form of *S. invicta* was first introduced to the United States around 1940 and spread quickly in the following decades throughout the southeast, where it now attains higher densities than it does in its South American native range (Porter et al. 1992). The success of this species has been attributed to a variety of causes including human modifications of the landscape, community simplifications resulting from prior invasions of other ants (e.g., *L.*

humile and *S. richteri*), and the use of pesticides (Morrison 2000). Escape from coevolved competitors, pathogens, and parasites probably also allowed increases in the density of its populations (Jouvenaz 1990; Porter et al. 1992, 1997). For example, the presence of host-specific parasitoids, such as phorid flies, can inhibit foraging in *S. invicta* colonies (Porter et al. 1995, Orr et al. 1995). Although the population-level effects of phorids on fire ants are unknown at present, the absence of phorids specialized on *S. invicta* throughout its introduced range may allow colonies there to grow faster and to attain larger sizes.

In the 1970s, polygyne colonies of *S. invicta* were first reported from the United States (Glancey et al. 1973), and the polygyne form is now widespread (but discontinuously distributed) in the southeast, particularly in Texas and Florida (Porter et al. 1991, 1992). As in the United States, the distribution of polygynous colonies of *S. invicta* varies in South America; polygyny is present in Argentina (Jouvenaz et al. 1989, Ross et al. 1996) but apparently absent in Brazil (Jouvenaz et al. 1989, Porter et al. 1992). Although the monogyne form of *S. invicta* is considered highly invasive (Wilson & Brown 1958, Apperson & Powell 1984, Tschinkel 1993), the polygyne form may be a bigger problem (Porter & Savignano 1990). In Florida, population densities (estimated using biomass) of polygyne *S. invicta* are two times higher than those of the monogyne form (Macom & Porter 1996). Likewise, in Texas, colonies of the polygyne form can recruit up to twice as many workers to baits compared to colonies of the monogyne form (MacKay et al. 1994). The higher densities of the polygyne form, compared to the monogyne form, may result from diminished intraspecific aggression and a concomitant reduction in intraspecific territoriality, an increased ability to monopolize resources from interspecific competitors, or differences in patterns of sex allocation (Macom & Porter 1996). Because polygyne and monogyne populations presently occur together in parts of the southeastern United States, an interesting opportunity exists to test in more detail how colony structure variation alone influences the ecological effects of these invasions.

The transition from monogyny to polygyny described above has been a topic of much interest. Ross & Keller (1995) and Ross et al. (1996) have hypothesized that polygyny became prevalent in introduced populations in response to ecological constraints, as proposed for the evolution of cooperative breeding in other taxa (Emlen 1982). Following introduction to the United States, population densities of monogyne *S. invicta* increased. With increasing colony densities, suitable nest sites became saturated, reducing the fitness of queens that attempted to found colonies independently and favoring queens seeking adoption into established colonies (Nonacs 1993, Ross & Keller 1995). Frequent queen adoption may have then led to an erosion of nestmate recognition abilities as levels of genetic diversity increased within polygyne colonies (Hölldobler & Michener 1980). This loss of nestmate recognition could then have further increased polygyny as colonies accepted more foreign queens (Hölldobler & Wilson 1977, Ross et al. 1996), leading to a runaway process of ever-increasing polygyny (Pamilo 1991, Ross et al. 1996).

In light of recent studies that illuminate the genetic machinery underlying monogyny and polygyny in *S. invicta* (Keller & Ross 1999, Krieger & Ross 2002), it appears unlikely that polygyny arose in introduced populations as a result of ecological constraints alone. These studies implicate genetic differences between the social forms at the *Gp-9* locus as the primary, and perhaps exclusive, determinant of colony queen number in *S. invicta*. Queens from the two social forms can be distinguished by their genotype at the *Gp-9* allozyme locus. Virtually all egg-laying queens in introduced polygyne colonies are *Bb* heterozygotes at *Gp-9* (Ross 1997) and possess a distinctive polygyne queen phenotype: physically small queens with fewer fat reserves and more gradual oogenesis than monogyne queens (Keller & Ross 1993). Workers in polygyne colonies accept additional queens based on their genotype at *Gp-9* (Keller & Ross 1998). Monogyne queens, on the other hand, are heavy, possess the large fat reserves necessary for independent colony founding, exhibit rapid oogenesis (Keller & Ross 1993), and possess the *BB* genotype at *Gp-9* in both the native and introduced ranges (Ross 1997). Workers in monogyne colonies do not permit additional queens to join (Keller & Ross 1998). The *bb* genotype is thought to be lethal, and is almost completely absent in adult workers and queens in both ranges (Ross 1997).

Thus, the reproductive strategies of monogyne and polygyne queens appear to be fixed by genetics, regardless of the ecological context. Previous work on the infiltration of mature colonies by dispersing monogyne queens further supports the idea that genetic factors limit queen number in *S. invicta* colonies. In populations of monogyne *S. invicta*, for example, ecological constraints on independent colony founding are so strong that some overwintered, monogyne queens attempt to infiltrate previously established colonies (Tschinkel 1996, DeHeer & Tschinkel 1998). When these monogyne queens, which presumably possess the BB genotype at *Gp-9*, attempt to enter colonies that contain queens, they are likely killed by workers, (Tschinkel 1996, DeHeer & Tschinkel 1998, VanderMeer & Alonso 2002) as described above. Thus, it appears that these dispersing queens are only able to successfully enter colonies that have lost their queen (DeHeer & Tschinkel 1998, VanderMeer & Alonso 2002). These studies illustrate that newly produced queens appear to be caught, on one hand, between ecological constraints that reduce opportunities for independent colony founding and, on the other hand, strong genetic constraints that prevent monogyne queens from successfully entering established colonies.

EFFECTS OF ANT INVASIONS ON OTHER TAXA

Although the displacement of native ants by invasive ants is the most obvious effect of ant invasions, many additional effects occur following invasion. Given the variety of ecological roles filled by native ants, it seems likely that reductions in native ant diversity and abundance would indirectly affect many different taxa. Moreover, because invasive ants are widespread, abundant, aggressive, and omnivorous, one would predict that they would disrupt invaded communities (Diamond & Case 1986, Pimm 1991, Parker et al. 1999). These effects might be most noticeable on island ecosystems that lack native ants. Numerous lines of empirical evidence support these predictions and illustrate the diversity of ecological effects that result. Nonetheless, the effects of ant invasions remain incompletely studied, reflecting the inadequate state of knowledge concerning the consequences of species introductions generally (Parker et al. 1999, Pimentel et al. 2000). Much evidence bearing on the effects of ant invasions, for example, is either anecdotal or correlative, and few studies determine if effects of invasive ants differ from those of native ants that are displaced. Experimental, long-term and large-scale studies are therefore needed to develop a more quantitative understanding of the impacts caused by invasive ants. Below we discuss the known ecological effects of ant invasions but recognize that current information on this topic is incomplete.

Competition and Predation

A clear understanding of the ecological role of invasive ants as predators and competitors is hampered by a poor understanding of their diets, studies that confuse predation and scavenging, and those that fail to distinguish between predation and competition. Although many lines of evidence illustrate the role of invasive ants as predators of at least certain taxa, the same cannot be said of competition. Some evidence exists for ants generally competing with both vertebrates (Brown & Davidson 1977, Aho et al. 1999) and non-ant invertebrates (Halaj et al. 1997), but there is, at present, little unequivocal evidence demonstrating the existence of competition (especially exploitative competition) between invasive ants and non-ant taxa. For these reasons, we discuss predation and competition together.

IMPACTS OF INVASIVE ANTS ON INVERTEBRATES   Every species of invasive ant listed in Table 1 has been implicated in the decline of non-ant invertebrates, but the effects of *L. humile* and *S. invicta* have been examined in the most detail (Table 3). Some of the best evidence illustrating the role of invasive ants as predators of invertebrates comes from studies conducted in agricultural settings that document invasive ants preying on insect herbivores (Table 3). As in agro-ecosystems, invasive ants occupying less manipulated environments also prey opportunistically on invertebrate eggs, larvae, and certain adult forms (Table 3). Such predation may jeopardize populations of some invertebrates, especially those on oceanic islands, such as Hawaii, which evolved in the absence of predaceous ants (Zimmerman 1970, Gillespie & Reimer 1993, Cole et al. 1992). Zimmerman (1970), for example, recounts the disappearance (and apparent extinction) of a once abundant wingless and ground-dwelling fly from forests on Oahu shortly after introduction of *P. megacephala*. In addition to affecting other invertebrates through predation, invasive ants may also compete with non-ant invertebrates.

Reported impacts of invasive ants on invertebrates range from qualitative observations, such as the absence of a species from an invaded area, to studies that estimate changes in diversity, abundance, or biomass between invaded and uninvaded areas (Table 3). *Solenopsis invicta* appears to cause declines in a variety of invertebrate groups including ground-dwelling arthropods (Nichols & Sites 1989,

TABLE 3 Reported effects of invasive ants on non-ant taxa

Species	Ant	Location	Reported effect/mechanism	Study
BIRDS				
Sooty tern (*Sterna fruscata*)	*A. gracilipes*	Bird Island, Seychelles	Failure to nest in invaded areas	Feare 1999
White tern (*Gygis alba*)	*A. gracilipes*	Bird Island, Seychelles	Death of chicks	Feare 1999
California gnatcatcher (*Polioptila melanura*)	*L. humile*	CA, USA	Nest failure	Sockman 1997
Northern bobwhite (*Colinus virginianus*)	*S. richteri*	TX, USA	Mortality of pipping chicks	Johnson 1961
Least tern (*Sterna antillarum*)	*S. invicta*	MS, USA	Dead chicks covered with ants, ant suppression increased chick survival	Lockley 1995
Colonial waterbirds (7 spp.)	*S. invicta*	TX, USA	Ant suppression increased nesting success	Drees 1994
Crested caracara (*Caracara plancus*)	*S. invicta*	TX, USA	Dead chicks covered in ants	Dickinson 1995
Northern bobwhite (*Colinus virginianus*)	*S. invicta*	TX, USA	Bobwhite densities decline after invasion, and increase after ant suppression	Allen et al. 1995
Northern bobwhite (*Colinus virginianus*)	*S. invicta*	GA/FL/SC, USA	Bobwhite densities decline after invasion	Allen et al. 2000
Northern bobwhite (*Colinus virginianus*)	*S. invicta*	TX, USA	Ant suppression increased chick survival	Mueller et al. 1999
Northern bobwhite (*Colinus virginianus*)	*S. invicta*	Laboratory study	Ant exposure reduced chick survival	Giuliano et al. 1996
Cliff swallow (*Hirundo pyrrhonota*)	*S. invicta*	TX, USA	Mortality of pipping chicks, nesting success reduced in invaded areas	Sikes & Arnold 1986
Barn swallow (*Hirundo rustica*)	*S. invicta*	TX, USA	Nest predation	Kopachena et al. 2000
Black rail (*Laterallus jamaicensis*)	*S. invicta*	FL, USA	Mortality of pipping chick	Legare & Eddleman 2001
Wood duck (*Aix sponsa*)	*S. invicta*	TX, USA	Predation on nestlings and pipped eggs	Ridlehuber 1982
Loggerhead shrike (*Lanius ludovicianus*)	*S. invicta*	FL, USA	Reduced food supply	Yosef & Lohrer 1995
REPTILES & AMPHIBIANS				
Skink (*Mabuya seychellensis*)	*A. gracilipes*	Bird Island, Seychelles	Disappearance of skinks in invaded areas	Feare 1999
Coast horned lizard (*Phrynosoma coronatum*)	*L. humile*	CA, USA	Displacement of ant prey	Suarez et al. 2000
Coast horned lizard (*Phrynosoma coronatum*)	*L. humile*	Laboratory study	Lower growth rate on *L. humile* diet	Suarez & Case 2002
Coast horned lizard (*Phrynosoma coronatum*)	*L. humile*	CA, USA	Lack of overlap between lizards and ants	Fisher et al. 2002

(Continued)

TABLE 3 (Continued)

Species	Ant	Location	Reported effect/mechanism	Study
Texas horned lizard (*Phrynosoma cornutum*)	*S. invicta*	TX, USA	Displacement of ant prey	Donaldson et al. 1994
Six-lined racerunner (*Cnemidophorus sexlineatus*)	*S. invicta*	AL, USA	Predation of eggs	Mount et al. 1981
Alligator (*Alligator mississippiensis*)	*S. invicta*	FL, USA	Killed hatchlings, lower hatchling mass	Allen et al. 1997a
Alligator (*Alligator mississippiensis*)	*S. invicta*	LA, USA	Predation of nestlings, pipped eggs	Reagan et al. 2000
Gopher tortoise (*Gopherus polyphemus*)	*S. invicta*	GA, USA	Predation of hatchlings	Landers et al. 1980
Loggerhead sea turtle (*Caretta caretta*)	*S. invicta*	GA/FL, USA	Predation of eggs and hatchlings	Moulis 1997, Wilmers et al. 1996
Green turtle (*Chelonia mydas*)	*S. invicta*	FL, USA	Predation of pipping hatchlings	Wilmers et al. 1996
Florida red-bellied turtle (*Pseudemys nelsoni*)	*S. invicta*	FL, USA	Predation of pipping hatchlings in lab	Allen et al. 2001
Slider turtle (*Trachemys scripta*)	*S. invicta*	SC, USA	Predation of eggs and hatchlings	Buhlmann & Coffman 2001
Snapping turtle (*Chelydra serpentina*)	*S. invicta*	AL, USA	Ants on dead hatchlings	Conners 1998a
Box turtle (*Terrapene carolina*)	*S. invicta*		Mortality from stings	Montgomery 1996
Rough green snake (*Opheodrys aestivus*)	*S. invicta*	AL, USA	Ants punctured eggs	Conners 1998b
Hognose snake (*Heterodon simus*)	*S. invicta*	SE, USA	"Responsible for decline"	Tuberville et al. 2000
Houston toad (*Bufo houstonensis*)	*S. invicta*	TX, USA	Mortality from stings	Freed & Neitman 1988
Reptile populations	*W. auropunctata*	New Caledonia	Decrease in populations in invaded areas	Jourdan et al. 2001
FISH				
Redear sunfish (*Leponis microlophus*)	*S. invicta*	Laboratory study	Ingestion of fire ants	Green & Hutchins 1960
Rainbow trout (*Oncorhynchus mykiss*)	*S. invicta*	TX, USA	Ingestion of fire ants	Contreras & Labay 1999
MAMMALS				
Christmas Island shrew (*Crocidura attenuata*)	*A. gracilipes*	Indian Ocean	Shrew disappeared from island	Meek 2000
Gray shrew (*Notiosorex crawfordi*)	*L. humile*	CA, USA	Negative relationship between number of shrew captures and ant density	Laakkonen et al. 2001
White-tailed deer (*Odocoileus virginianus*)	*S. invicta*	TX, USA	Fawn recruitment higher in treated areas	Allen et al. 1997b
White-tailed deer (*Odocoileus virginianus*)	*S. invicta*	TX, USA	Ants increase deer movement	Mueller et al. 2001
Northern pygmy mice (*Baiomys taylori*)	*S. invicta*	TX, USA	Ants influence seasonal microhabitat use	Smith et al. 1990
Northern pygmy mice (*Baiomys taylori*)	*S. invicta*	TX, USA	Capture rate higher in treated areas	Killion et al. 1995
Deer mice (*Peromyscus maniculatus*)	*S. invicta*	Laboratory study	Mice change foraging in presence of ants	Holtcamp et al. 1997

Taxon	Ant species	Location	Effect	Reference
Cotton rats (*Sigmodon hispidus*)	*S. invicta*	TX, USA	Trapped rats killed by ants	Flickinger 1989
Small mammals (4 spp.)	*S. invicta*	TX, USA	Negative relationship between number of mammal captures and ant mound density	Ferris et al. 1998, Killion & Grant 1993
Small mammals (3 spp.)	*S. invicta*	TX, USA	Trapped mammals killed by ants	Masser & Grant 1986
INVERTEBRATES (GENERAL)				
Land crab (*Cardisoma* sp.)	*A. gracilipes*	Bird Island, Seychelles	Dead crabs in invaded areas	Feare 1999
Insects (especially large beetles)	*A. gracilipes*	Bird Island, Seychelles	Ants seen killing insects	Feare 1999
Invertebrates	*A. gracilipes*	Seychelles	"Responsible for decline"	Haines & Haines 1978b
Araneae (Tetragnatha)	*A. gracilipes*	HI, USA	Spiders absent from invaded areas	Gillespie & Reimer 1993
Collembola, flies, spiders	*L. humile*	CA, USA	Reduced or absent in invaded areas	Human & Gordon 1997
Flies and beetles	*L. humile*	CA, USA	Reduced abundance in invaded areas	Bolger et al. 2000
Longhorn beetle (*Desmocerus californicus*)	*L. humile*	CA, USA	Negative association between beetle and ant	Huxel 2000
Arthropods (8 orders)	*L. humile*	HI, USA	Reduced abundance in invaded areas	Cole et al. 1992
Yellowjackets (4 spp.)	*L. humile*	CA, USA	Ants attack colonies	Gambino 1990
Honeybees (*Apis mellifera*)	*L. humile*	W. Cape, S. Africa	Ants collect 42% of nectar before bees forage	Buys 1987
Invertebrates	*P. megacephala*	N. Territory, Austrailia	42–85% decrease in abundance	Hoffman et al. 1999
Araneae (Tetragnatha)	*P. megacephala*	HI, USA	Spiders absent from invaded areas	Gillespie & Reimer 1993
Araneae (Tetragnatha)	*S. papuana*	HI, USA	Spider abundance reduced in invaded areas	Gillespie & Reimer 1993
Lone star tick (*Amblyomma americanum*)	*S. richteri*	LA, USA	Predation of ticks in non-treated areas	Harris & Burns 1972
Mosquito (*Psorophora columbiae*)	*S. invicta*	Laboratory study	Ants prey upon mosquito eggs	Lee et al. 1994
Tree snail (*Orthalicus reses reses*)	*S. invicta*	FL, USA	Predation of snails	Forys et al. 2001a
Apple snails (*Pomacea paludosa*)	*S. invicta*	FL, USA	Ants attack and kill snails in dry tanks	Stevens et al. 1999
Swallowtail butterfly (*Papilio cresphontes*)	*S. invicta*	FL, USA	Eggs, pupae, and most larvae consumed	Forys et al. 2001b
Ceratoma catalpae (Lepidoptera)	*S. invicta*	GA, USA	Predation on prepupae & pupae greater	Ness 2001
Cotesia congregata (Hymenoptera)	*S. invicta*	GA USA	Pupal survival lower relative to native ants	Ness 2001
Monarch butterfly (*Danaus plexippus*)	*S. invicta*	TX, USA	Larvae and eggs absent in invaded areas	Calvert 1996
Mites (Erythraeidae), Scarab (*Canthon*	*S. invicta*	TX, USA	Absent from invaded areas	Porter & Savignano 1990

(Continued)

TABLE 3 *(Continued)*

Species	Ant	Location	Reported effect/mechanism	Study
Gryllus sp., Trombidiidae, Linyphiidae	*S. invicta*	TX, USA	Reduced abundance in invaded areas	Nichols & Sites 1989
Canopy arthropods	*S. invicta*	TX, USA	Fewer in ant-infested trees	Kaspari 2000
Carrion decomposers (6 families)	*S. invicta*	TX, USA	Ant presence reduces arthropod abundance	Stoker et al. 1995
Dung-breeding Diptera (5 spp.)	*S. invicta*	TX, USA	Fly production increases when ants excluded	Schmidt 1984
Coprophagous Scarab beetles	*S. invicta*	TX, USA	2 of 8 spp. lower abundance with ant	Summerlin et al. 1984a
Horn fly (*Haematobia irritans*) and *Orthellia caesarion*	*S. invicta*	TX, USA	Ants prey upon eggs and larvae	Summerlin et al. 1984b
Plant-decomposers	*S. invicta*	TX, USA	Reduced abundance and diversity, eat fruit	Vinson 1991
Dung-inhabiting spp. (Diptera, Coleoptera)	*S. invicta*	FL, USA	Abundance increases after ants excluded; ants prey on larvae, pupae and adult flies	Hu & Frank 1996
Scorpions, some spiders, invert abund.	*W. auropunctata*	Galapagos, Ecuador	Eliminated or reduced	Lubin 1984
Pseudoscorpions	*W. auropunctata*	New Caledonia	Excluded from invaded areas	Jourdan 1997
Terrestrial invertebrates	*W. auropunctata*	New Caledonia	Partial exclusion from invaded areas	Jourdan 1997
Crickets	Many species	HI, USA	Crickets reduced or absent when ants present	La Polla et al. 2000
INVERTEBRATES (HERBIVORES IN AGRICULTURAL SYSTEMS)				
Cacao weevil (*Pantorhytes szentivanyi*)	*A. gracilipes*	Papua New Guinea (cacao)	Harasses adults, causes dispersion	Baker 1972, Room & Smith 1975
Mirid bugs (2 spp.)	*A. gracilipes*	Papua New Guinea (cacao)	Disturbance of feeding and egg laying	Entwistle 1972
Araucaria looper (*Millionia isodoxa*)	*A. gracilipes*	Papua New Guinea (hoop pine)	Late instar larvae attacked	Wylie 1974
Beetle pests (3 spp.)	*A. gracilipes*	Seychelles (coconut)	Thought to attack and control pests	Lewis et al. 1976
Coconut bug (*Amblypelta cocophaga*)	*A. gracilipes*	Solomon Islands (palms)	Observations of nutfall changes	Greenslade 1971
Eucalyptus borer (*Phoracantha aemipunctata*)	*L. humile*	Central/south Portugal	Predation on eggs	Way et al. 1992
Chrysoperla carnea	*L. humile*	CA, USA (tulip trees)	98% of eggs removed on infested trees	Dreistadt et al. 1986
Citrus red mite (*Panonychus citri*)	*L. humile*	CA, USA (citrus)	Interferes with spider mite destroyer	Haney et al. 1987
Liothrips urichi	*P. megacephala*	HI, USA (*Clidemia hirta*)	Higher predation in ants exposed areas	Reimer 1988
Hemipteran and lepidopteran pests	*S. geminata*	Philippines (rice)	Ant-collected larvae, nymphs and eggs	Way et al. 1998

Organism	Ant species	Location (crop)	Effect	Reference
Golden apple snail (*Pomacea canaliculata*)	*S. geminata*	Philippines (rice)	Ants observed collecting eggs and young	Way et al. 1998
Cabbage webworm	*S. geminata*	Malaysia (cabbage)	Ants prey on prepupal and pupal stages	Sivapragasam & Chua 1997
Fall armyworm (*Spodoptera frugiperda*)	*S. geminata*	Honduras (maize)	Negative association with ant density	Cañas & O'Neil 1998
Diaprepes abbreviatus	*S. geminata*	Caribbean (citrus)	Able to prey on all life stages	Jaffe et al. 1990
Soybean looper (*Pseudoplusia includens*)	*S. geminata*	FL, USA (soybeans)	Egg predation higher when ants present	Nickerson et al. 1977
Boll weevils (*Anthonomus grandis*)	*S. invicta*	TX, USA (cotton)	Ant removal increased emergence, survival	Jones & Sterling 1979
Cotton leafworm (*Alabama argillacea*)	*S. invicta*	TX, USA (cotton)	Egg removal	Gravena & Sterling 1983
Helicoverpa zea	*S. invicta*	TX, USA (cotton)	Egg removal	Nuessly & Sterling 1994
Southern green stinkbug eggs	*S. invicta*	LA, USA (soybean)	Dominant egg predator in young soybean	Stam et al. 1987
Sugarcane borer (*Diatrea saccharalis*)	*S. invicta*	LA, USA	Predation of larvae	Negm & Hensley 1969
Fall armyworm (*Spodoptera frugiperda*)	*S. invicta*	LA, USA (sugarcane)	Higher infestations in ant-suppressed plots	Fuller et al. 1997
Weevil (*Hypera postica*)	*S. invicta*	GA, USA (greenhouse)	Larvae removal	Morrill 1978
Lacewing (*Chrysoperla rufilabris*)	*S. invicta*	GA, USA (pecans)	Eggs, larvae, and pupae decline	Tedders et al. 1990
Syrphid (*Allograpta oblique*)	*S. invicta*	GA, USA (pecans)	Ants preyed on puparia	Tedders et al. 1990
Lepidopteran larvae	*S. invicta*	OK, USA (peanuts)	Ants carried Lepidoptera larvae	Vogt et al. 2001
Herbivores (16 of 16 taxa)	*S. invicta*	AL, USA (cotton)	Negative association with ant density	Eubanks 2001
Herbivores (13 of 16 taxa)	*S. invicta*	AL, USA (cotton)	Negative association with ant density	Eubanks 2001
Fleahopper (*Pseudatomoscelis seriatus*)	*S. invicta*	Laboratory study	100% mortality when exposed to ants	Breene et al. 1990
Capsid bug (*Sahlbergella singularis*)	*W. auropunctata*	W. Africa	Reduces mirids and other insects	Entwistle 1972
INVERTEBRATES (HOMOPTERA)				
Scale (*Coccus viridis*)	*A. gracilipes*	Indonesia (coffee)	Reduced parasitism, increased growth	Van Der Goot 1916
Ceroplastes rubens, Coccus viridis	*A. gracilipes*	Seychelles (cinnamon)	5–160× more abundant when ants present	Haines & Haines 1978b
Black scale, *Saissetia oleae*	*L. humile*	Laboratory study	Decreased parasitism	Barzman & Daane 2001
Scale (*Coccus hesperidium*)	*L. humile*	Laboratory study	Decreased parasitization by 27–98%	Bartlett 1961
Red scale (*Aonidiella aurantii*)	*L. humile*	South Africa (citrus)	Associated with outbreaks	Samways et al. 1982
Mealybug (*Planococcus ficus*)	*L. humile*	South Africa (grapes)	Associated with outbreaks	Addison & Samways 2000

(Continued)

TABLE 3 *(Continued)*

Species	Ant	Location	Reported effect/mechanism	Study
Mealybug, (*Pseudococcus adonidum*)	*L. humile*	CA, USA (Cherimoyas)	Infestation correlated with ant trail intensity	Phillips et al. 1987
Walnut aphid (*Chromaphis juglandicola*)	*L. humile*	CA, USA (walnuts)	Removes parasitized aphids	Frazer & Van den Bosch 1973
Scale (*Coccus viridis*)	*P. megacephala*	HI, USA (*Pluchea indica*)	Removes predators, increases reproduction	Bach 1991
Mealybug (*Dysmicoccus neobrevipes*)	*P. megacephala*	HI, USA (pineapple)	Infestation increases with ant abundance	Beardsley et al. 1982
Mealybug (*Dysmicoccus brevipes*)	*P. megacephala*	HI, USA (pineapple)	Interferes with natural enemy behavior	Gonzalez-Hernandez et al. 1999
Scale (*Coccus viridis*)	*P. megacephala*	HI, USA (coffee)	Interferes with or preys on natural enemies	Reimer et al. 1993
Soft scale, mealybug	*P. megacephala*	South Africa (citrus)	Associated with outbreaks	Samways et al. 1982
Cassava mealybug (*Phenacoccus manihoti*)	*P. megacephala*	Ghana (cassava)	Creates tent shelter, deters natural enemies	Cudjoe et al. 1993
Mealybugs (*Planococcus njalensis*, *P. Citri*)	*P. megacephala*	Ghana (cocoa)	2–4× more abundant when ants tending	Campbell 1994
Brown citrus aphid (*Toxoptera citricida*)	*S. invicta*	Puerto Rico (citrus)	Decreases natural enemy larvae	Michaud & Browning 1999
Corn leaf aphid (*Rhopalosiphum maidis*)	*S. invicta*	Laboratory study	Increases parasitoid search time, destroys mummies	Vinson & Scarborough 1991
Cotton aphid (*Aphis gossypii*)	*S. invicta*	Laboratory study	Aphid predators ineffective when ants present	Vinson & Scarborough 1989
Cowpea aphid (*Aphis craccivora*)	*S. invicta*	Laboratory study	Deters coccinellid aphid predators	Dutcher et al. 1999
Mealybug (*Planococcus citri*)	*W. auropunctata*	Bahia, Brazil (cocoa)	4× more abundant when ants present	De Souza et al. 1998

Porter & Savignano 1990 [but see Morrison 2002]), canopy arthropods (Kaspari 2000), and decomposers (Vinson 1991, Summerlin et al. 1984a, Stoker et al. 1995, Hu & Frank 1996). Argentine ants also affect other invertebrates negatively, but studies report mixed results (Table 3), ranging from little apparent impact (Holway 1998a) to declines in the abundance of two orders (Human & Gordon 1997), three orders (Bolger et al. 2000), and eight orders (Cole et al. 1992).

Much of the evidence from which we draw conclusions about community-level effects on invertebrates comes from studies that compare faunas of invaded areas with those of comparable, uninvaded areas. If carefully designed and replicated, such comparisons can yield insights into a wide variety of impacts associated with ant invasions. This approach, however, has a number of important limitations. First, invaded and uninvaded sites may differ inherently with respect to environmental variables such as soil moisture, soil type, elevation, disturbance history, distance to edge, or the presence of other invaders. In such cases, the effects of invasive ants can be difficult to tease apart from the effects of covarying environmental variation (but see Bolger et al. 2000). Second, in community-level comparisons, the effects of invasive ants on a single species may be hard to detect or to test for statistically; this is especially true for rare species. Moreover, it is often difficult to generalize about effects on a given taxon from one community to another. For example, whereas some studies report negative effects of invasive ants on spiders (Lubin 1978, Haines & Haines 1978b, Cole et al. 1992, Gillespie & Reimer 1993), others either failed to detect significant effects (Porter & Savignano 1990, Human & Gordon 1997, Holway 1998a) or found positive associations (Bolger et al. 2000). Taken together, these results permit little insight into the kinds of interactions that take place between spiders and invasive ants. It seems likely that certain spiders (e.g., ground-dwelling species, Hawaiian endemics) are more affected by invasive ants than others (e.g., web-building species, species that evolved with ants), but even these generalizations are not well supported by data.

IMPACTS OF INVASIVE ANTS ON VERTEBRATES   Considerable correlative and limited experimental evidence suggests that vertebrate populations may also decline as a result of ant invasions. Putatively affected taxa (Table 3) include mammals (Killion & Grant 1993, Ferris et al. 1998, Meek 2000, Laakkonen et al. 2001), lizards (Donaldson et al. 1994, Feare 1999, Jourdan et al. 2001, Fisher et al. 2002), and birds (Allen et al. 1995, Feare 1999). Decreases in vertebrate populations have been attributed to most invasive ant species, including *L. humile* (Laakkonen et al. 2001, Fisher et al. 2002), *W. auropunctata* (Jourdan et al. 2001), and *A. gracilipes* (Feare 1999, Meek 2000), although most reports involve *S. invicta* (Wojcik et al. 2001).

Whereas a causal relationship between ant invasion and vertebrate population decline is commonly suggested, the specific mechanisms responsible are often obscure. Predation is frequently argued to be important; this is especially true for studies on *S. invicta*. But accounts are often limited to situations in which animals cannot escape attack (Table 3). Small mammals in cages or traps, for example, are vulnerable to predation by *S. invicta* (Masser & Grant 1986, Flickinger 1989) (Table 3). Vertebrates also suffer nest predation by invasive ants. *Solenopsis invicta* has been implicated in nest failure for at least seven species of birds and nine species of reptiles (Table 3). The impacts of other invasive ants on the nesting success of vertebrates remain understudied, but both *A. gracilipes* and *L. humile* may cause nest failure in some bird species (Table 3). In some cases, it is unclear if ants caused nest failure or simply recruited to dead or dying nestlings. Moreover, it is often uncertain whether invasive ants reduce nesting success more than native ants do (Travis 1938, Chalcraft & Andrews 1999).

In addition to predation, invasive ants may affect vertebrates through other means. High densities of invasive ants may reduce the suitability of nest sites (Ridlehuber 1982) and alter behavioral patterns (Pedersen et al. 1996, Holtcamp et al. 1997), possibly increasing susceptibility to predation (Mueller et al. 2001). Changes in arthropod communities associated with ant invasions may also contribute to declines of insectivorous vertebrates including loggerhead shrikes [Lymn & Temple 1991 (but see Yosef & Lohrer 1995)], northern bobwhites (Allen et al. 1995), and horned lizards (Donaldson et al. 1994, Suarez et al. 2000).

The link between ant invasions and vertebrate declines is also supported by limited experimental evidence. Research on least terns *Sterna antillarum* (Lockley 1995), northern bobwhites *Colinus virginianus* (Mueller et al. 1999), and colonial waterbirds (Drees 1994) demonstrates that suppression of *S. invicta* can enhance nesting success between 27% (Lockley 1995) and 92% (Drees 1994). Similarly, captures of northern pygmy mice *Baiomys taylori* increased by over 50% after six months of fire ant suppression (Killion et al. 1995). It should be noted, however, that large-scale ant suppression using pesticides probably results in changes beyond the reduction of invasive ants (Yosef & Lohrer 1995, Hill & Dent 1985). Pesticides can negatively affect species directly through poisoning (Collins et al. 1974, Hill & Dent 1985, Williams et al. 2001) and indirectly through the reduction of arthropod prey (Lymn & Temple 1991).

To illustrate better the diversity of mechanisms by which invasive ants may affect vertebrate populations, we describe in more detail two relatively well-studied examples: the northern bobwhite (*C. virginianus*) and the coastal horned lizard (*Phrynosoma coronatum*).

Northern bobwhites and red imported fire ants   Several lines of evidence link declining northern bobwhite populations in the southeastern United States to invasion by *S. invicta*. First, significant correlations exist between the timing of fire ant infestation and drops in bobwhite density estimated from Christmas bird counts in Texas (Allen et al. 1995) and in Florida and South Carolina (Allen et al. 2000). Second, exposure to red imported fire ants decreases the growth rates and survival of chicks (Giuliano et al. 1996) and alters their time budgets, reducing time available for sleeping and foraging (Pedersen et al. 1996). Lastly, suppression of *S. invicta* leads to increases in both chick survival (Mueller et al. 1999) and adult density (Allen et al. 1995). It should be noted, however, that native fire ants

(*S. geminata*) also decrease nesting success in northern bobwhites (Travis 1938). The role of *S. invicta* in the decline of bobwhites has been debated in part for this reason (Brennan 1993).

Coastal horned lizards and Argentine ants   The invasion of *L. humile* into southern California is also correlated with the decline of a vertebrate species, the coastal horned lizard. This reptile has disappeared from up to 50% of its former range as a result of habitat destruction and collection for the pet trade (Fisher et al. 2002). However, portions of its remaining range, particularly in coastal California, may be unsuitable owing to invasion by Argentine ants. Like other *Phrynosoma* (Pianka & Parker 1975), the diet of *P. coronatum* consists primarily of ants, particularly large harvester species (e.g., *Messor* and *Pogonomyrmex*) that can constitute over 50% of their prey (and ≫50% of prey mass) (Suarez et al. 2000). Like many aboveground foraging ants, harvester ants are vulnerable to ant invasions (Hook & Porter 1990; Human & Gordon 1996; Suarez et al. 1998, 2000). Moreover, Argentine ants are unsuitable nutritional surrogates for native ants (Suarez & Case 2002). Hatchling horned lizards lose weight when raised on either Argentine ants or arthropods typical of invaded communities, whereas hatchlings raised on *Crematogster californica*, a common native ant, were able to maintain growth rates comparable to those of wild lizards (Suarez & Case 2002). Horned lizards avoid eating Argentine ants in the field, possibly because of their small size, noxious chemical defenses, or aggressive mobbing behavior (Suarez et al. 2000). It is also possible that Argentine ants can cause nest failure in horned lizards, although this remains to be tested. As a consequence of these factors, horned lizards are either absent from or occur at low densities in areas occupied by Argentine ants in coastal southern California (Fisher et al. 2002). The effects of invasive ants may extend to other horned lizard species as well. In Texas, the red imported fire ant has been implicated in the decline of the Texas horned lizard (*Phrynosoma cornutum*) (Donaldson et al. 1994).

Effects on Mutualistic Interactions

Ants enter into a variety of mutualistic interactions with plants and other insects. These interactions may be obligate or facultative, loose associations or species-specific, and may not always be mutually positive. How these interactions change in the context of ant invasions is a largely unexplored line of research teeming with questions of evolutionary, behavioral, and ecological significance. Much of what is known comes from agricultural settings. The extent to which patterns observed in agro-ecosystems occur in less manipulated settings remains to be documented in detail.

HOMOPTERA   The relationship between ants and honeydew-excreting Homoptera including scale insects, mealybugs, aphids, and treehoppers, is well known (Way 1963, Buckley 1987, Hölldobler & Wilson 1990). Among the benefits Homoptera derive are protection from natural enemies, removal of exudates that may otherwise foul the immediate environment, increased feeding potential, and relocation to more favorable parts of the host plant (Way 1963). In exchange, ants acquire a reliable, defendable source of carbohydrate-rich food (Sudd 1987, Hölldobler & Wilson 1990). Whereas partnerships are often facultative and non-species specific (Way 1963), certain pairings may yield mutually higher benefits than others (Greenslade 1972, Bristow 1984, Gaume et al. 1998).

Although there have been few direct comparisons among invasive and other ants, the presence of invasive ants is frequently associated with local increases in homopteran abundance, both in the introduced and native ranges (Table 3). It should be noted, however, that *Solenopsis* may be an exception to this general pattern (Adams 1986, Tedders et al. 1990, Clarke & DeBarr 1996, Dutcher et al. 1999; but see Vinson & Scarborough 1989, Michaud & Browning 1999). In spite of these numerous reports, it is often unclear why invasive ants are exceptional tenders relative to native ants, and research examining the dynamics of these interactions in nonagricultural settings is almost completely lacking. Ants with modified crops, such as *L. humile* and *A. gracilipes*, can ingest relatively large quantities of liquid food, allowing them to excel at collecting honeydew (Eisner 1957, Davidson 1998). In addition, because tending ants can be a limiting resource for Homoptera (Sudd 1987, Cushman & Whitham 1991, Breton & Addicott 1992), the high abundance achieved by invasive ants may remove this limitation and allow Homoptera to thrive. Invasive ants may be especially effective at deterring natural enemies of Homoptera (Table 3).

Why is sustaining high densities of Homoptera beneficial for ants? As discussed above, access to carbohydrate-rich resources may be related to ecological dominance in ant communities (Davidson 1997, 1998). Empirical evidence on this point is scarce, correlative, and limited to agricultural settings but generally supports the hypothesis. The presence of Homoptera appears necessary for the maintenance of *A. gracilipes* in cocoa plantations in Papua New Guinea (Baker 1972). Likewise, in cocoa in Ghana, *P. megacephala* achieves dominance only when the Homoptera with which it is most closely associated are present (Campbell 1994). Similarly, *L. humile* does not appear to become dominant in South African vineyards with low levels of Homoptera (Addison & Samways 2000).

High densities of Homoptera associated with invasive ants may have repercussions for the host plant. Since Homoptera feed on plant phloem, large aggregations can lead to direct damage, fouling from mold, and higher susceptibility and exposure to phytopathogens (Way 1963, Buckley 1987). Evidence is plentiful. The ants listed in Table 1 have all variously been classified as pests because of their tending ability (Table 3). Alternatively, the ant-Homoptera mutualism may be beneficial to the host plant in cases where ants attack other herbivores (Messina 1981, Compton & Robertson 1988). For example, in Portugal, aphids attract *L. humile* to pines where the ants in turn prey upon larvae of the pine processionary moth, a major defoliator (Way et al. 1999). Three-cornered alfalfa nymphs attract *S. geminata* to soybean plants where they remove 77% of soybean looper eggs; only 37% of eggs are removed from plants lacking nymphs (Nickerson et al. 1977). In addition, ants benefit plants when they remove enough honeydew to prevent sooty mold (Bach 1991).

PLANTS   Ant-plant mutualisms range from obligate interactions involving specialized domatia or food structures characteristic of true myrmecophytes to more facultative, nonspecific interactions (Buckley 1982, Keeler 1989, Hölldobler & Wilson 1990, Huxley & Cutler 1991). Not surprisingly, invasive ant-plant relationships fall in the facultative, nonspecific end of the spectrum and include tending, seed dispersal, and interactions in flowers. When native ants are displaced, invaders may usurp their roles and alter the dynamics of the interaction, or they may fail to replace natives functionally, in some cases disrupting relationships beneficial to the plant (Lach 2002). Invasive ants may also interact with plants in ways that native ants do not, to the potential detriment of the plant. Moreover, positive and negative effects of the same ant on the same plant may counteract or combine; such variability makes it difficult to generalize about the effects of invasive ants on plants (Lach 2002).

Flowers and pollination   If the ability to capitalize on carbohydrate-rich resources is important to becoming invasive, we might expect invasive ants to be attracted to floral nectar. However, few studies have examined associations among invasive ants and flowers. The acceptability of floral nectar to ants generally has been debated (Janzen 1977, Baker & Baker 1978, Feinsinger & Swarm 1978), and while it is clear that ants are repelled by the chemical or mechanical defenses of the flowers of some species (Willmer & Stone 1997, Ghazoul 2000), they readily consume nectar from the flowers of others (Haber et al. 1981, Koptur & Truong 1998).

Since ants are notoriously poor pollinators (Beattie et al. 1984, Hölldobler & Wilson 1990, Peakall et al. 1991), ants that are able to use floral nectar may be doing so at a cost to both the plant and legitimate pollinators. Data supporting these hypotheses are scant, but Buys (1987) found that Argentine ants exploited 42% of black ironbark nectar before honeybees began foraging, and Visser et al. (1996) documented a decline in arthropod visitors to *Protea nitida* flowers when *L. humile* was present in high numbers. Argentine-ant associated declines in seed set have been suspected (Potgieter 1937, Durr 1952), but unequivocal evidence is so far lacking (Buys 1990). Alternatively, the presence of invasive ants in flowers may enhance pollination, if it results in increased repositioning frequency of pollinators (Lach 2002). Other invasive ants are also known to visit flowers, sometimes to the observed detriment of the plant (Knight 1944, Adams 1986, Lofgren 1986, Hara & Hata 1992, Hata et al. 1995). Detrimental effects on pollinators may also occur through interactions away from the plant (e.g., Cole et al. 1992) (Table 3).

Extrafloral nectaries   As with flowers and honeydew-excreting Homoptera, we would expect that invasive ants exploit extrafloral nectaries (EFNs) as a carbohydrate-rich resource. EFNs are generally attractive to ants (Carroll & Janzen 1973), and a number of hypotheses exist to account for the association. Plants with EFNs may attract ants in order to deter herbivores (Bentley 1977, Buckley 1982) or to distract ants from potentially detrimental activities such as tending Homoptera (Becerra & Venable 1989) or visiting flowers (Zachariades & Midgley 1999).

The paucity of research on invasive ant-EFN interactions precludes concluding whether invasive ants are more or less likely than native ants to be attracted to EFNs and to fulfill any of the roles played by native ants. There are scattered reports of invasive ants, both in their native and introduced ranges, visiting EFNs (Meier 1994), and a few of these studies measure the effects of the ants on the plant (Koptur 1979, Agnew et al. 1982, de la Fuente & Marquis 1999, Fleet & Young 2000) compared to other ants (Horvitz & Schemske 1984, Freitas et al. 2000, Hoffmann et al. 1999, Ness 2001). Although EFNs attract ants generally, the invasive ant-plant interaction may differ from the native ant-plant interaction if the invaders and native ants diverge in their nutritional preferences, periods of activity, foraging behavior, interactions with herbivores, or abundance (Lach 2002). For example, *S. invicta* visits *Catalpa bignonioides* EFNs less frequently than native ants because of differences in seasonal diet preferences; however, it is exceptionally intolerant of herbivores, so plant protection is not diminished (Ness 2001).

Seed dispersal   Seed dispersal by ants, or myrmecochory, is another type of mutualism between ants and plants. Ants transport seeds away from a parent plant, often in exchange for an elaiosome, a lipid-rich attachment to the seed (Buckley 1982, Beattie 1985). The few studies to date suggest that invasive ants may be poor seed dispersers relative to at least some ants they displace. In South Africa, Argentine ants displace most native ants that are effective seed dispersers, but they fail to disperse or to bury seeds, instead eating the elaiosome and leaving the seed aboveground where it is susceptible to rodent predation and fire (Bond & Slingsby 1984). Some of the smaller native ants are able to coexist with *L. humile* and continue dispersing small seeds, but displacement of the larger native ants may lead to declines in large-seeded plant species (Christian 2001). Similarly, in Corsica, *L. humile* appears less effective than *Aphaenogaster spinosa*, a dominant native ant, at dispersing the seeds of a rare endemic plant, but the consequences for the plant's population dynamics are unclear (Quilichini & Debussche 2000). Other invasive ants also may affect seed dispersal. Red imported fire ants, for example, collect the seeds of eliaosome-bearing herbaceous plants in South Carolina and leave them scarified and exposed on their trash piles (Zettler et al. 2001). In Australia, *P. megacephala* outcompetes native ants on some rehabilitated sand mines where it takes seeds of the elaiosome bearing *Acacia concurrens* (Majer 1985). Whereas both *S. geminata* and *W. auropunctata* interfere with seed dispersal of a myrmecochorous herb within their presumed native range in Mexico (Horvitz & Schemske 1986), any effect on seed dispersal in their introduced ranges is undocumented. Additional experimental studies will clarify the extent to which invasive ants affect plant communities through the disruption of ant-mediated seed dispersal. Experiments that control for the effects of invasive ants on plant vigor and reproduction (e.g., pollination) and other confounding variables will be particularly valuable.

OTHER MUTUALISMS   Hölldobler & Wilson (1990) review symbiotic relationships between ants and other arthropods, a small subset of which can be considered mutualistic. Clearly, those species involved in obligate and species-specific relationships with native ant species that are vulnerable to displacement by invasive ants may themselves succumb to local extinction following invasion. Despite this concern, little research has been published on how ant invasions affect the ecology of myrmecophilic arthropods. Argentine ants, for example, may imperil lycaenid butterflies in South Africa because they displace the native ants that tend them, but probably do not fulfill their tending roles (A. Heath, unpublished observation). Whereas *A. gracilipes* participates in mutualistic relationships with a coprophagous reduviid bug in India (Ambrose & Livingstone 1979) and two species of coreid bugs on Malaysian bamboo (Maschwitz et al. 1987), and tends larvae of a lycaenid butterfly in Sulawesi (Kitching 1987), the origins of this ant are disputed, and it is unclear whether these constitute new associations or coevolved relationships.

Other Ant-Plant Interactions

DIRECT IMPACTS ON PLANTS   Direct effects of ants on plants include soil excavation around root systems, herbivory, and seed predation. Fire ants, for example, damage plants (Taber 2000) and frequently incorporate plant materials in their diet (Risch & Carroll 1986, Trabanino et al. 1989, Tennant & Porter 1991). In India, *S. geminata* attacks cucumber, tomato, cotton, and potato crops (Lakshmikantha et al. 1996). *Solenopsis invicta* also damages seeds, seedlings, and root systems of a variety of agricultural crops (Adams 1986, Banks et al. 1991, Drees et al. 1991, Vinson 1997, Shatters & Vander Meer 2000). Effects are likely not limited to agricultural systems; only 18 of 96 crop and noncrop seed species tested with *S. invicta* colonies in a laboratory experiment were resistant to damage (Ready & Vinson 1995). It is important to note, however, that in the southeastern United States, *S. invicta* commonly displaces *S. geminata*, a species that exhibits an even greater preference for seeds (Tennant & Porter 1991). Therefore, the impacts of *S. invicta* as a seed predator (e.g., Zettler et al. 2001) must be considered in the context of declining populations of *S. geminata*. Although most reports of invasive ants damaging plants focus on fire ants, *A. gracilipes* undermines the roots of several agricultural plants (Haines & Haines 1978b, Veeresh 1990), and *L. humile* damages figs and orange blossoms (Newell & Barber 1913) and spreads avocado stem canker (El Hamalawi & Menge 1996). Because of their high abundance, invasive ants may damage plants to a greater extent than do native ant species, but few direct comparisons exist.

EFFECTS ON HERBIVORES AND HERBIVORE ENEMIES   As discussed above, invasive ants prey upon a wide variety of invertebrates including herbivores that are important plant pests. Supportive evidence includes direct observations, correlations in abundance between ants and herbivores, and controlled cage experiments (Table 3). As yet, there is no evidence for a hierarchy of prey desirable or acceptable to predaceous ants (Way & Khoo 1992). As for studies examining the predatory habits of invasive ants generally, the majority of studies on herbivore predation focus on *S. invicta*, which preys upon or drives off numerous species of insect herbivores, sometimes to the benefit of the plant, in a diversity of agricultural systems (Table 3; reviewed in Taber 2000). Similarly, populations of *A. gracilipes* in cacao (Baker 1972, Room & Smith 1975) and planted hoop pine (Wylie 1974), and *W. auropunctata* in cacao in West Africa (Entwistle 1972) are encouraged for their detrimental effects on economically important herbivores.

The predatory habits of invasive ants may also harm beneficial insects, resulting in negative impacts for the plants on which the interactions occur. In Zanzibar, for example, neither *A. gracilipes* nor *P. megacephala* affects the coconut bug, *Pseudotheraptus wayi*, but both species displace the native weaver ant, *Oecophylla longinoda*, an effective predator of this pest (Way 1953, Zerhusen & Rashid 1992). *Pheidole megacephala* also displaces beneficial ants from coconut palms in the Solomon Islands but fails to fulfill their roles as predators of the coconut bug, *Amblypelta cocophaga* (Greenslade 1971). In Malaysia and Indonesia, *A. gracilipes* eliminates the native ants (*Oecophylla smaragdina* and *Dolichoderus* spp.) that protect cacao against mirid *Helopeltis* spp., but it in turn fails to prey on these pests (Way & Khoo 1989). Invasive ants may also prey on or otherwise displace non-ant enemies of herbivores (Table 3).

Other Effects

OBLIGATE ASSOCIATES AND VISUAL MIMICS   Ants generally support a rich fauna of associates, many of which are other insects (Hölldobler & Wilson 1990). Some of these taxa form obligate, species-specific associations with ants; examples include mymecophilic beetles as well as dipteran and hymenopteran parasitoids. Although little studied from the perspective of ant invasions, such taxa would seem highly vulnerable, especially because many obligate associates of ants are rare and local to begin with (Hölldobler & Wilson 1990). For example, *S. geminata* supports species-specific phorid fly parasitoids (Morrison et al. 1999) that almost certainly decline in abundance as their host is displaced by *S. invicta* throughout the southeastern United States. Arthropods (mostly insects and spiders) are also visual mimics of ants. Through their superficial resemblance to ants, some of these mimics must enjoy safety from predators uninterested in ants as prey. To the extent that native ants serve as models to support the existence of such mimicry, visual mimics also seem in jeopardy from ant invasions.

SOIL CHEMISTRY, TURNOVER AND EROSION   Because the nesting activities of ants turn over large quantities of soil and alter its chemistry and physical structure (Hölldobler & Wilson 1990, Jolivet 1996, Folgarait 1998), the replacement of native ants by invasive ants might generate ecosystem-level effects. Such changes might be especially important in situations where the nesting behaviors of native and invasive ants differ greatly. For example, in coastal California, Argentine ants displace *Messor* and *Pogonomyrmex* harvester ants (Erickson 1971, Human & Gordon 1996, Suarez et al. 1998). These harvester ants construct deep, long-lived nests in which seeds are cached and refuse (rich in organic matter and often including uneaten seeds) is discarded in a midden surrounding the nest entrance (MacMahon et al. 2000). Argentine ants, in contrast, typically occupy short-lived nest sites and usually fail to penetrate very deeply underground. It thus seems likely that the replacement of harvester ants by Argentine ants alters soil characteristics. The importance of such effects is unknown but deserves further scrutiny.

SYNERGISTIC EFFECTS   In some circumstances, the success of invasive ants may be facilitated by other invaders. Such mutually positive interactions may be a common feature of invasions (Simberloff & Von Holle 1999), but their frequency and importance with respect to ant invasions are not well known. Such interactions probably do occur in this context and deserve closer scrutiny. In Australia, for example, *P. megacephala* tends an EFN-bearing weed and may encourage its spread by deterring herbivores; the ant presumably benefits as well through the acquisition of food resources (Hoffmann et al. 1999). See Koptur (1979) for a similar example involing *L. humile* and a weedy vetch in California. The evidence linking ant dominance to availability of carbohydrate-rich resources suggests that honeydew-producing Homoptera too might facilitate the spread of invasive ants and vice versa. Bach (1991) describes such an interaction between *P. megacephala*, a non-native homopteran, and an introduced plant in Hawaii. Although not interpreted in terms of the spread of these non-native organisms, Bach's study illustrates that complexes of non-native species could invade in concert.

DIRECTIONS FOR FUTURE RESEARCH

In this review, we have attempted to synthesize a wealth of published information concerning the causes and consequences of ant invasions. Whereas many recent studies have enhanced a general understanding of these invasions, at the same time, they point to large gaps in knowledge. Given the focus on *L. humile* and *S. invicta*, there is an obvious need for research on additional species of currently or potentially invasive ants, especially those invading tropical environments. Below, we outline what we consider to be other key research needs.

Comparisons of Native and Introduced Populations

It is remarkable that almost all of what is known about the biology of invasive ants comes from studies of introduced populations. Although the same could be said of other invasive species (Steneck & Carlton 2001), this bias seems especially prominent for invasive ants. For example, accurate information about the location and boundaries of native ranges for most of the species listed in Table 1 either remains incomplete or is lacking all together. Two recent studies (Ross et al. 1996, Tsutsui et al. 2000) demonstrate the extent to which introduced populations can differ from native populations and in doing so make it clear that native populations should serve as an essential benchmark for any evolutionary inference. Given the likelihood of differences between native and introduced populations, between-range comparisons have great potential to add to what is known about ant invasions. Such comparisons will aid in the identification of geographic origins (Ross & Trager 1990, Tsutsui et al. 2001), clarify poorly resolved taxonomic and phylogenetic relationships, and elucidate the forces responsible for transitions in social organization (Ross et al. 1996, Tsutsui et al. 2000). Comparisons of native and introduced populations of invasive ants may also shed light on the relative importance of competitive release from native ants (Buren 1983), escape from natural enemies (Porter et al. 1997), and shifts in colony organization (Holway et al. 1998, Tsutsui et al. 2000) as factors influencing invasion success.

More Experimental, Large-Scale and Long-Term Studies

As is true for invasion biology generally, ecological research on invasive ants has largely been correlative or observational. A more comprehensive understanding of the causes and consequences of ant invasions will be achieved only through the implementation of manipulative experiments and studies conducted at larger spatial or temporal scales.

The value of manipulative field experiments in invasion biology is constrained somewhat by ethical concerns associated with introducing known invaders into new areas as an experimental treatment, but experiments nonetheless hold promise as a means to clarify both the causes and the consequences of ant invasions. Both short-term removals (Morrison 1996, Holway 1999) and short-term introductions (Bhatkar et al. 1972; Roubik 1978, 1980; Schaffer et al. 1983; Torres 1984; Human & Gordon 1996, 1999; Holway 1999) allow the study of behavioral interactions between native and introduced species of social insects. Although longer-term introduction experiments are often ethically untenable, a greater number of long-term removals (or partial removals), especially ones conducted at the leading edge of invasion fronts, should be attempted. Examples include studies that use pesticides to lower the density of red imported fire ants (Howard & Oliver 1978, Sterling et al. 1979, Allen et al. 1995, Adams & Tschinkel 2001). Although the confounding effects of pesticide treatment need to be carefully considered, experiments that lower the density of invasive ants can be highly informative.

Increasing the spatio-temporal scale of invasive ant research is also important. Some of the most dramatic examples of the impacts associated with ant invasions come from long-term (Erickson 1971, Greenslade 1971) or large-scale (Gotelli & Arnett 2000) studies. Long-term studies are an especially powerful means by which to study the ecological effects of ant invasions, as they allow explicit before-and-after comparisons of the same physical areas. A recent study by Morrison (2002) exemplifies this approach and illustrates how the ecological effects of ant invasions can vary greatly through time. Although few examples exist, studies conducted across large spatial scales can clarify the determinants of geographic variation in invasion success, the factors governing range limits, and the extent to which the effects of ant invasions exhibit scale dependency.

Better Estimates of Density and Biomass

A comprehensive understanding of the ecological effects of ant invasions is also hindered by a lack of quantitative comparisons of the density and biomass of invasive and native ants. Such comparisons are needed to gauge the impacts associated with ant invasions but are also of more general interest in that they provide information on how the density and biomass of an important group of consumers are related to diversity.

The red imported fire ant is perhaps unique among invasive ants in that colony density and biomass can be estimated from mound number and mound volume, respectively (Tschinkel 1992, Porter et al. 1992). Interesting and informative comparisons of density and biomass thus exist for native and introduced populations (Porter et al. 1992, 1997) and areas occupied by the two social forms in the introduced range (Macom & Porter 1996). It is also possible to relate measures of colony size and biomass in *S. invicta* to territory area (Tschinkel et al. 1995).

Less progress has been made in trying to measure the density or biomass of other invasive ants or to compare such measures with those of native ants. Such comparisons are difficult to make in part because, unlike *S. invicta*, other invasive ants maintain diffuse supercolonies composed of ephemeral and poorly defined nests that may differ greatly from one another in size.

Prevention and Control

As information accumulates concerning the ecology of invasive ants, a framework for identifying potential invaders will hopefully be constructed from common features and knowledge of mechanisms. In the interim, a "guilty until proven innocent" policy (Ruesink et al. 1995) seems warranted given the great difficulty involved in eradicating established populations and the numerous problems that can result from successful invasion. Although ants have been advocated as agents of biological control in agricultural settings and may in fact be useful in such circumstances (Way & Khoo 1992), introducing any ant species into a new location seems unwise.

As with invasions of other organisms, identifying new infestations of invasive ants as rapidly as possible must greatly increase opportunities for eradication. Established populations present difficult challenges in that eradication over large areas is unfeasible and even local management, at present, remains difficult (Davidson & Stone 1989, Williams et al. 2001). Nonetheless, the development of integrated pest management strategies that incorporate both time-honored approaches and innovative ideas should remain an important goal. For example, if unicolonial-like colony structures are an important determinant of the high densities of some invasive ants, then tactics that lead to the dissolution of supercolonies through increased intraspecific aggression could be profitable lines of attack (Suarez et al. 1999). Such strategies, while not resulting in eradication, could be used in concert with more traditional approaches to decrease both the magnitude and variety of negative ecological effects associated with these invasions.

ACKNOWLEDGMENTS

The authors are grateful to the numerous reviewers who provided many detailed and insightful comments: L.E. Alonso, A.N. Andersen, C.J. De Heer, D.H. Feener, L. Greenberg, M. Kaspari, P. Krushelnycky, T.P. McGlynn, L.W. Morrison, J.H. Ness, H. Robertson, D. Simberloff, and P.S. Ward. Research support was provided by the following: USDA NRICGP 99-35302-8675 (DAH), EPA Science to Achieve Results fellowship and NSF dissertation enhancement grant (LL), USDA NRICGP 00-35302-9417 and Miller Institute for Basic Research in Science (AVS), NSF OCE 99-06741 (to R.K. Grosberg) (NDT), and NSF DEB 96-10306 and Metropolitan Water District of Southern California (TJC).

The *Annual Review of Ecology and Systematics* is online at
http://ecolsys.annualreviews.org

LITERATURE CITED

Adams CT. 1986. Agricultural and medical impact of the imported fire ants. In *Fire and Leaf Cutting Ants: Biology and Management*, ed. CS Lofgren, RK Vander Meer, pp. 48–57. Boulder, CO: Westview

Adams CT, Banks WA, Plumley JK. 1976. Polygyny in the tropical fire ant, *Solenopsis geminata*, with notes on the red imported fire ant, *Solenopsis invicta*. *Fla. Entomol.* 59:411–16

Adams ES. 1990. Boundary disputes in the territorial ant *Azteca trigona*: effects of asymmetries in colony size. *Anim. Behav.* 39:321–28

Adams ES. 1998. Territory size and shape in fire ants: a model based on neighborhood interactions. *Ecology* 79:1125–34

Adams ES, Traniello JFA. 1981. Chemical interference competition by *Monomorium minimum* (Hymenoptera: Formicidae). *Oecologia* 51:265–70

Adams ES, Tschinkel WR. 1995a. Density-dependent competition in fire ants: effects

on colony survivorship and size variation. *J. Anim. Ecol.* 64:315–24

Adams ES, Tschinkel WR. 1995b. Spatial dynamics of colony interactions in young populations of the fire ant *Solenopsis invicta*. *Oecologia* 102:156–63

Adams ES, Tschinkel WR. 2001. Mechanisms of population regulation in the fire ant *Solenopsis invicta*: an experimental study. *J. Anim. Ecol.* 70:355–69

Addison P, Samways MJ. 2000. A survey of ants (Hymenoptera: Formicidae) that forage in vineyards in the Western Cape Province, South Africa. *Afr. Entomol.* 8:251–60

Adler FR. 1999. The balance of terror: an alternative mechanism for competitive trade-offs and its implications for invading species. *Am. Nat.* 154:497–509

Agnew CW, Sterling WL, Dean DA. 1982. Influence of cotton nectar on red imported fire ants and other predators. *Environ. Entomol.* 11:629–34

Agosti D, Majer JD, Alonso LE, Schultz TR,

eds. 2000. *Ants—Standard Methods for Measuring and Monitoring Biodiversity.* Washington, DC: Smithsonian Inst.

Aho T, Kuitunen M, Suhonen J, Jantti A, Hakkari T. 1999. Reproductive success of Eurasion tree creepers, *Certhia familiaris*, lower in territories with wood ants. *Ecology* 80:998–1007

Allen CR, Demarais S, Lutz RS. 1997a. Effects of red imported fire ants on recruitment of white-tailed deer fawns. *J. Wildl. Manag.* 6: 911–16

Allen CR, Forys EA, Rice KG, Wojcik DP. 2001. Effects of fire ants (Hymenoptera: Formicidae) on hatching turtles and prevalence of fire ants on sea turtle nesting beaches in Florida. *Fla. Entomol.* 84:250–53

Allen CR, Lutz SR, Demarais S. 1995. Red imported fire ant impacts on Northern Bobwhite populations. *Ecol. Appl.* 5:632–38

Allen CR, Rice KG, Wojcik DP, Percival HF. 1997b. Effect of red imported fire ant envenomization on neonatal American alligators. *J. Herpetol.* 31:318–21

Allen CR, Willey RD, Myers PE, Horton PM, Buffa J. 2000. Impact of red imported fire ant infestation on northern bobwhite quail abundance trends in southeastern United States. *J. Agric. Urban Entomol.* 17:43–51

Ambrose DP, Livingstone D. 1979. On the bioecology of *Lophocephala guerini* (Reduviidae: Harpactorinae) a coprophagous reduviid from the Palghat Gap, India. *J. Nat. Hist.* 13:581–88

Andersen AN. 1992. Regulation of "momentary" diversity by dominant species in exceptionally rich ant communities of the Australian seasonal tropics. *Am. Nat.* 140: 401–20

Andersen AN. 1997. Functional groups and patterns of organization in North American ant communities: a comparison with Australia. *J. Biogeogr.* 24:433–60

Andersen AN, Blum MS, Jones TH. 1991. Venom alkaloids in *Monomorium "rothsteini"* Forel repel other ants: Is this the secret to success by *Monomorium* in Australian ant communities? *Oecologia* 88:157–60

Andersen AN, Patel AD. 1994. Meat ants as dominant members of Australian ant communities: an experimental test of their influence on the foraging success and forager abundance of other species. *Oecologia* 98: 15–24

Apperson CS, Powell EE. 1984. Foraging activity of ants (Hymenoptera: Formicidae) in a pasture inhabited by the red imported fire ant. *Fla. Entomol.* 67:383–93

Aron S. 2001. Reproductive strategy: an essential component in the success of incipient colonies of the invasive Argentine ant. *Insect Soc.* 48:25–27

Bach CE. 1991. Direct and indirect interactions between ants (*Pheidole megacephala*), scales (*Coccus viridis*) and plants (*Pluchea indica*). *Oecologia* 87:233–39

Baker G. 1972. The role of *Anoplolepis longipes* Jerdon (Hymenoptera: Formicidae) in the entomology of cacao in the northern district of Papua New Guinea. *14th Int. Entomol. Congr., Aug. 22–30, Canberra, Aust.*

Baker HG, Baker I. 1978. Ants and flowers. *Biotropica* 10:80

Banks WA, Adams CT, Lofgren CS. 1991. Damage to young citrus trees by the red imported fire ant (Hymenoptera: Formicidae). *J. Econ. Entomol.* 84:241–46

Banks WA, Plumley JK, Hicks DM. 1973. Polygyny in a colony of the fire ant *Solenopsis geminata*. *Ann. Entomol. Soc. Am.* 66:234–35

Banks WA, Williams DF. 1989. Competitive displacement of *Paratrechina longicornis* (Latreille) (Hymenoptera: Formicidae) from baits by fire ants in Mato Grosso, Brazil. *J. Entomol. Sci.* 24:381–91

Barber ER. 1916. The Argentine ant: distribution and control in the United States. *Bur. Entomol. Bull.* 377:1–23

Baroni Urbani C, Kannowski PB. 1974. Patterns in the red imported fire ant settlement of a Louisiana pasture: some demographic parameters, interspecific competition and food sharing. *Environ. Entomol.* 3:755–60

Bartlett BR. 1961. The influence of ants upon

parasites, predators, and scale insects. *Ann. Entomol. Soc. Am.* 54:543–51

Barzman MS, Daane KM. 2001. Host-handling behaviours in parasitoids of the black scale: a case for ant-mediated evolution. *J. Anim. Ecol.* 70:237–47

Beardsley JW, Su TH, McEwen FL, Gerling D. 1982. Field investigations of the interrelationships of the big-headed ant, *Pheidole megacephala*, the gray pineapple mealybug, *Dysmicoccus neobrevipes*, and the pineapple mealybug wilt disease in Hawaii, USA. *Proc. Hawaii. Entomol. Soc.* 24:51–58

Beattie AJ. 1985. *The Evolutionary Ecology of Ant-Plant Mutualisms.* Cambridge: Cambridge Univ. Press

Beattie AJ, Turnbull CL, Knox RB, Williams EG. 1984. Ant inhibition of pollen function: a possible reason why ant pollination is rare. *Am. J. Bot.* 71:421–26

Becerra JXI, Venable DL. 1989. Extrafloral nectaries: a defense against ant-Homoptera mutualisms? *Oikos* 55:276–80

Bentley BL. 1977. Extrafloral nectaries and protection by pugnacious bodyguards. *Annu. Rev. Ecol. Syst.* 8:407–27

Bernasconi G, Strassmann JE. 1999. Cooperation among unrelated individuals: the ant foundress case. *Trends Ecol. Evol.* 14:477–82

Beye M, Neumann P, Chapuisat M, Pamilo P, Moritz RFA. 1998. Nestmate recognition and the genetic relatedness of nests in the ant *Formica pratensis. Behav. Ecol. Sociobiol.* 43:67–72

Bhatkar AP. 1988. Confrontation behavior between *Solenopsis invicta* and *S. geminata*, and competitiveness of certain Florida ant species against *S. invicta*. See Trager 1988, pp. 445–64

Bhatkar AP, Vinson SB. 1987. Colony limits in *Solenopsis invicta* Buren. In *Chemistry and Biology of Social Insects*, ed. J Eder, H Rembold, pp. 599–600. Munich: Paperny

Bhatkar AP, Whitcomb WH, Buren WF, Callahan P, Carlysle T. 1972. Confrontation behavior between *Lasius neoniger* (Hymenoptera:

Formicidae) and the imported fire ant. *Environ. Entomol.* 1:274–79

Bolger DT, Suarez AV, Crooks KR, Morrison SA, Case TJ. 2000. Arthropods in urban habitat fragments in southern California: area, age and edge effects. *Ecol. Appl.* 10:1230–48

Bond W, Slingsby P. 1984. Collapse of an ant-plant mutualism—the Argentine Ant (*Iridomyrmex humilis*) and myrmecochorous Proteaceae. *Ecology* 65:1031–37

Boomsma JJ, Brouwer AH, Van Loon AJ. 1990. A new polygynous *Lasius* species (Hymenoptera: Formicidae) from central-Europe. 2. Allozymatic confirmation of species status and social structure. *Insect Soc.* 37:363–75

Bourke AFG, Franks NR. 1995. *Social Evolution in Ants.* Princeton, NJ: Princeton Univ. Press

Brandao CRF, Paiva RVS. 1994. The Galapagos ant fauna and the attributes of colonizing ant species. See Williams 1994, pp. 1–10

Breene RG, Sterling WL, Nyffeler M. 1990. Efficacy of spider and ant predators on the cotton fleahopper (Hemiptera: Miridae). *Entomophaga* 35:393–401

Brennan LA. 1993. Fire ants and northern bobwhites: a real problem or a red herring? *Wildl. Soc. Bull.* 21:351–55

Breton LM, Addicott JF. 1992. Density-dependent mutualism in an aphid-ant interaction. *Ecology* 73:2175–80

Bristow CM. 1984. Differential benefits from ant attendance to two species of Homoptera on New York ironweed. *J. Anim. Ecol.* 53:715–26

Brown JH, Davidson DW. 1977. Competition between seed-eating rodents and ants in desert ecosystems. *Science* 196:880–82

Brown W. 2001. Diversity of ants. In *Ants—Standard Methods For Measuring and Monitoring Biodiversity*, ed. D Agosti, JD Majer, LE Alonso, TR Schultz, pp. 45–79. Washington, DC: Smithsonian Inst.

Buckley RC. 1982. Ant-plant interactions: a world review. In *Ant-Plant Interactions in*

Australia, ed. RC Buckley, pp. 111–42. The Hague: Dr W Junk

Buckley RC. 1987. Interactions involving plants, Homoptera and ants. *Annu. Rev. Ecol. Syst.* 18:111–35

Buhlmann KA, Coffman G. 2001. Fire ant predation of turtle nestlings and implications for the strategy of delayed emergence. *J. Elisha Mitchell Sci. Soc.* 117:94–100

Buren WF. 1983. Artificial faunal replacement for imported fire ant control. *Fla. Entomol.* 66:93–100

Buys B. 1987. Competition for nectar between Argentine ants (*Iridomyrmex humilis*) and honeybees (*Apis mellifera*) on black ironbark (*Eucalyptus sideroxylon*). *S. Afr. J. Zool.* 22:173–74

Buys B. 1990. Relationships between Argentine ants and honeybees in South Africa. See Vander Meer et al. 1990a, pp. 519–24

Calvert WH. 1996. Fire ant predation on monarch larvae (Nymphalidae: Danainae) in a central Texas prairie. *J. Lepid. Soc.* 50:149–51

Camilo GR, Philips SA. 1990. Evolution of ant communities in response to invasion by the fire ant *Solenopsis invicta*. See Vander Meer et al. 1990a, pp. 190–98

Cammell ME, Way MJ, Paiva MR. 1996. Diversity and structure of ant communities associated with oak, pine, eucalyptus and arable habitats in Portugal. *Insect Soc.* 43:37–46

Campbell CAM. 1994. Homoptera associated with the ants *Crematogaster clariventri*, *Pheidole megacephala*, and *Tetramorium aculeatum* (Hymenoptera: Formicidae) on cocoa in Ghana. *Bull. Entomol. Res.* 84:313–18

Cañas LA, O'Neil RJ. 1998. Applications of sugar solutions to maize, and the impact of natural enemies on fall armyworm. *Int. J. Pest Manag.* 44:59–64

Carlin NF, Hölldobler B. 1986. The kin recognition system of carpenter ants (*Camponotus* spp.): I. Hierarchical cues in small colonies. *Behav. Ecol. Sociobiol.* 19:123–34

Carlin NF, Hölldobler B. 1987. The kin recognition system of carpenter ants (*Camponotus* spp.): II. Larger colonies. *Behav. Ecol. Sociobiol.* 20:209–17

Carroll CR, Janzen DH. 1973. Ecology of foraging by ants. *Annu. Rev. Ecol. Syst.* 4:231–57

Chalcraft DR, Andrews RM. 1999. Predation on lizard eggs by ants: species interactions in a variable physical environment. *Oecologia* 119:285–92

Chang VCS. 1985. Colony revival and notes on rearing and life history of the big-headed ant. *Proc. Hawaii. Entomol. Soc.* 25:53–58

Chapman RE, Bourke AFG. 2001. The influence of sociality on the conservation biology of social insects. *Ecol. Lett.* 4:650–62

Christian CE. 2001. Consequences of a biological invasion reveal the importance of mutualism for plant communities. *Nature* 413:635–39

Clark DB, Guayasamin C, Pazmino O, Donoso C, de Villacis YP. 1982. The tramp ant *Wasmannia auropunctata*: autecology and effects on ant diversity and distribution on Santa Cruz Island, Galapagos. *Biotropica* 14:196–207

Clarke SR, DeBarr GL. 1996. Impacts of red imported fire ants (Hymenoptera: Formicidae) on striped pine scale (Homoptera: Coccidae) populations. *J. Entomol. Sci.* 31:229–39

Cole FR, Medeiros AC, Loope LL, Zuehlke WW. 1992. Effects of the Argentine ant on arthropod fauna of Hawaiian high-elevation shrubland. *Ecology* 73:1313–22

Collins HL, Markin GP, Davis J. 1974. Residue accumulation in selected vertebrates following a single aerial application of mirex bait, Louisiana 1971–72. *Pestic. Monit. J.* 8:125–30

Compton SG, Robertson HG. 1988. Complex interactions between mutualisms: ants tending homopterans protect fig seeds and pollinators. *Ecology* 69:1302–5

Conners JS. 1998a. Testudines: *Chelydra serpentina* (common snapping turtle). Predation. *Herpatol. Rev.* 29:235

Conners JS. 1998b. Serpentes: *Opheodrys aestivus* (Rough green snake). Egg predation. *Herpatol. Rev.* 29:243

Contreras C, Labay A. 1999. Rainbow trout kills induced by fire ant ingestion. *Tex. J. Sci.* 51:199–200

Crozier RH, Pamilo P. 1996. *Evolution of Social Insect Colonies.* Oxford: Oxford Univ. Press

Cudjoe AR, Neuenschwander P, Copland MJW. 1993. Interference by ants in biological control of the cassava mealybug *Phenacoccus manihoti* (Hemiptera: Pseudococcidae) in Ghana. *Bull. Entomol. Res.* 83:15–22

Cushman JH, Whitham TG. 1991. Competition mediating the outcome of a mutualism: protective services of ants as a limiting resource for membracids. *Am. Nat.* 138:851–65

Davidson DW. 1997. The role of resource imbalances in the evolutionary ecology of tropical arboreal ants. *Biol. J. Linn. Soc.* 61:153–81

Davidson DW. 1998. Resource discovery versus resource domination in ants: a functional mechanism for breaking the trade-off. *Ecol. Entomol.* 23:484–90

Davidson NA, Stone ND. 1989. Imported fire ants. In *Eradication of Exotic Pests: Analysis With Case Histories*, ed. DL Dahlsten, R Garcia, pp. 196–217. New Haven, CT: Yale Univ. Press

DeHeer CJ, Goodisman MAD, Ross KG. 1999. Queen dispersal strategies in the multiple-queen form of the fire ant *Solenopsis invicta. Am. Nat.* 153:660–75

DeHeer CJ, Tschinkel WR. 1998. The success of alternative reproductive tactics in monogyne populations of the ant *Solenopsis invicta:* signficance for transitions in social organization. *Behav. Ecol.* 9:130–35

De Kock AE. 1990. Interactions between the introduced Argentine ant, *Iridomyrmex humilis* Mayr, and two indigenous fynbos ant species. *J. Entomol. Soc. S. Afr.* 53:107–8

de la Fuente MAS, Marquis RJ. 1999. The role of ant-tended extrafloral nectaries in the protection and benefit of a Neotropical rainforest tree. *Oecologia* 118:192–202

De Souza ALB, Delabie JHC, Fowler HG. 1998. *Wasmannia* spp. (Hymenoptera: Formicidae) and insect damages to cocoa in Brazilian farms. *J. Appl. Entomol.* 122:339–41

Deyrup M, Davis L, Cover S. 2000. Exotic ants in Florida. *Trans. Am. Entomol. Soc.* 126:293–326

Diamond J, Case TJ. 1986. Overview: introductions, extinctions, exterminations, and invasions. In *Community Ecology,* ed. J Diamond, TJ Case, pp. 65–79. New York: Harper & Row

Dickinson VM. 1995. Red imported fire ant predation on Crested Caracara nestlings in south Texas. *Wilson Bull.* 107:761–62

Donaldson W, Price AH, Morse J. 1994. The current status and future prospects of the Texas horned lizard (*Phrynosoma cornutum*) in Texas. *Tex. J. Sci.* 46:97–113

Drees BM. 1994. Red imported fire ant predation on nestling of colonial waterbirds. *Southwest. Entomol.* 19:355–59

Drees BM, Berger LA, Cavazos R, Vinson SB. 1991. Factors affecting sorghum and corn seed predation by foraging red imported fire ants (Hymenoptera: Formicidae). *J. Econ. Entomol.* 84:285–89

Dreistadt SH, Hagen KD, Dahlsten DL. 1986. Predation by *Iridomyrmex humilis* (Hymenoptera: Formicidae) on eggs of *Chrysoperla carnea* (Neuroptera: Chrysopidae) released for inundative control of *Illinoia liriodendri* (Homoptera: Aphididae) infesting *Liriodendron tulipifera. Entomophaga* 31:397–400

Durr HJR. 1952. The Argentine ant, *Iridomyrmex humilis* Mayr. *Farming S. Afr.* 27:381–84

Dutcher JD, Estes PM, Dutcher MJ. 1999. Interactions in entomology: aphids, aphidophaga and ants in pecan orchards. *J. Entomol. Sci.* 34:40–56

Eisner T. 1957. A comparative morphological study of the proventriculus of ants (Hymenoptera: Formicidae). *Bull. Mus. Comp. Zool.* 116:429–90

El Hamalawi ZA, Menge JA. 1996. The role of snails and ants in transmitting the avocado stem canker pathogen, *Phytophthora citricola. J. Am. Soc. Hortic. Sci.* 121:973–77

Emlen ST. 1982. The evolution of helping. I. An ecological constraints model. *Am. Nat.* 119:29–39

Entwistle PF. 1972. *Pests of Cocoa.* London: Longman Group

Erickson JM. 1971. The displacement of native ant species by the introduced Argentine ant *Iridomyrmex humilis* Mayr. *Psyche* 78:257–66

Eubanks MD. 2001. Estimates of the direct and indirect effects of red imported fire ants on biological control in field crops. *Biol. Control* 21:35–43

Feare C. 1999. Ants take over from rats on Bird Island, Seychelles. *Bird Conserv. Int.* 9:95–96

Feener DH. 2000. Is the assembly of ant communities mediated by parasitoids? *Oikos* 90:79–88

Feener DH, Brown BV. 1997. Diptera as parasitoids. *Annu. Rev. Entomol.* 42:73–97

Feinsinger P, Swarm LA. 1978. How common are ant-repellent nectars? *Biotropica* 10:238–39

Fellers JH. 1987. Interference and exploitation in a guild of woodland ants. *Ecology* 68:1466–78

Ferris DK, Killion MJ, Ferris KP, Grant WE, Vinson SB. 1998. Influence of relative abundance of red imported fire ants (*Solenopsis invicta*) on small mammal captures. *Southwest. Nat.* 43:97–100

Fisher RN, Suarez AV, Case TJ. 2002. Spatial patterns in the abundance of the coastal horned lizard. *Conserv. Biol.* 16:205–15

Fleet RR, Young BL. 2000. Facultative mutualism between imported fire ants (*Solenopsis invicta*) and a legume (*Senna occidentalis*). *Southwest. Nat.* 45:289–98

Flickinger EL. 1989. Observation of predation by red imported fire ants on live-trapped wild cotton rats. *Tex. J. Sci.* 41:223–24

Fluker SS, Beardsley JW. 1970. Sympatric associations of three ants: *Iridomyrmex humilis, Pheidole megacephala,* and *Anoplolepis longipes. Ann. Entomol. Soc. Am.* 63:1290–96

Folgarait PJ. 1998. Ant biodiversity and its rela-

tionship to ecosystem functioning: a review. *Biodivers. Conserv.* 7:1221–44

Fontenla Rizo JL. 1995. Un comentario sobre las "hormigas locas" (*Paratrechina*) cubanas, con enfasis en *P. fulva. Cocuyo (Havana)* 2:6–7

Forys EA, Quistorff A, Allen CR. 2001a. Potential fire ant (Hymenoptera: Formicidae) impact on the endangered Schaus swallowtail (Lepidoptera: Papilionidae). *Fla. Entomol.* 84:254–58

Forys EA, Quistorff A, Allen CR, Wojcik DP. 2001b. The likely cause of extinction of the tree snail *Orthalicus reses reses* (Say). *J. Molluscan Stud.* 67:369–76

Frazer BD, Van den Bosch R. 1973. Biological control of the walnut aphid in California: the interrelationship of the aphid and its parasite. *Environ. Entomol.* 2:561–68

Freed PS, Neitman K. 1988. Notes on predation on the endangered Houston toad, *Bufo houstonensis. Tex. J. Sci.* 40:454–56

Freitas L, Galetto L, Berndall G, Paoli AAS. 2000. Ant exclusion and reproduction of *Croton sarcopetalus* (Euphorbiaceae). *Flora* 195:398–402

Fuller W, Reagan TE, Flynn JL, Boetel MA. 1997. Predation on fall armyworm (Lepidoptera: Noctuidae) in sweet sorghum. *J. Agric. Entomol.* 14:151–55

Gambino P. 1990. Argentine ant *Iridomyrmex humilis* (Hymenoptera: Formicidae) predation on yellowjackets (Hymenoptera: Vespidae) in California. *Sociobiology* 17:287–98

Gaume L, McKey D, Terrin S. 1998. Ant-plant-homopteran mutualism: how the third partner affects the interaction between a plant-specialist ant and its myrmecophyte host. *Proc. R. Soc. London Ser. B* 265:569–75

Ghazoul J. 2000. Repellent flowers keep plants attractive. *Bull. Br. Ecol. Soc.* 31:4

Gillespie RG, Reimer N. 1993. The effect of alien predatory ants (Hymenoptera: Formicidae) on Hawaiian endemic spiders (Araneae: Tetragnathidae). *Pac. Sci.* 47:21–33

Giraud T, Pedersen JS, Keller L. 2002. Evolution of supercolonies: the Argentine ants of southern Europe. *Proc. Natl. Acad. Sci. USA* 99:6075–99

Giuliano WM, Allen CR, Lutz RS, Demarais S. 1996. Effects of red imported fire ants on northern bobwhite chicks. *J. Wildl. Manag.* 60:309–13

Glancey BM, Craig CH, Stringer CE, Bishop PM. 1973. Multiple fertile queens in colonies of the imported fire ant, *Solenopsis invicta. J. Ga. Entomol. Soc.* 8:237–38

Glancey BM, Wojcik DP, Craig CH, Mitchell JA. 1976. Ants of Mobile County, AL, as monitored by bait transects. *J. Ga. Entomol. Soc.* 11:191–97

Gonzalez-Hernandez H, Johnson MW, Reimer NJ. 1999. Impact of *Pheidole megacephala* (F.) (Hymenoptera: Formicidae) on the biological control of *Dysmicoccus brevipes* (Cockerell) (Homoptera: Pseudococcidae). *Biol. Contr.* 15:145–52

Goodisman MAD, De Heer CJ, Ross KG. 2000. Unusual behavior of polygyne fire ant queens on nuptial flights. *J. Insect Behav.* 13:455–68

Gordon DM. 1995. The expandable network of ant exploration. *Anim. Behav.* 50:995–1007

Gordon DM, Moses L, Falkovitz-Halpern M, Wong EH. 2001. Effect of weather on infestation of buildings by the invasive Argentine ant, *Linepithema humile* (Hymenoptera: Formicidae). *Am. Midl. Nat.* 146:321–28

Gotelli NJ. 1996. Ant community structure: effects of predatory ant lions. *Ecology* 77:630–38

Gotelli NJ, Arnett AE. 2000. Biogeographic effects of red fire ant invasion. *Ecol. Lett.* 3:257–61

Gravena S, Sterling WL. 1983. Natural predation on the cotton leafworm (Lepidoptera: Noctuidae). *J. Econ. Entomol.* 76:779–84

Green HB, Hutchins RE. 1960. Laboratory study of toxicity of imported fire ants to bluegill fish. *J. Econ. Entomol.* 53:1137–38

Greenberg L, Vinson SB, Ellison S. 1992. Nine-year study of a field containing both monogyne and polygyne red imported fire ants (Hymenoptera: Formicidae). *Ann. Entomol. Soc. Am.* 85:686–95

Greenslade PJM. 1971. Interspecific competition and frequency changes among ants in Solomon Islands coconut plantations. *J. Appl. Ecol.* 8:323–52

Greenslade PJM. 1972. Comparative ecology of four tropical ant species. *Insect Soc.* 19:195–212

Grosholz ED. 1996. Contrasting rates of spread for introduced species in terrestrial and marine systems. *Ecology* 77:1680–86

Haber WA, Frankie GW, Baker HG, Baker I, Koptur S. 1981. Ants like flower nectar. *Biotropica* 13:211–14

Haines IH, Haines JB. 1978a. Colony structure, seasonality and food requirements of the crazy ant, *Anoplolepis longipes* (Jerd.), in the Seychelles. *Ecol. Entomol.* 3:109–18

Haines IH, Haines JB. 1978b. Pest status of the crazy ant, *Anoplolepis longipes* (Jerdon) (Hymenoptera: Formicidae), in the Seychelles. *Bull. Entomol. Res.* 68:627–38

Halaj J, Ross DW, Moldenke AR. 1997. Negative effects of ant foraging on spiders in Douglas-fir canopies. *Oecologia* 109:313–22

Haney PB, Luck RF, Moreno DS. 1987. Increases in densities of the red citrus mite, *Panonychus citri* (Acarina: Tetranychidae), in association with the Argentine ant, *Iridomyrmex humilis* (Hymenoptera: Formicidae), in southern California (USA) citrus. *Entomophaga* 32:49–58

Hara AH, Hata TY. 1992. Ant control on protea in Hawaii. *Sci. Hortic.* 51:155–63

Harris WG, Burns EC. 1972. Predation on the lone star tick by the imported fire ant. *Environ. Entomol.* 1:362–65

Haskins CP, Haskins EF. 1965. *Pheidole megacephala* and *Iridomyrmex humilis* in Bermuda: equilibrium or slow replacement? *Ecology* 46:736–40

Hata TY, Hara AH, Hu BKS, Kaneko RT, Tenbrink VL. 1995. Excluding pests from red ginger flowers with insecticides and pollinating, polyester, or polyethlyene bags. *J. Econ. Entomol.* 88:393–97

Hee J, Holway DA, Suarez AV, Case TJ. 2000. Role of propagule size in the success of incipient colonies of the invasive Argentine ant. *Conserv. Biol.* 14:559–63

Heraty JM. 1994. Biology and importance of two eucharitid parasites of *Wasmannia* and *Solenopsis*. See Williams 1994, pp. 104–20

Heterick B. 1997. The interaction between the coastal brown ant, *Pheidole megacephala* (Fabricius), and other invertebrate fauna of Mt Coot-tha (Brisbane Australia). *Aust. J. Ecol.* 22:218–21

Hill EP, Dent DM. 1985. Mirex residues in 7 groups of aquatic and terrestrial mammals. *Arch. Environ. Contam. Toxicol.* 14:7–12

Hoffmann BD. 1998. The big-headed ant *Pheidole megacephala*: a new threat to monsoonal northwestern Australia. *Pac. Conserv. Biol.* 4:250–55

Hoffmann BD, Andersen AN, Hill GJE. 1999. Impact of an introduced ant on native forest invertebrates: *Pheidole megacephala* in monsoonal Australia. *Oecologia* 120:595–604

Hölldobler B, Lumsden CJ. 1980. Territorial strategies in ants. *Science* 210:732–39

Hölldobler B, Michener CD. 1980. Mechanisms of identification and discrimination in social Hymenoptera. In *Evolution of Social Behavior: Hypotheses and Empirical Tests*, ed. H Markl, pp. 35–58. Weinheim: Verlag Chem.

Hölldobler B, Wilson EO. 1977. The number of queens: an important trait in ant evolution. *Naturwissenschaften* 64:8–15

Hölldobler B, Wilson EO. 1990. *The Ants*. Cambridge, MA: Belknap

Holtcamp WN, Grant WE, Vinson SB. 1997. Patch use under predation hazard: effect of the red imported fire ant on deer mouse foraging behavior. *Ecology* 78:308–17

Holway DA. 1995. Distribution of the Argentine ant (*Linepithema humile*) in northern California. *Conserv. Biol.* 9:1634–37

Holway DA. 1998a. Effect of Argentine ant invasions on ground-dwelling arthropods in northern California riparian woodlands. *Oecologia* 116:252–58

Holway DA. 1998b. Factors governing rate of invasion: a natural experiment using Argentine ants. *Oecologia* 115:206–12

Holway DA. 1999. Competitive mechanisms underlying the displacement of native ants by the invasive Argentine ant. *Ecology* 80:238–51

Holway DA, Case TJ. 2000. Mechanisms of dispersed central-place foraging in polydomous colonies of the Argentine ant. *Anim. Behav.* 59:433–41

Holway DA, Case TJ. 2001. Effects of colony-level variation on competitive ability in the invasive Argentine ant. *Anim. Behav.* 61:1181–92

Holway DA, Suarez AV, Case TJ. 1998. Loss of intraspecific aggression in the success of a widespread invasive social insect. *Science* 282:949–52

Holway DA, Suarez AV, Case TJ. 2002. Role of abiotic factors in governing susceptibility to invasion: a test with Argentine ants. *Ecology* 83:1610–19

Hook AW, Porter SD. 1990. Destruction of harvester ant colonies by invading fire ants in south-central Texas (Hymenoptera: Formicidae). *Southwest. Nat.* 35:477–78

Horvitz CC, Schemske DW. 1984. Effects of ants and an ant-tended herbivore on seed production of a neotropical herb. *Ecology* 65:1369–78

Horvitz CC, Schemske DW. 1986. Seed dispersal of a neotropical myrmecochore: variation in removal rates and dispersal distance. *Biotropica* 18:319–23

Howard FW, Oliver AD. 1978. Arthropod populations in permanent pastures treated and untreated with mirex for red imported fire ant control. *Environ. Entomol.* 7:901–3

Howard FW, Oliver AD. 1979. Field observations of ants (Hymenoptera: Formicidae) associated with red imported fire ants, *Solenopsis invicta* Buren, in Louisiana pastures. *J. Ga. Entomol. Soc.* 14:259–63

Hu GY, Frank JH. 1996. Effect of the red imported fire ant (Hymenoptera: Formicidae) on dung-inhabiting arthropods in Florida. *Environ. Entomol.* 25:1290–96

Human KG, Gordon DM. 1996. Exploitation and interference competition between the invasive Argentine ant, *Linepithema humile*, and native ant species. *Oecologia* 105:405–12

Human KG, Gordon DM. 1997. Effects of Argentine ants on invertebrate biodiversity in northern California. *Conserv. Biol.* 11:1242–48

Human KG, Gordon DM. 1999. Behavioral interactions of the invasive Argentine ant with native ant species. *Insect Soc.* 46:159–63

Human KG, Weiss S, Sandler B, Gordon DM. 1998. Effects of abiotic factors on the distribution and activity of the invasive Argentine ant (Hymenoptera: Formicidae). *Environ. Entomol.* 27:822–33

Huxel GR. 2000. The effect of the Argentine ant on the threatened valley elderberry longhorn beetle. *Biol. Invas.* 2:81–85

Huxley CR, Cutler DF, eds. 1991. *Ant-Plant Interactions*. Oxford, UK: Oxford Univ. Press

Jaffe K, Mauleon H, Kermarrec A. 1990. Qualitative evaluation of ants as biological control agents with special reference to predators on *Diaprepes* spp. (Coleoptera: Curculionidae) on citrus groves in Martinique and Guadeloupe. In *Colloques de l'INRA; Caribbean Meetings on Biological Control*, ed. C Pavis, A Kermarrec, pp. 405–16. Route de St-Cyr: Inst. Natl. Rech. Agron. (INRA)

Janzen DH. 1977. Why don't ants visit flowers? *Biotropica* 9:252

Johnson AS. 1961. Antagonistic relationships between ants and wildlife with special reference to imported fire ants and bobwhite quail in the southeast. *Proc. Ann. Conf. Southeast. Assoc. Game Fish Comm.* 15:88–107

Johnson LK. 1981. Effect of flower clumping on defense of artificial flowers by aggressive stingless bees. *Biotropica* 13:151–57

Johnson LK, Hubbell SP, Feener DH. 1987. Defense of food supply by eusocial colonies. *Am. Zool.* 27:347–58

Jolivet P. 1996. *Ants and Plants: An Example of Coevolution* (enlarged edition). Leiden, Netherlands: Backhuys

Jones D, Sterling WL. 1979. Manipulation of red imported fire ants in a trap crop for boll weevil suppression. *Environ. Entomol.* 8:1073–77

Jones SR, Phillips SA. 1987. Aggressive and defensive propensities of *Solenopsis invicta* (Hymenoptera: Formicidae) and three indigenous ant species in Texas. *Tex. J. Sci.* 39:107–15

Jones SR, Phillips SA. 1990. Resource collecting abilities of *Solenopsis invicta* (Hymenoptera: Formicidae) compared with those of three sympatric Texas ants. *Southwest. Nat.* 35:416–22

Jourdan H. 1997. Threats on Pacific islands: The spread of the tramp ant *Wasmannia auropunctata* (Hymenoptera: Formicidae). *Pac. Conserv. Biol.* 3:61–64

Jourdan H, Sadlier RA, Bauer AM. 2001. Little fire ant invasion (*Wasmannia auropunctata*) as a threat to New Caledonian lizards: evidences from a Sclerophyll forest (Hymenoptera: Formicidae). *Sociobiology* 38:283–301

Jouvenaz DP. 1990. Approaches to biological control of fire ants in the United States. In *Exotic Ants: Impact and Control of Introduced Species*, ed. RK Vander Meer, A Dendeno, pp. 620–27. Oxford: Westview

Jouvenaz DP, Wojcik DP, Vander Meer RK. 1989. First observation of polygyny in fire ants, *Solenopsis* spp., in South America. *Psyche* 96:161–65

Kaspari M. 2000. Do imported fire ants impact canopy arthropods? Evidence from simple arboreal pitfall traps. *Southwest. Nat.* 45:118–22

Kaspari M, Vargo EL. 1995. Colony size as a buffer against seasonality: Bergmann's rule in social insects. *Am. Nat.* 145:610–32

Keeler KH. 1989. Ant-plant interactions. In *Plant-Animal Interactions*, ed. WG Abrahamson, pp. 207–42. New York: McGraw-Hill

Keller L, ed. 1993. *Queen Number and Sociality in Insects*. Oxford, UK: Oxford Univ. Press

Keller L, Passera L. 1989. Size and fat content of gynes in relation to the mode of colony founding in ants (Hymenoptera: Formicidae). *Oecologia* 80:236–40

Keller L, Ross KG. 1993. Phenotypic basis of reproductive success in a social insect:

genetic and social determinants. *Science* 260: 1107–10

Keller L, Ross KG. 1998. Selfish genes: a green beard in the red fire ant. *Nature* 394:573–75

Keller L, Ross KG. 1999. Major gene effects on phenotype and fitness: the relative roles of Pgm-3 and Gp-9 in introduced populations of the fire ant *Solenopsis invicta. J. Evol. Biol.* 12:672–80

Kennedy TA. 1998. Patterns of an invasion by Argentine ants (*Linepithema humile*) in a riparian corridor and its effects on ant diversity. *Am. Midl. Nat.* 140:343–50

Killion MJ, Grant WE. 1993. Scale effects in assessing the impact of imported fire ants on small mammals. *Southwest. Nat.* 38:393–96

Killion MJ, Grant WE, Vinson SB. 1995. Response of *Baiomys taylori* to changes in density of imported fire ants. *J. Mammal.* 76:141–47

King TG, Phillips SA. 1992. Destruction of young colonies of the red imported fire ant by the pavement ant (Hymenoptera: Formicidae). *Entomol. News* 103:72–77

Kitching RL. 1987. Aspects of the natural history of the Lycaenid butterfly *Allotinus major* in Sulawesi. *J. Nat. Hist.* 21:535–44

Knight P. 1944. Insects associated with the Palay rubber vine in Haiti. *J. Econ. Entomol.* 37:100–2

Kopachena JG, Buckley AJ, Potts GA. 2000. Effects of the red imported fire ant (*Solenopsis invicta*) on reproductive success of barn swallows (*Hirundo rustica*) in northeast Texas. *Southwest. Nat.* 45:477–82

Koptur S. 1979. Facultative mutualism between weedy vetches bearing extrafloral nectaries and weedy ants in California. *Am. J. Bot.* 66:1016–20

Koptur S, Truong N. 1998. Facultative ant-plant interactions: nectar sugar preferences of introduced pest ant species in South Florida. *Biotropica* 30:179–89

Korzukhin MD, Porter SD, Thompson LC, Wiley S. 2001. Modeling temperature-dependent range limits for the fire ant

Solenopsis invicta (Hymenoptera: Formicidae) in the United States. *Environ. Entomol.* 30:645–55

Krieger MJB, Keller L. 2000. Mating frequency and genetic structure of the Argentine ant *Linepithema humile. Mol. Ecol.* 9:119–26

Krieger MJB, Ross KG. 2002. Identification of a major gene regulating complex social behavior. *Science* 295:328–32

Laakkonen J, Fisher RN, Case TJ. 2001. Effect of land cover, habitat fragmentation, and ant colonies on the distribution and abundance of shrews in southern California. *J. Anim. Ecol.* 70:776–88

Lach L. 2002. Invasive ants: unwanted partners in ant-plant interactions? *J. Mo. Bot. Gard.* In press

Lakshmikantha BP, Lakshminarayan NG, Ali TMM, Veeresh GK. 1996. Fire-ant damage to potato in Bangalore. *J. Indian Potato Assoc.* 23:75–76

Landers JL, Garner JA, McRae WA. 1980. Reproduction of gopher tortoises (*Gopherus polyphemus*) in southwestern Georgia. *Herpetologica* 36:353–61

La Polla JS, Otte D, Spearman LA. 2000. Assessment of the effects of ants on Hawaiian crickets. *J. Orthoptera Res.* 9:139–48

Lawton JH, Brown KC. 1986. The population and community ecology of invading insects. *Philos. Trans. R. Soc. London Ser. B* 314:607–17

Lee DK, Bhatkar AP, Vinson SB, Olson JK. 1994. Impact of foraging red imported fire ants (*Solenopsis invicta*) (Hymenoptera: Formicidae) on *Psorophora columbiae* eggs. *J. Am. Mosq. Control Assoc.* 10:163–73

Legare ML, Eddleman WR. 2001. Home range size, nest-site selection and nesting success of Black Rails in Florida. *J. Field Ornithol.* 72:170–77

Levins R, Pressick ML, Heatwole H. 1973. Coexistence patterns in insular ants. *Am. Sci.* 61:463–72

Lewis T, Cherrett JM, Haines I, Haines JB, Mathias PL. 1976. The crazy ant (*Anoplolepis longipes*) (Jerd.) (Hymenoptera: Formicidae)

in Seychelles, and its chemical control. *Bull. Entomol. Res.* 66:97–111

Lieberburg I, Kranz PM, Seip A. 1975. Bermudian ants revisited: the status and interaction of *Pheidole megacephala* and *Iridomyrmex humilis. Ecology* 56:473–78

Lockley TC. 1995. Effect of imported fire ant predation on a population of the least tern: an endangered species. *Southwest. Entomol.* 20:517–19

Lofgren CS. 1986. The economic importance and control of imported fire ants in the United States. See Vinson 1986, pp. 227–56

Lubin YD. 1984. Changes in the native fauna of the Galapagos Islands following invasion by the little red fire ant, *Wasmannia auropunctata. Biol. J. Linn. Soc.* 21:229–42

Lymn N, Temple SA. 1991. Land-use changes in the Gulf Coast region: links to declines in Midwestern Loggerhead Shrike populations. *Passenger Pigeon* 53:315–25

MacKay WP, Greenberg L, Vinson SB. 1994. A comparison of bait recruitment in monogynous and polygynous forms of the red imported fire ant, *Solenopsis invicta* Buren. *J. Kans. Entomol. Soc.* 67:133–36

MacKay WP, Porter S, Gonzalez D, Rodriguez A, Armendedo H, et al. 1990. A comparison of monogyne and polygyne populations of the tropical fire ant, *Solenopsis geminata* (Hymenoptera: Formicidae) in Mexico. *J. Kans. Entomol. Soc.* 63:611–15

MacMahon JA, Mull JF, Crist TO. 2000. Harvester ants (*Pogonomyrmex* spp.): their community and ecosystem influences. *Annu. Rev. Ecol. Syst.* 31:265–91

Macom TE, Porter SD. 1996. Comparison of polygyne and monogyne red imported fire ant (Hymenoptera: Formicidae) population densities. *Ann. Entomol. Soc. Am.* 89:535–43

Majer JD. 1985. Recolonization by ants of rehabilitated mineral sand mines on North Stradbroke Island, Queensland, with particular reference to seed removal. *Aust. J. Ecol.* 10:31–48

Majer JD. 1994. Spread of Argentine ants (*Linepithema humile*) with special reference to Western Australia. See Williams 1994, pp. 163–73

Markin GP. 1970a. Foraging behavior of the Argentine ant in a California citrus grove. *J. Econ. Entomol.* 63:740–44

Markin GP. 1970b. The seasonal life cycle of the Argentine ant, *Iridomyrmex humilis* (Hymenoptera: Formicidae), in southern California. *Ann. Entomol. Soc. Am.* 63:1238–42

Markin GP, Collins HL, Dillier JH. 1972. Colony founding by queens of the red imported fire ant, *Solenopsis invicta. Ann. Entomol. Soc. Am.* 65:1053–58

Markin GP, Dillier JH, Collins HL. 1973. Growth and development of colonies of the red imported fire ant, *Solenopsis invicta. Ann. Entomol. Soc. Am.* 66:803–8

Maschwitz U, Fiala B, Dolling WR. 1987. New trophobiotic symbioses of ants with South East Asian bugs. *J. Nat. Hist.* 21:1097–108

Masser MP, Grant WE. 1986. Fire ant-induced trap mortality of small mammals in east-central Texas, USA. *Southwest. Nat.* 31:540–42

McGlynn TP. 1999a. The worldwide transfer of ants: geographical distribution and ecological invasions. *J. Biogeogr.* 26:535–48

McGlynn TP. 1999b. Non-native ants are smaller than related native ants. *Am. Nat.* 154:690–99

Meek PD. 2000. The decline and current status of the Christmas Island shrew *Crocidura attenuata trichura* on Christmas Island, Indian Ocean. *Aust. Mammal.* 22:43–49

Meier RE. 1994. Coexisting patterns and foraging behavior of introduced and native ants (Hymenoptera: Formicidae) in the Galapagos Islands (Ecuador). See Williams 1994, pp. 44–61

Messina FJ. 1981. Plant protection as a consequence of an ant-membracid mutualism: interactions on goldenrod (*Solidago sp*). *Ecology* 62:1433–40

Michaud JP, Browning HW. 1999. Seasonal abundance of the brown citrus aphid, *Toxoptera citricida*, (Homoptera: Aphididae) and its natural enemies in Puerto Rico. *Fla. Entomol.* 82:424–47

Montgomery WB. 1996. Predation by the fire ant, *Solenopsis invicta*, on the three-toed box turtle, *Terapene carolina triunguis. Bull. Chicago Herpetol. Soc.* 31:105–6

Morel L, Vander Meer RK, Lofgren CS. 1990. Comparison of nestmate recognition between monogyne and polygyne populations of *Solenopsis invicta* (Hymenoptera: Formicidae). *Ann. Entomol. Soc. Am.* 83:642–47

Morrill WL. 1974. Dispersal of red imported fire ants by water. *Fla. Entomol.* 57:39–42

Morrill WL. 1978. Red imported fire ant predation on the alfalfa weevil and pea aphid. *J. Econ. Entomol.* 71:867–68

Morris JR, Steigman KL. 1993. Effects of polygyne fire ant invasion on native ants of a blackland prairie in Texas. *Southwest. Nat.* 38:136–40

Morrison LW. 1996. Community organization in a recently assembled fauna: the case of Polynesian ants. *Oecologia* 107:243–56

Morrison LW. 1999. Indirect effects of phorid fly parasitoids on the mechanisms of interspecific competition among ants. *Oecologia* 121:113–22

Morrison LW. 2000. Mechanisms of interspecific competition among an invasive and two native fire ants. *Oikos* 90:238–52

Morrison LW. 2002. Long-term impacts of the invasion of an arthropod community by the red imported fire ant, *Solenopsis invicta. Ecology* 83:In press

Morrison LW, Kawazoe EA, Guerra R, Gilbert LE. 1999. Phonology and dispersal in *Pseudacteon* flies (Diptera: Phoridae), parasitoids of *Solenopsis* fire ants (Hymenoptera: Formicidae). *Ann. Entomol. Soc. Am.* 92:198–207

Moulis RA. 1997. Predation by the imported fire ant (*Solenopsis invicta*) on loggerhead sea turtle (*Caretta caretta*) nests on Wassaw National Wildlife Refuge, Georgia. *Chelon. Conserv. Biol.* 2:433–36

Mount RH, Trauth SE, Mason WH. 1981. Predation by the red imported fire ant, *Solenopsis invicta* (Hymenoptera: Formicidae), on eggs of the lizard *Cnemidophorus sexlineatus* (Squamata: Teiidae). *J. Ala. Acad. Sci.* 52:66–70

Mueller JM, Dabbert CB, Demarais S, Forbes AR. 1999. Northern bobwhite chick mortality caused by red imported fire ants. *J. Wildl. Manag.* 63:1291–98

Mueller JM, Dabbert CB, Forbes AR. 2001. Negative effects of imported fire ants on deer: the "increased movement" hypothesis. *Tex. J. Sci.* 53:87–90

Nagamitsu T, Inoue T. 1997. Aggressive foraging of social bees as a mechanism of floral resource partitioning in an Asian tropical rainforest. *Oecologia* 110:432–39

Negm AA, Hensley SD. 1969. Evaluation of certain biological control agents of the sugarcane borer in Louisiana. *J. Econ. Entomol.* 62:1008–13

Ness JH. 2001. *The* Catalpa bignonioides *food web: implications of variable interactions among four trophic levels.* PhD thesis. Univ. Ga., Athens. 138 pp.

Newell W. 1908. Notes on the habits of the Argentine ant or "New Orleans" ant, *Iridomyrmex humilis* Mayr. *J. Econ. Entomol.* 1:21–34

Newell W, Barber TC. 1913. The Argentine ant. *Bur. Entomol. Bull.* 122:1–98

Nichols BJ, Sites RW. 1989. A comparison of arthropod species within and outside the range of *Solenopsis invicta* Buren in central Texas. *Southwest. Entomol.* 14:345–50

Nickerson JC, Rolph Kay CA, Buschman LL, Whitcomb WH. 1977. The presence of *Spissistilus festinus* as a factor affecting egg predation by ants in soybeans. *Fla. Entomol.* 60:193–99

Nonacs P. 1993. The effects of polygyny and colony life history on optimal sex investment. In *Queen Number and Sociality in Insects*, ed. L Keller, pp. 110–31. Oxford, UK: Oxford Univ. Press

Nuessly GS, Sterling WL. 1994. Mortality of *Helicoverpa zea* (Lepidoptera: Noctuidae) eggs in cotton as a function of oviposition sites, predator species and desiccation. *Environ. Entomol.* 23:1189–202

Obin MS, Vander Meer RK. 1985. Gaster flagging by fire ants (*Solenopsis* spp.): functional significance of venom dispersal behavior. *J. Chem. Ecol.* 11:1757–68

Orr MR, Seike SH. 1998. Parasitoids deter foraging by Argentine ants (*Linepithema humile*) in their native habitat in Brazil. *Oecologia* 117:420–25

Orr MR, Seike SH, Benson WW, Dahlsten DL. 2001. Host specificity of *Pseudacteon* (Diptera: Phoridae) parasitoids that attack *Linepithema* (Hymenoptera: Formicidae) in South America. *Environ. Entomol.* 30:742–47

Orr MR, Seike SH, Benson WW, Gilbert LE. 1995. Flies suppress fire ants. *Nature* 373:292–93

Pamilo P. 1991. Evolution of colony characteristics in social insects. II. Number of reproductive individuals. *Am. Nat.* 138:412–33

Parker IM, Simberloff D, Lonsdale WM, Goodell K, Wonham M, et al. 1999. Impact: toward a framework for understanding the ecological effects of invaders. *Biol. Invasions* 1:3–19

Passera L. 1994. Characteristics of tramp species. See Williams 1994, pp. 23–43

Passera L, Keller L, Suzzoni JP. 1988. Queen replacement in dequeened colonies of the Argentine ant *Iridomyrmex humilis* (Mayr). *Psyche* 95:59–66

Peakall R, Handel SN, Beattie AJ. 1991. The evidence for, and importance of, ant pollination. See Huxley & Cutler 1991, pp. 421–29

Pedersen EK, Grant WE, Longnecker MT. 1996. Effects of red imported fire ants on newly-hatched northern bobwhite. *J. Wildl. Manag.* 60:164–69

Perfecto I. 1994. Foraging behavior as a determinant of asymmetric competitive interaction between two ant species in a tropical agroecosystem. *Oecologia* 98:184–92

Phillips PA, Bekey RS, Goodall GE. 1987. Argentine ant management in cherimoyas. *Cal. Agric.* March–April:8–9

Pianka ER, Parker WS. 1975. Ecology of horned lizards: a review with special reference to *Phrynosoma platyrhinos*. *Copeia* 1975:141–62

Pimentel D, Lach L, Zuniga R, Morrison D. 2000. Environmental and economic costs of nonindigenous species in the United States. *BioScience* 50:53–65

Pimm SL. 1991. *The Balance of Nature: Ecological Issues in the Conservation of Species and Communities.* Chicago, IL: Univ. Chicago Press

Porter SD, Bhatkar A, Mulder R, Vinson SB, Clair DJ. 1991. Distribution and density of polygyne fire ants (Hymenoptera: Formicidae) in Texas. *J. Econ. Entomol.* 84:866–74

Porter SD, Eastmond DA. 1982. *Euryopis coki* (Theridiidae), a spider that preys upon *Pogonomyrmex* ants. *J. Arachnol.* 10:275–77

Porter SD, Fowler HG, MacKay WP. 1992. Fire ant mound densities in the United States and Brazil (Hymenoptera:Formicidae). *J. Econ. Entomol.* 85:1154–61

Porter SD, Savignano DA. 1990. Invasion of polygyne fire ants decimates native ants and disrupts arthropod community. *Ecology* 71:2095–106

Porter SD, Vander Meer RK, Pesquero MA, Campiolo S, Fowler HG. 1995. *Solenopsis* (Hymenoptera: Formicidae) fire ant reactions to attacks of *Pseudacteon* flies (Diptera: Phoridae) in southeastern Brazil. *Ann. Entomol. Soc. Am.* 88:570–75

Porter SD, Van Eimeren B, Gilbert LE. 1988. Invasion of red imported fire ants (Hymenoptera: Formicidae): microgeography of competitive displacement. *Ann. Entomol. Soc. Am.* 81:913–18

Porter SD, Williams DF, Patterson RS, Fowler HG. 1997. Intercontinental differences in the abundance of *Solenopsis* fire ants (Hymenoptera: Formicidae): escape from natural enemies. *Environ. Entomol.* 26:373–84

Potgieter JT. 1937. The Argentine ant. *Farming S. Afr.* 12:160

Quilichini A, Debussche M. 2000. Seed dispersal and germination patterns in a rare Mediterranean island endemic (*Anchusa crispa* Viv., Boraginaceae). *Acta Oecol.* 21:303–13

Ready CC, Vinson SB. 1995. Seed selection by the red imported fire ant (Hymenoptera: Formicidae) in the laboratory. *Environ. Entomol.* 24:1422–31

Reagan SR, Ertel JM, Wright VL. 2000. David and Goliath retold: fire ants and alligators. *J Herpetol.* 34:475–78

Redford KH. 1987. Ants and termites as food: patterns of mammalian myrmecophagy. In *Current Mammalogy*, ed. HH Genoways, pp. 349–99. New York: Plenum

Reimer NJ. 1988. Predation on *Liothrips urichi* (Karney) (Thysanoptera: Phlaeothripidae) a case of biotic interference. *Environ. Entomol.* 17:132–34

Reimer NJ. 1994. Distribution and impact of alien ants in vulnerable Hawaiian ecosystems. See Williams 1994, pp. 11–22

Reimer NJ, Cope M-L, Yasuda G. 1993. Interference of *Pheidole megacephala* (Hymenoptera: Formicidae) with biological control of *Coccus viridis* (Homoptera: Coccidae) in coffee. *Environ. Entomol.* 22:483–88

Ridlehuber KT. 1982. Fire ant predation on wood duck ducklings and pipped eggs. *Southwest. Nat.* 27:222

Risch SJ, Carroll CR. 1986. Effects of seed predation by a tropical ant on competition among weeds. *Ecology* 67:1319–27

Room PM, Smith ESC. 1975. Relative abundance and distribution of insect pests, and other components of the cocoa ecosystem in Papua New Guinea. *J. Appl. Ecol.* 12:31–46

Ross KG. 1997. Multilocus evolution in fire ants: effects of selection, gene flow and recombination. *Genetics* 145:961–74

Ross KG, Keller L. 1995. Ecology and evolution of social organization: insights from fire ants and other highly eusocial insects. *Annu. Rev. Ecol. Syst.* 26:631–56

Ross KG, Trager JC. 1990. Systematics and population genetics of fire ants (*Solenopsis saevissima* complex) from Argentina. *Evolution* 44:2113–34

Ross KG, Vargo EL, Fletcher DJC. 1987. Comparative biochemical genetics of three fire ant species in North America, with special reference to the two social forms of *Solenopsis invicta* (Hymenoptera: Formicidae). *Evolution* 41:979–90

Ross KG, Vargo EL, Keller L. 1996. Social evolution in a new environment: the case of in-

troduced fire ants. *Proc. Natl. Acad. Sci. USA* 93:3021–25

Roubik DW. 1978. Competitive interactions between neotropical pollinators and africanized honey bees. *Science* 201:1030–32

Roubik DW. 1980. Foraging behavior of competing africanized honey bees and stingless bees. *Ecology* 61:836–45

Ruesink JL, Parker IM, Groom MJ, Kareiva PM. 1995. Reducing risks of nonidigenous species introductions. *BioScience* 45:465–77

Samways MJ, Magda N, Prins AJ. 1982. Ants (Hymenoptera: Formicidae) foraging in citrus trees and attending honeydew-producing Homoptera. *Phytophylactica* 14:155–57

Sanders NJ, Barton KE, Gordon DM. 2001. Long-term dynamics of the distribution of the invasive Argentine ant, *Linepithema humile*, and native ant taxa in northern California *Oecologia* 127:123–30

Schaffer WM, Zeh DW, Buchmann SL, Kleinhans S, Schaffer MV, Antrim J. 1983. Competition for nectar between introduced honey bees and native North American bees and ants. *Ecology* 64:564–77

Schmidt CD. 1984. Influence of fire ants (*Solenopsis invicta*) on horn flies and other dung-breeding Diptera in Bexar County, Texas (USA). *Southwest. Entomol.* 9:174–77

Seifert B. 2000. Rapid range expansion in *Lasius neglectus* (Hymenoptera: Formicidae)— an Asian invader swamps Europe. *Dtsch. Entomol. Z.* 47:173–79

Shatters RG, Vander Meer RK. 2000. Characterizing the interaction between fire ants (Hymenoptera: Formicidae) and developing soybean plants. *J. Econ. Entomol.* 93:1680–87

Shoemaker DD, Ross KG, Arnold ML. 1996. Genetic structure and evolution of a fire ant hybrid zone. *Evolution* 50:1958–76

Sikes PJ, Arnold KA. 1986. Red imported fire ant *Solenopsis invicta* predation on cliff swallow *Hirundo pyrrhonota* nestlings in east-central Texas USA. *Southwest. Nat.* 31:105–6

Silverman J, Nsimba B. 2000. Soil-free collection of Argentine ants (Hymenoptera:

Formicidae) based on food-directed brood and queen movement. *Fla. Entomol.* 83:10–16

Simberloff D. 1989. Which insect introductions succeed and which fail? In *Biological Invasions: A Global Perspective*, ed. JA Drake, pp. 61–75. New York: Wiley

Simberloff D, Von Holle B. 1999. Positive interactions of nonindigenous species: invasional meltdown? *Biol. Invasions* 1:21–32

Sivapragasam A, Chua TH. 1997. Natural enemies for the cabbage webworm, *Hellula undalis* (Lepidoptera: Pyralidae) in Malaysia. *Res. Popul. Ecol.* 39:3–10

Smith TS, Smith SA, Schmidly DJ. 1990. Impact of fire ant (*Solenopsis invicta*) density on northern pygmy mice (*Baiomys taylori*). *Southwest. Nat.* 35:158–62

Sockman KW. 1997. Variation in life-history traits and nest-site selection affects risk of nest predation in the California gnatcatcher. *Auk* 114:324–32

Stam PA, Newsom LD, Lambremont EN. 1987. Predation and food as factors affecting survival of *Nezara viridula* (L.) (Hemiptera: Pentatomidae) in a soybean ecosystem. *Environ. Entomol.* 16:1211–16

Stein MB, Thorvilson HG. 1989. Ant species sympatric with the red imported fire ant in southeastern Texas. *Southwest. Entomol.* 14:225–31

Steneck RS, Carlton JT. 2001. Human alterations of marine communities: Students beware! In *Marine Community Ecology*, ed. MD Bertness, SD Gaines, ME Hay, pp. 445–68. Sunderland, MA: Sinauer

Sterling WL, Jones D, Dean DA. 1979. Failure of the red imported fire ant to reduce entomophagous insect and spider abundance in a cotton agroecosystem. *Environ. Entomol.* 8:976–81

Stevens AJ, Stevens NM, Darby PC, Percival HF. 1999. Observations of fire ants (*Solenopsis invicta* Buren) attacking apple snails (*Pomacea paludosa* Say) exposed during dry down conditions. *J. Molluscan Stud.* 65:507–10

Stoker RL, Grant WE, Vinson SB. 1995.

Solenopsis invicta (Hymenoptera: Formicidae) effect on invertebrate decomposers of carrion in central Texas. *Environ. Entomol.* 24:817–22

Stuart RJ. 1987. Transient nestmate recognition cues contribute to a multicolonial population structure in the ant, *Leptothorax curvispinosus*. *Behav. Ecol. Sociobiol.* 21:229–35

Suarez AV, Bolger DT, Case TJ. 1998. Effects of fragmentation and invasion on native ant communities in coastal southern California. *Ecology* 79:2041–56

Suarez AV, Case TJ. 2002. Bottom-up effects on persistence of a specialist predator: ant invasions and horned lizards. *Ecol. Appl.* 12:291–98

Suarez AV, Holway DA, Case TJ. 2001. Patterns of spread in biological invasions dominated by long-distance jump dispersal: insights from Argentine ants. *Proc. Natl. Acad. Sci. USA* 98:1095–100

Suarez AV, Richmond JQ, Case TJ. 2000. Prey selection in horned lizards following the invasion of Argentine ants in southern California. *Ecol. Appl.* 10:711–25

Suarez AV, Tsutsui ND, Holway DA, Case TJ. 1999. Behavioral and genetic differentiation between native and introduced populations of the Argentine ant. *Biol. Invasion* 1:43–53

Sudd JH. 1987. Ant aphid mutualism. In *Aphids: Their Biology, Natural Enemies and Control A*, ed. AK Minks, P Harrewijn, pp. 335–65. Amsterdam: Elsevier

Summerlin JW, Harris RL, Petersen HD. 1984a. Red imported fire ant (*Solenopsis invicta*) (Hymenoptera: Formicidae): frequency and intensity of invasion of fresh cattle droppings. *Environ. Entomol.* 13:1161–63

Summerlin JW, Hung ACF, Vinson SB. 1977. Residues in non-target ants, species simplification and recovery of populations following aerial application of mirex. *Environ. Entomol.* 6:193–97

Summerlin JW, Petersen HD, Harris RL. 1984b. Red imported fire ant (*Solenopsis invicta*) (Hymenoptera: Formicidae): effects on the horn fly (*Haematobia irritans*) (Diptera:

Muscidae) and coprophagous scarabs. *Environ. Entomol.* 13:1405–10

Taber SW. 2000. *Fire Ants.* College Station: Tex. A&M Univ. Press

Tedders WL, Reilly CC, Wood BW, Morrison RK, Lofgren CS. 1990. Behavior of *Solenopsis invicta* (Hymenoptera: Formicidae) in pecan orchards. *Environ. Entomol.* 19:44–53

Tennant LE, Porter SD. 1991. Comparison of diets of two fire ant species (Hymenoptera: Formicidae): solid and liquid components. *J. Entomol. Sci.* 26:450–65

Tilman D. 1999. The ecological consequences of changes in biodiversity: a search for general principles. *Ecology* 80:1455–74

Tobin JE. 1994. Ants as primary consumers: diet and abundance in the Formicidae. In *Nourishment and Evolution in Insect Societies*, ed. JH Hunt, CA Nalepa, pp. 279–308. Boulder, CO: Westview

Torres JA. 1984. Niches and coexistence of ant communities in Puerto Rico: repeated patterns. *Biotropica* 16:284–95

Trabanino CR, Pitre HN, Andrews KL, Meckenstock DH. 1989. Effect of seed size, color and number of seeds per hill and depth of planting on sorghum seed survival and stand establishment: relationship to phytophagous insects. *Trop. Agric.* 66:225–29

Trager JC, ed. 1988. *Advances in Myrmecology.* New York: EJ Brill

Travis BV. 1938. The fire ant (*Solenopsis* spp.) as a pest of quail. *J. Econ. Entomol.* 31:649–52

Tremper BS. 1976. *Distribution of the Argentine ant,* Iridomyrmex humilis *Mayr, in relation to certain native ants of California: ecological, physiological, and behavioral aspects.* PhD thesis. Univ. Calif., at Berkeley

Tschinkel WR. 1988. Distribution of the fire ants *Solenopsis invicta* and *S. geminata* (Hymenoptera: Formicidae) in northern Florida in relation to habitat and disturbance. *Ann. Entomol. Soc. Am.* 81:76–81

Tschinkel WR. 1992. Sociometry and sociogenesis of colonies of the fire ant Solenopsis invicta during one annual cycle. *Ecol. Monogr.* 63:425–57

Tschinkel WR. 1993. The fire ant (*Solenopsis invicta*): still unvanquished. In *Biological Pollution: The Control and Impact of Invasive Exotic Species*, ed. BN McKnight, pp. 121–36. Indianapolis: Indiana Acad. Sci.

Tschinkel WR. 1996. A newly-discovered mode of colony founding among fire ants. *Insect. Soc.* 43:267–76

Tschinkel WR. 1998. The reproductive biology of fire ant societies. *BioScience* 48:593–605

Tschinkel WR, Adams ES, Macom T. 1995. Territory area and colony size in the fire ant, *Solenopsis invicta. J. Anim. Ecol.* 64:473–80

Tschinkel WR, Howard DF. 1978. Queen replacement in orphaned colonies of the fire ant, *Solenopsis invicta. Behav. Ecol. Sociobiol.* 3:297–310

Tschinkel WR, Howard DF. 1983. Colony founding by pleometrosis in the fire ant, *Solenopsis invicta. Behav. Ecol. Sociobiol.* 12:103–13

Tsutsui ND, Case TJ. 2001. Population genetics and colony structure of the Argentine ant (*Linepithema humile*) in its native and introduced ranges. *Evolution* 55:976–85

Tsutsui ND, Suarez AV, Holway DA, Case TJ. 2000. Reduced genetic variation and the success of an invasive species. *Proc. Natl. Acad. Sci. USA* 97:5948–53

Tsutsui ND, Suarez AV, Holway DA, Case TJ. 2001. Relationships among native and introduced populations of the Argentine ant (*Linepithema humile*) and the source of introduced populations. *Mol. Ecol.* 10:2151–61

Tuberville TD, Bodie JR, Jensen JB, Laclaire L, Whitfield GJ. 2000. Apparent decline of the southern hog-nosed snake, *Heterodon simus. J. Elisha Mitchell Sci. Soc.* 116:19–40

Ulloa-Chacon P, Cherix D. 1990. The little fire ant *Wasmannia auropunctata* (R.) (Hymenoptera: Formicidae). See Vander Meer et al. 1990a, pp. 281–89

Van Der Goot P. 1916. Further investigations regarding the economic importance of the Gramang-ant. *Rev. Appl. Entomol.* 5:273–395

Vander Meer RK, Alonso LE. 2002. Queen primer pheromone affects conspecific fire ant

(*Solenopsis invicta*) aggression. *Behav. Ecol. Sociobiol.* 51:122–30

Vander Meer RK, Jaffe K, Cedeno A, eds. 1990a. *Applied Myrmecology: A World Perspective.* Boulder, CO: Westview

Vander Meer RK, Obin MS, Morel L. 1990b. Nestmate recognition in fire ants: monogyne and polygyne populations. See Vander Meer et al. 1990a, pp. 322–28

Vanderwoude C, Lobry De Bruyn LA, House PN. 2000. Response of an open-forest ant community to invasion by the introduced ant, *Pheidole megacephala. Aust. Ecol.* 25:253–59

Van Loon AJ, Boomsma JJ, Andrasfalvy A. 1990. A new polygynous *Lasius* species (Hymenoptera: Formicidae) from central-Europe. 1. Description and general biology. *Insect Soc.* 37:348–62

Vargo EL, Passera L. 1991. Pheromonal and behavioural queen control over the production of gynes in the Argentine ant *Iridomyrmex humilis* (Mayr). *Behav. Ecol. Sociobiol.* 28: 161–69

Veeresh GK. 1990. Pest ants of India. See Vander Meer et al. 1990a, pp. 15–24

Vinson SB, ed. 1986. *Economic Impact and Control of Social Insects.* Westport, CT: Greenwood

Vinson SB. 1991. Effect of the red imported fire ant on a small plant-decomposing arthropod community. *Environ. Entomol.* 20:98–103

Vinson SB. 1997. Invasion of the red imported fire ant (Hymenoptera: Formicidae): spread, biology, and impact. *Am. Entomol.* 43:23–39

Vinson SB, Greenberg L. 1986. The biology, physiology, and ecology of imported fire ants. See Vinson 1986, pp. 193–226

Vinson SB, Scarborough TA. 1989. Impact of the imported fire ant on laboratory populations of cotton aphid (*Aphis gossypii*) predators. *Fla. Entomol.* 72:107–11

Vinson SB, Scarborough TA. 1991. Interactions between *Solenopsis invicta* (Hymenoptera: Formicidae), *Rhopalsiphum maidis* (Homoptera: Aphididae) and the parasitoid *Lysiphlebus testaceipes* Cresson (Hymenoptera: Aphidiidae). *Ann. Entomol. Soc. Am.* 84:158–64

Visser D, Wright MG, Giliomee JH. 1996. The effect of the Argentine ant, *Linepithema humile* (Mayr) (Hymenoptera: Formicidae), on flower-visiting insects of *Protea nitida* Mill. (Proteaceae). *Afr. Entomol.* 4:285–87

Vogt JT, Grantham RA, Smith WA, Arnold DC. 2001. Prey of the red imported fire ant (Hymenoptea: Formicidae) in Oklahoma peanuts. *Environ. Entomol.* 30:123–28

Ward PS. 1987. Distribution of the introduced Argentine ant (*Iridomyrmex humilis*) in natural habitats of the lower Sacramento Valley and its effects on the indigenous ant fauna. *Hilgardia* 55(2):1–16

Washburn JO. 1984. Mutualism between a cynipid gall wasp and ants. *Ecology* 65:654–56

Way MJ. 1953. The relationship between certain ant species with particular reference to biological control of the coreid, *Theraptus sp. Bull. Entomol. Res.* 45:669–91

Way MJ. 1963. Mutualism between ants and honeydew producing Homoptera. *Annu. Rev. Entomol.* 8:307–44

Way MJ, Cammell ME, Paiva MR. 1992. Studies on egg predation by ants (Hymenoptera: Formicidae) especially on the eucalyptus borer *Phoracantha semipunctata* (Coleoptera: Cerambycidae) in Portugal. *Bull. Entomol. Res.* 82:425–32

Way MJ, Cammell ME, Paiva MR, Collingwood CA. 1997. Distribution and dynamics of the Argentine ant *Linepithema* (*Iridomyrmex*) *humile* (Mayr) in relation to vegetation, soil conditions, topography and native competitor ants in Portugal. *Insect Soc.* 44:415–33

Way MJ, Islam Z, Heong KL, Joshi RC. 1998. Ants in tropical irrigated rice: distribution and abundance, especially of *Solenopsis geminata* (Hymenoptera: Formicidae). *Bull. Entomol. Res.* 88:467–76

Way MJ, Khoo KC. 1989. Relationships between *Helopeltis theobromae* damage and ants with special reference to Malaysian cocoa smallholdings. *J. Plant. Prot. Trop.* 6:1–12

Way MJ, Khoo KC. 1992. Role of ants in pest management. *Annu. Rev. Entomol.* 37:479–503

Way MJ, Paiva MR, Cammell ME. 1999. Natural biological control of the pine processionary moth *Thaumetopoea pityocampa* (Den & Schiff) by the Argentine ant *Linepithema humile* (Mayr) in Portugal. *Agric. Forest. Entomol.* 1:27–31

Wetterer JK, Miller SE, Wheeler DE, Olson CA, Polhemus DA, et al. 1999. Ecological dominance by *Paratrechina longicornis* (Hymenoptera: Formicidae) an invasive tramp ant, in Biosphere 2. *Fla. Entomol.* 82:381–88

Wheeler WM. 1910. *Ants: Their Structure, Development and Behaviour.* New York: Columbia Univ. Press

Williams DF, ed. 1994. *Exotic Ants: Biology, Impact, and Control of Introduced Species.* Boulder, CO: Westview

Williams DF, Collins HL, Oi DH. 2001. The red imported fire ant (Hymenoptera: Formicidae): an historical perspective of treatment programs and the development of chemical baits for control. *Am. Entomol.* 47:146–59

Williams DF, Whelan P. 1991. Polygynous colonies of *Solenopsis geminata* (Hymenoptera: Formicidae) in the Galapagos Islands, Ecuador. *Fla. Entomol.* 74:368–71

Willmer PG, Stone GN. 1997. How aggressive ant-guards assist seed-set in *Acacia* flowers. *Nature* 388:165–67

Wilson EO. 1951. Variation and adaptation in the imported fire ant. *Evolution* 5:68–79

Wilson EO. 1971. *The Insect Societies.* Cambridge, MA: Harvard Univ. Press

Wilson EO, Brown WL. 1958. Recent changes in the introduced population of the fire ant *Solenopsis saevissima* (Fr. Smith). *Evolution* 12:211–18

Wilson EO, Taylor RW. 1967. The ants of Polynesia (Hymenoptera: Formicidae). *Pac. Insects Monogr.* 14:1–109

Witt ABR, Giliomee JH. 1999. Soil-surface temperatures at which six species of ants (Hymenoptera: Formicidae) are active. *Afr. Entomol.* 7:161–64

Wojcik DP, Allen CR, Brenner RJ, Forys EA, Jouvenaz DP, Lutz SR. 2001. Red imported fire ants: impact on biodiversity. *Am. Entomol.* 47:16–23

Wylie FR. 1974. The distribution and life history of *Milionia isodoxa* (Lepidoptera: Geometridae) a pest of planted hoop pine in Papua New Guinea. *Bull. Entomol. Res.* 63:649–59

Yosef R, Lohrer FE. 1995. Loggerhead shrikes, red fire ants and red herrings. *Condor* 97:1053–56

Zachariades C, Midgley JJ. 1999. Extrafloral nectaries of South African Proteaceae attract insects but do not reduce herbivory. *Afr. Entomol.* 7:67–76

Zenner-Polania I. 1994. Impact of *Paratrechina fulva* on other ant species. See Williams 1994, pp. 121–32

Zerhusen D, Rashid M. 1992. Control of the big-headed ant *Pheidole megacephala* Mayr (Hymenoptera: Formicidae) with the fire ant bait 'AMDRO' and its secondary effect on the population of the African weaver ant *Oecophylla longinoda* Latreille (Hymenoptera: Formicidae). *J. Appl. Entomol.* 113:258–64

Zettler JA, Spira TP, Allen CR. 2001. Ant-seed mutualisms: Can red imported fire ants sour the relationship? *Biol. Conserv.* 101:249–53

Zimmerman EC. 1970. Adaptive radiation in Hawaii with special reference to insects. *Biotropica* 2:32–38

Annu. Rev. Ecol. Syst. 2002. 33:235–63
doi: 10.1146/annurev.ecolsys.33.010802.150513
First published online as a Review in Advance on August 14, 2002

Gulf of Mexico Hypoxia, a.k.a. "The Dead Zone"

Nancy N. Rabalais,[1] R. Eugene Turner,[2] and
William J. Wiseman, Jr.[3]

[1]*Louisiana Universities Marine Consortium, 8124 Hwy. 56, Chauvin,
Louisiana 70344; email: nrabalais@lumcon.edu*
[2]*Coastal Ecology Institute, and Department of Oceanography & Coastal Sciences,
Louisiana State University, Baton Rouge, Louisiana 70803; email: euturne@lsu.edu*
[3]*Coastal Studies Institute, and Department of Oceanography & Coastal Sciences,
Louisiana State University, Baton Rouge, Louisiana 70803; email: wwiseman@lsu.edu*

Key Words anoxia, eutrophication, nutrient enrichment, Mississippi River,
coastal, estuary

■ **Abstract** The second largest zone of coastal hypoxia (oxygen-depleted waters) in the world is found on the northern Gulf of Mexico continental shelf adjacent to the outflows of the Mississippi and Atchafalaya Rivers. The combination of high freshwater discharge, wind mixing, regional circulation, and summer warming controls the strength of stratification that goes through a well-defined seasonal cycle. The physical structure of the water column and high nutrient loads that enhance primary production lead to an annual formation of the hypoxic water mass that is dominant from spring through late summer. Paleoindicators in dated sediment cores indicate that hypoxic conditions likely began to appear around the turn of the last century and became more severe since the 1950s as the nitrate flux from the Mississippi River to the Gulf of Mexico tripled. Whereas increased nutrients enhance the production of some organisms, others are eliminated from water masses (they either emigrate from the area or die) where the oxygen level falls below 2 mg l^{-1} or lower for a prolonged period. A hypoxia-stressed benthos is typified by short-lived, smaller surface deposit-feeding polychaetes and the absence of marine invertebrates such as pericaridean crustaceans, bivalves, gastropods, and ophiuroids. The changes in benthic communities, along with the low dissolved oxygen, result in altered sediment structure and sediment biogeochemical cycles. Important fisheries are variably affected by increased or decreased food supplies, mortality, forced migration, reduction in suitable habitat, increased susceptibility to predation, and disruption of life cycles.

INTRODUCTION

Waters with less than full oxygen saturation occur in many parts of the world's oceans (Kamykowski & Zentara 1990). Hypoxic (low oxygen) and anoxic (no oxygen) waters have existed throughout geologic time and presently occur in many

0066-4162/02/1215-0235$14.00

235

of the ocean's deeper environs, such as oxygen minimum layers, deep basins, and fjords. The occurrence of hypoxia and anoxia in shallow, coastal and estuarine areas, however, appears to be increasing, most likely accelerated by human activities (Diaz & Rosenberg 1995). One of the world's largest zones of estuarine and coastal hypoxia is located in the northern Gulf of Mexico on the Louisiana/Texas continental shelf. Referred to as the "Dead Zone" in the popular press and literature, the areal extent of the severe oxygen deficiency reached a record size of 20,700 km^2 in mid-summer 2001 (Rabalais et al. 2001). The term dead zone refers to the failure to capture fish, shrimp, and crabs in bottom-dragging trawls when the oxygen concentration falls below a critical level in water near the seabed (Renaud 1986). The numbers of stressed or dying benthic infaunal organisms within the sediments increase substantially when the oxygen levels remain low for prolonged periods, biodiversity is diminished, and community structure and ecosystem functioning are altered (Rabalais et al. 2001a,b). Higher in the water column and in the surface mixed layer, however, there is sufficient oxygen to support sizeable populations of fish and swimming crabs. Also, there are anaerobic or hypoxia-adapted organisms that survive in sediments overlain by hypoxic or anoxic waters, so that the term dead zone is not entirely accurate (Rabalais & Turner 2001a). Still, the area is large enough to garner a high level of public attention, if only because of the lack of catchable demersal fish and shrimp.

There is no universally accepted terminology or concentration limits to describe oxygen-deficient conditions (Tyson & Pearson 1991). Hypoxia is a term long used by physiologists to describe conditions or responses produced by stressful levels of oxygen deficiency. The use of the term to describe oxygen-deficient environments became common in the mid- to late 1970s, and was especially attributed to those working in the Gulf of Mexico by Tyson & Pearson (1991). Based on laboratory or field-observed responses of organisms to oxygen stress, hypoxia has been variously defined as corresponding to a range of 3.0–0.2 ml l^{-1}, with the consensus in favor of 1.4 ml l^{-1} (= 2 mg l^{-1} or ppm). Another convenient threshold for effects that was adopted by Breitburg (2002) was dissolved oxygen concentration <50% saturation owing to avoidance behavior, reduced growth, or other signs of physiological stress in sensitive fish. For this review, we define hypoxia for the northern Gulf of Mexico as dissolved oxygen levels below 2 mg l^{-1}. This is the level below which bottom-dragging trawls usually do not capture any shrimp or demersal fish (Renaud 1986). This concentration approximates 20% oxygen saturation in 25°C and salinity of 35 in summertime bottom waters of the Gulf of Mexico (Rabalais et al. 1999); in less saline, cooler Chesapeake Bay waters 2 mg l^{-1} equals 24% saturation at 20°C and salinity of 15 (Breitburg 2002). The range of organismal responses depends on the severity of the hypoxia, the length of exposure, and the periodicity and frequency of exposure. The hypoxic zone of the northern Gulf of Mexico, one of hundreds globally, will be the focal point of this review with similarities and differences drawn from other coastal and estuarine areas.

The coastal areas of the Baltic Sea, northern Gulf of Mexico, and northwestern shelf of the Black Sea are the largest such coastal hypoxic zones in the world, reaching 84,000 km^2, 21,000 km^2, and 40,000 km^2 (until recently), respectively

(Rosenberg 1985, Rabalais et al. 2002, Mee 2001). Smaller and less frequent zones of hypoxia occur in the northern Adriatic Sea (Justić et al. 1987), the southern bight of the North Sea (Fransz & Verhagen 1985) and in many U.S. coastal and estuarine areas, for example, New York Bight (Garside & Malone 1978, Swanson & Sindermann 1979, Swanson & Parker 1988), Chesapeake Bay (Officer et al. 1984, Malone 1991), Long Island Sound (Welsh & Eller 1991, Welsh et al. 1994, Parker & O'Reilly 1991), Mobile Bay (Turner et al. 1987), and the Neuse River estuary (Paerl et al. 1998). In a review of 47 known anthropogenic hypoxic zones where benthic community effects were documented, Diaz & Rosenberg (1995) noted that no other environmental variable of such ecological importance to estuarine and coastal marine ecosystems around the world as dissolved oxygen has changed so drastically and in such a short period of time. For those zones reviewed, they found that there was a consistent trend of increasing severity (either duration, intensity, or size) where hypoxia occurred historically, or hypoxia existed presently when it did not occur before. The occurrence of hypoxia in estuarine and coastal areas is increasing and the trend is consistent with the increase in human activities that result in nutrient over-enrichment.

Hypoxia is but one of the symptoms of eutrophication, defined by Nixon (1995) as an increase in the rate of production and accumulation of carbon in aquatic systems. Eutrophication may be a natural process, but very often results from an increase in nutrient loading, particularly by forms of nitrogen and phosphorus. Coastal eutrophication, often accompanied by hypoxia, tracks increases in population, a focusing of that populace in coastal regions, agricultural expansion in river basins, and increasing food and energy consumption (Nixon 1995; Howarth et al. 1995, 1996; Vitousek et al. 1997; Caraco & Cole 1999; Bennett et al. 2001). Nutrient over-enrichment from anthropogenic sources is one of the major stressors impacting estuarine and coastal ecosystems (Bricker et al. 1999, Howarth et al. 2000, National Research Council 2000, Cloern 2001). There is increasing concern in many areas around the world that an oversupply of nutrients from multiple sources is having pervasive ecological effects on shallow coastal waters. These effects include reduced light penetration, increased abundance of nuisance macroalgae, loss of aquatic habitat such as seagrass or macroalgal beds, noxious and toxic algal blooms, hypoxia and anoxia, shifts in trophic interactions and food webs, and impacts on living resources (Vitousek et al. 1997, Schramm 1999, Anderson et al. 2002, Rabalais 2002). Whereas many of these primary and secondary responses of ecosystems to nutrient over-enrichment are interrelated, this review will focus on aspects of hypoxia, emphasizing the Gulf of Mexico, in describing causes, temporal and spatial variability, effects on living resources, historical changes, and management challenges.

CAUSES OF HYPOXIA—FRESH WATER AND NUTRIENTS

Two principal factors lead to the development and maintenance of hypoxia. First, the water column must be stratified so that the bottom layer is isolated from the surface layer and the diffusion of oxygen from surface to bottom. The physical structure is dictated by water masses that differ in temperature or salinity or both. Fresher waters derived from rivers and seasonally-warmed surface waters are less dense than, and reside above, the saltier, cooler and more dense water masses near the bottom. Both salinity and temperature are important in influencing the strength of stratification in the northern Gulf of Mexico (Rabalais et al. 1991), Chesapeake Bay (Officer et al. 1984), and Kiel Bay (Arntz 1981), whereas stratification is controlled primarily by temperature in the Gulf of Trieste (Stachowitsch 1986), New York Bight (Falkowski et al. 1980), and Long Island Sound (Welsh & Eller 1991). Stratification goes through a well-defined seasonal cycle that generally exhibits maximum stratification during summer and weakest stratification during winter months (Figure 1, see color insert). This cycle is due to the strength and phasing of river discharge, wind mixing, regional circulation and air-sea heat exchange processes.

The second factor is decomposition of organic matter that leads to reduced oxygen levels in the bottom waters. The source of the organic matter settling to the seabed in the northern Gulf of Mexico is mostly from phytoplankton growth stimulated by riverine-delivered nutrients, although some river-borne organic matter is delivered by the Mississippi River (Eadie et al. 1994, Turner & Rabalais 1994a, Committee on Environment and Natural Resources 2000). The concentrations and total loads of nitrogen, phosphorus, and silica to the coastal ocean influence the productivity of the phytoplankton community, the types of phytoplankton that are most likely to grow, and ultimately the flux of phytoplankton-derived organic matter (Turner & Rabalais 1994b, Lohrenz et al. 1997, Dortch et al. 2001). Phytoplankton not incorporated into the food web and fecal material generated by the food web sink into bottom waters where they are decomposed by aerobic bacteria, causing oxygen depletion. The relative influence of the physical features of the system and the progression of biological processes varies spatially and over an annual cycle. Both are inter-related in the northern Gulf of Mexico and directly linked with the dynamics of the Mississippi and Atchafalaya River discharge. (The Atchafalaya River carries up to one-third of diverted Mississippi River discharge to a second delta 180 km west of the birdfoot delta.)

Some coastal areas and estuaries are more susceptible to eutrophication, including the formation of hypoxia, from nutrient enrichment based on two key physical features of the water body—the dilution capacity of the water column and its flushing/retention time (National Research Council 2000). Systems with relatively large volumes and short flushing times are less susceptible to eutrophication; systems with longer flushing times are more susceptible. Superimposed upon this continuum of susceptibility are the nutrient loadings to the water body, the timing of the loads, and the relative proportion of the nutrients. Light availability can play a critical role in determining the response of estuarine systems to nutrient loading (Cloern 1999), and grazing of phytoplankton by benthic filter feeders or water column zooplankton can limit the accumulation of algal biomass (Alpine & Cloern 1992).

SOURCES OF NUTRIENTS

Human activity has dramatically increased the flux of phosphorus and nitrogen to the world's oceans. The fluvial drainage of phosphorus from terrestrial and freshwater ecosystems is currently nearly three times greater than pre-industrial levels (Bennett et al. 2001). The change in nitrogen flux is even more dramatic, for example fourfold in the Mississippi River, eightfold in the rivers of the northeastern United States, and tenfold in the rivers draining to the North Sea (Howarth et al. 1996). The human activities and sources of increased nitrogen and phosphorus vary across watersheds, but their relative proportion and rate of change identify potential foci for eventual attempts at reduction. Nonpoint sources of nutrients are the dominant and least easily controlled inputs into coastal waters from large watersheds, and especially from watersheds with extensive agricultural activity (e.g., Mississippi River system) or atmospheric nitrogen pollution from fossil-fuel combustion (e.g., dominant in the northeastern United States).

Some estuaries receive nutrients across their boundary with the ocean along with inputs from land, rivers, and atmospheric deposition (Boynton et al. 1995, Nixon et al. 1995). Offshore waters on continental shelves can similarly receive nutrients from river inputs from land, direct deposition from the atmosphere, and advection of deeper, nutrient-rich oceanic waters. The relative proportion of these inputs varies among coastal waters of the U.S., but can be a dominant nutrient input (Nixon et al. 1996). In the northern Gulf of Mexico, however, the Mississippi River drainage is by far the dominant source of nutrients fueling the hypoxia zone.

Dunn (1996) calculated the nutrient inflows from 37 U.S. streams discharging into the Gulf of Mexico for water years 1972–1993. The combined flows of the Mississippi and Atchafalaya Rivers account for 91% of the estimated total nitrogen load. If only streams between Galveston Bay (Texas) and the Mississippi River delta are considered, i.e., those most likely to influence the zone of hypoxia, the combined flows of the Mississippi and Atchafalaya Rivers account for 96% of the annual freshwater discharge and 98.5% of the total annual nitrogen load. Similar calculations for the annual total phosphorus load are 88% of the total 37 streams and 98% of the streams between Galveston Bay and the Mississippi River delta. The relative contribution of direct atmospheric deposition of nitrogen to the total nitrogen load for an area twice the size of the hypoxic zone is 1% (Goolsby et al. 1999). Groundwater sources to the area affected by hypoxia are unlikely to be important because of the lack of shallow aquifers along the Louisiana coast and the low potential for transfer in a cross-shelf direction to the area where hypoxia develops (Rabalais et al. 1999). The relative contribution of offshore sources of nutrients from upwelled waters of the continental slope is unknown but expected to be minimal considering the alongshore current regime. The Mississippi River system is, by far, the major source of nutrients to the northern Gulf of Mexico where hypoxia is likely to develop, and thereby influences the primary production on the shelf and eventual flux of organic matter to the lower water column below the pycnocline.

Away from the discharges of the Mississippi and Atchafalaya Rivers, there is a consistent westward and downstream transition from lower to higher salinities, higher to lower nutrients, and higher to lower surface chlorophyll concentrations (Rabalais et al. 1996, 1999). Ultimately nutrients become limiting to phytoplankton productivity. These gradients away from the riverine sources are further reflected in the flux of organic material as seen in surface-to-bottom pigment ratios and accumulation of phaeopigments in the lower water column. Respiration rates are related to chlorophyll a concentrations (Turner & Allen 1982a). Therefore, there is a consistent transition away from the river discharges along the coastal plume in flux of organic material, respiration rates and incidence of bottom water hypoxia. The variability in freshwater discharge and related flux of nutrients on seasonal, annual, decadal, and longer scales underlies many important physical and biological processes affecting coastal productivity and food webs.

DIMENSIONS OF GULF HYPOXIA

The shelfwide distribution of hypoxia is determined annually from a five-day mapping cruise in mid-summer, usually between mid-July and mid-August (Rabalais et al. 1999). The hypoxic water mass extends west from the Mississippi River bird-foot delta across the Louisiana shelf and onto the upper Texas coast, from within 1 km of the barrier shoreface to as much as 125 km offshore. Hypoxia is found in water depths up to 60 m, but more typically between 5 and 30 m. Hypoxia is not found just in a thin lens overlying bottom sediments, but occurs well up into the water column depending on the location of the pycnocline(s). Hypoxia may encompass from 10% to over 80% of the lower water column, but normally affects 20–50% (Figure 1).

Mid-summer hypoxic zones, between 1985 and 1992, generally formed in two distinct areas west of the Mississippi and Atchafalaya River deltas (Figure 2), with the total area averaging 8000–9000 km^2 (Figure 3). The size of the hypoxic zone doubled in response to the Great Mississippi River Flood of 1993, forming a single continuous zone across the Louisiana shelf (Figures 2 and 3). The examples from 1986 and 1993 (Figure 2) illustrate the typical pre- and post-flood distribution and doubling in average size between the two periods (Rabalais et al. 2002). Persistent currents from the west to the east across the Louisiana shelf often force the hypoxic water mass onto the southeastern shelf, which is represented by the 1998 example (Figure 2). Although the extent of hypoxia across the bottom in 1998 was less than 1997 (Figure 3), the volume of the hypoxic water mass was greater because more of the total depth of the water column was hypoxic in 1998 compared to 1997 (N. Rabalais et al. unpublished data). Low river discharge and nutrient flux in 1988 and 2000 resulted in smaller areas of hypoxia. Hypoxia developed in the spring of 1988 when discharge was normal, but was confined to a single station off Terrebonne Bay in July when flow reached a record low. The smaller area of

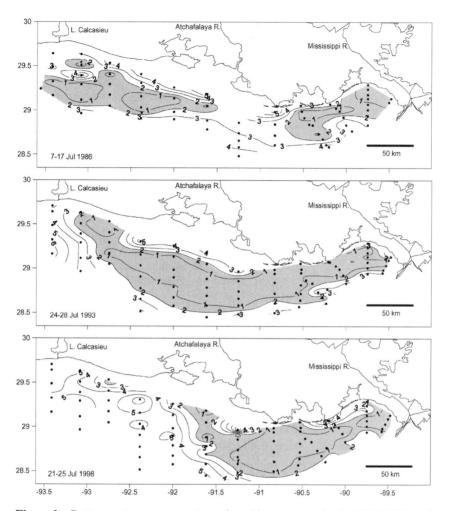

Figure 2   Bottom-water oxygen contours for mid-summer cruises in 1986, 1993, and 1998. The area of dissolved oxygen less than 2 mg l^{-1} is indicated by shading.

hypoxia in 2000 was proportional to the reduced Mississippi River discharge and nutrient flux that spring.

A compilation of sixteen mid-summer shelfwide surveys (1985–2001, Figure 4) shows that the frequency of >50% occurrence of hypoxia is highest down current (west) from the freshwater and nutrient discharges of the Mississippi and Atchafalaya Rivers. There are strong statistical relationships among nitrate flux, primary production, net production, and hypoxia in the area between the Mississippi River and transect C off Terrebonne Bay (Justić et al. 1993, 1997; Lohrenz et al. 1997). Similar statistical relationships exist between Atchafalaya River

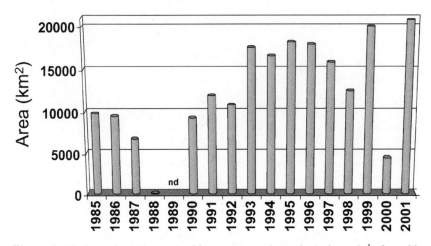

Figure 3   Estimated areal extent of bottom-water hypoxia (≤ 2 mg l^{-1}) for mid-summer cruises in 1985–2001 (updated from Rabalais et al. 1999, Rabalais & Turner 2001b). nd = no data collected.

discharge and hypoxia on the southwestern Louisiana shelf (Pokryfki & Randall 1987).

More frequent sampling along transect C on the southeastern Louisiana coast (location shown in Figure 4) indicates that critically depressed dissolved oxygen concentrations occur below the pycnocline from as early as late February through early October and nearly continuously from mid-May through mid-September (Rabalais et al. 1999). Data from fisheries' independent trawl surveys in the Mississippi River bight, the area immediately west of the birdfoot delta, indicate that hypoxia occurs in that area in 6- to 10-m water depth as late as November (T. Romaire,

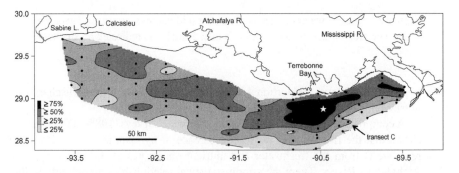

Figure 4   Distribution of frequency of occurrence of mid-summer bottom-water hypoxia over the 60- to 80-station grid from 1985–2001 (updated from Rabalais et al. 1999, Rabalais & Turner 2001b). Star indicates general location of stations C6A and C6B; transect C identified.

personal communication). Hypoxia is rare in late fall and winter. The monthly average value of bottom oxygen for each station along transect C is shown along a depth gradient from onshore to offshore by month (Figure 5, see color insert), and illustrates the seasonal progression of worsening hypoxia along an increasingly greater portion of the seabed in May through August. The persistence of extensive and severe hypoxia into September and October depends on the timing of the breakdown of vertical stratification by winds from either tropical storms or passage of cold fronts.

Once hypoxia becomes well established in mid-summer, much of the onshore-offshore variability in distribution can be attributed to wind-induced cross-shelf advection (Rabalais et al. 1991) (Figure 6). The low oxygen water mass is displaced into deeper water following winds that produce downwelling-favorable conditions. Upwelling-favorable conditions push the hypoxic water mass closer to the barrier island shore. Similar advection of saline, hypoxic water onto the flanks of Chesapeake Bay and into the lower reaches of the Choptank River, an adjoining tributary estuary, is driven by large-amplitude wind and tidally forced lateral internal oscillations of the pycnocline in the mainstem of the Bay (Breitburg 1990, Sanford et al. 1990). It is the impingement of these hypoxic water masses close to shore that results in "jubilees" along Louisiana barrier islands. Jubilees are events that usually follow a north wind and movement of the hypoxic water mass onto shore where stunned or stressed fish, shrimp and crabs are concentrated in the shallow waters along the beach and can be easily harvested. A jubilee "gone bad" occurs when the water mass is extremely low in oxygen or contains hydrogen sulfide, and the trapped fish cannot escape, resulting in massive fish kills. Similar jubilees occur along the eastern shore of Mobile Bay in response to intrusion of upwelled hypoxic water (Loesch 1960, May 1973).

Continuously recording (15-min interval) oxygen meters have been deployed near the bottom (20-m water depth) at Stations C6A or C6B off Terrebonne Bay during spring-fall since 1990 (Rabalais et al. 1999). There is variability within the year and between years, but the pattern generally depicted is (*a*) gradual decline of bottom oxygen concentrations through the spring and summer, with periodic reoxygenation from wind-mixing events, (*b*) persistent hypoxia and often anoxia for extended periods of the record in May-September, (*c*) isolated intrusion of higher oxygen content waters from deeper water during upwelling-favorable wind conditions, and (*d*) persistent wind-mixing events in the late summer and fall that mix the water column sufficiently to prevent prolonged instances of bottom-water hypoxia. The illustrated bottom oxygen series from station C6A from mid-June through mid-September 1990 (Figure 7) shows prolonged periods of hypoxia and anoxia, a short period of elevated dissolved oxygen in late August, and persistent reaeration of the water column beginning in late September. For comparison, a recording oxygen meter was deployed during the same period near the bottom in 20-m water depth but 77 km to the east and closer to the Mississippi River delta (Figure 7, Rabalais et al. 1994). At that station, hypoxia occurred for only

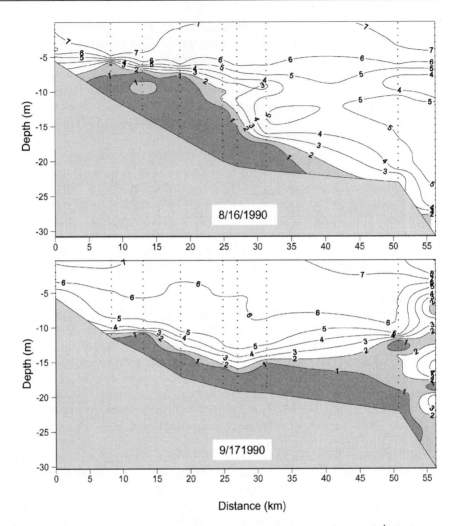

Figure 6 Cross-shelf contours for dissolved oxygen concentration (mg l^{-1}) during an upwelling-favorable wind condition (16 August 1990) and a downwelling-favorable wind condition (17 September 1990).

44% of the record from mid-June through mid-October (compared to 75% at station C6A), and there was a strong diurnal pattern in the oxygen time-series data for the former and not for the latter. The dominant coherence at the diurnal peaks of the oxygen record, from the site in the Mississippi River bight, with the bottom pressure record from a gauge located just offshore of Terrebonne Bay suggests that the dissolved oxygen signal was due principally to advection of the interface between hypoxic and normoxic water by tidal currents. These two bottom

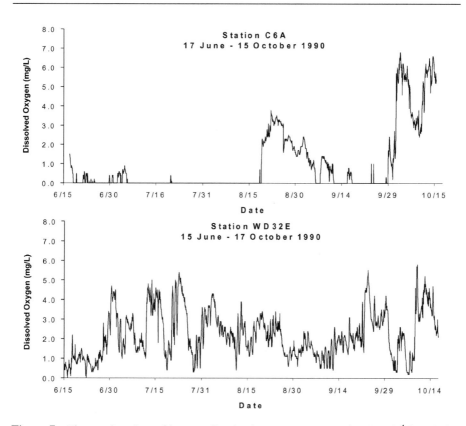

Figure 7   Time-series plots of bottom dissolved oxygen concentration (mg l^{-1} in 1-hr intervals) at stations C6A on transect C and WD32E, 77 km to the east in the Mississippi River bight (from Rabalais et al. 1994; reproduced with permission of the Estuarine Research Federation).

oxygen records illustrate the differences of exposure of benthic and demersal organisms to persistent, severe low oxygen versus intermittent periods of hypoxia and normoxia.

Most instances of hypoxia elsewhere in the northern Gulf of Mexico along the Texas, Mississippi, Alabama, and Florida coasts are infrequent, short-lived, and limited in extent (N.N. Rabalais 1992, unpublished data). Hypoxia on the upper Texas coast is usually an extension of the hypoxic zone off Louisiana (Harper et al. 1991, Pokryfki & Randall 1987), although isolated areas may be found farther south (Gulf States Fisheries Commission 1982–91). Hypoxia east of the Mississippi River is also isolated and ephemeral, but occurs more frequently during high stages of the Mississippi River in flood years or when summer currents move more Mississippi River water to the east of the birdfoot delta. Hypoxia occurs off Mobile Bay in bathymetric low areas (Rabalais 1992). From limited data where both sides of the delta were surveyed (Turner & Allen 1982b), there was no evidence that the area of low oxygen formed a continuous band around the delta.

HISTORY OF HYPOXIA AND RELATED FACTORS

Whereas Gulf of Mexico coastal hypoxia has been systematically mapped only since 1985, there are data that document its presence since the early 1970s (Rabalais 1992, Rabalais et al. 1999, 2002). [Several references to low oxygen in the Gulf during the mid-1930s are clearly referring to data from the oxygen minimum layer in deeper Gulf waters that are not connected to the continental shelf hypoxia (Rabalais et al. 1999, 2002).] A series of environmental assessments of oil and gas production activities, strategic petroleum reserve sites, and locations for offshore oil lightering operations and groundfish fishery assessments put an increasing number of scientists onto the Louisiana continental shelf with conductivity/temperature/depth/oxygen meters in the period 1972–1984. These studies were usually site-specific and rarely assessed the entire Louisiana continental shelf where hypoxia has been documented since 1985 (Figure 2), but still showed that hypoxia occurred in many locations in waters 5–30 m deep in the spring to fall, with most occurrences in the summer. In 1975–1976, Louisiana shelfwide surveys and surveys between Mobile Bay and Atchafalaya Bay (Ragan et al. 1978, Turner & Allen 1982b, respectively) found hypoxic areas that were less extensive than those mapped since 1985; however, the Turner & Allen (1982b) study area was smaller than the present-day grid. Whereas it appears from this compilation of data that hypoxia may have been increasing in frequency or extent, the history of hypoxia over decadal and century scales remained poorly resolved.

One might expect a propensity for the northern Gulf of Mexico ecosystem to develop hypoxia naturally because of the high volume of fresh water and nutrients delivered by the Mississippi River and the nature of the stratified coastal system. The question is often posed: hasn't hypoxia always been a feature of this system? Because relevant water column data do not exist before 1972, and systematic surveys did not start until 1985, we turned to the sediment record for paleoindicators of long-term transitions related to eutrophication and oxygen conditions beneath the Mississippi River plume. Sediment cores from inside and outside the current hypoxic region contain chemical and biological remains that reflect conditions extant in surface and bottom waters at the time the sediment was deposited and thus provide clues to decadal and century-long changes.

Sediment core indicators clearly document recent eutrophication and increased organic sedimentation in bottom waters, with the changes more apparent in areas of chronic hypoxia and coincident with the increasing nitrogen loads from the Mississippi River system in the 1950s (Eadie et al. 1994, Turner & Rabalais 1994a). This evidence comes as an increased accumulation of diatom remains (biologically bound silica) and marine-origin carbon (stable carbon isotope analysis) accumulation in the sediments. The increases in diatom-based productivity were also indicated in an analysis of changing dissolved silicate mixing diagrams along the Mississippi River plume (Turner & Rabalais 1994b). There have been no significant increases in either the riverine organic carbon or silica loads (Goolsby et al. 1999). The suspended sediment loads carried by the Mississippi River to the Gulf of Mexico have decreased by one-half since the Mississippi valley was first settled by European colonists (Meade et al. 1990). Alterations occurred as a result of deforestation and agriculture, changes in land management, and construction of dams, diversions, and levees. The decrease in suspended sediments has happened mostly since 1950 when the largest natural sources of sediments in the drainage basin were cut off from the Mississippi River mainstem by the construction of large reservoirs on the Missouri and Arkansas Rivers (Meade & Parker 1995). This large decrease in sediments from the western tributaries was counterbalanced somewhat by a five- to tenfold increase in sediment loads in the Ohio River as a result of deforestation and row-crop farming (Keown et al. 1986). A decrease in sediment load could potentially affect both the particulate and dissolved organic nitrogen flux. Because Mississippi River suspended sediment load has decreased since the 1950s (Meade 1995) and the dissolved inorganic nitrogen pool has increased from anthropogenic influences (Turner & Rabalais 1991), the relative importance of the organic nitrogen associated with the suspended sediment load might be less for the Mississippi River than for other large world rivers (Mayer et al. 1998). It is reasonable to infer that the increases in paleoindicators of phytoplankton productivity in the sediment record since the 1950s are related to in situ production of marine algae stimulated by dissolved inorganic nutrients.

Time courses for several surrogates for oxygen conditions are shown in Figure 8 (see color insert). The mineral glauconite forms under reducing conditions in sediments, and its abundance is an indication of low oxygen conditions. (Glauconite also forms in reducing sediments whose overlying waters are >2 mg l^{-1} dissolved oxygen.) The average glauconite abundance in the coarse fraction of sediments was $\sim5.8\%$ from 1900 to a transition period between 1940 and 1950, when it increased to $\sim13.4\%$ (Figure 8; Nelsen et al. 1994), suggesting that hypoxia may have existed at some level before the 1940–1950 time period, but that it worsened since then. Benthic foraminiferans and ostracods are also useful indicators of reduced oxygen levels because oxygen stress decreases their overall diversity as measured by the Shannon-Wiener Diversity Index (SWDI). Foraminiferal and ostracod diversity decreased since the 1940s and early 1950s, respectively (Figure 8; Nelsen et al. 1994, TA Nelsen, unpublished data). While present-day foraminiferal diversity is generally low in the Mississippi River bight, comparisons among assemblages from areas of different oxygen depletion indicate that the dominance of *Ammonia parkinsoniana* over *Elphidium* spp. (A-E index) was much more pronounced in oxygen-depleted compared to well-oxygenated waters (Rabalais et al. 1996, Sen Gupta et al. 1996). The A-E index has also proven to be a strong, consistent oxygen-stress signal in other coastal areas (Chesapeake Bay, Karlsen et al. 2000; Long Island Sound, Thomas et al. 2000). The A-E index from sediment cores increased significantly after the 1950s, suggesting increased oxygen stress (in intensity or duration) in the last half century (Figure 8). *Buliminella morgani*, a hypoxia-tolerant species, known only from the Gulf of Mexico, dominates the present-day population (>50%) within areas of chronic seasonal hypoxia, and has also increased markedly in recent decades (Sen Gupta et al. 1996). *Quinqueloculina* sp. (a hypoxia-intolerant foraminiferan) was a conspicuous member of the fauna from 1700 to 1900 (not illustrated; Rabalais et al. 1996), indicating that oxygen stress was not a problem prior to 1900. The trend of hypoxia, beginning at some level at the turn of the century with a significant increase after the 1950s when nitrate loads began to increase in the Mississippi River, has been further corroborated by analysis of bacterial chloropigments in dated sediment cores from the Mississippi River bight (Chen et al. 2001).

The long-term changes in hypoxia indicators can be compared with other long-term data sets and conditions in the watershed and Mississippi River to devise a scenario of what factors have changed through time that are most relevant to the changing hypoxia indicated in the sediment cores. The surrogates for oxygen conditions indicate an overall increase in continental shelf oxygen stress (in intensity or duration or both) in the last 100 years that seems especially severe since the 1950s. The indicators of worsening oxygen conditions parallel the increase in indicators of surface water primary production that accumulate in the sediments, i.e., diatom remains and phytoplankton-derived carbon, and the increasing flux of dissolved nitrate in the Mississippi River discharge (Turner & Rabalais 1991, 1994a; Goolsby et al. 1999).

In addition to a steady population increase within the Mississippi basin with related inputs of nitrogen through municipal wastewater systems, human activities have changed the natural functioning of the Mississippi River system. Navigation channelization and flood control through leveeing along the length of the river are clearly important watershed alterations, but most of these activities occurred well before the 1950s. Other significant alterations in landscape (e.g., deforestation, conversion of wetlands to cropland, loss of riparian zones, expansion of artificial agricultural drainage) removed most of the buffer for removing nutrients from runoff into the Mississippi tributaries and main stem. There was an increase in the area of land artificially drained between 1900 and 1920, and another significant burst in drainage during 1945–1960 (Mitsch et al. 2001). There was a dramatic increase in nitrogen input into the Mississippi River drainage basin, primarily from fertilizer application, between the 1950s and 1980s (Goolsby et al. 1999). These important alterations in land use and nitrogen input led to significant increases in riverine nitrate concentrations and flux to the Gulf (Turner & Rabalais 1991, Goolsby et al. 1999).

Because the amount of fresh water delivered to the northern Gulf of Mexico influences both the nitrogen load and the strength of salinity stratification on the shelf, climate-induced variability in river discharge will influence the extent and severity of hypoxia. Annual river discharge since the early 1900s has been highly variable, but total annual discharge increased by only 15% since 1900 and 30% since the 1950s (Bratkovich et al. 1994). This is in contrast to the 300% increase in nitrate load since the 1950s (Turner & Rabalais 1991, Goolsby et al. 1999). Clearly, the most significant driver in the change in nitrate load is the increase in nitrate river concentration, not freshwater discharge (Justić et al. 2002). In addition, the change in annual freshwater discharge appears to be due mostly to increased discharge in September–December (Bratkovich et al. 1994), a period less important to both the physical and biological processes that lead to hypoxia development and maintenance.

Organic carbon supplied by the Mississippi River was proposed as a cause of the formation of hypoxia in the Gulf (Carey et al. 1999). This view has been largely discounted (Committee on Environment and Natural Resources 2001) because: (*a*) the suspended sediment load in the river has declined by about half since the 1950s, (*b*) the particulate organic load that could settle on the Louisiana shelf has also most likely decreased since then, (*c*) the distances to which particulate organic carbon would need to be transported to affect the large area of hypoxia are too great for the riverine particles to provide a significant load, and (*d*) stable carbon isotope data for sediment cores from the hypoxic zone indicate that 80% of the carbon accumulated in the sediments is of marine, not terrestrial origin. A similar argument for carbon associated with wetland loss in coastal Louisiana can be applied. Wetland loss rates in coastal Louisiana peaked at 1200 ha yr^{-1} in the period 1955 to 1978, but the loss rate subsequently declined (Turner 1997), and mass flux calculations clearly indicate that this is a relatively small source of organic carbon to the hypoxic region. In addition, stable carbon isotope analyses indicate that wetland organic carbon accumulation is confined to a narrow band next to shore (Turner & Rabalais 1994a).

Where sufficient long-term data exist, e.g., Chesapeake Bay, the northern Adriatic Sea, the Baltic and the Black Seas, there is clear evidence for increases in nutrient flux, increased primary production, and worsening hypoxia (Cloern 2001). Thorough analyses of multiple indicators in sediment cores from the Chesapeake Bay indicate that sedimentation rates and eutrophication of the waters of the Bay have increased dramatically since the time of European settlement of the watershed (Cooper & Brush 1993, Cooper 1995, Karlsen et al. 2000). In addition, results indicate that hypoxia and anoxia may have been more severe and of longer duration in the last 50 years, particularly since the 1970s. The sediment core findings corroborate long-term changes in Chesapeake Bay water column chlorophyll biomass since the 1950s (Harding & Perry 1997). The parallels of the Chesapeake Bay eutrophication and hypoxia to those of the Mississippi River watershed and Gulf of Mexico hypoxia are striking, in particular those of the last half-century.

EFFECTS OF HYPOXIA ON LIVING RESOURCES

Increases in loads of nitrogen and phosphorus, sometimes accompanied by reductions in dissolved silicate, to estuarine and coastal systems can result in enhancement of primary and often secondary production (Nixon 1995, Cloern 2001,

Rabalais & Turner 2001a, Rabalais 2002). The shifts in nutrient ratios may also shift the composition of the phytoplankton base and thus affect trophic interactions, the transfer of energy through marine food webs, and the flux of carbon that causes the development of hypoxia (Officer & Ryther 1980, Turner et al. 1998, Dortch et al. 2001, Turner 2001). Whereas these multiple effects of nutrient enhancement certainly deserve attention (Rabalais 2002), we focus here on the effects of hypoxia, or the reduction in available oxygen, on living resources in the Gulf of Mexico with comparative information from elsewhere, especially where information on Gulf resources is more limited.

As the oxygen concentration falls from saturated or optimal levels toward depletion, a variety of physiological impairments, including survivability, affect animals residing in the lower water column, in the sediments, or attached to hard substrates. The obvious effects of hypoxia/anoxia are displacement of pelagic organisms and selective loss of demersal and benthic organisms (Figure 9) (Rabalais et al. 2001a,b). These impacts may be aperiodic or recurring with some or full

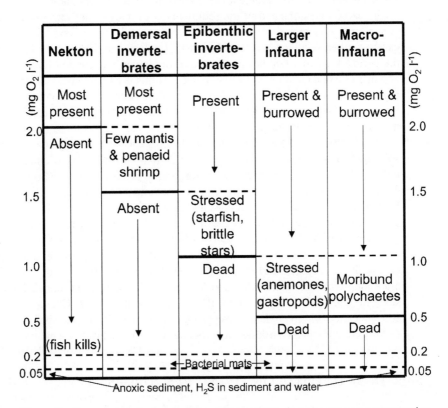

Figure 9 Progressive changes in fish and invertebrate fauna as oxygen concentration decreases from 2 mg l^{-1} to anoxia (from Rabalais et al. 2001a; reproduced with permission of the Americal Geophysical Union).

recovery, or permanent so that long-term ecosystem structure and function shift. Some organisms are adapted physiologically or behaviorally to persist in or avoid hypoxic environments (Burnett & Stickle 2001, Marcus 2001, Purcell et al. 2001). Behavioral responses include active avoidance, a response only available for swimming organisms versus those attached to substrates or associated with the bottom, and reduced feeding. Physiological adaptations include dormancy, reduction in molting and growth rates, increased ventilation rate, increased circulation of blood past respiratory organs, increased production of hemoglobin or other oxygen binding proteins, and anaerobic metabolism.

The distribution, abundance, and community composition of micro- and macrozooplankton are affected by episodic hypoxic/anoxic conditions in estuarine and coastal areas (Marcus 2001, Qureshi & Rabalais 2001, Purcell et al. 2001, Powers et al. 2001). Stages of the copepods *Acartia tonsa* and *Oithona colcarva* that normally migrated to bottom waters in Chesapeake Bay during daylight hours were not found there when oxygen levels were <1.0 mg 1^{-1}, but were concentrated at the pycnocline (Roman et al. 1993). *A. tonsa* also did not occur in anoxic bottom waters of a stratified lagoon in France, with the depth distribution varying with the location of the oxycline above the bottom (Cervetto et al. 1995). Qureshi & Rabalais (2001) found a similar exclusion of copepod nauplii, copepodites and adults when the oxygen concentration was less than 1 mg 1^{-1} in the Gulf of Mexico hypoxic zone, and vertical migration appeared to be disrupted at some stations. They also found a concentration of copepods and copepod nauplii below the pycnocline-oxycline but above the bottom where the oxygen concentration was reduced. Meroplankton (a water column life stage that alternates with a benthic life stage; usually larvae of benthic invertebrates) were also concentrated above oxygen-deficient bottom waters in summer and were either delaying metamorphosis or were unable to recruit to the seabed. Among polychaete larvae, however, only *Paraprionospio pinnata* appeared to be delaying metamorphosis by remaining in the water column above the seabed until the oxygen concentration returned to a level above 2 mg 1^{-1}; other polychaete larvae were distributed throughout the water column regardless of oxygen concentration (Powers et al. 2001). Eventual recruitment of macroinfaunal benthos to mostly defaunated sediments following long periods of hypoxia/anoxia was characterized by the predominance of juvenile *P. pinnata* owing to their delayed metamorphosis in combination with the inability of other recruits to survive conditions at the bottom (low oxygen and possibly hydrogen sulfide).

Most pelagic cnidarians and ctenophores (commonly grouped as jellyfish) do not live in hypoxic waters, although some species occur in high densities at very low oxygen concentrations or concentrate at the pycnocline/oxycline above severely hypoxic waters like the other zooplankton described above (Purcell et al. 2001). Experimental work with the schyphomedusan *Chrysaura quinquecirrha* and the ctenophore *Mnemiopsi leidyi* from Chesapeake Bay shows that the two jellyfish have prolonged survival at dissolved oxygen concentrations less than 2 mg 1^{-1}. They also prey on fish larvae that are less tolerant of low oxygen conditions, giving these jellyfish a competitive advantage over fish in hypoxic waters such as Chesapeake Bay, the Black Sea, and potentially the Gulf of Mexico. The abundance of *C. quinquecirrha* has increased over the Louisiana-Texas continental shelf from 5% of all SEAMAP (Southeast Area Mapping and Assessment Program for fisheries-independent groundfish surveys) stations sampled between 1987 and 1991 to 13–18% since 1992. At the same time, the mean oxygen concentration in the bottom water of stations where this jellyfish was collected dropped from 6.3 mg l^{-1} in 1987 to 3.4 mg l^{-1} in 1997 (Purcell et al. 2001). Summertime *C. quinquecirrha* medusa populations increased numerically and expanded away from shore across the portion of the Louisiana shelf where hypoxia is most frequent in summer, and there was a significant overlap between summer *C. quinquecirrha* abundance and lower water column hypoxia (Graham 2001). Changes in abundance of these jellyfish, however, as noted by Graham (2001), could also be related to overfishing of predators such as menhaden, expansion of hard substrates available for benthic polyp stages, and long-term climate variations, although eutrophication and increasing hypoxia appear to be a more credible explanation.

The responses of the fauna associated with the lower water column or seabed vary, depending on the concentration of dissolved oxygen, but there is a fairly consistent pattern of progressive stress and mortality as the oxygen concentration decreases from 2 mg l^{-1} to anoxia (Rabalais et al. 2001a,b). Motile organisms (fish, portunid crabs, stomatopods, penaeid shrimp, and squid) are seldom found in bottom waters with oxygen concentrations less than 2 mg l^{-1}. Below 1.5 to 1 mg l^{-1} oxygen concentration, less motile and burrowing invertebrates exhibit stress behavior, such as emergence from the sediments. They eventually die if the oxygen remains low for an extended period. As oxygen levels fall from 0.5 mg l^{-1} toward 0 mg l^{-1}, there is a fairly linear decrease in benthic infaunal diversity, abundance, and biomass. The sediments are never completely azoic, however. Some macroinfauna, such as the polychaetes *Ampharete* and *Magelona* and the sipunculan *Aspidosiphon*, are capable of surviving extremely low dissolved oxygen concentrations and/or high hydrogen sulfide concentrations. At minimal concentrations just above anoxia, sulfur-oxidizing bacteria form white mats on the sediment surface, and at 0 mg l^{-1}, there is no sign of aerobic life, just black anoxic sediments.

The severely stressed seasonal hypoxic/anoxic zone of the Louisiana inner shelf in the northern Gulf of Mexico loses many higher taxa during the peak of hypoxia (Rabalais et al. 2001b). Larger, longer-lived burrowing infauna are replaced by short-lived, smaller surface deposit–feeding polychaetes. Certain typical marine invertebrates are absent from the fauna, for example, pericaridean crustaceans, bivalves, gastropods, and ophiuroids. The hypoxia-affected fauna in Chesapeake Bay is characterized by a lower proportion of deeper-burrowing equilibrium species such as long-lived bivalves and a greater dominance of short-lived surface-dwelling forms (Dauer et al. 1992). Long-term trends for the Skagerrak coast of western Sweden in semi-enclosed fjordic areas experiencing increased oxygen stress (Rosenberg 1990) show declines in the total abundance and biomass of macroinfauna, abundance and biomass of mollusks, and abundance of suspension feeders and carnivores. These changes in benthic communities result in an impoverished diet for bottom-feeding fish and crustaceans and contribute, along with low dissolved oxygen, to altered sediment structure and sediment biogeochemical cycles.

Although there is a consistent decline in the macroinfauna as the oxygen concentration approaches 0 mg l^{-1}, many of the meiofaunal populations maintain similar numbers to those found under normoxic conditions. Other populations, however, go through a definite decline. Within a given habitat, certain species of foraminiferans and nematodes are typically most tolerant to hypoxia and anoxia while crustacean meiofauna, primarily the harpacticoid copepods, are least tolerant (Murrell & Fleeger 1989, Wetzel et al. 2001). Even apparent declines in the number of nematodes might as easily be explained by their apparent ability to emigrate into the water column in high numbers where they survive hypoxic conditions until normoxic conditions are re-established (Wetzel et al. 2001). While the numbers of macroinfauna become sufficiently depleted to no longer serve as a suitable food resource for demersal feeders such as shrimp, nematode numbers remain surprisingly high. Nematodes, although reduced in abundance at more severely affected stations (15-m depth) than inshore stations (8-m depth) on the southeastern Louisiana shelf, average about 1200 individuals per 10 cm^2 over an annual cycle (Murrell & Fleeger 1989). The insensitivity of nematode densities to oxygen deficiency or their occasional increase under severe hypoxia (Josefson & Widbom 1988, Cook et al. 2000) may make these remaining meiofaunal organisms potential food for foraging fish. The relative suitability of this potential nematode food to demersal feeders on the shelf compared to harpacticoid copepods and macroinfauna is not known. Fish and shrimp would not be potential predators during mid-summer severe hypoxia, but nematodes may be suitable prey for some foragers during the fall after hypoxia dissipates.

Important fishery resources are variably affected by direct mortality, forced migration, reduction in suitable habitat, increased susceptibility to predation, changes in food resources, and disruption of life cycles (Rabalais et al. 2001a). The effects of eutrophication, including hypoxia, are well known for some systems and include the increase in production of some species with a loss of others, including commercially important fisheries. The multi-level impacts of increased nutrient inputs and worsening hypoxia are not known for many components of productivity in the Gulf of Mexico, including pelagic and benthic, primary and secondary, food web linkages, and ultimately, fisheries yield. Comparisons of ecosystems along a gradient of increasing nutrient enrichment and eutrophication, or changes of a specific ecosystem over time through a gradient toward increasing eutrophication, provide information on how nutrient enrichment affects coastal communities. Caddy (1993) suggested a unimodal distribution of fishery yield in semi-enclosed seas in response to increasing eutrophication. The fishery yield is low in waters with low nutrients, but the yield increases as the nutrients increase, up to a point. As the ecosystem becomes increasingly eutrophied, there is a drop in fishery yield, but the decreases vary by habitat and feeding mode. The benthos are the first resources to be reduced by increasing frequency of seasonal hypoxia and eventually anoxia; bottom-feeding fishes then decline. The loss of a planktivorous fishery follows as eutrophication increases, with eventually a change in the zooplankton community composition. Where the current Gulf of Mexico fisheries lie along the gradient of increasing eutrophication is not known. It is well documented, however, that there is a negative relationship between the catch of brown shrimp (the largest economic fishery in the northern Gulf of Mexico) and the relative size of the mid-summer hypoxic zone (Zimmerman & Nance 2001). The catch per unit effort of brown shrimp has also declined significantly during the recent interval in which hypoxia was known to expand. As noted above for alternative hypotheses for increasing jellyfish abundances in areas of hypoxia off Louisiana, there may be additional or alternative reasons for shrimp decline in fisheries-dependent data, such as a trend of decreasing salinity in the coastal marshes, loss of habitat, or changes in fishing pressure.

Documenting the loss of fisheries related to the secondary effects of eutrophication, such as the loss of seabed vegetation and extensive bottom-water oxygen depletion, is complicated by poor fisheries data, inadequate economic indicators, increased fishing effort during the period of habitat degradation, natural variability of fish populations, shifts in harvestable populations, and climatic variability (Caddy 2000, Jackson et al. 2001, Boesch et al. 2001). Eutrophication of surface waters accompanied by oxygen deficient bottom waters can shift dominant fish stocks from demersal to pelagic. In the Baltic Sea and Kattegatt, where eutrophication-related ecological changes occurred mainly after World War II (reviewed by Elmgren & Larsson 2001a), some fish stocks increased because of increased food supply (e.g., pike perch in Baltic archipelagos) while others decreased (e.g., from oxygen deficiency that reduced Baltic cod recruitment and eventual harvest). Similar shifts are suggested by limited data on the Mississippi River–nfluenced shelf, with the increase in selected pelagic species in bycatch from trawls and a decrease in certain demersal species (Chesney & Baltz 2001). Still another analysis, however, indicated that the pelagic fishery yield may be impacted by extensive hypoxia on the northern Gulf of Mexico continental shelf (Smith 2001; J. Smith, unpublished data). In the case of commercial fisheries in the Black Sea, it is difficult to distinguish the impacts of eutrophication (e.g., the loss of macroalgal habitat and oxygen deficiency or a shift from benthic to pelagic production), from the possibility of overfishing. After the mid-1970s, benthic fish populations (e.g., turbot) collapsed, and pelagic fish populations (e.g., anchovy and sprat) started to increase. The commercial fisheries diversity declined from some 25 fished species to about 5 in 20 years (1960s to 1980s), while anchovy stocks and fisheries increased rapidly (Mee 2001). The point remains unclear on the gradient of increasing nutrients as to where benefits of higher overall yields become subsumed by environmental problems that lead to decreased landings quantity or quality.

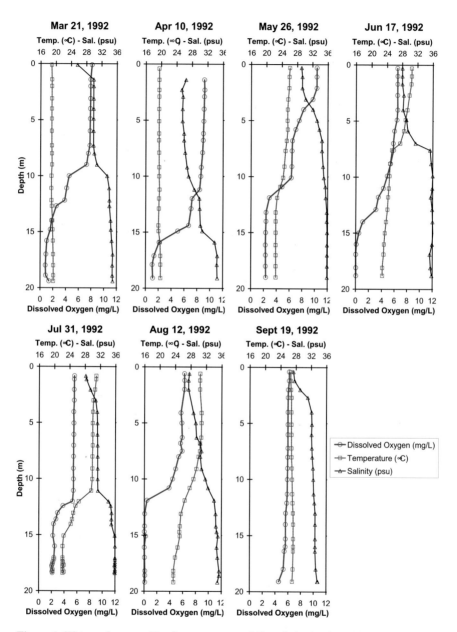

Figure 1 Water column profiles for temperature (°C), salinity (psu) and dissolved oxygen (mg l^{-1}) for representative dates of monthly sampling at station C6B off Terrebonne Bay on the southeastern Louisiana shelf in 1992 (modified from Rabalais & Turner 2001b, reprinted with permission of the American Geophysical Union). The station location is identified in Figure 4.

Average Monthly Bottom Dissolved Oxygen (mg/L)
1985 - 2001

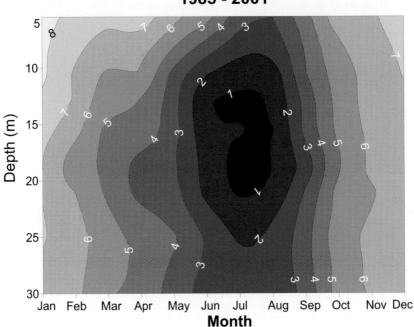

Figure 5 Contours of average values of bottom-water dissolved oxygen (mg l[-1]) by month for the eight stations along transect C (see Figure 4) by depth across the continental shelf from inshore to offshore and from January through December for the period 1985-2001. [Note, the number of values for each month varies, with fewer data in the winter than in spring-fall.]

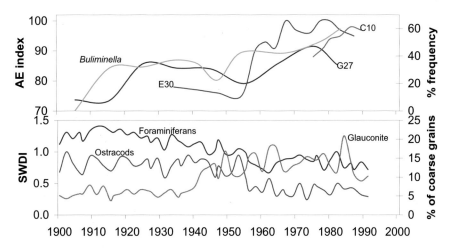

Figure 8 (*Top*) A-E index for cores C10 (3-yr running average), E30, G27 (Sen Gupta et al. 1996); % frequency of *Buliminella* in core G27 (Rabalais et al. 1996), % frequency of *Quenquiloculina* in core G27 (Rabalais et al. 1996). (*Bottom*) SWDI (Shannon-Wiener Diversity Index) for foraminiferans and ostracods (Nelsen et al. 1994, TA Nelson unpublished data); % glauconite of coarse grains (Nelsen et al. 1994).

NUTRIENT MANAGEMENT AND RECOVERY

The accelerated time course of coastal eutrophication in the northern Gulf of Mexico since the 1950s was typical for most temperate coastal regions at the terminus of modified rivers flowing through developed countries. In the northern Gulf of Mexico, the time course of eutrophication and hypoxia followed most closely the exponential growth of fertilizer use beginning in the 1950s. Elsewhere in the world, the relative proportion of agriculture-source nutrients may not be as high as in the Mississippi River basin, but other sources of nutrients, such as municipal and industrial wastewater and atmospheric deposition, also increased substantially since the 1950s. The consumption of fertilizers has plateaued in many developed countries, but continues to increase in developing countries (Seitzinger et al. 2002). There is no indication that fertilizer use will decrease, and controlling the non-point sources of nutrient pollution has proven much more difficult than controls emplaced for point sources (Boesch 2002). Without the curtailing of nutrient loads, the trajectory of coastal water quality degradation in the northern Gulf of Mexico will likely continue, or perhaps worsen under scenarios of increased precipitation in climate change models (Justić et al. 1996). Elsewhere in the world, eutrophication, with sometimes accompanying hypoxia, will continue without a reduction in nutrient loads and will certainly accelerate in areas where nutrient loads are on the rise. There are, however, several good examples of nutrient management schemes that have resulted in reversal of symptoms of eutrophication and economically driven experiments where similar results have occurred. However, many attempts to curtail nutrient export to estuarine and coastal waters have met with limited success in meeting reduction goals or environmental goals.

Hypoxia and anoxia have existed in the open waters of the Black Sea for millennia, but hypoxia on the northwestern Black Sea shelf became more frequent and widespread in the 1970s and 1980s (Tolmazin 1985, Zaitsev 1992, Mee 2001), affecting up to 40,000 km^2 in depths of 8 to 40 m. There is also evidence that the suboxic zone of the open Black Sea enlarged toward the surface by about 10 m since 1970. As a result of the economic collapse of the former Soviet Union and declines in subsidies for fertilizers, the decade of the 1990s witnessed a substantially decreased input of nutrients via the Danube River, which was accompanied by signs of recovery in the pelagic and benthic ecosystems of the Black Sea (Mee 2001). There is a recovery in zoobenthos species diversity, phytoplankton biomass has declined by about 30% from the 1990 maxima, there is some recovery of the diatoms, the phytoplankton are more diverse, the incidence of intense blooms has declined, and there is a limited recovery of some zooplankton stocks and diversity in limited geographic areas. There has been no recovery of benthic macroalgal beds. For the first time in several decades, oxygen deficiency was absent from the northwestern shelf of the Black Sea in 1996, and receded to an area less than 1000 km^2 in 1999. Most fish stocks in the northwestern Black Sea are still depleted (Mee 2001), however. There should be little doubt of the strong relationships among human activities, Danube River nutrient loads, Black Sea eutrophication, and the demise of pelagic and benthic coastal ecosystems, as well as similar linkages in the partial recovery of those systems following reduced nutrient loading.

The severe degradation of water quality in Tampa Bay and loss of valuable habitat, particularly seagrass beds, was followed by nutrient management schemes that reduced the nitrogen and phosphorus inputs to the bay (Johansson & Lewis 1992). Four years after improved sewage treatment, ambient chlorophyll *a* concentrations decreased in Hillsborough Bay, and the noxious filamentous cyanobacteria *Schizothrix calcicola* also decreased. Modest seagrass recovery followed. An aggressive nutrient management program with broad-scale public, institutional, and private participation continues under the Tampa Bay Estuary Program (National Research Council 2000), with the current goal to hold gains made in nutrient load reductions by making additional reductions in point and diffuse sources to offset the effects of anticipated high rate of population growth in the region.

On a larger scale, with the causes of hypoxia more clearly defined and the sources of nutrients fairly well understood, a Task Force forwarded to Congress in 2001 an Action Plan for Reducing, Mitigating, and Controlling Hypoxia in the Northern Gulf of Mexico (Mississippi River/Gulf of Mexico Watershed Nutrient Task Force 2001, Rabalais et al. 2002). Reaching this step of nutrient policy development required a scientific consensus that the increase in nitrogen loading was the primary factor in the worsening of hypoxia in the northern Gulf of Mexico. Finding common ground on the hypoxia assessment and a general action plan that calls for reducing hypoxia in the northern Gulf of Mexico by two-thirds was not reached without debate or controversy (Rabalais et al. 2002). The next step, development of a detailed resource management plan that will probably require the reduction of nitrogen inputs by 30% for a watershed covering 41% of the contiguous United States, remains a daunting task.

Within estuaries and coastal systems, a decrease in external nutrient loads does not produce an immediate shift in the eutrophic condition of the system, in part because of the continued remineralization of labile carbon and releases of regenerated nutrients. The response of the Black Sea coastal ecosystem to lower nutrient loading took almost a decade. Seagrass recovery in Tampa Bay lagged nutrient load reductions by about 8 years. Boynton & Kemp (2000) suggested a nutrient memory over timescales of a year rather than seasonal periods as suggested by Chesapeake Bay water residence times. Assessing the recovery of Chesapeake Bay in response to the nutrient load reductions achieved so far is complicated by numerous factors. One indicator that could be attributed to reduced nutrients is a return of sea grasses to some regions, although the present coverage is only a small portion of the habitat occupied in the 1950s (Boesch 2001). Justić et al. (1997) suggested that at least a year of continued carbon respiration following high deposition of carbon in a flood year contributed to oxygen demand on the Louisiana continental shelf in the following summer. Researchers within the narrow, coastal inlets of the Bodden are less optimistic about system recovery where nutrients to that sector of the Baltic were reduced greatly during the last decade of the twentieth century, but the expected improvement of water quality has not been demonstrated

(Meyer-Reil & Köster 2000). In addition, there is an inertia in terrestrial systems and rivers and streams with regard to losses from land to sea following nutrient reductions actually achieved or planned (Grimvall et al. 2000). On the timescale of a few years, changes in the anthropogenic impact on water quality may easily be overshadowed by natural fluctuations in climate. These facts are relevant to management strategies to mitigate nutrient loads to estuaries and coastal waters, and the perceived projection for recovery.

For many environmental problems, including excess nutrients, scientists, political institutions, and society must often agree on policy and intervention measures in the face of scientific uncertainty. Fortunately, around the globe, measures to decrease excess nutrient loads proceeded without complete scientific consensus on which nutrients to reduce, how much, and by what methods. Knowledge gaps will remain and new ones will surface, but most action plans for reducing and mitigating the effects of nutrients are conceived within a long-term adaptive management strategy that links management actions with monitoring, research, modeling, and a commitment to reassess conditions periodically (Elmgren 2001, Elmgren & Larsson 2001b, Rabalais et al. 2002). This can only improve the long-term prognosis for improved water quality resulting from robust actions while improving the science.

ACKNOWLEDGMENTS

The data presented were collected beginning in 1985 with support from the National Oceanic and Atmospheric Administration (NOAA), National Ocean Service, and subsequently from the Louisiana Board of Regents Support Fund, NOAA Nutrient Enhanced Coastal Ocean Productivity program, the Louisiana-Texas Physical Oceanography Program sponsored by Minerals Management Service (MMS), the Louisiana Sea Grant College Program, the NOAA National Undersea Research Program, the MMS University Research Initiative, the Gulf of Mexico Program, and the NOAA Coastal Ocean Program. We thank these organizations for their support. We also thank Ben Cole for maintaining our data in a manner helpful to analysis of long-term trends and for assistance with the preparation of figures.

The *Annual Review of Ecology and Systematics* is online at
http://ecolsys.annualreviews.org

LITERATURE CITED

Alpine AE, Cloern JE. 1992. Trophic interactions and direct physical effects control phytoplankton biomass and production in an estuary. *Limnol. Oceanogr.* 37:946–55

Anderson DM, Glibert PM, Burkholder JM.

2002. Harmful algal blooms and eutrophication: nutrient sources, composition, and consequences. *Estuaries.* 25: In press

Arntz WE. 1981. Zonation and dynamics of macrobenthos in an area stressed by oxygen deficiency. In *Stress Effects on Natural*

Ecosystems, ed. GW Barrett, R Rosenberg, pp. 215–25. New York: Wiley

Bennett EM, Carpenter SR, Caraco NF. 2001. Human impact on erodable phosphorus and eutrophication: a global perspective. *BioScience* 51:227–34

Boesch DF. 2001. Science and integrated drainage basin coastal management. Chesapeake Bay and Mississippi Delta. See von Bodungen & Turner 2001, pp. 37–50

Boesch DF. 2002. Challenges and opportunities for science in reducing nutrient overenrichment of coastal ecosystems. *Estuaries*. 25: In press

Boesch DF, Burreson E, Dennison W, Houde E, Kemp M, et al. 2001. Factors in the decline of coastal ecosystems. *Science* 293:1589–90

Boynton WR, Garber JH, Summers R, Kemp WM. 1995. Inputs, transformations, and transport of nitrogen and phosphorus in Chesapeake Bay and selected tributaries. *Estuaries* 18:285–314

Boynton WR, Kemp WM. 2000. Influence of river flow and nutrient loads on selected ecosystem processes. A synthesis of Chesapeake Bay data. In *Estuarine Science: A Synthetic Approach to Research and Practice*, ed. JE Hobbie, pp. 269–98. Washington, DC: Island

Bratkovich A, Dinnel SP, Goolsby DA. 1994. Variability and prediction of freshwater and nitrate fluxes for the Louisiana-Texas shelf: Mississippi and Atchafalaya River source functions. *Estuaries* 17:766–78

Breitburg DL. 1990. Near-shore hypoxia in the Chesapeake Bay: patterns and relationships among physical factors. *Estuar. Coast. Shelf Sci.* 30:593–609

Breitburg DL. 2002. Effects of hypoxia, and the balance between hypoxia and enrichment, on coastal fishes and fisheries. *Estuaries*. 25: In press

Bricker SB, Clement CG, Pirhalla DE, Orlando SP, Farrow DRG. 1999. *National estuarine Eutrophication Assessment: Effects of Nutrient Enrichment in the Nation's Estuaries*. Silver Spring, MD: NOAA 71 pp.

Burnett LE, Stickle WB. 2001. Physiological

responses to hypoxia. See Rabalais & Turner 2001a, pp. 101–14

Caddy JF. 1993. Toward a comparative evaluation of human impacts on fishery ecosystems of enclosed and semi-enclosed seas. *Rev. Fish. Sci.* 1:57–95

Caddy JF. 2000. Marine catchment basin effects versus impacts of fisheries on semi-enclosed seas. *ICES J. Mar. Sci.* 57:628–40

Caraco NF, Cole JJ. 1999. Human impact on nitrate export: an analysis using major world rivers. *Ambio* 28:167–70

Carey AC, Pennock JR, Hehrter JC, Lyons WB, Schroeder WW, et al. 1999. The role of the Mississippi River in Gulf of Mexico hypoxia. *Rep. Fertilizer Inst., Environ. Inst. Publ. No. 70*. Univ. Ala., Tuscaloosa 79 pp.

Cervetto G, Pagano M, Gaudy R. 1995. Feeding behavior and migrations in a natural population of the copepod *Acartia tonsa*. *Hydrobiologia* 300/301:237–48

Chen N, Bianchi TS, McKee BA, Bland JM. 2001. Historical trends of hypoxia on the Louisiana shelf: application of pigments as biomarkers. *Org. Geochem.* 32:543–61

Chesney EJ, Baltz DM. 2001. The effects of hypoxia on the northern Gulf of Mexico coastal ecosystem: a fisheries perspective. See Rabalais & Turner 2001a, pp. 321–54

Cloern JE. 1999. The relative importance of light and nutrient limitation of phytoplankton growth: a simple index of coastal ecosystems sensitivity to nutrient enrichment. *Aquat. Ecol.* 33:3–16

Cloern JE. 2001. Review. Our evolving conceptual model of the coastal eutrophication problem. *Mar. Ecol. Prog. Ser.* 210:223–53

Comm. Environ. Nat. Resourc. (CENR). 2000. *Integrated Assessment of Hypoxia in the Northern Gulf of Mexico*. Washington, DC: Nat. Sci. Tech. Counc. 58 pp.

Cook AA, Lambshead PJD, Hawkins LE, Mitchell N, Levin LA. 2000. Nematode abundance at the oxygen minimum zone in the Arabian Sea. *Deep-Sea Res. II* 47:75–85

Cooper SR. 1995. Chesapeake Bay watershed

historical land use: impact on water quality and diatom communities. *Ecol. Appl.* 5:703–23

Cooper SR, Brush GS. 1993. A 2500-year history of anoxia and eutrophication in Chesapeake Bay. *Estuaries* 16:617–26

Dauer DM, Rodi AJ Jr, Ranasinghe JA. 1992. Effects of low dissolved oxygen events on the macrobenthos of the lower Chesapeake Bay. *Estuaries* 15:384–91

Diaz RJ, Rosenberg R. 1995. Marine benthic hypoxia: a review of its ecological effects and the behavioural responses of benthic macrofauna. *Oceanogr. Mar. Biol. Ann. Rev.* 33:245–303

Dortch Q, Rabalais NN, Turner RE, Qureshi NA. 2001. Impacts of changing Si/N ratios and phytoplankton species composition. See Rabalais & Turner 2001a, pp. 37–48

Dunn DD. 1996. Trends in nutrient inflows to the Gulf of Mexico from streams draining the conterminous United States 1972–1993. *US Geol. Surv., Water-Res. Invest. Rep. 96–4113.* Austin, TX: US Geol. Surv. 60 pp.

Dyer KR, Orth RJ, eds. 1994. *Changes in Fluxes in Estuaries: Implications from Science to Management, Proc. ECSA22/ERF Symp., Int. Symp. Ser.* Fredensborg, Denmark: Olsen & Olsen. 483 pp.

Eadie BJ, McKee BA, Lansing MB, Robbins JA, Metz S, et al. 1994. Records of nutrient-enhanced coastal ocean productivity in sediments from the Louisiana continental shelf. *Estuaries* 17:754–65

Elmgren R. 2001. Understanding human impact on the Baltic ecosystem: changing views in recent decades. *Ambio* 30:222–31

Elmgren R, Larsson U. 2001a. Eutrophication in the Baltic Sea area. Integrated coastal management issues. See von Bodungen & Turner 2001, pp. 15–35

Elmgren R, Larsson U. 2001b. Nitrogen and the Baltic Sea: managing nitrogen in relation to phosphorus. In *Optimizing Nitrogen Management in Food And Energy Production and Environmental Protection: Proc.2nd Int. Nitrogen Conf. Sci. Pol. Sci. World* 1(S2):371–77

Falkowski PG, Hopkins TS, Walsh JJ. 1980. An analysis of factors affecting oxygen depletion in the New York bight. *J. Mar. Res.* 38:479–506

Fransz HG, Verhagen JHG. 1985. Modeling research on the production cycle of phytoplankton in the southern bight of the North Sea in relation to riverborne nutrient loads. *Netherlands J. Sea Res.* 19:241–50

Garside C, Malone TC. 1978. Monthly oxygen and carbon budgets of the New York bight apex. *Estuar. Coast. Shelf Sci.* 6:93–104

Goolsby DA, Battaglin WA, Lawrence GB, Artz RS, Aulenbach BT, et al. 1999. Flux and sources of nutrients in the Mississippi-Atchafalaya River Basin, Topic 3 Rep. Integrated Assessment of Hypoxia in the Gulf of Mexico. *NOAA Coast. Ocean Prog. Dec. Anal. Ser. 17.* Silver Spring, MD: NOAA. 130 pp.

Graham WM. 2001. Numerical increases and distributional shifts of *Chrysaora quinquecirrha* (Desor) and *Aurelia aurita* (Linné) (Cnidaria: Scyphozoa) in the northern Gulf of Mexico. *Hydrobiologia* 451:97–111

Grimvall A, Stålnacke P, Tonderski A. 2000. Time scales of nutrient losses from land to sea—a European perspective. *Ecol. Eng.* 14:363–71

Gulf States Mar. Fish. Comm. 1982–91. *SEAMAP Environmental and Biological Atlas of the Gulf of Mexico, 1982.* Ocean Springs, MS: Gulf States Mar. Fish. Comm.

Harding LW Jr, Perry ES. 1997. Long-term increase of phytoplankton biomass in Chesapeake Bay, 1950–1994. *Mar. Ecol. Prog. Ser.* 157:39–52

Harper DE Jr, McKinney LD, Nance JM, Salzer RR. 1991. Recovery responses of two benthic assemblages following an acute hypoxic event on the Texas continental shelf, northwestern Gulf of Mexico. See Tyson & Pearson 1991a, pp. 49–64

Howarth R, Anderson D, Cloern J, Elfring C, Hopkinson C, et al. 2000. Nutrient pollution of coastal rivers, bays, and seas. *Issues Ecol.* 7:1–15

Howarth RW, Billen G, Swaney D, Townsend A, Jaworski N, et al. 1996. Regional nitrogen budgets and riverine N & P fluxes for the drainages to the North Atlantic Ocean: natural and human influences. *Biogeochemistry* 35:75–79

Howarth RW, Jensen H, Marino R, Postma H. 1995. Transport to and processing of P in near-shore and oceanic waters. In *Phosphorus in the Global Environment: Transfers, Cycles and Management*, ed. H Tiessen, pp. 323–45. Chichester, UK: Wiley

Jackson JBC, Kirby MX, Berger WH, Bjorndal KA, Botsford LW, et al. 2001. Historical overfishing and the recent collapse of coastal ecosystems. *Science* 293:629–38

Johansson JOR, Lewis RR III. 1992. Recent improvements of water quality and biological indicators in Hillsborough Bay, a highly impacted subdivision of Tampa Bay, Florida, USA. In *Marine Coastal Eutrophication. The Response of Marine Transitional Systems to Human Impact: Problems and Perspectives for Restoration. Proc. Int. Conf., Bologna, Italy, March 1990*, ed. RA Vollenweider, R Marchetti, R Viviani, pp. 1199–1215. Amsterdam: Elsevier

Josefson AB, Widbom B. 1988. Differential response of benthic macrofauna and meiofauna to hypoxia in the Gullmar Fjord basin. *Mar. Biol.* 100:31–40

Justić D, Legovic T, Rottini-Sandrini L. 1987. Trends in oxygen content 1911–1984 and occurrence of benthic mortality in the northern Adriatic Sea. *Estuar. Coast. Shelf Sci.* 25:435–45

Justić D, Rabalais NN, Turner RE. 1996. Effects of climate change on hypoxia in coastal waters: a doubled CO_2 scenario for the northern Gulf of Mexico. *Limnol. Oceanogr.* 41:992–1003

Justić D, Rabalais NN, Turner RE. 1997. Impacts of climate change on net productivity of coastal waters: implications for carbon budget and hypoxia. *Clim. Res.* 8:225–37

Justić D, Rabalais NN, Turner RE. 2002. Modeling the impacts of decadal changes in riverine nutrient fluxes on coastal eutrophication near the Mississippi River Delta. *Ecol. Model.* 152:33–46

Justić D, Rabalais NN, Turner RE, Wiseman WJ Jr. 1993. Seasonal coupling between riverborne nutrients, net productivity and hypoxia. *Mar. Pollut. Bull.* 26:184–89

Kamykowski D, Zentara S-J. 1990. Hypoxia in the world ocean as recorded in the historical data set. *Deep-Sea Res.* 37:1861–74

Karlsen AW, Cronin TM, Ishman SE, Willard DA, Kerhin R, et al. 2000. Historical trends in Chesapeake Bay dissolved oxygen based on benthic Foraminifera from sediment cores. *Estuaries* 23:488–508

Keown MP, Dardeau EA Jr, Causey EM. 1986. Historic trends in the sediment flow regime of the Mississippi River. *Water Resourc. Res.* 22:1555–64

Loesch H. 1960. Sporadic mass shoreward migrations of demersal fish and crustaceans in Mobile Bay, Alabama. *Ecology* 41:292–98

Lohrenz SE, Fahnenstiel GL, Redalje DG, Lang GA, Chen X, et al. 1997. Variations in primary production of northern Gulf of Mexico continental shelf waters linked to nutrient inputs from the Mississippi River. *Mar. Ecol. Prog. Ser.* 155:45–54

Malone TC. 1991. River flow, phytoplankton production and oxygen depletion in Chesapeake Bay. See Tyson & Pearson 1991a, pp. 83–93

Marcus NH. 2001. Zooplankton: responses to and consequences of hypoxia. See Rabalais & Turner 2001a, pp. 49–60

May EB. 1973. Extensive oxygen depletion in Mobile Bay, Alabama. *Limnol. Oceanogr.* 18:353–66

Mayer LM, Keil RG, Macko SA, Joye SB, Ruttenberg KC, et al. 1998. Importance of suspended particulates in riverine delivery of bioavailable nitrogen to coastal zones. *Glob. Biogeochem. Cycles* 12(4):573–79

Meade RH, ed. 1995. Contaminants in the Mississippi River, 1987–1992. *US Geol. Surv. Circ. 1133*. Denver, CO: US Dep. Int. 140 pp.

Meade RH, Parker R. 1985. Sediment in rivers of the United States. *US Geol. Surv. Water Supply Pap. 2275*, pp. 49–60. Washington, DC: GPO

Meade RH, Yuzyk TR, Day TJ. 1990. Movement and storage of sediment in rivers of the United States and Canada. In *The Geology of North America, Vol. O-1*, ed. MG Wolman, HC Riggs, pp. 255–80. Boulder, CO: Geol. Soc. Am.

Mee LD. 2001. Eutrophication in the Black Sea and a basin-wide approach to its control. See von Bodungen & Turner 2001, pp. 71–91

Meyer-Reil L-A, Köster M. 2000. Eutrophication of marine waters: effects on benthic microbial communities. *Mar. Pollut. Bull.* 41:255–63

Mississippi River/Gulf of Mexico Watershed Nutrient Task Force. 2001. *Action Plan for Reducing, Mitigating, and Controlling Hypoxia in the Northern Gulf of Mexico*. Washington, DC: EPA

Mitsch WJ, Day JW Jr, Gilliam JW, Groffman PM, Hey DL, et al. 2001. Reducing nitrogen loading to the Gulf of Mexico from the Mississippi River basin: strategies to counter a persistent ecological problem. *BioScience* 15:373–88

Murrell MC, Fleeger JW. 1989. Meiofauna abundance on the Gulf of Mexico continental shelf affected by hypoxia. *Cont. Shelf Res.* 9:1049–62

Natl. Res. Counc. 2000. *Clean Coastal Waters—Understanding and Reducing the Effects of Nutrient Pollution*. Washington, DC: Natl. Acad. Press

Nelsen TA, Blackwelder P, Hood T, McKee B, Romer N, et al. 1994. Time-based correlation of biogenic, lithogenic and authigenic sediment components with anthropogenic inputs in the Gulf of Mexico NECOP study area. *Estuaries* 17:873–85

Nixon SW. 1995. Coastal marine eutrophication: a definition, social causes, and future concerns. *Ophelia* 41:199–219

Nixon SW, Ammerman JW, Atkinson LP, Berounsky VM, Billen G, et al. 1996. The fate of nitrogen and phosphorus at the land-sea margin of the North Atlantic Ocean. *Biogeochemistry* 35:141–80

Nixon SW, Granger SL, Nowicki BL. 1995. An assessment of the annual mass balance of carbon, nitrogen, and phosphorus in Narragansett Bay. *Biogeochemistry* 31:15–61

Officer CB, Biggs RB, Taft JL, Cronin LE, Tyler MA, et al. 1984. Chesapeake Bay anoxia: origin, development and significance. *Science* 223:22–27

Officer CB, Ryther JH. 1980. The possible importance of silicon in marine eutrophication. *Mar. Ecol. Prog. Ser.* 3:83–91

Paerl H, Pinckney J, Fear J, Peierls B. 1998. Ecosystem responses to internal and watershed organic matter loading: consequences for hypoxia in the eutrophying Neuse River Estuary, NC, USA. *Mar. Ecol. Prog. Ser.* 166:17–25

Parker CA, O'Reilly JE. 1991. Oxygen depletion in Long Island Sound: a historical perspective. *Estuaries* 14:248–64

Pokryfki L, Randall RE. 1987. Nearshore hypoxia in the bottom water of the northwestern Gulf of Mexico from 1981 to 1984. *Mar. Environ. Res.* 22:75–90

Powers SP, Harper DE Jr, Rabalais NN. 2001. Effect of hypoxia/anoxia on the supply and settlement of benthic invertebrate larvae. See Rabalais & Turner 2001a, pp. 185–210

Purcell JE, Breitburg DL, Decker MB, Graham WM, Youngbluth MJ, et al. 2001. Pelagic cnidarians and ctenophores in low dissolved oxygen environments: a review. See Rabalais & Turner 2001a, pp. 77–100

Qureshi NA, Rabalais NN. 2001. Distribution of zooplankton on a seasonally hypoxic continental shelf. See Rabalais & Turner 2001a, pp. 61–76

Rabalais NN. 1992. *An updated summary of status and trends in indicators of nutrient enrichment in the Gulf of Mexico. EPA Publ. EPA/800–R–92–004*. Stennis Space Cent., MS: Gulf Mexico Program. 421 pp.

Rabalais NN. 2002. Nitrogen in aquatic ecosystems. *Ambio.* 31:102–12

Rabalais NN, Harper DE Jr, Turner RE. 2001a. Responses of nekton and demersal and benthic fauna to decreasing oxygen concentrations. See Rabalais & Turner 2001a, pp. 115–28

Rabalais NN, Smith LE, Harper DE Jr, Justić D. 2001b. Effects of seasonal hypoxia on continental shelf benthos. See Rabalais & Turner 2001a, pp. 211–40

Rabalais NN, Turner RE, eds. 2001a. *Coastal Hypoxia: Consequences for Living Resources and Ecosystems. Coastal and Estuarine Stud. 58.* Washington, DC: Am. Geophys. Union

Rabalais NN, Turner RE. 2001b. Hypoxia in the Northern Gulf of Mexico: description, causes and change. See Rabalais & Turner 2001a, pp. 1–36

Rabalais NN, Turner RE, Justić D, Dortch Q, Wiseman WJ Jr. 1999. Characterization of hypoxia: topic 1 report for the integrated assessment of hypoxia in the Gulf of Mexico. *NOAA Coast. Ocean Prog. Decis. Anal. Ser. 15.* Silver Spring, MD: NOAA. 167 pp.

Rabalais NN, Turner RE, Justić D, Dortch Q, Wiseman WJ Jr, et al. 1996. Nutrient changes in the Mississippi River and system responses on the adjacent continental shelf. *Estuaries* 19:386–407

Rabalais NN, Turner RE, Scavia D. 2002. Beyond science into policy: Gulf of Mexico hypoxia and the Mississippi River. *BioScience* 52:129–42

Rabalais NN, Turner RE, Wiseman WJ Jr, Boesch DF. 1991. A brief summary of hypoxia on the northern Gulf of Mexico continental shelf: 1985–1988. See Tyson & Pearson 1991a, pp. 35–46

Rabalais NN, Wiseman WJ Jr, Turner RE. 1994. Comparison of continuous records of near-bottom dissolved oxygen from the hypoxia zone along the Louisiana coast. *Estuaries* 17:850–61

Ragan JG, Harris AH, Green JH. 1978. Temperature, salinity and oxygen measurements of surface and bottom waters on the continental shelf off Louisiana during portions of 1975 and 1976. *Prof. Pap. Ser. Biol.* 3:1–29. Thibodaux, LA: Nicholls State Univ.

Renaud M. 1986. Hypoxia in Louisiana coastal waters during 1983: implications for fisheries. *Fish. Bull.* 84:19–26

Roman MR, Gauzens AL, Rhinehart WK, White JR. 1993. Effects of low oxygen waters on Chesapeake Bay zooplankton. *Limnol. Oceanogr.* 38:1603–14

Rosenberg R. 1985. Eutrophication—the future marine coastal nuisance? *Mar. Pollut. Bull.* 16:227–31

Rosenberg R. 1990. Negative oxygen trends in Swedish coastal bottom waters. *Mar. Pollut. Bull.* 21:335–39

Sanford LP, Sellner KG, Breitburg DL. 1990. Covariability of dissolved oxygen with physical processes in the summertime Chesapeake Bay. *J. Mar. Res.* 48:567–90

Schramm W. 1999. Factors influencing seaweed responses to eutrophication: some results from EU-project EUMAC. *J. Appl. Phycol.* 11:69–78

Seitzinger SP, Koreze C, Bouwman AF, Caraco N, Dentener F, et al. 2002. Global patterns of dissolved inorganic and particulate nitrogen inputs to coastal systems: recent conditions and future projections. *Estuaries.* 25: In press

Sen Gupta BK, Turner RE, Rabalais NN. 1996. Seasonal oxygen depletion in continental-shelf waters of Louisiana: historical record of benthic foraminifers. *Geology* 24:227–30

Smith JW. 2001. Distribution of catch in the Gulf menhaden, *Brevoortia patronus*, purse seine fishery in the northern Gulf of Mexico from logbook information: Are there relationships to the hypoxic zone? See Rabalais & Turner 2001a, pp. 311–20

Stachowitsch M. 1986. The Gulf of Trieste: a sensitive ecosystem. *Nova Thalass.* 8(Suppl. 3):221–35

Swanson RL, Parker CA. 1988. Physical environmental factors contributing to recurring hypoxia in the New York bight. *Trans. Am. Fish. Soc.* 117:37–47

Swanson RL, Sindermann CJ, eds. 1979. Oxygen depletion and associated benthic mortalities in New York bight, 1976. *Natl. Ocean. Atmos. Admin. Prof. Pap. 11.* Rockville, MD: Natl. Ocean. Atmos. Admin.

Thomas E, Gapotchenko T, Varekamp JC,

Mecray EL, ter Brink MRB. 2000. Benthic Foraminifera and environmental changes in Long Island Sound. *J. Coast. Res.* 16:641–45

Tolmazin R. 1985. Changing coastal oceanography of the Black Sea. I. Northwestern shelf. *Prog. Oceanogr.* 15:217–76

Turner RE. 1997. Wetland loss in the northern Gulf of Mexico: multiple working hypotheses. *Estuaries* 20:1–13

Turner RE. 2001. Some effects of eutrophication on pelagic and demersal marine food webs. See Rabalais & Turner 2001a, pp. 371–98

Turner RE, Allen RL. 1982a. Plankton respiration rates in the bottom waters of the Mississippi River Delta bight. *Contr. Mar. Sci.* 25:173–79

Turner RE, Allen RL. 1982b. Bottom water oxygen concentration in the Mississippi River Delta bight. *Contr. Mar. Sci.* 25:161–72

Turner RE, Qureshi N, Rabalais NN, Dortch Q, Justić D, et al. 1998. Fluctuating silicate:nitrate ratios and coastal plankton food webs. *Proc. Natl. Acad. Sci. USA* 95:13048–51

Turner RE, Rabalais NN. 1991. Changes in Mississippi River water quality this century: implications for coastal food webs. *Bio-Science* 41: 140–48

Turner RE, Rabalais NN. 1994a. Coastal eutrophication near the Mississippi river delta. *Nature* 368:619–21

Turner RE, Rabalais NN. 1994b. Changes in the Mississippi River nutrient supply and offshore silicate-based phytoplankton community responses. See Dyer & Orth 1994, pp. 147–50

Turner RE, Schroeder WW, Wiseman WJ Jr. 1987. The role of stratification in the deoxygenation of Mobile Bay and adjacent shelf bottom waters. *Estuaries* 10:13–19

Tyson RV, Pearson TH, eds. 1991a. *Modern and Ancient Continental Shelf Anoxia, Geol. Soc. Spec. Publ. 58.* London: Geol. Soc. 460 pp.

Tyson RV, Pearson TH. 1991b. Modern and ancient continental shelf anoxia: an overview. See Tyson & Pearson 1991a, pp. 1–24

Vitousek PM, Aber JD, Howarth RW, Likens GE, Matson PA, et al. 1997. Human alteration of the global nitrogen cycle: sources and consequences. *Ecol. Appl.* 7:737–50

von Bodungen B, Turner RK, eds. 2001. *Science and Integrated Coastal Management.* Berlin: Dahlem Univ. Press 364 pp.

Welsh BL, Eller FC. 1991. Mechanisms controlling summertime oxygen depletion in western Long Island Sound. *Estuaries* 14:265–78

Welsh BL, Welsh RI, DiGiacomo-Cohen ML. 1994. Quantifying hypoxia and anoxia in Long Island Sound. See Dyer & Orth 1994, pp. 131–37

Wetzel MA, Fleeger JW, Powers SP. 2001. Effects of hypoxia and anoxia on meiofauna: a review with new data from the Gulf of Mexico. See Rabalais & Turner 2001a, pp. 165–84

Zaitsev YP. 1992. Recent changes in the trophic structure of the Black Sea. *Fish. Oceanogr.* 1:180–89

Zimmerman RJ, Nance JM. 2001. Effects of hypoxia on the shrimp fishery of Louisiana and Texas. See Rabalais & Turner 2001a, pp. 293–310

Annu. Rev. Ecol. Syst. 2002. 33:265–89
doi: 10.1146/annurev.ecolsys.33.010802.150511
First published online as a Review in Advance on August 6, 2002

THE (SUPER)TREE OF LIFE:
Procedures, Problems, and Prospects

Olaf R. P. Bininda-Emonds,[1,2] John L. Gittleman,[3] and Mike A. Steel[4]

[1]*Current address: Lehrstuhl für Tierzucht, Technical University of Munich, D-85354 Freising-Weihenstephan, Germany; email: Olaf.Bininda@tierzucht.tum.de*
[2]*Institute of Evolutionary and Ecological Sciences, Leiden University, Kaiserstraat 63, 2300 RA Leiden, The Netherlands*
[3]*Department of Biology, Gilmer Hall, University of Virginia, Charlottesville, Virginia 22904-4328; email: JLGittleman@virginia.edu*
[4]*Biomathematics Research Center, Department of Mathematics and Statistics, University of Canterbury, Christchurch, New Zealand; email: m.steel@math.canterbury.ac.nz*

Key Words   matrix representation, consensus techniques, total evidence, trees, algorithms, macroevolution, biodiversity

■ **Abstract**   Supertree construction is a new, rigorous approach for combining phylogenetic information to produce more inclusive phylogenies. It has been used to provide some of the largest, most complete phylogenies for diverse groups (e.g., mammals, flowering plants, and dinosaurs) at a variety of taxonomic levels. We critically review methods for assembling supertrees, discuss some of their more interesting mathematical properties, and describe the strengths and limitations of the supertree approach. To document the need for supertrees in biology, we identify how supertrees have already been used beyond the systematic information they provide to examine models of evolution, test rates of cladogenesis, detect patterns of trait evolution, and extend phylogenetic information to biodiversity conservation.

INTRODUCTION

The scope of phylogenetic analyses has increased tremendously over the past decade. The seed for this trend was sown by Chase et al. (1993) in an analysis of ∼500 chloroplast *rbcL* sequences sampled across angiosperms, which advanced the size of phylogenetic studies far beyond previous attempts and almost beyond the computational power then available. Now phylogenetic studies of hundreds of organisms are becoming routine (e.g., Van de Peer & de Wachter 1997, Bush et al. 1999, Soltis et al. 1999, Savolainen et al. 2000), and even those with thousands of organisms are being conducted (Källersjö et al. 1998).

Our ability to infer such large phylogenies derives from two factors. First, the molecular revolution, combined with on-line databases such as GenBank or

SwissProt, has afforded more phylogenetic data in a readily accessible format. Encouragingly, accurate answers to large phylogenetic problems may require much less data than previously thought (i.e., <10,000 bp) (Hillis 1996, Bininda-Emonds et al. 2000). Many studies now use sequence data on this order of magnitude (e.g., Madsen et al. 2001, Murphy et al. 2001a). Second, methodological advances are overcoming the basic limitation of phylogenetic inference, namely that the number of possible solutions to be examined ("tree space") increases super-exponentially with the number of taxa (Felsenstein 1978b). Continued advances in computer technology, in concert with algorithmic shortcuts and search strategies, will result in ever-larger phylogenetic problems becoming tractable, even if optimal solutions cannot be guaranteed (see Sanderson & Shaffer 2002, this volume).

The primary constraint to building complete phylogenies is still data accumulation. With a few notable exceptions such as the plant systematic community (e.g., Källersjö et al. 1998), data collection is largely uncoordinated and opportunistic, resulting in a patchwork of coverage for a given taxonomic group. Some species are overrepresented, whereas others are drastically underrepresented, if sampled at all. Moreover, the molecular sampling effort is confined to only a few genes, even for the more coordinated efforts. It remains unclear whether the entire extended history of a large group can be reconstructed adequately using only a few genes (but see Källersjö et al. 1998). Therefore, considerable motivation exists for developing methods that combine existing phylogenetic data—either the raw data themselves ("total evidence"; sensu Kluge 1989) or the tree topologies derived from them ("taxonomic congruence"; sensu Mickevich 1978)—to produce more inclusive phylogenies.

Recently a new approach for combining source trees—supertree construction—has gained popularity for its ability to produce phylogenies based on all data sources (i.e., morphological and molecular), even if the trees only overlap partially in the taxa they contain. Supertree construction has yielded comprehensive phylogenies for all extant members of the mammalian orders Primates (Purvis 1995a), Carnivora (Bininda-Emonds et al. 1999), Chiroptera (Jones et al. 2002b), and Lagomorpha (Stoner et al. 2003); for all extant families of mammal (Liu et al. 2001); for all species of procellariiform seabirds (Kennedy & Page 2002); for the major extant clades within the legume subfamily Papilionoideae (Wojciechowski et al. 2000); for 403 genera of the grass family Poaceae (Salamin et al. 2002); and for all genera of Dinosauria (Pisani et al. 2002).

We review the procedure of supertree construction, including its mathematical properties and potential utility to the biological community. Our discussion focuses on recent, formal supertree techniques in contrast to informal ones such as subjective syntheses of all available information (e.g., Novacek 1992) or pasting together individual hierarchically nested phylogenies (e.g., Weiblen et al. 2000) in a form of "taxonomic substitution" (sensu Wilkinson et al. 2001). Although informal methods have a long history, they lack an objective analytical methodology (Wilkinson et al. 2001); thus, our emphasis is on the more formal techniques.

DEFINITIONS AND TYPES OF SUPERTREES

Since being introduced formally by Gordon (1986), supertree construction has taken on a looser, less mathematical definition (e.g., Sanderson et al. 1998). We follow the latter to define supertree construction as the generation of one or more output trees (the supertrees) from a set of source trees that possess fully or partially overlapping sets of taxa. Because the source trees need only overlap—minimally each source tree must share at least two taxa with the rest of the set of source trees—the supertree can be more inclusive than any individual source tree contributing to it. The supertree ordinarily contains all taxa found in the set of source trees. Our definition distinguishes between supertree and consensus techniques, the latter of which we hold to combine fully overlapping source trees only (following Neumann 1983). We recognize that this distinction is arbitrary in that many consensus techniques can be adapted for a supertree setting.

Supertree techniques can be classified broadly as either "direct" or "indirect" (sensu Wilkinson et al. 2001). Direct supertree methods are akin to classical consensus techniques whereby the output tree is derived directly from source trees without a discrete intermediate step (Figure 1). Examples include strict consensus supertrees (Gordon 1986, Steel 1992), their generalization as MinCutSupertrees (Semple & Steel 2000), and Lanyon's (1993) modification of the semi-strict consensus algorithm. Informal supertrees could also be included here.

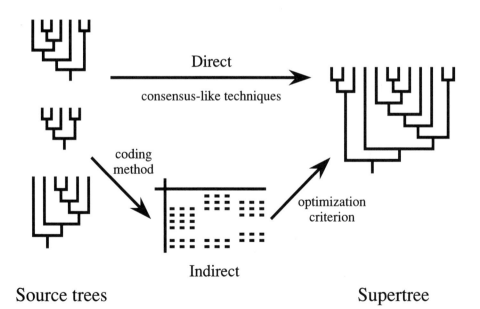

Figure 1 Diagrammatic representation of supertree construction, illustrating both direct and indirect methods.

Indirect supertree construction uses some form of matrix representation (Ponstein 1966, Ragan 1992) to encode individual source tree topologies as matrices that are then combined and analyzed using an optimization criterion (Figure 1). The best-known technique is matrix representation with parsimony (MRP) (Baum 1992, Ragan 1992; also Brooks 1981, Doyle 1992). In MRP the nodes of each source tree are encoded as follows: taxa descended from the focal node score 1; those that do not but that are present elsewhere in the source tree score 0; all other taxa are scored as missing (?). A fictitious all-zero outgroup is added to the matrix to polarize the subsequent parsimony analysis. The outgroup is then pruned to derive the supertree (Figure 2). Variants on this basic form of MRP include modifying the coding procedure (Purvis 1995b, Wilkinson et al. 2001, Semple & Steel 2002); transforming individual cells in the matrix to remove homoplasy ("flip" supertrees) (Chen et al. 2001); or using irreversible parsimony (Bininda-Emonds & Bryant 1998) or compatibility (Rodrigo 1993) to analyze the matrix. Another indirect technique is the average consensus procedure (Lapointe & Cucumel 1997). This method encodes the topology and branch lengths of source trees in individual path-length matrices, computes the average of these matrices,

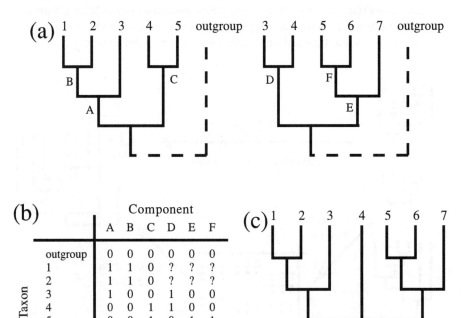

Figure 2 Matrix representation with parsimony (MRP) supertree construction. (*a*) Two source trees for the taxa 1 to 7. A hypothetical outgroup has been added to each source tree. (*b*) The matrix representations of the source trees. (*c*) The MRP supertree.

and then applies a least-squares algorithm to the average matrix to obtain the supertree.

THE THEORY OF SUPERTREES: MATHEMATICAL AND COMPUTATIONAL ASPECTS

In this section we look at some of the theory that underlies the supertree approach. In particular, we describe how concepts from discrete mathematics and computer science can illuminate three aspects of the supertree problem. First, mathematics helps formalize otherwise vague concepts (e.g., what does it mean to say that trees "fit together" consistently or that one tree "contains" another). Only once these ideas have been defined precisely is it possible to establish clear statements (theorems) concerning the properties of different supertree methods. Second, mathematical assessment can determine the limits of current or future supertree methods ("impossibility theorems") (see Arrow 1963). Third, computational techniques can help develop and refine algorithms to construct large-scale supertrees more efficiently.

We deal with rooted trees for two reasons. First, most biologists use rooted trees. Second, and more fundamentally, the supertree problem has no satisfactory solution with unrooted source trees. For example, no supertree method that returns a single output tree can guarantee to simultaneously (*a*) treat each species equally and (*b*) display the relationships present in the unrooted source trees whenever they can be combined without conflict (Steel et al. 2000). An earlier result by McMorris (1985) also places limitations on what can be achieved with unrooted trees. Finally, even determining whether unrooted source trees can be combined without conflict is computationally intractable, or "NP-hard" in mathematical terms (Steel 1992, Böcker et al. 2000).

Formally, two equivalent ways exist to describe a (rooted) phylogenetic tree. The more common method represents the tree visually as a collection of nodes from which three or more branches lead [often called (internal) vertices and edges, respectively, in the language of graph theory]. With rooted trees, the branches inherit a natural direction; if we orient them away from the ancestral root, they point from the past to the future. This is an example of what is often called a (directed) graph. For phylogenetic trees, branches may be either internal and connect two nodes or terminal and lead to a terminal taxon. If all nodes have exactly two outgoing branches, the tree is said to be binary.

An alternative way to describe a phylogenetic tree is simply to specify its clusters (or clades) (Figure 3). If we let X denote the entire set of species under study, a cluster is the collection of species from X that are descended from the most recent common ancestor of some pair of species from X. For example, Figure 3*b* shows the clusters associated with the phylogenetic tree in Figure 3*a*. In this example the most recent common ancestor of 1 and 3 identifies the cluster $\{1,2,3,4\}$, which corresponds to the set of species that are descended from that ancestor.

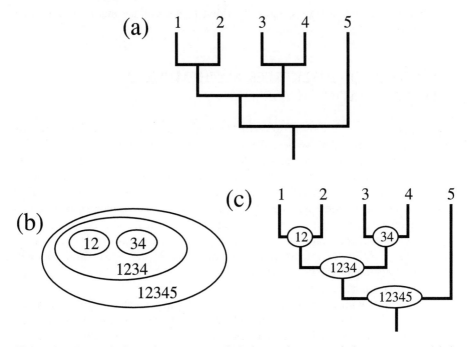

Figure 3 The equivalence between rooted phylogenetic trees and cluster systems. (*a*) A rooted phylogenetic tree *T* as a graph. (*b*) Its associated set of clusters. (*c*) The graph corresponding to this cluster system recovers *T*.

Given a collection of clusters, two fundamental questions arise that are central for many approaches to the supertree problem. Do the clusters come from a phylogenetic tree? If so, do the clusters determine this tree uniquely, and can we construct this tree easily? Regarding the first question, consider a collection *C*, composed of subsets of the entire set *X* of species. *C* forms the clusters of a phylogenetic tree for *X* precisely if all the species in *X* are contained in *C* and any two clusters *A*, *B* in *C* have the following "nesting" property: *A* and *B* either have no species in common, or all the species in one cluster are contained in the other (in set-theoretic notation, $A \cap B \in \{A, B, \phi\}$).

From such a set of the clusters, reconstructing the associated phylogeny is straightforward (Figure 3*c*). To do this, we simply place a directed branch from any cluster *A* (e.g., {1,2,3,4}) to any other cluster *B* (e.g., {1,2}) if all the species in *B* are contained in *A*, provided *A* is the smallest such cluster with this property. Thus, *A* could not be the cluster {1,2,3,4,5} in this example even though this cluster contains *B*. This process provides an equivalence between rooted phylogenetic trees and nested cluster systems that holds even when polytomies are present. Cluster representation provides the most convenient way to define certain consensus and many indirect (and informal) supertree methods.

The Ideal World: Compatible Source Trees

Consider two phylogenetic trees, T and T'. We say that T resolves T' if T' can be obtained from T by collapsing branches. In terms of clusters, this means that each of the clusters of T' is a cluster of T. Additionally, T contains T' if the phylogenetic tree that one obtains from T by deleting all taxa (and connecting branches) that do not appear in T' is equal to or resolves T'. In allowing for resolution, polytomies are held to be "soft" (Maddison 1989), reflecting uncertainty as to the exact order of speciation. Finally, a collection C of phylogenetic trees is compatible if a phylogenetic tree T exists that contains each tree in C. In this case, we call T the parent tree for the collection C. These concepts are illustrated in Figure 4. The trees in Figure 4a are compatible because they are contained in the parent tree in Figure 4b. Informally, a collection of trees is compatible if the trees tell a consistent evolutionary story and can be combined without having to suppress or alter any of the details.

For a set of compatible source trees, Gordon's (1986) strict consensus supertree method provides the strict consensus of all parent trees. Under these conditions the strict consensus and MRP supertree methods are equivalent because the MRP spectrum (i.e., the collection of all equally most parsimonious trees obtained using standard MRP) coincides exactly with the set of parent trees for the source trees. Consequently, the strict consensus MRP supertree

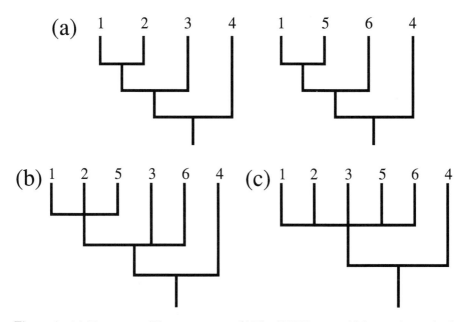

Figure 4 (a) Two compatible source trees. (b) The *BUILD* tree, which contains each of the source trees. (c) The matrix representation with parsimony supertree.

is identical to the strict consensus supertree of the source trees (Thorley 2000).

This finding has an important practical implication. Even for a compatible collection of source trees, computing the MRP supertree exactly may be problematic because the MRP spectrum may contain a potentially huge number of (parent) trees. However, the MRP supertree can be found quickly because it coincides with the strict consensus supertree, which can be constructed exactly by an efficient polynomial time algorithm (Steel 1992).

For compatible source trees, although each parent tree necessarily contains each of the source trees, their strict consensus (i.e., the MRP supertree or strict consensus supertree) may fail to do so. This is because there may simply be too much "slackness" in the way the source trees can fit together. For example, for the two source trees in Figure 4a, we may "attach" species 5 and 6 at many places within or leading to the cluster $\{1,2,3\}$ of the left source tree to obtain a parent tree. The tree in Figure 4b represents only one such parent tree; many more are possible, and their strict consensus is shown in Figure 4c. In this simple example, neither of the two source trees is contained in the MRP/strict consensus supertree.

In contrast, the *BUILD* algorithm (Aho et al. 1981) will decide whether a collection of rooted trees is compatible and, if so, construct a parent tree that contains each of the source trees. The *BUILD* algorithm builds the clusters of a tree as follows. Begin with the cluster $C = X$ (i.e., all species under study). Then repeatedly apply the following rule:

> Place an undirected branch between any two species i and j in C, provided there is some species k in C, and some source tree T for which i and j both lie in some cluster of T that does not contain k. Form new clusters by combining together species that can be connected by a sequence of branches. If this generates just one cluster consisting of all of C (and C has more than one element), then the source trees are incompatible; otherwise this step is repeated on each new cluster.

For example, again consider the two source trees in Figure 4a. The graph that we get for $C = X$ is shown on the left in Figure 5a and gives the cluster $\{1,2,3,5,6\}$ and the isolated species 4. This produces the initial tree shown on the right of Figure 5a. Repeating this procedure on the new cluster $C = \{1,2,3,5,6\}$ gives the graph on the left in Figure 5b, which produces the new cluster $\{1,2,5\}$ and the isolated species 3 and 6. This resolves the initial tree, as shown on the right of Figure 5b. Repeating once more for the cluster $\{1,2,5\}$ we obtain the three remaining isolated species 1, 2, and 5. Combining the three clusters obtained by this process recovers the tree shown in Figure 4b, which contains each of the source trees.

The *BUILD* algorithm has recently been refined. Ng & Wormald (1996) and Constantinescu & Sankoff (1995) independently showed how to output all the parent trees; Ng & Wormald also allowed hard polytomies in the source trees. As there may be many parent trees, it is useful to have an efficient algorithm that outputs just the minimally resolved parent trees (e.g., Semple 2002). Semple also

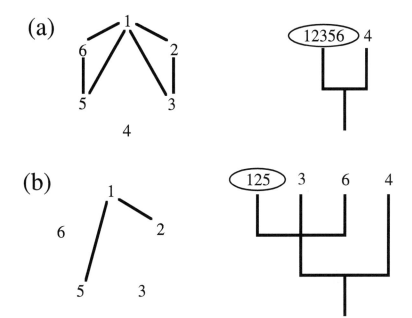

Figure 5   Diagrammatic representation of how the *BUILD* algorithm works using the source trees in Figure 4*a*. The *BUILD* supertree is given in Figure 4*b*.

characterized the *BUILD* tree in terms of a certain clustering property. It follows from this characterization that a collection of source trees has a unique parent tree if and only if the *BUILD* algorithm constructs a binary tree.

In summary, the two natural supertree approaches for compatible source trees are the *BUILD* algorithm and MRP/strict consensus supertrees. The choice of which is preferable depends, respectively, on whether the output tree should display all the source trees or only those relationships supported explicitly in the source trees. This is no longer a mathematical question; rather it represents a judgment of the biologist. For instance, it may be argued that the extra resolution the *BUILD* tree possesses over the MRP supertree in Figure 4 is misleading because neither source tree supports the cluster {1,2,5}.

The Real World: Incompatible Source Trees

In practice, some incompatibility is usually present among the source trees. Two source trees are incompatible if and only if they contain contradictory trees on the same subset of species. However, this need not be the case for three or more trees: One can easily construct examples in which every pair of trees is compatible but the entire collection is not.

Given incompatibility, two general approaches are appropriate: (*a*) try to resolve the incompatibilities by "correcting" the source trees to produce a compatible collection of source trees or (*b*) use an algorithm that does not require compatible source trees or modify an existing algorithm for this purpose.

Two strategies exist for option (*a*), both of which attempt to produce a set of compatible source trees through analyses of subtrees. The first strategy directly corrects incompatibility among trees based on quartet methods for phylogeny reconstruction (e.g., Willson 2001 and references therein). This approach has not been examined in detail but is useful in principle in a supertree framework.

The second strategy prunes "troublesome" taxa from the source trees. One approach for a small number of trees that have a considerable overlap of species is to assess if the subtrees on the species they share agree. If the reduced trees do not agree, then the source trees are incompatible. In that case, one might apply a consensus method to the subtrees and then possibly reattach the remaining species to this consensus tree (Gordon 1986, Bininda-Emonds et al. 1999). Another approach is to look for any species that show widely differing placements in the induced trees. By removing these few species, better resolution of the output tree may result (Wilkinson & Thorley 1998, Wilkinson et al. 2001), and it may be still considered a supertree because it may be more inclusive than any single source tree. However, reduced tree methods have not found wide acceptance in the biological community.

MRP is the most commonly used method under option (*b*), which we discuss in detail below. Another recent option is MinCutSupertrees, in which the *BUILD* algorithm has been adapted to handle any input of rooted trees while still preserving its desirable properties (e.g., retaining clusterings and relationships that are present in all the source trees). Essentially, whenever the *BUILD* algorithm gets "stuck" on some unbreakable cluster (see Figure 6), one deletes branches so that the cluster

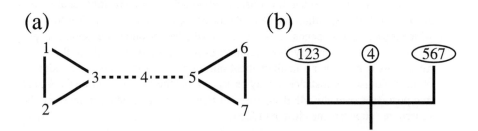

Figure 6   Unsticking the *BUILD* algorithm for incompatible source trees. Applying the *BUILD* algorithm to the source trees in Figure 2a (with the hypothetical outgroup taxon removed) results in the unbreakable cluster in (*a*). However, the *BUILD* algorithm can proceed if a single edge is cut, either that connecting taxa 3 and 4 or taxa 4 and 5 (*dashed lines*). Because there is no reason to cut either edge preferentially, both are cut to obtain the clusters in (*b*). Further application of the *BUILD* algorithm yields the MinCutSupertree, which in this case is identical to the matrix representation with parsimony supertree in Figure 2c.

breaks in a certain minimal way (for details see Semple & Steel 2000). This process can also be directed according to differential support among the source trees.

Any method of combining incompatible source trees should preserve phylogenetic relationships between species when these relationships are present in at least one source tree and are not contradicted by any other source tree. If we write $(IJ)K$ to denote the binary rooted tree in which taxa I and J form a clade to the exclusion of taxon K (the tree can also include other taxa), this desired property, P_1, is more formally written as: If at least one source tree contains $(IJ)K$ and no source tree contains the contradictory $(IK)J$ or $(JK)I$, then the output tree contains $(IJ)K$.

However, no method can satisfy property P_1 in general, even in the consensus setting. Consider four rooted trees for the five taxa labeled 1, 2, ..., 5 (see Figure 2 in Steel et al. 2000). Each tree has one nontrivial cluster, thereby grouping together the following pairs of taxa: {1,2}, {2,3}, {3,4}, and {4,5}. These trees contain, respectively, the subtrees (12)5, (23)5, (34)1, and (45)1; none of the source trees contains $(IK)J$ or $(JK)I$ for any subtree $(IJ)K$ in this list. Any consensus method satisfying P_1 would have to output (12)5, (23)5, (34)1, and (45)1. However, it is obvious that no rooted phylogenetic tree can contain these four trees simultaneously.

It might be objected that the trees in this example are nonbinary. If the source trees are binary, then property P_1 is equivalent to the (otherwise weaker) condition P_2: if all the source trees contain $(IJ)K$ then the output tree contains $(IJ)K$. In the supertree setting, the MinCutSupertree algorithm has been shown to satisfy property P_2 (Semple & Steel 2000). In contrast, MRP can fail to satisfy P_2 on sufficiently contrived data, even in the consensus setting (results not shown).

A CRITICAL LOOK AT SUPERTREE CONSTRUCTION

General Criticisms

Supertree construction has been criticized strongly because, like taxonomic congruence, it loses contact with the primary data (Rodrigo 1993, 1996; Novacek 2001; Springer & de Jong 2001; Gatesy et al. 2002). Thus, Springer & de Jong have argued that supertrees present only a useful summary of the source trees, rather than an accurate phylogenetic reconstruction. Novacek added that supertrees do not provide strong tests of previous phylogenetic hypotheses because they are not based on new data sets. Slowinski & Page (1999) have also questioned how supertree analyses should be interpreted, particularly the biological meaning of any "homoplasy" (which actually only represents source tree incongruence).

However, the inherent loss of information from using the source trees is a necessary trade-off to be able to combine all possible sources of phylogenetic information. Simulation studies have demonstrated that the cost of this trade-off is not very high, at least for MRP (Bininda-Emonds & Sanderson 2001). Over a wide range of conditions, MRP performed about on a par with total evidence at reconstructing a known model tree. Also, through the use of character weighting, many supertree methods, particularly the indirect ones, can incorporate information about differential levels of evidential support both within and among source studies. The use of weighting in this manner has been demonstrated to improve the fit between the primary data and the MRP representation of their associated source tree (Ronquist 1996). Further, the simulation study of Bininda-Emonds & Sanderson (2001) showed that "weighted MRP" outperforms total evidence under most of the conditions they examined.

Another potential problem with supertree construction is data set nonindependence. For example, Springer & de Jong (2001) pointed out that the family-level mammal supertree of Liu et al. (2001) includes five different source trees that all use the same transferrin immunology data set for bats. Some nonindependence is inevitable, despite steps to minimize it (e.g., Purvis 1995a, Bininda-Emonds et al. 1999, Liu et al. 2001, Jones et al. 2002b). The effect of this problem on supertree construction is unknown. However, the interaction between the repeated data and a different suite of data sets in each source study may minimize the influence of any single repeated data set. Moreover, the different assumptions, models, and methods of analysis used mean that even source studies based on virtually the same data set can present different phylogenetic estimates. Data set nonindependence nevertheless remains a concern for supertree and taxonomic congruence approaches (unlike total evidence studies). It is likely to become more of an issue in the future as the popularity of total evidence increases and primary analyses include more and more previously published data.

Matrix Representation Techniques

Matrix representation methods have attracted the most attention in the biological community. The popularity of MRP derives both from its universal applicability (i.e., it can combine all source trees) and its ease of use. Coding of even large source trees is a trivial, if involved, process. Several programs now exist to automate this process, [e.g., PAUP* (Swofford 2002), RadCon (Thorley & Page 2000), SuperTree (Salamin et al. 2002), r8s (available from http://ginger.ucdavis.edu/r8s/)], and many parsimony programs are available for the subsequent analysis. However, in addition to occasionally failing to satisfy property P_2 above, MRP, and by extension most matrix representation methods, have other potentially undesirable characteristics.

Although matrix representation is a well-grounded technique in basic graph and network theory, an exact one-to-one correspondence between a tree and its matrix representation exists only for single source trees (Ragan 1992). The derivation of a single supertree from an optimality analysis of the combined matrix representations must be viewed as a heuristic (Baum & Ragan 1993, Lapointe & Cucumel 1997). The behavior of matrix representation methods, and MRP in particular, is poorly characterized. MRP has a complex size bias in which the more inclusive of two competing, analogous clades is favored in the supertree because it contributes more "characters" to the matrix (Purvis 1995b, Bininda-Emonds & Bryant 1998).

MRP may also favor source trees that are more unbalanced (Wilkinson et al. 2001).

Attempts to correct for MRP's inherent size bias have been unsuccessful (Bininda-Emonds & Bryant 1998). Convex multistate coding (Semple & Steel 2002), which encodes binary source trees of any size by five multistate characters, eliminates the bias at the cost of losing the one-to-one correspondence between nodes and matrix elements. Therefore, it is difficult to incorporate information about differential signal strength within source trees using this technique.

It is debatable, however, whether MRP's size bias is problematic. If larger trees are held to be more accurate and contain more information, then the bias is appropriate (see Bininda-Emonds & Bryant 1998). More importantly, the impact of the size and balance biases appears to be minimal in practice. Simulations comparing MRP to total evidence show that both techniques produce similar phylogenies and behave similarly with respect to a number of variables (e.g., source tree size, number of source trees, and degree of taxon overlap between studies) (Bininda-Emonds & Sanderson 2001).

Direct Supertree Methods

Owing to their common usage of cluster representation, many direct supertree methods share the inability of classical consensus techniques (and, by extension, taxonomic congruence) to accommodate information regarding differential support. This is in terms of both (*a*) encoding such information within and among source studies and (*b*) summarizing such information for the supertree itself. The MinCutSupertree algorithm is uniquely able to account for differential source tree support among direct methods. It does so through a simple weighting function that dictates which of multiple possible branches to delete first to break apart a cluster (Figure 6) (Semple & Steel 2000). However, all direct methods can only provide conservative support information for the supertree. For example, the most that can be said for those clades present in a strict consensus supertree is that they are not contradicted by any source trees. At best, differential information is limited to the frequency of the clades within the set of source trees. In contrast, the clades of an indirect supertree can be characterized by various support metrics (e.g., Bremer support or bootstrap frequencies modified to account for the nonindependence within source trees). The use of such metrics for direct methods seems unlikely. Similarly, no direct method currently provides support information for the entire supertree, whereas indirect methods can through various ensemble goodness-of-fit statistics (e.g., consistency index).

APPLICATIONS OF SUPERTREES

Compared to conventional phylogenies, supertrees provide a greater potential for complete taxonomic coverage based on a consensus of all phylogenetic information (e.g., morphological, molecular, and other phenotypic traits). This feature permits broad-scale ecological and evolutionary analyses that are rarely, if ever, tractable using conventional phylogenies. Here, we review applications of supertrees (see Table 1), including novel and prospective ones. Our examples are not exhaustive and deal mainly with mammals because many supertrees are now available for them.

TABLE 1   Examples of supertrees and their applications

Supertree	Taxon	Level of tips (size)	Application	Reference[a]
Purvis (1995a)	Primates	Species (201)	Evolutionary rates	
			(Life histories)	12
			(Brain size)	6
			Extinction risk	11
			Extinction rates	10
			Immune system function	9
			Extinction and speciation	1, 5
			Brain size and behavior	2
			Allometry and home range size	8
Bininda-Emonds et al. (1999)	Carnivora	Species (271)	Evolutionary rates (Life histories)	12
			Body size and species richness	3
			Taxonomic differences	7
			Extinction risk	11
			Extinction rates	10
			Species richness	supertree study
Kirsch et al. (1997)	Marsupialia	Species (81)	Extinction and range size	4
Jones et al. (2002b)	Chiroptera	Species (925)	Extinction risk	13
Liu et al. (2001)	Mammalia	Families (90)	Phylogeny	supertree study
Weiblen et al. (2000)	Monocotyledons	Species (918)	Evolution of breeding systems	supertree study
Webb (2000)	Trees	Species (324)	Structure of ecological communities	supertree study
Linder (2000)	Restionaceae	Genera (62)	Convergent evolution	supertree study

[a]1, Purvis et al. (1995); 2, Barton (1996); 3, Gittleman & Purvis (1998); 4, Johnson (1998); 5, Paradis (1998); 6, Deaner & Nunn (1999); 7, Bininda-Emonds & Gittleman (2000); 8, Nunn & Barton (2000); 9, Nunn et al. (2000); 10, Purvis et al. (2000a); 11, Purvis et al. (2000b); 12, Purvis et al. (2003); 13, Jones et al. (2002a).

Descriptive Systematics

Phylogenetic reconstruction involves many contentious issues (Hull 1980, Felsenstein 2001). Supertrees, if anything, have added more fuel to the fire. However, the process of culling all phylogenetic information for supertree construction is extremely useful for descriptive systematics. It helps assess what information is available, differences in relative research effort among taxa, and degree of phylogenetic congruence among studies. At the very least, supertrees highlight groups that have received little systematic attention.

For example, the MRP supertree of the higher-level relationships across eutherian (placental) mammals used 430 molecular and morphological source trees from 315 research articles (Liu et al. 2001). Taxonomic coverage was unequal among the 90 families and orders, reflecting which groups are viewed as charismatic and economically valued. Of the 1965 MRP "characters," Solenodontidae (Insectivora) was represented in only 342, whereas Bovidae (Artiodactyla) was present in 1520. Taxonomic coverage was also generally poor. In addition to bovids, only 9 other terminal taxa were represented in more than 45% of the characters. Despite this, the apparent accuracy of available phylogenies is encouraging. Except for Artiodactyla and Insectivora, which are now widely agreed to be paraphyletic, the monophyly of the remaining orders was strongly supported. Most accepted interfamilial relationships were also corroborated. The molecular and morphological trees reconstructed the same interfamilial patterns with only five exceptions, most relating to recent molecular findings involving nonmonophyly. In essence, Liu et al.'s (2001) supertree indicates that the vast majority of molecular and morphological trees are congruent for eutherian phylogeny. Any disagreement seems to stem simply from the lack of information for orders such as Insectivora and Xenarthra.

Similar patterns were found in species-level MRP supertrees of the mammalian orders Primates (Purvis 1995a), Carnivora (Bininda-Emonds et al. 1999) and Chiroptera (Jones et al. 2002b). In particular, taxonomic coverage was extremely patchy, with relationships in poorly researched groups showing less resolution in the supertree. Nevertheless, the overall resolution is high for carnivores and primates (78.1% and 79.0%, respectively, compared with a fully resolved, binary tree), reflecting both sufficient phylogenetic information to construct a supertree and generally strong agreement among the source trees. Bats reveal poorer resolution (46.4%) owing to comparatively little treatment.

When coupled with descriptive statistics, the summary of phylogenetic information assembled for supertree construction can reveal what factors influence disagreement among phylogenies. Using the carnivore supertree, Bininda-Emonds (2000) examined the effect of tree selection criteria, data source, study size, and date of study on phylogenetic reconstructions of the order. Two interesting conclusions emerged. First, widespread disparity in study effort exists. Certain groups were confirmed as receiving disproportionately more attention, discrete characters in concert with parsimony analysis were favored owing to their simplicity, and molecular trees were more abundant but covered fewer species. Second, significant differences were rare among test variables as to whether they generated differences in phylogenetic topologies. From this finding, Bininda-Emonds (2000) inferred that most estimates of carnivore phylogeny are pointing generally at the same solution.

Another way of visualizing differences between phylogenies using supertrees is through a "sliding window" form of time-series analysis in which source studies are ordered chronologically and then combined in contiguous overlapping sets (e.g., Bininda-Emonds 2000, 2003). This approach is particularly amenable to historical study, in which phylogenetic statements without an explicit data matrix characterize much of the literature. Sliding window analyses often reveal trends that are not immediately obvious. For example, the giant panda (*Ailuropoda melanoleuca*) has received tremendous systematic attention recently, partly because of its conservation importance but also because molecular analysis was held to be able to pinpoint its phylogenetic position, whereas earlier morphological studies could not (O'Brien et al. 1985). A sliding window analysis of phylogenies published from 1869 to 1999 for giant pandas and putative sister taxa showed that giant pandas are unequivocally bears and that this relationship has held prior to the advent of molecular analysis (Bininda-Emonds 2003).

Adding branch length information reflecting divergence times can enhance the descriptive utility of supertrees. Times of divergence are also of critical importance for testing macroevolutionary hypotheses (Purvis 1996, Mooers & Heard 1997; see below). Assembling fossil and molecular dates for nodes in a supertree allows analyses of the level at which independent estimates of divergence times agree. In the carnivore supertree 73 nodes had at least one date estimate from both molecular and fossil sources. In contrast to an earlier, smaller study (Wayne et al. 1991), a slight but significant difference between fossil and molecular dates was found (Bininda-Emonds et al. 1999). Additional supertrees will be valuable in verifying divergence times, measuring the tempo of macroevolutionary change, and testing whether the ages of younger or older lineages are consistently over- or underestimated between different sources of dating.

Evolutionary Models

In addition to testing rates of evolutionary change, models of evolution are inherent to applying comparative phylogenetic methods for tests of adaptation and character evolution. The commonly accepted null model of evolution is Brownian motion (Felsenstein 1985), in which the amount of character evolution scales to branch length. Under this model closely related taxa are more similar to one another than to more distantly related ones. Although this is an effective null model for various macroevolutionary tests (Purvis et al. 1994), we do not know whether it is correct. Non-Brownian motion models also exist and are theoretically viable (e.g., Price 1997, Harvey & Rambaut 2000; see also Losos 1999). This is where supertrees should come into the picture. With a complete tree and associated branch lengths, characters could be correlated with tree structure to examine how species are evolving relative to each model and whether morphological, life history, or behavioral traits are actually following non-Brownian motion models.

Rates of Cladogenesis

Despite considerable effort, at least two fundamental questions about the tempo and mode of macroevolutionary change remain. First, do different lineages show different patterns of speciation and extinction? Second, how are any differential patterns related to rates of evolution? Investigating these questions hinges on complete taxonomic coverage (Nee et al. 1992, Pybus & Harvey 2000), which supertrees can provide. Based on the primate supertree, it was shown that speciation and extinction probably do not vary through time in this group (Purvis et al. 1995, Paradis 1998), although there may be variable rates at lower taxonomic levels using some statistical criteria (Pybus & Harvey 2000). Analysis of the carnivore supertree revealed that eight lineages contain significantly more species than expected by chance if all lineages had equal probabilities of diversifying (Bininda-Emonds et al. 1999).

A future trend in supertree applications will be comparing patterns and processes among trees. For example, whereas primates have had few explosive radiations for certain isolated clades only, carnivores have had massive radiations occurring many times among independent clades (Purvis et al. 1995, Bininda-Emonds et al. 1999). Analogous analyses on an informal supertree of flowering plants show that 10 clades are unexpectedly rich in species, whereas 13 clades have a lower than expected number (Magallón & Sanderson 2001). The distribution of these clades indicates that specific characteristics are unlikely to have promoted diversification within angiosperms. These patterns not only raise interesting questions about what factors lead to high diversification rates in some taxa, but also pertain to important consequences for conserving future biodiversity (see "Biodiversity and Conservation" below).

With an incomplete phylogeny, it is difficult to test for differential patterns of species richness (Pybus & Harvey 2000). The completeness of supertrees thus allows us to test null models of diversification (see Simberloff et al. 1981, Purvis et al. 1995). Even without knowing actual speciation and extinction parameters, simulation can be used to find the richness patterns generated using different values, which can be compared against the supertree. For example, such a simulation analysis compared against the carnivore supertree revealed that a massive extinction event likely occurred within the past five million years that killed off two thirds to five sixths of all species (Purvis et al. 2001). Even though these numbers are rough approximations, they are not the result of perceived losses from species not included in the tree.

Evolutionary Patterns

The surprisingly modern realization that hierarchical relationships among taxa are represented by phylogenies revolutionized comparative evolutionary biology (Felsenstein 1985, Harvey & Pagel 1991). The resultant interest in simply diagnosing the macroevolutionary patterns between traits and trees (Gittleman & Kot 1990, Martins 1996, Losos 1999) was limited by a lack of comprehensive phylogenies. Initially, researchers turned to taxonomies, which are often poor reflections of phylogenies. Supertrees, owing to their increased completeness, allow us to address questions of a broader scope with increased power. They also mitigate the known adverse effects in comparative analyses associated with incomplete taxon sampling (see Gittleman 1989).

The use of a supertree allows proper tests of functional relationships. For example, using the primate supertree with an independent contrasts analysis, Nunn & Barton (2000) found that home-range size scales to body mass with an exponent of 0.75 following Kleiber's law, not 1.0 as suggested by other comparative studies. Similar comparative studies have benefited from using supertrees to investigate brain size evolution in primates (Barton 1996), convergent evolution in vascular plants (Linder 2000), abundance patterns in Australian marsupials (Johnson 1998), and immune system functions in primates (Nunn et al. 2000). All these studies required supertrees because of their scope; without supertrees, sample sizes would often have been halved.

Comparative tests to associate differences in species richness with changes in a given trait likewise require complete phylogenetic information. For example, frequency distributions of body sizes in a clade are generally right-skewed. Most species tend to be small-bodied, perhaps because small-bodied lineages have the biological properties to speciate at higher rates. Previous attempts to test this hypothesis have been hampered by missing taxa, inadequate comparative statistics, and no null model for comparisons. Using the carnivore and primate supertrees, Gittleman & Purvis (1998) showed for the first time that species richness is not related to body size in primates and only partially so in carnivores, appearing mainly in the "dog-like" carnivores (i.e., canids, procyonids, pinnipeds, ursids, and mustelids).

Finally, supertrees have been used to test the "evolutionary lag" phenomenon (see Harvey & Pagel 1991), an oft-cited reason for failing to find macroevolutionary patterns among traits that are expected to be related. For example, observed slopes of brain size on body size in mammals are usually less than isometric, with taxon-level effects revealing shallower slopes at higher taxonomic levels. This is consistent with evolutionary lag in brain size evolution (Lande 1979, Pagel & Harvey 1988). Although this explanation is prevalent in the literature, inadequate samples across independent clades prevented direct tests of it. Deaner & Nunn (1999), using the primate supertree together with a method they developed to measure the relative change in two traits along the same time axis (branch), found no evidence for evolutionary lag in brain size. Because the method can only be used on sister species, supertrees are essential; sample sizes would otherwise be prohibitively small. In this example, only 22–25 contrasts were available across the order, even with the complete supertree.

Biodiversity and Conservation

Although appreciated for some time (Vane-Wright et al. 1991; see Vázquez & Gittleman 1998 for additional references), the importance of phylogenies to conservation was highlighted by Nee & May (1997). Using simulation and analytical modeling, they showed that evolutionary history need not necessarily be lost at a profound rate if (*a*) extinction is random, (*b*) branch lengths are used to represent a measure of "phylogenetic diversity" (PD), and (*c*) the topology of the tree is relatively balanced. The realism of these simulations could be tested using supertrees. For instance, Purvis et al. (2000a) used the primate and carnivore supertrees to compare the amount of PD lost if species classified as threatened went extinct relative to random extinction. Greater amounts of PD would indeed be lost. Interestingly, only the primates showed a significant loss. This is consistent with expectation (Nee & May 1997, Heard & Mooers 2000): The primate supertree is more unbalanced than the carnivore supertree (compare Purvis 1995a, Bininda-Emonds et al. 1999), revealing the underlying effects of random extinction. If incomplete trees were used for these groups, it would not be possible to separate the effects of extinction models relative to phylogenetic incompleteness. Similarly, Sechrest et al. (2002) used the carnivore and primate supertrees to show that PD is not randomly distributed, in that the 25 designated global biodiversity hotspots harbor significant amounts of PD.

Supertrees have also helped in investigations of the factors that may contribute to extinction risk. Despite many reviews of these factors (see McKinney 1997), empirical tests have been lacking because of too few comparative databases and comprehensive phylogenetic estimates. Purvis et al. (2000b) assembled trait data on carnivores and primates and performed contrast analyses using the supertrees of each group. Of 12 possible variables, high trophic level, low population density, slow life history (particularly gestation length), and especially small geographic range were significantly and independently correlated with the International Union for Conservation of Nature and Natural Resources (IUCN) *Red List* extinction risk designations. The traits combined explained nearly 50% of the variation across species. Supertrees were again essential. Without them, a multivariate analysis would not have been possible. Even with the complete supertrees, slightly less than 50% of all species were represented in the complete multivariate model. In the face of not having complete molecular or morphological phylogenies in the near future, supertrees are important tools for contributing information about the biological past in order to conserve present and future evolutionary history.

CONCLUSION: THE FUTURE OF SUPERTREE CONSTRUCTION

An unspoken sentiment is that supertree construction merely represents a stopgap measure, possibly based on economic grounds (see Sanderson et al. 1998), to infer the tree of life until there is sufficient molecular data. However, there are three strong arguments against this viewpoint.

The first relates to sampling issues. Despite the tremendous increase in sequencing effort, most species will continue to be poorly characterized at the molecular level for the foreseeable future. For example, two recent large-scale molecular studies across mammals (Madsen et al. 2001, Murphy et al. 2001a) together sequenced only 146 species, 27 of which were in common; a combined and expanded third study by both groups increased the number of common species sampled to 42 (Murphy et al. 2001b). Although prodigious, this represents a fraction of the approximately 4500 extant species of mammal, which in turn represent a fraction of the earth's total biota, much of which is not studied as intensively. Future sampling effort will continue to be nonrandom, with priority being given to species that are of economic or conservation importance, or that are simply more appealing to us (O'Brien et al. 2001). Thus, phylogenetic inference of molecular data will be prone to problems from taxon sampling (Lecointre et al. 1993, Bininda-Emonds et al. 1998, Hillis 1998), long-branch attraction (Felsenstein 1978a, Huelsenbeck 1995), and missing data (Sanderson et al. 1998, Wiens 1998).

The second argument arises from analytical problems inherent to molecular data, particularly in a maximum likelihood framework. Especially for noncoding DNA, assessing homology and therefore aligning the sequences becomes more difficult as the included taxa become increasingly diverse (Sanderson et al. 1998). There is also the computational demand of incorporating appropriate models of molecular evolution, models that are becoming increasingly sophisticated and complex (Sanderson & Shaffer 2002, this volume). Different genes are often analyzed most appropriately using different evolutionary models. However, current computer programs are limited in their ability to incorporate multiple models. The computational demand of such analyses is probably prohibitive in the foreseeable future (Sanderson & Kim 2000), necessitating the use of overly simplistic analyses (e.g., Gatesy et al. 2002). Instead, it seems more efficient to analyze each gene separately under an appropriate evolutionary model and then combine the results as a supertree (see Doyle 1992).

Finally, there is the need to include nonmolecular data. The inclusion of fossil taxa, for which molecular data are normally unavailable, can overturn phylogenetic hypotheses based on extant species (Donoghue et al. 1989). Moreover, the principle of total evidence dictates that the best hypothesis is that derived from as many independent data sources as possible. The nonindependence of molecular data owing to linkage associations between genes is often underestimated. For example, it is often argued that mitochondrial DNA with its many genes constitutes a single phylogenetic data source because it forms a single heritable unit that is not normally subject to recombination (Cummings et al. 1995). Although extremely valuable, molecular data represent only one tool to reconstruct the tree of life.

In summary, we maintain that supertree construction has a valid and continuing role in phylogenetic systematics. Although simultaneous analysis of the primary data is preferable owing to the greater information content retained, its applicability is limited by incompatible data types and the requirement of a single optimization criterion. Current supertree methods are not perfect. However, they generally show good performance in simulation, suggesting that we can be reasonably confident in the phylogenetic estimates derived from them. Supertree techniques continue to be developed, and advances such as MinCutSupertrees and flip supertrees are likely to improve the performance of supertree construction in the future.

ACKNOWLEDGMENTS

We thank Kate Jones, Arne Mooers, Sam Price, and Brad Shaffer for their constructive and helpful comments. OBE was supported by the van der Leeuw Fonds. JLG thanks NSF for support (grant #DEB 0129009) and also for supporting the NCEAS Working Group "Phylogeny and Conservation" (grant #DEB-94-21535). MAS was supported by the New Zealand Marsden Fund.

The *Annual Review of Ecology and Systematics* is online at
http://ecolsys.annualreviews.org

LITERATURE CITED

Aho AV, Sagiv Y, Szymanski TG, Ullman JD. 1981. Inferring a tree from lowest common ancestors with an application to the optimization of relational expressions. *SIAM J. Comput.* 10:405–21

Arrow KJ. 1963. *Social Choice and Individual Values.* New York: Wiley. 124 pp.

Barton RA. 1996. Neocortex size and behavioural ecology in primates. *Proc. R. Soc. London Ser. B* 263:173–77

Baum BR. 1992. Combining trees as a way of combining data sets for phylogenetic inference, and the desirability of combining gene trees. *Taxon* 41:3–10

Baum BR, Ragan MA. 1993. Reply to A.G. Rodrigo's "A comment on Baum's method for combining phylogenetic trees." *Taxon* 42: 637–40

Bininda-Emonds ORP. 2000. Factors influencing phylogenetic inference: a case study using the mammalian carnivores. *Mol. Phylogenet. Evol.* 16:113–26

Bininda-Emonds ORP. 2003. The phylogenetic position of the giant panda (*Ailuropoda melanoleuca*): a historical consensus through supertree analysis. In *Biology and Conservation of the Giant Panda*, ed. DG Lindburg, K Baragona. Berkeley: Univ. Calif. Press. In press

Bininda-Emonds ORP, Brady SG, Sanderson

MJ, Kim J. 2000. Scaling of accuracy in extremely large phylogenetic trees. In *Pacific Symposium on Biocomputing 2001*, ed. RB Altman, AK Dunker, L Hunter, K Lauderdale, TE Klein, pp. 547–58. River Edge, NJ: World Scientific

Bininda-Emonds ORP, Bryant HN. 1998. Properties of matrix representation with parsimony analyses. *Syst. Biol.* 47:497–508

Bininda-Emonds ORP, Bryant HN, Russell AP. 1998. Supraspecific taxa as terminals in cladistic analysis: implicit assumptions of monophyly and a comparison of methods. *Biol. J. Linn. Soc.* 64:101–33

Bininda-Emonds ORP, Gittleman JL. 2000. Are pinnipeds functionally different from fissiped carnivores? The importance of phylogenetic comparative analyses. *Evolution* 54: 1011–23

Bininda-Emonds ORP, Gittleman JL, Purvis A. 1999. Building large trees by combining phylogenetic information: a complete phylogeny of the extant Carnivora (Mammalia). *Biol. Rev.* 74:143–75

Bininda-Emonds ORP, Sanderson MJ. 2001. Assessment of the accuracy of matrix representation with parsimony supertree construction. *Syst. Biol.* 50:565–79

Böcker S, Bryant D, Dress AWM, Steel MA.

2000. Algorithmic aspects of tree amalgamation. *J. Algorithms* 37:522–37

Brooks DR. 1981. Hennig's parasitological method: a proposed solution. *Syst. Zool.* 30: 229–49

Bush RM, Fitch WM, Bender CA, Cox NJ. 1999. Positive selection on the H3 hemagglutinin gene of human influenza virus A. *Mol. Biol. Evol.* 16:1457–65

Chase MW, Soltis DE, Olmstead RG, Morgan D, Les DH, et al. 1993. Phylogenetics of seed plants: an analysis of nucleotide sequences from the plastid gene *rbc*L. *Ann. Mo. Bot. Gard.* 80:528–80

Chen D, Eulenstein O, Fernández-Baca D, Sanderson MJ. 2001. *Supertrees by flipping. Tech. Rep. TR02-01.* Iowa State Univ., Dept. Comput. Sci.

Constantinescu M, Sankoff D. 1995. An efficient algorithm for supertrees. *J. Classif.* 12: 101–12

Cummings MP, Otto SP, Wakeley J. 1995. Sampling properties of DNA sequence data in phylogenetic analysis. *Mol. Biol. Evol.* 12: 814–22

Deaner RO, Nunn CL. 1999. How quickly do brains catch up with bodies? A comparative method for detecting evolutionary lag. *Proc. R. Soc. London Ser. B* 266:687–94

Donoghue MJ, Doyle JA, Gauthier J, Kluge AG, Rowe T. 1989. The importance of fossils in phylogeny reconstruction. *Annu. Rev. Ecol. Syst.* 20:431–60

Doyle JJ. 1992. Gene trees and species trees: molecular systematics as one-character taxonomy. *Syst. Bot.* 17:144–63

Felsenstein J. 1978a. Cases in which parsimony or compatibility methods will be positively misleading. *Syst. Zool.* 27:401–10

Felsenstein J. 1978b. The number of evolutionary trees. *Syst. Zool.* 27:27–33

Felsenstein J. 1985. Phylogenies and the comparative method. *Am. Nat.* 125:1–15

Felsenstein J. 2001. The troubled growth of statistical phylogenetics. *Syst. Biol.* 50:465–67

Gatesy J, Matthee C, DeSalle R, Hayashi C. 2002. Resolution of a supertree/supermatrix paradox. *Syst. Biol.* 51:652–64

Gittleman JL. 1989. The comparative approach in ethology: aims and limitations. In *Perspectives in Ethology,* ed. PPG Bateson, PH Klopfer, pp. 55–83. New York: Plenum

Gittleman JL, Kot M. 1990. Adaptation: statistics and a null model for estimating phylogenetic effects. *Syst. Zool.* 39:227–41

Gittleman JL, Purvis A. 1998. Body size and species-richness in carnivores and primates. *Proc. R. Soc. London Ser. B* 265:113–19

Gordon AD. 1986. Consensus supertrees: the synthesis of rooted trees containing overlapping sets of labeled leaves. *J. Classif.* 3:31–39

Harvey PH, Pagel MD. 1991. *The Comparative Method in Evolutionary Biology.* Oxford: Oxford Univ. Press. 239 pp.

Harvey PH, Rambaut A. 2000. Comparative analyses for adaptive radiations. *Philos. Trans. R. Soc. London Ser. B* 355:1599–605

Heard SB, Mooers AØ. 2000. Phylogenetically patterned speciation rates and extinction risks change the loss of evolutionary history during extinctions. *Proc. R. Soc. London Ser. B* 267:613–20

Hillis DM. 1996. Inferring complex phylogenies. *Nature* 383:130–31

Hillis DM. 1998. Taxonomic sampling, phylogenetic accuracy, and investigator bias. *Syst. Biol.* 47:3–8

Huelsenbeck JP. 1995. Performance of phylogenetic methods in simulation. *Syst. Biol.* 44:17–48

Hull DL. 1980. *Science as a Process.* Chicago: Univ. Chicago Press

Johnson CN. 1998. Species extinction and the relationship between distribution and abundance. *Nature* 394:272–74

Jones KE, Gittleman JL, Purvis A. 2002a. Bat extinction risk. *J. Anim. Ecol.* Submitted

Jones KE, Purvis A, MacLarnon A, Bininda-Emonds ORP, Simmons NB. 2002b. A phylogenetic supertree of the bats (Mammalia: Chiroptera). *Biol. Rev.* 77:223–59

Källersjö M, Farris JS, Chase MW, Bremer B, Fay MF, et al. 1998. Simultaneous parsimony jackknife analysis of 2538 *rbc*L DNA sequences reveals support for major clades of green plants, land plants, seed plants and flowering plants. *Plant. Syst. Evol.* 213:259–87

Kennedy M, Page RDM. 2002. Seabird supertrees: combining partial estimates of Procellariiform phylogeny. *The Auk* 119:88–108

Kirsch JAW, Lapointe F-J, Springer MS. 1997. DNA-hybridisation studies of marsupials and their implications for metatherian classification. *Aust. J. Zool.* 45:211–80

Kluge AG. 1989. A concern for evidence and a phylogenetic hypothesis of relationships among *Epicrates* (Boidae, Serpentes). *Syst. Zool.* 38:7–25

Lande R. 1979. Quantitative genetic analysis of multivariate evolution, applied to brain:body size allometry. *Evolution* 33:402–16

Lanyon SM. 1993. Phylogenetic frameworks: towards a firmer foundation for the comparative approach. *Biol. J. Linn. Soc.* 49:45–61

Lapointe F-J, Cucumel G. 1997. The average consensus procedure: combination of weighted trees containing identical or overlapping sets of taxa. *Syst. Biol.* 46:306–12

Lecointre G, Philippe H, Vân Lê HL, Le Guyader H. 1993. Species sampling has a major impact on phylogenetic inference. *Mol. Phylogenet. Evol.* 2:205–24

Linder HP. 2000. Vicariance, climate change, anatomy and phylogeny of Restionaceae. *Bot. J. Linn. Soc.* 134:159–77

Liu F-GR, Miyamoto MM, Freire NP, Ong PQ, Tennant MR, et al. 2001. Molecular and morphological supertrees for eutherian (placental) mammals. *Science* 291:1786–89

Losos JB. 1999. Uncertainty in the reconstruction of ancestral character states and limitations on the use of phylogenetic comparative methods. *Anim. Behav.* 58:1319–24

Maddison WP. 1989. Reconstructing character evolution on polytomous cladograms. *Cladistics* 5:365–77

Madsen O, Scally M, Douady CJ, Kao DJ, DeBry RW, et al. 2001. Parallel adaptive radiations in two major clades of placental mammals. *Nature* 409:610–14

Magallón S, Sanderson MJ. 2001. Absolute diversification rates in angiosperm clades. *Evolution* 55:1762–80

Martins EP. 1996. Phylogenies, spatial autoregression, and the comparative method: a computer simulation test. *Evolution* 50:1750–65

McKinney ML. 1997. Extinction vulnerability and selectivity: combining ecological and paleontological views. *Annu. Rev. Ecol. Syst.* 28:495–516

McMorris FR. 1985. Axioms for consensus functions on undirected phylogenetic trees. *Math. Biosci.* 74:17–21

Mickevich MF. 1978. Taxonomic congruence. *Syst. Zool.* 27:143–58

Mooers AØ, Heard SJ. 1997. Inferring evolutionary process from phylogenetic tree shape. *Q. Rev. Biol.* 72:31–54

Murphy WJ, Eizirik E, Johnson WE, Zhang YP, Ryder OA, O'Brien SJ. 2001a. Molecular phylogenetics and the origins of placental mammals. *Nature* 409:614–18

Murphy WJ, Eizirik E, O'Brien SJ, Madsen O, Scally M, et al. 2001b. Resolution of the early placental mammal radiation using Bayesian phylogenetics. *Science* 294:2348–51

Nee S, May RM. 1997. Extinction and the loss of evolutionary history. *Science* 278:692–95

Nee S, Mooers AØ, Harvey PH. 1992. Tempo and mode of evolution revealed from molecular phylogenies. *Proc. Natl. Acad. Sci. USA* 89:8322–26

Neumann DA. 1983. Faithful consensus methods for *n*-trees. *Math. Biosci.* 63:271–87

Ng MP, Wormald NC. 1996. Reconstruction of rooted trees from subtrees. *Discrete Appl. Math.* 69:19–31

Novacek MJ. 1992. Mammalian phylogeny: shaking the tree. *Nature* 356:121–25

Novacek MJ. 2001. Mammalian phylogeny: genes and supertrees. *Curr. Biol.* 11:R573–75

Nunn CL, Barton RA. 2000. Allometric slopes and independent contrasts: a comparative test of Kleiber's law in primate ranging patterns. *Am. Nat.* 156:519–33

Nunn CL, Gittleman JL, Antonovics J. 2000. Promiscuity and the primate immune system. *Science* 290:1168–70

O'Brien SJ, Eizirik E, Murphy WJ. 2001.

Genomics. On choosing mammalian genomes for sequencing. *Science* 292:2264–66

O'Brien SJ, Nash WG, Wildt DE, Bush ME, Benveniste RE. 1985. A molecular solution to the riddle of the giant panda's phylogeny. *Nature* 317:140–44

Pagel MD, Harvey PH. 1988. The taxon level problem in mammalian brain size evolution: facts and artifacts. *Am. Nat.* 132:344–59

Paradis E. 1998. Detecting shifts in diversification rates without fossils. *Am. Nat.* 152:176–88

Pisani D, Yates AM, Langer MC, Benton MJ. 2002. A genus-level supertree of the Dinosauria. *Proc. R. Soc. London Ser. B.* 269:915–21

Ponstein J. 1966. *Matrices in Graph and Network Theory.* Assen, The Netherlands: Van Gorcum

Price T. 1997. Correlated evolution and independent contrasts. *Philos. Trans. R. Soc. London Ser. B* 352:519–29

Purvis A. 1995a. A composite estimate of primate phylogeny. *Philos. Trans. R. Soc. London Ser. B* 348:405–21

Purvis A. 1995b. A modification to Baum and Ragan's method for combining phylogenetic trees. *Syst. Biol.* 44:251–55

Purvis A. 1996. Using interspecies phylogenies to test macroevolutionary hypotheses. In *New Uses for New Phylogenies*, ed. PH Harvey, AJ Leigh Brown, J Maynard Smith, S Nee, pp. 153–68. Oxford: Oxford Univ. Press

Purvis A, Agapow P-M, Gittleman JL, Mace GM. 2000a. Nonrandom extinction and the loss of evolutionary history. *Science* 288:328–30

Purvis A, Gittleman JL, Cowlishaw G, Mace GM. 2000b. Predicting extinction risk in declining species. *Proc. R. Soc. London Ser. B* 267:1947–52

Purvis A, Gittleman JL, Luh H-K. 1994. Truth or consequences: effects of phylogenetic accuracy on two comparative methods. *J. Theor. Biol.* 167:293–300

Purvis A, Mace GM, Gittleman JL. 2001. Extinction risk in carnivores: a phylogenetic

approach. In *Carnivore Conservation*, ed. JL Gittleman, S Funk, D Macdonald, RK Wayne, pp. 11–34. Cambridge: Cambridge Univ. Press

Purvis A, Nee S, Harvey PH. 1995. Macroevolutionary inferences from primate phylogeny. *Proc. R. Soc. London Ser. B* 260:329–33

Purvis A, Webster AJ, Agapow P-M, Jones KE, Isaac NJG. 2003. Primate life histories and phylogeny. In *Primate Life History*, ed. PM Keppeler, M Pereira. Cambridge: Cambridge Univ. Press. In press

Pybus OG, Harvey PH. 2000. Testing macroevolutionary models using incomplete molecular phylogenies. *Proc. R. Soc. London Ser. B* 267:2267–72

Ragan MA. 1992. Phylogenetic inference based on matrix representation of trees. *Mol. Phylogenet. Evol.* 1:53–58

Rodrigo AG. 1993. A comment on Baum's method for combining phylogenetic trees. *Taxon* 42:631–66

Rodrigo AG. 1996. On combining cladograms. *Taxon* 45:267–74

Ronquist F. 1996. Matrix representation of trees, redundancy, and weighting. *Syst. Biol.* 45:247–53

Salamin N, Hodkinson TR, Savolainen V. 2002. Building supertrees: an empirical assessment using the grass family (Poaceae). *Syst. Biol.* 51:134–50

Sanderson MJ, Kim J. 2000. Parametric phylogenetics? *Syst. Biol.* 49:817–29

Sanderson MJ, Purvis A, Henze C. 1998. Phylogenetic supertrees: assembling the trees of life. *Trends Ecol. Evol.* 13:105–9

Sanderson MJ, Shaffer HB. 2002. *Annu. Rev. Ecol. Syst.* 33:49–72

Savolainen V, Chase MW, Hoot SB, Morton CM, Soltis DE, et al. 2000. Phylogenetics of flowering plants based on combined analysis of plastid *atpB* and *rbcL* gene sequences. *Syst. Biol.* 49:306–62

Sechrest W, Brooks TM, da Fonseca GAB, Konstant WR, Mittermeier RA, et al. 2002. Hotspots and the conservation of evolutionary history. *Proc. Natl. Acad. Sci. USA* 99: 2067–71

Semple C. 2002. Reconstructing minimal rooted trees. *Discrete Appl. Math.* In press

Semple C, Steel M. 2000. A supertree method for rooted trees. *Discrete Appl. Math.* 105: 147–58

Semple C, Steel M. 2002. Tree reconstruction from multi-state characters. *Adv. Appl. Math.* 28:169–84

Simberloff D, Hecht KL, McCoy ED, Connor EF. 1981. There have been no statistical tests of cladistic biogeographical hypotheses. In *Vicariance Biogeography: a Critique*, ed. G Nelson, DE Rosen, pp. 40–63. New York: Columbia Univ. Press

Slowinski JB, Page RDM. 1999. How should species phylogenies be inferred from sequence data? *Syst. Biol.* 48:814–25

Soltis PS, Soltis DE, Chase MW. 1999. Angiosperm phylogeny inferred from multiple genes as a tool for comparative biology. *Nature* 402:402–4

Springer MS, de Jong WW. 2001. Phylogenetics. Which mammalian supertree to bark up? *Science* 291:1709–11

Steel M. 1992. The complexity of reconstructing trees from qualitative characters and subtrees. *J. Classif.* 9:91–116

Steel M, Dress AWM, Böcker S. 2000. Simple but fundamental limitations on supertree and consensus tree methods. *Syst. Biol.* 49:363–68

Stoner CJ, Bininda-Emonds ORP, Caro TM. 2003. The adaptive significance of colouration in lagomorphs. *Biol. J. Linn. Soc.* In press

Swofford DL. 2002. *PAUP*. Phylogenetic Analysis Using Parsimony (*and Other Methods). Version 4.* Sunderland, MA: Sinauer

Thorley JL. 2000. *Cladistic information, leaf stability and supertree construction.* PhD thesis, Univ. Bristol, UK

Thorley JL, Page RD. 2000. RadCon: phylogenetic tree comparison and consensus. *Bioinformatics* 16:486–87

Van de Peer Y, de Wachter R. 1997. Evolutionary relationships among the eukaryotic crown taxa taking into account site-to-site rate variation in 18S rRNA. *J. Mol. Evol.* 45: 619–30

Vane-Wright RI, Humphries CJ, Williams PH. 1991. What to protect? Systematics and the agony of choice. *Biol. Conserv.* 55:235–54

Vázquez DP, Gittleman JL. 1998. Biodiversity conservation: Does phylogeny matter? *Curr. Biol.* 8:R379–81

Wayne RK, Van Valkenburgh B, O'Brien SJ. 1991. Molecular distance and divergence time in carnivores and primates. *Mol. Biol. Evol.* 8:297–319

Webb CO. 2000. Exploring the phylogenetic structure of ecological communities: an example for rain forest trees. *Am. Nat.* 156:145–55

Weiblen GD, Oyama RK, Donoghue MJ. 2000. Phylogenetic analysis of dioecy in monocotyledons. *Am. Nat.* 155:46–58

Wiens JJ. 1998. Does adding characters with missing data increase or decrease phylogenetic accuracy? *Syst. Biol.* 47:625–40

Wilkinson M, Thorley JL. 1998. Reduced supertrees. *Trends Ecol. Evol.* 13:283

Wilkinson M, Thorley JL, Littlewood DTJ, Bray RA. 2001. Towards a phylogenetic supertree of Platyhelminthes? In *Interrelationships of the Platyhelminthes*, ed. DTJ Littlewood, RA Bray, pp. 292–301. London: Taylor & Francis

Willson SJ. 2001. An error-correcting map for quartets can improve the signals for phylogenetic trees. *Mol. Biol. Evol.* 18:344–51

Wojciechowski MF, Sanderson MJ, Steele KP, Liston A. 2000. Molecular phylogeny of the "temperate herbaceous tribes" of papilionoid legumes: a supertree approach. In *Advances in Legume Systematics*, ed. P Herendeen, A Bruneau, pp. 277–98 (full tree available at: http://loco.la.asu.edu/plantbiology/faculty/wojciechowski/htm). Kew, UK: Royal Botanic Gardens

Annu. Rev. Ecol. Syst. 2002. 33:291–315
doi: 10.1146/annurev.ecolsys.33.010802.150429
First published online as a Review in Advance on August 6, 2002

HOMOGENIZATION OF FRESHWATER FAUNAS

Frank J. Rahel

*Department of Zoology and Physiology, University of Wyoming, Laramie, Wyoming
82071; email: frahel@uwyo.edu*

Key Words   biotic homogenization, introductions, extinctions, nonindigenous,
species, invasions

■ **Abstract**   Biotic homogenization is the increased similarity of biotas over time
caused by the replacement of native species with nonindigenous species, usually as a
result of introductions by humans. Homogenization is the outcome of three interacting
processes: introductions of nonnative species, extirpation of native species, and habitat
alterations that facilitate these two processes. A central aspect of the homogenization process is the ability of species to overcome natural biogeographic barriers either
through intentional transport by humans or through colonization routes created by human activities. Habitat homogenization through reservoir construction contributes to
biotic homogenization as local riverine faunas are replaced with cosmopolitan lentic
species. The homogenization process has generally increased biodiversity in most
freshwater faunas, as the establishment of new species has outpaced the extinction
of native species. There are important exceptions, however, where the establishment
of nonindigenous species has had devastating impacts on endemic species. The homogenization process appears likely to continue, although it could be slowed through
reductions in the rate of invasions and extirpations and by rehabilitating aquatic habitats
so as to favor native species.

INTRODUCTION

Biotic homogenization is the increased similarity of biotas over time caused by the
replacement of native species with nonindigenous species, usually as a result of
introductions by humans. It is an accelerating phenomenon that is a consequence
of human domination of Earth's ecosystems (Vitousek et al. 1997, McKinney
& Lockwood 2001). Biologists are concerned about homogenization because it
often results in a decline in biodiversity (McKinney & Lockwood 1999). Even
when biodiversity is enhanced through species introductions, the enhancement
often includes taxa that are already widespread, tolerant of degraded habitats, and
considered a nuisance by humans (Angermeier 1994, Pimentel et al. 2000, Scott
& Helfman 2001).

Homogenization is a complicated process because it integrates many aspects
of the biodiversity crisis such as species introductions, extirpations, and habitat alteration. The introduction of cosmopolitan species will, by itself, increase

0066-4162/02/1215-0291$14.00

291

homogenization, but this effect will be magnified if the introduced species also cause extinction of endemic species that make the aquatic system unique. An unfortunate example is the loss of almost 200 species of endemic cichlids following introduction of the predatory Nile perch *Lates nilotica* into Lake Victoria in Africa (Kaufman 1992). Habitat alteration may directly cause the extirpation of native species that cannot tolerate the new abiotic conditions. A well known example is the loss of mussel species following reservoir creation on rivers in the southeastern United States (Williams et al. 1993). But sometimes the effect of habitat alteration in causing extirpations is indirect. For example, Baltz & Moyle (1993) reported that altered abiotic conditions allow nonnative fishes to become established in California streams, and these new species then eliminate native species through competition or predation.

Much of the literature on biotic homogenization has focused on quantifying how the species composition of disjunct regions has become more similar. However, the process of homogenization extends across all levels of biological organization. For example, habitat homogenization has resulted in similar habitats across North America such as urban business districts, golf courses, canals, and warmwater reservoirs. In these human-created habitats, endemic species typically are replaced by cosmopolitan species with the result that entire ecosystems resembling each other now occur in disparate parts of the country. Blair (2001) reported that bird and butterfly assemblages from urban areas in California and Ohio were more similar to each other than they were to the native assemblages they replaced. The same phenomenon occurs in urban lakes across North America that tend to be dominated by a suite of organisms tolerant of degraded water quality, such as aquatic oligochaetes, common carp *Cyprinus carpio*, and goldfish *Carassius auratus*. Below the species level, genetic homogenization is of concern for taxa subject to artificial propagation and widespread stocking, such as many species of salmonids.

In the following sections I discuss how biotic homogenization is measured and examine three interacting mechanisms that result in homogenization: introductions, extirpations, and habitat alterations. Next, I discuss the consequences of homogenization on aquatic systems such as the replacement of unique regional biotas with cosmopolitan species, and the loss of genetic diversity for some widely cultivated species. I end with future scenarios of biotic homogenization and how the process may be slowed, although it is not likely to be stopped altogether. My review is focused on North America because much of the work on homogenization of aquatic biotas has been done there, but homogenization is a global phenomenon (Arthington 1991, Holcik 1991, Ogutu-Ohwayo & Hecky 1991, Contreras-Balderas 1999, Lodge et al. 2000a).

HOW HOMOGENIZATION IS MEASURED

Homogenization is defined as an increase in the similarity of biotas over time, and an obvious way to measure this increase is through similarity indices. A widely used index is Jaccard's coefficient of similarity, calculated as percent similarity $= [a/(a+b+c)] \times 100$ where a = number of species present in both biotas, b = number of species present only in the first biota and c = number of species present only in the second biota (Radomski & Goeman 1995, Marchetti et al. 2001). Values can range from 0% (biotas have no species in common) to 100% (biotas have identical species composition). A common approach for assessing temporal changes in homogenization is to calculate the similarity between a pair of biotas at two times, typically pre- and post-alteration by humans. If similarity has increased, then the biotas have become more homogeneous. Rahel (2000) used this approach to quantify the change in similarity of fish faunas among the 48 coterminous United States from pre-European settlement to the present.

Ecologists distinguish between α diversity (the number of species in a specific habitat) and β diversity (the turnover of species across habitats). A decrease in β diversity indicates that homogenization has occurred. Spatial turnover or β diversity can be measured by counting the number of species lost or gained as one moves from one site to another site along a spatial continuum (Russell 1999). The resultant value is then scaled to the size of the combined species pool at both sites. One formula for measuring β diversity is $T = (G + L)/\alpha$ where T is spatial turnover, G is the number of species found in the first site but not in the second, L is the number of species found in the second site but not the first, and α is the total number of species found within both. Duncan & Lockwood (2001) used this approach to examine the change in spatial turnover of fish, amphibian, and mussel species across zoogeographic zones in Tennessee. Other measures of species turnover along spatial gradients are discussed by Sheldon (1988) and Russell (1999).

Cluster analysis and ordinations are multivariate approaches used to assess similarity among a group of sampling sites. In a cluster analysis, sites with similar faunal assemblages are sorted into hierarchical groups showing a progressive increase in their similarity. This approach allows one to see relationships among all sites, not just one pair at a time. Blair (2001) used cluster analysis to show that the physical habitat and biotic communities of urban areas in California and Ohio were more similar than the original ecosystems present in these areas. Thus, these areas have experienced homogenization of both habitats and biota. In an ordination, sites are projected onto a reduced set of axes that represent gradients of community composition. Sites close together in the ordination plot are more similar in their faunal composition than sites located far apart. Jackson (2002) used this approach to examine homogenization of fish faunas in lakes following the addition of a piscivore.

CAUSES OF BIOTIC HOMOGENIZATION

Biotic homogenization is the outcome of three interacting processes: introductions of nonnative species, extirpation of native species, and habitat alterations that facilitate these two processes. Although invasions have always been a part of nature, they now occur at an accelerated rate as a result of human activities (Vermeij 1991,

Benson & Boydstun 1999, Fuller et al. 1999). Ecologists often describe the species assemblage of a local area as the result of filters that reduce the regional species pool to a subset of species that have had the opportunity to colonize the habitat, are physiologically adapted to the abiotic conditions, and have the ecological characteristics needed to interact successfully with the other species present. In such a model of community assembly, the first filter is represented by glacial events and biogeographic barriers that prevent many species from colonizing a region (Figure 1). Some of these species are physiologically and ecologically suited to the region but have not had the opportunity to realize this potential. A major effect of humans is to move species across barriers to colonization and thus to eliminate the biogeographic filter as a factor in determining the species composition of local assemblages (Figure 1). Bypassing of the biogeographic filter can occur when species are transplanted across basin divides within a region (Brown & Moyle 1997) or when species are transported to new continents or oceans (Baltz 1991, Ricciardi & MacIsaac 2000). In some cases, humans may not actively transport species but may create colonization routes that did not exist naturally. For example, shipping canals near Chicago, Illinois that link the Great Lakes with the Mississippi River have allowed the exchange of 15 species of fish and invertebrates formerly confined to just one of the basins (Kolar & Lodge 2000).

Introductions can either increase or decrease the similarity among biotas. Increased similarity occurs when the same group of species is introduced into two biotas that originally had few species in common (Scenario 1 in Figure 2). The widespread introduction of a group of common sport fish such as largemouth bass *Micropterus salmoides* and rainbow trout *Oncorhynchus mykiss* across the United States is a good example of this phenomenon (Rahel 2000). By contrast, decreased similarity occurs when different species are introduced into initially similar biotas (Scenario 2 in Figure 2). Marchetti et al. (2001) provided an example of this phenomenon for watersheds in California. Within a given ecoregion, watershed fish faunas that were initially similar because of zoogeographic and historical reasons have diverged as a result of haphazard introductions among different watersheds.

Extirpations also can increase or decrease the similarity of biotas. An increase in similarity occurs when each biota loses its unique species but retains widespread species (Scenario 3 in Figure 2). Duncan & Lockwood (2001) argue that this scenario will result in the future homogenization of amphibian, fish, and mussel faunas in Tennessee. Their rationale is that different ecoregions currently have a large number of unique species that are localized with small population sizes and thus highly vulnerable to extinction. As these species become extinct in the future, only relatively widespread and abundant species will remain. Hence the ecoregions will become more similar, even without introductions of new, cosmopolitan species. A decrease in similarity due to extirpation would happen if two biotas lose the species they have in common and retain their unique species (Scenario 4 in Figure 2). It is difficult to imagine this phenomenon occurring on a large scale because widespread species tend to be abundant and have many populations that facilitate recolonization if local extinction occurs. Thus, widespread species

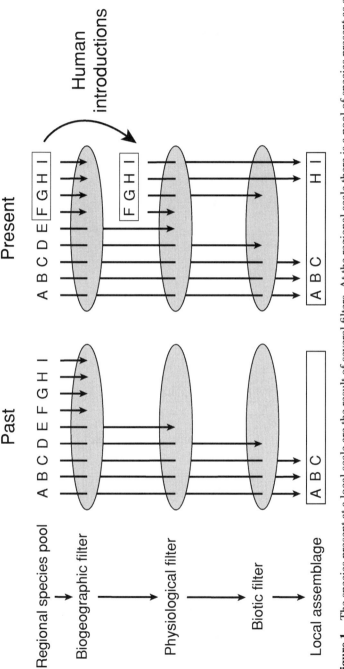

Figure 1 The species present at a local scale are the result of several filters. At the regional scale there is a pool of species present as a result of continental movement patterns and evolutionary events. Biogeographic filters such as glaciation and geographic barriers prevent some species from colonizing certain water bodies or drainage systems. Species that make it through this filter must be able to tolerate the abiotic conditions (physiological filter) and then interact successfully with the other species present (biotic filter). In the past, species such as A, B, and C that made it through all three filters comprised the local assemblage. Humans act to circumvent biogeographic filters by introducing species into areas they would not be able to colonize on their own. Introduced species such as H and I that subsequently pass through the physiological and biotic filters then become members of the local assemblage.

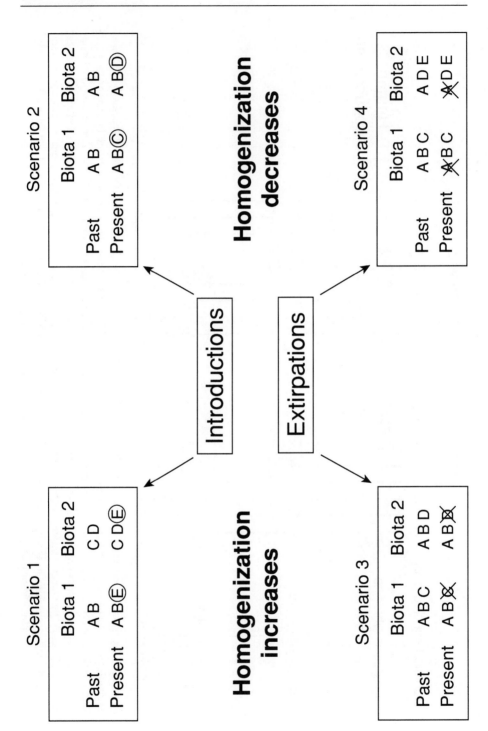

generally are not vulnerable to extinction. However, at a local scale, it is possible for two water bodies to lose some species they originally had in common and thus experience a decrease in their similarity. This scenario might be important for rehabilitation efforts involving degraded urban water bodies that currently share the same group of nonnative, pollution-tolerant species. As habitat conditions improve, cosmopolitan species such as common carp, goldfish, and bullheads may be replaced by more distinctive species native to the region (Kanehl et al. 1997).

Introductions and extirpations can facilitate each other and habitat alterations can strongly influence both (Figure 3). For example, the introduction of a predator or strong competitor can eliminate native species even when habitat remains relatively intact (Path A in Figure 3). Examples include the replacement of native trout by introduced trout in many pristine habitats in western North America (Harig et al. 2000), loss of amphibians in high elevation lakes to predation from introduced fish (Bradford et al. 1993), and replacement of native crayfish by introduced crayfish in undisturbed habitats in the midwestern United States (Lodge et al. 2000a). Conversely, loss of a native species may allow nonnatives to invade as resources and living space are freed up (Path B in Figure 3). An example is the changed fish assemblage in Lake Michigan. Overfishing and predation by sea lampreys *Petromyzon marinus* devastated the native top predators in the system, especially lake trout *Salvelinus namaycush*. The absence of predators provided an opportunity for introduction of nonnative predators such as Pacific salmon and steelhead trout *Oncorhynchus mykiss* that would use nuisance prey fish populations such as alewife *Alosa pseudoharengus* (Jude & Leach 1999). The predators were intentionally introduced by management agencies, but their success was due partly to the loss of native piscivores from the system. In a similar vein, loss of native crayfish populations in Europe as a result of introduced diseases has facilitated invasion by species introduced from North America (Lodge et al. 2000a).

Habitat alterations can lead to establishment of nonnative species (Path C in Figure 3) or extirpation of native species (Path D in Figure 3). As an example of Path C, flow regulation favors nonnative over native fishes in California streams (Marchetti & Moyle 2001). Under natural flow regimes, high flows in winter and spring prevent the establishment of nonnative species. But when flows are stabilized by dams, nonnatives dominate and displace natives through competition and predation. The role of habitat alteration in causing extirpations (Path D) is

←

Figure 2   Introductions and extirpations can either decrease or increase the similarity of biotas. Introduced species are indicated by circles and extirpated species are indicated by being crossed out. Introductions increase homogenization when the same cosmopolitan species are introduced to sites with initially distinct biotas (Scenario 1) or decrease homogenization if disparate species are introduced into sites with initially similar biotas (Scenario 2). Extirpations increase homogenization when disparate species are eliminated from biotas (Scenario 3) and decrease homogenization when common species are eliminated (Scenario 4).

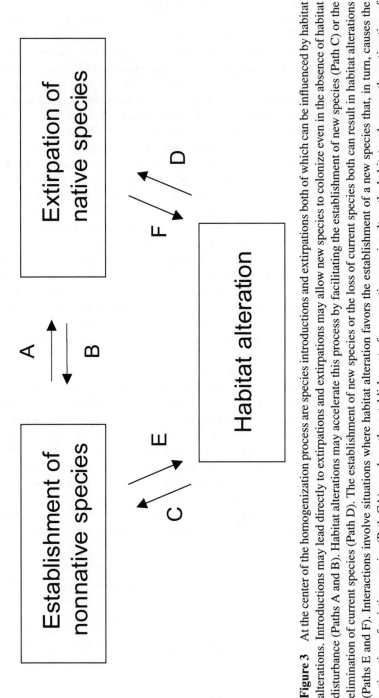

Figure 3 At the center of the homogenization process are species introductions and extirpations both of which can be influenced by habitat alterations. Introductions may lead directly to extirpations and extirpations may allow new species to colonize even in the absence of habitat disturbance (Paths A and B). Habitat alterations may accelerate this process by facilitating the establishment of new species (Path C) or the elimination of current species (Path D). The establishment of new species or the loss of current species both can result in habitat alterations (Paths E and F). Interactions involve situations where habitat alteration favors the establishment of a new species that, in turn, causes the extirpation of existing species (Path CA) or where the establishment of a nonnative species alters the habitat and causes the extirpation of native species (Path ED).

clearly documented for freshwater mussels. The extirpation of many populations in the southeastern United States is attributed to reservoir construction that eliminated the riffle-habitats required by many species (Williams et al. 1993, Bogan 1993).

An interesting situation is when establishment of nonnative species leads to a major alteration of the habitat (Path E in Figure 3). Establishment of the zebra mussel *Dreissena polymorpha* in the Great Lakes has substantially altered water clarity, primary productivity, and benthic substrates (Dermott & Kerec 1997). Because of their tremendous capacity for filtering algae and detritus from the water, energy is shifted away from the pelagic zone to the benthic zone. The result is less energy for pelagic organisms including larval fish. Because of their high numbers, zebra mussels replace complex benthic substrates with a more uniform carpet of mussel shells. This change has negative consequences for other benthic fauna, especially native mussels (Nalepa et al. 1996). The establishment of the Asian clam *Potamocorbula amurensis* in San Francisco Bay also has redirected energy pathways to the benthic zone with negative consequences for many pelagic species in this ecosystem (Grosholz 2002).

Sometimes extirpation of a native species can lead to major habitat alterations (Path F in Figure 3). An example is extirpation of the sea otter *Enhydra lutris* off the Pacific coast of North America (Estes & Palmisano 1974). Loss of otters resulted in loss of kelp forests in the near shore region. This was a major habitat alteration because the kelp forests reduced the effects of waves and provided structure for an entire fish assemblage. The otters preyed on the sea urchins and the loss of otters through overhunting allowed sea urchin populations to expand and decimate the kelp forests. Cessation of hunting allowed sea otter populations to recover, which led to a reduction in sea urchin populations and, subsequently, expansion of kelp forests. Recently, sea otter populations have declined again in some areas because of increased predation by killer whales *Orcinus orca*. The result has been another cascade of trophic interactions ultimately leading to reduction of kelp beds. (Estes et al. 1998). Beaver *Castor canadensis* provide another example where extirpation of a species causes significant habitat alteration. Beaver create small impoundments that provide habitat for a different assemblage of fishes than normally occurs in free-flowing stream reaches (Snodgrass & Meffe 1998, Schlosser & Kallemeyn 2000). Loss of beaver results in a reversion to flowing water habitat and loss of many pond-dependent taxa. Ecologists refer to species such as sea otters and beavers as ecosystem engineers because of their dominant role in determining the structure of the habitats they inhabit (Jones et al. 1994).

Introductions, extirpations and habitat alterations often will interact in complex ways. For example, habitat alteration in the California streams discussed above leads to establishment of nonnative fish species that then extirpate native species. This process corresponds to a combination of Paths C and A (Path CA in Figure 3). Loss of sea otters or beavers leads to loss of many other species that depend on habitat conditions maintained by these ecosystem engineers. This process would correspond to community changes along Path FD in Figure 3. Introduction of zebra mussels causes major habitat changes that have negative effects on many native species; hence this case would exemplify Path ED in Figure 3. A particularly insidious mechanism altering biotas is represented by Path EC in Figure 3 where the introduction of a nonnative species alters the environment to favor introductions of additional nonnative species. Simberloff & Von Holle (1999) referred to this positive feedback loop as "invasional meltdown." A recent series of invasions by euryhaline organisms from the Black and Caspian seas might be the start of an invasional meltdown in the North American Great Lakes. Ricciardi (2001) reported that establishment of large zebra mussel populations likely facilitated the rapid invasion of the round goby *Neogobius melanostomus*, a major predator of the mussel in the Caspian Sea basin. Additionally, *Echinogammarus*, a deposit-feeder associated with zebra mussels in Europe, has replaced other amphipods in zebra mussel beds in Lake Erie and Lake Ontario. And a hydroid from the Black and Caspian seas, *Cordylophora caspia*, that feeds on zebra mussel larvae and uses mussel shells as a substrate has produced luxuriant colonies on newly formed mussel beds in Lake Michigan. It seems likely that zebra mussels have altered the benthic environment of the Great Lakes so as to facilitate invasions by additional exotic species.

EVIDENCE OF BIOTIC HOMOGENIZATION IN AQUATIC SYSTEMS

Rahel (2000) reported that fish faunas across the United States have become increasingly homogenized, largely as the result of introductions of sport and food fishes. Jaccard's coefficient was used to calculate the similarity between each pair of states based on species presence and absence data. Similarity was calculated for two time periods, past and present, with the latter accounting for species extirpations and introductions that have occurred since European settlement. When the change in similarity (present minus past similarity) was determined for all 1,128 possible pairwise combinations of the 48 coterminous states, it was clear that fish faunas had become more similar (Figure 4a). Also, pairs of states averaged 15.4 more species in common now than they did in the past. The 89 pairs of states that historically had zero similarity (no species in common) now have an average similarity of 12.2% and an average of 25 species in common. The cause of homogenization was examined by calculating the change in similarity assuming only extirpations had occurred (i.e., omitting all introduced species from current state fish faunas) or that only introductions had occurred (i.e., assuming no species had been extirpated). Extirpations accounted for little of the observed change in similarity between past and present fish faunas (Figure 4b). By contrast, introductions produced big increases in similarity and were clearly the key factor homogenizing fish faunas (Figure 4c). Most of the introductions were of species intentionally introduced for sport or aquaculture purposes by management agencies. This example parallels the situation for birds, where most introductions result from intentional transport by humans (Lockwood et al. 2000).

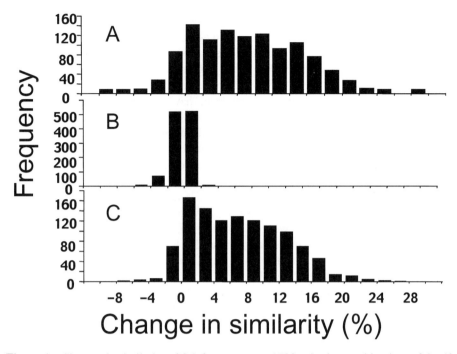

Figure 4 Changes in similarity of fish faunas among 1128 pairwise combinations of the 48 coterminous United States. (*A*) Change in similarity based on combined effects of species extirpations and introductions. Distribution is skewed toward positive values indicating fish faunas have become more similar by an average of 7.2%. (*B*) Change in similarity based on species extirpations only. Extirpations have caused a negligible change in the similarity among state fish faunas. (*C*) Change in similarity based on introductions only. Distribution resembles that in *A*, indicating most of the increased similarity in fish faunas is due to introduction of a group of cosmopolitan species. (From Rahel 2000).

Radomski & Goeman (1995) found an increase in homogenization of fish assemblages among lakes in Minnesota. They quantified the similarity of past and present fish assemblages among pairs of lakes using Jaccard's coefficient of similarity and presence and absence data from fisheries surveys. Changes in similarity were contrasted between a group of lakes subject to extensive stocking and a group with little stocking activity. Fish assemblages had become more similar among the stocked lakes but not the unstocked lakes, again implicating the role of intentional introductions in homogenizing fish faunas. Most of the stocked lakes increased their richness by one to three species. As a result of stocking activities, six species had become significantly more widespread including five game fish species and common carp. One species was less widespread, the bowfin *Amia calva*, which is piscivorous and a likely competitor with the introduced game fish species.

Marchetti et al. (2001) studied homogenization of California fish assemblages, using Jaccard's coefficient to measure the similarity of past (pre-1850) and current assemblages. A unique aspect of this study was examination of changes in similarity at three spatial scales. At the largest scale, there was an increase in the similarity of fish faunas across the six zoogeographic provinces of California. This increase was largely due to all six provinces' having gained a similar set of species not native to California. At the middle spatial scale, similarity either decreased or showed no clear pattern for fish assemblages in watersheds within each of the zoogeographic provinces. This result was attributed to the haphazard nature of introductions within individual watersheds that diversified the historically uniform within-province faunas. At the finest spatial scale, within-province comparisons were made among the fish faunas of reservoirs and the river reaches they replaced. For one province, the reservoir fish faunas were more similar today than the corresponding river reaches were in the past. No pattern was evident for the other province for which adequate data were available. To explore the relationship between habitat alteration and faunal change, Marchetti et al. (2001) examined the effects of environmental variables on the change in composition between each watershed's original (pre-1850) and current fish fauna. The degree of change between historical and current fish faunas was quantified using Jaccard's coefficient of similarity and species presence/absence data. A low similarity indicated extensive faunal change within the watershed. Extent of water development, degree of urban development, watershed area and mean elevation were positively associated with the degree of faunal change in a watershed. Only mean rainfall was negatively associated with the degree of faunal change. These results supported the idea that habitat alterations drove alteration of fish faunas at the watershed scale.

Duncan & Lockwood (2001) examined homogenization of amphibian, fish, and mussel faunas by quantifying the spatial turnover of species (beta diversity) across ecoregions in Tennessee. They first estimated spatial turnover for pre-settlement faunas prior to extirpations or introductions. This analysis provided a baseline for spatial turnover prior to significant human disturbance. Then they estimated turnover when introduced species were included but extirpated species as well as those classified as threatened, endangered, or vulnerable were omitted from the data set. The idea was to simulate spatial turnover in the future following species losses. Under the future scenarios, regional differences in species composition (spatial turnover) declined an average of 16% for fish, 21% for mussels, and 30% for amphibians. This represents a significant homogenization of aquatic faunas among ecoregions in Tennessee. In contrast to the studies discussed previously (Radomski & Goeman 1995, Rahel 2000, Marchetti et al. 2001), the future homogenization of aquatic faunas in Tennessee was predicted to result from extirpations rather than introductions. This prediction was based on the assumptions that no new species would invade and all threatened, endangered, and vulnerable species would be extirpated. Whereas these assumptions may be somewhat extreme, they do indicate that in regions with diverse faunas characterized by many endemic species with limited geographic ranges, extirpation rather than introductions will be the major homogenizing factor (Warren et al. 2000).

Scott & Helfman (2001) discussed the influence of native species invasions on the homogenization of fish faunas in the southeastern United States where headwater streams have many endemic fishes adapted to cool, clear, nutrient-poor conditions, and low sediment loads. As these streams flow into lower elevations, they are inhabited by widespread, generalist fish species adapted to warm, turbid, sediment-rich, and nutrient-rich conditions. Headwater species occur, on average, in only 3.3 drainages, whereas downstream species occur in an average of 29.5 drainages. Land use practices such as deforestation degrade stream habitats and reduce habitat diversity. As a result of this habitat homogenization, endemic headwater species are lost and the streams are invaded by generalist, downstream species. Even though the invaders are native to the drainage, they are not native to the headwater reaches. Although this process was not quantified, it seems plausible that replacement of unique headwater species with widespread downstream species will homogenize fish faunas across the region. Scott & Helfman (2001) cautioned that assessments of the integrity or conservation value of aquatic habitats should consider the effects of native invasions, as well as invasions by exotic species.

Jackson (2002) found that Ontario lakes containing bass (*Micropterus* spp.) were grouped together more closely in an ordination plot than lakes without bass. Thus, as a group, lakes with bass had more similar fish assemblages than lakes without bass. Bass were not native to the region, and all lakes were assumed to have had the same initial suite of species. The implication, therefore, was that bass had homogenized the fish assemblages by eliminating a diverse group of small-bodied prey species.

To date, much of the effort in quantifying faunal homogenization in freshwaters has focused on fish. Two other taxa that appear vulnerable to large-scale homogenization in North America are crayfish and freshwater mussels. In both cases, extirpation of many endemic species and their replacement by a few generalist species is considered likely. Lodge et al. (2000a) reviewed the status of crayfish and concluded that "the most important threat to native North American crayfish biodiversity is nonindigenous crayfishes (many from within North America)." Two species in particular, the signal crayfish (*Pacifastacus leniusculus*) and the rusty crayfish (*Orconectes rusticus*), have proven to be aggressive invaders that have contributed to the extinction of one crayfish species, the endangerment of another, and the loss of many populations of other species. There is concern that introduction of the rusty crayfish into the southeastern United States would be devastating to the many endemic crayfish found there (Taylor et al. 1996). Many of these species have small native ranges and thus are highly vulnerable to extirpation by competitively superior species such as the rusty crayfish (Hill & Lodge 1999).

Freshwater mussels in the families Margaritiferidae and Unionidae are worldwide in distribution but reach their greatest diversity in North America with 281 species and 16 subspecies (Williams et al. 1993). They are the most imperiled freshwater fauna with 12% of the taxa listed as probably extinct and 60% considered endangered, threatened, or of special concern (Ricciardi et al. 1998). During the twentieth century, the most important source of imperilment was destruction of habitat by damming, dredging, and channelization of rivers, especially in the southeastern United States. Dams were particularly harmful because of the change from riverine to reservoir habitat and the disruption of the reproductive cycle by eliminating host fish species needed to harbor the larval stages (Bogan 1993). Over-exploitation by commercial harvest also has been a concern (Anthony & Downing 2001). In the 1990s, a significant new threat was added with the establishment of the Eurasian zebra mussel, a suspension feeding organism that smothers the shells of other mollusks and competes with them for food (Nalepa et al. 1996). This species has rapidly spread throughout the Great Lakes and the Mississippi River basin, often reaching densities in excess of 3000 individuals m^{-2} and ex-tirpating native mussel species within 4–8 years. Ricciardi et al. (1998) projected that if the current rate of spread continues, the zebra mussel will accelerate the regional extinction rates of North American freshwater mussels by ten-fold and will threaten the existence of over 60 endemic species in the Mississippi River basin. The replacement of so many native mussel species by a single nonnative species would constitute an extreme case of biotic homogenization.

Another example of biotic homogenization is the establishment of over 145 nonnative species in the North American Great Lakes (Mills et al. 1993). Espe-cially noteworthy is that since 1985, 70% of the invading species are native to fresh and brackish waters of the Ponto-Caspian region (Black, Caspian, and Azov seas) (Ricciardi & Maclsaac 2000). Some of these species have achieved high abun-dance and now play a major role in Great Lake food webs. Examples include ruffe *Gymnocephalus cernuus*, round goby *Neogobius melanostomus*, tubenose goby *Proterorhinus marmoratus*, zebra mussel, quagga mussel *Dreissena bugensis*, and the amphipod *Echinogammarus ischnus*. Most of the invasions result from ship ballast water release. This invasion is decidedly one-sided, as few North Ameri-can species have invaded the Ponto-Caspian region. Still, it illustrates a situation where the introduction of diverse taxonomic groups is contributing to an increasing similarity between major water bodies located half a world apart.

CONSEQUENCES OF BIOTIC HOMOGENIZATION

Biotic homogenization results in the paradox of gaining species but losing diversity. This is because local richness often increases with the introduction of cosmopolitan species while, at the same time, regional and global diversity decrease as endemic species are driven to extinction. Consider Clear Lake, California, which originally contained 12 native fish taxa, including three endemic to the lake (Hunter 1996). As a result of efforts to increase the fish diversity in the lake, 16 species have been introduced and become established. Most are common sport fishes such as sunfishes, basses, and catfishes, and their establishment has made the fish fauna in Clear Lake more similar to fish faunas across the United States. But these introductions along with habitat alterations caused the global extinction of two native species, the Clear Lake splittail *Pogonichthys ciscoides* and the thicktail chub *Gila crassicauda*. Thus, although Clear Lake has gained 16 species, the earth's fish fauna has declined by two species. The gain in species and loss of diversity is evident at the scale of the United States where 39 species have been added to the fish fauna, all of which were already common in other parts of the world, and 19 species found nowhere else in the world have become extinct (Miller et al. 1989, Rahel 2000).

Because biotic homogenization is the combined result of introductions and extirpations, the negative consequences of both processes also apply to the process of homogenization (Tilman 1999, Mack et al. 2000). Concerns about introductions typically center on species that become pests or reduce the abundance of more desirable species. Most introduced species do not have major, detectable effects on native species, and in some cases they provide economic benefits related to sport fishing or aquaculture (Horak 1995). Mills et al. (1993) estimated that only 10% of the 139 introduced species in the North American Great Lakes have had demonstrably substantial impacts. In a review of the literature on invasive species, Lodge (1993) noted that between 2 and 40% of introduced species had an impact large enough to be detected. But the minority of introduced species that prove harmful can have major ecological and economic impacts (Pimental et al. 2000). Zebra mussels, for example, have had negative economic effects by clogging water intake structures, and they appear poised to have major ecological effects by causing the extinction of native mussel species in North America (Ricciardi et al. 1998). Common carp reduce the abundance of native fish species, and there is a long history of expensive and generally unsuccessful efforts to control them (Cooper 1987). Nile perch have contributed to the global extinction of hundreds of endemic cichlid species in the Great Lakes of Africa (Kaufman 1992). Most natural resource managers today are aware of the dangers of introducing nonnative species, and the rate of official introductions has declined (Townsend & Winterbourn 1992, Rahel 1997). However, unofficial and often illegal introductions continue to be a problem. There have been 210 instances of illegal fish species introductions within the state of Montana (Vashro 1995). And in the Pacific northwest of the United States, illegal introductions of northern pike *Esox lucius* and walleye *Stizestidion vitreum* are a concern (McMahon & Bennett 1996).

The consequences of species extinctions include lost opportunities for human use and potential negative effects on ecosystem services (Tilman 1999). Even the loss of seemingly unimportant species can be detrimental because such species may play a hidden role in supporting other species and because biodiversity provides insurance that ecosystem functions will be maintained in the face of environmental change (Yachi & Loreau 1999). As Hector et al. (2001) noted, we are far from being able to identify which subset of species is most important for the long-term health of ecosystems and whose conservation would preclude the need for other species. From a homogenization viewpoint, introductions are especially important when they cause extirpations of native species either through direct interactions (Path A in Figure 3) or by making habitats unsuitable for native species (Path ED in Figure 3).

An interesting question is whether aquatic biotas are more vulnerable to homogenization than terrestrial biotas. This question has three components: Are aquatic habitats more invasible than other habitats, are extinctions more likely for aquatic than terrestrial organisms, and are aquatic habitats more altered by human activities than terrestrial habitats? Moyle (1999) noted that many freshwater and estuarine ecosystems have been extensively invaded by nonnative species, but he did not attribute this pattern to aquatic systems being innately more invasible. Rather, Moyle felt the high frequency of successful invasions indicated (*a*) most aquatic environments have been altered by human activity, (*b*) there has been a high frequency of introductions into aquatic systems, both intentionally (sport fish) and as byproduct introductions (e.g., canal building, ballast water discharge, aquaculture operations) and (*c*) people have been highly successful introducing aquatic organisms in matching the organism to the local environment. There does appear to be a difference in extinction rates for aquatic taxa compared to terrestrial taxa. Ricciardi & Rasmussen (1999) reported that extinction rates for North American freshwater fauna (fish, crayfish, mussels, gastropods, and amphibians) were five times higher than extinction rates for terrestrial and marine fauna (birds, reptiles, land mammals, and marine mammals). Extinction rates for all taxa were projected to increase, but rates for aquatic fauna would remain higher. The high extinction rate of aquatic organisms has been attributed to the extensive deterioration of aquatic ecosystems (Richter et al. 1997, Ricciardi & Rasmussen 1999, Pringle et al. 2000). Moyle (1999) indicated that freshwater and estuarine habitats are among the most altered ecosystems on Earth because they are the ultimate sumps for watershed pollutants, their water is increasingly diverted for human use through dams and reservoirs that alter flow regimes and fragment drainages, they are the focus of most human activity, especially large cities and agriculture that degrade water quality, and their biota is subject to intense exploitation.

Homogenization of terrestrial faunas is in large part driven by human habitat alteration. This is because the biodiversity of a region is strongly linked to the diversity of habitats, and human activities tend to replace diverse natural habitats with already common agricultural and urban landscapes (Blair 2001, McKinney & Lockwood 2001). Thus, homogenization of terrestrial faunas is a consequence of spatial homogenization followed by range expansions of species adapted to human-created habitats. McKinney & Lockwood (1999) referred to this process as a few winners replacing many losers. Habitat homogenization also plays a role in homogenizing aquatic biotas. The effects of urbanization are similar in streams across North America and include flashier hydrographs, increased nutrients and toxicants, absence of woody debris, and warmer temperatures (Paul & Meyer 2001). As a result, urban streams tend to be dominated by the same suite of pollution tolerant taxa. Also, diverse riverine habitats have been replaced by reservoirs dominated by the same suite of introduced game fishes (Marchetti et al. 2001). Thus the physical convergence of habitats (abiotic homogenization) has facilitated an increase in faunal similarity (biotic homogenization) in many aquatic systems. However, habitat alteration is not always the major cause of biotic homogenization in aquatic systems. Even relatively undisturbed aquatic habitats can harbor many nonnative species as a result of human introductions (Drake & Naiman 2000, Findlay et al. 2000, Lodge et al. 2000a).

Loss of genetic diversity is another aspect of biotic homogenization. Genetic homogenization reduces the ability of species to adapt to changing environmental conditions or new diseases (Allendorf et al. 2001). In many species, there is a large spatial component to genetic variation that is lost when one or a few hatchery stocks are used to replace extirpated populations or supplement declining populations (Allendorf & Leary 1988, Nehlsen et al. 1991). Genetic homogenization is especially a problem for cultured species that are widely distributed into the natural environment through intentional stocking or escapes from aquaculture facilities (Hindar et al. 1991, Beveridge et al. 1994). Also, introduction of genotypes outside of their native range can disrupt native gene pools (Philipp et al. 1993, Bulak et al. 1995). Allendorf et al. (2001) cautioned against the genetic homogenization of trout populations that would occur if a single strain resistant to whirling disease was developed and adopted for widespread stocking throughout North America. In addition to genetic homogenization at the intraspecific level, a similar process can reduce biotic diversity at the species level. Hybridization with introduced species is thought to have been a major factor in the extinction of several fish species in North America and hybridization with introduced rainbow trout currently threatens several native trout species in the southwestern United States (Miller et al. 1989, Rhymer & Simberloff 1996).

FUTURE TRENDS IN BIOTIC HOMOGENIZATION

The increase in similarity among aquatic ecosystems will continue because the three main drivers of homogenization—species introductions, species extirpations, and habitat alteration—are likely to continue (Vitousek et al. 1997, Ricciardi & Rasmussen 1999, Fuller et al. 1999). An important issue is whether the primary consequence of homogenization will be the addition of new species with relatively minor effects on ecosystems, or mass extinctions of current species and substantial alteration of ecosystems. To date, most studies have shown an increase in biodiversity because the introduction of new species has outpaced extirpation of native species (Hobbs & Mooney 1998, Gido & Brown 1999, Rahel 2000). Additionally, most invaders become integrated without major negative effects (e.g., extirpations) on the communities being invaded. Moyle & Light (1996) felt this was true for most fish assemblages, although they noted important exceptions when the invading species was an effective piscivore such as bass or Nile perch (Kaufman 1992, Jackson 2002). Gido & Brown (1999) found that, in 80% of the drainages across North America, the number of introduced species exceeded the number of extirpated species, indicating a net increase in species richness due to species introductions. Rahel (2000) reported 4.6 introductions for every extirpation among fish faunas of the 48 coterminous United States, again supporting the view that most introductions do not result in extirpations. Griffiths (1997) found that local fish species richness was proportional to regional species richness in North American lakes. He noted that lakes do not appear to be species-saturated, suggesting that new species could be added without eliminating existing species.

Some of the increased biodiversity from homogenization may prove to be transient. Most introductions are relatively recent, and there simply may not have been enough time for the extinction process to complete its course. McKinney (2002) argued this would be the scenario for fish assemblages across the United States because extirpations would outpace introductions in the future. Scott & Helfman (2001) presented a similar scenario for upland streams in the southeastern United States, where biodiversity would initially increase because of invasions by native, downstream species but would eventually decline as endemic upland species were extirpated. These scenarios resemble the process of faunal relaxation that is hypothesized to occur when islands are created from formerly contiguous habitat (Brooks et al. 1999). The islands initially contain more species than predicted by species-areas curves but then slowly lose species until they reach a number more in line with other islands of similar size. We may have inflated the species richness in some habitats through introductions, and the result could be a return to lower levels of richness, albeit with a biota containing many nonindigenous species. Continued habitat alteration also may cause extinctions to outpace introductions as the extinction debt is paid off (Tilman et al. 1994). This seems to be the case for plants on Staten Island, New York. Robinson et al. (1994) documented a net gain of several hundred plant species through 1930, but most of this gain was erased by 1991 by the extinction of hundreds of native species attributable, in part, to habitat loss from urbanization.

Even if most introduced species are relatively harmless, a minority have disastrous effects, and we often lack the ecological knowledge to forecast whether a species will be relatively benign or will cause major environmental harm (Moyle et al. 1986). An important research agenda for conservation biology is determining whether future biodiversity will decrease or increase and how invasion/extinction processes are influenced by the initial species composition of communities, characteristics of the invading species, and the degree of habitat alteration (Moyle & Light 1996, Kolar & Lodge 2001).

WHAT CAN BE DONE TO REDUCE THE RATE OF BIOTIC HOMOGENIZATION?

What can be done to reduce the rate of biotic homogenization in freshwater habitats? Solutions involve controlling the three factors promoting homogenization: introductions, extinctions, and habitat homogenization. Agency-sponsored introductions have declined in recent years as natural resource managers have gained an awareness of the problems nonnative species can cause (Rahel 1997, Moyle 1999). Sterile hybrids are increasingly used for aquaculture or stocking, which should reduce the risk of nonnative species becoming established (Hindar et al. 1991). We are beginning to take steps to curb inadvertent introductions such as those associated with ballast water release or the pet industry, although regulatory statutes remain weak and enforcement is problematic (Locke et al. 1993, Dextrase & Coscarelli 1999). Most states and provinces regulate the use of crayfish or fish as bait, and there is an increased awareness of problems caused by bait bucket releases (Litvak & Mandrak 1999, Lodge et al. 2000b). However, illegal introductions remain a problem, and public education about the harmful effects of introduced species (Vashro 1995) or rewards for the identification of violators could help (Kaeding et al. 1996). The U.S. Geological Survey maintains a web site that offers ideas on how to dispose of unwanted pet fish humanely instead of releasing them into local waters (http://nas.er.usgs.gov/fishes/).

In addition to preventing new introductions, resource managers also are working to remove naturalized populations of nonnative species prior to reestablishing native species (Bradford et al. 1993, Thompson & Rahel 1996, Young & Harig 2001). In big river systems or large lakes, elimination of nonnatives may not be possible, but controlling their abundance helps in recovery efforts for native species (Tyus & Saunders 2000).

Preventing species extinctions is an ongoing effort (Rahel et al. 1999, Abell et al. 2000). Unfortunately, no fish species have been removed from the U.S. endangered species list because of successful recovery, although several have been removed because they became extinct (Williams et al. 1989, Young & Harig 2001). Because people often do not discriminate between native and nonnative diversity (Brown et al. 1979), resource managers must find ways to promote the appreciation of native taxa. This may be easier to accomplish with bird and mammal species that have more charisma with the public than do most aquatic taxa (Meffe & Carroll 1997). Still, efforts could be made to capitalize on regionally unique organisms. One example is the Cutt-Slam Program of the Wyoming Game and Fish Department. The program is designed to have anglers learn about the four subspecies of cutthroat trout that occur in Wyoming and to develop more appreciation and support for conservation efforts. When an angler catches all four subspecies, s/he is awarded a certificate featuring color pictures of all four subspecies. The program also promotes catch-and-release fishing to prevent mortality of the fish. Such programs encourage the public to discriminate between native and introduced fishes and to appreciate biodiversity even at the subspecies level. Partnering with the fishing public could be an important tool for preservation of native species because anglers can be an important political force.

The third approach for reducing biotic homogenization is to minimize habitat alteration and homogenization. Habitat preservation and rehabilitation is a cornerstone of most species recovery efforts and there is an increasing recognition of the need to deal with habitat issues at larger spatial scales than we have in the past (Angermeier & Schlosser 1995, Poff et al. 1997). For example, the natural-flow paradigm has been proposed as a landscape-level approach for restoring native assemblages in streams and rivers. The idea is that restoring natural flow regimes will favor native species that are better adapted to these conditions than introduced species (Minckley & Meffe 1987, Marchetti & Moyle 2001, Valdez et al. 2001). Removing dams and their associated reservoirs is another ecosystem-level approach to restoring native biodiversity (Fahlund 2000). Removal of a dam from the Milwaukee River in Wisconsin resulted in a switch from a fish assemblage dominated by pollution-tolerant nonnative species that are common in urban environments across North America to a fish assemblage characterized by regionally distinctive darter and sucker species (Kanehl et al. 1997). Cleaning up pollution also can reverse the effects of biotic homogenization. For example, a reversal of eutrophication in Lake Erie has resulted in a decline in pollution-tolerant exotic species such as common carp and goldfish; recovery of native species such as burbot *Lota lota*, lake whitefish *Coregonus clupeaformis*, and several minnow species and the reappearance of nine aquatic plants species thought to have been extirpated (Stuckey & Moore 1995, Ludsin et al. 2001). Thus habitat rehabilitation through restoration of natural flows, removal of dams, and elimination of pollution can reverse biotic homogenization. Of course, there can never be a complete reversal of biotic homogenization if endemic taxa, such as the blue pike *Stizostedion vitreum glaucum* in Lake Erie, have become globally extinct or nonnatives such as the round goby have established widespread reproducing populations.

CONCLUSION

Despite progress in dealing with the above issues, it is apparent that during the next 100 years the earth will lose many more species through human-related extinctions than it will gain through the evolutionary creation of new ones. Thus on a global scale, species richness will continue to decline (McKinney & Lockwood 1999). Continued introductions of cosmopolitan species both intentionally and inadvertently likely will continue and may even accelerate with climate change (Leach 1999). The prognosis then is for the continued homogenization of aquatic systems. With greater awareness of the problems, however, and more attention to reducing the rate of both extirpations and introductions, there is hope that the rate of biotic homogenization in freshwater systems can be lessened.

ACKNOWLEDGMENTS

I thank Amy J. Schrank, Seth M. White, Laura A. Thel, Carlos Martinez del Rio, Daniel Simberloff, and Wayne A. Hubert for comments on the manuscript. Elizabeth Ono Rahel prepared the figures.

The *Annual Review of Ecology and Systematics* is online at
http://ecolsys.annualreviews.org

LITERATURE CITED

Abell RA, Olson DM, Dinerstein E, Hurley PT, Diggs JT, et al. 2000. *Freshwater Ecoregions of North America: A Conservation Assessment*. Washington, DC: Island. 319 pp.

Allendorf FW, Leary RF. 1988. Conservation and distribution of genetic variation in a poly-typic species: the cutthroat trout. *Conserv. Biol.* 2:170–84

Allendorf FW, Spruell P, Utter FM. 2001. Whirling disease and wild trout: Darwinian fisheries management. *Fisheries* 26(5):27–29

Angermeier PL. 1994. Does biodiversity include artificial diversity? *Conserv. Biol.* 8: 600–2

Angermeier PL, Schlosser IJ. 1995. Conserving aquatic biodiversity: beyond species and populations. *Am. Fish. Soc. Symp.* 17:402–14

Anthony JL, Downing JA. 2001. Exploitation trajectory of a declining fauna: a century of freshwater mussel fisheries in North America. *Can. J. Fish. Aquat. Sci.* 58:2071–90

Arthington AH. 1991. Ecological and genetic impacts of introduced and translocated freshwater fishes in Australia. *Can. J. Fish. Aquat. Sci.* 48(Suppl. 1):33–43

Baltz DM. 1991. Introduced fishes in marine ecosystems and seas. *Biol. Conserv.* 56:151–78

Baltz DM, Moyle PB. 1993. Invasion resistance to introduced species by a native assemblage of California stream fishes. *Ecol. Appl.* 3:246–55

Benson AJ, Boydstun CP. 1999. Documenting over a century of aquatic introductions in the U.S. See Claudi & Leach 1999, pp. 1–31

Beveridge MCM, Lindsay GR, Kelly LA. 1994. Aquaculture and biodiversity. *Ambio* 23: 497–502

Blair BB. 2001. Birds and butterflies along urban gradients in two ecoregions of the United States: Is urbanization creating a homogeneous fauna? See Lockwood & McKinney 2001, pp. 33–56

Bogan AE. 1993. Freshwater bivalve extinctions (Mollusca:Unionoida): a search for causes. *Am. Zool.* 33:599–609

Bradford DF, Tabatabai F, Graber DM. 1993. Isolation of remaining populations of the native frog, *Rana muscosa*, by introduced fishes in Sequoia and Kings Canyon National Parks, California. *Conserv. Biol.* 7:882–88

Brooks TM, Pimm SL, Oyugi JO. 1999. Time lag between deforestation and bird extinction in tropical forest fragments. *Conserv. Biol.* 13:1140–50

Brown TC, Dawson C, Miller R. 1979. Interests and attitudes of metropolitan New York residents about wildlife. *N. Am. Wildl. Nat. Res. Conf.* 44:289–97

Brown LR, Moyle PB. 1997. Invading species in the Eel River, California: successes, failures, and relationships with resident species. *Environ. Biol. Fish.* 49:271–91

Bulak J, Leitner J, Hilbish T, Dunham RA. 1995. Distribution of largemouth bass genotypes in South Carolina: initial implications. *Am. Fish. Soc. Symp.* 15:226–35

Claudi R, Leach JH, eds. 1999. *Nonindigenous Freshwater Organisms*. Boca Raton, FL: Lewis

Contreras-Balderas S. 1999. Annotated checklist of introduced invasive fishes in Mexico, with examples of some recent introductions. See Claudi & Leach 1999, pp. 33–54

Cooper EL, ed. 1987. *Carp in North America*. Bethesda, MD: Am. Fish. Soc.

Dermott R, Kerec D. 1997. Changes to the deepwater benthos of eastern Lake Erie since the invasion of *Dreissena*: 1979–1993. *Can. J. Fish. Aquat. Sci.* 54:922–30

Dextrase AJ, Coscarelli MA. 1999. Intentional introductions of nonindigenous freshwater organisms in North America. See Claudi & Leach 1999, pp. 61–98

Drake DC, Naiman RJ. 2000. An evaluation of restoration efforts in fishless lakes stocked with exotic trout. *Conserv. Biol.* 14:1807–20

Duncan JR, Lockwood JL. 2001. Spatial homogenization of the aquatic fauna of Tennessee: extinction and invasion following land use change and habitat alteration. See Lockwood & McKinney 2001, pp. 245–57

Estes JA, Palmisano JF. 1974. Sea otters: their role in structuring nearshore communities. *Science* 185:1058–60

Estes JA, Tinker MT, Williams TM, Doak DF. 1998. Killer whale predation on sea otters linking oceanic and nearshore ecosystems. *Science* 282:473–76

Fahlund A. 2000. Reoperation and decommission of hydropower dams: an opportunity for river rehabilitation. See Abell et al. 2000, pp. 117–19

Findlay CS, Bert DG, Zheng L. 2000. Effect of introduced piscivores on native minnow communities in Adirondack lakes. *Can. J. Fish. Aquat. Sci.* 57:570–80

Fuller PL, Nico LG, Williams JD. 1999. Non-indigenous fishes introduced into inland waters of the United States. *Am. Fish. Soc. Spec. Publ. 27*

Gido KB, Brown JH. 1999. Invasion of North American drainages by alien fish species. *Freshw. Biol.* 42:387–99

Griffiths D. 1997. Local and regional species richness in North American lacustrine fish. *J. Anim. Ecol.* 66:49–56

Grosholz E. 2002. Ecological and evolutionary consequences of coastal invasions. *Trends Ecol. Evol.* 17:22–27

Harig AL, Fausch KD, Young MK. 2000. Factors influencing success of greenback cutthroat trout translocations. *N. Am. J. Fish. Manag.* 20:994–1004

Hector A, Joshi J, Lawler SP, Spehn EM, Wilby A. 2001. Conservation implications of the link between biodiversity and ecosystem functioning. *Oecologia* 129:624–28

Hill AM, Lodge DM. 1999. Replacement of resident crayfishes by an exotic crayfish: the roles of competition and predation. *Ecol. Appl.* 9:678–90

Hindar K, Ryman N, Utter F. 1991. Genetic effects of cultured fish on natural fish populations. *Can. J. Fish. Aquat. Sci.* 48:945–57

Hobbs RJ, Mooney HA. 1998. Broadening the extinction debate: population deletions and additions in California and western Australia. *Conserv. Biol.* 1 2:271–83

Holcik J. 1991. Fish introductions in Europe with particular reference to its central and eastern part. *Can. J. Fish. Aquat. Sci.* 48(Suppl. 1):13–23

Horak D. 1995. Native and nonnative fish species used in state fisheries management programs in the United States. *Am. Fish. Soc. Symp.* 15:61–67

Hunter ML Jr. 1996. *Fundamentals of Conservation Biology.* Cambridge, MA: Blackwell Sci. 482 pp.

Jackson DA. 2002. Ecological effects of *Micropterus* introductions: the dark side of black bass. In *Proc. Black Bass 2000 Symp.* ed. DP Philipp, M Ridgeway. Bethesda, MD: Am. Fish. Soc. In press

Jones CG, Lawton JH, Shachak M. 1994. Organisms as ecosystem engineers. *Oikos* 69:373–86

Jude DJ, Leach J. 1999. Great Lakes Fisheries. In *Inland Fisheries Management in North America*, ed. CC Kohler, WA Hubert, pp. 623–64. Bethesda, MD: Am. Fish. Soc. 2nd ed.

Kaeding LR, Boltz GD, Carty DG. 1996. Lake trout discovered in Yellowstone Lake threaten native cutthroat trout. *Fisheries* 21(3):16–20

Kanehl PD, Lyons J, Nelson JE. 1997. Changes in the habitat and fish community of the Milwaukee River, Wisconsin, following removal of the Woolen Mills dam. *N. Am. J. Fish. Manag.* 17:387–400

Kaufman L. 1992. Catastrophic change in species-rich freshwater ecosystems: the lessons of Lake Victoria. *BioScience* 42:846–58

Kolar CS, Lodge DM. 2000. Freshwater non-indigenous species: interactions with other global changes. In *Invasive Species in a Changing World*, ed. HA Mooney, RJ Hobbs, pp. 3–30. Washington, DC: Island

Kolar CS, Lodge DM. 2001. Progress in invasion biology: predicting invaders. *Trends Ecol. Evol.* 16:199–204

Leach JH. 1999. Climate change and the future distribution of aquatic organisms in North America. See Claudi & Leach 1999, pp. 399–400

Litvak MK, Mandrak NE. 1999. Baitfish trade as a vector of aquatic introductions. See Claudi & Leach 1999, pp. 163–180

Locke A, Reid DM, van Leeuwen HC, Sprules WG, Carlton JT. 1993. Ballast water exchange as a means of controlling dispersal of freshwater organisms by ships. *Can. J. Fish. Aquat. Sci.* 50:2086–93

Lockwood JL, Brooks TM, McKinney ML. 2000. Taxonomic homogenization of the global avifauna. *Anim. Conserv.* 3:27–35

Lockwood JL, McKinney ML, eds. 2001. *Biotic Homogenization.* New York: Kluwer

Lodge DM. 1993. Biological invasions: lessons for ecology. *Trends Ecol. Evol.* 8:133–37

Lodge DM, Taylor CA, Holdich DM, Skurdal J. 2000a. Nonindigenous crayfishes threaten North American freshwater biodiversity: lessons from Europe. *Fisheries* 25(8):7–20

Lodge DM, Taylor CA, Holdich DM, Skurdal J. 2000b. Reducing impacts of exotic crayfishes: new policies needed. *Fisheries* 25(8):21–23

Ludsin SA, Kershner MW, Blocksom KA, Knight RL, Stein RA. 2001. Life after death in Lake Erie: nutrient controls drive fish species richness, rehabilitation. *Ecol. Appl.* 11:731–46

Mack RN, Simberloff D, Lonsdale WM, Evans H, Clout M, Bazzaz FA. 2000. Biotic invasions: causes, epidemiology, global consequences, and control. *Ecol. Appl.* 10:689–710

Marchetti MP, Light T, Feliciano J, Armstrong T, Hogan Z, et al. 2001. Homogenization of California's fish fauna through abiotic change. See Lockwood & McKinney 2001, pp. 259–78

Marchetti MP, Moyle PB. 2001. Effects of flow regime on fish assemblages in a regulated California stream. *Ecol. Appl.* 11:530–39

McKinney ML. 2002. Do human activities raise species richness? Contrasting patterns in United States plants and fishes. *Glob. Ecol. Biogeogr. Lett.* 11:343–48

McKinney ML, Lockwood JL. 1999. Biotic homogenization: a few winners replacing many losers in the next mass extinction. *Trends Ecol. Evol.* 14:450–53

McKinney ML, Lockwood JL. 2001. Biotic homogenization: a sequential and selective process. See Lockwood & McKinney 2001, pp. 1–17

McMahon TE, Bennett DH. 1996. Walleye and northern pike: boost or bane to northwest fisheries? *Fisheries* 21(8):6–13

Meffe GK, Carroll CR. 1997. The species in conservation. In *Principles of Conservation Biology*, ed. GK Meffe, CR Carroll, pp. 57–86. Sunderland, MA: Sinauer

Miller RR, Williams JD, Williams JE. 1989. Extinctions of North American fishes during the past century. *Fisheries* 14(6):22–38

Mills EL, Leach JH, Carlton JT, Secor CL. 1993. Exotic species in the Great Lakes: A history of biotic crisis and anthropogenic introductions. *J. Great Lakes Res.* 19:1–54

Minckley WL, Meffe GK. 1987. Differential selection by flooding in stream fish communities of the arid American Southwest. In *Community and Evolutionary Ecology of North American Stream Fishes*, ed. WJ Matthews, DC Heins, pp. 93–104. Norman, OK: Univ. Okla. Press

Moyle PB. 1999. Effects of invading species on freshwater and estuarine ecosystems. In *Invasive Species and Biodiversity Management*, ed. T Sandlund, PJ Schei, A Viken, pp. 177–91. Dordrecht, The Netherlands: Kluwer

Moyle PB, Li HW, Barton BW. 1986. The Frankenstein effect: impact of introduced fishes on native fishes of North America. In *The Role of Fish Culture in Fisheries Management*, ed. R Stroud, pp. 415–26. Bethesda MD: Am. Fish. Soc.

Moyle PB, Light T. 1996. Biological invasions of fresh water: empirical rules and assembly theory. *Biol. Conserv.* 78:149–61

Nalepa TF, Hartson DJ, Gostenik GW, Fanslow DL, Lang GA. 1996. Changes in the freshwater mussel community of Lake St. Clair: from Unionoida to *Dreissena polymorpha* in eight years. *J. Great Lakes Res.* 22:354–69

Nehlsen W, Williams JE, Lichatowich JA. 1991. Pacific salmon at the crossroads: stocks at risk from California, Oregon, Idaho and Washington. *Fisheries* 16(2):4–21

Ogutu-Ohwayo R, Hecky RE. 1991. Fish introductions in Africa and some of their implications. *Can. J. Fish. Aquat. Sci.* 48(Suppl. 1): 8–12

Paul MJ, Meyer JL. 2001. Streams in the urban landscape. *Annu. Rev. Ecol. Syst.* 32:333–65

Philipp DP, Epifanio JM, Jennings MJ. 1993. Conservation genetics and current stocking practices: Are they compatible? *Fisheries* 18(12):14–16

Pimental D, Lach L, Zuniga R, Morrison D. 2000. Environmental and economic costs of nonindigenous species in the United States. *BioScience* 50:53–65

Poff NL, Allan JD, Bain MB, Karr JR, Prestegaard KL, et al. 1997. The natural flow regime: a paradigm for river conservation and restoration. *BioScience* 47:769–84

Pringle CM, Freeman MC, Freeman BJ. 2000. Regional effects of hydrologic alterations on riverine macrobiota in the new world: tropical-temperate comparisons. *BioScience* 50:807–23

Radomski PJ, Goeman TJ. 1995. The homogenizing of Minnesota lake fish assemblages. *Fisheries* 20(7):20–23

Rahel FJ. 1997. From Johnny Appleseed to Dr. Frankenstein: changing values and the legacy of fisheries management. *Fisheries* 22(8):8–9

Rahel FJ. 2000. Homogenization of fish faunas across the United States. *Science* 288:854–56

Rahel FJ, Muth RT, Carlson CA. 1999. Endangered species management. In *Inland Fisheries Management in North America*, ed. CC Kohler, WA Hubert, pp. 345–74. Bethesda, MD: Am. Fish. Soc.

Rhymer JM, Simberloff D. 1996. Extinction by hybridization and introgresion. *Annu. Rev. Ecol. Syst.* 27:83–109

Ricciardi A. 2001. Facilitative interactions among aquatic invaders: Is an "invasional meltdown" occurring in the Great Lakes? *Can. J. Fish. Aquat. Sci.* 58:2513–25

Ricciardi A, MacIsaac HJ. 2000. Recent mass invasion of the North American Great Lakes by Ponto-Caspian species. *Trends Ecol. Evol.* 15:62–65

Ricciardi A, Neves RJ, Rasmussen JB. 1998. Impending extinctions of North American freshwater mussels (Unionoida) following the zebra mussel (*Dreissena polymorpha*) invasion. *J. Anim. Ecol.* 67:613–19

Ricciardi A, Rasmussen JB. 1999. Extinction rates of North American freshwater fauna. *Conserv. Biol.* 13:1220–22

Richter BD, Braun DP, Mendelson MA, Master LL. 1997. Threats to imperiled freshwater fauna. *Conserv. Biol.* 11:1081–93

Robinson GR, Yurlina ME, Handel SN. 1994. A century of change in the Staten Island flora: ecological correlates of species losses and invasions. *Bull. Torrey Bot. Club* 121:119–29

Russell GJ. 1999. Turnover dynamics across ecological and geological scales. In *Biodiversity Dynamics: Turnover of Populations, Taxa, and Communities*, ed. ML McKinney, JA Drake, pp. 377–404. New York: Columbia Univ. Press

Schlosser IJ, Kallemeyn LW. 2000. Spatial variation in fish assemblages across a beaver-influenced successional landscape. *Ecology* 81:1371–82

Scott MC, Helfman GS. 2001. Native invasions, homogenization, and the mismeasure of integrity of fish assemblages. *Fisheries* 26(11):6–15

Sheldon AL. 1988. Conservation of stream fishes: patterns of diversity, rarity, and risk. *Conserv. Biol.* 2:149–56

Simberloff D, Von Holle B. 1999. Positive interactions of nonindigenous species: invasional meltdown? *Biol. Invasions* 1:21–32

Snodgrass JW, Meffe GK. 1998. Influence of beavers on stream fish assemblages: effects of pond age and watershed position. *Ecology* 79:928–42

Stuckey RL, Moore DL. 1995. Return and increase in abundance of aquatic flowering plants in Put-In-Bay Harbor, Lake Erie, Ohio. *Ohio J. Sci.* 95:261–66

Taylor CA, Warren ML Jr, Fitzpatrick JF Jr, Hobbs HH III, Jezerinac RF, et al. 1996. Conservation status of crayfishes of the United States and Canada. *Fisheries* 21(4):25–38

Thompson PD, Rahel FJ. 1996. Evaluation of depletion-removal electrofishing of brook trout in small Rocky Mountain streams. *N. Am. J. Fish. Manag.* 16:332–39

Tilman D. 1999. The ecological consequences of changes in biodiversity: a search for general principles. *Ecology* 80:1455–74

Tilman D, May RM, Lehman CL, Nowak MA. 1994. Habitat destruction and the extinction debt. *Nature* 371:65–66

Townsend CR, Winterbourn MJ. 1992. Assessment of the environmental risk posed by an exotic fish: the proposed introduction of channel catfish (Ictalurus punctatus) to New Zealand. *Conserv. Biol.* 6:273–82

Tyus HM, Saunders JF III. 2000. Nonnative fish control and endangered fish recovery: lessons for the Colorado River. *Fisheries* 25(9):17–24

Valdez RA, Hoffnagle TL, McIvor CC, McKinney T, Leibfried WC. 2001. Effects of a test flood on fishes of the Colorado River in Grand Canyon, Arizona. *Ecol. Appl.* 11:686–700

Vashro J. 1995. The "bucket brigade" is ruining our fisheries. *Montana Outdoors* 26(5):34–37

Vermeij GJ. 1991. When biotas meet: understanding biotic interchange. *Science* 253:1099–1104

Vitousek PM, Mooney HA, Lubchenco J, Melillo JM. 1997. Human domination of Earth's ecosystems. *Science* 277:494–499

Warren ML Jr, Burr BM, Walsh SJ, Bart HL Jr, Cashner RC, et al. 2000. Diversity, distribution, and conservation status of the native freshwater fishes of the southern United States. *Fisheries* 25(10):7–31

Williams JD, Warren ML Jr, Cummings KS, Harris JL, Neves RJ. 1993. Conservation status of freshwater mussels of the United States and Canada. *Fisheries* 18(9):6–22

Williams JE, Johnson JE, Hendrickson DA, Contreras-Balderas S, Williams JD, et al. 1989. Fishes of North America endangered, threatened, or of special concern: 1989. *Fisheries* 14(6):2–20

Yachi S, Loreau M. 1999. Biodiversity and ecosystem productivity in a fluctuating environment: the insurance hypothesis. *Proc. Natl. Acad. Sci. USA* 96:1463–68

Young MK, Harig AL. 2001. A critique of the recovery of greenback cutthroat trout. *Conserv. Biol.* 15:1575–84

Annu. Rev. Ecol. Syst. 2002. 33:317–40
doi: 10.1146/annurev.ecolsys.33.010802.150524
First published online as a Review in Advance on August 6, 2002

THE RENAISSANCE OF COMMUNITY-BASED MARINE RESOURCE MANAGEMENT IN OCEANIA

R. E. Johannes

R. E. Johannes Pty Ltd., 8 Tyndall Court, Bonnet Hill, Tasmania 7053, Australia;
email: bobjoh@netspace.net.au

Key Words marine protected areas, customary marine tenure, Pacific Island conservation

■ **Abstract** Twenty-five years ago, the centuries-old Pacific Island practice of community-based marine resource management (CBMRM) was in decline, the victim of various impacts of westernization. During the past two decades, however, this decline has reversed in various island countries. Today CBMRM continues to grow, refuting the claim that traditional non-Western attitudes toward nature cannot provide a sound foundation for contemporary natural resource management. Limited entry, marine protected areas, closed areas, closed seasons, and restrictions on damaging or overly efficient fishing methods are some of the methods being used. Factors contributing to the upsurge include a growing perception of scarcity, the restrengthening of traditional village-based authority, and marine tenure by means of legal recognition and government support, better conservation education, and increasingly effective assistance, and advice from regional and national governments and NGOs. Today's CBMRM is thus a form of cooperative management, but one in which the community still makes and acts upon most of the management decisions.

INTRODUCTION

Twenty-four years ago I published a paper in this *Annual Review* series entitled "Traditional Marine Resource Management in Oceania and Its Demise" (Johannes 1978). In it, I used historical and anthropological information to demonstrate that some tropical Pacific Island cultures invented and employed marine resource management measures centuries before the west did. These included limited entry, closed seasons, closed areas, size limits, and (albeit rarely) gear restrictions. I described how the impacts on these cultures of cash economies, export markets, new technology, and other concomitants of westernization were eroding these practices. As the article's title indicates, I thought their demise was not far off. As the title of the present article reveals, my pessimism was unwarranted.

In Oceania, said Fa'asili & Kelokolo (1999, p. 10), "regardless of legislation or enforcement, the responsible management of marine resources will only be

0066-4162/02/1215-0317$14.00 **317**

achieved when fishing communities see it as their responsibility." Today, communities spread widely throughout the region are rising to this challenge and adapting their traditional practices to fit contemporary circumstances. "Community" is used here in the broadest sense of a functional social unit; at different times and in different cultures, the most relevant social unit in connection with local marine resource management may be a group of villages, a single village, a clan, a family, or a chief or other influential individual in the community.

Judging by the literature, community-based marine resource management (CBMRM) may be more widespread in Oceania today than in any other tropical region in the world. And as Hviding & Ruddle (1991, p. 1) have said, the Pacific Island region "has much to contribute to innovative thinking about small scale fisheries management worldwide." Here I describe the revitalization of CBMRM in Oceania in the past two decades, some of the factors that led to it, and some of the lessons that are emerging.

VANUATU

A striking upsurge in CBMRM occurred in Vanuatu fishing villages beginning in 1990. In 27 villages surveyed in 1993, only 1 had not introduced MRM measures in the previous four years (Johannes 1998a). (I define an MRM measure as a measure employed consciously to reduce or eliminate overfishing or other damaging human impacts on marine resources). Enforcement was by village authorities, not the Fisheries Department.

Johannes & Hickey (2002) did a follow-up study of 21 of these villages in 2001 to gauge the success of these initiatives. Direct before-and-after measurements of the health of the reef communities and reef fisheries involved were far beyond our resources. So we determined how many of MRM measures operating in 1993 had lapsed and how many new ones had been initiated. Our reasoning was that maintaining or increasing MRM measures, which all entail short or medium-term sacrifice to fishers, would occur only if the fishers thought they were worth the longer-term benefits. The results revealed that of a total of 40 MRM measures operating in 1993, 5 had lapsed but 51 new ones had been implemented (Table 1). In short, MRM measures had more than doubled.

TABLE 1   Total numbers of marine resource management measures in 21 Vanuatu villages, 1993 and 2001[a]

	1993	2001
Total MRM measures operating	40	86
Average number per village	1.9	4.1
Lapsed MRM measures since 1993	N/A	5

[a]From Johannes & Hickey 2002.

The most often used MRM measures in 2001 were:

- fishing ground closures (18)
- trochus[1] closures (11)
- ban on taking turtles or their eggs (11)
- bêche-de-mer (sea cucumber) closures (10)
- spearfishing controls (8)
- controls on using fishing nets (7)

The main initial impetus for these developments was the Vanuatu Fisheries Department's promotion of a voluntary village-based trochus management program starting in 1990. Initially the program involved only a few fishing villages out of a total of several hundred. Selecting villages that declared an interest in obtaining their advice, the department surveyed their trochus stocks and advised them that regular several-year closures of trochus harvesting, followed by brief openings, would generate far more profit than the usual practice of harvesting more or less continually. They left it to the villagers to decide whether or not to act on this advice.

My 1993 study (Johannes 1998a) revealed that communities that followed the Fisheries department's advice on trochus management found it so profitable that other communities quickly followed suit. Moreover, observing what conservation could do for trochus stocks, many communities decided to implement their own conservation measures to protect other marine animals, including finfishes, lobsters, clams, bêche-de-mer, and crabs, as well as to ban or restrict certain overly efficient fishing practices such as night spearfishing and the use of nets, especially gillnets. One of the surveyed communities set up a marine protected area and stocked it with giant clams (*Tridacna* spp.).

I (Johannes 1998a) described how this locally funded shoestring operation has enjoyed greater success than a foreign-aid-funded fisheries development project in Vanuatu costing tens of millions of dollars.

While the Fisheries Department continued its work in the villages and broadened its scope, another potent stimulus for CBMRM emerged in 1995—a locally renowned traveling theatre group called Wan Smolbag (WSB). Operating out of the capital, Port Vila, since 1989, this group has made many village tours, putting on plays that simultaneously entertain and inform villagers about important issues such as HIV/AIDS, malaria reduction through mosquito control, etc. In 1995, the theme of the main play presented in the villages was the plight of sea turtles and the need to conserve them. The villagers were apparently receptive to this message in part because, as many informants told us, they were already aware of a marked decline in turtle numbers in their waters over the previous several decades. Conserving sea turtles has proven to be one of the most difficult conservation

---

[1]Trochus is a large marine snail, the shell of which is sold for making buttons and inlay; it is rural Vanuatu's biggest commercial marine export.

measures to persuade fishers to adopt in most tropical Pacific Islands (see World Bank 1999). Nevertheless, out of the 21 villages we surveyed, 11 had banned or restricted harvesting of turtles and turtle eggs within the past several years. None of these villages, or any others, had controlled turtle harvesting in 1993 (Johannes & Hickey 2002).

WSB also encouraged many villages to select turtle monitors to help oversee the conservation of turtles and turtle eggs in their villages. By 2001, 150 turtle monitors had been appointed in about 80 communities throughout Vanuatu. The program was so successful that WSB is training the turtle monitors to expand their efforts to encompass natural resources in general, assisted by the Department of Fisheries and other conservation organizations and foreign-aid sources.

In addition, when national marine conservation regulations[2] were explained to villagers by the Fisheries Department and were perceived by them to coincide with village interests, the regulations were often incorporated into their own management. This adoption greatly enhanced the observance of these regulations according to many informants (Johannes & Hickey 2002). Ignorance of these laws and their purposes had previously been widespread in rural Vanuatu.

Customary marine tenure (CMT)—the right to control access to and actions on one's traditional nearshore fishing grounds—remains generally strong in Vanuatu's villages and is recognized in the country's constitution (Amos 1993). CMT provides villagers—here and elsewhere in Oceania (see below) with the critical incentive to make MRM measures and enforce them. This is because the enhanced resources that result cannot be harvested by outsiders without permission and payment or reciprocal resource-access agreements.

SAMOA

A rapid increase in CBMRM also occurred in Samoa (formerly Western Samoa) in the 1990s. The Samoa Fisheries Division provided the impetus. It began by helping to design and implement a legal device that allowed villagers to overcome their inability to prevent poaching on their fishing grounds.

Samoa once had a strong CMT system (von Bulow 1902). But ownership of marine waters was transferred to the state during colonial rule (Fairbairn 1992). In recent decades, as a consequence, problems arose in connection with enforcement of custom-based fishing regulations by village authorities. Whereas the chiefs could generally control the actions of their own villagers, it became increasingly hard to control the actions of outsiders, especially fishers from neighboring villages.

In earlier times when seafood stocks were abundant, reciprocal rights of access to tenured fishing grounds had often been accorded to neighboring villages. But

---

[2]These laws set size limits on trochus, crayfish, and green snail, and ban taking turtle eggs or crayfish with eggs, or using poisons or explosives for fishing.

in recent decades, the pressures of expanding populations and depleted marine resources prompted many villages to try to withdraw access to their fishing grounds by outside fishers (Fairbairn 1992).

Some of the latter argued successfully in court cases, however, that since the area from the high water out to sea was legally public domain, CMT was not legally enforceable. The incentive of villagers to manage their traditional fishing grounds was thus undercut. Because outsiders could come into their fishing grounds and fish at will under the protection of national law, villagers had little incentive to manage and conserve their marine stocks.

Legal steps were taken to address this problem beginning in the late 1980s. With the passage of the Fisheries Act (1988), any village regulation concerning its nearshore fishing grounds could now become a legally recognized bylaw after consultation with and acceptance by the Fisheries Division and gazetting by the Legislative Assembly. Traditional authority was further reinforced by the Village Fono (council of chiefs) Bill (1990), which amended the constitution to provide for the exercise of chiefly authority in accordance with Samoan custom and to recognize the primacy of village rights, including the right to manage nearshore fisheries (Ruddle 1994, Fa'asili & Kelokolo 1999). The incentive of villagers to manage their fishing grounds was thus restored. Now, in addition to imposing traditional fines of pigs or taro on their own village transgressors, they could take formal legal action against outsiders if traditional measures did not work.

Converting village regulations into formal bylaws was no small task, however. The implications of the new laws had to be explained to the villagers, and they had to be assisted, village by village, in suitably framing their village laws and getting them gazetted. The Fisheries Division made this a major focus of its activities through the 1990s. [See King & Faasili (1998a) for a description of the methods used and problems encountered.] The results were transforming; after decades of decay, CBMRM underwent a strong revival.

In addition, as described by King & Faasili (1998b, p. 14), "when a village had proposed a reserve in an unsuitable position (e.g., an area of bare sand or coral rubble), additional scientific information was provided to encourage the community to select a more appropriate site. Some villages initially elected to have very large reserves and a few wanted to ban fishing in their entire lagoon area. In such cases, extension staff was obliged to curb over-enthusiasm, and ask the community to balance the perceived fish production advantages of a large reserve against the sociological disadvantages of banning fishing in a large proportion of the village's fishing area. In the latter case, although young men would still be able to go fishing beyond the reef, women (who traditionally collect echinoderms and mollusks in subtidal areas) and the elderly would be particularly disadvantaged in losing access to shallow-water fishing areas."

By August 1998, a total of 51 villages had marine resource management plans in place compared to a design target of 30. Of these, 46 had established village fish reserves compared to three pilot reserves initially envisaged by the Fisheries

TABLE 2   Marine resource management measures implemented by various Samoan villages[a]

Action/ Regulation	Percentage
Ban use of chemicals and dynamite to kill fish	100
Ban use of traditional plant-derived fish poisons	96
Establish small protected areas in which fishing is banned	86
Ban other destructive fishing methods (e.g., smashing corals to extract seafood)	82
Enforce (national) mesh size limits on nets	73
Ban dumping rubbish in lagoon	75
Set minimum size limit for fish	39
Ban coral collection for export	39
Ban removal of mangroves	30
Restrict or ban use of flashlights for night spearfishing	16
Ban removal of beach sand	13
Control or limit numbers of fish fences or traps	7

[a]Figures in the right-hand column indicate the percentage of 62 villages in the Samoan Fisheries Project that implemented the measures listed in the left-hand column by mid-1999 (modified from King & Faasili, 1999).

Division (Fa'asili & Kelokolo 1999). As of early 2002, there were 64 villages with Village Fisheries Management Plans. Of these, 52 had community-owned fish reserves (marine protected areas) (M. King, personal communication). Some of the management actions are summarized in Table 2.

Interviews with fishers and fish management committees in 15 villages (Australian Government Overseas Aid Program 2000) indicated that

- virtually all villages supported the concept of conservation and the establishment of reserves; other coastal villages also wished to be included in the program;

- most villagers were proud of their reserves and highlighted their use as fish dormitories, with fish aggregating in the reserves to sleep and leaving to feed during the day (prompting some villagers to complain about "their fish" leaving their waters to be caught by neighboring villages); and

- many reserves were seen as effective in improving lagoon conditions.

King & Faasili (1999, p. 4) stated, "Because the Samoan Village Fish Reserves are being managed by communities with direct interest in their success, compliance with bans on fishing is high and there are not the enforcement costs associated with national reserves." The most recent assessment of community-based management (in 59 villages) suggests that 23 communities are managing their fisheries and marine environment very well (with a score greater than 85%) and 2 are doing poorly (with a score less than 55%) (M. King, personal communication). Some communities have built watch houses and routinely use watchmen in patrol canoes to monitor illegal activity in their fishing grounds and marine-protected areas (MPAs).

COOK ISLANDS

CMT and tradition-based MRM in the Cook Islands were once highly developed, but they were largely eroded by colonial regulations, major demographic changes, and other Western impacts (well-described by Sims 1989). These practices have since been revived, however, and adapted to fit contemporary conditions in a number of notable instances.

In 1989, legislation was passed that effectively gave island councils total control over management of the living marine resources in their lagoons. CMT was then reevaluated by Cook Islanders as a means of regulating aquaculture of pearl shell, giant clam (*Tridacna* spp.), and the seaweed *Eucheuma*. In addition, periodic closures of commercial trochus and pearl shell grounds patterned after traditional area closures known as *ra'ui* were implemented. Sims (1989, p. 343) reported that such tradition-based regulations were "fairly well accepted, in contrast with the less traditional concepts of size limits and harvest quotas."

In 1982, the Manihiki cultured pearl shell farming management was transferred to the island council. Initially management was poor and over harvesting was serious (Sims 1989). But by 1992, the fishery was described as being "tightly managed" by the council (South Pacific Commission 1992). In 1994, pearl shell farming was extended to Tongareva, where its island council also managed it under a modified tenure system. As in various other Pacific Islands (Table 3), some Cook Island councils have restricted or banned spearfishing or regulated gillnet mesh size and length (e.g., Sims 1989).

Sims (1989) said there appeared to be little incentive to reinstate tenure over subsistence fisheries resources because demographic and material culture revolutions decreased reliance on them for subsistence purposes. However, in the remote atoll of Pukapuka, where subsistence fishing remains vital, Munro (1996) described a variety of MRM measures in operation, most of them recently implemented. These include a ban on fishing using explosives, bans on spearfishing within the lagoon, a ban on hunting undersized turtles or harvesting turtle eggs, areal bans on using gillnets overnight, seasonal bans on taking milkfish, and seasonal and areal bans on taking coconut crabs and seabirds.

In 1998, the traditional chiefs of Rarotonga (the most populous island and site of the capital of the Cook Islands) designated five coastal areas as marine reserves. Patterned after the traditional *ra'ui*, which had not operated on Rarotonga for four decades, they were initiated entirely by local people, with no push from outside sources such as aid donors, although the latter and the government assisted in their establishment (K. Passfield, personal communication).

By 2000, a permanent sanctuary in the lagoon had also been designated, and three more *ra'ui* were operating. Different *ra'ui* have different management arrangements. Some are opened temporarily for limited types of harvesting, such as for trochus. The results yield information on the effects of what is, in essence, a form of experimental management and are expected to lead to gradual improvements in *ra'ui* design. The *ra'ui* are repopulating quickly with some species (B. Ponia, personal communication) and are proving to be a significant tourist attraction. There seems to be widespread community support for them, judging by various local newspaper reports, but poaching by islanders returning from overseas for the Christmas holidays has been a problem.

FIJI

In Fiji, stated Veitayaki (1998, p. 57), "It is becoming abundantly clear that customary fishing area owners are taking seriously their role in the proper management of the resources within their areas." He described gillnetting closures at Kaba Point, Verata, and Macuata, banning of all commercial fishing in Lau, and fishing ground closures elsewhere. He also described how the chief and the people of Kiuva repeatedly opposed the construction of a road to their village because it would have involved clearing and draining extensive mangrove areas that provide the people's main fisheries resources (Veitayaki 1998).

Fong (1994) described a variety of CBMRM measures instituted since 1989 in Macuata Province, including bêche-de-mer closures and restrictions on gillnet use and on spearfishing using scuba. She also described a number of statements by villagers and outside commercial fishermen indicating that they believed fishing had improved significantly as a consequence of these management measures. She noted the widespread opinion among villagers that banning gillnets has proven especially effective in increasing catch-per-unit effort and numbers and sizes of fish.

Anderson (1999) reported banning or controlling gillnetting, banning night spearfishing, fishing area closures, and total fishing bans in various Fijian communities. Cooke (1994a, p. 181) mentioned community-based reef closures of one year or more in the Ba area of Fiji "in direct response to declines in relevant stocks," and taboos on dynamite fishing.

Naqasima-Sobey & Vuki (2002) described recent taboos in some Fijian villages on commercial harvesting of various invertebrates, as well as rotating fishing ground closures. They also related how the establishment of community-based MPAs is increasing and discussed some special considerations that must be addressed in order to encourage their establishment. Much of the local interest in MPAs is being generated by the awareness-raising activities of nongovernmental organizations (e.g., Calamia 2000). Villagers are also recognizing the need to protect and enhance their fisheries resources and combat beach erosion by replanting mangroves (Veitayaki 2001).

The Ueunivanua community in the Verata area closed a 24-hectare area of seagrass and mudflats to harvesting of the blood cockle (*Anadara* sp.) in 1997. After being trained by a University of South Pacific team, the villagers did their own monitoring of the impact. Within two years cockle abundance increased by 1365% and mean size also increased. In the open habitat downstream of the harvest area, cockles increased by 523% owing to increased recruitment. Remonitoring by a university team revealed that the village monitoring data were sound (A. Tawake & W. Aalbersberg, in review). The results so impressed the community that they set up additional closed areas in mangrove and reef areas to protect other species. Word of the success spread to other villages in the area, and seven of them implemented their own tabooed fishing areas. By 2000 the total protected area in these waters had increased to 7 km^2. Following local media coverage of the Ueunivanua project, similar efforts began in four other sites across Fiji, and the Ueunivanua monitoring team was in high demand to serve as trainers (Tawake et al. 2001).

According to Tawake et al. (2001, p. 35), the Ueunivanua "team presented their results to fishery policy makers in the Fijian government. After they recovered from their surprise at being given scientific findings by community members, the government policy makers embraced the idea of adopting traditional Fijian customs to manage marine resources. As a result, the government recently developed a full-time project focusing on locally managed marine reserves within Fiji's coastal waters."

To oversee village fishing regulations and reduce poaching, honorary fish wardens patrol their fishing grounds (Fong 1994). Unpaid and part time, they see this "as a natural part of their traditional service to the community" (Adams 1993). Some receive training from the Fisheries Department or NGOs such as WWF (Naqasima-Sobey & Vuki 2002). Some of their costs may be subsidized by gifts to the community from outside commercial fishermen who must seek permission to use tenured fishing grounds annually.

Cooke (1994a, p. 180) reported from a survey in the Ba area that "most owners said they considered conservation for future generations of their own people more important than deriving revenue from the resource, and even placed it above the option of optimizing revenue while conserving." The seriousness with which some Fijians take marine conservation in recent years is illustrated by Fong's (1994) account of how some chiefs decided to go further than the government regulations in restricting the harvest of bêche-de-mer in their waters.

But despite all these promising developments, nearshore marine resource management in Fiji has some serious problems that are discussed in a later section.

PALAU

In the 20 years since its adoption of a constitution and seven years since independence, Palau has evolved an awkward, complex, three-tiered system of government. Although, in theory, its constitution grants more authority to customary law than most in Oceania, in practice the system has led to its erosion, including the decline of traditional marine tenure (but not marine tenure per se) and the decline of village-based (but not community-based) marine resource management. In an excellent treatment of a complex subject, Graham & Idechong (1998) explained these distinctions and (p. 143) described how these recent changes "may portend an important shift back toward decentralized, if not exactly traditional, control over the use of Palau's inshore resources."

Palau's 16 states have an average population of a few hundred people except for the state of Koror, which contains the capital. Physically, most states consist of small village clusters rather than what are usually thought of as states.

The line between traditional and modern governance is blurring; it is not uncommon for chiefs to be represented in state governments. But although traditional authority per se is being weakened by this new political arrangement according to Graham & Idechong (1998), CBMRM is not; local authority for purposes of CBMRM is simply shifting from village leaders to state governments and may be strengthening in the process according to these authors.

Seven or more states have established one or more marine reserves and some have placed seasonal closures on a number of important reef-fish spawning-aggregation sites (Graham & Idechong 1998, Johannes et al. 1999; T. Graham, personal communication).

HAWAI'I

Hawai'i is highly cosmopolitan and native Hawaiians have tended to be marginalized economically and numerically by Asians and Europeans. Customary marine tenure, once strong, ceased to function owing to colonial impacts several generations ago (Kosaki 1954).

Yet even there a modest revival of CBMRM has been occurring. There remain a few areas where native Hawaiian communities still dominate nearshore fishing, and local seafood remains an important source of subsistence. Recognizing this fact, as well as the need for better nearshore fisheries management in the state, the Hawai'i State Legislature created a process in 1994 for designating community-based subsistence fishing areas and providing local communities with some degree of management assistance and authority.

Using this opportunity, the community in the Ho'olehu Hawaiian Homesteads on the island of Moloka'i implemented a fisheries management plan, described by Friedlander et al. (2002), in order to revitalize a locally sanctioned code of fishing. They also established a marine resource monitoring program that integrates traditional observational methods and science-based technique. They devised a novel means of circumventing a problem they could not legally control—the behavior of outsiders who use these fishing grounds, which belong to the State. The only road to these fishing grounds goes through their lands. Outside fishers who use this road must observe community fishing rules and regulations. These include closures during the season when many food fish are known to spawn, a ban on night fishing, and size restrictions (Friedlander et al. 2002, Friedlander, personal communication).

The Hawaii State Legislature also established the Kahoòlawe Island Reserve consisting of the island and its surrounding ocean waters within a two-mile radius of shore. The island and its waters can be used only for native Hawaiian cultural, spiritual, and subsistence purposes, including fishing, for environmental restoration, and for historic preservation and education.

TUVALU

The tiny nation of Tuvalu may have been the first to use bylaws for the purpose of fisheries management in Oceania. I gathered the following information during a visit in 2000. In 1979, a bylaw was passed in Funafuti Atoll banning use of gillnets of less than 1-inch-stretched mesh for catching rabbitfish. In 1980, a bylaw was passed banning fishtraps and nets in designated areas of the reef and lagoon.

According to Nukulaelae Atoll's Control of *Faapuku* and *Kaumu* Bylaw of 1984, fishing with nets or spear for *faapuku*—two species of serranids—is banned when they aggregate during their spawning season. Nukulaelae is possibly the second place in the entire Indo-Pacific to pass a modern law to protect spawning aggregations. Only Palau, as far as I know, was earlier in this regard—1977 (Johannes 2000).

DISCUSSION

Why the Renaissance?

Over the two decades since I described the decline of CBMRM in the Pacific Islands (Johannes 1978), many conditions that led to that decline have intensified. These include the spread of the cash economy, new export markets, improved harvesting and transport technology, burgeoning populations, and the decline of traditional authority.

What, then, has led to the renaissance of CBMRM in the face of such obstacles? Some of the contributing factors are not hard to deduce. One obvious factor is the perception among islanders of the growing scarcity of their marine resources owing to the demands of growing export markets and local populations (see, e.g., Fong 1994). In Marovo Lagoon, Solomon Islands, said Hviding (1989, p. 36) "the conservation of resources is a key concern for most of today's leaders—more so than for previous generations, when population density was low and resources abundant." Another factor is the income that some communities can now earn from keeping their reefs healthy in order to attract tourists (see, e.g., Calamia 2000, Johannes & Hickey 2002).

In addition, pride in one's culture is growing among many indigenous peoples, including Pacific Islanders (see, e.g., Adams 1998). One manifestation of this is that islanders are rediscovering the value of some of their natural resource management practices, albeit often in altered forms to fit contemporary circumstances.

Recent political independence is an important related factor. Most of the island countries of the western and southern Pacific gained independence during the past three decades. And, as described above, many of the resulting constitutions have granted renewed authority to traditional leaders and customary laws and processes (e.g., Ghai 1988)[3].

STRENGHTENING CMT   The renewed status that independence has given CMT in various Pacific Island countries is an especially important incentive for CBMRM. Where the ability to exclude outsiders from ones' fishing grounds is absent or weak, as noted earlier, so is the incentive to conserve ones' marine resources because outsiders can expropriate the benefits.

Recent studies in the Pacific Islands support the conventional (but not uncontested) wisdom that marine tenure plays a vital role in nearshore CBMRM[4]. For example, in Samoa, as described above, CBMRM blossomed only after the legal stumbling block that prevented villagers from excluding outsiders from their fishing grounds was removed. Eight of the 21 Vanuatu villages surveyed by Johannes & Hickey (2002) had internal disputes over control of fishing ground tenure. The mean number of MRM measures operating in these villages was less than half the number found in the 13 villages that reported no such disputes, and the difference was highly statistically significant.

The data in Table 3 further reinforce the importance of CMT for CBMRM, providing a clear contrast between the varied CBMRM measures taken in some Pacific Island countries where community-based marine tenure is secure, and the dearth of such measures in two where it no longer exists.

Kiribati

Little information is available on contemporary community-based management in the nation as a whole. But Johannes & Yeeting (2001) described the decline and disappearance of CMT on Tarawa Atoll because of past colonial government actions. The incentive of fishing communities to conserve is minimal since the law does not recognize their traditional rights to prevent outsiders from taking what they leave unharvested so it can breed or grow. There is, however, a growing push from local communities to legally formalize rights to their surrounding fishing areas to facilitate CBMRM.

A first attempt at CBMRM was initiated recently in Buariki village in order to try to stop overexploitation of bonefish spawning runs in its waters. Although its villagers have no legal authority from the national government to do this, they were able to get some support through their island council by incorporating restrictions in the island council bylaws. Although this attempt is not binding legally, it is a start.

---

[3]As Ghai (1988) and Graham & Idechong (1998) pointed out, however, these constitutions are not always an umixed blessing for customary authority—they sometimes weaken its power even when nominally intending to strengthen it.

[4]The first published article in which the importance of CMT for fisheries management was recognized in the Pacific Islands is 25 years old (Johannes 1977). Pacific Island fisheries managers (who were, in those days, almost entirely colonial expatriates) considered CMT to be nuisance, an impediment to fisheries development—if they considered it at all.

TABLE 3   Customary marine tenure and some community-based fisheries management measures found in various Pacific Islands today

	Palau	Cook Islands	Solomon Islands	Fiji	Samoa	Vanuatu	Tonga	Tarawa
Customary marine tenure	•	•	•	•	•	•	—	—
Spearfishing restrictions	•	•	•	•	•	•	—	—
Netting restrictions	*	•	•	•	•	•	—	—
Destructive fishing methods ban	*	•	•	•	•	•	—	—
Marine protected areas	•	•	—	•	•	•	Δ	—
Periodic closures–species or areas	•	•	•	•	—	•	—	π

•—community law; *—national law enforced by the community; ◊—de facto (see text); Δ—national law–poorly observed; π—not legal but tolerated (see text).

Sources: Palau: Graham & Idechong 1998, Johannes 2002; Cook Islands: Sims 1989; Solomon Islands: Hviding 1998; Fiji: Veitayaki 1998, Fong 1994, Cooke & Moce 1995; Samoa: Fa'asili & Kelokolo 1999, King & Fa'asili 1998b; Vanuatu: Johannes & Hickey 2002; Tonga: World Bank 1999; Tarawa: Johannes & Yeeting 2001.

Moreover, it seems to be effective. The Kiribati Fisheries Department recognizes the need for some kind of management of this important but threatened fishery but has been unable to do it themselves. So they encourage this initiative as a first step before getting proper national legislative support, which can often be a long process (B. Yeeting, personal communication).

Tonga

CMT disappeared generations ago (Malm 2001). Tonga is also devoid of CBMRM (Table 3). Perminow (1996, cited in Malm 2001) provides an example of the effect of the absence of community fishing rights in Tonga. Although fishers on Kotu island in the Ha'apai group knew that the increasingly intense exploitation of lagoon species and invertebrates for sale might be too taxing on the lagoon resources to be sustainable, they felt that there was no point in reducing the intensity of exploitation because the resources could be exploited by fishermen from other islands in the district. Arguments have been advanced to introduce some legal form of community-based control over local fishing grounds in Tonga (e.g., Pelelo et al. 1995).

EDUCATION FOR CBMRM   In the past 20 years, government fisheries managers in Oceania have also come to recognize that the research required to manage complex fisheries on a rigorous scientific basis is far beyond their (or anyone else's) abilities. Centrally based government management is often too expensive to justify the cost in any event (Johannes 1998b). In many circumstances, therefore, nearshore fisheries resources must be managed largely by villagers.

There are, to be sure, important exceptions. For example, central governments can usually best monitor compliance with certain marine resource regulations (e.g., trochus size limits at collection or shipping points, species export bans at airports and ports) as well as license foreign fishing operations.

To foster greater reliance on CBMRM, some government fisheries departments have directed increasing effort into appropriate extension work, some of which has been described above. These efforts include education, which has clearly helped influence many communities to pursue CBMRM.

Those providing education for better natural resource management in the villages include not just fisheries departments, moreover, but importantly the SPC Coastal Fisheries Program, including its new Community Fisheries Section, various NGOs, and the University of the South Pacific. Better education is also taking the form of improved environmental curricula in schools and the increased access that promising island students have to overseas training in the University of the South Pacific Marine Studies Program, the University of Guam Marine Laboratory, and beyond. Without such support, the current growth in CBMRM would decelerate and probably even decline.

The public are not the only ones who can benefit from more education concerning nearshore fisheries in Oceania. National governments need to realize that, in almost every Pacific Island country, subsistence fisheries are worth more than nearshore commercial fisheries (Dalzell et al. 1996)[5]. On economic grounds, then, extension work in rural fishing communities deserves a larger proportion of fisheries funding than it usually gets. Commercial fisheries usually receive disproportionate attention when island politicians and aid donors decide on funding priorities.

COMPENSATING FOR MANAGEMENT RESTRICTIONS   King & Faasili (1998a, p. 37) asserted, it is "unreasonable to expect communities to adopt conservation measures, which would (at least in the short term) reduce present catches of seafood even further, without offering alternatives." This is not always so. But it is most likely to be true where inshore resources are severely overfished as in Samoa, to which the above authors were referring, or on atolls with depleted marine resources and extremely limited terrestrial resources.

Some of the cases described above do indicate a willingness to make short-term sacrifices even without outside assistance with alternative employment. This is more likely to happen when alternative fishing grounds, perhaps further from the village, or unused agricultural lands are available for cultivation in order to compensate for restricted fishing (e.g., Johannes & Hickey 2001). In short, some alternative source of income, either from underused community sources or external sources, is usually essential if fishing is to be effectively restricted for management purposes.

---

[5]The value of the subsistence catch was calculated by these authors as the price it would fetch if it were sold. In the absence of this catch, precious foreign exchange funds would often have to be used to replace it in island diets.

Many early attempts by fisheries departments to provide alternative income sources in Pacific fishing villages failed (see, e.g., Johannes 1998a, Veitayaki 2000). But as experience increased, there have been more successes. In Samoa, for example, the diversion of fishing pressure to areas immediately beyond the reefs through the introduction of medium-sized, low-cost boats has been useful. Encouraging aquaculture, although often fraught with problems (see, e.g., Veitayaki 2000), sometimes worked well, as did pearl culture in the Cook Islands. Fish aggregation devices (FADs) are used with increasing effectiveness to redirect some of the pressure on nearshore marine resources to less heavily exploited offshore stocks (SPC Fish Aggregation Device Information Bulletin).

MAINTAINING FLEXIBILITY OF TRADITIONAL MANAGEMENT   CBMRM in Oceania usually implies the operation of customary law. The articulation of customary law and modern, Western-based law is an exceedingly complex and troublesome issue in the region (see, e.g., Ghai 1988, Adams 1998, Graham & Idechong 1998). Among the strengths of traditional law are its culture-specific and locale-specific nature; needs and customs can differ greatly from village to village even on single islands[6]. The statement by Graham & Idechong (1998, p. 150) that Palauan custom is "anything but certain, general, fixed, and uniform" can be generalized to Oceania as a whole.

The ability of traditional laws to adapt quickly to meet changing circumstances is also especially useful in the fast-changing world of Oceania during the past few decades (e.g., Hviding 1998).

National Western-based laws typically possess neither of these attributes; their uniformity and slowness to change is based more on the need to grease the wheels of commerce rather than for community harmony or equity. Thus, explicit and detailed codification in national legislation of customary natural resource law runs the risk of homogenizing and freezing it (e.g., Ruddle et al. 1992, McKinnon 1993). Graham (1994, p. 6) argued, however, that "The question facing governments that want to keep traditional management systems intact is not whether or not to codify; it is to what degree to codify" (see also Aswani 1997). Recent developments have proven Graham right.

Despite the formal recognition of traditional authority and village-based customary laws in the constitutions of the region, the ability of some village leaders to enforce village laws is, in fact, weakening (e.g., Graham & Idechong 1998), especially around urban centers (Johannes & Hickey 2002). Many village leaders

---

[6]Adams (1998, p. 139) discussed the advantages of village-to-village differences in CBMRM: "Community management goals may not differ only from those of government, but from those of other communities. This diversity of approach is actually one of the main advantages of community-based decision making for artisanal fisheries in the Pacific Islands. Even if there are cases where excessive exploitation, or unwise leadership occurs, if responsibility is sufficiently fine-grained there was will always be converse cases. ... The effect over the fisheries of the nation as a whole will tend to be more stable than in cases where a centralized government attempts to experiment in maintaining a sustainable fishery by manipulating the rules across the country as a whole."

have, accordingly, moved from avoiding recourse to national law except in extreme circumstances (see, e.g., Johannes 1998a) to seeking ways of enlisting national law to shore up their authority and to back up specific village laws with formal legal recognition (see, e.g., Fa'asili & Kelokolo 1999, Johannes & Hickey 2002).

The introduction of bylaws is proving to be an especially valuable strategy in this connection. Bylaws are selected village regulations that are accorded legal recognition in national courts of law. They are a means of legally sanctioning these regulations without encasing the entire body of customary law in legislative concrete.

Nowhere have bylaws been employed with greater effect in the context of CBMRM in Oceania than in Samoa. There, as described above, it was the introduction in the mid-1990s of the bylaw system that more than any other action precipitated the upsurge in CBMRM. The flexibility of village laws is preserved; from time to time these bylaws may be altered or revoked as required. The heterogeneity of village regulations is also preserved; each village has its own set of bylaws to fit its particular needs.

The Vanuatu Government is also looking into enacting appropriate legislation to strengthen traditional authority. In some villages in the meantime, government police officers are informally backing up the chiefs when necessary (Johannes & Hickey 2002).

The Cook Islands has also adopted the bylaw approach to CBMRM (Adams 1998). Fong (1994) recommended a similar approach for Fiji. In Isabel Province, Solomon Islands, the use of ordinances (similar to bylaws) to provide legislative support for customary owners of natural resources who wish to manage them more effectively is discussed by Peart (1993).

No longer possessing CMT, the people of Kiribati are currently experimenting with bylaws in order to give fishing communities some new form of fishing ground tenure coupled with the ability to manage their marine resources. Bylaws are enabling the people of Abaiang, for example, to regulate a live reef food fish export operation (B. Yeeting, personal communication).

Graham & Idechong (1998) noted a precedent for the use of a similar strategy in Palau. Here a state legislature passed a law that recognized a marine reserve established by local chiefs, strengthening their authority to manage it, especially (as with the Samoan bylaws discussed above) in connection with their ability to control the actions of outsiders. But this law also reflects the new states' growing exercise of their authority to regulate local marine resource use, which, as described above, amounts to a new, nontraditional form of CBMRM. (Recall that most states in Palau consist of small village clusters.)

The differences between the Western free-enterprise concept of natural resource ownership and islanders' quite different concept of communal natural resource tenure can create difficulties. The sea (like the land) is a perpetual source of sustenance rather than a commodity among Pacific Islanders; it is an integral part of their culture, and the depth of their emotional attachment is much greater than many Westerners can easily comprehend. Under these circumstances the sale of the shallow marine areas, like the sale of land, is often virtually unthinkable, and many Pacific Island countries have laws against the sale of tenured land or nearshore waters. These laws have made it difficult for companies to invest in activities such as aquaculture in tenured waters and thus retail development. A solution increasingly used is long-term leasing. This approach makes available marine areas for commercial development while protecting communal tenure and allowing for the maintenance of those aspects of traditional management and use agreed upon by lessors and lessees. A lease agreement might stipulate, for example, that the leased waters surrounding an aquaculture facility would still be available for community fishing and subject to management restrictions such as closed seasons, exclusion of outside fishers, etc.

FIRMER CONTROL OVER OUTSIDE FISHERIES OPERATORS   In the 1990s, outside companies were increasingly required by governments to negotiate formal agreements with villagers, to guarantee their adherence to various measures to protect marine resources before operating in tenured waters. Often government fisheries personnel or NGOs assisted in these negotiations. In connection with live reef food fish export operations, for example, such agreements are used in Fiji (Yeeting 1999), Papua New Guinea (Gisawa & Lokani 2001), Solomon Islands (Johannes & Lam 1999), Vanuatu (Naviti & Hickey 2001), and Palau (Graham 2001).

Other communities have simply banned enterprises perceived as threats to their marine resources. In Solomon Islands, for example, some traditional leaders have closed their tenured lagoon waters to tuna-bait fishing by transnational companies (Hviding 1989). Some communities have also unceremoniously evicted companies that did not adhere to their fishing agreements (e.g., Naviti & Hickey 2001, Johannes & Riepen 1995).

SELF-MONITORING OF MANAGEMENT IMPACTS   One recent element to emerge in the moves to reinvigorate CBMRM is systematic data gathering by trained community members to determine the impacts their management measures are having on their marine resources. Above, I described the successful community-based monitoring of protected clam stocks in Fiji and the monitoring of marine resources by a Hawaiian community. Communities in some other Pacific Islands are planniing similar activities (see, e.g., Tawake et al. 2001).

BETTER CRITERIA FOR GOVERNMENT ASSISTANCE   Government fisheries personnel in Vanuatu and Samoa now accord disputatious villages low priority in providing assistance. These are relatively new policies; it will be instructive to see if they provide a useful incentive for such villages to settle their differences. It should, at a minimum, help reduce government money and effort spent on projects that fail.

The Downside

The trend toward reinvigorated CBMRM in various Pacific Islands is no cause for complacency. Indeed, many islanders themselves are under no illusion that the future of their inshore fisheries is secure. A survey of community fishes in Fiji,

Kiribati, Palau, Samoa, Solomon Islands, and Tonga (World Bank 1999) revealed that only Samoans believed that the condition of their coastal resources would improve in future.

One problem relates to the size of tenured fishing grounds and of marine reserves. Fiji's CMT units seem especially problematic in this regard. Many groups have exclusive rights to territories far from their adjacent waters and sometimes separated from the rights-holding communities by waters belonging to other social groups (Ruddle 1995). Moreover, Fiji's 411 legally recognized customary fishing rights areas range in area from 1 km^2 to a whopping 5000 km^2 (Cooke 1994b, cited in Ruddle 1995). Their sizes and the quantities of resources contained therein are, in addition, only weakly related to the sizes of the populations that depend upon them. Some readjustments in the boundaries, sizes, and locations of fishing rights areas, where politically feasible, might be desirable. The district of Sasa has already combined four of its fishing rights areas into one (Fong 1994) for greater equitability of access.

The problems of enforcement in large Fijian customary fishing grounds are exacerbated by the extensive fishing carried out in them by commercial fishermen with no cultural link to their owners, and thus with a lesser commitment to long-term sustainability of their resources. Many disputes arise relating to revenue distribution in connection with commercial fishing (see, e.g., Lagibalavu 1994, Ledua 1995). Policing is difficult; physical clashes are not uncommon. In one case, wardens who reported fishermen involved in blast-fishing had their own boats blown up (Cooke & Moce 1995). Because of these and other problems, said Ruddle (1995, p. 10), "this seemingly straightforward and modern management of traditional rights areas is, in reality, confused and emotionally charged."

The locations and sizes of tenured fishing grounds in Oceania are not based on biologically optimal management units but on historical developments and geographic features. [It must be stressed, however, that there is little consensus among biologists on how to select ideal unit sizes or locations for these extremely complex nearshore fisheries (e.g., Adams 1998).] Some flexibility is afforded by the collective action of owners of two or more adjacent fishing grounds. In Palau, for example, the chiefs of Kayangel and Ngerchelongl decided to share some of their communities' fishing grounds and jointly close to fishing certain reef channels known to be important spawning aggregation sites (Graham & Idechong 1998).

Even large tenured fishing grounds are too small to protect species that routinely migrate beyond their limits. The incentive to moderate harvest of such species on one's fishing grounds is reduced because they can simply be caught by fishers further along their migration path. This problem has led, for example, to serious depletion of mullet on their spawning migrations in various parts of Oceania (e.g., Johannes & Hickey 2002).

Community-based reserves are sometimes too small or poorly located to be very effective in conserving finfish (see, for example, Samoa, above). The sizes of marine reserves must usually be constrained by the size of the tenured fishing grounds in which they are to be placed (e.g., Adams 1998). Small reserves may suffice, however, for managing stocks of less mobile invertebrates if they are appropriately sited (e.g., A. Tawake & W. Aalbersberg, in review).

One obvious criterion of the suitability of a fishing rights area is how well it can be policed. Surveillance is difficult in fishing grounds that extend far from shore or far from the fishing communities to which they belong, as in parts of Fiji.

Natural resource rape by some leaders—from village leaders to those high up in national governments—is on the increase in parts of Oceania. Political "irregularities" in the selloff of some islands' natural resources especially to overseas interests, is widely discussed in the regional media. It is fairly well documented in relation to Melanesia's forest resources (e.g., Barlow & Winduo 1997). But it has not been adequately studied in connection with the regions's marine resources (e.g., Adams 1998, Johannes 1999). The live reef food fish trade in some island countries is ripe for this kind of of exploitation.

The secretary to Papua New Guinea's Department of Fisheries and Marine Resources reportedly told his law-enforcement officers that within three months of his appointment he had been offered a total of US$23,000 in bribes (which he did not accept) (Fisheries 'Bribes' in PNG. *South Seas Digest* 13(7): June 18, 1993).

There are also growing numbers of reports of leaders viewing the profits to be obtained from access fees to their villages' marine resources as theirs, not their village's, leading to the erosion in some areas of traditional principles of redistribution (e.g., Schug 1996).

Additional problems include the following.

- Although some island fisheries departments actively engage fishing communities and effectively catalyze CBMRM, as in Vanuatu and Samoa (see above), others do not (e.g., Adams 1998). A World Bank study (1999) revealed that only about one fourth of the staff time of national fisheries agencies in Oceania is spent on coastal management matters. Only about 40% of the villages in the Bank study team survey had been visited by a government official to discuss coastal resource management issues during the previous 10 years. Many villages do not as yet participate in the new CBMRM

- As Fong (1994) pointed out, good community leadership is critical to good CBMRM (see also World Bank 1999). I have seen more than one CBMRM regime decline when an ineffective leader replaced a good one. But the process works both ways; moribund CBMRM can be rejuvenated when strong leadership succeeds weak.

CONCLUSIONS

Young professionals in Oceania are sometimes discouraged when faced with fisheries management systems that are less than perfect, unaware, perhaps, that there is no other kind. Given the parlous state of fisheries management around the world, the growth of CBMRM in Oceania, despite its imperfections, should be a source of optimism.

The enthusiasm for the CBMRM in Oceania may surprise some tropical small-scale fisheries experts working elsewhere. For those who are fighting uphill battles to introduce community-based controls, such as MPAs, in various parts of the world, the situation in Samoa, for example, can only be greeted with envy. Not only, as noted above, did villagers establish 46 small fishing reserves instead of the 3 that were initially hoped for by the Fisheries Department, but also the Department was "often obliged to curb overenthusiasm for (impractically) large MPAs" (King & Faasili 1999, p. 37).

Space does not permit an adequate appraisal of why these CBMRM successes seem more common today in Oceania than in many other tropical regions. But clearly one important factor is the widespread existence of CMT and its formal recognition by various Pacific Island governments.

RESEARCH POTENTIALS OF CBMRM   Adams (1998, p. 139) said "given the current state of tropical fisheries ecology, let alone tropical fisheries management, small-island governments are obliged to experiment wildly when asked to balance cash-economy development against the hope of sustainability." In the small, often well-demarcated fishing grounds that characterize many nearshore fisheries in Oceania, there is much scope for experimental management research as proposed by Walters & Hilborn (1976). As described above, many such experiments are being carried out today, but few are being rigorously monitored (the clam closure in Fiji, discussed above, is one of the exceptions).

Adams (1996, p. 342) pointed out, "Local communities observe the results of their management experiments and adapt accordingly, whilst researchers observe the results of their own management experiments and amend theory accordingly. Scientists however, rarely observe the results of local community adaptations. Whilst these adaptations are not scientific, they can provide a very cost-effective source of information for the formulation of hypotheses to be tested scientifically."

To learn from these management experiments all researchers need to do is identify experimenting villages where village authority is strong and respected (so the experiment is more likely to proceed smoothly)—and monitor the results or train community members to monitor them (see, e.g., Tawake & Aalbersberg 2002) (after obtaining permission from appropriate leaders using culturally appropriate approaches). Ready-made controls are available in some areas—that is, open or unregulated fishing grounds and closed or regulated fishing grounds with very similar environmental features may often be found adjacent to one another. Vanuatu, Samoa, and Fiji seem especially attractive for such research.

For the Doubters

As Hviding & Ruddle (1991, p. 10) have noted, "far from being overwhelmed by commercialization and resource scarcity, many CMT systems in Oceania appear to have considerable capacity for handling and adapting to new circumstances, thereby becoming potentially important tools in the contemporary management of fisheries and of the coastal zone in general."

A small but destructive group of anthropologists maintains, however, that building contemporary conservation on traditional natural resource management is bound to fail because of differences between Western and indigenous concepts of nature. This is an astonishing generalization, coming as it does from a profession that normally serves to restrain Western ethnocentrism, for it implies that only Westerners are capable of deducing the connection between harvesting pressure and natural resource availability. No one has been more outspoken on this issue than Dwyer (1994, p. 91) who has claimed that, "To represent indigenous management systems as being well-suited to the needs of modern conservation, or as founded on the same ethic, is both facile and wrong." This opinion arises from generalizing too freely from experience gained in certain cultures for which the statement may well be true.

But some Pacific Island fishing cultures have long recognized the relationship between fishing pressure and the state of their fish stocks and have regulated their fishing accordingly. Their traditional management systems not only predated Western ones by centuries (Johannes 1978, in press) but also today provide an invaluable, adaptive foundation for the renaissance in CBMRM in Oceania.

ACKNOWLEDGMENTS

I am grateful to Tom Graham and Tim Adams for very constructively reviewing an earlier draft of this paper. Thanks also to Kelvin Passfield, Ben Ponia, Mike King, Beeing Yeeting, and Alan Friedlander for information on various points.

The *Annual Review of Ecology and Systematics* is online at
http://ecolsys.annualreviews.org

LITERATURE CITED

Adams TJ. 1993. Forthcoming changes to the legal status of traditional fishing rights in Fiji. *SPC Tradit. Mar. Res. Manag. Knowledge Inf. Bull.* 2:21–22

Adams TJ. 1996. Modern institutional framework for reef fisheries management. In *Reef Fisheries*, ed. N Polunin, C Roberts, pp. 337–60. London: Chapman & Hall. 477 pp.

Adams TJ. 1998. The interface between traditional and modern methods of fishery management in the Pacific Islands. *Ocean Coast. Manag.* 40:127–42

Amos M. 1993. Traditionally based marine management systems in Vanuatu. *Tradit.*

Mar. Res. Manag. Knowledge Inf. Bull. 2:14–17

Anderson JA. 1999. Project background and research methods. In *The Performance of Customary Marine Tenure in the Management of Community Fishery Resources in Melanesia*, ed. JA Anderson, CC Mees. *Final Tech. Rep. UK Dep. Int. Dev. 1*. London: MRAG

Aswani S. 1997. Troubled water in SW New Georgia: Is codification of the commons a viable venue for resource use regularisation? *SPC Tradit. Mar. Res. Manag. Knowledge Inf. Bull.* 8:2–16

Aust. Gov. Overseas Aid Program. 2000.

Increasing rural incomes: an evaluation of the rural sector projects in Samoa. *Aust. Gov. Overseas Aid Program Qual. Assur. Ser.* 19:40–61

Barlow K, Winduo S, eds. 1997. *Logging the Southwestern Pacific: Perspective from Papua New Guinea, Solomon Islands, and Vanuatu. Contemp. Pac.* 9(1):1–193

Calamia MA. 2000. *Power, money, and tradition in the establishment of marine protected areas: lessons from the Fiji Islands.* Presented at Am. Anthropol. Assoc. Annu. Meet. San Francisco, CA, Nov. 15–19

Cooke A. 1994a. The qoliqoli of Fiji—some preliminary research findings in relation to management by customary owners. See South et al. 1994, pp. 179–82

Cooke A. 1994b. *The qoliqoli of Fiji: management of resources in traditional fishing grounds.* MSc. thesis. Dep. Mar. Sci. Coast. Manag., Univ. Newcastle-upon-Tyne. Cited in Ruddle 1995

Cooke A, Moce K. 1995. Current trends in the management of *qoliqoli* in Fiji. *SPC Tradit. Mar. Res. Manag. Knowledge Inf. Bull.* 5:2–6

Dalzell P, Adams TJH, Polunin NVC. 1996. Coastal fisheries in the Pacific Islands. *Oceanogr. Mar. Biol: Ann. Rev.* 34:395–51

Dwyer PD. 1994. Modern conservation and indigenous peoples: in search of wisdom. *Pac. Conserv. Biol.* 1:91–97

Fa'asili U, Kelokolo I. 1999. The use of village bylaws in marine conservation and fisheries management. *SPC Tradit. Mar. Res. Manag. Knowledge Inf. Bull.* 11:7–10

Fairbairn TIJ. 1992. Traditional reef and lagoon tenure in Western Samoa and its implications for giant clam mariculture in Giant Clams. In *The Sustainable Development of the South Pacific,* ed. C Tisdell, pp. 169–89. Canberra, Aust.: Aust. Cent. Int. Agric. Res. 275 pp.

Foale S, Macintyre M. 2000. Dynamic and flexible aspects of land and marine tenure West Nggela: implications for marine resource management. *Oceania* 71:30–45

Fong GM. 1994. Case study of a traditional marine management system: Sasa Village, Macuata Province, Fiji. *FAO Field Rep.* RAS/92/TO5. 94/1. Rome: FAO. 85 pp.

Friedlander AK, Poepoe K, Poepoe K, Helm P, Bartram PK, et al. 2002. Application of Hawaiian traditions to community-based fishery management. *Proc. 9th Int. Coral Reef Symp., Bali.* In press

Ghai Y. 1988. Constitution making and decolonisation. In *Law, Government and Politics in the Pacific Island States,* ed. Y Ghai, pp. 1–53. Suva, Fiji: Inst. Pac. Stud., Univ. S. Pac.

Gisawa L, Lokani P. 2001. Trial community fishing and management of live reef food fisheries in Papua New Guinea. *SPC Live Reef Fish Inf. Bull.* 8:3–6

Graham T. 1994. Flexibility and the codification of traditional fisheries management systems. *SPC Tradit. Mar. Manag. Knowledge Inf. Bull.* 4:2–6

Graham T. 2001. The live reef fisheries of Palau: history and prospects for management. *Asia Pac. Coastal Mar. Program. Rep. 0103, Nat. Conserv.,* Honolulu, HI

Graham T, Idechong N. 1998. Reconciling customary and constitutional law: managing marine resources in Palau, Micronesia. *Ocean Coast. Manag.* 40:143–64

Hviding E. 1989. Keeping the sea: aspects of marine tenure in Marovo Lagoon, Solomon Islands. In *Traditional Marine Resource Management in the Pacific Basin: An Anthology,* ed. K Ruddle, RE Johannes, pp. 7–44. Jakarta: UNESCO/Reg. Off. Sci. Technol. SE Asia. 410 pp.

Hviding E. 1998. Contextual flexibility: present status and future of customary marine tenure in Solomon Islands. *Ocean Coast. Manag.* 40:253–69

Hviding E, Ruddle K. 1991. A regional assessment of the potential role of customary marine tenure systems (cmt) in contemporary fisheries management in the South Pacific. *Forum Fish. Agency Rep.* 91/71. 20 pp.

Johannes RE. 1977. Traditional law of the sea in Micronesia. *Micronesia* 13:121–27

Johannes RE. 1978. Traditional marine conservation methods in Oceania and their demise. *Annu. Rev. Ecol. Syst.* 9:349–64

Johannes RE. 1998a. Government-supported, village-based management of marine resources in Vanuatu. *Ocean Coast. Manag.* 40:165–86

Johannes RE. 1998b. The case for data-less marine resource management: examples from tropical nearshore fisheries. *Trends Ecol. Evol.* 13:243–46

Johannes RE. 1999. Breaking environmental laws. *SPC Live Reef Fish. Inf. Bull.* 6:1–2

Johannes RE. 2000. Palau: protection of reef fish spawning aggregations. In *Marine and Coastal Protected Areas,* ed. RV Salm, JR Clark, pp. 320–23. Gland, Switz.: IUCN. 3rd ed.

Johannes RE. 2002. Did indigenous conservation ethics exist? *SPC Tradit. Mar. Res. Manag. Knowledge Inf. Bull.* In press

Johannes RE, Hickey F. 2002. Evolution of village-based marine resource management in Vanuatu between 1993 and 2001. *Rep. UNESCO, Sect. Environ. Dev. Coast. Reg. Small Islands.* In press

Johannes RE, Lam M. 1999. The live reef food fish trade in Solomon Islands. *SPC Live Reef Fish Inf. Bull.* 5:8–15

Johannes RE, Riepen M. 1995. Environmental, economic and social implications of the live reef fish trade in Asia and the Western Pacific. *Rep. The Nature Conservancy and the Forum Fisheries Agency.* 83 pp.

Johannes RE, Squire L, Graham T, Sadovy Y, Renguul H. 1999. Spawning aggregations of groupers (Serranidae) in Palau. *The Nature Conservancy Marine Conservation Research Rep. 1.* The Nature Conservancy & Forum Fisheries Agency. 144 pp. (available online at http://www.conserveonline.org)

Johannes RE, Yeeting B. 2001. I-Kiribati Knowledge and Management of Tarawa's Lagoon resources. *Atoll Res. Bull.* 489. 24 pp.

King M, Faasili U. 1998a. Village fisheries management and community-owned marine protected areas in Samoa. *Naga, The ICLARM Q.* April–June:34–38

King M, Faasili U. 1998b. A network of small, community-owned fish reserves in Samoa. *PARKS* 8:11–16

King M, Faasili U. 1999. Community-based management of subsistence fisheries in Samoa. *Fish. Manag. Ecol.* 6:133–44

Kosaki RH. 1954. Konohiki fishing rights. *Hawaii Legislative Ref. Bur. Rep. 1.* 35 pp.

Lagibalavu M. 1994. Traditional marine tenure and policy recommendations: the Fiji experience. See South et al. 1994, pp. 269–73

Ledua E. 1995. *Policies, problems, laws and regulations with regards to inshore fisheries resource management in Fiji.* Presented at the Jt. FFA/Workshop on the Management of South Pacific Inshore Fisheries. Noumea, New Caledonia: S. Pac. Comm. Doc. SPC/Inshore Fish. Manag./CP 7. 9 pp.

Malm T. 2001. The tragedy of the commoners: the decline of the customary marine tenure system of Tonga. *SPC Tradit. Mar. Res. Manag. Knowledge Inf. Bull.* 13:3–14

McKinnon J. 1993. Resource management under traditional tenure; the political ecology of a contemporary problem, New Georgia Islands, Solomon Islands. *S. Pac. Study* 14:96–117

Munro DM. 1996. A case study of traditional marine management systems in Pukapuka Atoll, Cook Islands. *FAO Field Rep.* RAS/92/TO5. No. 96/2. Rome: FAO

Naqasima-Sobey M, Vuki V. 2002. Customary marine tenureship and establishment of MPAs in Fiji. *Proc. 2nd Water Forum 2000.* The Hague: UNESCO

Naviti W, Hickey FR. 2001. Live reef food fishery trial generates problems in Vanuatu. *SPC Live Reef Fish Inf. Bull.* 9:3–4

Peart R. 1993. *Supporting traditional conservation laws through legislation—a case study, from Isabel Province, Solomon Island.* Presented at 5th S. Pac. Conf. Nat. Conserv. Protected Areas, Nuku'alofa, Tonga

Pelelo A, Matoto SV, Gillett R. 1995. The case for community-based fisheries management in Tonga. In *Manuscript Collection of Country Statements and Background Papers of* the SPC/FFA Workshop on the Management of South Pacific Inshore Fisheries. Vol. II: Integrated Coastal Fisheries Management Project Fisheries Doc. 12, ed. P Dalzell, T Adams, pp. 487–93. Noumea, New Caledonia: S. Pac. Comm.

Perminow AA. 1996. Moving things of love: an ethnography of constitutive motions on Kotu Island in Tonga. PhD thesis. Univ. Oslo, Oslo, Norway. Cited in Malm 2001

Ruddle K. 1994. A guide to the literature on traditional community-based fishery management in the Asia-Pacific tropics. FAO Fish. Circ. No. 869. 114 pp.

Ruddle K. 1995. A guide to the literature on traditional community-based fishery management in Fiji. SPC Tradit. Mar. Res. Manag. Knowledge Inf. Bull. 5:7–15

Ruddle K, Hviding E, Johannes RE. 1992. Marine resources management in the context of customary tenure. Mar. Res. Econ. 7:249–73

Schug DM. 1996. The revival of territorial use rights in Pacific island inshore fisheries. In Ocean Yearbook 12, ed. E Mann, E Borgese, N Ginsburg, J Morgan, pp. 235–46. Chicago: Univ. Chicago Press

Sims N. 1989. Adapting traditional marine tenure and management practices to the modern fisheries framework in the Cook Islands. In Traditional Marine Resource Management in the Pacific Basin: An Anthology, ed. K Ruddle, RE Johannes, pp. 323–58. Jakarta: UNESCO/Reg. Off. Sci. Technol. SE Asia. 410 pp.

S. Pac. Comm. 1992. Study of the Aitutaki trochus fishery. S. Pac. Comm. Fish. Newsl. 59:5–6

South R, Goulet D, Tuquiri S, Church M, eds. 1994. Traditional Marine Tenure and Sustainable Management of Marine Resources

in Asia and the Pacific. Suva, Fiji: Int. Ocean Inst./Univ. S. Pac.

Tawake A, Aalbersberg W. 2002. Community-based refugia management in Fiji. In Coastal Protection For and By the People of the Indo-Pacific: Learning from 13 Case Studies. Washington, DC: World Resour. Inst. In review

Tawake A, Parks J, Radikedike P, Aalbersberg B, Vuki V, et al. 2001. Harvesting clams and data. Conserv. Biol. Pract. 2(4):32–35

Veitayaki J. 1998. Traditional community-based marine resources management system in Fiji: an evolving integrative process. Coast. Manag. 26:47–60

Veitayaki J. 2000. Inshore fisheries development in Fiji. In Fiji Before the Storm: Elections and the Politics of Development, ed. BV Lal, pp. 135–48 Canberra, Aust.: Asia Pac. Press/Aust. Natl. Univ.

Veitayaki J. 2001. Customary marine tenure and the empowerment of resource owners in Fiji. In Oceans in the New Millennium: Challenges and Opportunities for the Islands, ed. GR South, G Cleave, PA Skelton, pp. 151–61. Romania: Publ. House Dada

von Bulow W. 1902. Fishing rights of the natives of German Samoa. Globus 82:40–41 (From German). In Giant Clams in the Sustainable Development of the South Pacific, ed. C Tisdel, pp. 188–89. Canberra: Aust. Cent. Int. Agric. Res. 275 pp.

Walters CJ, Hilborn R. 1976. Adaptive control of fishing systems. J. Fish. Res. Bd. Can. 33:145–59

World Bank. 1999. Voices from the Village: A Comparative Study of Coastal Resource Management in the Pacific Islands. Summary Rep. Washington, DC

Yeeting B. 1999. Live reef fish developments in Fiji. SPC Live Reef Fish Inf. Bull. 6:19–24

Annu. Rev. Ecol. Syst. 2002. 33:341–70
doi: 10.1146/annurev.ecolsys.33.010802.150519
First published online as a Review in Advance on August 14, 2002

NUTRIENT CYCLING BY ANIMALS IN FRESHWATER ECOSYSTEMS

Michael J. Vanni

Department of Zoology, Miami University, Oxford, Ohio 45056;
email: vannimj@muohio.edu

Key Words   lakes, nitrogen, nutrient excretion, phosphorus, streams

■ **Abstract**   Animals are important in nutrient cycling in freshwater ecosystems. Via excretory processes, animals can supply nutrients (nitrogen and phosphorus) at rates comparable to major nutrient sources, and nutrient cycling by animals can support a substantial proportion of the nutrient demands of primary producers. In addition, animals may exert strong impacts on the species composition of primary producers via effects on nutrient supply rates and ratios. Animals can either recycle nutrients within a habitat, or translocate nutrients across habitats or ecosystems. Nutrient translocation by relatively large animals may be particularly important for stimulating new primary production and for increasing nutrient standing stocks in recipient habitats. Animals also have numerous indirect effects on nutrient fluxes via effects on their prey or by modification of the physical environment. Future studies must quantify how the importance of animal-mediated nutrient cycling varies among taxa and along environmental gradients such as ecosystem size and productivity.

INTRODUCTION

The cycling of nutrients is critical for the sustenance of ecosystems (DeAngelis et al. 1989, DeAngelis 1992, Costanza et al. 1997, Chapin et al. 2000). Nutrient cycling may be defined as the transformation of nutrients from one chemical form to another, and/or the flux of nutrients between organisms, habitats, or ecosystems. In most ecosystems, microbes (bacteria and fungi) are important agents of nutrient cycling (Schlesinger 1997). Nutrient inputs from outside ecosystem boundaries (often referred to as allochthonous inputs) are also important in many ecosystems (Polis et al. 1997, Carpenter et al. 1998). However, over the past three decades, ecologists have shown that animals can be important in the cycling of nutrients in terrestrial, marine, and freshwater ecosystems (e.g., Kitchell et al. 1979; Meyer & Schultz 1985; Grimm 1988a,b; Pastor et al. 1993; Vanni 1996; McNaughton et al. 1997; Vanni et al. 1997; Sirotnak & Huntly 2000; Hjerne & Hansson 2002). In most aquatic ecosystems, attention has focused on the cycling of nitrogen (N) and phosphorus (P) because they are the nutrients most likely to limit primary

0066-4162/02/1215-0341$14.00   **341**

producers and perhaps heterotrophic microbes (Pace & Funke 1991, Suberkropp & Chauvet 1995, Smith 1998, Rosemond et al. 2002).

Animals have many strong effects on aquatic food webs and ecosystems, and it is necessary to place the role of animal-mediated nutrient cycling within this context. Predators such as fish can directly or indirectly control the biomass and species composition of trophic levels below them. One of the most well-studied effects is the trophic cascade (e.g., Carpenter et al. 1985), whereby predation by fish results in reduced biomass and altered species composition of herbivores, and in increased biomass and altered species composition of primary producers (usually algae). Several studies have shown that the trophic cascade also affects nutrient concentrations, the relative apportionment of nutrients to different ecosystem pools, and the extent and severity of nutrient limitation (e.g., Shapiro & Wright 1984, Andersson et al. 1988, Elser et al. 1988, Mazumder et al. 1989, Reinertsen et al. 1990, Carpenter et al. 1992, Rosemond 1993, Rosemond et al. 1993, Vanni et al. 1997, Drenner et al. 1998, Elser et al. 2000). These studies and others show that the increase in primary producers set in motion by carnivores cannot be completely explained by a reduction in herbivory, and they suggest that changes in nutrient cycling may at least partly explain the trophic cascade response of primary producers. Indeed, in referring to P-limited lakes, Carpenter et al. (1992) suggest that "changes in trophic structure that derive from trophic cascades can be viewed as changes in the phosphorus cycle driven by fishes."

Freshwater animals can affect nutrient cycling in many ways (Figure 1), which can be characterized as direct and indirect. I consider their direct effects to be those that emanate from the physiological transformation of nutrients from one form to another within their own bodies. This includes consumption of nutrients and their subsequent allocation to feces, growth, and nutrient excretion (Figure 1). Indirect effects occur when animals affect nutrient fluxes through impacts on their prey and/or on physical habitat structure (Figure 1). In this review I first consider direct effects animals have on nutrient cycling, starting with processes at the level of individual animals and then proceeding to effects on communities and ecosystems. Then I consider the indirect effects animals have on nutrients.

NUTRIENT CYCLING AT THE INDIVIDUAL LEVEL

Nutrient Mass Balance

The amount of nutrients ingested and released by an animal must follow principles of mass balance. Nutrients that are ingested but not assimilated through an animal's gut wall are released as feces, a process referred to as nutrient egestion. Fecal nutrients are not usually directly available to primary producers, which require nutrients in dissolved form. However, fecal nutrients may subsequently become available to primary producers via decomposition and remineralization by microbes (e.g., Hansson et al. 1987). Assimilated nutrients have two fates: They can be sequestered into animal tissues via growth, in which case the nutrients are

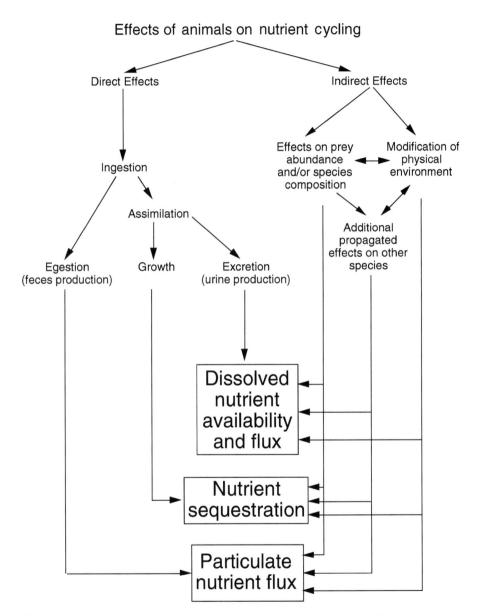

Figure 1   Schematic diagram of animal-mediated nutrient cycling via direct and indirect pathways.

not immediately available to other organisms; alternatively, they are released in dissolved form through kidneys or functionally similar organs, a process known as nutrient excretion.

Nutrient excretion is the most direct means by which animals can provide nutrients for primary producers (algae and vascular plants) and heterotrophic microbes (bacteria and fungi). Although some freshwater animals excrete certain nutrients in organic form (e.g., urea), most N and P is excreted in inorganic forms (e.g., ammonia, phosphate). Thus the rates at which animals excrete N and P are potentially important for primary producers and heterotrophic microbes. In addition, because either N or P can be limiting, the ratio at which animals excrete these nutrients (hereafter excretion N:P) is potentially important in determining the relative degree of N vs. P limitation (Elser et al. 1988, Sterner 1990, Sterner & Elser 2002) and algal species composition (Tilman et al. 1982, Smith 1983).

Nutrient Recycling versus Nutrient Translocation

Nutrient cycling through an animal's body can be divided into two functionally distinct processes: nutrient recycling and nutrient translocation. Nutrient recycling occurs when an animal releases nutrients within the same habitat in which food was ingested. For example, when zooplankton consume phytoplankon in the open water of a lake and excrete nutrients back into the water, they recycle nutrients already in that habitat. In contrast, nutrient translocation (or transport) refers to the process by which an animal physically moves nutrients between habitats or ecosystems, often accompanied by transformation of nutrients from one chemical form to another (Kitchell et al. 1979, Shapiro & Carlson 1982, Vanni 1996). For example, when an animal feeds on benthic prey and excretes nutrients into the water, it translocates nutrients from benthic to pelagic habitats and converts nutrients from particulate to dissolved forms. In this case, animals move nutrients between habitats within a single ecosystem, but nutrient translocation can also occur between different ecosystems, often at great spatial and temporal scales.

What makes nutrient translocation different from recycling is that in the former, nutrients are moved across physical boundaries or against physical processes that impede nutrient movement. Such impediments include the sediment-water interface, the thermocline that separates surface and deep water layers, and the downstream flow of water. In many aquatic systems, primary producers cannot utilize nutrients in deep waters because of inadequate light. Therefore, any process that brings nutrients to the euphotic zone (surface waters where light intensity is sufficient for photosynthesis) is potentially important. Nutrients delivered to the euphotic zone from deeper waters, littoral/benthic areas, or outside the ecosystem are often referred to as "new" nutrients (Dugdale & Goering 1967, Eppley & Peterson 1979, Caraco et al. 1992) because these nutrients have been recently translocated from another habitat or ecosystem. New (translocated) nutrients can stimulate "new primary production" (Dugdale & Goering 1967, Eppley &

Peterson 1979, Caraco et al. 1992, Vanni 1996) and increase the total mass of nutrients in the recipient habitat or ecosystem. In contrast, nutrient recycling cannot directly increase the mass of nutrients in a habitat or ecosystem, but rather it sustains "recycled production." The implications of whether animals recycle or translocate nutrients are considerable and are discussed throughout this review.

Measurement of Excretion Rates

Excretion rates can be estimated by using bioenergetics/mass balance models or by direct measurement. In the former approach, excretion rate is estimated as nutrients ingested minus nutrients allocated to egestion and growth (e.g., Kraft 1992, Schindler et al. 1993). Direct measurement usually entails capturing animals in the field and placing them in containers in which the accumulation of nutrients is quantified. Both methods have their advantages and disadvantages, but limited comparative data suggest that they yield similar excretion rates and ratios for fish (Mather et al. 1995, Vanni 1996, Schindler & Eby 1997, Hood 2000, Vanni et al. 2002). Therefore I treat rates obtained by both methods equally. However, many early attempts to measure excretion rates in the field used relatively long incubation times (length of time an animal is held without food after collection). This leads to underestimation of rates because animals are held without food during incubations, and excretion rates decline quickly after feeding ceases (Lehman 1980a, Devine & Vanni 2002). Therefore, I do not consider studies with long incubation times in evaluating nutrient excretion rates in nature.

FACTORS MEDIATING NUTRIENT EXCRETION BY ANIMALS

Body Size

Because of allometric constraints on metabolism (Peters 1983), mass-specific nutrient excretion rates of animals (i.e., nutrients excreted per unit body mass per unit time) usually decline with increasing body mass. Much of the evidence for allometric relationships derives from laboratory studies, in which animals were either not fed or fed a controlled ration, or from field studies with relatively long incubation times. These studies yield excretion rates that are not necessarily indicative of those in nature. Nevertheless, they show convincing allometric effects for fish (e.g., Gerking 1955), zooplankton (e.g., Wen & Peters 1994), and benthic invertebrates (e.g., Lauritsen & Mosley 1989), as do many field studies (Brabrand et al. 1990, Schaus et al. 1997, Davis et al. 2000, Gido 2002).

Temperature

Nutrient excretion rates of aquatic animals also increase with temperature due to the dependence of metabolic rates on temperature. Estimates of Q_{10} (i.e., the factor by which a rate increases for every 10°C increase in temperature) for nutrient excretion rates of animals are similar to Q_{10} values for other metabolic processes, and such values are generally between 1.5 and 2.5 (e.g., Gardner et al. 1981, Wen & Peters 1994, Schaus et al. 1997, Devine & Vanni 2002).

Ecological Stoichiometry: Body and Food Nutrient Composition

THEORY   Because nutrient excretion is constrained by mass balance, rates must be affected by the nutrient composition of the animal's body and its food (Olsen et al. 1986, Sterner 1990, Sterner et al. 1992, Elser & Urabe 1999, Sterner & Elser 2002). Ecological stoichiometry theory proposes that individual animal species maintain relatively constant body nutrient contents per unit body mass. Thus, during growth, an animal will incorporate nutrients at a rate needed to maintain constant body nutrient composition and will excrete nutrients that are assimilated but not needed for growth. Therefore, an animal feeding on a nutrient-rich food source will excrete more nutrients than one feeding on a nutrient-poor source, all else being equal. Similarly, an animal with a relatively low nutrient content in its body will allocate fewer nutrients to growth and will hence excrete more nutrients than an animal with a high body nutrient composition. Stoichiometry theory also predicts that the excretion N:P of animals is a function of the imbalance between the N:P ratios in its body and its food. An individual with a low body N:P ratio should release nutrients at a relatively high N:P ratio compared to an individual with a high body N:P ratio, if the two are feeding on the same food. More generally, the N:P ratio released by an animal should be negatively correlated with the N:P of its body tissues and positively correlated with the N:P of its food.

EVIDENCE   Relatively few field studies have explicitly tested the hypothesis that nutrient excretion rates are functions of the nutrient composition of animals and their food. Elser & Urabe (1999) suggested that for herbivorous zooplankton, food N:P is more important than body N:P in determining excretion N:P, perhaps because the N:P of their food (phytoplankton) is much more variable than the N:P of zooplankton body tissues. Schindler & Eby (1997) used bioenergetics/mass balance models to show that the excretion N:P of 18 species of fish is relatively invariant and low (generally <15:1 molar) as long as fish growth is not limited by P. The excretion N:P can be much higher if fish growth is limited by P because fish need to sequester a greater proportion of assimilated P, but fish growth rates are apparently rarely limited by P (Schindler & Eby 1997). Similarly, Sterner & George (2000) showed that four species of cyprinid fish (minnows) differed only slightly in body N and P contents and assimilation rates, and by implication, excretion rates of these species were probably similar.

In contrast, Vanni et al. (2002) found > tenfold variation in excretion rates and ratios among 26 fish and 2 amphibian species in a tropical stream in Venezuela, and ecological stoichiometry explained much of this variation (body size was important also). Mass-specific P excretion rate was negatively correlated with body P content, and excretion N:P was negatively correlated with body N:P, as predicted by stoichiometry theory. Body nutrient content may have had a greater effect on excretion rates and ratios than in other studies because of relatively large interspecific variation in body P content. Some fish, particularly the armored catfish (Loricariidae), have very high P contents in their bodies compared to other taxa, apparently because they need to sequester P to make their armor, which is modified bone (Vanni et al. 2002). Loricariids excrete P at very low rates and have high excretion N:P ratios. The relative abundance of loricariids in neotropical streams (Lowe-McConnell 1987, Power 1990) may be very important in determining variation among taxa in excretion rates and ratios.

IMPORTANCE OF NUTRIENT EXCRETION BY ANIMALS FOR COMMUNITIES AND ECOSYSTEMS

Approaches to Quantifying the Importance of Nutrient Excretion by Animals

There are three basic approaches to estimating the importance of animal-mediated nutrient excretion. One is to compare animals' nutrient excretion rates to the rates at which nutrients are supplied by other sources. While this may seem straightforward, in reality it is often very difficult to quantify, and sometimes even to identify, all nutrient fluxes in an ecosystem (e.g., Caraco et al. 1992). Therefore, nutrient excretion by animals is often compared to other sources known to be important in many ecosystems, such as inputs from watersheds and release of nutrients from sediments via microbial processes.

An alternative, or surrogate, approach is to compare nutrient excretion rates by animals to nutrient demand by producers in the ecosystem. If nutrient excretion by animals supports a substantial proportion of nutrient demand, it can be concluded that animals are important in overall nutrient supply, without actually measuring other nutrient fluxes. Nutrient demand is often estimated as the rate at which primary producers utilize nutrients (e.g., Grimm 1988a,b; Schindler et al. 1993). However, this may underestimate total ecosystem demand because heterotrophic microbes also utilize, and may compete with algae for, limiting nutrients (e.g., Sterner et al. 1995, Suberkropp & Chauvet 1995). One assumption behind the supply/demand approach is that total nutrient demand equals nutrient supply from all sources; therefore the proportion of demand supported by any one source can be equated with the proportion of total nutrients supplied. This approach is valid only for the limiting nutrient because the total supply rate of a nonlimiting nutrient may be much higher than demand for that nutrient.

The third way to evaluate the importance of nutrient excretion by animals is to experimentally isolate and quantify the effects of nutrient excretion on recipients of nutrients (primary producers or heterotrophic microbes). This approach seeks mainly to quantify population- or community-level effects, and it involves experimentally separating effects of consumption and excretion for the animals in question.

Evidence for the Importance of Nutrient Excretion by Animals: Nutrient Supply and Demand Studies

Small animals such as zooplankton are well known as an important potential source of nutrients in lakes and oceans. The role of larger animals such as fish is less clear, and some investigators have argued that large animals play only a minor role in supplying nutrients compared to small animals (e.g., Nakashima & Leggett 1980, Hudson et al. 1999). This may seem logical because large animals excrete nutrients at lower mass-specific rates and often have lower population biomass than do small animals. However, the biomass of large animals can sometimes be quite high, and the available data suggest that ecosystem-wide excretion rates of large animals can be as high as those of small animals (Table 1). In addition, because of greater mobility, large animals are more likely than small animals to translocate nutrients (Vanni 1996). Because body size has been considered a possible mediating factor in regulating nutrient cycling rates, I have organized this section according to body size.

ZOOPLANKTON   For decades, ecologists have realized that nutrient excretion by zooplankton can be important in sustaining primary production in lakes and in the sea (Barlow & Bishop 1965, Goldman et al. 1979, Lehman 1980a, Lehman & Sandgren 1985, Sterner 1989). Unfortunately, many early estimates of nutrient excretion by zooplankton may have produced biased rates due to methodological problems (as discussed by Lehman 1980a,b), and on a per-mass basis, nutrient excretion rates vary greatly among studies (e.g., Lehman 1980b, Gulati et al. 1995). Nevertheless, recent estimates using refined methods suggest that excretion by zooplankton can supply substantial amounts of nutrients and support a substantial fraction of phytoplankton primary production (Table 1). In many lakes, nutrient excretion by zooplankton represents mostly recycled nutrients, rather than translocated nutrients, because zooplankton feed and excrete in the euphotic zone. Zooplankton can undergo daily vertical migrations and, in the process, translocate nutrients between deep water and the euphotic zone. However, the net effect of vertical migration is probably a loss of nutrients from the euphotic zone because zooplankton feed and excrete there, but usually do not feed (but excrete) in deeper waters (Wright & Shapiro 1984).

BENTHIC INVERTEBRATES   Nutrient excretion by benthic invertebrates can also be important in lakes and streams (Table 1; Gardner et al. 1981, Grimm 1988a, Arnott & Vanni 1996, Devine & Vanni 2002). For example, benthic insects and snails supplied 15% to 70% of algal N demand in a desert stream (Grimm 1988a), and P excretion by unionid mussels exceeded direct P release from sediments in a mesotrophic lake (Nalepa et al. 1991). Benthic invertebrates can either recycle or translocate nutrients. Burrowing invertebrates (e.g., chironomids, worms) mostly

TABLE 1 N and P excretion rates and excretion N:P of various animals, and the percentage of primary producer N and P demand supported by excretion

Lake or stream	Taxonomic identity	N excretion rate[a] (mg N/m²/d)	P excretion rate[a] (mg P/m²/d)	Excretion N:P[a] (molar)	N demand supported[a] (%)	P demand supported[a] (%)	Source
Zooplankton							
Lake Washington	Assemblage[b]	20.5	4.00	13	24	33	Lehman 1980b
Castle Lake	Assemblage	10.9 (1.5–24.5)			57 (6–160)		Axler et al. 1981
Lake Tahoe	Assemblage				0.5		Carney & Elser 1990
Lake Titicaca	Assemblage				11.5		Carney & Elser 1990
Lake Michigan	Assemblage					58	Carney & Elser 1990
Latvian lakes (8 lakes), May–Oct	Assemblage		3.03 (0.69–4.99)			16 (2–34)	Gutelmacher and Makartseva 1990
Latvian lakes (8 lakes), winter	Assemblage		0.05 (0.01–0.12)				Gutelmacher & Makartseva 1990
West Long Lake	Assemblage		0.83			26	Schindler et al. 1993
Peter Lake	Assemblage		0.16			4	Schindler et al. 1993
Lake Biwa	Assemblage	1.4–11.7	0.07–0.61	19–23	43 (3–104)	15 (1–36)	Urabe et al. 1995
Lake Pend Oreille	Opposum shrimp (*Mysis relicta*)		0.07				Chipps & Bennett 2000
Benthic invertebrates							
Sycamore Creek	Insects and snails	85.0 (33.0–137.0)			15–70		Grimm 1988a
Lake St. Clair	*Lampsilis radiata siliquoidea* (unionid)		0.16				Nalepa et al. 1991
Lake Suwa	Dipterans, tubificids		1.20				Fukuhara & Yasuda 1985
Lake Erie, western basin	Zebra mussel (*Dreissena polymorpha*)	222.3	58.50	8 (3–18)			Arnott & Vanni 1996
Snowflake Lake	*Gammarus lacustris* (amphipod)		1.11				Wilhelm et al. 1999
Acton Lake	Dipterans, tubificids	10.8	1.45	16	8	10	Devine & Vanni 2002; Knoll et al. in review
Fish							
Sycamore Creek	Longfin dace (*Agosia chrysogaster*)	19.1 (13.2–24.9)			5–10		Grimm 1988b
Rio Las Marías	Assemblage	63.9	7.17	20	49	126	Hood 2000, Vanni et al. 2002

(Continued)

TABLE 1 *(Continued)*

Lake or stream	Taxonomic identity	N excretion rate[a] (mg N/m²/d)	P excretion rate[a] (mg P/m²/d)	Excretion N:P[a] (molar)	N demand supported[a] (%)	P demand supported[a] (%)	Source
Lake Gjersjøen (May–Oct)	Roach (*Rutilus rutilus*)		1.21				Brabrand et al. 1990
Lake Gjersjøen (June–July)	Roach (*Rutilus rutilus*)		2.95				Brabrand et al. 1990
Lake Memphremagog	Young-of-year Yellow perch (*Perca flavescens*)	3.5–7.4	0.35–2.07	13–47			Kraft 1992
Lake Michigan[c]	Alewife (*Alosa pseudoharengus*)		2.19				Kraft 1993
West Long Lake	Assemblage		0.15			5	Schindler et al. 1993
Peter Lake	Assemblage		2.30			36	Schindler et al. 1993
Lake Finjasjön	Roach (*Rutilus rutilus*), Bream (*Abramis brama*)		0.53				Persson 1997a
Acton Lake	Gizzard shad (*Dorosoma cepedianum*)	35.3	5.46	17	25	36	Schaus et al. 1997, Knoll et al. in review
Bautzen Reservoir	Young-of-year fish		0.01–0.05				Mehner et al. 1998
Lake Pend Oreille	Kokanee salmon (*Oncorhynchus nerka*)		0.02				Chipps & Bennett 2000
Sierra Nevada lakes (5 lakes, stocked)	Trout		0.02				Schindler et al. 2001
Sierra Nevada lakes (7 lakes, unstocked)	Trout		0.01				Schindler et al. 2001
Lake Texoma	Benthivorous fish assemblage	35.0	1.00	82			Gido 2002
Birds							
Bosque del Apache wetland[d]	Lesser snow geese (*Chen caerulescens caerulescens*)	43.1	5.42	18			Post et al. 1998

[a]Values outside parentheses are means, whereas those in parentheses are ranges.

[b]Assemblage means that parameters were quantified for the entire assemblage (all taxa) within that group.

[c]Includes P egestion as well as excretion.

[d]Includes only new nutrients, i.e., nutrients translocated from outside the wetland.

consume benthic food and translocate nutrients into the water column. In contrast, taxa such as unionid and zebra mussels, which filter phytoplankton from the water column, mostly recycle nutrients (Nalepa et al. 1991, Arnott & Vanni 1996). Note that many earlier studies may have underestimated excretion rates of benthic invertebrates because incubation times were too long, thus producing rates not reflective of natural feeding conditions (Devine & Vanni 2002).

FISH   Several recent studies show the importance of nutrient excretion by fish (Table 1). Nutrient excretion rates of fish assemblages can be comparable to, or exceed, nutrient input rates from external sources in lakes (inflow streams: Brabrand et al. 1990, Persson 1997a; atmosphere: Schindler et al. 2001) and can support a substantial fraction of algal nutrient demand in lakes (Schindler et al. 1993) and streams (Grimm 1988b, Hood 2000, Vanni et al. 2002). P excretion rates of fish can exceed watershed inputs even in reservoirs that are located in highly agricultural watersheds and thus receive large quantities of allochthonous nutrients (Schaus et al. 1997, Vanni et al. 2001). However, the importance of fish may be most pronounced during dry periods when external inputs are reduced (Gido 2002). In many lakes, most nutrients excreted by fish are derived from benthic/littoral food sources, indicating that fish translocate nutrients to pelagic habitats (Brabrand et al. 1990; Schindler et al. 1993, 2001; Schaus et al. 1997; Gido 2002). Even fish referred to as "planktivores" often rely heavily on littoral/benthic prey (Schindler et al. 1993). In contrast, Kraft (1992, 1993) and Persson (1997a) found that fish fed mainly on plankton and thus provided primarily recycled nutrients. The extent to which fish provide new or recycled nutrients will depend on fish species as well as variation in diet, which can be great even within a species. Stable isotope studies show that most freshwater fish obtain a substantial fraction of their food from benthic sources (Hecky & Hesslein 1995, Schindler & Scheuerell 2002), so in many ecosystems, a substantial proportion of nutrients excreted by fish are likely to be translocated from benthic to pelagic habitats.

The relative roles of different animal taxa in nutrient cycling are likely to depend on food web configuration. For example, Schindler et al. (1993) found that fish and zooplankton provided 5% and 26%, respectively, of phytoplankton P demand in a lake dominated by piscivorous fish and with few small fish. However, in a lake dominated by small fish (which fed mostly on littoral prey but also zooplankton), nutrient excretion by fish supported 36% of P demand by phytoplankton, and excretion by zooplankton supported only 4% (Schindler et al. 1993). Fish were more important in the latter lake because small fish suppress zooplankton populations, rendering their excretion less important, and because small fish have much higher mass-specific excretion rates than piscivores (Schindler et al. 1993). Interestingly, primary production was about 30% higher in the lake with small fish, corresponding to a 40% higher excretion rate by fish and zooplankton combined.

EXCRETION N:P   Surprisingly few field studies have quantified excretion N:P and its impacts on nutrient recipients. Seasonal or interannual increases in *Daphnia* are associated with more severe P limitation and less severe N limitation of phytoplankton (Elser et al. 1988, 2000; Urabe et al. 1995; MacKay & Elser 1998), presumably because *Daphnia* excretes nutrients at a high N:P ratio (Sterner et al. 1992). Most direct measurements of excretion rates of fish and benthic invertebrates reveal relatively low N:P excretion ratios (usually <20 molar; Table 1), as do stoichiometric models for fish (Schindler & Eby 1997, Sterner & George 2000). There is some evidence that unionid mussels have a relatively high N:P excretion ratio (often >20; Nalepa et al. 1991, Davis et al. 2000). However, excretion ratios can be quite variable both among and within species of invertebrates and fish (Nalepa et al. 1991, Arnott & Vanni 1996, Davis et al. 2000, Devine & Vanni 2002, Gido 2002, Vanni et al. 2002). Although Vanni et al. (2002) found that much of the interspecific variation in N:P excretion ratio can be explained by body nutrient ratios and size, clearly more studies are needed that explore interspecific variation in excretion N:P, its relationship to stoichiometry, and its significance for nutrient limitation.

Evidence for the Importance of Nutrient Excretion by Animals: Experimental Studies

HERBIVORES   Many studies provide experimental evidence for the importance of herbivores in nutrient cycling. For example, some phytoplankton taxa increase when zooplankton biomass is increased experimentally. These taxa are usually large (hence relatively inedible) and nutrient-limited, suggesting that they are enhanced by nutrient recycling by zooplankton (e.g., Lehman & Sandgren 1985, Elser et al. 1987, Vanni & Temte 1990).

A few investigators have employed a nested design in which herbivory and nutrient cycling processes are experimentally separated. Natural algal assemblages are placed in enclosures with the animals of interest where they are exposed to both grazing and nutrient cycling. In addition, algae are incubated in nutrient-permeable chambers placed inside the enclosures, or sections of the enclosures, that allow passage of nutrients but not animals or algae. Algae incubated in chambers are exposed only to nutrient cycling by animals, and not to direct herbivory. Cuker (1983) used this approach and found that nutrient cycling by snails had no effect on algae in an arctic lake. He proposed that nutrient transfer between snails and algae occurs at small scales, e.g., within snail guts or feces. In contrast, Sterner (1986) found that nutrient regeneration by *Daphnia* increased total phytoplankton growth rate as well as that of several taxa; further, the taxon responding most positively to nutrient recycling by *Daphnia* (pennate diatoms) was also the most nutrient-limited.

CARNIVORES AND OMNIVORES   Recently, the nested design has been expanded to include effects of higher trophic levels. Vanni & Layne (1997) and Attayde & Hansson (2001a) conducted experiments in lakes. Their enclosures contained fish, zooplankton, and phytoplankton, while nutrient-permeable chambers contained only phytoplankton. Both studies found that some algal taxa responded positively in nutrient-permeable chambers when fish were present in the surrounding enclosures, showing that these taxa are positively affected by increased nutrient excretion in the presence of fish. Some phytoplankton taxa responded more to nutrient excretion than to direct herbivory, while grazing was more important for others. The extent to which grazing or nutrient cycling has a greater effect on phytoplankton taxa is probably a function of edibility. For example, large (and presumably relatively inedible) taxa such as cyanobacteria and large dinoflagellates are affected much more by nutrient cycling than by grazing (Vanni & Layne 1997, Attayde & Hansson 2001a). However, some relatively edible taxa (e.g., cryptomonads) also responded positively to increased nutrient cycling by animals (Attayde & Hansson 2001a). Geddes (1999) found that algae responded positively to increased nutrient cycling by animals in some nested experiments with a benthic food web containing omnivorous fish and/or shrimp, herbivores, and attached algae.

Thus, in all three of the studies using the nested design with higher trophic levels, increased nutrient availability accounted for some of the observed trophic cascade response, i.e., increased algal biomass stimulated by the top predator. Increased nutrient availability can be mediated by nutrient excretion by top predators themselves (Vanni & Layne 1997), and/or increased nutrient excretion by herbivores (Attayde & Hansson 2001a). Most likely, the relative importance of nutrient excretion by fish and zooplankton will depend on fish biomass, availability of non-planktonic prey for fish, and herbivore size-structure and biomass.

Another experimental approach is to confine fish (the potential source of nutrients) instead of algae (the potential recipients of nutrients) (Schindler 1992, Persson 1997b, Attayde & Hansson 2001b). These experiments have treatments that attempt to isolate the direct effects of fish excretion from other processes, but they differ from the experiments described above in that algae are concurrently exposed to direct herbivory in all treatments. All of these studies showed that phytoplankton abundance and/or productivity was enhanced to some extent by nutrient excretion by fish. However, net effects of excretion by fish on phytoplankton communities may be manifested only when herbivory rates are low (Persson 1997b, Attayde & Hansson 2001b).

TRANSLOCATION EFFECTS OF BENTHIC-FEEDING FISH   Havens (1991, 1993), and Schaus & Vanni (2000) experimentally separated nutrient translocation effects from trophic cascade effects of fish on phytoplankton in eutrophic lakes. Some enclosures were fitted with screens placed just above the sediments to prevent fish from feeding on sediments (but allowed them to feed on zooplankton), while in other enclosures fish had access to sediments. Both studies found that when fish had access to sediments, they increased phytoplankton biomass and total P in the water column three- to fivefold and altered phytoplankton community composition (nutrient translocation effect). In contrast, fish had weak or no effects on phytoplankton and water column nutrients when screens prevented them from feeding on sediments (trophic cascade effect). Benthic-feeding fish can also increase water column nutrients and phytoplankton by resuspending sediments (bioturbation, see below), and it is very difficult to separate effects via nutrient excretion and bioturbation. However, in the experiments described above as well as others with carp (Lamarra 1975), it appears that nutrient translocation (excretion) can account for most of the effects of fish on phytoplankton and nutrients.

Animals as a Source of New Nutrients: a Simple Model

As mentioned above, some authors have argued that smaller animals such as zooplankton are much more important than fish as nutrient sources (e.g., Hudson et al. 1999.) However, small animals such as zooplankton are more likely to recycle nutrients, whereas large animals like fish often translocate nutrients. Supply of new nutrients may play a critical role in ecosystems (Dugdale & Goering 1967, Caraco et al. 1992). To explore the potential impact of nutrient translocation, here I develop a simple model based on data on nutrient translocation by gizzard shad (*Dorosoma cepedianum*) in Acton Lake, a eutrophic reservoir in Ohio, USA. Gizzard shad is the dominant fish species in this lake (Schaus et al. 1997) and many other lakes in the eastern United States (Stein et al. 1995, Bachmann et al. 1996, Vanni & Headworth in press). In this lake and in most reservoirs, adult gizzard shad obtain most of their energy and nutrients from sediment detritus (Schaus et al. 2002).

In this model (Figure 2), I simulate the dynamics of water column total phosphorus concentration and primary production from May to October, when most production occurs. Gizzard shad excretion rate (P translocation) was set to 0.97 μg P L^{-1} d^{-1}, based on data from 1994 to 1999 (Schaus et al. 1997, M.J. Vanni unpublished data). Two other sources of new P were watershed inputs (1.23 μg P L^{-1} d^{-1}, the mean rate of PO_4 inputs from May–October 1994–1998; Vanni et al. 2001), and release of P from sediments (0.12 μg P L^{-1} d^{-1}, based on data from Evarts 1997). Following Smith (1979), primary production in the euphotic zone (PPR, mg C m^{-2} d^{-1}) was assumed to be a function of water column total phosphorus concentration (TP, μg P L^{-1}) based on a relationship for Ohio reservoirs: PPR $= 8.36$TP $- 115.9$ (Knoll et al. 2002). Sedimentation of P was assumed to be the major loss process and was modeled using the relationship between export ratio (ER, proportion of primary production lost via sedimentation) and PPR: ER $= -0.000163$PPR $+ 0.459$ (from Table 6 in Baines & Pace 1994). PPR was then multiplied by ER to obtain the loss of C via sedimentation, which was then multiplied by the P:C ratio of sedimenting material (set equal to 0.018 by mass, based on data from sediment traps in Acton Lake) to obtain sedimentation of P from the water column.

Simulations show that nutrient translocation by gizzard shad has major impacts on pelagic P and primary production (Figure 2). With nutrient translocation by gizzard shad, TP and PPR increased gradually, but when translocation was not

→

Figure 2 Simulation model illustrating effects of nutrient translocation by sediment-feeding fish (gizzard shad, *Dorosoma cepedianum*). *A*: Diagram showing fluxes of phosphorus (P) modeled. *B*, *C*, and *D*: Simulated water column total P, phytoplankton primary production, and planktonic P regeneration with and without nutrient translocation by gizzard shad.

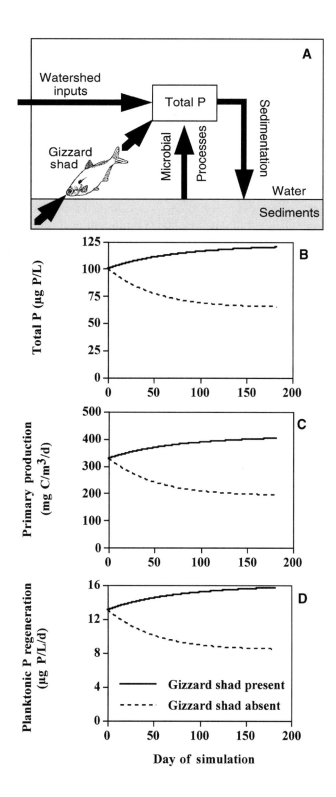

included, TP and PPR declined. Averaged over the simulation period (approximately equal to the length of a growing season), exclusion of gizzard shad excretion resulted in a 35% reduction in TP, which is similar to the effect of shad exclusion in a field experiment (Schaus & Vanni 2000), and a 40% reduction in PPR. Interestingly, this model also showed that regeneration of nutrients by plankton also depends on inputs of new (translocated) nutrients by gizzard shad. I calculated planktonic P regeneration (REG) according to the equation provided by Hudson et al. (1999): $\log REG = 1.0077 \log TP + 0.7206$. In the model, planktonic P regeneration declines by 35% when gizzard shad excretion is excluded. In essence, gizzard shad provide new P to the water column that plankton can recycle. Note that even though predicted P excretion by gizzard shad is an order of magnitude less than P excretion by plankton, gizzard shad have major impacts on water column P and primary production because they provide new P. In contrast, because plankton recycle P but do not provide new P, they have no direct impact on water column P concentration.

ANIMALS AS NUTRIENT SINKS

The processes of nutrient consumption and nutrient release by animals may be temporally uncoupled, and this has implications for whether animals function as a nutrient source or sink. Relatively long-lived animals, such as fish and unionid mussels, can sequester large amounts of nutrients in their bodies over timescales relevant to aquatic primary producers. These animals function as a major nutrient sink rather than a source (Kitchell et al. 1979, Kraft 1992, Vaughn & Havenkamp 2001). Kitchell et al. (1975) found that most of the water column phosphorus in highly productive Lake Wingra is stored in fish biomass. Animals are most likely to be important nutrient sinks when their population biomass is expanding and thus sequestering nutrients (Kraft 1992). Zooplankton can also act as nutrient sinks over timescales relevant for phytoplankton. For example, Urabe et al. (1995) found that sequesteration of P in *Daphnia* bodies can lead to low rates of P recycling and increased P limitation of phytoplankton.

Nutrients sequestered in animal bodies may be made available as animals die and decompose, thereby liberating nutrients. Kitchell et al. (1979) suggested that remineralization of nutrients from fish bodies following postspawning mortality could be an important P source. Alternating periods of storage and supply through a single population can occur within an ecosystem or among ecosystems. An excellent example of the latter is the migration of Pacific salmon, discussed below.

NUTRIENT TRANSPORT ACROSS ECOSYSTEM BOUNDARIES

In addition to translocating nutrients between habitats within an ecosystem, animals can transport nutrients among ecosystems, often over great distances and long timescales. For example, geese often forage in terrestrial areas and roost in wetlands, transporting nutrients in the process. Geese in the Bosque del Apache

National Wildlife Refuge in New Mexico feed on land but excrete much of their nutrients in wetlands, thereby providing nearly 40% of the N and 75% of the P entering their primary roosting wetland (Post et al. 1998, Kitchell et al. 1999). Stable isotope data also show that these nutrients move up the wetland food web (Kitchell et al. 1999).

Perhaps the most spectacular example of how animals can transport nutrients long distances is that of anadromous Pacific salmon, which can transport nutrients hundreds of kilometers (reviewed by Naiman et al. 2002). Pacific salmon are born and spend their early life in freshwaters, but live most of their life in the ocean where they grow and accumulate the vast majority of their body nutrients. They return to freshwater ecosystems as adults, spawn once, and then die. Remineralization of nutrients from decomposition of adult salmon bodies can represent a major nutrient source to streams and lakes in which salmon spawn (Richey et al. 1975, Kline et al. 1993, Bilby et al. 1996, Finney et al. 2000). Salmon carcasses provide up to 70% of total N inputs (25% on average) into salmon nursery lakes in Alaska (Finney et al. 2000). Because salmon bodies have a relatively low N:P ratio, the contribution of salmon to P inputs is likely to be even greater but is not well quantified (Naiman et al. 2002). Marine-derived nutrients from salmon can also have impacts on riparian terrestrial vegetation, via either direct uptake of nutrients released from carcasses or via urine produced by salmon-eating animals such as bears (Ben-David et al. 1998, Hilderbrand et al. 1999).

Downstream migration of young salmon from freshwaters to the oceans also translocates nutrients, but the quantity of nutrients moved upstream by adult salmon greatly exceeds downstream transport by young fish (Naiman et al. 2002). Thus, anadromous salmon are a sink for marine-derived nutrients, but a nutrient source for freshwaters. Nutrient translocation by salmon historically occurred in thousands of lakes and streams but is being reduced greatly by human-caused declines in salmon populations. This reduction has decreased the productivity of freshwater nursery lakes (Naiman et al. 2002).

INDIRECT EFFECTS OF ANIMALS ON NUTRIENT CYCLING

In addition to effects mediated by the physiological processing of nutrients, animals may have indirect effects on nutrient fluxes. Indirect effects are propagated through direct effects on prey assemblages and/or physical properties of ecosystems. Note that many indirect effects emanate from consumption activities of animals, as do direct effects. However, unlike direct effects, the major pathways for indirect effects are mechanisms other than storage or release of nutrients by the animal initially propagating the effects (Figure 1).

Effects Via Size-Selective Predation

Predation by fish and invertebrates can shift the size distributions of prey assemblages (Brooks & Dodson 1965, Blumenshine et al. 2000). Because mass-specific nutrient excretion rates of animals decline with body size, size-selective predation can therefore affect nutrient excretion rates of prey assemblages. For example, predation by fish on large zooplankton can shift the zooplankton assemblage to smaller species and therefore increase the rate at which zooplankton recycle nutrients (Bartell & Kitchell 1978, Bartell 1981). Similar effects probably result from size-selective predation on benthic invertebrates or fish. Thus, predators can have important indirect effects on nutrient excretion by animals, even if the predators themselves are not important as direct nutrient sources (Schindler et al. 1993).

Effects on Horizontal Nutrient Transport

EFFECTS ON NUTRIENT UPTAKE LENGTH   In running water ecosystems, dissolved nutrients are transported downstream with flow, taken up by organisms on the stream bottom, and then released back into dissolved form for further downstream transport, a process referred to as nutrient spiraling (Webster & Patten 1979). The nutrient uptake length is the length of stream over which a dissolved nutrient molecule travels before being taken up by biota (or abiotic processes) on the stream bottom (Newbold et al. 1981). In general, nutrient uptake lengths are shorter when algae or heterotrophic microbes are nutrient limited and when the biomass of these organisms is high. Animals can affect nutrient uptake length in several ways (Mulholland 1996). Grazing by herbivorous snails increases P uptake length by reducing algal biomass and thus total nutrient demand (Mulholland et al. 1983, 1994; Steinman et al. 1991). Grazers can also alter the relative importance of nutrient sources. For example, when periphyton biomass is high (i.e., when grazing is low), the algal mats represent transient storage zones for nutrients, and most algal nutrient demand is met via recycling within the mat. In contrast, when algal biomass is low, a greater fraction of nutrient demand is met by nutrients flowing by in overlying water (Mulholland et al. 1994). In some ways these two nutrient pools are analogous to recycled and new nutrient categories discussed above for lakes and oceans.

PARTICLE PROCESSING EFFECTS   Animals can also affect the size distribution, standing stocks, deposition rates, and transport rates of particles via consumption and egestion, and by "sloppy feeding" (whereby particles are broken up into smaller pieces but not ingested), with subsequent effects on nutrients attached to particles. This is perhaps most important in streams, where many animals ("shredders") feed on large detrital particles (e.g., leaves) and convert them into smaller particles that become available to other consumers such as filter-feeding "collectors" (Cummins & Klug 1979, Webster & Wallace 1996). Via this process, animals can increase the downstream transport of particulate nutrients. For example, experimental removal of macroinvertebrates caused a large reduction in the concentration and downstream transport of fine particulate matter (Wallace et al. 1991). In some tropical streams, sediment-feeding fish can also affect the rate at which sediments (and associated particulate nutrients) accumulate, and thus the degree to which it is transported downstream (Flecker 1996). Other studies show the impacts of consumption of large particulate organic matter. In tropical Puerto Rico streams, experimental exclusion of shrimp caused decreased leaf decay rates; increased accrual of organic matter, particulate C and particulate N; and increased C:N ratio in material accumulating on the stream bottom (Pringle et al. 1999, March et al. 2001). However, effects depend on the species composition of the shrimp assemblage. By processing leaves into smaller particles one shrimp genus (*Xiphocaris*) increased leaf decay rates, downstream transport of suspended particulate organic matter, and concentrations of dissolved organic carbon and nitrogen (Crowl et al. 2001). Another genus (*Atya*) increased leaf decay rate slightly but had no effect on downstream transport (because they consumed the fine particulates) or dissolved nutrients (Crowl et al. 2001). In contrast to these effects, Rosemond et al. (1998) found that exclusion of fish and shrimp elicited an increase in small invertebrates, but no effects on detrital processing rates in a Costa Rica stream.

BEAVER DAM CONSTRUCTION   Dams constructed by beavers (*Castor canadensis*) can greatly affect the downstream transport of nutrients (Naiman et al. 1988, 1994; Correll et al. 2000). In general, the decrease in flow in beaver ponds enhances uptake of dissolved nutrients by algae, sedimentation and retention of particulate nutrients, and denitrification rates (Naiman et al. 1988). Correll et al. (2000) found that a single beaver pond retained or volatilized (i.e., prevented downstream transport of) 18%, 21%, 32%, and 27%, respectively, of total N, P, Si, and suspended solids entering the pond over a six-year period. Over long timescales, nutrients are sequestered in meadows that develop after beaver ponds are abandoned, resulting in long-term reductions in downstream nutrient transport (Naiman et al. 1988, Correll et al. 2000).

Effects on Vertical Fluxes of Nutrients

EFFECTS ON SEDIMENTATION   In aquatic ecosystems, particulate matter, including phytoplankton and detritus, can sink from the euphotic zone into the sediments. Sedimentation can represent a major loss of nutrients and productivity from the euphotic zone (e.g., Figure 2; Caraco et al. 1992, Guy et al. 1994), although it may represent a source of nutrients for benthic organisms.

Many studies show that animals can affect the sedimentation of nutrients in lakes, but the effects are quite variable in terms of magnitude, mechanism, and even direction (Uehlinger & Bloesch 1987; Bloesch & Bürgi 1989; Mazumder et al. 1989; Sarnelle 1992, 1999; Elser et al. 1995; Larocque et al. 1996; Houser et al. 2000). Grazing by zooplankton can either increase or decrease nutrient sedimentation rates (Elser et al. 1995, Sarnelle 1999, Houser et al. 2000). Grazing can increase sedimentation rate by shifting size distributions of phytoplankton towards larger phytoplankton taxa, which sink at faster rates than smaller taxa (Mazumder et al. 1989, Larocque et al. 1996) and by production of fecal pellets (Bloesch & Bürgi 1989), which tend to be larger and hence sink faster than phytoplankton.

However, grazing can also decrease net nutrient sedimentation rate by reducing the amount of algal particles, and hence total nutrient mass, in the water column. Thus, nutrient sedimentation rate may decline under high grazing pressure simply because there is less nutrient mass in the water column available for sedimentation (Sarnelle 1992, 1999).

Grazing by zooplankton may increase nutrient sedimentation rate when productivity (nutrient concentration) is high, but decrease sedimentation rate when productivity is low (Houser et al. 2000). However, in some cases sedimentation rate may be maximal at intermediate zooplankton abundance (Sarnelle 1999). The net effects of zooplankton on nutrient sedimentation rate will depend on the direct rate of phytoplankton sinking (a function of phytoplankton size-structure and swimming ability), the proportion of zooplankton fecal material exiting the euphotic zone (a function of zooplankton taxonomic composition, as well as the rate at which fecal nutrients are remineralized, which in turn depends on the depth of the euphotic zone and turbulence), zooplankton assimilation efficiency (which determines fecal production rates), and ecosystem productivity (Elser et al. 1995, Sarnelle 1999).

Some bivalves can also greatly increase the rate at which particles are transported from the water column to the sediments (Strayer et al. 1999, Vaughn & Havenkamp 2001). These benthic organisms filter particles and subsequently deposit some of these (feces as well as particles captured but not consumed) onto the sediments. This can represent a significant loss of energy and nutrients from the water column and shift an ecosystem toward more benthic production and less pelagic production (Strayer et al. 1999).

BIOTURBATION   Many aquatic animals, including fish and invertebrates, physically disturb sediments via feeding or other activities, a process known as bioturbation, and this can greatly affect exchange of nutrients between sediments and overlying water. Nutrients often accumulate in sediment porewaters (water in between sediment particles). Physical mixing of sediments by benthic invertebrates can increase the rate at which porewater nutrients are released to overlying waters, and this process can be as important as nutrient excretion by these animals (Gallepp 1979, Graneli 1979, Matisoff et al. 1985, Starkel 1985, Tatrai 1986, Fukuhara & Sakamoto 1987). However, bioturbation can also decrease the rate at which nutrients are released from sediments and therefore can counteract excretion. For example, bioturbation can oxygenate near-surface porewaters by increasing diffusion of oxygen from overlying water. This can lead to increased rates of nitrification of excreted ammonium, increased denitrification rates because of increased nitrate flux to anoxic sediment layers, and/or precipitation of excreted phosphorus (Svensson 1997, Tuominen et al. 1999). The net effects of benthic invertebrates may be taxon-specific. Chironomid excretion products are more likely to be released into the overlying water due to the construction of burrows, which facilitate water movement (Fukuhara & Sakamoto, 1987). In contrast, excretion products of oligochaetes may be trapped in sediments because nutrients are excreted directly into the sediment (Fukuhara & Sakamoto, 1987). Fish can also affect nutrient flux via bioturbation, either directly by their own activities (e.g., Cline et al. 1994) or by predatory effects on benthic invertebrates (e.g., Svensson et al. 1999).

Atmosphere-Water Fluxes

Effects of animals on nutrient cycling can extend even to fluxes of nutrients between freshwaters and the atmosphere. For example, grazing by stream fishes can increase the abundance of grazing-resistant cyanobacteria, which could result in increased N-fixation rates and thus the total flux of N from the atmosphere (Power et al. 1988). In contrast, MacKay & Elser (1998) showed that nutrient excretion by *Daphnia* in a eutrophic lake reduced the abundance of cyanobacteria and N-fixation rates. Schindler et al. (1997) and Cole et al. (2000) showed that food web structure can affect the flux of CO_2 to the atmosphere. When lakes have relatively low nutrient inputs and food webs dominated by piscivores, then phytoplankton biomass is low, microbial respiration exceeds primary production, lake water is supersaturated with CO_2, and there is a net evacuation of CO_2 to the atmosphere. In contrast, when lakes are dominated by planktivorous fish and have high nutrient inputs, phytoplankton biomass and productivity is higher, and higher algal productivity draws CO_2 into the lake from the atmosphere. These lakes are more likely to be net sinks for CO_2 (Schindler et al. 1997, Cole et al. 2000).

FUTURE DIRECTIONS

It is clear that animals can have strong effects on nutrient cycling in some freshwater ecosystems, but further research is needed to assess generality. Future studies need to take a more holistic approach in terms of nutrient sources, nutrient recipients, and community-level consequences of animal-mediated nutrient cycling. While several studies have compared the importance of nutrient excretion by animals to other nutrient fluxes or to algal nutrient demand, or experimentally explored the role of animal-mediated nutrient cycling, no published studies have included all these approaches. In addition, nutrient demand by heterotrophic microbes as well as algae must be included in supply/demand approaches to achieve an ecosystem scale assessment of the role of animals. While inclusion of all these elements is labor-intensive, it is certainly feasible, and holistic studies will help facilitate the integration of species-based and biogeochemical approaches to ecology (Jones & Lawton 1995).

Future research must determine how often animals play a key role in nutrient cycling and what factors mediate this role. Mediating factors must include characteristics of the animals themselves (e.g., taxonomic affiliation, feeding guild, body size), and the ecosystems in which they reside. With regard to the former, there is considerable interest in ascertaining the roles of species identity and biodiversity in mediating ecosystem processes (Loreau et al. 2001). Exploration of variation among freshwater animals in how they mediate nutrient cycling may offer an excellent means of addressing this question, particularly in tropical ecosystems where the diversity of species and guilds is very high (Lowe-McConnell 1987, Covich et al. 1999, Vanni et al. 2002). Furthermore, ecological stoichiometry provides a sound theoretical basis for generating predictions about how animals may vary in mediating nutrient cycling (Sterner & Elser 2002, Vanni et al. 2002).

Ecosystem factors probably also affect the importance of animal-mediated nutrient cycling. Ecosystem size is likely to be important. For example, as lake size (surface area) decreases, the ratio of littoral to pelagic habitats increases (Schindler & Scheuerell 2002). Therefore, the role of animals in translocating nutrients from littoral to pelagic habitats is probably greater in small lakes, and most lakes world-wide are small (Wetzel 1990). In contrast, the importance of nutrient recycling within the water column is likely to increase with lake size (Fee et al. 1994). Ecosystem size (e.g., lake surface area, watershed area, stream channel width) may also affect the magnitude of nutrient fluxes from abiotic sources such as watershed runoff, ground water inputs, and wind-mediated nutrient resuspension. The magnitude of these inputs will determine the relative role of animals in supporting the nutrient demands of autotrophs and heterotrophic microbes. Productivity may also be important in mediating the role of animals in nutrient cycling. Attayde & Hansson (2001b) suggested that excretion by animals is more important (relative to herbivory) in unproductive lakes than in productive lakes because nutrient availability is lower in the former. In contrast, Drenner et al. (1998) and Vanni & Headworth (in press) suggest that the importance of nutrient translocation by benthic-feeding fish is likely to increase with lake productivity. No studies have explicitly quantified how nutrient cycling by animals varies with productivity. In addition, factors related to the animals themselves and to ecosystems may interact. For example, species richness and food chain length both increase with lake size, and the former is also related to productivity (Dodson et al. 2000, Post et al. 2001). Species richness and food chain length can mediate the role of animals in nutrient cycling in a number of ways.

Finally, we need to know how the role of animals differs in streams and lakes. Essington & Carpenter (2000) suggested that consumers in streams are most likely to affect nutrient cycling by controlling the rate at which dissolved nutrients are taken up by biota. For example, animals can affect transient storage zones and nutrient uptake lengths via consumption of periphyton mats or detritus. In contrast, Essington & Carpenter (2000) suggest that in lakes, animals are most likely to play a role by converting particulate nutrients into dissolved nutrients via excretion. Certainly, stream ecologists have focused much attention on nutrient uptake length and spiraling (Mulholland 1996, Meyer et al. 1988), and the role of nutrient excretion by animals has been explored much more in lakes than in streams (Table 1). However, it is not clear if the apparently different roles of animals in lakes and streams are real or simply due to different approaches taken by stream and lake ecologists. Webster & Wallace (1996) suggest that the role of nutrient excretion by stream animals warrants much more study, and the few studies that have quantified this in streams demonstrate its importance (Grimm 1988a,b; Hood 2000; Vanni et al. 2002). Future studies need to assess how animal-mediated nutrient cycling varies between lotic and lentic ecosystems, and along gradients such as ecosystem size, productivity, and species composition.

ACKNOWLEDGMENTS

I thank A.M. Bowling, S.P. Glaholt, S.J. Harper, M.J. Horgan, D.M. Post, A.D. Rosemond, D.E. Schindler, K.A. Sigler, H.M.H. Stevens, and an anonymous reviewer for comments on an earlier draft, and the National Science Foundation (DEB 9615620, 9726877 and 9982124) for supporting my research on nutrient cycling.

The *Annual Review of Ecology and Systematics* is online at
http://ecolsys.annualreviews.org

LITERATURE CITED

Andersson G, Granéli W, Stenson J. 1988. The influence of animals on phosphorus cycling in lake ecosystems. *Hydrobiologia* 170:267–84

Arnott DL, Vanni MJ. 1996. Nitrogen and phosphorus recycling by zebra mussels (*Dreissena polymorpha*) in the western basin of Lake Erie. *Can. J. Fish. Aquat. Sci.* 53:646–59

Attayde JL, Hansson LA. 2001a. Fish-mediated nutrient recycling and the trophic cascade in lakes. *Can. J. Fish. Aquat. Sci.* 58:1924–31

Attayde JL, Hansson LA. 2001b. The relative importance of fish predation and excretion effects on planktonic communities. *Limnol. Oceanogr.* 46:1001–12

Axler RP, Redfield GW, Goldman CR. 1981. The importance of regenerated nitrogen to phytoplankton productivity in a subalpine lake. *Ecology* 62:345–54

Bachmann RW, Jones BL, Fox DD, Hoyer M, Bull LA, Canfield DE. 1996. Relations between trophic state indicators and fish in Florida (USA) lakes. *Can. J. Fish. Aquat. Sci.* 53:842–55

Baines SB, Pace ML. 1994. Relationships between suspended particulate matter and sinking flux along a trophic gradient and implications for the fate of planktonic primary production. *Can. J. Fish. Aquat. Sci.* 51:25–36

Barlow JP, Bishop JW. 1965. Phosphate regeneration by zooplankton in Cayuga Lake. *Limnol. Oceanogr.* 10(Suppl.):R15–24

Bartell SM. 1981. Potential impact of size-selective planktivory on P release by zooplankton. *Hydrobiologia* 80:139–46

Bartell SM, Kitchell JF. 1978. Seasonal impact of planktivory on phosphorus release by Lake Wingra zooplankton. *Verh. Int. Verein. Limnol.* 20:466–74

Ben-David M, Hanley TA, Schell SM. 1998. Fertilization of terrestrial vegetation by spawning Pacific salmon. *Can. J. Zool.* 75:376–82

Bilby RE, Fransen BR, Bisson PA. 1996. Incorporation of nitrogen and carbon from spawning coho salmon into the trophic system of small streams: evidence from stable isotopes. *Can. J. Fish. Aquat. Sci.* 53:164–73

Bloesch J, Bürgi H-R. 1989. Changes in phytoplankton and zooplankton biomass and composition reflected by sedimentation. *Limnol. Oceanogr.* 34:1048–61

Blumenshine SC, Lodge DM, Hodgson JR. 2000. Gradient of fish predation alters body size distributions of lake benthos. *Ecology* 81:374–86

Brabrand A, Faafeng BA, Nilssen JPM. 1990. Relative importance of phosphorus supply to phytoplankton production: fish excretion versus external loading. *Can. J. Fish. Aquat. Sci.* 47:364–72

Brooks JL, Dodson SI. 1965. Predation, body size, and composition of plankton. *Science* 150:28–35

Caraco NF, Cole JJ, Likens GE. 1992. New and recycled primary production in an oligotrophic lake: insights for summer phosphorus dynamics. *Limnol. Oceanogr.* 37:590–602

Carney JH, Elser JJ. 1990. Strength of zooplankton-phytoplankton coupling in relation to lake trophic state. In *Ecological Structure and Function in Large Lakes*, ed. MM Tilzer, C Serruya, pp. 616–31. Berlin: Springer

Carpenter SR, Caraco NF, Correll DL, Howarth RW, Sharpley AN, Smith VH. 1998. Nonpoint pollution of surface waters with phosphorus and nitrogen. *Ecol. Appl.* 8:559–68

Carpenter SR, Kitchell JF, Hodgson JR. 1985. Cascading trophic interactions and lake productivity. *BioScience* 35:634–39

Carpenter SR, Kraft CE, Wright R, He X, Soranno PA, Hodgson JR. 1992. Resilience and resistance of a lake phosphorus cycle before and after food web manipulation. *Am. Nat.* 140:781–98

Chapin FS III, Zavaleta ES, Eviners VT, Naylor RL, Vitousek PM, et al. 2000. Consequences of changing biodiversity. *Nature* 405:234–42

Chipps SR, Bennett DH. 2000. Zooplanktivory and nutrient regeneration by invertebrate (*Mysis relicta*) and vertebrate (*Oncorhynchus nerka*) planktivores: implications for trophic interactions in oligotrophic lakes. *Trans. Am. Fish. Soc.* 129: 569–83

Cline JM, East TL, Threlkeld ST. 1994. Fish interactions with the sediment-water interface, *Hydrobiologia* 276:301–11

Cole JJ, Pace ML, Carpenter SR, Kitchell JF. 2000. Persistence of net heterotrophy in lakes during nutrient addition and food web manipulations. *Limnol. Oceanogr.* 45:1718–30

Correll DL, Jordan TE, Weller DE. 2000. Beaver pond biogeochemical effects in the Maryland coastal plain. *Biogeochemistry* 49:217–39

Costanza R, d'Arge R, deGroot R, Farber S, Grasso M, et al. 1997. The value of the world's ecosystem services and natural capital. *Nature* 387:253–60

Covich AP, Palmer MA, Crowl TA. 1999. The role of benthic invertebrate species in freshwater ecosystems. *BioScience* 49:119–28

Crowl TA, McDowell WH, Covich AP, Johnson SL. 2001. Freshwater shrimp effects on detrital processing and nutrients in a tropical headwater stream. *Ecology* 82:775–73

Cuker BE. 1983. Grazing and nutrient interactions in controlling the activity and composition of the epilithic algal community of an arctic lake. *Limnol. Oceanogr.* 28:133–41

Cummins KW, Klug MJ. 1979. Feeding ecology of stream invertebrates. *Annu. Rev. Ecol. Syst.* 10:147–72

Davis WR, Christian AD, Berg DJ. 2000. Seasonal nitrogen and phosphorus cycling by three unionid bivalves (Unionidae: Bivalvia) in headwater streams. In *Freshwater Mollusk Symposium Proceedings*, ed. RS Tankersley, DO Warmolts, GT Watters, BJ Armitage, PD Johnson, RS Butler, pp. 1–10. Columbus, OH: Ohio Biol. Surv.

DeAngelis DL. 1992. *Dynamics of Nutrient Cycling and Food Webs*. London: Chapman & Hall

DeAngelis DL, Mulholland PJ, Palumbo AV, Steinman AD, Huston MA, Elwood JW. 1989. Nutrient dynamics and food-web stability. *Annu. Rev. Ecol. Syst.* 20:71–95

Devine JA, Vanni MJ. 2002. Spatial and seasonal variation in nutrient excretion by benthic invertebrates in a eutrophic reservoir. *Freshwater Biol.* 47:1107–21

Dodson SI, Arnott SE, Cottingham KL. 2000. The relationship in lake communities between primary productivity and species richness. *Ecology* 81:2662–79

Drenner RW, Gallo KL, Baca RM, Smith JD. 1998. Synergistic effects of nutrient loading and omnivorous fish on phytoplankton biomass. *Can. J. Fish. Aquat. Sci.* 55:2087–96

Dugdale RC, Goering JJ. 1967. Uptake of new and regenerated forms of nitrogen in primary productivity. *Limnol. Oceanogr.* 12:196–206

Elser JJ, Elser MM, MacKay NA, Carpenter SR. 1988. Zooplankton-mediated transitions between N- and P-limited algal growth. *Limnol. Oceanogr.* 33:1–14

Elser JJ, Foster DK, Hecky RE. 1995. Effects of zooplankton on sedimentation in pelagic ecosystems: theory and test in two lakes of the Canadian shield. *Biogeochemistry* 30:143–70

Elser JJ, Goff NC, MacKay NA, St. Amand AL, Elser MM, Carpenter SR. 1987. Species-specific algal responses to zooplankton-experimental and field observations in 3 nutrient-limited lakes. *J. Plankton Res.* 9:699–717

Elser JJ, Sterner RW, Galford AE, Chrzanowski TH, Findlay DL, et al. 2000. Pelagic C:N:P stoichiometry in a eutrophied lake: responses to a whole-lake manipulation. *Ecosystems* 3:293–307

Elser JJ, Urabe J. 1999. The stoichiometry of consumer-driven nutrient recycling: theory, observations, and consequences. *Ecology* 80:735–51

Eppley RW, Peterson BJ. 1979. Particulate organic matter flux and planktonic new production in the deep ocean. *Nature* 282:677–80

Essington TE, Carpenter SR. 2000. Nutrient cycling in lakes and streams: insights from a comparative analysis. *Ecosystems* 3:131–43

Evarts JE. 1997. *Nutrient release from the sediments in a midwest eutrophic reservoir.* Masters thesis. Miami Univ., Oxford, Ohio. 55 pp.

Fee EJ, Hecky RE, Regehr GW, Hendzel LL, Wilkinson P. 1994. Effects of lake size on nutrient availability in the mixed-layer during summer stratification. *Can. J. Fish. Aquat. Sci.* 51:2756–68

Finney BP, Gregory-Eaves I, Sweetman J, Douglas MSV, Smol JP. 2000. Impacts of climatic change and fishing on Pacific salmon abundance over the past 300 years. *Science* 290:795–99

Flecker AS. 1996. Ecosystem engineering by a dominant detritivore in a diverse tropical stream. *Ecology* 77:1845–54

Fukuhara H, Sakamoto M. 1987. Enhancement of inorganic nitrogen and phosphate release from lake sediment by tubificid worms and chironomid larvae. *Oikos* 38:312–20

Fukuhara H, Yasuda K. 1985. Phosphorus excretion by some zoobenthos in a eutrophic freshwater lake and its temperature dependency. *Jpn. J. Limnol.* 46:287–96

Gallepp GW. 1979. Chironomid influence on phosphorus release in sediment-water microcosms. *Ecology* 60:547–56

Gardner WS, Nalepa TF, Quigley MA, Malcyk JM. 1981. Release of phosphorus by certain benthic invertebrates. *Can. J. Fish. Aquat. Sci.* 38:978–81

Geddes P. 1999. *Omnivory and periphyton mats: uncoupling and quantifying consumer effects in the Florida Everglades.* Masters Thesis. Fla. Int. Univ., Miami, FL. 100 pp.

Gerking SD. 1955. Endogenous nitrogen excretion of bluegill sunfish. *Physiol. Zool.* 28:283–89

Gido KB. 2002. Interspecific comparisons and the potential importance of nutrient excretion by benthic fishes in a large reservoir. *Trans. Am. Fish. Soc.* 131:260–70

Goldman JC, McCarthy JJ, Peavey DG. 1979. Growth rate influence on the chemical composition of phytoplankton in oceanic waters. *Nature* 279:210–15

Granéli W. 1979. The influence of *Chironomus plumosa* larvae on the exchange of dissolved substances between sediments and water. *Hydrobiologia* 66:149–59

Grimm NB. 1988a. Role of macroinvertebrates in nitrogen dynamics of a desert stream. *Ecology* 69:1884–93

Grimm NB. 1988b. Feeding dynamics, nitrogen budgets, and ecosystem role of a desert stream omnivore, *Agosia chrysogaster* (Pisces: Cyprinidae). *Environ. Biol. Fish.* 21:143–52

Gulati RD, Martinez CP, Siewertsen K. 1995. Zooplankton as a compound mineralizing and synthesizing system: phosphorus excretion. *Hydrobiologia* 315:25–37

Gutelmakher BL, Makartseva. 1990. The significance of zooplankton in the cycling of phosphorus in lakes of different trophic categories. *Int. Rev. Gesamten Hydrobiol.* 75:143–51

Guy M, Taylor WD, Carter JCH. 1994. Decline in total phosphorus in the surface waters of lakes during summer stratification, and its relationship to size distribution of particles and sedimentation. *Can. J. Fish. Aquat. Sci.* 51:1330–37

Hansson LA, Johansson L, Persson L. 1987. Effects of fish grazing on nutrient release and succession of primary producers. *Limnol. Oceanogr.* 32:723–29

Havens KE. 1991. Fish-induced sediment resuspension: effects on phtyoplankton biomass and community structure in a shallow hypereutrophic lake. *J. Plankton Res.* 13:1163–76

Havens KE. 1993. Response to experimental fish manipulations in a shallow, hypereutrophic lake: the relative importance of benthic nutrient recycling and trophic cascade. *Hydrobiologia* 254:73–80

Hecky RE, Hesslein RH. 1995. Contributions of benthic algae to lake food webs as revealed by stable isotope analysis. *J. N. Am. Benth. Soc.* 14:631–53

Hilderbrand GV, Hanley TA, Robbins CT, Schwartz CC. 1999. Role of brown bears (*Ursa arctos*) in the flow of marine nitrogen into a terrestrial ecosystem. *Oecologia* 121:546–50

Hjerne O, Hansson S. 2002. The role of fish and fisheries in Baltic Sea nutrient dynamics. *Limnol. Oceanogr.* 47:1023–32

Hood JM. 2000. *The potential importance of nutrient regeneration by fish in a neotropical stream.* Thesis, Miami Univ., Oxford, OH. 62 pp.

Houser JN, Carpenter SR, Cole JJ. 2000. Food web structure and nutrient enrichment: effects on sediment phosphorus retention in whole-lake experiments. *Can. J. Fish. Aquat. Sci.* 57:1524–33

Hudson JJ, Taylor WD, Schindler DW. 1999. Planktonic nutrient regeneration and cycling efficiency in temperate lakes. *Nature* 400:659–61

Jones CG, Lawton JH, ed. 1995. *Linking Species and Ecosystems.* London: Chapman & Hall

Kitchell JF, Koonce JF, Tennis PS. 1975. Phosphorus flux through fishes. *Verh. Internat. Verein. Limnol.* 19:2478–84

Kitchell JF, O'Neill RV, Webb D, Gallepp GW, Bartell SM, et al. 1979. Consumer regulation of nutrient cycling. *BioScience* 29:28–34

Kitchell JF, Schindler DE, Herwig BR, Post DM, Olson MH, Oldham M. 1999. Nutrient cycling at the landscape scale: the role of diel foraging migrations by geese at the Bosque del Apache National Wildlife Refuge, New Mexico. *Limnol. Oceanogr.* 44:828–36

Kline TC Jr, Goering JJ, Mathisen OA, Poe PH, Parker PL. 1993. Recycling of elements transported upstream by runs of Pacific salmon: I. ^{15}N and ^{13}C evidence in Sashin Creek, southeastern Alaska. *Can. J. Fish. Aquat. Sci.* 47:136–44

Knoll LB, Vanni MJ, Renwick WH. 2002. Phytoplankton primary production and photosynthetic parameters in reservoirs along a gradient of watershed land use. *Limnol. Oceangr.* In press

Kraft CE. 1992. Estimates of phosphorus cycling by fishes using a bioenergetics model. *Can. J. Fish. Aquat. Sci.* 49:2596–2604

Kraft CE. 1993. Phosphorus regeneration by Lake Michigan alewives in the mid-1970s. *Trans. Am. Fish. Soc.* 122:749–55

Lamarra VA Jr. 1975. Digestive activities of carp as a major contributor to the nutrient loading of lakes. *Verh. Int. Verein. Limnol.* 19:2461–68

Larocque I, Mazumder A, Proulx M, Lean DRS, Pick FR. 1996. Sedimentation of algae: relationships with biomass and size distribution. *Can J. Fish. Aquat. Sci.* 53:1133–42

Lauritsen DD, Mosley SC. 1989. Nutrient excretion by the Asiatic clam *Corbicula fluminea. J. N. Am. Benthol. Soc.* 8:134–39

Lehman JT. 1980a. Release and cycling of nutrients between planktonic algae and herbivores. *Limnol. Oceanogr.* 25:620–32

Lehman JT. 1980b. Nutrient recycling as an interface between algae and grazers in freshwater communities. In *Evolution and Ecology of Zooplankton Communities*, ed. WC Kerfoot, pp. 251–63. Hanover, NH: New England

Lehman JT, Sandgren CD. 1985. Species-specific rates of growth and grazing loss among freshwater algae. *Limnol. Oceanogr.* 30:34–46

Loreau M, Naeem S, Inchausti P, Bengtsson J, Grime JP, et al. 2001. Ecology-biodiversity and ecosystem functioning: current knowledge and future challenges. *Science* 294:804–88

Lowe-McConnell RH. 1987. *Ecological Studies in Tropical Fish Communities.* Cambridge, UK: Cambridge Univ.

MacKay NA, Elser JJ. 1998. Nutrient recycling by *Daphnia* reduces N_2 fixation by cyanobacteria. *Limnol. Oceanogr.* 43:347–54

March JG, Benstead JP, Pringle CM, Ruebel MW. 2001. Linking shrimp assemblages with rates of detrital processing along an elevational gradient in a tropical stream. *Can. J. Fish. Aquat. Sci.* 58:470–78

Mather ME, Vanni MJ, Wissing TE, Davis SA, Schaus MH. 1995. Regeneration of nitrogen and phosphorus by bluegill and gizzard shad: effect of feeding history. *Can. J. Fish. Aquat. Sci.* 52:2327–38

Matisoff G, Fisher JB, Matis S. 1985. Effects of benthic macroinvertebrates on the exchange of solutes between sediments and freshwater. *Hydrobiologia* 122:19–33

Mazumder A, Taylor WD, McQueen DJ, Lean DRS. 1989. Effects of fertilization and planktivorous fish on epilimnetic phosphorus and phosphorus sedimentation in large enclosures. *Can. J. Fish. Aquat. Sci.* 46:1735–42

McNaughton SJ, Banyikwa FF, McNaughton MM. 1997. Promotion of the cycling of diet-enhancing nutrients by African grazers. *Science* 278:1798–900

Mehner T, Mattukat F, Bauer D, Voigt H, Benndorf J. 1998. Influence of diet shifts in underyearling fish on phosphorus recycling in

a hypereutrophic biomanipulated reservoir. *Freshwater Biol.* 40:759–69

Meyer JL, McDowell WH, Bott TL, Elwood JW, Ishizaki C, et al. 1988. Elemental dynamics in streams. *J. N. Am. Benthol. Soc.* 7:410–32

Meyer JL, Schultz ET. 1985. Migrating haemulid fishes as a source of nutrients and organic matter on coral reefs. *Limnol. Oceanogr.* 30:146–56

Mulholland PJ. 1996. Role in nutrient cycling in streams. In *Algal Ecology: Freshwater Benthic Ecosystems*, ed. RJ Stevenson, ML Bothwell, RL Lowe, pp. 609–39. San Diego, CA: Academic

Mulholland PJ, Newbold JD, Elwood JW, Hom CL. 1983. The effect of grazing intensity on phosphorus spiralling in autotrophic streams. *Oecologia* 58:358–66

Mulholland PJ, Steinman AD, Marxolf ER, Hart DR, DeAngelis DL. 1994. Effect of periphyton biomass on hydraulic characteristics and nutrient cycling in streams. *Oecologia* 98:40–47

Naiman RJ, Bilby RE, Schindler DE, Helfield JM. 2002. Pacific salmon, nutrients, and the dynamics of freshwater and riparian ecosystems. *Ecosystems* 5:399–417

Naiman RJ, Johnston CA, Kelley JC. 1988. Alteration of North American streams by beaver. *BioScience* 38:753–62

Naiman RJ, Pinay G, Johnston CA, Pastor J. 1994. Beaver influences on the long-term biogeochemical characteristics of boreal forest drainage networks. *Ecology* 75:905–21

Nakashima BS, Leggett WC. 1980. The role of fishes in the regulation of phosphorus availability in lakes. *Can. J. Fish. Aquat. Sci.* 37:1540–49

Nalepa TF, Gardner WS, Malcyk JM. 1991. Phosphorus cycling by mussels (Unionidae: Bivalvia) in Lake St. Clair. *Hydrobiolgia* 219:239–50

Newbold JD, Elwood JW, O'Neill RV, Van Winkle W. 1981. Measuring nutrient spiralling in streams. *Can. J. Fish. Aquat. Sci.* 38:860–63

Olsen Y, Varum KM, Jensen A. 1996. Dependence of the rate of release of phosphorus by zooplankton on the C:P ratio in the food supply, as calculated by a recycling model. *Limnol. Oceanogr.* 31:34–44

Pace ML, Funke E. 1991. Regulation of planktonic microbial communities by nutrients and herbivores. *Ecology* 72:904–14

Pastor J, Dewey B, Naiman RJ, McInnes PF, Cohen Y. 1993. Moose browsing and soil fertility in the boreal forest of Isle Royale National Park. *Ecology* 74:467–80

Persson A. 1997a. Phosphorus release by fish in relation to external and internal load in a eutrophic lake. *Limnol. Oceanogr.* 42:577–83

Persson A. 1997b. Effect of fish predation and excretion on the configuration of aquatic food webs. *Oikos* 79:137–46

Peters RH. 1983. *The Ecological Consequences of Body Size.* New York: Cambridge Univ. Press

Polis GA, Anderson WB, Holt RD. 1997. Toward an integration of landscape and food web ecology: the dynamics of spatially subsidized food webs. *Annu. Rev. Ecol. Syst.* 28:289–316

Post DM, Pace ML, Hairston NG Jr. 2000. Ecosystem size determines food-chain length in lakes. *Nature* 405:1047–49

Post DM, Taylor JP, Kitchell JF, Olson MH, Schindler DE, Herwig BR. 1998. The role of migratory waterfowl as nutrient vectors in a managed wetland. *Cons. Biol.* 12:910–20

Power ME. 1990. Resource enhancement by indirect effects of grazers—armored catfish, algae, and sediment. *Ecology* 71:897–904

Power ME, Stewart AJ, Matthews WJ. 1988. Grazer control of algae in an Ozark mountain stream—effects of short-term exclusion. *Ecology* 69:1894–98

Pringle CM, Hemphill N, McDowell WH, Bednarek A, March JG. 1999. Linking species and ecosystems: different biotic assemblages cause interstream differences in organic matter. *Ecology* 80:1860–72

Reinertsen H, Jensen A, Koksvik JL, Langel A, Olsen Y. 1990. Effects of fish removal on the limnetic ecosystem of a eutrophic lake. *Can. J. Fish. Aquat. Sci.* 47:166–73

Richey JE, Perkins AM, Goldman CR. 1975.

Effects of kokanee salmon (*Oncorhynchus nerka*) decomposition on the ecology of a subalpine stream. *J. Fish. Res. Board. Can.* 32:817–20

Rosemond AD. 1993. Interactions among irradiance, nutrients, and herbivores constrain a stream algal community. *Oecologia* 94:585–94

Rosemond AD, Mulholland PJ, Elwood JW. 1993. Top-down and bottom-up control of stream periphyton: effects of nutrients and herbivores. *Ecology* 74:1264–80

Rosemond AD, Pringle CM, Ramírez A. 1998. Macroconsumer effects on insect detritivores and detritus processing in a tropical stream. *Freshwater Biol.* 39:515–23

Rosemond AD, Pringle CM, Ramírez A, Paul MJ, Meyer JL. 2002. Landscape variation in phosphorus concentration and effects on detritus-based tropical streams. *Limnol. Oceanogr.* 47:278–89

Sarnelle O. 1992. Contrasting effects of *Daphnia* on ratios of nitrogen to phosphorus in a eutrophic, hard-water lake. *Limnol. Oceanogr.* 37:1527–42

Sarnelle O. 1999. Zooplankton effects on vertical particulate flux: testable models and experimental results. *Limnol. Oceanogr.* 44:357–70

Schaus MH, Vanni MJ. 2000. Effects of omnivorous gizzard shad on phytoplankton and nutrient dynamics: effect of sediment-feeding and fish size. *Ecology* 81:1701–19

Schaus MH, Vanni MJ, Wissing TE. 2002. Biomass-dependent diet shifts in omnivorous gizzard shad: implications for growth, food webs and ecosystem effects. *Trans. Am. Fish. Soc.* 131:40–54

Schaus MH, Vanni MJ, Wissing TE, Bremigan MT, Garvey JA, Stein RA. 1997. Nitrogen and phosphorus excretion by detritivorous gizzard shad in a reservoir ecosystem. *Limnol. Oceanogr.* 42:1386–97

Schindler DE. 1992. Nutrient regeneration by sockeye salmon (*Oncorhynchus nerka*) fry and subsequent effects on zooplankton and phytoplankton. *Can. J. Fish. Aquat. Sci.* 49:2498–2506

Schindler DE, Carpenter SR, Cole JJ, Kitchell JF, Pace ML. 1997. Influence of food web structure on carbon exchange between lakes and the atmosphere. *Science* 277:248–51

Schindler DE, Eby LA. 1997. Stoichiometry of fishes and their prey: implications for nutrient cycling. *Ecology* 78:1816–31

Schindler DE, Kitchell JF, He X, Hodgson JR, Carpenter SR. 1993. Food web structure and phosphorus recycling in lakes. *Trans. Am. Fish. Soc.* 122:756–72

Schindler DE, Knapp RA, Leavitt PR. 2001. Alteration of nutrient cycles and algal production resulting from fish introduction into mountain lakes. *Ecosystems* 4:308–21

Schindler DE, Scheuerell MD. 2002. Habitat coupling in lake ecosystems. *Oikos.* 98:177–89

Schlesinger WH. 1997. *Biogeochemistry: An Analysis of Global Change.* San Diego, CA: Academic. 2nd ed.

Shapiro J, Carlson RE. 1982. Comment on the role of fishes in the regulation of phosphorus availability in lakes. *Can. J. Fish. Aquat. Sci.* 39:364

Shapiro J, Wright DI. 1984. Lake restoration by biomanipulation—Round Lake, Minnesota, the 1st 2 years. *Freshwater Biol.* 14:371–83

Sirotnak JM, Huntly NJ. 2000. Direct and indirect effects of herbivores on nitrogen dynamics: voles in riparian areas. *Ecology* 81:78–87

Smith VH. 1979. Nutrient dependence of primary productivity in lakes. *Limnol. Oceanogr.* 24:1051–64

Smith VH. 1983. Low nitrogen to phosphorus ratios favor dominance by blue-green algae in lake phytoplankton. *Science* 221:669–71

Smith VH. 1998. Cultural eutrophication of inland, estuarine, and coastal waters. In *Successes, Limitations, and Frontiers in Ecosystem Science,* ed. ML Pace, PM Groffman, pp. 7–49. New York: Springer

Starkel WM. 1985. Predicting the effect of macrobenthos on the sediment-water flux of metals and phosphorus. *Can. J. Fish. Aquat. Sci.* 42:95–100

Stein RA, DeVries DR, Dettmers JM. 1995. Food-web regulation by a planktivore: exploring the generality of the trophic cascade hypothesis. *Can. J. Fish. Aquat. Sci.* 52:2518–26

Steinman AD, Mulholland PJ, Kirschtel DB. 1991. Interactive effects of nutrient reduction and herbivory on biomass, taxonomic structure, and P uptake in lotic periphyton communities. *Can. J. Fish. Aquat. Sci.* 48:1951–59

Sterner RW. 1986. Herbivores' direct and indirect effects on algal populations. *Science* 231:605–7

Sterner RW. 1989. The role of grazers in phytoplankton succession. In *Plankton Ecology: Succession in Plankton Communities,* ed. U Sommer, pp. 107–70. Berlin: Springer

Sterner RW. 1990. The ratio of nitrogen to phosphorus resupplied by herbivores: zooplankton and the algal competitive arena. *Am. Nat.* 136:209–29

Sterner RW, Chrzanowski TH, Elser JJ, George NB. 1995. Sources of nitrogen and phosphorus supporting the growth of bacterioplankton and phytoplankton in an oligotrophic Canadian shield lake. *Limnol. Oceanogr.* 40:242–49

Sterner RW, Elser JJ. 2002. *Ecological Stoichiometry: The Biology of Elements from Molecules to the Biosphere.* Princeton, NJ: Princeton Univ. Press

Sterner RW, Elser JJ, Hessen DO. 1992. Stoichiometric relationships among producers, consumers and nutrient cycling in pelagic ecosystems. *Biogeochemistry* 17:49–67

Sterner RW, George NB. 2000. Carbon, nitrogen, and phosphorus stoichiometry of cyprinid fishes. *Ecology* 81:127–40

Strayer DL, Caraco NF, Cole JJ, Findlay S, Pace ML. 1999. Transformation of freshwater ecosystems by bivalves—a case study of zebra mussels in the Hudson River. *BioScience* 49:19–27

Suberkropp K, Chauvet E. 1995. Regulation of leaf breakdown by fungi in streams: influences of water chemistry. *Ecology* 76:1433–45

Svensson JM. 1997. Influence of *Chironomus plumosa* larvae on ammonium flux and denitrification (measured by the acetylene blockage- and the isotope pairing-techniques) in eutrophic lake sediment. *Hydrobiologia* 346:157–68

Svensson JM, Bergman E, Andersson G. 1999. Impact of cyprinid reduction on the benthic macroinvertebrate community and implications for increased nitrogen retention. *Hydrobiologia* 404:99–112

Tatrai I. 1986. Rates of ammonia release from sediments by chironomid larvae. *Freshwater Biol.* 16:61–66

Tilman D, Kilham SS, Kilham P. 1982. Phytoplankton community ecology—the role of limiting nutrients. *Annu. Rev. Ecol. Syst.* 13:349–72

Tuominen L, Mäkelä K, Lehtonen KK, Haahti H, Hietanen S, Kuparinen J. 1999. Nutrient fluxes, porewater profiles and denitrification in sediment influenced by algal sedimentation and bioturbation by *Monoporeia affinis*. *Estuar. Coast. Shelf Sci.* 49:83–97

Uehlinger U, Bloesch J. 1987. The influence of crustacean zooplankton on the size structure of algal biomass and suspended and settling seston (biomanipulation in limnocorrals II). *Int. Rev. Gesamten Hydrobiol.* 72:473–86

Urabe J, Nakashini M, Kawabata K. 1995. Contribution of metazoan plankton to the cycling of nitrogen and phosphorus in Lake Biwa. *Limnol. Oceanogr.* 40:232–41

Vanni MJ. 1996. Nutrient transport and recycling by consumers in lake food webs: implications for algal communities. In *Food Webs: Integration of Patterns and Dynamics*, ed. GA Polis, KO Winemiller, pp. 81–95. New York: Chapman & Hall

Vanni MJ, Flecker AS, Hood JM, Headworth JL. 2002. Stoichiometry of nutrient cycling by vertebrates in a tropical stream: linking species identity and ecosystem processes. *Ecol. Lett.* 5:285–93

Vanni MJ, Headworth JL. 2003. Cross-habitat transport of nutrients by omnivorous fish along a productivity gradient: integrating watersheds and reservoir food webs. In *Food Webs at the Landscape Level*, ed. GA Po-

lis, ME Power, GL Huxel. Chicago: Univ. Chicago Press. In press

Vanni MJ, Layne CD. 1997. Nutrient recycling and herbivory as mechanisms in the "top-down" effects of fish on phytoplankton in lakes. *Ecology* 78:21–41

Vanni MJ, Layne CD, Arnott SE. 1997. "Top-down" trophic interactions in lakes: effects of fish on plankton and nutrient dynamics. *Ecology* 78:1–20

Vanni MJ, Renwick WH, Headworth JL, Auch JD, Schaus MH. 2001. Dissolved and particulate flux from three adjacent agricultural watersheds: a five-year study. *Biogeochemistry* 54:85–114

Vanni MJ, Temte J. 1990. Seasonal patterns of grazing and nutrient limitation of phytoplankton in a eutrophic lake. *Limnol. Oceanogr.* 35:697–709

Vaughn CC, Havenkamp CC. 2001. The functional role of burrowing bivalves in freshwater ecosystems. *Freshwater Biol.* 46:1431–46

Wallace JB, Cuffney TF, Webster JR, Lugthart GJ, Chung K, Goldowitz BS. 1991. Export of fine organic particles from headwater streams: effects of season, extreme discharges, and invertebrate manipulations. *Limnol. Oceanogr.* 36:670–82

Webster JB, Patten BC. 1979. Effects of watershed perturbation on stream potassium and calcium dynamics. *Ecol. Monogr.* 19:51–72

Webster JB, Wallace JR. 1996. The roles of macroinvertebrates in stream ecosystem function. *Annu. Rev. Entomol.* 41:115–39

Wen YH, Peters RH. 1994. Empirical-models of phosphorus and nitrogen excretion rates by zooplankton. *Limnol. Oceanogr.* 39:1669–79

Wetzel RG. 1990. Land-water interfaces: metabolic and limnological regulators. *Verh. Int. Verein. Limnol.* 24:6–24

Wilhelm FM, Hudson JJ, Schindler DW. 1999. Contribution of *Gammarus lacustris* to phosphorus recycling in a fishless alpine lake. *Can. J. Fish. Aquat. Sci.* 56:1679–86

Wright DI, Shapiro J. 1984. Nutrient reduction by biomanipulation: an unexpected phenomenon and its possible cause. *Verh. Int. Verein. Limnol.* 22:518–24

Annu. Rev. Ecol. Syst. 2002. 33:371–96
doi: 10.1146/annurev.ecolsys.33.010802.150434
First published online as a Review in Advance on August 14, 2002

Effects of UV-B Radiation on Terrestrial and Aquatic Primary Producers*

Thomas A. Day[1] and Patrick J. Neale[2]

[1]*Department of Plant Biology, Arizona State University, Tempe, Arizona 85287-1601;*
email: tadday@asu.edu
[2]*Smithsonian Environmental Research Center, PO Box 28, Edgewater, Maryland 21037;*
email: neale@serc.si.edu

Key Words   biological weighting functions, ozone depletion, photosynthesis, phytoplankton, UV screening

■ **Abstract**   Ozone depletion by anthropogenic gases has increased the atmospheric transmission of solar ultraviolet-B radiation (UV-B, 280–315 nm). Our understanding of the consequencences of enhanced UV-B levels on primary producers has grown dramatically over the past 20 years, but it has been hampered by how realistically experimental UV-B exposures mimic ozone-depletion scenarios. Overcoming these shortcomings will require sophisticated and creative approaches. Biological weighting functions and solar spectral irradiance estimates are critical in evaluating effects and require more attention. Whereas UV screening compounds in terrestrial and aquatic producers commonly increase with UV-B exposure, the implications, while potentially far reaching, are unclear. Photosynthesis is more sensitive to UV-B in phytoplankton than in terrestrial plants, probably owing to less effective screening in phytoplankton. Productivity of terrestrial plants is usually unaffected by enhanced UV-B, although reduced growth has been observed and may increase in magnitude over successive years. Aquatic productivity is often compromised by short-term exposures to enhanced UV-B, and long-term assessments are complicated by the dynamic nature of aquatic systems and by nonlinear responses. Recent work examining UV-B effects on multiple trophic levels suggests that outcomes will be diverse and difficult to predict. Such effects may lead to feedbacks on primary producers.

INTRODUCTION

Concerns that anthropogenic emissions might deplete stratospheric ozone, leading to greater atmospheric transmission of ultraviolet-B radiation (UV-B) and higher surface fluxes, emerged in the early 1970s, with a focus on nitrogen oxide emissions from high-altitude aircraft. This threat proved short lived, as

---

*The US Government has the right to retain a nonexclusive, royalty-free license in and to any copyright covering this paper.

improved atmospheric models showed this risk to be low. However, at about this time, Molina & Rowland (1974) proposed that chlorofluorocarbons (CFCs), used mainly as refrigerants and spray propellants, might photodegrade in the stratosphere, releasing chlorine that could catalyze ozone depletion. At that time, understanding of the influence of solar UV-B on primary producers was limited, although pioneering work of Caldwell (1968, 1971) on terrestrial plants provided a solid framework for future research. A few early marine studies had observed that solar UV could reduce phytoplankton production (Steemann-Nielsen 1964, Jitts et al. 1976). This effect was subsequently found to be widespread among aquatic ecosystems (Harris 1978). By the early 1980s ozone depletion by CFCs was perceived as a significant threat (National Academy of Sciences 1979), and research addressing UV-B effects began in earnest. Many issues deemed important in terms of UV-B effects on primary producers were defined at several international workshops during the early 1980s, resulting in, among others, landmark reviews on the role of solar UV in terrestrial (Caldwell 1981) and marine (Calkins 1982) ecosystems and a special issue of *Physiologia Plantarum* (Vol. 58, No. 3, 1983).

The discovery of thinning ozone over Antarctica in the mid-1980s, and its subsequent link to CFCs, sparked a large research effort on UV-B effects that continues today. Whereas ozone depletion is most pronounced over polar regions, particularly Antarctica, where about one-half of the ozone column is depleted each spring (Jones & Shanklin 1995, Solomon 1999), these ozone-depleted air masses disperse and ultimately dilute concentrations over more temperate latitudes. Since 1980, biologically effective erythemal levels of UV-B have increased ~4–7% at northern mid-latitudes, 6% at southern mid-latitudes, and 130 and 22% during springtime in the Antarctic and arctic, respectively (Madronich et al. 1998). Ozone-depleting chemicals are thought to have reached peak loads over the past few years. Experts suggest that maximal ozone depletion and peak UV-B levels will occur over the next decade, with a return to pre-1980 levels of stratospheric ozone and UV-B by the middle of this century. However, many factors, including feedbacks from rising concentrations of greenhouse gases, could delay this return (Shindell et al. 1998, Montzka et al. 1999, Newman et al. 2001). A further complication for aquatic environments is that climate change affects several factors determining UV-B penetration (Hargreaves 2002).

The large amount of research addressing UV-B effects on primary producers and ecosystems over the past 20 years has prompted many reviews and edited volumes including those of Lumsden (1997), Rozema et al. (1997), Franklin & Forster (1997), Häder (1997), Björn et al. (1998), Caldwell et al. (1998a,b), de Mora et al. (2000), Hester & Harrison (2000), Cockell & Blaustein (2001), and Helbling & Zagarese (2002). Rather than provide an exhaustive review, we examine a handful of issues that appear particularly important in predicting effects of enhanced UV-B on primary producers. By comparing existing UV-B paradigms in the terrestrial vascular plant versus the aquatic plant (mainly phytoplankton) research fields, we hope to provide some new perspectives.

EXPERIMENTAL PROTOCOLS

Exposure Approaches

Much of the early work examining UV-B effects on terrestrial plants was conducted indoors using growth chambers or greenhouses, in which plants were exposed to UV-B from lamps (sometimes at unnaturally high levels) against a background of low UV-A (ultraviolet-A radiation, 315–400 nm) and PAR (photosynthetically active radiation, 400–700 nm). By the 1990s it was widely accepted that UV-B effects on terrestrial plants were typically exaggerated under the unnaturally high ratios of UV-B:UV-A and UV-B:PAR found in most indoor studies, and extrapolation to field responses was questioned (Caldwell & Flint 1994, 1997). Suggestions made in the late 1980s that one-third to one-half of the terrestrial plant species studied were sensitive to enhanced UV-B associated with ozone depletion may have been exaggerated because they rested on findings from indoor studies (Fiscus & Booker 1995).

As in terrestrial studies, much of the early work examining UV-B effects on phytoplankton was conducted indoors with UV lamps (Calkins & Thordardottir 1980, Thomson et al. 1980, Worrest et al. 1981). Again, because of unnaturally low UV-B:PAR ratios, these results were not thought appropriate for predicting the effects of enhanced UV-B outdoors (Cullen & Neale 1994, Prézelin et al. 1994). Moreover, other limitations were recognized, such as assumptions about the linearity of time responses and how vertical mixing affects the residence time of phytoplankton in the near-surface zone (Smith & Baker 1982, Worrest 1983, Cullen & Neale 1994).

These concerns spurred an emphasis on outdoor studies, which took advantage of the UV and visible background irradiance provided by sunlight. The two most widely used approaches in these outdoor studies involve either attenuating some of the solar UV-B reaching plants with filters ("attenuation" studies), or supplementing solar UV-B levels with UV lamps ("enhancement" studies).

FILTER ATTENUATION OF SOLAR UV-B   The attenuation approach uses sheet filters (terrestrial plants) or incubation containers (phytoplankton) constructed of materials that either absorb or transmit large percentages of UV-B, typically providing a sub-ambient and near-ambient UV-B treatment, respectively.

Some of the earliest experiments on the influence of solar UV-B exposure on terrestrial plants employed the filter attenuation approach (Caldwell 1968). More recent attenuation studies have documented that solar UV-B exposure can reduce the biomass production of some plant species by 10–35% (Krizek et al. 1997, 1998; Mazza et al. 1999a; Xiong & Day 2001; Day et al. 2001; Day 2001a).

During the 1980s and early 1990s experiments assessing UV effects on phytoplankton relied heavily on the attenuation approach in combination with in situ incubation (Smith et al. 1980, Maske 1984, Bühlmann et al. 1987). Several groups used this approach in Antarctica to assess the effects of enhanced solar UV-B during ozone-depletion events. They found that reductions in productivity were generally much lower than earlier estimates from indoor lamp studies, ranging from 2–10% of daily production (Lubin et al. 1992; Vernet et al. 1994; Smith et al. 1992; Helbling et al. 1992, 1994; Prézelin et al. 1994). Such modest effects result because most of the surface inhibition is due to UV-A, and much of the daily productivity occurs deep in the water column, where UV exposure is low (Holm-Hansen et al. 1993, Cullen & Neale 1993).

Advantages of the attenuation approach are its low cost, its simplicity, and the fact that it does not require power. A disadvantage is that it typically only tests for the effects of ambient solar UV-B exposure, rather than above-ambient levels associated with ozone depletion. The exception is when experiments are conducted during ozone depletion events when solar UV-B is naturally enhanced (e.g., Antarctica). Regardless of whether or not solar UVB is naturally enhanced during an attenuation study, a common shortcoming of the attenuation approach is its use of filter designs that remove a large percentage of ambient UV-B. For example, UV-B–absorbing filters are usually arranged on frames such that the biologically effective UV-B dose received by terrestrial plants is ≈20% of ambient (Day 2001a). The control treatment involves placing UV-B–transparent filters on frames, but reflection and absorption invariably reduce UV-B doses under these frames to ≈85% of ambient (Day 2001a). This difference in doses is far larger than the enhancements in solar UV-B usually predicted with continued ozone depletion. It does provide a test of the effects of solar UV-B on plant performance, which is a worthy question in itself. However, in the context of ozone depletion, greater realism in attenuation experiments could be achieved with filter designs that provide smaller reductions in the ambient UV-B (e.g., Ballaré et al. 1996).

A related disadvantage of attenuation studies is that in order for the results to be used to predict the consequences of ozone depletion one must assume that responses are linear over a wide range of UV-B levels. In other words, plants will respond to a given increase in UV-B similarly, regardless of whether this increase is applied to a subambient or ambient level of UV-B. In the case of terrestrial plants this assumption has not been well tested, although responses are likely to be nonlinear, based on the variability in responses observed in studies employing several above-ambient UV-B levels. In the case of phytoplankton productivity responses are clearly nonlinear and tend to saturate at ambient levels (Neale 2000). These nonlinear responses make the extrapolation of findings from attenuation studies to ozone depletion scenarios problematic.

LAMP ENHANCEMENTS TO SOLAR UV-B   In contrast to filter attenuation studies, enhancement studies supplement ambient UV-B with fluorescent lamps to mimic future UV-B enhancements associated with ozone depletion. This approach has been used since the 1970s (Fox & Caldwell 1978) and most outdoor UV-B terrestrial plant studies have employed this method.

This approach has also been used with phytoplankton (Smith et al. 1980, Worrest et al. 1981, El-Sayed et al. 1990, Behrenfeld et al. 1993). Although the UV-B:PAR ratios in these experiments were much closer to those of natural sunlight than those in indoor studies, their predictive value was still questioned by the phytoplankton research community, based on the spectral difference between fluorescent lamps and solar UV (Cullen & Neale 1994, Prézelin et al. 1994). In contrast to terrestrial plant research, in which outdoor fluorescent lamp enhancements are the norm in UV assessments, many aquatic researchers are using solar simulators, whereby the output from a filtered xenon-arc lamp is used indoors to simulate natural sunlight. Although not a perfect simulator of sunlight, the spectral output from these lamps, when appropriately filtered, resembles that of the solar spectrum through the UV and visible wavebands and allows more control and resolution than can be obtained outdoors. This approach has been used by terrestrial plant researchers, but has generally been confined to experiments aimed at developing biological weighting functions (see below).

Because of their strong reliance on the outdoor enhancement approach, terrestrial plant researchers have refined this approach. When the UV-B supplement is held constant over the day, large levels of UV-B can be supplemented to small background levels of PAR (e.g., on overcast days or at low solar angles). This supplementation results in unnaturally high ratios of UV-B:PAR, and plant responses to these so-called squarewave lamp supplements may be exaggerated (Sullivan et al. 1994). Modulated systems (Caldwell et al. 1983a), which measure ambient solar irradiance and adjust lamp output to compensate for low background UV-B, have largely overcome this problem, although often at a prohibitive cost. The problem can be addressed to some extent by using a step-wise delivery system, in which UV-B output from lamps increases through the morning and then declines after solar noon. In combination with switching lamps off during completely overcast days, this step-wise approach can provide UV enhancements similar to those provided by more costly modulated systems (Musil et al. 2002).

Whereas the outdoor enhancement approach is attractive because one can examine responses to above-ambient UV-B levels, the spectral differences between fluorescent lamps and sunlight introduce uncertainties into their predictive value. For example, the fluorescent lamps in these systems have an emission peak around 313 nm. Hence, the UV spectral output from these lamps does not match the spectral enhancement found in sunlight with ozone depletion. Attenuation studies are not immune to uncertainties about spectral differences, as the filters used to reduce UV-B also absorb some UV-A.

Biological Weighting Functions

Biological weighting functions (BWFs) were developed to calculate UV dose and compare doses under different spectral regimes (e.g., lampbanks versus naturally enhanced sunlight). For example, BWFs are used to weight the UV spectral irradiance predicted by a model under a desired ozone-depletion level, and the daily biologically effective UV (UV_{be}) dose is calculated. The daily dose under lampbanks is then matched to the dose predicted by the model under the desired ozone depletion scenario.

Several BWFs have been proposed for terrestrial plants based on a diverse array of responses (Rundel 1983, Caldwell et al. 1986, Caldwell & Flint 1997). BWFs are typically developed indoors using a solar simulator (Caldwell et al. 1986). Probably the most commonly used in terrestrial plant studies is Caldwell's generalized plant damage BWF (Plant BWF; Caldwell 1971). While all BWFs give greater effectiveness to shorter wavelengths of UV, the rate at which they decline with higher UV wavelengths and how far their tails extend into these higher wavelengths vary appreciably. This variation can lead to large differences among BWFs in their radiation amplification factor (RAF). The RAF is the proportional increase in UV_{be} produced by a given decrease in ozone column (Caldwell et al. 1986). Steep BWFs that give relatively more weight to UV-B than UV-A, such as the Plant BWF, have a greater RAF and tend to require the least UV output from lamps to achieve a desired ozone depletion dose, because UV fluorescent sunlamps supply more short- than long-wave UV-B compared with sunlight.

Flint & Caldwell (2002a,b) recently proposed a UV photomorphogenic BWF (Morph BWF) based on terrestrial plant growth. This BWF gives considerable weight to UV-A wavelengths and therefore has a lower RAF than many earlier BWFs. However, while this lower RAF provides a smaller change in UV_{be} per increment of ozone depletion, it also means that previous enhancement experiments may have supplied milder doses of additional UV than intended, because they were usually based on the Plant BWF, which may have overemphasized UV-B and ignored UV-A (Flint & Caldwell 2002a,b).

Biological weighting functions were also identified as key unknowns in predicting the impact of ozone depletion on marine phytoplankton (Smith et al. 1980, Lubin et al. 1992). Many BWFs for UV inhibition of phytoplankton photosynthesis have been developed over the past decade using solar simulators (Cullen et al. 1992, Neale et al. 1994, Neale 2000, Neale & Kieber 2000). The spectral responses of phytoplankton to UV are highly variable, although some general patterns have emerged (Banaszak & Neale 2001). The spectral shape of BWFs in the UV-B is similar to that suggested by Behrenfeld et al. (1993) when averaged over all assemblages for a specific region (e.g., Chesapeake Bay or Southern Ocean), but specific assemblages can diverge widely from the average. Current work on this subject is focusing on how to better relate phytoplankton BWFs to environmental conditions (Litchman et al. 2002) to achieve more accurate assessments of UV effects on global aquatic primary productivity. In the meantime, an examination of UV spectral responses at some level is needed in any study purporting to estimate the effects of UV-B and ozone depletion on aquatic productivity.

In many ways, phytoplankton lend themselves to solar simulator experiments and development of BWFs better than terrestrial plants. Their small size allows simultaneous exposure of many individuals, and their response times are relatively fast (Neale & Fritz 2001). In contrast, the small dose area provided by solar simulators makes adequate replication of terrestrial plants problematic, and response times of processes most indicative of overall performance, such as growth, are slow. Because BWFs have a large influence on the lamp output used in UV-B experiments, continued assessments and refinements in BWFs are well warranted.

Estimating Solar UV-B Levels

Another area of uncertainty concerning UV-B doses is the precision of models that predict solar UV-B irradiance under ambient and depleted ozone levels. Booker et al. (1992) found that the semi-empirical model of Green (1983), computer encoded in the program of Bjorn & Murphy (1985), showed relatively close correspondence with broad-band dosimeter measurements. This program is widely used by terrestrial plant researchers. However, Musil et al. (2002) recently found that this computer-encoded model consistently underestimated clear-sky background UV-B irradiance in South Africa. These underestimates could have a significant effect on the doses employed in enhancement experiments, although their prevalence needs to be more fully evaluated.

Radiative transfer models may provide more precise estimates of UV spectral irradiance (Koepke et al. 1998) and are increasingly used by phytoplankton researchers (Neale et al. 2001, Neale 2001). Examples include models available from the Norwegian Institute for Air Research (via internet interface at http://zardoz.nilu.no/~olaeng/fastrt/fastrt.html) and the University of Munich (as a software package accessed at http://www.meteo.physik.uni-muenchen.de/strahlung/ uvrad/Star/STARinfo.htm) (see Koepke et al. 1998 for more examples).

Recommendations on Experimental Protocols

With an aim to improve realism of UV experiments and facilitate a synthesis of findings, we make the following recommendations: (*a*) Express doses using several published BWFs, as recommended by Cullen & Neale (1997) and Flint & Caldwell (2002b), and cite sources for parameterizations. It would be useful to include contrasting BWFs in terms of their spectral slopes or RAFs. (*b*) In outdoor enhancement studies, in the absence of a modulated system, use a stepwise approach and switch lamps off on overcast days. (*c*) Consider using radiative transfer models to estimate solar spectral irradiance. (*d*) In attenuation studies, include treatments that provide relatively small reductions in ambient UV-B. (*e*) Consider how the temporal realism of UV-B manipulations may be improved, for example, by compensating for seasonal trends in ozone levels or UV-B irradiance.

RESPONSES TO UV-B

The complexities involved in manipulating and quantifying UV-B levels present many challenges for investigators assessing UV-B effects. Nevertheless, our understanding of UV-B effects on primary producers has improved dramatically over the past 20 years. Investigators have assessed these effects on an impressive breadth of processes across large spatial and temporal scales. We review the effects on a handful of these processes that appear particularly important for developing a framework to predict how UV-B impacts primary productivity.

UV-Screening Compounds

Concentrations of UV-absorbing compounds commonly increase in the leaves of terrestrial plants with greater UV-B exposure. For example, in a meta-analysis of outdoor lamp enhancement studies, Searles et al. (2001a) found increases in bulk leaf concentrations of soluble methanol-extractable UV-B–absorbing compounds to be the most consistent response to UV-B supplements. Corroborating this analysis, a qualitative survey of recent outdoor studies employing filter attenuations and modulated enhancements found that about half of these studies detected increases in concentrations of these compounds (Day 2001a). In higher plants these compounds include a large number of phenylpropanoids, with flavonoids and hydroxycinnamic acids likely to be most important in terms of UV-B screening (Beggs & Wellmann 1994, Cockell & Knowland 1999). Several key enzymes in the pathway leading to synthesis of these compounds are stimulated by UV-B, as well as UV-A, PAR, and several other environmental stress factors (Caldwell et al. 1998b).

Higher bulk concentrations of these compounds are usually inferred to reduce epidermal transmittance and provide greater protection to putative targets in the leaf mesophyll. Increases in concentrations of these compounds, particularly the subtle increases seen in response to outdoor UV-B enhancements (average increases of $\approx 10\%$) (Searles et al. 2001a), may not necessarily lead to appreciable reductions in epidermal UV-B transmittance and improved screening. For example, the relationship between concentrations and epidermal transmittance is complicated because these compounds are not uniformly distributed across the epidermis (leading to sieve effects), they are not confined to the epidermis, and some are wall bound and therefore insoluble in standard extraction procedures (Day 2001a). In spite of these caveats, some strong negative correlations between bulk concentrations of soluble absorbing compounds and epidermal transmittance have been documented. Upon UV-B exposure of plants in growth chambers, increases in soluble compound concentrations have been correlated with reductions in transmittance (Cen & Bornman 1993, Reuber et al. 1996). More recent work has often found significant correlations between soluble concentrations and transmittance among leaves that differ in age or UV-B exposure (Barnes et al. 2000, Mazza et al. 2000, Bilger et al. 2001, Kolb et al. 2001, Markstädter et al. 2001). With refinements in extraction protocols, these correlations can be particularly strong (Burchard et al. 2000, Bilger et al. 2001). Further work in this area can provide us with a sense of whether the increases in concentrations seen in acclimated plants in response to UV-B enhancements lead to appreciable reductions in epidermal transmittance and improved screening.

Do these UV-B–induced increases in screening-compound concentrations and reductions in epidermal transmittance in turn reduce damage in plants? In a growth-chamber study Tevini et al. (1991) found that increases in these compounds, brought about by preexposure of rye plants to UV-B, appeared to reduce the severity of photosystem II (PSII) inhibition when they were subsequently challenged with UV-B. Kolb et al. (2001) presented greenhouse-grown grape plants to a UV-B challenge outdoors and found that PSII inhibition declined in parallel with increased epidermal screening. Mazza et al. (2000) found that soybean leaves that had higher constitutive concentrations of screening compounds, and lower epidermal transmittance, experienced less DNA damage when subsequently challenged with a short exposure to solar UV-B. Hence, improved epidermal screening correlated with higher concentrations of screening compounds can afford protection of DNA and photosynthetic machinery, at least for nonacclimated or young leaves presented with a UV-B challenge. Whether or not the increases in concentrations found in outdoor plants in response to higher UV-B levels also afford greater protection is unclear and will require further study. In several cases higher UV-B levels do not elicit increases in concentrations of these compounds, and plant performance is not impaired, suggesting that constitutive concentrations, together with internal repair and protection mechanisms, provide adequate protection.

In phytoplankton the primary UV screening compounds are mycosporine-like amino acids (MAAs). The biochemistry, occurrence and possible functions of the compounds have recently been reviewed (Dunlap & Shick 1998, Roy 2000, Banaszak & Trench 2001, Shick & Dunlap 2002). MAAs have absorption maxima from 310 to 340 nm and occur in the highest concentrations in marine plants inhabiting high-light environments (Shick & Dunlap 2002). When MAAs accumulate, cellular UV absorption (per unit chlorophyll) exceeds PAR absorption by as much as an order of magnitude. This observation led to early suggestions of their role as UV sunscreens, but they appear to have other metabolic functions as well (Shick & Dunlap 2002).

Owing to very short pathlengths, significant UV protection of single cells is obtained only when cell diameters are greater than ≈ 20 μm (Garcia-Pichel 1994). Much of the ocean is dominated by phytoplankton with smaller cell diameters. Some species of phytoplankton (particularly dinoflagellates) and macroalgae (particularly red macroalgae) accumulate significant amounts of MAAs yet are still sensitive to UV-B, suggesting that MAAs provide only partial protection (Franklin et al. 2001; Lesser 1996a,b). Effective, albeit partial, optical screening of UV-B by MAAs has been inferred from the negative correlation between concentrations and sensitivity to UV-B in the MAA absorption waveband (Neale et al. 1998a, Litchman et al. 2002). Although MAAs screen UV-B, induction of MAAs in marine algae mainly depends on UV-A and PAR exposure (Carreto et al. 1990; Riegger & Robinson 1997; Neale et al. 1998a; Moisan & Mitchell 2001; Franklin et al. 1999, 2001). Concentrations of MAAs usually do not increase dramatically in response to UV-B supplements when background levels of UV-A and PAR are already high (Hannach & Sigleo 1998). Since UV-A and PAR penetrate deeper into aquatic environments, irradiances in these wavebands are stronger cues for induction of MAAs than UV-B, per se (Neale et al. 2002). This regulation of MAAs primarily by longer wavelengths typically leads to only modest increases in

screening compounds in response to increases in solar UV-B owing to ozone depletion. In summary, concentrations of UV-screening compounds commonly increase
in terrestrial plants and phytoplankton in response to UV-B supplements, but these
increases are usually subtle, and it is unclear how much additional protection they
provide under field situations.

Photosynthesis

Many indoor studies have shown that UV-B can impair all of the main processes of
leaf photosynthesis in terrestrial plants (Allen et al. 1998), with PSII appearing to
be particularly sensitive (Bornman 1989, Tevini 1993, Teramura & Sullivan 1994).
However, photosynthetic gas-exchange rates per unit leaf area in terrestrial plants
are usually unaffected by UV-B when realistic doses are applied to acclimated
plants outdoors (Fiscus & Booker 1995, Allen et al. 1998, Searles et al. 2001a,
Day 2001a).

Nevertheless, there have been cases in which some photosynthetic processes
appear compromised by UV-B exposure in the field. Upon exposure of nonacclimated plants to solar UV-B, CO_2 assimilation rates can be reduced (Kolb
et al. 2001). Xiong & Day (2001) found that gas-exchange rates per unit leaf area
were unaffected by solar UV-B exposure during ozone depletion events. Chlorophyll fluorescence signals from the upper leaf surface showed that photosynthetic
function in the upper mesophyll was impaired, but this impairment was apparently
compensated for by thicker leaves with higher pigment concentrations, such that
gas-exchange rates per unit leaf area were not compromised. Some of the strongest
evidence for reductions of assimilation rates in acclimated plants outdoors comes
from Keiller & Holmes (2001), who found that UV-B enhancements over several
years led to lower CO_2 assimilation rates in five tree species. Reductions in photosynthetic function in these three studies appeared attributable to impairments in
light-independent enzyme activity, rather than PSII damage per se, corroborating
the conclusion of Allen et al. (1998). Nonetheless, photosynthetic gas-exchange
rates per unit leaf area are usually not reduced in acclimated plants in response to
UV-B supplements outdoors.

In contrast to terrestrial plants, UV-B exposure commonly reduces photosynthetic rates in phytoplankton under field situations. The primary mechanism by
which UV-B inhibits photosynthesis is uncertain, but there are many possible targets (Vincent & Neale 2000). As for terrestrial plants, several laboratory studies
using high UV-B exposures have demonstrated damage to PSII, but field studies
with natural assemblages have found few or no effects on PSII electron transport
capacity (Vincent & Neale 2000). An integral part of algal defense against UV-B
is rapid repair of damaged photosynthetic complexes. This mechanism can be
inferred from the time course of photosynthesis under UV-B exposure: After an
initial decline in photosynthesis, a steady-state rate is attained and typically sustained over many hours of exposure (Cullen & Lesser 1991, Lesser et al. 1994,
Neale et al. 1998a, Heraud & Beardall 2000). This steady state is interpreted as

an equilibrium between damage and ongoing repair processes (Lesser et al. 1994, Neale 2000). Because of repair, UV-B inhibition of phytoplankton photosynthesis does not typically obey reciprocity, i.e., damage is not solely a function of cumulative exposure. Under some conditions repair is slowed, increasing sensitivity to UV-B. This slowing occurs when protein synthesis is inhibited (Cullen & Lesser 1991, Lesser et al. 1994, Neale et al. 1998a) or under low nitrogen availability (Cullen & Lesser 1991, Litchman et al. 2002). For phytoplankton in deeply mixed layers of the Southern Ocean, repair is so slow that equilibrium is not attained under high UV-B exposure conditions and inhibition does depend on cumulative exposure (Neale et al. 1998b).

In general, terrestrial plants receive appreciably higher UV-B doses than aquatic algae because UV-B penetration of the water column is limited by the presence of chromophoric dissolved organic matter (Hargreaves 2002). However, because leaves have a relatively thick epidermal cell layer, containing relatively high concentrations of screening compounds, UV-B transmittance to internal targets is likely lower in leaves than in single-celled algae. Leaf thickening in response to UV-B (Ballaré et al. 1996, Xiong & Day 2001, Phoenix et al. 2001) may also provide protection by increasing the pathlength to targets. The greater UV-B pathlength found in leaves, together with multiple layers of photosynthetic cells that allows compensatory adjustments (Day & Vogelmann 1995), may explain why photosynthetic rates of leaves are less sensitive to UV-B than those of single-celled free-living algae.

What Traits Confer Tolerance?

Over two decades ago substantial variability was noted in UV-B responsiveness among terrestrial plant species, although the reasons for this variability were unclear (Caldwell 1981). Explanations remain elusive. One possibility is differences in leaf UV-B–screening effectiveness. Epidermal UV-B transmittance varies substantially among species, ranging from ≈ 0 to >40% (Robberecht & Caldwell 1978, Robberecht et al. 1980, Caldwell et al. 1983b; Day et al. 1992, Day 1993). In general, foliage of evergreens is particularly effective at screening [UV-B (300 nm) transmittance <10%, and <2% in most conifers], whereas foliage of deciduous species, especially herbaceous species, is usually >15% (Day et al. 1992, 1994; Day 1993). However, evergreens generally appear no less sensitive to UV-B than other plants (Johanson et al. 1995a,b; Sullivan et al. 1996; Björn et al. 1998). Attempts to relate differences in the UV-B sensitivity of growth or photosynthesis among species to absorbing-compound concentrations (Barnes et al. 1987, Musil 1995, Sullivan et al. 1996) or epidermal transmittance (Caldwell et al. 1982, Sullivan et al. 1996) have generally failed. Monocots such as grasses tend to respond more to enhanced UV-B than dicots (Barnes et al. 1990, Caldwell et al. 1998b; but see Musil 1995), yet grasses tend to screen UV-B more effectively (Day et al. 1992). Differences in the relative sensitivity of plants, especially in terms of long-term processes such as growth, appear to depend on more than simply leaf screening effectiveness. Additional factors, such as the efficiency of DNA and oxidative damage repair systems, may well play a role.

Phytoplankton appear to use differing combinations of screening, antioxidant defense, and repair in defending against UV-B. Large-cell, MAA-accumulating species (such as dinoflagellates) acclimated to high light seem to be the most resistant to solar UV-B exposure (Neale et al. 1998), but relatively resistant species are found in all groups including those that are small or do not accumulate MAAs (Xiong et al. 1996, Laurion & Vincent 1998a). Variations in sensitivity of phytoplankton to UV-B parallel their ability to acclimate photosynthetically to their irradiance environment (Neale 1987, Falkowski & Raven 1997, MacIntyre et al. 2000, Neale & Kieber 2000). These differences in the ability to acclimate to UV-B appear to be best explained by differences in resource allocation to light harvesting versus defense/repair (Ivanov et al. 2000). In Antarctica the average light intensity can be low, owing to a deep surface mixed layer, so that survival may depend more on the efficiency of photosynthesis at low irradiance than on resistance to UV-B (Neale et al. 1998b). In other areas, phytoplankton growth is strongly limited by available inorganic nitrogen. Dinoflagellates in nitrogen-limited culture allocate less nitrogen to UV-B resistance (via synthesis of MAAs and enzymes needed for antioxidant defense and repair of damaged proteins) than nitrogen-sufficient cultures (Litchman et al. 2002). However, assemblages with low sensitivity are encountered in selected environments such as shallow surface mixed layers (Vernet et al. 1994, Neale & Kieber 2000) and in long-term incubations of natural assemblages under near-surface intensities (Neale et al. 1994, Karentz 1994). It appears that in these environments inhibition is sufficiently severe and chronic that only species capable of substantial allocation to UV defense survive.

Productivity and Timescales

Exposure to solar UV-B can reduce biomass production of some terrestrial plant species. For example, half of recent attenuation studies observed reductions in biomass production of plants exposed to solar UV-B (Day 2001a), and some of these reductions are substantial, ranging from 10 to 35% (Krizek et al. 1997, 1998; Mazza et al. 1999a; Xiong & Day 2001; Day et al. 2001). These reductions illustrate that solar UV-B constrains the performance of some terrestrial plants, although extrapolation of these findings to enhanced UV-B/ozone-depletion scenarios may not be warranted because of the nonlinear nature of responses. These reductions in total biomass are usually attributable to reduced aboveground biomass, which is often found in conjunction with slower leaf elongation rates, smaller leaves, and smaller whole-plant leaf area (Searles et al. 1995, Ballaré et al. 1996, 2001; Krizek et al. 1997, 1998; Mazza et al. 1999a; Xiong & Day 2001; Day et al. 2001). The mechanisms responsible for leaf stunting are unclear, although several have been proffered (Caldwell et al. 1998a,b). As alluded to earlier, these reductions in production typically occur in the absence of reductions in photosynthetic rates per unit leaf area.

In contrast to filter attenuation studies, reductions in plant biomass production or leaf area are usually not observed in outdoor lamp enhancement studies (Caldwell et al. 1998a,b; Day 2001a; Searles et al. 2001a). Hence, it appears that productivity of most terrestrial plants under enhanced UV-B regimes associated with ozone depletion is unlikely to be compromised. However, several recent studies have found that outdoor UV-B enhancements can reduce various plant growth parameters in some species. This includes subarctic heath species (Johanson et al. 1995a; Björn et al. 1996; Phoenix et al. 2000, 2001), ephemerals of South African deserts (Musil 1995), and the dominant grass in a Dutch grassland (Oudejans et al. 2001). Changes in plant morphology and canopy architecture are commonly observed under outdoor UV-B supplements, rather than reductions in overall growth or production per se. These changes might indirectly alter productivity of species by shifting the competitive balance, for example, through altering interspecific competition for visible light (Barnes et al. 1995, Caldwell et al. 1998b). Nevertheless, plants are generally more sensitive to UV-B in attenuation studies in which solar UV-B is reduced, than in enhancement studies in which solar UV-B is supplemented (Day 2001a). One explanation is that targets may become saturated at ambient solar UV-B levels, such that above-ambient supplements elicit negligible, smaller, or more elusive reductions in growth (Day 2001a), which would explain why dose responses tend to be nonlinear.

Whereas most UV-B field studies examining growth and production of terrestrial plants have been limited to single growing seasons, some have been conducted over several years. Results from these long-term studies suggest that the reductions in some growth parameters attributable to ambient or enhanced UV-B can increase or accumulate over successive years in some species, including trees at mid-northern latitudes (Sullivan & Teramura 1992), shrubs at high-northern latitudes (Johanson et al. 1995a, Björn et al. 1998, Phoenix et al. 2000), and herbaceous perennials in Antarctica (Day et al. 2001). These cumulative reductions in growth may result from reductions in leaf area such that plant carbon assimilation is more reduced over successive years (Day et al. 2001), or from increased leaf dark respiration rates, which may be required for protection or repair at the expense of photoassimilate storage in roots (Gwynn-Jones 2001). It is unclear how prevalent these cumulative growth reductions are among terrestrial plants, but they could lead to significant long-term changes in plant communities.

The time of the response relative to the duration of treatment has been a recurring theme in our discussion of the response of phytoplankton to UV-B. In general, characteristic response times of phytoplankton are quite short, compared with those of terrestrial plants. Changes in UV-B exposure alter photosynthetic rates within minutes to hours (Neale 2000) and reproduction (cell division) within 1–2 days, facilitating measurement of short-term effects. However, unlike sessile terrestrial plants, phytoplankton occupying a given patch in the surface layer of an ocean or lake are constantly changing because of horizontal and vertical mixing. Thus, it is not reasonable to conduct long-term UV-B manipulations over an aquatic patch to assess integrated effects. Long-term UV-B exposures have been conducted on aquatic ecosystems in enclosures (mesocosms). This is a useful approach to study UV-B effects on trophic interactions (see below), but can usually only be conducted for 1–2 weeks before wall growth and other effects render mesocosms substantially different from their surroundings. Therefore, estimating the effect of UV exposure on seasonal or annual aquatic productivity usually requires extrapolation of short-term measurements to longer timescales. Initial estimates of UV-B effects on Antarctic marine primary production were based on relatively simple proportionality relationships to extend the results of partial-day incubations to full days and seasons. Using this approach, estimates of reductions in annual Southern Ocean productivity owing to increased UV-B caused by seasonal ozone depletion ranged from $\leq 0.2\%$ (Holm-Hansen et al. 1993) to 2–4% (Smith et al. 1992).

These simple approaches assume a linear relationship between UV-B exposure and decreased production, but responses to UV-B are inherently nonlinear, either because ongoing repair results in failure of reciprocity or, when repair is low, damage can build up to the point that the scarcity of undamaged target sites limits further damage (Neale 2000). Strategies to cope with these complexities have varied. One approach is to use short-term measurements to define various exposure response relationships and then combine these into a model forced by longer-term changes. When this has been attempted, interacting complexities have resulted in some surprising results. For example, in a model of phytoplankton response that accounts for light variation owing to vertical mixing, the effect of UV-B enhancements associated with ozone depletion was greatest not under static conditions (which result in the longest residence time near the surface), but when there was a moderate amount of mixing (which minimized the effect of nonlinear responses) (Neale et al. 1998c). This model estimated that springtime phytoplankton production in the Weddell-Scotia confluence in Antarctica was reduced by 0.7–8.5% by enhanced UV-B associated with ozone depletion, depending on the depth and strength of mixing as well as natural variability in sensitivity. More information on the timescales of mixing in the Southern Ocean is needed to refine these estimates further (Neale et al. 2002). The results of another model (Arrigo 1994), which did not include mixing effects but did account for nonlinear responses, suggested that the negative effects of enhanced UV-B caused by ozone depletion on Southern Ocean productivity could largely be offset by gains in productivity resulting from increased transmission of PAR through the ozone visible-absorbance bands early in the season, when algae growing beneath ice are strongly light limited. The scope and accuracy of such modeling-based estimates of UV-B effects will improve as parallel progress is made on a number of relevant issues including global estimates of UV-B penetration (Vasilkov et al. 2001), variation in BWFs (Neale & Kieber 2000), and algorithms for calculation of UV-B–dependent decreases in production (Lehmann et al. 2000).

Feedback from Other Trophic Levels

Ultimately, the response of primary producers to enhanced UV-B should be viewed in an ecosystem context, taking into account feedback from other trophic levels.

Because of the often subtle nature of UV-B effects on terrestrial plants, it has been suggested that effects of UV-B on other trophic levels are likely to have a greater impact on terrestrial plant performance, via feedback, than direct effects on plants (Björn et al. 1996, Caldwell et al. 1998b).

Effects of UV-B on other trophic levels are quite variable and specific to the system studied, making generalizations difficult (Day 2001b). For example, direct exposure of leaf litter to enhanced UV-B can accelerate (Rozema et al. 1997) or slow litter decomposition (Gehrke et al. 1995, Newsham et al. 1997, Moody et al. 2001) and increase litter C:N ratios (Moody et al. 2001). Plant UV-B exposure also variably affects subsequent litter decomposition. These so-called indirect effects of UV-B accelerate decomposition rates in some systems (Newsham et al. 1997, Newsham et al. 2001) but slow them in others (Rozema et al. 1997). Several seasons of enhanced UV-B exposure can decrease C:N ratios in soil microbial pools and alter microbial species composition (Johnson et al. 2002), suggesting large differences in UV-B sensitivity among microbes (Moody et al. 1999). Similar opposing results have been found for mycorrhizal infection (Ballaré et al. 2001, Van de Staaij et al. 2001). Exposure to naturally enhanced UV-B over several growing seasons increased numbers of testate amoebae in Tierra del Fuego (Searles et al. 2001b) and reduced numbers of microarthropods in Antarctica (Convey et al. 2003), primarily through indirect effects. UV-B exposure of plants tends to increase the risk of fungal infection to aboveground organisms, but direct exposure can damage fungi and reduce disease (Paul 2000). Exposure to higher UV-B levels typically reduces insect herbivory, abundance, and performance (Ballaré et al. 1996, Rousseaux et al. 1998, Mazza et al. 1999b, Zavala et al. 2001).

The variability in responses among other trophic levels to UV-B may stem from differences in plant phenolic chemistry among species, as well as constitutive differences in the effectiveness of microbial communities to adjust to changes in litter quality. In addition to leaf UV-B screening, many other roles have been proposed for leaf phenylpropanoids and related phenolics such as deterrents to insects, fungal pathogens, and microbial decomposers (Lumsden 1997; Caldwell et al. 1998b; Day 2001a,b). However, evidence of how upregulation of some of these compounds in turn affects responses in other trophic levels is generally lacking. The diversity of these compounds in plants might explain why the responses of organisms in other trophic levels tend to be specific to the plant system studied. How effects on these trophic levels may in turn feed back to affect plant performance remains speculative, but such feedback could be significant. For example, long-term UV-B enhancements in a subarctic heath resulted in a large decrease in C:N ratios in the soil microbial biomass that might ultimately lead to strong feedback on plants in this N-limited system (Johnson et al. 2002).

In aquatic systems it is essential to consider trophic interactions in UV-B assessments of primary producers because production, grazing, and decomposition are so tightly coupled. Effects of UV-B on grazing or nutrient remineralization should rapidly feed back to primary producers. In theory at least, UV-B could affect all these processes because they are carried out by microorganisms (i.e., bacteria, phytoplankton, and zooplankton), which are inherently vulnerable to UV-B because of their small size, as previously discussed. In fact, sensitivity to UV-B varies among microorganisms depending on how their optical vulnerability is counteracted by other defenses (Helbling & Zagarese 2002). Thus, it is difficult to generalize as to which trophic levels in aquatic ecosystems will be most affected by enhanced UV-B.

Some indication of how ecosystems respond can be obtained by experimental manipulation of UV-B in model ecosystems that enclose organisms in mesocosms (using tanks, bags, or flumes) that receive attenuated and/or enhanced UV-B. These treatments are limited for practical reasons to 1–3 week exposures. One of the first such studies found that zooplankton were the most UV-B–sensitive component of a stream ecosystem (Bothwell et al. 1994). These zooplankton exerted strong top-down control on the primary producers in this system, such that reduced grazing under higher UV-B exposures led to higher algal biomass. A study of a planktonic system (tanks filled with St. Lawrence Estuary water) found an analogous result (Mostajir et al. 1999): Enhanced UV-B exposure had a disproportionate effect on large organisms (>5 μm), which included most of the grazers (ciliates), so UV-B exposure led to a greater proportion of small organisms (mainly bacteria and small phytoplankton normally grazed by the ciliates). UV-B induced a shift in species composition but little change in overall biomass of phytoplankton in another set of mesocosms filled with coastal water (Wangberg et al. 2001). Other studies in lakes and coastal marine waters have found little effect from UV-B attenuation or enhancement (Bergeron & Vincent 1997; Halac et al. 1997; Keller et al. 1997a,b; Laurion et al. 1998b). Wangberg et al. (1999) found that UV-B had a positive effect on phytoplankton activity under low nutrient availability, which they attributed to the stimulation of mineralizing bacteria via increased organic carbon from chromophoric dissolved organic matter photolysis. Mesocosm results have led to a better appreciation of the complexities of aquatic producers responses to UV-B, but no general approach has emerged to predict how systems as a whole will respond. In part, this is because mesocosm studies have focused more on trends in biomass and activity of various components and less on the mechanisms responsible for the dynamics. In the future, the best strategy seems to be to use a combination of detailed studies of specific responses, whole-system manipulation and identification of model structure to develop and test ecosystem models that include specific descriptions of UV-B effects.

CONCLUSIONS

Stratospheric ozone depletion by anthropogenic gases has led to increases in levels of solar UV-B reaching the Earth's surface. Our understanding of the consequences of these enhanced UV-B levels on primary producers has improved dramatically over the past 20 years but has been hampered by the complexities involved in realistically simulating UV-B enhancements. Further development and assessment is needed for BWFs, which are critical in determining UV-B doses and evaluating effects. For terrestrial plants, their large size and the long response times of integrative measures of performance (e.g., growth) make development of BWFs particularly challenging. Estimates of solar spectral irradiance under ambient and depleted ozone scenarios, which are also critical in determining UV-B doses, might be improved through the use of radiative transfer models, and further ground truthing of these models is needed. Whereas concentrations of UV screening compounds in terrestrial and marine producers commonly increase with UV-B exposure, the implications, while potentially far-reaching, are poorly understood. The subtle increases observed in response to higher UV-B levels may translate into only modest improvements in protection but may influence many other processes because of the diverse roles suspected of these compounds. Photosynthetic rates (per unit leaf area) of terrestrial plants are usually not affected by UV-B exposure, whereas they are reduced in many phytoplankton, likely because of less effective UV-B screening. Levels of ambient solar UV-B reduce the productivity of some terrestrial plants, but UV-B enhancements simulating ozone depletion do not elicit further reductions in productivity in most cases. However, reductions in some growth parameters have been observed in some terrestrial plants in response to these enhancements, particularly in long-term studies, and these growth reductions can sometimes increase over successive years. Aquatic productivity is often compromised by short-term exposures to ambient as well as enhanced UV-B. However, longer-term assessments of UV-B effects on aquatic producers are complicated because of difficulties involved in accounting for the horizontal and vertical mixing, and the nonlinear nature of responses. Recent work examining UV-B effects on multiple trophic levels suggests that effects on other trophic levels can be diverse and hard to predict but have the potential to feedback to primary producers.

ACKNOWLEDGMENTS

TAD thanks Martyn M. Caldwell for his insights and perspectives on the history of UV plant research, along with support from NSF, Office of Polar Programs (OPP-9596188 and 9615268) and USDA, Plant Response to the Environment Program (NRICGP-93-371008875 and 97-351004212). PJN acknowledges support from NSF, Office of Polar Programs (OPP-9615342), Biological Oceanography (OCE-9812036), Environmental Biology (DEB-9973938), and the US EPA (CISNET grant R826943).

The *Annual Review of Ecology and Systematics* is online at
http://ecolsys.annualreviews.org

LITERATURE CITED

Allen DJ, Nogués S, Baker NR. 1998. Ozone depletion and increased UV-B radiation: Is there a real threat to photosynthesis? *J. Exp. Bot.* 49:1775–88

Arrigo KR. 1994. Impact of ozone depletion on phytoplankton growth in the southern ocean: large-scale spatial and temporal variability. *Mar. Ecol. Prog. Ser.* 114:1–12

Ballaré CL, Rousseaux MC, Searles PS, Zaller JG, Giordano CV, et al. 2001. Impacts of solar ultraviolet-B radiation on terrestrial ecosystems of Tierra del Fuego (southern Argentina): an overview of progress. *J. Photochem. Photobiol. B: Biol.* 62:67–77

Ballaré CL, Scopel AL, Stapleton AE, Yanovsky MJ. 1996. Solar ultraviolet-B radiation affects seedling emergence, DNA integrity, plant morphology, growth rate, and attractiveness to herbivore insects in *Datura ferox. Plant Physiol.* 112:161–70

Banaszak AT, Neale PJ. 2001. Ultraviolet radiation sensitivity of photosynthesis in phytoplankton from an estuarine environment. *Limnol. Oceanogr.* 46:592–603

Banaszak AT, Trench RK. 2001. Ultraviolet sunscreens in dinoflagellates. *Protist* 152: 93–101

Barnes PW, Flint SD, Caldwell MM. 1987. Photosynthesis damage and protective pigments in plants from a latitudinal arctic/alpine gradient exposed to supplemental UV-B radiation in the field. *Arct. Alp. Res.* 19:21–27

Barnes PW, Flint SD, Caldwell MM. 1990. Morphological responses of crop and weed species of different growth forms to ultraviolet-B radiation. *Am. J. Bot.* 77:1354–60

Barnes PW, Flint SD, Caldwell MM. 1995. Early-season effects of supplemented solar UV-B radiation on seedling emergence, canopy structure, simulated stand photosynthesis and competition for light. *Glob. Change Biol.* 1:43–53

Barnes PW, Searles PS, Ballaré CL, Ryel RJ, Caldwell MM. 2000. Non-invasive measurements of leaf epidermal transmittance of UV radiation using chlorophyll fluorescence: field and laboratory studies. *Physiol. Plant.* 109:274–83

Beggs CJ, Wellmann E. 1994. Photocontrol of flavonoid biosynthesis. In *Photomorphogenesis in Plants*, ed. RE Kendrick, GHM Kronenberg, pp. 733–51. Dordrecht, The Netherlands: Kluwer Academic

Behrenfeld MJ, Chapman JW, Hardy JT, Lee HI. 1993. Is there a common response to ultraviolet-B radiation by marine phytoplankton? *Mar. Ecol. Progr. Ser.* 102:59–68

Bergeron M, Vincent WF. 1997. Microbial food web responses to phosphorus supply and solar UV radiation in a subarctic lake. *Aquat. Microbial Ecol.* 12:239–49

Bilger W, Johnsen T, Schreiber U. 2001. UV-excited chlorophyll fluorescence as a tool for the assessment of UV-protection by the epidermis of plants. *J. Exp. Bot.* 52:2007–14

Björn LO, Callaghan TV, Gehrke C, Gwynn-Jones D, Holmgren B, et al. 1996. Effects of UV-B radiation of subarctic vegetation. In *Ecology of Arctic Environments*, ed. SJ Woodin, M Marquiss, pp. 241–53. Spec. publ. no. 13, Br. Ecol. Soc. London: Blackwell Sci.

Björn LO, Callaghan TV, Gehrke C, Johanson U, Sonesson M, et al. 1998. The problem of ozone depletion in northern Europe. *Ambio* 27:275–79

Björn LO, Murphy TM. 1985. Computer calculation of solar ultraviolet radiation at ground level. *Physiol. Vég.* 23:555–61

Booker FL, Fiscus EL, Philbeck AS, Miller JE, Heck WW. 1992. A supplemental ultraviolet-B radiation system using open-top field chambers. *J. Environ. Qual.* 21:56–61

Bornman JF. 1989. Target sites of UV-B radiation in photosynthesis of higher plants. *J. Photochem. Photobiol. B: Biol.* 4:145–58

Bothwell ML, Sherbot DMJ, Pollock CM. 1994. Ecosystem response to solar ultraviolet-B radiation: influence of trophic-level interactions. *Science* 265:97–100

Bühlmann B, Bossard P, Uehlinger U. 1987. The influence of longwave ultraviolet radiation (u.v. A) on the photosynthetic activity (^{14}C-assimilation) of phytoplankton. *J. Plankton Res.* 9:935–43

Burchard P, Bilger W, Weissenböck G. 2000. Contribution of hydroxycinnamates and flavonoids to epidermal shielding of UV-A and UV-B radiation in developing rye primary leaves as assessed by ultraviolet-induced chlorophyll fluorescence measurements. *Plant Cell Environ.* 23:1373–80

Caldwell MM. 1968. Solar ultraviolet radiation as an ecological factor for alpine plants. *Ecol. Monogr.* 38:243–68

Caldwell MM. 1971. Solar UV radiation and the growth and development of higher plants. In *Photophysiology*, ed. AC Giese, 6:131–77. New York: Academic

Caldwell MM. 1981. Plant response to solar ultraviolet radiation. In *Encyclopedia of Plant Physiology, New Series, Physiological Ecology I*, ed. OL Lange, CB Osmond, H Ziegler, 12A:170–97. Berlin: Springer-Verlag

Caldwell MM, Björn LO, Bornman JF, Flint SD, Kulandaivelu G, et al. 1998a. Effects of increased solar ultraviolet radiation on terrestrial ecosystems. *J. Photochem. Photobiol. B: Biol.* 46:40–52

Caldwell MM, Camp LB, Warner CW, Flint SD. 1986. Action spectra and their key role in assessing biological consequences of solar UV-B radiation change. In *Stratospheric Ozone Reduction, Solar Ultraviolet Radiation and Plant Life*, ed. RC Worrest, MM Caldwell, pp. 87–111. Berlin: Springer-Verlag

Caldwell MM, Flint SD. 1994. Stratospheric ozone reduction, solar UV-B radiation and terrestrial ecosystems. *Clim. Change* 28:375–94

Caldwell MM, Flint SD. 1997. Uses of biological spectral weighting functions and the need of scaling for the ozone reduction problem. *Plant Ecol.* 128:66–76

Caldwell MM, Gold WG, Harris G, Ashurst CW. 1983a. A modulated lamp system for solar UV-B (280–320 nm). Supplementation studies in the field. *Photochem. Photobiol.* 37:479–85

Caldwell MM, Robberecht R, Billings WD. 1980. A steep latitudinal gradient of solar ultraviolet-B radiation in the arctic-alpine life zone. *Ecology* 61:600–11

Caldwell MM, Robberecht R, Flint SD. 1983b. Internal filters: prospects of UV-acclimation in higher plants. *Physiol. Plant.* 37:479–85

Caldwell MM, Robberecht R, Nowak RS. 1982. Differential photosynthetic inhibition by ultraviolet radiation in species from the arctic-alpine life zone. *Arct. Alp. Res.* 14:195–202

Caldwell MM, Searles PS, Flint SD, Barnes. PW Barnes. PW 1998b. Terrestrial ecosystem responses to solar UV-B mediated by vegetation, microbes and abiotic photochemistry. In *Physiological Plant Ecology*, ed. MC Press, JD Scholes, MG Barker, pp. 241–62. Oxford: Blackwell

Calkins J, ed. 1982. *The Role of Solar UV Radiation in Marine Ecosystems*. New York: Plenum

Calkins J, Thordardottir T. 1980. The ecological significance of solar UV radiation on aquatic organisms. *Nature* 283:563–66

Carreto JI, Lutz VA, De Marco SG, Carignan MO. 1990. Fluence and wavelength dependence of mycosporine-like amino acid synthesis in the dinoflagellate *Alexandrium excavatum*. In *Toxic Marine Phytoplankton*, ed. E Graneli, L Edler, B Sundström, DM Anderson, pp. 275–79. Amsterdam: Elsevier

Cen Y-P, Bornman JF. 1993. The effect of exposure to enhanced UV-B radiation on the penetration of monochromatic and polychromatic UV-B radiation in leaves of *Brassica napus*. *Physiol. Plant.* 87:249–55

Cockell CS, Blaustein AR, eds. 2001. *Ecosystems, Evolution, and Ultraviolet Radiation*. New York: Springer-Verlag

Cockell CS, Knowland J. 1999. Ultraviolet radiation screening compounds. *Biol. Rev.* 74:311–45

Convey P, Pugh PJA, Jackson C, Murray AW, Ruhland CT, et al. 2003. Response of Antarctic terrestrial microarthropods to multifactorial climate manipulation over a four year period. *Ecology.* In press

Cullen JJ, Lesser MP. 1991. Inhibition of photosynthesis by ultraviolet radiation as a function of dose and dosage rate: results for a marine diatom. *Mar. Biol.* 111:183–90

Cullen JJ, Neale PJ. 1993. Quantifying the effects of ultraviolet radiation on aquatic photosynthesis. In *Photosynthetic Responses to the Environment*, ed. H Yamamoto, CM Smith, pp. 45–60. Current Topics in Plant Physiology. Vol. 8. Washington, DC: Am. Soc. Plant Physiol.

Cullen JJ, Neale PJ. 1994. Ultraviolet radiation, ozone depletion, and marine photosynthesis. *Photosynth. Res.* 39:303–20

Cullen JJ, Neale PJ. 1997. Biological weighting functions for describing the effects of ultraviolet radiation on aquatic systems. In *Effects of Ozone Depletion on Aquatic Ecosystems*, ed. D-P Häder, pp. 97–118. Austin, TX: Landes

Cullen JJ, Neale PJ, Lesser MP. 1992. Biological weighting function for the inhibition of phytoplankton photosynthesis by ultraviolet radiation. *Science* 258:646–50

Dahlback A, Stamnes K. 1991. A new spherical model for computing the radiation field available for photolysis and heating at twilight. *Planet. Space Sci.* 39:671–83

Day TA. 1993. Relating UV-B radiation screening effectiveness of foliage to absorbing-compound concentration and anatomical characteristics in a diverse group of plants. *Oecologia* 95:542–50

Day TA. 2001a. Ultraviolet radiation and plant ecosystems. See Cockell & Blaustein 2001, pp. 80–117

Day TA. 2001b. Multiple trophic levels in UV-B assessments: completing the ecosystem. *New Phytol.* 152:183–86

Day TA, Howells BW, Rice WJ. 1994. Ultraviolet absorption and epidermal-transmittance spectra in foliage. *Physiol. Plant.* 92: 207–18

Day TA, Ruhland CT, Xiong FS. 2001. Influence of solar ultraviolet-B radiation on Antarctic terrestrial plants: results from a 4-year field study. *J. Photochem. Photobiol. B: Biol.* 62:78–87

Day TA, Vogelmann TC, DeLucia EH. 1992. Are some plant life forms more effective at screening UV-B radiation? *Oecologia* 92: 513–19

de Mora SJ, Demers S, Vernet M, eds. 2000. *The Effects of UV Radiation on Marine Ecosystems*. Environ. Chem. Ser. Cambridge: Cambridge Univ. Press

Dunlap WC, Shick JM. 1998. Ultraviolet radiation-absorbing mycosporine-like amino acids in coral reef organisms: a biochemical and environmental perspective. *J. Phycol.* 34:418–30

El-Sayed SZ, Stephens FC, Bidigare RR, Ondrusek ME. 1990. Effect of ultraviolet radiation on Antarctic marine phytoplankton. In *Antarctic Ecosystems: Ecological Change and Conservation*, ed. KR Kerry, G Hempel, pp. 379–85. Berlin/Heidelberg: Springer-Verlag

Falkowski PG, Raven JA. 1997. *Aquatic Photosynthesis*. London: Blackwell

Fiscus EL, Booker FL. 1995. Is UV-B a hazard to crop photosynthesis and productivity? Results of an ozone-UV-B interaction study and model predictions. *Photosynth. Res.* 43:81–92

Flint SD, Caldwell MM. 2002a. A biological spectral weighting function for ozone depletion research with higher plants. *Physiol. Plant.* In press

Flint SD, Caldwell MM. 2002b. Field testing of UV biological spectral weighting functions for higher plants. *Physiol. Plant.* In press

Fox FM, Caldwell MM. 1978. Competitive interaction in plant populations exposed to supplementary ultraviolet-B radiation. *Oecologia* 36:173–90

Franklin LA, Forster RM. 1997. The changing irradiance environment: consequences for marine macrophyte physiology, productivity and ecology. *Eur. J. Phycol.* 32:207–32

Franklin LA, Kräbs G, Kuhlenkamp R. 2001. Blue light and UV-A radiation control the synthesis of mycosporine-like amino acids in *Chondrus crispus* (Florideophyceae). *J. Phycol.* 37:257–70

Franklin LA, Yakovleva I, Karsten U, Lüning K. 1999. Synthesis of mycosporine-like amino acids in *Chondrus crispus* (Florideophyceae) and the consequences for sensitivity to ultraviolet-B radiation. *J. Phycol.* 35: 682–93

Garcia-Pichel F. 1994. A model for self-shading in planktonic organisms and its implications for the usefulness of ultraviolet sunscreens. *Limnol. Oceanogr.* 39:1704–17

Gehrke C, Johanson U, Callaghan TV, Chadwick D, Robinson CH. 1995. The impact of enhanced ultraviolet-B radiation on litter quality and decomposition processes in *Vaccinium* leaves from the Subarctic. *Oikos* 72:213–22

Green AES. 1983. The penetration of ultraviolet radiation to the ground. *Physiol. Plant.* 58:351–59

Green AES, Cross KR, Smith LA. 1980. Improved characterization of skylight. *Photochem. Photobiol.* 31:59–65

Green AES, Sawada T, Shettle EP. 1974. The middle ultraviolet reaching the ground. *Photochem. Photobiol.* 19:251–59

Gwynn-Jones D. 2001. Short-term impacts of enhanced UV-B radiation on photoassimilate allocation and metabolism: a possible interpretation for time-dependent inhibition of growth. *Plant Ecol.* 154:67–73

Häder D-P. 1997. *The Effects of Ozone Depletion on Aquatic Ecosystems.* Georgetown, TX: Landes

Halac S, Felip M, Camarero L, Sommaruga-Wögrath S, Psenner R, et al. 1997. An in situ enclosure experiment to test the solar UVB impact on plankton in a high-altitude mountain lake. I. Lack of effect on phytoplankton species composition and growth. *J. Plankton Res.* 19:167–86

Hannach G, Sigleo AC. 1998. Photoinduction of UV-absorbing compounds in six species of marine phytoplankton. *Mar. Ecol. Prog. Ser.* 174:207–22

Hargreaves BR. 2002. Water column optics and penetration of UVR. See Helbling & Zagarese 2002, In press

Harris GP. 1978. Photosynthesis, productivity and growth: the physiological ecology of phytoplankton. *Arch. Hydrobiol. Beih. Ergebn. Limnol.* 10:1–171

Helbling EW, Villafañe V, Ferrario M, Holm-Hansen O. 1992. Impact of natural ultraviolet radiation on rates of photosynthesis and on specific marine phytoplankton species. *Mar. Ecol. Prog. Ser.* 80:89–100

Helbling EW, Villafañe V, Holm-Hansen O. 1994. Effects of ultraviolet radiation on Antarctic marine phytoplankton photosynthesis with particular attention to the influ-

ence of mixing. See Weiler & Penhale 1994, pp. 207–27

Helbling EW, Zagarese HE, eds. 2002. *UV Effects in Aquatic Organisms and Ecosystems.* In press

Heraud P, Beardall J. 2000. Changes in chlorophyll fluorescence during exposure of *Dunaliella tertiolecta* to UV radiation indicate a dynamic interaction between damage and repair processes. *Photosynth. Res.* 63:123–34

Hester RE, Harrison RM, eds. 2000. *Causes and Environmental Implications of Increased U.V.-B. Radiation*, Vol. 14. Cambridge: R. Soc. Chem.

Holm-Hansen O, Helbling EW, Lubin D. 1993. Ultraviolet radiation in Antarctica: inhibition of primary production. *Photochem. Photobiol.* 58:567–70

Ivanov AG, Miskiewicz E, Clarke AK, Greenberg BM, Huner NP. 2000. Protection of photosystem II against UV-A and UV-B radiation in the cyanobacterium *Plectonema boryanum*: the role of growth temperature and growth irradiance. *Photochem. Photobiol.* 72:772–79

Jitts HR, Morel A, Saijo Y. 1976. The relation of oceanic primary production to available photosynthetic irradiance. *Aust. J. Mar. Freshw. Res.* 27:441–54

Johanson U, Gehrke C, Björn LO, Callaghan TV. 1995a. The effects of enhanced UV-B radiation on the growth of dwarf shrubs in a subarctic heathland. *Funct. Ecol.* 9:713–19

Johanson U, Gehrke C, Björn LO, Callaghan TV. 1995b. The effects of enhanced UV-B radiation on a subarctic heath ecosystem. *Ambio* 24:106–11

Johnson D, Campbell CD, Lee JA, Callaghan TV, Gwynn-Jones D. 2002. Arctic microorganisms respond more to elevated UV-B radiation than CO_2. *Nature* 416:82–83

Jones AE, Shanklin JD. 1995. Continued decline of total ozone over Halley, Antarctica, since 1985. *Nature* 376:409–11

Jones LW, Kok B. 1966. Photoinhibition of chloroplast reactions. I. Kinetics and action spectra. *Plant Physiol.* 41:1037–43

Karentz D. 1994. Ultraviolet tolerance mechanisms in Antarctic marine organisms. See Weiler & Penhale 1994, pp. 93–110

Keiller DR, Holmes MG. 2001. Effects of long-term exposure to elevated UV-B radiation on the photosynthetic performance of five broad-leaved tree species. *Photosynth. Res.* 67:229–40

Keller A, Hargraves P, Jeon H, Klein-Macphee G, Klos E, et al. 1997a. Effects of ultraviolet-B radiation enhancement on marine trophic levels in a stratified coastal system. *Mar. Biol.* 130:277–87

Keller A, Hargraves P, Jeon H, Klein-Macphee G, Klos E, et al. 1997b. Ultraviolet-B radiation enhancement does not affect marine trophic levels during a winter-spring bloom. *Ecoscience* 4:129–39

Koepke P, Bais A, Balis D, Buchwitz M, De Backer H, et al. 1998. Comparison of models used for UV index calculations. *Photochem. Photobiol.* 67:657–62

Kolb CA, Käser MA, Kopecký J, Zotz G, Riederer M, Pfündel EE. 2001. Effects of natural intensities of visible and ultraviolet radiation on epidermal ultraviolet screening and photosynthesis in grape leaves. *Plant Physiol.* 127:863–75

Krizek DT, Britz SJ, Mirecki RM. 1998. Inhibitory effects of ambient levels of solar UV-A and UV-B radiation on growth of cv. New Red Fire lettuce. *Physiol. Plant.* 103:1–7

Krizek DT, Mirecki RM, Britz SJ. 1997. Inhibitory effects of ambient levels of solar UV-A and UV-B radiation on growth of cucumber. *Physiol. Plant.* 100:886–93

Laurion I, Lean DRS, Vincent WF. 1998. UVB effects on a plankton community: results from a large-scale enclosure assay. *Aquat. Microb. Ecol.* 16:189–98

Laurion I, Vincent WF. 1998. Cell size versus taxonomic composition as determinants of UV-sensitivity in natural phytoplankton communities. *Limnol. Oceanogr.* 43:1774–79

Lehmann MK, Davis RF, Huot Y, Cullen JJ. 2000. Biologically weighted transparency: A predictor for water-column photosynthesis

and its inhibition by ultraviolet radiation. *Proc. Ocean Opt. XV [CDROM], Musee Oceanogr., Monaco.* Washington, DC: ONR—Ocean, Atmos. Space Sci. Tech. Dept.

Lesser MP. 1996a. Acclimation of phytoplankton to UV-B radiation: oxidative stress and photoinhibition of photosynthesis are not prevented by UV-absorbing compounds in the dinoflagellate *Prorocentrum micans*. *Mar. Ecol. Prog. Ser.* 132:287–97

Lesser MP. 1996b. Acclimation of phytoplankton to UV-B radiation: oxidative stress and photoinhibition of photosynthesis are not prevented by UV-absorbing compounds in the dinoflagellate *Prorocentrum micans* (Correction). *Mar. Ecol. Prog. Ser.* 141:312

Lesser MP, Cullen JJ, Neale PJ. 1994. Carbon uptake in a marine diatom during acute exposure to ultraviolet B radiation: relative importance of damage and repair. *J. Phycol.* 30:183–92

Litchman E, Neale PJ, Banaszak AT. 2002. Increased sensitivity to ultraviolet radiation in nitrogen-limited dinoflagellates: photoprotection and repair. *Limnol. Oceanogr.* 47:86–94

Lubin D, Mitchell BG, Frederick JE, Alberts AD, Booth CR, et al. 1992. A contribution toward understanding the biospherical significance of antarctic ozone depletion. *J. Geophys. Res.* 97:7817–28

Lumsden P, ed. 1997. *Plants and UV-B: Responses to Environmental Change.* New York: Cambridge Univ. Press

MacIntyre HL, Kana TM, Geider RJ. 2000. The effect of water motion on short-term rates of photosynthesis by marine phytoplankton. *Trends Plant Sci.* 5:12–17

Madronich S, McKenzie RL, Björn LO, Caldwell MM. 1998. Changes in biologically active ultraviolet radiation reaching the Earth's surface. *J. Photochem. Photobiol. B: Biol.* 46:5–19

Markstädter C, Queck I, Baumeister J, Riederer M, Schreiber U, et al. 2001. Epidermal transmittance of leaves of *Vicia faba* for UV radiation as determined by two different methods. *Photosynth. Res.* 67:17–25

Maske H. 1984. Daylight ultraviolet radiation and the photoinhibition of phytoplankton carbon uptake. *J. Plankton Res.* 6:351–57

Mazza CA, Battista D, Zima AM, Szwarcberg-Bracchitta M, Giordano CV, et al. 1999a. The effects of solar ultraviolet-B radiation on the growth and yield of barley are accompanied by increased DNA damage and antioxidant responses. *Plant Cell Environ.* 22:61–70

Mazza CA, Boccalandro HE, Giordano CV, Battista D, Scopel AL, et al. 2000. Functional significance and induction by solar radiation of ultraviolet-absorbing sunscreens in field-grown soybean crops. *Plant Physiol.* 122:117–25

Mazza CA, Zavala J, Scopel AL, Ballaré CL. 1999b. Perception of solar UVB radiation by phytophagous insects: behavioral responses and ecosystem implications. *Proc. Natl. Acad. Sci. USA* 96:980–85

Moisan TA, Mitchell BG. 2001. UV absorption by mycosporine-like amino acids in *Phaeocystis antarctica* Karsten induced by photosynthetically available radiation. *Mar. Biol.* 138:217–27

Molina JM, Rowland FS. 1974. Stratospheric sink for chlorofluoromethanes: chlorine atom-catalysed destruction of ozone. *Nature* 249:810–12

Montzka SA, Butler JH, Elkins JW, Thompson TM, Clarke AD, et al. 1999. Present and future trends in the atmospheric burden of ozone-depleting halogens. *Nature* 398:690–94

Moody SA, Newsham KK, Ayres PG, Paul ND. 1999. Variation in the responses of litter and phyllophane fungi to UV-B radiation (290–315 nm). *Mycol. Res.* 103:1469–77

Moody SA, Paul ND, Björn LO, Callaghan TV, Lee JA, et al. 2001. The direct effects of UV-B radiation on *Betula pubescens* litter decomposing at four European field sites. *Plant Ecol.* 154:29–36

Mostajir B, Demers S, de Mora S, Belzile C, Chanut JP, et al. 1999. Experimental test of the effect of ultraviolet-B radiation in a planktonic community. *Limnol. Oceanogr.* 44:586–96

Musil CF. 1995. Differential effects of elevated ultraviolet-B radiation on the photochemical and reproductive performances of dicotyledonous and monocotyledonous arid-environment ephemerals. *Plant Cell Environ.* 18:844–54

Musil CF, Björn LO, Scourfield MWJ, Bodeker GE. 2002. How substantial are ultraviolet-B supplementation inaccuracies in experimental square-wave delivery systems? *Environ. Exp. Bot.* 47:25–38

Musil CF, Rutherford MC, Powrie LW, Björn LO, McDonald DJ. 1999. Spatial and temporal changes in South African solar ultraviolet-B exposure: implications for threatened taxa. *Ambio* 28:450–56

National Academy of Sciences. 1979. *Protection Against Depletion of Stratospheric Ozone by Chlorofluorocarbons.* Washington, DC: Natl. Acad. Sci.

Neale PJ. 1987. Algal photoinhibition and photosynthesis in the aquatic environment. In *Photoinhibition*, ed. DJ Kyle, CB Osmond, CJ Arntzen. pp. 35–65. Amsterdam: Elsevier

Neale PJ. 2000. Spectral weighting functions for quantifying the effects of ultraviolet radiation in marine ecosystems. See de Mora et al. 2000, pp. 73–100

Neale PJ. 2001. Effects of ultraviolet radiation on estuarine phytoplankton production: impact of variations in exposure and sensitivity to inhibition. *J. Photochem. Photobiol. B: Biol.* 62:1–8

Neale PJ, Banaszak AT, Jarriel CR. 1998a. Ultraviolet sunscreens in dinoflagellates: Mycosporine-like amino acids protect against inhibition of photosynthesis. *J. Phycol.* 34:928–38

Neale PJ, Bossard P, Huot Y, Sommaruga R. 2001. Incident and *in situ* irradiance in Lakes Cadagno and Lucerne: a comparison of methods and models. *Aquat. Sci.* 63:250–64

Neale PJ, Cullen JJ, Davis RF. 1998b. Inhibition of marine photosynthesis by ultraviolet radiation: variable sensitivity of phytoplankton in the Weddell-Scotia Sea during the austral spring. *Limnol. Oceanogr.* 43:433–48

Neale PJ, Davis RF, Cullen JJ. 1998c. Interactive effects of ozone depletion and vertical mixing on photosynthesis of Antarctic phytoplankton. *Nature* 392:585–89

Neale PJ, Fritz JJ. 2001. Experimental exposure of plankton suspensions to polychromatic ultraviolet radiation for determination of spectral weighting functions. In *Ultraviolet Ground- and Space-Based Measurements, Models, and Effects*, ed. J Slusser, JR Herman, W Gao, pp. 291–96. Aerospace, Remote Sensing, and Astronomy. Vol. 4482. San Diego, CA: SPIE—Int. Soc. Optical Eng.

Neale PJ, Helbling EW, Zagarese HE. 2002. Modulation of UV exposure and effects by vertical mixing and advection. See Helbling & Zagarese 2002. In press

Neale PJ, Kieber DJ. 2000. Assessing biological and chemical effects of UV in the marine environment: spectral weighting functions. See Hester & Harrison 2000, pp. 61–83

Neale PJ, Lesser MP, Cullen JJ. 1994. Effects of ultraviolet radiation on the photosynthesis of phytoplankton in the vicinity of McMurdo Station (78°S). See Weiler & Penhale 1994, pp. 125–42

Newman PA, Nash ER, Rosenfield JE. 2001. What controls the temperature of the Arctic stratosphere during the spring? *J. Geophys. Res. Atmos.* 106:19999–20010

Newsham KK, Anderson JM, Sparks TH, Splatt P, Woods C, McLeod AR. 2001. UV-B effect on *Quercus robur* leaf litter decomposition persists over four years. *Glob. Change Biol.* 7:479–83

Newsham KK, McLeod AR, Roberts JD, Greenslade PD, Emmett BA. 1997. Direct effects of elevated UV-B radiation on the decomposition of *Quercus robur* leaf litter. *Oikos* 79:592–602

Oudejans AMC, Nijssen A, Huls JS, Rozema J. 2001. The reduction of aboveground *Calamagrostis epigeios* mass and tiller number by enhanced UV-B in a dune-grassland ecosystem. *Plant Ecol.* 154:39–48

Paul ND. 2000. Stratospheric ozone depletion, UV-B radiation and crop disease. *Environ. Pollut.* 108:343–55

Phoenix GK, Gwynn-Jones D, Callaghan TV, Sleep D, Lee JA. 2001. Effects of global change on a sub-Arctic heath: effects of enhanced UV-B radiation and increased summer precipitation. *J. Ecol.* 89:256–67

Phoenix GK, Gwynn-Jones D, Lee JA, Callaghan TV. 2000. The impacts of UV-B radiation on the regeneration of a sub-arctic heath community. *Plant Ecol.* 146:67–75

Prézelin BB, Boucher NP, Schofield O. 1994. Evaluation of field studies of UVB radiation effects on Antarctic marine primary productivity. In *Stratospheric Ozone Depletion/UV-B Radiation in the Biosphere*, ed. H Biggs, M Joyner, pp. 181–94. Berlin: Springer-Verlag

Quaite FE, Sutherland BM, Sutherland JC. 1992. Action spectrum for DNA damage in alfalfa lowers predicted impact of ozone depletion. *Nature* 358:576–78

Reuber S, Bornman JF, Weissenböck G. 1996. A flavonoid mutant of barley (*Hordeum vulgare* L.) exhibits increased sensitivity to UV-B radiation in the primary leaf. *Plant Cell Environ.* 19:593–601

Riegger L, Robinson D. 1997. Photoinduction of UV-absorbing compounds in Antarctic diatoms and *Phaeocystis antarctica*. *Mar. Ecol. Prog. Ser.* 160:13–25

Robberecht R, Caldwell MM. 1978. Leaf epidermal transmittance of ultraviolet radiation and its implications for plant sensitivity to ultraviolet-radiation induced injury. *Oecologia* 32:277–87

Robberecht R, Caldwell MM, Billings WD. 1980. Leaf ultraviolet optical properties along a latitudinal gradient in the arctic-alpine life zone. *Ecology* 61:612–19

Roy S. 2000. Strategies for minimisation of UV-induced damage. See de Mora et al. 2000, pp. 177–205

Rousseaux MC, Ballaré CL, Scopel AL, Searles PS, Caldwell MM. 1998. Solar ultraviolet-B radiation affects plant-insect interactions in a natural ecosystem of Tierra del Fuego (southern Argentina). *Oecologia* 116:528–35

Rozema J, van de Staaij J, Björn LO, Caldwell MM. 1997. UV-B as an environmental factor in plant life: stress and regulation. *Trends Ecol. Evol.* 12:22–28

Rundel RD. 1983. Action spectra and estimation of biologically effective UV radiation. *Physiol. Plant.* 58:360–66

Searles PS, Caldwell MM, Winter K. 1995. The response of five tropical dicotyledon species to solar ultraviolet-B radiation. *Am. J. Bot.* 82:445–53

Searles PS, Flint SD, Caldwell MM. 2001a. A meta-analysis of plant field studies stimulating stratospheric ozone depletion. *Oecologia* 127:1–10

Searles PS, Kropp BR, Flint SD, Caldwell MM. 2001b. Influence of solar UV-B radiation on peatland microbial communities of southern Argentina. *New Phytol.* 152:213–21

Shick JM, Dunlap W. 2002. Mycosporine-like amino acids and related gadusols: biosynthesis, accumulation, and UV-protective functions in aquatic organisms. *Annu. Rev. Physiol.* 64:223–62

Shindell DT, Rind D, Lonergan P. 1998. Increased polar stratospheric ozone losses and delayed eventual recovery owing to increasing greenhouse-gas concentrations. *Nature* 392:589–92

Smith RC, Baker KS. 1982. Assessment of the influence of enhanced UV-B on marine primary productivity. In *The Role of Solar Ultraviolet Radiation in Marine Ecosystems*, ed. J Calkins, pp. 509–37. New York: Plenum

Smith RC, Baker KS, Holm-Hansen O, Olson RS. 1980. Photoinhibition of photosynthesis in natural waters. *Photochem. Photobiol.* 31:585–92

Smith RC, Prézelin BB, Baker KS, Bidigare RR, Boucher NP, et al. 1992. Ozone depletion: ultraviolet radiation and phytoplankton biology in Antarctic waters. *Science* 255:952–59

Solomon S. 1999. Stratospheric ozone depletion: a review of concepts and history. *Rev. Geophys.* 37:275–316

Steemann-Nielsen E. 1964. On a complication in marine productivity work due to the influence of ultraviolet light. *J. Cons. Perm. Int. Explor. Mer.* 22:130–5

Sullivan JH, Howells BW, Ruhland CT, Day TA. 1996. Changes in leaf expansion and epidermal screening effectiveness in *Liquidambar styraciflua* and *Pinus taeda* in response to UV-B radiation. *Physiol. Plant.* 98:349–57

Sullivan JH, Teramura AH. 1992. The effects of ultraviolet-B radiation on loblolly pine. *Trees* 6:115–20

Sullivan JH, Teramura AH, Adamse P, Kramer GF, Upadhyaya A, et al. 1994. Comparison of the response of soybean to supplemental UV-B radiation supplied by either square-wave or modulated irradiation systems. In *Stratospheric Ozone Depletion/UV-B Radiation in the Biosphere*, ed. RH Biggs, MEB Joyner, pp. 211–20. Berlin: Springer-Verlag

Teramura AH, Sullivan JH. 1994. Effects of UV-B radiation on photosynthesis and growth of terrestrial plants. *Photosynth. Res.* 39:463–73

Tevini M, ed. 1993. *UV-B Radiation and Ozone Depletion: Effects on Humans, Animals, Plants, Microorganisms, and Materials*. Boca Raton, Florida: Lewis

Tevini M, Braun J, Fieser G. 1991. The protective function of the epidermal layer of rye seedlings against ultraviolet-B radiation. *Photochem. Photobiol.* 53:329–33

Thomson BE, Worrest RC, Van Dyke H. 1980. The growth response of an estuarine diatom (*Melosira nummuloides* Dillw. Ag.) to UV-B (290–320 nm) radiation. *Estuaries* 3:69–72

Van de Staaij J, Rozema J, van Beem A, Aerts R. 2001. Increased solar UV-B radiation may reduce infection by arbuscular mycorrhizal fungi (AMF) in dune grassland plants: evidence from five years of field exposure. *Plant Ecol.* 154:171–77

Vasilkov A, Krotkov N, Herman J, McClain J, Arrigo K, Robinson W. 2001. Global mapping of underwater UV fluxes and DNA-weighted exposures using TOMS and SeaWifs data products. *J. Geophys. Res.* 106:27205–219

Vernet M, Brody EA, Holm-Hansen O, Mitchell BG. 1994. The response of Antarctic phytoplankton to ultraviolet radiation: absorption, photosynthesis, and taxonomic composition. See Weiler & Penhale 1994, pp. 143–58

Vincent WF, Neale PJ. 2000. Mechanisms of UV damage to aquatic organisms. See de Mora et al. 2000, pp. 73–100

Wangberg S, Garde K, Gustavson K, Selmer J. 1999. Effects of UVB radiation on marine phytoplankton communities. *J. Plankton Res.* 21:147–66

Wangberg SA, Wulff A, Nilsson C, Stagell U. 2001. Impact of UV-B radiation on microalgae and bacteria: a mesocosm study with computer modulated UV-B radiation addition. *Aquat. Microb. Ecol.* 25:75–86

Weiler CS, Penhale PA, eds. 1994. *Ultraviolet Radiation in Antarctica: Measurements and Biological Effects*. Antarctic Res. Ser., Vol. 62. Washington, DC: Am. Geophys. Union

Worrest RC. 1983. Impact of solar ultraviolet-B radiation (290–320 nm) upon marine microalgae. *Physiol. Plant.* 58:428–34

Worrest RC, Wolniakowski KU, Scott JD, Brooker DL, Thomson BE, Van Dyke H. 1981. Sensitivity of marine phytoplankton to UV-B radiation: impact upon a model ecosystem. *Photochem. Photobiol.* 33:223–27

Xiong FS, Day TA. 2001. Effect of solar ultraviolet-B radiation during springtime ozone depletion on photosynthesis and biomass production of Antarctic vascular plants. *Plant Physiol.* 125:738–51

Xiong FS, Lederer F, Lukavsky J, Nedbal L. 1996. Screening of freshwater algae (Chlorophyta, Chromophyta) for ultraviolet-B sensitivity of the photosynthetic apparatus. *J. Plant Physiol.* 148:42–48

Zavala AL, Scopel AL, Ballaré CL. 2001. Effects of ambient UV-B radiation on soybean crops: impact on leaf herbivory by *Anticarsia gemmatialis*. *Plant Ecol.* 156:121–30

Annu. Rev. Ecol. Syst. 2002. 33:397–425
doi: 10.1146/annurev.ecolsys.33.010802.150419
First published online as a Review in Advance on August 14, 2002

THE EVOLUTION AND MAINTENANCE OF ANDRODIOECY

John R. Pannell

*Department of Plant Sciences, University of Oxford, Oxford OX1 3RB, United Kingdom;
email: john.pannell@plants.ox.ac.uk*

Key Words   dioecy, gynodioecy, hermaphroditism, metapopulation, fitness-gain curves

■ **Abstract**   Examples of androdioecy, the coexistence of males and hermaphrodites, was unknown when the subject was last reviewed about two decades ago. Since then, several examples have been discovered in both plants and animals, and we are now in a position to reappraise theoretical work on the subject. Whereas early ideas were framed largely in terms of the invasion of males into hermaphroditic populations, all of the clearest examples of androdioecy now known appear to have evolved from dioecy. There are strong indications that this has occurred repeatedly as a result of the selection of self-fertile hermaphroditism for reproductive assurance during colonization. Male frequencies in these species are highly variable, self-fertilization in hermaphrodites is delayed, and mating opportunities appear to depend strongly on population density. Results from theoretical work on the evolution and maintenance of androdioecy in single populations and in metapopulations are summarized, and several case studies of androdioecious plants and animals are reviewed.

INTRODUCTION

It is not often that research on an evolutionary topic carried out independently by botanists and zoologists produces conclusions which are virtually identical. When this does happen, one cannot restrain a feeling that a principle of more than superficial importance has been uncovered.

H.G. Baker (1955)

Darwin (1877) devoted considerable space in his book *The Different Forms of Flowers* to a discussion of gynodioecy, the coexistence of females and hermaphrodites in a population. Much less attention was given to the apparently analogous case of androdioecy, where males co-occur with hermaphrodites. Apart from speculating about the possible role that androdioecy may have played as a pathway in the evolution of dioecy, Darwin was aware of no actual examples of androdioecy in nature and regarded it as being not worthy of further consideration. The issue of why it appeared to be so rare was left unexplained.

0066-4162/02/1215-0397$14.00

In the century following Darwin's work, occasional instances of androdioecy were reported in the literature (see especially Yampolsky & Yampolsky 1922). These were based largely on morphological observations of males and hermaphrodite forms in single populations. In 1984 D. Charlesworth published a comprehensive review of these cases in the light of mathematical models for the evolution and maintenance of androdioecy (Lloyd 1975, Charlesworth & Charlesworth 1978, Charlesworth 1984). Based on the predictions made by these models, Charlesworth was unable to confirm any cases of true functional androdioecy. Where adequate data were available, recorded observations of pollen morphology or patterns of sex allocation strongly suggested that individuals with hermaphrodite morphology were in fact functional females with dysfunctional pollen or anthers.

In the two decades since Charlesworth's review, the empirical base for our understanding of androdioecy has changed substantially. Several instances of androdioecy have now come to the fore in both plants and animals, and observations of morphology, estimates of mating-system parameters, and measures of sex allocation in these populations have confirmed its functional status. Although androdioecy must still be regarded as exceedingly rare, and the low number of species that display it places obvious limitations on the generalizations we can make, several interesting patterns have begun to emerge. Perhaps most importantly, the clearest cases of androdioecy all appear to have evolved from dioecy as a result of selection for self-fertile hermaphroditism, rather than from hermaphroditism. The highly fragmented nature of androdioecious populations has prompted metapopulation models of mating-system evolution that make predictions largely consistent with empirical observations.

In this review, I draw together the knowledge we have gained about androdioecy since Charlesworth's (1984) review. First, I summarize the key predictions of early theoretical studies as well as the results emerging from more recent theoretical work on the effects of fluctuating local population sizes and mating opportunities in a metapopulation. I then review cases of androdioecy from the empirical literature and conclude with a discussion of what these cases have taught us about androdioecy and the selection of combined versus separate sexes. Throughout I refer to individuals that contribute genetically through both sexual functions (not necessarily in equal measure) as "hermaphrodites," without implying that, in the case of plants, they have perfect flowers. I use the term androdioecy in its functional rather than morphological sense, unless otherwise stated.

THEORETICAL BACKGROUND

Comparisons Between Androdioecy and Gynodioecy

Theoretical models of androdioecy have largely concerned its evolution and maintenance in terms of patterns of sex allocation, selfing rates, levels of inbreeding depression, self-incompatibility, and life-history variation (Lloyd 1975, Charnov et al. 1976, Ross & Weir 1976, Charlesworth & Charlesworth 1978, Ross 1982, Charlesworth 1984, Maurice & Fleming 1995, Vassiliadis et al. 2000b). More recently, the effects of population structure and metapopulation dynamics on its evolution and maintenance have also been considered (Pannell 1997a, 2001). All these studies fall into a broader class of models that ask general questions concerning the selection of combined versus separate sexes (Lloyd 1982, 1988). Within this context a great deal of effort has been directed toward understanding the evolution and maintenance of gynodioecy, both through theoretical (reviewed in Charlesworth 1999) and empirical (reviewed in Webb 1999) studies. Gynodioecious populations have long presented us with an opportunity to address the fitness implications of combined versus separate sexes from the female perspective, by comparing hermaphrodite and female traits within the same population. The discovery of several cases of androdioecy offers similar possibilities from the male point of view.

At first glance, androdioecy might seem to be the symmetrical male analogue of gynodioecy. The striking rarity of androdioecy relative to gynodioecy indicates empirically that this is not the case, and there are also good theoretical reasons for this. It can readily be seen, for example, that fitness gains accrued through male and female functions are commonly limited in different ways. According to Bateman's Principle (Bateman 1948, Wilson et al. 1994), female fitness tends to be limited by resources needed to fill seeds and fruits or to nourish and care for offspring, whereas male fitness is more likely to be limited by mating opportunities (although see Elle & Meagher 2000). Models comparing androdioecy and gynodioecy show that self-fertilization by hermaphrodites has very different implications for the ability of males versus females to invade and spread in populations (Lloyd 1975; Ross & Weir 1976; Charlesworth & Charlesworth 1978, 1981; Charlesworth 1984). Whereas selfing by hermaphrodites reduces the availability of ovules and thus limits male mating opportunities, it does not directly compromise mating opportunities for females (Figure 1).

These critical differences between males and females have been quantified in the models of Lloyd (1975), Charlesworth & Charlesworth (1978), and Charlesworth (1984). In the case of androdioecy, these models show that because males lack a female component of fitness, they must make more than twice the genetic contribution through their male function as do hermaphrodites if they are to spread in a population (see Figure 1). In fully outcrossing populations this means that males must sire more than twice the number of progeny as do hermaphrodites, either through the dispersal of more pollen or through increased siring success of the pollen they disperse. If resources formerly allocated to female function are diverted without loss to male function, one might expect a maximum of a twofold increase in male fertility in unisexual males. This is because in an outcrossing hermaphroditic population the costs of investment through each sexual function are expected to be equal (Fisher 1930, Charnov 1982, Lloyd 1983). In the absence of full "compensation," as Darwin (1877) referred to the male-female allocation trade-off, males should not achieve the required male fertility threshold unless

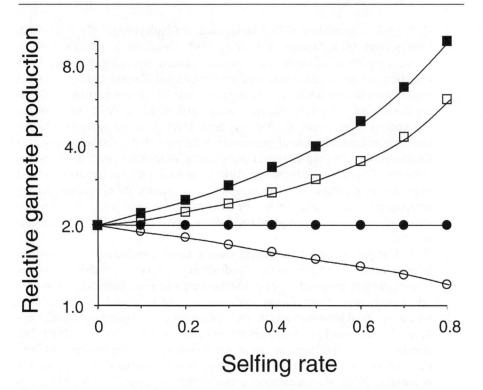

Figure 1   The minimum relative pollen or seed production of males (*squares*) and females (*circles*), respectively, for their invasion into a population of hermaphrodites as a function of the hermaphrodite selfing rate. Curves are shown for the case in which there is no inbreeding depression (*open symbols*) and with inbreeding depression $= 0.5$ (*closed symbols*). Drawn according to equations modified from Charlesworth & Charlesworth (1978).

their siring ability is increased for reasons other than simple increases in pollen production. Specifically, this implies that males should be able to spread only under ecological conditions in which male fitness is an accelerating function of allocation (Charnov et al. 1976, Charnov 1979).

In outcrossing hermaphroditic populations, similar reasoning also applies to the evolution of gynodioecy; for females to invade and spread, their seed production must exceed twice that of hermaphrodites (Lloyd 1975; Ross & Weir 1976; Charlesworth & Charlesworth 1978, 1981). However, this symmetry breaks down if hermaphrodites self-fertilize. In the case of androdioecy, partial selfing by hermaphrodites increases the fertility threshold above which males can invade the population, and even in the presence of strong inbreeding depression, males must always successfully disperse more than twice the pollen that hermaphrodites do. By contrast, if the inbred progeny of hermaphrodites suffer sufficient inbreeding depression, partial selfing reduces the relative female fertility threshold for the invasion of females and thus may favor the evolution of gynodioecy (Lloyd 1975, Charlesworth & Charlesworth 1978). These predictions suggest that whereas gynodioecy may frequently evolve from hermaphroditism as a response to selection for inbreeding avoidance, androdioecy is most likely to do so in outcrossing populations in which ecological factors in some way give rise to an accelerating male fitness-gain curve (Lloyd 1975, Charlesworth & Charlesworth 1978, Charlesworth 1984). As Charlesworth (1984) noted, such circumstances are generally favorable for the evolution of full dioecy, so that androdioecy might evolve as an evolutionarily unstable state in the transition from hermaphroditism to dioecy (see also Seger & Eckhart 1996, Charlesworth 1999).

Androdioecy as an Evolutionarily Stable Strategy

Charnov et al. (1976) described the evolutionary stability of androdioecy in graphical terms. They suggested that a rather specific type of curve describing the trade-off between male and female fitness is required (see Figure 2). Whereas concave-out and concave-in curves in general are necessary for the stability of dioecy and hermaphroditism, respectively, the stability of androdioecy may require a curve that is concave-out at the male end (favoring pure males) and concave-in at the female end (favoring female-biased hermaphrodites) (Charnov et al. 1976). If this is true, it would appear to place a severe restriction on the stability of androdioecy, although no suggestions have yet been made as to what biological circumstances might give rise to such a concave-convex fitness set.

The gain-curve analysis proposed by Charnov et al. (1976) assumes complete outcrossing and ignores the possibility that a shortage of mates might reduce female reproductive success. Charlesworth & Charlesworth (1981) extended this analysis to partially inbred populations and showed that the shape of the fitness-gain curve is no longer a good predictor of which breeding system will be stable. However, their study did not identify situations in which androdioecy might be more easily maintained. Maurice & Fleming (1995) considered the effect of pollen-limited seed production on the stability of combined versus separate sexes. In their model, hermaphrodites could overcome the problem of pollen limitation through selfing and could thus invade and spread in dioecious populations (see also Després & Maurice 1995). However, this was more likely to lead to trioecy, with males, females, and hermaphrodites maintained together, rather than to stable androdioecy (Maurice & Fleming 1995). According to this analysis, androdioecy is therefore not likely to evolve in response to increased pollen limitation within populations, contrary to speculations concerning reasons for the breakdown of dioecy in *Datisca glomerata* (for example, Fritsch & Rieseberg 1992, Charlesworth 1993) (see below).

Androdioecy in a Metapopulation

One specific circumstance under which pollen or mate limitation may be severe is during colonization, when self-fertile hermaphrodites, which can produce progeny

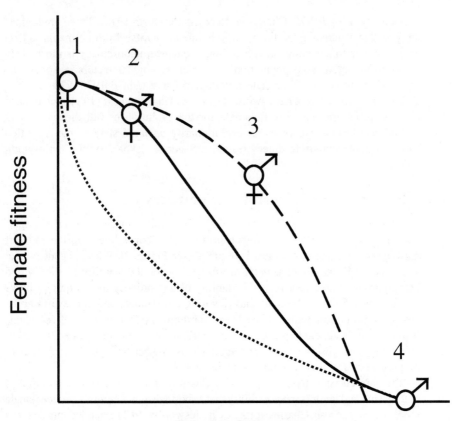

Figure 2 Theoretical curves depicting the trade-off between male and female fitness for a population with a dioecious, a hermaphroditic, and an androdioecious evolutionarily stable strategy. See Charnov et al. (1976) for details. (1) Pure female strategy; (2) female-biased hermaphrodite strategy; (3) hermaphrodite strategy; (4) pure male strategy. Monomorphic hermaphroditic populations are expected to be at (3); dimorphic dioecious populations are expected to be at (1) and (4); androdioecious populations are expected to be at (2) and (4).

on their own, have a distinct advantage over self-incompatible hermaphrodites or unisexuals (Darwin 1876). The importance of the selection of hermaphroditism in the context of colonization by long-distance dispersal, first emphasized by Baker (1955, 1967) and termed "Baker's Law" by Stebbins (1957), has been invoked in explanations of the breakdown of self-incompatibility or dioecy in a wide range of species (reviewed in Pannell & Barrett 1998), as well as in the peripheral distribution of selfing variants in otherwise potentially outcrossing species (e.g., Barrett & Shore 1987, Barrett et al. 1989, Vogel et al. 1999). It has also been suggested as having played a particularly important role in the evolution of androdioecy in a metapopulation with frequent local extinction and recolonization (Pannell 1997a, 2001).

The selection of combined versus separate sexes under pollen or mate limitation in a metapopulation differs from that in a single, demographically stable population. This is because the fluctuations in population size and density, which are involved in colonization and subsequent population growth, may frequently cause local populations to deviate from their evolutionarily stable sex allocation (Pannell 2001). This applies particularly to species with density-dependent rates of self-fertilization, in which low density causes high selfing rates and outcrossing becomes increasingly frequent as populations grow. This situation is likely to be common in wind-pollinated species, in which mating occurs by mass action (e.g., Farris & Mitton 1984, Schoen & Stewart 1986, Burczyk & Prat 1997), but there is growing evidence for similar patterns in insect-pollinated plants (e.g., van Treuren et al. 1993, Kohn & Barrett 1994, Routley et al. 1999; and see Pannell 2001). It is also more likely to be found where hermaphrodites are protogynous, a situation that may allow delayed selfing in pollen-limited conditions (Lloyd & Schoen 1992) but outcrossing in the presence of mates.

In the extreme event that a population is founded by a single self-fertile hermaphrodite, all progeny must be the result of self-fertilization if reproduction is to occur. If such colonization is frequent, female-biased sex allocation will be selected in hermaphrodites at the metapopulation level as a result of episodic local mate competition (Hamilton 1967, Charnov 1982). However, if the growth of population size and density following colonization gives rise to increased outcrossing opportunities, then the female-biased sex allocation of the founding population will no longer be evolutionarily stable, and the population will be open to the invasion of males. Male invasion is thus made possible not necessarily because of accelerating fitness gains within local populations, but rather because pure males achieve a greater than twofold male fertility advantage over hermaphrodites because the latter invest only lightly in male function (Pannell 2001).

It is not yet clear whether this type of scenario will give rise to the long-term maintenance of androdioecy, and further theoretical work is needed. Because the process that selects for female-biased sex allocation in hermaphrodites also selects against males at the metapopulation level, it seems likely that for the long-term evolutionary persistence of androdioecy conditions for an accelerating male fitness-gain curve within populations may still also be required. Nevertheless, it is possible that androdioecy may be maintained along these lines for long periods while selection slowly adjusts patterns of sex allocation toward either dioecy or hermaphroditism. It is also important to note that, even if the polymorphic state is transient, it is androdioecy that should be maintained rather than gynodioecy or trioecy. This is because females are disadvantaged not only as a result of selection for reproductive assurance, but also because they cannot invade female-biased local populations, whereas males can (Charnov 1982, Pannell 2001; compare with Sarkissian et al. 2001).

If we assume a female-biased sex allocation in hermaphrodites, it is possible to investigate the conditions under which males are maintained with hermaphrodites in a metapopulation. Analysis of an androdioecious metapopulation has shown that the frequency of males is reduced as a result of selection for reproductive assurance by increases in the extinction rate or decreases in the mean number of immigrants into habitat patches each generation (Pannell 1997a). As long as androdioecy is maintained at the metapopulation level, male frequencies are predicted to vary greatly among populations, with newly founded populations comprising only hermaphrodites, and older, larger populations more likely to be androdioecious (Pannell 1997a, 2001). With sufficiently high migration rates among populations and the reduced importance of population turnover, sex-ratio variation is diminished, and all populations are then expected to be androdioecious (Figure 3) (see below). At the other extreme, rapid population turnover with sufficiently small colonizing propagules and low migration rates cause males to be lost from the metapopulation entirely, leaving a monomorphic metapopulation of female-biased hermaphrodites (Pannell 1997a). These models therefore suggest that androdioecy can be maintained only within relatively narrow margins of metapopulation parameters (Figure 3). The rarity of androdioecy, even among species with a metapopulation dynamic, may thus reflect its frequent loss as much as its infrequent evolution.

Androdioecious Sex Ratios

The models of Lloyd (1975), Charlesworth & Charlesworth (1978), and Charlesworth (1984) make clear predictions concerning the frequency of males that we should expect to find in androdioecious populations. Using Lloyd's (1975) approach with a slight modification of the parameters used, an expression for the equilibrium proportion of males at reproductive maturity (q_{rep}) can be derived in terms of the relative pollen fecundity of males versus hermaphrodites (r), the selfing rate (s), the level of inbreeding depression suffered by selfed progeny (δ), and the relative probability that hermaphrodite versus male zygotes reach reproductive maturity (v):

$$q_{rep} = \frac{r(1-s) + 2v(s\delta - 1)}{r\{1 - s(1 - v + 2\delta v) + v\} + 2v(s\delta - 1)}.$$

Figure 3 The proportion of males maintained in a metapopulation under recurrent population turnover as a function of the average number of individuals immigrating into local populations per generation. Curves are given for an extinction rate of 0.2 (*open squares*) and 0.025 (*closed circles*) assuming (*a*) dominant sex determination and (*b*) recessive sex determination of maleness. Data are from simulations assuming that males disperse five times more pollen than hermaphrodites (as found, for example, in *Mercurialis annua*). Modified from Pannell (1997a).

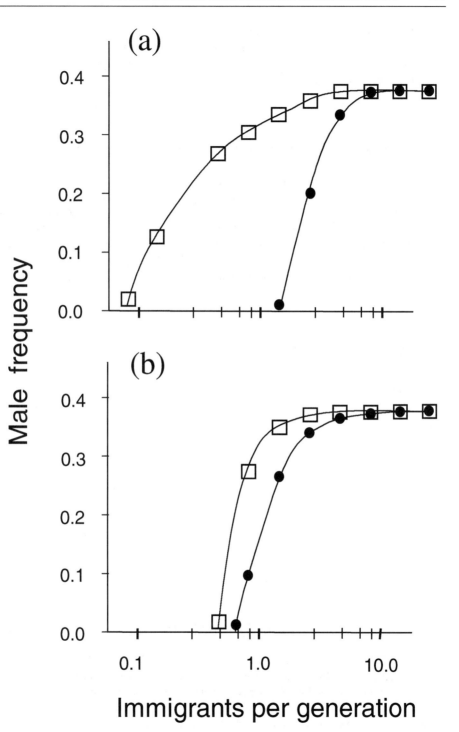

Recognizing that the outcrossing opportunities of a male or hermaphrodite zygote depend also on the viability of other hermaphrodites in the population, Lloyd's (1975) analysis can be extended to derive an expression for the primary sex ratio (q_{zyg}) in an androdioecious population at equilibrium

$$q_{zyg} = \frac{1 - s(v + 2\delta - 1) + v + 1 - r(1 - s)}{1 - s(v + 2\delta - 1) + v + 1 - 2r(1 - s)},$$

where clearly $v > 0$ for populaton persistence. These expressions for q_{rep} and q_{zyg} are plotted in Figure 4 for a range of parameter values. In the absence of differential mortality between males and hermaphrodites (i.e., when $v = 1$), q_{rep} and q_{zyg} are of course equal. In this case, male frequencies in an androdioecious population are always <0.5, with the population sex ratio approaching equality only when the male function of hermaphrodites approaches zero, i.e., when the population is functionally dioecious. Thus, in the absence of evidence for differential mortality between the sexes, observed male frequencies of 0.5 or greater are difficult to explain in terms of androdioecy and are more likely to be cases of

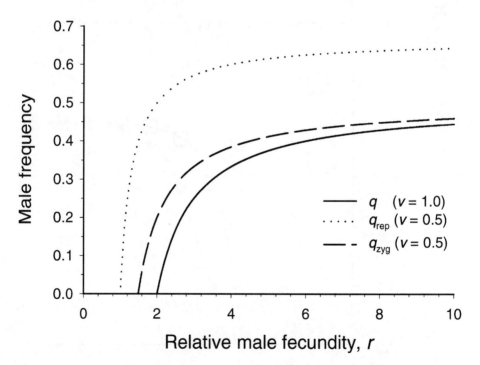

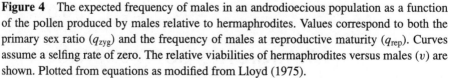

Figure 4 The expected frequency of males in an androdioecious population as a function of the pollen produced by males relative to hermaphrodites. Values correspond to both the primary sex ratio (q_{zyg}) and the frequency of males at reproductive maturity (q_{rep}). Curves assume a selfing rate of zero. The relative viabilities of hermaphrodites versus males (v) are shown. Plotted from equations as modified from Lloyd (1975).

cryptic functional dioecy (Charlesworth 1984, Mayer & Charlesworth 1991, Pannell 2002).

Of course, field tallies of males and hermaphrodites may not correspond to the primary sex ratio, and it is clear that q_{rep} may rise above 0.5 if hermaphrodites suffer higher mortality than males ($v < 1$ in Figure 4). Lloyd's model applies implicitly to a semelparous life history, but Charlesworth's (1984) analysis includes the possibility of iteroparity and of differential mortality in years following first reproduction. Given the costs of fruit production, this seems plausible. It is also possible that hermaphrodites flower less frequently than males, so that if only flowering individuals are counted, the apparent sex ratio may be greater than equality (e.g., Pannell & Ojeda 2000). Finally, it is important to note that if survivorship is reduced sufficiently in hermaphrodites, androdioecy can in fact evolve and be maintained even when the relative pollen fertility of males is below the twofold threshold (Figure 4). These processes have been suggested as possible explanations for the maintenance of putative androdioecy in several perennial plant species (e.g., Lepart & Dommée 1992, Molau 1992, Pannell & Ojeda 2000).

Mode of Sex Determination

The population-based models of androdioecy make predictions in terms of allocation and sex expression strategies that make few assumptions about how they are governed mechanistically. In fact, the evolutionarily stable strategy models discussed above assume only genic determination of the allocation strategy, in the sense that the evolutionarily stable strategy is determined by the transmission interests of nuclear genes in organisms with Mendelian inheritance. Just as maternally inherited cytoplasmic genes have played an important role in the evolution and maintenance of gynodioecy (Kaul 1988, Charlesworth 1999), so genes with non-Mendelian paternal inheritance might in principle give rise to an androdioecious breeding system. The invasion criteria and conditions for evolutionary maintenance of males with hermaphrodites would be different with such inheritance, but no such case is known, and no theoretical models have yet dealt with this issue (although see Vassiliadis et al. 2000b).

Under straightforward genetic sex determination at a nuclear locus, an androdioecious evolutionarily stable strategy does not depend on the dominance relations of the genes involved. This is not the case, however, for the maintenance of androdioecy in a metapopulation. In particular, the range of extinction rates and colonizing propagule sizes under which androdioecy can be maintained is greater when maleness is determined by a dominant allele than when the male-determining allele is recessive (Pannell 1997a). This difference is due to an interaction between processes occurring at the local population and the metapopulation levels. For example, if males enter populations that have grown substantially in size following an initial hermaphrodite colonization, their rise in frequency will be very much slower under recessive than under dominant male determination as a result of the action, within local populations, of Haldane's sieve (Haldane 1927, Crow &

Kimura 1970, Orr & Betancourt 2001). By contrast, if recessive male-determining genes immigrate into small populations, e.g., at or soon after their establishment, frequency-dependent selection will favor their spread much more quickly (Pannell 1997a). This suggests that immigration rates should be higher for the maintenance of recessive maleness, but we have noted that high migration rates may be more conducive to the maintenance of dioecy than androdioecy. Thus, if metapopulation dynamics have been important in the maintenance of specific cases of androdioecy, we might expect maleness to be determined by a dominant allele.

Androdioecy when Hermaphrodites Cannot Outcross as Males

The models reviewed above assume that hermaphrodites can use their pollen both to self-fertilize their own ovules and to cross with other hermaphrodites in the population. As I review in detail below, however, androdioecy has evolved in several animal taxa in which hermaphrodites are unable to cross with each other. In populations lacking males, hermaphrodites are therefore completely selfing. The conditions under which males can be maintained in such populations has been studied in a population genetic model by Otto et al. (1993). Using a slightly modified nomenclature, in accordance with terms defined in Equations 1 and 2 above, the conditions for the maintenance of androdioecy in their model are given by

$$\alpha > 2\beta(1 - \delta)v$$

where males fertilize a total of α times as many eggs as are produced by a single hermaphrodite, a proportion β of the eggs not fertilized by male sperm are self-fertilized, and v and δ are the relative viability of hermaphrodites and the inbreeding depression suffered by selfed progeny, respectively, as defined above. For males to spread from rarity in a population, they must clearly fertilize a sufficiently large number of eggs in the population. The likelihood of their spread is increased if progeny produced by hermaphrodites through selfing suffer high inbreeding depression, if not all hermaphrodite eggs are fertilized, and if hermaphrodites suffer reduced viability relative to males. In the absence of inbreeding depression and viability differences, and when all eggs in the population are fertilized either by males or through selfing, males can invade a population if they fertilize a number of eggs equivalent to that produced by at least two hermaphrodites (Otto et al. 1993).

These conditions for the maintenance of androdioecy are different from those predicted by the models conceived with plant examples in mind, in which males cannot invade in response to selection for inbreeding avoidance. Perhaps an even more important difference between those models and the model of Otto et al. (1993) is that only in the latter do conditions exist for males to invade a population with a selfing rate of one. It is important to note, however, that this model assumes delayed selfing, whereas the models of Lloyd (1975), Charlesworth & Charlesworth (1978), and Charlesworth (1984), for example, assume that a fixed proportion of ovules are selfed prior to any opportunities that males might have to mate. The assumption of delayed selfing by Otto et al. (1993) is similar to that implicit in metapopulation models, in which selfing rates change dynamically with the changing composition of local populations.

Evolutionary Pathways to Androdioecy

Models of androdioecy have tended to be framed in terms of conditions that would allow the invasion of males into hermaphroditic populations, partly reflecting interest in evolutionary pathways to dioecy (Ross 1982, Charlesworth 1999). One suggested evolutionary path to dioecy from hermaphroditism is via andromonoecy (Ross 1982), in which individuals possess both male and perfect (hermaphroditic) flowers. In the plant genus *Solanum*, there is a strong phylogenetic association between andromonoecy and cryptic dioecy, in which females produce inaperturate pollen (Anderson 1979, Knapp 1991, Knapp et al. 1998). Ross (1982) conjectured that androdioecy may have been an intermediate stage in the evolution of cryptic dioecy from andromonoecy in *Solanum*, but there is no direct evidence for this transition. As in other species with cryptic dioecy, *Solanum* flowers are nectarless, and male function in the females is probably maintained as a reward for pollinators (Charlesworth 1984, Mayer & Charlesworth 1991, Knapp et al. 1998). In these cases the evolution to dioecy via a gynodioecious state may be more likely (cf., Sarkissian et al. 2001).

Androdioecy may also be derived from dioecy, with the replacement of females by hermaphrodites (Liston et al. 1990, Rieseberg et al. 1992). We have already seen that the breakdown of dioecy under pollen limitation may more likely lead to trioecy than to androdioecy in a single population (Maurice & Fleming 1995). However, Pannell (2001) has recently considered hypotheses for the evolution of androdioecy in a metapopulation from both an hermaphroditic and a dioecious ancestral state and has argued that a dioecious precursor is more probable. This is because, if hermaphroditism is maintained through selection for reproductive assurance, the immigration of males from populations with a dioecious history would be an easier evolutionary step than their de novo generation through mutation, especially because natural selection under dioecy may have led to specialist males with secondary sexual characteristics that yield accelerating male fitness-gain curves (see Eckhart 1999, Geber 1999). The evolution of androdioecy in a hermaphroditic metapopulation as a result of a relaxation of metapopulation dynamics thus seems more probable in lineages with a dioecious history for much the same reasons. We might also predict that, under a regime of increased migration and reduced extinction rates, the classic signature of population turnover would be absent, i.e., high variability in sex ratios between populations.

EMPIRICAL OBSERVATIONS

The cases of androdioecy that have been confirmed for both plant and animal taxa since Charlesworth's (1984) review are listed in Table 1. Although they are few in number and cannot provide a statistically firm comparative test of the theory,

TABLE 1   Taxa within which androdioecy has either been confirmed, or for which limited evidence is consistent with androdioecy. Characteristics predicted to be important on theoretical grounds are also shown. See text for explanation and references

	Plant/ animal	Male frequencies <0.5	Sex ratios variable	Mating context– dependent	Male sex determination	2° sexual characters
Androdioecy derived from dioecy						
Datisca glomerata	Plant	+	+	e	Dominant	?
Mercurialis annua	Plant	+	+	d	Dominant	+g, h
Schizopepon bryoniaefolius	Plant	+	+	e	?	+g
Spinifex littoreus	Plant	?	+?	?	?	+g
Castilla elastica	Plant	+?	?	d	?	+g
Branchiopod crustaceans[a]	Animal	+	+	f	Recessive	+i
Rhabtidid nematodes[b]	Animal	+?	+?	e	Dominant[k]	+i
Androdioecy derived from hermaphroditism						
Sagittaria lancifolia	Plant	+	+	d	Dominant	?
Oleaceae[c]	Plant	+/−	−	d	?	+j

?, more information required.

[a]Several species of Notostracan (Sassaman 1991) and Conchostracan branchiopods (Sassaman 1995).

[b]Several species of rhabtidid nematode (Fitch 1997), especially in the Elegans group (Sudhaus & Kiontke 1996).

[c]Several species in the Oleaceae family, particularly in the general *Phillyrea* and *Fraxinus*.

[d]Suspected density-dependent rates of self-fertilization as a result of wind pollination or pollination by generalist insect pollinators.

[e]Negative correlations found between inbreeding coefficient and male frequencies (see Figure 5).

[f]Hermaphrodites favor mating with males over self-fertilization (see text).

[g]Males and hermaphrodites have distinct inflorescence architecture.

[h]Males are taller for their biomass than hermaphrodites (see text).

[i]Hermaphrodites lack organs for outcrossing as males; life-history differences between males and hermaphrodites (see text).

[j]Differences in life history found, either in flowering frequency or rates of mortality (see text).

[k]Effective dominance through an XO chromosomal system.

it is satisfying to note that the predictions are borne out well. In this section I review what is known about androdioecy in the species in Table 1, beginning with examples that appear to have been derived from dioecy and which are now quite well understood. I then discuss more equivocal cases of putative androdioecy derived from hermaphroditism and conclude the section with comments on several species that are almost certainly not androdioecious, although claims in favor of it have been made.

Datisca Glomerata (Datiscaceae)

The first species to be confirmed as displaying androdioecy was *Datisca glomerata* (Datiscaceae), an herbaceous, wind-pollinated perennial occupying small "islands" of disturbed riparian habitat in south-western North America (Liston et al. 1990). Liston et al. showed that hermaphrodites were self-compatible and produced fully competent pollen, that male flowers produced about 3.8 times as many anthers as did those of hermaphrodites, that population male frequencies were highly variable and always much lower than 0.5, and that male individuals were not simply small hermaphrodites. Males of *D. glomerata* are determined by a dominant allele (Wolf et al. 2001).

Philbrick & Rieseberg (1994) confirmed that the high relative anther count in male flowers in *D. glomerata* translated into sufficiently high relative male fertilities. Moreover, high outcrossing rates estimated for two populations of the species (Fritsch & Rieseberg 1992), as well as moderate levels of inbreeding depression (Rieseberg et al. 1993), are also consistent with the maintenance of androdioecy. Hermaphrodite flowers are protogynous (Rieseberg et al. 1993), suggesting the likelihood of delayed selfing, and a single comparison of outcrossing rates between a pair of populations with different plant densities (Fritsch & Rieseberg 1992) further supports the idea that mating patterns are likely to be density dependent. Although recent work on the polarity of breeding-system transitions in the clade to which *D. glomerata* belongs (Swensen et al. 1998) has shown that patterns are less clear than they once seemed (cf., Rieseberg et al. 1992), the phylogenetic evidence strongly suggests that androdioecy has been derived from dioecy in the Datiscaceae (see Weller & Sakai 1999).

Mercurialis annua (Euphorbiaceae)

Androdioecy has also been documented in hexaploid populations of the annual plant *Mercurialis annua* (Euphorbiaceae), a wind-pollinated weed of anthropogenic habitat with a wide distribution across Europe and around the Mediterranean (Durand 1963, Durand & Durand 1992, Pannell 1997d). Diploid populations of *M. annua* are dioecious, as are six of the other seven species in the genus, and it is clear that monoecy (and thus androdioecy) is a derived character in *M. annua* hexaploids (Durand 1963, Durand & Durand 1992).

Male frequencies in androdioecious populations in southern Spain and northern Morocco are highly variable, with males always found at frequencies significantly lower than 0.5 (Durand 1963; Durand & Durand 1992; Pannell 1997c,d). The hermaphrodites are protogynous at the level of the inflorescence and are self-compatible, and although rates of selfing in this species are not yet known, isolated plants set full seed upon selfing (Pannell 1997d). Males disperse between 4 and 10 times more pollen than do hermaphrodites, well above the twofold threshold required for the maintenance of androdioecy (Pannell 1997c,d). Moreover, male inflorescence morphology differs qualitatively from that of the hermaphrodites. On males, staminate flowers are produced on erect axillary peduncles that are held above the plant, whereas they form tight axillary clusters around single subsessile pistillate flowers on hermaphrodites (Durand 1963, Pannell 1997d). This would suggest that pollen produced by males should enjoy increased siring success compared with that produced by hermaphrodites, potentially giving rise to accelerating male fitness gains. Maleness in *M. annua* appears to be determined by a dominant allele, although the sex expression of genetic males may also be influenced by plant density (Pannell 1997b).

Schizopepon bryoniaefolius (Cucurbitaceae)

Androdioecy has also been confirmed in *Schizopepon bryoniaefolius* (Cucurbitaceae), a slender annual vine that occupies roadside habitats in East Asia (Akimoto et al. 1999). *S. bryoniaefolius* possesses small white flowers and is pollinated by generalist insect pollinators. As in *Mercurialis annua*, inflorescence architecture of *S. bryoniaefolius* differs between males and hermaphrodites: males produce racemose or paniculate inflorescences with many staminate flowers, whereas hermaphrodites produce solitary perfect flowers in leaf axils that are capable of autonomous self-fertilization (Akimoto et al. 1999). Although estimates of total pollen production by males and hermaphrodites have not been published, staminate and perfect flowers each possess three stamens of similar size with identical pollen, so that the substantial difference in flower number between the sexes suggests that males produce a great deal more pollen than do hermaphrodites (Akimoto et al. 1999). In common with *D. glomerata* and *M. annua*, male frequencies vary widely between zero and about 0.3, and inbreeding coefficients correlate negatively with male frequency across populations, suggesting context-dependent outcrossing rates (Figure 5).

Castilla elastica (Moraceae)

A new case of putative androdioecy has recently been described in *Castilla elastica*, a tropical tree in the fig family (Moraceae) that is pollinated by thrips (Sakai 2001). There are no detailed sex-allocation data for this species, but males in the small population studied were in the minority and were not just small hermaphrodites (Sakai 2001). Hermaphrodites of *C. elastica* are monoecious and possess distinct male and female inflorescences. Particularly interesting is the fact that the staminate inflorescences produced by males differ morphologically from those on hermaphrodites, adding weight to the interpretation of true sexual dimorphism for this species (Sakai 2001).

Notwithstanding the dimorphism in staminate inflorescences, males and hermaphrodites of *C. elastica* produce identical and fertile pollen. A close relative, *C. tuna*, is dioecious, and Sakai (2001) has suggested that androdioecy may have evolved from dioecy in the genus as a response to selection for reproductive assurance during colonization. Apparently, the rate of thrips visitation to the staminate inflorescences on hermaphrodites is much lower than that to males, suggesting that hermaphrodites do not produce pollen merely as a pollinator reward. It

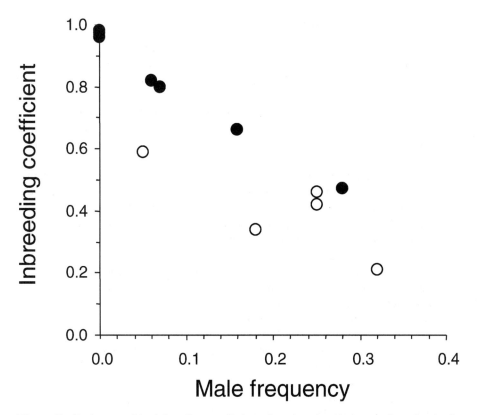

Figure 5   Estimates of the inbreeding coefficient plotted against the proportion of males in each population for *Schizopepon bryoniaefolius* (*open circles*) (from Akimoto et al. 1999) and *Eulimnadia texana* (*closed circles*) (from Sassaman 1989).

is tempting to consider that the unusual inflorescence differentiation in *C. elastica* may have evolved as a strategy to maintain high outcrossing rates in androdioecious populations, with hermaphrodites functioning almost as females, but with selfing possible when populations are small.

Spinifex littoreus (Poaceae)

The wind-pollinated dune grass *Spinifex littoreus*, distributed in southern and south-east Asia, is found mainly in dioecious populations, with male and female inflorescences being markedly distinct (Connor 1996). In the south-eastern extent of the species' distribution, however, populations are smaller and are either monomorphic, comprising modified females with some staminate flowers, or apparently androdioecious and comprising these same modified females along with males. There are as yet no data on sex allocation, nor are there any indications of the extent to which males actually sire progeny in this clonal species, but observations show some interesting similarities to the case of *M. annua* (see above).

Branchiopod Crustaceans

Androdioecy has been documented in several species of branchiopod crustaceans that inhabit freshwater pools or ditches and thus possess intrinsically structured populations (Sassaman 1991, 1995). As in the plant examples cited above, male frequencies in the androdioecious populations are highly variable, ranging between zero and about 0.4 (Sassaman 1989, 1991, 1995). Hermaphroditism has clearly been derived from dioecy on several independent occasions within the group (Sassaman 1995), with females producing testicular lobes in the ovaries (Longhurst 1955, Zucker et al. 1997). In dioecious species females are the heterogametic sex (see Sassaman 1995), and this genetic mode of sex determination has been retained in the androdioecious species, so that males are determined by a recessive allele. All of the androdioecious species are found as either androdioecious or purely hermaphroditic populations, but the tadpole shrimp *Triops longicaudatus* has also been found in dioecious populations in the eastern extreme of its range in North America (Sassaman et al. 1997).

A particularly detailed analysis of androdioecy has been conducted for the conchostracan shrimp *Eulimnadia texana*, which inhabits ephemeral freshwater ponds and ditches in the southwestern United States (Sassaman 1989). In common with other branchiopods, hermaphrodites of *E. texana* are unable to outcross with one another (Sassaman & Weeks 1993), and in the absence of males they self-fertilize all their eggs (Knoll & Zucker 1995, Weeks et al. 2001b). It seems likely that negative frequency-dependent selection on males is crucial to the maintenance of androdioecy in *E. texana*. Males are fittest at low frequency when they encounter only hermaphrodites (Hollenbeck et al. 2002), which show a preference for outcrossing by swimming more slowly, spending longer periods of time in the vicinity of males, and postponing self fertilization when males are not present (Medland et al. 2000, Zucker et al. 2002). Selection for such behavior may have been mediated by high levels of inbreeding depression found in these populations, in which selfed progeny are up to 68% less fit than their outcrossed counterparts (Weeks et al. 1999, 2000). Male fitness may be lower at high frequencies, at which their mortality is increased, partly as a result of antagonistic male-male interactions (Zucker et al. 2001). As in *S. bryoniaefolius* (see above), the inbreeding coefficient correlates negatively with the frequency of males (Sassaman 1989, Weeks & Zucker 1999) (see Figure 5). However, in contrast to *D. glomerata* and *M. annua*, maleness in *E. texana* is determined by a recessive allele, so that hermaphrodites are either homozygous dominant (monogenic) or heterozygous (amphigenic) at the sex-determining locus (Sassaman & Weeks 1993). The fact that amphigenic hermaphrodites are about 13% fitter than monogenics is therefore another factor contributing to the maintenance of males in androdioecious populations (Weeks et al. 2001a).

Rhabditid Nematodes

Androdioecy is hypothesized to have arisen multiple times within the rhabditid nematodes (Fitch 1997, Haag & Kimble 2000) and is found in two of the seven species in the Elegans group, *C. elegans* and *C. briggsae* (Sudhaus & Kiontke

1996). Although there is some uncertainty about the polarity of the transition between dioecy and androdioecy along specific branches of the phylogeny (see discussion by Haag & Kimble 2000), there is no doubt that dioecy is ancestral in the Rhabditidae generally, even though reversions from androdioecy to dioecy are known (Haag & Kimble 2000). As in the branchiopods, self-fertile rhabditid hermaphrodites are modified females that produce some sperm. Also in common with the branchiopods is the inability of hermaphrodites to outcross through male function (Wood 1988). In contrast with the branchiopods, however, sex determination is an XO system, which effectively gives rise to dominant male determination (Wood 1988), as in *D. glomerata* and *M. annua*.

Unfortunately, sex ratio variation from natural populations of these species has not yet been reported. However, the recurrent generation of males through nondisjunction of the sex-determining chromosome suggests that it may be functionally significant. Reproduction by hermaphrodites upon selfing is limited by the low numbers of sperm they produce, and mating with males can increase life-time fecundity fourfold (Hodgekin & Barnes 1991). Because male sperm also displace hermaphrodite sperm prior to fertilization (Ward & Carrel 1979), it seems possible that males might be selected under conditions in the wild in which male-hermaphrodite encounters are frequent. Hermaphrodites might thus benefit from reproductive assurance in the absence of males, and male frequencies would be expected to rise when populations become dense. Recently, male frequencies of experimental *C. elegans* populations were tracked over several generations, but their decline toward zero demonstrated that males cannot be maintained with hermaphrodites under typical laboratory conditions (Stewart & Phillips 2002). Field data or data from experiments under more natural conditions than typical laboratory cultures would be valuable.

Putative Androdioecy in the Oleaceae

In recent years the breeding system of several species in the olive family (Oleaceae) has been studied. Patterns of sex allocation in these long-lived woody species are still poorly understood, but the evolution of dioecy from hermaphroditism via an androdioecious path seems possible. In *Phillyrea angustifolia*, a woody perennial that grows in fire-prone vegetation in the western Mediterranean Basin, hermaphrodites with perfect flowers coexist with female-sterile males (Lepart & Dommée 1992). There is good evidence that individuals do not change sex between seasons (Lepart & Dommée 1992, Vassiliadis et al. 2000a), but plants that flower one year may not flower the next, and populations may thus comprise a high proportion of nonflowering plants in any one season (Traveset 1994, Pannell & Ojeda 2000).

P. angustifolia is at least partially self-incompatible, and male-hermaphrodite matings tend to be more successful than matings between hermaphrodites (Lepart & Dommée 1992, Traveset 1994, Vassiliadis et al. 2000a). Although pollen from males and hermaphrodites is equally viable (Lepart & Dommée 1992), pollen grains differ morphologically between the sexes, with those from males being more porous than those from hermaphrodites (Traveset 1994). These facts suggest differing pollen function between males and hermaphrodites in *P. angustifolia* and argue against a straightforward androdioecious breeding system. Lepart & Dommée (1992) documented male frequencies that were not significantly lower than 0.5, an observation consistent with functional cryptic dioecy. This is very likely to be the case for the related species *P. latifolia*, which has also been described as morphologically androdioecious (Aronne & Wilcock 1992). Patterns of sex allocation in western Mediterranean populations of *P. latifolia* need to be studied in more detail, but limited data from Israeli populations suggest a 1:1 sex ratio, consistent with functional dioecy (Rottenberg 1998).

Lepart & Dommée (1992) have interpreted androdioecy in *P. angustifolia* as being an example of "leaky dioecy" (see, e.g., Baker & Cox 1984), with the evolution of self-fertility through selection for reproductive assurance during colonization processes. However, if colonization by hermaphrodites were common, we would find high levels of sex-ratio variation among populations, with some all-hermaphrodite populations; in contrast, male frequencies are uniformly quite high (Pannell & Ojeda 2000). Although some populations of *P. angustifolia* appear to be cryptically dioecious, others may indeed be functionally androdioecious. In a larger-scale survey of sex-ratio variation in the species in Spain and Portugal, for example, Pannell & Ojeda (2000) found that male frequencies were consistently lower than 0.5 when the proportion of nonflowering individuals was taken into account. This appears to be supported by recent paternity analysis in one population of the species, which indicates that hermaphrodites may sire a sizable proportion of seeds (Vassiliadis et al. 2002).

Wallander (2001) recently studied the evolution of pollination syndromes and breeding systems in the Oleaceae. According to her analysis, no fewer than 37 species in this family are morphologically androdioecious, most of which occur in a single subtribe. These species possess small white to yellow-green flowers and are pollinated either by wind or by both wind and insects. Apart from *P. angustifolia*, discussed above, Wallander (2001) has cited three further species from this list for which androdioecy is claimed to be functional: *Fraxinus ornus* (Dommée et al. 1999, Wallander 2001), *F. lanuginosa* (Ishida & Hiura 1998), and *F. longicuspis* (Wallander 2001). As in *P. angustifolia*, hermaphrodites in each of these species are partially self-incompatible. Moreover, although hermaphrodite pollen is generally viable, it is less germinable and substantially less successful in fertilizing ovules produced by other hermaphrodites than is the pollen of males, at least in some populations. In *F. lanuginosa* and *F. ornus* there are no differences in pollen size between males and hermaphrodites (Ishida & Hiura 1998, Dommée et al. 1999), but the pollen of hermaphrodites is substantially smaller in *F. longicuspis* (Wallander 2001). It is interesting to note that the hermaphrodites of several other morphologically androdioecious species also appear to possess "smaller and seemingly non-functional anthers" (Wallander 2001). These species lack petals and are presumably pollinated by wind, whereas those with petals, which are probably pollinated by insects, show no obvious abortion of male function (Wallander 2001).

Thus, the apparent trend toward the abortion of male function in hermaphrodites in *F. ornus*, *F. lanuginosa*, and *F. longicuspis* is most extreme in those species that are cryptically dioecious (Charlesworth 1984, Mayer & Charlesworth 1991).

Another common feature in the putatively androdioecious *Fraxinus* species are their high male frequencies. In *F. ornus* (Dommée et al. 1999) and *F. longicuspis* (Wallander 2001), males are either as frequent as or more frequent than hermaphrodites (although the sample size for the latter species was small). Unless males have lower mortality rates than hermaphrodites (there is no evidence either way), these sex ratios also strongly suggest functional dioecy rather than androdioecy. In *F. lanuginosa*, male frequencies vary between about 0.1 and 0.5 and are thus similar to those found for *P. angustifolia* in southern Spain and Portugal (Pannell & Ojeda 2000).

Sagittaria lancifolia (Alismataceae)

An interesting case of androdioecy has been documented in populations of the aquatic perennial herb *Sagittaria lancifolia* (Muenchow 1998). *S. lancifolia* occupies freshwater wetlands from the United States to northern South America and is visited by generalist insect pollinators. Males, which are determined by a dominant allele, co-occur with monoecious individuals at frequencies of about 0.16. Although they sometimes produce fruits, there is a distinct bimodality in the distribution of gender within the population (Muenchow 1998). Almost all the close relatives of *S. lancifolia* are hermaphroditic, suggesting that androdioecy is derived from hermaphroditism, as in the Oleaceae (Muenchow 1998, Sarkissian et al. 2001).

Inflorescences of *S. lancifolia* are predated by the weevil *Listronotus appendiculatus*, which eats the uppermost flowers first (Muenchow 1998). Because pistillate flowers are basal and staminate flowers apical within a protogynous inflorescence, this pattern of predation tends to reduce the male fertility of the hermaphrodites, with the result that males ultimately disperse pollen from about 4.5 times more staminate flowers than hermaphrodites do (Muenchow 1998). Given a moderate amount of self-fertilization, this value of relative realized male fecundity is consistent with the maintenance of males at a frequency of 0.16, as observed.

According to Muenchow (1998), males of *S. lancifolia* would appear to be maintained through accelerating male fitness gains imposed by an interaction with the species' weevil predator, whereas the occurrence of some all-hermaphrodite populations suggests that reproductive assurance during colonization may have contributed to the selection of male function in the hermaphrodites. In the related species *S. latifolia*, Sarkissian et al. (2001) recently found wide variation in sex allocation across monoecious populations, and they interpreted apparent androdioecy in some of these as being the result of size-dependent sex allocation. *S. lancifolia* appears to be different, but sex-allocation data from a wider sample of populations would be valuable. For the moment, the interpretation of androdioecy seems plausible.

Further Putative Cases of Androdioecy

There are several other possible cases of androdioecy for which we have too little data to make an adequate assessment. The very limited sex-allocation data available for *Neobuxbaumia mezcalaensis* (Cactaceae), a bat-pollinated columnar cactus in Mexico, appear to be consistent with an androdioecious breeding system (Valiente-Banuet et al. 1997). Few columnar cactus species have separate sexes, and it seems possible that androdioecy in this species, if it is confirmed, has arisen from self-incompatible hermaphroditism (Valiente-Banuet et al. 1997). The case merits further study.

The coexistence of morphological males and hermaphrodites has been documented in *Saxifraga cernua* and *S. foliolosa* (Saxifragaceae), two species with circumpolar arctic distributions (Molau 1992, Molau & Prentice 1992). Populations appear to be a complex mix of rare sexuality with predominant asexuality owing to meiotic irregularities (Brochmann & Hapnes 2001), and although fruits are very occasionally set, it is not clear that males play a significant part in the sexual process, even when it does occur. A similar pattern of reproduction, with males and hermaphrodites co-occurring in largely asexual populations, has been found in filamentous fungi (Leslie & Klein 1996) and the field elm *Ulmus minor* (J.C. Lopez-Almansa, J.R. Pannell & L. Gil, unpublished data).

A recent study of the distribution of gender expression in the maple, *Acer japonicum*, pointed to the possible role of androdioecy as an evolutionary step toward dioecy from heterodichogamy, the coexistence in a population of hermaphrodites with either a male-first (protandrous) or a female-first (protogynous) flowering sequence within each reproductive season (Sato 2002). In one study population, males were found coexisting with protogynous and protandrous hermaphrodites. Whereas some individuals were male in both of the two consecutive years of study, some individuals oscillated in phenotype between male and protandrous hermaphrodite, and so it is possible that there is in fact no distinct all-male strategy in the species. Further study would be valuable.

Reports have also been made of possible androdioecy in *Oxalis suksdorfii* (Oxalidaceae) (Ornduff 1964, 1972) and *Nivenia corymbosa* (Iridaceae) (Ornduff 1983). However, the former case is more likely an example of the breakdown of tristyly towards distyly, whereas mating patterns in the second are consistent with the evolution of dioecy from distyly. Finally, a breeding-system analysis of the short-lived Mediterranean perennial *Anagallis monelli* has produced data that may be consistent with subandrodioecy (Gibbs & Talavera 2001). In this species several individuals produced many flowers and were fully competent as male parents, but they set very few seeds. The case merits further attention.

CONCLUSIONS

Although androdioecy must still be regarded as exceedingly rare, it has now been found in several species of plants and animals with diverse origins. The theoretical predictions made by the seminal models of Lloyd (1975), Charlesworth

& Charlesworth (1978, 1981), and Charlesworth (1984) are borne out by these cases, inasmuch as the relevant mating-system and sex-allocation parameters have been measured. The clearest examples of androdioecy are those that appear to be derived from dioecy. In all cases males are the minority gender. These species tend to be short-lived colonizers, and the wide between-population variance in their sex ratios is consistent with metapopulation models of androdioecy (Pannell 1997d, 2001). The importance of selection for reproductive assurance during colony establishment provides a plausible explanation for the evolution of hermaphroditism in these species, and attributes that lead to delayed selfing, preferences for male-hermaphrodite mating, and density- and/or frequency-dependent selfing rates explain how males might be maintained in the face of selection for uniparental reproduction.

The few putative examples of the evolution of androdioecy from hermaphroditism are more difficult to interpret. They tend to occur in long-lived plants for which life-history differences between males and hermaphrodites are difficult to estimate, although they may be important. Androdioecy in these species tends to verge on cryptic dioecy, particularly in the Oleaceae, in which several possible cases have been found. The retention of male function by "hermaphrodites" may have a simple function in providing a reward to pollen-collecting pollinators (Mayer & Charlesworth 1991). However, in species of the Oleaceae pollinated by both wind and insects, hermaphrodite pollen is not completely infertile (Wallander 2001), and hermaphrodites may sire some progeny. Where hermaphrodites can be shown to contribute genetically through their male function, androdioecy in such species might best be viewed as the outcome selection for dioecy that has not yet run its course, combined with a need to reward pollinators. It seems fairly clear that in these cases selection to avoid selfing has led to a reduction in hermaphrodite siring ability, as suggested by Charlesworth (1984); otherwise one might have expected pollen to have retained its function as a fully fertile gametophyte in addition to serving as a reward for pollinators.

Interesting questions remain. Although androdioecy has evolved independently in several distantly related groups, what is it about the Oleaceae, the branchiopod crustaceans, or the nematodes, for example, that has led to its repeated evolution within these groups? These are not the only groups of animals or plants in which colonization dynamics or pollination by both wind and insects, for example, may have been important, and it seems likely that other taxonomically more specific factors will have contributed to the evolution of androdioecy. More generally, is androdioecy always just an unstable intermediate step in the transition between hermaphroditism and dioecy? This seems most likely for the path from hermaphroditism. However, the high incidence of cases derived from dioecy within two taxonomic orders of the branchiopods, for example, suggests that its evolutionary maintenance along this path is possible. Further theoretical work needs to be done to establish this.

Finally, to what extent might an influence of environmental factors on sex allocation play a role in maintaining androdioecy? Pannell (1997b) found that, although maleness in *M. annua* is apparently determined by a dominant allele, at least some males can respond to reduced plant density by expressing hermaphroditic function. This sort of plasticity should enjoy a large advantage where densities and mating opportunities fluctuate during colony establishment, but its evolution will be constrained by the ability of individuals to assess their environment accurately (Charnov & Bull 1977, Schlichting & Pigliucci 1998). It is interesting that a presumably quite accurate means of assessing local mating opportunities has evolved in the homosporous ferns (Haig & Westoby 1988). In these plants, hermaphroditic gametophyte thalli disperse a hormone that causes germinating spores to develop as males (Hickok et al. 1998). Males are often the majority gender in these populations (e.g., Hamilton & Lloyd 1991), indicating that they must enjoy limited siring success, and they should thus not be viewed as androdioecious (Pannell 2002). Nevertheless, it appears that hermaphroditism has evolved from dioecy in these species as a result of selection for reproductive assurance during colonization (Haig & Westoby 1988; see Vogel et al. 1999). It thus seems significant that the evolution of an effective assessment of local mating opportunities under such circumstances has led to a male-hermaphrodite rather than a female-hermaphrodite strategy.

ACKNOWLEDGMENTS

I am grateful to S.C.H. Barrett, D. Charlesworth, L.H. Rieseberg, D.W. Schemske, C. Vassiliadis, E. Wallander, and S.C. Weeks for valuable comments on the manuscript, and to M. Verdú and E. Wallander for discussions.

The *Annual Review of Ecology and Systematics* is online at
http://ecolsys.annualreviews.org

LITERATURE CITED

Akimoto J, Fukuhara T, Kikuzawa K. 1999. Sex ratios and genetic variation in a functionally androdioecious species, *Schizopepon bryoniaefolius* (Cucurbitaceae). *Am. J. Bot.* 86:880–86

Anderson GJ. 1979. Dioecious *Solanum* species of hermaphroditic origin is an example of a broad convergence. *Nature* 282:836–38

Aronne G, Wilcock CC. 1992. Breeding system of *Phillyrea latifolia* L. and *Phillyrea angustifolia* L: evidence for androdioecy. *G. Bot. Ital.* 126:263

Baker HG. 1955. Self-compatibility and establishment after "long-distance" dispersal. *Evolution* 9:347–48

Baker HG. 1967. Support for Baker's Law—as a rule. *Evolution* 21:853–56

Baker HG, Cox PA. 1984. Further thoughts on dioecism and islands. *Ann. Mo. Bot. Gard.* 71:244–53

Barrett SCH, Morgan MT, Husband BC. 1989. The dissolution of a complex genetic polymorphism: the evolution of self-fertilization in tristylous *Eichhornia paniculata* (Pontederiaceae). *Evolution* 43:1398–416

Barrett SCH, Shore JS. 1987. Variation and evolution of breeding systems in the *Turnera ulmifolia* complex (Turneraceae). *Evolution* 41:340–54

Bateman AJ. 1948. Intra-sexual selection in *Drosophila. Heredity* 2:349–68

Brochmann C, Hapnes A. 2001. Reproductive strategies in some arctic *Saxifraga* (Saxifragaceae), with emphasis on the narrow endemic *S. svalbardensis* and its parental species. *Bot. J. Linn. Soc.* 137:31–49

Burczyk JL, Prat D. 1997. Male reproductive success in *Pseudotsuga menziesii* (Mirb.) Franco: the effects of spatial structure and flowering characteristics. *Heredity* 79:638–47

Charlesworth D. 1984. Androdioecy and the evolution of dioecy. *Biol. J. Linn. Soc.* 22: 333–48

Charlesworth D. 1993. Save the male. *Curr. Biol.* 3:155–57

Charlesworth D. 1999. Theories of the evolution of dioecy. See Geber et al. 1999, pp. 33–60

Charlesworth D, Charlesworth B. 1978. A model for the evolution of dioecy and gynodioecy. *Am. Nat.* 112:975–97

Charlesworth D, Charlesworth B. 1981. Allocation of resources to male and female functions in hermaphrodites. *Biol. J. Linn. Soc.* 15:57–74

Charnov EL. 1979. Simultaneous hermaphroditism and sexual selection. *Proc. Natl. Acad. Sci. USA* 76:2480–82

Charnov EL. 1982. *The Theory of Sex Allocation.* Princeton, NJ: Princeton Univ. Press

Charnov EL, Bull J. 1977. When is sex environmentally determined? *Nature* 266:228–30

Charnov EL, Maynard Smith J, Bull JJ. 1976. Why be an hermaphrodite? *Nature* 263:125–26

Connor HE. 1996. Breeding systems in Indomalesian *Spinifex* (Paniceae: Gramineae). *Blumea* 41:445–54

Crow JF, Kimura M. 1970. *An Introduction to Population Genetics Theory.* New York: Harper & Row

Darwin C. 1876. *The Effects of Cross- and Self-Fertilization in the Vegetable Kingdom.* London: Murray

Darwin C. 1877. *The Different Forms of Flowers on Plants of the Same Species.* New York: Appleton

Després L, Maurice S. 1995. The evolution of dimorphism and separate sexes in schistosomes. *Proc. R. Soc. London Ser. B* 262:175–80

Dommée B, Geslot A, Thompson JD, Reille M, Denelle N. 1999. Androdioecy in the entomophilous tree *Fraxinus ornus* (Oleaceae). *New Phytol.* 143:419–26

Durand B. 1963. Le complexe *Mercurialis annua* L. *s.l.*: une étude biosystématique. *Ann. Sci. Nat. Bot. Paris* 12:579–736

Durand R, Durand B. 1992. Dioecy, monoecy, polyploidy and speciation in annual Mercuries. *Bull. Soc. Bot. France Lett. Bot.* 139: 377–99

Eckhart VM. 1999. Sexual dimorphism in flowers and inflorescences. See Geber et al. 1999, pp. 123–48

Elle E, Meagher TR. 2000. Sex allocation and reproductive success in the andromonoecious perennial *Solanum carolinense* (Solanaceae). II. Paternity and functional gender. *Am. Nat.* 156:622–36

Farris MA, Mitton JW. 1984. Population density, outcrossing rate, and heterozygous superiority in Ponderosa pine. *Evolution* 38: 1151–54

Fisher RA. 1930. *The Genetical Theory of Natural Selection.* Oxford: Oxford Univ. Press

Fitch DHA. 1997. Evolution of male tail morphology and development in rhabditid nematodes related to *Caenorhabditis elegans. Syst. Biol.* 46:145–79

Fritsch P, Rieseberg LH. 1992. High outcrossing rates maintain male and hermaphrodite individuals in populations of the flowering plant *Datisca glomerata. Nature* 359: 633–36

Geber MA. 1999. Theories of the evolution of sexual dimorphism. See Geber et al. 1999, pp. 97–122

Geber MA, Dawson TE, Delph LF, eds. 1999. *Gender and Sexual Dimorphism in Flowering Plants.* Heidelberg: Springer

Gibbs PE, Talavera S. 2001. Breeding system studies with three species of *Anagallis* (Primulaceae): self-incompatibility and reduced female fertility in *A. monelli* L. *Ann. Bot.* 88:139–44

Haag ES, Kimble J. 2000. Regulatory elements required for development of *Caenorhabditis elegans* hermaphrodites are conserved in the tra-2 homologue of *C. remanei*, a male/female sister species. *Genetics* 155:105–16

Haig D, Westoby M. 1988. Sex expression in homosporous ferns: an evolutionary perspective. *Evol. Trends Plants* 2:111–20

Haldane JBS. 1927. A mathematical theory of natural and artificial selection. Part V. Selection and mutation. *Proc. Cambridge Philos. Soc.* 28:838–44

Hamilton RG, Lloyd RM. 1991. Antheridiogen in the wild: the development of fern gametophyte communities. *Funct. Ecol.* 5:804–9

Hamilton WD. 1967. Extraordinary sex ratios. *Science* 156:477–88

Hickok LG, Warne TR, Baxter SL, Melear CT. 1998. Sex and the C-fern: not just another life cycle. *Bioscience* 48:1031–37

Hodgekin J, Barnes TM. 1991. More is not better: brood size and population growth in a self-fertilizing hermaphrodite. *Proc. R. Soc. London Ser. B* 246:19–24

Hollenbeck VG, Weeks SC, Gould W, Zucker N. 2002. Maintenance of androdioecy in the freshwater shrimp *Eulimnadia texana*: sexual encounter rates and outcrossing success. *Behav. Ecol.* 13:561–70

Ishida K, Hiura T. 1998. Pollen fertility and flowering phenology in an androdioecious tree, *Fraxinus lanuginosa* (Oleaceae), in Hokkaido, Japan. *Int. J. Plant Sci.* 159:941–47

Kaul MLH. 1988. *Male Sterility in Higher Plants*. Berlin: Springer Verlag

Knapp S. 1991. A cladistic analysis of the *Solanum* sessile species group (Section *Geminata pro parte*, Solanaceae). *Bot. J. Linn. Soc.* 106:73–89

Knapp S, Persson V, Blackmore S. 1998. Pollen morphology and functional dioecy in *Solanum* (Solanaceae). *Plant Syst. Evol.* 210:113–39

Knoll L, Zucker N. 1995. Selfing versus outcrossing in the androdioecious clam shrimp, *Eulimnadia texana* (Crustacea, Conchostraca). *Hydrobiologia* 298:83–86

Kohn JR, Barrett SCH. 1994. Pollen discounting and the spread of a selfing variant in tristylous *Eichhornia paniculata*: evidence from experimental populations. *Evolution* 48:1576–94

Lepart J, Dommée B. 1992. Is *Phillyrea angustifolia* L. (*Oleaceae*) an androdioecious species? *Bot. J. Linn. Soc.* 108:375–87

Leslie JF, Klein KK. 1996. Female fertility and mating type effects on effective population size and evolution in filamentous fungi. *Genetics* 144:557–67

Liston A, Rieseberg LH, Elias TS. 1990. Functional androdioecy in the flowering plant *Datisca glomerata*. *Nature* 343:641–42

Lloyd D. 1975. The maintenance of gynodioecy and androdioecy in angiosperms. *Genetica* 45:325–39

Lloyd DG. 1982. Selection of combined versus separate sexes in seed plants. *Am. Nat.* 120:571–85

Lloyd DG. 1983. Evolutionary stable sex ratios and sex allocations. *J. Theor. Biol.* 105:525–39

Lloyd DG. 1988. Benefits and costs of biparental and uniparental reproduction in plants. In *The Evolution of Sex: An Examination of Current Ideas*, ed. BR Levin, RE Michod, pp. 233–52. Sunderland, MA: Sinauer

Lloyd DG, Schoen DJ. 1992. Self- and cross-fertilization in plants. I. Functional dimensions. *Int. J. Plant Sci.* 153:358–69

Longhurst AR. 1955. Evolution in the Notostraca. *Evolution* 9:84–86

Maurice S, Fleming TH. 1995. The effect of pollen limitation on plant reproductive systems and the maintenance of sexual polymorphisms. *Oikos* 74:55–60

Mayer SS, Charlesworth D. 1991. Cryptic dioecy in flowering plants. *Trends Ecol. Evol.* 6:320–25

Medland VL, Zucker N, Weeks SC. 2000. Implications for the maintenance of androdioecy in the freshwater shrimp, *Eulimnadia texana* Packard: encounters between males and hermaphrodites are not random. *Ethology* 106:839–48

Molau U. 1992. On the occurrence of sexual reproduction in *Saxifraga cernua* and

Saxifraga foliolosa (Saxifragaceae). *Nord. J. Bot.* 12:197–203

Molau U, Prentice HC. 1992. Reproductive system and population structure in three arctic *Saxifraga* spp. *J. Ecol.* 80:149–61

Muenchow GE. 1998. Subandrodioecy and male fitness in *Sagittaria lancifolia* subsp. *lancifolia* (Alismataceae). *Am. J. Bot.* 85:513–20

Ornduff R. 1964. The breeding system of *Oxalis suksdorfii*. *Am. J. Bot.* 51:307–14

Ornduff R. 1972. The breakdown of trimorphic incompatibility in *Oxalis* section Corniculatae. *Evolution* 26:52–65

Ornduff R. 1983. Studies on the reproductive system of *Nivenia corymbosa* (Iridaceae), an apparently androdioecious species. *Ann. Mo. Bot. Gard.* 70:146–48

Orr HA, Betancourt AJ. 2001. Haldane's sieve and adaptation from the standing genetic variation. *Genetics* 157:875–84

Otto SP, Sassaman C, Feldman MW. 1993. Evolution of sex determination in the conchostracan shrimp *Eulimnadia texana*. *Am. Nat.* 141:327–37

Pannell J. 1997a. The maintenance of gynodioecy and androdioecy in a metapopulation. *Evolution* 51:10–20

Pannell J. 1997b. Mixed genetic and environmental sex determination in an androdioecious population of *Mercurialis annua*. *Heredity* 78:50–56

Pannell J. 1997c. Variation in sex ratios and sex allocation in androdioecious *Mercurialis annua*. *J. Ecol.* 85:57–69

Pannell J. 1997d. Widespread functional androdioecy in *Mercurialis annua* L. (*Euphorbiaceae*). *Biol. J. Linn. Soc.* 61:95–116

Pannell JR. 2001. A hypothesis for the evolution of androdioecy: the joint influence of reproductive assurance and local mate competition in a metapopulation. *Evol. Ecol.* 14:195–211

Pannell JR. 2002. What is functional androdioecy? *Funct. Ecol.* In press

Pannell JR, Barrett SCH. 1998. Baker's Law revisited: reproductive assurance in a metapopulation. *Evolution* 52:657–68

Pannell JR, Ojeda F. 2000. Patterns of flowering and sex-ratio variation in the Mediterranean shrub *Phillyrea angustifolia* (Oleaceae): implications for the maintenance of males with hermaphrodites. *Ecol. Lett.* 3:495–502

Philbrick CT, Rieseberg LH. 1994. Pollen production in the androdioecious *Datisca glomerata* (Datiscaceae): implications for breeding system equilibrium. *Plant Species Biol.* 9:43–46

Rieseberg LH, Hanson MA, Philbrick CT. 1992. Androdioecy is derived from dioecy in Datiscaceae: evidence from restriction site mapping of PCR-amplified chloroplast DNA fragments. *Syst. Bot.* 17:324–36

Rieseberg LH, Philbrick CT, Pack PE, Hanson MA, Fritsch P. 1993. Inbreeding depression in androdioecious populations of *Datisca glomerata* (Datiscaeae). *Am. J. Bot.* 80:757–62

Ross MD. 1982. Five evolutionary pathways to subdioecy. *Am. Nat.* 119:297–318

Ross MD, Weir BS. 1976. Maintenance of males and females in hermaphrodite populations and the evolution of dioecy. *Evolution* 30:425–41

Rottenberg A. 1998. Sex ratio and gender stability in the dioecious plants of Israel. *Bot. J. Linn. Soc.* 128:137–48

Routley MB, Mavraganis K, Eckert CG. 1999. Effect of population size on the mating system in a self-compatible, autogamous plant, *Aquilegia canadensis* (Ranunculaceae). *Heredity* 82:518–28

Sakai S. 2001. Thrips pollination of androdioecious *Castilla elastica* (Moraceae) in a seasonal tropical forest. *Am. J. Bot.* 88:1527–34

Sarkissian TS, Barrett SCH, Harder LD. 2001. Gender variation in *Sagittaria latifolia* (Alismataceae): Is size all that matters? *Ecology* 82:360–73

Sassaman C. 1989. Inbreeding and sex ratio variation in female-biased populations of a clam shrimp, *Eulimnada texana*. *Bull. Mar. Sci.* 45:425–32

Sassaman C. 1991. Sex ratio variation in female-biased populations of Notostracans. *Hydrobiologia* 212:169–79

Sassaman C. 1995. Sex determination and the evolution of unisexuality in the Conchostraca. *Hydrobiologia* 298:45–65

Sassaman C, Simovich MA, Fugate M. 1997. Reproductive isolation and genetic differentiation in North American species of *Triops* (Crustacea: Branchiopoda: Notostraca). *Hydrobiologia* 359:125–47

Sassaman C, Weeks SC. 1993. The genetic mechanism of sex determination in the conchostracan shrimp *Eulimnadia texana*. *Am. Nat.* 141:314–28

Sato T. 2002. Phenology of sex expression and gender variation in heterodichogamous maple, *Acer japonicum*. *Ecology* 85:1226–38

Schlichting CD, Pigliucci M. 1998. *Phenotypic Evolution: A Reaction Norm Perspective.* Sunderland, MA: Sinauer

Schoen DJ, Stewart SC. 1986. Variation in male reproductive investment and male reproductive success in white spruce. *Evolution* 40:1109–20

Seger J, Eckhart VM. 1996. Evolution of sexual systems and sex allocation in plants when growth and reproduction overlap. *Proc. R. Soc. London Ser. B* 263:833–41

Stebbins GL. 1957. Self-fertilization and population variability in the higher plants. *Am. Nat.* 91:337–54

Stewart AD, Phillips PC. 2002. Selection and maintenance of androdioecy in *Caenorhabditis elegans*. *Genetics* 160:975–82

Sudhaus W, Kiontke K. 1996. Phylogeny of *Rhabditis* subgenus *Caenorhabditis* (Rhabditidae, Nematoda). *J. Zool. Syst. Evol. Res.* 34:217–33

Swensen SM, Luthi JN, Rieseberg LH. 1998. Datiscaceae revisited: monophyly and the sequence of breeding system evolution. *Syst. Bot.* 23:157–69

Traveset A. 1994. Reproductive biology of *Phillyrea angustifolia* L. (*Oleaceae*) and effect of galling-insects on its reproductive output. *Bot. J. Linn. Soc.* 114:153–66

Valiente-Banuet A, Rojas-Martinez A, del Coro-Arizmendi M, Davila P. 1997. Pollination biology of two columnar cacti

(*Neobuxbaumia mezcalaensis* and *Neobuxbaumia macrocephala*) in the Tehuacan Valley, Central Mexico. *Am. J. Bot.* 84:452–55

van Treuren R, Bijlsma R, Ouborg NJ, van Delden W. 1993. The effects of population size and plant density on outcrossing rates in locally endangered *Salvia pratensis*. *Evolution* 47:1094–104

Vassiliadis C, Lepart J, Saumitou-Laprade P, Vernet P. 2000a. Self-incompatibility and male fertilization success in *Phillyrea angustifolia* (Oleaceae). *Int. J. Plant Sci.* 161:393–402

Vassiliadis C, Saumitou-Laprade P, Lepart J, Viard F. 2002. High male reproductive success of hermaphrodites in the androdioecious *Phillyrea angustifolia*. *Evolution*. In press

Vassiliadis C, Valero M, Saumitou-Laprade P, Godelle B. 2000b. A model for the evolution of high frequencies of males in an androdioecious plant based on a cross-compatibility advantage of males. *Heredity* 85:413–22

Vogel JC, Rumsay FJ, Schneller JJ, Barrett JA, Gibby M. 1999. Where are the glacial refugia in Europe? Evidence from pteridophytes. *Biol. J. Linn. Soc.* 66:23–37

Wallander E. 2001. *Evolution of wind pollination in Fraxinus (Oleaceae): an ecophylogenetic approach.* PhD thesis. Göteborg Univ., Göteborg, Sweden

Ward S, Carrel JS. 1979. Fertilization and sperm competition in the nematode *Caenorhabditis elegans*. *Dev. Biol.* 73:304–21

Webb CJ. 1999. Empirical studies: evolution and maintenance of dimorphic breeding systems. See Geber et al. 1999, pp. 61–95

Weeks SC, Crosser BR, Bennett R, Gray M, Zucker N. 2000. Maintenance of androdioecy in the freshwater shrimp, *Eulimnadia texana*: estimates of inbreeding depression in two populations. *Evolution* 54:878–87

Weeks SC, Crosser BR, Gray MM. 2001a. Relative fitness of two hermaphroditic mating types in the androdioecious clam shrimp, *Eulimnadia texana. J. Evol. Biol.* 14:83–94

Weeks SC, Hutchison J, Zucker N. 2001b. Maintenance of androdioecy in the freshwater shrimp, *Eulimnadia texana*: Do hermaphrodites need males for complete fertilization? *Evol. Ecol.* 15:205–21

Weeks SC, Marcus V, Crosser BR. 1999. Inbreeding depression in a self-compatible, androdioecious crustacean, *Eulimnadia texana. Evolution* 53:472–83

Weeks SC, Zucker N. 1999. Rates of inbreeding in the androdioecious clam shrimp *Eulimnadia texana. Can. J. Zool.* 77:1402–8

Weller SG, Sakai AK. 1999. Using phylogenetic approaches for the analysis of plant breeding system evolution. *Annu. Rev. Ecol. Syst.* 30:167–99

Wilson P, Thomson JD, Stanton ML, Rigney LP. 1994. Beyond floral Batemania: gender biases in selection for pollination success. *Am. Nat.* 143:283–96

Wolf DE, Satkoski JA, White K, Rieseberg LH. 2001. Sex determination in the androdioecious plant *Datisca glomerata* and its dioecious sister species *D. cannabina. Genetics* 159:1243–57

Wood WB. 1988. Introduction to *C. elegans* biology. In *The Nematode Caenorhabditis Elegans*, ed. WB Wood, pp. 1–16. Cold Spring Harbor, NY: Cold Spring Harbor Lab. Press

Yampolsky C, Yampolsky H. 1922. Distribution of sex forms in the phanerogamic flora. *Bibliogr. Genet.* 3:4–62

Zucker N, Aguilar GA, Weeks SC, McCandless LG. 2002. Variation in reproductive cycle between selfing and outcrossing hermaphrodites in an androdioecious desert shrimp. *Invertebr. Biol.* 121:66–72

Zucker N, Cunningham M, Adams HP. 1997. Anatomical evidence for androdioecy in the clam shrimp *Eulimnadia texana. Hydrobiologia* 359:171–75

Zucker N, Stafki B, Weeks SC. 2001. Maintenance of androdioecy in the freshwater clam shrimp *Eulimnadia texana*: longevity of males relative to hermaphrodites. *Can. J. Zool.* 79:393–401

Annu. Rev. Ecol. Syst. 2002. 33:427–47
doi: 10.1146/annurev.ecolsys.33.020602.095433

MAST SEEDING IN PERENNIAL PLANTS:
Why, How, Where?

Dave Kelly[1] and Victoria L. Sork[2]

[1]*Plant and Microbial Sciences, University of Canterbury, Private Bag 4800, Christchurch
8001, New Zealand; email: d.kelly@botn.canterbury.ac.nz*
[2]*Department of Organismic Biology, Ecology, and Evolution; and Institute of the
Environment, University of California Los Angeles, Los Angeles, California 90095-1786;
email: vlsork@ucla.edu*

Key Words   dispersal, economies of scale, mass flowering, predator satiation, wind pollination

■ **Abstract**   For many years biologists have debated whether mast seeding (the synchronous intermittent production of large seed crops in perennial plants) results from weather conditions or is an evolved plant reproductive strategy. In this review, we analyze the evidence for the underlying causes of masting. In the absence of selection for higher or lower variability, plants will vary in tandem with the environment (resource matching). Two selective factors often favor the evolution of masting: increased pollination efficiency in wind-pollinated species, and satiation of seed predators. Other factors select against masting, including animal pollination and frugivore dispersal. A survey of 570 masting datasets shows that wind-pollinated species had higher seed production coefficients of variation (CVs) than biotically pollinated ones. Frugivore-dispersed species had low CVs whereas predator-dispersed plants had high CVs, consistent with gaining benefits from predator satiation rather than dispersal. The global pattern of masting shows highest seed crop variability at mid latitudes and in the Southern Hemisphere, which are similar to the patterns in variability of rainfall. We conclude that masting is often an adaptive reproductive trait overlaid on the direct influence of weather.

INTRODUCTION

A reproductive episode that results in a superabundance of seeds can be a remarkable phenomenon. Anecdotes about entire forests being swamped with seeds, such as Malaysian forests dominated by Dipterocarpaceae (Janzen 1974, Ashton et al. 1988, Curran & Leighton 2000), or an entire area of a bamboo species suddenly coming into flower (Janzen 1976), draw attention. Many years ago, this pattern of reproduction became known as mast seeding, from the German word for fattening livestock on abundant seed crops, and hence years of high abundance are called mast years (Janzen 1971, Silvertown 1980, Kelly 1994). The pulse of resources through masting can have effects throughout the ecosystem. For example, in the eastern United States masting triggers interacting density fluctuations in rodents, deer, Lyme disease, and gypsy moths (Ostfeld et al. 1996). Such community effects have been recently reviewed elsewhere (Ostfeld & Keesing 2000, Vander Wall 2001), so the focus of this review will be the evolution of masting from the plant's perspective. We ask whether variable seeding is simply a plant's reproductive response to variable weather conditions, or is a reproductive trait that has evolved through natural selection despite the costs of lost opportunities for reproduction (Waller 1979) and probable higher density-dependent seedling mortality in mast years (e.g., Hett 1971).

The most parsimonious hypothesis for variable seed production is variable weather conditions: the resource matching hypothesis (Norton & Kelly 1988, Sork 1993, Kelly 1994, Houle 1999). A recent study by Koenig & Knops (2000) provides evidence both for and against this hypothesis. They found that the scale of autocorrelation in seed production of Northern Hemisphere tree species occurred at the same spatial scale as autocorrelation in rainfall and temperature data, consistent with the underlying effect of climatic factors on masting. However, they also found that seed production has much higher variability than the weather factors, and the temporal patterns of autocorrelation in the climate variables do not match those of seed production. Thus, they concluded that weather alone could not be responsible for masting. In this review, we discuss the role of weather and list criteria for refuting the weather hypothesis as the sole explanation. For many species, selective factors are the ultimate cause of masting, but weather and resources must be involved as proximate causes. One recent advance is the development of new models to explain the mechanism of producing variable, synchronized seed crops. Our discussion will outline these models.

For plants with masting as a reproductive strategy, the question is why has it evolved? Two of the most prominent hypotheses are the pollination efficiency hypothesis, which states that synchronized, occasional flowering increases pollination success in wind-pollinated plants (Nilsson & Wästljung 1987, Norton & Kelly 1988, Smith et al. 1990, Kelly et al. 2001), and the predator satiation hypothesis, which states that large intermittent seed crops reduce losses to seed predators (Janzen 1971, Silvertown 1980). In animal-dispersed plants, masting may be favored or selected against, depending on the response of the disperser to mast crops (Janzen 1969, 1971, 1974, 1978; Silvertown 1980). For all of these hypotheses, the common element is that the selective advantage occurs through an economy of scale (Janzen 1978, Norton & Kelly 1988) whereby large reproductive efforts are more efficient than small ones, so plants reproducing in step with mast years will have higher fitness. In this analysis we will identify the criteria for demonstrating selection and review studies meeting these criteria. Masting has three key elements: variability, synchrony, and periodicity. The population level variation in seed crops (CV_p), as measured by the coefficient of variation ($CV = SD/mean$), results from the interaction of individual variability (CV_i, the average CV of individual plants over time) and synchrony among plants (Janzen

1971, Herrera 1998a). We will consider how economies of scale affect all of these elements.

One little-studied question is the global distribution of masting. Koenig & Knops (2000) reported higher interannual variability in seed production among species (based on CV) at lower latitudes, which suggests that weather conditions associated with latitude may influence the variability of seed production. Herrera et al. (1998) showed that CVs differ among species with different pollination and seed dispersal modes, in a pattern consistent with the selection hypotheses discussed above. However, Herrera et al.'s study did not control for the global distribution of those modes or the underlying variability in weather. One way to examine the evidence for weather versus selection is to study the global patterns of CV taking into account simultaneously the effects of latitude, pollination, and dispersal modes. In this paper, we present just such an analysis.

We have organized our ideas into three main sections. In the first section, we evaluate the roles of weather and the resource-matching hypothesis. Next, we review the evidence for the key selective pressures that might favor the evolution of masting. The third section analyzes patterns of among-year variation from a global perspective. Here we examine the latitudinal trend in variability in a key weather variable (rainfall). We also examine the CVs of 570 seedfall datasets to evaluate the extent to which latitude, pollination mode and dispersal mode can account for global patterns of CV. We conclude our paper by outlining productive areas of future research.

RESOURCES AND WEATHER

Weather and resources are clearly involved in mast seeding, but the exact nature of their involvement is less clear. One alternative is that plants simply respond to variable weather by flowering more in good years. The other alternative is that there are selective advantages to masting (ultimate factors) which modify plant responses to weather and internal resource levels (proximate factors) in order to enhance the interannual variation in seedfall. This section will review data on the links between weather and masting for evidence supporting either the weather or selection arguments. We will then examine cases that involve selective factors to see how weather and resources act as proximate factors.

Resource Matching: Weather Affecting Resources

The resource matching hypothesis states that *in the absence of selection for (or against) masting, seed crops will vary in response to environmental variation* (Kelly 1994).

The resource matching (RM) or weather tracking hypothesis is the oldest hypothesis for mast seeding. It postulated that each plant's available resources vary each year, being higher during favorable conditions, and reproductive effort mirrors this variation (Büsgen & Münch 1929, Norton & Kelly 1988, Sork 1993, Kelly 1994). Hence, RM should produce a positive correlation between growth and reproduction within years (good years are good for both). Synchrony among plants arises incidentally because they experience similar weather. This nonadaptive hypothesis is the most parsimonious explanation for variability in reproduction.

Support for resource matching comes from various lines of evidence. Adverse weather (e.g., frost or drought) sometimes prevents reproduction, causing synchronization of low years (Sharp & Sprague 1967, Ågren 1988, Allen & Platt 1990, Fenner 1991, Sork et al. 1993, Houle 1999, Selas 2000, Inouye 2000). RM may therefore apply in highly variable environments, where reproduction may be impossible when conditions are bad. A possible example is restricted flowering and growth of perennials in semideserts in frequent dry years (e.g., reproduction in Western Australian shrubs, Davies 1976). Of more interest is how often RM explains high seed years (the core of the RM hypothesis). One good example is *Pinus banksiana* in Quebec, which shows positive correlations within plants across years among reproduction, growth, and good growing conditions (Despland & Houle 1997). Significantly, this species is strongly serotinous and the lifetime seed crop is retained on the tree until fire occurs, making the exact year of seed production irrelevant, so the plant is free to follow resource matching. Other studies claimed as supporting RM (Nienstaedt 1985, Byram et al. 1986, Willson 1986, Oyama 1990, Cremer 1992, Lord 1998) do not directly test it, usually because they compare multi-year averages of growth and reproduction across different plants.

The key evidence that disproves the RM hypothesis is the presence of "switching," where in successive years plants move resources into, then away from, reproduction (Norton & Kelly 1988). A negative correlation between growth and reproduction (good years for reproduction are bad for growth) demonstrates switching and refutes the RM hypothesis. Such data are common for trees including *Abies*, *Acer*, *Betula*, *Dacrydium*, *Fagus*, *Picea*, *Pinus*, and *Pseudotsuga* (Morris 1951; Holmsgaard 1956, in Silvertown 1980; Eis et al. 1965; Tappeiner 1969; Gross 1972; Harper 1977, p. 654; Norton & Kelly 1988; Fenner 1991; Silvertown & Dodd 1999; Houle 1999; and references therein). Koenig & Knops (1998) in a review of 298 datasets found that negative correlations were widespread in Northern Hemisphere conifers. Negative correlations are also reported for herbaceous plants (e.g., Payton & Mark 1979, Muir 1995, Greer & McCarthy 2000).

Patterns of internal resource allocation in masting plants provide two other tests of RM. First, negative autocorrelation (current reproduction negatively related to previous reproduction) suggests the presence of switching. Negative autocorrelations are widespread (Norton & Kelly 1988; Allen & Platt 1990; Sork 1993; Sork et al. 1993; Sork & Bramble 1993; Koenig et al. 1994; Herrera et al. 1998; Stevenson & Shackel 1998; Koenig & Knops 1998, 2000; Selas 2000). Another test is the existence of bimodality (high and low years) in reproduction, which also suggests switching since environmental conditions are not bimodal (Koenig & Knops 2000). Bimodality is found in many species (Norton & Kelly 1988, Herrera et al. 1998), but not all (Kelly 1994, Koenig & Knops 2000).

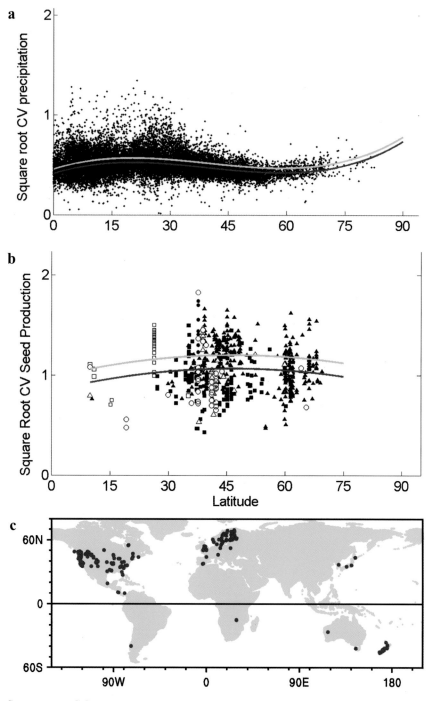

See text page C-2

Figure 1 Global patterns of mast seeding. (*a*) Variability of annual precipitation (CV) from 19,279 weather stations versus absolute latitude (both hemispheres combined). The equation for Northern and Southern hemispheres is: square root $CV_{precipitation} = \beta_o + 0.0128$ (latitude) - 0.0004 (latitude^2) + 3.5×10^{-6} (latitude^3) where $\beta_o = 0.4551$ for Southern hemisphere (green line, upper) and 0.4098 for Northern hemisphere (blue line, lower; $R^2 = 11.5\%$, $P < 0.0001$). (*b*) Seedfall variability (CV_p) in 570 datasets versus latitude. Symbols indicate mode of pollination (open symbols = biotic, filled = abiotic) and dispersal (abiotic = triangle, frugivore = square, predator = circle). The equation for both hemispheres is: square root $CV_p = \beta_o + 0.0095$ (latitude) $- 0.0001$ (latitude^2) where $\beta_o = 0.9790$ for Southern hemisphere (green line, upper) and $\beta_o = 0.8444$ for Northern hemisphere (blue line, lower). See Table 1 for summary of ANCOVA statistics. (*c*) Distribution of the 570 seedfall datasets. Note the abundance of data from North America (250 datasets), Europe (216) and Australasia (79), and the dearth of data for Asia (12), Central and South America (11), Africa (2), and the tropics generally.

Our conclusion is that switching (and hence selection for masting) is much more common than pure resource matching in plants with high interannual variability.

Cues for Masting: Weather as a Signal

Masting requires synchrony among plants, usually reliant on entrainment to a weather cue (Janzen 1971). Weather cues that are correlated with resource abundance offer the advantage of minimizing storage costs (Norton & Kelly 1988, Koenig et al. 1996). Therefore, correlations between high seed crops and years of abundant resources (e.g., Wright & van Schaik 1994, Fenner 1998, Schauber et al. 2002) do not separate RM from the hypothesis that there has been selection for masting.

Cues are often not associated with increased resources, consistent with selection but not with RM. For example, masting in dipterocarps is triggered by night temperatures dropping 2°C over three nights (Ashton et al. 1988). In *Fraxinus excelsior*, the weather cues for heavy flowering were different from the weather variables associated with greater growth (Tapper 1996). In *Chionochloa pallens*, a simple binary weather cue (January temperatures over 11.5°C) affects flowering more than whole-season growing degree-days (Rees et al. 2002). Flowering may be triggered by drought (van Schaik et al. 1993, Wright et al. 1999) or fire (Payton & Mark 1979, Kelly 1994), both of which decrease resources. Therefore, evidence of the weather signals that trigger masting is universally consistent with selection, but frequently inconsistent with RM.

Resource Models: The Mechanisms of Masting

Ultimately, masting as a reproductive strategy requires some kind of resource allocation mechanism that exaggerates variation among years. The underlying question is, if a plant has a physiological mechanism that alters flowering effort in relation to an environmental signal, why does it do so in just this fashion? Janzen (1971) predicted that in masting species, selection should favor flowering to be hypersensitive to weather variables and to levels of reserves in the plant. Recent work supports this prediction. The first step was developing models for how a plant's internal resource levels influence flowering patterns. Several recent models generate intermittent reproduction by each plant, driven by resource thresholds for reproduction coupled with large expenditure when reproduction occurs (Isagi et al. 1997, Satake & Iwasa 2000). This internal process would lead to strong negative autocorrelations and increase individual variation. The exact level of individual variation depends on how heavily the plant invests in reproduction during years when it does reproduce (the depletion coefficient); with a large depletion coefficient, reserves are exhausted and the plant cannot flower again for some time. Importantly, the model predicts stable flowering provided that the depletion coefficient is below a critical value, otherwise chaotic masting dynamics result (Satake & Iwasa 2000).

A version of this resource threshold model has now been applied to a 12-year dataset for *Chionochloa pallens* in New Zealand (Rees et al. 2002). The population had a high CV_p (1.88) and very high synchrony among plants (mean pairwise $r = 0.77$). Rees et al. showed that RM models fit the data poorly, whereas the resource threshold model fit excellently. The depletion coefficient calculated by the *C. pallens* model was in the region giving chaotic masting dynamics. Moreover, they determined that this value of the depletion coefficent minimizes seed predation by specialist invertebrates. This case illustrates how seed predators can select for a hypersensitive internal mechanism that controls and synchronizes reproduction among individuals.

Thus, the proximate mechanisms that integrate weather cues and resource utilization can produce more variable patterns of reproduction than expected from simple resource matching. Next, we ask why this would be advantageous.

SELECTIVE FORCES AND ECONOMIES OF SCALE

Types of Selective Advantage

An economy of scale (EOS) is required for mast seeding to be selectively advantageous (Janzen 1978, Norton & Kelly 1988). Here we review the three EOS hypotheses with the most experimental testing—wind pollination, predator satiation, and animal dispersal—and comment briefly on the animal pollination hypothesis (that large flower crops attract disproportionately more pollinators). Six other published hypotheses are not reviewed for the following reasons. The outcrossing hypothesis (that mast years facilitate outbreeding: Janzen 1978, Tisch & Kelly 1998) awaits experimental testing. The accessory costs hypothesis (high fixed costs of reproduction favor fewer, larger reproductive episodes: Kelly 1994), and the large seed hypothesis (selection for larger seeds increases the recovery time between seed crops: Sork 1993) do not require synchrony among plants. Three hypotheses apply only in specialized situations: the environmental prediction hypothesis (that plants reproduce heavily in years that will be favorable for seedling establishment: Kelly 1994), the bamboo fire cycle hypothesis (synchronized death of bamboos after masting encourages fire, which prevents trees from out-competing the bamboo: Keeley & Bond 1999), and the predator cleansing hypothesis (the synchronized death of bamboos reduces densities of herbivores feeding on adult leaves: Pearson et al. 1994).

Wind Pollination

The pollination efficiency hypothesis states that *masting should be strongly selected in species that can achieve greater pollination efficiency through synchronized above-average flowering effort.* This efficiency is most likely to be seen in wind pollinated plants because they are at less risk of saturating the pollinators (Janzen 1978, Smith et al. 1990, Sork 1993, Kelly et al. 2001).

To support this hypothesis, one must show that percent fruit set is higher when flower density is higher. This has been frequently shown in wind-pollinated species (Nilsson & Wästljung 1987, Norton & Kelly 1988, Smith et al. 1990, Allen & Platt 1990, Burrows & Allen 1991, Kelly 1994, Shibata et al. 1998, Houle 1999, Kelly et al. 2001), although some species show no effect (Sork 1993, Kelly & Sullivan 1997). The species most likely to gain benefits are those whose pollination success at the long-term average flowering effort is low compared to the maximum achievable with superabundant pollen (Kelly et al. 2001). Pollination is likely to be more sensitive to flowering density in obligate outcrossers (e.g., dioecious or self-incompatible species). The consequences of inefficient pollination should be more severe among species with expensive unpollinated female structures (Smith et al. 1990), so these species should gain more from masting, but empirical tests of this prediction have been equivocal (Kelly et al. 2001).

The pollination efficiency hypothesis depends on the size of the current flowering effort, but the sequence of reproductive efforts among years is irrelevant; for example, having successive mast years would not be disadvantageous (Norton & Kelly 1988). Plants in productive habitats could reach a high enough level of reproductive effort for efficient pollination every year, so they could reproduce constantly and avoid the negative consequences of masting (Hett 1971, Waller 1979, Kelly 1994). Plants in unproductive habitats are unable to reach this level every year, so they accumulate reserves to be expended in occasional large efforts.

Pollination efficiency could apply to animal pollinators if they are attracted to large flowering displays (Kelly 1994). Supporting data are scarce; the insect-pollinated *Acer saccharum* shows higher fruit set in mast years (Curtis 1959, p. 105; Graber & Leak 1992), although its pollen can also be wind-dispersed. In general, animals are more likely to be saturated by large crops, providing diseconomies of scale (Herrera et al. 1998, and see "Global Patterns of Variability in Seed Production," below). We conclude that wind pollination often provides an EOS but animal pollination does not.

Predator Satiation

The predator satiation hypothesis states that *seed predators cause selection for masting when larger seed crops synchronized among individuals experience lower percentage seed predation* (Janzen 1971).

Predator satiation favors masting when variation in seed crops satiates seed predators in high-seed years. Salisbury (1942, p. 2) pointed out that in beech and oaks the only seeds to escape predation were produced in mast years, and that if a species had a constant seed crop its natural enemies could increase in number until all seeds were destroyed every year. Janzen (1971, 1974, 1976, 1978) rekindled interest in mast seeding when he explored predator satiation at length. Predator satiation requires interannual variation in seed crops, but it is unclear whether selection acts directly to favor gaps between mast years by starving predators in low-seed years, or whether gaps are an inevitable consequence of selection for larger crop size (Janzen 1978, Silvertown 1980).

The simplest evidence for predator satiation is lower seed predation in high-seed years. The ideal data are a long time-series for both seed production and seed predation at a site. Variation in space within a year is sometimes used (e.g., Nilsson & Wästljung 1987) but may not be a good analog for temporal variation. Numerous studies provide evidence of lower predation during mast years, both for insect predators (e.g., McQuilkin & Musbach 1977, DeSteven 1983, Sork 1983, Schupp 1990, Crawley & Long 1995, Sullivan & Kelly 2000, Shibata et al. 1998) and vertebrates (e.g., Boucher 1981, Nilsson 1985, Smith et al. 1990, Crawley & Long 1995, Hart 1995, Wolff 1996, Forget et al. 1999, Vander Wall 2001, Theimer 2001), although some counter-examples exist (Ågren 1988, Hart 1995, Sperens 1997). Ideally, data should record losses separately for different predator species (e.g., Hedlin 1964, McKone et al. 2001). Different seed predators may vary in their responses to masting: Insects may eat less but vertebrates eat more of a large seed crop (Nilsson & Wästljung 1987, Graber & Leak 1992). Different responses can occur even within a guild (e.g., two specialist dipteran predators of *Chionochloa pallens*, McKone et al. 2001).

A thorough analysis for predator satiation requires testing for a numerical response from the predator (reduced predator densities following low seed years), detected by lower seed losses in years following a small seed crop (Silvertown 1980, Kelly & Sullivan 1997). Because predation is often sequence-dependent, two similarly sized flowering efforts could have very different levels of predation if one followed a high seed year and the other followed a low seed year (e.g., Hedlin 1964, McDonald 1992). If the seedfall time series shows simple alternation, then current crop can be a reasonable proxy for the change in crop. It will not be a good proxy where the time series is more complicated, in which case predator satiation may be unrelated to current reproduction but significantly related to the change in reproductive effort (e.g., *Chionochloa pallens*: Kelly & Sullivan 1997). The numerical response should be present in many, though not all, cases of predator satiation. When insect predators have extended diapause (e.g., Hedlin 1964, Kelly et al. 2000), effects from previous years will be more complex.

Selection by predators for or against synchrony is strongly affected by the mobility of the predator (Janzen 1978). If the predator can move easily between plants, selection will favor synchrony among plants at a scale comparable to the predator's mobility. The best large-scale examples involved specialist birds like the passenger pigeon, Carolina parakeet, and the Javanese finch *Serinus estherae* (Janzen 1971, 1976). The mammalian predators of dipterocarp species also select for synchrony on a large scale (Curran & Leighton 2000). In contrast some invertebrates may be satiated at the level of a few trees or even a single tree (e.g., *Cydia fagiglandana*: Nilsson & Wästljung 1987), necessitating very local, or no, synchrony. If the predator is a mobile generalist, it may preferentially forage on an abundant seed crop, causing higher seed losses in mast years (Janzen 1971, Nilsson & Wästljung 1987). Therefore, a plant with several different types of seed predator may experience contradictory selection pressures. For example in *Betula alleghaniensis* larger seed crops experienced lower percentage losses to insects but higher percentage losses to birds (Graber & Leak 1992). Because of the difference in mobility, these contradictory selection pressures could be resolved; *Betula alleghaniensis* might reduce losses to invertebrates by increasing individual seed crop variability and simultaneously reduce losses to birds by reducing synchrony to reduce the degree of masting (Kelly et al. 2001). However, the dramatic examples of masting clearly require both high interannual variability and high synchrony.

The predator satiation hypothesis is widely studied and widely accepted, perhaps beyond what is warranted by the data. The challenge for future studies is to estimate selective impacts, both by modeling seed survival at the population level, and by using long-term data on individual plants to document the relationship between the selective pressures and individual fitness, while simultaneously estimating the selective impact of multiple seed predators.

Animal Dispersal

The animal dispersal hypothesis states that *masting should be selected against in plants dispersed by frugivores that are saturated with large fruit production, creating diseconomies of scale* (Janzen 1971, Silvertown 1980, Herrera et al. 1998).

The effect of masting on seed dispersal varies among dispersal modes. Abiotic dispersal and dispersal on the outside of animals (ectozoochory) are unaffected by masting, whereas frugivore dispersal (endozoochory) would be negatively affected by masting, and predator dispersal (scatterhoarding or dyszoochory, Janzen 1971) may be affected positively. Dispersal could provide an EOS if large fruit crops attract a generalist frugivore (Bawa 1980, Kelly 1994) or result in wider dispersal by scatterhoarders (Smith et al. 1990). The key evidence supporting this hypothesis would be that in high seed years, either a higher fraction of the seed crop is dispersed, or the mean dispersal distance is greater. However, such evidence is rare (e.g., Christensen & Whitham 1991).

Several studies have shown that dispersal by frugivores is negatively affected by masting (Ballardie & Whelan 1986, Koenig et al. 1994, Herrera et al. 1994, Levey & Benkman 1999). The percentage of fruit crop removed is usually either independent of crop size (Davidar & Morton 1986) or is smaller in large crops due to satiation (Jordano 1987; Herrera et al. 1994, 1998; Herrera 1998b). Therefore frugivore-dispersed plants should be less likely to mast than plants with other kinds of dispersal, just as Silvertown (1980) and Herrera et al. (1998) reported from literature reviews (see also "Global Patterns of Variability in Seed Production," below).

For predator dispersal Vander Wall (2001) claims, with little evidence, that masting enhances dispersal. Dispersal is typically either unaffected or worsened in high seed years. For example, in *Pinus monophylla*, scatterhoarding birds collected 89% of seeds from the canopy in a low seed year but only 43% in a high seed year (Vander Wall 1997). Thus, less dispersal took place in a mast year. Also, seeds may be moved shorter distances in a mast year if animals recache seeds less often when seeds are abundant (Vander Wall & Joyner 1998). Masting benefits predator-dispersed plants, but wholly by improving the chances of seeds escaping postdispersal predation (Janzen 1971, Vander Wall & Balda 1977, Christensen & Whitham 1991, Herrera et al. 1998, Vander Wall 2001, Theimer 2001), consistent with the predator satiation hypothesis.

Conclusions: Site Productivity and Contrasts with Animals

We have shown that both wind-pollination efficiency and predator satiation often select for masting, and that dispersal by frugivores may sometimes select against it. In this concluding section, we consider the implications of these factors and make two predictions: Masting should be more common in both unproductive habitats and dominant plant species. Secondly, we consider why the equivalent of masting appears to be rarely selected for in animals.

Plants in less productive habitats should show more pronounced masting. Lower productivity increases the time required to recover between high seed crops, whether those seed crops are large to gain wind pollination benefits or to satiate predators. Factors that reduce plant productivity and are associated with more pronounced masting within or among species include altitude (Allen & Platt 1990, Webb & Kelly 1993, Mencuccini et al. 1995, Sullivan & Kelly 2000, Kelly et al. 2001), latitude (Hagner 1965, in Harper 1977), and soil infertility (Janzen 1974). For example, Gysel & Lemmien (1964) measured fruit output of *Elaeagnus angustifolia* and *Lonicera tartarica* in Michigan on normal and impoverished soils. In both cases reproductive output decreased on poorer soils and varied more among years.

Plants that dominate their communities are more likely to show masting (Janzen 1978, Boucher 1981). First, wind pollination tends to be found in dominant rather than sparse species; and second, dominant species are more vulnerable to seed predation, as they cannot escape by having low apparency. Predator satiation is also easier where one species (or group of related species, e.g., the Dipterocarpaceae) dominate the local vegetation, so masting should be especially well-developed in low-diversity communities, e.g., temperate forests (Janzen 1971).

These trends lead to an interesting prediction—that the tropics should have few masting species. Because the tropics are typified by biotic pollination and dispersal (which select against masting, see above), high plant species diversity, high site productivity, and relatively low year-to-year variation in climate, mast seeding should be uncommon there. The spectacular exception of dipterocarp forests may prove the rule (Wright et al. 1999) because these forests are dominated by closely related species that require synchrony and large crops to satiate shared vertebrate seed predators (Janzen 1969).

Finally, we comment briefly on the contrast with animals, within which "masting" is extremely rare. There are a few species of synchronized semelparous insects, most famously six species of cicadas (Heliövaara et al. 1994). Synchronized iteroparity is almost unknown. We know of only two cases where a long-lived iteroparous animal will synchronously breed in some years and refrain in others even given an adequate food supply. Two parrots, the kaka (Wilson et al.

1998), and the kakapo (Clout & Merton 1998) breed only in response to a masting food crop, regardless of supplementary artificial feeding. Other possible cases include the passenger pigeon (Bucher 1992), New Zealand parakeets (Moorhouse 1997), Australian banded stilt (Flannery 1994, p. 89), and New Zealand pigeon (Clout et al. 1995). We conclude that such behavior is so rare in animals yet so common in plants because animals are shorter-lived (increasing the costs of lost reproductive opportunities) and use mobility to solve the problems (finding mates, avoiding predators) that plants respond to by masting. Thus, mast seeding may be one of the "evolutionary consequences of being a plant" (Bradshaw 1972). Massed reproduction in immobile animals such as corals may be analogous to mast seeding.

GLOBAL PATTERNS OF VARIABILITY IN SEED PRODUCTION

One way to evaluate the relative impacts of weather and selection on variability of seed production is to test the two hypotheses simultaneously. Two recent reviews have tested them one at a time. Koenig & Knops (2000) used 443 datasets on reproduction of temperate Northern Hemisphere trees to test predictions of the weather hypothesis. As discussed previously (in the "Introduction," above) they found that weather influences seedfall CV (seedfall variability decreases at higher latitudes, in parallel with decreases in the CV of rainfall), but that weather could not account fully for the observed patterns. In a separate review of 296 datasets for woody plants, Herrera et al. (1998) compared interannual variability in seedfall against postulated selective factors (pollination efficiency, predator satiation, and seed dispersal), while controlling for phylogenetic constraints. They predicted that wind-pollinated species should have higher CV_ps than biotically pollinated species but the differences were not quite significant. For dispersal mode they predicted, and found, that frugivore-dispersed plants had significantly lower CV_ps than abiotically and predator-dispersed species. Herrera et al. (1998) conclude that their analysis supports the hypotheses that these factors selected for masting. Here, we integrate the approaches of these two studies to simultaneously test predictions about differences in CV_ps across pollination and dispersal modes while controlling for, and estimating, the effect of latitude.

To understand the influence of weather, we will first report a global analysis of the CV of annual precipitation, a key weather factor for plants, using 30 years of data from 19,279 weather stations (downloaded from the Global Historical Climatology Network, ftp://ftp.ncdc.noaa.gov/pub/data/ghcn/), which was analyzed with a regression of the CV for rainfall versus latitude, with hemisphere as a class variable. Next, we will examine seed variability data using an Analysis of Covariance (ANCOVA) to evaluate the impact of pollen vector, seed dispersal mode, growth form, and hemisphere as factors, and latitude and length of study as covariates, on the CV of seed production. Our analysis employs 570 studies of ≥6 years duration compiled from reviews (Silvertown 1980, Webb & Kelly 1993, Kelly 1994, Herrera et al. 1998, Kelly et al. 2000, Koenig

& Knops 2000, Schauber et al. 2002) and the primary literature (Appendix A, http://www.annualreviews.org/MastingAppendix.html). We particularly sought extra datasets from herbaceous plants and from tropical latitudes. We categorized pollination and dispersal modes in the same way as Herrera et al. (1998).

In our global weather analysis, the CV of rainfall shows a nonlinear association with latitude (Figure 1a, color insert) with a peak at about 20–25° and decreasing variability nearer the equator and toward 60° latitude. The regression model had significant first, second, and third order terms (Figure 1a), with a significant effect of hemisphere (df = 1 and 19,265; F = 671.8, $P < 0.0001$). The mean CV for the Southern Hemisphere (back-transformed CV = 0.303, n = 7225) was greater than that of the Northern Hemisphere (CV = 0.248, n = 12,045), but both means are small compared to seedfall mean CVs. The conclusion is that the annual variability of a key weather variable, rainfall, changes significantly with latitude. If rainfall affects seed crops, then we predict that CV_p should peak at mid latitudes and be greater in the Southern Hemisphere.

Our analysis of seed production CV_p shows that both latitudinal effects and differences among pollen and seed vectors are significant (Table 1). The analysis supports the hypothesis that climatic factors influence variability in seed production. Latitude had significant first and second order terms (Table 1). The curvilinear relationship between CV_p and latitude is consistent with the precipitation/latitude relationship (Figure 1b, color insert), but here the peak is around 45°. Due to an imperfect spread of seedfall data across latitudes, we cannot determine why the latitudinal curves for rainfall and seed production do not concur precisely. Both the precipitation and seedfall data indicate lower variability in the tropics (consistent with the prediction in "Conclusions: Site Productivity, and Contrasts with Animals," above) but we have very few tropical seedfall datasets. In contrast the prediction from "Conclusions: Site Productivity, and Contrasts with Animals," above, of higher CV_p at high latitudes is not supported. Similar to the rainfall data, seed production variability for the Southern Hemisphere (back-transformed least squares mean CV_p = 1.369, n = 84) is greater than that of the Northern Hemisphere (CV_p = 1.083, n = 488). The estimates of CV_p are almost an order of magnitude greater than precipitation CVs. Thus, one or more factors are causing greater annual variability in seed production than occurs in precipitation.

In addition to the significant effect of latitude the ANCOVA model supports the predictions of the EOS hypotheses. By using the least square mean estimates that remove the effect of the other parameters, we find that CVs differ significantly between pollen vectors and among seed dispersal modes (Table 1) in a pattern consistent with selective economies of scale. Abiotic pollination has a higher mean CV_p than biotic pollination (pollination efficiency hypothesis). Species with seed dispersal by predators had a higher mean CV than species dispersed abiotically or by frugivores, supporting the predator satiation hypothesis. We also found a significant interaction between pollen and seed vector, which is not surprising because selection on pollen mode will affect seeding variability and vice versa. Consistent with the pollination efficiency and seed dispersal

TABLE 1   ANCOVA of factors affecting square root transformed CV of yearly seed production based on 570 datasets from 175 species (For sources of data, see Appendix A, http://www. annualreviews.org/MastingAppendix.html)

Source	DF	SS	MS	F Value	Pr > F
A ANCOVA for square root transformed CV of yearly seed production ($R^2 = 0.289$)					
Model	18	9.051	0.502	12.45	<.0001
Pollen vector	1	0.363	0.363	8.99	0.0028
Seed vector	2	0.501	0.251	6.20	0.0022
Pollen × seed	2	0.410	0.205	5.08	0.0065
Growth form	4	0.657	0.164	4.07	0.0029
Hemisphere	1	0.441	0.441	10.91	0.0010
Pollen × hemisphere	1	0.089	0.089	2.20	0.1229
Seed × hemisphere	2	0.708	0.354	8.76	0.0002
Latitude	1	0.414	0.414	10.26	0.0014
Latitude^2	1	0.179	0.179	4.43	0.0359
Latitude × hemisphere	1	0.259	0.259	6.42	0.0116
Latitude^2 × hemisphere	1	0.096	0.096	2.39	0.1229
Length of study	1	0.654	0.654	16.20	<.0001
Error	551	22.253	0.040		

B Least square means of CV, using back-transformed values of square root transformed CV, by pollination and dispersal mode. Sample sizes are given in parentheses. Overall mean CV = 1.050

	Abiotic pollination	Biotic pollination	Combined
Predator-dispersed	1.672 (117)	1.861 (28)	1.765 (145)
Frugivore-dispersed	1.443 (15)	0.692 (40)	1.034 (55)
Abiotically dispersed	1.162 (347)	0.741(23)	0.940 (370)
Combined	1.418 (479)	1.039 (91)	

hypotheses, biotic pollination and frugivore dispersal are associated with the lowest mean CV.

The geographic distribution of pollen and dispersal modes is clearly not random with respect to latitude (Figure 1*b*). Wind-pollinated, predator-dispersed species (filled triangles) are common above 45° latitude while animal-pollinated, frugivore-dispersed species (hollow squares) are common below 45° latitude. However, the ANCOVA model suggests that latitude has an effect even when pollination and dispersal modes are controlled for and vice versa.

We tested the effect of growth form because masting should be favored in long-lived plants, not just in trees (Waller 1979, Silvertown 1980). The mean CV_ps differed significantly among growth forms from a maximum in herbaceous dicotyledons (mean $CV_p = 1.683$, n = 17 datasets), through herbaceous monocots (1.403, n = 20), trees (1.253, n = 474), and shrubs (1.177, n = 58), to a minimum in woody monocots (0.700, n = 3 palm datasets). The fact that trees do not have the highest value suggests the need for better study of masting in other kinds of perennials.

Length of study accounted for significant variation in our model (Table 1). Although the average length of study was 11.1 years and we had 86 datasets with at least 15 years of data, the range (6 to 35 years) is considerable. The fact that longer studies had higher CVs of seed production (in contrast to Herrera et al. 1998) suggests either that as number of years increases, the CV becomes larger (the red shift often seen in ecological data; Pimm & Redfearn 1988), or that investigators are more likely to prolong studies of more variable species. Our data support the latter alternative. Datasets that were continued for 20 years or more were already significantly more variable at the 10-year stage (mean CV = 1.53, n = 39) than datasets that were terminated after exactly 10 years (mean CV = 1.17, n = 69) according to a one-way ANOVA ($F_{1, 106}$ = 14.11, $P < 0.001$). In contrast, in these longer datasets the mean CVs after 10 years (1.53) did not increase significantly after 15 years (1.52) or after the full duration (mean duration 25.7 years, mean CV = 1.57, $F_{2, 114}$ = 0.14, NS). These results indicate that plant variability affects study duration rather than study duration affecting the measured variability.

Although we collated many datasets, there are limitations in the spread of data. The 570 datasets spanned 175 species, 74 genera, and 37 families (Table 2) but, as in previous reviews (Herrera et al. 1998, Koenig & Knops 2000), were skewed toward north temperate trees. For example, *Pinus* contributed 135 studies and 14 species, and *Quercus* 58 studies and 23 species (Table 2). Geographically, the datasets span 19 countries, skewed toward the temperate Northern Hemisphere (Figure 1c, color insert) with the USA and Finland contributing 50% of all studies.

Finally, we comment briefly on the unavoidable confounding effects in our model. First, we noted earlier that pollen and seed dispersal modes were not distributed randomly with respect to latitude nor are they independent of each other. Statistically our samples should ideally be more balanced across latitude, but the problem is more one of evolution than sampling error. Second, genera and species are very unevenly represented. We decided to include every available dataset despite replication of taxa. When we compared CVs for different sites within a taxon, we found a great deal of variation. In fact, if latitude and local weather conditions as well as pollen/dispersal traits jointly influence CV, then data from separate sites are partially independent and offer additional information. To check that our results were not driven by over-representation of *Pinus*, we ran the ANCOVA excluding this genus (n = 435) and the conclusions were identical. Third, we omitted taxonomic status from our final model. If it was included, we lost the latitude effect because different taxa occur in different parts of the world. For this particular analysis, we were more interested in how the CVs vary globally with latitude than in how CVs vary with taxon. Reassuringly, Herrera et al. (1998) found that controlling for phylogeny did not alter the conclusions from their analysis of pollination and dispersal modes.

TABLE 2 Taxa involved in the 570 studies of seed production for 6 or more years (for details of sources see Appendix A)

Family	# studies	Family	# studies	Family	# studies
A List of all 37 families and number of studies per family ranked by frequency of study					
Pinaceae	279	Caprifoliaceae	4	Tiliaceae	2
Fagaceae	95	Arecaceae	3	Bombacaceae	1
Betulaceae	46	Elaeocarpaceae	3	Burseraceae	1
Fabaceae	29	Anacardiaceae	2	Hippocastanaceae	1
Poaceae	20	Cornaceae	2	Labiateae	1
Cupressaceae	9	Corylaceae	2	Meliaceae	1
Oleaceae	9	Elaeagnaceae	2	Nyssaceae	1
Rosaceae	9	Ericaceae	2	Rubiaceae	1
Agavaceae	8	Eucryphiaceae	2	Smilacaceae	1
Juglandaceae	7	Liliaceae	2	Violaceae	1
Podocarpaceae	7	Monimiaceae	2	Winteraceae	1
Gentianaceae	6	Taxodiaceae	2		
Aceraceae	4	Thymelaeaceae	2		

B List of genera (with family) out of 74 sampled genera with more than 10 studies, ranked by frequency

Genus (Family)	Number of studies	Genus (Family)	Number of studies
Pinus (Pinaceae)	135	*Chionochloa* (Poaceae)	17
Picea (Pinaceae)	66	*Fagus* (Fagaceae)	17
Quercus (Fagaceae)	58	*Acacia* (Fabaceae)	16
Abies (Pinaceae)	43	*Pseudotsuga* (Pinaceae)	12
Betula (Betulaceae)	35	*Tsuga* (Pinaceae)	12
Nothofagus (Fagaceae)	18	*Larix* (Pinaceae)	11

Our ANCOVA analysis supports the notion that selection favors the evolution of masting behavior for species with certain pollen and seed attributes. The large values of CV_p relative to the precipitation CVs and the significant effects of pollen and dispersal modes are strong evidence. Ideally, a global analysis would include more data points for tropical, herbaceous, and Southern Hemisphere populations. However, generalizations about the global patterns of CV are now reasonably robust. Our conclusions are consistent with those of Herrera et al. (1998) and Koenig & Knops (2000). We now need studies with a different approach, particularly long-term studies with data for individual plants, to study the selective pressures directly and to see whether they are affecting variability, synchrony, or masting interval. In other words, we need microevolutionary studies, especially from outside the Pinaceae and Fagaceae. It would be helpful to study different species at the same site because they would be experiencing the same weather. In addition, we could benefit from macroevolutionary studies that compare seeding schedules in related species with different pollen or seed dispersal modes. These approaches will answer questions that broad surveys cannot.

FUTURE DIRECTIONS

We highlight eight promising directions for future research. (*a*) We need more tests of predator satiation that model the selective benefits of masting and test for numerical responses. (*b*) Physiological work on resource acquisition and depletion will help to quantify the underlying mechanisms of masting. Long-term observational studies on individual plants and manipulative experiments that alter resources would both be worthwhile. (*c*) We need to study the relationship between site productivity and degree of masting, including more data on tropical plants and plants of arid environments. (*d*) We should study the selective forces and physiological mechanisms of plants with very constant reproduction. (*f*) Study of anthropogenic impacts on masting is desirable. For example, global warming may disrupt the temperature cues that synchronize plants, causing masting to fail (McKone et al. 1998), or fragmentation may devalue the benefits of masting (Curran & Leighton 2000, Kelly et al. 2001). (*g*) We need to model the impact of pulsed resources on animal communities (Ostfeld & Keesing 2000), particularly if global warming may disrupt masting. (*h*) We need explicit study of the spatial dimension in masting. Studies of local and meso-scale spatial autocorrelation similar to those done with animal populations (Bjørnstad et al. 1999) can clarify the scale of biotic versus abiotic interactions. (*i*) Time-series analysis of periodicity in masting species (e.g., Bonferroni series) can elucidate both the dynamics of seedfall patterns and interactions with predator population responses if data sets of adequate duration can be assembled.

The history of masting has had several phases. Janzen and others initiated many evolutionary hypotheses over a short period, which were followed by gradual empirical testing of these ideas. Now we are entering an era of mechanistic understanding of the spatial and temporal patterns, the causes, the physiological dynamics, and the cascading consequences of mast seeding.

ACKNOWLEDGMENTS

This paper benefited greatly from discussions in the Masting Dynamics Working Group (Ottar Bjørnstad, John Buonaccorsi, Richard Duncan, Joe Elkinton, Walt Koenig, Bill Kuhn, Andrew Liebhold, Mikko Peltonen, Mark Rees, Chris Smith, and Bob Westfall) supported by the National Center for Ecological Analysis and Synthesis, a Center funded by NSF (Grant #DEB-0072909), the University of California, and UC Santa Barbara. VLS acknowledges support from the Missouri Department of Conservation and thanks Judy Bramble for years of wonderful discussion on masting. Bill Kuhn helped collate datasets, Juan Fernandez and Rodney Dyer helped analyze data and prepare figures, and Bob Westfall provided statistical advice. Rob Allen, Bob Brockie, Phil Cowan, Diane DeSteven, Carlos Herrera, David Inouye, Walt Koenig, and W. Joseph Wright kindly provided access to unpublished data. Carol Augspurger, Pedro Jordano, Linda Newstrom-Lloyd, Laura Sessions, Jon Sullivan, and Don Waller made helpful comments on the manuscript.

The *Annual Review of Ecology and Systematics* is online at
http://ecolsys.annualreviews.org

LITERATURE CITED

Ågren J. 1988. Between-year variation in flowering and fruit set in frost-prone and frost-sheltered populations of dioecious Rubus chamaemorus. *Oecologia* 76:175–83

Allen RB, Platt KH. 1990. Annual seedfall variation in *Nothofagus solandri* (Fagaceae), Canterbury, New Zealand. *Oikos* 57:199–206

Ashton PS, Givnish TJ, Appanah S. 1988. Staggered flowering in the Dipterocarpaceae: new insights into floral induction and the evolution of mast fruiting in the aseasonal tropics. *Am. Nat.* 132:44–66

Ballardie RT, Whelan RJ. 1986. Masting, seed dispersal and seed predation in the cycad *Macrozamia communis*. *Oecologia* 70:100–5

Bawa KS. 1980. Evolution of dioecy in flowering plants. *Annu. Rev. Ecol. Syst.* 11:15–39

Bjørnstad ON, Ims RA, Lambin X. 1999. Spatial population dynamics: analyzing patterns and processes of population synchrony. *Trends Ecol. Evol.* 14:427–32

Boucher DH. 1981. Seed predation by mammals and forest dominance by *Quercus oleoides*, a tropical lowland oak. *Oecologia* 49:409–19

Bradshaw AD. 1972. Some of the evolutionary consequences of being a plant. *Evol. Biol.* 5:25–47

Bucher EH. 1992. The causes of extinction of the passenger pigeon. In *Current Ornithology*, ed. DM Power pp. 1–36. New York: Plenum

Büsgen M, Münch E. 1929. *The Structure and Life of Forest Trees*. London: Chapman & Hall, Transl. T. Thompson, 3rd ed.

Burrows LE, Allen RB. 1991. Silver beech (*Nothofagus menziesii* (Hook. f.) Oerst.) seedfall patterns in the Takitimu Range, South Island, New Zealand. *NZ J. Bot.* 29:361–65

Byram TD, Lowe WJ, McGriff JA. 1986. Clonal and annual variation in cone production in loblolly pine seed orchards. *For. Sci.* 32:1067–73

Christensen KM, Whitham TG. 1991. Indirect herbivore mediation of avian seed dispersal in pinyon pine. *Ecology* 72:534–42

Clout MN, Karl BJ, Pierce RJ, Robertson HA. 1995. Breeding and survival of New Zealand pigeons *Hemiphaga novaeseelandiae*. *Ibis* 137:264–71

Clout MN, Merton DV. 1998. Saving the Kakapo: the conservation of the world's most peculiar parrot. *Bird Conserv. Int.* 8:281–96

Crawley MJ, Long CR. 1995. Alternate bearing, predator satiation and seedling recruitment in *Quercus robur* L. *J. Ecol.* 83:683–96

Cremer KW. 1992. Relations between reproductive growth and vegetative growth of *Pinus radiata*. *For. Ecol. Manage.* 52:179–99

Curran LM, Leighton M. 2000. Vertebrate responses to spatiotemporal variation in production of mast-fruiting Dipterocarpaceae. *Ecol. Monogr.* 70:101–28

Curtis JT. 1959. *The Vegetation of Wisconsin*. Madison: Univ. Wisc. Press 657 pp.

Davidar P, Morton ES. 1986. The relationship between fruit crop sizes and fruit removal rates by birds. *Ecology* 67:262–65

Davies SJJF. 1976. Studies of the flowering season and fruit production of some arid zone shrubs and trees in western Australia. *J. Ecol.* 64:665–87

Despland E, Houle G. 1997. Climate influences on growth and reproduction of *Pinus banksiana* (Pinaceae) at the limit of the species distribution in eastern North America. *Am. J. Bot.* 84:928–37

DeSteven D. 1983. Reproductive consequences of insect seed predation in *Hamamelis virginiana*. *Ecology* 64:89–98

Eis S, Garman EH, Ebell LF. 1965. Relation between cone production and diameter increment of Douglas Fir (*Pseudotsuga menziesii* (Mirb.) Franco), Grand Fir (*Abies grandis* (Dougl.) Lindl.) and Western White Pine (*Pinus monticola* Dougl.). *Can. J. Bot.* 43:1553–59

Fenner M. 1991. Irregular seed crops in forest trees. *Q. J. For.* 85:166–72

Fenner M. 1998. The phenology of growth and reproduction in plants. *Perspect. Plant Ecol. Evol. Syst.* 1:78–91

Flannery TF. 1994. *The Future Eaters.* Kew, Victoria, Aust.: Reed. 423 pp.

Forget P-M, Kitajima K, Foster RB. 1999. Pre- and post-dispersal seed predation in *Tachigali versicolor* (Caesalpiniaceae): effects of timing of fruiting and variation among trees. *J. Trop. Ecol.* 15:61–81

Graber RE, Leak WB. 1992. *Seed fall in an old-growth northern hardwood forest.* Radnor, Pa.: USDA, Research paper NE-663, 11 pp.

Greer GK, McCarthy BC. 2000. Patterns of growth and reproduction in a natural population of the fern *Polystichum acrostichoides*. *Am. Fern J.* 90:60–76

Gross HL. 1972. Crown deterioration and reduced growth associated with excessive seed production by birch. *Can. J. Bot.* 50:2431–37

Gysel LW, Lemmien WA. 1964. An eight-year record of fruit production. *J. Wildl. Manage.* 28:175–77

Harper JL. 1977. *Population Biology of Plants.* London: Academic. 892 pp.

Hart TB. 1995. Seed, seedling and subcanopy survival in monodominant and mixed forests of the Ituri forest, Africa. *J. Trop. Ecol.* 11:443–59

Hedlin AF. 1964. A six-year plot study on douglas-fir cone insect population fluctuations. *For. Sci.* 10:124–28

Heliövaara K, Väisänen R, Simon C. 1994. Evolutionary ecology of periodical insects. *Trends Ecol. Evol.* 9:475–80

Herrera CM. 1998a. Population-level estimates of interannual variability in seed production:

What do they actually tell us? *Oikos* 82:612–16

Herrera CM. 1998b. Long-term dynamics of Mediterranean frugivorous birds and fleshy fruits: a 12-yr study. *Ecol. Monogr.* 68:511–38

Herrera CM, Jordano P, Guitián J, Traveset A. 1998. Annual variability in seed production by woody plants and the masting concept: reassessment of principles and relationship to pollination and seed dispersal. *Am. Nat.* 152:576–94

Herrera CM, Jordano P, López-Soria L, Amat JA. 1994. Recruitment of a mast-fruiting, bird-dispersed tree: bridging frugivore activity and seedling establishment. *Ecol. Monogr.* 64:315–44

Hett JM. 1971. A dynamic analysis of age in sugar maple seedlings. *Ecology* 52:1071–74

Houle G. 1999. Mast seeding in *Abies balsamea, Acer saccharum* and *Betula alleghaniensis* in an old growth, cold temperate forest of north-eastern North America. *J. Ecol.* 87:413–22

Inouye DW. 2000. The ecological and evolutionary significance of frost in the context of climate change. *Ecol. Lett.* 3:457–63

Isagi Y, Sugimura K, Sumida A, Ito H. 1997. How does masting happen and synchronize? *J. Theor. Biol.* 187:231–39

Janzen DH. 1969. Seed-eaters versus seed size, number, toxicity and dispersal. *Evolution* 23:1–27

Janzen DH. 1971. Seed predation by animals. *Annu. Rev. Ecol. Syst.* 2:465–92

Janzen DH. 1974. Tropical blackwater rivers, animals, and mast fruiting by the Dipterocarpaceae. *Biotropica* 6:69–103

Janzen DH. 1976. Why bamboos wait so long to flower. *Annu. Rev. Ecol. Syst.* 7:347–91

Janzen DH. 1978. Seeding patterns of tropical trees. In *Tropical Trees as Living Systems,* ed. PB Tomlinson, MH Zimmermann, pp. 83–128. Cambridge, UK: Cambridge Univ. Press

Jordano P. 1987. Patterns of mutualistic interactions in pollination and seed dispersal:

connectance, dependence asymmetries, and coevolution. *Am. Nat.* 129:657–77

Keeley JE, Bond WJ. 1999. Mast flowering and semelparity in bamboos: the bamboo fire cycle hypothesis. *Am. Nat.* 154:383–91

Kelly D. 1994. The evolutionary ecology of mast seeding. *Trends Ecol. Evol.* 9: 465–70

Kelly D, Sullivan JJ. 1997. Quantifying the benefits of mast seeding on predator satiation and wind pollination in *Chionochloa pallens* (Poaceae). *Oikos* 78:143–50

Kelly D, Harrison AL, Lee WG, Payton IJ, Wilson PR, Schauber EM. 2000. Predator satiation and extreme mast seeding in 11 species of *Chionochloa* (Poaceae). *Oikos* 90:477–88

Kelly D, Hart DE, Allen RB. 2001. Evaluating the wind-pollination benefits of mast seeding. *Ecology* 82:117–26

Koenig WD, Knops JMH. 1998. Scale of mast-seeding and tree-ring growth. *Nature* 396:225–26

Koenig WD, Knops JMH. 2000. Patterns of annual seed production by Northern Hemisphere trees: a global perspective. *Am. Nat.* 155:59–69

Koenig WD, Knops JMH, Carmen WJ, Stanback MT, Mumme RL. 1996. Acorn production by oaks in central coastal California: influence of weather at three levels. *Can. J. For. Res.* 26:1677–83

Koenig WD, Mumme RL, Carmen WJ, Stanback MT. 1994. Acorn production by oaks in central coastal California: variation within and among years. *Ecology* 75:99–109

Levey DJ, Benkman CW. 1999. Fruit-seed disperser interactions: timely insights from a long-term perspective. *Trends Ecol. Evol.* 14:41–43

Lord J. 1998. Effect of flowering on vegetative growth and further reproduction in *Festuca novae-zelandiae*. *NZ J. Ecol.* 22:25–31

McDonald PM. 1992. Estimating seed crops of conifer and hardwood species. *Can. J. For. Res.* 22:832–38

McKone MJ, Kelly D, Harrison AL, Sullivan JJ, Cone AJ. 2001. Biology of insects that feed in the inflorescences of *Chionochloa*

pallens (Poaceae) in New Zealand and their relevance to mast seeding. *NZ J. Zool.* 28:89–101

McKone MJ, Kelly D, Lee WG. 1998. Effect of climate change on masting species: frequency of mass flowering and escape from specialist insect seed predators. *Glob. Change Biol.* 4:591–96

McQuilkin RA, Musbach RA. 1977. Pin oak acorn production on green tree reservoirs in southeastern Missouri. *J. Wildl. Manage.* 41:218–25

Mencuccini M, Piussi P, Sulli AZ. 1995. 30 years of seed production in a subalpine Norway spruce forest. *For. Ecol. Manage.* 76:109–25

Moorhouse R. 1997. The diet of the North Island kaka (*Nestor meridionalis septentrionalis*) on Kapiti Island. *NZ J. Ecol.* 21:141–52

Morris RF. 1951. The effects of flowering on the foliage production and growth of balsam fir. *For. Chron.* 27:40–57

Muir AN. 1995. The cost of reproduction to the clonal herb *Asarum canadense* (wild ginger). *Can. J. Bot.* 73:1683–86

Nienstaedt H. 1985. Inheritance and correlation of frost injury, growth, flowering and cone characteristics in white spruce. *Picea glauca* (Moench) Voss. *Can. J. For. Res.* 15:498–504

Nilsson SG. 1985. Ecological and evolutionary interactions between reproduction of beech *Fagus sylvatica* and seed eating animals. *Oikos* 44:157–64

Nilsson SG, Wästljung U. 1987. Seed predation and cross-pollination in mast-seeding beech (*Fagus sylvatica*) patches. *Ecology* 68:260–65

Norton DA, Kelly D. 1988. Mast seeding over 33 years by *Dacrydium cupressinum* Lamb. (rimu) (Podocarpaceae) in New Zealand: the importance of economies of scale. *Funct. Ecol.* 2:399–408

Ostfeld RS, Jones CG, Wolff JO. 1996. Of mice and mast: ecological connections in eastern deciduous forests. *Bioscience* 46:323–30

Ostfeld R, Keesing F. 2000. Pulsed resources and consumer community dynamics in terrestrial ecosystems. *Trends Ecol. Evol.* 15:232–37

Oyama K. 1990. Variation in the growth and reproduction in the neotropical dioecious palm *Chamaedorea tepejilote. J. Ecol.* 78:648–63

Payton IJ, Mark AF. 1979. Long-term effects of burning on growth, flowering, and carbohydrate reserves in narrow-leaved snow tussock (*Chionochloa rigida*). *NZ J. Bot.* 17:43–54

Pearson AK, Pearson OP, Gomez IA. 1994. Biology of the bamboo *Chusquea culeou* (Poaceae: Bambusoideae) in southern Argentina. *Vegetatio* 111:93–126

Pimm SL, Redfearn A. 1988. The variability of population densities. *Nature* 334:613–14

Rees M, Kelly D, Bjørnstad O. 2002. Snow tussocks, chaos, and the evolution of mast seeding. *Am. Nat.* 160:44–59

Salisbury EJ. 1942. *The Reproductive Capacity of Plants.* London: Bell. 244 pp.

Satake A, Iwasa Y. 2000. Pollen coupling of forest trees: forming synchronized and periodic reproduction out of chaos. *J. Theor. Biol.* 203:63–84

Schauber EM, Kelly D, Turchin P, Simon C, Lee WG, et al. 2002. Synchronous and asynchronous masting by 18 New Zealand plant species: the role of temperature cues and implications for climate change. *Ecology* 83:1214–25

Schupp EW. 1990. Annual variation in seedfall, postdispersal predation, and recruitment of a neotropical tree. *Ecology* 71:504–15

Selas V. 2000. Seed production of a masting dwarf shrub, *Vaccinium myrtillus*, in relation to previous reproduction and weather. *Can. J. Bot.* 78:423–29

Sharp WM, Sprague VG. 1967. Flowering and fruiting in the white oaks. Pistillate flowering, acorn development, weather and yields. *Ecology* 48:243–51

Shibata M, Tanaka H, Nakashizuka T. 1998. Causes and consequences of mast seed production of four co-occurring *Carpinus* species in Japan. *Ecology* 79:54–64

Silvertown JW. 1980. The evolutionary ecology of mast seeding in trees. *Biol. J. Linn. Soc.* 14:235–50

Silvertown JW, Dodd M. 1999. The demographic cost of reproduction and its consequences in balsam fir (*Abies balsamea*). *Am. Nat.* 154:321–32

Smith CC, Hamrick JL, Kramer CL. 1990. The advantage of mast years for wind pollination. *Am. Nat.* 136:154–66

Sork VL. 1983. Mast-fruiting in hickories and availability of nuts. *Am. Midl. Nat.* 109:81–88

Sork VL. 1993. Evolutionary ecology of mast-seeding in temperate and tropical oaks (*Quercus* spp.). *Vegetatio* 107/108:133–47

Sork VL, Bramble JE. 1993. Prediction of acorn crops in three species of North American oaks: *Quercus alba, Q. rubra and Q. velutina. Ann. Sci. For.* 50:128S–36

Sork VL, Bramble J, Sexton O. 1993. Ecology of mast-fruiting in three species of North American deciduous oaks. *Ecology* 74:528–41

Sperens U. 1997. Fruit production in *Sorbus aucuparia* L (Rosaceae) and predispersal seed predation by the apple fruit moth (*Argyresthia conjugella* Zell.). *Oecologia* 110:368–73

Stevenson MT, Shackel KA. 1998. Alternate bearing in pistachio as a masting phenomenon: construction cost of reproduction versus vegetative growth and storage. *J. Am. Soc. Hortic. Sci.* 123:1069–75

Sullivan JJ, Kelly D. 2000. Why is mast seeding in *Chionochloa rubra* (Poaceae) most extreme where seed predation is lowest? *NZ J. Bot.* 38:221–33

Tappeiner JC. 1969. Effect of cone production on branch, needle, and xylem ring growth of Sierra Nevada Douglas Fir. *For. Sci.* 15:171–74

Tapper P-G. 1996. Long-term patterns of mast fruiting in *Fraxinus excelsior. Ecology* 77:2567–72

Theimer TC. 2001. Seed scatterhoarding by white-tailed rats: consequences for seedling recruitment by an Australian rain forest tree. *J. Trop. Ecol.* 17:177–89

Tisch PA, Kelly D. 1998. Can wind pollination provide a selective benefit to mast seeding?

Chionochloa macra (Poaceae) at Mt Hutt, New Zealand. *NZ J. Bot.* 36:635–41

Vander Wall SB. 1997. Dispersal of singleleaf piñon pine (*Pinus monophylla*) by seed-caching rodents. *J. Mammal.* 78:181–91

Vander Wall SB. 2001. The evolutionary ecology of nut dispersal. *Bot. Rev.* 67:74–117

Vander Wall SB, Balda RP. 1977. Coadaptations of the Clark's Nutcracker and the piñon pine for efficient seed harvest and dispersal. *Ecol. Monogr.* 47:89–111

Vander Wall SB, Joyner JW. 1998. Recaching of Jeffrey pine (*Pinus jeffreyi*) seeds by yellow pine chipmunks (*Tamias amoenus*): potential effects on plant reproductive success. *Can. J. Zool.* 76:154–62

van Schaik CJ, Terborgh JW, Wright SJ. 1993. The phenology of tropical forests: adaptive significance and consequences for primary consumers. *Annu. Rev. Ecol. Syst.* 24:353–77

Waller DM. 1979. Models of mast fruiting in trees. *J. Theor. Biol.* 80:223–32

Webb CJ, Kelly D. 1993. The reproductive biology of the New Zealand flora. *Trends Ecol. Evol.* 8:442–47

Willson MF. 1986. On the costs of reproduction in plants: *Acer negundo. Am. Midl. Nat.* 115:204–7

Wilson PR, Karl BJ, Toft RJ, Beggs JR, Taylor RH. 1998. The role of introduced predators and competitors in the decline of kaka (*Nestor meridionalis*) populations in New Zealand. *Biol. Conserv.* 83:175–85

Wolff JO. 1996. Population fluctuations of mast-eating rodents are correlated with production of acorns. *J. Mammal.* 77:850–56

Wright SJ, Carrasco C, Calderon O, Paton S. 1999. The El Niño Southern Oscillation, variable fruit production, and famine in a tropical forest. *Ecology* 80:1632–47

Wright SJ, van Schaik CP. 1994. Light and the phenology of tropical trees. *Am. Nat.* 143:192–99

APPENDIX A: SOURCES OF DATASETS

See http://www.annualreviews.org/MastingAppendix.html

Annu. Rev. Ecol. Syst. 2002. 33:449–73
doi: 10.1146/annurev.ecolsys.33.010802.150515

DISTURBANCE TO MARINE BENTHIC HABITATS BY TRAWLING AND DREDGING: Implications for Marine Biodiversity

Simon F. Thrush[1] and Paul K. Dayton[2]

[1]National Institute of Water and Atmospheric Research, PO Box 11-115, Hamilton, New Zealand; email: s.thrush@niwa.cri.nz
[2]Scripps Institution of Oceanography, University of California, San Diego, La Jolla, California 92093-0201; email: pdayton@ucsd.edu

Key Words fishing, resilience, habitat structure, bioturbation, benthic-pelagic coupling, biocomplexity

■ Abstract   The direct effects of marine habitat disturbance by commercial fishing have been well documented. However, the potential ramifications to the ecological function of seafloor communities and ecosystems have yet to be considered. Soft-sediment organisms create much of their habitat's structure and also have crucial roles in many population, community, and ecosystem processes. Many of these roles are filled by species that are sensitive to habitat disturbance. Functional extinction refers to the situation in which species become so rare that they do not fulfill the ecosystem roles that have evolved in the system. This loss to the ecosystem occurs when there are restrictions in the size, density, and distribution of organisms that threaten the biodiversity, resilience, or provision of ecosystem services. Once the functionally important components of an ecosystem are missing, it is extremely difficult to identify and understand ecological thresholds. The extent and intensity of human disturbance to oceanic ecosystems is a significant threat to both structural and functional biodiversity and in many cases this has virtually eliminated natural systems that might serve as baselines to evaluate these impacts.

INTRODUCTION

The marine biota is remarkably diverse. There are well over 50 phyla and only one is strictly terrestrial; all the rest have marine representatives. Interestingly, all these phyla had differentiated by the dawn of the Cambrian, almost 600 million years ago, and all evolved in the sea. Since that time the sea has been frozen, experienced extensive anaerobic conditions, been blasted by meteorites, and undergone extensive variations in sea level. Also during this time the continental shelves have been fragmented, reshuffled, and coalesced in such a way that the biotic communities have been exposed to a wide array of environmental conditions. The present diverse

0066-4162/02/1215-0449$14.00

biota reflects the combination of historical events and present physical, chemical, and biological dynamics. We review structural patterns and functional processes important to the perseverance of this ecosystem and discuss the impact of fishing on these patterns and processes.

Whereas great attention has been paid to the decline in species diversity in terrestrial ecosystems, it is apparent that there have also been substantial changes in diversity in aquatic systems, albeit changes that may not be so readily detected. A common perception of marine seafloor biodiversity reflects a disproportionate interest in hard bottoms such as coral reefs, kelp forests, and the rocky intertidal. This bias is understandable because these hard-bottom communities lend themselves to terrestrial comparisons and ecological studies. However, about 70% of the earth's seafloor is composed of marine soft sediments (Wilson 1991, Snelgrove 1999). These soft-sediment habitats can be highly heterogeneous owing to interactions between broad-scale factors (e.g., hydrodynamic and nutrient regimes) and smaller-scale physical and biological features; nonetheless, the apparent three-dimensional habitat structure imposed by this heterogeneity may not be as obvious as that observed on hard-bottom habitats. Although these habitats do not always appear as highly structured as some terrestrial or marine reef habitats, they do support extremely high species diversity (Etter & Grassle 1992, Grassle & Maciolek 1992, Coleman et al. 1997, Gray et al. 1997, Snelgrove 1999). In fact, the organisms that inhabit the sediments create much of the structure in soft-sediment habitats, ranging from the micro-scale changes around individual animal burrows to the formation of extensive biogenic reefs.

As well as adding substantively to the variety of species found on earth, soft-sediment marine organisms have functional roles crucial to many ecosystem processes. The provision of protein for human consumption and the ecosystems that sustain fisheries are clear examples of the products and functions of marine ecosystems that benefit humankind. Other processes in which marine benthos play important roles include their influence on sediment stability, water column turbidity, nutrient and carbon processing, and contaminant sequestering, as well as the provision of pharmaceuticals and nutraceuticals and recreational and amenity values.

There is now good evidence that commercial fishing has a profound effect on marine ecosystems. Although there is a long history of concern about the environmental effects of fishing (Graham 1953, de Groot 1984), it is really only in the past decade or two that ecological research efforts have focused in this arena. In turn this focus has spawned a number of review articles and books that summarize and synthesize the environmental effects of fishing (Dayton et al. 1995, Auster & Langton 1999, Jennings & Kaiser 1998, Watling & Norse 1998, Hall 1999, Kaiser & de Groot 2000). This information has informed the debate over fisheries management and marine conservation, but it also highlights both the challenges and opportunities to test our current understanding of interactions between broad-scale habitat disturbance to seafloor communities and the functioning of benthic ecosystems.

In this review we place studies of the environmental effects of fishing into the context of direct and indirect effects on marine biodiversity. We consider biodiversity to have both structural and functional components. The distribution and abundance patterns of landscapes, habitats, communities, populations, and genotypes form the structural component of biodiversity. Functional components involve mechanisms that drive interactions between species themselves and between them and other components of the environment, as well as other processes generating fluxes of energy and matter. We consider impacts on community structure and physical changes in habitat structure along with functional changes to seafloor ecosystems (benthic-pelagic coupling, nutrient recycling, and biogeochemical processes). Our aim is not to review the literature that addresses the issues of fisheries, marine biodiversity, and the spatial and temporal scales of ecosystem resilience, but to draw together ecological processes and fishing impacts to focus attention on new and integrative research directions.

Scale-Dependant Disturbance

Disturbance regimes play a key role in influencing biodiversity (Connell 1977, Huston 1994). In marine benthic habitats small-scale natural disturbance plays an important role in influencing communities by generating patchiness (Dayton 1994, Hall et al. 1994, Sousa 1984). Many of the small-scale disturbances that impact benthic communities and generate heterogeneity result from the biological activities of organisms that live in or feed on the seafloor. The spatial heterogeneity created by local disturbance events can account for resource patchiness (Thistle 1981, Van Blaricom 1982), and ubiquity of opportunistic species in soft-sediment habitats. Such heterogeneity is an important component of the functioning of ecological systems (Kolasa & Pickett 1991, Legendre 1993, Giller et al. 1994) and has implications for the maintenance of diversity and stability at the population, community, and ecosystem levels (e.g., De Angelis & Waterhouse 1987, Pimm 1991, Loehle & Li 1996).

The fact that natural disturbance is important to soft-sediment communities has led to the suggestion that fishing disturbance can positively effect biodiversity. This application of Connell's (1977) intermediate disturbance hypothesis is not appropriate because this hypothesis is predicated on disturbance as a means of reducing resource monopolization such that diversity is enhanced. Direct competition for food or space, however, has been difficult to demonstrate as an important process in soft sediments, especially over broad spatial scales (Olafsson et al. 1994, Peterson 1979, Wilson 1991). Theoretical consideration of the intermediate disturbance hypothesis demonstrates that the effects of disturbance on multitrophic level systems can, in many situations, have no effect on the coexistence of competitors, as necessitated by the hypothesis, or may even cause a monotonic decline in diversity (Wootton 1998). Furthermore, the intermediate disturbance hypothesis has not been adequately tested over broad spatial scales relevant to fishing disturbance. Thus its application across species, community and habitat types, and over various scales of disturbance and recovery is unfounded.

Spatial heterogeneity in community structure is related to the spatial extent and/or the frequency of disturbance events; for disturbance to create patchiness it must be small relative to the colonization potential of the benthic community, but not so small as to enable the adjacent assemblage to quickly infill the disturbed patch. This concept is encapsulated in a simple ratio-based model of the effect of disturbance on landscapes (Turner et al. 1993). The temporal dimension is considered by the ratio of disturbance interval (time between events) to recovery time, and the spatial dimension is considered by the ratio of size of the disturbed area to size of the habitat. The model simplifies many of the complexities of disturbance-recovery dynamics and the potential for recovery processes to change with scale in a nonlinear fashion. Nevertheless, consideration of these ratios indicates disturbance regimes that, through their frequency, extent, or intensity, could result in catastrophic change across the seafloor landscape. Even such a simple model emphasizes the need to understand the scales of mobility and the processes affecting successful establishment and growth of potential colonists. Typically in soft sediments a wide array of species and life stages are involved in recovery processes within a disturbed patch (Zajac et al. 1998, Zajac 1999, Thrush & Whitlatch 2001, Whitlatch et al. 2002). The ratio model implies that significant threats to the integrity and resilience of marine benthic communities arise when the rate of human-induced change exceeds the rate at which nature can respond. This is particularly likely to occur where habitat structure and heterogeneity are reduced and large areas of habitat have been modified. Slow-growing and -reproducing species will be strongly affected, vastly reducing the potential for such species to reestablish themselves or colonize new areas. The homogenization of habitats and the loss of small-scale patchiness result in the risk of the loss of ecological function and natural heritage values in marine ecosystems.

Fishing as a Disturbance Agent on the Seafloor

Many types of trawls, dredges, and traps are dragged over or sit on the seafloor (Jennings & Kaiser 1998). The type of physical impact the fishing gear has on the seafloor depends on its mass, degree of contact with the seafloor, and the speed at which it is dragged. The way the gear is designed and operated also influences how it interacts with the seafloor and how many species other than the target species the gear removes from the seafloor or damages in situ (i.e., by-catch). Not all types of gear are used in all locations, and the impact of the gear depends on the habitat in which it is used.

Unfortunately there are limited data on the location and frequency of the area of the seafloor swept by fishing gear. The data available usually are based on broad-scale fisheries management units and not necessarily related to the spatial variation in seafloor habitats or biodiversity. For example, Churchill (1989) summarized trawling effort off the Middle Atlantic Bight, an area of intensive fishing pressure. The range of effort was quite variable along the coast because fishers do not work where there are no fish, but some areas off southern New England were on average exposed to 200% effort. Another typical fishery in northern California trawled across the same section of seafloor an average of 1.5 times per year, with selected areas trawled as often as 3 times per year (Friedlander et al. 1999). For a scampi fishery on the continental slope off New Zealand (200–600 m water depth), Cryer et al. (2002) calculated that on average about 2100 km^2 of the seafloor was swept each year by trawlers. The statistics suggest that within the study area trawlers swept about 20% of the upper continental slope each year, although about 80% of all scampi trawls were made in an area of about 1200 km^2. In some areas the extent and frequency of disturbance can be extreme, Pitcher et al. (2000) identifies one harbor in Hong Kong where every square meter of the seafloor was trawled three times a day. The ecological intensity of response is also determined by the resident species; even low-intensity disturbance can significantly affect sensitive species (Jenkins et al. 2001). The spatial distribution of fishing effort on the seafloor is patchy, reflecting the relative availability of the target species. In some cases refining the scale of measurement reveals higher levels of aggregation of fishing effort (e.g., Pitcher et al. 2000). Actually, we really do not know the global extent of disturbance to the seafloor by fishing. However, the magnitude of exploitation of global fishery resources provides some important clues as to the general extent of disturbance, with about 25–30% of the world's fishery populations overexploited or depleted and a further 40% considered heavily to fully exploited (Pauly et al. 1998). Often the scales of measurement of fishing effort (e.g., tens to hundreds of square kilometers) are difficult to match with ecological effects, as they do not match well with the scales of variability in seafloor ecological communities.

Evaluating the Direct Effects of Habitat Disturbance by Fishing

Many studies have been conducted to assess the direct effects of habitat disturbance by trawling or dredging on benthic communities (Table 1). These studies have been conducted in a variety of habitats and locations, generally in shallow water. We have reviewed a large number of these studies to offer examples from a variety of habitats and locations, rather than attempt a complete list. Our aim is to provide a brief summary of the range of effects observed. We hope this list, as well as previously mentioned reviews, offer the reader an entrée into this literature. There are a number of important issues to consider when summarizing such a diverse array of studies because they encompass a range of intensities and spatial and temporal scales of fishing disturbance. Study designs and assessment approaches are also widely different. We have summarized statistically significant or nonsignificant effects described in the individual papers but refer the reader to Loftis et al. (1991) and Nelder (1999) for comments on "significance." We recommend that readers study papers of interest in detail to assess for themselves the magnitude of ecological effects.

Marine benthic ecosystems are often challenging systems to study, and precise data are rare. Furthermore, the interpretation of results is frequently difficult. For

TABLE 1 Summary of effects on subtidal benthic communities reported in recent assessments of fishing impacts

Effects	Habitats/depths/location	Source
Removal of biogenic and physical habitat structure	Sand, 30 m depth; Northwest Atlantic	Auster et al. 1996
Decreased diversity in trawled plots	Mud, 75 m depth and heavily trawled; 35 m less frequently trawled; Irish Sea	Ball et al. 2000
Decreased density of common echinoderms, polychaetes, and molluscs	Sand, 30 m depth; North Sea	Bergman & Hup 1992
Direct mortality of 5–60% for species following single passage of trawl	Sand/mud, 10–45 m depth; North Sea	Bergman & van Santbrink 2000
Decreased number of organisms, biomass, species richness, species diversity, and biogenic habitat structure	Gravel pavement, 40–80 m depth; Georges Bank; Northwest Atlantic	Collie et al. 1997
6 of 10 common species decreased in abundance	Sand/coarse silt, 10–20 m depth; Port Phillip Bay, Vic., Australia	Currie & Parry 1996
No consistent trends in epifauna or infauna; high site-to-site variability	Sand/coarse silt, 10–20 m depth; Port Phillip Bay, Vic., Australia	Currie & Parry 1999
Decreased abundance of large epifauna and infaunal species abundance	Sand, 10 m depth; Loch Ewe, Scotland	Eleftheriou & Robertson 1992
Higher densities of epifauna in lightly trawled area; higher densities of predator/scavenger worms in heavily trawled area	Sand, 180 m depth; central California	Engel & Kvitek 1998
Large epifauna removed and damaged; boulders displaced	Pebble/cobble/boulder, 200 m depth; Gulf of Alaska	Freese et al. 1999
Temporal trends in community composition differentiate under heavy fishing pressure	Mud, 80 m depth fished and 50 m unfished sites; North Sea	Frid et al. 2000
No detectable changes in macrobenthic fauna	Sand, 10 m depth; Botany Bay, NSW, Australia	Gibbs et al. 1980
Overall decreases in biomass and abundance, but site and time interaction terms make detection of effect difficult	Mud, 73–96 m depth; Gullmarsfjord, Sweden	Hansson et al. 2000
70% reduction of mearl[a] habitat over 4 years	Sand/mud, 6–15 m depth; Firth of Clyde, Scotland	Hall-Spencer & Moore 2000

(Continued)

TABLE 1   (*Continued*)

Effects	Habitats/depths/location	Source
No significant effects on biomass or production in area of low fishing pressure; under high fishing pressure, significant decrease in biomass and production	High impact site muddy sand 55–75 m depth, low impact site sand 40–65 m; North Sea	Jennings et al. 2001a
Decrease in infaunal and epifaunal biomass, particularly bivalves and burrowing urchins, only detected at high impact site	High impact site muddy sand 55–75 m depth, low impact site sandy 40–65 m; North Sea	Jennings et al. 2001b
Decrease in density of epifauna and diversity in stable sand habitat; no effects detected in unstable sand habitat	2 habitats: one stable sand with rich epifauna, the other mobile sand, 26–34 m depth; Anglesey Bay, Irish Sea	Kaiser & Spencer 1996
Slight changes in community composition in stable habitat; no detectable effects on a number of species or diversity indices in either habitat	2 habitats: one stable sand with rich epifauna, the other mobile sand, 26–34 m depth; Anglesey Bay, Irish Sea	Kaiser et al. 1998
Larger individuals and increased density of epifauna in unfished area	Sand, 18–69 m depth; English Channel	Kaiser et al. 1999
Loss of sessile, emergent, high biomass species, increase in small-bodied infauna	Gravel/sand; Isle of Man, Irish Sea[b]	Kaiser et al. 2000
No effect detected	Sand, 120–146 m depth; Grand Banks of Newfoundland	Kenchington et al. 2001
Trawl reduced density of large epifauna about 15% on each pass; trawl flown 15 cm above seafloor had no detectable impact on large epifauna	50 m depth[c]; north-west shelf, Australia	Moran & Stephenson 2000
Higher diversity in unfished area; sedentary macrofauna more abundant in unfished sites; mixed response by motile species and infaunal bivalves	Sand, 44–53 m depth; eastern Bering Sea	McConnaughey et al. 2000
Trawls typically removed 5–20% of large benthic fauna	Sand[b], Great Barrier Reef Region, Australia	Pitcher et al. 2000
Short-term decreases in biomass and abundance of macrofauna; number of taxa showed no immediate effect but increased in trawled area after 7 days	Sand, 24 m depth; Adriatic Sea	Pranovi et al. 2000

(*Continued*)

TABLE 1 (*Continued*)

Effects	Habitats/depths/location	Source
Decrease in epifaunal biomass following disturbance; no significant impacts on dominant molluscs	Sand, 120–146 m depth; Grand Banks of Newfoundland	Prena et al. 1999
Decrease in species richness and diversity	Mud, 30–40 m depth; Catalan coast	Sanchez et al. 2000
Decrease in small-scale heterogeneity of sediment texture after trawling	Sand, 120–146 m depth; Grand Banks of Newfoundland	Schwinghamer et al. 1996
Higher numbers of epifauna and diversity, abundance, and biomass of macrofauna outside trawled area	Mud, 200 m depth; Crete	Smith et al. 2000
Changes in community structure, decreased density of common bivalves and polychaetes, increased density of nemerteans	Mud, 60 m depth; Penobscot Bay, Maine	Sparks-McConkey & Watling 2001
Density decreased, effects on number of taxa detected at one site only	Sand, 24 m depth. Mercury Bay, New Zealand	Thrush et al. 1995
Fishing decreased density of burrowing urchins, long-lived surface dwellers, and diversity and increased density of deposit feeders and small opportunists	Varied sediments 1–48% mud, 17–35 m depth; Hauraki Gulf, N.Z.	Thrush et al. 1998
Numbers of species, individuals, and various diversity indices increased in fished area	Mud, 30–35 m depth; Gareloch, Scotland	Tuck et al. 1998
Barrel sponges (*Cliona*) significantly reduced in abundance but recovered in 12 months	Low relief hard-bottom habitat, 20 m depth; Georgia	Van Dolah et al. 1987
No detectable effect of trawling	Sand, Port Royal (8 m depth) and St. Helena (30 m depth) sounds, S. Carolina	Van Dolah et al. 1991
Species diversity, richness, total number of species decreased with increased fishing effort	Sand, 20–67 m depth; north Irish Sea	Veale et al. 2000
Decrease in numerical dominants and changes in sediment food quality	Silty sand 15 m depth. Damariscotta River, Maine	Watling et al. 2001

[a]sediments with a surface layer of slow-growing unattached coralline algae.

[b]depth not given.

[c]sediment type not given.

example, Pitcher et al. (2000) documented the removal of 7 tonnes of epifaunal biomass during experimental trawling, but were unable to detect significant changes by surveying the density of epifauna on the seafloor. Whereas the direct effect of such an impact on benthic communities appears obvious, its magnitude has been difficult to evaluate with regard to other components of the ecosystem. There is often a failure to detect the effect of experimental fishing disturbance in areas exposed to extreme natural disturbance to the seafloor (e.g., storms or very strong tidal flows) (Hall et al. 1990, Brylinsky et al. 1994, Kaiser & Spencer 1996, DeAlteris et al. 1999). Effects are not always consistent across sites, even within studies. Given the variety of experimental designs and habitats studied, these variations in response are far from surprising. Nevertheless these studies emphasize a number of changes in benthic communities including loss of habitat-structuring species, changes in species richness, and loss of large and long-lived organisms.

One of the most conspicuous long-term physical effects of bottom fishing is the homogenization of the substratum and reduced species diversity (Veale et al. 2000). Thus, apart from all the usual difficulties of study design, the history of fishing disturbance can make it impossible to control experimental studies. Some European and Adriatic waters have a long history of fishing by bottom trawling (de Groot 1984, Kaiser & de Groot 2000, Pranovi et al. 2000). Frid et al. (2000) considered reports from the North Sea going back to the 1920s and concluded that fishing practices have changed benthic communities in some parts of the North Sea; in other areas fishing impacts could not be evaluated without a longer time series of data. Aronson (1989, 1990) argued that overfishing has virtually eliminated the evolutionarily new teleost predators, resulting in a rebirth of the Mesozoic-like system dominated by echinoderms and crustacea. Recent analyses of fish predation in the North Sea provide some support for this view (Frid et al. 1999). There is also a long history of transformation of marine coastal ecosystems in the western Atlantic (Steele & Schumacher 2000, Jackson 2001) and eastern Pacific (Dayton & Tegner 1984, Dayton et al. 1998). There are few, if any, unfished habitats with economically exploitable stocks outside the Antarctic region (Dayton et al. 2000). Human disturbances that exceed the rate of natural recovery dynamics have been underway for decades and possibly centuries. Some marine organisms have been driven to extinction by human activities (e.g., the Atlantic gray whale, the great auk). Others are probably close to extinction, for example, the Irish ray (Brander 1981), the barndoor skate (Casey & Myers 1998), and the white abalone (Tegner et al. 1996). A recent review listed 82 species and subspecies of endangered fish in the United States (Musick et al. 2000). Over time, repeated intense disturbance will select for species with appropriate facultative responses, and communities are likely to become dominated by juvenile stages, mobile species, and rapid colonists. Such broad-scale descriptions point to the problem of identifying effects in ecological systems that are potentially already affected. The important point is that the potential for both direct and indirect effects on biodiversity cannot be ignored because of variability in ecological response across such a diverse array of studies. As always, manipulative experiments are useful but difficult to control properly and limited in scope.

The broader-scale implications of fishing impacts have been inferred from benthic surveys and time-series data (Reise 1982, Holme 1983, Langton & Robinson 1990, Cranfield et al. 1999, Kaiser et al. 2000). One valuable approach to integrating experimental results into broader-scale patterns is to develop iterative procedures and test a priori hypotheses with data collected over broad scales. This process can provide good evidence of large-scale change (Thrush et al. 1998). The use of effort data in the rapidly expanding trawl fisheries of the eastern Bering Sea enabled McConnaughey et al. (2000) to contrast areas with different exposure to fishing disturbance and investigate the long-term consequences for benthic communities. They found that the numbers of sedentary epifauna (animals such as anemones, soft corals, sponges, bryozoans, etc.) and diversity and the niche breadth of sedentary taxa decreased with fishing. Mixed patterns of response were apparent for mobile epifauna and infaunal bivalves, suggesting species-specific responses based on life-history characteristics. It is important to note that the clear changes recorded in this study were documented in an area with a high potential for storm-generated wave disturbance, emphasizing the value of carefully designed and analyzed assessments.

The Potential for Functional Changes in Biodiversity

The studies described above provide evidence for direct changes in response to habitat disturbance by fishing, but we must also consider the potential for changes in the functional roles played by organisms, communities, and ecosystems. Usually, as density declines the size of the individuals and their spatial distribution also change; thus, although not biologically extinct, they may be functionally so, being unable to fulfill their natural roles in community and ecosystem function (Dayton et al. 1998).

In soft-sediment habitats the creation of small-scale habitat structure by biogenic features can play key roles in influencing diversity and resilience. Some benthic fishes, such as rays, have an important influence on small-scale habitat structure (Van Blaricom 1982, Levin 1984, Thrush et al. 1994). Organisms that live at the sediment surface or create mounds, tubes, and burrows within it also provide habitat structure and frequently have important roles in the sequestering and recycling processes essential to ecosystem function. Studies listed in Table 1 provide evidence of direct effects on the density and distribution of such organisms; in some cases they are the most susceptible to habitat disturbance by dredging and bottom trawling. As yet, however, there have been no direct assessments of the implications of the loss of these functionally important species to ecosystem function and resilience.

Alteration to marine food webs through changes in the abundance and size distribution of fish populations could have important consequences for benthic communities. Many types of fish prey upon and disturb the seafloor; all rays and some sharks and bony fishes that make up many important demersal fisheries (e.g., sparids, scorpaenids, labrids, gadoids, pleronectids) are important benthic predators, although their wider ecological role is rarely studied (but see Sala & Ballesteros 1997). Many fishes have life-history characteristics that make them extremely vulnerable to over exploitation, thus highlighting the potential for the role of these animals to be diminished. Jennings et al. (1999) examined long-term trends in the abundance of North Sea demersal fishes and demonstrated that those species that decreased in abundance compared with their nearest relative matured later, grew larger, and had a lower potential for rapid population increase. Many important demersal fish stocks (gadoids, sparids, pleronectids, and scorpaenids) appear to show limited recoverability after over exploitation (Hutchings 2000). We do not have a good understanding of the role of broader community- and ecosystem-level processes in the resilience of fish stocks.

Declining density of a species is usually associated with reductions in both the geographic distribution and the size of individuals. Steele & Schumacher (2000) discussed the implications for marine food webs of historic fish stocks in the Northwest Atlantic possibly being an order of magnitude higher than stocks in the last half of the twentieth century. Density changes of this magnitude could result in a profoundly altered ecosystem. For example, in the Gulf of Maine, the removal of top fish predators through intensive fishing apparently released other predators such as crabs and starfish, thus changing the benthic communities (Witman & Sebens 1992, Steneck 1997). Frid et al. (1999) also provide an example that links changes in the abundance of fish to changes in benthos: Changes in fish biomass in the North Sea appear to have resulted in changes in the taxonomic composition of benthos consumed by fish and an overall increase in predation pressure on benthos. However, such effects are difficult to identify without extensive study; tracking effects through marine foodwebs is difficult because of the inherently complex interactions and weak and indirect effects (Micheli 1999).

Fishing can also directly alter the physical habitat by influencing sediment particle size, resuspension regimes, and biogeochemical flux rates (Churchill 1989, Currie & Parry 1999, Mayer et al. 1991, Palanques et al. 2001). Sediment quality is important because of the intimate relationship between particle size and benthic community structure and function (Gray 1974, Rhoads 1974, Whitlatch 1980, Snelgrove & Butman 1994). One study found significant declines in some organisms (70% for scallops and 20–30% for burrowing anemones and fan worms) owing to a scallop-fishing–induced shift in sediment (organic-rich silty sand to sandy gravel with shell hash) (Langton & Robinson 1990). Caddy (1973) also documented the smothering of suspension feeders as a result of sediment resuspended by fishing. Other effects include modifications to microbial activity (Mayer et al. 1991, Watling et al. 2001), resuspension of contaminants, and increases in benthic/pelagic nutrient flux (Krost 1990). Trawling-induced resuspension of sediments in the Gulf of Maine has been hypothesized to have changed the nature of nutrient supply from the seafloor with potentially ecosystem-wide consequences on phytoplankton growth (Pilskaln et al. 1998).

Habitat-Structuring Organisms, Functions, and Biodiversity

Interactions between hydrodynamic conditions and the benthic habitat drive many of the important processes occurring at the sediment-water interface. The increased drag created by structures protruding into the near-bed water flow and active feeding currents generated by suspension feeders influence localized rates of erosion and deposition (Eckman & Nowell 1984, Frechette et al. 1989, Shimeta & Jumars 1991, Dame 1993, Wildish & Kristmanson 1997). Common benthic suspension feeders include a diverse array of corals, bryozoans, sponges, gorgonians, seapens, echinoderms, brachiopods, and bivalves. Patches of these organisms can further modify hydrodynamics over a wide range of spatial scales, significantly influencing both the vertical and horizontal flux of food and larvae at the seafloor. Both the size of the organisms and the patch are important factors influencing these interactions (Green et al. 1998, Nikora et al. 2002). Bivalves expend high levels of energy drawing water over their gills to feed (Rhodes & Thompson 1993). Thus, suspension-feeding bivalves are capable of actively removing 60–90% of the suspended matter from the horizontal particle flux (Loo & Rosenberg 1989). These bivalves package any particles that are unsuitable for ingestion in mucous and eject them as pseudofeces, thus appreciably influencing the rate of particle deposition to the seafloor (Graf & Rosenberg 1997). These processes create variation in seafloor ecosystems and add to their biodiversity (Cummings et al. 2001, Norkko et al. 2001).

Organisms that live at the sediment surface, as well as the small-scale disturbances created by benthic-feeding predators, can also increase the three-dimensional structure of the habitat. For example, small heterogeneities in sediment topography (e.g., tubes and burrows) and even sparsely distributed epifauna characterize most soft-sediment habitats. Such structure at the sediment-water interface, along with variations in sediment particle size, is positively related to macrobenthic diversity (Thrush et al. 2001). Spatially these small-scale features are often highly variable (e.g., Schneider et al. 1987) and can be important to commercially valuable species (Auster et al. 1995). The shear vastness of the area covered by such habitats results in an important role for these small-scale features in biogeochemical processes and species and habitat diversity, and it is these elements that are most susceptible to habitat disturbance by dredging and bottom trawling.

Highly structured habitats can provide refuges for both predators and prey. Many studies show significant variations in predator-prey interactions associated with variations in habitat complexity (e.g., Woodin 1978, Ruiz et al. 1993, Irlandi 1994, Skilleter 1994). Habitat structure influences predation rates on fish, particularly juvenile life stages (e.g., Heck & Thoman 1981, Persson & Eklov 1995, Rooker et al. 1998). Topographic complexity can have a significant and positive influence on the growth and survivorship of juvenile life stages of commercially valuable species, often as a result of reduced risk of predation (Tupper & Boutilier 1995, Lindholm et al. 1999). On the Australian northwest shelf, Sainsbury (1988) showed a decrease in the number and variety of epifauna, particularly sponges, collected over time as by-catch. This reduction was associated with shifts in the fishery from high- to low-value species. A probable explanation was a positive role for epifauna in affecting the survivorship of the commercially valuable fish species. In many cases the habitats damaged by trawling probably constitute very important nursery areas for many species, often including some of the target species of fisheries (Turner et al. 1999).

The importance of habitat-structuring organisms is not restricted to shallow water because shelf-break and seamount habitats can exhibit marvelous levels of habitat complexity generated by biogenic structure. Even the deep-sea basins once considered constant and uniform exhibit high levels of both local biogenic complexity (e.g., Jumars & Eckman 1983, Levin & Gooday 1992) and regional diversity (Levin et al. 2001). Improved technologies and the demand for fish are opening up deep-water habitats to exploitation. Cryer et al. (2002) provide empirical evidence of the large-scale effects of trawling on a deep-water soft-sediment system by demonstrating substantive decreases in the diversity of large benthic invertebrates associated with a continental slope (a depth of 200–600-m) scampi (*Metanephrops challengeri*) fishery. This result emphasizes that the impacts on seafloor communities that have been more readily documented in shallower water are also occurring in deeper water. In these environments effects on biodiversity are likely to be exacerbated because deep-sea communities are generally characterized by life-history adaptations such as slow growth, extreme longevity, delayed age of maturation, and low natural adult mortality, and they exhibit slow rates of recovery from disturbance.

Deep-water corals occur in the upper bathyl zone throughout the world and are under threat from human activity, particularly fishing and oil exploration (Roberts et al. 2000, Rogers 1999). The biology of most of these deep-water coral species is unknown, but they appear to have exceptionally slow growth and low reproductive rates, with individual colonies being hundreds to thousands of years old (e.g., Druffel et al. 1995). These thicket-forming corals are often associated with a diverse fauna, and levels of diversity appear to be similar to those of shallow water tropical coral reefs. Squires (1965) reports the first detection of a deep-water coral structure in the Pacific, at a depth of 320 m on the Campbell Plateau. These coral structures generated about 40-m-high relief on the seafloor. *Lophelia pertusa* is a deep-water coral that occurs in discrete patches hundreds of meters to several kilometers in diameter, and up to 45 m high. Off Norway and the Faeroe Islands *Lophelia* reefs have several hundred species in association, and with the exception of small areas off Norway, most have been heavily damaged (Roberts 1997).

Seamounts too have been the focus of intensive deep-water fisheries. In the southern hemisphere these fisheries were usually initiated to capture orange roughy (*Hoplostethus atlanticus*). When the New Zealand fishery targeting spawning aggregations of orange roughy began, the trawls brought up a great deal of benthic by-catch, but these levels decreased with repeated trawling (Probert et al. 1997). For the orange roughy fishery on seamounts off Tasmania, Koslow et al. (2001)

report tonnes of coralline material brought to the surface in a single trawl when fishing a new area. Surveys confirmed characterizations of changes in benthic communities based on by-catch (de Forges et al. 2000, Koslow et al. 2001). These authors report at least 299 species from a single short seamount cruise near Tasmania; 24–43% of these were new to science. The benthic biomass from fished seamounts was 83% less than from lightly fished or unfished habitats.

Fluxes, Ecosystem Effects, and Biodiversity

Sediments play important roles in transformation and exchange processes of organic matter and nutrients. For example, sediments on marine continental shelves, while occupying only 7% of the area of the planet covered by marine sediments, are responsible for 52% of global organic matter mineralization. Slope sediments (i.e., 200–2000-m depth, 9% area) remineralize another 30% (Middleburg et al. 1997). These disproportionately large contributions to organic matter mineralization reflect the importance of biological activity within marine sediments influencing solute and particle transport. By enhancing the transport of labile particulate organic carbon to subsurface layers within the sediment, organisms stimulate anaerobic degradation and so affect the form and rate at which metabolites are returned to the water column (Herman et al. 1999). The seas above continental shelf environments typically receive one third to half their nutrients for primary production from sediment (Pilskaln et al. 1998). These nutrients are derived from organic matter decay and nutrient remineralization within the sediments, followed by molecular diffusion or biological irrigation back into the water column. The process of sediment manipulation by resident animals (i.e., bioturbation) is the dominant mode of sediment transport in the upper centimeters of oceanic sediments (Middleburg et al. 1997). Solute pumping, burrowing, and feeding increase the area of the sediment-water interface (Aller 1982). Bioturbation affects the stability and composition of marine sediments and influences their role as geochemical sources and sinks (McCall & Tevesz 1982, Marinelli 1994, Bird et al. 1999). Thayer (1983) extensively reviewed estimates of the rate of bioturbation for a wide variety of marine organisms, with rates of sediment reworking ranging from 5×10^{-5} to 2.1×10^6 cm^3 day^{-1} per individual and depths to which sediment was reworked ranging from 0.1 to 400 cm. Larger organisms play a particularly important role in influencing sediment-reworking rates (Thayer 1983, Sandnes et al. 2000). Typically animals increase the particle exchange between water and sediment by a factor of 2 to 10 (Graf 1999).

Direct disturbance of the seafloor enhances the upward flux of nutrients by releasing pore-water nutrients as a pulse, rather than a more steady release controlled by bioturbation (Pilskaln et al. 1998). Fanning et al. (1982) estimated that a storm that imposed sufficient energy on the seafloor to resuspend the top 1 mm of sediment could intermittently augment overlying productivity by as much as 100–200%. This depth of sediment disturbance is much less than what occurs as a result of many types of trawling and dredging. Dredges usually disturb the top 2–6 cm of sediment, while the doors that hold open trawl nets in the water can plough furrows from 0.2–2 m wide and up to 30 cm deep in the sediment (e.g., Caddy 1973, Jennings & Kaiser 1998, Krost 1990). However, it is not appropriate to equate storm disturbances with fishing because the latter may involve a much higher intensity of disturbance, although its frequency and extent will be highly location dependent.

The rate and efficiency of bioturbation processes are determined by interactions between organisms and between the organisms and their environment. The degree of particle flux enhancement varies with faunal composition and density (Cadee 1979, Thayer 1983) in conjunction with organic carbon flux to the sediment (Legeleux et al. 1994). Interactions between bioturbation and mineralization processes in sediments are highly nonlinear and are characterized by the presence of strong feedback loops between deposit feeders, their food, and their chemical environment (Herman et al. 1999). Variation in large burrow structures and animal activity can result in markedly different biogeochemical fluxes, in terms of both rates and chemicals (Hughes et al. 2000). As well as influencing water column production, bioturbation can also affect the growth of benthic species that use this resource (Weinberg & Whitlatch 1983). Deposit-feeding typically controls the biological mixing of near-shore and, probably, deep-sea sediments. Soetaert et al. (1996) measured and modeled the total flux of ^{210}Pb entering the sediment (used as a marker of particle transport) along a transect from 208–4500 m over the Goban Spur in the northeast Atlantic. Their analysis showed that between 8 and 86% of the particle flux was derived from nonlocal exchange processes (i.e., active pumping/flow through burrows), with these processes most important in shallower waters where trawling is most intense.

Apart from burrowing and actively pumping water and particles through burrows, animals can also influence fluxes by modifying surface sediment topography, which then interacts with sediment boundary water flows. Huettel et al. (1996) demonstrated how shrimp mounds protruding from sandy sediments can alter flow patterns to increase the flux of fine particulate matter into the sediment. Small pressure gradients generated by boundary flow–topography interactions also increase the flux of oxygenated water into the pore waters of sandy sediments, thus increasing the oxic volume of the sediments and affecting biogeochemical processes (Forster et al. 1996, Ziebis et al. 1996).

From these studies it is clear that bioturbation is important in ecosystem functioning. There is some evidence that bioturbation has been a significant factor influencing the evolution and enhancement of marine biodiversity over geological time scales. Regeneration of nitrogen from the seafloor may exceed inputs from freshwater in the coastal zone (Rowe et al. 1975). The faster recycling of nutrients by increased rates of bioturbation over evolutionary time scales may have contributed to the diversity of phytoplankton and zooplankton in the Mesozoic (Thayer 1983). Martin (1996) contends that elevated nutrient levels associated with increased rates of ocean circulation, continental erosion, and bioturbation have played a key role in enhancing the productivity and diversity of marine systems over the Phanerozoic (i.e., essentially post-Precambrian).

Implications for Research and Biodiversity Management

We have tried to synthesize the effects of fishing disturbance on the biodiversity of the seafloor by reviewing studies of direct effects and discussing the functional roles of soft-sediment organisms. In doing this we have summarized immediate effects that include changes in species diversity and ecosystem processes, habitat modification, and loss of predators. Because of the high variability within and between studies, there is no single definitive study that adequately describes the range of impacts of fishing disturbance. Nevertheless, there is evidence of effects on seafloor communities that have important ramifications for ecosystem function and resilience.

Whereas gathering good data on the magnitude and extent of disturbance at appropriate spatial scales is important, from both a scientific and management perspective, we have moved beyond the question of what the immediate effect of habitat disturbance by fishing is and now need to focus on the implications of loss of structural and functional biodiversity over various space and timescales. Pitcher (2001) argues that the only hope for fisheries themselves is to move their management to a focus on ecosystem rebuilding. He contends that the goal of sustainable yield of single species in a fishery is a fundamental mistake. The potential for the change in functionally important ecosystem processes leads us to ask very broad questions and test the generality of many fundamental theories. The themes of scale, complexity, resilience, and strong coupling of physical and biological processes in marine benthic ecosystems are pervasive.

We have emphasized the functional roles of marine benthic organisms. To develop this theme further we need a much better understanding of the basic ecology of these species, how their ecosystem roles may be modified by their size, density, and spatial arrangement, and how these characteristics enable them to cope with disturbance. The integration of small-scale variation into broader patterns is important because we should expect threshold effects and nonlinearities in the multispecies biotic and environmental processes that create biodiversity. For example, differences in density and species among a functionally similar group of bioturbators can result in different effects on biodiversity (Widdicombe & Austen 1999, Widdicombe et al. 2000). The potential for different responses of macrobenthic assemblages to the presence of a large epifaunal bivalve under a number of different physical regimes and local species pools has also been demonstrated (Cummings et al. 2001). We must use natural history and environmental information to both design and interpret mechanistic studies because responses are usually scale dependent (Thrush et al. 1999). Standing back from the detail and looking for more general and abstract patterns also provides a basis for revealing emergent phenomena (Brown 1995).

Natural systems have a great deal of structure in time and space, and it is important to identify thresholds of change in this structure and the processes it influences to gauge ecosystem resilience. This means that our predictions of the ecological consequences of human activity in the marine environment require an understanding of broad-scale forcing functions as well as knowledge of the natural and life-history characteristics of individual species. The removal of small-scale heterogeneity associated with the homogenization of habitats is, by definition, loss of biodiversity. We argue this is important even over the extensive sand- and mud-flats of the seafloor that are often considered "featureless." Spatial mosaics that result from local biological disturbance events, as well as the organisms that create them, can be obliterated by intense broad-scale disturbances. There are winners and losers in the ecological response to any disturbance, but a key issue is understanding ecological heterogeneity and its role in modifying the consequences of habitat disturbance to ecosystem processes. We need to better understand the implications of natural and anthropogenic habitat fragmentation and how it relates to the intensity and frequency of disturbance. Empirical and theoretical studies addressing this issue will need to integrate biogeochemistry, hydrodynamics, and ecology. Site history and the effect of multiple stresses will be particularly important in many areas in order to address the role of the resident species assemblage and environmental context in affecting disturbance-recovery dynamics in fragmented habitats. Once the functionally important components of an ecosystem are missing, it is extremely difficult to identify and understand ecological thresholds that are violated beyond the point of recovery, at which point the anthropogenic disturbances are less obvious. Some knowledge of these issues will be necessary to address the more fundamental question, "At what point are there ecosystem thresholds beyond which recovery is unlikely?" Ecological systems can shift into alternative states as a result of the loss of ecosystem functions, and we must be able to assess the consequences of these shifts in terms of loss of diversity and ecological services (e.g., Carpenter et al. 1999, Scheffer et al. 2001).

ACKNOWLEDGMENTS

We dedicate this review to the memory of Mike Mullin and Mia Tegner for their deep concern for fisheries and related conservation issues. We thank Judi Hewitt, Vonda Cummings, and Bob Whitlatch for comments on the manuscript and Mary Powers and Dan Simberloff for their insightful reviews. FRST CO1804 supported this research.

The *Annual Review of Ecology and Systematics* is online at
http://ecolsys.annualreviews.org

LITERATURE CITED

Aller RC. 1982. The effects of macrobenthos on chemical properties of marine sediment and overlying water. See McCall & Tevesz, pp. 53–102

Aronson RB. 1989. Brittlestar beds: low preda-tion anachronisms in the British Isles. *Ecology* 70:856–65

Aronson RB. 1990. Onshore-offshore patterns of human fishing activity. *Palaios* 5:88–93

Auster PJ, Langton RW. 1999. The effects of

fishing on fish habitat. *Am. Fish Soc. Symp.* 22:150–87

Auster PJ, Malatesta RJ, Langton RW, Watling L, Valentine PC, et al. 1996. The impact of mobile fishing gear on seafloor habitats in the Gulf of Maine (Northwest Atlantic): implications for conservation of fish populations. *Rev. Fish Sci* 4:185–202

Auster PJ, Malatesta RJ, LaRosa SC. 1995. Patterns of microhabitat utilization by mobile megafauna on the southern New England (USA) continental shelf and slope. *Mar. Ecol. Prog. Ser.* 127:77–85

Ball BJ, Fox G, Munday BW. 2000. Long- and short-term consequences of a *Nephrops* trawl fishery on the benthos and environment of the Irish Sea. *ICES J. Mar. Sci.* 57:1315–20

Bergman MJN, Hup M. 1992. Direct effects of beamtrawling in macrofauna in a sandy sediment in the southern North Sea. *ICES J. Mar. Sci.* 49:5–11

Bergman MJN, vanSantbrink JW. 2000. Mortality in megafaunal benthic populations caused by trawl fisheries on the Dutch continental shelf in the North Sea in 1994. *ICES J. Mar. Sci.* 57:1321–31

Bird FL, Ford PW, Hancock GJ. 1999. Effects of burrowing macrobenthos on the flux of dissolved substances across the water-sediment interface. *Mar. Freshw. Res.* 50:523–32

Brander KM. 1981. Disappearance of the common skate, *Raja batis* from the Irish Sea. *Nature* 270:48–49

Brown JH. 1995. *Macroecology*. Chicago: Univ. Chicago Press. 269 pp.

Brylinsky M, Gibson J, Gordon DC Jr. 1994. Impacts of flounder trawls on the intertidal habitat and community of the Minas Basin, Bay of Fundy. *Can. J. Fish Aquat. Sci.* 51:650–61

Caddy JF. 1973. Underwater observations on tracks of dredges and trawls and some effects of dredging on a scallop ground. *J. Fish. Res. Bd. Can.* 30:173–80

Cadee GC. 1979. Sediment reworking by the polychaete *Heteromastus filiformis* on a tidal flat in the Dutch Wadden Sea. *Neth. J. Sea Res.* 13:441–54

Carpenter SR, Ludwig D, Brock WA. 1999. Management of eutrophication for lakes subject to potentially irreversible change. *Ecol. Appl.* 9:751–71

Casey JM, Myers RA. 1998. Near extinction of a large, widely distributed fish. *Science* 281:690–92

Churchill JH. 1989. The effect of commercial trawling on sediment resuspension and transport over the Middle Atlantic Bight Continental Shelf. *Cont. Shelf Res.* 9:841–64

Coleman N, Gason ASH, Poore GCB. 1997. High species richness in the shallow marine waters of southeast Australia. *Mar. Ecol. Prog. Ser.* 154:17–26

Collie JS, Escanero GA, Valentine PC. 1997. Effects of bottom fishing on the benthic megafauna of Georges Bank. *Mar. Ecol. Prog. Ser.* 155:159–72

Connell JH. 1977. Diversity in tropical rainforests and coral reefs. *Science* 199:1302–10

Cranfield HJ, Michael KP, Doonan IJ. 1999. Changes in the distribution of epifaunal reefs and oysters during 130 years of dredging for oysters in Foveaux Strait, southern New Zealand. *Aquat. Conserv. Mar. Freshw. Ecosyst.* 9:461–83

Cryer M, Hartill B, O'Shea S. 2002. Modification of marine benthos by trawling: towards a generalization for the deep ocean? *Ecol. Appl.* 12: In press

Cummings VJ, Thrush SF, Hewitt JE, Funnell GA. 2001. Variable effect of a large suspension-feeding bivalve on infauna: experimenting in a complex system. *Mar. Ecol. Prog. Ser.* 209:159–75

Currie DR, Parry GD. 1996. Effects of scallop dredging on a soft sediment community: a large-scale experimental study. *Mar. Ecol. Prog. Ser.* 134:131–50

Currie DR, Parry GD. 1999. Impacts and efficiency of scallop dredging on different soft substrates. *Can. J. Fish Aquat. Sci.* 56:539–50

Dame RF, ed. 1993. *Bivalve Filter Feeders in Estuarine and Coastal Ecosystem Processes*, Vol. G 33. Berlin: Springer-Verlag. 579 pp.

Dayton PK. 1994. Community landscape: scale

and stability in hard bottom marine communities. See Giller et al. 1994, pp. 289–332

Dayton PK, Sala E, Tegner MJ, Thrush S. 2000. Marine reserves: parks, baselines and fishery enhancement. *Bull. Mar. Sci.* 66:617–34

Dayton PK, Tegner MJ. 1984. The importance of scale in community ecology: a kelp forest example with terrestrial analogs. In *A New Ecology: Novel Approaches to Interactive Systems*, ed. PW Price, CN Slobodchikoff, WS Gaud, pp. 457–83. New York: Wiley & Sons. 515 pp.

Dayton PK, Tegner MJ, Edwards PB, Riser KL. 1998. Sliding baselines, ghosts, and reduced expectations in kelp forest communities. *Ecol. Appl.* 8:309–22

Dayton PK, Thrush SF, Agardy TM, Hofman RJ. 1995. Environmental effects of fishing. *Aquat. Conserv. Mar. Freshw. Ecosyst.* 5:205–32

DeAlteris J, Skrobe L, Lipsky C. 1999. The significance of seafloor disturbance by mobile fishing gear relative to natural processes: a case study in Narragansett Bay, Rhode Island. *Am. Fish Soc. Symp.* 22:224–37

De Angelis DL, Waterhouse JC. 1987. Equilibrium and nonequilibrium concepts in ecological models. *Ecol. Monogr.* 57:1–21

de Forges BR, Koslow JA, Poore GCB. 2000. Diversity and endemism of the benthic seamount fauna in the southwest Pacific. *Nature* 405:944–47

de Groot SJ. 1984. The impact of bottom trawling on benthic fauna of the North Sea. *Ocean Manag.* 9:177–90

Druffel ERM, Griffin S, Witter A, Nelson E, Southon J, et al. 1995. *Gerardia*: bristlecone pine of the deep-sea? *Geochem. Cosmoschim. Acta* 59:5031–36

Eckman JE, Nowell AR. 1984. Boundary skin friction and sediment transport about an animal-tube mimic. *Sedimentology* 31:851–62

Eleftheriou A, Robertson MR. 1992. The effects of experimental scallop dredging on the fauna and physical environment of a shallow sandy community. *Neth. J. Sea Res.* 30:289–99

Engel J, Kvitek R. 1998. Impacts of otter trawling on a benthic community in Monterey Bay National Marine Sanctuary. *Conserv. Biol.* 12:1204–14

Etter RJ, Grassle F. 1992. Patterns of species diversity in the deep sea as a function of sediment particle size diversity. *Nature* 360:576–78

Fanning KA, Carder KL, Betzer PR. 1982. Sediment resuspension by coastal waters: a potential mechanism for nutrient re-cycling on the ocean's margins. *Deep-Sea Res.* 29:953–65

Forster S, Huettel M, Ziebis W. 1996. Impact of boundary layer flow velocity on oxygen utilisation in coastal sediments. *Mar. Ecol. Prog. Ser.* 143:173–85

Frechette M, Butman CA, Geyer WR. 1989. The importance of boundary-layer flows in supplying phytoplankton to the benthic suspension feeder, *Mytilus edulis* L. *Limnol. Oceanogr.* 34:19–36

Freese L, Auster PJ, Heifetz J, Wing BL. 1999. Effects of trawling on seafloor habitat and associated invertebrate taxa in the Gulf of Alaska. *Mar. Ecol. Prog. Ser.* 182:119–26

Frid CLJ, Hansson S, Ragnarsson SA, Rijnsdorp A, Steingrimsson SA. 1999. Changing levels of predation on benthos as a result of exploitation of fish populations. *Ambio* 28:578–82

Frid CLJ, Harwood KG, Hall SJ, Hall JA. 2000. Long-term changes in the benthic communities on North Sea fishing grounds. *ICES J. Mar. Sci.* 57:1303–9

Friedlander AM, Boehlert GW, Field ME, Mason JE, Gardner JV, Dartnell P. 1999. Side-scan sonar mapping of benthic trawl marks on the shelf and slope off Eureka, California. *Fish. Bull.* 97:786–801

Gibbs PJ, Collins AJ, Collett LC. 1980. Effect of otter prawn trawling on the macrobenthos of a sandy substratum in a New South Wales estuary. *Aust. J. Mar. Freshw. Res.* 31:509–16

Giller PS, Hildrew AG, Raffaelli D. 1994. *Aquatic Ecology: Scale, Pattern and Process.* Oxford: Blackwell Sci. 649 pp.

Graf G. 1999. Do benthic animals control the particle exchange between bioturbated sediments and benthic turbidity zones? See Gray et al. 1999, pp. 153–59

Graf G, Rosenberg R. 1997. Bioresuspension and biodeposition: a review. *J. Mar. Syst.* 11:269–78

Graham M. 1953. Effect of trawling on animals of the sea bed. *Deep-Sea Res.* 3:1–6

Grassle JF, Maciolek NJ. 1992. Deep-sea species richness: regional and local diversity estimate from quantitative bottom samples. *Am. Nat.* 1992:313–41

Gray JS. 1974. Animal-sediment relationships. *Oceanogr. Mar. Biol. Annu. Rev.* 12:707–22

Gray JS, Ambrose W Jr, Szaniawska A, eds. 1999. *Biogeochemical Cycling and Sediment Ecology.* Dordrecht, The Netherlands: Kluwer Academic. 236 pp.

Gray JS, Poore GCB, Ugland KI, Wilson RS, Olsgard F, Johannessen O. 1997. Coastal and deep-sea benthic diversities compared. *Mar. Ecol. Prog. Ser.* 159:97–103

Green MO, Hewitt JE, Thrush SF. 1998. Seafloor drag coefficients over natural beds of horse mussels (*Atrina zelanadica*). *J. Mar. Res.* 56:613–37

Hall SJ. 1999. *The Effects of Fisheries on Ecosystems and Communities.* Oxford: Blackwell Sci. 274 pp.

Hall SJ, Basford DJ, Robertson MR. 1990. The impacts of hydraulic dredging for razor clams *Ensis* sp. on an infaunal community. *Neth. J. Sea Res.* 27:119–25

Hall SJ, Raffaelli D, Thrush SF. 1994. Patchiness and disturbance in shallow water benthic assemblages. See Hildrew et al. 1994, pp. 333–75

Hall-Spencer JM, Moore PG. 2000. Scallop dredging has profound, long-term impacts on maerl habitats. *ICES J. Mar. Sci.* 57:1407–15

Hansson M, Lindegarth M, Valentinsson D, Ulmestrand M. 2000. Effects of shrimp-trawling on abundance of benthic macrofauna in Gullmarsfjorden, Sweden. *Mar. Ecol. Prog. Ser.* 198:191–201

Heck KLJ, Thoman TA. 1981. Experiments on predator-prey interactions in vegetated aquatic habitats. *J. Exp. Mar. Biol. Ecol.* 53:125–34

Herman PMJ, Middelburg JJ, VandeKoppel J, Heip CHR. 1999. Ecology of estuarine macrobenthos. *Adv. Ecol. Res.* 29:195–231

Holme NA. 1983. Fluctuations in the benthos of the Western Channel. *Oceanol. Acta, Proc. 17th Eur. Mar. Biol. Symp.* pp. 121–24. Elsevier. 225 pp.

Huettel M, Ziebis W, Forster S. 1996. Flow-induced uptake of particulate matter in permeable sediments. *Limnol. Oceanogr.* 41:309–22

Hughes DJ, Atkinson RJA, Ansell AD. 2000. A field test of the effects of megafaunal burrows on benthic chamber measurements of sediment-water solute fluxes. *Mar. Ecol. Prog. Ser.* 195:189–99

Huston MA. 1994. *Biological Diversity: The Coexistence of Species on Changing Landscapes.* Cambridge: Cambridge Univ. Press. 681 pp.

Hutchings JA. 2000. Collapse and recovery of marine fishes. *Nature* 406:882–85

Irlandi EA. 1994. Large- and small-scale effects of habitat structure on rates of predation: how percent coverage of seagrass affects rates of predation and siphon nipping on an infaunal bivalve. *Oecologia* 98:176–83

Jackson JBC. 2001. What was natural in the coastal oceans? *Proc. Natl. Acad. Sci. USA* 98:5411–18

Jenkins SR, Beukers-Stewart BD, Brand AR. 2001. Impact of scallop dredging on benthic megafauna: a comparison of damage levels in captured and non-captured organisms. *Mar. Ecol. Prog. Ser.* 215:297–301

Jennings S, Dinmore TA, Duplisea DE, Warr KJ, Lancaster JE. 2001a. Trawling disturbance can modify benthic production processes. *J. Anim. Ecol.* 70:459–75

Jennings S, Greenstreet SPR, Reynolds JD. 1999. Structural change in an exploited fish community: a consequence of differential fishing effects on species with contrasting life histories. *J. Anim. Ecol.* 68:617–27

Jennings S, Kaiser MJ. 1998. The effects of fishing on marine ecosystems. *Adv. Mar. Biol.* 34:203–314

Jennings S, Pinnegar JK, Polunin NVC, Warr KJ. 2001b. Impacts of trawling disturbance on the trophic structure of benthic invertebrate communities. *Mar. Ecol. Prog. Ser.* 213:127–42

Jumars PA, Eckman JE. 1983. Spatial structure within deep-sea benthic communities. In *The Sea*, ed. GT Rowe, pp. 399–451. New York: Wiley & Sons. 569 pp.

Kaiser MJ, de Groot SJ, eds. 2000. *The Effects of Fishing on Non-Target Species and Habitats*. Oxford: Blackwell Sci. 416 pp.

Kaiser MJ, Edwards DB, Armstrong PJ, Radford K, Lough NEL, et al. 1998. Changes in megafaunal benthic communities in different habitats after trawling disturbance. *ICES J. Mar. Sci.* 55:353–61

Kaiser MJ, Ramsay K, Richardson CA, Spence FE, Brand AR. 2000. Chronic fishing disturbance has changed shelf sea benthic community structure. *J. Anim. Ecol.* 69:494–503

Kaiser MJ, Spence FE, Hart PJB. 1999. Fishing-gear restrictions and conservation of benthic habitat complexity. *Conserv. Biol.* 14:1512–25

Kaiser MJ, Spencer BE. 1996. The effects of beam-trawl disturbance on infaunal communities in different habitats. *J. Anim. Ecol.* 65:348–58

Kenchington ELR, Prena J, Gilkinson KD, Gordon DC, MacIsaac K, et al. 2001. Effects of experimental otter trawling on the macrofauna of a sandy bottom ecosystem on the Grand Banks of Newfoundland. *Can. J. Fish Aquat. Sci.* 58:1043–57

Kolasa J, Pickett STA. 1991. *Ecological Heterogeneity*. New York: Springer-Verlag. 332 pp.

Koslow JA, Gowelett-Holmes K, Lowry JK, O'Hara T, Poore GCB, Williams A. 2001. Seamount benthic macrofauna off southern Tasmania: community structure and impacts of trawling. *Mar. Ecol. Prog. Ser.* 213:111–25

Krost P. 1990. The impact of otter-trawl fishery on nutrient release from the sediment

and macrofauna of Kieler Bucht (Western Baltic). *Ber. Inst. Meereskd. Nr.* 200:167

Langton RW, Robinson WE. 1990. Faunal association on scallop grounds in the Western Gulf of Maine. *J. Exp. Mar. Biol. Ecol.* 144:157–71

Legeleux F, Reyss J-L, Schmidt S. 1994. Particle mixing rates in sediments of the northeast tropical Atlantic: evidence from $^{210}Pb_{xs}$, ^{137}Cs, $^{228}Th_{xs}$ and $^{234}Th_{xs}$ downcore distributions. *Earth Planet Sci. Lett.* 128:545–62

Legendre P. 1993. Spatial autocorrelation: trouble or new paradigm? *Ecology* 74:1659–73

Levin LA. 1984. Life history and dispersal patterns in a dense infaunal polychaete assemblage: community structure and response to disturbance. *Ecology* 65:1185–200

Levin LA, Etter RJ, Rex MA, Cooday AJ, Smith CR, et al. 2001. Environmental influences on regional deep-sea species diversity. *Annu. Rev. Ecol. Syst.* 32:51–93

Levin LA, Gooday AJ. 1992. Possible roles for xenophyophores in deep-sea carbon cycling. In *Deep-Sea Food Chains and the Global Carbon Cycle*, ed. GT Rowe, V Pariente, pp. 93–104. The Netherlands: Kluwer Academic. 410 pp.

Lindholm JB, Auster PJ, Kaufman LS. 1999. Habitat-mediated survivorship of juvenile (0-year) Atlantic cod *Gadus morhua*. *Mar. Ecol. Prog. Ser.* 180:247–55

Loehle C, Li B-L. 1996. Habitat destruction and the extinction debt revisited. *Ecol. Appl.* 6:784–89

Loftis JC, McBride GB, Ellis JC. 1991. Considerations of scale in water quality monitoring and data analysis. *Water Res. Bull.* 27:255–64

Loo L-O, Rosenberg R. 1989. Bivalve suspension-feeding dynamics and benthic-pelagic coupling in an eutrophicated marine bay. *J. Exp. Mar. Biol. Ecol.* 130:253–76

Marinelli RL. 1994. Effects of burrow ventilation on activities of a terebellid polychaete and silicate removal from sediment pore waters. *Limnol. Oceanogr.* 39:303–17

Martin RE. 1996. Secular increase in nutrient levels through the phanerozoic: implications for productivity, biomass and diversity of the marine biosphere. *Palaios* 11:209–19

Mayer LM, Schick DF, Findlay RH, Rice DL. 1991. Effects of commercial dragging on sedimentary organic matter. *Mar. Environ. Res.* 31:249–61

McCall PL, Tevesz MJS, eds. 1982. *Animal-Sediment Relations: The Biogenic Alteration of Sediments*. New York: Plenum. 351 pp.

McConnaughey RA, Mier KL, Dew CB. 2000. An examination of chronic trawling effects on soft-bottom benthos of the eastern Bering Sea. *ICES J. Mar. Sci.* 57:1377–88

Micheli F. 1999. Eutrophication, fisheries and consumer-resource dynamics in marine pelagic ecosystems. *Science* 285:325–26

Middleburg JJ, Soetaert K, Herman PMJ. 1997. Empirical relationships for use in global diagenetic models. *Deep-Sea Res.* 44:327–44

Moran MJ, Stephenson PC. 2000. Effects of otter trawling on macrobenthos and management of demersal scalefish fisheries on the continental shelf of north-western Australia. *ICES J. Mar. Sci.* 57:510–16

Musick JA, Harbin MM, Berkeley SA, Burgess GH, Eklund AM, et al. 2000. Marine, estuarine, and diadromous fish stocks at risk of extinction in North America (exclusive of Pacific salmonids). *Fisheries* 25:6–30

Nelder JA. 1999. Statistics for the millennium: from statistics to statistical science. *Statistician* 48:257–69

Nikora V, Green MO, Thrush SF, Hume TM, Goring DG. 2002. Structure of the internal boundary layer over a patch of horse mussels (*Atrina zelandica*) in an estuary. *J. Mar. Res.* 60:121–50

Norkko A, Hewitt JE, Thrush SF, Funnell GA. 2001. Benthic-pelagic coupling and suspension feeding bivalves: linking site-specific sediment flux and biodeposition to benthic community structure. *Limnol. Oceanogr.* 46:2067–72

Olafsson EB, Peterson CH, Ambrose WGJ. 1994. Does recruitment limitation structure populations and communities of macroinvertebrates in marine soft sediments: the relative significance of pre- and post-settlement

processes. *Oceangr. Mar. Biol. Annu. Rev.* 32:65–109

Palanques A, Guillen J, Puig P. 2001. Impact of bottom trawling on water turbidity and muddy sediment of an unfished continental shelf. *Limnol. Oceanogr.* 46:1100–10

Pauly D, Christensen V, Dalsgaard J, Froese R, Torres F Jr. 1998. Fishing down marine food-webs. *Science* 279:860–3

Persson L, Eklov P. 1995. Prey refuges affecting interactions between piscivorous perch and juvenile perch and roach. *Ecology* 76:70–81

Peterson CH. 1979. The importance of predation and competition in organising the intertidal epifaunal communities of Barnegat Inlet, New Jersey. *Oecologia* 39:1–24

Pilskaln CH, Churchill JH, Mayer LM. 1998. Resuspension of sediments by bottom trawling in the Gulf of Maine and potential geochemical consequences. *Conserv. Biol.* 12:1223–24

Pimm SL. 1991. *The Balance of Nature: Ecological Issues in the Conservation of Species and Communities*. Chicago/London: Univ. Chicago Press. 434 pp.

Pitcher CR, Poiner IR, Hill BJ, Burridge CY. 2000. Implications of the effects of trawling on sessile megazoobenthos on a tropical shelf in northeastern Australia. *ICES J. Mar. Sci.* 57:1359–68

Pitcher TJ. 2001. Fisheries managed to rebuild ecosystems? Reconstructing the past to salvage the future. *Ecol. Appl.* 11:601–17

Pitcher TJ, Watson R, Haggan N, Guenette S, Kennish R, et al. 2000. Marine reserves and the restoration of fisheries and marine ecosystems in the South China Sea. *Bull. Mar. Sci.* 66:543–66

Pranovi F, Raicevich S, Franceschini G, Farrace MG, Giovanardi O. 2000. Rapido trawling in the northern Adriatic Sea: effects on benthic communities in an experimental area. *ICES J. Mar. Sci.* 57:517–24

Prena J, Schwinghammer P, Rowell TW, Gordon DC Jr, Gilkinson KD, et al. 1999. Experimental otter trawling on a sandy bottom ecosystem of the Grand Banks of

Newfoundland: analysis of the trawl bycatch and effects on epifauna. *Mar. Ecol. Prog. Ser.* 181:107–24

Probert PK, McKnight DG, Grove SL. 1997. Benthic invertebrate bycatch from a deep-water trawl fishery, Chatham Rise, New Zealand. *Aquat. Conserv. Mar. Freshw. Ecosyst.* 7:27–40

Reise K. 1982. Long-term changes in the macrobenthic invertebrate fauna of the Wadden sea. *Neth. J. Sea Res.* 16:29–36

Rhoads DC. 1974. Organism-sediment relations on the muddy seafloor. *Oceanogr. Mar. Biol. Annu. Rev.* 12:263–300

Rhodes MC, Thompson RJ. 1993. Comparative physiology of suspension feeding in living brachiopods and bivalves: evolutionary implications. *Paleobiology* 19:322–34

Roberts JM, Harvey SM, Lamont PA, Gage JD, Humphery JD. 2000. Seafloor photography, environmental assessment and evidence for deep-water trawling on the continental margin west of the Hebrides. *Hydrobiologia* 441:173–83

Roberts M. 1997. Coral in deep water. *New Sci.* 155:40–43

Rogers AD. 1999. The biology of *Lophelia pertusa* (Linnaeus 1758) and other deep-water reef forming corals and impacts from human activities. *Int. Rev. Hydrobiol.* 84:315–406

Rooker JR, Holt GJ, Holt SA. 1998. Vulnerability of newly settled red drum (*Scianops ocellatus*) to predatory fish: Is early-life survival enhanced by seagrass meadows? *Mar. Biol.* 131:145–51

Rowe GT, Clifford CH, Smith KL Jr, Hamilton PL. 1975. Benthic nutrient regeneration and its coupling to primary productivity in coastal waters. *Nature* 255:215–17

Ruiz GM, Hines AH, Posey MH. 1993. Shallow water as a refuge habitat for fish and crustaceans in non-vegetated estuaries: an example from Chesapeake Bay. *Mar. Ecol. Prog. Ser.* 99:1–16

Sainsbury KJ. 1988. The ecological basis of multispecies fisheries and management of a demersal fishery in tropical Australia. In *Fish Population Dynamics: The Implications for Management*, ed. JA Gulland, pp. 349–82. New York: Wiley. 351 pp.

Sala E, Ballesteros E. 1997. Partitioning of space and food resources by three fish of the genus *Diplodus* (Sparidae) in a Mediterranean rocky infralittoral ecosystem. *Mar. Ecol. Prog. Ser.* 152:273–83

Sanchez P, Demestre M, Ramon M, Kaiser MJ. 2000. The impact of otter trawling on mud communities in the northwestern Mediterranean. *ICES J. Mar. Sci.* 57:1352–58

Sandnes J, Forbes T, Hansen R, Sandnes B, Rygg B. 2000. Bioturbation and irrigation in natural sediments, described by animal-community parameters. *Mar. Ecol. Prog. Ser.* 197:169–79

Scheffer M, Carpenter S, Foley JA, Folke C, Walker B. 2001. Catastrophic shifts in ecosystems. *Nature* 413:591–96

Schneider DC, Gagnon JM, Gilkinson KD. 1987. Patchiness of epibenthic megafauna on the outer Grand Banks of Newfoundland. *Mar. Ecol. Prog. Ser.* 39:1–13

Schwinghamer P, Guigne JY, Siu WC. 1996. Quantifying the impact of trawling on benthic habitat structure using high resolution acoustics and chaos theory. *Can. J. Fish Aquat. Sci.* 53:288–96

Shimeta JS, Jumars PA. 1991. Physical mechanisms and rates of particle capture by suspension feeders. *Oceanogr. Mar. Biol. Annu. Rev.* 29:191–257

Skilleter GA. 1994. Refuges from predation and the persistence of estuarine clam populations. *Mar. Ecol. Prog. Ser.* 109:29–42

Smith CJ, Papadopoulou KN, Diliberto S. 2000. Impact of otter trawling on an eastern Mediterranean commercial trawl fishing ground. *ICES J. Mar. Sci.* 57:1340–51

Snelgrove PVR. 1999. Getting to the bottom of marine biodiversity: sedimentary habitats—ocean bottoms are the most widespread habitat on Earth and support high biodiversity and key ecosystem services. *Bioscience* 49:129–38

Snelgrove PVR, Butman CA. 1994. Animal-sediment relationships revisited: cause versus effect. *Oceanogr. Mar. Biol. Annu. Rev.* 32:111–77

Soetaert K, Herman PMJ, Middleburg JJ, Heip C, deStigter HS, et al. 1996. Modeling ^{210}Pb-derived mixing activity in ocean margin sediments: diffusive verses nonlocal mixing. *J. Mar. Res.* 54:1207–27

Sousa WP. 1984. The role of disturbance in natural communities. *Annu. Rev. Ecol. Syst.* 15:353–91

Sparks -McConkey PJ, Watling L. 2001. Effects on the ecological integrity of a soft-bottom habitat from a trawling disturbance. *Hydrobiologia* 456:73–85

Squires DF. 1965. Deep-water coral structure on the Cambell Plateau, New Zealand. *Deep-Sea Res.* 12:785–88

Steele JH, Schumacher M. 2000. Ecosystem structure before fishing. *Fish. Res.* 44:201–5

Steneck RS. 1997. Fisheries-induced biological changes to the structure and function of the Gulf of Maine. In *Proc. Gulf of Maine Ecosys. Dynamics Scientific Symp. and Workshop. RARGOM Rep., 91-1. Regional Assoc. for Res. on the Gulf of Maine*, ed. GT Wallace, EF Braasch, pp. 151–65. Hanover, NH: Maine/New Hamphsire Sea Grant Progr.

Tegner MJ, Basch LV, Dayton PK. 1996. Near extinction of an exploited marine invertebrate. *TREE* 11:278–80

Thayer CW. 1983. Sediment-mediated biological disturbance and the evolution of marine benthos. In *Biotic Interactions in Recent and Fossil Benthic Communities*, ed. MJS Tevesz, pp. 479–625. New York/London: Plenum

Thistle D. 1981. Natural physical disturbances and the communities of marine soft bottoms. *Mar. Ecol. Prog. Ser.* 6:223–28

Thrush SF, Hewitt JE, Cummings VJ, Dayton PK. 1995. The impact of habitat disturbance by scallop dredging on marine benthic communities: What can be predicted from the results of experiments? *Mar. Ecol. Prog. Ser.* 129:141–50

Thrush SF, Hewitt JE, Cummings VJ, Dayton PK, Cryer M, et al. 1998. Disturbance of the marine benthic habitat by commercial fishing: impacts at the scale of the fishery. *Ecol. Appl.* 8:866–79

Thrush SF, Hewitt JE, Funnell GA, Cummings VJ, Ellis J, et al. 2001. Fishing disturbance and marine biodiversity: the role of habitat structure in simple soft-sediment systems. *Mar. Ecol. Prog. Ser.* 221:255–64

Thrush SF, Lawrie SM, Hewitt JE, Cummings VJ. 1999. The problem of scale: uncertainties and implications for soft-bottom marine communities and the assessment of human impacts. See Gray et al. 1999, pp. 195–210

Thrush SF, Pridmore RD, Hewitt JE, Cummings VJ. 1994. The importance of predators on a sandflat: interplay between seasonal changes in prey densities and predator effects. *Mar. Ecol. Prog. Ser.* 107:211–22

Thrush SF, Whitlatch RB. 2001. Recovery dynamics in benthic communities: balancing detail with simplification. In *Ecological Comparisons of Sedimentary Shores*, ed. K Reise, pp. 297–316. Berlin: Springer-Verlag. 384 pp.

Tuck ID, Hall SJ, Robertson MR, Armstrong E, Basford DJ. 1998. Effects of physical trawling disturbance in a previously unfished sheltered Scottish sea loch. *Mar. Ecol. Progr. Ser.* 162:227–42

Tupper M, Boutilier RG. 1995. Effects of habitat on settlement, growth and postsettlement survival of Atlantic cod (*Gadus morhua*). *Can. J. Fish Aquat. Sci.* 52:1834–41

Turner MG, Romme WH, Gardner RH, O'Neill RV, Kratz TK. 1993. A revised concept of landscape equilibrium: disturbance and stability on scaled landscapes. *Landscape Ecol.* 8:213–27

Turner SJ, Thrush SF, Hewitt JE, Cummings VJ, Funnell G. 1999. Fishing impacts and the degradation or loss of habitat structure. *Fish Manag. Ecol.* 6:401–20

Van Blaricom GR. 1982. Experimental analysis of structural regulation in a marine sand community exposed to oceanic swell. *Ecol. Monogr.* 52:283–305

Van Dolah RF, Wendt PH, Levisen MV. 1991. A study of the effects of shrimp trawling on benthic communities in two South Carolina sounds. *Fish Res.* 12:139–56

Van Dolah RF, Wendt PH, Nichloson N. 1987. Effects of a research trawl on a hard-bottom assembalge of sponges and corals. *Fish Res.* 5:39–54

Veale LO, Hill AS, Hawkins SJ, Brand AR. 2000. Effects of long-term physical disturbance by commercial scallop fishing on subtidal epifaunal assemblages and habitats. *Mar. Biol.* 137:325–37

Watling L, Findlay RH, Mayer LM, Schick DF. 2001. Impact of a scallop drag on the sediment chemistry, microbiota, and faunal assemblages of a shallow subtidal marine benthic community. *J. Sea Res.* 46:309–24

Watling L, Norse EA. 1998. Disturbance of the seafloor by mobile fishing gear: a comparison to forest clear cutting. *Conserv. Biol.* 12:1180–97

Weinberg JR, Whitlatch RB. 1983. Enhanced growth of a filter-feeding bivalve by a deposit-feeding polychaete by means of nutrient regeneration. *J. Mar. Res.* 41:557–69

Whitlatch RB. 1980. Patterns of resource utilization and co-existence in marine intertidal deposit-feeding communities. *J. Mar. Res.* 38:743–65

Whitlatch RB, Lohrer AM, Thrush SF. 2002. Scale-dependent recovery of the benthos: effects of larval and post-larval stages. In *Organism-Sediment Workshop*, ed. R Aller, J Aller, SA Woodin, pp. 181–98 Columbia: Univ. SC. 403 pp.

Widdicombe S, Austen MC. 1999. Mesocosm investigation into the effects of bioturbation on the diversity and structure of a subtidal macrobenthic community. *Mar. Ecol. Prog. Ser.* 189:181–93

Widdicombe S, Austen MC, Kendall MA, Warwick RM, Jones MB. 2000. Bioturbation as a mechanism for setting and maintaining levels of diversity in subtidal macrobenthic communities. *Hydrobiologia* 440:369–77

Wildish D, Kristmanson D. 1997. *Benthic Suspension Feeders and Flow*. Cambridge: Cambridge Univ. Press. 422 pp.

Wilson WH. 1991. Competition and predation in marine soft-sediment communities. *Annu. Rev. Ecol. Syst.* 21:221–41

Witman JD, Sebens K. 1992. Regional variation in fish predation intensity: a historical perspective in the Gulf of Maine. *Oecologia* 90:305–15

Woodin SA. 1978. Refuges, disturbance and community structure: a marine soft-bottom example. *Ecology* 59:274–84

Wootton JT. 1998. Effects of disturbance on species diversity: a multitrophic perspective. *Am. Nat.* 152:803–25

Zajac RN. 1999. Understanding the sea floor landscape in relation to impact assessment and environmental management in coastal marine sediments. See Gray et al. 1999, pp. 211–27

Zajac RN, Whitlatch RB, Thrush SF. 1998. Recolonisation and succession in soft-sediment infaunal communities: the spatial scale of controlling factors. *Hydrobiologia* 376:227–40

Ziebis W, Huettel M, Forster S. 1996. Impact of biogenic sediment topography on oxygen fluxes in permeable seafloors. *Mar. Ecol. Prog. Ser.* 140:227–37

Annu. Rev. Ecol. Syst. 2002. 33:475–505
doi: 10.1146/annurev.ecolsys.33.010802.150448
Copyright © 2002 by Annual Reviews. All rights reserved
First published online as a Review in Advance on August 14, 2002

PHYLOGENIES AND COMMUNITY ECOLOGY

Campbell O. Webb[1], David D. Ackerly[2], Mark A. McPeek[3], and Michael J. Donoghue[1]

[1]*Department of Ecology and Evolutionary Biology, Yale University, New Haven, Connecticut 06511; email: campbell.webb@yale.edu, michael.donoghue@yale.edu*
[2]*Department of Biological Sciences, Stanford University, Stanford, California 94305; email: dackerly@stanford.edu*
[3]*Department of Biology, Dartmouth College, Hanover, New Hampshire 03755; email: mark.mcpeek@dartmouth.edu*

Key Words community assembly and organization, phylogenetic conservatism, biogeography, species diversity, niche differentiation

■ **Abstract** As better phylogenetic hypotheses become available for many groups of organisms, studies in community ecology can be informed by knowledge of the evolutionary relationships among coexisting species. We note three primary approaches to integrating phylogenetic information into studies of community organization: 1. examining the phylogenetic structure of community assemblages, 2. exploring the phylogenetic basis of community niche structure, and 3. adding a community context to studies of trait evolution and biogeography. We recognize a common pattern of phylogenetic conservatism in ecological character and highlight the challenges of using phylogenies of partial lineages. We also review phylogenetic approaches to three emergent properties of communities: species diversity, relative abundance distributions, and range sizes. Methodological advances in phylogenetic supertree construction, character reconstruction, null models for community assembly and character evolution, and metrics of community phylogenetic structure underlie the recent progress in these areas. We highlight the potential for community ecologists to benefit from phylogenetic knowledge and suggest several avenues for future research.

INTRODUCTION

The differences among species that co-occur in an ecological community are the result of modifications to a common ancestor that the species all ultimately share. As molecular and analytical methods make the elucidation of phylogenetic relationships easier and more reliable, ecologists have an invaluable new dimension of information available with which to make sense of these differences among species. However, despite recognition of the potential for using phylogenies in community ecology (Brooks & McLennan 1991, Losos 1996, Thompson et al. 2001), and increasing interest in the role of history in ecology (Ricklefs 1987, Ricklefs & Schluter 1993a), integration of evolutionary biology and community ecology remains elusive. This is due partly to the conceptual and methodological difficulties of bridging gaps of temporal and spatial scale and partly to poor communication: many ecologists are either unaware of the potential benefits of knowing about the phylogenetic relationships in their communities or are deterred by the unfamiliarity of molecular techniques and phylogenetic methods and the accompanying terminology. Similarly, many systematists are unaware of the fascinating ecological questions that can be addressed using the phylogenies they produce or the ways in which knowledge of community composition might bear on studies of character evolution, diversification rate, and historical biogeography. Our intention in this review is to introduce to both parties the various approaches that have already been taken to incorporate phylogenetic information into community ecology.

Phylogenies are being used extensively in the larger field of evolutionary ecology (see Miles & Dunham 1993, Miller & Wenzel 1995, Ackerly et al. 2000), so we limit our review to studies and concepts explicitly relating to the phylogenetic and taxonomic structure of local communities. We do not explicitly review character displacement in species pairs (Schluter 2000a), adaptive radiation in particular clades (Schluter 2000b), "host-client" coevolution (host-parasite, plant-herbivore, and host-pathogen), general historical biogeography, or the uses of microbial phylogenies. Previous reviews and discussions of the interaction of phylogeny with community ecology include Wanntorp et al. (1990), Brooks & McLennan (1991, 2002), Eggleton & Vane-Wright (1994), Losos (1996), McPeek & Miller (1996), Grandcolas (1998), and Nel et al. (1998).

Empirically, phylogenies and community ecology have been put together predominantly in studies of community assembly, organization, and species co-occurrence, and we identify in this literature three major approaches (Figure 1). Other questions of community ecology, relating to relative abundance, range size distributions, and species richness have received less attention from a phylogenetic perspective, but we cover the work that has been done so far. We then review recent methodological advances and conclude with suggested directions for further work.

COMMUNITY STRUCTURE AND COEXISTENCE

Even though phylogenetic methods were developed fairly recently, a connection between taxonomy and community ecology has long been recognized:

> As species of the same genus have usually, though by no means invariably, some similarity in habits and constitution, and always in structure, the struggle will generally be more severe between species of the same genus, when they come into competition with each other, than between species of distinct genera (Darwin 1859).

Darwin's statement already contains what we see to be the essential elements of an evolutionary understanding of community organization: that species interact

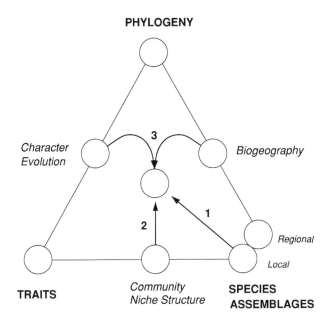

Figure 1   Schematic summary of various approaches to the integration of phylogeny, traits, and communities. (1) Examining the phylogenetic structure of community assemblages; (2) exploring the phylogenetic basis of community niche structure; (3) adding community context to studies of trait evolution and biogeography.

in communities, that species interact based on their phenotypic differences and similarities, and that phenotypic variation has a basis in evolutionary history. In a synthetic understanding of the origin and maintenance of community composition, three elements are drawn together: phylogeny, community composition, and trait information (Figure 1). Researchers have tended to approach this synthesis using one (or more) of three methods: 1. analyzing community taxonomic or phylogenetic structure, 2. exploring the phylogenetic basis of niche differentiation, and 3. adding community context to character evolution and biogeography. We discuss these approaches below, in order of increasing information requirements and increasing potential to reveal both ecology and evolution in the past and present.

The Phylogenetic Structure of Community Assemblages

Key question: Is the distribution of species among habitats (or samples) in a community nonrandom with respect to phylogeny? (Using: species list of local community + distribution of species among community samples + phylogeny of community species list)

Since the advent of formal classification, natural historians have asked why different areas are dominated by different species, genera, and families (e.g., Gentry

1982). The quantitative taxonomic structure of communities was first addressed by Elton (1946), who reasoned that the lower number of species per genus observed in local areas than in the whole of Britain was evidence for competitive exclusion of ecologically similar congeners in local habitats. Interest continued in species/genus ratios for a number of years (Moreau 1948, Williams 1964, Simberloff 1970, Tokeshi 1991) and was notable as the context for the first use of null models in ecology (Gotelli & Graves 1996). Implicit in these analyses was the same three-part interaction discussed above (Figure 1): community organization (i.e., the role of competition) can be deduced from the (assumed) ecological similarity within a genus, and the taxonomic structure of a community (i.e., the significant departure of species/genus ratios in community samples relative to a regional species pool). More recently, the global consistency of taxonomic structure in forest communities has been examined by Enquist et al. (2002), who compared the species/genus and species/family ratios across many standardized 0.1 ha plots. They found an exponential relationship between numbers of genera or families and the numbers of species, across two orders of magnitude of species number, and suggest that this result indicates the existence of forces acting to constrain phylogenetic structure.

The availability of phylogenies, along with methods for the construction of supertrees and for assembling the phylogenies of communities, now permits community structure to be assessed phylogenetically. A simple logical framework can then be employed to infer mechanisms of contemporary coexistence (Table 1, and see Figure 2 for terminology). A clumped phylogenetic distribution of taxa ("phylogenetic attraction") indicates that habitat-use is a conserved trait within the pool of species in the community, and that phenotypic attraction dominates over repulsion. However, phylogenetic overdispersion (repulsion) can result either when closely related taxa with the most similar niche-use are being locally excluded (phenotypically repulsed), such that there is minimum niche overlap of coexisting species, or when distantly related taxa have converged on similar niche-use and are phenotypically attracted. Note that the fourth possible interaction, phenotypic repulsion of traits that are convergent, will not tend to recreate phylogenetically clustered communities, but phylogenetically random ones.

For example, Webb (2000) found that the tree taxa that co-occurred in 0.16 ha plots in Indonesian Borneo were more closely related than expected from a random sampling of the local species pool. Assuming that conservatism dominates in the phylogenetic distribution of ecological character, he interpreted this as evidence for the predominant role of habitat filtering (and phenotypic attraction), as opposed to local competitive exclusion (and phenotypic repulsion) of similar species. In a similar study, H. Steers (personal communication) determined that a measure of the frequency of co-occurrence of tree species pairs in a Mexican dry tropical forest was positively correlated with their phylogenetic proximity, again interpreting this as evidence of habitat selection for ecologically similar, phylogenetically related species. Kelly (1999) found that British plant taxa in extreme environments were more closely related than expected by chance, which was seen as evidence that these

TABLE 1   The expected distribution of sample taxa on the phylogeny of a pool at a larger spatial scale, given various combinations of phylogenetic trait distribution and ecological process

	Ecological traits phylogenetically	
	Conserved	Convergent
Dominant ecological force:		
Habitat filtering (phenotypic attraction)	Clustered	Overdispersed
Competitive exclusion (phenotypic repulsion)	Overdispersed	Random

species were ecologically similar. Conversely, Graves & Gotelli (1993) showed that congeners seldom co-occur in the same mixed-species foraging flock in the Amazon, but that this "checkerboard" pattern breaks down at higher taxonomic levels. They interpreted this finding as the effect of intra-community competitive exclusion among ecologically similar species (i.e., phenotypic repulsion), with congeners being most similar. In Florida woodland communities dominated by oaks, J. Cavender-Bares (personal communication) also found that close relatives co-occurred less than expected by chance. In this case each plot generally had one species from each of three major *Quercus* clades (sections).

The spatial scale of samples used in studies of community phylogenetic structure is of great importance to the interpretation of the patterns found because the biological nature of phenotypic and phylogenetic attraction and repulsion depends upon the scale involved. At the largest, continental scales (e.g., 1,000–10,000 km), phylogenetic clustering of members of a regional sample on a global phylogeny reflects biogeographic rather than ecological processes, as clades diversify within the sample region, and cause many taxa in the region to be, on average, more related to each other than to taxa outside the region. Within a region (e.g., 10–1,000 km), phenotypic sorting might occur among communities that differ environmentally from one another (e.g., wetlands versus montane). Such phenotypic attraction might lead to phylogenetic attraction or repulsion of the community sample on the regional pool, depending on the phylogenetic distribution of important traits. Sustained phenotypic repulsion within a community might also lead to semipermanent exclusion of too-similar taxa from individual communities, with taxa maintained in the regional pool by low rates of dispersal among communities (e.g., Tilman 1994). At the community scale (e.g., 100 m–10 km), species should segregate into habitats based on the relative strengths of habitat filtering versus competition among similar species (see Figure 2). Finally, at the smallest, neighborhood scales (e.g., <100 m), one might observe the effect of individual-based interactions that lead to within-habitat filtering or "neighborhood exclusion." Hence, a spatially nested analysis of community phylogenetic structure may detect different patterns of phylogenetic clustering or over-dispersion at different scales, providing more information about community processes than an analysis at just a single scale.

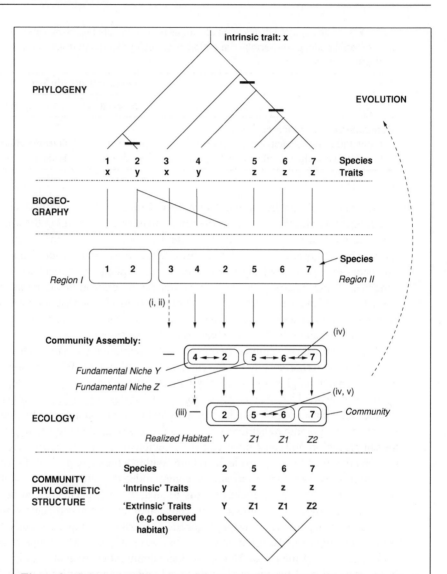

Figure 2 Schematic of the general framework employed in this review, with associated terms. A lineage may diversify by the division of its ancestral range and allopatric speciation, such that sister clades are no longer coregional (ancestor of species 1 and 2 versus ancestor of species 3–7; *BIOGEOGRAPHY*). Alternatively, sympatric and parapatric or even allopatric speciation mechanisms may lead to the origination of new species that are coregional with their sister species (species 3–7). A phylogeny can be reconstructed for the lineage (*PHYLOGENY*) using molecular and morphological species traits. Species may appear in regions either through the geographical division of their area (vicariance) or by subsequent dispersal (species 2

Figure 2   (*Continued*) dispersed into region II). The phylogenies of lineages and the distribution of taxa among regions can be used to infer the historical patterns of movement in the taxa and associated reconstruction of area connectedness (area cladograms), using biogeographic methods. Trait change occurs as the lineage diversifies, and ancestral state changes can be reconstructed (or traced) on a phylogeny using, for example, parsimony or maximum likelihood (x to y, y to z). Traits can usefully be divided into intrinsic (morphological or physiological traits that can be assessed validly when an organism is removed from its environment, e.g., skeletal structure, beak size, body size, plant sexual system) and extrinsic (traits that only have meaning in an external abiotic and biotic environment, e.g., swimming speed, maximum growth rate, drought tolerance, shade tolerance, prey choice, diet breadth). The "ecological character" of an organism is usually a complex set of correlated characters, but can often be directly related to simpler intrinsic morphological characters (e.g., Losos 1995).

The pattern of evolution of any trait can be characterized as conservative (more closely related taxa are more similar) or convergent (homoplasious, the independent evolution of similarity). Trait divergence is not intrinsically conservative or convergent, but because the greater the change in a trait, the more likely it is to resemble the value of a species in an independent lineage, divergence often results in convergence. Additionally, the rate of homoplasy increases with decreasing number of potential trait states (Donoghue & Ree 2000) and increasing number of ways for species to be functionally similar while intrinsically different (e.g., fleshy-fruitedness arises via a number of anatomical paths). When refering to ecological traits, the term "phylogenetic niche conservatism" has been used (Harvey & Pagel 1991, Lord et al. 1995). This conservatism may be due to active, stabilizing selection (Lord et al. 1995) or to a reduction in the potential range of ecological character evolution caused by the fixation of ancestral traits (i.e., developmental constraints; Westoby et al. 1995).

Species are assembled into communities from a regional species pool (the list of all species in an area at the next-highest spatial scale from the scale under consideration; e.g., species 2–7 form a species pool relative to the community) (e.g., Liebold 1998, Fox 1999, Blackburn & Gaston 2001) (*ECOLOGY*). In this review we consider a community to be more than two species in the same trophic level and the same guild (e.g. meadow plants, desert granivores) co-occurring spatially at a scale over which species might disperse within a few generations. Some species present in a region may fail to meet the ecological requirements to survive in any of the niches in a community, that is, they are filtered out, or simply fail to reach a particular community by chance (species 3 and process i or ii, respectively). Community species richness is often correlated with regional pool richness, suggesting that communities seldom saturate (Cornell & Lawton 1992). Species may persist (coexist for long periods) in the same community by occupying different niches, thus minimizing competition for resources

Figure 2   (*Continued*) ("habitat partitioning"; species 2 and 4 versus species 5–7; Wisheu 1998). However, with enough species in the community, several species with similar niche requirements will tend to be filtered into the same niche (a phenotypic attraction). Simultaneously, negative interindividual interaction (phenotypic repulsion) is expected among species that occupy the same niche and/or habitat (process iv). This negative interaction is expected to be stronger among ecologically more similar species and may alter the realized niche/habitat distribution of taxa into sub-niches (or localities) within the fundamental niche (but never beyond the bounds of the fundamental niche; species 7 versus 5 and 6), or may lead to classic competitive exclusion (process iii on species 4). It may also lead to the exclusion of competing species on the most proximate, neighborhood scale (e.g., at the scale of individual interactions), without leading to community exclusion. Species may coexist within a habitat by non-niche-partitioning, equalizing (sensu Chesson 2000) processes (species 5, 6, and process v), or may be in the slow, nonequilibrial process of being excluded from the community. We use habitat to mean the spatial location where a combination of resource levels come together.

A phylogeny can be reconstructed for species sampled in a spatially defined area (region, community), rather than for all species in a lineage (*COMMUNITY PHYLOGENETIC STRUCTURE*). The occurrence of taxa in areas at a smaller spatial scale (community and habitat, respectively) can be indicated on the phylogeny of the larger pool of species. The distribution of these taxa can be phylogenetically clumped, random, or over-dispersed on the phylogeny of the entire pool (e.g., a sample containing species 6 and 7 is clustered on the phylogeny of the community of species 2, 5, 6, and 7).

Sustained selective pressure from individual competitors or environmental changes leads eventually to change in ecological character over evolutionary time, usually in the direction of reduced niche overlap (*EVOLUTION*). The community context may also cause stabilizing selection, especially in diverse systems in which interspecific interactions are unpredictable. Trait change may also be the result of drift, founder effects, or pleiotropy. Trait change resulting from ecological interactions may itself lead to ecological speciation within a region (McPeek 1996, Schluter 2001).

The Phylogenetic Basis of Community Niche Structure

Key question: How are niche differences in communities arrayed on a phylogeny? (Using: species list of local community + distribution of species within community + phylogeny of community species list + ecological character data for those species)

Ecologists have long studied the distribution of ecological characters of species in communities to understand community organization (e.g., MacArthur & Levins 1967, Bowers & Brown 1982). An even dispersion of trait values along some gradient (e.g., constant body-size ratios) has been held to be evidence for both contemporary competitive exclusion and the long-term evolutionary effect of such competition (Connell 1980), although the establishment of the significance of this overdispersion has been controversial (reviewed by Gotelli & Graves 1996).

The availability of data on the taxonomic or phylogenetic distribution of niche differences enables the allocation of contemporary niche structure to either contemporary ecological or historic evolutionary causes, or a combination thereof. For example, because Cavender-Bares (see above, personal communication) had assessed water-use trait data as well as water availability in sample plots (Cavender-Bares & Holbrook 2001), she was able to interpret the phylogenetic overdispersion of local plots as the result of contemporary habitat filtering mediated by characters that were phylogenetically convergent. Similarly, Webb & Peart (2000) assessed the species, genus, and family associations of rain forest trees with three habitat types and found both genera and families that had all of their species associated with a particular habitat. Although not phylogenetic, this pattern of ecological conservatism supports Webb's (2000) interpretation of the role of habitat filtering. In the same forests, Ashton (1988) has argued that too much emphasis has been placed on differences in habitat use among species and that in the Dipterocarpaceae, genera and sections differ significantly in ecological character, with species within them differing mainly by minor morphological (particularly floral) changes. At a deeper phylogenetic level, Dimichele & Phillips (1996) showed consistent occupation of various habitats in Pennsylvanian-age fossil plant communities by different lineages (lycopsids, seed plants, ferns, and sphenopsids), and T. Feild and colleagues (personal communication) have inferred the maintenance of an ecological niche (disturbed understory) in several early angiosperm lineages.

Whereas clear cases of simple conservatism occur, it is likely that with a large community the phylogenetic distribution of traits is a complex mix of conservatism and convergence. A different phylogenetic scope of a study (e.g., a community of *Quercus* versus all angiosperms) might thus have a strong influence on the community phylogenetic structure observed. Silvertown et al. (1999) demonstrated for meadow plants that mean pairwise co-occurrence of species in a hydrologically defined niche-space was less than expected, indicating significant divergences in habitat use among species. Using the same methods, comparing just the species within a genus, they found a few cases of segregation of species, some cases with random overlap, and some examples of higher than expected niche overlap (Silvertown et al. 2001). This mixture of overlap and segregation was also seen at higher taxonomic levels, although the broadest comparison showed eudicots and monocots to be more segregated than expected. These results indicate that the underlying ecological traits exhibited varying patterns of divergence and stasis (i.e., convergence and conservatism) corresponding to the observation of segregation and overlap, respectively. In a related study of community assembly, Tofts & Silvertown (2000) assessed the effect of environmental filtering on grassland community membership, finding that trait values for species in a local community were less variable than those in a regional pool, independent of phylogenetic effects; they used phylogenetic independent contrasts to remove the effect of phylogenetic trait conservatism.

The trait-phylogeny-community relationship may also appear to be random: Winston (1995) found no difference in co-occurrence rate between groups of phylogenetically closely related and more distantly related stream fish, even though co-occurrence was less among morphologically similar species than among less similar ones, a result he attributed to the effects of competition. Barraclough et al. (1999) tested whether species of tiger beetle that co-occurred in the same habitat at a locality had lower than expected similarity in various ecomorphological characters, using a phylogenetically based null model. They found no evidence for character divergence between co-regional or co-occurring species, or for habitat divergence in co-regional species. Divergence patterns were indistinguishable from random character change (see also McCallum et al. 2001).

Some studies perform the equivalent of phylogenetic independent contrast analysis (Felsenstein 1985) on the association between particular traits and community membership. For example, Chazdon et al. (2002) asked whether there were associations between reproductive traits and forest types for Costa Rican trees, and found that the significant associations found could be explained by the different phylogenetic composition of the forest types and the generally phylogenetically conservative nature of the reproductive traits. Ibarra-Manriquez et al. (2001) showed the same outcome for differences in seedling germination type among different forests. Other studies have looked primarily at the distribution of ecologically related traits of species from a species pool, without reference to community-level variation in species composition (Grandcolas 1993, Brandl et al. 1994). For example, Böhning-Gaese & Oberrath (1999) found little evidence of conservatism in ecological traits for 151 co-occurring bird species.

Biogeographic History, Character Evolution, and Community Assembly

Key question: Where and under what community conditions did patterns of niche-use originate? [Using: species list of local communities + distribution of species within communities + phylogeny of full lineage (and of other lineages) + ecological trait data for lineage members + biogeographic reconstruction of regional distribution of ancestral taxa]

When phylogenetic and ecological information is available for all extant taxa in a lineage, including species in and outside of a particular community, ancestral character reconstruction of ecological traits and niche use can be examined (within limits of reconstruction methods; Cunningham et al. 1998). The questions of niche-use evolution described above can thus be answered more accurately (e.g., McPeek & Brown 2000).

Including all taxa in a lineage permits assessment of the contemporary geographic distribution of clades. Where all the members of a clade are currently co-regional, and no members of other (closely related) clades are present, the region may be considered to be closed (without immigration), and character evolution methods can also be used to reconstruct the evolution of community structure. Furthermore, where there is evidence that (*a*) intra-community interactions occurred among all taxa in the region (often assumed) and (*b*) no major interactors in the historical communities came from other clades (e.g., ants versus rodents), trait changes can be interpreted as both cause and effect: The ancestral communities can be reconstructed and change in characters can be interpreted as the response to character states in other ancestral species. Work by Losos and colleagues (Losos 1992, 1995; Losos et al. 1998) on Greater Antillean *Anolis* lizards exemplifies this approach: the majority of species on each island result from intra-island radiations, and a similar pattern of habitat use (e.g., crown, trunk, twig) has evolved on each island within each radiation. This method also allowed the researchers to reconstruct the composition of ancestral communities and to infer that diversification in habitat use was a response to the habitats already occupied.

Most regions are, however, open (Losos 1996), i.e., they contain members of clades that can be inferred (using biogeographic methods) to have originated elsewhere and to have arrived in a region by dispersal. For example, McPeek & Brown (2000) reconstructed the phylogeny of all *Enallagma* (damselfly) species in eastern North America and were able to differentiate between a recent radiation in New England and an older diversification in the southeast United States; most eastern lakes from the Gulf of Mexico to Canada, however, contain members of both clades. Combining such biogeographic data with trait reconstructions permits the determination of whether traits and current niche-use evolved in situ or whether the traits were established elsewhere, outside the current region (Brown & Zeng 1989). Cadle & Greene (1993) analyzed the clade composition and size distribution of 15 Neotropical colubrid snake communities to test whether the overall distribution in sizes in any community was independent of the clade composition. They found that (*a*) the overall size distribution of a community bore a direct relationship to the clade composition, (*b*) different clades had different mean sizes (and size-associated differences in diet), and (*c*) the lineages appear to have originated and diversified in isolated areas. The work thus indicated the role of history in shaping the regional pool of species and did not support the hypothesis that snake communities are organized by processes that tend to maintain some overall distribution in size. In his study of Lesser Antillean *Anolis* species, Losos (1992) noted that the one or two species on each island are not generalists, as the ancestors of the Greater Antilles lizards are inferred to be, but show conservatism of niche-use and appear to have been ecologically sorted onto the islands according to the available habitats (see also Vitt & Zani 1996, Zimmerman & Simberloff 1996, Sturmbauer 1998, Barker & Mayhill 1999, Craig et al. 2001, Galassi 2001). Richman (1996) used phylogenies to suggest that Japanese and European assemblages of *Phylloscopus* warblers were formed by multiple invasions from mainland Asia, with only limited subsequent speciation and little associated morphological diversification. There was strong conservatism in body size throughout the genus, irrespective of region, and no evidence of convergence in morphology or habitat-use in Europe and Japan. In addition (in contrast to previous analyses, e.g., Richman & Price 1992), habitat-use (low-elevation conifer versus high-elevation deciduous) divergence was shown to have mainly occurred early in the diversification of the lineage (although habitat-use was only inferred from morphology, not recorded directly).

It is often noted that the ecological character of species in today's plant communities reflects the biogeographic history of the species and their recent ancestors. For example, Lechowicz (1984) asked why temperate tree communities show a wide range in the time of leaf appearance in spring, using physiological, phylogenetic, and biogeographic data. There was no significant phylogenetic clustering of early versus late leafers (based on taxonomies available at the time), but early leafers were generally derived from cool temperate lineages, whereas late leafers were primarily of tropical lineage origin, reflecting evolutionary conservatism in physiological traits. Similarly, in subtropical montane forests of Mexico the evergreen understory is composed of tropical elements, while the deciduous overstory is composed primarily of species from temperate clades (Williams-Linera 1997; see also ter Steege & Hammond 2001).

Mediterranean-type ecosystems have been the subject of considerable study owing to the apparent convergences in plant and animal communities on five continents (Cody & Mooney 1978). Community-level convergence in distributions of ecological traits may arise by a combination of recent adaptive responses of the organisms to current environments and by sorting or assembly of lineages drawn from the regional species pools. Phylogenetic, biogeographic, and historical studies are key to evaluating these alternatives. In the plants of Andalusia, Spain, Herrera (1992) demonstrated that character syndromes differed between "old" and "new" lineages (distinguished by fossil records and intercontinental disjunctions). The persistence of ecological character in old lineages again emphasizes the importance of conservatism in traits contributing to community structure. Expanded and better resolved phylogenies now make it possible to evaluate such hypotheses with detailed studies of ancestral states in relation to community assembly. In California chaparral, D.D. Ackerly (unpublished) found that plants with similar leaf characteristics in the contemporary flora were derived from both subtropical and north-temperate lineages. In this case the subtropical lineages maintained ancestral traits that predated the mediterranean-type climate, while the temperate lineages exhibited more recent evolutionary shifts, presumably representing adaptations to changing climatic conditions compared with their biogeographic origins (see also Verdu et al. 2002).

Reconstructing ancestral communities and determining the causes of trait change is harder in open regions than in closed ones, but methods have been developed for this. When several lineages share the same area cladogram and can be assumed to have evolved in the same time frame, co-regionality can be inferred for ancestral species (co-speciation sensu Brooks & McLennan 1993). If community co-occurrence and biotic interaction can be reasonably assumed between members of different lineages, patterns of trait change can be interpreted as co-adaptations (sensu Brooks & McLennan 1993). Co-adaptation may most reasonably be inferred where trait change in one lineage is associated with putative dispersal into a region by a species in another lineage (see figure 3 in Losos 1996). Brooks & McLennan (1991) reconstructed the influence of pre-adaptation and co-adaptation in helminth parasite communities in Neotropical stingrays, considering different host species as different niches at six different sites in South America. They showed that three contemporary communities came about primarily by vicariant division of ancestral communities, two originated by dispersal of parasites into a region without switching host species (or niches), whereas one represented a complicated assembly of vicariance and dispersal, with some host conservatism and some host switching (see also Poulin 1999, Sasal et al. 1999, Morand & Guegan 2000). Mayden (1987) used geological and fish phylogeny information to reconstruct the history of vicariance in Ozark river drainages. Gorman (1992) then added information about habitat use (position in water column) of contemporary species to reconstruct ancestral community composition and habitat occupancy and was able to infer which competitive interactions were leading to contemporary niche displacement, which he tested experimentally. Losos et al. (1997) also applied an experimental approach to test hypotheses derived from biogeographic and trait evolution data, finding that *Anolis* limb length evolved adaptively in populations established on small islands.

Community Organization: Conclusions

A phylogenetic approach to studying community organization provides a new perspective on the perennial questions of the role of competition and the maintenance of diversity in communities, by highlighting the similarities of co-occurring species as well as the differences. A dominant perception in evolutionary ecology is that co-existing species must differ significantly and that most variation between closely related species is the adaptive response to past competition when species did not differ (e.g., Harvey & Rambaut 2000). The general prediction from this model of evolution is that divergence (and therefore homoplasy) in ecological character should be widespread. However, phylogenetic analysis reveals that many (possibly the majority of) lineages studied show evidence for conservatism of dominant ecological character, in both animal lineages (e.g., Richman 1996, Barraclough et al. 1999, Peterson et al. 1999, Lindeman 2000, McPeek & Brown 2000, Price et al. 2000, Forstmeier et al. 2001) and in plants (Grime 1984, Peat & Fitter 1994, Lord et al. 1995, Ackerly 1999, Prinzing et al. 2001). For plants this runs counter to a long-standing belief that functional and ecological strategies evolve rapidly, leading to widespread convergence and little correspondence between taxonomic and ecological groupings (e.g., Warming 1909, Cronquist 1988).

An associated insight arising from a phylogenetic perspective is that even if convergent evolution has occurred in a single trait (e.g., diet), other axes of ecological similarity (e.g., forest-type use) are often conserved. The more traits involved, the more likely it is that a composite measure of "net ecological similarity" will be conserved in a lineage, especially if there are life-history trade-offs among traits. Change in such a composite measure would perhaps be best modeled by evolutionary drift. Ecologically, difference on one niche axis alone may be sufficient to reduce competition with other species under stable conditions, but it is more likely that the competitive environment shifts over time, with species experiencing intermittant interactions with other species on different niche axes (food, shelter, water, nutrients, space). Hence, when viewed across many traits, interactions that challenge the co-existence of closely related, ecologically similar species may be more frequent than recognized.

If conservatism of ecological traits is widespread, then the effect of contemporary competition can be assessed by the extent to which phylogenetically related, ecologically similar species co-occur: avoidance of closely related species by one another (e.g., Graves & Gotelli 1993) would be indicative of strong competition among similar species. Because the strength of negative interaction should generally be proportional to the phylogenetic proximity, a likely community outcome might be a hierarchical pattern of both phylogenetic clumping and overdispersion: Some conserved characters will determine the ability of taxa to occupy a fundamental niche (leading possibly to overall phylogenetic attraction of co-occurring taxa, e.g., taxa 5, 6, and 7 in Figure 2), whereas others will cause local competitive exclusion (leading to phylogenetic repulsion of co-occurring taxa within those attracted clades, e.g., taxa 6 and 7 in Figure 2). Where related, similar species do co-occur, attention must be given to mechanisms that permit the co-existence of similar rather than different organisms (e.g., Chesson & Warner 1981, McPeek & Brown 2000, Hubbell 2001).

The study of adaptive radiations on islands (e.g., Givnish 1998) may have led to an overemphasis on evolutionary character displacement and ecological niche partitioning. In diverse, continental communities (in which most species live), interactions may be more unpredictable over time, both on the timescale of individuals (because of the diversity of neighbors) (Connell 1980), and because species ranges may change more often. This would lead to weaker pairwise interactions among sister taxa, and species radiations might occur with little ecological differentiation (change being nonadaptive) (sensu McPeek & Brown 2000, e.g., Richardson et al. 2001). Local communities on continents may then represent a selective sampling of the regional pool to minimize ecological similarity or may only be able to contain species that are similar (over long periods of time, if not indefinitely). On islands, species must change or go extinct. A systematic review of the prevalence of conservatism and convergence in both island and continental systems would be very valuable.

Incorporating phylogenetic information offers important new perspectives but also brings additional challenges. Perhaps the most immediate challenge is to explicitly define the ecological, spatial, and taxonomic scales in a study because the processes that structure the assembly of regions, communities, and habitats differ (see The Phylogenetic Structure of Community Assemblages, above), and using phylogenies only of taxa that co-occur at a particular spatial scale can confound inferences about ecological and evolutionary processes (see Jablonski & Sepkoski 1996). Although it is valid to assess the correlation between relatedness and similarity (i.e., the effective degree of conservatism) in any sample of species, it must be remembered that partial (community) phylogenies may not provide accurate inferences about character evolution or lineage-wide assessment of conservatism and homoplasy (the issue of "taxon sampling"; Ackerly 2000). An example is given in Figure 2, in which the taxa included in the community phylogeny would lead to an incorrect reconstruction of the evolution of traits y and z. We recognize that all phylogenies of extant taxa are only partial samples, owing to extinction, but phylogenies for co-occurring species are particularly egregious samples and are likely to introduce systematic bias in the study of trait evolution owing to the distribution of ecological characters in different communities.

Obtaining estimates of the absolute lengths of branches in a partial (community) phylogeny greatly increases the accuracy of inference about the correlation between relatedness and similarity. Estimates of relative branch lengths often come from rates of molecular evolution but ultimately depend upon the dating and phylogenetic placement of fossils. Age estimates are also vital for the accurate reconstruction of the species composition of ancestral communities. Congruence of the topologies of different lineages without age information cannot determine the order of arrival of taxa in an ancestral region or the temporal order of trait change in the lineages. Disentangling cause from effect requires temporal information (see Donoghue et al. 2001, Hunn & Upchurch 2001, Sanmartin et al. 2001).

A related challenge when using regional-scale biogeographic methods to reconstruct historical intra-community interactions is that community co-occurrence of ancestral taxa must be assumed. The spatial scale of the areas in an area cladogram is usually far larger than a single community (Grandcolas 1998), and coregional ancestral species may never have interacted, because they occurred only in different types of communities or because they were spatially segregated by chance while still sharing a region. However, phylogeographic methods (Avise 2000) may permit the use of intra-species genetic variation to reconstruct historical population movement and to indicate patterns of intra-region historical co-occurrence (e.g., Zink 1996, Comes & Kadereit 1998, Taberlet et al. 1998). Fossil communities (e.g., Wing et al. 1993, Hadly & Maurer 2001) might also be used to confirm community co-occurrence.

EMERGENT PROPERTIES OF COMMUNITIES

Most work to date at the intersection of phylogenetics and community ecology has dealt with the presence and absence of species in a community, associated differences in traits, and the assembly of communities through time. Community ecology also deals with species diversity, the relative abundance of co-occurring species, and the distribution of range sizes. There have been a number of creative uses of phylogeny in these areas.

Species Diversity

Key question: Why do different areas vary in the species richness of particular clades? (Using: species lists of local communities + local and global estimates of species richness of clades + phylogenetic relationships of local and global clades)

One answer to the question "Why do similar habitats in different regions have different numbers of species?" is that differing histories of the areas have led to occupancy by different clades (Latham & Ricklefs 1993, Schluter & Ricklefs 1993, Qian & Ricklefs 1999, Ricklefs 2002) and that different clades have different potentials for diversification (Farrell et al. 1991, Sanderson & Donoghue 1996, Dodd et al. 1999, Gardezi & da Silva 1999) and thus different numbers of extant species. A second answer is that the areas differ in the length of time they have been occupied (Brown et al. 2000, Ricklefs 2002). For example, a latitudinal gradient in diversity may reflect the relative ages of major climate regions (and the challenges of adapting to temperate climates) as much as the effect of local ecological processes (Blondel & Vigne 1993, Latham & Ricklefs 1993). Phylogenetic analyses and age inferences are central to testing such hypotheses: For example, Ricklefs & Schluter (1993b) found that the clades of passerine birds in Panamanian forest were on average 2.6 times the age of the clades in forests in Illinois.

The dominant hypothesis for why islands differ in species richness is MacArthur & Wilson's (1967) equilibrium theory of island biogeography. Phylogenies are playing a major role in testing the applicability of this hypothesis. Using molecular estimates of species age, Ricklefs & Bermingham (2001) were able to reject the assumption of constant rates of immigration and extinction in Antillean birds. Using island species lists and a phylogeny, Losos & Schluter (2000) were able to differentiate between in situ speciation of *Anolis* lizards on Caribbean islands, and immigrants from other islands. They showed that on islands larger than 3,000 km^2, in situ speciation overtakes immigration as a source.

Relative Abundance

Key question: How does the relative abundance of taxa vary across a community phylogeny? (Using: species list of local community + relative abundance structure of local community + phylogeny of species in local community)

The distribution of relative abundance in communities has long been the subject of attention by ecologists. Whereas there are many ways to statistically characterize the shape of abundance-distribution curves (e.g., log-series, log-normal), satisfactory explanatory models have been few (but see MacArthur 1960). Fewer still make predictions about the relation of phylogeny to relative abundance. The most comprehensive model of community abundance (Hubbell 2001) predicts that older species should be more abundant and widespread than younger species (in itself, an old idea; Willis 1922). Methods exist to date nodes in a phylogeny, but the pruning of phylogenies by extinction means the time since divergence from the most closely related extant taxon may often be a poor predictor of species age (Box 2 in Chown & Gaston 2000).

An increasing number of studies have asked if there is a relation between characters of a taxonomic group and the mean abundance of its members (Farrell et al. 1991; see also Heard & Hauser 1995, Edwards & Westoby 2000, Murray & Westoby 2000). Schwartz & Simberloff (2001) found that vascular plant families with few species tended to have fewer than expected rare species. Such analyses will benefit from an explicit phylogenetic framework: Webb & Pitman (2002) found that a rank-based association between common species and diverse families of rain forest trees disappeared when considered phylogenetically.

Another approach to relative abundance is illustrated by the work on bird size and abundance (e.g., Cotgreave & Harvey 1994, Harvey & Nee 1994). In Britain, overall bird population size is negatively related to body size (Nee et al. 1991), but within a tribe the relationship is often positive. If ecological similarity and competition are correlated, and if larger bodied birds attain higher densities under competition than smaller birds, then those clades that contain the most ecologically similar species should show the strongest positive association of body size and abundance. Nee et al. (1991) found that the branch length (using taxonomic levels) from the base of a tribe's clade to the rest of the birds was a good predictor of the strength of the positive relationship, longer branches being associated with more complete guilds.

Examining relative abundance structure from a phylogenetic standpoint will surely be an exciting avenue of research. Finding any association between abundance and relatedness could indicate that local abundance is actively influenced by phylogenetically conserved characters.

Geographical Range

Key question: How do the sizes and spatial arrangement of species ranges vary across a phylogeny? (Using: species lists of local community or region + range information for these species + phylogeny for these species)

Range size can be treated as a continuous character, and its evolution and association with other characters assessed. The community context of the species in such analyses is generally not addressed. In the few cases studied, closely related species tend to have more similar range sizes than distantly related species (Jablonski 1987, Ricklefs & Latham 1992, Brown 1995). This might appear to imply that (*a*) some ecological traits are responsible for range size and (*b*) there is some conservatism in these traits. However, an alternative explanation for such a pattern is that related species tend to be of a more similar age than less related species, and if age is correlated with area (Willis 1922, Fjeldsa & Lovett 1997), then area would appear as a phylogenetically conserved attribute of species. Kelly & Woodward (1996) investigated the correlation between life-form and range size in British plants, using a phylogeny, and found that trees have larger ranges than nontrees, and that wind-pollinated species have larger ranges than related non–wind-pollinated species. Gregory (1995) found that phylogenetic conservatism did not explain the relationship between range size and body size in British birds. Gotelli & Taylor (1999) used phylogenetic independent contrasts in an analysis correlating the probability of stream colonization by fish with body size, population size, range size, and distance from range center. They found that the importance of removing phylogenetic effects to detecting the effect of distance-from-range-center was substantial.

Range information has also been used to investigate modes of speciation, reasoning in the "opposite direction" from most work reviewed here (Lynch 1989, Barraclough & Nee 2001, Gimaret-Carpentier et al. 2002). Species range size is likely to be closely linked to the probability of further diversification under any model of allopatric speciation (Barraclough & Vogler 2000, Chown & Gaston 2000). Barraclough et al. (1998) used the range overlap of clades as an indicator of the mode of speciation in tiger beetles. If allopatric speciation and subsequent range movement dominated, the degree of range overlap between sister clades should start low for two sister species and increase as more taxa are included in both clades. However, sympatric speciation and subsequent range movement should start with high range overlap of sister species and decrease as more taxa are included. The authors found that overlap started low and increased with increasing clade inclusiveness and inferred allopatric speciation.

METHODOLOGICAL ADVANCES

Community Phylogenies

Phylogenetic methods are in a sustained phase of rapid development, with new maximum-likelihood (Lewis 2001) and Bayesian (Huelsenbeck et al. 2001) approaches being explored. Phylogenies from separate studies can now be joined to form "supertrees" (Sanderson et al. 1998, Bininda-Emonds et al. 1999), either being assembled "by hand" (e.g., Donoghue et al. 1998) or using algorithms to resolve conflict among trees (Semple & Steel 2000, Salamin et al. 2002, Bininda-Emonds et al. 2002). From these supertrees, phylogenies of community species lists can be prepared (e.g., Tofts & Silvertown 2000, Webb 2000); tools are now available to facilitate this process (Webb & Donoghue 2002). Community phylogenies constructed from supertrees usually lack information about branch length, but there are supertree methods that yield branch lengths (Lapointe & Cucumel 1997), and we anticipate that branch lengths based on absolute age estimates will soon be available for many groups (e.g., Magallon & Sanderson 2001).

Tests for Phylogenetic Conservatism

As we have emphasized, predictions and interpretation of patterns of phylogenetic community structure depend on patterns of ecological similarity and divergence among related species. Using taxonomic information, these patterns have been evaluated with hierarchical analysis of variance, partitioning interspecific variation into different levels: species within genera, genera within families, families within orders, etc. (e.g., Mazer & Wheelwright 1993, Peat & Fitter 1994, Lord et al. 1995). With the development of well-resolved phylogenies, a variety of quantitative methods have been proposed to examine the extent to which ecological traits are conserved or convergent on a phylogeny. For discrete characters, randomization-based tests of the number of reconstructed character changes have been used (Maddison & Slatkin 1991, Barraclough et al. 1999); a conserved character has fewer changes, or "steps," than expected based on the number of occurrences in the terminal taxa. Such tests are easily modified to handle ordered or continuous characters (the Quantitative Convergence Index; Ackerly & Donoghue 1998, Prinzing et al. 2001). An alternative method for continuous traits is based on regressions of trait differences versus phylogenetic distance (Legendre et al. 1994); a positive correlation indicates that traits are conserved. The phylogenetic "neighborhood" over which trait conservatism is evident can be assessed with phylogenetic "autocorrelation" (Cheverud et al. 1985, Gittleman et al. 1998, Böhning-Gaese & Oberrath 1999). Owing to the nonparametric structure of phylogenetic data, significance testing often requires randomization methods or null model simulations (Legendre et al. 1994, Lapointe & Garland 2001).

Despite the proliferation of such tests, few comparisons have been conducted to evaluate their performance on common data sets. Morales (2000) conducted one such comparison, and Ackerly has reanalyzed his data with several additional methods (D. Ackerly, unpublished data). Ackerly found that hierarchical ANOVA, the quantitative convergence index, and phylogenetic correlation of distance matrices give parallel results across different traits; however, there was little correspondence between these methods and phylogenetic autocorrelation or eigenvector analyses. For applications to community data sets, we favor methods based on phylogenetic distance (regression or autocorrelation) rather than parsimony-based trait mapping, to avoid the suggestion of inferring patterns of historical trait evolution from just the community-based taxon sample. More work is needed to examine the statistical power of these methods and their sensitivity to different kinds of deviation from random patterns.

Null Models for Community Phylogenies and for Community Assembly

There has been extensive work on the generation of random phylogenies (e.g., Raup et al. 1973, Losos & Adler 1995, Heard & Mooers 2000), but few studies have employed an explicitly ecological model (but see Maley 1998, Doebeli & Dieckmann 2000). Hubbell (2001) has claimed that a realistic null model for phylogenies must include information on a region's biota (his "meta-community"), because the probability of a taxon's extinction is inversely related to its population size, and the sum of all populations of all extant taxa is often limited (e.g., for canopy trees). The probability of extinction is therefore dependent on the number of species in the region because increasing species richness will tend to increase the number of species with small populations. Hubbell's (2001) null models for phylogenies generate patterns of hierarchical diversity (e.g., frequency distributions of species per family) that fit observed data well. Jansen & Mulder (1999) incorporated speciation into a patch-dynamic model to simulate the evolution of lineages in an explicitly competitive environment.

The simplest null models for community phylogenies are generated by subsampling the taxa in a larger area, using existing phylogenies for the relationships among those taxa (e.g., Webb 2000). The large literature on null models for the assembly of communities used to detect nonrandom co-occurrence patterns and assembly rules (usually independent of phylogeny/taxonomy) is relevant here (e.g., Diamond 1975, Connor & Simberloff 1979; reviewed by Gotelli & Graves 1996). Null models also exist for the distribution of ecological traits expected in communities where competition is important (Colwell & Winkler 1984, Leibold 1998, Stevens & Willig 2000) and for the evolution of traits in lineages without effects of competition (e.g., Ackerly 2000); these approaches need to be combined in future models.

Metrics of Community Phylogenetic Structure

Metrics that quantify the distribution of taxa in a sample relative to a pool have been developed by Webb (2000). The net relatedness index (NRI) is a standardized measure of the mean pairwise phylogenetic distance of taxa in a sample, relative to a phylogeny of an appropriate species pool, and quantifies overall clustering of taxa on a tree (similar to Clarke & Warwick's 1999 and von Euler & Svensson's 2001 metrics). It is calculated as $-1 \cdot ((\text{mn}(X_{\text{obs}}) - \text{mnX}(n))/\text{sdX}(n))$, where X_{obs} is the phylogenetic distance between two taxa (the sum of all intervening branch lengths) in the phylogeny of the pool, $\text{mn}(X_{\text{obs}})$ is the mean of all possible pairs of n taxa, and $\text{mnX}(n)$ and $\text{sdX}(n)$ are the mean and standard deviation expected for n taxa randomly distributed on the phylogeny of the pool (found by multiple iteration; note that this formulation is slightly modified from Webb 2000). Where continuous branch length estimates are not available, phylogenetic distances can be based on the number of nodes separating two taxa (Farris 1969, Gittleman & Kot 1990). The nearest taxon index (NTI) is a standardized measure of the phylogenetic distance to the nearest taxon for each taxon in the sample and quantifies the extent of terminal clustering, independent of deep level clustering. NTI is calculated as $-1 \cdot ((\text{mn}(Y_{\text{obs}}) - \text{mnY}(n))/\text{sdY}(n))$, where Y_{obs} is the phylogenetic distance to the nearest taxon in the phylogeny of the pool; $\text{mn}(Y_{\text{obs}})$, $\text{mnY}(n)$, and $\text{sdY}(n)$ are calculated as for X.

These metrics share much in common with those developed to assess the phylogenetic uniqueness of taxa in a conservation area (e.g., Williams et al. 1991, Faith 1996, Crozier 1997, Nee & May 1997, Clarke & Warwick 1999, Sechrest et al. 2002). Both NRI and NTI increase with increasing clustering and become negative with overdispersion. The precise response of NRI and NTI in communities formed by phenotypic attraction (Table 1) depends upon the form the trait conservatism takes. Maximum conservatism in traits, at a deep level (leading to a high consistency index), yields both high NRI and NTI. Conservatism at more terminal levels in the phylogeny causes NTI to increase in significance relative to NRI. Both NRI and NTI depend on the particular species pool, and further study is required to determine when and how these measures can be compared across different studies. A suitable null model of community assembly (see above) can be used to generate expectations for the distribution of relatedness indices with which the observed values can be compared. An alternative approach to assessing whether the taxa that co-occur in samples are more related than expected by chance is to correlate a metric of co-occurrence with phylogenetic distance for all possible pairs of taxa (H. Steers, personal communication).

DIRECTIONS FOR FUTURE WORK

Beyond the directions already taken and reviewed in this paper, we have identified a number of areas that might be profitable to pursue.

Dynamics of Community Phylogenetic Structure

The static patterns of community phylogenetic structure described above (Webb 2000; H. Steers personal communication, J. Cavender-Bares personal communication) result from differential mortality of species that vary in phylogenetic relatedness and ecological characteristics. Changes in phylogenetic structure could also be observed directly over time in the many existing long-term datasets of community composition. In age- and size-structured populations, comparing the community structure of different age- or size-classes at a single time can provide a (limited) proxy for the direct observation of change over time (e.g., Webb & Peart 1999). For example, increasing size classes of seedlings and trees in small plots in Bornean rain forest shows a monotonic increase in phylogenetic clustering (C.O. Webb, unpublished data). This pattern, at a single time, is consistent with the cumulative mortality of locally ill-suited species over time (if ecological suitability is phylogenetically conserved).

Using Phylogenetic Information in Models of Neighborhood Performance

Most models of the performance response (growth, survival) of focal individuals to neighborhood density classify neighbors either as conspecifics or heterospecifics. This dichotomy hides a great range of ecological similarity between species and an expected range of magnitude of effect. Weighting the interaction by a measure of phylogenetic relatedness should greatly improve the performance of such models, if important parameters of ecological similarity are phylogenetically conserved. For instance, if (*a*) negative neighborhood interactions are mediated by pathogens or herbivores (e.g., Gilbert et al. 1994), (*b*) at least some of the pathogen species infect multiple host species, and (*c*) at least some of the polyphagous pathogens have a phylogenetically restricted set of host species (Futuyma & Mitter 1996, Farrell 2001, Frenzel & Brandl 2001, Novotny et al. 2002), then the expected effect of neighbor density will be greater the more closely related it is to the focal individual. Analytical models of community stability, based on modified Lotka-Voltera competition models with phylogenetically structured interaction coefficients; may also be possible and would be expected to predict the maintenance of a phylogenetically diverse (or overdispersed) set of species.

Comparative Surveys of Community Phylogenetic Structure

Understanding variation in community phylogenetic structure across known gradients (e.g., moisture regime and species richness) may yield important insights into community organization (Thompson et al. 2001). Including gradients that integrate change in both resources and predation (e.g., Leibold 1996) will be especially revealing. The basic analysis of community phylogenetic structure requires only plot-based samples and a species list (which can be converted into a community supertree), and can thus be rapidly conducted on preexisting data. Where phylogenies can be constructed for fossils (e.g., Upchurch 1995, Vermeij & Carlson 2000) and a stratigraphic turnover of communities can be reconstructed (e.g., Olszewski & Patzkowsky 2001, Jackson & Overpeck 2000), change in community phylogenetic structure could be assessed over time.

Phylogenetic Ordination and Classification

Basing ordination and classification methods on intersample distances that reflect net phylogenetic dissimilarity rather than Euclidean distance in N-dimensional species space offers a means to display the phylogenetic relations among sample-plots. Such methods can reveal meaningful ecological relationships hidden by standard, nonphylogenetic methods: e.g., plots sharing many genera should still cluster even if they share none of the same species.

Balance of Community Phylogenies

Tree balance (the degree to which sister clades differ in their number of taxa) provides another way to quantify the complex branching structure of community phylogenies (e.g., Heard & Mooers 2000). Models relating the phylogenetic distribution of niche space among taxa in a regional pool and the niche structure of local communities should generate predictions about the balance of local community phylogenies.

CONCLUSIONS

We resist the temptation to declare that "phylogenetic community ecology" is a new field. Rather, we view phylogenetic information as a "glue" that can stick ecological and evolutionary studies together, where often they have slid past each other, their practitioners speaking different languages. We want to emphasize, however, that despite its great utility, there is no simple, single way to apply phylogenetic information in community ecology, as is highlighted by the diversity of approaches reviewed here. Phylogenies must also be used with full knowledge of the assumptions and uncertainties that underlie them. There is a real danger that with the increasing ease of obtaining phylogenetic information, ecologists will forget that phylogenies are hypotheses to be further tested, and not the truth. This said, we genuinely believe that no ecological study can fail to benefit in some way from an understanding of the phylogenetic relationships of its taxa. Community ecologists and phylogenetic biologists should continue to engage in a discussion that will surely enrich and hopefully unite both disciplines.

ACKNOWLEDGMENTS

The synthesis presented here has grown out of discussions with many people over the past four years, to whom we are very grateful. We especially thank M. Ashton, P. Ashton, D. Baum, A. Blundell, J. Cavender-Bares, R. Chazdon, F. Cohen, J. Connell, R. Colwell, S. Davies, B. Enquist, T. Givnish, S. Heard, S. Hubbell, J. Losos, M. Martinez-Ramos, D. Peart, N. Pitman, R. Ree, M. Silman, P. Stevens, D. Wagner, M. Westoby, S. Zens, the Donoghue labs at Harvard and Yale, a 1999 graduate seminar at Harvard, and an NCEAS working group on Life History Strategies of Neotropical Trees. As the manuscript was being finalized, a meeting was held at NCEAS on "Phylogenies and Community Ecology" (organized by Donoghue & Webb), and we valued the opportunity to discuss many of the ideas in this paper with those who attended. J. Losos, E. Palkovacs, and especially B. Shaffer gave useful feedback on a draft of this manuscript. C.O.W. has been funded by a Mercer Fellowship of the Arnold Arboretum and a Donnelley Fellowship of the Yale Institute for Biospheric Studies.

The *Annual Review of Ecology and Systematics* is online at
http://ecolsys.annualreviews.org

LITERATURE CITED

Ackerly DD. 1999. Comparative plant ecology and the role of phylogenetic information. In *Physiological Plant Ecology*, ed. M Press, JD Scholes, MG Barker, pp. 391–413. Oxford, UK: Blackwell Sci.

Ackerly DD. 2000. Taxon sampling, correlated evolution, and independent contrasts. *Evolution* 54:1480–92

Ackerly DD, Donoghue M. 1998. Leaf size, sapling allometry, and Corner's rules: phylogeny and correlated evolution in maples (*Acer*). *Am. Nat.* 152:767–91

Ackerly DD, Dudley SA, Sultan SE, Schmitt J, Coleman JS, et al. 2000. The evolution of plant ecophysiological traits: recent advances and future directions. *Bioscience* 50:979–95

Ashton PS. 1988. Dipterocarp biology as a window to the understanding of tropical forest structure. *Annu. Rev. Ecol. Syst.* 19:347–70

Avise JC. 2000. *Phylogeography: the History and Formation of Species*. Cambridge, MA: Harvard Univ. Press

Barker GM, Mayhill PC. 1999. Patterns of

diversity and habitat relationships in terrestrial mollusc communities of the Pukeamaru ecological district, northeastern New Zealand. *J. Biogeogr.* 26:215–38

Barraclough TG, Hogan JE, Vogler AP. 1999. Testing whether ecological factors promote cladogenesis in a group of tiger beetles (Coleoptera: Cicindelidae). *Proc. R. Soc. London Ser. B* 266:1061–67

Barraclough TG, Nee S. 2001. Phylogentics and speciation. *Trends Ecol. Evol.* 16:391–99

Barraclough TG, Vogler AP. 2000. Detecting the geographical pattern of speciation from species-level phylogenies. *Am. Nat.* 155:419–34

Barraclough TG, Vogler AP, Harvey PH. 1998. Revealing the factors that promote speciation. *Philos. Trans. R. Soc. London Ser. B* 353:241–49

Bininda-Emonds ORP, Gittleman JL, Purvis A. 1999. Building large trees by combining phylogenetic information: a complete phylogeny of the extant Carnivora (Mammalia). *Biol. Rev.* 74:143–75

Bininda-Emonds ORP, Gittleman JL, Steel M. 2002. The (super)tree of life: procedures, problems, and prospects. *Annu. Rev. Ecol. Syst.* 33:265–89

Blackburn TM, Gaston KJ. 2001. Local avian assemblages as random draws from regional pools. *Ecography* 24:50–58

Blondel J, Vigne J-D. 1993. Space, time and man as determinants of diversity of birds and mammals in the Mediterranean region. See Ricklefs & Schluter 1993a, pp. 135–46

Böhning-Gaese K, Oberrath R. 1999. Phylogenetic effects on morphological, life-history, behavioural and ecological traits of birds. *Evol. Ecol. Res.* 1:347–64

Bowers MA, Brown JH. 1982. Body size and coexistence in desert rodents: chance or community structure? *Ecology* 63:391–400

Brandl R, Kristin A, Leisler B. 1994. Dietary niche breadth in a local-community of passerine birds, an analysis using phylogenetic contrasts. *Oecologia* 98:109–16

Brooks DR, McLennan DA. 1991. *Phylogeny, Ecology, and Behavior: A Research Program*

in Comparative Biology. Chicago: Univ. Chicago Press

Brooks DR, McLennan DA. 1993. Historical ecology: examining phylogenetic components of community evolution. See Ricklefs & Schluter 1993a, pp. 267–80

Brooks DR, McLennan DA. 2002. *The Nature of Diversity.* Chicago: Univ. Chicago Press

Brown JH. 1995. *Macroecology.* Chicago: Univ. Chicago Press

Brown JH, Zeng Z. 1989. Comparative population ecology of eleven species of rodents in the Chihuahuan desert. *Ecology* 70:1507–25

Brown JM, McPeek MA, May ML. 2000. A phylogenetic perspective on habitat shifts and diversity in the North American *Enallagma* damselflies. *Syst. Biol.* 49:697–712

Cadle JE, Greene HW. 1993. Phylogenetic patterns, biogeography, and the ecological structure of Neotropical snake assemblages. See Ricklefs & Schluter 1993a, pp. 281–93

Cavender-Bares J, Holbrook NM. 2001. Hydraulic properties and freezing-induced xylem cavitation in evergreen and deciduous oaks with contrasting habitats. *Plant, Cell Environ.* 24:1243–56

Chazdon RL, Careaga S, Webb CO, Vargas O. 2002. Community and phylogenetic structure of reproductive traits of woody species in wet tropical forests. *Ecol. Monogr.* In press

Chesson P. 2000. Mechanisms of maintenance of species diversity. *Annu. Rev. Ecol. Syst.* 31:343–66

Chesson PL, Warner RR. 1981. Environmental variability promotes coexistence in lottery competitive systems. *Am. Nat.* 117:923–43

Cheverud JM, Dow MM, Leutnegger W. 1985. The quantitative assessment of phylogenetic constraints in comparative analyses: sexual dimorphism in body weight among primates. *Evolution* 39:1335–51

Chown SL, Gaston KJ. 2000. Areas, cradles and museums: the latitudinal gradient in species richness. *Trends Ecol. Evol.* 15:311–15

Clarke KR, Warwick RM. 1999. The taxonomic distinctness measure of biodiversity: weighting of step lengths between hierarchical levels. *Mar. Ecol. Prog. Ser.* 184:21–29

Cody ML, Mooney HA. 1978. Convergence versus non-convergence in Mediterranean-climate ecosystems. *Annu. Rev. Ecol. Syst.* 9:265–321

Colwell RK, Winkler DW. 1984. A null model for null models in biogeography. In *Ecological Communities: Conceptual Issues and the Evidence*, ed. DR Strong, D Simberloff, LG Abele, AB Thistle, pp. 344–59. Princeton, NJ: Princeton Univ. Press

Comes HP, Kadereit JW. 1998. The effects of Quaternary climatic changes on plant distribution and evolution. *Trends Plant Sci.* 3:432–38

Connell JH. 1980. Diversity and the coevolution of competitors, or the ghost of competition past. *Oikos* 35:131–38

Connor EF, Simberloff D. 1979. The assembly of species communities: chance or competition? *Ecology* 60:1132–40

Cornell HV, Lawton JH. 1992. Species interactions, local and regional processes, and limits to the richness of ecological communities: a theoretical perspective. *J. Anim. Ecol.* 61:1–12

Cotgreave P, Harvey PH. 1994. Phylogeny and the relationship between body size and abundance in bird communities. *Funct. Ecol.* 8:219–28

Craig DA, Currie DC, Joy DA. 2001. Geographical history of the central-western pacific black fly subgenus *Inseliellum* (Diptera: Simuliidae: *Simulium*) based on a reconstructed phylogeny of the species, hot-spot archipelagoes and hydrological considerations. *J. Biogeogr.* 28:1101–27

Cronquist A. 1988. *The Evolution and Classification of Flowering Plants.* New York: NY Bot. Gard.

Crozier RH. 1997. Preserving the information content of species: genetic diversity, phylogeny, and conservation worth. *Annu. Rev. Ecol. Syst.* 28:243–68

Cunningham CW, Omland KE, Oakley TH. 1998. Reconstructing ancestral character states: a critical reappraisal. *Trends Ecol. Evol.* 13:361–66

Darwin C. 1859. *The Origin of Species by Means of Natural Selection.* London: Murray

Diamond JM. 1975. Assembly of species communities. In *Ecology and Evolution of Communities*, ed. ML Cody, JM Diamond, pp. 342–444. Cambridge, MA: Harvard Univ. Press

Dimichele WA, Phillips TL. 1996. Clades, ecological amplitudes, and ecomorphs: phylogenetic effects and persistence of primitive plant communities in the Pennsylvanian-age tropical wetlands. *Palaeogeogr. Palaeoclim. Palaeoecol.* 127:83–105

Dodd M, Silvertown J, Chase M. 1999. Phylogenetic analysis of trait evolution and species diversity variation among Angiosperm families. *Evolution* 53:732–44

Doebeli M, Dieckmann U. 2000. Evolutionary branching and sympatric speciation caused by different types of ecological interactions. *Am. Nat.* 156:S77–101

Donoghue MJ, Bell CD, Li J. 2001. Phylogenetic patterns in Northern Hemisphere plant geography. *Int. J. Plant Sci.* 162:S41–52

Donoghue MJ, Ree RH. 2000. Homoplasy and developmental constraint: a model and example from plants. *Am. Zool.* 40:759–69

Donoghue MJ, Ree RH, Baum DA. 1998. Phylogeny and the evolution of flower symmetry in the Asteridae. *Trends Plant Sci.* 3:311–17

Edwards W, Westoby M. 2000. Families with highest proportions of rare species are not consistent between floras. *J. Biogeogr.* 27:733–40

Eggleton P, Vane-Wright RI, eds. 1994. *Phylogenetics and Ecology.* London: Academic

Elton C. 1946. Competition and the structure of ecological communities. *J. Anim. Ecol.* 15:54–68

Enquist BJ, Haskell JP, Tiffney BH. 2002. General patterns of taxonomic structure and biomass division within extant and fossil woody plant communities. *Nature.* In press

Faith DP. 1996. Conservation priorities and phylogenetic pattern. *Conserv. Biol.* 10:1286–89

Farrell BD. 2001. Evolutionary assembly of the milkweed fauna: cytochrome oxidase I and the age of *Tetraopes* beetles. *Mol. Phylogenet. Evol.* 18:467–78

Farrell BD, Dussourd DE, Mitter C. 1991. Escalation of plant defense: Do latex and resin canals spur plant diversification? *Am. Nat.* 138:881–900

Farris JS. 1969. A successive approximations approach to character weighting. *Syst. Zool.* 18:374–85

Felsenstein J. 1985. Phylogenies and the comparative method. *Am. Nat.* 125:1–15

Fjeldsa J, Lovett JC. 1997. Geographical patterns of old and young species in African forest biota: the significance of specific montane areas as evolutionary centres. *Biodivers. Conserv.* 6:325–46

Forstmeier W, Bourski OV, Leisler B. 2001. Habitat choice in *Phylloscopus* warblers: the role of morphology, phylogeny and competition. *Oecologia* 128:566–76

Fox BJ. 1999. The genesis and devlopment of guild assembly rules. In *Ecological Assembly Rules: Perspectives, Advances, Retreats*, ed. E Weiher, P Keddy, pp. 23–57. Cambridge, UK: Cambridge Univ. Press

Frenzel M, Brandl R. 2001. Hosts as habitats: faunal similarity of phytophagous insects between host plants. *Ecol. Entomol.* 26:594–601

Futuyma DJ, Mitter C. 1996. Insect-plant interactions: the evolution of component communities. *Philos. Trans. R. Soc. London Ser. B* 351:1361–66

Galassi DMP. 2001. Groundwater copepods: diversity patterns over ecological and evolutionary scales. *Hydrobiologia* 453:227–53

Gardezi T, da Silva J. 1999. Diversity in relation to body size in mammals: a comparative study. *Am. Nat.* 153:110–23

Gentry AH. 1982. Patterns of Neotropical plant species diversity. *Evol. Biol.* 15:1–84

Gilbert GS, Hubbell SP, Foster RB. 1994. Density and distance-to-adult effects of a canker disease of trees in a moist tropical forest. *Oecologia* 98:100–8

Gimaret-Carpentier C, Dray S, Pascal J-P. 2002. Broad-scale biodiversity pattern analyses of the endemic tree flora of the Western Ghats

(India) using canonical correlation analysis of herbarium records. *Ecography.* In press

Gittleman JL, Anderson CG, Cates SE, Luh H-K, Smith JD. 1998. Detecting ecological patterns in phylogenies. In *Biodiversity Dynamics: Turnover of Populations, Taxa, and Communities*, ed. ML McKinney, JA Drake, pp. 51–69. New York: Columbia Univ. Press

Gittleman JL, Kot M. 1990. Adaptation: statistics and a null model for estimating phylogenetic effects. *Syst. Zool.* 39:227–41

Givnish TJ. 1998. Adaptive radiation of plants on oceanic islands: classical patterns, molecular data, new insights. In *Evolution on Islands*, ed. P Grant, pp. 281–304. Oxford, UK: Oxford Univ. Press

Gorman OT. 1992. Evolutionary ecology and historical ecology: assembly, structure, and organization of stream fish communities. In *Systematics, Historical Ecology, and North American Fresh Water Fishes*, ed. RL Mayden, pp. 659–88. Stanford, CA: Stanford Univ. Press

Gotelli NJ, Graves GR. 1996. *Null Models in Ecology.* Washington, DC: Smithsonian Inst. Press

Gotelli NJ, Taylor CM. 1999. Testing macroecological models with stream-fish assemblages. *Evol. Ecol. Res.* 1:847–58

Grandcolas P. 1993. The origin of biological diversity in a tropical cockroach lineage: a phylogenetic analysis of habitat choice and biome occupancy. *Acta Oecol.* 14:259–70

Grandcolas P. 1998. Phylogenetic analysis and the study of community structure. *Oikos* 82:397–400

Graves GR, Gotelli NJ. 1993. Assembly of avian mixed-species flocks in Amazonia. *Proc. Natl. Acad. Sci. USA* 90:1388–91

Gregory RD. 1995. Phylogeny and relations among abundance, geographical range and body size of British breeding birds. *Philos. Trans. R. Soc. London Ser. B* 349:345–51

Grime JP. 1984. The ecology of species, families and communities of the contemporary British flora. *New Phytol.* 98:15–33

Hadly EA, Maurer BA. 2001. Spatial and temporal patterns of species diversity in montane mammal communities of western North America. *Evol. Ecol. Res.* 3:477–86

Harvey PH, Nee S. 1994. Comparing real with expected patterns from molecular phylogenies. See Eggleton & Vane-Wright 1994, pp. 659–88

Harvey PH, Pagel MD. 1991. *The Comparative Method in Evolutionary Biology.* Oxford, UK: Oxford Univ. Press

Harvey PH, Rambaut A. 2000. Comparative analyses for adaptive radiations. *Philos. Trans. R. Soc. London Ser. B* 355:1599–605

Heard SB, Hauser DL. 1995. Key evolutionary innovations and their ecological mechanisms. *Hist. Biol.* 10:151–73

Heard SB, Mooers AO. 2000. Phylogenetically patterned speciation rates and extinction risks change the loss of evolutionary history during extinctions. *Proc. R. Soc. London Ser. B* 267:613–20

Herrera CM. 1992. Historical effects and sorting processes as explanations for contemporary ecological patterns: character syndromes in Mediterranean woody plants. *Am. Nat.* 140:421–46

Hubbell SP. 2001. *The Unified Neutral Theory of Biodiversity and Biogeography.* Princeton, NJ: Princeton Univ. Press

Huelsenbeck JP, Ronquist F, Nielsen R, Bollback JP. 2001. Bayesian inference of phylogeny and its impact on evolutionary biology. *Science* 294:2310–14

Hunn CA, Upchurch P. 2001. The importance of time/space in diagnosing the causality of phylogenetic events: towards a "chronobiogeographical" paradigm. *Syst. Biol.* 50:391–407

Ibarra-Manríquez G, Martínez-Ramos M, Oyama K. 2001. Seedling functional types in a lowland rain forest in Mexico. *Am. J. Bot.* 88:1801–12

Jablonski D. 1987. Heritability at the species level: analysis of geographic ranges of cretaceous mollusks. *Science* 238:360–63

Jablonski D, Sepkoski JJ. 1996. Paleobiology, community ecology, and scales of ecological pattern. *Ecology* 77:1367–78

Jackson ST, Overpeck JT. 2000. Responses of plant populations and communities to environmental changes of the late Quaternary. *Paleobiol.* 26:194–220

Jansen VAA, Mulder GSEE. 1999. Evolving biodiversity. *Ecol. Lett.* 2:379–86

Kelly CK. 1999. On the relationship between function and phylogenetic relatedness: environmental severity and community structure. *XVI Intl. Botan. Congr.* (Abstr.)

Kelly CK, Woodward FI. 1996. Ecological correlates of plant range size: taxonomies and phylogenies in the study of plant commonness and rarity in great britain. *Philos. Trans. R. Soc. London Ser. B* 351:1261–69

Lapointe F-J, Cucumel G. 1997. The average consensus procedure: combination of weighted trees containing identical or overlapping sets of taxa. *Syst. Biol.* 46:306–12

Lapointe F-J, Garland T. 2001. A generalized permutation model for the analysis of cross-species data. *J. Classif.* 18:109–27

Latham RE, Ricklefs RE. 1993. Continental comparisons of temperate-zone tree species diversity. See Ricklefs & Schluter 1993a, pp. 294–314

Lechowicz MJ. 1984. Why do temperate deciduous trees leaf out at different times? Adaptations and ecology of forest communities. *Am. Nat.* 124:821–42

Legendre P, Lapointe F, Casgrain P. 1994. Modeling brain evolution from behavior: a permutational regression approach. *Evolution* 48:1487–99

Leibold MA. 1996. A graphical model of keystone predators in food webs: trophic regulation of abundance, incidence and diversity patterns in communities. *Am. Nat.* 147:784–812

Leibold MA. 1998. Similarity and local coexistence of species in regional biotas. *Evol. Ecol.* 12:95–110

Lewis PO. 2001. Phylogenetic systematics turns over a new leaf. *Trends Ecol. Evol.* 18:30–37

Lindeman PV. 2000. Resource use of five sympatric turtle species: effects of competition, phylogeny, and morphology. *Can. J. Zool.* 78:992–1008

Lord J, Westoby M, Leishman M. 1995. Seed size and phylogeny in six temperate floras: constraints, niche conservatism, and adaptation. *Am. Nat.* 146:349–64

Losos JB. 1992. The evolution of convergent structure in Caribbean *Anolis* communities. *Syst. Biol.* 41:403–20

Losos JB. 1995. Community evolution in Greater Antillean *Anolis* lizards: phylogenetic patterns and experimental tests. *Philos. Trans. R. Soc. London Ser. B* 349:69–75

Losos JB. 1996. Phylogenetic perspectives on community ecology. *Ecology* 77:1344–54

Losos JB, Adler FR. 1995. Stumped by trees: a generalized null model for patterns of organismal diversity. *Am. Nat.* 145:329–42

Losos JB, Jackman TR, Larson A, de Queiroz K, Rogriguez-Schettino L. 1998. Contingency and determinism in replicated adaptive radiations of island lizards. *Science* 279:2115–18

Losos JB, Schluter D. 2000. Analysis of an evolutionary species-area relationship. *Nature* 408:847–50

Losos JB, Warheit KI, Schoener TW. 1997. Adaptive differentiation following experimental island colonization in *Anolis* lizards. *Nature* 387:70–73

Lynch JD. 1989. The gauge of speciation: on the frequency of modes of speciation. In *Speciation and Its Consequences*, ed. D Otte, JA Endler, pp. 527–53. Sunderland, MA: Sinauer

MacArthur RH. 1960. On the relative abundance of species. *Am. Nat.* 94:25–36

MacArthur RH, Levins R. 1967. The limiting similarity, convergence and divergence of coexisting species. *Am. Nat.* 101:377–85

MacArthur RH, Wilson EO. 1967. *The Theory of Island Biogeography*. Princeton, NJ: Princeton Univ. Press

Maddison WP, Slatkin M. 1991. Null models for the number of evolutionary steps in a character on a phylogenetic tree. *Evolution* 45:1184–97

Magallon S, Sanderson MJ. 2001. Absolute diversification rates in angiosperm clades. *Evolution* 55:1762–80

Maley CC. 1998. *The evolution of biodiversity: a simulation approach*. PhD thesis. Cambridge, MA: MIT

Mayden RL. 1987. Historical ecology and North American highland fishes: a research program in community ecology. In *Community and Evolutionary Ecology of North American Stream Fishes*, ed. WJ Matthews, DC Heins, pp. 210–22. Norman, OK: Univ. Okla. Press

Mazer SJ, Wheelwright NT. 1993. Fruit size and shape: allometry at different taxonomic levels in bird-dispersed plants. *Evol. Ecol.* 7:556–75

McCallum DA, Gill FB, Gaunt SLL. 2001. Community assembly patterns of parids along an elevational gradient in western China. *Wilson Bull.* 113:53–64

McPeek MA. 1996. Linking local species interactions to rates of speciation in communities. *Ecology* 77:1355–66

McPeek MA, Brown JM. 2000. Building a regional species pool: diversification of the *Enallagma* damselflies in eastern North American waters. *Ecology* 81:904–20

McPeek MA, Miller TE. 1996. Evolutionary biology and community ecology. *Ecology* 77:1319–20

Miles DB, Dunham AE. 1993. Historical perspectives in ecology and evolutionary biology—the use of phylogenetic comparative analyses. *Annu. Rev. Ecol. Syst.* 25:587–619

Miller JS, Wenzel JW. 1995. Ecological characters and phylogeny. *Annu. Rev. Entomol.* 40:389–415

Morales E. 2000. Estimating phylogenetic inertia in *Tithonia* (Asteraceae): a comparative approach. *Evolution* 54:475–84

Morand S, Guegan JF. 2000. Distribution and abundance of parasite nematodes: ecological specialisation, phylogenetic constraint or simply epidemiology? *Oikos* 88:563–73

Moreau RE. 1948. Ecological isolation in a rich tropical avifauna. *J. Anim. Ecol.* 17:113–26

Murray BR, Westoby M. 2000. Properties of species in the tail of rank-abundance curves:

the potential for increase in abundance. *Evol. Ecol. Res.* 2:583–92

Nee S, May RM. 1997. Extinction and loss of evolutionary history. *Science* 278:692–95

Nee S, Read AF, Greenwood JJD, Harvey PH. 1991. Phylogeny and the relationship between abundance and body size in British birds. *Nature* 351:312–13

Nel A, Nel J, Masselot G, Thomas A. 1998. An investigation into the application of the Wagner parsimony method in synecology. *Biol. J. Linn. Soc.* 65:165–89

Novotny V, Basset Y, Miller SE, Weiblen G, Bremer B, et al. 2002. Low host specificity of herbivorous insects in a tropical forest. *Nature* 416:841–44

Olszewski TD, Patzkowsky ME. 2001. Evaluating taxonomic turnover: Pennsylvanian-Permian brachiopods and bivalves of the North American midcontinent. *Paleobiology* 27:646–68

Peat HJ, Fitter AH. 1994. Comparative analyses of ecological characteristics of British angiosperms. *Biol. Rev.* 69:95–115

Peterson AT, Soberon J, Sanchez-Cordero V. 1999. Conservatism of ecological niches in evolutionary time. *Science* 285:1265–67

Poulin R. 1999. Speciation and diversification of parasite lineages: an analysis of congeneric parasite species in vertebrates. *Evol. Ecol.* 13:455–67

Price T, Lovette IJ, Bermingham E, Gibbs HL, Richman AD. 2000. The imprint of history on communities of North American and Asian warblers. *Am. Nat.* 156:354–67

Prinzing A, Durka W, Klotz S, Brandl R. 2001. The niche of higher plants: evidence for phylogenetic conservatism. *Proc. R. Soc. London Ser. B* 268:2383–89

Qian H, Ricklefs RE. 1999. A comparison of the taxonomic richness of vascular plants in China and the United States. *Am. Nat.* 154:160–81

Raup DM, Gould SJ, Schopf TJM, Simberloff DS. 1973. Stochastic models of phylogeny and the evolution of diversity. *J. Geol.* 81:525–42

Richardson JE, Pennington RT, Pennington TD,

Hollingsworth PM. 2001. Rapid diversification of a species-rich genus of Neotropical rain forest trees. *Science* 293:2242–45

Richman AD. 1996. Ecological diversification and community structure in the Old World leaf warblers(genus *Phylloscopus*): a phylogenetic perspective. *Evolution* 50:2461–70

Richman AD, Price T. 1992. Evolution of ecological differences in the Old World leaf warblers. *Nature* 355:817–21

Ricklefs RE. 1987. Community diversity: relative roles of local and regional processes. *Science* 235:167–71

Ricklefs RE. 2002. Phylogenetic perspectives on patterns of regional and local species richness. In *Tropical Rainforests: Past and Future*, ed. C Moritz, E Bermingham. In press

Ricklefs RE, Bermingham E. 2001. Nonequilibrium diversity dynamics of the Lesser Antillean avifauna. *Science* 294:1522–24

Ricklefs RE, Latham RE. 1992. Intercontinental correlation of geographical ranges suggests stasis in ecological traits of relict genera of temperate perennial herbs. *Am. Nat.* 139:1305–21

Ricklefs RE, Schluter D, eds. 1993a. *Species Diversity in Ecological Communities: Historical and Geographical Perspectives.* Chicago: Univ. Chicago Press

Ricklefs RE, Schluter D. 1993b. Species diversity: regional and historical influences. See Ricklefs & Schluter 1993a, pp. 350–63

Salamin N, Hodkinson TR, Savolainen V. 2002. Building supertrees: an empirical assessment using the grass family (Poaceae). *Syst. Biol.* 51:136–50

Sanderson MJ, Donoghue MJ. 1996. Reconstructing shifts in diversification rates on phylogenetic trees. *Trends Ecol. Evol.* 11:15–20

Sanderson MJ, Purvis A, Henze C. 1998. Phylogenetic supertrees: assembling the trees of life. *Trends Ecol. Evol.* 13:105–9

Sanmartin I, Enghof H, Ronquist F. 2001. Patterns of animal dispersal, vicariance and diversification in the Holarctic. *Biol. J. Linn. Soc.* 73:345–90

Sasal P, Niquil N, Bartoli P. 1999. Community structure of digenean parasites of sparid and labrid fishes of the Mediterranean sea: a new approach. *Parasitology* 119:635–48

Schluter D. 2000a. Ecological character displacement in adaptive radiation. *Am. Nat.* 156:S4–16

Schluter D. 2000b. *The Ecology of Adaptive Radiation*. Oxford, UK: Oxford Univ. Press

Schluter D. 2001. Ecology and the origin of species. *Trends Ecol. Evol.* 16:372–80

Schluter D, Ricklefs RE. 1993. Convergence and the regional component of species diversity. See Ricklefs & Schluter 1993a, pp. 230–41

Schwarz MW, Simberloff D. 2001. Taxon size predicts rates of rarity in vascular plants. *Ecol. Lett.* 4:464–69

Sechrest W, Brooks TM, da Fonseca GAB, Konstant WR, Mittermeier RA. 2002. Hotspots and the conservation of evolutionary history. *Proc. Natl. Acad. Sci. USA* 99: 2067–71

Semple C, Steel M. 2000. A supertree method for rooted trees. *Discrete Appl. Math.* 105: 147–58

Silvertown J, Dodd ME, Gowing DJG. 2001. Phylogeny and the niche structure of meadow plant communities. *J. Ecol.* 89:428–35

Silvertown J, Dodd ME, Gowing DJG, Mountford JO. 1999. Hydrologically defined niches reveal a basis for species richness in communites. *Nature* 400:61–63

Simberloff DS. 1970. Taxonomic diversity of island biotas. *Evolution* 24:23–47

Stevens RD, Willig MR. 2000. Community structure, abundance, and morphology. *Oikos* 88:48–56

Sturmbauer C. 1998. Explosive speciation in cichlid fishes of the African great lakes: a dynamic model of adaptive radiation. *J. Fish Biol.* 53:18–36

Taberlet P, Fumagalli L, Wust-Saucy A, Cosson J. 1998. Comparative phylogeography and postglacial colonization routes in Europe. *Mol. Ecol.* 7:453–64

ter Steege H, Hammond DS. 2001. Character convergence, diversity, and disturbance in tropical rain forest in Guyana. *Ecology* 82:3197–212

Thompson JN, Reichman OJ, Morin PJ, Polis GA, Power ME, et al. 2001. Frontiers of ecology. *Bioscience* 51:15–24

Tilman D. 1994. Competition and biodiversity in spatially structured habitats. *Ecology* 75:2–16

Tofts R, Silvertown J. 2000. A phylogenetic approach to community assembly from a local species pool. *Proc. R. Soc. London Ser. B* 267:363–69

Tokeshi M. 1991. Faunal assembly in chironomids (Diptera): generic association and spread. *Biol. J. Linn. Soc.* 44:353–67

Upchurch P. 1995. The evolutionary history of sauropod dinosaurs. *Philos. Trans. R. Soc. London Ser. B* 349:365–90

Verdu M, Barron-Sevilla JA, Valiente-Banuet A, Flores-Hernandez N, Garcia-Fayos P. 2002. Mexican plant phenology: Is it similar to Mediterranean communities? *Bot. J. Linn. Soc.* 138:297–303

Vermeij GJ, Carlson SJ. 2000. The muricid gastropod subfamily Rapaninae: phylogeny and ecological history. *Paleobiology* 26:19–46

Vitt LJ, Zani PA. 1996. Organization of a taxonomically diverse lizard assemblage in Amazonian Ecuador. *Can. J. Zool.* 74:1313–35

von Euler F, Svensson S. 2001. Taxonomic distinctness and species richness as measures of functional structure in bird assemblages. *Oecologia* 129:304–11

Wanntorp HE, Brooks DR, Nilsson T, Nylin S, Ronquist F. 1990. Phylogenetic approaches in ecology. *Oikos* 57:119–32

Warming E. 1909. *Oecology of Plants*. London: Oxford Univ. Press

Webb CO. 2000. Exploring the phylogenetic structure of ecological communities: an example for rain forest trees. *Am. Nat.* 156:145–55

Webb CO, Donoghue MJ. 2002. *Phylomatic: A Database for Applied Phylogenetics*. http://www.phylodiversity.net/phylomatic

Webb CO, Peart DR. 1999. Seedling density dependence promotes coexistence of Bornean rain forest trees. *Ecology* 80:2006–17

Webb CO, Peart DR. 2000. Habitat associations of trees and seedlings in a Bornean rain forest. *J. Ecol.* 88:464–78

Webb CO, Pitman NC. 2002. Phylogenetic balance and ecological evenness. *Syst. Biol.* In press

Westoby M, Leishman M, Lord J. 1995. On misinterpreting the "phylogenetic correction." *J. Ecol.* 83:531–34

Williams CB. 1964. *Patterns in the Balance of Nature. New York: Academic*

Williams PH, Humphries CJ, Vane-Wright RI. 1991. Measuring biodiversity: taxonomic relatedness for conservation priorities. *Austr. Syst. Bot.* 4:665–79

Williams-Linera G. 1997. Phenology of deciduous and broadleaved-evergreen tree species in a Mexican tropical lower montane forest. *Glob. Ecol. Biogeogr. Lett.* 6:115–27

Willis JC. 1922. *Area and Age: A Study of Geographical Distribution and Origin in Species.* Cambridge, UK: Cambridge Univ. Press

Wing SL, Hickey LJ, Swisher CC. 1993. Late Cretaceous vegetation. *Nature* 363:342–44

Winston MR. 1995. Cooccurrence of morphologically similar species of stream fishes. *Am. Nat.* 145:527–45

Wisheu IC. 1998. How organisms partition habitats: different types of community organization can produce identical patterns. *Oikos* 83:246–58

Zimmerman BL, Simberloff D. 1996. An historical interpretation of habitat use by frogs in a Central Amazonian forest. *J. Biogeogr.* 23:27–46

Zinc RM. 1996. Comparative phylogeography in North American birds. *Evolution* 50:308–17

Annu. Rev. Ecol. Syst. 2002. 33:507–59
doi: 10.1146/annurev.ecolsys.33.020602.095451
Copyright © 2002 by Annual Reviews. All rights reserved
First published online as a Review in Advance on August 14, 2002

Stable Isotopes in Plant Ecology

Todd E. Dawson[1], Stefania Mambelli[1],
Agneta H. Plamboeck[1], Pamela H. Templer[2],
and Kevin P. Tu[1]

[1]*Center for Stable Isotope Biogeochemistry and the Department of Integrative
Biology, University of California, Berkeley, California 94720;
email: tdawson@socrates.berkeley.edu, mambelli@socrates.berkeley.edu,
ktu@socrates.berkeley.edu, agneta@socrates.berkeley.edu*
[2]*Ecosystem Sciences Division, Department of Environmental Science,
Policy and Management, University of California, Berkeley, California 94720;
email: ptempler@nature.berkeley.edu*

Key Words   plant-environment interactions, resources, tracers, integrators, scaling, methods

■ **Abstract**   The use of stable isotope techniques in plant ecological research has grown steadily during the past two decades. This trend will continue as investigators realize that stable isotopes can serve as valuable nonradioactive tracers and nondestructive integrators of how plants today and in the past have interacted with and responded to their abiotic and biotic environments. At the center of nearly all plant ecological research which has made use of stable isotope methods are the notions of interactions and the resources that mediate or influence them. Our review, therefore, highlights recent advances in plant ecology that have embraced these notions, particularly at different spatial and temporal scales. Specifically, we review how isotope measurements associated with the critical plant resources carbon, water, and nitrogen have helped deepen our understanding of plant-resource acquisition, plant interactions with other organisms, and the role of plants in ecosystem studies. Where possible we also introduce how stable isotope information has provided insights into plant ecological research being done in a paleontological context. Progress in our understanding of plants in natural environments has shown that the future of plant ecological research will continue to see some of its greatest advances when stable isotope methods are applied.

INTRODUCTION AND SCOPE

At the core of all plant ecological investigations is the notion of plant-environment interactions, be these with the physical environment or with other organisms. Investigations that seek to understand the nature of these interactions commonly focus on particular resources such as light, water, carbon dioxide, or nutrients and how the interaction is influenced or mediated by that resource. Understanding the importance of a particular resource in a plant ecological context requires the acquisition of observational and experimental data; the collection of such data, in turn, requires suitable methods and measurements.

As with most areas of science our ability to obtain appropriate measurements that aid us with addressing the unanswered questions in plant ecology have often been limited or constrained by available tools. In this regard, stable isotope methods have recently emerged as one of the more powerful tools for advancing understanding of relationships between plants and their environment. Stable isotope techniques have permitted plant ecologists to address issues that have seemed intractable using other methods. They have therefore had a significant and positive impact on the science of plant ecology much like modern molecular techniques have for the fields of genetics, biochemistry, and evolutionary biology. Stable isotope information has provided insights across a range of spatial scales from the cell to the plant community, ecosystem, or region and over temporal scales from seconds to centuries. The elegance of stable isotope methods derives from the fact that it is generally easy to learn and the behavior of stable isotopes in ecological systems and biogeochemical cycles is reasonably well understood owing to the pioneering work of isotope chemists and geochemists. Areas in need of a deeper understanding seem well within our reach as stable isotope investigations and methods become more refined. Arguably, stable isotope methods are now among the most important empirical tools in modern plant ecological research, and the information they provide has yielded some of the newest and most important insights about plants in natural environments since the advent of the common garden experiment.

Our review highlights recent advances in plant ecology that have used stable isotope data to address questions at a variety of scales. We focus on carbon (C), water (H_2O), and nitrogen (N) because they are three of the most important resources influencing plant function, growth, distribution, and the biogeochemical cycles in which plants participate. We begin by briefly reviewing stable isotope terminology. Some essential principles for understanding how and why isotopes vary in nature and how isotope values alone or in mixtures are calculated are found in special topic boxes within the text. The sections that follow review studies that illustrate particular issues and identify where emerging trends or patterns exist, where areas of controversy and/or disagreement remain, and where promising areas of future research lie. For the newcomer, our hope is that this review enhances understanding of how stable isotopes might be used in plant ecological research. For the seasoned user, this review serves as a place to retrieve information on what we know, do not know, and need to know. Because space is limited, our review is not comprehensive but we hope that we have not been superficial and we apologize to scientists whose work we could not include. Wherever we feel it is helpful and possible we direct the reader toward other literature that provides more in-depth discussions of particularly important issues or areas of research. Last, we restrict our review to H, C, N, and O isotopes and terrestrial systems and attempt to show how both natural abundance and enrichment studies can be used.

STABLE ISOTOPES: CONCEPTS, TERMINOLOGY AND EXPRESSIONS

As noted previously (Peterson & Fry 1987), ecological studies have been informed by using stable isotopes at naturally occurring levels (called "natural abundance"; Table 1B) and at levels well outside the natural range of values (called "enriched" levels); enriched isotope studies therefore use "labeled" substances. Isotope abundance in any sample, enriched or not, is measured using a mass spectrometer. The details of how these measurements are made and how the mass spectrometer works

TABLE 1   The (*A*) isotope abundance ratios measured and their internationally accepted reference standards, and (*B*) the elements, their isotopes, their percent abundance, common form, relative molecular mass difference, and range (in ‰) measured in terrestrial environments of the principle stable isotopes discussed in this review

Isotope	Ratio measured	Standard	Abundance ratio of reference standard
A			
^{2}H (D)[a]	^{2}H/^{1}H (D/H)	V-SMOW[b]	1.5575×10^{-4}
^{13}C	^{13}C/^{12}C	V-PDB[c]	1.1237×10^{-2}
^{15}N	^{15}N/^{14}N	N$_2$-atm.[d]	3.6764×10^{-3}
^{18}O	^{18}O/^{16}O	V-SMOW,	2.0052×10^{-3}
		V-PDB	2.0672×10^{-3}

Element	Isotope	Percent abundance	Form & relative molecular mass difference	Terrestrial[e] range
B				
Hydrogen	^{1}H	99.984	^{1}HD/^{1}H^{1}H	~700‰
	^{2}H (D)	0.0156	(3/2), 50%	
Carbon	^{12}C	98.982	^{13}C^{16}O^{16}O/^{12}C^{16}O^{16}O	~100‰
	^{13}C	1.108	(45/44), 2.3%	
Nitrogen	^{14}N	99.63	^{15}N^{14}N/^{14}N^{14}N	~50‰
	^{15}N	0.3663	(29/28), 3.6%	
Oxygen	^{16}O	99.759	^{12}C^{16}O^{18}O/^{12}C^{16}O^{16}O	~100‰
	^{17}O	0.037	(46/44), 4.5%	
	^{18}O	0.204		

[a]The hydrogen stable isotope with mass two is also called deuterium, D.

[b]The original standard was standard mean ocean water (SMOW) which is no longer available; however, Vienna-SMOW is available from the IAEA.

[c]The original standard was a belemnite from the PeeDee formation (PDB) which is no longer available; however, "Vienna"-PDB is available from the IAEA.

[d]atm. = atmospheric gas.

[e]Approximate range measured in all analyzed substances from Earth (gasses, solids, biological materials).

can be found in reviews by Criss (1999), Ehleringer et al. (2000b), and Dawson & Brooks (2001).

Natural abundance stable isotopes are used as both natural integrators and tracers of ecological processes. As integrators, they permit ecologists to evaluate the net outcome of many processes that vary both spatially and temporally, while not disrupting the natural activity or behavior of the element in that system (Handley & Raven 1992, Högberg 1997, Robinson 2001). As tracers, they allow ecologists to follow the fates and transformations of a resource. Using natural abundance isotopes as tracers requires that the different potential sources have repeatable and distinct δ values (Equation 1) that are broader than the natural range of plant δ values measured. Furthermore, for tracers to be most useful there must not be significant fractionation (Text Box 1) or mixing of sources during the steps that move the resource from its source to the plant. Because it can be very difficult to fulfill all these requirements, many plant ecological investigations cannot use natural abundance isotope data to determine sources or process rates (Handley & Scrimgeour 1997); these studies rely on enriched isotope approaches.

BOX 1 Natural Abundance Stable Isotope Fractionation

Changes in the partitioning of heavy and light isotopes between a source substrate and the product(s) is termed isotope fractionation. Fractionation occurs because physical and chemical processes that influence the representation of each isotope in a particular phase (e.g., liquid vs. vapor) are proportional to their mass. In plant ecological investigations, though these fractionations are typically quite small, they are nonetheless important and must be understood for proper data interpretation. Isotope fractionations are categorized as primarily of two types; equilibrium fractionation and kinetic fractionation. *Equilibrium isotope fractionation* occurs during isotope exchange reactions that convert one phase (e.g., liquid) to another phase (e.g., vapor). The forward and back reaction rates of the rare isotope that leads to isotope redistribution are identical to each other. These reactions are often incomplete or take place in an open system that result in unequal (or nonequilibrium) representation of all of the isotope species in the mixture in all of the phases. If the system in which the isotope exchange reaction is taking place is closed and/or the reaction is allowed to go to completion (full exchange has taken place), there will be no net fractionation. This can occur under natural conditions, for example, when all of a particular substrate is consumed, but under many circumstances these reactions are incomplete and therefore net fractionation does exist. *Kinetic isotope fractionation* occurs when the reaction is unidirectional and the reaction rates are mass-dependent. In biological systems, kinetic fractionations are often catalyzed by an enzyme that discriminates among the isotopes in the mixture such that the substrate and product end up isotopically distinct from one another. Biologically mediated isotope fractionation is also called isotope discrimination. Fractionations exist because the lighter isotope (with a lower atomic mass) forms bonds that are more easily broken. Therefore, the lighter isotope is more reactive and likely to be concentrated in the product faster and more easily than the heavier isotope (Kendall & McDonnell 1998, Dawson & Brooks 2001). Many biochemical and biogeochemical processes discriminate against the heavier isotopic species in a mixture (e.g., against $^{13}CO_2$ more than $^{12}CO_2$ during C3 photosynthetic C fixation). This discrimination leads to marked variation in the isotopic ratios of source and product pools at different stages of a chemical reaction or biogeochemical cycle and of the different resources used by the organisms from these pools. Fractionations involved in biogeochemical reactions can provide information about processes. The resulting isotope ratio of any substance that is part of the reaction can act also as a fingerprint for that resource or transitional form. It can therefore be used as a tracer to follow the reaction products through complex cycles or into diets (e.g., you are what you eat ±0.1 to 1‰ for $\delta^{13}C$, and ±3 to 4+‰ for $\delta^{15}N$; see Griffiths 1998, Robinson 2001) or along an isotope gradient of continuum such as when water moves from soils through plants and into the atmosphere (Gat 1996, Dawson et al. 1998).

Enriched isotope methods involve applying trace amounts of a labeled substance. This procedure permits one to follow the flows and fates of an element without altering its natural behavior (Schimel 1993; go to Text Box 3). Because the substances are enriched, relative to the background, tracer studies remove or minimize problems of interpretation brought about by fractionation (Text Box 1) among pools that mix because the signal (the label) is amplified relative to the noise (variation caused by fractionation). Thus, the addition of an enriched substance acts as a powerful tracer for following a specific element's cohort through a system as well as for determining rates of biological process within the system (see Nadelhoffer & Fry 1994).

For natural abundance work we express the stable isotope composition of a particular material or substance as a ratio relative to an internationally accepted reference standard (Table 1A) as,

$$\delta^{XX}E = 1000 \cdot \left(\frac{R_{sample}}{R_{standard}} - 1 \right), ‰ \qquad 1.$$

where E is the element of interest (e.g., ^2H or D, ^{13}C, ^{15}N or ^{18}O), "^{XX}" is the mass of the rarest (and heavier) isotope in the abundance ratio, and R is the abundance ratio of those isotopes (e.g., $^{18}O/^{16}O$). Absolute abundance ratios are often very small (on the order of a few parts per thousand; Table 1B), so expressing isotope values relative to a standard and multiplying these by 1000 simply expresses the very small fractional differences in convenient "per mil" (or ppt) notation shown as ‰. The final δ value is expressed as the amount of the rarest to commonest (heavy to light) isotope in the sample with higher values indicating greater amounts of the heavier isotope. By definition, standards have a δ value of 0‰. A positive δ value therefore indicates that the sample contains more of the heavy isotope than the standard whereas a negative δ value indicates that the sample contains less than the standard.

For studies using enriched materials, the labeled substance added (e.g., $^{15}NO_3$) has an isotopic composition that significantly differs (usually exceeds) from any natural occurring level. The expression of the isotopic composition of this type of material is referred to in "atom %" (A_b) which is defined as,

$$A_b = \frac{X_{heavy}}{X_{heavy} + X_{light}} = 100 \cdot \left(\frac{R_{sample}}{R_{sample} + 1} \right), \% \qquad 2.$$

where X_{heavy} and X_{light} are the numbers of heavy and light atoms present in the sample and R_{sample} is the isotope ratio (as above). Equation 2 is most commonly used when values of A_b exceed ~ 0.5 atom % (or 500‰). Atom % is thus the percentage contribution of the heavy isotope to the total number of atoms of that element in the sample.

STABLE ISOTOPES AND PLANT ECOLOGICAL RESEARCH

The sections that follow are arranged hierarchically. We begin with the individual plant and how it interacts with the environment to acquire the resources it needs. We then review examples of how isotopes have informed us about interactions that occur between plants and other organisms; this section therefore focuses on the population and community scales. The final section reviews ecosystem studies where plants play a central role. We conclude with our views on future directions in plant ecology where we believe stable isotopes can have a positive impact.

Plant Level Studies

The acquisition of resources is the dominant theme that encapsulates stable isotope research at the individual plant level. Stable isotope information has been most informative in studies focused on water, carbon, and nitrogen—three resources that influence and limit plant growth, survival, and distribution. The sections that follow are organized around these resources.

CARBON   The utility of using C isotopes as an ecological index of plant function stems from the correlation between habitat quality and the biochemical discrimination (Text Box 1) against $^{13}CO_2$ during gas exchange, noted here as Δ. In C3 plants, discrimination (Δ) against ^{13}C by the carboxylating enzyme, Rubisco ($\sim 27‰$), is linked to photosynthesis via c_i/c_a, the ratio of intercellular to atmospheric CO_2 concentrations (Farquhar et al. 1982, Brugnoli et al. 1988). This ratio reflects the relative magnitudes of net assimilation (A) and stomatal conductance (g) that relate to demand and supply of CO_2, respectively. Carbon-13 data are thus a useful index for assessing intrinsic water use efficiency (A/g; the ratio of carbon acquired to water vapor losses via stomatal conductance, g) and may even provide information on actual water use efficiency (the ratio of assimilation to transpiration) when the leaf-to-air vapor pressure difference is known (Farquhar et al. 1989). In C4 plants, variations in c_i/c_a and Δ are relatively small despite variation in A and g (Farquhar 1983, Henderson et al. 1992, Buchmann et al. 1996a). In CAM plants, Δ values generally lie between that for C3 ($\sim$15 to 25‰ or -20 to -35‰ using $\delta^{13}C$ notation) and C4 ($\sim$2.5 to 5‰ or -11 to -15‰ using $\delta^{13}C$ notation) plants (Griffiths 1992). Variation in Δ of nonvascular plants is similar to that in C3 plants (Rundel et al. 1979, Teeri 1981, Proctor et al. 1992).

In contrast to gas exchange techniques that provide measurements of photosynthetic rates at a single time, $\delta^{13}C$ integrates photosynthetic activity throughout the period the leaf tissue was synthesized. Moreover, leaf $\delta^{13}C$ values reflect the interplay among all aspects of plant carbon and water relations and are thereby more useful than plant gas exchange measurements as integrators of whole plant function (Figure 1). As such, the earliest observations of $\delta^{13}C$ values in plant tissues (Wickman 1952, Craig 1953, Baertschi 1953) quickly established that C-isotope analyses were an important tool for integrating photosynthetic performance across ecological gradients in both space and time. As reviewed by Ehleringer (1988, 1993a,b), C isotopes have also been instrumental in revealing how species adjust their gas exchange metabolism, strategies of resource acquisition and use, and life-history patterns to ensure competitiveness and survival in a given habitat. Variation in Δ is caused by genetic and environmental factors that combine to influence gas exchange through morphological and functional plant responses. Discrimination has been observed to vary in response to soil moisture (Ehleringer & Cooper 1988; Ehleringer 1993a, 1993b; Stewart et al. 1995; Korol et al. 1999), low humidity (Madhavan et al. 1991, Comstock & Ehleringer 1992, Panek & Waring 1997), irradiance (Ehleringer et al. 1986, Zimmerman & Ehleringer 1990), temperature (Welker et al. 1993, Panek & Waring 1995), nitrogen availability (Condon et al. 1992, Högberg et al. 1993, Guehl et al. 1995), salinity (Bowman et al. 1989, Sandquist & Ehleringer 1995, Poss et al. 2000), and atmospheric CO_2 concentration (Bettarini et al. 1995, Ehleringer & Cerling 1995, Williams et al. 2001). Furthermore, morphological features also impose constraints on the physiological response to these various conditions through their influence on such factors as leaf boundary layer resistance, hydraulic conductivity through xylem, and leaf internal resistance to CO_2 and H_2O. Accordingly, variation in Δ has been found in relation to leaf size (Geber & Dawson 1990) and thickness (Vitousek et al. 1990, Hanba et al. 1999, Hultine & Marshall 2000), stomatal density (Hultine & Marshall 2000), branch length (Waring & Silvester 1994, Panek & Waring 1995, Panek 1996, Walcroft et al. 1996, Warren & Adams 2000), and canopy height (Yoder et al. 1994, Martinelli et al. 1998). Finally, Δ is, to a large extent, genetically determined, as relative rankings within and among genotypes are maintained irrespective of variations in the environment or plant condition (Farquhar et al. 1989, Ehleringer 1990, Ehleringer et al. 1990, Geber & Dawson 1990, Johnson et al. 1990, Schuster et al. 1992a, Dawson & Ehleringer 1993, Zhang et al. 1993, Donovan & Ehleringer 1994, Johnsen & Flanagan 1995, Zhang & Marshall 1995, Damesin et al. 1998, Johnsen et al. 1999). In contrast to vascular plants, genetic variation among nonvascular plants appears to have little effect on Δ whereas environmental factors

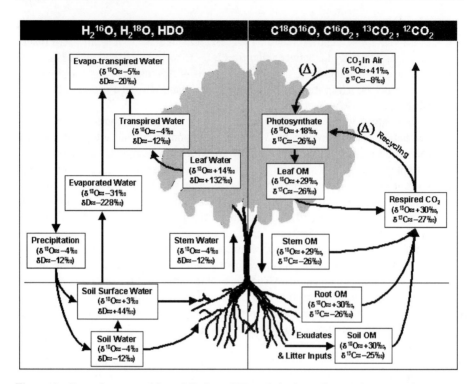

Figure 1   Isotopic composition of C, O, and H pools in the carbon and water cycles. Boxes represent pools and arrows represent processes. The values are rough approximations and can vary greatly with geographical location and environmental conditions. For demonstration purposes, we include data based on an example from Israel; δD values were estimated from $\delta^{18}O$ using the local meteoric water line for the same region following Gat & Carni (1970). The main advantages of the isotopic approach lie in the unique labeling of flux components; leaf transpiration and soil evaporation are isotopically very different; root and soil respiration can have distinct ^{13}C labeling; photosynthesis (depleted uptake) tends to enrich the atmosphere, while respiration (depleted release) tends to deplete the atmosphere in ^{18}O and ^{13}C. OM refers to organic matter. Δ indicates discrimination occurs during photosynthetic assimilation. Values are on the SMOW and PDB scales for $\delta^{18}O$ and $\delta^{13}C$ values, respectively. [Modified from Yakir & Sternberg (2000).]

such as moisture availability and water content are most important (Rice & Giles 1996, Williams & Flanagan 1996, Rice 2000). But unlike in vascular plants, Δ tends to increase with water limitation in nonvascular plant taxa (Williams & Flanagan 1996, 1998).

The factors cited above explain much of the variation in Δ observed with respect to phenology (Lowden & Dyck 1974, Smedley et al. 1991, Ehleringer et al. 1992, Damesin et al. 1998), development (Geber & Dawson 1990), age (Yoder et al. 1994, Fessenden & Ehleringer 2002), and gender (Dawson & Ehleringer 1993, Kohorn et al. 1994, Retuerto et al. 2000, Ward et al. 2002). In addition, the aforementioned factors also explain spatial gradients in Δ found within canopies (Medina & Minchin 1980, Garten & Taylor 1992, Buchmann et al. 1997a, Hanba et al. 1997, Le Roux et al. 2001), across landscapes (Williams & Ehleringer 1996, Moore et al. 1999), and with altitude (Körner et al. 1991, Morecroft & Woodward 1990, Hultine & Marshall 2000). It should be noted that the causes of variation in Δ are clearly complex and are, at times, not straightforward. This complexity can make correlations between Δ and a single factor such as hydraulic conductivity (Cernusak & Marshall 2001), water availability (Warren et al. 2001), or rainfall (Miller et al. 2001) problematic. Further, variations in $\delta^{13}CO_2$ of source-air owing to recycling of respired CO_2 within canopies (Figure 1) may confound the ecological interpretation of $\delta^{13}C$ or Δ in leaf tissues (Schleser & Jayasekera 1985, Sternberg et al. 1989, Broadmeadow et al. 1992, Buchmann et al. 1997b, Yakir & Sternberg 2000).

Because of the integrative response of Δ to multiple eco-physiological constraints through time, C isotopes can be used to assess traits that co-vary with gas exchange, C gain, and water relations, including water use efficiency (WUE) (Farquhar & Richards 1984, Henderson et al. 1998), photosynthetic capacity (Virgona & Farquhar 1996), stomatal conductance (Condon et al. 1987, Ehleringer 1990, Ehleringer et al. 1990, Virgona et al. 1990, Meinzer et al. 1992), leaf nitrogen content (Sparks & Ehleringer 1997, Schulze et al. 1998), leaf mass per area (Vitousek et al. 1990, Hultine & Marshall 2000, Williams & Ehleringer 2000), longevity (DeLucia et al. 1988, Schuster et al. 1992b), and relative growth rate (Ehleringer 1993b, Poorter & Farquhar 1994). For example, working in boreal ecosystems, Brooks et al. (1997) used $\delta^{13}C$ as a surrogate for physiological characteristics and found that life form (deciduous or evergreen trees, shrubs, forbs, and mosses) can be a robust indicator of functional group membership related to carbon and water fluxes (see also Flanagan et al. 1997a). Whereas these data are consistent with those gathered by Marshall & Zhang (1994), Kelly & Woodward (1995) found that life form had no effect on Δ among three altitude catagories. In another example, Smedley et al. (1991) examined the seasonal time-course of $\delta^{13}C$ among grassland species and found lower WUE among the taxa active during the initial, less stressful months of the growing season. Further, WUE increased with evaporative demand as soil moisture declined. In a related fashion, Kloeppel et al. (1998) used $\delta^{13}C$ of leaf tissue to assess WUE and determined that, in general, larches (*Larix* spp.) use water less efficiently and maintain higher photosynthetic capacity (based on foliar N concentration) than co-occurring evergreen conifers from 20 locations in the northern hemisphere. Their results suggest that water is not the most limiting resource at these high elevation (3000–4000 m) sites. Finally, Flanagan et al. (1992) and Valentini et al. (1992) used $\delta^{13}C$ of leaves to assess WUE of species from different functional groups in a Pinyon-Juniper Woodland and the Mediterranean macchia, respectively. By concurrently measuring δD in xylem water to distinguish surface versus groundwater sources (see next section), they found that species with more negative $\delta^{13}C$ values and therefore lower WUE had deeper rooting depths and a more reliable water supply than species that relied on rain water in the upper soil layers (also see Lajtha & Marshall 1994).

To separate the independent effects of photosynthetic capacity and stomatal conductance on c_i/c_a, Scheidegger et al. (2000) have recently proposed measuring both $\delta^{13}C$ and $\delta^{18}O$ in leaf organic matter (Figure 1). Whereas $\delta^{13}C$ reflects c_i/c_a, $\delta^{18}O$ generally varies with ambient humidity, which in turn reflects changes in water use [g] (Ball et al. 1987, Grantz 1990; but see also Mott & Parkhurst 1991, Monteith 1995). The $\delta^{18}O$ of leaf and tree ring cellulose are largely determined by the integrated leaf-to-air vapor pressure gradient during photosynthetic gas exchange (Farquhar et al. 1998). This leaf-air vapor pressure gradient changes with environmental conditions (atmospheric humidity, soil moisture, air temperature) and plant response to these environmental changes (e.g., g, leaf temperature, A). So measurement of the ^{18}O composition of plant tissues aids with the interpretation of differences in $\delta^{13}C$ among individual plants growing in the same location and among species in different environments. Moreover, the determination of WUE is greatly improved by the simultaneous use of $\delta^{13}C$ and $\delta^{18}O$ in plant tissues (Saurer et al. 1997). By considering concurrent variations $\delta^{13}C$ and $\delta^{18}O$, one can distinguish between biochemical and stomatal limitations to photosynthesis in response to a change in environmental conditions. Alternative methods to distinguish such effects rely on instantaneous gas exchange measurements (Farquhar & Sharkey 1982) that are more difficult to extend through time or to apply simultaneously on a large number of samples. Although further research is needed to develop a quantitative dual (C and O) isotope model, this approach should improve our ability to relate gas exchange characteristics to $\delta^{13}C$ and $\delta^{18}O$ signals in plant leaves and tree rings.

In fact, bulk wood or purified cellulose obtained from tree rings has provided some of the best samples for isotope analyses because the $\delta^{13}C$, δD, $\delta^{18}O$, and even $\delta^{15}N$ isotopes in the wood can record a great deal about the ecophysiology of the plants (Leavitt & Long 1986, 1988, 1989, 1991; Livingston & Spittehouse 1993; Bert et al. 1997; Saurer et al. 1997; Borella & Saurer 1999; Roden & Ehleringer 1999a,b), the resources they use (Roden et al. 2000, Ward et al. 2002), and the environments they inhabit, both now (Barbour et al. 2000, Roden et al. 2000) and in the past (Edwards et al. 1985; Leavitt 1993; Lipp et al. 1996; Switsur et al. 1996; Feng 1999; Hemming et al. 1998, 2000; Monserud & Marshall 2001). Recent modeling efforts that use $\delta^{13}C$ (Hemming et al. 2000) and δD and $\delta^{18}O$ (Roden & Ehleringer 2000; Barbour et al. 2000, 2001) have very much improved our interpretations of isotope variation in tree rings. In these models, fractionations are better understood and accounted for, allowing one to make more precise inferences about environmental conditions (temperature and humidity), plant resource status (water-use efficiency and/or sources of water and nitrogen), and environmental change (Epstein & Krishnamurthy 1990, Duquesnay et al. 1998).

In addition to the use of natural abundance $\delta^{13}C$ to deepen our understanding of plant gas exchange, stable isotope techniques based on 13-C labeling (see Text

Box 2) offer a means to trace and quantify the fate of C allocation to various plant organs, tissues, and even specific compounds or the surrounding soil that would have been difficult to study using other methods. For example, Simard et al. (1997a) used enriched ^{13}C-tracers (see Text Box 2) to examine the fate of C fixed by paper birch (*Betula papyrifera*) and Douglas fir (*Psuedotsuga menziesii*) seedlings and showed that half of the labeled ^{13}C that was assimilated was quickly lost via root and shoot respiration, root exudation, and tissue death. Interestingly, paper birch had higher assimilation rates and higher allocation of its remaining C to roots than Douglas fir, suggesting that it may have a competitive advantage when grown in mixed communities. In another study, the effect of seasonality on the temporal pattern of C partitioning to the root system was investigated (Mordacq et al. 1986). Here it was shown that very little C was allocated to roots during shoot and leaf elongation, whereas later in the season, the stem and root system were the major sinks. Working in grasslands, Niklaus et al. (2001) used labeled ^{13}C at elevated CO_2 concentrations to determine that an increased supply of CO_2 enhanced C assimilation but not root exudation and turnover.

BOX 2 Isotope additions: Tracer and Isotope Dilution Methods

The addition to a system (plant, soil, etc.) of a quantity of material with a δ value significantly different from any natural background level, followed by the observation of its fate, allows the study of its fluxes and/or transformations in undisturbed conditions. The amount of material required depends on detection limits for measurements, size of the pool to be labeled, rates of production or consumption, and duration of labeling. In tracer experiments, a known quantity of isotope is added to a pool and then recovered in a recipient pool after a known amount of time. The total amount of an element (tracer plus background isotope, M_{AB}) that moved from the labeled source pool (A) to the sink (B) can be calculated knowing the atom % excess (I_A) of pool A, and the mass (P_B) and the atom % excess (I_B) of pool B at the end of the experiment (from Stark 2000):

$$M_{AB} = \frac{P_B \cdot I_B}{I_A}.$$ B2.1.

Dividing M_{AB} by the length of the experiment gives the flux rate from A to B. Calculation of % recovery of the label within a sink pool requires the measurement of the mass of the labeled source pool (P_A):

$$\% \text{ recovery} = 100 \cdot \frac{P_B \cdot I_B}{P_A \cdot I_A}.$$ B2.2.

This technique is usually applied in mass balance studies. The assumptions underlying the tracer method are: (*a*) Nonlabeled and labeled materials have the same chemical behavior (fractionations, except for small amounts caused by differences in heavy and light isotopes, do not occur), (*b*) the source pool is uniformly labeled by the isotope, (*c*) the addition of the isotope does not stimulate rates of transformations and (*d*) material that is lost from sink pool(s) is accounted for (see Stark 2000, p. 224 for additional equation). The calculation of the rates of flow will be inaccurate if material flows into the source pool because of dilution or if there is outflow from the sink pool(s).

Isotope dilution techniques, in general, include the labeling of a pool that is the product of the transformation of interest (for example, labeling the soil ammonium pool to estimate the gross rate of mineralization) and the measuring of how rapidly the added isotope is diluted by influx of the natural isotope into the pool over time. This approach allows one to measure flux rates through different compartments that have simultaneous inputs and outputs. Some of the same requirements mentioned above, in particular a–c, must be met. In addition, the rates of transformation must be constant and unidirectional over the timescale of the experiment. When these assumptions are satisfied, the gross production rate (GPR) for a particular process can be calculated knowing the size (P) and atom % excess (I) of the labeled pool at the beginning (P_0 and I_0) and at the end (P_t and I_t) of the experiment (from Hart et al. 1994, modified by Stark 2000):

$$GPR = \frac{P_0 - P_t}{t} \cdot \frac{\log(I_0/I_t)}{\log(P_0/P_t)}.$$
B2.3.

Many different formulations for resolving specific applications of tracer and isotope dilution experiments exist and they are reviewed by Schimel (1993), Stark (2000), and Palta (2001).

WATER   Stable isotope analyses of both H and O have significantly improved our understanding of water source acquisition by plants because the "pools" of water used by plants can easily be distinguished (Figure 1; Gat 1996). There are now examples from a wide range of ecosystems showing how different plant species use water resources in time and in space. Because much of this work has been reviewed recently (Ehleringer & Dawson 1992, Dawson 1993a, Dawson & Ehleringer 1998, Dawson et al. 1998, Ehleringer et al. 2000b, Walker et al. 2001), we mention only a few examples.

It is easy to apply δD and $\delta^{18}O$ data to water acquisition studies because there is no isotopic fractionation during water uptake by terrestrial plants (Wershaw et al. 1966), although Lin & Sternberg (1993) provide an example for a marine, salt-excluding plant species that does seem to fractionate H (but not O) during water uptake. For terrestrial plants, if samples of the different water sources can be obtained and the water within the plant's xylem sap is also extracted, it is possible to assess which sources of water are being used (Ehleringer et al. 2000b and Turner et al. 2001 provide detailed methods). Applying isotope mixing models (Text Box 3) coupled with other ecological or physiological measurements then becomes a particularly powerful way to link the water sources used by plants to other aspects of their water relations (Dawson 1993b, Flanagan et al. 1992, Jackson et al. 1995, Williams & Ehleringer 2000b).

BOX 3 Isotope Mixing Models

In situations where the isotope values of the sources can be determined, one can partition the use of each source using what have been called end-member mixing models. In the simplest case, proportional use of two known sources using a single isotope involves using a two-source mixing model that can be written as,

$$\delta_t = f_A\delta_A + (1 - f_A)\delta_B, \qquad \text{B3.1.}$$

where δ_t is the total δ value (e.g., plant xylem water $\delta^{18}O$), δ_A, and δ_B are the isotope values of sources A and B, and f_A is the fraction of the total contributed by source A. Rearranging for f_A gives

$$f_A = \frac{(\delta_t - \delta_B)}{(\delta_A - \delta_B)}. \qquad \text{B3.2.}$$

Brunel and co-workers (1995) proposed a variation on the model presented above that uses both the δD and $\delta^{18}O$ of source and plant water to determine proportional use of each source. This may be necessary when it is difficult to resolve the zone of uptake using only one isotope as reported in several plant water uptake studies in saline soils (Thorburn & Walker 1994, Thorburn et al. 1994, Mensforth et al. 1994, Mensforth & Walker 1996, Walker et al. 2001).

When more than two different sources are present, assessing the proportional use of each becomes much more challenging. However, a recent set of papers has provided guidelines for approaching this issue in a more robust fashion (see Phillips 2001, Ben-David & Schell 2001, Phillips & Gregg 2001, Phillips & Koch 2002, Zencich et al. 2002). The authors discuss sampling issues and develop significant improvements in the application of mixing models that involve de-convoluting the use of multiple-sources using stable isotope data. Moreover and perhaps more importantly, guidelines are provided for statistical error analysis associated with mixing model applications.

In our view, these are extremely important improvements over previous approaches because they not only provide a means to assess where, for example, the water being used by particular plants comes from when roots are found throughout a soil profile, but they standardize and improve upon the mathematical evaluation of the data obtained. Whereas these more complex models require additional information about the spatial and temporal variation of the resources within a system, they may be the only way to quantitatively access resource use and partitioning among multiple end-members and species, within an ecosystem. In some instances, there may be several different interpretations, and simple two- or even three-source models cannot and should not be applied. In these cases one may still be able to arrive at a best-case interpretation (as shown by Phillips & Gregg 2000).

Hydrogen and O isotope analyses have been used effectively to determine the reliance of a species on shallow versus deep soil water (White et al. 1985, White 1989, Dawson 1993b, Brunel et al. 1995), surface runoff or streamwater versus soil water (Dawson & Ehleringer 1991, Busch et al. 1992, Thorburn et al. 1994, Phillips & Ehleringer 1995, Kolb et al. 1997), and winter precipitation versus fog (Dawson

1998) or monsoonal rain (Lin et al. 1996, Williams & Ehleringer 2000b). Other studies have used δD or $\delta^{18}O$ to investigate differential water resource use among different species within a community (Ehleringer et al. 1991a; Flanagan et al. 1992; Dawson & Ehleringer 1991, 1998; Dawson 1993a,b; Jackson et al. 1995, 1999; Meinzer et al. 1999) and to examine the relationship between plant distribution along natural gradients of water availability and the depth at which plants obtain their water (Sternberg & Swart 1987; Thorburn et al. 1993a,b; Mensforth et al. 1994). Still others have used isotope analyses of source and plant waters to look at changes in the zone of water uptake over time when soil moisture at different depths varies with season (Ehleringer et al. 1991a, Dawson & Pate 1996, Lin et al. 1996, Dawson 1998) or in relation to life history stages (Feild & Dawson 1998), life form differences (Williams & Ehleringer 2000b), functional group classifications (Ehleringer et al. 1991a, Dawson 1993a), or changes in plant size (Dawson 1996, Meinzer et al. 1999).

Thorburn & Ehleringer (1995) provide an important caution regarding simple interpretations from isotope data used in tracing water source use. By measuring soil water isotope profiles, xylem water, and root water from three different systems they showed that plants may extract water from several zones simultaneously. Their data showed how the application of simple two-source (shallow versus deep) mixing models (Text Box 3) can lead to erroneous interpretations. Whereas this complication means that additional sampling is needed, this sampling can be done and adds additional power to the interpretation (also see Cramer et al. 1999).

The aforementioned examples highlight the importance of using the isotope data to determine the zones in soils where plant roots are actively extracting moisture (Ehleringer & Dawson 1992). This is critical information because roots can often be found throughout the soil profile. Therefore, isotope data show that the presence of roots does not always mean these roots are active in water uptake; we know no other method that can provide this perspective. Armed with this information, investigators may be better able to link species characteristics and species diversity to ecosystem functions (Dawson 1993b, 1996; Jackson et al. 2000; Meinzer et al. 2001). There may also be practical applications here such as the design of agroforestry systems (Pate & Dawson 1999).

At times, combining enriched and natural abundance tracers in water relations studies has helped in the interpretation of dynamic plant water uptake (Lin et al. 1996, Plamboeck et al. 1999) as well as water transfer behavior among ramets on the same plant (De Kroon et al. 1996). In these studies, pulses of D_2O-labeled water were added to individual plants or entire plots (see Plamboeck et al. 1999, Lin et al. 1996, Schwinning et al. 2002, for examples) in order to distinguish the precise zone in the soil (Moreira et al. 2000) or time of day or season that plants use their water in relation to a transition between conditions of low soil moisture availability and short episodes of high soil moisture availability (Plamboeck et al. 1999, Williams & Ehleringer 2000b). Pulse labeling has also helped reveal how different life forms within a community may partition water resources (Lin et al. 1996, Schwinning et al. 2002). In all these cases, whether using natural abundance or enriched tracers of water, stable isotope methods have helped quantify the absolute and relative use of shallow versus deep soil water sources in different species. When it is difficult or impossible to use two- or three-source mixing models (Text Box 2), tracer-pulse methods, coupled with measurements of plant transpiration (T) and the exchangeable water volume (V) can provide a more precise quantification of water source utilization from different sources. Schwinning and coworkers (2002) developed a dynamic mixing model that uses measures of T and V as well as a labeled (f) and an unlabeled ($1\text{-}f$) water source to better understand water use behavior in a mixed shrub-grass community. In their study, the concentration of the D_2O-label, in a fixed V, depends on the relative contribution of f and T/V. However, because water uptake rates and ratios may not remain constant during or after a pulse, the pulse use must be estimated by integrating parameters over time. From these experimental and modeling efforts, Schwinning et al. (2002) suggested that relative pulse utilization by plants with contrasting life history strategies can be estimated with fair accuracy and the species-specific water use strategies estimated. This information in turn permitted Schwinning and her coworkers to better evaluate the validity of the assertion that plants partition water resources in desert communities (also see Schwinning & Ehleringer 2001).

NITROGEN   Plant $\delta^{15}N$ values can vary as much as 10‰ in co-occurring species (Handley & Scrimgeour 1997). Early studies assumed that the $\delta^{15}N$ of whole plant or leaf tissues reflected the $\delta^{15}N$ of the form (source) of N most used by that plant. This assumption was the conceptual basis for predicting, for example, the ability of co-occurring plants to partition acquired N in time, space, and by form. It is now clear that such an interpretation is incorrect. Natural abundance $\delta^{15}N$ of plants we now know reflects the net effect of a range of processes (reviewed by Handley & Scrimgeour 1997, Robinson 2001, Evans 2001, Stewart 2001). The presence of multiple N-sources with distinct isotopic values (Figure 2), mycorrhizal associations, temporal and spatial variation in N availability, and changes in plant demand can all influence plant $\delta^{15}N$. To determine the importance of these effects, we must develop a more complete understanding of the controls over $\delta^{15}N$ in the plant-soil system (Handley et al. 1998). Even if the N source is the most important factor in determining plant $\delta^{15}N$, at the moment there is no easy technique for isolating and analyzing $\delta^{15}N$ in soil N pools that are available for plant uptake. Promising techniques that are currently being used, such as dual-isotope techniques (e.g., $\delta^{15}N$ and $\delta^{18}O$ in NO_3; Durka et al. 1994), need to be further explored so that we can more successfully distinguish the dominant N sources to plants. Better understanding of plant N sources will also be aided by the production of a mechanistic model demonstrating how ^{15}N behaves during N-acquisition by plants.

Recent investigations have begun to study the causal relationships between uptake, assimilation, and allocation of N and plant $\delta^{15}N$ values. To date, there is no evidence of fractionation of either ^{14}N or ^{15}N during its physical movement across living membranes (passive and active uptake; Handley & Raven 1992). Differences in $\delta^{15}N$ between the N source(s) and the plant are generally due to

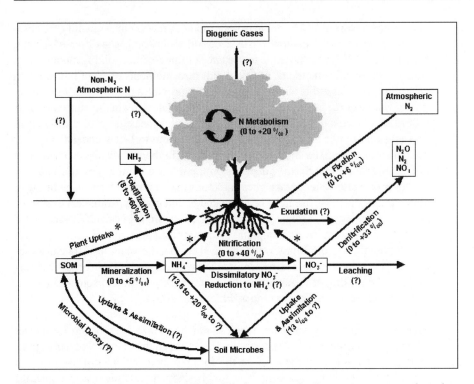

Figure 2 Major pathways and pools in the nitrogen cycle. Boxes represent pools and arrows represent processes. Figure illustrates that different N transformations lead to varying amounts of ^{15}N enrichment or depletion. Values in parentheses represent potential range of discriminations observed in controlled laboratory, greenhouse, and field experiments [values from Shearer & Kohl (1986) and Handley et al. (1999)]. Similar to the water and carbon cycle, these can vary greatly with geographical location and environmental conditions. The "*" indicates that plant N uptake could be via mycorrhizae and "?" indicates that the specific fractionation associated with that process is not fully characterized. SOM includes soil organic matter derived from plant, microbial, and faunal organic tissues. This figure does not include all of the possible N fractionations that occur within the N cycle (e.g., does not include NH_4^+ exchange in soils). [See Shearer & Kohl (1986) and Handley et al. (1999) for more details.]

enzyme-mediated reactions that commonly discriminate against the heavier ^{15}N (for example, assimilation of NH_4 mediated by glutamine synthetase or assimilation of NO_3 mediated by nitrate reductase) (Handley & Raven 1992). Whole plant δ^{15}N appears similar to the δ^{15}N value of the external source (implying no discrimination) only if the entire N pool is converted from source to product. But net isotope discrimination can occur if fractionated plant parts or fractionated N compounds are lost from roots (exudation) or leaves (loss of gaseous N; Figure 2).

New insights about the net discrimination that occurs during N-assimilation have been obtained by growing plants hydroponically in the presence of a single inorganic N-source. When NH_4 is the sole source of N there is little, if any, discrimination when the N concentration is limiting (Evans et al. 1996). But there may be large whole plant depletion in ^{15}N at high N concentrations (Yoneyama et al. 2001). The reasons why depletion occurs, including exudation of ^{15}N enriched NH_4 or organic N from roots, requires more explicit testing. External NO_3 concentrations, instead, appear to have a small effect on whole plant $\delta^{15}N$ (Mariotti et al. 1982, Kohl & Shearer 1980, Bergersen et al. 1988, Yoneyama & Kaneko 1989, Evans et al. 1996, Yoneyama et al. 2001). Plants can be either slightly enriched or slightly depleted in ^{15}N compared to the NO_3 source (Yoneyama et al. 2001). Release of ^{15}N fractionated compounds could account for this observation (Robinson et al. 1998). Very high external NO_3 concentrations, osmotic stress, or drought may induce N isotopic fractionation, although the mechanisms for this fractionation are still unknown (Handley et al. 1994, 1997). Remarkably little is known about whether fractionation occurs during the uptake and assimilation of organic forms of N.

Different plant parts, organs, and compounds can differ from the whole plant $\delta^{15}N$ values. Within-plant variation is typically small, between 2 and 3‰, but it can reach 7‰ in desert plants (Evans 2001). Understanding what causes this variation may help elucidate the metabolic events responsible for it because differences between plant parts or compounds are determined by differences in assimilation, reallocation, and/or N losses. If NO_3, for example, is the only source of N and is partially assimilated in the roots, there can be an isotopic difference between the root and shoot (Evans et al. 1996, Yoneyama et al. 2001). Kolb & Evans (in review) observed no discrimination during N resorption, but they did observe discrimination during re-allocation of N from storage.

Recently, it has been suggested that natural isotopic variation of phloem and xylem saps may provide insight into resource acquisition and use (Yoneyama et al. 1997). Further development of new technologies, such as a chromatographic separation of specific compounds coupled to the isotope ratio mass spectrometer, could provide powerful insights for resolving metabolic processes and making inferences about plants in relation to N.

Providing an enriched ^{15}N-label (Text Box 2) to plants and following its fate can be an effective way to differentiate between external and internal plant sources and to quantify the importance of the internal cycling of N to support new growth (Proe et al. 2000). Dual-isotope (^{13}C and ^{15}N) labeling may be the most powerful tool for demonstrating the relationship between internal N stores and recently fixed photosynthate (Dyckmans & Flessa 2001). Enriched N tracers can also be used to determine the interaction between N supply and atmospheric CO_2. For example, oak seedlings responded positively to elevated CO_2 concentration when N availability was high, whereas N deficiency created a growth imbalance, enhancing biomass accumulation and partitioning of newly assimilated C and N to the root system (Vivin et al. 1996, Maillard et al. 2001).

Population and Community Studies

Stable isotope methods have only very recently been featured in studies of plants at the population and community levels. These studies have employed H, C, O, and N isotope data to investigate competition, facilitation, parasitism, herbivory, and symbiotic relationships formed between fungi and N-fixing bacteria and their host plants. The sections below highlight some of this work and are organized by theme and then by resource.

Competition and Facilitation

Both positive (facilitation) and negative (competition) interactions among plants have been investigated using stable isotopes. We review some examples below, by resource.

CARBON   A few recent studies are now using C isotope data to investigate competition. For example, based on the early work by Svejcar & Boutton (1985) that used $\delta^{13}C$ for investigating rooting patterns in mixed plant stands, Polley et al. (1992) provided some of the first C (and N) stable isotope evidence from root and shoot tissue analyses that competition structured mixed grass-shrubland communities. This approach was extended in two recent studies. Williams et al. (1991) used ^{13}C-analyses of two grasses in the presence and the absence of a co-occurring shrub. A second study (Rice et al. 1993) examined blue oak–grassland mixtures and used $\delta^{13}C$ values to infer the efficiency of water use in the presence or absence of either annual grass or perennial bunchgrass neighborhoods. In both studies, $\delta^{13}C$ values were useful in showing how the efficiency of resource use varied in the presence or absence of different neighbors. The work of Rice et al. (1993) on tree-grass interactions has recently been expanded in a study by Archer (1995). Working in savanna parklands in the southwestern United States, Archer (1995) used soil $\delta^{13}C$ data to show how this community, once dominated by C4 grasses, has been largely replaced by C3 shrubs during the past 100–200 years; these shifts appear to have been facilitated by changes in soil N.

WATER   Our understanding of competition or facilitation among plants inhabiting the same population or community has been enhanced in a limited way through H and O isotope work (but see Schwinning et al. 2002 and above). Dawson (1993a) provided some of the first isotopic evidence that trees can have a positive influence (facilitation) on their neighbors via the process of hydraulic lift; the redistribution of water by plant root systems from deep moist layers in the soil to drier shallow layers (reviewed by Caldwell et al. 1998). Dawson (1993a) showed that water used by plants growing nearest sugar maple trees that showed hydraulic lift obtained a significant fraction of their water in summer from the trees that lifted and released this water into the soil and rhizosphere (see also Burgess et al. 2000). These same plants were less water-stressed, had greater water use, and grew more than plants inhabiting areas farther from trees. This study demonstrated that the long-held assumption that plants living in close proximity will show negative, competitive, effects may not always be true. The δD data used in conjunction with measures of plant water use (Dawson 1996, 1998) and soil water transport (Emerman & Dawson 1996), showed that trees can have a positive influence on neighboring plants and on community diversity. This idea is supported by recent work of Caldeira et al. (2001) on positive biodiversity-production relationships; these workers used $\delta^{13}C$ data to establish a relationship between species richness and productivity. Their data indicated that there was higher water use and productivity when plants grew in species-rich mixtures compared to monocultures. Work by Ludwig (2001) and coworkers (in review) in *Acacia*-woodland/savannas in East Africa has shown how both competition between trees and grasses and facilitation brought about by hydraulic lift from the *Acacia* trees influences the interannual dynamics of tree-grass interactions. Using the $\delta^{18}O$ of plant and source waters coupled with trenching experiments and measures of productivity, Ludwig and his colleagues show that in dry years competition dominates and in wetter years facilitation dominates. Lastly, studies of the facilitation effects by plants on each other have now been discussed in the context of designing intercropping systems in agroforestry (Emerman & Dawson 1996) that in some cases mimic the natural ecosystems they are embedded within or replace (Pate & Dawson 1999). In this context, Smith et al. (1997, 1998) recently used isotope data to help design useful tree-crop systems in drought-prone Africa.

Parasitism

Plant parasites are important for shaping plant-plant interactions; here stable isotopes have been particularly powerful tools for understanding these interactions.

CARBON AND NITROGEN   Variation in $\delta^{13}C$, when combined with anatomical, morphological, and physiological information, has provided important insights into the complex physiological relationships between host plants and their associated plant parasites. Obligate parasites of higher plants have specialized absorptive organs called haustoria that penetrate the tissues of their hosts and acquire resources from them. Mistletoes, which are epiphytic parasites and root hemiparasites, are connected to the host via the xylem but are also photosynthetic. Estimates of WUE based on the $\delta^{13}C$ in host and mistletoe pairs have shown that not only do they differ but that mistletoes have lower WUE and higher overall water use, especially under conditions of N shortage (Schulze & Ehleringer 1984, Ehleringer et al. 1985). These findings strengthened the hypothesis that mistletoes regulate their transpiration (upwards) to obtain N from their hosts. Press et al. (1987) were the first to compare the $\delta^{13}C$ of a C4 host and its C3 root hemiparasites with predicted values based on foliar gas-exchange data; this information was used to demonstrate that substantial C transfer from host to parasite can occur even when the parasite can photosynthesize. The same approach was expanded by Marshall & Ehleringer (1990) to study host-mistletoe unions and subsequently formalized into a model

(Marshall et al. 1994) that calculated the C-contribution from heterotrophic carbon to the total C-budget of the parasite. The Marshall et al. (1994) model showed that the heterotrophic C contribution ranged from 15% to 60%. They also found a correlation between the total C gained by the mistletoe [via A plus heterotrophic C (acquisition) obtained in the transpiration stream] and the host's A, suggesting a convergence in parasite and host growth rates, which can provide a basis for the N-parasitism hypothesis. In contrast, Bannister & Strong (2001) recently found that the estimation of mistletoe heterotrophy calculated from $\delta^{13}C$ data can be confounded in moist temperate environments like New Zealand, where very small differences between host and parasite $\delta^{13}C$ were observed.

The movement of N from putative hosts to a woody root hemiparasite has been investigated in a coastand heathland in southwest Australia (Tennakoon et al. 1997). The similarity between the $\delta^{15}N$ values of the parasite and those of the N_2-fixing hosts was interpreted as a qualitative indicator of the parasite dependence on fixed N as opposed to soil inorganic N. These findings might explain why woody root hemiparasites can achieve such high biomass and become the dominant growth form in this ecosystem deficient in soil-available N.

Trophic Interactions—Herbivory/Grazing

The use of stable isotopes has recently helped inform research on trophic interactions. We review some important case studies below.

CARBON   Several recent studies have illustrated that stable isotopes coupled with mixing models (Text Box 3) can inform us about the nature of trophic interactions. For example, using the $\delta^{13}C$ and $\delta^{15}N$ of ants in the genus *Philidris* and their host, the tropical epiphyte *Dischidia major*, Treseder et al. (1995) showed that ants provide C (as respired CO_2) and N in exchange for shelter. Because the epiphyte is an obligate CAM plant with higher $\delta^{13}C$ than a C3 tree, and the ants feed on Homoptera that ingest the phloem sap of C3 trees, unexpectedly low $\delta^{13}C$ values in the epiphyte showed the extent to which the plant was assimilating C derived from ant respiration. *D. major* leaves were also more enriched in ^{15}N than those of another epiphyte that grew close by but did not host ants, suggesting that *D. major* uses ant debris as an N source. Sagers et al. (2000) also documented that, within the specialized mutualism of *Azteca* ants and the plant *Cecropia peltata* there is exchange of resources between the ants and the trees, where the ants use some plant C products and provide N to their host. Earlier studies of carnivorous plants had used both the $\delta^{13}C$ and $\delta^{15}N$ of the predominant prey (ants or termites) to determine their contribution to the plants' resource requirements (Schulze et al. 1991, 1997; Moran et al. 2001). In these sorts of trophic investigations, the application of mixing models for determining proportional contributions of different sources is critical.

Approaches to isotope mixing have recently been critiqued (Phillips 2001, Phillips & Gregg 2001, Ben-David & Schell 2001, Phillips & Koch 2002). These authors recommend measuring the isotopic composition of each component in the study system, its elemental concentration (Phillips & Koch 2002), and its variation, in order to achieve robust estimates of partitioning. Replication over time and space is critical for enhancing statistical power. Moreover, the interpretations should be based on good a priori knowledge of the system itself (Handley & Scrimgeour 1997).

Very few studies have used data on natural abundance δ^{13}C to assess herbivore impacts on plant performance. Fry et al. (1978) were among the first to use δ^{13}C data to look at effects of grasshopper feeding on plant C-balance. Recently, Alstad and coworkers (1999) showed that plant WUE improved in *Salix* with elk browsing, and Kielland & Bryant (1998) used δ^{13}C (and δ^{15}N) of animals, vegetation, and soil to elucidate the important role of moose in shaping vegetation dynamics in a taiga forest (see additional examples in Lajtha & Michener 1994).

Tracer techniques have also proven useful in elucidating mechanisms of plant tolerance to herbivory, particularly when these mechanisms are based on patterns of C (or nutrient) allocation within the plant. For a group of C4 perennial prairie grasses, tolerance was associated with the capacity of the plant to reallocate C rapidly from roots to shoots in response to defoliation (Briske et al. 1996). Olson & Wallander (1999), however, showed that the amount of C allocated to the root system by the invasive leafy spurge (*Euphorbia esula*) was unaffected by defoliation, suggesting that the grazing tolerance of this species is instead linked to its ability to maintain an extensive root system that may store C and allow regrowth after tissue loss.

NITROGEN   Despite the limitations in applying natural abundance δ^{15}N of plants to understanding N dynamics at the population and community levels (Handley & Scrimgeour 1997, Högberg 1997, Evans 2001, Robinson 2001, Stewart 2001), there is some consensus that under certain conditions N isotopes provide useful insights. This may be particularly true when pools within a system have distinct δ^{15}N values and these have been measured. Here, one may be able to use δ^{15}N to explore broad patterns of N use at the population and/or community level. For example, in sites that receive a high input of nitrogen from animals (animals are highly enriched compared to plants and soils; see Robinson 2001), tracing N inputs and movement within a plant community may be possible (Erskine et al. 1998, Stewart 2001). Under these conditions, δ^{15}N can serve as a nonenriched tracer, as recently demonstrated by Frank & Evans (1997) and Frank et al. (2000) in an elegant example of how N inputs by large herbivores (buffalo and elk) can alter both plant and soil δ^{15}N in the plant communities of Yellowstone National Park in the United States.

Under conditions where the study of natural abundance ^{15}N poses limitations, enriched ^{15}N-tracers can act as indispensable tools to evaluate the importance of N availability in shaping natural communities. Differences in uptake of N tracers, both spatially and temporally, have been used to infer what maintains species diversity (or niche diversification) in old-field (McKane et al. 1990) and desert

(Gebauer & Ehleringer 2000) plant communities. ^{15}N-tracers also have been used to show how plants can enhance rates of N uptake following defoliation (Wallace & Macko 1993). Tracers also showed that after simulated mammalian browsing the performance of several evergreen and deciduous tree saplings depended upon both the capacity to store N and the site of storage during the previous winter (Millard et al. 2001).

Symbioses with Plants—Mycorrhizal Fungi and N-Fixing Bacteria

The importance of plant symbioses have recently been studied using stable isotope methods. We review some important recent studies below.

CARBON   Mycorrhizal fungi are known to influence plant acquisition of C, nutrients and water. Generally, in plant-mycorrhizal symbioses C moves from the plant to the fungus, whereas nutrients derived from the soil are passed from the fungus to the plant (Smith & Read 1997). Because the connection between plant roots and the fungus is not easily observed, and because many fungi are not host-specific, fungi might receive C from several hosts, whereas the plant in turn could receive N from many different fungal symbionts.

Several laboratory-based ^{13}C labeling studies indicate that C is transported between plants connected via ectomycorrhizal (ECM) and arbuscular-mycorrhizal (AM) networks (Simard et al. 1997b, Watkins et al. 1996, Fitter et al. 1998). This type of plant-to-plant C transport occurs via a mycorrhizal intermediate and may influence both plant C-balance and competition within a population or community. For example, this relationship may be of considerable ecological and physiological importance to the establishment of seedlings living in the shade of overstory trees where C-income of these very small plants may be severely limited. If the root system of the developing seedling were to be colonized by fungi that are connected to canopy trees then the shaded seedlings could potentially receive assimilates synthesized by large trees, thereby enhancing their growth and survival. Simard et al. (1997b) recently used both ^{13}C and radioactive ^{14}C to trace the movement of C through ectomycorrhizal *Betula papyrifera* and *Pseudotsuga menziesii*. The authors asserted that bidirectional transfer of C between plants through the fungi had occurred. However, when the fungal mycelia that linked the plants were severed, variation among replicates increased, making it difficult to demonstrate statistically the significance of the C-transfer between plants. Earlier, Watkins et al. (1996) used only natural abundance ^{13}C to quantify transfer of C between *Plantago lanceolata*, a C3 plant, connected by an AM network to *Cyndon dactylon*, a C4 plant. Their approach indicated that the gross, unidirectional transfer of C, via the mycorrhizae, from *P. lanceolata* into *C. dactylon* averaged 10% of the total C in the roots of *C. dactylon*. Fitter and coworkers (1998) used the same approach and showed that C transferred from the C3 plant to the C4 plant via the fungi remained in the roots and was never transferred to the plant shoots.

Because direct analysis of fungal mycelia is very difficult, most studies that have used isotopes to interpret the trophic status of fungi are based on measures of ^{13}C in the sporocarps. However, fungi may receive their C via hyphal connections to plants, decomposing organic matter, or both. Despite these potential limitations, $\delta^{13}C$ data have provided some general insights. For example, ECM fungi are generally more depleted in ^{13}C than saprotrophic fungi and both groups are enriched relative to the source they take up (Högberg et al. 1999b; Hobbie et al. 1999a, 2001; Kohzu et al. 1999). If there is net C-transport between a donor and recipient plant via mycorrhizae, it is critical to understand if fractionation occurs, otherwise estimates of the quantity of C-transferred made from isotope data may be in error. To date, uncertainties as to why and how $\delta^{13}C$ is altered when C is transported from the plant to the fungi remain. However, recent evidence presented by Henn & Chapela (2000) suggests there is an enrichment of ^{13}C in the fungal biomass that could occur because of selective uptake by the fungi of ^{13}C-enriched carbohydrates. They also suggest that the degree of enrichment caused by fractionation varies because of imbalances between respiratory physiology and fermentative physiology. This interpretation provides a challenge for previous interpretations, despite the fact that some investigators had used a dual-isotope ($\delta^{13}C$ and $\delta^{15}N$) approach. Clearly, more research is needed and compound-specific isotope analyses of plants, fungi, and the compounds they contain will be the method of choice.

WATER   An early study that used the radioactive isotope, ^{3}H, reported that water could be transported through the mycorrhizal mycelium to plants (Duddridge et al. 1980). This investigation has stimulated three new studies (A. Plamboeck, E. Lilleskov, J. Querejeta, personal communication) using δD or $\delta^{18}O$ to look at water transport between mycorrhizae and plants. Having access to a greater soil water volume than plant roots and being able to literally act as a bridge among plant taxa in the same community means that these fungi have the potential to influence and/or mediate an array of individual- and community-level phenomena.

NITROGEN   Because plants cannot directly access atmospheric N_2 and because soil N is not always available because of strong adsorption to soil particles, competition with soil microbes and processes that lead to N-losses, the majority of plants have evolved symbiotic relationships with mycorrhizal fungi and/or N-fixing bacteria (Figure 2). As stated already, these organisms supply the plant with N and in return receive C (Newman & Reddell 1987, Smith & Read 1997). Moreover, the mycorrhizae may enable host plants to use forms of N, such as labile organic N, that the host cannot assimilate directly. Abuzinadah & Read (1986a,b) and Finlay et al. (1992) demonstrated that many ECM fungi can use soluble peptides and proteins of animal and plant origin. The application of an enriched ^{15}N-tracer has been especially useful for demonstrating that ECM fungi are able to use, as N sources, amino acids and proteins that are not accessible for direct uptake by *Eucalyptus* species (Turnbull et al. 1995), and that the uptake of ^{15}N-labeled alanine and ammonium by *Pinus sylvestris* associated with a ECM fungi can be much higher than the uptake of ^{15}N-labeled nitrate (Wallander et al. 1997). In a field study, Näsholm et al. (1998) showed for the first time that 91%, 64%, and 42% of the N added to the organic layer as the double-labeled (^{13}C, ^{15}N) amino acid glycine was taken up by dwarf shrubs, a grass, and the trees, respectively. They also showed that the plants, irrespective of their different types of mycorrhizal symbiosis, used organic N and thereby bypassed N mineralization. Michelsen et al. (1996) suggested that mycorrhizal species in the subarctic are specialized in using organic N from the litter and that this N source is depleted in ^{15}N relative to soil mineral N.

It is still unclear how, and under what conditions, fungi alter the δ^{15}N when N is taken up and transferred to the plant (Högberg 1997; Hobbie et al. 2000, 2001; Evans 2001). Data from Michelsen et al. (1998) and Emmerton et al. (2001) show that the ^{15}N abundance in plants is closely correlated with the presence and type of mycorrhizae. Sporocarps (Gebauer & Dietrich 1993, Taylor et al. 1997, Hobbie et al. 1999a) and sheaths of ECM fungi (Högberg et al. 1996) are enriched in ^{15}N compared to their host plants. Also, mycorrhizal fungi are generally more enriched in ^{15}N than saprotrophic fungi (Gebauer & Dietrich 1993, Taylor et al. 1997, Hobbie et al. 1999a, Kohzu et al. 1999).

The observed variation in δ^{15}N among plant species has led researchers to hypothesize that this could arise during the transfer of N from the fungus to the host (Högberg et al. 1999a, Hobbie et al. 1999a, Kohzu et al. 2000, Emmerton et al. 2001). Using N-sources with known δ^{15}N values, Högberg et al. (1999a) showed that plant uptake and assimilation of N discriminates against ^{15}N for both NH$_4$ and NO$_3$. The magnitude of the fractionation appeared to decrease with increased uptake of plant N and was largest when NH$_4$ was the N-source. Interestingly, these authors observed no apparent differences in fractionation between ECM and nonmycorrhizal plants. In field and modeling studies aimed at relating variation in the δ^{15}N of mineral N, plants, soils, and mycorrhizal fungi to N availability along a successional gradient in Glacier Bay, Alaska, the best fit model included net fractionation during mycorrhizal transfer (Hobbie et al. 1999b). The results for this study highlighted the importance of designing good experiments if we are to learn how different sources of N affect the ^{15}N value of plants associated with various mycorrhizae.

Studies using both enriched and natural abundance ^{15}N have advanced the measurement of biological N fixation (BNF) tremendously, as they avoid many of the problems associated with earlier methods based on acetylene reduction assay (Hardy et al. 1968) and plant and ecosystem N budgeting (see reviews by Boddey et al. 2000; Shearer & Kohl 1986, 1991; Unkovich & Pate 2001; Warembourg 1993). Isotopic measurements have been used to estimate the proportion of plant N derived from the atmosphere. Using labeled ^{15}N tracers, one creates an enriched ^{15}N$_2$ gas atmosphere and follows the movement of the ^{15}N tracer into the plant (Warembourg 1993). Alternatively, one can add enriched ^{15}N to the soil and observe the plant as soil 15-N becomes diluted over time because of uptake of depleted-^{15}N from BNF sources. Using natural abundance ^{15}N takes advantage of the fact that soil commonly has a $\delta^{15}N$ signature relatively distinct from the atmosphere, which has a $\delta^{15}N$ signature equal to zero. Thus, plant $\delta^{15}N$ can be used to determine whether the source of its N was predominantly atmosphere or the soil pool. Nitrogen derived from fixation is calculated by comparing the isotopic composition of an N_2-fixing plant with that of plant available soil N (usually not directly measured, but derived by the $\delta^{15}N$ of a nonfixing reference plant that relies solely on soil-derived N). Problems can arise if the N_2-fixing plant and the nonfixing reference plant differ in root distribution, temporal N uptake patterns, or preferences for soil N-forms (organic versus inorganic). Furthermore, this approach assumes that only two sources of N are available (N_2 and soil N) and that there is no movement of N from the N-fixing plant to the nonfixing reference plant (Shearer & Kohl 1991). Violation of these assumptions can cloud the interpretation of the observed variation in plant $\delta^{15}N$. Because all of these assumptions are often not met, the natural abundance ^{15}N method usually provides a qualitative, rather than completely quantitative measure of BNF. It is important to note though that each method for calculating BNF has its own advantages and disadvantages (Shearer & Kohl 1986, Warembourg 1993).

Isotopes, Plants and Ecosystem Studies

Stable isotopes have provided key insights into biogeochemical interactions between plants, soils, and the atmosphere. Through the exchange of gases and the uptake of water and nutrients by roots, plants mediate the influx of energy and the gain and loss of materials from ecosystems. The subsequent effects on the metabolism and resource status of the soil ultimately feed back to influence plant function via the dynamics of soil nutrient and water availability. The application of stable isotopes in the plant-ecosystem context is rapidly increasing as improved methods are developed to integrate plant function over large spatial scales and in response to recent changes in the global cycles of C, water, and N.

CARBON   Spatially and temporally integrated values of ecosystem C-isotope discrimination (Δ) can be obtained from measurements of the $\delta^{13}C$ of whole ecosystem respiration (Figure 1). The so-called Keeling plot approach (Text Box 4) is used to obtain these data. This ecosystem Δ should convey analogous information to leaf level discrimination (see "Plant Level Studies" above). However, the comprehensive data sets needed to understand the biophysical processes that control its variation are only recently becoming available (Buchmann et al. 1998a, Pataki et al., in press). At present, ecosystem Δ is known to exhibit a high degree of spatial and temporal variability, with precipitation (Pataki et al., in press), water availability (Ometto et al., personal communication), vapor pressure deficit (Bowling et al. 2002), stand age (Fessenden & Ehleringer 2002), and species composition (Buchmann et al. 1997b) as the major drivers. These factors will affect canopy discrimination (Lloyd et al. 1996, Bowling et al. 2001) and the magnitude of leaf and root respiration, as well as respiration by rhizosphere organisms using carbon exudates from plant roots (Buchmann et al. 1998b; JPH Ometto et al., personal communication). As shifts in $\delta^{13}C$ of ecosystem respiration are likely to be dominated by fast-cycling carbon fixed from the atmosphere during the previous few days (Ekblad & Högberg 2001, Bowling et al. 2002), ecosystem discrimination should serve as an important tool for assessing the integrated response of the ecosystem to recent environmental changes (Pataki et al. 2002). A key issue in this approach is the distinction between whole ecosystem Δ and that of the plant and canopy, which will differ owing to the contribution of respiration from roots and soil organic matter decomposition, respectively. Over time, the $\delta^{13}C$ of the soil organic matter will approach that of leaf litter itself (Balesdent et al. 1993, Ehleringer et al. 2000a).

BOX 4 Keeling Plot Technique for Determining Source Signatures

Keeling (1958, 1961) developed a simple technique to determine the isotope ratio of respired CO_2 based on diurnal changes in the concentration and isotopic ratio of atmospheric CO_2 within a vegetation canopy. At night, the CO_2 concentration within the forest boundary layer increases owing to the input of respiratory CO_2. The CO_2 released from plant and soil respiration is depleted in ^{13}C and so causes a decline in $^{13}C/^{12}C$ ratio of atmospheric CO_2 within the forest boundary layer.

Keeling (1958, 1961) showed that by plotting the isotopic composition of the air (δ_{air}) against the inverse of its concentration ($1/[CO_2]_{air}$), a linear relationship was obtained and the intercept (b) of a linear regression provided an estimate of the isotope ratio of the ecosystem respiration (Flanagan & Ehleringer 1998, Yakir & Sternberg 2000),

$$\delta_{air} = m \cdot \frac{1}{[CO_2]_{air}} + b,$$ B4.1.

where m and b are determined empirically as the slope and intercept of the regression. Conceptually, as the ecosystem respires, the CO_2 concentration approaches infinity ($1/[CO_2]_{air} \rightarrow 0$) and the isotopic composition of the air (δ_{air}) approaches that of the ecosystem itself ($\delta_{air} \rightarrow b$).

At the canopy scale, the intercept represents a spatially integrated measure of the $\delta^{13}C$ of respiration from aboveground vegetation and soil components. The spatial area integrated by the calculation depends on the height at which air samples are collected, or the footprint of the air sample mast. The intercept also represents a temporal integration because it includes contributions from different aged carbon pools in plants and soil that have different turnover times and different $\delta^{13}C$ values.

This approach has proven useful when applied to other gases (e.g., H_2O) and isotopes (e.g., ^{18}O; Flanagan et al. 1997b, Moreira et al. 1997). (Modified from Flanagan & Ehleringer 1998).

As tracers, C and O isotopes in CO_2 provide the means to trace the flow of CO_2 among plants, soil, and the atmosphere (Flanagan & Ehleringer 1991, 1998). However, this method requires contrasting isotopic signatures of the sources and sinks in question. As the largest difference in plant $\delta^{13}C$ exists between plants with the C3 ($\sim -27‰$) and C4 ($\sim -12‰$) photosynthetic pathways, the principle application has been to distinguish the origin of respired CO_2 as being from either of these plant types (Miranda et al. 1997, Rochette & Flanagan 1997). This approach is limited to systems where both photosynthetic pathways are represented or that have experienced recent changes from C3 to C4 vegetation (or vice versa), as might occur after land use conversion (Trumbore et al. 1995) or agricultural cultivation (Robinson & Scrimgeour 1995, Schubler et al. 2000). Working in a mixed C3/C4 grassland, Still et al. (C. Still, submitted) used the Keeling plot approach (Text Box 4) and a two-source mixing model (Text Box 2) to determine the relative contribution of C3 and C4 sources to total ecosystem respiration by comparing $\delta^{13}C$ of ecosystem respiration to that of C3 ($\sim -28‰$) and C4 ($\sim -12‰$) plants. Rochette et al. (1999) used the $\delta^{13}C$ of C4 corn ($-12‰$) currently growing in a field where C3 wheat ($-26‰$) previously dominated to calculate the contribution of rhizosphere respiration (C4) to the total soil surface CO_2 flux (a mixture of respiration from both C3 and C4 sources). They found that it contributed up to 45% of total soil respiration and overall comprised 17% of crop net assimilation. Because these results were comparable to those made by root exclusion and ^{14}C labeling techniques, respectively, the natural abundance ^{13}C approach has the advantage of allowing for in situ measurements throughout the growing season. To date this approach has been limited to partitioning root from soil respiration. However, in any system where the isotopic difference between ecosystem components (roots, soils, stems, leaves, etc.) can be resolved, this approach can be applied to partition respiration sources (Tu & Dawson 2003 and unpublished manuscript). In a related approach, Hungate et al. (1997) and Andrews et al. (1999) separated root and microbial respiration from total soil surface CO_2 flux based on their different $\delta^{13}C$ values that resulted from exposing the vegetation to CO_2 depleted in ^{13}C ($\delta^{13}C$ of $-35‰$ versus $-8‰$ for CO_2 in air).

In a related approach, Still et al. (C. Still, submitted) determined the relative contribution of C3 and C4 vegetation to whole canopy photosynthesis by combining leaf-level C3 and C4 Δ measurements (see Evans et al. 1986) with estimates of canopy Δ derived from vertical gradients of $\delta^{13}CO_2$ (see Lloyd et al. 1996) and estimates of $\delta^{13}C$ in CO_2 respired during soil organic matter decomposition. By analyzing the isotopic composition of CO_2 rather than plant biomass, they could determine the relative flux-weighted physiological activity of C3 and C4 plants rather than simply their relative biomass abundance (e.g., Tieszen et al. 1997). Extending this approach further, Yakir & Wang (1996) used $\delta^{13}CO_2$ above crop canopies to partition measurements of net ecosystem CO_2 exchange between rates of photosynthesis and respiration. Bowling et al. (1999a, 1999b, 2001) provide a theoretical framework and experimental evidence to apply this approach in forests ecosystems. Combining isotope and micrometeorological measurements in this fashion will ultimately provide the continuous long-term observations necessary to understand the ecology and dynamics of both carbon production and storage in ecosystems. As noted by Bowling et al. (2001), the greatest uncertainty in this approach lies in the determination of canopy photosynthetic discrimination (Δ_{canopy}), which cannot presently be measured directly (see also Lloyd et al. 1996). As the collection of CO_2 without fractionation is technically demanding (Ehleringer &

Cook 1998, Bowling et al. 1999b, Tu et al. 2001), the development of sensors capable of fast response in situ measurements of isotope ratios or concentrations will greatly expand the application and potential of stable isotopes in the study of plant-atmosphere interactions.

It should be noted that in the above examples, O isotopes in CO_2 (Figure 1) provide a similar but alternative approach to trace the flow of CO_2 between plants, soils, and the atmosphere (Flanagan 1998, Flanagan et al. 1999). A major advantage of O isotopes is that large variations in $\delta^{18}O$ of CO_2 exchanged between these pools can occur even when the differences in the $\delta^{13}C$ signals are small (Yakir 1992, Flanagan & Varney 1995). Although substantial progress has been made in understanding the mechanistic basis for O isotope effects during plant-soil-atmosphere exchange (Farquhar et al. 1993, Farquhar & Lloyd 1993, Flanagan et al. 1994, Flanagan & Varney 1995, Tans 1998, Miller et al. 1999, Angert et al. 2001, Stern et al. 2001), the large variability and spatial heterogeneity in the $\delta^{18}O$ signal of CO_2 can at times make Keeling plot approaches or ecological interpretations difficult (Flanagan et al. 1997b, Bowling et al. 1999b).

The application of $\delta^{13}C$ as a tracer of CO_2 fluxes generally relies on the assumption that respired CO_2 has the same isotopic composition as the bulk organic C from which the CO_2 presumably originated. However, the results of several recent studies indicate that CO_2 evolved during dark respiration might be enriched relative to bulk leaf material by up to 6‰ (Duranceau et al. 1999, 2001; Ghashghaie et al. 2001). Earlier studies both support (Park & Epstein 1961, Troughton et al. 1974) and contradict (Troughton et al. 1974, Cheng 1996) these findings (see, also, O'Leary 1981). Some of this variation may be caused by differences between the C substrate for respiration and that of the bulk material (Duranceau et al. 1999, Cernusak et al. 2001). Further evidence suggests that apparent fractionation (Text Box 1) may occur when respired CO_2 reflects recent photosynthates whereas $\delta^{13}C$ of bulk tissues reflects C fixed earlier (Pate 2001). Discrimination during bark photosynthesis may also contribute to apparent differences between respired CO_2 and whole tissue samples (Cernusak et al. 2001). Whereas Lin & Ehleringer (1997) demonstrated that fractionation does not occur during dark respiration, fractionation can occur during synthesis of secondary metabolites (Park & Epstein 1961, Winkler et al. 1978, O'Leary 1981, Schmidt & Gleixner 1998). As noted by Park & Epstein (1961), any depletion (or enrichment) of ^{13}C in a compound must necessarily complement an enrichment (or depletion) of ^{13}C in some other compound such as respired CO_2. As there are isotopic differences among different plant tissues, such as leaves and roots (Gleixner et al. 1993), and among different compounds, such as starch and lipids (DeNiro & Epstein 1977, Ghashghaie et al. 2001), there may be differences in respired CO_2 during both biosynthesis and decomposition by soil microorganisms of these various tissues and metabolites. Further research is needed to develop a predictive understanding of the fractionations that occur during C metabolism in both plants and microbes and their effect on the ecological interpretation of isotopes in plants, soils, and air.

Lastly, research aimed at elucidating patterns of ecosystem (vegetation) and faunal change over hundreds to millions of years has also used stable C and O isotope information preserved in fossilized materials (Cerling et al. 1993, 1998; Tu et al. 1999; Eshetu & Högberg 2000b; MacFadden 2000). The analysis of $\delta^{13}C$ in pedogenic carbonates (Quade et al. 1992) as well as the C and O isotope composition preserved in tooth enamel (see MacFadden 2000), fossil seeds, or bone collagen (DeNiro & Epstein 1979) has helped show when transitions occurred between vegetation dominated by C3 plants versus C4 plants (which were largely grasses), and how this timing may have had an impact on diet choice in animals. In addition, these data have been used to make arguments about the evolution of new photosynthetic pathways (e.g., C4; Ehleringer et al. 1991b), the rise and fall of atmospheric CO_2 (Cerling et al. 1998, Ehleringer et al. 1998, Arens et al. 2000) and other environmental changes (e.g., the origin and patterns of precipitation). In a related approach, Ehleringer & Cerling (1995) inferred c_i/c_a ratios from the $\delta^{13}C$ of current and fossil leaf material from glacial-interglacial periods to the present. Based on this isotopic evidence, they found strong regulation of c_i/c_a ratios over the range of conditions expected on evolutionary timescales (several million years) and predicted less control by plants exposed to current atmospheric CO_2 concentrations that are outside the range under which they evolved.

WATER   Plants have an important influence on the magnitude and speed of water moving in ecosystems (Jackson et al. 2000, Yakir & Sternberg 2000, Feddes et al. 2001). In this context, δD and $\delta^{18}O$ analyses in soil, plant and water vapor have been used to explore the role of plants in catchment-scale processes (e.g., fluxes and runoff; Brunel et al. 1991; Busch et al. 1992; Thorburn et al. 1993a,b; Dawson & Ehleringer 1998; Harwood et al. 1998; Walker et al. 2001) and in the hydrological cycle itself (Gat 1996, 1998; Dawson et al. 1998; Figure 1). Much of this work involves understanding baseline hydrology and the isotope variation in water sources and precipitation within a region (Mazor 1991, Gat 1996, Ingraham 1998, Kendall & McDonnell 1998), how water vapor over the vegetation behaves (Bariac et al. 1989, Brunel et al. 1991, Harwood et al. 1998), and how isotope values in water change along the soil-plant-atmospheric continuum (Dawson et al. 1998). For example, Bariac and his coworkers (1983, 1987, 1989) established the utility of isotope analyses of leaf water to assess evapotranspirational flux from an alfalfa field (also see Wang & Yakir 1995). This work has been extended to forested ecosystems where a handful of investigations use direct measurements of water transpired from canopies (Harwood et al. 1998) plus other isotope data to link water-source uptake with ecosystem-level water loss; these data are used to make inferences about stand-level hydrologic processes (Walker & Brunel 1990, Brunel et al. 1991, Thorburn et al. 1993b) and to determine the role of the trees in these processes (Dawson 1996, Dawson & Ehleringer 1998).

Most recently, the information contained in the isotopic values of leaf water, the water vapor leaving leaf surfaces, atmospheric water vapor, and the sources of water taken up by plants has been used to estimate the proportion of water vapor flux leaving an ecosystem that comes from plant canopies versus from evaporation. Using the Keeling plot approach (Text Box 4), applied to the $\delta^{18}O$ of water vapor leaving an Amazonian forest (instead of CO_2), Moreira and coworkers

(1997) attempted to partition evapotranspiration between soil evaporation and plant transpiration (Figure 1). Because this approach has limitations (Mathieu & Bariac 1996a,b), the authors concluded that canopy transpiration was the dominant path for water vapor flux from this forested ecosystem (also, see Yakir & Sternberg 2000 for additional examples). A related approach was employed by Dawson (1996) to determine whether soil or ground water was being transpired by different age classes of sugar maple trees in a northeastern North American temperate forest ecosystem. Here, it was shown that small trees contribute water to the ecosystem flux from shallow layers in the soil, whereas larger trees transpire mostly deeper ground water (Dawson 1996). The findings from both of these investigations indicate that water vapor loss from forested ecosystems is dominated by transpiration and may vary with stand age and thus successional status, with season, and perhaps with tree species.

Related to the examples discussed above are ongoing efforts that use H or O isotope variation in leaf water either to determine the atmospheric conditions during plant transpiration or to directly estimate transpiration rate itself (Figure 1). As already critiqued in some detail elsewhere (Dawson et al. 1998, Yakir & Sternberg 2000) these types of investigation are still being refined and need further development before they can be used reliably. However, the incorporation of new empirical methods, such as relaxed eddy accumulation (REA; Pattey et al. 1993) and laser- or spectroscopic-based measurements of water vapor (and other gases) also need to be explored more fully. Such techniques allow the $\delta^{18}O$ of water vapor flux leaving an ecosystem to be directly measured (D. Hollinger, T. Dawson, K.P. Tu, unpublished manuscript). In addition, the development of predictive models that use these data is needed (W. Riley, C. Still, personal communication). Observations show that many assumptions in our current models may not be valid or may be valid only under certain circumstances (Wang & Yakir 1995, Yakir 1998, Dawson et al. 1998). If refined, this area of research holds a great deal of promise in allowing us to link plant ecophysiological behavior to local, regional, and perhaps even global hydrological processes.

NITROGEN   Within the N-cycle, soil microbial processes fractionate ^{15}N leading to a large degree of variation in the $\delta^{15}N$ of N within soil pools (Figure 2). Therefore, understanding the processes that lead to changes in the soil $\delta^{15}N$ is imperative for determining the N sources used by plants because they obtain much of their N from the soil (Handley & Raven 1992). The major soil N transformations mediated by microbes, such as mineralization (conversion of organic to inorganic forms of N; Nadelhoffer & Fry 1994), nitrification (conversion of NH_4 to NO_3; Nadelhoffer & Fry 1994, Högberg 1997, Handley & Raven 1992), and denitrification (conversion of NO_3 to atmospheric NO, N_2O, or N_2 during microbial respiration; Piccolo et al. 1996) lead to N products that are depleted in ^{15}N relative to the substrates from which they were produced (Peterson & Fry 1987, Yoneyama 1996). For example, the residual organic matter pool may become increasingly ^{15}N enriched if the NO_3 produced during nitrification exceeds soil and plant demand and leaches out of a forest (Piccolo et al. 1994, Nadelhoffer & Fry 1988). Further ^{15}N enrichment can occur during microbial decomposition of organic matter. A study in Wisconsin forests showed that the δ^{15}N of deeper soil reflected the presence of more enriched ^{15}N products from decomposition, whereas surface soils reflected leaf litter inputs that were depleted in ^{15}N (Nadelhoffer & Fry 1988). The pattern of ^{15}N enrichment with depth has been observed elsewhere (Rennie et al. 1976, Shearer et al. 1978, Shearer & Kohl 1986) and is important because it means that roots taking up the same form of N (e.g., NH_4) could have a different δ^{15}N.

Nitrogen-15 pool dilution (Text Box 2) techniques can help determine what forms of N are available and what forms plants and soil microbes may compete for as shown in Alaska birch forests (Van Cleve & White 1980). The development of the pool dilution technique allows ecologists to examine gross rates of mineralization and nitrification in natural settings and to better understand the forms of inorganic N present in an ecosystem, rather than the net transformations over time. Davidson and coworkers (1992) employed this method to demonstrate that young and old forests of California have detectable rates of gross nitrification, a process previously thought not to occur in these mature forests. Schimel et al. (1989) used this method to compare plant and microbial competition for N in a California grassland.

There are many applications for adding enriched ^{15}N tracers to ecosystems and following the fate of N (Text Box 2). For example, Knowles (1975) showed that in many studies less than 30% of applied fertilizer N ended up in crop trees. Other research has used ^{15}N as a way to determine plant, microbial, and soil sinks within evergreen and deciduous forests of Europe (Koopmans et al. 1996, Buchmann et al. 1996b, Schleppi et al. 1999) and North America (Nadelhoffer et al. 1995, 1999; Groffman et al. 1993; Zak et al. 1990; Zogg et al. 2000; Templer 2001). Nitrogen-15 tracers have also been used in greenhouse experiments to determine if foliar uptake and assimilation of wet-deposited N takes place and to quantify its contribution to the N requirements of each plant species (Bowden et al. 1989, Garten & Hanson 1990, Wilson & Tiley 1998).

Measurements of natural abundance ^{15}N have been used increasingly as an indicator of change in the N cycling of forests. Humans have doubled the amount of N naturally fixed in the environment (Galloway et al. 1995, Vitousek et al. 1997). When the amount of N deposited exceeds biological demand, an ecosystem can reach N-saturation and increasing amounts of N may leave either through leaching or gas loss (Aber et al. 1989, Agren & Bosatta 1988, Peterjohn et al. 1996, Stoddard 1994). Elevated levels of N availability can lead to increased rates of N cycling. This increase in turn results in ^{15}N enrichment of each soil pool as the lighter ^{14}N isotopes are preferentially lost through leaching and denitrification (Figure 2). Plants accessing this soil N pool can then become relatively ^{15}N-enriched over time. Because plant biomass turns over at a faster rate than the total soil pool, plants themselves can also be used as indicators of anthropogenically caused environmental change (Johannisson & Högberg 1994). In this way, the measurement of foliar δ^{15}N has the potential to be used to indicate that

N cycling rates have increased. For example, several studies have found a strong correlation between enriched levels of ^{15}N in foliage and increased soil N (Emmett et al. 1998, Meints et al. 1975, Högberg 1990), increased rates of N-cycling (Garten 1993, Garten & Van Miegroet 1994), and increased loss of N (Högberg & Johannisson 1993). Other studies have used natural abundance ^{15}N and [18]O within NO_3^- simultaneously to examine whether N deposited onto a forest is cycled within plants and microbes or directly passes through the forest into nearby streams without biological processing, an indicator that the system may have reached N saturation (Durka et al. 1994). Other studies have linked land use change to a decline in N fixing cryptobiotic crusts in aridlands, which in turn has led to changes in N availability and subsequent enrichment in plant and soil $\delta^{15}N$ (Evans & Belnap 1999; Evans & Ehleringer 1993, 1994). Additionally, some studies have measured N pools and natural abundance ^{15}N simultaneously to indicate the relative degree of openness of the N-cycle, which relates how much N is lost from an ecosystem relative to its internal pool size N (Austin & Vitousek 1998, Chang & Handley 2000, Eshetu & Högberg 2000a, Brenner et al. 2001).

From a longer-term perspective, the analysis of ^{15}N in tree-ring cellulose and in peat deposits has been used as an indicator of ongoing environmental change. The work of Poulson et al. (1995) shows that ^{15}N in the wood of two eastern hemlock trees decreased from the early 1960s to 1992. The authors attribute this decrease to either a decrease in ^{15}N of available N over time or to isotope fractionation during translocation within the tree. Untangling which mechanism is responsible for these changes is the next step. In this same context, Bergstrom et al. (2002) examined the ^{15}N in fossil peat; this work allowed the authors to not only look at the long-term (8500 years) trend in N dynamics but also allowed them to show that some of this N-input is from animal sources.

FUTURE DIRECTIONS

This review has attempted to highlight how our knowledge of plant-environment interactions has been advanced by the application of stable isotope methods. We believe that the literature clearly shows that plant ecology has made very significant progress as a result of collecting and interpreting isotope data. Several important areas are in need of further research.

We believe that isotope methods, when merged with other information from modeling, molecular, and/or genetic data, have the potential to deepen our understanding of population- and community-level processes. For example, we need to determine if and how mycorrhizae may contribute to the plant water uptake in areas where water limits growth and survival. Additionally, further study is needed to determine what happens to C and N at the plant-mycorrhizae-interface. This study is important because it will enhance our understanding of what contributes to the overall variation in soil and plant $\delta^{15}N$ values. Finally, if we want to refine our estimates of the quantities of C transferred between plants via mycorrhizae, ^{13}C methods hold great promise.

The use of stable isotopes as integrators of plant C and water relations at spatial scales of individual leaves, whole plants, and entire ecosystems and temporal scales ranging from instantaneous gas exchange to paleontological tree-ring studies would be greatly enhanced by developing a quantitative linkage between C and O isotopes in leaf and stem tissues and physiological characteristics such as photosynthetic capacity and stomatal conductance. In this way, an improved measure of plant physiological status could be obtained from stable isotope signals recorded in tree rings under past environmental conditions and in plant tissues collected from remote locations or where gas exchange is not practical. Current single-isotope or single resource (e.g., isotopes in water) approaches rely on knowledge of ancillary climate or environmental data. In addition, further research is needed to develop a predictive understanding of C isotope fractionations during plant and microbial metabolism. The magnitude of such fractionations and their influence on the C isotope composition of respired CO_2 is not known, but they could have important implications for partitioning canopy photosynthesis, root respiration, and soil organic matter decomposition (see Tu & Dawson 2003). Then, for both C and N it is clear that further development of compound-specific $\delta^{13}C$ and $\delta^{15}N$ isotope analyses and dual-isotope approaches (e.g., Robinson et al. 2000) has the potential to enhance our understanding of plant ecological phenomena at all spatial and temporal scales.

Future advances in the use of ^{15}N isotopes in plant ecology will result from combining modeling and empirical studies that address the mechanisms underlying the variation in $\delta^{15}N$ of plant and ecosystem pools. A complete mechanistic model of ^{15}N behavior in plants is still lacking. Fractionations that may occur during direct uptake of inorganic N forms, during N uptake that is mediated by mycorrhizae, during N exudation, as well as within-plant transformations all need deeper understanding. Our current interpretation of plant-soil interactions is also constrained by the technical difficulties of measuring the $\delta^{15}N$ of soil N pools that is available for plant uptake. At the ecosystem scale, modeling may be the best tool for using $\delta^{15}N$ as an integrator of N cycling processes. For example, Hobbie et al. (1999b) developed the NIFTE model to predict the relationship between N cycling and natural abundance ^{15}N within ecosystem pools. Another model has been developed to examine the movement of enriched ^{15}N tracers throughout ecosystems (called TRACE; Currie et al. 1999, Currie & Nadelhoffer 1999). Future work needs to expand on these models and incorporate results from mechanistic studies.

ACKNOWLEDGMENTS

Foremost, we thank Jim Ehleringer; Jim has not only been a leading advocate for using stable isotopes in plant ecology but he has provided all of us, and many others, with the opportunity and inspiration to involve ourselves in using isotope methods in our own research. We thank Rick Harrison as well for initially encouraging this review and to our many colleagues who provided us with important feedback and literature related to it. Finally, we thank Larry Flanagan, Thorsten Grams, Dave Evans, Murray Unkovich, Dave Williams, and John Roden for their comments and suggestions that improved the manuscript.

The *Annual Review of Ecology and Systematics* is online at
http://ecolsys.annualreviews.org

LITERATURE CITED

Aber JD, Nadelhoffer KJ, Steudler P, Melillo JM. 1989. Nitrogen saturation in northern forest ecosystems. *BioScience* 39:378–86

Abuzinadah RA, Finlay RD, Read DJ. 1986a. The role of proteins in the nitrogen nutrition of ectomycorrhizal plants. III. Protein utilisation by *Betula, Picea,* and *Pinus* in mycorrhizal association with *Hebeloma crustuliniforme. New Phytol.* 103:507–14

Abuzinadah RA, Finlay RD, Read DJ. 1986b. The role of proteins in the nitrogen nutrition of ectomycorrhizal plants. II. Utilisation of protein by mycorrhizal plants of Pinus contorta. *New Phytol.* 103:495–506

Ågren GI, Bosatta E. 1988. Nitrogen saturation of terrestrial ecosystems. *Environ. Pollut.* 54:185–98

Alstad KP, Welker JM, Williams SA, Trlica MJ. 1999. Carbon and water relations of *Salix monticola* in response to winter browsing and changes in surface water hydrology: an isotopic study using $\delta^{13}C$ and $\delta^{18}O$. *Oecologia* 120:375–85

Amundson R, Stern L, Baisden T, Wang Y. 1998. The isotopic composition of soil and soil-respired CO_2. *Geoderma* 82:83–114

Andrews JA, Harrison KG, Matamala R, Schlesinger WH. 1999. Separation of rot respiration from total soil respiration using carbon-13 labeling during free-air carbon dioxide enrichment (FACE). *Soil Sci. Soc. Am. J.* 63:1429–35

Angert A, Luz B, Yakir D. 2001. Fractionation of oxygen isotopes by respiration and diffusion in soils and its implications for the isotopic composition of atmospheric CO_2. *Glob. Biogeochem. Cycles* 15:871–80

Archer S. 1995. Tree-grass dynamics in a Prosopis-thornscrub savanna parkland: reconstructing the past and predicting the future. *Ecoscience* 2:83–99

Arens NC, Jahren AH, Amundson R. 2000. Can C3 plants faithfully record the carbon isotopic composition of atmospheric carbon dioxide? *Paleobiology* 26:137–64

Austin AT, Vitousek PM. 1998. Nutrient dynamics on a precipitation gradient in Hawai'i. *Oecologia* 113:519–29

Baertschi P. 1953. Die Fraktionierung der naturlichen kohlenstoffisotopen im kohlendioxydstoffenwechsel gruner pflanzen. *Helv. Chim. Acta* 36(4):773–81

Balesdent J, Girardin C, Mariotti A. 1993. Site-related ^{13}C of tree leaves and soil organic matter in a temperate forest. *Ecology* 74:1713–21

Ball JT, Woodrow IE, Berry JA. 1987. A model predicting stomatal conductance and its contribution to the control of photosynthesis under different environmental conditions. In *Progress in Photosynthesis Research,* ed. I Biggins, pp. 221–24. Netherlands: Martinus Nijoff

Bannister P, Strong GL. 2001. Carbon and nitrogen isotope ratios, nitrogen content and heterotrophy in New Zealand mistletoes. *Oecologia* 126:10–20

Barbour MM, Andrews TJ, Farquhar GD. 2001. Correlations between oxygen isotope ratios of wood constituents of Quercus and Pinus samples from around the world. *Aust. J. Plant Physiol.* 28:335–48

Barbour MM, Fischer RA, Sayre KD, Farquhar GD. 2000. Oxygen isotope ratio of leaf and grain material correlates with stomatal conductance and grain yield in irrigated wheat. *Aust. J. Plant Physiol.* 27:625–37

Bariac T, Ferhi A, Jusserand C, Létolle R. 1983. Sol-plante-atmosphère: contribution à l'étude dé la composition isotopique de l'eau des differentes composantes de ce systeme. *Proc. Symp. Isotope Radiation Tech. Soil Phys. Irrig. Stud. IAEA, Vienna*, pp. 561–76

Bariac T, Klamecki S, Jusserand C, Létolle R. 1987. Evolution de la composition isotopique de l'eau (^{18}O) dans le continuum sol-plante-atmosphère (example d'une parcelle cultivée en blé, Versailles, France, Juin 1984). *Catena* 14:55–72

Bariac T, Rambal S, Jusserand C, Berger A. 1989. Evaluating water fluxes of field-grown alfalfa from diurnal observations of natural isotope concentrations, energy budget and ecophysiological parameters. *Agric. For. Meteorol.* 48:263–84

Ben-David M, Schell DM. 2001. Mixing models in analyses of diet using multiple stable isotopes: a response. *Oecologia* 127:180–84

Bergersen FJ, Peoples MB, Turner GL. 1988. Isotopic discriminations during the accumulation of nitrogen by soybeans. *Aust. J. Plant Physiol.* 15:407–20

Bergstrom DM, Stewart GR, Selkirk PM, Schmidt S. 2002. ^{15}N natural abundance of fossil peat reflects the influence of animal-derived nitrogen on vegetation. *Oecologia* 130:309–14

Bert D, Leavitt S, Dupouey J-L. 1997. Variations of wood $\delta^{13}C$ and water-use efficiency of *Abies alba* during the last century. *Ecology* 78:1588–96

Bettarini I, Calderoni G, Miglietta F, Raschi A, Ehleringer J. 1995. Isotopic carbon discrimination and leaf nitrogen content of *Erica arborea* L. along a CO_2 concentration gradient in a CO_2 spring in Italy. *Tree Physiol.* 15:327–32

Boddey RM, Peoples MB, Palmer B, Dart PJ. 2000. Use of the ^{15}N natural abundance technique to quantify biological nitrogen fixation by woody perennials. *Nutr. Cycl. Agroecosys.* 57:235–70

Borella S, Leuenberger M, Saurer M. 1999.

Analysis of $\delta^{18}O$ in tree rings: wood-cellulose comparison and method dependent sensitivity. *J. Geophys. Res.* 104:19267–73

Bowden RD, Geballe GT, Bowden WB. 1989. Foliar uptake of ^{15}N from simulated cloud water by red spruce (*Picea rubens*) seedlings. *Can. J. For. Res.* 19:382–86

Bowling DR, Baldocchi DD, Monson RK. 1999a. Dynamics of isotopic exchange of carbon dioxide in a Tennessee deciduous forest. *Glob. Biogeochem. Cycles* 13:903–22

Bowling DR, Delany AC, Turnipseed AA, Baldocchi DD, Monson RK. 1999b. Modification of the relaxed eddy accumulation technique to maximize measured scalar mixing ratio differences in updrafts and downdrafts. *J. Geophys. Res.* 104:9121–33

Bowling DR, McDowell NG, Bond BJ, Law BE, Ehleringer JR. 2002. ^{13}C content of ecosystem respiration is linked to precipitation and vapor pressure deficit. *Oecologia* 131:113–24

Bowling DR, Tans PP, Monson RK. 2001. Partitioning net ecosystem carbon exchange with isotopic fluxes of CO_2. *Glob. Change Biol.* 7(2):127–45

Bowman WD, Hubick KT, von Caemmerer S. 1989. Short-term changes in leaf carbon isotope discrimination in salt- and water-stressed C4 grasses. *Plant Physiol.* 90(1):162–66

Brenner DL, Amundson R, Baisden WT, Kendall C, Harden J. 2001. Soil N and ^{15}N variation with time in a California annual grassland ecosystem. *Geochim. Cosmochim. Acta* 65:4171–86

Briske DD, Boutton TW, Wang Z. 1996. Contribution of flexible allocation priorities to herbivory tolerance in C4 perennial grasses: an evaluation with ^{13}C labeling. *Oecologia* 105:151–59

Broadmeadow MSJ, Griffiths H, Maxwell C, Borland AM. 1992. The carbon isotope ratio of plant organic material reflects temporal and spatial variations in CO_2 within tropical forest formations in Trinidad. *Oecologia* 89:435–41

Brooks JR, Flanagan LB, Buchmann N,

Ehleringer JR. 1997. Carbon isotope composition of boreal plants: functional grouping of life forms. *Oecologia* 110:301–11

Brugnoli E, Hubick KT, von Caemmerer S. 1988. Correlation between the carbon isotope discrimination in leaf starch and sugars of C3 plants and the ratio of intercellular and atmospheric partial pressures of carbon dioxide. *Plant Physiol.* 88:1418–24

Brunel J-P, Walker GR, Kennett-Smith AK. 1995. Field validation of isotopic procedures for determining sources of water used by plants in a semi-arid environment. *J. Hydrol.* 167:351–68

Brunel J-P, Walker GR, Walker CD, Dighton JC, Kennett-Smith A. 1991. Using stable isotopes of water to trace plant water uptake. In *Stable Isotopes in Plant Nutrition, Soil Fertility and Environmental Studies*, pp. 543–51. *Proc. Int. Symp., Oct. 1990.* Vienna: IAEA/FAO

Buchmann N, Brooks JR, Flanagan LB, Ehleringer JR. 1998a. Carbon isotope discrimination of terrestrial ecosystems. See Griffiths 1998, pp. 203–21

Buchmann N, Brooks JR, Rapp KD, Ehleringer JR. 1996a. Carbon isotope composition of C_4 grasses is influenced by light and water supply. *Plant Cell Environ.* 19:392–402

Buchmann N, Gebauer G, Schulze E-D. 1996b. Partitioning of ^{15}N-labeled ammonium and nitrate among soil, litter, below- and aboveground biomass of trees and understory in a 15-year-old *Picea abies* plantation. *Biogeochemistry* 33:1–23

Buchmann N, Guehl J-M, Barigah TS, Ehleringer JR. 1997a. Interseasonal comparison of CO_2 concentrations, isotopic composition, and carbon dynamics in an Amazonian rainforest (French Guiana). *Oecologia* 110:120–31

Buchmann N, Hinckely TM, Ehleringer JR. 1998b. Carbon isotope dynamics in *Abies amabilis* stands in the Cascades. *Can. J. For. Res.* 28:808–19

Buchmann N, Kao W-Y, Ehleringer JR. 1997b. Influence of stand structure on carbon-13 of vegetation, soils, and canopy air within deciduous and evergreen forests in Utah, United States. *Oecologia* 110:109–19

Burgess SO, Pate JS, Adams MA, Dawson T. 2000. Seasonal water acquisition and redistribution in the Australian woody phreatophyte. *Banksia prionotes. Ann. Bot.* 85:215–24

Busch DE, Ingraham NL, Smith SD. 1992. Water uptake in woody riparian phreatophytes of the Southwestern United States a stable isotope study. *Ecol. Appl.* 2:450–59

Caldeira MC, Ryel RJ, Lawton JH, Pereira JS. 2001. Mechanisms of positive biodiversity-production relationships: insights provided by δ^{13}C analysis in experimental Mediterranean grassland plots. *Ecol. Lett.* 4:439–43

Caldwell MM, Dawson TE, Richards JH. 1998. Hydraulic lift: consequences of water efflux from the roots of plants. *Oecologia* 113:151–61

Cerling TE, Ehleringer JR, Harris JM. 1998. Carbon dioxide starvation, the development of the C4 ecosystem, and mammalian evolution. *Philos. Trans. R. Soc. London Ser. B* 353:159–71

Cerling TE, Wang Y, Quade J. 1993. Expansion of C4 ecosystems as an indicator of global ecological change in the late Miocene. *Nature* 361:344–45

Cernusak LA, Marshall JD. 2001. Responses of foliar δ^{13}C gas exchange and leaf morphology to reduced hydraulic conductivity in *Pinus monticola* branches. *Tree Physiol.* 21:1215–22

Cernusak LA, Marshall JD, Comstock JP, Balster NJ. 2001. Carbon isotope discrimination in photosynthetic bark. *Oecologia* 128:24–35

Chang SX, Handley LL. 2000. Site history affects soil and plant ^{15}N natural abundances (δ^{15}N) in forests of northern Vancouver Island, British Columbia. *Funct. Ecol.* 14:273–80

Cheng W. 1996. Measurement of rhizosphere respiration and organic matter decomposition using natural ^{13}C. *Plant Soil.* 183:263–68

Comstock JP, Ehleringer JR. 1992. Correlating

genetic variation in carbon isotopic composition with complex climatic gradients. *Proc. Natl. Acad. Sci. USA* 89:7747–51

Condon AG, Richards RA, Farquhar GD. 1987. Carbon isotope discrimination is positively correlated with grain yield and dry matter production in field-grown wheat. *Crop Sci.* 27:996–1001

Condon AG, Richards RA, Farquhar GD. 1992. The effect of variation in soil water availability, vapour pressure deficit and nitrogen nutrition on carbon isotope discrimination in wheat. *Aust. J. Agric. Res.* 43:935–48

Craig H. 1953. The geochemistry of the stable carbon isotopes. *Geochim. Cosmochim. Acta* 3:53–92

Cramer VA, Thorburn PJ, Fraser GW. 1999. Transpiration and groundwater uptake from farm forest plots of *Casuarina glauca* and *Eucalyptus camaldulensis* in saline areas of southeast Queensland, Australia. *Agric. Water Manag.* 39:187–204

Criss RE. 1999. *Principles of Stable Isotope Distribution.* New York: Oxford Univ. Press. 254 pp.

Currie WS, Nadelhoffer KJ. 1999. Dynamic redistribution of isotopically labeled cohorts of nitrogen inputs in two temperate forests. *Ecosystems* 2:4–18

Currie WS, Nadelhoffer KJ, Aber JD. 1999. Soil detrital processes controlling the movement of ^{15}N tracers to forest vegetation. *Ecol. Appl.* 9:87–102

Damesin C, Rambal S, Joffre R. 1998. Seasonal drought and annual changes in leaf $\delta^{13}C$ in two co-occurring Mediterranean oaks: relations to leaf growth and drought progression. *Funct. Ecol.* 12:778–85

Davidson EA, Hart SC, Firestone MK. 1992. Internal cycling of nitrate in soils of a mature coniferous forest. *Ecology* 73:1148–56

Dawson TE. 1993a. Hydraulic lift and water use by plants: implications for water balance, performance and plant-plant interactions. *Oecologia* 95:565–74

Dawson TE. 1993b. Water sources of plants as determined from xylem-water isotopic composition: perspectives on plant competition, distribution, and water relations. See Ehleringer et al. 1993, pp. 465–96

Dawson TE. 1996. Determining water use by trees and forests from isotopic, energy balance and transpiration analyses: the roles of tree size and hydraulic lift. *Tree Physiol.* 16:263–72

Dawson TE. 1998. Fog in the California redwood forest: ecosystem inputs and use by plants. *Oecologia* 117:476–85

Dawson TE, Brooks PD. 2001. Fundamentals of stable isotope chemistry and measurement. See Unkovich et al. 2001, pp. 1–18

Dawson TE, Ehleringer JR. 1991. Streamside trees that do not use stream water. *Nature* 350:335–37

Dawson TE, Ehleringer JR. 1993. Gender-specific physiology, carbon isotope discrimination, and habitat distribution in boxelder. *Acer negundo. Ecology* 74:798–815

Dawson TE, Ehleringer JR. 1998. Plants, isotopes, and water use: a catchment-level perspective. See Kendall & McDonnell 1998, pp. 165–202

Dawson TE, Pate JS. 1996. Seasonal water uptake and movement in root systems of Australian phraeatophytic plants of dimorphic root morphology: a stable isotope investigation. *Oecologia* 107:13–20

Dawson TE, Pausch RC, Parker HM. 1998. The role of H and O stable isotopes in understanding water movement along the soil-plant-atmospheric continuum. See Griffiths 1998, pp. 169–83

De Kroon H, Fransen B, Van Rheenen JWA, Van Dijk A, Kreulen R. 1996. High levels of inter-ramet water translocation in two rhizomatous Carex species, as quantified by deuterium labelling. *Oecologia* 106:73–84

DeLucia EH, Schlesinger WH, Billings WD. 1988. Water relations and the maintenance of Sierran conifers on hydrothermally altered rock. *Ecology* 69:303–11

DeNiro MJ, Epstein S. 1977. Mechanism of carbon isotope fractionation associated with lipid synthesis. *Science* 197:261–63

DeNiro MJ, Epstein S. 1979. Relationship between the oxygen isotope ratios of terrestrial plant cellulose, carbon dioxide and water. *Science* 204:51–53

Donovan LA, Ehleringer JR. 1994. Carbon isotope discrimination, water-use efficiency, growth, and mortality in a natural shrub population. *Oecologia* 100:347–54

Duddridge JA, Malibari A, Read DJ. 1980. Structure and function of mycorrhizal rhizomorphs with special reference to their role in water transport. *Nature* 287:834–36

Duquesnay A, Breda N, Stievenard M, Dupouey JL. 1998. Changes of tree-ring $\delta^{13}C$ and water-use efficiency of beech (*Fagus sylvatica* L.) in north-eastern France during the past century. *Plant Cell Environ.* 21:565–72

Duranceau M, Ghashghaie J, Badeck F, Deleens E, Cornic G. 1999. $\delta^{13}C$ of CO_2 respired in the dark in relation to $\delta^{13}C$ of leaf carbohydrates in *Phaseolus vulgaris* L. under progressive drought. *Plant Cell Environ.* 22:515–23

Duranceau M, Ghashghaie J, Brugnoli E. 2001. Carbon isotope discrimination during photosynthesis and dark respiration in intact leaves of *Nicotiana sylvestris*: comparisons between wild type and mitochondrial mutant plants. *Aust. J. Plant Physiol.* 28:65–71

Durka WE, Schultze DS, Gebauer G, Voerkelius S. 1994. Effects of forest decline on uptake and leaching of deposited nitrate determined from ^{15}N and ^{18}O measurements. *Nature* 372:765–69

Dyckmans J, Flessa H. 2001. Influence of tree internal N status on uptake and translocation of C and N in beech: a dual ^{13}C and ^{15}N labeling approach. *Tree Physiol.* 21:395–401

Edwards TWD, Aravena RO, Fritz P, Morgan AV. 1985. Interpreting paleoclimate from ^{18}O and ^{2}H in wood cellulose: paleoclimatic implications for southern Ontario. *Can. J. Earth Sci.* 22:1720–26

Ehleringer JR. 1989. Carbon isotope ratios and physiological processes in aridland plants. See Rundel et al.1989, pp. 41–54

Ehleringer JR. 1990. Correlations between carbon isotope discrimination and leaf conductance to water vapor in common beans. *Plant Physiol.* 93:1422–25

Ehleringer JR. 1993a. Carbon and water relations in desert plants: an isotopic perspective. See Ehleringer et al. 1993, pp. 155–72

Ehleringer JR. 1993b. Variation in leaf carbon isotope discrimination in *Encelia farinosa*: implications for growth, competition, and drought survival. *Oecologia* 95:340–46

Ehleringer JR, Buchmann N, Flanagan LB. 2000a. Carbon isotope ratios in belowground carbon cycle processes. *Ecol. Appl.* 10:412–22

Ehleringer JR, Cerling TE. 1995. Atmospheric CO_2 and the ratio of intercellular to ambient CO_2 levels in plants. *Tree Physiol.* 15:105–11

Ehleringer JR, Cook CS. 1998. Carbon and oxygen isotope ratios of ecosystem respiration along an Oregon conifer transect: preliminary observations based on small-flask sampling. *Tree Physiol.* 18:513–19

Ehleringer JR, Cooper TA. 1988. Correlations between carbon isotope ratio and microhabitat in desert plants. *Oecologia* 76:562–66

Ehleringer JR, Dawson TE. 1992. Water uptake by plants: perspectives from stable isotope composition. *Plant Cell Environ.* 15:1073–82

Ehleringer JR, Evans RD, Williams D. 1998. Assessing sensitivity to change in desert ecosystem—a stable isotope approach. See Griffiths 1998, pp. 223–37

Ehleringer JR, Field CB, Lin Z-F, Kuo C-Y. 1986. Leaf carbon isotope and mineral composition in subtropical plants along an irradiance cline. *Oecologia* 70:520–26

Ehleringer JR, Hall AE, Farquhar GD, eds. 1993. *Stable Isotopes and Plant Carbon-Water Relations*. San Diego, CA: Academic

Ehleringer JR, Phillips SL, Comstock JP. 1992. Seasonal variation in the carbon isotopic composition of desert plants. *Funct. Ecol.* 6:396–404

Ehleringer JR, Phillips SL, Schuster WSF, Sandquist DR. 1991a. Differential utilization of summer rains by desert plants. *Oecologia* 88:430–34

Ehleringer JR, Roden J, Dawson TE. 2000b. Assessing ecosystem-level water relations through stable isotope ratio analyses. See Sala et al. 2000, pp. 181–98

Ehleringer JR, Sage RF, Flanagan LB, Pearcy RW. 1991b. Climate change and the evolution of C4 photosynthesis. *Trends Ecol. Evol.* 6:95–99

Ehleringer JR, Schulze ED, Ziegler H, Lange OL, Farquhar GD, et al. 1985. Xylem-tapping mistletoes: water or nutrient parasites? *Science* 227:1479–81

Ehleringer JR, White JW, Johnson DA, Brick M. 1990. Carbon isotope discrimination, photosynthetic gas exchange, and water-use efficiency in common bean and range grasses. *Acta Oecol.* 11:611–25

Ekblad A, Högberg P. 2001. Natural abundance of ^{13}C in CO_2 respired from forest soils reveals speed of link between tree photosynthesis and root respiration. *Oecologia* 127:305–8

Emerman SH, Dawson TE. 1996. The role of macropores in the cultivation of bell pepper in salinized soil. *Plant Soil* 181:241–49

Emmerton KS, Callaghan TV, Jones HE, Leake JR, Michelsen A, et al. 2001. Assimilation and isotopic fractionation of nitrogen by mycorrhizal and nonmycorrhizal subarctic plants. *New Phytol.* 151:513–24

Emmett BA, Kjonaas OJ, Gundersen P, Koopmans C, Tietema A, et al. 1998. Natural abundance of ^{15}N in forests across a nitrogen deposition gradient. *For. Ecol. Manag.* 101:9–18

Epstein S, Krishnamurthy RV. 1990. Environmental information in the isotopic record in trees. *Philos. Trans. R. Soc. London Ser. B* 330:427–39

Erskine PD, Bergstrom DM, Schmidt S, Stewart GR, Tweedie CE, et al. 1998. Subantarctic Macquarie Island: a model ecosystem for studying animal-derived nitrogen sources using ^{15}N natural abundance. *Oecologia* 117:187–93

Eshetu Z, Högberg P. 2000a. Effects of land use on ^{15}N natural abundance of soils in Ethiopian highlands. *Plant Soil* 222:109–17

Eshetu Z, Högberg P. 2000b. Reconstruction of forest site history in Ethiopian highlands based on ^{13}C natural abundance of soils. *Ambio* 29:83–89

Evans JR, Sharkey TD, Berry JA, Farquhar GD. 1986. Carbon isotope discrimination measured concurrently with gas exchange to investigate CO_2 diffusion in leaves of higher plants. *Aust. J. Plant Physiol.* 13:281–92

Evans RD. 2001. Physiological mechanisms influencing plant nitrogen isotope composition. *Trends Plant Sci.* 6:121–26

Evans RD, Belnap J. 1999. Long-term consequences of disturbance on nitrogen dynamics in an arid grassland ecosystem. *Ecology* 80:150–60

Evans RD, Bloom AJ, Sukrapanna SS, Ehleringer JR. 1996. Nitrogen isotope composition of tomato (*Lycopersicon esculentum* Mill, cv. *T-5*) *grown under ammonium or nitrate nutrition*. *Plant Cell Environ.* 19:1317–23

Evans RD, Ehleringer JR. 1993. A break in the nitrogen cycle of aridlands: evidence from delta ^{15}N of soils. *Oecologia* 94:314–17

Evans RD, Ehleringer JR. 1994. Water and nitrogen dynamics in an arid woodland. *Oecologia* 99:233–42

Farquhar GD. 1983. On the nature of carbon isotope discrimination in C4 species. *Aust. J. Plant Physiol.* 10:205–26

Farquhar GD, Barbour MM, Henry BK. 1998. Interpretation of oxygen isotope composition of leaf material. See Griffiths 1998, pp. 27–62

Farquhar GD, Ehleringer JR, Hubick KT. 1989. Carbon isotope discrimination and photosynthesis. *Annu. Rev. Plant Physiol. Plant Mol. Biol.* 40:503–37

Farquhar GD, Lloyd J. 1993. Carbon and oxygen isotope effects in the exchange of carbon dioxide between terrestrial ecosystems and the atmosphere. See Ehleringer et al. 1993, pp. 47–70

Farquhar GD, Lloyd J, Taylor JA, Flanagan LB, Syvertsen JP, et al. 1993. Vegetation effects on the isotope composition of oxygen in atmospheric carbon dioxide. *Nature* 363:439–43

Farquhar GD, O'Leary MH, Berry JA. 1982. On the relationship between carbon isotope discrimination and intercellular carbon dioxide concentration in leaves. *Aust. J. Plant Physiol.* 9:121–37

Farquhar GD, Richards RA. 1984. Isotopic composition of plant carbon correlates with water-use efficiency of wheat genotypes. *Aust. J. Plant Physiol.* 11:539–52

Farquhar GD, Sharkey TD. 1982. Stomatal conductance and photosynthesis. *Annu. Rev. Plant Physiol.* 33:317–45

Feddes RA, Hoff H, Bruen M, Dawson T, de Rosnay P, et al. 2001. Modeling root water uptake in hydrological and climate models. *B. Am. Meteorol. Soc.* 82:2797–810

Feild TS, Dawson TE. 1998. Water sources used by *Didymopanax pittieri* at different life stages in a tropical cloud forest. *Ecology* 79:1448–52

Feng X. 1999. Trends in intrinsic water-use efficiency of natural trees for past 100–200 years: a response to atmospheric CO_2 concentration. *Geochim. Cosmochim. Acta* 63:1891–903

Fessenden JE, Ehleringer JR. 2002. Age-related variations in ^{13}C of ecosystem respiration across a coniferous forest chronosequence in the Pacific Northwest. *Tree Physiol.* 22:159–67

Finlay RD, Frostegard A, Sonnerfeldt A-M. 1992. Utilization of organic and inorganic nitrogen sources by ectomycorrhizal fungi in pure culture and in symbiosis with *Pinus contorta* Dougl. ex Loud. *New Phytol.* 120:105–15

Fitter AH, Graves JD, Watkins NK, Robinson D, Scrimgeour C. 1998. Carbon transfer between plants and its control in network of arbuscular mycorrhizas. *Funct. Ecol.* 12:406–12

Flanagan LB. 1998. Oxygen isotope effects during CO_2 exchange: from leaf to ecosystem processes. See Griffiths 1998, pp. 185–201

Flanagan LB, Brooks JR, Ehleringer JR. 1997a. Photosynthesis and carbon isotope discrimination in boreal forest ecosystems: a comparison of functional characteristics in plants from three mature forest types. *J. Geophys. Res.* 102:28861–69

Flanagan LB, Brooks JR, Varney GT, Ehleringer JR. 1997b. Discrimination against $C^{18}O^{16}O$ during photosynthesis and the oxygen isotope ratio of respired CO_2 in boreal forest ecosystems. *Global Biogeochem. Cycles* 11:83–98

Flanagan LB, Ehleringer JR. 1991. Stable isotope composition of stem and leaf water: applications to the study of plant water use. *Funct. Ecol.* 5:270–77

Flanagan LB, Ehleringer JR. 1998. Ecosystem-atmosphere CO_2 exchange: interpreting signals of change using stable isotope ratios. *Trends Ecol. Evol.* 13:10–14

Flanagan LB, Ehleringer JR, Marshall JD. 1992. Differential uptake of summer precipitation among co-occurring trees and shrubs in a pinyon-juniper woodland. *Plant Cell Environ.* 15:831–36

Flanagan LB, Kubien DS, Ehleringer JR. 1999. Spatial and temporal variation in the carbon and oxygen stable isotope ratio of respired CO_2 in a boreal forest ecosystem. *Tellus B* 51:367–84

Flanagan LB, Phillips SL, Ehleringer JR, Lloyd J, Farquhar GD. 1994. Effect of changes in leaf water oxygen isotopic composition on discrimination against C18O16O during photosynthetic gas exchange. *Aust. J. Plant Physiol.* 21:221–34

Flanagan LB, Varney GT. 1995. Influence of vegetation and soil CO_2 exchange on the concentration and stable oxygen isotope ratio of atmospheric CO_2 within a *Pinus resinosa* canopy. *Oecologia* 101:37–44

Frank DA, Evans RD. 1997. Effects of native grazers on grassland N cycling in Yellowstone National Park. *Ecology* 78:2238–48

Frank DA, Groffman PM, Evans RD, Tracy BF. 2000. Ungulate stimulation of nitrogen cycling and retention in Yellowstone Park grasslands. *Oecologia* 123:116–21

Fry B, Joern A, Parker PL. 1978. Grasshopper food web analysis: use of carbon isotope ratios to examine feeding relationships among terrestrial herbivores. *Ecology* 59:498–506

Galloway JN, Schlesinger WH, Levy H, Michaels A, Schnoor JL. 1995. Nitrogen fixation: anthropogenic enhancement-environmental response. *Glob. Biogeochem. Cycles* 9:235–52

Garten CT. 1993. Variation in foliar ^{15}N abundance and the availability of soil nitrogen on Walker Branch Watershed. *Ecology* 74:2098–113

Garten CT Jr, Hanson PJ. 1990. Foliar retention of ^{15}N-nitrate and ^{15}N-ammonium by red maple (*Acer rubrum*) and white oak (*Quercus alba*) leaves from simulated rain. *Environ. Exp. Bot.* 30:333–42

Garten CT Jr, Taylor GE Jr. 1992. Foliar $\delta^{13}C$ within a temperate deciduous forest: spatial, temporal, and species sources of variation. *Oecologia* 90:1–7

Garten CT Jr, Van Miegroet H. 1994. Relationships between soil nitrogen dynamics and natural ^{15}N abundance in plant foliage from Great Smoky Mountains National Park. *Can. J. For. Res.* 24:1636–45

Gat JR. 1996. Oxygen and hydrogen stable isotopes in the hydrologic cycle. *Annu. Rev. Earth Plant. Sci.* 24:225–62

Gat JR. 1998. Stable isotopes, the hydrological cycle and the terrestrial biosphere. See Griffiths 1998, pp. 397–407

Gebauer G, Dietrich P. 1993. Nitrogen isotope ratios in different compartments of a mixed stand of spruce, larch and beech trees and of understorey vegetation including fungi. *Isotopenpraxis* 29:35–44

Gebauer G, Schulze ED. 1991. Carbon and nitrogen isotope ratios in different compartments of a healthy and a declining *Picea abies* forest in the Fichtelgebirge, northeastern Bavaria Germany. *Oecologia* 87:198–207

Gebauer RLE, Ehleringer JR. 2000. Water and nitrogen uptake patterns following moisture pulses in a cold desert community. *Ecology* 81:1415–24

Geber MA, Dawson TE. 1990. Genetic variation in and covariation between leaf gas exchange, morphology, and development in *Polygonum arenastrum*, an annual plant. *Oecologia.* 85:53–58

Ghashghaie J, Duranceau M, Badeck FW, Cor-

nic G, Adeline MT, et al. 2001. ^{13}C of CO_2 respired in the dark in relation to ^{13}C of leaf metabolites: comparison between *Nicotiana sylvestris* and *Helianthus annuus* under drought. *Plant Cell Environ.* 24:505–15

Gleixner G, Danier H-J, Werner RA, Schmidt H-L. 1993. Correlation between the ^{13}C content of primary and secondary plant products in different cell compartments and that in decomposing basidiomycetes. *Plant Physiol.* 102:1287–90

Grantz DA. 1990. Plant responses to atmospheric humidity. *Plant Cell Environ.* 13:667–79

Griffiths H. 1992. Carbon isotope discrimination and the integration of carbon assimilation pathways in terrestrial CAM plants: commissioned review. *Plant Cell Environ.* 15:1051–62

Griffiths H, ed. 1998. *Stable Isotopes: Integration of Biological, Ecological and Geochemical Processes.* Oxford: BIOS Sci.

Griffiths H, Smith J, eds. 1993. *Plant Responses to Water Deficits.* London, UK: BIOS Sci.

Groffman PM, Zak DR, Christensin S, Mosier A, Tiedje JM. 1993. Early spring nitrogen dynamics in a temperate forest landscape. *Ecology* 74:1579–85

Guehl J-M, Fort C, Ferhi A. 1995. Differential response of leaf conductance, carbon isotope discrimination and water-use efficiency to nitrogen deficiency in maritime pine and pedunculate oak plants. *New Phytol.* 131:149–57

Hanba YT, Miyazawa SI, Terashima I. 1999. The influence of leaf thickness on the CO_2 transfer conductance and leaf stable carbon isotope ratio for some evergreen tree species in Japanese warm–temperate forests. *Funct. Ecol.* 13:632–39

Hanba YT, Mori S, Lei TT, Koike T, Wada E. 1997. Variations in leaf $\delta^{13}C$ along a vertical profile of irradiance in a temperate Japanese forest. *Oecologia* 110:253–61

Handley LL, Austin AT, Robinson D, Scrimgeour CM, Raven JA, et al. 1999. The ^{15}N natural abundance ($\delta^{15}N$) of ecosystem samples reflects measures of water availability. *Aust. J. Plant Physiol.* 26:185–99

Handley LL, Odee D, Scrimgeour CM. 1994. $\delta^{15}N$ and $\delta^{13}C$ patterns in savanna vegetation: dependence on water availability and disturbance. *Funct. Ecol.* 8:306–14

Handley LL, Raven JA. 1992. The use of natural abundance of nitrogen isotopes in plant physiology and ecology. *Plant Cell Environ.* 15:965–85

Handley LL, Robinson D, Forster BP, Ellis RP, Scrimgeour CM, et al. 1997. Shoot delta-^{15}N correlates with genotype and salt stress in barley. *Planta* 201:100–2

Handley LL, Scrimgeour CM. 1997. Terrestrial plant ecology and ^{15}N natural abundance: The present limits to interpretation for uncultivated systems with original data from a Scottish old field. *Adv. Ecol. Res.* 27:133–212

Handley LL, Scrimgeour CM, Raven JA. 1998. ^{15}N at natural abundance levels in terrestrial vascular plants: a precis. See Griffiths 1998, pp. 89–98

Hardy RWF, Holsten RD, Jackson EK, Burns RC. 1968. The acetylene-ethylene assay for N2 fixation: laboratory and field evaluation. *Plant Physiol.* 43:1185–207

Hart S, Stark JM, Davidson EA, Firestone MK. 1994. Nitrogen mineralization, immobilization, and nitrification. In *Methods of Soil Analysis. Part 2. Microbiological and Biochemicam Properties*; ed. RW Weaver, S Angle, P Bottomley, D Bedzicek, S Smith, et al., pp. 985–1018. Madison, WI: Soil Sci. Soc. Am.

Harwood KG, Gillon JS, Griffiths H, Broadmeadow MSJ. 1998. Diurnal variation of $\delta^{13}CO_2$, $\delta C^{18}O^{16}O$ and evaporative site enrichment of $\delta H_2^{18}O$ in *Piper aduncum* under field conditions in Trinidad. *Plant Cell Environ.* 21:269–83

Hemming D, Fritts HC, Leavitt SW, Wright WE, Long A, Shashkin A. 2001. Modelling tree-ring $\delta^{13}C$. *Dendrochronologia* 19:23–38

Hemming DL, Switsur VR, Waterhouse JS, Heaton THE, Carter AHC. 1998. Climate variation and the stable carbon isotope composition of tree ring cellulose: an intercomparison of Quercus robur, Fagus silvatica and Pinus silvestris. *Tellus B* 50:25–33

Henderson S, von Caemmerer S, Farquhar GD, Wade L, Hammer G. 1998. Correlation between carbon isotope discrimination and transpiration efficiency in lines of the C4 species *Sorghum bicolor* in the glasshouse and the field. *Aust. J. Plant Physiol.* 25:111–23

Henderson SA, von Caemmerer S, Farquhar GD. 1992. Short-term measurements of carbon isotope discrimination in several C4 species. *Aust. J. Plant Physiol.* 19:263–85

Henn RM, Chapela IH. 2000. Differential C isotope discrimination by fungi during decomposition of C3- and C4-derived sucrose. *Appl. Environ. Microbiol.* 66:4180–86

Hobbie EA, Macko SA, Shugart HH. 1999a. Insights into nitrogen and carbon dynamics of ectomycorrhizal and saprotrophic fungi from isotopic evidence. *Oecologia* 118:353–60

Hobbie EA, Macko SA, Shugart HH. 1999b. Interpretation of nitrogen isotope signatures using the NIFTE model. *Oecologia* 120:405–15

Hobbie EA, Macko SA, Williams M. 2000. Correlation between foliar $\delta^{15}N$ and nitrogen concentrations may indicate plant-mycorrhizal interactions. *Oecologia* 122:273–83

Hobbie EA, Weber NS, Trappe NS. 2001. Mycorrhizal vs saprotrophic status of fungi: the isotopic evidence. *New Phytol.* 150:601–10

Högberg P. 1990. Forests losing large quantities of nitrogen have elevated nitrogen $^{15}N/^{14}N$ ratios. *Oecologia* 84:229–31

Högberg P. 1997. Tansley review no. 95 15N natural abundance in soil-plant systems. *New Phytol.* 137:179–203

Högberg P, Högberg MN, Quist ME, Ekblad A, Näsholm T. 1999a. Nitrogen isotope fractionation during nitrogen uptake by ectomycorrhizal and non-mycorrhizal *Pinus sylvestris*. *New Phytol.* 142:569–76

Högberg P, Högbom L, Schinkel H, Högberg M, Johannison C, et al. 1996. ^{15}N abundance of surface soils, roots and mycorrhizas in profiles of European forest soils. *Oecologia* 108:207–14

Högberg P, Johannisson C. 1993. [15]N abundance of forests is correlated with losses of nitrogen. *Plant Soil* 157:147–50

Högberg P, Johannisson C, Hällgren J-E. 1993. Studies of ^{13}C in the foliage reveal interactions between nutrients and water in fertilization experiments. *Plant Soil* 152:207–14

Högberg P, Plamboeck AH, Taylor AFS, Fransson PMA. 1999b. Natural ^{13}C abundance reveals trophic status of fungi and host-origin of carbon in mycorrhizal fungi in mixed forests. *Proc. Natl. Acad. Sci. USA* 96:8534–39

Hultine KR, Marshall JD. 2000. Altitude trends in conifer leaf morphology and stable carbon isotope composition. *Oecologia* 123:32–40

Hungate BA, Jackson RB, Chapin FS III, Mooney HA, Field CB. 1997. The fate of carbon in grasslands under carbon dioxide enrichment. *Nature* 388:576–79

Ingraham NL. 1998. Isotopic variations in precipitation. See Kendall & McDonell 1998, pp. 87–118

Jackson PC, Cavelier J, Goldstein G, Meinzer FC, Holbrook NM. 1995. Partitioning of water resources among plants of a lowland tropical forest. *Oecologia* 101:197–203

Jackson PC, Meinzer FC, Bustamante M, Goldstein G, Franco A, et al. 1999. Partitioning of soil water among tree species in a Brazilian Cerrado ecosystem. *Tree Physiol.* 19:717–24

Jackson RB, Sperry JS, Dawson TE. 2000. Root water uptake and transport: using physiological processes in global predictions. *Trends Plant Sci.* 5:482–88

Johannisson C, Högberg P. 1994. [15]N abundance of soils and plants along an experimentally induced forest nitrogen supply gradient. *Oecologia* 97:322–25

Johnsen KH, Flanagan LB. 1995. Genetic variation in carbon isotope discrimination and its relationship to growth under field conditions in full-sib families of *Picea mariana*. *Can. J. For. Res.* 25:39–47

Johnsen KH, Flanagan LB, Huber DA, Major JE. 1999. Genetic variation in growth, carbon isotope discrimination, and foliar N concentration in *Picea mariana*: analyses from a half-diallel mating design using field-grown trees. *Can. J. For. Res.* 29:1727–35

Johnson DA, Asay KH, Tieszen LL, Ehleringer JR, Jefferson PG. 1990. Carbon isotope discrimination—potential in screening cool-season grasses for water-limited environments. *Crop Sci.* 30:338–43

Keeling CD. 1958. The concentration and isotopic abundances of atmospheric carbon dioxide in rural areas. *Geochim. Cosmochim. Acta* 13:322–34

Keeling CD. 1961. The concentration and isotopic abundances of atmospheric carbon dioxide in rural and marine areas. *Geochim. Cosmochim. Acta* 24:277–98

Kelly FI, Woodward FI. 1995. Ecological correlates of carbon isotope composition of leaves: a comparative analysis testing for the effects of temperature, CO_2 and O_2 partial pressures and taxonomic relatedness on $\delta^{13}C/C$. *J. Ecol.* 83:509–15

Kendall C, McDonnell JJ, eds. 1998. *Isotope Tracers in Catchment Hydrology.* Amsterdam: Elsevier Science. 839 pp.

Kielland K, Bryant JP. 1998. Moose herbivory in taiga: effects on biogeochemistry and vegetation dynamics in primary succession. *Oikos* 82:377–83

Kloeppel BD, Gower ST, Treichel IW, Kharuk S. 1998. Foliar carbon isotope discrimination in *Larix* species and sympatric evergreen conifers: a global comparison. *Oecologia* 114:153–59

Knowles R. 1975. Interpretation of recent ^{15}N studies of nitrogen in forest systems. In *Forest Soils and Forest Land Management. Proc. 4th North Am. For. Soil Conf.*, ed. B Bernier, CH Winget, pp. 53–65. Quebec: Laval Univ. Press

Knowles R, Henry Blackburn T, eds. 1993. *Nitrogen Isotope Techniques.* San Diego, CA: Academic

Kohl DH, Shearer G. 1980. Isotope fractionation associated with symbiotic N_2 fixation and uptake of NO_3. *Plant Physiol.* 66:51–56

Kohorn LU, Goldstein G, Rundel PW. 1994.

Morphological and isotopic indicators of growth environment: variability in delta ^{13}C in *Simmondsia chinensis*, a dioecious desert shrub. *J. Exp. Bot.* 45:1817–22

Kohzu A, Tateishi T, Yamada K, Koba K, Wada E. 2000. Nitrogen isotope fractionation during nitrogen transport from ectomycorrhizal fungi. *Suillus granulatus*, to the host plant, *Pinus densiflora. Soil Sci. Plant Nutr.* 46:733–39

Kohzu A, Yoshioka T, Ando T, Takahashi M, Koba K, et al. 1999. Natural ^{13}C and ^{15}N abundance of field-collected fungi and their ecological implications. *New Phytol.* 144:323–30

Kolb TE, Hart SC, Amundson R. 1997. Boxelder water sources and physiology at perennial and ephemeral stream sites in Arizona. *Tree Physiol.* 17:151–60

Koopmans CJ, Tietema A, Boxman AW. 1996. The fate of ^{15}N enriched throughfall in two coniferous forest stands at different nitrogen deposition levels. *Biogeochemistry* 34:19–44

Körner C, Farquhar GD, Wong SC. 1991. Carbon isotope discrimination by follows latitudinal and altitudinal trends. *Oecologia* 88:30–40

Korol RL, Kirschbaum MUF, Farquhar GD, Jeffreys M. 1999. Effects of water status and soil fertility on the C-isotope signature in *Pinus radiata. Tree Physiol.* 19:551–62

Lajtha K, Marshall JD. 1994. Sources of variation in the stable isotopic composition of plants. See Lajtha & Michener 1994, pp. 1–21

Lajtha K, Michener RH, eds. 1994. *Stable Isotopes in Ecology and Environmental Science.* Oxford: Blackwell Sci.

Leavitt SW. 1993. Environmental information from 13C/12C ratios in wood. In *Climate Change in Continental Isotope Records*, ed. PK Swart, KC Lohmann, JA McKenzie, S Savin, 78:325–31. Am. Geophys. Union Monogr.

Leavitt SW, Long A. 1986. Stable-carbon isotope variability in tree foliage and wood. *Ecology* 67:1002–10

Leavitt SW, Long A. 1988. Stable carbon isotope chronologies from trees in the south western United States. *Glob. Biogeochem. Cycles* 2:189–98

Leavitt SW, Long A. 1989. Drought indicated in ^{13}C/^{12}C ratios of south western tree rings. *Water Resour. Bull.* 25:341–47

Leavitt SW, Long A. 1991. Seasonal stable carbon isotope variability in tree rings: possible paleoenvironmental signals. *Chem. Geol.* 87:59–70

Le Roux X, Bariac T, Sinoquet H, Genty B, Piel C, et al. 2001. Spatial distribution of leaf water-use efficiency and carbon isotope discrimination within an isolated tree crown. *Plant Cell Environ.* 24:1021–32

Lin G, Ehleringer JR. 1997. Carbon isotopic fractionation does not occur during dark respiration in C3 and C4 plants. *Plant Physiol.* 114:391–94

Lin G, Phillips SL, Ehleringer JR. 1996. Monsoonal precipitation responses of shrubs in a cold desert community on the Colorado Plateau. *Oecologia* 106:8–17

Lin G, Sternberg LDL. 1993. Hydrogen isotopic fractionation by plant roots during water uptake in coastal wetland plants. See Ehleringer et al. 1993, pp. 497–510

Lipp J, Timborn P, Edwards T, Waisel Y, Yakir D. 1996. Climatic effects on the $\delta^{18}O$ and $\delta^{13}C$ of cellulose in the desert tree *Tamarix jordanis. Geochim. Cosmochim. Acta* 60:3305–9

Livingston NJ, Spittlehouse DL. 1993. Carbon isotope fractionation in tree rings in relation to the growing season water balance. See Ehleringer et al. 1993, pp. 141–53

Lloyd J, Kruijt B, Hollinger DY, Grace J, Francey RJ, et al. 1996. Vegetation effects on the isotopic composition of atmospheric CO_2 at local and regional scales: theoretical aspects and a comparison between rain forest in Amazonia and a boreal forest in Siberia. *Aust. J. Plant Physiol.* 23:371–99

Lowden JA, Dyck W. 1974. Seasonal variations in the isotope ratios of carbon in maple leaves and other plants. *Can. J. Earth Sci.* 11:79–88

Ludwig F. 2001. *Tree-grass interactions on an*

east African savanna: the effects of competition, facilitation and hydraulic lift. PhD Diss., Wageningen Univ., The Netherlands

MacFadden BJ. 2000. Cenozoic mammalian herbivores from the Americas: reconstructing ancient diets and terrestrial communities. *Annu. Rev. Ecol. Syst.* 31:33–59

Madhavan S, Treichel I, O'Leary MH. 1991. Effects of relative humidity on carbon isotope fractionation in plants. *Bot. Acta* 104:292–94

Maillard P, Guehl JM, Muller JF, Gross P. 2001. Interactive effects of elevated CO_2 concentration and nitrogen supply on partitioning of newly fixed ^{13}C and ^{15}N between shoot and roots of pedunculate oak seedlings (*Quercus robur*). *Tree Physiol.* 21:163–72

Mariotti A, Mariotti F, Champigny ML, Amargar N, Moyse A. 1982. Nitrogen isotope fractionation associated with nitrate reductase activity and uptake of NO_3^- by pearl millet. *Plant Physiol.* 69:880–84

Marshall JD, Ehleringer JR. 1990. Are xylem-tapping mistletoes partially heterotrophic? *Oecologia* 84:244–48

Marshall JD, Ehleringer JR, Schulze ED, Farquhar G. 1994. Carbon isotope composition, gas exchange and heterotrophy in Australian mistletoes. *Funct. Ecol.* 8:237–41

Marshall JD, Zhang J. 1994. Carbon isotope discrimination and water-use efficiency in native plants of the north-central Rockies. *Ecology* 75:1887–95

Martinelli LA, Almeida S, Brown IF, Moreira MZ, Victoria RL, et al. 1998. Stable carbon isotope ratio of tree leaves, boles and fine litter in a tropical forest in Rondonia, Brazil. *Oecologia* 114:170–79

Mathieu R, Bariac T. 1996a. An isotopic study (2H and 18O) of water movements in clayey soils under a semi-arid climate. *Water Resour. Res.* 32:779–89

Mathieu R, Bariac T. 1996b. A numerical model for the simulation of stable isotope profiles in drying soils. *J. Geophys. Res.* 101:12585–696

Mazor E. 1991. Stable hydrogen and oxygen isotopes. *In Applied Chemical and Isotopic Groundwater Hydrology*, pp. 122–46. London: Halsted

McKane RB, Grigal DF, Russelle MP. 1990. Spatiotemporal differences in ^{15}N uptake and the organization of an old-field plant community. *Ecology* 71:1126–32

Medina E, Minchin P. 1980. Stratification of $\delta^{13}C$ values of leaves in Amazonian rain forests. *Oecologia* 45:377–78

Meints VW, Boone LV, Kurtz LT. 1975. Natural ^{15}N abundance in soil, leaves, and grain as influenced by long term additions of fertilizer N at several rates. *J. Environ. Qual.* 4:486–90

Meinzer FC, Andrade JL, Goldstein G, Holbrook NM, Cavelier J, et al. 1999. Partitioning of soil water among canopy trees in a seasonally dry tropical forest. *Oecologia* 121:293–301

Meinzer FC, Clearwater MJ, Goldstein G. 2001. Water transport in trees: current perspectives, new insights and some controversies. *Environ. Exp. Bot.* 45:239–62

Meinzer FC, Rundel PW, Goldstein G, Sharifi MR. 1992. Carbon isotope composition in relation to leaf gas exchange and environmental conditions in Hawaiian *Metrosideros polymorpha* populations. *Oecologia* 91:305–11

Mensforth LJ, Thorburn PJ, Tyerman SD, Walker GR. 1994. Sources of water used by riparian *Eucalyptus camaldulensis* overlying highly saline groundwater. *Oecologia* 100:21–28

Michelsen A, Quarmby C, Sleep D, Jonasson S. 1998. Vascular plant ^{15}N natural abundance in heath and forest tundra ecosystems is closely correlated with presence and type of mycorrhizal fungi in roots. *Oecologia* 115:406–18

Michelsen A, Schmidt IK, Jonasson S, Quarmby C, Sleep D. 1996. Leaf ^{15}N abundance of subarctic plants provides field evidence that ericoid, ectomycorrhizal and non- and arbuscular mycorrhizal species access different sources of soil nitrogen. *Oecologia* 105:53–63

Millard P, Hester A, Wendler R, Baillie G. 2001. Interspecific defoliation responses of trees depend on sites of winter nitrogen storage. *Funct. Ecol.* 15:535–43

Miller JB, Yakir D, White JWC, Tans PP. 1999. Measurement of $^{18}O/^{16}O$ in the soil-atmosphere CO_2-flux. *Glob. Biogeochem. Cycles* 13:761–74

Miller JM, Williams RJ, Farquhar GD. 2001. Carbon isotope discrimination by a sequence of *Eucalyptus* species along a subcontinental rainfall gradient in Australia. *Funct. Ecol.* 15:222–32

Miranda AC, Miranda HS, Lloyd J, Grace J, Francey RJ, et al. 1997. Fluxes of carbon, water and energy over Brazilian cerrado: an analysis using eddy covariance and stable isotopes. *Plant Cell Environ.* 20:315–28

Monserud RA, Marshall JD. 2001. Time-series analysis of $\delta^{13}C$ from tree rings. I. Time trends and autocorrelation. *Tree Physiol.* 21:1087–102

Monteith JL. 1995. A reinterpretation of stomatal responses to humidity. *Plant Cell Environ.* 18:357–64

Moore DJ, Nowak RS, Tausch RJ. 1999. Gas exchange and carbon isotope discrimination of *Juniperus osteosperma* and *Juniperus occidentalis* across environmental gradients in the great Basin of western North America. *Tree Physiol.* 19:421–33

Moran JA, Merbach MA, Livingston NJ, Clarke CM, Booth WE. 2001. Termite prey specialization in the pitcher plant *Nepenthes albomarginata*: evidence from stable isotope analysis. *Ann. Bot.* 88:307–11

Mordacq L, Mousseau M, Deleens E. 1986. A ^{13}C method of estimation of carbon allocation to roots in a young chestnut coppice. *Plant Cell Environ.* 9:735–40

Morecroft MD, Woodward FI. 1990. Experimental investigation on the environmental determination of $\delta^{13}C$ at different altitudes. *J. Exp. Bot.* 31:1303–8

Moreira MZ, Sternberg LDL, Martinelli LA, Victoria RL, Barbosa EM, et al. 1997. Contribution of transpiration to forest ambient vapor based on isotopic measurements. *Glob. Change Biol.* 3:439–50

Moreira MZ, Sternberg LDL, Nepstad DC.

2000. Vertical patterns of soil water uptake by plants in a primary forest and an abandoned pasture in the eastern Amazon: an isotopic approach. *Plant Soil* 222:95–107

Mott KA, Parkhurst DF. 1991. Stomatal responses to humidity in air and helox. *Plant Cell Environ.* 14:509–15

Nadelhoffer K, Downs M, Fry B, Magill A, Aber J. 1999. Controls on N retention and exports in a forested watershed. *Environ. Monit. Assess.* 55:187–210

Nadelhoffer K, Fry B. 1994. Nitrogen isotope studies in forest ecosystems. See Lajtha & Michener 1994, pp. 22–44

Nadelhoffer KJ, Downs MR, Fry B, Aber JD, Magill AH, et al. 1995. The fate of ^{15}N-labelled nitrate additions to a northern hardwood forest in eastern Maine, USA. *Oecologia* 103:292–301

Nadelhoffer KJ, Fry B. 1988. Controls on natural ^{15}N and ^{13}C abundances in forest soil organic matter. *Soil Sci. Soc. Am. J.* 52:1633–40

Näsholm T, Ekblad A, Nordin A, Giesler R, Högberg M, et al. 1998. Boreal forest plants take up organic nitrogen. *Nature* 392:914–16

Newman EI, Reddell P. 1987. The distribution of mycorrhizas among families of vascular plants. *New Phytol.* 106:745–51

Niklaus PA, Glockler E, Siegwolf R, Korner C. 2001. Carbon allocation in calcareous grassland under elevated CO_2: a combined ^{13}C pulse-labelling/soil physical fractionation study. *Funct. Ecol.* 15:43–50

O'Leary MH. 1981. Carbon isotope fractionation in plants. *Phytochemistry* 20:553–67

Olson BE, Wallander RT. 1999. Carbon allocation in *Euphorbia esula* and neighbours after defoliation. *Can. J. Bot.* 77:1641–47

Palta JA. 2001. Source/sink interactions in crop plants. See Unkovich et al. 2001, pp. 145–65

Panek JA. 1996. Correlations between stable carbon-isotope abundance and hydraulic conductivity in Douglas-fir across a climate gradient in Oregon, USA. *Tree Physiol.* 16:747–55

Panek JA, Waring RH. 1995. Carbon isotope variation in Douglas-fir foliage: improving the δ^{13}C-climate relationship. *Tree Physiol.* 15:657–63

Panek JA, Waring RH. 1997. Stable carbon isotopes as indicators of limitations to forest growth imposed by climate stress. *Ecol. Appl.* 7:854–63

Park R, Epstein S. 1961. Metabolic fractionation of ^{13}C and ^{12}C in plants. *Plant Physiol.* 36:133–38

Pataki DE, Ehleringer JR, Flanagan LB, Yakir D, Bowling DR, et al. 2002. The application and interpretation of Keeling plots in terrestrial carbon cycle research. *Glob. Biogeochem. Cycles.* In press

Pate J. 2001. Carbon isotope discrimination and plant water-use efficiency: case scenarios for C3 plants. See Unkovich et al. 2001, pp. 19–36

Pate JS, Dawson TE. 1999. Novel techniques for assessing the performance in uptake and utilisation of carbon, water and nutrients by woody plants: implications for designing agricultural mimics. *Agrofor. Syst.* 45:245–76

Pattey E, Desjardins RL, Rochette P. 1993. Accuracy of the relaxed eddy-accumulation technique, evaluated using CO_2 flux measurements. *Bound.-Lay. Meteorol.* 66:341–55

Peterjohn WT, Adams MB, Gilliam FS. 1996. Symptoms of nitrogen saturation in two central Appalachian hardwood forest ecosystems. *Biogeochemistry* 35:507–22

Peterson BJ, Fry B. 1987. Stable isotopes in ecosystem studies. *Annu. Rev. Ecol. Syst.* 18:293–320

Phillips DL. 2001. Mixing models in analyses of diet using multiple stable isotopes: a critique. *Oecologia* 127:166–70

Phillips DL, Gregg JW. 2001. Uncertainty in source partitioning using stable isotopes. *Oecologia* 127:171–79

Phillips DL, Koch PL. 2002. Incorporating concentration dependence in stable isotope mixing models. *Oecologia* 130:114–25

Phillips SL, Ehleringer JR. 1995. Limited uptake of summer precipitation by bigtooth maple (*Acer grandidentatum* Nutt) and

Gambel's oak (*Quercus gambelii* Nutt). *Trees* 9:214–19

Piccolo MC, Neill C, Cerri CC. 1994. Natural abundance of ^{15}N in soils along forest-to-pasture chronosequences in the western Brazilian Amazon Basin. *Oecologia* 99:112–17

Piccolo MC, Neill C, Melillo JM, Cerri CC, Steudler PA. 1996. ^{15}N natural abundance in forest and pasture soils of the Brazilian Amazon Basin. *Plant Soil* 182:249–58

Plamboeck AH, Grip H, Nygren U. 1999. A hydrological tracer study of water uptake depth in a Scots pine forest under two different water regimes. *Oecologia* 119:452–60

Polley HW, Johnson HB, Mayeux HS. 1992. Determination of root biomasses of three species grown in a mixture using stable isotopes of carbon and nitrogen. *Plant Soil* 142:97–106

Poorter H, Farquhar GD. 1994. Transpiration, intercellular carbon dioxide concentration and carbon-isotope discrimination of 24 wild species differing in relative growth rate. *Aust. J. Plant Physiol.* 21:507–17

Poss JA, Grattan SR, Suarez DL, Grieve CM. 2000. Stable carbon isotope discrimination: an indicator of cumulative salinity and boron stress in *Eucalyptus camaldulensis*. *Tree Physiol.* 20:1121–27

Poulson SR, Page Chamberlain C, Friedland AJ. 1995. Nitrogen isotope variation of tree rings as a potential indicator of environmental change. *Chem. Geol.* 125:307–15

Press MC, Shah N, Tuohy JM, Stewart GR. 1987. Carbon isotope ratios demonstrate carbon flux from C4 host to C3 parasite. *Plant Physiol.* 85:1143–45

Proctor MCF, Raven JA, Rice SK. 1992. Stable carbon isotope discrimination measurements in *Sphagnum* and other bryophytes: physiological and ecological implications. *J. Bryol.* 17:193–202

Proe MF, Midwood AJ, Craig J. 2000. Use of stable isotopes to quantify nitrogen, potassium and magnesium dynamics in young Scots pine (*Pinus sylvestris*). *New Phytol.* 146:461–69

Quade J, Cerling TE, Barry JC, Morgan ME, Pilbeam DR, et al. 1992. A 16 million year record of paleodiet from Pakistan using carbon isotopes in fossil teeth. *Chem. Geol.* 94:183–92

Rennie DA, Paul EA, Johns LE. 1976. Natural ^{15}N abundance of soil and plant samples. *Can. J. For. Res.* 56:43–50

Retuerto R, Lema BF, Roiloa SR, Obeso JR. 2000. Gender, light and water effects in carbon isotope discrimination, and growth rates in the dioecious tree *Ilex aquifolium*. *Funct. Ecol.* 14:529–37

Rice KJ, Gordon DR, Hardison JL, Welker JM. 1993. Phenotypic variation in seedlings of a "keystone" tree species (*Quercus douglasii*): The interactive effects of acorn source and competitive environment. *Oecologia* 96:537–47

Rice SK. 2000. Variation in carbon isotope discrimination within and among *Sphagnum* species in a temperate wetland. *Oecologia* 123:1–8

Rice SK, Giles L. 1996. The influence of water content and leaf anatomy on carbon isotope discrimination and photosynthesis in *Sphagnum*. *Plant Cell Environ.* 19:118–24

Robinson D. 2001. δ^{15}N as an integrator of the nitrogen cycle. *Trends Ecol. Evol.* 16:153–62

Robinson D, Handley LL, Scrimgeour CM. 1998. A theory for ^{15}N/^{14}N fractionation in nitrate-grown vascular plants. *Planta* 205:397–406

Robinson D, Handley LL, Scrimgeour CM, Gordon DC, Forster BP, et al. 2000. Using stable isotope natural abundances (δ^{15}N and δ^{13}C) to integrate the stress responses of wild barley (*Hordeum spontaneum* C. *Koch.*) genotypes. *J. Exp. Bot.* 51:41–50

Robinson D, Scrimgeour CM. 1995. The contribution of plant C to soil CO_2 measured using δ^{13}C. *Soil Biol. Biochem.* 27:1653–56

Rochette P, Flanagan LB. 1997. Quantifying rhizosphere respiration in a corn crop under field conditions. *Soil Sci. Soc. Am. J.* 61:466–74

Rochette P, Flanagan LB, Gregorich EG. 1999. Separating soil respiration into plant and soil

components using analyses of the natural abundance of carbon-13. *Soil Sci. Soc. Am. J.* 63:1207–13

Roden JS, Ehleringer JR. 1999a. Observations of hydrogen and oxygen isotopes in leaf water confirm the Craig-Gordon model under wide-ranging environmental conditions. *Plant Physiol.* 120:1165–73

Roden JS, Ehleringer JR. 1999b. Hydrogen and oxygen isotope ratios of tree-ring cellulose for riparian trees grown long-term under hydroponically controlled environments. *Oecologia* 121:467–77

Roden JS, Ehleringer JR. 2000. Hydrogen and oxygen isotope ratios of tree-ring cellulose for field grown riparian trees. *Oecologia* 123:481–89

Roden JS, Lin G, Ehleringer JR. 2000. A mechanistic model for interpretation of hydrogen and oxygen isotope ratios in tree-ring cellulose. *Geochim. Cosmochim. Acta* 64:21–35

Rundel PW, Ehleringer JR, Nagy KA, eds. 1989. *Stable Isotopes in Ecological Research. Ecological Studies*, Vol. 68. Heidelberg: Springer-Verlag. 525 pp.

Rundel PW, Stichler W, Zander RH, Ziegler H. 1979. Carbon and hydrogen isotope ratios of bryophytes from arid and humid regions. *Oecologia* 44:91–94

Sagers CL, Ginger SM, Evans RD. 2000. Carbon and nitrogen isotopes trace nutrient exchange in an ant-plant mutualism. *Oecologia* 123:582–86

Sala O, Jackson R, Mooney HA, eds. 2000. *Methods in Ecosystem Science*. San Diego, CA: Academic

Sandquist DR, Ehleringer JR. 1995. Carbon isotope discrimination in the C4 shrub *Atriplex confertifolia* along a salinity gradient. *Great Basin Nat.* 55:135–41

Saurer M, Aellen K, Siegwolf R. 1997. Correlating delta[13]C and delta[18]O in cellulose of trees. *Plant Cell Environ.* 20:1543–50

Scheidegger Y, Saurer M, Bahn M, Siegwolf R. 2000. Linking stable oxygen and carbon isotopes with stomatal conductance and photosynthetic capacity: a conceptual model. *Oecologia* 125:350–57

Schimel DS. 1993. *Theory and Application of Tracers*. San Diego, CA: Academic

Schimel JP, Jackson LE, Firestone MK. 1989. Spatial and temporal effects of plant-microbial competition for inorganic nitrogen in a California annual grassland USA. *Soil Biol. Biochem.* 21:1059–66

Schleppi P, Bucher-Wallin I, Siegwolf R, Saurer M, Muller N, et al. 1999. Simulation of increased nitrogen deposition to a montane forest ecosystem: partitioning of the added ^{15}N. *Water Air Soil Pollut.* 116:129–34

Schleser GH, Jayasekera R. 1985. δ^{13}C variations of leaves in forests as an indication of reassimilated CO_2 from the soil. *Oecologia* 65:536–42

Schmidt H-L, Gleixner G. 1998. Carbon isotope effects on key reactions in plant metabolism and ^{13}C-patterns in natural compounds. See Griffiths 1998, pp. 13–25

Schubler W, Neubert R, Levin I, Fischer N, Sonntag C. 2000. Determination of microbial versus root-produced CO_2 in an agricultural ecosystem by means of $\delta^{13}CO_2$ measurements in soil air. *Tellus* B 52:909–18

Schulze ED, Ehleringer JR. 1984. The effect of nitrogen supply on growth and water-use efficiency of xylem-tapping mistletoes. *Planta* 162:268–75

Schulze ED, Gebauer G, Schulze W, Pate JS. 1991. The utilization of nitrogen from insect capture by different growth forms of *Drosera* from Southwest Australia. *Oecologia* 87:240–46

Schulze ED, Williams RJ, Farquhar GD, Schulze W, Langridge J, et al. 1998. Carbon and nitrogen isotope discrimination and nitrogen nutrition of trees along a rainfall gradient in northern Australia. *Aust. J. Plant Physiol.* 25:413–25

Schulze W, Schulze ED, Pate JS, Gillison AN. 1997. The nitrogen supply from soils and insects during growth of the pitcher plants *Nepenthes mirabilis, Cephalotus follicularis* and *Darlingtonia californica*. *Oecologia* 112:464–71

Schuster WSF, Sandquist DR, Phillips SL, Ehleringer JR. 1992a. Comparisons of carbon isotope discrimination in populations of arid land plant species differing in lifespan. *Oecologia* 91:332–37

Schuster WSF, Sandquist DR, Phillips SL, Ehleringer JR. 1992b. Heritability of carbon isotope discrimination in *Gutierrezia microcephala* (Asteraceae). *Am. J. Bot.* 79:216–21

Schwinning S, Davis K, Richardson L, Ehleringer JR. 2002. Deuterium enriched irrigation indicates different forms of rain use in shrub/grass species of the Colorado Plateau. *Oecologia* 130:345–55

Schwinning S, Ehleringer JR. 2001. Water use trade-offs and optimal adaptations to pulse-driven arid ecosystems. *J. Ecol.* 89:464–80

Shearer G, Kohl DH. 1986. Nitrogen fixation in field settings estimations based on natural ^{15}N abundance. *Aust. J. Plant Physiol.* 13:699–756

Shearer G, Kohl DH. 1991. The ^{15}N natural abundance method for measuring biological nitrogen fixation: practicalities and possibilities. In *Stable Isotopes in Plant Nutrition, Soil Fertility and Environmental Studies*, pp. 103–15. Vienna: IAEA

Shearer G, Kohl DH. 1993. Natural abundance of ^{15}N: fractional contribution of two sources to a common sink and use of isotope discrimination. See Knowles & Henry Blackburn 1993, pp. 89–125

Shearer G, Kohl DH, Chien S. 1978. The ^{15}N abundance in a wide variety of soils. *Soil Sci. Soc. Am. J.* 42:899–902

Simard SW, Durall DM, Jones MD. 1997a. Carbon allocation and carbon transfer between *Betula papyrifera* with *Pseudotsuga menziesii* seedlings using a ^{13}C pulse-labeling method. *Plant Soil* 191:41–55

Simard SW, Jones MD, Durall DM, Perry DA, Myrold DD, et al. 1997b. Reciprocal transfer of carbon isotopes between ectomycorrhizal *Betula papyrifera* and *Pseudotsuga menziesii*. *New Phytol.* 137:529–42

Smedley MP, Dawson TE, Comstock JP, Donovan LA, Sherril DE, et al. 1991. Seasonal carbon isotope discrimination in a grassland community. *Oecologia* 85:314–20

Smith DM, Jarvis PG, Odongo JCW. 1997.

Sources of water used by trees and millet in Sahelian windbreak systems. *J. Hydrol.* 198:140–53

Smith DM, Jarvis PG, Odongo JCW. 1998. Management of windbreaks in the Sahel: the strategic implications of tree water use. *Agrofor. Syst.* 40:83–96

Smith SE, Read DJ. 1997. *Mycorrhizal Symbiosis.* London, UK: Academic. 2nd ed.

Sparks J, Ehleringer JR. 1997. Leaf carbon isotope discrimination and nitrogen content for riparian trees along elevational transect. *Oecologia* 109:362–67

Stark JM. 2000. Nutrient transformations. See Sala et al. 2000, pp. 215–34

Stern LA, Amundson R, Baisden WT. 2001. Influence of soils on oxygen isotope ratio of atmospheric CO_2. *Glob. Biogeochem. Cycles* 15:753–60

Sternberg LDL. 1989. Oxygen and hydrogen isotope ratios in plant cellulose: Mechanisms and applications. See Rundel et al. 1989, pp. 124–41

Sternberg LDL, Mulkey SS, Wright SJ. 1989. Ecological interpretation of leaf carbon isotope ratios: influence of respired carbon dioxide. *Ecology* 70:1317–24

Sternberg LDL, Swart PK. 1987. Utilization of freshwater and ocean water by coastal plants of southern Florida USA. *Ecology* 68:1898–1905

Stewart GR. 2001. What do $\delta^{15}N$ signatures tell us about nitrogen relations in natural ecosystems? See Unkovich et al. 2001, pp. 91–101

Stewart GR, Turnbull MH, Schmidt S, Erskine PD. 1995. ^{13}C natural abundance in plant communities along a rainfall gradient: a biological integrator of water availability. *Aust. J. Plant Physiol.* 22:51–55

Stoddard JL. 1994. Long-term changes in watershed retention of nitrogen. In *Environmental Chemistry of Lakes and Reservoirs, Advances in Chemistry Series,* ed. LA Baker, 237:223–84. Washington, DC: Am. Chem. Soc.

Svejcar TJ, Boutton TW. 1985. The use of stable carbon isotope analysis in rooting studies. *Oecologia* 67:205–8

Switsur VR, Waterhouse JS, Field EM, Carter AHC. 1996. Climatic signals from stable isotopes in oak trees from East Anglia, Great Britain. In *Tree Rings, Environment and Humanity,* ed. JS Dean, DM Meko, TW Swetnam, *Radiocarbon,* pp. 637–45

Tans PP. 1998. Oxygen isotopic equilibrium between carbon dioxide and water in soils. *Tellus* B 50:162–78

Taylor AFS, Högbom L, Högberg M, Lyon AJE, Näsholm T, et al. 1997. Natural ^{15}N abundance in fruit bodies of ectomycorrhizal fungi from boreal forests. *New Phytol.* 136:713–20

Teeri JA. 1981. Stable carbon isotopes analysis of mosses and lichens growing in xeric and moist habitats. *Bryologist* 84:82

Templer P. 2001. *Direct and indirect effects of tree species on forest nitrogen retention in the Catskill Mountains, NY.* PhD Diss., Cornell Univ., Ithaca, NY

Tennakoon KU, Pate JS, Arthur D. 1997. Ecophysiological aspects of the woody root hemiparasite *Santalum acuminatum* (R. Br.) A. DC and its common hosts in South Western Australia. *Ann. Bot.* 80:245–56

Thorburn PJ, Ehleringer JR. 1995. Root water uptake of field-growing plants indicated by measurements of natural-abundance deuterium. *Plant Soil* 177:225–33

Thorburn PJ, Hatton TJ, Walker GR. 1993. Combining measurements of transpiration and stable isotopes of water to determine groundwater discharge from forests. *J. Hydrol.* 150:563–87

Thorburn PJ, Mensforth LJ, Walker GR. 1994. Reliance of creek-side river red gums on creek water. *Aust. J. Mar. Freshwater Res.* 45:1439–43

Thorburn PJ, Walker GR. 1993. The source of water transpired by *Eucalyptus camaldulensis*: soil, groundwater, or streams? See Ehleringer et al. 1993, pp. 511–27

Tieszen LL, Reed BC, Bliss NB, Wylie BK, DeJong DD. 1997. NDVI, C3 and C4 production and distribution in Great Plains grassland land cover classes. *Ecol. Appl.* 7:59–78

Treseder KK, Davidson DW, Ehleringer JR.

1995. Absorption of ant-provided carbon dioxide and nitrogen by a tropical epiphyte. *Nature* 375:137–39

Troughton JH, Card KA, Hendy CH. 1974. Photosynthetic pathways and carbon isotope discrimination by plants. *Carnegie Inst. Washington Yearb.* 73:768–80

Trumbore SE, Davidson EA, De Camargo PB, Nepstad DC, Martinelli LA. 1995. Belowground cycling of carbon in forests and pastures of Eastern Amazonia. *Glob. Biogeochem. Cycles* 9:515–28

Tu KP, Brooks PD, Dawson TE. 2001. Using septum-capped vials with continuous-flow isotope ratio mass spectrometric analysis of atmospheric CO_2 for Keeling plot applications. *Rapid Commun. Mass Spectrom.* 15:952–56

Tu KP, Dawson TE. 2003. Partitioning ecosystem respiration using stable carbon isotopes. In *Stable Isotopes and Biosphere-Atmosphere Interactions*, ed. LB Flanagan, JR Ehleringer. San Diego: Academic. In press

Tu TTN, Bocherens H, Mariotti A, Baudin F, Pons D, Broutin J, et al. 1999. Ecological distribution of Cenomanian terrestrial plants based on $^{13}C/^{12}C$ ratios. *Palaeogeogr. Palaeoclimatol. Palaeoecol.* 145:79–93

Turnbull MH, Goodall R, Stewart GR. 1995. The impact of mycorrhization on nitrogen source utilisation in *Eucalyptus grandis* and *Eucalyptus maculata*. *Plant Cell Environ.* 18:1386–94

Turner JV, Farrington P, Gailitis V. 2001. Extraction and analysis of plant water for deuterium isotope measurement and application to field experiments. See Unkovich et al. 2001, pp. 37–55

Unkovich M, Pate J. 2001. Assessing N_2 fixation in annual legumes using ^{15}N natural abundance. See Unkovich et al. 2001, pp. 103–18

Unkovich M, Pate J, McNeill A, Gibbs JD, eds. 2001. *Stable Isotope Techniques in the Study of Biological Processes and Functioning of Ecosystems*. Dordrecht: Kluwer Academic

Valentini R, Mugnozza GES, Ehleringer JR.

1992. Hydrogen and carbon isotope ratios of selected species of a mediterranean macchia ecosystem. *Funct. Ecol.* 6:627–31

Van Cleve K, White R. 1980. Forest-floor nitrogen dynamics in a 60-year old paper birch ecosystem in interior Alaska. *Plant Soil* 54:359–81

Virgona JM, Farquhar GD. 1996. Genotypic variation in relative growth rate and carbon isotope discrimination in sunflower is related to photosynthetic capacity. *Aust. J. Plant Physiol.* 23:227–44

Virgona JM, Hubick KT, Rawson HM. 1990. Genotypic variation in transpiration efficiency, carbon-isotope discrimination and carbon allocation during early growth in sunflower. *Aust. J. Plant Physiol.* 17:207–14

Vitousek PM, Aber JD, Howarth RH, Likens GE, Matson PA, et al. 1997. Human alteration of the global nitrogen cycle: source and consequences. *Ecol. Appl.* 7:737–50

Vitousek PM, Field CB, Matson PA. 1990. Variation in foliar $\delta^{13}C$ in Hawaiian *Metrosideros polymorpha*: a case of internal resistance? *Oecologia* 84:362–70

Vivin P, Martin F, Guehl JM. 1996. Acquisition and within-plant allocation of ^{13}C and ^{15}N in CO_2-enriched *Quercus robur* plants. *Physiol. Plant.* 98:89–96

Walcroft AS, Silvester WB, Grace JC, Carson SD, Waring RH. 1996. Effects of branch length on carbon isotope discrimination in *Pinus radiata*. *Tree Physiol.* 16:281–86

Walker CD, Brunel JP. 1990. Examining evapotranspiration in a semi-arid region using stable isotopes of hydrogen and oxygen. *J. Hydrol.* 118:55–76

Walker G, Brunel J-P, Dighton J, Holland K, Leaney F, et al. 2001. The use of stable isotopes of water for determining sources of water for plant transpiration. See Unkovich et al. 2001, pp. 57–89

Wallace LL, Macko SA. 1993. Nutrient acquisition by clipped plants as a measure of competitive success: the effects of compensation. *Funct. Ecol.* 7:326–31

Wallander H, Arnebrant K, Östrand F, Kårén O. 1997. Uptake of ^{15}N-labelled alanine, ammonium and nitrate in *Pinus sylvestris* L. ectomycorrhiza growing in forest soil treated with nitrogen, sulphur or lime. *Plant Soil* 195:329–38

Wang XF, Yakir D. 1995. Temporal and spatial variations in the oxygen-18 content of leaf water in different plant species. *Plant Cell Environ.* 18:1377–85

Ward JK, Dawson TE, Ehleringer JR. 2002. Responses of *Acer negundo* genders to inter-annual differences in water availability determined from carbon isotope ratios of tree ring cellulose. *Tree Physiol.* 22:339–46

Warembourg FR. 1993. Nitrogen fixation in soil and plant systems. See Knowles & Henry Blackburn 1993, pp. 127–56

Waring RH, Silvester WB. 1994. Variation in foliar $\delta^{13}C$ values within the crowns of *Pinus radiata* trees. *Tree Physiol.* 14:1203–13

Warren CR, Adams MA. 2000. Water availability and branch length determine $\delta^{13}C$ in foliage of *Pinus pinaster*. *Tree Physiol.* 10:637–44

Warren CR, McGrath JF, Adams MA. 2001. Water availability and carbon isotope discrimination in conifers. *Oecologia* 127:426–86

Watkins NK, Fitter AH, Graves JD, Robinsson D. 1996. Carbon transfer between C3 and C4 plants linked by a common mycorrhizal network, quantified using stable carbon isotopes. *Soil Biol. Biochem.* 28:471–77

Welker JM, Wookey PA, Parsons AN. 1993. Leaf carbon isotope discrimination and vegatative responses of *Dryas octopetala* to temperature and water manipulations in a High Arctic polar semi-desert, Svalbard. *Oecologia* 95:463–69

Wershaw RL, Friedman I, Heller SJ. 1966. Hydrogen isotope fractionation of water passing through trees. In *Advances in Organic Geochemistry*, ed. F Hobson, M Speers, pp. 55–67. New York: Pergamon

White JWC. 1989. Stable hydrogen isotope ratios in plants: a review of current theory and some potential applications. See Rundel et al. 1989, pp. 142–62

White JWC, Cook ER, Lawrence JR, Broecker

WS. 1985. The deuterium to hydrogen ratios of sap in trees: implications for water sources and tree ring deuterium to hydrogen ratios. *Geochim. Cosmochim. Acta* 49:237–46

Wickman FE. 1952. Variations in the relative abundance of the carbon isotopes in plants. *Geochim. Cosmochim. Acta* 2:243–54

Williams DG, Ehleringer JR. 1996. Carbon isotope discrimination in three semi-arid woodland species along a monsoon gradient. *Oecologia* 106:455–60

Williams DG, Ehleringer JR. 2000a. Carbon isotope discrimination and water relations of oak hybrid populations in southwestern Utah. *West. N. Am. Nat.* 60:121–29

Williams DG, Ehleringer JR. 2000b. Intra- and interspecific variation for summer precipitation use in pinyon-juniper woodlands. *Ecol. Monogr.* 70:517–37

Williams DG, Gempko V, Fravolini A, Leavitt SW, Wall GW, et al. 2001. Carbon isotope discrimination by *Sorghum bicolor* under CO_2 enrichment and drought. *New Phytol.* 150:285–93

Williams K, Richards JH, Caldwell MM. 1991. Effect of competition on stable carbon isotope ratios of two tussock grass species. *Oecologida* 88:148–51

Williams TG, Flanagan LB. 1996. Effect of changes in water content on photosynthesis, transpiration and discrimination against $^{13}CO_2$ and $C^{18}O^{16}O$ in *Pleurozium* and *Sphagnum*. *Oecologia* 108:38–46

Williams TG, Flanagan LB. 1998. Measuring and modelling environmental influences on photosynthetic gas exchange in *Spagnum* and *Pleurozium*. *Plant Cell Environ.* 21:555–64

Wilson EJ, Tiley C. 1998. Foliar uptake of wet-deposited nitrogen by Norway spruce: an experiment using ^{15}N. *Atmos. Environ.* 32:513–18

Winkler FJ, Wirth E, Latzko E, Schmidt HL, Hoppe W, Wimmer P. 1978. Influence of growth conditions and development on ^{13}C values in different organs and constituents of wheat, oat, and maize. *Z. Pflanzenphysiol.* 87:255–63

Yakir D. 1992. Variations in the natural

abundance of oxygen-18 and deuterium in plant carbohydrates. *Plant Cell Environ.* 15: 1005–20

Yakir D. 1998. Oxygen-18 of leaf water: a crossroad for plant-associated isotopic signals. See Griffiths 1998, pp. 147–68

Yakir D, Sternberg LDL. 2000. The use of stable isotopes to study ecosystem gas exchange. *Oecologia* 123:297–311

Yakir D, Wang X-F. 1996. Fluxes of CO_2 and water between terrestrial vegetation and the atmosphere estimated from isotope measurements. *Nature* 380:515–17

Yoder BJ, Ryan MG, Waring RH, Schoettle AW, Kaufmann MR. 1994. Evidence of reduced photosynthetic rates in old trees. *For. Sci.* 40:513–27

Yoneyama T. 1996. Characterization of natural ^{15}N abundance of soils. In *Mass Spectrometry of Soils*, ed. TW Boutton, S Yamasaki, pp. 205–23. New York: Dekker

Yoneyama T, Handley LL, Scrimgeour CM, Fisher DB, Raven JA. 1997. Variations of the natural abundances of nitrogen and carbon isotopes in *Triticum aestivum*, with special reference to phloem and xylem exudates. *New Phytol.* 137:205–13

Yoneyama T, Kaneko A. 1989. Variations in the natural abundance of ^{15}N in nitrogenous fractions of Komatsuna plants supplied with nitrate. *Plant Cell Physiol.* 30:957–62

Yoneyama T, Matsumaru T, Usui K, Engelaar WMHG. 2001. Discrimination of nitrogen isotopes during absorption of ammonium and nitrate at different nitrogen concentrations by rice (*Oryza sativa* L.) plants. *Plant Cell Environ.* 24:133–39

Zak DR, Groffman PM, Pregitzer KS, Christensen S, Tiedje JM. 1990. The vernal dam: plant-microbe competition for nitrogen in northern Michigan, USA hardwood forests. *Ecology* 71:651–56

Zhang JW, Marshall JD. 1995. Variation in carbon isotope discrimination and photosynthetic gas exchange among populations of *Pseudotsuga mensiesii* and *Pinus ponderosa* in different environments. *Funct. Ecol.* 9:402–12

Zhang JW, Marshall JD, Jaquish BC. 1993. Genetic differentiation in carbon isotope discrimination and gas exchange in *Pseudotsuga mensiesii*: a common garden experiment. *Oecologia* 93:80–87

Zencich SJ, Froend RH, Turner JV, Gailitis. 2002. Influence of groundwater depth on the seasonal sources of water access by *Banksia* tree species on a shallow, sandy coastal aquifer. *Oecologia* 131:8–19

Zimmerman JK, Ehleringer JR. 1990. Carbon isotope ratios are correlated with irradiance levels in the Panamanian orchid *Catasetum viridiflavum*. *Oecologia* 83:247–49

Zogg GP, Zak DR, Pregitzer KS, Burton AJ. 2000. Microbial immobilization and the retention of anthropogenic nitrate in a northern hardwood forest. *Ecology* 81:1858–66

Annu. Rev. Ecol. Syst. 2002. 33:561–88
doi: 10.1146/annurev.ecolsys.33.030602.152151

THE QUALITY OF THE FOSSIL RECORD:
Implications for Evolutionary Analyses

Susan M. Kidwell
*Department of Geophysical Sciences, University of Chicago, 5734 South Ellis Avenue,
Chicago, Illinois 60637; email: skidwell@uchicago.edu*

Steven M. Holland
*Department of Geology, University of Georgia, Athens, Georgia 30602-2501;
email: stratum@gly.uga.edu*

Key Words paleobiology, taphonomy, stratigraphy, speciation, extinction,
phylogeny

■ **Abstract** Advances in taphonomy and stratigraphy over the past two decades
have dramatically improved our understanding of the causes, effects, and remedies of
incompleteness in the fossil record for the study of evolution. Taphonomic research
has focused on quantifying probabilities of preservation across taxonomic groups, the
temporal and spatial resolution of fossil deposits, and secular changes in preservation
over the course of the Phanerozoic. Stratigraphic research has elucidated systematic
trends in the formation of sedimentary gaps and permanent stratigraphic records, the
quantitative consequences of environmental change and variable rock accumulation
rates over short and long timescales, and has benefited from greatly improved methods of correlation and absolute age determination. We provide examples of how these
advances are transforming paleontologic investigations of the tempo and mode of morphologic change, phylogenetic analysis, and the environmental and temporal analysis
of macroevolutionary patterns.

INTRODUCTION

Ever since Darwin first raised concerns about the completeness of the fossil
record as an evolutionary archive, paleontologists have devoted considerable attention to the causes, recognition, and mitigation of gaps in the record (Paul
1982, Donovan & Paul 1998, McKinney 1991, Kidwell & Flessa 1996, Behrensmeyer et al. 2000, Holland 2000). This work has elaborated the many ways in
which the record can be an imperfect document of history, including gaps in
paleontologic time series from failures in fossil or rock preservation, and distortion of biological trends owing to variable environments and rates of sedimentary accumulation. However imperfect it may be, this record is a unique
window into life on earth and provides, at the very least, data on the minimum

0066-4162/02/1215-0561$14.00

possible ages of morphologies and taxa, on taxonomic richness and morphological variety through time, and on the environmental and geographic distribution
of life through time. The immense value of such information for evolutionary
analysis drives the exploration of the pitfalls of fossil preservation and the development of protocols to maximize the quality and quantity of data retrieved from the
record.

Here we summarize what paleontologists and geologists have learned over the
past 20 years about the nature of the stratigraphic record as an archive of biological information, highlight some of the successful strategies that have been developed to deal with incompleteness, and recommend directions for future research
(for a comparable treatment of ecological questions, see Kidwell & Flessa 1996).
Owing to the breadth of evolutionary questions, paleontologists are concerned
with the quality of the record at many scales, from that of populations sampled
at a single geological instant, to species-level traits such as geographic extent,
evolutionary duration, and interpopulation variation, to clade- and higher-level
dynamics that require time series spanning even greater stratigraphic intervals
and accurate determination of the relative geologic ages of widely spaced deposits (Figure 1). The natural processes that structure the available record in time
and space—the selective postmortem preservation of organic remains (taphonomy) and the selective archiving of the sedimentary deposits that entomb those
remains (stratigraphy)—are the subject of this review. These factors are examined at the scale of individual beds (single samples of the finest temporal resolution), stratigraphic sections (time series captured at a single point on earth)
(Figure 1A), and geographic regions whose natural boundaries are usually determined by climate and tectonics (Figure 1B,C). Because of our expertise, we
emphasize benthic macroinvertebrates whose biomineralized skeletons, together
with those of biomineralizing microfossils (foraminiferans, radiolarians, ostra-
codes, etc.), dominate the marine fossil record. However, many of these issues and
solutions devised for them also apply to the fossil record of vertebrates, plants,
and soft-bodied organisms.

In general, research into the nature of the fossil record has brought a long-needed
shift away from a search for total completeness, which is never achieved, even in
neontological sampling. Instead, paleontologists now test for the adequacy of a
particular segment of the fossil record, that is, whether data at hand are sufficient to
address a specific evolutionary question (Paul 1982). Research has also broadened
from a concern with gaps in evolutionary time series—the concept of completeness
in Darwin's sense (and see Sadler 1981)—to include additional aspects of data
quality (Behrensmeyer et al. 2000, Kowalewski & Bambach 2002). Resolution
refers to the level of detail that can be recovered from the fossil record, such as
organelle-versus cell-versus tissue-level preservation of anatomy, high versus low
degrees of time-averaging (mixing) of successive generations of a species into a
single bed, or high versus low resolution on the original spatial distribution of a
taxon. Bias refers to the distortion of the underlying biological signal by selective

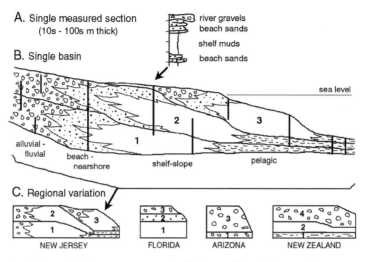

Figure 1   Paleontologists are concerned with the quality of the record at many scales. (*A*) Within a single measured section, that is, the succession of beds visible at a single locale that usually spans only part of a species' full evolutionary duration. The primary concerns are the extent of mixing of generations (time-averaging) within each sedimentary bed that is a few millimeters to a few tens of centimeters thick, the magnitude of time represented by each of the planes that separates beds (stratigraphic gaps) and the potential to mistake a morphological trend that is linked to local environmental change (geographic shifting of a cline) as true evolutionary change. Environments differ in the steadiness of sediment accumulation (frequency of gaps) and in the rate of sedimentation (both affecting time-averaging), and they also differ in their favorableness to the preservation of organic remains. (*B*) More complete time series are constructed by combining data from many local sections of varying completeness and quality. The *solid lines* running across this cross section separate three temporally distinct depositional units ("sequences"), each comprising a lateral array of depositional environments that have migrated over time first toward and then away from the continent. Each lithologically distinct body within a sequence may be formalized as a stratigraphic formation. Sequence boundaries are recognized by strong offsets in environment and signify larger gaps than those between beds within a sequence. Each of these major gaps diminishes in magnitude seaward. Paleontologic information on the tempo and mode of species evolution and on lineages usually requires compiling data from multiple sequences within a region because species have average durations of ∼5 million years (my), and depositional sequences each typically represent ∼1–3 my. (*C*) Paleontologic data from many regional composite time series must be correlated to build global perspectives on biotic change. Regions differ in the completeness of their record and in their dominant environments, owing to differences in their plate-tectonic and climatic histories.

preservation, sediment accumulation, and paleontological collection, including, for example, modification of the relative abundances of morphs because fragile forms have been lost, underestimation of the relative species-richness of clades because of differential preservation potentials of their body plans or of their preferred habitats, or skewing of temporal trends because successive samples could not be collected from comparable habitats or taphonomic conditions. Thus, a collection of fossils from a single bed might contain a high-resolution but highly biased sample of the original morphologic and species composition of the original fauna, and a data set generated by sampling through a series of beds might comprise a fairly complete time series composed of individual high-resolution assemblages but nonetheless might be biased by changes in habitat among successive beds (Figure 2).

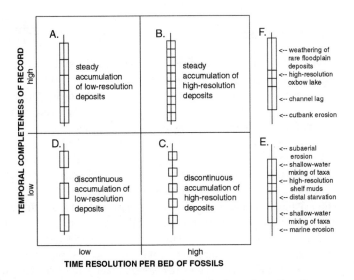

Figure 2 The quality of the fossil record is scale dependent. Fossils extracted from a single bed may constitute a virtual biological census and have a high degree of time resolution (*short blocks in cartoons*) or be time-averaged to some degree, providing a lower time-resolution sample (*long blocks*). The historical time series provided by these segments may be complete, containing fossils from each elapsed increment of time of a particular scale, or may contain significant gaps. End-member combinations of per-bed time resolution and section completeness are illustrated by timelines (*A–D*) (see text). Real patterns are generally more heterogenous, especially in shallow-marine macrobenthic records (*E*) (timeline to complement Figure 1*A*) and continental vertebrate records (*F*), owing to lateral shifts in environments over time. Compositing paleontologic data from multiple sections and sedimentary basins attempts to compensate for this variability within single sections.

GEOLOGIC CONTROLS ON THE QUALITY OF THE FOSSIL RECORD

Time-series data vary in completeness, temporal resolution, and compositional bias as a function of depositional environment and geologic history and as a function of the taxonomic group being targeted. These stratigraphic and taphonomic factors are key to acquiring high-quality paleontologic time series at any scale, that is, in acquiring data that are of comparable quality (isotaphonomic samples; see Behrensmeyer & Hook 1992).

Taphonomy: The Fossilization of Biological Remains

Much taphonomic research has been concerned with the rates and variability of postmortem processes, and the challenge now is to move from a phenomenology of the modification and accumulation of organic remains to quantitative models of bias in paleobiological data (for reviews see Behrensmeyer et al. 2000, Briggs & Crowther 2001). Four themes of this research are essential components of any quantitative model of preservation and are directly relevant to evolutionary analysis using the fossil record.

PROBABILITY OF PRESERVATION   Taxa do not have equal probabilities of preservation, with the most obvious demonstration being the poor fossil record of most organisms that lack mineralized skeletons. Preservation of entirely soft-bodied organisms requires unusual environmental conditions that are geologically rare, such as anoxia or catastrophic burial with rapid mineral replacement by specialist microbial communities (Briggs & Crowther 2001). When it occurs, such preservation provides valuable windows into the anatomy and habitats of these groups and can be important simply by virtue of being the earliest record of taxa and morphologic characters. However, stratigraphic horizons with comparable preservation are generally so widely spaced or environment specific that evolutionary time series for these groups are highly incomplete. Similarly, DNA is rarely, if ever, preserved in rocks older than 100,000 years (Bada et al. 1999, Cooper & Poinar 2000), despite a number of early claims.

In contrast, biomineralizing taxa are comparatively well represented and are consequently the focus of most evolutionary analyses by marine paleontologists. For example, $\sim$50% of scleractinian coral, $\sim$75% of echinoid, and $\sim$90% of shelled mollusk species found alive today are represented by dead skeletal material in locally accumulating sediments (reviewed by Kidwell & Flessa 1996). Postdepositional processes can reduce the percentage of taxa preserved; for example, only $\sim$80% of modern shelled mollusk species in the Californian province are preserved in the local Pleistocene record (Valentine 1989). However, at higher taxonomic levels and larger geographic scales, recovery is improved. For example, 100% of Californian molluscan genera have a local fossil record (Valentine 1989), and from a single intertidal zone, Schopf (1978) found that 75% of lightly mineral-ized and 100% of well-mineralized genera have a known fossil record somewhere in the world.

Within major biomineralizing groups, taxa lost in the fossilization process are often predictable from body size (small-bodied forms are most susceptible to loss), tissue microstructure (hardparts constructed of aragonite, rather than cal-cite, or of high organic content, rather than low organic content), population size, and other factors intrinsic to each taxon (Kidwell & Flessa 1996). These basic patterns determine the taxa that are best suited for testing a given evolutionary question. Alternatively, if the study is targeted to a particular taxonomic group, then these same guidelines can be used to identify a "taphonomic control taxon" to determine when absences of the target are biologically, rather than taphonomically, determined (Jablonski et al. 1997). Cyclostome bryozoans, for example, have been used as control taxa for the early evolution of cheilostome bryozoans. The control taxon has similar ecological requirements to the target and a preservation potential that is comparable or less. Thus, if the control taxon is present in a deposit, then the target taxon would be expected to have been preserved and collected if it actually co-occurred in that unit.

TIME RESOLUTION WITHIN SINGLE BEDS   Because sediment accumulation at the scale of beds is highly episodic in most environments (timed with storms and floods, for example, rather than occurring steadily), net rates of stratigraphic accumulation are commonly slow relative to the life span of individual organisms. Consequently, the skeletal remains of multiple generations typically become mixed within a single bed. This time-averaging of local but noncontemporaneous remains is pervasive in the fossil record: It explains why fossilized individuals are rarely found in their original life orientations, but instead are rotated and disarticulated (Walker & Bambach 1971). When conditions of slow net sediment accumulation are so prolonged that they encompass a period of environmental change, then remains from more than one habitat can become mixed into a single bed, leading to faunal or environmental condensation (Fürsich 1978, Kidwell & Bosence 1991). Thus, a sample from a time-averaged fossil assemblage is not expected to be comparable to a modern biological census; instead, it is a selective summation of individuals that were alive at some point during an extended period of time.

The absolute duration of time-averaging in fossil assemblages can be inferred from Recent sediments in which the organic remains of several different groups have been dated using radiocarbon decay and amino-acid racemization. These studies indicate that in many nearshore marine environments, presently accumu-lating sediments include mollusk and foraminiferan shells up to several hundreds or a few thousands of years old (Flessa & Kowalewski 1994, Meldahl et al. 1997, Martin et al. 1996, Kidwell & Best 2001). Shell ages range up to 20,000 or 30,000 years in the thin shelly sands that cover continental shelves and record the entire postglacial rise in sea level (Flessa & Kowalewski 1994, Flessa 1998, Anderson et al. 1997).

Areas of more rapid sediment accumulation, such as lagoons, bays, deltas, and lakes, permit less time averaging per bed and approach decadal time resolution for mollusks and pollen over significant periods of time (e.g., Brewster-Wingard et al. 2001, Webb 1993). To achieve a finer, e.g., annual or seasonal level of time resolution within time series, the paleontologist must focus on those cases where deposition was either continuous, such as deep-sea sediments composed of planktonic debris, or frequent, such as varved lake sediments (e.g., Bell et al. 1987), but permanent in either case. Due to episodic sedimentation, high-resolution assemblages also occur scattered within intervals dominated by lower resolution assemblages.

Paleontologists are thus able to generate time series with annual or decadal time resolution in only a few settings, but even in relatively strongly time-averaged records (hundreds or thousands of years per sampled bed), the level of temporal resolution is often many orders of magnitude finer than the average duration of species and high-resolution samples will be present (Jablonski 2000). Moreover, decadal and centennial time averaging is not necessarily disadvantageous to biological analysis because time averaging can smooth the noise of seasonal and annual fluctuations in populations (Peterson 1977). One encouraging result from recent research is the growing evidence that although time-averaged assemblages contain old specimens, they are numerically dominated by taphonomically young shells, that is, by individuals that died during the final phases of time averaging (Meldahl et al. 1997, Olszewski 1999, Kidwell 2002). Such samples should thus have an effective time resolution that is much finer than the full duration of time averaging.

SPATIAL RESOLUTION   Because population patches—even those of sessile colonies—commonly migrate or shift over time, time averaging generally also entails a certain amount of spatial averaging (Behrensmeyer & Chapman 1993, Miller & Cummins 1990). Spatial mixing from postmortem transport, in which species are preserved in sediments outside their life habitat, does not appear to be a significant bias for many taxonomic groups, including marine macrobenthos, benthic foraminifera, and land mammals not living near major rivers (Behrensmeyer & Dechant Boaz 1980, Kidwell & Flessa 1996, Anderson et al. 1997). Where out-of-habitat transport is significant, it is increasingly well understood (Jackson & Cheetham 1994, Davis 2000) and can even be advantageous from the perspective of gathering spatially coarse occurrence data, such as confirming the regional presence of a taxon. Postmortem transport of a taxon to areas outside its original biogeographic range appears to be negligible in all but a very few predictable groups, such as shelled cephalopods, which occasionally drift for vast distances. Species assemblages atypical of modern conditions are thus more likely to arise from individualistic behaviors of species during life (e.g., Webb 1993, Graham 1993) or from time averaging of individuals from provinces whose boundaries have shifted over time (e.g., Roy et al. 1996).

SECULAR CHANGES IN PRESERVATION   The quality of the fossil record as a biological archive has almost certainly changed over the course of earth's history for many reasons (Allison & Briggs 1993a, Kidwell & Brenchley 1994, Kowalewski 1996, Taylor & Allison 1999, Behrensmeyer et al. 2000, Behrensmeyer 1999). For example, plate tectonics has shifted many continents—particularly North America and Europe, whose fossils dominate the described fossil record—from tropical to nontropical positions over the Phanerozoic (past ~550 million years). Over the same time, Earth has undergone multiple cycles of greenhouse and icehouse conditions as well as possible changes in seawater chemistry that affect the postmortem solubility of calcium carbonate and other key biominerals. In addition, the acquisition via natural selection of different biominerals, skeletal microstructures, environmental preferences, and life habits by clades affects their preservation potential, as does the evolutionary appearance and expansion of organisms that modify or destroy the remains of other taxa (e.g., shell and bone-crushing predators, scavengers, and burrowing and boring organisms).

Thus arguably, one cannot assume that "typical" samples of the fossil record from one geologic age are taphonomically comparable to those from another, nor can one assume that a major clade has a constant preservation potential over its entire evolutionary duration. However, such bias will impinge mainly on attempts to compare patterns over very large stretches of geologic time (e.g., Paleozoic versus post-Paleozoic).

Stratal Architecture: The Accumulation of Fossil-Bearing Strata

The past two decades have seen an intense effort by stratigraphers to characterize and understand the architecture of the sedimentary record, including both (*a*) the origins of individual beds and sets of beds, each representing time intervals of minutes to tens of thousands of years, and (*b*) their organization into larger packages of strata, with the largest representing a few million, tens of millions, or hundreds of millions of years (third-, second-, and first-order depositional sequences, respectively) (Figure 1*B*). Much of this stratigraphic work is directly relevant to the quality of evolutionary data, in that stratal architecture determines the duration, timing, and geographic extent of gaps in the sedimentary record as well as the shifting over time of habitats suitable for life or for fossil preservation (Brett 1995, Holland 1995). Six themes of this research are particularly relevant to understanding the quality of the fossil record for evolutionary analysis.

NATURE AND DISTRIBUTION OF SEDIMENTARY GAPS   Gaps in the stratigraphic record are produced by local episodes of non-deposition, which can be caused by the failure of sediment supply, the bypassing of sediment to other areas, or the net erosion of previously deposited sediment. These gaps are signified by the planar to irregular surfaces that separate one bed from the next, as observed in outcrops. The durations recorded by these individual bedding planes and discontinuity surfaces can be estimated in various ways, and range from a few minutes (e.g., the pause between one avalanche of sediment down a migrating dune face to the next) up to hundreds of millions of years (e.g., the erosion surface created by tectonic uplift of a formerly largely submerged continental block; Sadler 1981, McKinney 1986, Anders et al. 1987). During these episodes, organisms may continue to produce skeletal and other organic remains (aquatic organisms if the area remains submerged, terrestrial organisms if it is exposed as land). However, these remains are especially prone to destruction through breakage, abrasion, boring, and dissolution because permanent burial is delayed, and erosional exhumation itself can also be a process of selective hardpart destruction. Thus, depending on their duration and the environment in which they form, individual gaps in the sedimentary record can be signified by discontinuity surfaces overlain by no skeletal hardparts, by an abundance of time-averaged hardparts, or by a thin lag of highly damaged hardparts, which will be preferentially biased toward the most physically robust, chemically resistant, ecologically abundant, and stratigraphically youngest forms (Kidwell 1991, 1993; Brett 1995; Rogers & Kidwell 2000). The time resolution and inherent bias of individual fossil assemblages in a stratigraphic section therefore depends on the context of their occurrence: Those that are physically associated with discontinuity surfaces must be approached with caution and are generally best suited for coarse-scale analyses (e.g., presence of a taxon within a geologic period and region).

Some environments are more prone to producing sedimentary gaps than others. Land is primarily an area of erosion and weathering, lakes and shallow-marine environments are primarily sites of sediment accumulation, and the deep-sea is starved of most sediment input other than the deposition of airborne dust, suspended clay, and pelagic organisms. Continental records thus commonly contain many discontinuity surfaces of long duration, so paleontologists commonly target the particular regions (e.g., areas undergoing sustained tectonic subsidence, such as foreland basins) and environments (e.g., lacustrine deltas, alluvial plains, wetlands) that favor more continuous sediment accumulation and postmortem conditions favorable to fossil preservation (Figure 2D,F). Lakes provide exceptionally continuous high time-resolution archives within the generally poor continental record (Figure 2A or B) but tend to be spatially limited and geologically short lived, persisting for a few thousand or tens of thousands of years at most (the Eocene Green River lake system of Wyoming is one of the most famous exceptions). Deposition is also relatively continuous but localized and short lived in estuaries and bays (Figure 2A or C). Shallow marine shelf and slope (hemipelagic) settings accumulate the bulk of land-derived sediment and in the tropics can be areas of major biogenic carbonate accumulation including reefs. These environments are both widespread and relatively long lived geologically, but deposition is highly variable and episodic on timescales ranging from seasons to millions of years (cycles of sea-level rise and fall), thus generating local sections of heterogeneous quality (Figure 2A,B,E). In many instances, this variability is still rapid, indeed, cyclic, relative to the duration of species, and thus has fewer consequences for some evolutionary analyses than for others. Finally, deep-sea records receive slight but highly continuous sediment input. Thus, they typically have few large gaps but can suffer from considerable time averaging and condensation of skeletal input because net stratigraphic accumulation is so slow (Figure 2*A* or *C*). In abyssal depths, calcareous hardparts are also subject to dissolution. Workers consequently focus on oceanic zones of relatively high productivity, where the rain of microplankton and other detritus was most intensive, and concentrate on the most dissolution-resistant taxa where possible. Unfortunately, because these deep-sea environments are prone to burial, deformation, and destruction by plate tectonic processes, nearly everything we know about marine life in the past (especially before the past ~150 million years) is from the shallowest water environments ($\leq$~200 m).

Within those constraints, however, the international Ocean Drilling Project (ODP; formerly Deep Sea Drilling Project, DSDP) has provided a wealth of richly fossiliferous sedimentary cores for evolutionary analysis at many scales (Jackson & Cheetham 1999, Chapman 2000, MacLeod et al. 2000, Norris 1991, Kucera & Malmgren 1998).

Thus, in any given local section, the number and type of gaps depend heavily on how environments shifted across the area during the time interval under study (Figure 1*B*). The longer the geologic interval is under study, the greater the potential for temporal variation in environments, which changes the quality of the record that is available to be sampled locally (Figure 1; 2*E* & *F*). This is a difficulty for local studies: The paleontologist must broaden the study area in order to be able to track a single habitat over time, and thus studies of long evolutionary time typically are regional in scope. Increasing the length of time encompassed by a study also increases the number of major gaps one can expect to encounter. In general, temporally prolonged gaps such as created by significant relative falls in sea level from ice-sheet formation or tectonic uplift affect larger geographic areas than do short gaps. Large gaps are thus more difficult to overcome by compositing data from multiple sections than are small gaps, such as those created by migrating tidal channels, and so interpretations of paleontologic patterns must generally take the fact of these larger breaks into account.

RELATIONSHIP BETWEEN THICKNESS AND TIME   The amount of time represented by a given bed varies widely: Ten centimeters of sediment can reflect nearly instantaneous deposition during a storm or flood, several decades of accumulation in a lake or bay, or several thousand years in an offshore habitat. Moreover, a local section can include any number and magnitude of erosional and nondepositional breaks. Consequently, stratigraphic or rock accumulation rates (thickness per unit time) range over 11 orders of magnitude and vary as a function of depositional setting (Sadler 1981). In lakes and the deep sea where sediments accumulate steadily, the relative spacing of fossils through a stratigraphic record can frequently be treated as a simple time series in which rock thickness can be a good proxy for elapsed geologic time between paleontologic events (but see MacLeod 1991 and MacLeod & Keller 1991 for exceptions). This is not feasible, however, in riverine and shallow marine settings where sediment accumulation is much less steady on scales often of thousands to tens of millions of years; apparent rates of evolution will be highly distorted if one takes rock thickness as a proxy for geologic time (Holland & Patzkowsky 1999).

Sadler (1981) proposed a practical method of evaluating the stratigraphic completeness of individual sections, based on the ratio of the rock accumulation rate of the target stratigraphic section over a span of time to the average accumulation rate expected for that sedimentary environment in such a period of time. Thus, if the stratigraphic record needs to be relatively free of gaps of a given length for a particular evolutionary study to be conducted, one can calculate the percent completeness of various candidate local sections at that requisite scale before paleontological sampling is begun.

CHANGES IN ENVIRONMENTAL CONDITIONS   Due to sea-level changes, climatic changes, and the latitudinal shifts of continents, depositional environments shift laterally over time, causing local stratigraphic sections to preserve a series of different depositional environments. In addition to influencing the completeness and accumulation rate of the rock record, such environmental shifts also exert an ecological and taphonomic influence on the occurrence of taxa in local sections, for example, delaying the first occurrence of a taxon after its actual time of evolutionary origination (because populations cannot immigrate until the appropriate environment exists locally), or causing the last local occurrence of a taxon to predate the actual time of extinction (local ecological extirpation usually precedes ultimate evolutionary exinction). In cases where environmental changes are minimal or where taxa are not strongly controlled by environment, numerical modeling indicates that this difference in age, known as range offset, is likely to be on the order of tens to hundreds of thousands of years on average (Holland & Patzkowsky 2002), an estimate in agreement with that seen in Neogene planktonic microfossils (Spencer-Cervato et al. 1994). Given average species durations of 4 million years, such values of range offset would represent on the order of 1% to 10% of the species duration. In contrast, values of range offset in shallow marine shelf are typically on the order of 1–2 million years in shallow marine shelf settings, where environments migrate strongly in response to third-order sea-level fluctuations (Holland & Patzkowsky 2002) (Figure 1). In such circumstances, different amounts of range offset for taxa that overlap in time can cause their relative times of apparent origination (or extinction) in a local section to be reversed. Latitudinal shifts of continents and longer-term sea-level changes could likewise cause local values of range offset of higher taxa to be in the tens of millions of years, so that, for example, the presence of tropical corals in a region will change according to its latitudinal position and sedimentary environment.

Although only recently quantified, this problem of environmental control on species occurrences within their evolutionary range is long recognized by paleontologists. The main method of reducing the artifacts in timing is also long recognized, namely, compositing data from stratigraphic sections over a broad geographic range so that key environments can be collected for each increment of geologic time. In practice, correcting for such environmental control has not been done as routinely as it should be. Furthermore, it is simply not possible in some instances: Geologic processes may have selectively removed the record of particular environments from throughout the known geographic range of the target taxa, making the removal of such bias in fossil ranges difficult unless areas with very different tectonic and depositional histories are compared (Smith et al. 2001). This is a particular problem for endemic taxa because rises and falls in sea level are likely to affect a relatively small area synchronously, leaving few local environmental refuges for taxa to occupy—and be preserved in—during some intervals of time. Environmental homogeneity of the record and endemicity of the biota are thus factors in determining whether a given fossil record is likely to be adequate for testing a particular evolutionary question.

FORMATION OF A PERMANENT ROCK RECORD   The net accumulation of sediments into a permanent stratigraphic record requires subsidence of Earth's crust, such that deposition exceeds erosion over evolutionarily meaningful intervals of time. The subsidence that generates sedimentary basins is localized both spatially and temporally by plate tectonics. The typical durations of subsidence episodes range from a few millions to a few tens of millions of years in relatively short-lived basins (e.g., rift systems of East Africa and the Gulf of California, small intermontane basins of the Rocky Mountains, basins along strike-slip faults such as the San Andreas) to several hundred million years in more slowly subsiding regions (e.g., continental margins of Atlantic-type ocean basins). As a result, individual biogeographic provinces vary in the completeness or even existence of their stratigraphic records (e.g., Foote et al. 1999) or may undergo windows of preservation and non-preservation, depending on their history of subsidence. Thus, comparative analyses of the diversity dynamics and biogeographic histories of co-occurring clades, all subject to similar natural sampling regimes, will often be more robust than comparisons among widely separated regions or among quite different time intervals. Furthermore, sedimentary basins must survive tectonic destruction during subduction and continental collision. For example, deep-sea environments have a minimal geologic record prior to the Jurassic because oceanic lithosphere undergoes subduction, resulting in the destruction of most accumulated sedimentary record.

RESOLUTION IN CORRELATION   The resolution of stratigraphic correlation, that is, of establishing the age equivalency of rocks in different areas, can be very high (<1 year in exceptional cases) over short distances, but generally decreases as geographic distance increases. The refinement of geophysical methods of correlation using the remnant magnetism of rocks, which record ancient reversals in earth's magnetic poles, has greatly improved the correlation of continental fossil records within some regions and of Cenozoic age rocks in general. However, it is difficult to establish a global chronology of biological events at the same fine resolution that can be achieved locally and regionally, and this difficulty becomes more acute deeper into the geological past. In Paleozoic and Mesozoic rocks, for example, global correlation generally cannot be achieved with higher resolution than a geologic stage, which has an average duration of ~6 million years. There is thus a trade-off between temporal resolution and spatial coverage in species-level and other evolutionary analyses (Jablonski 2000). Fortunately, recently developed geochemical correlation methods based on stratigraphic variation in oxygen, carbon, and strontium stable isotopes, as well as methods that tune pelagic lithologic cycles to Milankovitch astronomical cycles, are overcoming the usual tradeoff between temporal resolution and geographic distance and are finding widespread application, especially in the marine record of the last 60–80 million years (Hinnov 2000). For some intervals, these methods allow resolution in global correlation to approach tens to hundreds of thousands of years (Berggren et al. 1995). In addition, major improvements in the precision of radiometric dating are permitting higher resolution correlation throughout the fossil record, as well as better information on the absolute ages of events (Bowring et al. 1993, Bowring & Erwin 1998).

Thus both regional- and global-scale time series for lineages and clades can now be constructed with greater precision and confidence in the relative timing of events and in the rates of evolutionary processes. Advances in radiometric dating have also improved the calculation of evolutionary rates and, combined with a suite of new correlation methods, permit the fine structure of mass-extinction events and major taxonomic radiations to be resolved with much greater confidence (Bowring & Erwin 1998).

SECULAR CHANGES IN THE STRATIGRAPHIC RECORD   Some of the largest-scale patterns in evolutionary paleobiology, such as Phanerozoic trends in biodiversity, are vulnerable to broad changes in the nature of the stratigraphic record, wrought by physical evolution of the earth. These potential biasing factors include some of the same factors that are thought to cause large-scale trends in the quality of fossil preservation (see section above), and are the subject of intense current research (e.g., Peters & Foote 2001, Powell & Kowalewski 2002, Alroy et al. 2000, Smith 2001). All workers presently express considerable uncertainty about the actual magnitude of artifactual effects, and most of the same issues were hotly debated during the earliest modern work on the subject (e.g., Sepkoski et al. 1981). A new generation of analyses is now possible because of improved geological information, new approaches to modeling, and increasingly powerful database options.

Three factors appear to have greatest potential to bias diversity trends at this scale.

1. The shift of North American and European continents from exclusively tropical latitudes in the Paleozoic to largely temperate latitudes in the Mesozoic and Cenozoic (e.g., Allison & Briggs 1993b). Because most global trends are extrapolated from information drawn primarily from these two regions, some workers have estimated that global post-Paleozoic diversity might be underestimated by two- to fivefold at the genus and species levels given the strong diversity gradients documented for most major groups in modern and earlier times (Allison & Briggs 1993b, Jackson & Johnson 2001).

2. The Pull of the Recent (Raup 1979) inflates late Cenozoic taxonomic richness relative to earlier geologic intervals owing to our knowledge of living biota,

which extends the known stratigraphic ranges of fossil taxa up to the Recent from isolated fossil occurrences. Because no part of the history of a group in the deeper past is ever as well sampled as the modern fauna, the stratigraphic ranges of relatively young taxonomic groups are lengthened artificially by their present-day occurrences. The magnitude of the Pull of the Recent has historically been difficult to quantify (but see Foote 2001, 2002; Jablonski et al. 2002). Pull of the Recent would be expected to be most severe for taxa with poor preservation potential, rare taxa, stenotopic taxa, and taxa confined to poorly sampled geographic provinces; a concerted effort on compiling published and new data on Late Cenozoic occurrences will allow the actual magnitude of this effect to be tested rigorously. Similar effects to the Pull of the Recent occur to a lesser degree elsewhere in the stratigraphic record where the biota of one time interval is much better known than adjacent time intervals.

3. Fluctuations in the volume and environmental composition of the marine rock record that are available to sample track closely two ~300 million year cycles of sea-level rise and fall: Shallow-marine deposition was at a minimum in the late Proterozoic to earliest Cambrian Periods, peaked in the Ordovician, and returned to low levels in the Permian; the second cycle peaked in the Cretaceous and has since undergone a net decline (for some new metrics of available rock record, see Peters & Foote 2001). This first-order sea-level history resembles that of Phanerozoic marine biodiversity at the family level (Sepkoski 1984) except that diversity continues to rise from the Cretaceous to present-day, and new analyses find a close correspondence in stage-by-stage genus diversity (Sepkoski 1997) and available rock (Peters & Foote 2001, Smith 2001), again except for a final rise to present-day levels. The long-standing (Sepkoski et al. 1981) controversy concerns the biological reality of the "excess" diversity in the late Cenozoic. The Paleobiology Database project (http://paleodb.org) began as an effort to develop genus-level diversity data that were standardized to sampling intensity. Data are still insufficient to be conclusive (Alroy et al. 2000, Jackson & Johnson 2001), but the design of the database will ultimately allow a standardized sampling of Phanerozoic diversity, including tests of the effects of extrapolation from particular regions or latitudinal spread.

STUDYING EVOLUTION IN AN INCOMPLETE FOSSIL RECORD

Taphonomy and stratal architecture can undoubtedly distort evolutionary patterns in the fossil record and can even create apparent evolutionary patterns. However, in many cases it is possible to compensate and correct for the imperfections of local sections by tailoring sampling strategies and analytical techniques and by using taxonomic groups suited to the particular question at hand.

Tempo and Mode of Morphologic Change in Species and Lineages

Morphometric data are collected from the fossil record for a variety of evolutionary studies, ranging from the separation of ecophenotypic variation from evolutionary morphologic change between and during speciation events, to the tracking of clade trajectories through a multivariate morphospace (Foote 1997). A sample of morphologies collected from a single bed of fossils can usually be assumed to be time-averaged to some extent and thus will not be equivalent to a census of a living population, but the consequences for morphometric data are not yet clear. In theory, we expect that time averaging will tend to increase variance and even shift the mode relative to a live census owing to the mixing of populations, but under some circumstances, it might decrease variance by the selective destruction of fragile morphs (Kidwell 1986). In contrast, most empirical tests find neither effect (Bell et al. 1987; Cohen 1989; Bush et al. 1999, 2002; Hunt 2001). One possible explanation is that these particular test taxa are morphologically stable from census to census over the temporal and spatial scales captured by time averaging. Alternatively, time-averaged assemblages might be overwhelmingly dominated numerically by specimens added during the most recent interval of time and thus be closer in temporal resolution to a biological sample than previously thought (Kidwell 2002). Of course, for large-scale morphometric studies focused on the average form of a species rather than on intraspecific population-level variation, the effects of time averaging are unimportant or, at least, will generally be nullified by comparing among similarly time-averaged fossil collections, (for criteria to estimate time averaging, see Behrensmeyer & Hook 1992, Kidwell & Flessa 1996).

Local stratigraphic series of intraspecific morphologies reflect changes over time, but this can be interpreted in direct evolutionary terms only if the record is uncomplicated by environmental change over the same interval. In some settings, most notably the shallow marine records that contain the largest numbers of macrobenthic fossils, up-section changes in morphology reveal more about species' responses to environmental change than their evolutionary histories (Bayer & McGhee 1985; Cisne et al. 1980, 1982; Daley 1999; Ludvigsen et al. 1986; McGhee et al. 1991). Similar ambiguity can arise in nonmarine records, even when data from multiple sections have been composited if, for example, all sections are from a single basin and thus share a single climatic or other environmental history. Many reported cases of gradualism in the fossil record (e.g., Sheldon 1987) are compromised by up-section environmental change, which can be subtle for some environments.

Up-section variation in sedimentation rate can also distort the record of evolutionary patterns within single sections. In particular, intervals of low net sedimentation rates (condensed beds), or gaps formed by erosion or complete nondeposition of sediment, tend to increase the perceived rate of morphologic change (MacLeod 1991) and can generate apparent jumps in morphology. Again, the answer is to acquire good sedimentologic and taphonomic information on the context of fossil occurrences in local sections (in order to identify sharp changes or events that might be preservational noise), and then composite data from multiple sections, particularly sections with different histories of sedimentation and environmental change (in order to build a time series with the most useful segments of record). Using multiple sections from more than a single basin can permit geographic variation to be isolated successfully from true evolutionary change (Jackson & Cheetham 1994, Lieberman et al. 1995). It is also possible to play one lineage against another: For example, if one shows saltation in the same section where another is evolving gradually at a constant rate, it is difficult to argue that the saltation is an artifact of a gap (e.g., Fortey 1985).

In contrast to gradual and abrupt up-section changes in morphology, morphologic stasis is not produced by any known stratigraphic or taphonomic process and must represent a true evolutionary pattern. In fact, the presence of stasis in numerous lineages over long intervals of geologic time (Jackson & Cheetham 1999) is all the more remarkable given the pervasiveness of time averaging, changes in sedimentation rates, the presence of stratigraphic gaps, and the ubiquity of changing sedimentary environments in the fossil record, all of which would be expected to be reflected by changes in morphology.

Phylogenetic Analysis

The fossil record is the only direct evidence of the history of life on Earth. As such, fossils have played a variety of roles in phylogenetic analysis. One underlying debate is the extent to which the relative timing of first and last fossil occurrences in the stratigraphic record should be used, given long-standing concerns about the quality of occurrence data, new molecular lines of evidence for determining ancestor-descendent relationships, and widespread adaptation of cladistic methods, which were originally devised to rank relationships on biological evidence alone. At one extreme, investigators have argued that occurrence data are too flawed to play any but a secondary role in phylogenetic reconstruction. At the other extreme, researchers have argued that the fossil record is adequate to play a role equal to morphologic or molecular data (e.g., stratophenetics of Gingerich 1979, stratocladistics of Fisher 1994).

Given the highly variable quality of local and regional records (Figure 1), the feasibility of using fossil occurrence data depends on the taxonomic group and geologic setting of the analysis. For example, where fossil occurrences are based on relatively complete and environmentally homogeneous sections, the succession of closely related taxa probably can be inferred with confidence from the stratigraphic record. This does not eliminate the need to have strong spatial coverage or the need to employ statistical analysis to find the most robust occurrence patterns (e.g., Dowsett 1988). On probabilistic grounds, it has been argued that ancestor-descendant pairs are sufficiently common in the fossil record that phylogenetic methods must be modified to account for their presence or such methods will otherwise produce erroneous results (Foote 1996).

A variety of phylogenetic approaches have been developed to address the history of life as captured by the fossil record. Stratigraphic data can be used to choose among otherwise equivalent phylogenetic trees, such that the tree with the lowest "stratigraphic debt" or implied missing fossil record is the preferred tree. Similarly, stratigraphic debt and morphologic data can be minimized simultaneously, as in the method of stratocladistics (Fisher 1994). Fossils may also be used as tests of phylogenetic estimation through consistency metrics, which compare branching order with appearance in the fossil record, and gap metrics, which quantify the length of nonpreservational gaps implied by a phylogeny (Wagner & Sidor 2000). The stratigraphic consistency index (SCI), perhaps the best known of these metrics, is simply the proportion of phylogenetic nodes that is consistent with the fossil record (Huelsenbeck 1994). Conversely, some have used phylogenies to estimate the incompleteness of the fossil record. Statistical models that assume an incomplete fossil record have been used to test the plausibility of hypothesized ancestor-descendant relationships (Marshall 1995). Many of these models start with the working assumption that stratigraphic gaps follow a Dirichlet or similar distribution; however, the demonstration that gaps commonly do not follow such distributions requires a refinement of these methods. A promising lead is the realization that gaps occur in stratigraphically predictable positions, such that advances in stratigraphic modeling will lead to improved paleontological models of fossil distributions (e.g., Holland 1995, Marshall 1997).

Some recent phylogenetic analyses based on molecular clocks have questioned the veracity of the fossil record in reflecting evolutionary pathways (Heckman et al. 2001, Kumar & Hedges 1998, Murphy et al. 2001, Wray et al. 1996). All of these studies have postulated divergence times between clades that are far in excess of those observed in the fossil record. Few tests of the quality of the fossil record relative to the specific claims of these studies have been made, but in at least one case, investigators have concluded that the quality of the global fossil record would have to be at least an order of magnitude worse in the disputed time interval than in later times for divergence times to be significantly older than the fossil record indicates (Foote et al. 1999). The discrepancy between the fossil record and molecular phylogenies may indicate that older crown group taxa in the fossil record have been unrecognized as such, that older members of crown groups have anomalously low preservation rates, that the early histories of crown groups are hidden in areas with no known fossil record, or that rates of molecular evolution are heterogeneous across groups or vary through a group's history (Foote et al. 1999). All of these are possible solutions to this conflict.

Environmental Analysis of Macroevolutionary Patterns

At the largest temporal and taxonomic scales, evolutionary patterns have been analyzed most commonly at the global scale, in part owing to the compendia of genus- and family-level stratigraphic ranges by Sepkoski (1984, 1993, 1997). However, it is becoming increasingly clear that novel patterns can be detected when global data are decomposed into regional or habitat-specific time series. Analyses at these finer scales are vulnerable to artifacts generated by taphonomic and stratigraphic biases, so care must be taken either to seek patterns that are opposed to the predicted effects of the biases or to control for such biases, such as by using taphonomic control taxa to verify taxon absences. For example, post-Paleozoic sampling is more complete in northern temperate regions than in other regions (e.g., Allison & Briggs 1993b), but by controlling for sampling intensity using an inventory of species records across latitude, Jablonski (1993) found that a disproportionate number of marine invertebrate orders first appeared in the tropics. Other evolutionary patterns that appear to be robust to spatial or environmental biases include (*a*) onshore origination and subsequent offshore expansion of marine invertebrate orders (Sepkoski & Miller 1985, Jablonski & Bottjer 1991, Jablonski et al. 1997) and intercontinental variation, (*b*) the Ordovician marine radiation (Miller 1997, 1998), (*c*) late Permian plant extinctions (Rees 2002), and (*d*) molluscan recovery from the end-Cretaceous mass extinction (Jablonski 1998).

Great caution must be used when extrapolating from such environmental and regional differences to global-scale patterns. For example, Rees (2002) found that earlier global compilations of land-plant diversity were heavily biased toward data from a single paleo-continent, leading to interpretations of worldwide catastrophic die-off of vegetation at the Permian-Triassic boundary when the timing and taxonomic focus of extinction actually varied strongly among regions. In the marine realm, workers are well aware of, but rarely adjust for, a strong collection bias toward North America and Europe, which lay primarily in tropical latitudes in the Paleozoic but were extratropical in the post-Paleozoic. It is possible to compensate for these biases in several ways, including seeking more standardized sampling effort, but also by testing for effects that are disproportionate to known differences in sampling or that are contrary to expected taphonomic and stratigraphic biases.

Temporal Analysis of Macroevolutionary Patterns

One of the most extensive applications of the fossil record to evolutionary biology concerns patterns of origination and extinction at regional and global scales. Given the broad geographic coverage of these studies, they have necessarily adopted coarse, stage-level temporal resolution (i.e., average ∼6 million-year-long time bins) because of the difficulty of high-resolution stratigraphic correlation over long distances. A tacit assumption of these studies is that short-term taphonomic and stratigraphic complications (e.g., within single sections or sets of sections) are eliminated or at least minimized by sampling over broad expanses of time and space. For example, stratigraphic variation in time averaging is not thought to affect Phanerozoic trends in biodiversity. The extent to which such first-order assumptions are true is now being tested in a series of ongoing research projects.

One particular focus of these macroevolutionary studies is the tendency for taxonomic originations and extinctions to occur in brief time intervals. Such clustering may record real evolutionary dynamics, that is, times when life went through genuine bottlenecks or pulses of elevated extinction and diversification. But it is clear that such episodes demand taphonomic and stratigraphic evaluation because similar patterns can be generated by gaps in the record, which truncate the stratigraphic ranges of taxa that actually became extinct or first appeared in the unrecorded interval of time. Last and first appearances are thus artificially concentrated along major discontinuity surfaces. Rapid up-section changes in environment can produce a similar effect. For example, a rapid switch from species-rich-water marine environments to the species-poor deep-water marine environments at the Cenomanian-Turonian stage boundary in Europe has led to overestimating the actual magnitude of mass extinction in this region (Smith et al. 2001). Many other mass extinction events, such as the repeated regional trilobite extinctions within the Cambrian Period known as biomeres, the end-Ordovician extinction, and the end-Permian extinction were accompanied by rapid changes in sedimentary environment and preserved rock volume that have almost certainly exaggerated to varying degrees the taxonomic breadth and coordinated timing of these extinctions (Osleger & Read 1993, Palmer 1965, Saltzman 1999, Peters & Foote 2001, Smith 2001). However, numerical optimization techniques that simultaneously estimate origination, extinction, and preservation rates (Foote 2003) indicate that most of the "Big Five" mass extinctions in the Phanerozoic marine metazoan record (Raup & Sepkoski 1982) still show elevated extinction above background levels and elevated origination during subsequent recovery periods, even after correcting for changes in fossil preservation rates. Taphonomic and stratigraphic evaluation are thus leading to more conservative yet more confident estimates of evolutionary rates. Moreover, the persistence of such pulses in extinction and origination, even after correcting for unevenness in the fossil record, suggests that the relative changes in sea level that drive major changes in the quality of the fossil record may to some degree drive changes in origination and extinction rate. For example, the sea-level changes that produce stratigraphic gaps and surfaces of abrupt environmental change may also alter the area or nature of habitable shallow marine shelves, climatic changes, and shifts in ocean circulation, all of which may increase speciation and extinction rates (Copper 2001, Wignall & Hallam 1992).

In empirical studies, deconvolving the effects of discontinuous sedimentation and changing environmental conditions from true evolutionary dynamics requires (*a*) paying close attention to the bed-level and environmental context of sampled horizons and of up-section changes in these qualities so that one can test for homogeneity in data quality and (*b*) compositing data from a sufficient geographic area to escape any shared similarities among sections in the timing of gaps and in the nature of environmental change. Some workers have used the stereotypic patterns of gaps and environmental change produced by sea-level cycles (Figure 1) to design sampling strategies that can isolate such artifacts from evolutionarily important extinction and biotic invasion events (Patzkowsky & Holland 1997, 1999).

As an example of the difficulty of distinguishing stratigraphic artifact from paleobiological events, a pattern named coordinated stasis has been described from mid-Paleozoic strata of North America based on the persistence over several million years of marine macrobenthic faunas of relatively static ecological structure, taxonomic composition, and fossil morphologies (Brett & Baird 1995). These long blocks of stability are separated by short periods (10,000 to 100,000 years) of ecological restructuring with pulsed origination, extinction, migration, and morphologic change. In the original and subsequent reported examples, the pattern of coordinated stasis has not withstood scrutiny: Turnover is commonly timed with discontinuity surfaces and surfaces of rapid environmental change so that the turnover event may have been much more protracted than a face-value reading of the stratigraphic record would suggest (Baumiller 1996, Goldman et al. 1999, Holland 1996). One study that accounted for taphonomic and stratigraphic effects recognized a much more limited degree of faunal stability (Jackson et al. 1996), and another found no evidence of faunal stability (Patzkowsky & Holland 1997). The "turnover pulse hypothesis" interprets mammalian faunal change in the Cenozoic of Africa in a similar manner (Vrba 1985); subsequent taphonomic and stratigraphic analyses with broader spatial scope have suggested that this too is an artifact of discontinuous stratigraphic accumulation in the original study area (Behrensmeyer et al. 1997).

Although stratigraphic processes are capable of generating artifactual peaks in origination and extinction, limited sampling or the rarity of fossils can have the opposite effect on fossil occurrence data, causing a genuine pulse in origination or extinction to appear as a more gradual pattern in what has been called the Signor-Lipps Effect (Signor & Lipps 1982). When the actual probabilities of finding fossils at any given horizon are taken into account, some apparently gradual extinction records at the Cretaceous-Tertiary boundary are consistent with an extinction pulse that has been degraded by the Signor-Lipps Effect (Marshall & Ward 1996, Sheehan et al. 1991). Rarefaction and related sampling-standardization methods can also be used to correct for the distorting effects of variable sampling size (Miller & Foote 1996, Alroy et al. 2001). Confidence limits on the timing of origination and extinction have been an important recent advance in evaluating the quality of the fossil record (Strauss & Sadler 1989; Marshall 1990, 1997). Early confidence-limit methods made the simplifying assumptions that gaps are randomly distributed and that fossil preservation rates are constant through time, but more recent approaches have replaced these assumptions with more realistic ones, namely, that gaps are nonrandomly distributed and that fossil preservation rates vary through time (Marshall 1997, Holland 2001).

At the broader scale of reconstructing origination and extinction rates over geologic time, statistical methods that evaluate the probability of recovering fossils have allowed paleontologists to evaluate the robustness of observed paleobiological patterns. For example, capture-mark-recapture methods borrowed from field ecology have been used to simultaneously calculate extinction, origination, and preservation probabilities among taxa during the Ordovician radiation (Connolly & Miller 2001). Similarly, numerical optimization techniques have been used to remove the distorting effects of variable preservation probabilities on the record of

Phanerozoic origination and extinction rates (Foote 2001, 2003). These methods explore possible combinations of probabilities of preservation, origination, and extinction through time until a maximal fit to the fossil record is found. Although computationally expensive, they represent the first attempt to address quantitatively the effects of variable preservation on the fossil record of biodiversity and turnover rates.

CONCLUSIONS

The fossil record is highly variable in quality from place to place as well as over time at a single location. This variation—in the preservation potential of major groups, in the gappiness of the sedimentary record, in environments represented, and in the temporal and spatial mixing of fossils—is systematic and thus has potential to bias paleontological data. However, the variation is quantifiable and is increasingly well understood. Thus progress will continue in the development and application of sampling protocols and statistical methods that compensate for these effects.

In devising strategies, it is important to realize that (*a*) no single bias applies to all scales of evolutionary analysis or to all taxonomic groups and (*b*) no scale or type of evolutionary analysis or taxonomic group is completely free of taphonomic and stratigraphic bias. Thus, there is no single "simple fix": The specific problems faced by the paleontologist depend on the desired temporal and spatial scale of study and the taxonomic group at hand. Recovery of high-resolution evolutionary time series is inherently difficult for certain groups and environments (e.g., nonmineralized groups preserved only in lagerstätten, riverine versus lake environments, and shallow-versus deep-marine habitats). Arguably, the easiest strategies are to shift the target of the high-resolution analysis to conducive clades and settings or to change the focus of the analysis to better match the quality of targeted material: Many taphonomic artifacts at the finest scale, for example, the mixing of morphs from multiple generations or habitats into a single bed by time averaging, are less relevant when working at coarser spatial, temporal, and taxonomic scales. Other taphonomic and stratigraphic effects remain relevant at coarser scales, e.g., the lower preservation potential of taxa with smaller, more gracile, or less fully mineralized skeletons, and some stratigraphic biases come more fully to the foreground in coarser analyses, e.g., when encountering large geographically widespread gaps in the sedimentary record, changes in the proportions of environments available to sample, and declining confidence in the age-correlations necessary for accurate binning of data into composite time series. Some of the worst taphonomic and stratigraphic biases that arise in local sections can be compensated for by compositing data from multiple sections over biologically relevant regions, but in the future, our collection of even coarse-scale data needs to be far more sophisticated: We need to demonstrate, rather than assume, that environmental and taphonomic variation and local incompleteness have been compensated for by coarse binning, and this requires that fossil occurrences be scored according to their context in the record, rather than entered free of any encumbering information as to environment or quality of source.

Electronic databasing now permits the requisite level of taphonomic and stratigraphic bookkeeping on data quality that was not practical in the past. Basic taphonomic research in modern environments and in the stratigraphic record needs to continue its focus on ranking and quantifying taxa in terms of their preservation potential, both to better parameterize paleobiologic models and to inform empirical analysis. Comparisons of patterns among taxa or environments with different preservation potential may be another underexploited approach to separating taphonomic bias from biological signal in situations where simpler null hypotheses are not adequate. Thus, over the past twenty years, paleontologists have become much better positioned to deal rigorously with their long-standing concerns about the quality of the fossil record. That record is and will remain incomplete, but ongoing advances in taphonomy and stratigraphy—and of course continued work to increase the known record, especially in regions undersampled so far–will ensure that it continues to be a valuable source of insight into the evolution of life on Earth.

ACKNOWLEDGMENTS

We are grateful to D. Jablonski, M. Foote, and A.I. Miller for helpful reviews. We thank M. Foote for suggesting the title for the latter half of the paper.

The *Annual Review of Ecology and Systematics* is online at
http://ecolsys.annualreviews.org

LITERATURE CITED

Allison PA, Briggs DEG, eds. 1991. *Taphonomy: Releasing the Data Locked in the Fossil Record*. New York: Plenum

Allison PA, Briggs DEG. 1993a. Exceptional fossil record: distribution of soft-tissue preservation through the Phanerozoic. *Geology* 21:527–30

Allison PA, Briggs DEG. 1993b. Paleolatitudinal sampling bias, Phanerozoic species diversity, and the end-Permian extinction. *Geology* 21:65–68

Alroy J, Marshall CR, Bambach RK, Bezusko K, Foote M, et al. 2001. Effects of sampling standardization on estimates of Phanerozoic marine diversification. *Proc. Natl. Acad. Sci. USA* 98:6261–66

Anders MH, Krueger SW, Sadler PM. 1987. A new look at sedimentation rates and the completeness of the stratigraphic record. *J. Geol.* 95:1–14

Anderson LC, Sen Gupta BK, McBride RA, Byrnes MR. 1997. Reduced seasonality of Holocene climate and pervasive mixing of Holocene marine section: northeastern Gulf of Mexico shelf. *Geology* 25:127–30

Bada JL, Wang XS, Hamilton H. 1999. Preservation of key biomolecules in the fossil record: current knowledge and future challenges. *Philos. Trans. R. Soc. London Ser. B* 354:77–87

Baumiller TK. 1996. Exploring the pattern of coordinated stasis: simulations and extinction scenarios. *Palaeogeogr. Palaeoclimatol. Palaeoecol.* 127:135–46

Bayer U, McGhee GR. 1985. Evolution in marginal epicontinental basins: the role of phylogenetic and ecologic factors (Ammonite replacements in the German Lower and Middle Jurassic). In *Sedimentary and Evolutionary Cycles*, ed. U Bayer, A Seilacher, pp. 164–220. New York: Springer-Verlag

Behrensmeyer AK. 1999. Bonebeds through geologic time. *J. Vertebr. Paleontol.* 19 (Suppl. to 3):31–32A

Behrensmeyer AK, Chapman RE. 1993. Models and simulations of time-averaging in terrestrial vertebrate accumulations. See Kidwell & Behrensmeyer 1993, pp. 125–49

Behrensmeyer AK, Dechant Boaz DE. 1980. The recent bones of Amboseli Park, Kenya, in relation to East African paleoecology. In *Fossils in the Making*, ed. AK Behrensmeyer, AP Hill, pp. 72–92. Chicago, IL: Univ. Chicago Press

Behrensmeyer AK, Hook RW. 1992. Paleoenvironmental contexts and taphonomic modes in the terrestrial fossil record. In *Terrestrial Ecosystems Through Time*, ed. AK Behrensmeyer, JD Damuth, WA DiMichele, R Potts, H-D Sues, SL Wing, pp. 15–136. Chicago, IL: Univ. Chicago Press

Behrensmeyer AK, Kidwell SM, Gastaldo RA. 2000. Taphonomy and paleobiology. See Erwin & Wing 2000, pp. 103–47

Behrensmeyer AK, Todd NE, Potts R, McBrinn GE. 1997. Late Pliocene faunal turnover in the Turkana Basin, Kenya and Ethiopia. *Science* 278:1589–94

Bell MA, Sadagursky MS, Baumgartner JV. 1987. Utility of lacustrine deposits for the study of variation within fossil samples. *Palaios* 2:455–66

Berggren WA, Kent DV, Swisher III CC, Aubry MP. 1995. A revised Cenozoic geochronology and chronostratigraphy. In *Geochronology, Time Scales and Global Stratigraphic Correlation*, ed. WA Berggren, DV Kent, MP Aubry, J Hardenbol, pp. 129–212. Tulsa, OK: SEPM Spec. Publ. No. 54

Bowring SA, Erwin DH. 1998. A new look at evolutionary rates in deep time: uniting paleontology and high-precision geochronology. *Geol. Soc. Am. Today* 8:1–8

Bowring SA, Grotzinger JP, Isachsen CE, Knoll AH, Pelechaty SM, Kolosov P. 1993. Calibrating rates of early Cambrian evolution. *Science* 261:1293–98

Brett CE. 1995. Sequence stratigraphy, biostratigraphy, and taphonomy in shallow marine environments. *Palaios* 10:597–616

Brett CE, Baird GC. 1995. Coordinated stasis and evolutionary ecology of Silurian to Middle Devonian faunas in the Appalachian Basin. See Erwin & Anstey 1995, pp. 285–315

Brewster-Wingard GL, Stone JR, Holmes CW. 2001. Molluscan faunal distribution in Florida Bay, past and present: an integration of down-core and modern data. *Bull. Am. Paleontol.* 361:199–231

Briggs DEG, Crowther PR, eds. 2001. *Paleobiology II*. Oxford: Blackwell Sci. 583 pp.

Bush AM, Powell MG, Arnold WS, Bert TM, Daley GM. 2002. Time-averaging, evolution, and morphologic variation. *Paleobiology* 28:9–250

Chapman MR. 2000. The response of planktonic foraminifera to the Late Pliocene intensification of Northern Hemisphere glaciation. See Culver & Rawson, pp. 79–96

Cisne JL, Chandlee GO, Rabe BD, Cohen JA. 1982. Clinal variation, episodic evolution, and possible parapatric speciation: the trilobite *Flexicalymene senaria* along an Ordovician depth gradient. *Lethaia* 15:325–41

Cisne JL, Molenock J, Rabe BD. 1980. Evolution in a cline: the trilobite *Triarthrus* along an Ordovician depth gradient. *Lethaia* 13:47–59

Cohen AS. 1989. The taphonomy of gastropod shell accumulations in large lakes: an example from Lake Tanganyika, Africa. *Paleobiology* 15:26–45

Connolly SR, Miller AI. 2001. Joint estimation of sampling and turnover rates from fossil databases: capture-mark-recapture methods revisited. *Paleobiology* 27:751–67

Cooper A, Poinar HN. 2000. Ancient DNA: do it right or not at all. *Science* 289:1139

Copper P. 2001. Reefs during the multiple crises towards the Ordovician-Silurian boundary: Anticosti Island, eastern Canada, and worldwide. *Can. J. Earth Sci.* 38:153–71

Culver SJ, Rawson PF. 2000. *Biotic Response to Global Change, the Last 145 Million Years.* 501 pages. Cambridge, UK: Cambridge Univ. Press

Daley GM. 1999. Environmentally controlled variation in shell size of *Ambonychia* Hall (Mollusca: Bivalvia) in the type Cincinnatian (Upper Ordovician). *Palaios* 14:520–29

Davis MB. 2000. Palynology after Y2K—understanding the source area of pollen in sediments. *Annu. Rev. Ecol. Syst.* 28:1–18

Donovan SK, Paul CRC. 1998. *The Adequacy of the Fossil Record.* Chichester: Wiley & Sons. 312 pp.

Dowsett HJ. 1988. Diachrony of late Neogene microfossils in the southwest Pacific Ocean: application off the graphic correlation technique. *Paleoceanography* 3:209–22

Erwin DH, Anstey RL, eds. 1995. *New Approaches to Speciation in the Fossil Record.* New York: Columbia Univ. Press

Erwin DH, Wing SL, eds. 2000. *Deep Time: Paleobiology's Perspective.* Lawrence, KS: Paleontol. Soc.

Fisher DC. 1994. Stratocladistics: morphological and temporal patterns and their relation to phylogenetic process. In *Interpreting the Hierarchy of Nature*, ed. L Grande, O Rieppel, pp. 133–71. San Diego, CA: Academic

Flessa KW. 1998. Well-traveled cockles: shell transport during the Holocene transgression of the southern North Sea. *Geology* 26:187–90

Flessa KW, Kowalewski M. 1994. Shell survival and time-averaging in nearshore and shelf environments: estimates from the radiocarbon literature. *Lethaia* 27:153–65

Foote M. 1996. On the probability of ancestors in the fossil record. *Paleobiology* 22:141–51

Foote M. 1997. The evolution of morphological diversity. *Annu. Rev. Ecol. Syst.* 28:129–52

Foote M. 2001. Inferring temporal patterns of preservation, origination, and extinction from taxonomic survivorship analysis. *Paleobiology* 27:602–30

Foote M. 2003. Origination and extinction through the Phanerozoic: a new approach. *J. Geol.* 111: In press

Foote M, Hunter JP, Janis CM, Sepkoski JJ Jr. 1999. Evolutionary and preservational constraints on origins of biologic groups: divergence times of eutherian mammals. *Science* 283:1310–14

Fortey RA. 1985. Gradualism and punctuated equilibria as competing and complementary series. *Spec. Pap. Palaeontol.* 33:17–28

Fürsich FT. 1978. The influence of faunal condensation and mixing on the preservation of fossil benthic communities. *Lethaia* 11:243–50

Gingerich PD. 1979. The stratophenetic approach to phylogeny reconstruction in vertebrate paleontology. In *Phylogenetic Analysis and Paleontology*, ed. J Cracraft, N Eldredge, pp. 41–77. New York: Columbia Univ. Press

Goldman D, Mitchell CE, Joy MP. 1999. The stratigraphic distribution of graptolites in the classic upper Middle Ordovician Utica Shale of New York State: an evolutionary succession or a response to relative sea-level change? *Paleobiology* 25:273–94

Graham RW. 1993. Processes of time-averaging in the terrestrial vertebrate record. See Kidwell & Behrensmeyer 1993, pp. 102–24

Heckman DS, Geiser DM, Eidell BR, Stauffer RL, Kardos NL, Hedges SB. 2001. Molecular evidence for the early colonization of land by fungi and plants. *Science* 293:1129–33

Hinnov LA. 2000. New perspectives on orbitally forced stratigraphy. *Annu. Rev. Earth Planet. Sci.* 28:419–75

Holland SM. 1995. The stratigraphic distribution of fossils. *Paleobiology* 21:92–109

Holland SM. 1996. Recognizing artifactually generated coordinated stasis: implications of numerical models and strategies for field tests. *Palaeogeogr. Palaeoclimatol. Palaeoecol.* 127:147–56

Holland SM. 2000. The quality of the fossil

record—a sequence stratigraphic perspective. See Erwin & Wing 2000, pp. 148–68

Holland SM. 2001. Confidence limits on fossil ranges that account for sequence architecture. *Geol. Soc. Am. Abstr. Program* 33:A31

Holland SM, Patzkowsky ME. 1999. Models for simulating the fossil record. *Geology* 27:491–94

Holland SM, Patzkowsky ME. 2002. Stratigraphic variation in the timing of first and last occurrences. *Palaios.* 17:134–46

Huelsenbeck JP. 1994. Comparing the stratigraphic record to estimates of phylogeny. *Paleobiology* 20:470–83

Hunt G. 2001. Time averaging and morphometric data: Do fossil samples accurately reflect population-level variability? *Paleobios* 21:69

Jablonski D. 1993. The tropics as a source of evolutionary novelty: the post-Palaeozoic fossil record of marine invertebrates. *Nature* 364:142–44

Jablonski D. 1998. Geographic variation in the molluscan recovery from the end-Cretaceous extinction. *Science* 279:1327–30

Jablonski D. 2000. Micro- and macroevolution: scale and hierarchy in evolutionary biology and paleobiology. See Erwin & Wing 2000, p. 15–52

Jablonski D, Bottjer DJ. 1991. Environmental patterns in the origins of higher taxa: the post-Paleozoic fossil record. *Science* 252:1831–33

Jablonski D, Lidgard S, Taylor PD. 1997. Comparative ecology of bryozoan radiations: origin of novelties in cyclostomes and cheilostomes. *Palaios* 12:505–23

Jablonski D, Roy K, Valentine JW, Price RM, Anderson PS. 2002. Pull of the Recent? What pull of the Recent? An analysis of the marine Bivalvia. *Abstr. Progr. Geol. Soc. Amer.* 34(7): In press

Jackson JBC, Cheetham AH. 1994. Phylogeny reconstruction and the tempo of speciation in cheilostome Bryozoa. *Paleobiology* 20:407–23

Jackson JBC, Budd AF, Pandolfi JM. 1996. The shifting balance of natural communities? In *Evolutionary Paleobiology*, ed. D Jablonski, DH Erwin, JH Lipps, pp. 89–122. Chicago, IL: Univ. Chicago Press

Jackson JBC, Cheetham AH. 1999. Tempo and mode of speciation in the sea. *Trends Ecol. Evol.* 14:72–77

Kidwell SM. 1986. Models for fossil concentrations: Paleobiologic implications. *Paleobiology* 12:6–24

Kidwell SM. 1991. The stratigraphy of shell concentrations. See Allison & Briggs 1991, pp. 211–90

Kidwell SM. 1993. Taphonomic expressions of sedimentary hiatus: field observations on bioclastic concentrations and sequence anatomy in low, moderate and high subsidence settings. *Geol. Rundsch.* 82:189–202

Kidwell SM. 2002. Time-averaged molluscan death assemblages: palimpsests of richness, snapshots of abundance. *Geology* 30:803–6

Kidwell SM, Behrensmeyer AK, eds. 1993. *Taphonomic Approaches to Time Resolution in Fossil Assemblages*. Knoxville, TN: Paleontol. Soc. Short Courses Paleontol

Kidwell SM, Best MMR. 2001. Tropical time-averaging: disparate scales and taphonomic clocks in bivalve assemblages from modern subtidal siliciclastic and carbonate facies. North Am. Paleontol. Conv. 2001, Program Abstr. *PaleoBios* 21:79

Kidwell S, Bosence DWJ. 1991. Taphonomy and time-averaging of marine shelly faunas. See Allison & Briggs 1991, pp. 115–209

Kidwell SM, Brenchley PJ. 1994. Patterns in bioclastic accumulation through the Phanerozoic: changes in input or in destruction? *Geology* 22:1139–43

Kidwell SM, Flessa KW. 1996. The quality of the fossil record: populations, species, and communities. *Annu. Rev. Earth. Planet. Sci.* 24:433–64

Kowalewski M. 1996. Taphonomy of a living fossil: the lingulide brachiopod *Glottidia palmeri* Dall from Baja California, Mexico. *Palaios* 11:244–65

Kowalewski M, Bambach RK. 2002. The limits of paleontological resolution. In *High*

Resolution Approaches in Paleontology, ed. PJ Harries, DH Geary. New York: Kluwer/ Plenum. In press

Kucera M, Malmgren BA, 1998. Differences between evolution of mean form and evolution of new morphotypes; an example from Late Cretaceous planktonic Foraminifera. *Paleobiology* 24:49–63

Kumar S, Hedges SB. 1998. A molecular timescale for vertebrate evolution. *Nature* 392:917–20

Lieberman BS, Brett CE, Eldredge N. 1995. A study of stasis and change in two species lineages from the Middle Devonian of New York state. *Paleobiology* 21:15–27

Ludvigsen R, Westrop SR, Pratt BR, Tuffnell PA, Young GA. 1986. Dual biostratigraphy: zones and biofacies. *Geosci. Can.* 13:139–54

MacLeod N. 1991. Punctuated anagenesis and the importance of stratigraphy to paleobiology. *Paleobiology* 17:167–88

MacLeod N, Keller G. 1991. Hiatus distributions and mass extinctions at the Cretaceous/ Tertiary boundary. *Geology* 19:497–501

MacLeod N, Ortiz N, Fefferman N, Clyde W, Schulter C, MacLean J. 2000. Phenotypic response of foraminifera to episodes of global environmental change. See Culver & Rawson 2000, pp. 51–78

Marshall CR. 1990. Confidence intervals on stratigraphic ranges. *Paleobiology* 16:1–10

Marshall CR. 1995. Stratigraphy, the true order of species originations and extinctions, and testing ancestor-descendant hypotheses among Caribbean Neogene bryozoans. See Erwin & Anstey 1995, pp. 208–35

Marshall CR. 1997. Confidence intervals on stratigraphic ranges with nonrandom distributions of fossil horizons. *Paleobiology* 23:165–73

Marshall CR, Ward PD. 1996. Sudden and gradual molluscan extinctions in the latest Cretaceous of Western European Tethys. *Science* 274:1360–63

Martin RE, Wehmiller JF, Harris MS, Liddell WD. 1996. Comparative taphonomy of foraminifera and bivalves in Holocene shallowwater carbonate and siliciclastic regimes: taphonomic grades and temporal resolution. *Paleobiology* 22:80–90

McGhee GR Jr, Bayer U, Seilacher A. 1991. Biological and evolutionary responses to transgressive-regressive cycles. In *Cycles and Events in Stratigraphy*, ed. W Ricken, A Seilacher, pp. 696–708. Berlin: Springer-Verlag

McKinney ML. 1986. How biostratigraphic gaps form. *J. Geol.* 94:875–84

McKinney ML. 1991. Completeness of the fossil record: an overview. In *The Processes of Fossilization*, ed. SK Donovan, pp. 66–83. New York: Columbia Univ. Press

Meldahl KH, Flessa KW, Cutler AH. 1997. Time-averaging and postmortem skeletal survival in benthic fossil assemblages: quantitative comparisons among Holocene environments. *Paleobiology* 23:209–29

Miller AI. 1997. Dissecting global diversity patterns: examples from the Ordovician Radiation. *Annu. Rev. Ecol. Syst.* 28:85–104

Miller AI. 1998. Biotic transitions in global marine diversity. *Science* 281:1157–60

Miller AI, Cummins H. 1990. A numerical model for the formation of fossil assemblages: estimating the amount of post- mortem transport along environmental gradient. *Palaios* 5:303–16

Miller AI, Foote M. 1996. Calibrating the Ordovician radiation of marine life: implications for Phanerozoic diversity trends. *Paleobiology* 22:304–9

Murphy WJ, Eizirik E, O'Brien SJ, Madsen O, Scally M, et al. 2001. Resolution of the early placental mammal radiation using Bayesian phylogenetics. *Science* 294:2348–51

Norris RD. 1991. Biased extinction and evolutionary trends. *Paleobiology* 17:388–99

Olszewski TD. 1999. Taking advantage of time-averaging. *Paleobiology* 25:226–38

Osleger D, Read JF. 1993. Comparative analysis of methods used to define eustatic variations in outcrop: late Cambrian interbasinal sequence development. *Am. J. Sci.* 293:157–216

Palmer AR. 1965. Biomere—a new kind of biostratigraphic unit. *J. Paleontol.* 39:149–53

Patzkowsky ME, Holland SM. 1997. Patterns of turnover in Middle and Upper Ordovician brachiopods of the eastern United States: a test of coordinated stasis. *Paleobiology* 23: 420–43

Patzkowsky ME, Holland SM. 1999. Biofacies replacement in a sequence stratigraphic framework: Middle and Upper Ordovician of the Nashville Dome, Tennessee, USA. *Palaios* 14:301–23

Paul CRC. 1982. The adequacy of the fossil record. In *Problems of Phylogenetic Reconstruction*, ed. KA Joysey, AE Friday, pp. 75–117. New York: Academic

Peters SE, Foote M. 2001. Biodiversity in the Phanerozoic: a reinterpretation. *Paleobiology* 27:583–601

Peterson CH. 1977. The paleoecological significance of undetected short-term temporal variability. *J. Paleontol.* 51:976–81

Powell MG, Kowalewski M. 2002. Increase in evenness and sampled alpha diversity through the Phanerozoic: comparison of early Paleozoic and Cenozoic marine fossil assumblages. *Geology* 30:331–34

Raup DM. 1979. Biases in the fossil record of species and genera. *Bull. Carnegie Mus. Nat. Hist.* 13:85–91

Raup DM, Sepkoski JJ Jr. 1982. Mass extinctions in the marine fossil record. *Science* 215: 1501–2

Rees PM. 2002. Land-plant diversity and the end-Permian mass extinction. *Geology* 30: 827–30

Rogers RR, Kidwell SM. 2000. Associations of vertebrate skeletal concentrations and discontinuity surfaces in continental and shallow marine records: a test in the Cretaceous of Montana. *J. Geol.* 108:131–54

Roy K, Valentine JW, Jablonski D, Kidwell S. 1996. Scales of climatic variability and time averaging in Pleistocene biotas: implications for ecology and evolution. *Trends Ecol. Evol.* 11:458–63

Sadler PM. 1981. Sediment accumulation rates and the completeness of stratigraphic sections. *J. Geol.* 89:569–84

Saltzman MR. 1999. Upper Cambrian carbonate platform evolution, *Elvinia* and *Taenicephalus* Zones (Pterocephaliid-Ptychaspid biomere boundary), northwestern Wyoming. *J. Sediment. Res.* 69:926–38

Schopf TJM. 1978. Fossilization potential of an intertidal fauna: Friday Harbor, Washington. *Paleobiology* 4:261–70

Sepkoski JJ Jr. 1984. A kinetic-model of Phanerozoic taxonomic diversity. III. Post-Paleozoic families and mass extinctions. *Paleobiology* 10:246–67

Sepkoski JJ Jr. 1993. Ten years in the library: new data confirm paleontological patterns. *Paleobiology* 19:43–51

Sepkoski JJ Jr. 1997. Biodiversity: past, present, and future. *J. Paleontol.* 71:533–39

Sepkoski JJ Jr, Bambach RK, Raup DM, Valentine JW. 1981. Phanerozoic marine diversity and the fossil record. *Nature* 293:435–37

Sepkoski JJ Jr, Miller AI. 1985. Evolutionary faunas and the distribution of Paleozoic benthic communities in space and time. In *Phanerozoic Diversity Patterns*, ed JW Valentine, pp. 153–90. Princeton, NJ: Princeton Univ. Press

Sheehan PM, Fastovsky DE, Hoffman RG, Berghaus CB, Gabriel DL. 1991. Sudden extinction of the dinosaurs: latest Cretaceous, Upper Great Plains, U.S.A. *Science* 254:835–39

Sheldon PR. 1987. Parallel gradualistic evolution of Ordovician trilobites. *Nature* 330: 561–63

Signor PW, Lipps JH. 1982. Sampling bias, gradual extinction patterns, and catastrophes in the fossil record. *Geol. Soc. Am. Spec. Pap.* 190:291–96

Smith AB. 2001. Large-scale heterogeneity of the fossil record: implications for Phanerozoic biodiversity studies. *Philos. Trans. R. Soc. London Ser. B* 356:351–67

Smith AB, Gale AS, Monks NEA. 2001. Sea-level change and rock record bias in the Cretaceous: a problem for extinction and biodiversity studies. *Paleobiology* 27:241–53

Spencer-Cervato C, Thierstein HR, Lazarus DB, Beckmann JP. 1994. How synchronous are Neogene marine plankton events? *Paleoceanography* 9:739–63

Strauss D, Sadler PM. 1989. Classical confidence intervals and the Bayesian probability estimates for the ends of local taxon ranges. *Math. Geol.* 21:411–27

Taylor PD, Allison P. 1999. Bryozoan carbonates through time and space. *Geology* 26:459–62

Valentine JW. 1989. How good was the fossil record? Clues from the Californian Pleistocene. *Paleobiology* 15:83–94

Vrba E. 1985. Environment and evolution: alternative causes of the temporal distribution of evolutionary events. *S. Afr. J. Sci.* 81:229–36

Wagner PJ, Sidor CA. 2000. Age rank/clade rank metrics—sampling, taxonomy, and the meaning of "stratigraphic consistency." *Syst. Biol.* 49:463–79

Walker KR, Bambach RK. 1971. The significance of fossil assemblages from fine-grained sediments: time-averaged communities. *Geol. Soc. Am. Abstr. Programs* 3:783–84

Webb T III. 1993. Constructing the past from late Quaternary pollen data: temporal resolution and a zoom lens space-time perspective. See Kidwell & Behrensmeyer 1993, pp. 79–101

Wignall PB, Hallam A. 1992. Anoxia as a cause of the Permian/Triassic extinction: facies evidence from northern Italy and the western United States. *Palaeogeogr. Palaeoclimatol. Palaeoecol.* 93:21–46

Wray GA, Levinton JS, Shapiro LH. 1996. Molecular evidence for deep Precambrian divergences among metazoan phyla. *Science* 274:568–73

Annu. Rev. Ecol. Syst. 2002. 33:589–639
doi: 10.1146/annurev.ecolsys.33.010802.150437
Copyright © 2002 by Annual Reviews. All rights reserved
First published online as a Review in Advance on August 14, 2002

NEOPOLYPLOIDY IN FLOWERING PLANTS

Justin Ramsey[1,2] and Douglas W. Schemske[3]

[1]Department of Botany, Box 355325, University of Washington, Seattle,
Washington 98195-5325; email: jramsey@u.washington.edu
[2]Present address: Department of Botany, University of Guelph, Guelph,
Ontario N1G 2W1, Canada; email: jramsey@uoguelph.ca
[3]Department of Plant Biology and Kellogg Biological Station, Michigan State
University, East Lansing, Michigan 48824-1312; email: schem@msu.edu

Key Words adaptation, aneuploidy, cytogenetics, polyploidy, speciation

■ **Abstract** Here we review the biology of early generation neopolyploids and discuss the profound changes that accompany their formation. Newly formed auto- and allopolyploids exhibit considerable meiotic complexity, including multivalent pairing, multisomic inheritance, and the production of unbalanced gametes. The cytogenetic behavior of allopolyploids and autopolyploids differ statistically, but are more similar than commonly believed. The progeny of neopolyploids include a high frequency of aneuploids, pseudoeuploids and homeologue-recombinant genotypes that may contribute to the phenotypic variability observed in early generation polyploids. We find no evidence to support the traditional view that autopolyploids possess lower fertility than allopolyploids, casting doubt on the paradigm that allopolyploids should be more frequent due to their inherent fertility. The fertility of early generation polyploids increases rapidly, owing largely to selection against meiotic configurations that generate unbalanced gametes. Neopolyploids are commonly differentiated from progenitors by a combination of morphological, phenological and life-history characteristics. Further progress toward understanding polyploid evolution will require studies in natural populations that can evaluate the demographic and larger ecological significance of the cytogenetic and phenotypic character of neopolyploids.

INTRODUCTION

Polyploidy, the genome-wide multiplication of chromosome number, is a key feature in plant evolution. It is estimated that between 47 and 70% of flowering plants are the descendants of polyploid ancestors (Masterson 1994). Differences in ploidy are commonly observed among closely related plant species and among populations within species (Lewis 1980a), and recent molecular studies have revealed that polyploid taxa often have multiple origins (Soltis & Soltis 1993, 1999). These observations demonstrate that polyploidy in plants is a dynamic process. Polyploids generally differ markedly from their progenitors in morphological, ecological, physiological and cytological characteristics (Levin 1983, 2002; Lumaret 1988)

0066-4162/02/1215-0589$14.00

that can contribute both to exploitation of a new niche and to reproductive isolation. Thus, polyploidy is a major mechanism of adaptation and speciation in plants (Clausen et al. 1945, Stebbins 1950, Grant 1981, Otto & Whitton 2000, Levin 2002).

In spite of the importance of polyploidy, the factors contributing to polyploid evolution are poorly understood (Thompson & Lumaret 1992). There are two early stages of polyploid evolution: formation of new cytotypes and their demographic establishment. To understand the process of polyploid formation requires information on the pathways, cytological mechanisms, and rates of polyploid formation. To assess the likelihood that a new polyploid will successfully establish requires information on the viability and fertility of new cytotypes, as well as their phenotypic characteristics and fitness in different environments. A review of polyploid formation is provided in a companion *Annual Review of Ecology and Systematics* chapter (Ramsey & Schemske 1998). Here we review the literature regarding newly formed polyploids to answer the following questions: 1. What are the cytogenetic characteristics of neopolyploids, and how do these relate to the viability, fertility, and stability of polyploids? 2. What are the phenotypic consequences of polyploidy, and by what genetic means are they induced?

INFERENCE IN POLYPLOID RESEARCH

Despite an enormous literature concerning the biological characteristics of polyploids and their progenitors, most investigations compare naturally occuring established cytotypes. This approach may confound phenotypic differences attributable to ploidy per se with those that result from evolution since the time of polyploid formation (Bretagnolle & Lumaret 1995, De Kovel & De Jong 2000). For example, Smith (1946) documented substantial differences in the morphology, size, flowering phenology, and drought tolerance of diploid, tetraploid, and hexaploid "races" of *Sedum pulchellum*. Without comparative information from closely related homoploid taxa, we have no way of ascribing the divergence observed among cytotypes to polyploidy versus genic differentiation via natural selection, genetic drift, interspecific hybridization, or other mechanisms. It is also common practice to compare the geographic distribution of different cytotypes in an effort to identify the ecological consequences of polyploidy (Lewis 1980b). For example, Mosquin (1966) showed that diploids, tetraploids, and hexaploids of *Epilobium angustifolium* occupy very different geographic regions, an observation consistent with the hypothesis that polyploidy promotes ecological diversification. Yet, in most plants, profound ecological and geographic differentiation is perhaps just as frequent without changes in ploidy (Clausen et al. 1940).

This chapter is focused on the origins and demographic establishment of polyploids, phenomenon dependent on the characteristics of polyploids at the time of origin. One approach that minimizes the confounding effects of postformation evolution involves the comparison of diploid progenitors with newly formed polyploids (neopolyploids). For example, Müntzing (1951) induced autotetraploids in three varieties of rye and directly compared the growth, tillering, phenology, yield, and baking properties of tetraploids and progenitor diploids. In this way, the biological attributes of the early generation polyploids can be compared directly to those of their progenitor cytotypes. As a rule, we include only neopolyploids in our analyses here, though the characteristics of naturally established and cultivated polyploids are sometimes discussed.

In addition to elucidating the process of polyploid establishment, studies of neopolyploids provide insights into the nature of polyploidy as an adaptation. The difficulty of determining the role of polyploidy per se in creating cytotype differences hampers investigations of naturally occurring polyploids: What fraction of the phenotypic differences between related cytotypes are a direct consequence of polyploidy rather than incidental differentiation accrued via genic differentiation in allopatry? Three hypotheses may be posed to explain observed cytotype adaptation in natural populations. First, phenotypic differentiation may be driven by polyploidy per se, through the combined effects of increased cell size, gene dosage effects, allelic diversity and other mechanisms (Levin 1983, 2002; Lumaret 1988). Second, postzygotic reproductive isolation of cytotypes may facilitate ecotypic differentiation, either through reinforcement (Petit et al. 1999) or liberation from gene flow that may slow the process of local adaptation and range expansion (Kirkpatrick & Barton 1997). Finally, cytotype differences may simply represent allelic differences accumulated via conventional evolutionary processes that operate in both diploid and polyploid populations. Clearly, comparisons of neopolyploids and established cytotypes in natural polyploid complexes will play an important role in evaluating the role of polyploidy in adaptation and speciation. Where relevant, we discuss the implications of our results for both demographic establishment and the larger evolutionary significance of polyploidy.

APPROACH AND TERMINOLOGY

A major motivation for this review is to synthesize the diverse literature on early generation polyploids and thereby provide a resource for the development of future research on polyploid establishment. To this end, we have tabulated data from ~250 published studies and made this information available on the Annual Reviews web site (http://www.annualreviews.org; see Supplementary Materials). We summarize these data throughout the text and identify the location of each database on the web site. Textual citations generally include only the most comprehensive studies.

By necessity, most of the plants considered in this review are agricultural or horticultural cultivars, as well as classic genetic systems (e.g., *Datura* and *Nicotiana*). These studies provide insights into neopolyploids in natural populations, but we caution that further investigations are needed to test our results. First, polyploids generated in domesticated, highly modified and inbred cultivars may not always be representative of neopolyploids in natural populations. Second, many surveyed polyploids were induced by experimental treatments such as heat shock and colchicine. These methods provide a practical advantage in allowing researchers to generate large numbers of neopolyploids from diverse plant material, but also induce occasional chromosomal rearrangements, chromosome substitutions, aneuploidy and, in some cases, apparent genic mutations (Randolph 1932, Bergner et al. 1940, Smith 1943, Sanders & Franzke 1964, Salanki & Parameswarappa 1968). Induced neopolyploids, which are formed by somatic doubling, may also be less heterozygous than spontaneous neopolyploids, which are hypothesized to arise primarily via unreduced gametes (see Bretagnolle & Thompson 1995). The origin of all surveyed polyploids is documented in Web Tables. Finally, naturally occurring polyploids are unknown in some of the taxa included here. Inasmuch as some species (and species hybrids) may be predisposed to generate demographically successful neopolyploids, these data represent a random sample of newly formed polyploids rather than a sample of successful polyploids. As discussed below, there is a general need for studies of neopolyploidy in natural polyploid complexes.

Our surveys include a wide range of plant families, genera, and species, but because of limited sample size we do not interpret our results in a phylogenetic framework. As a general rule, data collected from related congeneric and conspecific varieties were as variable as those observed in distant relatives. Hence, data from related congeners, hybrid combinations, and different strains or populations of a single species are treated as independent. Most datasets incorporate data from numerous taxa, including one to several species or accessions per genus. Complete datasets are online in Supplementary Materials.

In this chapter, $2n$ refers to the somatic chromosome number and n to the gametic chromosome number regardless of the degree of polyploidy, while x is the most probable base number. This gives the following cytological designations: diploids ($2n = 2x$), triploids ($2n = 3x$), tetraploids ($2n = 4x$), etc. As summarized in Ramsey & Schemske (1998), the terms autopolyploid and allopolyploid have a complex history. We believe that the primary criterion for classifying a polyploid is its mode of origin. In this chapter we use the term autopolyploid to denote a polyploid arising within or between populations of a single species, and allopolyploid to indicate polyploids derived from hybrids between species, where species are defined according to their degree of pre- and post-zygotic reproductive isolation (biological species concept). Alternate definitions of polyploid types will be discussed in later sections.

MEIOTIC BEHAVIOR OF NEOPOLYPLOIDS

Chromosomes that are structurally similar and pair normally at meiosis are termed homologous. Parental diploid species have two sets of homologous chromosomes, and an autotetraploid has four sets. Allopolyploids possess chromosome complements from two or more evolutionary lineages. These chromosomes are differentiated to some degree by (*a*) DNA sequence (e.g., allelic differences caused by nucleotide substitutions or indels); (*b*) structure and gene order (e.g., chromosomal rearrangements); and/or (*c*) total number. Partially homologous chromosomes are called homeologous. For example, the chromosomes of the diploid species *Collinsia concolor* and *C. sparsiflora* ($x = 7$) are distinguished by reciprocal translocations and paracentric inversions as well as numerous gene differences (Garber & Gorsic 1956, Garber & Dhillon 1962). The diploid F1 hybrid has two sets of chromosomes that are homeologous, and can be assigned the genomic formula CS; at meiosis, the F1 exhibits interchange complexes, dicentric bridges and fragments, and a high frequency of unpaired chromosomes (univalents). Allotetraploid *Collinsia concolor* × *sparsiflora* (CCSS) has both homologous (C-C, S-S) and homeologous (C-S) chromosomes. At meiosis the allotetraploid exhibits no complexes, bridges or fragments, and few univalents.

At a basic level, meiosis can be understood as three sequential processes: chromosome pairing (synapsis), crossing over (chiasma formation), and chromosome distribution (Sybenga 1975; Singh 1993). Pairing occurs during prophase I. In polyploids, associations between homologous or homeologous chromosomes are determined at synapsis; hence, chromosome pairing determines broad patterns of genetic recombination and chromosome distribution at anaphase. At metaphase I, synapsed homologues and homeologues usually cross over and form chiasmata, exchanges of chromatid segments, leading to genetic recombination. The frequency and distribution of chiasmata determine the configurations of chromosomes during metaphase and anaphase. Finally, pairing partners, whether homologous or homeologous, are distributed to opposite poles of mother cells during anaphase I.

Autopolyploids possess only homologous chromosomes, while allotetraploids possess two or more sets of homeologous chromosomes. This has led to the prediction that meiosis and inheritance should differ between the two types of polyploids (Müntzing 1932, 1936; Winge 1932; Darlington 1937; Stebbins 1950; Soltis & Rieseberg 1986). In autopolyploids, the four or more sets of homologous chromosomes of autopolyploids should pair randomly during prophase (autosyndesis). These homologues should form groups of two (bivalent configuration), three (trivalent configuration), four (quadrivalent configuration) or more during metaphase. Hence, alleles at a given locus on the homologous chromosomes of autopolyploids should segregate at random; unlike diploids, autopolyploids should be characterized by multisomic inheritance. In allopolyploids, chromosomes are expected to pair nonrandomly, with homologous pairing (autosyndesis) occurring to the exclusion of homeologous pairing (allosyndesis). Homologous chromosomes of allopolyploids should then form only balanced pairs (bivalent configuration) during metaphase. Hence, alleles at a given locus on homeologous chromosomes of allopolyploids should segregate independently; like diploids, allopolyploids should be characterized by disomic inheritance. Examples of typical meiotic configurations in diploids and polyploids are shown in Figure 1.

In effect, autopolyploids are considered to represent cytogenetically complex, multisomic parental diploids, while allopolyploids represent fixed, disomic species hybrids. To test these hypotheses, we surveyed the literature regarding the meiotic behavior of neoauto- and neoallopolyploids. In these sections, we consider only

Cytotype

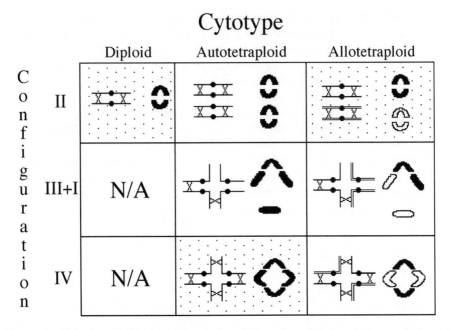

Figure 1   Univalent (I), bivalent (II), trivalent (III), and quadrivalent (IV) associations of homologous and homeologous metacentric chromosomes in diploids, autotetraploids and allotetraploids. Theoretical derivations (*left*) and resulting meiotic figures (*right*) are illustrated for each configuration and cytotype. Crossover events are shown by an "X," and centromeres by a dot (•). Configurations commonly hypothesized to be "normal" for a given cytotype are highlighted with stippled background. The formation of configurations in diploids and polyploids is dependent upon the distribution and number of chiasmata formed between homologous/homeologous chromosomes, hence illustrated associations represent a subset of all possible configurations. A comprehensive review of meiotic configurations is provided by Sybenga (1975).

even-ploidy cytotypes, i.e., tetraploids, hexaploids, and octoploids. The meiosis and progeny of triploids are evaluated in Ramsey & Schemske 1998.

Auto- and Allosyndesis

The major hypotheses regarding the character of hybrid and nonhybrid polyploids are predicated on the assumptions that autopolyploids exhibit random pairing of homologues and allopolyploids lack homeologous pairing. There are, however, few direct observations of auto- and allosyndetic pairing. Markers for distinguishing homologs at meiotic prophase and metaphase are often unavailable, preempting evaluation of preferential pairing in autopolyploids. Morphological differences between homeologous chromosomes of some taxa enable determinations of heteromorphic pairing in allopolyploids (e.g., Lindstrom & Humphrey 1933), but not systematic estimates of the frequency of auto- and allosyndesis. Homologous and homeologous pairing can be inferred from meiotic configurations and from inheritance data (see below), though indirect estimates do not assess pairing that is gamete- or sporophyte-lethal. This factor may explain the phenotypic stability of allopolyploids that have multivalent pairing and other meiotic irregularities (e.g., Brown 1951). In this scenario, allosyndesis contributes to the lowered fertility of polyploids.

Meiotic Configurations

In a simple model, one might predict that autopolyploids would exhibit 100% multivalent configurations, while allopolyploids would have 100% bivalent pairings. To evaluate the chromosome associations of auto- and allopolyploids we surveyed the occurrence of univalents, bivalents, trivalents and quadrivalents during late prophase (diakinesis) and metaphase I of neopolyploids. It should be noted that chiasma formation and configurations at diakenesis and metaphase are not always indicative of synapsis during prophase, because paired chromosomes sometimes fail to cross over (Sybenga 1975, Jackson 1976). Hence, metaphase configurations represent successful pairings that result in genetic recombination. Metaphase associations are standard cytological inference, and the only widely available index of meiotic pairing in polyploids.

In our survey, the mean percent occurrence of multivalents (trivalents and quadrivalents) is significantly higher in autopolyploids (28.8%, N = 93 studies) than in allopolyploids (8.0%, N = 78 studies) (Mann-Whitney U test, $P < 0.001$) (Figure 2; see Web Table 1). This result supports the hypothesis of differential chromosome behavior of auto- and allopolyploids, but also raises several issues. First, in autopolyploids, the occurrence of bivalent pairing is higher (mean 63.7%, range 12 to 98.2), and quadrivalent pairing lower (mean 26.8%, range 1.8 to 69.1), than what theoretically might be expected from homologous chromosomes (e.g., 0% bivalents and 100% quadrivalents). For example, induced autotetraploids of *Lolium perenne* exhibited 1% trivalent and 20% quadrivalent configurations at metaphase I (Simonsen 1973), while autotetraploid tomato had almost no trivalents and 19% quadrivalents (Upcott 1935).

Two mechanisms could limit bivalent pairing in first and early generation autopolyploids that have not been fertility-selected: (*a*) nonrandom chromosome associations among some homologues, and (*b*) existence of physical limitations or genetic factors limiting multivalent pairing between randomly associating homologous chromosomes. The former hypothesis is unlikely. "Diploidization" of autopolyploids is hypothesized to occur by the gradual accumulation of structural differences in homologous chromosomes (Doyle 1963, Sybenga 1969) as well as the evolution of genic factors that enforce preferential pairing in established polyploid populations. Moreover, the numbers of multivalents occurring per mother cell (0, 1, 2, 3, ... *x*) in new autopolyploids generally follows a binomial distribution, suggesting that the probability of quadrivalent formation is the same for each

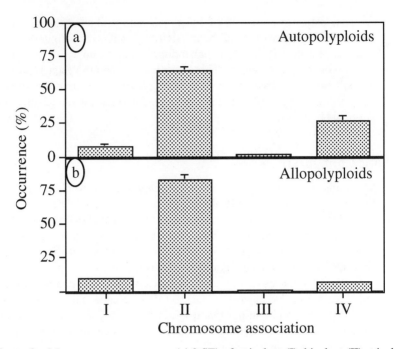

Figure 2   Mean percent occurrence (±2 SE) of univalent (I), bivalent (II), trivalent (III) and quadrivalent (IV) chromosome associations during metaphase I of newly formed (*a*) autopolyploids (N = 93) and (*b*) allopolyploids (N = 78). Data from Web Table 1.

set of homologous chromosomes (McCollum 1958, Morrison & Rajhathy 1960a, Simonsen 1973, Sybenga 1996).

Most cytogeneticists hypothesize that the frequency and distribution of chiasmata dictate the occurrence of multivalents in autopolyploids (Kostoff 1940, McCollum 1958, Hazarika & Rees 1967, Sybenga 1975). As an illustration, consider a single parameter, number of chiasma initiation sites per chromosome. Pairing behavior in diploids is robust to the manner of pairing initiation because 100% bivalent pairing will result if chiasma can initiate at one or more sites, and if each chromosome has at least one crossover. In contrast, configurations of polyploids are strongly influenced by chiasma initiation. In an autotetraploid, if pairing is initiated at only one chromosomal location, no more than two bivalents will form from each set of four homologues. If pairing is initiated at two chromosomal locations, two thirds of resulting configurations will be quadrivalents, while one third will be bivalents (John & Henderson 1962, Sved 1966). Assuming pairing initiation at two or more sites, other factors could decrease or increase multivalent frequencies (Sybenga 1975, Lavania 1986). For example, polyploids may tend to exhibit reduced chiasma formation compared to progenitor diploids (e.g., Hazarika & Rees 1967), perhaps because complex chromosome associations interfere with pairing and crossover. Polyploids may thus exhibit fewer or altered chiasma, and hence fewer multivalents, than observations of diploids would lead one to expect. Some chromosomes may be structurally incapable of forming all possible chiasmata between four or more homologues (Sybenga 1975).

The observation that the occurrence of two chiasma initiation points per chromosome leads to an expected 66% frequency of multivalents, coupled with early cytological surveys suggesting occurrences of multivalents range from 50% to 80% (Morrison & Rajhathy 1960a,b), led some cytogeneticists to hypothesize that two thirds of chromosome associations in "good" autopolyploids should be multivalents (Sved 1966; Jackson 1976). Our survey revealed that quadrivalent frequency of neoautotetraploids (mean 29%, range 2 to 69) is significantly different than 66% (One-Sample test, $P < 0.0001$). In the absence of cytological data regarding chiasma formation, there does not appear to be a reliable a priori expectation regarding the configurations of autopolyploids.

Another unexpected result of this survey is the common occurrence of multivalents in allopolyploids (see Web Table 1). Trivalents and quadrivalents were observed in 80% of surveyed allopolyploids, and percent occurrence is significantly different than zero (One-sample Sign Test, $P < 0.0001$). The mean frequency of multivalent pairing observed in allopolyploids (mean 8.0%, range 0 to 52) is approximately one quarter the occurrence in autopolyploids (mean 28.8%, range 2 to 69). Multivalents are reported in polyploids generated by wide crosses (e.g., Howard 1938, Stebbins & Vaarama 1954, Phillips 1964) and in many classic textbook polyploids. For example, *Primula kewensis*, the famous allopolyploid derived by somatic doubling of sterile diploid F1 *P. floribunda* × *verticillata*, exhibited 18% multivalent pairing at metaphase I (Upcott 1940).

Multivalent pairing in allopolyploids is biologically significant for two reasons. First, homeologous pairing in bivalents and multivalents will lead to the production of genically unbalanced euploid chromosomes, and hence reduced fertility (Howard 1938; see below). Second, the occurrence of multivalents provides evidence of intergenomic recombination in allopolyploids. Sved (1966) estimated the relationship between multivalent and recombination frequencies for allotetraploids with two pairing initiation sites per chromosome but differing degrees of preferential pairing. The relationship between quadrivalent occurrence and segregation frequency (indexed as the frequency of homozygous *aa* recessive gametes produced by the heterozygote *aaAA* parent) is approximately linear. For the mean frequency of quadrivalents in our survey, the expected segregation frequency is 1.2%. In comparison, 16.7% segregation would be expected in an autopolyploid (Sved 1966).

Patterns of Inheritance in Neopolyploids

It is commonly believed that strict autosyndesis in allopolyploids leads to independent segregation of alleles on homeologous chromosomes, and hence "fixed heterozygosity" inherited in a disomic manner (Winge 1932, Roose & Gottlieb 1976, Soltis & Soltis 1993). Consider an allotetraploid with the duplex heterozygote genotype *AAaa*, where one pair of homeologs carries the *A* allele, and the other has the *a* allele. Autosyndesis leads to a single arrangement of bivalent pairings, and all gametes will be heterozygous *Aa* (Sybenga 1969). In contrast, pairing of duplicate chromosomes in autopolyploids is hypothesized to occur at random during meiotic prophase I. Crossover may involve any homologue, and segregation at a given locus can involve as many alleles as there are homologous chromosomes. If the above-described heterozygote *AAaa* was an autopolyploid, 12 quadrivalent and bivalent associations could occur during meiosis. Although the heterozygous gamete *Aa* would be the most frequent product of meiosis, ~30–45% of the gametes would be completely homozygous. The exact outcome is dependent upon the frequency of recombination between the marker locus and the centromere and the occurrence of multivalents (Little 1945, Burnham 1962, Bever & Felber 1992).

The chromatid segregation model postulates that recombination can occur between a locus and the centromere. Hence, sister chromatids of homologs associated as multivalents will sometimes end up with different alleles, depending on the structure of the multivalent and the crossover frequency. By chance, anaphase II may reunite identical alleles initially held on sister chromatids but recombined to separate homologs by chiasma during metaphase I, a phenomenon termed double reduction. Assuming 100% quadrivalent configurations and the consistent presence of one chiasma between the *A* locus and the centromere, the chromatid segregation model would predict gametic ratios of 2 *aa*:5 *Aa*:2 *AA* from the *AAaa* sporophyte (Burnham 1962, Jackson & Jackson 1996). In the chromosome segregation model, there would be no crossing over between the *A* locus and the centromere, for example because the locus is located near the centromere. Chromatids of homologs associated as quadrivalents will possess the same allele, and double reduction is not possible. Segregation ratios are thus robust to the frequency of multivalents and total gamete ratios will be 1 *aa*:4 *Aa*:1 *AA* (Sybenga 1969, Jackson & Jackson 1996). Segregation rates of most loci in autopolyploids are probably intermediate between those predicted by chromatid and chromosome segregation models of multisomic inheritance. Also, the frequent occurrence of aneuploids in the progeny of neopolyploids alters realized segregation rates.

Available studies of spontaneous and induced autopolyploids are all consistent with the hypothesis of multisomic inheritance (see Web Table 2; N = 14 loci in 7 species), thus supporting the hypothesis of random association of homologues in nonhybrid polyploids. For example, flower color in diploid Jimsonweed (*Datura*) is determined by two alleles at a single locus, with purple (*P*) dominant to white (*p*). In autotetraploids, backcrosses of duplex heterozygotes (*PPpp*) to homozygous recessives (*pppp*) generated 905 purple-flowered individuals and 179 white-flowered individuals, frequencies that correspond to an expected 5:1 ratio. Crosses of induced autotetraploid *Tradescantia paludosa* that were duplex heterozygous for self-incompatibility alleles (i.e., S1133 × S2244) generated 10% di-allelic, 40% tri-allelic, and 50% tetra-allelic progeny, frequencies consistent with the chromosome segregation model of multisomic inheritance (Annerstedt & Lundqvist 1967).

Segregation ratios in allopolyploids vary dramatically across taxa, sometimes approaching disomic inheritance but more typically multisomic inheritance (see

Web Table 2; $N = 58$ loci in 23 species). The most comprehensive data involve *Gossypium* species. A variety of monogenic characters are known from the cultivated cottons, *G. hirsutum* and *G. barbadense*, and their wild relatives. Gerstel & Phillips (1958) used marker alleles to test segregation patterns in polyploids derived from congeners of varying phylogenetic affinity. *G. hirsutum* and *G. barbadense* are stable allopolyploids derived from distant relatives; extant accessions are characterized by bivalent pairing and disomic inheritance, and are assigned the genome formula AADD. Strains of tetraploid cotton that were homozygous for dominant marker alleles on the D or A genomes were crossed to a variety of natural diploid D and A genome species, and the converse. These crosses generated triploid hybrids (AD_1D_2 and A_1A_2D), which were treated with colchicine to induce allohexaploids ($AAD_1D_1D_2D_2$ and $A_1A_1A_2A_2DD$). Depending upon the occurrence of homeologous pairing of D and A chromosomes during meiosis, hexaploids would generate either (*a*) gametes that were always heterozygous for the dominant marker (AD_1D_2 and A_1A_2D), or (*b*) a combination of heterozygous and homozygous gametes (AD_1D_1, AD_1D_2, and AD_2D_2, or A_1A_1D, A_1A_2D, and A_2A_2D). The latter possibility is manifest as the occurrence of recessive (wild-type) individuals in a test cross, in which hexaploids were backcrossed to diploid parentals to form tetraploids. A multisomic chromosome segregation model would predict a 1:5 occurrence of recessives, which would never occur under conditions of strict disomic inheritance. Segregation rates varied widely between crosses, and were correlated with frequencies of multivalents at meiotic metaphase (Figure 3) (Phillips 1962, 1964). For example, segregation of recessives in *G. hirsutum* × *arboreum* averaged 5.1:1, approximately the frequency expected with polysomic inheritance, while *G. barbadense* × *gossypioides* averaged 72:1 (Figure 3). The latter value is less than that predicted by multisomic inheritance, but very different from the predicted value of 0 for disomic inheritance.

Because the diploid progenitors of most neoallopolyploids are morphologically distinct, it is possible to evaluate allosyndesis and the breakdown of disomic inheritance in a general way by comparing the phenotypic characteristics of parentals and primary allopolyploids with those of successive allopolyploid generations. For example, Grant (1954) described segregation of leaf width, pubescence, calyx lobe shape, corolla length, filament length, and other traits in progeny of the spontaneous allopolyploid *Gilia millefoliata* × *achilleaefolia*. We found that 31 of 42 studies of neopolyploid accessions identified substantial segregation for morphological characteristics, marker alleles and fertility (see Web Table 3). It should be noted that segregation in spontaneous neopolyploids may be caused by recombination in unreduced gametes produced by progenitor F1 hybrids and triploid intermediates in addition to allosyndesis in the neopolyploid itself (Howard 1938, Müntzing 1932). Also, aneuploidy and gene silencing can generate patterns of phenotypic variability that may be confused with intergenomic recombination.

Nonetheless, the combined data on inheritance and segregation in allopolyploids (see Web Tables 2, 3) clearly suggest that (*a*) variation in the frequency of allosyndesis is expected for different pairwise combinations of species, and

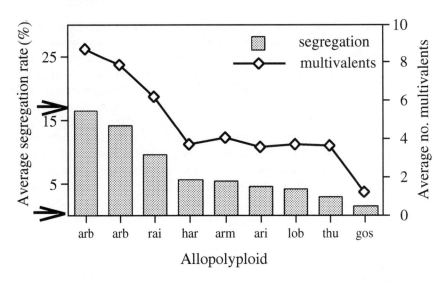

Figure 3 Segregation of recessive genotypes (*zzzz*) from crosses of duplex heterozygotes (*ZZzz*) to recessive testers (*zzzz*) (*left axis, columns*), and percent occurrence of multivalents at meiotic metaphase I (*right axis, line*), in allopolyploids involving cotton (*Gossypium hirsutum* or *G. barbadense*) and wild relatives (*G. arboreum, G. raimondii, G. harknessii, G. armourianum, G. aridum, G. lobatum, G. thurberi,* and *G. gossypioides*). Arrows show expected segregation for multisomic inheritance (chromosome segregation) and disomic inheritance. Data from Phillips (1962, 1964).

(*b*) except in wide crosses, the mean frequency of allosyndesis is considerably greater than zero (Goodspeed & Bradley 1942, Grant 1975). The long-term evolutionary consequences of intergenomic segregation are unknown (Sybenga 1969, 1996). Rapid evolution of chromosome structure and genic control of chromosome pairing (see below) may lead to strict homologue pairing and disomic inheritance, though possibly after considerable recombination between homeologues. Alternatively, allosyndesis may result in chance fixation of chromosome segments from one homeologue or the other, generating hybrid polyploids that are multisomic for individual chromosomes or chromosome regions (Stebbins 1950). Models suggest that in the absence of mitigating factors, even occasional multivalent pairing can rapidly deteriorate to a system of multisomic inheritance (Sybenga 1996). As discussed below, these possibilities challenge the perception of allopolyploids as "constant species hybrids" (Winge 1932).

ANEUPLOIDY

Aneuploidy, defined as the possession of chromosome numbers either greater or less than an exact multiple of the base chromosome number x, is common in flowering plants (Grant 1981). Aneuploidy is hypothesized to contribute to phenotypic evolution and speciation in some genera, and may, in some cases, enable transitions between euploid chromosome numbers. Some polyploid lineages have a high occurrence of aneuploid variation. For example, some taxa exhibit polyploid drop, aneuploid reduction at the polyploid chromosome number (Darlington 1963). Here we examine the origin and maintenance of aneuploid cytotypes in neopolyploid populations.

Polyploids Generate a High Frequency of Aneuploid Gametes

The occurrence of univalents and multivalents during polyploid meiosis complicates the orderly separation of homologs/homeologues. Univalents and trivalents by necessity divide unequally during anaphase I because there is no mechanism to evenly divide the chromosomes of an odd-number configuration (though by chance, unbalanced divisions may compensate each other, for example by a 2–1 separation of a trivalent, and a 0–1 division of a univalent). The divisions of tetraploids are more complicated to assess. Some ten types of quadrivalent configurations can be formed, depending on which homologous/homeologous chromosomes happen to cross over (Sybenga 1975, Singh 1993). Quadrivalent configurations can broadly be divided into ring configurations (each homolog/homeolog forming two chiasma) and chain configurations (each homolog/homeolog forms one or two chiasma), analogous to the ring and rod configurations of diploids. Among ring and chain configurations, one may distinguish alternate orientations (proximate homologs/homeologs oriented in opposite directions) and adjacent orientations (proximate homologs/homeologs oriented, to varying degrees, in the same direction). Alternate quadrivalent orientations, sometimes called zigzag orientations, are believed to nearly always generate equal (2–2) chromosome disjunctions (Garber 1955, McCollum 1958), whereas disjunctions from adjacent orientations will include both balanced and unbalanced separations. Orientation frequencies are dependent upon the initial likelihood of formation, and stability during the transition from metaphase to anaphase; reorientation of individual configurations is probably frequent (Sybenga 1975). Frequencies of alternate quadrivalent frequencies vary among polyploid species (Myers & Hill 1940, Garber 1954, Jones 1956, McCollum 1958), so the contributions of quadrivalents to unbalanced meiotic divisions may vary across taxa. The evolution of increased zigzag quadrivalent orientations may be an important mechanism increasing fertility of neopolyploid lineages.

To evaluate the extent of aneuploidy in the gametes of neopolyploids we summarized the literature on anaphase I and metaphase II chromosome distributions in pollen mother cells, as well as chromosome counts in maturing pollen grains. Chromosome compositions determined at later stages of development (e.g., pollen mitoses) more accurately reflect actual pollen cytotypes than determinations made early in development (e.g., anaphase I). There was no clear influence of stage of development on the frequency of aneuploidy in our dataset (see Web Table 4; Gulcan

& Sybenga 1967), but where more than one stage was investigated in a study we included later measurements. No data on megaspore mother cells and ovules were available.

Occurrences of aneuploid pollen in auto- and allopolyploids were not significantly different (mean 36.0 versus 43.1%, Mann Whitney U test, $P = 0.6173$; see Web Table 4), so we pooled polyploid types for consideration of specific cytotypes. Euploid (n) pollen were most frequent (mean 62.5%, range 31.1 to 78.5, $N = 26$ studies), followed by $n - 1$ and $n + 1$ cytotypes (mean 16.4 and 13.2%, respectively) (Figure 4a; see Web Table 4). Pollen lacking or gaining three or

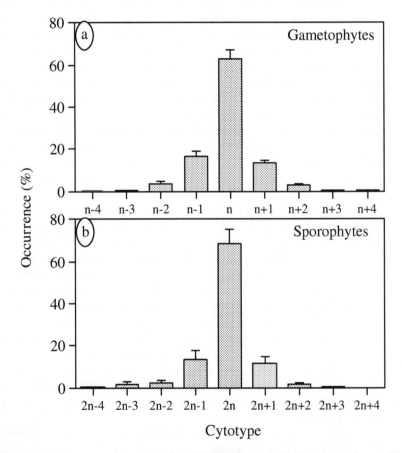

Figure 4   Mean percent occurrence (± 2 SE) of aneuploidy in gametophytes ($N = 26$) and sporophytes ($N = 33$) of neopolyploids (auto- and allopolyploids pooled). (a) Frequency distribution of pollen cytotypes produced by neopolyploids, as determined from pollen mitoses as well as anaphase I or metaphase II stages in pollen mother cells. (b) Frequency distribution of chromosome numbers in the progeny of neopolyploids. Data from Web Tables 4 and 5.

more chromosomes on average constituted <2% of cytotypes (Figure 4a; see Web Table 4). Hypoeuploid pollen cytotypes (i.e., aneuploids with less than the euploid chromosome number n) were significantly more common than hypereuploid pollen cytotypes (aneuploids with more than the euploid chromosome number n) (means 20.8 versus 16.7%; Wilcoxon Signed Rank test, $P = 0.0023$). The likely cause of this phenomenon is the occurrence of lagging univalents, which often fail to incorporate into either daughter nuclei. In essence, there are two mechanisms that lose chromosomes during meiotic divisions (unbalanced anaphase separations, lagging chromosomes), but only one way to gain chromosomes (unbalanced anaphase separations).

These analyses suggest that aneuploid cytotypes constitute a substantial portion of the gametes produced by neopolyploids. The mean frequency of euploid gametes in autoploids (64.0%, $N = 22$) closely matches the mean occurrence of bivalents (63.7%, $N = 93$) (see Web Tables 1, 4; Figures 2, 4a). Bivalent pairings involving homologous chromosomes, such as would be observed in parental diploid species and autopolyploids, rarely exhibit irregularities at anaphase. This implicates the remaining configurations (univalents, trivalents and quadrivalents) as important contributors to unbalanced meiotic divisions. In contrast, the mean frequency of euploid gametes in allopolyploids (56.9%, $N = 7$) is much less than the mean occurrence of bivalents (82.3%, $N = 78$) (see Web Tables 1, 4; Figure 2, 4a). This discrepancy may be a spurious result of the small number of anaphase I and metaphase II chromosome distributions determined in allopolyploids. However, this trend holds in individual allopolyploid studies with estimates of the occurrence of both meiotic configurations and aneuploid gametes (Kostoff 1938, Upcott 1940, Singh & Hymowitz 1985). A likely explanation is that a portion of bivalent configurations observed in allopolyploids in fact involve homeologous chromosomes (i.e., allosyndesis), which may lead to irregular separation at anaphase.

Variability in gamete cytotypes is considerably less in the even-ploidy cytotypes (tetraploids, hexaploids, and octoploids) considered here, than in triploids (Ramsey & Schemske 1998, Figure 1). The difference probably originates from the meiotic configurations that occur in odd and even ploidy cytotypes. In triploids, every chromosome will display either (a) two bivalents and one univalent; (b) one trivalent, or (c) three univalents. In each case, a high degree of irregularity is anticipated during anaphase. In tetraploids, univalents and trivalents may occur, but bivalents and quadrivalents are the most abundant associations (Figure 2).

Polyploids Generate a High Frequency of Aneuploid Progeny

The high frequency of aneuploidy in the gametes of neopolyploids has two possible outcomes. First, aneuploidy may be lethal at the gamete or embryo development stages. In this scenario the progeny of neopolyploids are euploid, but of limited number. Second, aneuploid gametes may function similarly to euploid gametes, and generate viable gametes. In this case, polyploid progeny will be more numerous but include a high percentage of aneuploid individuals. Here we summarize the frequency of aneuploids in the progeny of neopolyploids.

Occurrences of aneuploid progeny from auto- and allopolyploids were not significantly different (mean 29.0 versus 28.3%, Mann Whitney U test, $P = 0.3339$; see Web Table 5), so we consider the two polyploid types together in estimates of cytotype occurrences. Euploid ($2n$) sporophytes were most frequent (mean 68.6%, range 20 to 96.9, N = 33 studies), followed by $2n - 1$ and $2n + 1$ cytotypes (mean 13.4 and 11.5%, respectively) (Figure 4b; see Web Table 5). Sporophytes lacking or gaining three or more chromosomes on average constituted <3% of progeny cytotypes (Figure 4b; see Web Table 5). Mean occurrences of hypo- and hypereuploid progeny were not significantly different (means 17.6 and 13.5; Wilcoxon Signed Rank test, $P = 0.6402$). Asymmetries were observed in some individual species (see Web Table 5), but there was no consistent trend for a higher frequency of hypoeuploids in sporophyte cytotypes as observed in pollen cytotypes.

The most striking feature of this analysis is the close correspondence between cytotype distributions of the pollen and progeny of neopolyploids (Figure 4a,b). Euploids are somewhat more common among sporophytes than gametes, and the difference is marginally significant (68.6 versus 62.5%; Mann Whitney U test, $P = 0.0859$). The lack of cytological observations of megasporogenesis complicates comparisons of gametes and sporophytes. If chromosome separations during megasporogenesis are more balanced than those during microsporogenesis, little selection may be operating on gamete and sporophyte cytotypes in polyploids. Cytologists generally assume that chromosome behavior during micro- and megasporogenesis of polyploids is similar, although some studies of aneuploid chromosome transmission rates (see Riley & Kimber 1961) and analyses of reciprocal crosses involving triploids (Ramsey & Schemske 1998) suggest that there is stronger selection for euploid pollen than ovule cytotypes. Assuming that frequencies of pollen and ovule cytotypes are approximately equal, the observed frequency of euploid sporophytes (mean 62.6%) is significantly different than expected (43.6%) (One-Sample Sign test, $P < 0.0001$).

There is less selection to remove aneuploids from the progeny of even-ploidy cytotypes than from the progeny of odd-ploidy cytotypes. For example, cytotype distributions of the gametes and progeny of triploids are nearly inverted (Ramsey & Schemske 1998, Figures 1, 2, 3), whereas the distributions for even-ploidy cytotypes differ significantly, but are similar in overall shape. There are several explanations for the difference in selection. First, the gametes of triploids have a wider distribution of aneuploid types compared to the gametes of tetraploids, hexaploids, and octoploids. Selection may act strongly on more numerically deviating aneuploid cytotypes, skewing the distribution of sporophyte cytotypes away from that of the gamete cytotypes. Second, the gametes and progeny of triploids on average have a lower total chromosome number compared to the gametes and progeny of tetraploids, hexaploids, and octoploids.

In general, studies indicate that aneuploid cytotypes occur more often, exhibit less deviant phenotypes, and have higher reproductive fitness in polyploid than diploid populations (Khush 1973). For example, the mean occurrence of aneuploids

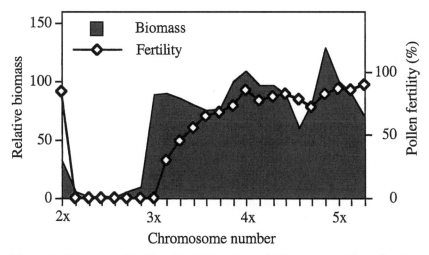

Figure 5   Biomass and fertility of euploid and aneuploid cytotypes of *Dactylis glomerata*, generated by crosses between diploid, triploid, tetraploid and pentaploid cytotypes. Data from Müntzing (1937).

in surveyed diploid systems is 1% (see Web Table 5), thirtyfold less than the mean occurrence in polyploid systems (37.4%). The difference is attributable not only to the complex meiosis of polyploids, but also to the low viability of aneuploids at the diploid level. In diploids, aneuploids are readily identifiable by their aberrant phenotypes and sterility (Avery et al. 1959, Ellis & Janick 1960, Khush & Rick 1966). For example, many of the *Oenothera* "species" and "mutants" described by Hugo de Vries were trisomics (Emerson 1935). Aneuploid-polyploids often appear somewhat distinctive, but as a rule survive and compete successfully with euploids (Kostoff 1938; O'Mara 1943; Clausen et al. 1945; Einset 1947; Bernstrom 1954; Rommel 1961; Bingham 1968; Ahloowalia 1971; McNaughton 1973; Simonsen 1973, 1975). Müntzing (1937) generated sporophytes with all chromosome numbers between $2n = 2x = 14$ and $2n = 5x = 35$ by crossing $2x$, $3x$, $4x$, and $5x$ cytotypes of established *Dactylis glomerata* (Figure 5). Biomass of aneuploid sporophytes was greatly reduced for $2n = 14$ to 21, but for higher chromosome numbers there was little difference between euploids and aneuploids. Pollen fertility of aneuploid cytotypes showed a similar pattern, but high fertility was regained at $2n = 4x = 28$ rather than $2n = 3x = 21$ (Figure 5). Similar, though less comprehensive, results are reported in neopolyploid *Beta*, *Lactuca*, *Lamium*, *Medicago*, *Nicotiana*, and *Raphanus-Brassica*. As concluded by Clausen et al. (1945) from studies of the spontaneous allohexaploid *Madia sativa* × *citriodora*, the balance of chromosomes and genes in polyploids is relatively flexible, "permitting survival of plants that deviate slightly from the hexaploid level."

To evaluate the dynamics of aneuploid formation in polyploid populations one must consider the formation of aneuploids both by euploids and by existing aneuploids. A survey of chromosome transmission rates suggests that aneuploid-polyploids tend to generate progeny with cytotypes somewhat closer to the euploid chromosome number than the parent's cytotype (Web Table 6). For example, in a survey of 26 aneuploid cytotypes in 7 species, 19 of the aneuploids generated progeny with a modal chromosome number one or several steps closer to euploid than the parent's chromosome number (Web Table 6). In spite of the tendency to revert to euploid chromosome numbers, aneuploids still generate large numbers of new aneuploids. In neopolyploid *Coix lacryma-jobi*, 70%–80% of the selfed progeny of $2n = 4x - 1$ and $2n = 4x + 1$ plants are euploid (Rao 1976). Euploids are the modal progeny of selfed aneuploid cytotypes ($2n = 4x - 2$ to $2n = 4x + 6$) in neotetraploid *Lamium amplexicaule*, representing on average 39% of offspring (Bernstrom 1954). An exception to this reversion-to-euploid trend relates to those plants that become di-, tetra- or hexasomic for specific homologues or homeologues via aneuploidy and/or chromosome substitution (Riley & Kimber 1961). Such plants are chromosomally balanced and will replicate themselves via selfing.

The frequency of aneuploidy in the gametes and sporophytes of neopolyploids (Figure 4*a*,*b*) and aneuploid-polyploids (Web Table 6) suggests that a component of polyploid populations are "continuously moving through aneuploid conditions" (Riley & Kimber 1961). A schematic of this process is illustrated for maize (Figure 6) (Randolph 1935, Shaver 1962). Newly formed autotetraploid *Zea perennis* is characterized by frequent (41.2%) multivalent associations at metaphase I. Subsequent anaphase divisions are often unequal, and 45% of maturing pollen grains are aneuploid, mostly $2x - 1$ and $2x + 1$ (Figure 6*a*). Progeny of $4x \times 4x$ crosses are only 50% euploid (Figure 6*b*), suggesting that aneuploid gametes function and generate viable aneuploid offspring. The aneuploid progeny of euploid crosses are semifertile, with ovules showing more inviability than pollen (Figure 6*c*). This difference in male and female fertility, which is typical of most polyploid systems (see below), is probably an artifact of measuring pollen fertility early in development (e.g., stainability at anthesis) but ovule fertility late (e.g., percent of ovules producing viable seed). Euploids constitute a large fraction of the progeny of aneuploids (Figure 6*d*,*f*), though a somewhat wider range of aneuploid cytotypes is also recovered. A numerical simulation of this system, including ovule fertility and progeny cytotype parameters and assuming self-fertilization, suggests that an entirely euploid population rapidly accrues aneuploid cytotypes until reaching an equilibrium value of 46% euploid cytotypes in 8–10 generations (data not shown).

The occurrence of aneuploidy in neopolyploid populations has several implications. First, aneuploids represent a form of genetic load (Doyle 1986). Although most polyploids accommodate the loss or gain of individual chromosomes, aneuploid crosses will generate increasingly deviant cytotypes as well as multiple deficiencies for a single chromosome. Second, considerable phenotypic variability can be introduced into a neopolyploid accession via aneuploidy (Kostoff

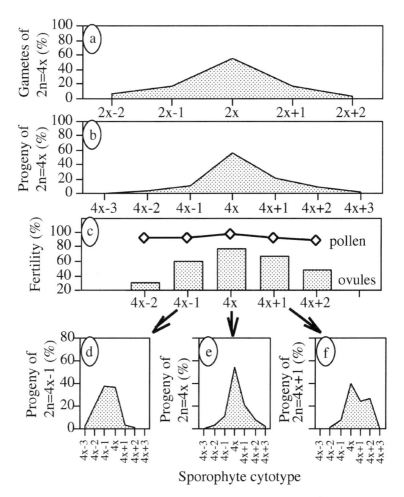

Figure 6   Production and maintenance of aneuploid variation in neoautotetraploid *Zea perennis*. (*a*) Distribution of pollen cytotypes produced by $2n = 4x$ individuals, based on chromosome counts in maturing pollen grains. (*b*) Chromosome numbers of progeny generated by euploid $4x \times 4x$ crosses. (*c*) Pollen and ovule fertility of euploid and commonly occurring aneuploid cytotypes. (*d,e,f*) Chromosome numbers of progeny of euploid and aneuploid individuals, generated by $4x \times 4x$, $4x-1 \times 4x-1$, and $4x + 1 \times 4x + 1$ crosses. Data from Randolph (1935) and Shaver (1962).

1938, Clausen et al. 1945), because of the chromosomal heterogeneity of individual aneuploid cytotypes. For an autotetraploid with $x = 7$, there will be 7 types of the hypereuploid $2n = 4x + 1$, and 28 types of $2n = 4x + 2$. In allopolyploids, the loss or gain of homeologs provides an additional mechanism of phenotypic alteration. Advantageous chromosome combinations in both auto- and alloploids can be stabilized by hexasomy and disomy. For example, in an autotetraploid, the loss of two copies of a certain chromosome would lead to stable chromosome transmission of the two remaining homologues. Resulting progeny would be disomic for loci on the affected chromosome.

Chromosome Substitution

In polyploids, some gametes with the euploid chromosome number are generated by unequal but numerically compensating divisions during meiosis, which in turn generates chromosomally unbalanced sporophytes. For example, in a tetraploid species, the four homologues (or homeologues) of a chromosome may split 3–1 during anaphase I in a micro- or megaspore mother cell, whereas the homologues (homeologues) of another chromosome may split 1–3 (i.e., double-opposed nondisjunction). If other chromosomes separate equally, the resulting gametes (s), while all $n = 2x$, will be monosomic for one chromosome and trisomic for another.

"Pseudoeuploid" sporophytes are often reported in the progeny of polyploids, but estimating their frequency is problematic because of the difficulty of morphologically distinguishing chromosomes. Simonsen (1973) was able to determine exact chromosome constitutions in one half of 62 "euploid" progeny of neotetraploid *Lolium perenne*, and identified 2 (6.5%) pseudoeuploids. Seven percent of the aneuploid and euploid progeny of neotetraploid *Cyrtanthus parviflorus* × *mackenii* were similarly numerically-compensating (Ising 1966). A general estimate of pseudoeuploid frequencies can be made under the assumptions that (*a*) most pseudoeuploids arise via double nondisjunction (quadruple and sextuple nondisjunctions are rare; see Web Table 4); and (*b*) chromosome nondisjunctions are randomly oriented (Belling & Blakeslee 1924, Beasley 1942). In this scenario, gametes with $n = 2x \pm 2$ (or $n = 3x \pm 2$, $n = 4x \pm 2$, etc.) should occur at the same frequency as unbalanced $N = 2x$ gametes. The mean frequency of $n = 2x - 2$ and $n = 2x + 2$ gametes in our survey is 3.1% (range 0 to 10, $N = 27$; see Web Table 4), suggesting an average occurrence of pseudoeuploids in polyploid sporophytes similar to those reported in *Lolium* and *Cyrtanthus* ($2 \times 3.1 = 6.2\%$). The actual mean frequency may be somewhat higher because pseudoaneuploids may be formed by other means (e.g., quadruple opposed nondisjunction) or, alternatively, lower, because selection favors balanced euploid gametes and sporophytes. In the latter case, chromosome substitution contributes to the reduced fertility of polyploids (see below).

Chromosome substitution in allopolyploids can occur via nonhomologous/nonhomeologous substitution, as well as through homeologous substitution, i.e., replacing a chromosome from one progenitor diploid species with a homeologue from another progenitor. Homeologue substitution can occur via compensating but unequal divisions of univalents and multivalents, but also via bivalent pairing of homeologues. Poole (1932) screened the progeny of a spontaneous *Crepis rubra* × *foetida* for homeologous substitution of a single chromosome that was differentiated by a morphological marker. Thirty-nine percent of the "euploid"

(4x) progeny were unbalanced for these homeologs. If rates of chromosome substitution are comparable across the genome, only 5.8% of the progeny of this allotetraploid would be numerically and genically balanced.

FERTILITY OF NEOPOLYPLOIDS

Reduced fertility is commonly thought to constrain the demographic success of newly formed polyploids, especially autopolyploids (Stebbins 1950, 1971; Darlington 1963; Sybenga 1969; Briggs & Walters 1997). For example, Stebbins (1950) refers to a sterility "bottleneck" in the demographic establishment of polyploids, and Darlington (1963) muses that "in light of the reduced fertility of auto-tetraploids it is surprising that autotetraploid races and species are by no means uncommon in nature." To assess the fertility of neopolyploids and its evolutionary significance, we surveyed the literature for information regarding the pollen and seed fertility of newly formed autopolyploids and allopolyploids, and of progenitor diploid species and interspecific hybrids (see Web Table 7). In all studies, pollen fertility is expressed as percent pollen viability, measured by standard assays like stainability in aniline blue or acetocarmine. In contrast, measures of seed fertility varied considerably between studies, and included such indices as number of seeds per fruit, number of seeds per inflorescence, seed mass, and proportion of ovules producing seeds (see Web Table 7). To produce a relative index of seed fertility, we include two types of data. First, for studies that measure seed production of neopolyploids and progenitor diploids, we report percent fertility (ratio of neopoly-ploid to progenitor fertilities). Second, where only neopolyploids are measured, we include an inherently relative measure of fertility such as the proportion of ovules or florets producing viable seeds, when it is implicit that diploid parentals would be fully fertile.

The fertilities of surveyed auto- and allopolyploids varied enormously, ranging from zero to nearly 100% (Figure 7). The mean pollen viability of new auto- and allopolyploids was not significantly different (mean 70.9 versus 72.2%, $N = 176$, Mann Whitney U test, $P = 0.5524$) (Figure 7). Similarly, mean seed fertilities did not differ significantly between polyploid types (mean 39.4 versus 46.6%, $N = 113$, Mann Whitney U test, $P = 0.1929$) (Figure 7). Overall, there are no clear differences in fertilities of newly formed auto- and allopolyploids.

In polyploids, measurements of percent pollen fertility are often much higher than percent seed fertility (see Web Table 7; Figure 7) (Eskilsson 1963). Several factors probably contribute to this trend. First, measures of pollen fertility are made early during development (e.g., stainability at anthesis), whereas seed fertility is measured late (e.g., production of viable seeds). Studies have found that pollen viability is substantially higher than pollen germinability, suggesting that early measures of pollen viability tend to overestimate pollen fertility (e.g., Tanaka 1940). Second, the reduced pollen fertilities of neopolyploids may lead to pollen limitation, because there is not enough functional pollen to fertilize all fertile ovules. Finally, reduced seed fertility may reflect differences in development and

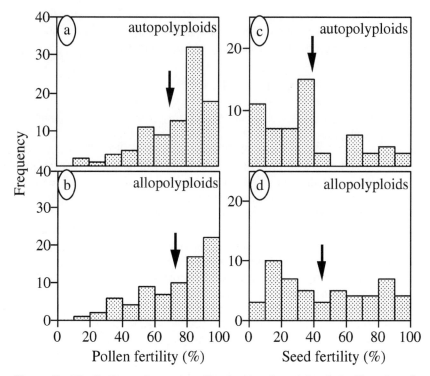

Figure 7   Distributions of percent pollen (*a, b*) and seed (*c, d*) fertility of newly formed autopolyploids and allopolyploids. Arrows show mean values. Data from Web Table 7.

allocation of neopolyploids. For example, neopolyploids often have larger floral organs made up of fewer constituent parts, and commonly generate fewer, but bigger, seeds (e.g., Howard 1939, Bretagnolle & Lumaret 1995).

Pollen and seed fertilities of neopolyploids are significantly reduced compared to progenitor diploid parentals (pollen mean 71.5 versus 91.4%, N = 94, Wilcoxon Signed Rank test, $P < 0.0001$; seed index mean 27.0 versus 68.1, N = 69, Wilcoxon Signed Rank test, $P < 0.0001$). Pollen fertility of new allopolyploids is significantly greater than their progenitor diploid interspecific F1 hybrids (pollen mean 70.5 versus 16.8, N = 34, Wilcoxon Signed Rank test, $P < 0.0001$). There are too few measures of F1 seed fertilities to allow statistical comparisons.

The results of this survey both support and challenge conventional wisdom regarding the fertility of polyploids. First, it is often assumed that autopolyploids have much lower fertility than allopolyploids, primarily due to the higher expected frequency of univalents and multivalents in autopolyploids (Stebbins 1950, 1971; Briggs & Walters 1997). Our analyses suggest that fertilities of early generation autopolyploids and allopolyploids are enormously variable and not significantly different. Reduced fertility appears to be a general characteristic of polyploids irrespective of origin, a conclusion also reached by Gottschalk (1978). Second, the reduced fertility of neopolyploids is commonly regarded as a major constraint on polyploid establishment and persistence (Stebbins 1950, Darlington 1963). These analyses do suggest a large reduction in neopolyploid fertility relative to progenitor diploids, with an average 20% reduction of pollen fertility, and a twofold difference in seed fertility. However, the reduction in some systems is modest (see Web Table 7), and even strong reductions could be compensated by increased survivorship and growth. Additional empirical and theoretical studies are needed to evaluate the impact of neopolyploid fertility on demographic establishment.

The Fertility of Polyploids Can Evolve

The meiotic behavior of plants is known to be under genic control and influenced by chromosome structure (Sybenga 1975, Jackson 1976). To determine if the fertility of polyploids may be improved by natural selection, we reviewed the literature on fertility in neopolyploid populations. Most studies were conducted in agricultural settings, where induced autotetraploids were evaluated for agronomic potential, though the fertility of natural allopolyploid lines has also been studied (see Web Table 8). Populations were selected for 2 to 19 generations, in all cases for seed fertility.

Of 12 studies that quantified pre- and postselection fertility, 11 reported increases. Fertility improvements were generally large, both in terms of rate of increase per generation and in total gains (see Web Table 8). For example, seed fertility of neopolyploid *Brassica campestris* increased from 1.5 to 16.8 over nineteen generations of selection, an average increase of 53.7% per generation (Swaminathan & Sulbha 1959). In four generations of selection, tetraploid *Nicotiana glauca* × *langsdorfii* increased pollen fertility from 59% to 99% (17.1% increase per generation), and seed fertility from 48 to 150 seeds per fruit (53.1% per generation) (Kostoff 1938). Overall, mean increases per generation in our survey were 14.0% for pollen viability (range −0.1 to 36.0%), and 39.7% for seed fertility (range −0.8 to 104.1%) (see Web Table 8). These results suggest that the infertility of polyploids may be a transient phenomenon.

With clear evidence that the fertility of neopolyploids can be increased rapidly by natural selection, we next ask if polyploidy sets an upper limit to fertility. To evaluate this question we surveyed the pollen viability of established, naturally occurring even-ploidy cytotypes and their diploid relatives (see Web Table 9). We sampled taxonomic species consisting of multiple cytotypes, as well as related congeners that differ in ploidy. Average fertilities of diploid, tetraploid, and hexaploid cytotypes were similar (diploid mean 89.1%, N = 33, range 64.9 to 100; tetraploid mean 86.4, N = 34, range 65.7 to 99.8; hexaploid mean 89.9%, N = 17, range 46.0 to 99.0) (see Web Table 9). To evaluate these data statistically, we computed mean values for each cytotype in genera where two or more ploidy levels were sampled. Comparisons of low versus high ploidy values are significantly different (mean

89.2 versus 83.5%; N = 14, Wilcoxon Signed Rank test, $P = 0.0238$). These results suggest that the fertility of polyploids ultimately reaches a level slightly lower than that of their progenitors.

CAUSES OF INFERTILITY IN POLYPLOIDS

Explanations of polyploid infertility are complex and fraught with controversy (Müntzing 1936; Howard 1938; Myers & Hill 1940; Randolph 1941; Sparrow et al. 1942; Stebbins 1950; Doggett 1964; Sybenga 1969, 1973a; Gottschalk 1978). Neopolyploids have simultaneous alterations of chromosome number, gene dosage, allele number, DNA content, cell size, growth and development, so it is difficult to distinguish contributions of each potential factor. Broadly, three causes of sterility have been identified. First, fertility may be reduced by meiotic abnormalities. In particular, univalent, multivalent, and homeologous bivalent pairings can lead to the production of inviable aneuploid gametes and sporophytes. Second, fertility may be reduced by genetic causes that are independent of obvious meiotic aberrations. The nature of these so-called genotypic effects are obscure, but a number of studies point to their importance. Finally, infertility may be related to incidental phenotypic effects of polyploidy. For example, the increased gene dosage, DNA content or cell size of polyploids may affect the development, growth, or physiology of sporophytes or gametophytes in such a way as to reduce the number or viability of gametes produced. Here we review each of the proposed mechanisms.

Meiotic Aberrations

Meiotic aberrations probably represent the most general factor affecting polyploid fertility. As described previously, polyploids exhibit frequent univalents, trivalents, and quadrivalents during metaphase (Figure 2) that lead to the production of chromosomally and genically unbalanced gametes (Figure 4a). Homeologue bivalent pairing in allopolyploids probably generates aneuploid gametes as well. The frequency distribution of sporophyte cytotypes contains significantly more euploids than would be expected on the basis of gamete cytotypes (Figure 4b). Moreover, aneuploid sporophytes are typically less fertile than euploid sporophytes (Figures 5, 6). Together, these observations indicate that the production of unbalanced gametes leads to aborted ovules, inviable seeds, and semisterile adults. Strain differences in the fertility of polyploid crops have been correlated with occurrences of aneuploidy (Aastveit 1968, Weimarck 1973). Moreover, established polyploid crops generate substantially fewer aneuploid progeny than neopolyploid cultivars (Riley & Kimber 1961, Bingham 1968, see Web Table 5).

In allopolyploids, multivalents and homeologue bivalent pairing will generate unbalanced gametes that are numerically euploid (Howard 1938). In the allotetraploid, $A_1A_1A_2A_2$, homeologue bivalents (A_1A_1) will separate A_1A_1/A_2A_2 one half the time, generating gametes that contain only one homeologue. Also, A_1A_2/A_1A_2

separations will sometimes contain recombined segments, which in turn generate unbalanced gametes. First generation allopolyploids derived from unreduced gametes may inherit unbalanced, recombined genomes from homeologous bivalent pairing in the progenitor F1 hybrid (Howard 1938). In a diploid F1, one half of the unreduced gametes produced by homeologous pairing and subsequent formation of a restitution nucleus will be unbalanced for chromosomal segments. Only one quarter of the neopolyploids generated by the unreduced gametes would be a balanced euploid. Hence, infertility caused by homeologous pairing would result both from the production of unbalanced, inviable gametes by balanced sporophytes, and the presence of semisterile unbalanced sporophytes in populations. Some early-generation allopolyploids exhibit surprising variability in fertility that is consistent with a homeologue pairing model of polyploid infertility (Clausen et al. 1945, Stebbins 1949, Gajewski 1953, Stebbins & Vaarama 1954).

Cytological observations of fertility-selected neopolyploid populations generally suggest that improvements to seed set occur concomitant to changes in meiotic pairing behavior. In 10 of the 12 surveyed studies examining meiotic behavior of pre- and postselected plants, an increase in bivalent and a decrease in quadrivalent pairing was documented (see Web Table 8). For example, the mean frequency of bivalents in *Pennisetum typhoides* increased from 5.8 to 10.2 per pollen mother cell during six generations of fertility selection (Jauhar 1970). Overall, the percent increase of bivalent configurations in our survey averaged 4.1% per generation (range 1.6 to 12.6%) in those studies observing increased bivalent pairing (see Web Table 8). These data demonstrate that increased fertility of auto- and allopolyploids is often accomplished by increased bivalent associations.

Other results suggest that increases of quadrivalent pairing may, in some autopolyploids, also improve fertility. In autotetraploid *Lolium perenne*, six generations of fertility selection led to a 30% increase in the frequency of quadrivalent configurations (Crowley & Rees 1968). Hazarika & Rees (1967) induced autotetraploids in inbred lines of *Secale cereale*. Polyploid seed fertility varied substantially among lines and was positively correlated with the occurrence of chiasmata and quadrivalents. The differential effects of quadrivalents on seed and pollen fertility (see Web Table 1) probably reflect variation in quadrivalent orientations. As described earlier, alternate (zigzag) orientations are thought to split evenly, whereas adjacent orientations often have unbalanced separations. Autopolyploids with primarily alternate orientations of quadrivalents may increase quadrivalent frequencies in response to fertility selection, perhaps by increasing the frequency of chiasma (Sybenga 1969, 1975).

Correlation analyses of meiotic configurations versus pollen fertility from surveyed neopolyploids (see Web Tables 1, 7) implicate univalent and trivalent configurations as substantial contributors to reduced fertility. The sum occurrence of odd-numbered configurations is termed the chromosome disjunction index (Hazarika & Rees 1967). In our dataset, the disjunction index of neopolyploids is negatively correlated with pollen fertility (Spearman Rank Correlation, $N = 115$, $r_s = -0.470$, $P < 0.0001$). In contrast, quadrivalent configurations are uncorrelated with pollen viability ($r_s = -0.073$, $P = 0.4296$), and bivalent pairing only weakly correlated ($r_s = +0.346, P = 0.0002$). Due to variability in the quadrivalent orientation among taxa (McCollum 1958), quadrivalents may be negatively correlated with fertility in some systems, but not in others. In contrast, univalent and trivalent configurations always lead to the formation of aneuploid or pseudoeuploid gametes. Although the correlation between the disjunction index and pollen fertility is highly significant, the magnitude of the correlation is less than 1.0. There are several possible explanations. First, the true fertility of polyploids is typically lower than measures of pollen viability indicate (Tanaka 1940, Eskilsson 1963), so the actual correlation between fertility and meiotic configurations may be greater than what is reported here. Second, fertility may be influenced by subtle meiotic features, such as orientation, which are not commonly reported. Finally, as outlined below, polyploid fertility may be affected by factors besides meiotic aberrations.

Genic Factors

Meiotic aberrations play a clear role in neopolyploid fertility. There is, however, evidence that genic factors with no obvious meiotic effects may also influence fertility. For example, neopolyploid fertility varies substantially between lines and varieties within some species (see Web Table 7). Doggett (1964) induced polyploids in nine varieties of *Sorghum*, and found that percent seed set ranged from 40% to 77%; varietal differences were statistically significant, and highly heritable. Also, substantial increases in seed set of autopolyploids have been obtained by crosses between induced or spontaneous polyploids derived from different lines, strains, and varieties (Randolph 1941; Müntzing 1948b, 1951; Bingham 1980). For example, Doggett (1964) induced autotetraploids in two lines of *Sorghum* and found each had 31% seed set. F1 crosses averaged 68% fertility, and the F2 averaged 74%. In *Antirrhinum majus*, tetraploids induced in intervarietal hybrids averaged 86% pollen fertility, while intravarietal tetraploids averaged 55.3% (Sparrow et al. 1942). Autotetraploids derived from inbred lines of maize set few seeds, while outcrossed plants and interline hybrids exhibit high seed set (Randolph 1941). In *Sorghum*, *Antirrhinum*, and *Zea*, the increased fertilities of hybrids were not associated with obvious changes in chromosome pairing. For example, the mean frequency of multivalents in sterile and fertile lines of autotetraploid maize is fairly constant, between 7.5 and 8 (Randolph 1941), whereas fertility-selected lines exhibited equivalent meiotic pairings to controls (Mastenbroek et al. 1982). The nature of so-called genic fertility effects are obscure, but may relate to the effects of allelic heterozygosity on overall plant fitness (see below).

In some allopolyploids, genic sterility factors impede the development of pollen, ovules, embryos or endosperm but not meiosis per se. In these cases, polyploids of hybrid origin may be partially or completely sterile, but exhibit few or no meiotic abnormalities. Allotetraploid *Nicotiana sylvestris* × *tomentosa* and *N. sylvestris* × *tomentosiformis* exhibited bivalent pairing and >90% pollen viability, but was completely female sterile due to abortion of developing embryos 1–3 days after pollination (Greenleaf 1941). Backcrosses to the related tetraploid *N. tabacum* (tobacco) generated semi-fertile progeny, while F2 progeny of *N. tobacco* × *N. sylvestris-tomentosiformis* included a few fully sterile individuals. The sterility of the primary allotetraploids were hypothesized to result from complementary sterility genes, possibly few in number, which in certain allelic combinations inhibited ovule development (Greenleaf 1941). Genic and chromosomal sterility is well described in hybrids of diploid species. Allopolyploids generated by semisterile diploid hybrids are generally much more fertile than their progenitors (Web Table 7). Much hybrid sterility in plants may thus be hypothesized to result from unpaired or mispaired homeologous chromosomes and subsequent production of unbalanced gametes (Stebbins 1950), a difficulty corrected in polyploids via the duplication of all chromosomes. However, the fertility discrepancy between parental diploids and new allopolyploids (see Web Table 7; Figure 7) may in part reflect genic incompatibilities that are independent of meiosis. Because the gametes of polyploids are (at least) diploid, incompatibilities may involve both intra- and interlocus interactions. Further research is needed to investigate the impacts of genic incompatibilities on both the viability and fertility of hybrid polyploids.

Incidental Phenotypic Effects

In some cases, the infertility of polyploids is attributable to specific changes in the development, growth, or physiology of sporophytes or gametophytes. In the trivial case, polyploids may have altered patterns of allocation that reduce fecundity, which is subsequently misinterpreted as reduced fertility. For example, polyploids often have fewer (but bigger) flowers, fewer (but bigger) pollen and ovules, fewer (but bigger) seeds, and delayed flowering (Müntzing 1951, Jaranowski & Kalasa 1971, Roy & Dutt 1972, see below). Also, when grown in competition with more numerous progenitor diploid cytotypes, polyploids may appear semisterile due to triploid block (Randolph 1935, Olsson 1948, Hagberg & Ellerström 1959). Munzting (1951) found that induced tetraploid *Secale cereale* exhibited uniformly low seed unless grown in isolation plots away from diploid progenitors. Reduced fecundity of neopolyploids is probably often attributed to physiological infertility, though clear developmental sterility has been identified in some systems. In induced autotetraploid *Lactuca sativa*, ~20% of developing ovules abort during megasporogenesis (Einset 1944). Moreover, most pollen grains fail to germinate, or burst while growing through the style.

Population Genetics of Fertility Improvement

Fertility selection on neopolyploid populations rapidly increases pollen viability and seed set, often with corresponding increases in bivalent (or quadrivalent) pairing (see Web Table 8). It is difficult to imagine how a small population of polyploids could possess sufficient genetic variation to produce such rapid evolutionary change. Several mechanisms have been discussed, mostly focused on meiotic aberrations (Kostoff 1938; Howard 1938; Hazarika & Rees 1967; Sybenga 1969, 1973a). For autopolyploids, perhaps the simplest explanation involves allele substitutions at loci influencing the frequency and distribution of chiasma. Chromosome synapsis appears to be a polygenic trait (Sybenga 1975, Singh 1993). For example, numerous mutations are known to affect chiasma frequency or distribution in diploid rye, and polygenic inheritance of chiasmata has been shown in crosses of inbred lines (Prakken 1943; Rees 1955, 1961; Koduru & Rao 1981). Bivalent pairing in diploids is somewhat robust to chiasma formation, so chiasma factors may be neutral or under balancing selection and hence segregating in diploid populations. In neopolyploid populations, such alleles may quickly reach fixation due to strong effects on the type and orientation of configurations during meiotic metaphase. For example, increases in occurrences of chiasma in neopolyploids would limit unpaired chromosomes and laggards, commonly associated with polyploid sterility (Web Tables 1, 7, 8). Localization of chiasmata to specific chromosome regions may limit multivalent associations and cause a breakdown of prophase multivalents into metaphase bivalents (Levan 1940, Shaver 1962). Chiasma frequency and distribution also probably affect orientation, for example alternate versus adjacent configurations (Sybenga 1975). Hazarika & Rees (1967) identified variation for the occurrence of chiasma in inbred lines of diploid rye. The frequency of chiasma in induced autotetraploids was correlated with the frequencies in progenitor diploids, and also with the occurrence of quadrivalent configurations and seed fertility.

Genetic control of the frequency and distribution of chiasma may also underlie fertility improvement in new allopolyploids. Alternatively, genic factors could enforce homologous pairing by either altering the premeiotic alignment of chromosomes such that pairing of homeologous chromosomes does not occur, or increasing the stringency of crossover such that recombination only occurs between homologous chromosomes (Luo et al. 1996, Vega & Feldman 1998). Genic control of homologue pairing is known from several established allopolyploid crop species (e.g., Riley & Chapman 1958, Jauhar 1975, Evans & Aung 1985). In allohexaploid wheat, bivalent pairing is controlled in large part by a single gene, *Ph1*, located on chromosome 5B (Sears 1976). The origins of genic factors which influence meiotic pairing in allopolyploid crops and their role in the initial establishment and domestication of cultivars are generally unknown. Recent research in *Triticum* suggests that allelic variation in *Ph1* may exist in wild polyploid species, including wild tetraploid relatives of cultivated bread wheat (Hakan & Feldman 2001, Martinez et al. 2001). Thus hexaploid wheat may have inherited one or more homeologue pairing suppressor alleles directly from their polyploid progenitors at the time of formation, rather than evolving de novo genic control of homologous pairing in the face of strong fertility selection. This scenario may be common for hexaploids, octoploids and other higher-ploidy cytotypes that are derived from long-established polyploids, but presumably cannot explain rapid evolution of meiotic behavior in neopolyploids derived from diploids.

Changes in chromosome structure could also alter chiasma formation among homologous or homeologous chromosomes in neopolyploids, leading to preferential pairing and "diploidization" (Sybenga 1969, Feldman et al. 1997). Introduction of a reciprocal inversion in autotetraploid maize significantly reduced quadrivalent pairing and segregation of recessive alleles (Doyle 1963), but reciprocal translocations had no effect in autotetraploid rye (Sybenga 1973b). Somewhat increased bivalent pairing of newly formed autopolyploids has sometimes been achieved by inducing chromosomal rearrangements with radiation and mutagens (Srinivasachar & Singh 1967, Gottschalk 1978). Beneficial chromosomal rearrangements are most likely to evolve in selfing species, because identical rearrangements would rarely come together as pairs in an outcrossing population. The accumulation of chromosomal differences probably contributes in part to the evolution of meiotic regularity in selfing autopolyploids (Sybenga 1969), but in isolation are probably inadequate to explain rapid transitions of pairing behavior and fertility. In contrast, chromosomal rearrangements can have a strong effect on chromosome pairing in allopolyploids (Shaver 1963, Sybenga 1973a). For example, inversions greatly reduced allosyndesis in allotetraploid *Zea mays* × *Euchlaena* perennis, but caused only slight preferential pairing in autotetraploid maize. Elimination of DNA sequences may similarly differentiate homeologous chromosomes and thus enforce homologous pairing in allopolyploids (Feldman et al. 1997, Liu et al. 1998a). Hence, the evolution of chromosome structure may strongly limit pairing of homeologues that already differ in the number and position of chiasma initiation points, or the timing of chromosomal processes (e.g., long distance attraction and condensation) (Sybenga 1969).

DEFINITIONS OF AUTO- AND ALLOPOLYPLOIDY

Definitions of auto- and allopolyploidy either emphasize mode of origin (hereafter, MO) or cytological criteria (hereafter, CC) as primary criteria. MO autopolyploids arise within single populations or between ecotypes of a single species, whereas allopolyploids are derived from interspecific hybrids (Müntzing 1936, Darlington 1937, Burnham 1962, Gottschalk 1978). This is the definition used in this chapter. CC allopolyploids are expected to display bivalent pairing, lack of allosyndesis and disomic inheritance, while CC autopolyploids will exhibit multivalent configurations, nonpreferential pairing at metaphase, and multisomic inheritance (Stebbins 1980, Jackson 1982, Jackson & Jackson 1996).

Our review of newly formed auto- and allopolyploids indicates that cytological behavior at the time of origin is more similar than might be expected. For example, 80% of newly formed MO allopolyploids displayed multivalent pairing (Figure 2; see Web Table 1) and are expected to undergo some degree of homeologous pairing. Intergenomic segregation in the progeny of MO allopolyploids was postulated to occur in approximately two thirds of surveyed studies (see Web Table 3). Hence, strict CC allopolyploid definitions would exclude many known

MO neopolyploids, including the famous "*Primula kewensis*" (Upcott 1940), "*Galeopsis tetrahit*" (Müntzing 1932), and possibly "*Raphanobrassica*" (Richharia 1937, Howard 1938). For their part, early generation MO autopolyploids correspond to most definitions of CC autopolyploids, because there is no evidence of preferential pairing or disomic inheritance (see Web Tables 1, 2). However, multivalent configurations can be much less common than is sometimes expected (Figure 2). In later generations, accumulation of chromosomal rearrangements, evolution of genic factors controlling meiotic pairings or hybridization may lead to diploidized MO autopolyploids that cytologically behave like CC allopolyploids.

In short, students of polyploidy are faced with a choice: either (*a*) recognize intuitive, mode of origin polyploid categories, but expect heterogeneity in the genetic characteristics of autopolyploids and allopolyploids; or (*b*) recognize categories that rely on posthoc cytological analyses of polyploids, but maintain homogeneity in the characteristics of polyploid types. The terms genomic and segmental allopolyploids were proposed to distinguish MO allopolyploids with strict versus weak preferential pairing of homologues (Stebbins 1947, 1950) but are now also used to distinguish degrees of CC allopolyploidy (Jackson & Jackson 1996). For the purposes of studies of ecology and systematics, we advocate the use of mode-of-origins definitions of polyploidy. Irrespective of cytological characteristics, the biology of polyploids derived from interspecific hybrids is quite different than the biology of intraspecific polyploids. MO allopolyploids are formed by the breakdown of reproductive isolation between species, followed by polyploidization in a meiotically or mitotically unstable, semifertile hybrids (Ramsey & Schemske 1998). In contast, MO autopolyploids form by the production of occasional meiotic or mitotic aberrations in single species populations. Although derivatives of multisomic allopolyploids will be more polymorphic than the progeny of fixed disomic allopolyploids, all allopolyploids retain a degree of hybrid character unlike anything found in an autopolyploid. Conversely, the evolution of preferential pairing in MO autopolyploids via chromosome rearrangements or genic control of pairing would not lead immediately to the hybridity of allopolyploids.

IS POLYPLOIDY REVERSIBLE?

Conventional wisdom holds that chromosome evolution via polyploidy is a one-directional process: In a given lineage, ploidy level either increases through time, or remains constant. The production of occasional haploid progeny by otherwise polyploid species challenges this assertion (Randolph & Fischer 1939, Raven & Thompson 1964). These so-called polyhaploids halve the chromosome number of their progenitors, but in the case of tetraploids or octoploids will have a balanced diploid or tetraploid chromosome number. Establishment of polyhaploid mutants in a polyploid lineage may lead to the evolution of diploidy (Raven & Thompson 1964).

Here we examine the mechanistic potential of ploidy reversal by reviewing the occurrence, cytological origins and characteristics of polyhaploids in established and neopolyploid systems. Out of 55 reviewed studies, 6 (10.9%) reported the occurrence of polyhaploids (see Web Table 10). In comparison, 52 studies (94.5%) reported aneuploids, while 5 studies (9.1%) found higher polyploids (e.g., neohexaploids in the progeny of tetraploids) (see Web Table 10). In the 6 investigations reporting spontaneous polyhaploidy, frequencies of polyhaploids were generally low (see Web Table 11). For example, Randolph & Fischer (1939) found 23 parthenogenetically produced diploids among 17,165 progeny of neotetraploid maize. Higher frequencies of polyhaploids have been observed in several colchicine-induced neopolyploids (Pundir et al. 1983, Singh 1986). In such circumstances, polyhaploids are often suspected to have originated via ploidy chimaerism (a common artifact of colchicine treatment) coupled with selfing. Excluding these cases, polyhaploids were usually attributed to apomixis (see Web Table 11). Apomictic polyhaploidy was observed to occur spontaneously in seven systems, while in three systems it was induced via interspecific pollinations (see Web Table 11). Polyembryony (twinning) is also associated with a high incidence of haploid progeny in plants, probably via reduced parthenogenesis (Webber 1940, Müntzing 1948a, Dewey 1961). However, the overall frequency of polyembryonic seeds in populations is low.

In evaluating the characteristics of polyhaploids, we pooled induced and spontaneous occurrences. In most systems, polyhaploids were found to have reduced viability, pollen viability, and seed fertility. Eight of ten studies described the polyhaploids as having reduced growth, survivorship and/or reproduction. The polyhaploids were often characterized as being small and slender in appearance (e.g., Lesins 1957, Dewey 1961), opposite the gigas characteristics of polyploids. Mean pollen viability was 6.5% ($N = 10$), as compared to 87.9% ($N = 4$) in diploid or polyploid controls. Mean seed fertility of polyhaploids was 9.4% ($N = 8$). Qualitative assessments of fertility generally paralleled these quantitative data (see Web Table 11). In several studies, diploid polyhaploids were selfed or crossed to related diploids and tetraploids. Tetraploids were the most common progeny reported, followed by triploids, diploids, and aneuploids (Müntzing 1948a, Gerstel & Mishanec 1950, Lesins 1957), suggesting polyhaploids are characterized by frequent asynapsis. Several factors may contribute to the overall reduced fitness of polyhaploids. Polyhaploids derived from allopolyploids (or hybridized autopolyploids) will harbor homeologous chromosomes of different progenitor species, and are expected to be somewhat asynaptic and sterile. In the case of autopolyploids, the presence of multiple chromosome sets may allow the maintenance of deleterious alleles and chromosomal deficiencies that are expressed more in polyhaploids than progenitor tetraploids (Müntzing 1948a, Dewey 1961).

This survey suggests that ploidy reduction via polyhaploidy is, on average, an unlikely event. Excluding studies of suspected chimeric colchiploids, we estimate the occurrence of polyhaploids to be 0.19% (6 of 55 studies reporting polyhaploids, times average frequency of 1.8% in those studies finding polyhaploids). Polyhaploids were reported to have low viability and fertility (average 7% fertility of related diploid and polyploids) and do not perpetuate themselves efficiently. The conditions for establishment of neopolyploids are generally regarded as restrictive

(Felber 1991, Rodríguez 1996). Polyhaploid mutations face the same frequency-dependent selection that limits polyploid establishment, and are characterized by lower average viability, fertility and cytological stability (see Web Table 11 versus Web Tables 4, 5, 6, 7). Our survey supports the view of Stebbins (1980) that polyhaploidy is unlikely to contribute to reversions to diploidy in most established polyploid species.

ARE ALLOPOLYPLOIDS "CONSTANT SPECIES HYBRIDS?"

In an influential paper, Winge (1932) summarized early investigations of experimental allopolyploids, and hypothesized that hybrid polyploids represent constant species hybrids with "the quality of a pure species." This description was intended to contrast with the character of interspecific diploid hybrids, which segregate extensively in F2 and backcross generations due to allosyndesis and subsequent random segregation of homeologs during meiosis. Ecologists and systematists took their cue from Winge's characterization, and naturally occurring established allopolyploids were expected to be fixed heterozygotes that are phenotypically and ecologically intermediate to their related diploid species (e.g., Clausen et al. 1945, Lewis & Lewis 1955). The primary criteria for identifying allopolyploidy include fixed intermediacy of morphology and genetic markers, bivalent configurations during meiotic metaphase, and disomic inheritance (Stebbins 1950, 1971; Soltis & Soltis 1993).

At a broad stroke, Winge's characterization is certainly correct. Because of preferential pairing of duplicate chromosomes, allopolyploids are more stable than diploid hybrids. However, cytogeneticists often concluded that newly formed allopolyploids were considerably less stable than Winge's initial description would indicate (e.g., Poole 1932, Lindstrom & Humphrey 1933, Richharia 1937, Howard 1938, Kostoff 1938, Davis 1943, Gajewski 1953, Grant 1954, Gerstel & Phillips 1958, Phillips 1964, Day 1965). First, neoallopolyploids display occasional to frequent allosyndesis and multivalent pairing during meiosis (Figure 2; see Web Table 1). Second, allopolyploids sometimes have multisomic rather than disomic inheritance (Figure 3; see Web Table 2) and segregate for parental characteristics in subsequent generations (see Web Table 3). Third, allopolyploids generate large numbers of aneuploid gametes and progeny (Figure 4a,b; see Web Tables 4, 5). Finally, fertility is observed to be low in first-generation allopolyploids (Figure 7; see Web Table 7), but improves over time (see Web Tables 8, 9).

Kostoff (1938) analyzed the origin, meiotic behavior, chromosome balance, fertility and phenotypic characteristics of early-generation allopolyploid *Nicotiana glauca* × *langsdorfii*. The diploid parental species, *N. langsdorfii* ($x = 9$) and *N. glauca* ($x = 12$), are phenotypically differentiated for a number of growth, leaf and floral characters, and can be crossed (with difficulty) to form a highly sterile F1 hybrid. An allopolyploid ($2n = 4x = 42$) F2 had the intermediate morphological character of the diploid F1, but had gigas characters, including larger and coarser leaves, stems and flowers. The F2 was semifertile (51% and 19% pollen and seed fertilities), exhibited some multivalent and homeologous bivalent pairing, and frequent unequal chromosome separations at anaphase I (see Web Tables 1, 4). With four generations of fertility selection, Kostoff (1938) substantially increased frequencies of bivalents, euploid gametes, euploid sporophytes, pollen and seed fertility (Figure 8a,b,c,d,e). Concomitantly, segregation of the morphological differences of the progenitor diploids was found. For example, the mean corolla length of allotetraploid families differed substantially and was more variable than that observed in parental species or F1 hybrids (Figure 8f,g,h,i,j). Kostoff

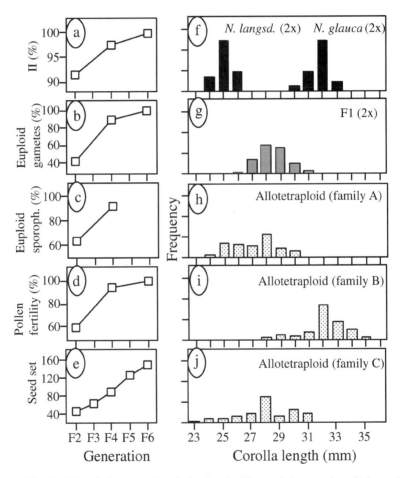

Figure 8 Rapid evolution of meiotic behavior, fertility, and phenotypic traits in neotetraploid *Nicotiana langsdorfii* × *glauca*. Mean percent bivalent pairing (a), aneuploid gametes (b), aneuploid sporophytes (c), pollen and seed fertility (d, e) of four fertility-selected generations derived by selfing a single spontaneous F2 allopolyploid. Corolla lengths of diploid *N. langsdorfii* and *N. glauca* (f), their diploid F1 hybrid (g), and three allotetraploid F4 families (h, i, j). Data from Kostoff (1938).

(1938) concluded that the combination of aneuploidy and recombination of genes on homeologous chromosomes of the allotetraploid created phenotypically heterogeneous progenies segregating for parental characteristics. The evolution of preferential pairing decreased the occurrence of aneuploids and homeologous recombination, so that by the F5 generation little intrafamily variation was observed. By the F6 generation there were some plants more or less reconstituting the phenotypes of the parental species, and many varying intermediates (Kostoff 1938).

Allopolyploids may sometimes give rise to polymorphic species consisting of varying amounts of the chromosomes of each parental species (Kostoff 1938; Sybenga 1969, 1996). Kostoff (1938) speculates "it seems that the constancy of the amphidiploids is very questionable ... the process of meiosis in the majority of the amphidiploids recorded by various authors suggests that they should not be constant, and most of them actually producing inconstant progeny." After similar experiences with allotetraploid *Oenothera*, Davis (1943) concludes "accounts of amphidiploids have frequently assumed that these plants even from hybrids would breed true ... but as more examples have been investigated it has become evident that irregularities of chromosome distribution at meiosis are common and that the pairing of sister chromosomes may not take place as often as might be expected."

The occurrence of unstable neopolyploids presents a challenge to systematists seeking to evaluate the phylogenetic origins of natural polyploids, and ecologists examining the adaptive significance of ploidy variation in species populations. Many criteria for distinguishing natural auto- and allopolyploids may be reliable for distinguishing only recently derived autopolyploids and genomic allopolyploids (Goodspeed & Bradley 1942, Grant 1975, Sybenga 1996), especially given the possibilities of the extinction of progenitors and interspecific crosses at the polyploid level. Moreover, it may be difficult to evaluate the ecological significance of some established allopolyploids when cytotype differences are created by a combination of polyploidy per se, homeologous recombination and segregation, and genic evolution.

POLYPLOIDY AND PHENOTYPIC EVOLUTION

Polyploids initially attracted attention because of their unique cytogenetics and their reproductive isolation from diploids (Blakeslee 1921, Jørgensen 1928). It was soon recognized that polyploids also exhibited distinctive phenotypic traits (Müntzing 1936, Randolph 1941). Comprehensive summaries of the characteristics of polyploids have been published previously (e.g., Tal 1980; Levin 1983, 2002; Lumaret 1988), but have not always differentiated between early generation spontaneous or induced polyploids and naturally occurring, established polyploids.

Neopolyploids are commonly differentiated from progenitor diploids by a combination of morphological, reproductive, phenological, life-history, and physiological traits (Table 1). For example, new polyploids commonly exhibit gigas characteristics, including sturdier foliage, thicker stems, and enlarged reproductive structures that are typically less numerous than in progenitor diploids. In comparison to the progenitor F1 hybrid, spontaneous allotetraploid *Gossypium davidsonii* × *anomalum* exhibit thickened leaves and enlarged flowers and seeds (Brown 1951). Growth and development is often slowed in neopolyploids, leading to a delayed and prolonged flowering phenology. For example, the flowering of induced autotetraploids of several *Medicago* and *Melilotus* spp. was extended weeks or months beyond diploid progenitors, due to a combination of lush growth and reduced fertility (Jaranowski & Kalasa 1971). The life-history of new polyploids is often distinctive. For example, induced autotetraploids of *Agropryon*, *Lamium*, *Pennisetum*, and *Secale* produced substantially fewer tillers than progenitor diploids (Table 1). Water relations, photosynthetic rates, and other physiological traits are known to differ between neopolyploids and progenitors. Induced tetraploid barley exhibited reduced respiration and transpiration rates compared to progenitor diploids (Chen & Tang 1945). Although there are no data on the ecological interactions of neopolyploids, they are likely to differ from those of diploids. For example, delayed phenology may alter associations with pollinators (Segraves & Thompson 1999), whereas changes in secondary chemistry (Kostoff 1938, Sullivan & Myers 1939) could influence plant-herbivore interactions. Most observations of neopolyploids are based in autopolyploid systems. First generation allopolyploids typically exhibit a phenotype intermediate to their diploid parents, though they differ from progenitor interspecific F1 hybrids in a manner similar to the difference between autopolyploids and progenitor diploid parents (e.g., Buxton & Darlington 1931, Davis 1943, Clausen et al. 1945, Gajewski 1953). Later generation allopolyploids may or may not segregate for parental characteristics (see Web Table 3).

Causes of Phenotypic Alterations

To develop a conceptual framework for evaluating the contributions of polyploidy to phenotypic evolution, we briefly summarize the mechanistic bases of phenotypic alteration.

INCREASED DNA CONTENT   Transitions to polyploidy are accompanied by physical alterations in the size and geometry of cells, which in turn may affect biochemistry, development, anatomy, and ultimately whole-plant morphology, growth, and physiology. Many of the gigas traits of neopolyploids, such as increased cell size, enlarged floral structures, sturdier foliage, and robust stems, may be a direct consequence of increased DNA content (Randolph 1941). Also, the characteristic slowed growth, altered phenology, and prolonged flowering of polyploids may, in part, result from slowed mitotic divisions and cell divisions of larger cells with more chromosomes (Noggle 1946, Stebbins 1971). The contributions of DNA content are perhaps the most consistent effects of polyploidy and may characterize both neo- and established polyploids.

TABLE 1   Common phenotypic characteristics of newly-formed auto- and allopolyploids

Trait type	Characteristic	Genus (reference)
Morphology	Leaves coarser, thicker, larger	*Amaranthus* (Pal & Khoshoo 1977), *Manihot* (Hahn et al. 1990), *Medicago* (Bingham & Binek 1969), *Solanum* (Jørgensen 1928)
Reproductive character	Flowers larger	*Allium* (Levan 1941), *Digitalis* (Buxton & Darlington 1931), *Gossypium* (Beasley 1940; Brown 1951), *Zea mays* (Randolph 1935)
	Seeds larger, but fewer	*Brassica* (Howard 1939), *Dactylis* (Bretagnolle & Lumaret 1995), *Triticum* (Kuspira et al. 1985), *Vicia* (Nordenskiöld 1953)
Phenology	Flowering initiated later	*Artemisia* (Clausen et al. 1940), *Crepis* (Navashin 1925), *Cucumis* (Shifriss 1942)
	Flowering duration longer	*Capsicum* (Raghuvanshi & Joshi 1964), *Luffa* (Roy & Dutt 1972), *Melilotus* (Jaranowski & Kalasa 1971), *Ocimum* (Bose & Choudhury 1962)
Life-history	Reduced tillering	*Agropyron* (Tai & Dewey 1966), *Lamium* (Bernstrom 1954), *Pennisetum* (Jauhar 1970), *Secale* (Müntzing 1951)
Physiology	Altered water relations	*Hordeum* (Chen & Tong 1945), *Solanum* (Tal & Gardi 1976)
	Altered photosynthesis	*Mimulus* (Hiesey et al. 1971), *Thalictrum* (Mooney & Johnson 1965)

ANEUPLOIDY, CHROMOSOME SUBSTITUTION, AND RECOMBINATION OF HOMEOLO-GOUS CHROMOSOMES   As described earlier, the cytological instability of neopolyploids generates substantial chromosomal and genic variation. On average, nearly 40% of the progeny of neopolyploids are expected to be aneuploid or pseudo-euploid (see Web Tables 4, 5). Poly-aneuploids are typically fertile and exhibit distinctive phenotypes that contribute to phenotypic variability of populations (Müntzing 1937, Bernstrom 1954). For example, Kostoff (1938) concludes that "aneuploidy augments the numbers of new forms in the progenies of the am-phidiploid *Nicotiana glauca × langsdorffii* and leads to more striking divergence in the new forms." In allopolyploids, segregation of parental traits is another fac-tor contributing to phenotypic variation (see Web Tables 2, 3). Derivatives of allopolyploids may thus be polymorphic, and include a variety of combinations of parental genomes (Kostoff 1938, Sybenga 1969). Molecular data may corroborate these hypotheses (Soltis & Soltis 1999). Song et al. (1993, 1995) created syn-thetic allopolyploids from diploid *Brassica* species and compared nuclear RFLP genotypes of parents and polyploid lines derived by selfing a single, homozygous progenitor. Genomic changes were frequent, with evidence for the gain and loss of parental alleles, as well as the appearance of unique DNA fragments. The degree of genetic change was high, with genetic distances between S6 plants and their parents ranging from 2.1% to 9.6%. Moreover, polyploid lines displayed substantial heritable variation for phenotypic traits (Schranz & Osbourne 2000). For example, days to flowering of S6 progeny varied from 39–75 days, with parent-offspring regression indicating high heritability of this trait. Hence, "de novo" genetic and phenotypic variation may be induced by allopolyploidy in *Brassica* (Schranz & Osbourne 2000). Intergenomic recombination was proposed to be the major cause of the observed genome changes (Song et al. 1995), but other factors could also be involved (see below).

GENE DOSAGE EFFECTS   Polyploidy causes an immediate increase in gene dosage throughout the genome. With more gene copies, polyploids may have increased (or altered) gene expression, which would potentially affect many phenotypic traits. Guo et al. (1996) compared gene expression in leaves of $1x$, $2x$, $3x$, and $4x$ maize, and found that most studied loci were expressed in proportion to dosage. In contrast, experiments using aneuploid dosage series revealed extensive alterations in expression (Guo & Bircher 1994). Guo et al. (1996) proposed that chromosomal imbalance in aneuploids would change the stoichiometry, resulting in altered gene expression, whereas in euploid series, dosage is changed proportionally. These results may explain the common observation that aneuploids, especially at the diploid level, exhibit more deviant phenotypes than polyploids (Khush 1973). Further research is needed to link dosage effects with specific phenotypic characters of polyploids and aneuploids.

A particular type of dosage effect relates to the loss of self-incompatibility (SI) in polyploids (Stout & Chandler 1941, Pandey 1968). Crossing experiments between self-compatible polyploids and self-incompatible diploids indicate that the breakdown in the SI reaction occurs in pollen. Hence, it is generally believed that interactions between incompatibility alleles in diploid pollen grains will induce self compatibility in systems where the SI reaction is determined by the genotype of pollen (i.e., gametophytic SI) (Lewis 1947). We surveyed the literature to determine the frequency of SI breakdown in neopolyploids (see Web Table 12). In sporophytic systems in which the SI reaction is determined by the sporophyte genotype rather than the pollen genotype, no breakdown was observed (N = 6 studies; see Web Table 12). Moreover, no loss of SI was observed in monocot species with either 1- or 2-locus gametophytic SI (N = 3 studies; Web Table 12). Among dicot taxa with 1-locus gametophytic SI, the induction of polyploidy generated self-compatible plants in 8 of 10 systems (Web Table 12).

In systems exhibiting a loss of SI, self-pollination generally resulted in variable seed set, which, on average, was considerably less than that observed from outcrossing. For example, mean seed set from self pollinations on neotetraploid *Trifolium hybridum* was 8% of that generated from outcrossed pollinations (Armstrong & Robertson 1956). In neopolyploid pear, seed set from self pollinations was about 60% that of outcross pollinations (Crane & Lewis 1942). Limited seed set observed from selfing self-compatible neopolyploids is probably explained by varying degrees of competition between alleles in heterozygotes, though inbreeding depression may also be a factor.

INFERTILITY   Neopolyploids generally exhibit reduced fertility compared to their diploid progenitors (see Web Table 7). Abortion of developing pollen grains and ovules, coupled with reduced seed development, alters growth and allocation late in the life cycle. The common observation that neopolyploids have prolonged flowering and luxurious growth late in the season (Table 1) may be a simple result of infertility.

ALLELIC DIVERSITY   For a given locus, diploids will possess a maximum of two alleles. Increased gene dosage allows polyploids to harbor three or more alleles per locus, and hence exhibit greater overdominance than a diploid (Bingham 1980, Bever & Felber 1992). The proposed advantage is that multi-allelic, overdominant loci will allow polyploid individuals to achieve higher fitness in a wider diversity of environments than diploids. There is strong evidence that the growth, fertility, and yield of polyploid crops is correlated with heterozygosity, as illustrated by the performance of intervarietal autopolyploid hybrids as well as neopolyploids derived from highly heterozygous unreduced gametes versus somatic doubling (Bingham 1980, Stebbins 1980, Werner & Peloquin 1991).

In spite of the clear importance of allelic heterozygosity in plant breeding, there are reasons to doubt the impact of allelic diversity in early-generation neopolyploids in natural populations. First, it is unlikely that a few individual diploid progenitors of first-generation polyploids will possess the required allelic diversity, i.e., three or more alleles per locus (e.g., Bretagnolle & Lumaret 1995). Moreover, it is improbable that an entire diploid progenitor population would possess three or more overdominant alleles at a single locus. As Futuyma (1998, p. 385) concludes, "only under exceptional circumstances can heterozygous advantage maintain three or more alleles as a stable polymorphism." Second, only strong overdominance will maintain substantial heterozygosity in self-fertilizing populations thus neopolyploids derived from self-fertilizing progenitors are unlikely to possess overdominant alleles. Finally, as noted by Bever & Felber (1992), allelic overdominance in diploids is rare.

GENETIC LOAD   With the potential for harboring three or more alleles per locus, neopolyploids may express deleterious codominant and recessive alleles less often than their progenitor diploids. Because the genetic load of early generation polyploids is the same as that of diploids, and the probability of producing a recessive homozygote is low, the mean fitness of neopolyploids may be consistently higher than that of diploids. Recent simulations show that new autopolyploids will, in fact, possess lower genetic load than diploids, but the advantage declines through time as a function of the genetic basis of inbreeding depression (Otto & Whitton 2000). The expected magnitude of inbreeding depression in polyploids has been estimated at mutation–selection equilibrium. Assuming that inbreeding depression is due to recessive deleterious alleles, Lande & Schemske (1985) proposed that autopolyploids possess half the inbreeding depression of diploids. In contrast, Ronfort (1999) concluded that under the model of complete recessivity, diploids and autopolyploids should have equivalent inbreeding depression. Estimating the expected inbreeding depression in the more plausible case where deleterious alleles are partially dominant is complicated by the fact that polyploids produce several heterozygous genotypes. Such an analysis revealed that inbreeding depression at equilibrium can decrease or increase with changes in the selfing rate (Ronfort 1999).

There are few studies comparing inbreeding depression of polyploids and related diploids. Results are mixed for populations of established cytotypes (Kalton et al. 1952, Dewey 1966, Johnston & Schoen 1996, Husband & Schemske 1997). The few studies of early generation polyploids indicate lower inbreeding depression in neopolyploids than diploids. Dewey (1969) found a 17% reduction in forage yield in selfed neotetraploids of *Agropyron cristatum* as compared to a 55% reduction for their selfed diploid progenitors. These results are consistent with observations of neopolyploids in *Lolium multiflorum*, maize, and clover (Dewey 1969). There are no empirical data examining inbreeding depression in newly formed allopolyploids, or theoretical models that evaluate the evolution of genetic load in allopolyploid populations. The genetic system of allopolyploids limits allosyndetic pairing that would lead to the production of homozygous genotypes, so inbreeding depression in new allopolyploids is expected to be very low (Sybenga 1969). The fixed heterozygosity of allopolyploids may be an important cause of their evolutionary success.

GENOMIC EVOLUTION   In addition to homeologous recombination (discussed above), genome evolution in polyploids may involve gene silencing, divergence of gene function, and other processes (Wendel 2000). Many aspects of genome evolution occur over long timescales, but some mechanisms could cause substantial genomic evolution in early-generation polyploids. Feldman et al. (1997) and Liu et al. (1998a) found that nonrandom elimination of noncoding DNA sequences in allotetrapoid and allohexaploid wheat contributes to molecular diploidization and the diploid-like meiotic behavior of early generation polyploids. A parallel study of coding sequences identified loss of parental fragments and the appearance of novel fragments, but no elimination of parental sequences (Liu et al. 1998b). These changes may cause gene inactivation, reduced expression, and functional diversification. Liu et al. (1998b) conclude that the changes in coding regions are a result of methylation, not intergenomic recombination.

Although gene silencing via mutation is thought to be a slow process, epigenetic control of gene expression can be immediate (Wendel 2000). The instability and infertility of early generation allopolyploids may reflect intergenomic incompatibilities, which can be resolved by gene silencing (Comai 2000). Comai et al. (2000)

produced synthetic allopolyploids of *Arabidopsis thaliana* × *Cardaminopsis suecica* and estimated that 0.4% of genes are silenced, some by epigenetic regulation. Epigenetic gene silencing is also observed in nonhybrid systems. Scheid et al. (1996) observed epigenetic regulation of a transgene in *Arabidopsis thaliana*, with reduced expression in triploids as compared to diploids. They suggest that epigenetic silencing in plants is a direct response to changes in chromosome number or nuclear DNA content that could have substantial effects on gene regulation.

POPULATION GENETICS   Autopolyploids exhibit multisomic inheritance that alters basic population genetic processes. It might be asked whether neopolyploids can respond more rapidly to selection and thus be more "adaptable" than progenitor diploids. Single-locus theory shows that polyploids may evolve faster or slower than diploids, depending on the dominance coefficients of advantageous alleles. For an additive, completely dominant allele, Hill (1971) found that the response to selection was always greater in diploid than in autotetraploid populations. Otto & Whitton (2000) showed that tetraploids with tetrasomic inheritance will evolve faster than diploids when $h_1 > h/2$, where h_1 and h are the dominance coefficients for the advantageous mutant allele in tetraploids (*AAAa*) and diploids (*Aa*), respectively. In tetraploids with disomic inheritance, an advantageous mutant allele will become fixed in only one of the two gene copies. In this case, tetraploids will evolve faster than diploids if $h_1 h_2 > h/2$, where h_2 is the dominance coefficient of the mutant allele in tetraploid genotype *AAaa* (Otto & Whitton 2000). See Bever & Felber (1992) for a comprehensive review of the population genetic consequences of polyploidy.

Ignoring epigenetic sources of genetic variation and that due to aneuploidy, chromosomal rearrangements, and homeologous recombination, neopolyploids depend on the genetic variation present in the founding population. In most systems, the effective population size of early generation neopolyploids will be exceedingly small, even in cases of multiple formation. This genetic bottleneck will reduce the genetic variation available to a new polyploid population, and the small population size will limit input of new variation by mutation. Adaptive evolution by the fixation of individual alleles may be relatively unimportant during the early phases of polyploid establishment.

FITNESS AND ADAPTATION

Polyploidy may contribute to local adaptation in several distinct ways. First, neopolyploid populations may exhibit increased phenotypic variability due to the combined effects of aneuploidy, chromosome substitution, homeologous recombination, and epigenetic changes. Although auto- and allopolyploids both tend to display more variability than progenitor diploids, allopolyploids probably vary more due to the greater phenotypic consequences of numerical representation and recombination of homeologous than homologous chromosomes. According to this model, polyploidy contributes to adaptation by making populations adaptable to a wider range of environmental conditions. Second, polyploidy may induce immediate phenotypic changes that incidentally preadapt plants to a new ecological niche. For example, increased DNA content, gene dosage effects, and masked genetic load could alter the traits of a species in such a way as to increase its survival and reproduction in a novel environment. According to this hypothesis, the important phenotypic effects of polyploidy are imbued immediately and are in some ways predictable.

Although polyploidy is widely believed to be a mechanism of local adaptation (Clausen et al. 1945, Levin 1983), little is known about the fitness of neopolyploids (or, in fact, established polyploids) in different environments. The seed yield of early-generation polyploid crops is generally reduced compared to progenitor diploids (e.g., Hagberg & Ellerström 1959, Tai & Dewey 1966, Jaranowski & Kalasa 1971). In experimental plantings of rye, induced autotetraploids exhibited fewer flowering tillers, fewer flowers per spike, and lower seed set than diploids (Müntzing 1951). However, tetraploid seeds were 50% larger than diploid seeds and germinated better. In general, yield differences are attributed to the reduced fertility of neopolyploids and do not reflect possible differences in survival and growth in natural field conditions involving environmental stress and competition. There is a general need for investigations of the fitness of neopolyploids in natural plant species, especially when grown in the field. Stebbins (1949, 1985) induced autotetraploids in the exotic annual grass *Ehrharta erecta* and planted seeds and seedlings in 22 environmentally contrasting sites in central California. Diploid plants were demographically successful in most sites, but tetraploids survived in only two. In these two transplant sites, tetraploids were less numerous than diploids and occurred in a narrower range of environmental conditions. Bretagnolle & Lumaret (1995) examined the phenotypic characteristics and reproductive fitness of several neotetraploid clones of *Dactylis glomerata* generated by unreduced gametes. Diploids and tetraploids were grown in a common garden with four environmental treatments. In all environments, tetraploids had lower seed set and similar total biomass compared to diploids.

CONCLUSIONS AND FUTURE DIRECTIONS

We find that early-generation autopolyploids are characterized by random associations of homologous chromosomes, variable degrees of multivalent pairing, multisomic inheritance, and frequent production of aneuploid and pseudoeuploid gametes and progeny. Although some newly formed allopoyploids exhibited characteristics of true genomic allopolyploids, most exhibited occasional allosyndesis and multivalent configurations, inheritance intermediate to disomic and multisomic models, and frequent production of aneuploid gametes and offspring. Hence, the cytogenetic character of newly formed auto- and allopolyploids differ statistically, but are not as distinct as might be expected. In particular, allopolyploids may not always represent constant species hybrids due to allosyndetic pairing and homeologous recombination.

Although aneuploids and pseudoeuploids are commonly observed in the progeny of neopolyploids, their evolutionary significance is poorly understood. Aneuploidy may contribute to phenotypic variability, and hence adaptability, of neopolyploid populations, but also represents a form of genetic load. In the case of allopolyploids, homeologous recombination may also contribute to genetic and phenotypic variation in populations. A critical evaluation of the evolutionary significance of aneuploidy and homeologous recombination will be difficult, requiring measurements of the occurrence, phenotypes, progeny, and fitness of aneuploid and segregant genotypes in different environments.

Autopolyploids are traditionally thought to be less fertile than allopolyploids, and hence less likely to establish and maintain viable populations. Although we find that neopolyploids exhibit significantly reduced pollen and seed fertility compared to their diploid progenitors, there are no differences between the fertilities of newly formed auto- and allopolyploids. The fertility of neopolyploids may thus be a general barrier to establishment of both polyploid types. Rapid evolution of pollen and seed fertility is often observed in fertility-selected neopolyploid populations, suggesting infertility of neopolyploids may be somewhat transient. However, comparisons of naturally established low and high ploidy cytotypes indicate that polyploidy may set an upper limit to the fertility of a cytotype. Additional studies are needed to evaluate the demographic consequences of fertility because neopolyploids may compensate for reduced fertility by their increased survival and growth.

The causes of polyploid infertility are complex and include meiotic aberrations, physiological effects of polyploidy, and genic factors. Meiotic aberrations related to univalent and multivalent pairings probably represent the primary cause of sterility in most polyploids. Fertility selected neopolyploids nearly always show an increase in bivalent pairing, or else quadrivalent pairing. In our datasets, pollen fertility is significantly correlated with the occurrence of univalents and trivalents, but not with quadrivalent configurations. Increased fertility of autopolyploids is hypothesized to occur primarily by the fixation of alleles controlling the frequency and distribution of chiasma. In contrast, fertility improvement of allopolyploids probably involves suppression of homeologue pairing via genic control of pre-meiotic chromosome alignment or the stringency of crossover. Evolution of chromosome structure (e.g., chromosomal rearrangements or sequence elimination) may also be an important factor in allopolyploids. Further cytological studies, perhaps in combination with linkage mapping approaches, will be necessary to unambiguously identify the genetic architecture of fertility improvement in neopolyploids.

Neopolyploids often differ from their diploid progenitors by a combination of morphological, reproductive, phenological, and life-history traits. The distinctive characteristics of neopolyploids are probably caused by a variety of factors, including increased DNA content and cell size, gene dosage effects, aneuploidy, homeologous recombination, masked genetic load, and epigenetic changes.

The combined dataset regarding the cytogenetics and phenotypes of neopoly-ploids is deep, and historical investigations of cultivars and model systems have elucidated many of the characteristics of newly formed auto- and allopolyploids. However, to evaluate questions related to the establishment of polyploid popula-tions and the nature of polyploidy as an evolutionary mechanism, it will be neces-sary to link classical cytogenetic approaches with investigations of neopolyploid formation in natural populations. In spite of the widespread belief that polyploidy is an agent of adaptation and speciation, there are no studies of neopolyploid fitness and reproductive isolation in the field, much less the impacts of specific chromo-somal and genic factors on the evolution of neopolyploid populations. Moreover, investigations of the dynamics of polyploid formation and establishment remain theoretical, even though published evaluations of neopolyploid crops demonstrate the feasibility of an empirical approach.

Hence, there is a need for experimental studies that compare the genetic and phenotypic characteristics of neopolyploids and early-generation polyploids with those of established polyploid and progenitor populations. Critical questions in-clude: What fraction of the total fitness differential between established polyploids and their progenitors is obtained in neopolyploids? Which characters of established polyploids are the products of genic evolution, and which are due to polyploidy per se? What are the genetic, developmental, and ecological attributes of neopoly-ploids, and how do these contribute to polyploid establishment?

ACKNOWLEDGMENTS

We thank Mary Bricker, Brenda Clifton, Tara Fletcher, Abby Sine, and Marlene Wagner for assistance with locating references and preparing the manuscript. Toby Bradshaw, Luca Comai, Carol Goodwillie, Chris Pires, Eric Schranz and John N. Thompson provided helpful comments on a draft of this manuscript. This material is based on work supported under a National Science Foundation fellowship.

The *Annual Review of Ecology and Systematics* is online at
http://ecolsys.annualreviews.org

LITERATURE CITED

Aastveit K. 1968. Variation and selection for seed set in tetraploid rye. *Hereditas* 60:294–315

Ahloowalia BS. 1971. Frequency, origin, and survival of aneuploids in the tetraploid rye-grass. *Genetica* 42:471–84

Annerstedt I, Lundqvist A. 1967. Genetics of self-incompatibility in *Tradescantia palu-dosa* (Commelinaceae). *Hereditas* 58:13–30

Armstrong JM, Robertson RW. 1956. Studies of colchicine-induced tetraploids of *Trifolium hybridum* L. *Can. J. Agric. Sci.* 36:255–66

Avery AG, Satina S, Rietsema J. 1959. *Blakes-lee: The Genus Datura*. New York: Ronald

Beasley JO. 1940. The production of polyploids in *Gossypium*. *J. Hered.* 31:39–48

Beasley JO. 1942. Meiotic chromosome behav-ior in species, species hybrids, haploids, and induced polyploids of *Gossypium. Genetics* 27:25–54

Belling J, Blakeslee AF. 1924. The distribution of chromosomes in tetraploid *Daturas. Am. Nat.* 58:60–70

Bergner AD, Avery AG, Blakeslee AF. 1940. Chromosomal deficiencies in *Datura stramonium* induced by colchicine treatment. *Am. J. Bot.* 27:676–83

Bernstrom P. 1954. Fertility and aneuploidy in new autotetraploids in *Lamium. Hereditas* 40:181–241

Bever JD, Felber F. 1992. The theoretical population genetics of autopolyploidy. In *Oxford Surveys in Evolutionary Biology*, ed. D Futuyma, J Antonovics, pp. 185–217. Oxford, UK: Oxford

Bingham ET. 1968. Aneuploids in seedling populations of tetraploid alfalfa *Medicago sativa* L. *Crop Sci.* 8:571–74

Bingham ET. 1980. Maximizing heterozygosity in autopolyploids. See Lewis 1980, pp. 471–89

Bingham ET, Binek A. 1969. Hexaploid alfalfa *Medicago sativa* L.: origin, fertility and cytology. *Can. J. Genet. Cytol.* 11:359–66

Blakeslee AF. 1921. Types of mutations and their possible significance in evolution. *Am. Nat.* 55:254–67

Bose RB, Choudhury JK. 1962. A comparative study of the cytotaxonomy, palynology, and physiology of diploid and polypoid plants of *Ocimum kilimandscharicum* Guerke and their yield of raw material and volatile contents. *Caryologia* 15:435–54

Bretagnolle F, Lumaret R. 1995. Bilateral polyploidization in *Dactylis glomerata* L. subsp. *lusitanica*: occurrence, morphological and genetic characteristics of first polyploids. *Euphytica* 84:197–207

Bretagnolle F, Thompson J. 1995. Gametes with somatic chromosome number: their mechanisms of formation and role in the evolution of autopolyploid plants. *New Phytol.* 129:1–22

Briggs D, Walters SM. 1997. *Plant Variation and Evolution.* Cambridge: Cambridge Univ. Press. 3rd ed.

Brown MS. 1951. The spontaneous occurrence of amphidiploidy in species hybrids of *Gossypium. Evolution* 5:25–41

Burnham CR. 1962. *Discussions in cytogenetics.* Minneapolis: Burgess

Buxton BH, Darlington CD. 1931. Behaviour of a new species, *Digitalis mertonensis. Nature* 127:94

Chen S, Tang PS. 1945. Studies on colchicine-induced autotetraploid barley. III. Physiological studies. *Am. J. Bot.* 32:103–76

Clausen J, Keck DD, Hiesey WM. 1940. Experimental studies on the nature of species. I. Effect of varied environments on western North American plants. *Carnegie Inst. Washington Publ.* 520

Clausen J, Keck DD, Hiesey WM. 1945. Experimental studies on the nature of species. II. Plant evolution through amphiploidy and autoploidy, with examples from the *Madiinae. Carnegie Inst. Washington Publ.* 564

Comai L. 2000. Genetic and epigenetic interactions in allopolyploid plants. *Plant Mol. Biol.* 43:387–99

Comai L, Tyagi AP, Winter K, Holmes-Davis R, Reynolds SH, Stevens Y, Byers B. 2000. Phenotypic instability and rapid gene silencing in newly formed *Arabidopsis* allotetraploids. *Plant Cell* 12:1551–1567

Crane MB, Lewis D. 1942. Genetical studies in pears. III. Incompatibility and sterility. *J. Genet.* 43:31–43

Crowley JG, Rees H. 1968. Fertility and selection in tetraploid *Lolium. Chromosoma* 24:300–8

Darlington CD. 1937. *Recent Advances in Cytology.* Philadelphia: Blakiston. 2nd ed.

Darlington CD. 1963. *Chromosome Botany and the Origins of Cultivated Plants.* New York: Hafner. 2nd ed.

Davis BM. 1943. An amphidiploid in the F_1 generation from the cross *Oenothera franciscana* × *Oenothera biennis* and its progeny. *Genetics* 28:275–85

Day A. 1965. The evolution of a pair of sibling allotetraploid species of Cobwebby *Gilias* (Polemoniaceae). *Aliso* 6:25–75

De Kovel CGF, De Jong G. 2000. Selection on apomictic lineages of *Taraxacum* at establishment in a mixed sexual-apomicitic population. *J. Evol. Biol.* 13:561–68

Dewey DR. 1961. Polyhaploids of crested wheatgrass. *Crop Sci.* 1:249–54

Dewey DR. 1966. Inbreeding depression in diploid, tetraploid and hexaploid crested wheatgrass. *Crop Sci.* 6:144–47

Dewey DR. 1969. Inbreeding depression in diploid and induced-autotetraploid crested wheatgrass. *Crop Sci.* 9:592–95

Doggett H. 1964. Fertility improvement in autotetraploid sorghum. *Heredity* 19:403–17

Doyle GG. 1963. Preferential pairing in structural heterozygotes of *Zea mays*. *Genetics* 48:1011–27

Doyle GG. 1986. Aneuploidy and inbreeding depression in random mating and self-fertilizing autotetraploid populations. *Theor. Appl. Genet.* 72:799–806

Einset J. 1944. Cytological basis for sterility in induced autotetraploid lettuce (*Lactuca sativa* L.). *Am. J. Bot.* 31:336–42

Einset J. 1947. Aneuploidy in relation to partial sterility in autotetraploid lettuce (*Lactuca sativa* L.). *Am. J. Bot.* 34:99–105

Ellis JR, Janick J. 1960. The chromosomes of *Spinacea oleracea*. *Am. J. Bot.* 47:210–14

Emerson S. 1935. The genetic nature of de Vries' mutations in *Oenothera lamarckiana*. *Am. Nat.* 69:545–59

Eskilsson L. 1963. A method for estimating pollen quality in autopolyploid plants. *Hereditas* 49:185–88

Evans GM, Aung T. 1985. Identification of a diploidizing genotype of *Lolium multiflorum*. *Can J. Genet. Cytol.* 27:498–505

Felber F. 1991. Establishment of a tetraploid cytotype in a diploid population: effect of relative fitness of the cytotypes. *J. Evol. Biol.* 4:195–207

Feldman M, Liu B, Segal G, Abbo S, Levy AA, Vega JM. 1997. Rapid elimination of low-copy DNA sequences in polyploid wheat: a possible mechanism for differentiation of homeologous chromosomes. *Genetics* 147:1381–87

Futuyma DJ. 1998. *Evolutionary Biology*. Sunderland, MA: Sinauer

Gajewski W. 1953. A fertile amphipolyploid hybrid of *Geum rivale* with *G. macrophyllum*. *Acta Soc. Bot. Pol.* 22:411–39

Garber ED. 1954. Cytotaxonomic studies in the genus *Sorghum*. III. The polyploid species of the subgenera *Para-Sorghum* and *Stiposorghum*. *Bot. Gaz.* 115:336–42

Garber ED. 1955. Cytogenetics of *Sorghum*. I. The orientation of interchange complexes and quadrivalents at metaphase I in *S. purpureo-sericeum*. *Bot. Gaz.* 116:369–72

Garber ED, Dhillon TS. 1962. The genus *Collinsia*. XVII. Preferential pairing in four amphidiploids and three triploid interspecific hybrids. *Can. J. Genet. Cytol.* 4:6–13

Garber ED, Gorsic J. 1956. The genus *Collinsia*. II. Interspecific hybrids involving *C. heterophylla*, *C. concolor*, and *C. sparsiflora*. *Bot. Gaz.* 118:73–77

Gerstel DU, Mishanec WM. 1950. On the inheritance of apomixis in *Parthenium argentatum*. *Bot. Gaz.* 111:96–106

Gerstel DU, Phillips LL. 1958. Segregation of synthetic amphiploids in *Gossypium* and *Nicotiana*. *Cold Spring Harbor Symp. Quant. Biol.* 23:225–37

Goodspeed TH, Bradley MV. 1942. Amphidiploidy. *Bot. Rev.* 8:271–316

Gottschalk W. 1978. Open problems in polyploidy research. *Nucleus* 21:99–112

Grant V. 1954. Genetic and taxonomic studies in *Gilia*. V. *Gilia clivorum*. *Aliso* 3:19–34

Grant V. 1975. *Genetics of Flowering Plants*. New York: Columbia

Grant V. 1981. *Plant Speciation*. New York: Columbia. 2nd ed.

Greenleaf WH. 1941. Sterile and fertile amphidiploids: their possible relation to the origin of *Nicotiana tabacum*. *Genetics* 26:301–24

Gulcan R, Sybenga J. 1967. Relative rate of development of aneuploid cells in euploid environment. *Genetica* 38:163–70

Guo M, Bircher JA. 1994. Trans-acting dosage effects on the expression of model gene systems in maize aneuploids. *Science* 266:1999–2002

Guo M, Davis D, Bircher JA. 1996. Dosage effects on gene expression in a maize ploidy series. *Genetics* 142:1349–55

Hagberg A, Ellerström S. 1959. The competition between diploid, tetraploid and aneuploid rye: theoretical and practical aspects. *Hereditas* 45:369–415

Hahn SK, Bai KV, Asiedu R. 1990. Tetraploids, triploids, and 2n pollen from diploid interspecific crosses with cassava. *Theor. Appl. Genet.* 79:433–39

Hazarika MH, Rees H. 1967. Genotypic control of chromosome behaviour in rye. *Heredity* 22:317–32

Hill RR. 1971. Selection in autotetraploids. *Theor. Appl. Genet.* 41:181–86

Howard HW. 1938. The fertility of amphidiploids from the cross *Raphanus sativus* × *Brassica oleracea*. *J. Genet.* 36:239–73

Howard HW. 1939. The size of seeds in diploid and autotetraploid *Brassica oleracea* L. *J. Genet.* 38:325–40

Husband BC, Schemske DW. 1997. The effect of inbreeding in diploid and tetraploid populations of *Epilobium angustifolium* (Onagraceae): implications for the genetic basis of inbreeding depression. *Evolution* 51:737–46

Ising G. 1966. Cytogenetic studies in *Cyrtanthus*. I. Segregation in an allotetraploid. *Hereditas* 56:27–53

Jackson RC. 1976. Evolution and systematic significance of polyploidy. *Annu. Rev. Ecol. Syst.* 7:209–34

Jackson RC. 1982. Polyploidy and diploidy: new perspectives on chromosome pairing and its evolutionary implications. *Am. J. Bot.* 69:1512–23

Jackson RC, Jackson JW. 1996. Gene segregation in autotetraploids: Prediction from meiotic configurations. *Am. J. Bot.* 83:673–78

Jaranowski J, Kalasa M. 1971. Comparative analysis of fertility in several *Trifolium, Melilotus, Medicago* and *Trigonella* species and forms on a di- and tetraploid level. *Genet. Pol.* 12:1–16

Jauhar PP. 1970. Chromosome behavior and fertility of the raw and evolved synthetic tetraploids of pearl millet *Pennisetum typhoides* Stapf et Hubb. *Genetica* 41:407–24

Jauhar PP. 1975. Genetic control of diploid like meiosis in hexaploid tall fescue. *Nature* 254:595–97

John B, Henderson SA. 1962. Asynapsis and polyploidy in *Schistocerca paranensis*. *Chromosoma* 13:111–47

Johnston MO, Schoen DJ. 1996. Correlated evolution of self-fertilization and inbreeding depression: an experimental study of nine populations of *Amsinckia*. *Evolution* 50:1478–91

Jones K. 1956. Species differentiation in *Agrostis*. I. Cytological relationships in *Agrostis canina* L. *J. Genet.* 54:370–76

Jørgensen CA. 1928. The experimental formation of heteroploid plants in the genus *Solanum*. *J. Genet.* 19:133–210

Kalton RR, Smit AG, Leffel RC. 1952. Parent–inbred progeny relationships of selected orchardgrass clones. *Agron. J.* 44:481–86

Khush GS. 1973. *Cytogenetics of Aneuploids*. New York: Academic

Khush GS, Rick CM. 1966. The origin, identification, and cytogenetic behavior of tomato monosomics. *Chromosoma* 18:407–20

Kirkpatrick M, Barton NH. 1997. Evolution of a species' range. *Am. Nat.* 150:1–23

Koduru PRK, Rao MK. 1981. Cytogenetics of synaptic mutants in higher plants. *Theor. Appl. Genet.* 59:197–214

Kostoff D. 1938. Studies on polyploid plants. XXI. Cytogenetic behaviour of the allopolyploid hybrids *Nicotiana glauca* Grah. × *Nicotiana langsdorffii* Weinm. and their evolutionary significance. *J. Genet.* 37:129–209

Kostoff D. 1940. Fertility and chromosome length. *J. Hered.* 31:33–34

Kuspira J, Bhambhani RN, Shimada T. 1985. Genetic and cytogenetic analyses of the A genome of *Tritium gonococcus*. I. Cytology, breeding behaviour, fertility, and morphology of induced autotetraploids. *Can. J. Genet. Cytol.* 27:51–63

Lande R, Schemske DW. 1985. The evolution of self-fertilization and inbreeding depression in plants. I. Genetic models. *Evolution* 39:24–40

Lavania UC. 1986. High bivalent frequencies in artificial autopolyploids of *Hyoscyamus muticus* L. *Can. J. Genet. Cytol.* 28:7–11

Lesins K. 1957. Cytogenetic study on a tetraploid plant at the diploid chromosome level. *Can. J. Bot.* 35:181–96

Levan A. 1940. Meiosis in *Allium porrum*, a tetraploid species with chiasma localization. *Hereditas* 26:454–62

Levin DA. 1975. Minority cytotype exclusion in local plant populations. *Taxon* 24:35–43

Levin DA. 1983. Polyploidy and novelty in flowering plants. *Am. Nat.* 122:1–25

Levin DA. 2002. *The Role of Chromosomal Change in Plant Evolution.* Oxford, UK: Oxford Univ. Press

Lewis D. 1947. Competition and dominance of incompatibility alleles in diploid pollen. *Heredity* 1:85–108

Lewis H, Lewis ME. 1955. The genus *Clarkia*. *Univ. Calif. Publ. Bot.* 20:241–392

Lewis WH, ed. 1980. *Polyploidy: Biological Relevance.* New York/London: Plenum

Lewis WH. 1980. Polyploidy in species populations. See Lewis 1980, pp. 103–44

Lindstrom EW, Humphrey LM. 1933. Comparative cyto-genetic studies of tetraploid tomatoes from different origins. *Genetics* 18:193–209

Little TM. 1945. Gene segregation in autotetraploids. *Bot. Rev.* 11:60–85

Liu B, Vega JM, Feldman M. 1998a. Rapid genomic changes in newly synthesized amphiploids of *Triticum* and *Aegilops*. II. Changes in low-copy coding DNA sequences. *Genome* 41:535–42

Liu B, Vega JM, Segal S, Abbo S, Rodova M, Feldman M. 1998b. Rapid genomic changes in newly synthesized amphiploids of *Triticum* and *Aegilops*. I. Changes in low-copy non-coding DNA sequences. *Genome* 41:272–77

Lumaret R. 1988. Adaptive strategies and

ploidy levels. *Acta Oecol. Oecol. Plant* 9:83–93

Luo MC, Dubcovsky J, Dvorak J. 1996. Recognition of homeology by the wheat *Ph1* locus. *Genetics* 144:1195–203

Martinez M, Naranjo T, Cuadrado C, Romero C. 2001. The synaptic behavior of the wild forms of *Triticum turgidum* and *T. timopheevi*. *Genome* 44:517–22

Mastenbroek I, Dewet JMJ, Lu C. 1982. Chromosome behaviour in early and advanced generations of tetraploid maize. *Caryologia* 35:463–70

Masterson J. 1994. Stomatal size in fossil plants: evidence for polyploidy in majority of angiosperms. *Science* 264:421–24

McCollum GD. 1958. Comparative studies of chromosome pairing in natural and induced tetraploid *Dactylis*. *Chromosoma* 9:571–605

McNaughton IH. 1973. Synthesis and sterility of *Raphanobrassica*. *Euphytica* 22:70–80

Mooney HA, Johnson W. 1965. Comparative physiological ecology of an arctic and alpine population of *Thalictrum alpinum* L. *Ecology* 46:721–27

Morrison JW, Rajhathy T. 1960a. Chromosome behaviour in autotetraploid cereals and grasses. *Chromosoma* 11:297–309

Morrison JW, Rajhathy T. 1960b. Frequency of quadrivalents in autotetraploid plants. *Nature* 187:528–30

Mosquin T. 1966. A new taxonomy for *Epilobium angustifolium* L. (*Onagraceae*). *Brittonia* 18:167–88

Müntzing A. 1932. Cyto-genetic investigations on synthetic *Galeopsis tetrahit*. *Hereditas* 16:105–54

Müntzing A. 1936. The evolutionary significance of autopolyploidy. *Hereditas* 21:263–378

Müntzing A. 1937. The effects of chromosomal variation in *Dactylis*. *Hereditas* 23:113–235

Müntzing A. 1948a. Characteristics of two haploid twins in *Dactylis glomerata*. *Hereditas* 29:135–40

Müntzing A. 1948b. Fertility improvement by recombination in autotetraploid *Galeopsis pubescens*. *Hereditas* 29:201–4

Müntzing A. 1951. Cyto-genetic properties and practical value of tetraploid rye. *Hereditas* 37:17–84

Myers WM, Hill HD. 1940. Studies of chromosomal association and behavior and occurrence of aneuploidy in autotetraploid grass species, orchard grass, tall oat grass, and crested wheatgrass. *Bot. Gaz.* 102:236–55

Navashin M. 1925. Polyploid mutations in *Crepis*. Triploid and pentaploid mutants of *Crepis capillaris*. *Genetics* 10:583–92

Noggle GR. 1946. The physiology of polyploidy in plants. I. Review of the literature. *Lloydia* 9:153–73

Nordenskiöld H. 1953. An investigation into two tetraploid strains of vetches (*Vicia sativa*) and their hybrid products. *Kungl. Lantbrukshögsk. Ann.* 19:209–25

Olsson G. 1948. Spontaneous crossing between diploid and tetraploid *Sinapis alba*. *Hereditas* 34:351–65

O'Mara JG. 1943. Meiosis in autotetraploid *Secale cereale*. *Bot. Gaz.* 104:563–75

Otto SP, Whitton J. 2000. Polyploid incidence and evolution. *Annu. Rev. Genet.* 34:401–37

Ozkan H, Feldman M. 2001. Genotypic variation in tetraploid wheat affecting homoeologous pairing in hybrids with *Aegilops peregrina*. *Genome* 44:1000–6

Pal M, Khoshoo TN. 1977. Evolution and improvement of cultivated amaranths. VIII. Induced autotetraploidy in grain types. *Z. Pflanzenzuecht.* 78:135–48

Pandey KK. 1968. Colchicine-induced changes in the self-incompatibility behaviour of *Nicotiana*. *Genetica* 39:257–71

Petit C, Bretagnolle F, Felber F. 1999. Evolutionary consequences of diploid-polyploid hybrid zones in wild species. *Trends Ecol. Evol.* 14:306–11

Phillips LL. 1962. Segregation in new allopolyploids of *Gossypium*. IV. Segregation in New World × Asiatic and New World × wild American hexaploids. *Am. J. Bot.* 49:51–57

Phillips LL. 1964. Segregation in new allopolyploids of *Gossypium*. V. Multivalent formation in New World × Asiatic and New World × wild American hexaploids. *Am. J. Bot.* 51:324–29

Poole CF. 1932. The interspecific hybrid, *Crepis rubra* × *C. foetida*, and some of its derivatives. II. Two selfed generations from an amphidiploid hybrid. *Univ. Calif. Publ. Agric. Sci.* 6:231–55

Prakken R. 1943. Studies of asynapsis in rye. *Hereditas* 29:475–95

Pundir RPS, Roa NK, van der Maeson LJG. 1983. Induced autotetraploidy in chickpea (*Cicer arietinum* L.). *Theor. Appl. Genet.* 65:119–22

Ramsey J, Schemske DW. 1998. Pathways, mechanisms, and rates of polyploid formation in flowering plants. *Annu. Rev. Ecol. Syst.* 29:467–501

Randolph LF. 1935. Cytogenetics of tetraploid maize. *J. Agric. Res. Washington, DC* 50:591–605

Randolph LR. 1932. Some effects of high temperature on polyploidy and other variations in maize. *Proc. Natl. Acad. Sci. USA* 18:222–29

Randolph LF. 1941. An evaluation of induced polyploidy as a method of breeding crop plants. *Am. Nat.* 75:347–63

Randolph LF, Fischer HE. 1939. The occurrence of pathenogenetic diploids in tetraploid maize. *Proc. Natl. Acad. Sci. USA* 25:161–64

Rao PN. 1976. Studies on the occurrence, cytology, fertility and breeding behaviour of aneuploids in induced autotetraploid Job's tears. *Cytologia* 41:142–52

Raven PH, Thompson HJ. 1964. Haploidy and angiosperm evolution. *Am. Nat.* 98:251–52

Rees H. 1955. Heterosis in chromosome behaviour. *Proc. R. Soc. Lond. B* 144:150–59

Rees H. 1961. Genotypic control of chromosome form and behaviour. *Bot. Rev.* 26:288–318

Richharia RH. 1937. Cytological investigation of *Raphanus sativus, Brassica oleracea*, and their F_1 and F_2 hybrids. *J. Genet.* 34:21–44

Riley R, Chapman V. 1958. Genetic control of cytologically diploid behaviour of hexaploid wheat. *Nature* 162:713–15

Riley R, Kimber G. 1961. Aneuploids and the cytogenetic structure of wheat varietal populations. *Heredity* 16:275–90

Rodríguez DJ. 1996. A model for the establishment of polyploidy in plants. *Am. Nat.* 147:33–46

Rommel M. 1961. Aneuploidy, seed set and sterility in artificially induced autotetraploid *Hordeum vulgare* L. *Can. J. Genet. Cytol.* 3:272–82

Ronfort J. 1999. The mutation load under tetrasomic inheritance and its consequences for the evolution of the selfing rate in autotetraploid species. *Genet. Res.* 74:31–42

Roose ML, Gottlieb LD. 1976. Genetic and biochemical consequences of polyploidy in *Tragopogon*. *Evolution* 30:818–30

Roy RP, Dutt B. 1972. Cytomorphological studies in induced polyploids of *Luffa acutangula* Roxb. *Nucleus* 15:171–80

Salanki MS, Parameswarappa R. 1968. Colchicine-induced mutant in cotton (*Gossypium hirsutum*, L.). *Curr. Sci.* 37:357–58

Sanders ME, Franzke CJ. 1964. A proposed explanation for the origin of colchicine-induced diploid mutants in *Sorghum*. *J. Arnold Arbor. Harv. Univ.* 45:36–37

Scheid OM, Jakovleva L, Afsar K, Maluszynska J, Paszkowski J. 1996. A change of ploidy can modify epigenetic silencing. *Proc. Natl. Acad. Sci. USA* 93:7114–19

Schranz ME, Osborn TC. 2000. Novel flowering time variation in the resynthesized polyploid *Brassica napus*. *J. Hered.* 91:242–46

Sears ER. 1976. Genetic control of chromosome pairing in wheat. *Annu. Rev. Genet.* 10:31–51

Segraves KA, Thompson JN. 1999. Plant polyploidy and pollination: floral traits and insect visits to diploid and tetraploid *Heuchera grossulariifolia*. *Evolution* 53:1114–27

Shaver DL. 1962. A study of aneuploidy in autotetraploid maize. *Can. J. Genet. Cytol.* 4:226–33

Shaver DL. 1963. The effect of structural heterozygosity on the degree of preferential pairing in allotetraploids of *Zea*. *Genetics* 48:515–24

Shifriss O. 1942. Polyploids in the genus *Cucumis*. *J. Hered.* 33:144–52

Simonsen Ø. 1973. Cytogenetic investigations in diploid and autotetraploid populations of *Lolium perenne* L. *Hereditas* 75:157–88

Simonsen Ø. 1975. Cytogenetic investigations in diploid and autotetraploid populations of *Festuca pratensis*. *Hereditas* 79:73–108

Singh AK. 1986. Utilization of wild relatives in the genetic improvement of *Arachis hypogaea* L. 7. Autotetraploid production and prospects of interspecific breeding. *Theor. Appl. Genet.* 72:164–69

Singh RJ. 1993. *Plant Cytogenetics.* Boca Raton, Fl: CRC

Singh RJ, Hymowitz T. 1985. Diploid-like meiotic behavior in synthesized amphidiploids of the genus *Glycine* Willd. subgenus *Glycine*. *Can. J. Genet. Cytol.* 27:655–60

Smith HE. 1946. *Sedum pulchellum*: a physiological and morphological comparison of diploid, tetraploid, and hexaploid races. *Bull. Torrey Bot. Club* 73:495–541

Smith HH. 1943. Studies on induced heteroploids of *Nicotiana*. *Am. J. Bot.* 30:121–30

Soltis DE, Rieseberg LH. 1986. Autopolyploidy in *Tolmiea menziesii* (Saxifragaceae): Genetic insights from enzyme electrophoresis. *Am. J. Bot.* 73:310–18

Soltis DE, Soltis PS. 1993. Molecular data and the dynamic nature of polyploidy. *Crit. Rev. Plant Sci.* 12:243–73

Soltis DE, Soltis PS. 1999. Polyploidy: recurrent formation and genome evolution. *Trends Ecol. Evol.* 14:348–52

Song K, Liu P, Tang K, Osborn TC. 1995. Rapid genome change in synthetic polyploids of *Brassica* and its implications for polyploid evolution. *Proc. Natl. Acad. Sci. USA* 92:7719–23

Song K, Tang K, Osborn TC. 1993. Development of synthetic *Brassica* amphidiploids by reciprocal hybridization and comparison to natural amphidiploids. *Theor. Appl. Genet.* 86:811–21

Sparrow AH, Ruttle ML, Nebel BR. 1942. Comparative cytology of sterile intra- and fertile inter-varietal tetraploids of *Antirrhinum majus* L. *Am. J. Bot.* 29:711–15

Srinivasachar D, Singh KP. 1967. Induced allopolyploidization of autotetraploid *Brassica campestris*. *Genetica* 38:375–80

Stebbins GL. 1947. Types of polyploids: their classification and significance. *Adv. Genet.* 1:403–29

Stebbins GL. 1949. The evolutionary significance of natural and artificial polyploids in the family Gramineae. *Proc. Int. Congr. Genet.* 8th pp. 461–85

Stebbins GL. 1950. *Variation and Evolution in Plants*. New York: Columbia

Stebbins GL. 1971. *Chromosomal Evolution in Higher Plants*. London: Addison-Wesley

Stebbins GL. 1980. Polyploidy in plants: unsolved problems and prospects. See Lewis 1980, pp. 495–520

Stebbins GL. 1985. Polyploidy, hybridization, and the invasion of new habitats. *Annu. Missouri Bot. Gard.* 72:824–32

Stebbins GL, Vaarama A. 1954. Artificial hybrids in the Gramineae, tribe Hordeae. VII. Hybrids and allopolyploids between *Elymus glaucus* and *Sitanion* spp. *Genetics* 39:378–95

Stout AB, Chandler C. 1941. Change from self-incompatibility to self-compatibility accompanying change from diploidy to tetraploidy. *Science* 94:118

Sullivan JT, Myers WM. 1939. Chemical composition of diploid and tetraploid *Lolium perenne* L. *J. Am. Soc. Agron.* 31:869–71

Sved JA. 1966. Telomere attachment of chromosomes. Some genetical and cytological consequences. *Genetics* 53:747–56

Swaminathan MS, Sulbha K. 1959. Multivalent frequency and seed frequency in raw and evolved tetraploids of *Brassica campestris* var. *toria*. *Z. Vererbungsl.* 90:385–92

Sybenga J. 1969. Allopolyploidization of autopolyploids. I. Possibilities and limitations. *Euphytica* 18:355–71

Sybenga J. 1973a. Allopolyploidization of autopolyploids. 2. Manipulation of the chromosome pairing system. *Euphytica* 22:433–44

Sybenga J. 1973b. The effect of reciprocal translocations on segregation and multivalent formation in autotetraploids of rye *Secale cereale*. *Genetica* 44:270–82

Sybenga J. 1975. *Meiotic Configurations: A Source of Information for Estimating Genetic Parameters*. Berlin: Springer-Verlag

Sybenga J. 1996. Chromosome pairing affinity and quadrivalent formation in polyploids: Do segmental allopolyploids exist? *Genome* 39:1176–84

Tai W, Dewey DR. 1966. Morphology, cytology and fertility of diploid and colchicine induced tetraploid crested wheatgrass. *Crop Sci.* 6:223–26

Tal M. 1980. Physiology of polyploids. See Lewis 1980, pp. 61–75

Tal M, Gardi I. 1976. Physiology of polyploid plants: water balance in autotetraploid and diploid tomato under low and high salinity. *Physiol. Pl.* 38:257–61

Tanaka N. 1940. Chromosome studies in Cyperaceae. VIII. Meiosis in diploid and tetraploid forms of *Carex siderostica* Hance. *Cytologia* 11:282–310

Thompson JD, Lumaret R. 1992. The evolutionary dynamics of polyploid plants: origins, establishment and persistence. *Trends Ecol. Evol.* 7:302–7

Upcott M. 1935. The cytology of triploid and tetraploid *Lycopersicum esulentum*. *J. Genet.* 31:1–19

Upcott M. 1940. The nature of tetraploidy in *Primula kewensis*. *J. Genet.* 39:79–100

Vega J, Feldman J. 1998. Effects of the pairing gene *Ph1* on centromere misdivision in common wheat. *Genetics* 148:1285–94

Webber JM. 1940. Polyembryony. *Bot. Rev.* 6:575–98

Weimarck A. 1973. Cytogenetic behaviour in

octoploid *Triticale*. I. Meiosis, aneuploidy and fertility. *Hereditas* 74:103–18

Wendel JF. 2000. Genome evolution in polyploids. *Plant Mol. Biol.* 42:225–49

Werner JE, Peloquin SJ. 1991. Significance of allelic diversity and 2N gametes for approaching maximum diversity in 4x potatoes. *Euphytica* 58:21–29

Winge Ö. 1932. On the origin of constant species-hybrids. *Sven. Bot. Tidskr.* 26:107–22

Annu. Rev. Ecol. Syst. 2002. 33:641–64
doi: 10.1146/annurev.ecolsys.33.010802.150443
First published online as a Review in Advance on August 14, 2002

HERBIVORE OFFENSE

Richard Karban

*Department of Entomology, University of California, Davis, California 95616;
email: rkarban@ucdavis.edu*

Anurag A. Agrawal

*Department of Botany, University of Toronto, 25 Willcocks Street, Toronto,
Ontario M5S 3B2, Canada; email: agrawal@botany.utoronto.ca*

Key Words   Plant-herbivore interactions, coevolution, choice,
preference-performance correlation, enzymatic metabolism, sequestration

■ **Abstract**   Herbivore offense describes traits that allow herbivores to increase their
feeding and other uses of host plants when these uses benefit the herbivores. We argue
that ecological interactions and coevolution between plants and herbivores cannot be
understood without an offense-defense framework. Thus far, plant defense theory and
data have far outpaced knowledge of herbivore offense. Offensive tactics include feeding and oviposition choices, enzymatic metabolism of plant compounds, sequestration,
morphological adaptations, symbionts, induction of plant galls, and induced plant susceptibility, trenching, and gregarious feeding. We propose that offensive tactics can be
categorized usefully depending upon when they are effective and whether they are plastic or fixed traits. The advantages of offensive traits have not been adequately described
in terms of herbivore fitness. Similarly, a more complete understanding of the costs
and limitations of offensive traits will help put the herbivore back in plant-herbivore
interactions and coevolution.

INTRODUCTION

The past 30 years have seen incredible advances in our understanding of plant
defenses, the mechanisms that plants use to protect themselves from being eaten
by their herbivores. By comparison, biologists have paid much less attention to
herbivore offense, the mechanisms that herbivores use to exploit their host plants.
Offensive traits are those that currently allow herbivores to increase their feeding
and use of a host when this use benefits the herbivores. These offensive traits
increase herbivore performance or fitness. Offensive traits are determined from
the herbivore's point of view; they do not necessarily cause more damage to plants
or reduce plant fitness.

We have devised a dichotomous key to define offensive traits and to help organize our thinking and research about them (Table 1). We start by asking whether
the herbivore's traits appear to match the challenges that its host plant presents. For

0066-4162/02/1215-0641$14.00   **641**

TABLE 1   A dichotomous key for offensive traits

	No	Yes
1. Do herbivores have traits that correspond to host traits?	The trait is unlikely to be involved in herbivore offense	Possible offensive trait (proceed to 2)
2. Does the trait produce a measurable increase in the rate or efficiency of the herbivore's feeding or other uses (e.g., oviposition) of the host?	The trait is not involved in herbivore offense	Probable offensive trait (proceed to 3)
3. Does the trait increase herbivore performance (fitness)?	The trait is not beneficial	The trait is currently offensive (proceed to 4)
4. Did the trait evolve as an adaptation to herbivory?	The trait is currently beneficial but was shaped or maintained by other selective pressures	The trait is currently beneficial and has been shaped or maintained by herbivory (adapted offensive trait)

example, do its mouthparts or ovipositor correspond to features of its host? This effort often takes the form of examining variation in herbivore traits when the herbivore feeds on different species. We compare the traits observed with alternative traits that are observed in other herbivores.

Our next step should be to examine whether the herbivore trait increases the rate or efficiency of feeding or other host use. The comparison is with herbivores that exhibit alternative traits. The fitness consequences of possibly offensive herbivore traits have rarely been quantified, although a connection between herbivore efficiency and fitness is often assumed. Ideally, fitness should be measured over the entire lifetime of the herbivore, as traits may increase performance during one stage but decrease performance during other stages (e.g., Scheirs et al. 2000). The relationship between traits that appear offensive (correspond to the host and increase herbivore efficiency) and herbivore fitness deserves much further study.

Offensive traits can include plastic traits that change over ecological time as well as fixed traits that change over evolutionary time. Our definition does not require an offensive trait to have evolved in response to host plant traits (adaptation sensu; Reeve & Sherman 2001). Demonstrating that a trait has been shaped by a particular selective factor is a very difficult task that requires information about the selective environment and how the trait performs, its genetic variability and heritability, and a comparison among related species (Endler 1986, Reznick & Travis 1996). We presently have an incomplete understanding of the selective pressures that shaped any of the adaptations that we observe in nature. Many traits that now appear offensive may have evolved for other reasons.

Expanding on the scheme proposed by Rhoades (1985), we envision three strategies that have been employed by herbivores to exploit hosts (Table 2). The first and

TABLE 2 Herbivore offensive strategies and their consequences[a]

	Strategy	Tactic	Employed by	Resulting population dynamics
Least aggressive	Choice	Avoidance, attraction	Opportunistic herbivores	Variable populations
	Change herbivore morphology, physiology	Metabolize, detoxify		
Most aggressive	Manipulate the host	Change host nutrition & defense	Stealthy herbivores	Low, invariant populations

[a]Modified from Rhoades 1985.

least aggressive of these is choice. Herbivores do not randomly eat the green world but can make sophisticated choices, avoiding many potential foods and selectively seeking others. We consider choice to be offensive when it increases herbivore efficiency and performance. The second strategy is altering the herbivore's morphology and physiology over ecological and evolutionary time to better exploit existing hosts and to begin to exploit novel ones. The third and most aggressive strategy is active manipulation of the host. Herbivore offense may occur prior to ingestion, for instance, by suppressing the host's ability to respond defensively or by inducing the host to create a highly nutritious gall. It may also occur after ingestion by metabolizing the plant's secondary chemicals (those that are not part of the plant's primary metabolism and presumed to be defensive) or by sequestering them to protect the herbivore against its own predators. Many specific tactics that exemplify each of these strategies will be elaborated below although in only a few cases have the tactics been linked to herbivore fitness.

BEHAVIORAL MECHANISMS—LEAST AGGRESSIVE

Feeding Choices

Herbivores are faced with a food supply that is not very nutritious. The tissues of all animals contain one to several orders of magnitude more nitrogen than their plant foods (Mattson 1980). Other nutrients and water are also commonly limiting to herbivores. In addition, plants contain many physical and chemical traits that reduce their digestibility or are directly toxic to most herbivores. Natural selection has favored herbivores that are selective about the foods they ingest. For several well-studied species of insect herbivores, individuals that were allowed to select their own diets grew faster than individuals that were fed different ratios of the same nutrients or diets that lacked a self-selected nutrient (Waldbauer & Friedman 1991). Self-selection of food is likely to be more important for herbivores that are more mobile and can gain access to varied diets. Of course, not all foraging decisions by herbivores are optimal. When herbivores fail to make appropriate decisions, these "mistakes" tend to be conservative, such that herbivores reject foods that are actually nutritious to them (Waldbauer & Fraenkel 1961).

Plant quality can vary greatly in time and space. In response, many herbivores compensate for suboptimal foods. For example, when faced with a diluted diet, herbivores may eat more, increasing meal size, eating meals more frequently, or altering their diets (Simpson & Simpson 1990, Slansky 1993). It appears that compensatory feeding among insect herbivores occurs as the result of nutritional feedback; herbivores respond to the osmolality and amino acid content of their hemolymph. Herbivores may also assess the bulk of digested food and its rate of passage through their guts and adjust their decisions about intake accordingly.

Insect herbivores can supplement their diets with food that is not living vegetation but is high in protein and other essential micronutrients. These supplemental foods include pollen, their own cast skins, and even other living animals. Cannibalism is very common among herbivorous insects, in fact more common than among predators (Fox 1975). Cannibalism and predation on other herbivores is associated with nutritional benefits and is more likely when plant foods are scarce or less nutritious (Fox 1975, Via 1999). For example, when omnivorous western flower thrips feed on plants with high levels of defense, they shift their diet to include more small arthropods (Agrawal et al. 1999). Similarly, adult Lepidoptera of many families feed from the margins of moist ground, as well as on carrion and excreta (Arms et al. 1974, Beck et al. 1999). Cafeteria choice experiments revealed that this "puddling" behavior in some butterfly species was associated with sources of sodium, whereas other butterflies were more interested in protein.

Generalist herbivores (e.g., caterpillars and grasshoppers) have been found to perform best when offered a variety of food plants rather than a single plant. This result has been explained by mixed diets balancing nutrient intake (Pulliam 1975) or by facilitating avoidance of high levels of any single secondary compound (Freeland & Janzen 1974). It has been suggested that these two different mechanisms will produce different signatures that can be used to distinguish the mechanisms causing the pattern (Pennings et al. 1993, Bernays et al. 1994). If the mechanism involves critical complementary nutrients, then an herbivore will grow more on a mixed diet without necessarily ingesting more. However, if a mixed diet allows ingestion of low doses of several different toxins, then an herbivore will eat more of a mixed diet and will perform better as a result of eating more food. However, as is often the case, inferring process from pattern in ecology is very tricky. To separate these mechanisms we must assume that no single diet will satisfy the nutritional requirements as well as a mixed diet and that nutrients and toxins are distinct categories (Hagele & Rowell-Rahier 1999). Not only are these assumptions questionable, but alternative selective pressures can also make it very difficult to see a correspondence between the outcome of selection (pattern) and the selective force (process). For example, an herbivore's choice of diet will affect many other aspects of its life such as risk of predation and disease, likelihood of mating, thermal balance, and so on.

Innate preferences are probably responsible for much of food selection by herbivores, although there is growing evidence that most insects are capable of learning (Papaj & Prokopy 1989). Prior experience alters the ability of many herbivores to acquire food. Some authors (e.g., Bernays 1993) argue that learning is responsible for the ability of herbivores to self-select their diets, although no specific data are offered to support this argument. It is well substantiated that grasshoppers can avoid nutritionally inadequate foods and foods associated with adverse physical responses (Bernays & Lee 1988, Bernays 1993). Grasshoppers can also learn to associate plant odors with limiting nutrients, and they actively seek these odors (Simpson & White 1990). Such associative learning allows grasshoppers to grow more rapidly when compared with individuals that are prevented from learning (Dukas & Bernays 2000).

Even without learning, herbivores can choose to avoid patchy plant defenses. Small insects with mouthparts that are smaller than the distance between defensive structures in their host plants can eat around those structures. Young caterpillars of several species "window feed" on maple leaves, removing discrete pieces of mesophyll and epidermis and avoiding sclerenchymous bundle sheaths (Hagan & Chabot 1986). *Trichoplusia ni* caterpillars feed between the veins of wild parsnip leaves and in so doing reduce their intake of furanocoumarins (Zangerl & Bazzaz 1992). Similarly, larvae of the cotton leaf perforator skeletonize the leaves of their wild cotton host plant, eating everything except the epidermis and the pigment glands, which contain defensive terpenoid aldehydes (R. Karban & A.A. Agrawal, personal observations).

Many of these examples of individual feeding choices should be classified as possible offensive traits (sensu Table 1). In the cases of window feeding and cannibalism, the observations are primarily behavioral and no benefit to the herbivores, as required for our definition of offense, has been demonstrated. Compensatory feeding increases consumption (probable offense), but lack of true comparisons with noncompensatory feeding individuals makes benefits and offense ambiguous in these cases. Increases in herbivore performance associated with self selection and diet mixing lack mechanistic detail but do demonstrate herbivore offense.

Oviposition Choices

Many herbivorous insects cannot move much during the course of their development. According to the prevailing paradigm, mobile ovipositing females assess different host plant species and place their eggs or offspring on those hosts that result in the highest performance. A review of the recent literature found that just slightly more than half of the studies supported this positive correlation between adult oviposition and offspring performance (Mayhew 1997). This conclusion echoes previous findings that the relationship between oviposition preference and offspring performance is weaker than expected (Thompson 1988). Many explanations have been proposed for the bad fit between oviposition preference and offspring performance (Thompson 1988, Mayhew 2001). The empirical studies reviewed by these authors consider the correlation between mean oviposition preference and survival and growth of larvae averaged across many females in an herbivore population. The relevant question from the perspective of natural selection is whether individuals select the best hosts for their own offspring (Thompson & Pellmyr 1991). Singer et al. (1988) considered this relationship at the individual level and found a strong correlation between the host species preferences of adult *Euphydryas editha* butterflies and the growth rates of their offspring. However, two studies that have looked for similar correlations among individuals have failed to find them for fruit flies and bruchid beetles (Jaenike 1989, Fox 1993). More studies of individual preferences and performance are required to resolve this issue.

Herbivores may select host plants that provide poor growth but good protection from their own natural enemies (Price et al. 1980, Bernays & Graham 1988). For example, a marine amphipod specializes on a seaweed that minimizes its risk of predation by fish (Hay et al. 1990). If host choice affects risk of predation then correlations between choice and growth in the laboratory may be much poorer than correlations between choice and fitness in the field.

Most investigators have looked for correlations between adult choices and performance of offspring. One recent study that considered the correlation between adult oviposition decisions and adult performance of the parents in leaf-mining flies found a very strong correlation between these two (Scheirs et al. 2000). Adult herbivores may be maximizing their own lifetime fitness, and this may come at the expense of some of the individual offspring. Future field studies of the preference-performance correlation should consider adult performance as distinct from offspring performance.

In general, female herbivores seem more adept at making appropriate intraspecific choices among host plants than they do at choosing among host species. This greater discrimination between plants within a species seems counterintuitive, although here too there are examples of less than optimal choices. *Pemphigus* aphids provide a good example of intraspecific preference that correlates very well with offspring success. *Pemphigus* colonists make appropriate choices about leaf size and location on a leaf. Large leaves are much preferred by colonists and these large leaves increase the weight of stem mothers and their progeny, the number of progeny, their development rate, and their fecundity (Whitham 1978). Colonists at the base of leaves were more likely to survive and produced more progeny than those at other leaf positions and this corresponded to choices of colonists and to their willingness to engage in aggressive territorial behaviors (Whitham 1980).

Preference for appropriate host plants is a possible offensive trait sensu Table 1. In most cases, when differences in performance have been observed (step 3, "herbivore offense" in Table 1) the causal mechanisms (step 2) have not been examined or were not well documented. Knowledge of mechanism is not essential for determining if a trait is offensive. However, failure to grasp the mechanisms limits our understanding of the conditions under which the trait will be effective and how it may have evolved.

PHYSIOLOGICAL AND MORPHOLOGICAL TRAITS
OF HERBIVORES—SOMETIMES AGGRESSIVE

Enzymes

Herbivores possess various physiological and morphological traits that enable them to exploit their host plants. For example, all herbivores must deal with chemicals that are potentially damaging to their cellular processes; these come from various sources including secondary chemicals of plants that can be toxic or antinutritive to them (Duffey & Stout 1996). One class of traits that allows them to deal with such threats is enzymes that reduce the detrimental effects of these plant compounds. Mixed function oxidases are widespread enzymes that are thought to function in detoxification by catalyzing oxidative reactions (Brattsten 1988, Feyereisen 1999). In particular, cytochrome P-450 enzymes have been causally linked to detoxification of plant secondary metabolites. Production of these enzymes is induced facultatively by exposure to plant metabolites and such induction is also correlated with diet choice. For example, *Manduca sexta* caterpillars only increase consumption of food containing nicotine after increasing activity of P-450 enzymes in their midguts (Snyder & Glendinning 1996). Conversely, consumption of inhibitors of P-450 enzymes caused caterpillars to reduce consumption of foods with nicotine. Spider mites also use P-450 enzymes (Mullin & Croft 1983). Spider mites with inducible enzyme systems have higher fecundity when feeding on toxic host plants compared with mites with their detoxification systems blocked (Agrawal et al. 2002). This advantage was not apparent on nontoxic hosts, providing evidence that inducible P-450 systems are offensive.

Plants also contain antinutritive compounds that make essential constituents unavailable for digestion by herbivores. Proteinase inhibitors in plants have received considerable attention as antinutritive agents that inhibit the action of digestive protease enzymes in herbivores and thus make protein unavailable (Ryan 1990). However, careful work has revealed that although proteinase inhibitors effectively reduced the function of particular proteases in the guts of six lepidopteran species, corresponding reductions in caterpillar growth were minimal (Broadway 1995, 1997). After feeding on diets with proteinase inhibitors, these caterpillars responded by producing other proteases that were relatively insensitive to the inhibitors in their diets. Similar induction of insensitive proteases has been reported for proteinase inhibitors expressed in high levels in transgenic plants (Jongsma et al. 1995, Cloutier et al. 2000) and for proteinase inhibitors induced by methyl jasmonate (Bolter & Jongsma 1995). Both P-450 mediated detoxification and production of alternative proteases increase feeding and growth by reducing effects of plant defenses that are induced following ingestion.

Herbivores also produce salivary enzymes constitutively, prior to ingestion, that minimize the effectiveness of plant defenses. Such enzymes are applied to leaf wounds as the herbivores chew and these may reduce the activation of induced defenses in plants. Glucose oxidase is a major constituent of caterpillar saliva that appears to suppress early signaling events in plant defense (Felton & Eichenseer 1999). Glucose oxidase inhibits lipoxygenase activity, a critical step in the biosynthesis of the induction signal, and thereby reduces induced resistance in plants. Caterpillars that were inhibited from producing glucose oxidase fed less and grew less, indicating that offense had been blocked (Musser et al. 2002).

Like caterpillars, other herbivores may also secrete saliva that interferes with plant defenses. Aphid saliva gels into a sheath at the interface between the aphid's stylet and the plant's cells. This sheath may prevent signals from diffusing from the site of penetration, reducing the plant's induced response (Felton & Eichenseer 1999). The evidence for this interaction and its possible mechanism are less well established for aphids compared with similar processes in caterpillars. It has also been suggested that constituents of the aphid sheath may oxidize defenses produced by the plant and reduce their spread (Miles 1990). Salivary constituents from grasshoppers have been implicated in enhanced growth of plants (Detling & Dyer 1981, Dyer et al. 1995, Dyer 2000). In this case, the offensive strategy may be to stimulate plant growth. However, a benefit to the individual herbivore has not been demonstrated in this example.

Herbivores that specialize on particular host plants often have very specific enzymatic systems that allow them to metabolize the secondary chemicals of those hosts. For example, some tropical legume seeds contain nonprotein amino acids that are potent defenses against most herbivores (Rosenthal et al. 1978, 1982). Specialized bruchid beetles are able to metabolize these compounds, synthesizing their own amino acids in the process. Similarly, specialized *Heliconius* caterpillars employ a novel enzymatic system that allows them to convert cyanogenic glycosides into thiols (Engler et al. 2000). This process prevents the plant from releasing cyanide and allows the caterpillars to use the normally toxic compounds as a source of nitrogen. These tactics result in increased feeding ability and may be offensive; however their effects on herbivore performance have not been widely measured.

Sequestering Host Chemicals

Many herbivorous insects incorporate or sequester the biologically active chemicals of their host plants into their own tissues or glands, and thereby gain protection from their predators and parasites (Duffey 1980). Sequestration as an offensive strategy involves several components (Dobler 2001). First, the herbivore must be willing to ingest the host's secondary chemicals. Second, the herbivore must be relatively tolerant of the chemicals. Third, the herbivore must ingest the chemicals without metabolizing them into biologically inactive products. Fourth, the herbivore must deposit the secondary chemicals in particular tissues, or otherwise store them. Some herbivores such as decorator crabs sequester protective chemicals behaviorally by attaching noxious plants to the outsides of their bodies and thereby reduce their risk of predation (Stachowicz & Hay 1999). Sequestration may be relatively expensive as an offensive strategy. Although the strategy allows herbivores to save the costs associated with synthesis of defensive chemicals, it often requires specialized structures to separate, selectively uptake, and/or modify plant compounds (Bowers 1992). Common sense suggests that sequestration may require a very specialized diet. Indeed, in a survey of 70 lepidopteran species from Costa Rica, narrow diet breadth was correlated with defense against predators that was acquired by consumption of plant-derived chemicals (Dyer 1995), although generalists are also known to sequester (Blum et al. 1990, Pennings & Paul 1993).

Morphological Adaptations

Morphological traits of herbivores affect their abilities to exploit particular host plants. A comparative approach revealed striking correspondence between feeding morphologies and the type of food ingested, indicating possible offense (Bernays 1991, Labandeira 1997). Within insect families and feeding guilds, particular mouthparts are associated with particular strategies of consumption. For example, within the Sphingidae, some species have typical incisors for tearing and chewing relatively softer leaves whereas other species have toothless snipping mandibles for cutting larger, tougher grasses or mature leaves of trees (Bernays & Janzen 1988).

Morphological traits of herbivores track their particular diets over evolutionary time as the result of natural selection and over short periods of time as the result of phenotypic plasticity. An example of rapid evolution to novel hosts is provided by soapberry bugs that have mouthparts of precisely the right length to reach the seeds of their native sapindaceous host plant (Carroll & Boyd 1992). In the past century, three new sapindaceous plants have been introduced to North America; soapberry bugs that feed on each of these three new hosts have evolved mouthparts of the appropriate length to feed on seeds of each new plant. These feeding adaptations give the bugs enhanced fitness on their new hosts and concomitant reductions in performance on the ancestral host, demonstrating offense (Carroll et al. 1997, 1998).

Within the lifetime of an individual herbivore, diet can determine its feeding morphology. For example, caterpillars and grasshoppers fed on hard diets developed larger heads with correspondingly greater mandibular power than individuals fed on soft diets (Bernays 1986, Thompson 1992). This morphological plasticity was associated with enhanced feeding abilities for each of the morphs. Mandibular power and wear can potentially limit herbivore fitness (Raupp 1985). Genetic variation in the ability to respond to variability supports the notion that this plasticity in head shape is subject to natural selection (Thompson 1999).

Many other herbivore traits besides mouthparts can be important and offensive. For bruchid seed beetles, for example, females adjust egg size and probably other traits in response to particular host plants. On hosts with hard seeds, fitness is maximized by laying large eggs whereas on hosts with softer seeds, fitness is higher when females lay more, but smaller, eggs (Fox et al. 1997).

Symbionts

Herbivores harbor and vector microbial symbionts that allow them to exploit their host plants more effectively. Symbionts allow herbivores to use foods that the herbivores cannot themselves digest, exploit foods that would otherwise be toxic to herbivores, and weaken the host plant's defenses to the point of submission.

No herbivores can digest cellulose, although several groups of herbivores house bacteria, fungi, and protozoa that carry out this task for them (Heath 1988, Leschine 1995). Some herbivores feed on xylem and phloem sap, diets that lack sufficient amino acids for insect survival (Buchner 1965, Douglas 1998). Phloem-feeding aphids harbor bacteria in their guts that produce the missing nutrients and individuals that have been "cured" of their symbionts perform very poorly, if at all. Leafhoppers, scales, and aphids that have shifted away from phloem diets over evolutionary time no longer have symbiotic bacteria (Douglas 1998).

Symbionts may also be responsible for some of the advantages that accrue to herbivores that eat mixed diets. Turtles that supplemented their duckweed diet with insect larvae became more efficient, especially at digesting the duckweed (Bjorndal 1991). Nutritional analyses suggested that much of this effect resulted from insect larvae in the diet enhancing the numbers or diversity of cellulolytic microbes.

Herbivores must overcome the defenses of the plants that they exploit. As mentioned above, enzymatic detoxification systems are widespread among herbivores and some of these services are provided by microbial symbionts (Jones 1984, Dowd 1991). For example, fungal symbionts in the guts of cigarette beetles use a variety of plant allelochemicals as a carbon source and produce detoxification enzymes (esterases) in response to other potential toxins. Thus, symbionts may increase feeding and exploitation of plants by herbivores by providing nutrients, digestion, and detoxification.

More aggressive strategies involve inoculating hosts with microbes that mobilize nutrients or diminish plant defenses. Herbivores that vector plant diseases (thrips, beetles, mites, homopterans) may manipulate plants to their own benefits through infection. Aphids that vector plant viruses have shorter developmental times, higher fecundity, and a greater intrinsic rate of increase on virus-infected hosts compared with uninfected hosts (Fereres et al. 1989, Blua et al. 1994). In these systems, changes in sugar, total protein, and specific amino acid concentrations following infection may partly explain the benefit to aphids (Blua et al. 1994). Other insects may benefit by inoculating plants with pathogens that diminish the plant's ability to defend against herbivores; recent evidence leads to the speculation that defenses against pathogens may inhibit defenses against herbivores (Felton et al. 1999, Preston et al. 1999, Thaler et al. 1999).

Some herbivores are far more successful on damaged and weakened host plants, and symbionts may facilitate these herbivore attacks. For example, many species of bark beetles would be killed by defensive reactions of their host trees. However, invading beetles introduce blue stain fungi that cause necrotic lesions and reduce the tree's ability to respond defensively (Whitney 1982, Nebeker et al. 1993). Although the causal mechanisms have not yet been elucidated, it is clear that the fungus causes many physiological changes in an attacked tree and these greatly facilitate beetle colonization (Klepzig et al. 1996).

HERBIVORES MANIPULATE THEIR
HOSTS—AGGRESSIVE

The examples discussed above mostly involve traits of herbivores that enable them to make efficient use of their host plants. In some cases (e.g., salivary enzymes, symbionts), herbivores manipulate the traits of their hosts rather than their own morphology, physiology, and behavior. The ability to manipulate the host's defensive reactions falls into this more aggressive category, as do other tactics discussed below.

Galls and Induced Plant Susceptibility

Galls are plant structures that are inhabited and fed upon by small herbivores (or by fungi or microorganisms). These structures are made of plant tissue, although gall morphology is controlled by the herbivores (Price et al. 1987). Variation in the morphology of the gall depends upon the phylogeny of the gall-making insect and is independent of the host phylogeny (Crespi & Worobey 1998, Nyman et al. 2000). Gall form is determined by substances secreted by the ovipositing female or by the spatial pattern of larval or adult feeding. In most cases, the interior of the gall where the herbivore feeds is made up of highly nutritious tissue. For example, observations using labeled ^{14}C in galled cottonwood leaves showed that aphid galls are strong physiological sinks that drew resources from surrounding plant tissues (Larson & Whitham 1991). In general, aphids that feed in galls are more efficient than free-feeding aphids (Llewellyn 1982). This evidence implicates the induction of galls as probable offensive traits. Data on the nutritional quality of the specific tissues fed on by gall makers rather than the entire gall would strengthen this case (Nyman & Julkunen-Tiitto 2000). Better comparisons of herbivore performance on galled and ungalled tissue would provide a more definitive test of offense. Galls provide the gall maker with a specific microhabitat that may include relative safety from pathogens, predators, and parasites, as well as a source of high quality nutrition (Cornell 1983, Weis & Kapelinski 1994).

Herbivores also modify the microhabitat that their host plants provide when they tie, fold, or glue foliage around themselves. These herbivore tactics can greatly reduce the effectiveness of photoactive plant defenses (Berenbaum 1987). Leaf rolling reduces light inside the roll and this behavior was associated with a reduction in leaf toughness and tannin concentration (Sagers 1992). Many herbivores that feed on plants containing photoactive hypericin reduce its activity by tying leaves and feeding in the dark (Sandberg & Berenbaum 1989). This tactic increases rates of development and decreases mortality of larvae and qualifies as offensive.

Damage by many herbivores causes their plants to become better hosts, a phenomenon we term induced plant susceptibility (Karban & Baldwin 1997, Table 4.2). Induced susceptibility is often caused by herbivores manipulating the ontogeny of their host plant to that stage that is most favorable for the herbivores. For example, feeding by green spruce aphids accelerated senescence of spruce needles (Fisher 1987). Senescent needles were superior hosts for aphids compared with unmanipulated needles that remained in the less mature green stage, making this an example of herbivore offense (sensu Table 1).

Trenching, Mowing, Haying, and Gardening

Herbivores exhibit a great diversity of behaviors associated with feeding that deactivate the defenses of their host plants. Plants can immobilize herbivore mouthparts and reduce feeding by releasing secretions from pressurized canals such as laticifers when herbivores bite. In response, many insects cut veins across the leaf blade prior to feeding. The herbivores then feed beyond the cuts, free from the plant exudates that would normally deter herbivory (Dussourd & Denno 1994, Dussourd 1999). This herbivore behavior precisely matches the architecture of secretory canals in each host plant (Dussourd & Denno 1991). For example, vein clipping is found on plants with arborescent canals, whereas continuous trenching is found on plants with net-like canals. Caterpillars that exhibited these offensive behaviors fed more efficiently and survived much more frequently than those lacking the behaviors on plants in many families that contained pressurized exudates (Dussourd 1993).

Herbivores manipulate plants in many other ways reminiscent of human agriculture. Many insect herbivores reduce the harmful effects of trichomes (plant hairs) by mowing them before they start to feed (e.g., Hulley 1988). Some caterpillars avoid the trichomes by laying down a silk covering or scaffolding over them (e.g., Rathcke & Poole 1975). Certain mammalian herbivores clip vegetation and cache it before consuming it, a behavior known as haying (Vander Wall 1990). For example, pikas cache food in summer that they consume in winter (Dearing 1997). Plants cached in "hay piles" facilitated food storage and also allowed toxins to degrade so that cached food became much more palatable after storage (probable offense).

Herbivores occasionally maintain elaborate gardens. For example, leaf-cutting attine ants gather pieces of fresh leaves and flowers, which they use to culture fungi, their ultimate source of nutrition (Holldobler & Wilson 1990). The ants prepare the pieces of vegetation, chewing, licking, and trimming the fragments before inoculating them with mycelia. The ants maintain a monoculture of their preferred fungus by carrying another mutualist, a filamentous bacterium that produces antibiotics that are particularly effective against parasites of their fungus garden (Currie et al. 1999). Phylogenetic evidence suggests that farming fungi on leaves arose from ants that collected fungi from insects and other substrates (Holldobler & Wilson 1990). More work on the phylogeny of this and other offensive traits could suggest important insights into how the traits evolved and if they have been responsible for diversification of herbivores.

Gregarious Feeding

Group feeding appears to enhance the ability of herbivores to exploit their host plants. There are other advantages to feeding in groups, such as enhanced host finding and reduced predation risk (Wilson 1975, Breton & Addicott 1992). Homopterans that suck phloem sap from plants create sinks that are preferentially supplied with nutrients (Way & Cammell 1970, Larson & Whitham 1991). The more aphids there are over a wide range of densities, the stronger the sink is. As a result, *Brevicoryne brassicae* showed increased per capita population growth in larger groups (Lopez et al. 1989). Benefits of gregarious feeding have also been reported for chewing insects. Gregarious caterpillars grew twice as fast in large groups (30 individuals) compared with smaller groups (1–10 individuals) (Denno & Benrey 1997). This effect was completely ameliorated when groups of these sizes were grown on excised foliage. Thus large groups either caused nutrient sinks or prevented induced defenses in intact plants compared with smaller groups of herbivores. Gregarious *Battus philenor* caterpillars showed a growth benefit that was positive and linearly associated with group size (J. Fordyce, unpublished manuscript). This effect was caused by a density-dependent induced plant response, perhaps through creation of a nutrient sink. Intraspecific competition and density dependent parasitism may be costs of herbivore aggregations. In most of the cases studied so far, evidence for a benefit to the herbivore preceded a mechanistic understanding of host manipulation.

SYNTHESIS

Models of Herbivore Offense

Herbivore offense has not received much consideration by theoreticians beyond general foraging theory and theories of optimal oviposition in different environments including host plants. These theories are not considered in this review. The small amount of specific herbivore offense theory that exists posits that different strategies will be used by different herbivores and that the strategies will be associated with particular suites of herbivore life-history traits and population dynamics (Rhoades 1985; Price 1990, 1992).

Rhoades described a dichotomy between opportunistic herbivores that avoid plants that are well defended and stealthy herbivores that metabolize plant defenses and manipulate their hosts (summarized in Table 2). Opportunistic herbivores are attracted to plants that provide abundant nutrients and repelled by secondary chemicals and other "deterrents." Feeding by opportunists often involuntarily causes induced resistance (not present prior to feeding), making their host plants more resistant to subsequent feeding. In contrast, stealthy herbivores metabolize or detoxify the secondary chemicals of their hosts. They manipulate their hosts, inducing them to provide more nutritious tissue for the herbivore to exploit. Stealthy herbivores co-opt the defenses of their hosts and use them for their own benefit, as feeding stimulants, mating rewards, or defenses against their own enemies. They also can inhibit the plant's induced responses by suppressing the recognition, deployment, or effect of induced resistance.

Offensive strategies may be associated with suites of herbivore life-history traits. For example, stealthy herbivores are posited to be solitary, territorial, and

TABLE 3   Price's (1990) model of herbivore offensive strategies and their consequences

Syndrome	Resources consumed	Choice?	Life history traits	Population dynamics
Latent herbivore	High quality	Ovipositing females choosy	High host specificity, emigration from high density	Relatively low Number, low variation
Eruptive herbivore	Lower quality	Females cannot assess quality or choose	Low host specificity, low dispersal	Can reach high numbers, high variation

have overdispersed damage, whereas opportunists are more likely to be gregarious, colonial, and clumped in space. These life-history traits produce very different patterns of population dynamics for opportunist and stealthy herbivores. Opportunists have a greater intrinsic growth rate and exhibit much more variable population dynamics. Rhoades argues that most outbreaking herbivores are opportunists although there are clearly exceptions.

Price (1990, 1992) also described a dichotomy in herbivore strategies but suggested very different relationships. Price was more explicit in assigning causality: the quality of the resource eaten by different herbivores determines how choosy ovipositing females are, which determines other life-history traits and ultimately population dynamics (Table 3). For "latent" species (whose populations do not outbreak), high-quality resources are scarce. This scarcity causes ovipositing females to be choosy, resulting in a suite of life-history traits including high specificity of host use, emigration from high density sites, and ultimately latent population dynamics. In contrast, eruptive species feed on lower quality resources and females lack the ability to assess or choose host quality. Low-quality resources and lack of choosiness result in the opposite set of life-history traits compared with those associated with latent herbivore species. These life-history traits allow populations to sometimes become very large. The key distinction according to this model is whether females assess and respond to resource quality.

The models of both Rhoades and Price consider host plant choice to be an important determinant of life histories and population dynamics, although they reach rather different conclusions and predictions. Rhoades is much more detailed about specific strategies and tactics. However, his conclusions about the consequences of those strategies are based on intuition rather than a thorough review of the literature. Price does not develop the distinction between the strategies, although his conclusions about their consequences are a first attempt to synthesize the existing literature, which is roughly supportive of his ideas (Price 1992). In a more recent treatment of this model, he and coworkers cite examples of latent herbivores that exhibit strong correlations between oviposition preference and larval performance (Pires et al. 2000). Similarly, eruptive species failed to exhibit choice or to demonstrate a positive correlation between preference for a host and performance on that host. These patterns from the literature support Price's conceptual model, although the review is clearly selective, as some herbivores with eruptive dynamics have been found to make appropriate choices (e.g., willow beetles; Rank et al. 1998) and some with latent dynamics fail to make appropriate choices (e.g., the pearly marblewing butterfly; Karban & Courtney 1987). However, these exceptions do not falsify the generality and a more complete meta-analysis of the many reported preference-performance studies would be a useful contribution.

The Importance of Choice

The theories of herbivore offense developed by Rhoades and especially by Price emphasized the importance of choice. We wish to underscore that observation and to note that many plant defenses will be effective only if herbivores make appropriate and rather sophisticated choices. A plant that poisons its herbivores will not necessarily have higher fitness than one that does not, unless poisoned herbivores reduce plant fitness less than herbivores that have not been poisoned. All else being equal, plant defenses that cause herbivores to avoid feeding on the plant will be most effective from the plant's perspective. Defenses that act prior to significant damage to the plant will be more effective than those that act after ingestion. Defended plants will probably benefit most from pre-ingestive herbivore strategies because herbivores will be deterred. In contrast to plants, herbivores can accrue benefits from both pre- and post-ingestive strategies.

Selection may favor plants that respond to their risk of herbivory as well as herbivores that assess plant quality and choose accordingly (Karban et al. 1997, 1999). Traits that allow herbivores to accurately assess plant quality may benefit herbivores as well as plants. Hamilton & Brown (2001) suggest that plants may provide reliable information to their herbivores indicating their level of defense. They interpret autumn coloration of tree leaves as a signal to colonizing herbivores that the brightly colored leaves are not very good hosts. There is some empirical support for this contention. For example, maple aphids were more likely to place overwintering offspring on maple leaves that were yellow or orange and less likely to oviposit in trees with red foliage (Furuta 1986). Those aphids that colonized the reddest trees experienced reduced fitness (Furuta 1990, in Hamilton & Brown 2001).

Herbivore Offense and Adaptive Plasticity

The models of Rhoades and Price were useful as broad generalizations, although their ability to offer a predictive framework was limited. As an alternative, we suggest that herbivore offense can be categorized along two axes: when they work and whether they are plastic or fixed (Table 4A). Offensive traits may work before the herbivore ingests the plant or after ingestion. The herbivore may alter its offensive traits depending upon the host and other conditions or the traits may be fixed. In theory, phenotypic plasticity can maximize fitness in variable environments (Levins 1968, Agrawal 2001). We consider two different ways for an herbivore to exhibit plasticity: (*a*) by choosing which resources to exploit or

TABLE 4   Herbivore offense categorized by plasticity and time of action

	Pre-ingestion	Post-ingestion
A Examples of offensive tactics		
When does the trait work?		
Plastic	Behavioral choice	Diet mixing, learning
	Trenching, haying of	Inducible enzymes,
	selected hosts	morphology
Not Plastic	No behavioral discrimination	Fixed physiology, morphology
	Galling and leaf rolling	Sequestration
	Salivary enzymes	Internal digestive symbionts
B Predictions		
When does the trait work?		
Plastic	Generalist	Generalist
	Unpredictable resource	Unpredictable resource
	Can be aggressive	Less aggressive
Not Plastic	More specialized	More specialized
	Predictable resource	Most predictable resource
	Most aggressive	Less aggressive

(b) by altering its morphology and physiology rather than its choice. There are examples of herbivores that are plastic in both ways (e.g., inducible enzymes of generalists), not plastic in either (e.g., many specialists), or plastic in one category but not the other.

Arranging offensive tactics along these two axes of when they work and whether they are plastic leads to several predictions (Table 4B). First, plastic offensive traits are often employed by generalist herbivores that feed on many plant species and adjust their traits accordingly. Conversely, more specialized herbivores are more likely to have offensive traits that are not plastic. Some offensive traits of generalists are nonplastic, such as the salivary enzymes or filter chambers of xylem-feeding homopterans. However, these traits have widespread utility against chemical and physical obstacles that are highly conserved across plant species. Similarly, xylem feeders are faced with a food source that is much more homogeneous across plant species than are other plant tissues. Second, offensive traits that are not plastic are most often associated with plant resources that are highly predictable to the herbivore. This seems particularly true for post-ingestive traits; once a food has entered an herbivore's gut, if the morphology and physiology of the gut are not plastic, the food had better be predictably palatable and nutritious. It is interesting to note that many of the offensive traits of specialist herbivores that are plastic involve unpredictable resources. For example, specialized monarch caterpillars do not cut veins consistently in their milkweed hosts. Vein cutting only occurs when latex volume and pressure in the milkweed host are high (A.A. Agrawal, personal observation). Third, most of the examples of aggressive offensive traits work prior to ingestion, especially for nonplastic traits. An herbivore has less opportunity to manipulate its host once it has ingested it. Rapidly growing, young, intact plants are generally the most inducible and most subject to manipulation by the herbivore (Karban & Baldwin 1997). In summary, the early attempts to categorize herbivore offense by Rhoades and Price have served as useful starting places for development of a theory for this field. However, we found their categories less than satisfying and have proposed that consideration of when a trait works and whether it is plastic may lead to more robust predictions. These categories are simplifications of a continuum and can never capture the full richness of species interactions. Although the categories have heuristic value, a systematic and objective analysis of many herbivore species is required to assess whether this framework adds clarity or just more arm waving.

Costs and Limitations to Herbivore Offense

A consideration of the costs and benefits has been very instructive in structuring our thinking about the ecology and evolution of plant defenses. Much of this review has been a compilation of the potential and realized benefits of traits that herbivores use to exploit their host plants. Generally, we have found only a limited number of studies demonstrating that these traits increase herbivore fitness. Similarly, there have been few tests of the notion that offensive traits are costly for herbivores.

Potentially, costs and limitations of offensive strategies can take many forms. One of the greatest costs of choosiness is that many suitable host plants will be rejected as unacceptable (Waldbauer & Fraenkel 1961). In such cases, herbivores are passing up good opportunities in order to minimize the likelihood of adversity. There are also likely to be limits to the processing abilities of herbivores; the ability to recognize some foods may come at the expense of recognition of other foods or potentially important stimuli (Bernays 2001). Similarly, enzyme systems that respond to food constituents carry potential allocation costs. In other words, resources that are allocated to enzyme systems cannot then be used by the herbivore for other purposes (growth, reproduction, etc.). Several workers have looked for allocation costs of enzyme systems that metabolize plant secondary chemicals and have failed to observe reductions in food utilization or growth parameters associated with the induction of enzymes (Neal 1987, Appel & Martin 1994). A few studies have found circumstantial evidence for measurable costs of enzymatic metabolism, although other costs such as autotoxicity or hormonal imbalances may be larger but unexamined (Berenbaum & Zangerl 1994, Agrawal et al. 2002).

Essentially, all the tactics considered in this review may be costly to the herbivores, although these costs have not been well documented. Symbionts may also be expensive for herbivores to maintain. Bacterial endosymbionts of aphids greatly increase aphid fecundity while reducing other aspects of aphid fitness on some hosts (Chen et al. 2000; T. Leonardo, unpublished manuscript). Clipping plant veins makes herbivores conspicuous to their predators and exposes them to physical challenges of sticky exudates. Mortality associated with this behavior is relatively high (Zalucki et al. 2002). Gregarious feeding increases levels of intraspecific competition, attracts predators, and may increase the likelihood of induced plant responses (e.g., Dyer & Gentry 1999, Williams & Whitham 1985). Sequestration of plant chemicals can come at the expense of growth. For example, specialist buckeye caterpillars, fed on diets high in iridoid glycosides, sequestered the glycosides but became less efficient at digesting their food (Camara 1997). Better measures of the costs and benefits of offensive tactics in terms of herbivore performance and especially fitness will sharpen our ability to predict the outcome of ecological interactions and coevolution.

CONCLUSIONS

We have outlined a list of criteria (questions) for herbivore offense. Do herbivore traits match their host plants? Have those traits been found to increase feeding and other uses? Do they increase herbivore performance? What is the evidence that herbivore traits have evolved to increase fitness on those hosts? For each of these questions, are alternative explanations possible? Answering these questions should allow us to be much more explicit about the nature and strength of the evidence for herbivore offense and allow more rapid progress in this subdiscipline. In the future, the theory and evidence for herbivore offense should match that accumulated for plant defense.

ACKNOWLEDGMENTS

We thank David Dussourd, Jim Fordyce, Mikaela Huntzinger, Marc Johnson, Andy McCall, Steve Pennings, Peter Price, Jennifer Thaler, Pete Van Zandt, and Louie Yang for comments on this manuscript. Our work was supported by NSF (DEB-0121050) and NSERC.

The *Annual Review of Ecology and Systematics* is online at
http://ecolsys.annualreviews.org

LITERATURE CITED

Agrawal AA. 2001. Phenotypic plasticity in the interactions and evolution of species. *Science* 294:321–26

Agrawal AA, Kobayashi C, Thaler JS. 1999. Influence of prey availability and induced host-plant resistance on omnivory by western flower thrips. *Ecology* 80:518–23

Agrawal AA, Vala F, Sabelis MW. 2002. Induction of preference and performance after acclimation to novel hosts in a phytophagous

spider mite: Adaptive plasticity? *Am. Nat.* 159:553–65

Appel HM, Martin MM. 1992. Significance of metabolic load in the evolution of host specificity of *Manduca sexta*. *Ecology* 73:216–28

Arms K, Feeny P, Lederhouse RC. 1974. Sodium: stimulus for puddling behavior by tiger swallowtail butterflies. *Papilio glaucus*. *Science* 185:372–74

Beck J, Muhlenberg E, Fiedler K. 1999. Mud-puddling behavior in tropical butterflies: In search of proteins or minerals? *Oecologia* 119:140–48

Berenbaum MR. 1987. Charge of the light brigade: phototoxicity as a defense against insects. In *Light-Activated Pesticides*, ed. JR Heitz, KR Downum, pp. 206–16. Washington, DC: Am. Chem. Soc.

Berenbaum MR, Zangerl AR. 1994. Costs of inducible defense: protein limitation, growth, and detoxification in parsnip webworms. *Ecology* 75:2311–17

Bernays EA. 1986. Diet-induced head allometry among foliage-chewing insects and its importance for graminivores. *Science* 231:495–97

Bernays EA, ed. 1990. *Insect-Plant Interactions. Vol. II*, Boca Raton, FL: CRC

Bernays EA. 1991. Evolution of insect morphology in relation to plants. *Philos. Trans. R. Soc. London Ser. B* 333:257–64

Bernays EA. 1993. Aversion learning and feeding. In *Insect Learning. Ecological and Evolutionary Perspectives*, ed. DR Papay, AC Lewis, pp. 1–17. New York: Chapman & Hall

Bernays EA. 2001. Neural limitations in phytophagous insects: implications for diet breadth and evolution of host affiliation. *Annu. Rev. Entomol.* 46:703–27

Bernays EA, Bright KL, Gonzalez N, Angel J. 1994. Dietary mixing in a generalist herbivore: tests of two hypotheses. *Ecology* 75: 1997–2006

Bernays EA, Graham M. 1988. On the evolution of host specificity in phytophagous arthropods. *Ecology* 69:886–92

Bernays EA, Janzen DH. 1988. Saturniid and sphingid caterpillars: two ways to eat leaves. *Ecology* 69:1153–60

Bernays EA, Lee JC. 1988. Food aversion learning in the polyphagous grasshopper *Schistocerca americana*. *Physiol. Entomol.* 13:131–37

Bjorndal KA. 1991. Diet mixing: nonadditive interactions of diet items in an omnivorous freshwater turtle. *Ecology* 72:1234–41

Blua MJ, Perring TM, Madore MA. 1994. Plant virus-induced changes in aphid population development and temporal fluctuations in plant nutrients. *J. Chem. Ecol.* 20:691–707

Blum MS, Severson RF, Arrendale RF, Whitman DW, Escoubas P, et al. 1990. A generalist herbivore in specialist mode. Metabolic sequestrative, and defensive consequences. *J. Chem. Ecol.* 16:223–44

Bolter CJ, Jongsma MA. 1995. Colorado potato beetles (*Leptinotarsa decemlineata*) adapt to proteinase inhibitors induced in potato leaves by methyl jasmonate. *J. Insect Physiol.* 41:1071–78

Bowers MD. 1992. The evolution of unpalatability and the costs of chemical defense in insects. In *Insect Chemical Ecology*, ed. BD Roitberg, MB Isman, pp. 216–44. New York, Chapman & Hall

Brattsten LB. 1988. Exzymatic adaptation in leaf-feeding insects to host-plant allelochemicals. *J. Chem. Ecol.* 14:1919–40

Breton LM, Addicott JF. 1992. Density-dependent mutualism in an aphid-ant interaction. *Ecology* 73:2175–80

Broadway RM. 1995. Are insects resistant to plant proteinase inhibitors? *J. Insect Physiol.* 41:107–16

Broadway RM. 1997. Dietary regulation of serine proteinases that are resistant to serine proteinase inhibitors. *J. Insect Physiol.* 43: 855–74

Buchner P. 1965. *Endosymbiosis of Animals with Plant Microorganisms*. New York: Intersci. 909 pp.

Camara MD. 1997. Physiological mechanisms underlying the costs of chemical defence in *Junonia coenia* Hubner (Nymphalidae): a gravimetric and quantitative genetic analysis. *Evol. Ecol.* 11:451–69

Carroll SP, Boyd C. 1992. Host race radiation in the soapberry bug: natural history with the history. *Evolution* 46:1052–69

Carroll SP, Dingle H, Klassen SP. 1997. Genetic differentiation of fitness-associated traits among rapidly evolving populations of the soapberry bug. *Evolution* 51:1182–88

Carroll SP, Klassen SP, Dingle H. 1998. Rapidly

evolving adaptations to host ecology and nutrition in the soapberry bug. *Evol. Ecol.* 12:955–68

Chen DQ, Montlor CB, Purcell AH. 2000. Fitness effects of two facultative endosymbiotic bacteria on the pea aphid. *Acyrthosiphon pisum*, and the blue alfalfa aphid, *A. kondoi.* *Entomol. Exp. Appl.* 95:315–23

Cloutier C, Jean C, Fournier M, Yelle S, Michaud D. 2000. Adult Colorado potato beetles, *Leptinotarsa decemlineata* compensate for nutritional stress on oryzacystatin I-transgenic potato plants by hypertrophic behavior and over-production of insensitive proteases. *Arch. Insect Biochem. Physiol.* 44:69–81

Cornell HV. 1983. The secondary chemistry and complex morphology of galls formed by the Cynipinae (Hymenoptera): Why and how? *Am. Midl. Nat.* 110:225–34

Crespi B, Worobey M. 1998. Comparative analysis of gall morphology in Australian gall thrips: the evolution of extended phenotypes. *Evolution* 52:1686–96

Currie CR, Scott JA, Summerbell RC, Malloch D. 1999. Fungus-growing ants use antibiotic-producing bacteria to control garden parasites. *Nature* 398:701–4

Dearing MD. 1997. The manipulation of plant toxins by a food-hoarding herbivore. *Ochotona princeps. Ecology* 78:774–81

Denno RF, Benrey B. 1997. Aggregation facilitates larval growth in the neotropical nymphalid butterfly *Chlosyne janais*. *Ecol. Entomol.* 22:133–41

Detling JK, Dyer MI. 1981. Evidence for potential plant growth regulators in grasshoppers. *Ecology* 62:485–88

Dobler S. 2001. Evolutionary aspects of defense by recycled plant compounds in herbivorous insects. *Basic Appl. Ecol.* 2:15–26

Douglas AE. 1998. Nutritional interactions in insect-microbial symbioses: aphids and their symbiotic bacteria *Buchnera*. *Annu. Rev. Entomol.* 43:17–37

Dowd P. 1991. Symbiont-mediated detoxification in insect herbivores. In *Microbial Mediation of Plant-Herbivore Interactions*, ed. P

Barbosa, VA Krischik, C Jones, pp. 411–40. New York: Wiley

Duffey SS. 1980. Sequestration of plant natural-products by insects. *Annu. Rev. Entomol.* 25:447–77

Duffey SS, Stout MJ. 1996. Antinutritive and toxic components of plant defense against insects. *Arch. Insect Biochem. Physiol.* 32:3–37

Dukas R, Bernays EA. 2000. Learning improves growth rate in grasshoppers. *Proc. Natl. Acad. Sci. USA* 97:2637–40

Dussourd DE. 1993. Foraging with finesse: caterpillar adaptations for circumventing plant defenses. In *Caterpillars: Ecological and Evolutionary Constraints on Foraging*, ed. NE Stamp, TM Casey, pp. 92–131. New York: Chapman & Hall

Dussourd DE. 1999. Behavioral sabotage of plant defense: Do vein cuts and trenches reduce insect exposure to exudates? *J. Insect Behav.* 12:501–15

Dussourd DE, Denno RF. 1991. Deactivation of plant defense: correspondence between insect behavior and secretory canal architecture. *Ecology* 72:1383–96

Dussourd DE, Denno RF. 1994. Host range of generalist caterpillars: Trenching permits feeding on plants with secretory canals. *Ecology* 75:69–78

Dyer LA. 1995. Tasty generalists and nasty specialists? Antipredator mechanisms in tropical lepidoptera larvae. *Ecology* 76:1483–96

Dyer LA, Gentry G. 1999. Predicting natural-enemy responses to herbivores in natural and managed systems. *Ecol. Appl.* 9:402–8

Dyer MI. 2000. Herbivores, biochemical messengers and plants: aspects of intertrophic transduction. In *Invertebrates as Webmasters in Ecosystems*, ed. DC Coleman, PF Hendrix, pp. 115–40. Wallingford, UK: CABI

Dyer MI, Moon AM, Brown MR, Crossley DA. 1995. Grasshopper crop and midgut extract effects on plants: an example of reward feedback. *Proc. Natl. Acad. Sci. USA* 92:5475–78

Endler JA. 1986. *Natural Selection in the Wild.* Princeton NJ: Princeton Univ. Press. 336 pp.

Engler HS, Spencer KC, Gilbert LE. 2000. Insect metabolism—preventing cyanide release from leaves. *Nature* 406:144–45

Felton GW, Eichenseer H. 1999. Herbivore saliva and its effect on plant defense against herbivores and pathogens. In *Induced Plant Defenses Against Pathogens and Herbivores*, ed. AA Agrawal, S Tuzun, E Bent, pp. 19–36. St. Paul, MN: Am. Phytopathol. Soc.

Felton GW, Korth KL, Bi JL, Wesley SV, Huhman DV, et al. 1999. Inverse relationship between systemic resistance of plants to microorganisms and to insect herbivory. *Curr. Biol.* 9:317–20

Fereres A, Lister RM, Araya JE, Foster JE. 1989. Development and reproduction of the English grain aphid (Homoptera: Aphididae) on wheat cultivars infected with barley yellow dwarf virus. *Environ. Entomol.* 18:388–93

Feyereisen R. 1999. Insect P450 enzymes. *Annu. Rev. Entomol.* 44:507–33

Fisher M. 1987. The effect of previously infested spruce needles on the growth of the green spruce aphid, *Elatobium abietinum*, and the effect of the aphid on the amino acid balance of the host plant. *Ann. Appl. Biol.* 111:33–41

Fox CW. 1993. A quantitative genetic analysis of oviposition preference and larval performance on two hosts in the bruchid beetle. *Callosobruchus maculatus. Evolution* 47:166–75

Fox CW, Thakar KJ, Mousseau TA. 1997. Egg size plasticity in a seed beetle: an adaptive maternal effect. *Am. Nat.* 149:149–63

Fox L. 1975. Cannibalism in natural populations. *Annu. Rev. Ecol. Syst.* 6:87–106

Freeland WJ, Janzen DH. 1974. Strategies in herbivory by mammals: the role of plant secondary compounds. *Am. Nat.* 108:269–89

Furuta K. 1986. Host preference and population dynamics in an autumnal population of the maple aphid. *Periphyllus californiensis* Shinji (Homoptera, Aphididae). *J. Appl. Entomol.* 102:93–100

Hagele BF, Rowell-Rahier M. 1999. Dietary

mixing in three generalist herbivores: Nutrient complementation or toxin dilution? *Oecologia* 119:521–33

Hagen RH, Chabot JF. 1986. Leaf anatomy of maples (*Acer*) and host use by Lepidoptera larvae. *Oikos* 47:335–45

Hamilton WD, Brown SP. 2001. Autumn tree colours as a handicap signal. *Proc. R. Soc. London Ser. B* 268:1489–93

Hay ME, Duffy JE, Fenical W. 1990. Host-plant specialization decreases predation on a marine amphipod: an herbivore in plant's clothing. *Ecology* 71:733–43

Heath IB. 1988. Gut fungi. *TREE* 3:167–71

Holldobler B, Wilson EO. 1990. *The Ants*. Cambridge MA: Harvard Univ. Press. 732 pp.

Hulley PE. 1988. Caterpillar attacks plant mechanical defence by mowing trichomes before feeding. *Ecol. Entomol.* 13:239–41

Jaenike J. 1989. Genetic population structure of *Drosophila tripunctata*: patterns of variation and covariation of traits affecting resource use. *Evolution* 43:1467–82

Jones CG. 1984. Microorganisms as mediators of plant resource exploitation by insect herbivores. In *A New Ecology: Novel Approaches to Interactive Systems*, ed. PW Price, CN Slobodchikoff, WS Gaud, pp. 53–99. New York: Wiley

Jongsma MA, Bakker PL, Peters J, Bosch D, Stiekema WJ. 1995. Adaptation of *Spodoptera exigua* larvae to plant proteinase inhibitors by induction of gut proteinase activity insensitive to inhibition. *Proc. Natl. Acad. Sci. USA* 92:8041–45

Karban R, Agrawal AA, Mangel M. 1997. The benefits of induced defenses against herbivores. *Ecology* 78:1351–55

Karban R, Agrawal AA, Thaler JS, Adler LS. 1999. Induced plant responses and information content about risk of herbivory. *TREE* 14:443–47

Karban R, Baldwin IT. 1997. *Induced Responses to Herbivory*. Chicago, IL: Univ. Chicago Press. 319 pp.

Karban R, Courtney S. 1987. Intraspecific host plant choice: Lack of consequences for *Streptanthus tortuosus* (Cruciferae) and

Eucloe hyantis (Lepidoptera: Pieridae). *Oikos* 48:243–48

Klepzig KD, Smalley EB, Raffa KF. 1996. Combined chemical defenses against an insect-fungal complex. *J. Chem. Ecol.* 22:1367–88

Labandeira CC. 1997. Insect mouthparts: ascertaining the paleobiology of insect feeding strategies. *Annu. Rev. Ecol. Syst.* 28:153–93

Larson KC, Whitham TG. 1991. Manipulation of food resources by a gall-forming aphid: the physiology of sink-source interactions. *Oecologia* 88:15–21

Leschine SB. 1995. Cellulose degradation in anaerobic environments. *Annu. Rev. Microbiol.* 49:399–426

Levins R. 1968. *Evolution in Changing Environments.* Princeton, NJ: Princeton Univ. Press

Llewwllyn M. 1982. The energy economy of fluid feeding herbivorous insects. In *International Symposium on Insect-Plant Relationships, Proc. 5th Int. Symp. Insect-Plant Relationships*, ed. JH Viller, AK Minks, pp. 243–51. Wageningen: Cent. Agric. Publ. Doc.

Lopez ER, Van Driesche RG, Elkinton JS. 1989. Influence of group size on daily per capita birth rates of the cabbage aphid (Homoptera: Aphididae) on collards. *Environ. Entomol.* 18:1086–89

Mattson WJ. 1980. Herbivory in relation to plant nitrogen content. *Annu. Rev. Ecol. Syst.* 11:119–61

Mayhew PJ. 1997. Adaptive patterns of host-plant selection by phytophagous insects. *Oikos* 79:417–28

Mayhew PJ. 2001. Herbivore host choice and optimal bad motherhood. *TREE* 16:165–67

Miles PW. 1990. Aphid salivary secretions and their involvement in plant toxicoses. In *Aphid-Plant Genotype Interactions*, ed. RK Campbell, RD Eikenbary, pp. 131–47. Amsterdam: Elsevier

Mullin C, Croft B. 1983. Host-related alterations of detoxification enzymes in *Tetranychus urticae* (Acari: Tetranychidae). *Environ. Entomol.* 12:1278–82

Musser RO, Hum-Musser SM, Eichenseer H,

Peiffer M, Ervin G, et al. 2002. Herbivory: caterpillar saliva beats plant defense—A new weapon emerges in the evolutionary arms race between plants and herbivores. *Nature* 416:599–600

Neal JJ. 1987. Metabolic costs of mixed-function oxidase induction in *Heliothis zea*. *Entomol. Exp. Appl.* 43:175–79

Nebeker TE, Hodges JD, Blanche CA. 1993. Host response to bark beetle and pathogen colonization. In *Beetle-Pathogen Interactions in Conifer Forests*, ed. T Schowalter, G Filip, pp. 157–73. New York, Academic

Nyman T, Julkunen-Tiitto R. 2000. Manipulation of the phenolic chemistry of willows by gall-inducing sawflies. *Proc. Natl. Acad. Sci. USA* 97:13184–87

Nyman T, Widmer A, Roinnen H. 2000. Evolution of gall morphology and host-plant relationships in willow-feeding sawflies (Hymenoptera: Tenthredinidae). *Evolution* 54:526–33

Papaj DR, Prokopy RJ. 1989. Ecological and evolutionary aspects of learning in phytophagous insects. *Annu. Rev. Entomol.* 34:315–50

Pennings SC, Nadeau MT, Paul VJ. 1993. Selectivity and growth of the generalist herbivore *Dolabella auricularia* feeding upon complimentary resources. *Ecology* 74:879–90

Pennings SC, Paul VJ. 1993. Sequestration of dietary secondary metabolites by three species of sea hares—location, specificity and dynamics. *Marine Biol.* 117:535–46

Pires CSS, Price PW, Fontes EG. 2000. Preference-performance linkage in the neotropical spittlebug *Deois flavopicta*, and its relation to the phylogenetic constraints hypothesis. *Ecol. Entomol.* 25:71–80

Preston CA, Lewandowski C, Enyedi AJ, Baldwin IT. 1999. Tobacco mosaic virus inoculation inhibits wound-induced jasmonic acid-mediated responses within but not between plants. *Planta* 209:87–95

Price PW. 1992. Plant resources as the mechanistic basis for insect herbivore population dynamics. In *Effects of Resource Distribution* on *Animal-Plant Interactions*, ed. MD Hunter, T Ohgushi, PW Price, pp. 139–73. San Diego: Academic

Price PW, Bouton CE, Gross P, McPheron BA, Thompson JN, Weis AE. 1980. Interactions among three trophic levels: influence of plants on interactions between insect herbivores and natural enemies. *Annu. Rev. Ecol. Syst.* 11:41–65

Price PW, Cobb N, Craig TP, Fernandes GW, Itami JK, et al. 1990. Insect herbivore population dynamics on trees and shrubs: new approaches relevant to latent and eruptive species and life table development. See Bernays 1990, pp. 1–38

Price PW, Fernandes GW, Waring GL. 1987. Adaptive nature of insect galls. *Environ. Entomol* 16:15–24

Pulliam HR. 1975. Diet optimization with nutrient constraints. *Am. Nat.* 109:765–68

Rank NE, Kopf A, Julkunen-Tiitto R, Tahvanainen J. 1998. Host preference and larval performance of the salicylate-using leaf beetle *Phratora vitellinae*. *Ecology* 79:618–31

Rathcke BJ, Poole RW. 1975. Coevolutionary race continues: butterfly larval adaptation to plant trichomes. *Science* 187:175–76

Raupp MJ. 1985. Effects of leaf toughness on mandibular wear of the leaf beetle. *Plagiodera versicolora*. *Ecol. Entomol.* 10:73–80

Reeve HK, Sherman PW. 2001. Optimality and phylogeny: a critique of current thought. In *Adaptation and Optimality*, ed. SH Orzack, E Sober, pp. 64–113. Cambridge, UK: Cambridge Univ. Press

Reznick D, Travis J. 1996. The empirical study of adaptation in natural populations. In *Adaptation*, ed. MR Rose, GV Lauder, pp. 243–89. San Diego, CA: Academic

Rhoades DF. 1985. Offensive-defensive interactions between herbivores and plants: Their relevance in herbivore population dynamics and ecological theory. *Am. Nat.* 125:205–38

Rosenthal GA, Dahlman DL, Janzen DH. 1978. L-canaline detoxification: A seed predator's biochemical mechanism. *Science* 202:528–29

Rosenthal GA, Hughes CG, Janzen DH. 1982. L-canavanine, a dietary nitrogen source for the seed predator *Caryedes brasiliensis* (Bruchidae). *Science* 217:353–55

Ryan CA. 1990. Proteinase inhibitors in plants: Genes for improving defenses against insects and pathogens. *Annu. Rev. Phytopathol.* 28:425–49

Sagers CL. 1992. Manipulation of host plant quality: Herbivores keep leaves in the dark. *Funct. Ecol.* 6:741–43

Sandberg SL, Berenbaum MR. 1989. Leaftying by tortricid larvae as an adaptation for feeding on phototoxic *Hypericum perforatum*. *J. Chem. Ecol.* 15:875–85

Scheirs J, De Bruyn L, Verhagen R. 2000. Optimization of adult performance determines host choice in a grass miner. *Proc. R. Soc. London Ser. B* 267:2065–69

Simpson SJ, Simpson CL. 1990. The mechanisms of nutritional compensation by phytophagous insects. See Bernays 1990, pp. 111–60

Simpson SJ, White PR. 1990. Associative learning and locust feeding: evidence for a 'learned hunger' for protein. *Anim. Behav.* 40:506–13

Singer MC, Ng D, Thomas CD. 1988. Heritability of oviposition preference and its relationship to offspring performance within a single insect population. *Evolution* 42:977–85

Slansky F. 1993. Nutritional ecology: the fundamental quest for nutrients. See Stamp & Casey 1993, pp. 29–91

Snyder MJ, Glendinning JI. 1996. Causal connection between detoxification enzyme activity and consumption of a toxic plant compound. *J. Comp. Physiol. A* 179:255–61

Stachowicz JJ, Hay ME. 1999. Reducing predation through chemically mediated camouflage: indirect effects of plant defenses on herbivores. *Ecology* 80:495–509

Stamp NE, Casey TM, eds. 1993. *Ecological and Evolutionary Constraints on Foraging*. New York: Chapman & Hall

Thaler JS, Fidantsef AL, Duffey SS, Bostock RM. 1999. Trade-offs in plant defense against pathogens and herbivores: a field demonstration of chemical elicitors of induced resistance. *J. Chem. Ecol.* 25:1597–1609

Thompson DB. 1992. Consumption rates and the evolution of diet-induced plasticity in the head morphology of *Melanoplus femurrubrum* (Othoptera: Acrididae). *Oecologia* 89:204–13

Thompson DB. 1999. Genotype-environment interaction and the ontogeny of diet-induced phenotypic plasticity in size and shape of *Melanoplus femurrubrum* (Orthoptera: Acrididae). *J. Evol. Biol.* 12:38–48

Thompson JN. 1988. Evolutionary ecology of the relationship between oviposition preference and performance of offspring in phytophagous insects. *Ent. Exp. Appl.* 47:3–14

Thompson JN, Pellmyr O. 1991. Evolution of oviposition behavior and host preference in Lepidoptera. *Annu. Rev. Entomol.* 36:65–89

Vander Wall SB. 1990. *Food Hoarding in Animals.* Chicago: Univ. Chicago Press. 445 pp.

Via S. 1999. Cannibalism facilitates the use of a novel environment in the flour beetle. *Tribolium castaneum. Heredity* 82:267–75

Waldbauer GP, Fraenkel G. 1961. Feeding on normally rejected plants by maxillectomized larvae of the tobacco hornworm. *Protoparce Sexta* (Johan.) *Ann. Ent. Soc. Am.* 54:477–85

Waldbauer GP, Friedman S. 1991. Self-selection of optimal diets by insects. *Annu. Rev. Entomol.* 36:43–63

Way MJ, Cammell M. 1970. Aggregation behaviour in relation to food utilization by aphids. In *Animal Populations in Relation to Their Food Resources,* ed. A Watson, pp. 229–47. Oxford: Blackwell

Weis AE, Kapelinski A. 1994. Variable selection on *Eurosta's* gall size. II. A path analysis of the ecological factors behind selection. *Evolution* 48:734–45

Whitham TG. 1978. Habitat selection by *Pemphigus* aphids in response to resource limitation and competition. *Ecology* 59:1164–76

Whitham TG. 1980. The theory of habitat selection: examined and extended using *Pemphigus* aphids. *Am. Nat.* 115:449–66

Whitney HS. 1982. Relationships between bark beetles and symbiotic organisms. In *Bark Beetles in North American Conifers,* ed. JB Mitton, KB Sturgeon, pp. 183–211. Austin, TX: Univ. Texas Press

Williams AG, Whitham TG. 1986. Premature leaf abscission: an induced plant defense against gall aphids. *Ecology* 67:1619–27

Wilson EO. 1975. *Sociobiology.* Cambridge MA: Harvard Univ. Press. 697 pp.

Zalucki MP, Clarke AR, Malcolm SB. 2002. Ecology and behavior of first instar larval lepidoptera. *Annu. Rev. Entomol.* 47:361–93

Zangerl AR, Bazzaz FA. 1992. Theory and pattern in plant defense allocation. In *Plant Resistance to Herbivores and Pathogens,* ed. RS Fritz, EL Simms, pp. 363–91. Chicago: Univ. Chicago Press

Annu. Rev. Ecol. Syst. 2002. 33:665–706
doi: 10.1146/annurev.ecolsys.33.010802.150504
First published online as a Review in Advance on August 14, 2002

THE PACIFIC SALMON WARS: What Science Brings to the Challenge of Recovering Species

Mary H. Ruckelshaus[1], Phil Levin[1], Jerald B. Johnson[1], and Peter M. Kareiva[2]

[1]NMFS Northwest Fisheries Science Center, 2725 Montlake Blvd. E., Seattle, Washington 98112; email: mary.ruckelshaus@noaa.gov, phil.levin@noaa.gov, jerry.johnson@noaa.gov
[2]The Nature Conservancy of Washington, 217 Pine Street, Ste. 1100, Seattle, Washington 98101; email: pkareiva@tnc.org

Key Words recovery planning, scenarios, conservation units, Endangered Species Act, science/policy interface

■ **Abstract** Politicians, scientists, government agencies, and the public are all engaged in recovery planning for Pacific salmon. In order for science to fulfill its potential in the arena of salmon recovery planning, several shortcomings of the science and its application to decision-making must be rectified. The definition of conservation units using genetic and phylogenetic inference needs to be sharpened. Ecological analyses must get beyond casting blame for past declines in salmon numbers and examine mixed strategies of management that consider interactions between hatcheries, harvest, hydropower, and habitat factors as well as background natural stresses and invasive species. Glib acceptance of expert opinion and extrapolated or inferred data should be tempered. To deal with uncertainty, recovery teams should engage in scenario analyses in which a wide variety of assumptions are played out. Finally, there is a pressing need for analyses aimed at determining what circumstances and communication strategies give science an effective voice in decision-making.

INTRODUCTION

As increasing numbers of species become threatened with extinction, ecologists and evolutionary biologists continue to devise strategies to rebuild depleted populations. Although scientists sometimes forget this point, it is not their job to recommend a recovery strategy to policy-makers. Rather, the job of science is to make clear the biological targets that constitute recovery and the biological consequences of different suites of actions (or inaction) aimed at recovering threatened and endangered species. The actual recovery policy adopted by public agencies will inevitably balance biological consequences with social, political, and economic values. Biological science cannot and should not have any special place in determining how biological value is traded off against an economic or social concern.

However, science can and should be the preeminent voice in describing targets and anticipating the likely ecological consequences of alternative management actions. In other words, science cannot make policy, but it can illuminate the consequences of policy for threatened and endangered species. Unfortunately, science often fails to play this idealized role of making clear the consequences of alternatives. In turn, the failure of many high-profile conservation-planning decisions to adequately incorporate ecology and evolutionary biology is partly due to the inadequate light shed by science on the consequences of decisions. In this paper, we offer an illustration of how science—the collection of biological information and translation of that information into useful forms—can be used fruitfully to inform conservation planning for Pacific salmon. Our experience with salmon illustrates both the impediments to and opportunities for incorporating biological criteria into recovery decisions.

First, it is important to emphasize that Pacific salmon are of enormous economic, cultural, and social value, and the extinction of any Pacific salmon species would be viewed as a tragedy by the public (Smith 1979, Beckham 1977, Brown 1983, Cone 1995, Lichatowich 1999, Connelly 2001, Frank 2002). Part of the fascination with salmon can be traced to their remarkable anadromous life history. The spectacular journeys of spring-run chinook and sockeye in the Snake River Basin—traveling thousands of kilometers to the ocean and back to small headwater streams and lakes in Idaho—have captured the imagination of Native Americans for centuries. More recently, local disappearances and declines in salmon populations have become fodder for the campaign speeches of politicians, who promise that these species will not vanish on their watch. Fortunately, compared to the state of other highly visible species in decline, most species of Pacific salmon are at a relatively low risk of immediate extinction. Because Pacific salmon are still abundant in places, and are unlikely to be rapidly extirpated, we have a unique opportunity to use science thoughtfully to design recovery strategies that include multiple options instead of last ditch efforts that often typify endangered species management (Gore 1992, NRC 1995, Doremus 2000).

There have been many reviews of salmon ecology, evolutionary biology, and even salmon management (e.g., Cone & Ridlington 1996, NRC 1996, Lichatowich 1999, ISG 1999). Here we adopt a different tack. We examine the role science plays in shaping recovery planning and highlight where science can be useful, and where it has not been as effective. By focusing on the role of science in salmon recovery, we hope to provide warnings and lessons that may help in future attempts to recover species using the best available science.

Conservation planning for salmon in the western United States is especially high-profile because nearly everything people do in the Pacific Northwest affects salmon in some way. Consequently, the potential remedies for the salmon problem have huge implications for the ways people extract water, develop industrial, commercial and private properties, generate electricity, transport goods, harvest fish, and conduct their daily lives. A benefit of the high stakes of salmon recovery is that scientists have a willing audience: politicians and decision-makers are eager for biological advice about how best to proceed in recovering salmon populations. Our charge as scientists is to respond to the intense pressure for biologically defensible answers in a way that clearly distinguishes scientific conclusions from policy choices. Recent crises in the public's confidence in biologists to contribute to policy decisions affecting high-profile species such as Canada lynx and grizzly bear (Mapes 2001, Boyle 2002, Dalton 2002, Mills 2002, *Seattle Times* 2002, Stokstad 2002, Strassel 2002, Weber 2002) illustrate the importance of our charge. Indeed, a recent controversial decision to cut off irrigation water to farmers in the Klamath River Basin to save listed coho salmon in Oregon and California prompted U. S. Secretary of Interior Gale Norton to request a National Academy of Sciences panel to review the scientific basis for the decision (Milstein 2002, NRC 2002). The heated policy debates surrounding the salmon recovery problem create an intensely challenging and exciting arena in which science is actively competing for influence.

Overview of Biological Declines and Legal Status

Seven species of anadromous Pacific salmonids in the genus *Oncorhynchus* occur in North America (*O. gorbuscha*—pink salmon, *O. keta*—chum salmon, *O. tshawytscha*—chinook salmon, *O. kisutch*—coho salmon, *O. nerka*—sockeye salmon, *O. mykiss*—steelhead, and *O. clarki clarki*—coastal cutthroat trout), with geographic ranges occurring throughout the north Pacific from Japan across to the west coast of the United States and Canada, from Alaska to southern California. Pacific salmonids generally follow a basic anadromous life history: they are born in freshwater streams, including such diverse locations as coastal lowlands and headwaters thousands of kilometers from the ocean. Juveniles grow in fresh water from several weeks up to two years, then migrate downstream to estuaries and oceans for one to six years before returning as adults to spawn in natal streams (Table 1, Groot & Margolis 1991). Five species of salmonids are obligately semelparous, and steelhead and coastal cutthroat trout can be iteroparous, making the fresh-to-saltwater migration two to three times before dying (Groot & Margolis 1991).

The myriad freshwater and saltwater habitats Pacific salmonids must traverse throughout their life cycle exposes them to a number of threats. Salmon declines have been blamed on many factors, including hydropower projects, harvest, reductions in freshwater habitat quantity and quality, predation by other fishes, birds and marine mammals, competition for rearing and spawning habitats, ocean conditions, and negative interactions between wild and hatchery-origin salmon. Combinations of such impacts have led to serious reductions in abundance, spatial distribution, and diversity of a number of Pacific salmon populations.

The National Marine Fisheries Service (NMFS) defines reproductively isolated groups of salmon with unique evolutionary legacies as Evolutionarily Significant Units (ESUs) under the Endangered Species Act (ESA) (NMFS 1991; Waples 1991, 1995). The NMFS has identified 51 ESUs within the 6 species of Pacific salmonids under its jurisdiction, and 26 of these currently are listed under the ESA

TABLE 1 Life history characteristics of Pacific salmon (modified from Pearcy 1992, Busby et al. 1996)

Species	Colloquial names	Ocean entry	Size at ocean entry (mm)	Freshwater residence	Marine residence	Age at maturity
Chinook salmon (*O. tshawytscha*)	King salmon; Tyee; Blackmouth; spring, summer, or fall chinook	May–Oct	40–110	0.5–2+ years	0.5–6 years	3–6 years
Coho salmon (*O. kisutch*)	Silver salmon	May–June	60–120	1 year	0.5–1.5 years	2–4 years; usually 3 years
Sockeye salmon (*O. nerka*)	Red salmon; Blueback	May–June	60–110	1–2 years	1–5 years	3–7 years; usually 3–4 years
Chum salmon (*O. keta*)	Dog salmon	March–June	30–40	1–3 months	2–4 years	2–7 years; usually 3–4 years
Pink salmon (*O. gorbuscha*)	Humpback salmon	May–June	30–40	1–3 weeks	1.6 years	2 years
Steelhead (*O. mykiss*)	Rainbow trout; redband trout; winter or summer steelhead	May–April	NA	2–7 years	0–4 years	2–7 years

TABLE 2   Conservation status of listed Evolutionarily Significant Units (ESUs) of Pacific salmonids*

Species	ESU	Listing status
Chinook salmon	Sacrameto River winter-run	Endangered
	Upper Columbia River spring-run	Endangered
	Snake River fall-run	Threatened
	Snake River spring/summer-run	Threatened
	Puget Sound	Threatened
	Lower Columbia River	Threatened
	Upper Willamette River	Threatened
	Central Valley spring-run	Threatened
	California coastal	Threatened
Coho salmon	Central California	Threatened
	Southern Oregon/northern California coasts	Threatened
	Oregon coast	Threatened
Chum salmon	Hood Canal summer-run	Threatened
	Columbia River	Threatened
Sockeye salmon	Snake River	Endangered
	Ozette Lake	Threatened
Steelhead	Southern California	Endangered
	Upper Columbia River	Endangered
	South-central California coast	Threatened
	Central California coast	Threatened
	Snake River Basin	Threatened
	Lower Columbia River	Threatened
	California Central Valley	Threatened
	Upper Willamette	Threatened
	Middle Columbia River	Threatened

*The listing status of each ESU under the Endangered Species Act is specified—ESUs are listed separately as "species" under the Act as distinct population segments

(Table 2). The listed ESUs are considered to be distinct population segments under the ESA, and for legal purposes they are treated as separate species that must be recovered.

Once an ESU is listed under the ESA, actions that might directly or indirectly negatively affect the species become restricted (Rohlf 1989). These so-called "take" prohibitions are designed to limit the deleterious effects of harvest, hydropower, and hatchery management practices, and of changes to habitat quantity and quality, on salmon populations. Recovery planning is complicated because, in addition to the ESA, the NMFS must also serve two other major legal mandates when making decisions about salmon. First, NMFS is charged with administering the Magnuson-Stevens Fishery Conservation and Management Act, as amended by the Sustainable Fisheries Act of 1996. This Act sets biological standards that must be met in salmon populations subjected to commercial and tribal harvest and describes methods by which harvest levels should be allocated (Magnuson-Stevens Act 16 U.S.C. § 1801 et seq.). In addition, the U.S. government is a trustee for the salmon resource for a number of Native American tribes in the Pacific Northwest. This relationship was originally established through historical treaties signed in the mid-1850s and was reaffirmed in 1974 in the *United States v. Washington* case (the Boldt decision) and in 1983 in the *United States v. Oregon* case between the tribes and United States government. Consequently, the NMFS works closely with state and tribal comanagers to review harvest and hatchery management plans to ensure compliance with statutory requirements in the ESA, the Magnuson-Stevens Act, and tribal treaty rights. For thriving populations of salmon, tensions among these three primary legal standards are minimal. However, for threatened and endangered populations, where the pie to be divided is smaller, the debate becomes more heated (e.g., *Alsea Valley Alliance v. Evans* 2001, Federal Register 2002, Verhovek 2002, *Washington Trout v. Lohn et al.* 2001).

NMFS's Design of Recovery Planning for Pacific Salmonids

The NMFS has implemented a geographically based approach to developing recovery plans for the 26 listed ESUs under its jurisdiction. Eight recovery-planning domains have been identified, each encompassing 3–6 salmon ESUs that are listed as threatened or endangered under the ESA (Figure 1). For each geographic region, a Technical Recovery Team has been (or will be) established; the team's charge is to develop biologically based delisting criteria and provide technical guidance in determining which actions are necessary for recovery of all ESUs within the domain (NMFS 2000a).

Recovery teams face three major scientific challenges: identifying the units of conservation, establishing recovery goals, and recommending management actions that will abate existing threats and ultimately meet those recovery goals. Approaches to establishing recovery goals fall largely under the umbrella of "population viability analysis," which recently has been thoroughly reviewed (Lande 1993, Morris et al. 1999, Brook et al. 2000, Holmes 2001, Beissinger & McCullough 2002, Holmes & Fagan 2002), including a detailed synthesis of ideas regarding what represents viability for salmonids (McElhany et al. 2000). In contrast, the scientific challenges of defining appropriate conservation units and then completing a threats analysis on those units have not been reviewed elsewhere. By focusing on these scientific issues, we illustrate how guidance from basic science and clear communication from conservation practitioners can help advance conservation planning for species such as salmon.

IDENTIFYING UNITS FOR CONSERVATION

Salmon species are composed of multiple lineages marked by varying levels of evolutionary divergence. This variation can be seen in neutral genetic markers (e.g., Utter et al. 1989, Wood et al. 1994, Nielsen et al. 1997, Ford 1998,

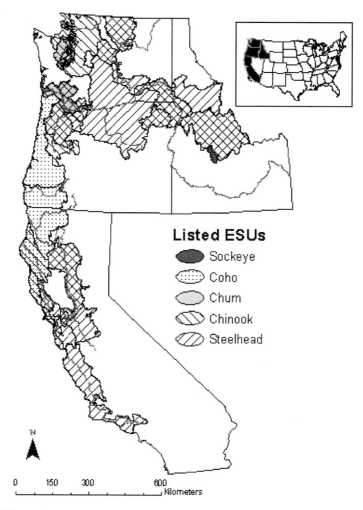

Figure 1   Map showing the geographic distribution of listed ESUs of Pacific salmonids.

McCusker et al. 2000, Smith et al. 2001), and also in the wide range of life history strategies employed within each species (Table 1; Groot & Margolis 1991, Taylor 1991, Waples et al. 2001). A demographically based viability analysis that simply counts fish—without regard to genetic or life history variants—could result in missed opportunities for future evolvability of a population. Yet, the reality is that not every salmon lineage can be saved; thus one of the first critical questions conservation planners must address is: What are the biologically significant units for conservation of the species?

There is a rich history to the question of how to identify and prioritize conservation units, and ongoing debates surrounding this issue for imperiled species continue in the biological literature (see discussion below). Advances in molecular genetic and statistical tools have led to impressive algorithms for discriminating among groups within species—in some cases, to the level of distinguishing between individuals. Nevertheless, resolution of distinct genetic lineages within a species of conservation concern does not help to identify which hierarchical level of organization is the appropriate one to save (Waples 1995). In this section, we offer a brief history of the debate about appropriate means for identifying conservation units and how they have been delineated for Pacific salmon under legal constraints of the ESA. We then highlight important omissions in formulating definitions of conservation units and how those definitions are applied in practice. We argue that incorporating a phylogenetic perspective into identification of conservation units will greatly improve assessment of their relevance for the conservation of imperiled species.

The Conservation Unit Debate: What is the Best Way to Identify Relevant Clusters of Diversity?

Delineating conservation unit boundaries and major evolutionary lineages within species has been the focus of considerable research over the past 15 years. Ryder (1986) introduced the concept of Evolutionarily Significant Units to categorize evolutionary diversity within rare mammal species. Waples (1991, 1995) later adopted the ESU concept as a way to identify stocks of Pacific salmon that could receive formal protection as "distinct population segments" under the ESA. Given their propensity to home to natal streams, salmon are naturally subdivided into a hierarchy of genetically distinct groups that tend to show local adaptation to varying selective environments (reviewed in Taylor 1991). Therefore, the biological framework used by NMFS to define an ESU includes: (a) substantial reproductive isolation and (b) assurance that the designated population or group of populations contains an important component of the evolutionary legacy of the species, which in application means that salmon ESUs show major divergence from one another (Waples 1991).

Although the ESU criteria for salmon are clear, approaches underlying ESU designations are not universally agreed upon in the broader conservation community (reviewed in Fraser & Bernatchez 2001). Recent controversy stems from disagreement about the relative importance of using neutral molecular data as opposed to ecological adaptations to delineate groups (Crandall et al. 2000). Moritz (1994) argued that ESUs should be distinct phylogenetic lineages and that the criterion to delineate such lineages should be reciprocal monophyly based on DNA sequence data. Strict reliance on neutral marker data to the exclusion of local adaptations can be troublesome (Barlow 1995, Crandall et al. 2000). Yet, the value of determining phylogenetic relationships among evolutionary lineages and integrating this information with knowledge of local adaptations remains underappreciated (Mayden & Wood 1995).

Recovering species must have continued opportunities to evolve. Waples (1995) envisioned this need by describing the evolutionary legacy of Pacific salmon as "a reservoir of genetic variability upon which future evolutionary potential depends." Yet most ESU designations, including those in Pacific salmon, are based largely on static views of evolution divergence. We seldom have a clear understanding of the multiple mechanisms by which patterns of evolutionary divergence (or convergence) could have occurred. In order to evaluate the significance of divergent lineages to species persistence, and when necessary to establish priorities among ESUs, the potential processes that gave rise to patterns of divergence must be understood. There are at least two problems with viewing reproductive isolation and evolutionary divergence as two independent criteria that offer unambiguous indications of conservation unit boundaries.

First, different evolutionary processes produce different rates of divergence between lineages for neutral markers (used to measure reproductive isolation) and adaptations. Applications of ESU concepts ignore these different processes, instead focusing on the degree of divergence at a single time. For example, rapid evolutionary shifts caused by selection can outpace neutral marker divergence and lineage sorting by orders of magnitude (Figure 2A). In some cases, evolution is so swift that distinct populations can be almost genetically indistinguishable at neutral alleles while showing significant adaptive divergence (Pascual et al. 2001). Salmon introduced to novel environments offer striking evidence for this phenomenon. Evolutionary radiations of chinook salmon in New Zealand (Quinn et al. 2001, Kinnison et al. 2002) and steelhead introduced to Argentina (Pascual et al. 2001) both show rapid life history divergence but very little neutral marker divergence. More troubling are cases where gene flow is high (Waples 1998), but selection is strong enough for populations to evolve local adaptations (Figure 2B; e.g., Tallman & Healy 1991, 1994). Such populations, although not reproductively isolated, constitute important evolutionary components of the biological species, but probably warrant no unique protection. Finally, populations can show high levels of reproductive isolation (measured by neutral markers) but show very little or no phenotypic divergence. Stabilizing selection or convergent evolution between geographically isolated populations could both lead to this pattern (Figure 2C; e.g., Teel et al. 2000, Craig & Foote 2001). Clearly, different evolutionary processes can lead to very different relationships between neutral markers and adaptive traits. In short, degree of divergence in traits per se is not a simple indicator of evolutionarily significant diversity that should be protected in a conservation plan.

Second, reproductive isolation and evolutionary divergence are not always independent events, and isolation does not always precede adaptive differentiation. In fact, in many cases natural selection itself causes reproductive isolation. This occurs when natural selection operates on traits that limit opportunities for groups of individuals to mate, or limits reproductive success of hybrids. This process, dubbed ecological speciation (Schluter 1998, Johannesson 2001), could be quite common in Pacific salmon, where selection on traits such as run timing or homing ability can result in spatial or temporal isolation among breeding groups. Recent evidence in sockeye salmon suggests that even subtle differences such as local morphological adaptation to spawning habitats can lead to early stages of reproductive isolation (Hendry 2000). If we are viewing life history lineages in the early stages

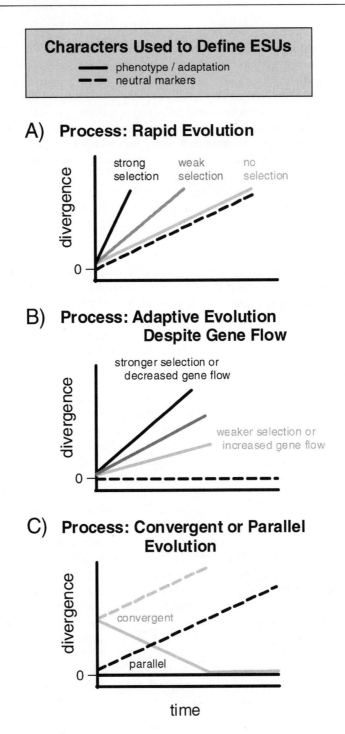

of ecological speciation, then there may be no genetic barriers to reproduction, and genetic differentiation could still be modest, yet the diverging lineages could be important from a conservation perspective because they represent evolution's answer to the most recent environmental challenges.

Finally, it is also important to recognize that reliance solely upon variability at neutral molecular markers as a way to assess evolutionary potential may fall short of what will actually be required to maintain a species. Bowen (1999) makes a strong case for consideration of additional criteria to identify potential sources of future evolutionary diversity, including spatial or temporal isolation, behavioral variation (Barlow 1995), unique niche use, morphological divergence, and preserving lineages with a recent history of evolutionary diversification. None of these ideas has been implemented in any species' recovery plan (Boersma et al. 2001).

In sum, we need to take care that our views of ESUs are not unduly typological, and instead accommodate the variety of processes that could generate superficially similar patterns. All genetically distinct populations do not warrant equal protection; it depends on the mix of adaptation, gene flow, uniqueness, or irreplaceability of the units. Part of the problem with applying ESU concepts in recovery planning grows out of what has become the standard interpretation of the ESA. The conventional view is that the primary intent of the ESA is to eliminate immediate extinction risks to biological species, or to distinct population segments within species (Federal Register 1996). Hence, recovery-planning efforts typically focus primarily on estimating short-term extinction risks and discussing ways to reduce those risks. However, conservation scientists recognize that one broader goal of conservation is to preserve the biological processes of life (Bowen 1999). For Pacific salmon, this means that it is first necessary to identify the phylogenetic

---

Figure 2 Expected patterns of divergence for neutral markers and adaptive traits over time between hypothetical lineages for three evolutionary processes. (*A*) depicts the case where there is no gene flow among lineages and diversifying selection is operating. Here, divergence at neutral markers increases linearly as a function of time; divergence for adaptive traits also increases as a function of time, with the rate dependent on the strength of diversifying selection. (*B*) depicts the case where gene flow is operating, but diversifying selection is strong enough to overcome the effects of gene flow. Here, lineages again show no differences at neutral markers, but show varying rates of divergence for adaptations, depending on the strength of selection and levels of gene flow. (*C*) depicts the case where there is no gene flow, but lineages experience identical selective pressures. Here, divergence at neutral markers increases linearly as a function of time, but adaptations converge or remain at common phenotypes, for convergent and parallel evolutionary processes, respectively. For simplicity, all models assume that there is adequate genetic variation for evolutionary adaptations to occur, and that neutral markers do not become saturated over time.

history and current evolutionary trajectories of salmon lineages in order to define the biological heritage we are charged to protect.

Identifying Conservation Units That are Commensurate with Conservation Goals

The ultimate goal of recovery planning is to foster natural persistence of imperiled species or conservation units. The difficulty in identifying relevant units arises from the fact that species continue to evolve, so it is not obvious whether it is better to conserve existing diversity or save units with maximum potential to adapt to changing environments. For example, if our underlying conservation goal is to preserve historical diversity, then a recently evolved lineage with closely related sister groups may have lower priority for protection than more distant evolutionarily distinct groups. Alternatively, if the goal is to foster future evolutionary radiation, then it may make more sense to protect rapidly diversifying lineages (despite their recent common ancestry) as opposed to lineages that are older but more evolutionarily inert.

Typically, data are not available to establish quantitative links between historical diversity or evolutionary potential and species viability. Instead, assertions about the importance of conserving a representative sample of historical diversity or a species' evolutionary potential are bolstered mainly by theoretical arguments (Lynch & Lande 1993, Lande & Shannon 1996, Storfer 1996). In practice, Pacific salmon ESUs have been defined by reproductive isolation based on allozyme frequency data or morphological, life history, or habitat characteristics (Myers et al. 1998; Busby et al. 1994, 1996; Weitkamp et al. 1995; Hard et al. 1996; Gustafson et al. 1997). As mentioned above, listed ESUs all are protected under the ESA, so recovery efforts must be made to ensure the persistence of any Pacific salmon ESU threatened with extinction, irrespective of its contribution to the long-term persistence of the species.

The difficult choices about which units of Pacific salmon to conserve therefore occur within ESUs, at the level of populations. For salmon ESUs, populations are defined based on their degree of relative demographic independence so that viability can be evaluated using single-population viability analysis (McElhany et al. 2000). The conservation goal underlying identification of independent populations is to determine how many and which viable populations are necessary for persistence of the ESU (Ruckelshaus et al. 2002). Work conducted to date suggests that the number of demographically independent populations per ESU ranges from 3 to 30, and life history and genetic diversity vary greatly among populations within an ESU (Ford et al. 2001, PS TRT 2001, Myers et al. 2002). Determining which populations must be saved for ESU persistence involves identifying critical components of ESU diversity that must be protected. We believe that estimating the population diversity necessary for ESU persistence will benefit from a phylogenetic perspective on the significance of the diversity we observe within salmon ESUs—a topic we address in the next section.

Incorporating Evolutionary Pattern and Process into Identifying Relevant Units for Conservation

Given political, economic, and even biological constraints, conservation planners are often forced to establish conservation priorities among evolutionary lineages. This necessity translates into difficult decisions about which traits should have primacy in establishing conservation priorities, or in extreme cases, deciding which groups are expendable. Conservation units—like most taxonomic designations—are best viewed as convenient ways of identifying and organizing phenotypic or genetic differences among diverging lineages. Advances in intraspecific phylogeography provide a historical framework to examine current patterns of evolutionary divergence in rare species and ESUs within species (Templeton 1998, Edwards & Beerli 2000, Avise 2001). Yet, in salmon recovery planning, very little attention has been paid to phylogeographic patterns and the underlying processes that give rise to distinct evolutionary lineages (see Bowen 1998, 1999; Moritz 2002).

A phylogenetic perspective could be used to assess the relative importance of salmon adaptations by revealing the order in which characters have evolved through time (Avise 1989). For example, Quinn et al. (2000) argue that timing of migration is a key life history trait in salmon because it provides a means by which populations become temporally isolated, thereby facilitating further divergence in secondary adaptations. Similar arguments could be made regarding the effects of natal homing on spatial isolation among breeding groups. By mapping the evolution of migratory timing to a phylogeny, we can formally test whether these adaptive traits consistently precede the evolution of secondary adaptations. Such information would help clarify the relative importance of conserving particular traits or populations containing such traits.

Phylogenies can also be used to identify the rarity of adaptations. For example, a recent reconstruction of relationships among rainbow trout populations in British Columbia showed a repeated evolutionary shift from anadromy to a freshwater life cycle and repeated evolutionary divergence in adult migration timing (McCusker et al. 2000). This pattern of independent evolution across multiple lineages suggests that on an evolutionary timescale, variation for both of these traits appears to be quite easy to recover. By comparison, sockeye salmon also show repeated evolution from anadromy to freshwater life cycles, but adult migration timing appears to be canalized on a very local scale (Wood 1995). Hence, in sockeye salmon, care should be taken to preserve populations containing adaptations associated with run timing because these traits may be difficult or impossible to recover once lost.

In practice, conducting analyses to illuminate the history of evolutionary processes occurring within salmon ESUs (i.e., processes generating diversity among populations) is difficult because of the ubiquitous presence of hatchery fish in most ESUs. Transfers of hatchery fish between streams within ESUs, and even from streams outside the geographic boundaries of an ESU, have been commonplace since the early to mid 1900s (e.g., Myers et al. 1998). Hatchery fish usually constitute an unknown fraction of the parents of any generation, so ascribing evolutionary mechanisms to observed genetic relationships or potential adaptive differences among stocks is very difficult.

Despite the difficulty in reconstructing patterns of and processes generating historical diversity among populations, recovery teams recognize the importance of describing major diversity groups within ESUs so that a representation of historical diversity can be included among those populations given highest priority for conservation efforts. In areas where patterns of extant diversity are not seriously tainted by the presence of hatchery fish, the approaches outlined in the previous section are a promising means of more clearly identifying those populations harboring significant sources of historical diversity or future evolutionary potential. Unfortunately, in order to characterize historical patterns of diversity among populations, recovery teams in most cases will be forced to use proxy information, such as historical diversity of habitat types or life history information from historical fisheries. Decisions about which populations should have highest priority for protection of diversity can then be based on what fraction of historical diversity across the ESU is deemed necessary for recovery. To the extent such characterizations of diversity can include a closer look at its likely evolutionary sources, we will improve our choices about which groups of salmon to protect.

THE ORTHODOX EXPLANATION FOR SALMON DECLINES: HARVEST, HABITAT, HATCHERIES, AND HYDROPOWER

Pacific salmon have been extirpated from nearly 40% of their historical habitat in the Pacific Northwest, and of the remaining salmon populations, almost one half are thought to be at risk of extinction (Nehlsen et al. 1991, Levin & Schiewe 2001). Recent quantitative estimates of extinction risk for 87 populations of 13 listed ESUs of chinook, chum, and steelhead suggest that 84% of the populations currently are not viable (as judged by persistence over 100 years or high risk of decline) (McClure et al. 2002; NMFS's recovery teams, unpublished data). However, contrary to the dismal views conjured up by these risk calculations, salmon are capable of generating remarkable rates of population growth and have successfully invaded many non-native ecosystems (Pascual et al. 2001, Quinn et al. 2001). Consequently, only modest changes in mortality may reverse salmon declines.

The key question for recovery then is, "What actions should we take to reverse population declines?" Unfortunately this question has too often been replaced by the question, "What caused the historic decline of salmon?" These two questions might seem similar, but they actually set up the scientific problem in very different ways. For example, it is clear that excessive harvest drove some salmon populations to low levels in the mid-twentieth century (NRC 1996). In response to these declines, harvest was reduced to low levels in several ESUs; some ESUs responded positively to harvest reductions—others did not (PSC 2001a,b). Therefore, further reductions in harvest are not likely to lead to recovery in all cases. In short, a factor could have caused a decline historically, but not be an effective place to look for recovery now.

This is a case of scientists posing a question (what historic factors caused the decline of salmon?) that does not shed as much direct light as one would hope on the consequences of decisions regarding alternative management actions that might be taken now to reverse salmon declines. History is important, but even more important is understanding the likely consequences of today's alternative actions. Hence, as we examine the risk factors in the next few pages, we focus on what might be done today to reduce these risks, as opposed to focusing on the history of ecological degradation and resource mismanagement that has gotten us to the current situation.

A Historical Depiction of the Salmon Crisis: The Single-Factor Approach

Salmon biologists have long held that four major human-induced factors have been responsible for salmon declines: (*a*) habitat degradation; (*b*) hydroelectric dams and other impoundments; (*c*) harvest practices; and (*d*) the proliferation of fish hatcheries. These factors (colloquially called the four H's) have been the primary focus of salmon recovery research and there is, indeed, considerable evidence that each can have direct negative impacts on salmon population status—sometimes dramatically so.

Habitat Degradation and Loss

With exponential human population growth, increases in timber harvest, intensive grazing, and urban development, habitat degradation is obviously a major factor underlying salmon declines (NRC 1996). Documentation of the threats posed by habitat loss and degradation generally has come in two forms: Large-scale, multi-factor correlative studies and smaller-scale explorations of the effects of a few habitat features on particular life stages of salmon. Results from broad-scale correlative studies suggest that human-influenced landscape features affecting salmon vary geographically. Human-induced characteristics such as percent of urbanized or agricultural lands, poor water quality, and road density are associated with reductions in population productivity, adult densities, and early life-stage production for chinook and coho salmon over large geographic areas (Bradford & Irvine 2000, Paulsen & Fisher 2001, Pess et al. 2002, Regetz 2002, Feist et al. 2002). At the other end of the geographic spectrum, there is a plethora of local-scale studies demonstrating the effects of stream habitat characteristics on salmon density and life history characteristics. Effects of such features as large woody debris in streams, water temperature and water flow, stream channel morphology, proportions of pools, riffles, and glides within streams, and nutrient budgets all can affect early life stages of salmon and their total population productivity (reviewed by Beechie et al. 2002).

Results from neither the large- nor small-scale studies alone are adequate to guide recovery planning in freshwater habitats. The large-scale crude classes of habitat variables give little guidance regarding what specific actions might be taken to rectify problems, and results from small-scale studies typically are not framed within the whole life cycle of salmon populations. Because of these realities, it has proven very difficult to identify robust relationships between habitat variables that can be managed or restored and salmon population dynamics (Collings et al. 1972, Regetz 2002). No one questions the importance of habitat to salmon; but no one has unequivocal recipes for how to alter habitat in a way that encourages the recovery of depleted salmon populations.

Management of habitat for salmon recovery reflects the difficulty articulated above—it largely has proceeded along two philosophically divergent tacks: (*a*) fix in-stream habitat characteristics so that they resemble those required by salmon, or (*b*) fix the landscape processes within watersheds so that in-stream habitat functions naturally, thus supporting all species in the ecosystem. The first approach has been criticized because of its focus on the symptoms, rather than the causes of habitat dysfunction (Frissell & Nawa 1992, Beechie et al. 1996, Beechie & Bolton 1999). A favorite example is in cases where thousands of dollars are spent in dumping high quality spawning gravels into a stream, only to have them be washed out of the system within a matter of months because the processes governing gravel recruitment have not been adequately repaired. Furthermore, rarely have either of the approaches to habitat management attempted to quantify fish population responses—the general assumption of habitat management is, "if you build it, they will come." A few good studies have shown how impossibly long lists of projects designed to fix processes or habitat features in a watershed can be prioritized with explicit comparison of the effects of alternative projects on fish population vital rates (e.g., Reeves et al. 1989, 1991; Beechie et al. 1994).

Hydroelectric Dams and Impoundments

Dams clearly harm anadromous salmon through their effects on the amount, timing, and characteristics of water flows and by blocking access to previously available habitat (Raymond 1988, Williams & Tuttle 1992, Bednarek 2001). The effects of hydropower development on salmon are vividly illustrated in the Columbia River Basin in the Pacific Northwest. For example, the construction of the Hell's Canyon dam resulted in the loss of 90% of the historic spawning habitat of fall-run chinook in the Snake River Basin (a major tributary of the Columbia River) in Idaho, landing this ESU squarely on the Endangered Species list (McClure et al. 2001). Even passable dams can be devastating to salmon—they convert rivers into reservoirs, change aquatic communities (Bednarek 2001), and also present passage challenges to down- and upstream migrants. For example, survival of juvenile chinook in the Columbia River during their seaward migration dramatically declined as hydropower projects were constructed along their migration corridor through the 1960s–1970s (Williams et al. 2001). This reduction in survival was associated with subsequent declines in adults returning to spawn (Schaller et al. 1999, Levin & Tolimieri 2001).

Much has been done to reduce the impacts of dams that completely block access to historical habitats. Logistically and financially demanding interventions such as truck and haul projects and creation of artificial spawning habitats have recreated whole salmon runs in areas completely isolated from the sea (WDFW & WWTIT 1993, Gustafson et al. 1997). Without constant human assistance, these populations would perish. The success of engineering solutions is even more striking in situations where hydropower projects provide for fish passage and upriver habitats are plentiful. Most noteworthy are the mitigation measures designed to reduce passage mortality in the Columbia Basin hydropower system after 1975— including fish-passage facilities, predator control, barging fish around dams, and flow augmentation. Taken together, these expensive efforts (costing billions of dollars) have increased the survival of spring/summer chinook salmon migrants to levels similar to those experienced before the hydropower construction boom of the 1960s–1970s (NRC 1996, Williams et al. 2001). Nevertheless, there are other ecological consequences of hydropower operations that are not captured by simply measuring the survival of chinook salmon while migrating through the river system (ISG 1996).

Harvest Practices

Commercial and recreational harvests have had clear negative impacts on Pacific salmon (NRC 1996). In addition to direct mortality caused by fishing, indirect mortality owing to hooking fish that are not ultimately landed or catch and release programs have been documented—the Pacific Salmon Commission uses an average hook and release mortality of 30% (PSC 1997). Furthermore, effects of fishing on life history traits—most notably, size-selectivity on adults and resultant reductions in fecundity (Healey 1986; Ricker 1995; Weitkamp et al. 1995; Ratner & Lande 2001; Hard 2002a,b)—can significantly impair population dynamics in ways that are difficult to quantify.

Often, one of the earliest steps taken to redress declines is to reduce harvest. In some cases, reduction in fishing pressure has coincided with a dramatic reversal of salmon declines. Such appears to have been the case for summer chum salmon returning to streams in Hood Canal, Washington (Figure 3). Although the recent rebound in numbers of summer chum salmon escaping to spawn might be attributed to harvest reductions, recent improvements in ocean survival probably also contribute to their increasing numbers. In contrast to the apparent success of harvest management in addressing declines in Hood Canal summer chum populations, reductions in harvest levels have not improved population status of chinook populations a few kilometers across Puget Sound (PSC 2001b). Clearly, other threats to the chinook salmon in northern Puget Sound must be addressed.

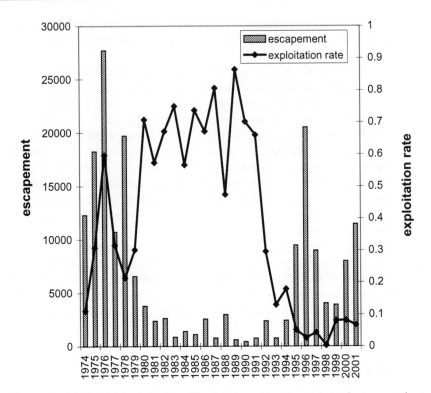

Figure 3   Total exploitation rates (the fraction of total number of summer chum returning to Hood Canal killed owing to direct and incidental harvest in Washington and Canadian waters) and numbers of summer chum salmon escaping to spawn in streams in Hood Canal, WA from 1974–2001 (data from WDFW & PNPTT 2000; T. Johnson, WDFW, personal communication).

Hatchery Propagation of Salmon

A common solution to the ills of salmon populations is to release sometimes enormous numbers of hatchery-reared fish (Lichatowich 1999). Each year hatcheries along the west coast of the United States release nearly 1.2 billion juvenile salmon (Mahnken et al. 1998), with 200 million salmon released annually into the Columbia River alone. There is extensive disagreement over the degree to which hatchery programs contribute to or detract from the viability of wild salmon populations. There are well-documented differences between hatchery and wild Pacific salmon in their survival, reproductive, foraging, and social behaviors, habitat preferences, response to predators, morphology, and physiology (reviewed by Flagg et al. 2000). These differences potentially can affect ecological and genetic interactions between wild and hatchery salmon, and it is clear that negative impacts of hatchery fish do occur in some cases. Misuse of hatcheries has the potential to reduce the viability of natural salmon populations through effects on vital rates or life history traits such as run timing (Nicholas & Hankin 1988; Fleming & Gross 1993; Hilborn 1992, 1999; Meffe 1992; NRC 1996; Reisenbichler & Rubin 1999; Waples 1999). Translocations of salmon through hatchery programs and subsequent introgression of non-native hatchery-origin genes into wild populations can reduce fitness of wild salmon (e.g., Zinn et al. 1977, Reisenbichler & Rubin 1999). Moreover, reduced smolt-to-adult returns in wild populations of chinook salmon are associated with exceptionally large releases of hatchery fish (Levin et al. 2001).

A risk/benefit analysis of hatchery fish must consider the goals of the hatchery program; generally, hatchery management is designed either to supplement declining wild populations (supplementation programs) or to provide extra salmon for harvest (production programs). Supplementation programs are designed to end— the goal is to produce a naturally self-sustaining population of salmon. Ideally, these programs maximize recovery of wild stocks while minimizing negative genetic or ecological impacts of the hatchery fish. In a recent review, Flagg et al. (2000) identified only three successful supplementation programs (i.e., a self-sustaining population of salmon existed after artificial propagation ended) among scores of candidates (see also Waples et al. 2002). In contrast to supplementation programs, production programs often are designed to exist in perpetuity so that fishers can be assured of catching salmon. Through the late 1980s, in spite of increased hatchery production of coho salmon along the Oregon coast and chinook salmon in Puget Sound, adult returns did not increase as hoped (Stouder et al. 1996; PSC 2001a,b). Risk evaluations on such hatcheries must evaluate the negative ecological and genetic interactions that can occur during juvenile rearing and outmigration stages, threats to adults in mixed hatchery-wild fisheries, competition for spawning sites, and interbreeding between wild and hatchery fish. These potential risks are not trivial—for example, in chinook populations where hatchery fish are marked, escaped hatchery salmon can constitute up to 60% of the spawning population in areas without planned supplementation programs (NMFS, unpublished data).

All Risk Factors Do Not Conveniently Fit into the 4-H Framework

Although scientists and federal agencies tend to emphasize the four H's as keys to salmon recovery (NRC 1996, NMFS 2000b), there are many other sources of salmon population declines. For example, the productivity of the Pacific Ocean varies in decadal oscillations, with sometimes tenfold changes in smolt-to-adult survival (Mantua et al. 1997, Hare et al. 1999, Hollowed et al. 2001). In addition, shorter-term events such as El Niño (occurring every two to seven years) are negatively related to sockeye population sizes in Alaska and British Columbia (Mysack 1986). During prolonged periods of bad ocean conditions, salmon may do so poorly that their populations plummet without any help from human factors (Beamish et al. 1999, McFarlane et al. 2000, Welch et al. 2000).

In addition to the stresses of bad ocean years, salmon face many other natural risks that threaten their persistence. Papers quantifying the effects of either predation or competition on Pacific salmon populations are plentiful (e.g., Reimers 1973, Glova 1984, Ruggerone 1986, Wood 1987, Gearin et al. 1989, Healey 1991, NOAA 1997, Collis et al. 2001). Notably high-profile stories include reports of Caspian terns eating an estimated 7% to 15% of the 100 million salmonid juveniles migrating downstream in the Columbia River (NMFS 1995, Collis et al. 2001) and sea lions eating up to 65% of steelhead adults returning to Lake Washington in Puget Sound, Washington (Gearin et al. 1988, 1989).

Finally, salmon are exposed to another form of environmental degradation that has recently come into the limelight as a global factor: the invasion of ecosystems by exotic species. In the western United States, exotic species now make up to 59% of the fish fauna (Moyle et al. 1986) and are common even in areas that are otherwise pristine (see, e.g., Achord et al. 1997), leading ecologists to suggest that nonindigenous species may rank highest among threats to imperiled species (Vitousek 1994, Simberloff 2000). Levin and colleagues (Levin et al. 2002) showed that survival of juvenile chinook in streams without exotic brook trout was nearly double that in streams with brook trout, even after differences in habitat were taken into account. The difference in juvenile chinook survival between sites with and without brook trout would increase population growth rate (λ) by about 2.5%—an increase sufficient to reverse the negative population growth observed in many chinook populations of the Snake River Basin (McClure et al. 2002).

Furthermore, the effects of nonindigenous species can be much less direct than that of brook trout. For example, one of the most important native predators of juvenile salmon in mainstem rivers and reservoirs as they make their seaward migration is the pike minnow *Ptychocheilus oregonensis* (Rieman et al. 1991). NMFS estimates that each year pike minnow consume 16.4 million salmonids in the Columbia River Basin alone (NMFS 2000c). It is interesting to note that the prominence of the pike minnow as a predator may result from interactions with exotic fishes (Poe et al. 1994). In the absence of nonindigenous fishes, such as small-mouth bass, walleye, or channel catfish, salmonids constituted only 2% of the diet of pike minnow (Buchanan et al. 1981). Instead, crayfish, insects, and sculpins dominated the diet. Poe and colleagues (1994) concluded that a shift in the pike minnow diet to salmon results from competition with exotic predators for sculpins and crayfish. Thus, in the absence of exotic piscivorous fishes, the impact of even native predators such as pike minnow may be reduced.

Management attention to the effects of ecological interactions and ocean conditions on salmon populations has been inconsistent at best. The lack of a systematic consideration of ecological and natural environmental conditions in salmon management plans can be attributed in part to lack of scientific information. For example, in the few cases where ecological impacts of predators on salmon were well documented (e.g., Caspian terns, sea lions, and pike minnow), management plans included—sometimes even featured—actions designed to reduce predation on salmon (Gearin et al. 1988, Beamesdurfer et al. 1996, Friesen et al. 1999, NWPPC 2001). The contents of salmon management plans thus appear to be guided by serendipitous treatment of these effects by scientists, rather than a more holistic biological vision of threats analysis.

RECOVERY REQUIRES ATTENTION TO INTERACTIONS AMONG RISK FACTORS AND A MIX OF OPTIONS FOR IMPROVEMENT

Although examples exist in which single factors appear to be primarily responsible for the decline of salmon populations, the demise of salmon nearly always results from a constellation of factors throughout their life history (Lawson 1993, NRC 1996, Brodeur et al. 2000, Finney et al. 2000, Levin & Schiewe 2001). The tidy compartmentalization of risks is convenient and analytically tractable, yet it neglects cumulative or interactive effects of risk factors. The alternative—comparing the relative value of a fix to one H to improvements accomplished by instead fixing more than one H (Kareiva et al. 2000)—has never been incorporated into a management plan to our knowledge. Because of the uncertainties inherent in predicting the effects of improvements in the H's on salmon populations, distributing efforts among several known threats increases the chances that hoped-for biological responses will occur. Thus, even if removing all hydropower projects or removing all hatcheries might recover salmon, the more scientifically prudent approach is to consider simultaneous alterations in many of the H's.

For some salmon ESUs, we actually have enough demographic detail in terms of stage-specific mortality rates that one can even use matrix models to examine the impacts of mixed-management activities. For example, Kareiva et al. (2000) found that a mix of modest reductions in juvenile mortality and estuarine mortality together could reverse the declining trends in Snake River spring-summer chinook salmon.

Ecologists are comfortable testing for significant effects. In salmon, all H's exert a significant effect, and the more pertinent question concerns relative importance, which is much harder to estimate. As a first approximation to this issue, currently we are gathering data for a large-scale comparative study designed to explore the combined effects of the threats to population status for salmon in California, Oregon, Idaho, and Washington. We have formulated an a priori set of candidate models that reflect a competing suite of biological hypotheses to explain salmon abundance and trends (Figure 4). Justification for these hypotheses is based on what we know about potential interactions among risk factors as illustrated by the examples below. A strength of this approach is that information theory can be used to select a best approximating model (or set of models) that can then be used to infer the relative importance of those factors thought to impact salmon population trends directly or indirectly (e.g., Hilborn & Mangel 1996, Burnham & Anderson

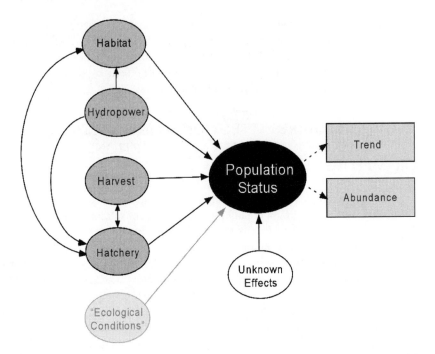

Figure 4   One of several possible model path diagrams depicting the direct and indirect effects of five putative causes of salmon declines (H's—see text) on salmon population status. The four H's (*gray shaded circles*) are modeled here as latent variables that can be estimated by factor analysis from a set of measured environmental parameters thought to collectively describe the factor of interest. A fifth factor (*light gray shaded circle*), defined here as natural ecological conditions (e.g., ocean conditions, interactions with nonindigenous species), could also shape salmon population dynamics, and although not modeled explicitly, could also interact with any of the four H's. The dependent variable, population status, is modeled as a latent variable (*blackened circle*) to be estimated from trend (i.e., λ) and change in abundance of salmon from historical levels. Residual variation (not explained by the combination of predictor variables) is depicted as an unmeasured factor (*open circle*). Arrows indicate the regressions of the H's on population status (direct effects) and on each other (indirect effects). Using model selection criteria, each candidate model (i.e., each combination of paths) represents a biological hypothesis that can be evaluated for fit to collected data to determine the most likely correlative causes of salmon decline.

1998, Anderson et al. 2000). Neither of the holistic analyses we are conducting informs us about how to improve population status, but they do reinforce the value of pursuing improvements simultaneously in multiple life stages and across threats—something that is often overlooked when assigning blame to each of the four H's.

Hatchery and Harvest Interactions

Recent harvest rates on federally listed Pacific salmon populations range from near 0 to 60% (PSC 2001b). As mentioned earlier, in many areas, a major reason fisheries persist is because of hatchery programs that provide fish to be caught. Depending on the extent of ecological and genetic interactions between hatchery and wild fish throughout their life cycle and the degree to which hatchery fish buffer wild fish from threats such as harvest, the net effect of hatchery programs on wild populations could be negative or positive (see "Hatchery Propagation of Salmon," above).

Aside from the potential direct and indirect negative effects of hatchery salmon on wild salmon, hatchery fish can mask the true status of wild fish in a population. In the Pacific Northwest, hatchery fish currently contribute 70% to 80% of the total salmon catch in coastal fisheries (Flagg et al. 2000). Some argue that the prevalence of hatchery fish in the catch should alleviate concerns about high harvest rates on listed salmon populations. The real test of whether harvest rates are significantly negatively affecting wild salmon populations is to estimate the status of the populations with and without harvest. This seemingly straightforward task is made very difficult because of the often unknown fraction of hatchery fish that are included in estimates of total potential returns to freshwater (recruits), catch, and spawning adults. For most populations of salmon, the proportion of hatchery fish mixed in with a count of total fish is not known—in some cases, estimates are available for a few years, and in very rare cases (e.g., where tagged fish can be automatically counted at dams), time series of hatchery fish in a run are available. In other words, in many cases, it is not possible to estimate the true status of wild salmon populations because of the presence of an unknown fraction of hatchery fish. In populations for which we have time series of estimates of total abundance and hatchery counts, the estimates of λ will always be lower when the presence of hatchery fish is factored out of the calculations in attempts to estimate the status of wild fish. Figure 5 illustrates variation in the effect hatchery fish have on estimates of λ for 15 populations of chinook salmon from Washington, Oregon, Idaho, and California.

Pacific salmon harvest managers must make annual decisions about levels of fishing mortality that do not constitute jeopardy to listed stocks under the ESA, defined as the case where an action does not "reduce appreciably the likelihood of both the survival and recovery of a listed species in the wild" (50 CFR S402.02). All the populations represented in Figure 5 are still harvested at rates greater than 30%—for example, the Nooksack chinook population in northern Puget Sound, Washington (in a listed ESU) had a mean harvest rate of over 52% in the years spanning 1989–1999 (PSC 2001b), and the population growth rate of wild chinook over that same time period was less than one. For populations with unknown fractions of hatchery fish included in total population counts, estimates of appreciable risk from harvest are fraught with uncertainty. This issue is especially pressing in the present political climate, where legal rulings such as the Hogan decision (e.g., *Alsea Valley Alliance v. Evans* 2001) and ongoing petitions to NMFS (Federal

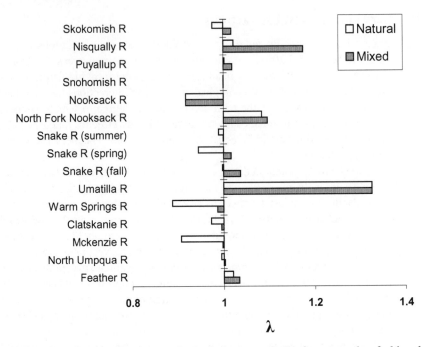

Figure 5 Average annual rate of population growth (λ) for a sample of chinook salmon spawning population counts from Washington, Oregon, Idaho, and California during 1980–1999. λ was calculated two different ways for each population: (*a*) including naturally spawning fish of hatchery and wild origin (mixed) and (*b*) counting only wild spawners whose parents were naturally produced (natural). (Source: NMFS, unpublished data.)

Register 2002) are questioning the means of accounting for hatchery fish in risk evaluations on listed wild salmon populations.

Hydropower and Hatchery Interactions

Much of the massive production of hatchery salmon throughout the Pacific Northwest was motivated by observed or anticipated declines in wild salmon (Lichatowich 1999). A prime example of how hatchery fish were used supposedly to mitigate losses in wild populations is where hatchery programs are associated with hydropower projects. In many cases, unanticipated negative effects of hatchery production on wild salmon species have been documented. For example, most hatchery and wild salmonids from the Snake River are captured at dams, loaded into barges, and transported to the estuary below the eight dams downstream in the Columbia River system they would otherwise have to traverse. Steelhead smolts from Snake River hatcheries average nearly 10 times the size of wild chinook smolts (Levin & Williams 2002), but they are loaded together into the same barges at densities up to 60 g/L. The result of the shared ride appears to be substantially reduced survival of the wild chinook (Levin & Williams 2002).

Another unintended consequence of hatchery practices is to concentrate salmonid juveniles after their release into freshwater, which can lead to higher susceptibility to predators who cue off of prey densities (reviewed by Flagg et al. 2000, Collis et al. 2001). Because of blockages such as dams and locks, increased predation rates on adult salmon also have been observed (Gearin et al. 1988, 1989; Nash et al. 2000).

Interactions Between Natural Processes and Human Activities

Not only can the effects of the four H's interact with each other, but they also interact with the environment. For example, human activities may interact with the environment by altering the strength of competitive interactions in the sea. Levin et al. (2001) showed that when releases from hatcheries are high and ocean productivity is low, the survival of chinook salmon in the marine phase declines. Similarly, Cooney & Brodeur (1998) found that hatchery effects on wild salmon depend on freshwater habitat capacity and ocean conditions experienced by the wild and hatchery fish. Recent improvements in rearing practices have resulted in hatchery-reared salmon that are now larger than their wild counterparts and exhibit greater marine survival than wild fish (Zabel & Williams 2002).

Another example of interactions between human threats and environmental sources of variability is found in the complicated effects of dams on salmon. The tendency to focus solely on direct mortality of salmon through dams ignores the possibility that factors affecting the fish before or after they pass through hydropower projects act synergistically with dams to reduce survival (Budy et al. 2002). For instance, recent work has highlighted the ecosystem importance of returning salmon that can have positive direct and indirect feedbacks on salmon population dynamics. Nutrients derived from decomposing salmon carcasses are now recognized to play an important role in the ecology of Pacific Northwest riparian habitats (Gresh et al. 2000). For one, the delivery of marine-derived nutrients by salmon returning from the sea appears to be crucial to the growth of juvenile salmon (Bilby et al. 1996, 1998; Larkin & Slaney 1997; Wipfli et al. 1999). Thus, the extreme reduction in salmon abundance caused by dams ostensibly has resulted in a nutrient deficit that could have contributed to reduced size of outmigrating juveniles (compare Bilby et al. 1998). Furthermore, hydropower projects slow the migration of salmon by several weeks, and this delay in migration appears to exacerbate size-selective marine mortality (Zabel & Williams 2002). Together, these independent results suggest that it is possible that the effects of nutrient deficiencies in the freshwater habitat are amplified by dams, resulting in increased marine mortality of juvenile salmon.

An Integrated Threats Analysis for Salmon Recovery

Scientists and managers engaged in salmon recovery increasingly are realizing the importance of whole-life-cycle, whole-ecosystem approaches to identifying significant threats to salmon populations. For example, a recent NRC panel leveled criticism against current practices of stock assessment and management for failing to account adequately for environmental fluctuations in their harvest decisions (NRC 1998), and a group of lead scientists from agencies involved in salmon management trumpeted the need for integrated research on classically isolated problems facing salmon (ISAB 2001a, ISST 2001).

Conducting more relevant research and using better models in making management decisions are relatively obvious things to change about the way risks to salmon populations are managed. Clearly, a widely agreed upon and supported strategic research plan is needed to organize guidance for salmon recovery efforts. More politically and logistically difficult to implement are manipulative, large-scale management experiments followed by thorough monitoring. Impressive experiments such as shutting down harvest or hatcheries or removing dams have been heralded as excellent ways to learn how salmon really would respond to improvements in their complex ecosystems (RSRP 2001). We agree with those pleas for experimentation—but we have not seen any undertaken yet. In the mean time, admitting sources of uncertainty, planning for alternative future states (e.g., Groves 2000), and generating technical analyses that offer a comparison of multiple options for recovering salmon will increase the chances that scientific input informs final recovery decisions designed to abate threats facing the fish.

GETTING REAL—SALMON RECOVERY PLANNING IN PRACTICE

The Art of Recovery Planning for Threatened and Endangered Salmonids: Creating Options

Two technical recovery teams established by the NMFS have been underway for about two years (Puget Sound and Willamette–Lower Columbia River domains), and three others were convened in the fall of 2001 (Interior Columbia Basin, southern Oregon–northern California, and California coastal domains; Figure 1). One of us (MR) is the chair of the Puget Sound recovery team. The technical challenges are a small part of the overall task of developing a recovery plan for listed chinook, chum, and sockeye ESUs within our geographic domain. The recovery teams are multi-stakeholder groups whose members are diverse, but all have the necessary technical backgrounds to satisfy recovery team selection criteria. On the Puget Sound team, we have representatives from federal, state, tribal, and local entities, each of whom views the problem from a distinct perspective. The perspectives of team members come from their experiences in biology spanning harvest management, local government, and basic science backgrounds. In addition, the member agencies represented on the team have been embroiled in salmon wars for decades—the recent listing under the ESA has merely brought their disputes into sharper focus. Nevertheless, having a diverse set of technical perspectives is the point of setting up multi-stakeholder recovery teams—the hope is that alternative views can be hashed out through the team dynamics, thus producing a relatively harmonized recovery plan that will be easier to communicate to parties on the policy end.

Nonscientists often are unaware of how contentious biological tasks such as those facing the recovery teams can be—truisms such as "numbers don't lie" lead to expectations that scientific disagreements are relatively easy to resolve. In fact, major technical disagreements stemming from philosophical differences that seem to run as deep as religious beliefs are commonplace in such technical teams. For example, one of our early jobs as a recovery team was to agree on viability criteria for the 21 demographically independent populations of chinook we have identified in Puget Sound (PSTRT 2001). Because of the previous thought and work that had gone into justifying the importance of four key population parameters— abundance, productivity or growth rate, spatial structure, and diversity (McElhany et al. 2000)—there was virtually no debate among team members on what the biological indicators of viability should be.

In contrast, the debate among recovery team members was heated on how best to model viability quantitatively. In particular, strong differences emerged for preferable model structure and complexity, the importance of incorporating density-dependence into population dynamic models, and whether representing uncertainty around model results was important. After a long period of debating the relative merits of alternative approaches, the team agreed to explore a number of different models aimed at addressing the same basic question: How many salmon are needed for a population to have a negligible risk of extinction over 100 years? Preliminary results from two of several planned alternative modeling approaches produced ranges of viable population size that are large and overlapping. Just the simple exercise of displaying these ranges to policy-makers has led to a better understanding of why different scientists produce results that do not agree. The technical team is working with the policy group organizing recovery planning in Puget Sound to decide how best to summarize and communicate these disparate technical results. This seems like such a simple solution to a seemingly unresolvable debate—but it is surprising how rarely a multi-model approach is used in conservation and management applications (Kareiva et al. 1998, Boersma et al. 2001).

Pacific salmon recovery teams are exploring another novel approach—scenarios analysis—to increase the chances that biological results will influence recovery-planning decisions. The idea behind scenario analysis is simple: When uncertainty is enormous, bundle it together in a small number of scenarios (usually three to five) that can then be more easily examined than a complicated model with hundreds or even thousands of unknown parameters. Each scenario should capture a coherent set of assumptions that is embraced by an important stakeholder group or scientific group. An example in the salmon world might be the notion that the major source of current salmon risk is ocean conditions, and that other risk factors such as hydropower, hatcheries, and habitat explain much less about the recent declines. Given a limited number of scenarios, scientists can then present the data that support or contradict each scenario, and policy-makers can examine the consequences of alternative management actions assuming different scenarios are apt reflections of reality.

Such an approach has been used in a few conservation applications—high-profile examples include the IPCC for global climate change actions and the Millenium Assessment project evaluating the value of global biodiversity under alternative future states—and its popularity is increasing (Reid 2000; Sala et al. 2000; Clark et al. 2001; MA 2001; G. Peterson et al., unpublished data).

Pacific salmon recovery teams are using scenarios to address two main questions: (*a*) How many and which populations within an ESU must be naturally self-sustaining for the ESU to be considered recovered (i.e., what does recovery look like)? and (*b*) What actions are necessary to recover populations and ESUs (i.e., how do we get there)? The first question is one that is likely to have multiple answers because most ESUs are comprised of many populations—for example, the Puget Sound chinook ESU has at least 21 demographically independent populations (PSTRT 2001), and it is plausible that not all will need to be viable for the ESU to have a negligible risk of extinction. The Puget Sound technical recovery team is looking at multiple ways of combining population characteristics to describe a recovered ESU. The key to generating multiple solutions is first to set desired ESU-wide targets for abundance, productivity/growth rate, spatial structure, and diversity. Tools such as siting algorithms used in reserve design can help produce biologically equivalent solutions to ESU-level recovery targets (Ruckelshaus et al. 2002). Which ESU viability scenario(s) will be chosen to satisfy regional recovery goals will be decided by a multi-stakeholder policy group in Puget Sound called the Shared Strategy Development Committee (http://www.sharedsalmonstrategy.org).

The second type of scenario analysis used in salmon recovery planning is proposing alternative sets of actions whose combined effects will suffice to recover the listed ESUs. Making predictions about the effects of a given action (e.g., reduction in harvest, change in hatchery management practices, protection or restoration of habitat, predator control) on the fate of salmon populations depends on assumptions about future environmental conditions. Previously, detailed simulation models typically have been used to explore different assumptions about the environment and its influence on salmon (ISAB 2001b). Scenario analyses have similar goals as such simulation models, but by collecting assumptions into only a handful of alternatives, they are easier for decision-makers to work with. If one were to consider only three scenarios for salmon recovery, they might be (*a*) oceanic conditions outside of human control are the dominant factor affecting salmon survival, thus only modest improvements in human-caused risk factors are needed; (*b*) risks due to human effects are severe, but technical solutions such as hatcheries, better-engineered dams, restored streams, and selective harvest are sufficient to achieve the required survival benefits; and last, (*c*) the degradation of the environment is so severe that only massive changes in environmental conditions through restoration of landscape-forming processes and revolutionary changes in harvest and hatchery management will recover salmon. Under each of these scenarios, alternative management actions will have different effects on salmon. The objective for science is to assess the data that support or refute each scenario, to evaluate the internal consistency of each scenario, and to provide an analytical framework that can use new data to resolve critical uncertainties. The ultimate decision about what to do will still come down to trade-offs between economics and ecology or social values and biology, but if scenario analysis is done artfully, those trade-offs will be clear.

It is impossible to predict fish population dynamics precisely and accurately twenty or even five years into the future, with or without management intervention. Uncertainty, lack of basic ecological understanding, and tremendous environmental variability make such an endeavor fruitless. But we can identify management actions that are likely to yield marked improvements in population status under a wide variety of scenarios. Based on what we have learned from recovery planning for Pacific salmonids, we offer the elements in Table 3 as a brief primer on how to incorporate science into a recovery plan effectively (see also Wondolleck & Yaffee 2000, Brick et al. 2001).

Nature and Limitations of Data Underlying Recovery Analyses

Compared to most other imperiled species, there is a plethora of information about salmon—they are counted on fishing boats, at weirs, in snorkel surveys, and at dams. In spite of notebooks and computers full of salmon counts, accessible and well-managed databases for salmon are lacking (Botkin et al. 2000, ISAB 2001a). Science advisory panels note that many questions about salmon recovery could be answered by contrasting populations that experience different risk factors, if only such broad geographic data were available (see, e.g., RSRP 2001). Inadequate monitoring and poor data accessibility are common deficiencies afflicting endangered species management (Kareiva et al. 1998; Campbell, unpublished data). In salmon management, however, there is a more insidious problem—the acceptance of coded expert opinion as a substitute for data. Using expert opinion as a basis for management decisions in the absence of data is a practical necessity—the problem arises when calls for empirical information are forgotten once a decision has been made.

Two expert opinion approaches have come to be widely embraced as scientific support tools for salmon management. The first, Ecosystem Diagnosis and Treatment (EDT), is used to evaluate habitat capacity for salmon. EDT uses 45 habitat descriptors to predict the capacity of different streams and watersheds to support salmon. The 45 habitat variables represent scores on a scale of zero to four that are typically filled in by experts or derived from written reports but almost never entail a scientist actually going to the sites to take repeatable, standardized measurements. The method thus involves making guesses about habitat attributes

TABLE 3   Elements of successful participation in a recovery planning exercise. We define "success" as creating a situation where biological information influences ultimate decisions made in designing a recovery plan for imperiled species

1. Work hard in advance to clearly write down the first principles of population and conservation biology that will guide your technical approaches to recovery planning. Codified conceptual guidance and the next step—well-documented, tested and transparent technical approaches to addressing key questions—go a long way toward settling disputes that inevitably arise in technical teams.

2. Be choosy about the level of technical detail you delve into in addressing questions posed to your group. Each of the questions posed to salmon recovery teams could involve several lifetimes of research and analyses to answer. Use guidance from policy staff involved in recovery planning to help you decide which questions are key to the decisions they will make. In the end, deadlines will be the ultimate check in deciding which analyses you undertake and how to balance a mix of quantitative assessments, proxy information, and biological intuition to justify your conclusions.

3. Plan to present technical results as scenarios rather than "the answer" to a particular question. There can be many potential solutions to questions such as "Which populations must be viable for an ESU to be recovered?" and "Which actions are necessary for ESU recovery?", and expecting a technical team to present a single answer is unwise for at least two reasons. First, presenting a single result makes it very difficult to see how sensitive results are to policy assumptions (e.g., probability of persistence, time frame used in viability models; degree of certainty in the effects of actions on population status). Second, single answers from a technical team box decision-makers into a corner, reducing the chances that biological input will influence ultimate decisions that must also consider political, social, and logistical factors.

4. Resist the temptation to walk away from the group when it gets frustrating. Abandoning ship is an option open to most academics—when one scientist disagrees with another on a technical matter, it is relatively easy to write the offender off as naïve, off-base, or just plain stupid. Attempting to provide useful scientific guidance for real conservation applications requires that you stay at the table to ensure that your opinions on analytical approaches and interpretation of results are included in the final analysis.

5. Invest the time to work with policy- or decision-makers who will influence how scientific results are used in the design of the final recovery plan. Interacting with policy-makers who typically lack a scientific background can appear to slow your progress, will almost surely reorder your priorities for technical analyses, and will generally try your patience. The reality is that political, social, and logistical constraints on recovery solutions are powerful drivers when a recovery plan is being designed—in the end, policy staff are your best hope for incorporating hard-won points of scientific agreement into recovery solutions.

and then running those "data" through a model that predicts population performance. The EDT model includes so many unknown parameters that it is prone to large error propagation (Burnham & Anderson 1998), and it is extremely difficult to validate. Although this approach has been criticized by a "blue ribbon" panel of ecologists (RSRP 2000), it remains popular as a management tool.

A similar expert opinion scientific tool is the "matrix of pathways and indicators" that the NMFS developed for conducting consultations under the ESA (NMFS 1996). The matrix consists of 18 habitat attributes such as sediment, chemical contaminations and nutrients, pool frequency, width/depth ratio, and floodplain connectivity, each of which is scored as properly functioning, at risk, or not properly functioning. Although these attributes could be directly measured, the more typical pattern is that a consultant or agency biologist visits the site and fills in the matrix using his or her professional judgement. It may be possible to fill in a matrix without even visiting the watershed (simply by looking at previous reports and maps). It is not surprising then that attempts to see how well the matrix scores relate to measures of salmon abundance or productivity have detected no relationships (T. Good et al., unpublished data). Both EDT and the NMFS habitat matrix are honest efforts to inform decisions pertaining to salmon habitat actions in the absence of data. The problem is that the public, and even some scientists, have ended up being too satisfied with an absence of data, and have been willing to accept the expert opinion systems as a permanent substitute for empirical information.

In fact, the largest and best-funded ecological database for the Columbia Basin, the Interior Columbia Basin Ecosystem Management Plan (ICBEMP) database, is replete with "data layers" that are only thinly related to anything that might be measured (Quigley & Arbelbide 1997, http://www.icbemp.gov). For example, one of the variables in the databases, "riparian integrity index," is derived using models of models, and cannot be interpreted as anything that could be directly measured. Even a variable like "road" in the ICBEMP data, which could conceivably be directly measured, is a categorical variable that results from two conditional inequalities (see metadata regarding field *ROAD in ICBEMP). ICBEMP also used Bayesian belief networks to summarize the quality of habitats in the Columbia basin for salmonids, and in turn the opportunities for aiding fish recovery via habitat restoration or protection (ISAB 2001b). Analyses of measured data were lacking from this exercise. Again, the derived nature of the data and the degree to which data are expert opinion or empirical are difficult to discern—these important details are often forgotten by those using the database for management.

Finally, even in those cases where measured data exist, they can be fraught with problems. For example, the most fundamental of all data for salmon are population counts, which are key to determining the status and risk of different ESUs. In most management models, these counts are befuddled by two major sources of error that are rarely given the attention they deserve. First, counts are usually made of redds, carcasses, or live fish in small index streams that may be chosen simply because they are easily accessed. These index counts are then used to estimate total population size along two to three dimensions of extrapolation: multiplying index stream length by a factor to get total anadromous stream miles, expanding the number of redds or carcasses into a number of live fish, and somehow translating counts made over a limited period to a number representing the entire returning population (by area-under-the-curve methods, or making assumptions about the longevity of a redd) (e.g., Smith & Castle 1994). All of these extrapolations are prone to huge errors. For example, total anadromous stream miles can err by 30% to 75% (Bahls & Ereth 1994) and even the simplest redd counts have sampling errors on the order of 25% to 75% (Holmes 2001). If these sampling errors are not accounted for, estimates of population risk could be greatly inflated (because variability owing to sampling error often gets translated as environmental variability and hence elevated extinction risk).

Data can be divided into four categories: measured data, extrapolated or inferred data, modeled data, and expert opinion. Salmon management relies primarily on expert opinion and extrapolated or inferred data. As a temporary solution this approach is sensible; yet excessive reliance on expert opinion relaxes the pressure to collect data.

CONCLUSIONS

The elegant experiment, clear hypothesis, and simple model are icons of good science. But when science enters the arena of endangered species recovery, the science is rarely elegant, clear, or simple. Although engaging in political and economic discussions can be frustrating to biologists, the much-maligned role of science in endangered species management has as much to do with the shortcomings of science as it does with the perils of politics. We have uncovered many shortcomings in the science underlying salmon recovery. Even the most basic question of identifying relevant units for conservation at the population level warrants more thoughtful consideration. For example, when populations are targeted for priority protection, multiple biological criteria need to be involved when defining those units, in concert with a compelling vision of exactly what one is attempting to preserve (e.g., future evolutionary potential or medium-range ecological persistence). Once conservation units have been delineated, tools for estimating their viability are relatively well developed and understood. What we lack are tools for translating different management scenarios into likely changes in viability. The difficulty in assessing the merits of different management scenarios emerges from salmon facing multiple sources of risk simultaneously; risks that interact with one another and with a backdrop of natural hazards and fluctuating environmental conditions. Finally, our data management simply is not sufficiently developed to support the numerical analyses we wish we could do. Salmon, especially Pacific salmon, are the subject of thousands of papers in scientific journals over the past decade. Nonetheless, the most fundamental data regarding abundance and distribution are not readily accessible. And for subtler data—such as habitat quality—managers rely all too often upon expert opinion and obscure extrapolations. Expediency dictates that management decisions are made based on insufficient data, nevertheless, calls for empirical data to redress the gaps in our understanding cannot be too insistent.

We are experimenting with approaches, such as scenario planning and multistakeholder technical recovery teams, for delivering scientific guidance in the face of data shortcomings and great uncertainty. There is a dire need for research aimed at determining what makes the infusion of science into decision-making sometimes work and sometimes fail miserably (Kinzig et al. 2000). One place to begin would be a systematic review of the effectiveness of science at influencing policy.

Finally, for salmon recovery, the most crucial questions are not purely scientific. Our experience has taught us that participants bring such different values to the table, that even when given exactly the same data and information, they come up with dramatically different solutions. The need for collaboration, consensus, compromises, and an accounting for values make the science/policy interface an uncomfortable environment for most scientists. Yet it is primarily in this environment that science can make the biggest difference—by making clear the likely biological consequences of actions, and stopping short of saying anything about what "should" be done.

ACKNOWLEDGMENTS

We would like to thank Tammy Harms and Juliet Fabbri for help with figures, data compiling, and endless logistics. Our thinking about salmon conservation problems has benefited greatly from discussions with Tim Beechie, Ted Case, Mike Ford, Fran James, Russ Lande, Simon Levin, Michelle McClure, Paul McElhany, Bill Murdoch, Bob Paine, Walt Reid, Bill Ruckelshaus, Mike Schiewe, Robin Waples, John Williams, Rich Zabel, and members of the Puget Sound Technical Recovery Team. Dan Simberloff provided insightful comments on an earlier version of this manuscript. The views expressed in this paper are those of the authors.

The *Annual Review of Ecology and Systematics* is online at
http://ecolsys.annualreviews.org

LITERATURE CITED

Achord S, Eppard MB, Hockersmith EE, Standford BP, Matthews GM. 1997. *Monitoring the migrations of wild Snake River spring/summer chinook salmon smolts 1996.* Bonneville Power Admin. Rep. Proj. 91–028, Portland, OR

Anderson DR, Burnham KP, Thompson WL. 2000. Null hypothesis testing: problems, prevalence, and an alternative. *J. Wildl. Manag.* 64:912–23

Avise JC. 1989. A role for molecular genetics in the recognition and conservation of endangered species. *Trends Ecol. Evol.* 4:29–281

Bahls P, Ereth M. 1994. *Stream typing error in Washington water type maps for watersheds of Hood Canal and Southwest Olympic Peninsula. Tech. Rep. 94–2.* Point-No-Point Treaty Counc., Kingston, Wash.

Barlow GW. 1995. The relevance of behavior and natural history to evolutionarily significant units. See Nielsen 1995, pp. 169–75

Beamesderfer RC, Ward DL, Nigro AA. 1996. Evaluation of the biological basis for a predator control program on northern squawfish (*Ptychocheilus oregonensis*) in the Columbia and Snake rivers. *Can. J. Fish. Aquat. Sci.* 53:2898–908

Beamish RJ, Noakes DJ, McFarlane GA, Klyashtorin L, Ivanov VV, Kurashov V. 1999. The regime concept and natural trends in the production of Pacific salmon. *Can. J. Fish. Aquat. Sci.* 56:516–26

Beckham SD. 1977. *The Indians of Western*

Oregon: This Land was Theirs. Coos Bay, OR: Arago Books

Bednarek AT. 2001. Undamming rivers: a review of the ecological impacts of dam removal. *Environ. Manag.* 27:803–14

Beechie TJ, Beamer E, Collins B, Benda L. 1996. Restoration of habitat-forming processes in Pacific Northwest watersheds: A locally adaptable approach to salmonid habitat restoration. In *The Role of Restoration in Ecosystem Management,* ed. DL Peterson, CV Klimas, pp. 48–67. Madison, WI: Soc. Ecol. Restor.

Beechie TJ, Beamer E, Wasserman L. 1994. Estimating coho salmon rearing habitat and smolt production losses in a large river basin, and implications for restoration. *N. Am. J. Fish. Manag.* 14:797–811

Beechie TJ, Bolton S. 1999. An approach to restoring salmonid habitat-forming processes in Pacific Northwest watersheds. *Fisheries* 24:6–15

Beechie TJ, Roni R, Steel EA. 2002. *Ecosystem recovery planning for listed salmon: an integrated assessment approach for salmon habitat. NOAA Tech. Memo. NMFS-NWFSC,* Northwest Fish. Sci. Cent., Seattle, WA. In press

Beissinger SR, McCullough DR, eds. 2002. *Population Viability Analysis.* Chicago, IL: Univ. Chicago Press

Bilby RE, Fransen BR, Bisson PA. 1996. Incorporation of nitrogen and carbon from spawning coho salmon into the trophic system of small streams: evidence from stable isotopes. *Can. J. Fish. Aquat. Sci.* 53(1):164–73

Bilby RE, Fransen BR, Bisson PA, Walter JK. 1998. Response of juvenile coho salmon (*Oncorhynchus kisutch*) and steelhead (*Oncorhynchus mykiss*) to the addition of salmon carcasses to two streams in southwestern Washington, U.S.A. *Can. J. Fish. Aquat. Sci.* 55:1090–118

Boersma PD, Kareiva PM, Fagan WF, Clark JA, Hoekstra JM. 2001. How good are endangered species recovery plans? *BioScience* 51:643–49

Botkin D, Peterson D, Calhoun J. 2000. *The scientific basis for validation monitoring of salmon for conservation and restoration plans.* Olympic Nat. Resource Cent. Tech. Rep., Forks, WA

Bowen BW. 1998. What is wrong with ESUs? *The gap between evolutionary theory and conservation principles. J. Shellfish Res.* 17:1355–58

Bowen BW. 1999. Preserving genes, species, or ecosystems? Healing the fractured foundations of conservation policy. *Mol. Ecol.* 8:S5–10

Boyle B. 2002. Restoring credibility to government science. *Seattle Times,* Feb. 27

Bradford MJ, Irvine JR. 2000. Land use, fishing, climate change, and the decline of Thompson River, British Columbia, coho salmon. *Can. J. Fish. Aquat. Sci.* 57:13–16

Brick P, Snow D, Van De Wetering S. 2001. *Across the Great Divide. Explorations in Collaborative Conservation and the American West.* Washington, DC: Island

Brodeur RD, Boehlert GW, Casillas E, Eldridge MB, Helle JH, et al. 2000. A coordinated research plan for estuarine and ocean research on Pacific Salmon. *Fisheries* 25(6):7–16

Brook BW, O'Grady J, Chapman A, Burgman M, Akcakaya HR, Frankham R. 2000. Predictive accuracy of population viability analysis in conservation biology. *Nature* 404:385–86

Brown B. 1983. *Mountain in the Clouds: A Search for the Wild Salmon.* Seattle: Univ. Wash. Press. 249 pp.

Buchanan DV, Hooton RM, Moring JR. 1981. Northern squawfish (*Ptychocheilus oregonensis*) predation on juvenile salmonids in sections of the Willamette River Basin, Oregon. *Can. J. Fish. Aquat. Sci.* 38:360–64

Budy P, Thiede GP, Bouwes N, Petrosky CE, Schaller H. 2002. Evidence linking delayed mortality of Snake River salmon to their earlier hydrosystem experience. *N. Am. J. Fish. Manag.* 22:35–51

Burnham KP, Anderson DR. 1998. *Model Selection and Inference: A Practical Information-Theoretic Approach.* New York: Springer-Verlag

Busby PJ, Wainwright TC, Bryant GJ, Lierheimer LJ, Waples RS, et al. 1996. *Status review of west coast steelhead from Washington, Idaho, Oregon, and California. NOAA Tech. Memo. NMFS-NWFSC-27.* Northwest Fish. Sci. Cent., Seattle, WA

Busby PJ, Wainwright TC, Waples RS. 1994. *Status review for Klamath Mountains Province steelhead. NOAA Tech. Memo. NMFS-NWFSC-19.* Northwest Fish. Sci. Cent., Seattle, WA

Clark JS, Carpenter SR, Barber M, Collins S, Dobson A, et al. 2001. Ecological forecasts: An emerging imperative. *Science* 293:657–60

Collings MR, Smith RW, Higgins GT. 1972. *Hydrology of four streams in western Washington as related to several Pacific salmon species: Humptulips, Elochoman, Green and Wynoochee rivers.* U.S. Geol. Serv. open file rep., Tacoma, Wash.

Collis K, Roby DD, Craig D, Ryan BA, Ledgerwood RD. 2001. Colonial waterbird predation on juvenile salmonids tagged with passive integrated transponders in the Columbia River estuary: vulnerability of different salmonid species, stocks, and rearing types. *Trans. Am. Fish. Soc.* 130:385–96

Cone J. 1995. *A Common Fate: Endangered Salmon and the People of the Pacific Northwest.* Corvallis: Oregon State Univ. Press

Cone J, Ridlington S, eds. 1996. *The Northwest Salmon Crisis: A Documented History.* Corvallis: Oregon State Univ. Press

Connelly M. 2001. Home is where they'll lay me down. *Orion* (Summer):17–25

Cooney RT, Brodeur RD. 1998. Carrying capacity and North Pacific salmon production: stock enhancement implications. *Bull. Mar. Sci.* 62:443–64

Craig JK, Foote CJ. 2001. Countergradient variation and secondary sexual color: phenotypic convergence promotes genetic divergence in carotenoid use between sympatric anadromous and nonanadromous morphs of sockeye salmon (*Oncorhynchus nerka*). *Evolution* 55:380–91

Crandall KA, Bininda-Emonds ORP, Mace GM, Wayne RK. 2000. Considering evolutionary processes in conservation biology. *Trends Ecol. Evol.* 15:290–95

Dalton R. 2002. Fur flies over lynx survey's suspect samples. *Nature* 415:107

Doremus H. 2000. Delisting endangered species: An aspirational goal, not a realistc expectation. *Environ. Law Rep.* 30:10434

Fed. Regist. 1996. Policy regarding the recognition of distinct vertebrate population segments under the Endangered Species Act. *Fed. Regist.* 61:4721–25

Fed. Regist. 2002. Endangered and threatened species: Findings on petitions to delist Pacific Salmonid ESUs. *CFR Fed. Regist.* 50:223–26. 67(28)

Feist BE, Steel EA, Pess GR, Bilby RE. 2002. What is the correct scale for habitat analyses aimed at prioritizing restoration efforts for salmon? *Anim. Conserv.* In review

Finney BP, Eaves G, Sweetman I, Douglas J, Smol MSV. 2000. Impacts of climatic change and fishing on Pacific salmon abundance over the past 300 years. *Science* 290:795–99

Flagg TA, Berejikian BA, Colt JE, Dickhoff WW, Harrell LW, et al., eds. 2000. *Ecological and behavioral impacts of artificial production strategies on the abundance of wild salmon populations—a review of practices in the Pacific Northwest.* NOAA Tech. Memo. NMFS-NWFSC-4, Northwest Fish. Sci. Cent., Seattle, WA

Fleming IA, Gross MR. 1993. Breeding success of hatchery and wild coho salmon (*Oncorhynchus kisutch*) in competition. *Ecol. Appl.* 3:230–45

Ford MJ. 1998. Testing models of migration and isolation among populations of Chinook salmon (*Oncorhynchus tschawytscha*). *Evolution* 52:539–57

Ford MJ, Budy P, Busack C, Chapman D, Cooney T, et al. 2001. *Upper Columbia River steelhead and spring chinook salmon population structure and biological requirements.* http://www.nwfsc.noaa.gov/

Frank BJR. 2002. Salmon need better habitat. *Seattle Post-Intelligencer*, Jan. 11

Fraser DJ, Bernatchez L. 2001. Adaptive evolutionary conservation: towards a unified concept for defining conservation units. *Mol. Ecol.* 10:2741–52

Friesen TA, Ward DL. 1999. Management of northern pikeminnow and implications for juvenile salmonid survival in the lower Columbia and Snake rivers. *N. Am. J. Fish. Manag.* 19:1008–19

Frissell CA, Nawa RK. 1992. Incidence and causes of physical failure of artificial habitat structures in streams of western Oregon and Washington. *N. Am. J. Fish. Manag.* 12:182–97

Gearin P, Pfeifer R, Jeffries S, Delong R. 1989. California sea lion predation depletes a steelhead run in Washington State. *Proc. Bien. Conf. Biol. Mar. Mammals, Dec. 7–11, Pacific Grove, CA* (Abstr.)

Gearin P, Pfeifer R, Jeffries S, Delong R, Johnson M. 1988. *Results of the 1986–87 California sea lion-steelhead trout predation control program at the Hiram M. Chittenden Locks.* Seattle, WA: NMFS Northwest Alaska Fish. Cent. Process. Rep. 88–30. 111 pp.

Glova GJ. 1984. Management implications of the distribution and diet of sympatric populations of juvenile coho salmon and coastal cutthroat trout in small streams in British Columbia, Canada. *Prog. Fish Cult.* 46:269–77

Gore AJR. 1992. *Earth in the Balance. Ecology and the Human Spirit.* New York: Penguin

Gresh T, Lichatowich J, Schoonmaker P. 2000. An estimation of historic and current levels of salmon production in the northeast Pacific ecosystem. *Fisheries* 25:15–21

Groot C, Margolis L. 1991. *Pacific Salmon Life Histories.* Vancouver: Univ. Br. Columbia Press

Groves C. 2000. *Designing a Geography of Hope: A Practitioner's Handbook to Ecoregional Conservation Planning.* Washington, DC: The Nat. Conserv. Press

Gustafson RG, Wainwright TC, Winans GA, Waknitz FW, Parker LT, Waples RS. 1997. *Status review of sockeye salmon from Washington and Oregon. NOAA Tech. Memo.*

NMFS-NWFSC-33. Northwest Fish. Sci. Cent., Seattle, WA

Hard JJ. 2002a. Case study of Pacific salmon. In *Fisheries-Induced Adaptive Change*, ed. U Dieckmann, OR Godø, M Heino, J Mork. Cambridge, UK. In press

Hard JJ. 2002b. Response of chinook salmon life history to selection under harvest. In *Salmonid Perspectives on Evolution*, ed. A Hendry, S Stearns. Oxford, UK: Oxford Univ. Press. In press

Hard JJ, Kope RG, Grant WS, Waknitz FS, Parker LT, Waples RS. 1996. *Status review of Pink salmon from Washington, Oregon, and California. NOAA Tech. Memo. NMFS-NWFSC-25.* Northwest Fish. Sci. Cent., Seattle, WA

Healey MC. 1986. Optimum size and age at maturity in Pacific salmon and effects of size-selective fisheries. *Can. Spec. Publ. Fish. Aquat. Sci.* 89:39–52

Healey MC. 1991. Diets and feeding rates of juvenile pink, chum and sockeye salmon in Hecate Strait, British Columbia. *Trans. Am. Fish. Soc.* 120:303–18

Hendry AP. 2000. Rapid evolution of reproductive isolation in the wild: evidence from introduced salmon. *Science* 290:516–18

Hilborn R. 1992. Hatcheries and the future of salmon in the Northwest. *Fisheries* 17:5–8

Hilborn R, Mangel M. 1997. *The Ecological Detective.* Princeton, NJ: Princeton Univ. Press

Hollowed AB, Hare SR, Wooster WS. 2001. Pacific Basin climate variability and patterns of Northeast Pacific Marine fish production. *Prog. Oceanogr.* 49:257–82

Holmes EE. 2001. Estimating risks in declining populations with poor data. *Proc. Natl. Acad. Sci. USA* 98:5072–77

Holmes EE, Fagan WF. 2002. Validating population viability analyses for corrupted data sets. *Ecology* In press

Independent Sci. Advis. Board (ISAB). 2001a. *A Review of Salmon Recovery Strategies for the Columbia River Basin, ISAB Rep. 2001–7.* Northwest Power Plan. Counc., Portland, OR

Independent Sci. Advis. Board (ISAB). 2001b. *ISAB Model Synthesis Report. An Analysis of Decision Support Tools Used in Columbia River Basin Salmon Management. ISAB Rep. 2001-1a.* Northwest Power Plan. Counc., Portland, OR

Independent Sci. Group (ISG). 1996. *Return to the river: restoration of salmonid fishes in the Columbia River ecosystem.* Northwest Power Plan. Counc., Portland, OR

Independent Sci. Group (ISG). 1999. Scientific issues in the restoration of salmonid fishes in the Columbia River. *Fisheries* 24:10–19

Interagency Sci. Salmon Team (ISST). 2001. *Identified Research Needs for Salmon Recovery.* 15 pp.

Johannesson K. 2001. Parallel speciation: a key to sympatric divergence. *Trends Ecol. Evol.* 16:148–53

Kareiva PM, Andelman S, Doak D, Elderd B, Groom M, et al. 1998. *Using science in habitat conservation plans.* http://www.nceas.ucsb.edu/projects/hcp

Kareiva PM, Marvier M, McClure M. 2000. Recovery and management options for spring/summer chinook salmon in the Columbia River Basin. *Science* 270:977–79

Kinnison MT, Bentzen P, Unwin MJ, Quinn TP. 2002. Reconstructing recent divergence: evaluating nonequilibrium population structure in New Zealand Chinook salmon. *Mol. Ecol.* 11:739–54

Kinzig AP, Carpenter S, Dove M, Heal G, Levin S, et al. 2000. *Nature and Society: An Imperative for Integrated Environmental Research.* http://lsweb.la.asu/akinzig/NSFReport.pdf

Lande R. 1993. Risks of population extinction from demographic and environmental stochasticity and random catastrophes. *Am. Nat.* 142:911–27

Lande R, Shannon S. 1996. The role of genetic variation in adaptation and population persistence in a changing environment. *Evolution* 50:434–37

Larkin G, Slaney PA. 1997. Implications of trends in marine-derived nutrient influx to south coastal British Columbia salmonid production. *Fisheries* 22(11):16–24

Lawson PW. 1993. Cycles in ocean productivity, trends in habitat quality, and the restoration of salmon runs in Oregon. *Fisheries* 18:6–10

Levin PS, Achord S, Feist B, Zabel RW. 2002. Non-indigenous brook trout and the demise of threatened Snake River salmon: a forgotten threat? *Proc. R. Soc. London Ser. B.* 269:1663–70

Levin PS, Schiewe MH. 2001. Preserving salmon biodiversity. *Am. Sci.* 89:220–27

Levin PS, Tolimieri N. 2001. Differences in the impacts of dams on the dynamics of salmon populations. *Anim. Conserv.* 4:291–99

Levin PS, Williams JG. 2002. Interspecific effects of artifically propagated fish: an additional conservation risk for salmon. *Conserv. Biol.* In press

Levin PS, Zabel RW, Williams JG. 2001. The road to extinction is paved with good intentions: negative associations of fish hatcheries with threatened salmon. *Proc. R. Soc. London Ser. B* 268:1–6

Lichatowich J. 1999. *Salmon Without Rivers.* Washington, DC: Island

Lynch M, Lande R. 1993. Evolution and extinction in response to environmental change. In *Biotic Interactions and Global Change*, ed. P Kareiva, JG Kingsolver, RB Huey, pp. 234–50. Sunderland, MA: Sinauer

Mahnken C, Ruggerone G, Waknitz W, Flagg T. 1998. A historical perspective on salmonid production from Pacific Rim hatcheries. *North Pac. Anadromous Fish. Comm. Bull.* 1:38–53

Mantua NJ, Hare SR, Zhang Y, Wallace JM, Francis RC. 1997. A Pacific interdecadal climate oscillation with impacts on salmon production. *Am. Meteor. Soc. Bull.* 78:1069–79

Mapes LV. 2001. Lynx-fur furor focuses on science role. *Seattle Times*, Dec. 30

Mayden RL, Wood RM. 1995. Systematics, species concepts, and the evolutionarily significant unit in biodiversity and conservation biology. See Nielsen 1995, pp. 58–113

McClure MM, Cooney T, Marvier M. 2001. Assessing the role of dams in salmon recovery. *Hydro Rev.* 20:36–45

McClure MM, Holmes EE, Sanderson BL, Jordan CE. 2002. A large-scale, multi-species assessment: Anadromous salmonids in the Columbia River Basin. *Ecol. Appl.* In press

McCusker MR, Parkinson E, Taylor EB. 2000. Mitochondrial DNA variation in rainbow trout (*Oncorhynchus mykiss*) across its native range: testing biogeographical hypotheses and their relevance to conservation. *Mol. Ecol.* 9:2089–108

McElhany P, Ruckelshaus MH, Ford MJ, Wainwright T, Bjorkstedt E. 2000. *Viable salmonid populations and the recovery of evolutionarily significant units. NOAA Tech. Memo. NMFS-NWFSC-42.* Washington, DC: US Dep. Comm. 156 pp.

McFarlane GA, King JR, Beamish RJ. 2000. Have there been recent changes in climate? Ask the fish. *Prog. Oceanogr.* 47:147–69

Meffe GK. 1992. Techno-arrogance and halfway technologies: salmon hatcheries on the Pacific coast of North America. *Conserv. Biol.* 6:350–54

Millenium Ecosyst. Assess. (MA). 2001. *Objectives, audience, process and conceptual framework.* Presented at MA Tech. Design Workshop, Bilthoven, The Netherlands

Mills LS. 2002. False samples are not the same as blind controls. *Nature* 415:471

Milstein M. 2002. Fish, lynx flubs hit agencies' credibility. *The Oregonian*, Feb. 5

Moritz C. 1994. Defining "evolutionarily significant units" for conservation. *Trends Ecol. Evol.* 9:373–75

Moritz C. 2002. Strategies to protect biodiversity and the evolutionary processes that sustain it. *Syst. Biol.* 51:238–54

Morris W, Doak D, Groom M, Kareiva PM, Fieberg J, et al. 1999. *A Practical Handbook for Population Viability Analysis.* Washington, DC: The Nat. Conserv.

Moyle PB, Li HW, Barton BA. 1986. *The Frankenstein Effect: Impact of Introduced Fishes on Native Fishes in North America,* pp. 415–26. Bethesda, MD: Am. Fish. Soc.

Myers J, Busack C, Rawding D. 2002. *Identifying historical populations of chinook and chum salmon and steelhead with the Lower Columbia River and Upper Willamette River Evolutionarily Significant Units.* Willamette/Lower Columbia River Tech. Recovery Team. Tech. Rev. Draft

Myers JM, Kope RG, Bryant GJ, Teel D, Lierheimer LJ, et al. 1998. *Status review of Chinook salmon from Washington, Idaho, Oregon, and California. NOAA Tech. Memo. NMFS-NWFSC-35.* Northwest Fish. Sci. Cent., Seattle, WA

Mysak LA. 1986. El Nino, interannual variability and fisheries in the northeast Pacific ocean. *Can. J. Fish. Aquat. Sci.* 43:464–97

Nash CE, Iwamoto RN, Mahnken CVW. 2000. Aquaculture risk management and marine mammal interactions in the Pacific Northwest. *Aquaculture* 183:307–23

Natl. Mar. Fish. Serv. (NMFS). 1991. Notice of policy: Policy on applying the definition of species under the Endangered Species Act to Pacific salmon. *Fed. Regist.* 56:58612–16

Natl. Mar. Fish. Serv. (NMFS). 1995. *Biological opinion for reinitiation of consultation on 1994–1998 operation of the Federal Columbia River Power System and juvenile transportation program in 1995 and future years.* NMFS Northwest Reg., Portland, OR

Natl. Mar. Fish. Serv. (NMFS). 1996. *Making ESA determinations of effect for individual or grouped actions at the watershed scale.* Environ. Tech. Serv Div., Habitat Conserv. Branch, Portland, OR

Natl. Mar. Fish. Serv. (NMFS). 2000a. *Recovery planning guidance for Technical Recovery Teams.* http://www.nwfsc.noaa.gov/cbd/trt/about.htm

Natl. Mar. Fish. Serv. (NMFS). 2000b. *Reinitiation of consultation on operation of the Federal Columbia River power system, including the juvenile fish transportation program, and 19 Bureau of Reclamation projects in the Columbia Basin.* Seattle, WA

Natl. Mar. Fish. Serv. (NMFS). 2000c. *Predation on salmonids relative to the Federal Columbia River Power System.* Northwest Fish. Sci. Cent., Seattle, WA

Natl. Oceanic Atmospheric Administration

(NOAA). 1997. *Impacts of California sea lions and Pacific harbor seals on salmonids and on the coastal ecosystems of Washington, Oregon and California. NOAA Tech. Memo. NMFS-NWFSC-28.* Northwest Fish. Sci. Cent., Seattle, WA

Natl. Res. Counc. (NRC), ed. 1995. *Science and the Endangered Species Act.* Washington, DC: Natl. Acad. Press

Natl. Res. Counc. (NRC), ed. 1996. *Upstream: Salmon and Society in the Pacific Northwest.* Washington, DC: Natl. Acad. Press

Natl. Res. Counc. (NRC), ed. 1998. *Improving Fish Stock Assessments.* Washington, DC: Natl. Acad. Press

Natl. Res. Counc. (NRC). 2002. *Scientific Evaluation of Biological Opinions on Endangered and Threatened Fishes in the Klamath River Basin: Interim Report.* Washington, DC: Natl. Acad. Press

Nehlsen W, Williams JE, Lichatowich JA. 1991. Pacific salmon at the crossroads: stocks at risk from California, Oregon, Idaho and Washington. *Fisheries* 16(2):4–21

Nicholas JW, Hankin DG. 1988. *Chinook salmon populations in Oregon coastal river basins: Description of life histories and assessment of recent trends in run strengths, Fish. Div. Inf. Rep., No. 88–1.* Portland, OR: Oreg. Dep. Fish Wildl. 359 pp.

Nielsen JL, ed. 1995. Evolution and the aquatic ecosystem: defining unique units in population conservation. *Proc. Am. Fish. Soc. Symp., 17th.* Bethesda, MD: Am. Fish. Soc.

Nielsen JL, Fountain MC, Wright JM. 1997. Biogeographical analysis of Pacific trout (*Oncorhynchus mykiss*) in California and Mexico based on mitochondrial DNA and nuclear microsatellites. In *Molecular Systematics of Fishes*, ed. TD Kocher, CA Stepien, pp. 53–69. New York: Academic

Northwest Power Plan. Counc. (NWPPC). 2001. *20th annual report of the Pacific Northwest Electric Power and Conservation Planning Council. Counc. Doc. 2001–3.* Northwest Power Plan. Counc., Portland, OR

Pac. Salmon Comm. (PSC). 1997. *Incidental fishing mortality of chinook salmon: mortal-*

ity rates applicable to Pacific salmon commission fisheries. Rep. TCCHINOOK (97)-1. Vancouver, BC, Can.: Pac. Salmon Comm.

Pac. Salmon Comm. (PSC). 2001a. *Catch and escapement of chinook salmon under Pacific Salmon Commission jurisdiction 1997–2000.* Rep. TCCHINOOK (01)-1. Vancouver, BC, Can.: Pac. Salmon Comm.

Pac. Salmon Comm. (PSC). 2001b. *Annual exploitation rate analysis and model calibration.* Rep. TCCHINOOK (01)-2. Vancouver, BC, Can.: Pac. Salmon Comm.

Pascual M, Bentzen P, Rossi CR, Mackey G, Kinnison MT, Walker R. 2001. First documented case of anadromy in a population of introduced rainbow trout in Patagonia, Argentina. *Trans. Am. Fish. Soc.* 130:53–67

Paulsen CM, Fisher TR. 2001. Statistical relationship between parr-to-smolt survival of Snake River spring-summer chinook salmon and indices of land use. *Trans. Am. Fish. Soc.* 130:347–58

Pearcy WG. 1992. *Ocean Ecology of North Pacific Salmonids.* Seattle, WA: Wash. Sea Grant

Pess GR, Montgomery DR, Steel EA, Bilby RE, Feist BE, Greenberg HM. 2002. Landscape characteristics, land use, and coho salmon (*Oncorhynchus kisutch*) abundance, Snohomish River, Wash. *Can. J. Fish. Aquat. Sci.* 59:613–23

Poe TP, Shively RS, Tabor RA. 1994. Ecological consequences of introduced piscivorous fishes in the lower Columbia and Snake Rivers. See Stouder et al. 1994, pp. 347–60

Puget Sound Tech. Recovery Team (PSTRT). 2001. *Independent populations of chinook salmon* (Oncorhynchus tshawytscha) *in Puget Sound.* Public Rev. Draft NMFS, Seattle, WA. http://www.nwfsc.noaa.gov/cbd/trt/popid.pdf

Quigley TM, Arbelbide SJ, eds. 1997. *An assessment of ecosystem components in the interior Columbia basin and portions of the Klamath and Great Basins. Gen. Tech. Rep. PNW-GTR-405.* Pac. Northwest Res. Stn., Portland, OR

Quinn TP, Kinnison MT, Unwin MJ. 2001. Evolution of Chinook salmon (*Oncorhynchus tshawytscha*) populations in New Zealand: pattern, rate, and process. *Genetica* 113:493–513

Quinn TP, Unwin MJ, Kinnison MT. 2000. Evolution of temporal isolation in the wild: genetic divergence in timing of migration and breeding by introduced Chinook salmon populations. *Evolution* 54:1372–85

Ratner S, Lande R. 2001. Demographic and evolutionary responses to selective harvesting in populations with discrete generations. *Ecology* 82(11):3093–104

Raymond HL. 1988. Effects of hydroelectric development and fisheries enhancement on spring and summer chinook salmon and steelhead in the Columbia River Basin. *N. Am. J. Fish. Manag.* 8:1–24

Recovery Sci. Rev. Panel (RSRP). 2000. *Report from the RSRP Meeting held Dec. 3–4.* http://www.nwfsc.noaa.gov/cbd/trt/rsrpdoc2.pdf

Recovery Sci. Rev. Panel (RSRP). 2001. *Report from the RSRP Meeting held March 13–14.* http://www.nwfsc.noaa.gov/cbd/trt/rsrp_mar01.pdf

Reeves GH, Everest FH, Nickelson TE. 1989. *Identification of physical habitats limiting the production of coho salmon in western Oregon and Washington. Gen. Tech. Rep. PNW-GTR-245,* Pac. Northwest Res. Stn., Portland, OR

Reeves GH, Hall JD, Roelofs TD, Hickman TL, Baker CO. 1991. Rehabilitating and modifying stream habitats. In *Influences of Forest and Rangeland Management on Salmonid Fishes and Their Habitats,* ed. WR Meehan, pp. 519–57. Bethesda, MD: Am. Fish. Soc.

Regetz J. 2002. Landscape-level constraints on recruitment of chinook salmon (*Oncorhynchus tshawytscha*) in the Columbia River basin, USA. *Aquat. Conserv.* In press

Reid WV. 2000. Ecosystem data to guide hard choices. *Issues Sci. Tech.* http://www.nap.edu/issues/

Reimers PE. 1973. *The length of residence of juvenile fall chinook salmon in Sixes River,* Oregon. Oregon Fish Comm. Res. Rep. 4(2). Portland: Oregon Dep. Fish Wildl. 43 pp.

Reisenbichler RR, Rubin SP. 1999. Genetic changes from artificial propagation of Pacific salmon affect the productivity and viability of supplemented populations. *ICES J. Mar. Sci.* 56:459–66

Ricker WE. 1995. Trends in the average size of Pacific salmon in Canadian catches. *Can. Spec. Publ. Fish. Aquat. Sci.* 121:593–602

Rieman BE, Beamesderfer RC, Vigg S, Poe TP. 1991. Estimated loss of juvenile salmonids to predation by Northern Squawfish, Walleyes, and Smallmouth Bass in John Day Reservoir, Columbia River. *Trans. Am. Fish. Soc.* 120:448–58

Rohlf DJ. 1989. *The Endangered Species Act: A Guide to Its Protections and Implementation.* Stanford, CA: Stanford Environ. Law Soc.

Ruckelshaus MH, McElhany P, Ford MJ. 2002. Recovering species of conservation concern: Are populations expendable? In *The Importance of Species: Setting Conservation Priorities,* pp. 305–29, ed. PM Kareiva, S Levin. Princeton, NJ: Princeton Univ. Press. In press

Ruggerone GT. 1986. Consumption of migrating juvenile salmonids by gulls foraging below a Columbia River dam. *Trans. Am. Fish. Soc.* 115:736–42

Ryder OA. 1986. Species conservation and systematics: the dilemma of subspecies. *Trends Ecol. Evol.* 1:9–10

Sala OE, Chapin FS III, Armesto JJ, Berlow E, Bloomfield J, et al. 2000. Global biodiversity scenarios for the year 2100. *Science* 287:1770–74

Schaller HA, Petrosky CE, Langness OP. 1999. Contrasting patterns of productivity and survival rates for stream-type chinook salmon (*Oncorhynchus tshawytscha*) populations of the Snake and Columbia Rivers. *Can. J. Fish. Aquat. Sci.* 56:1031–45

Schluter D. 1998. Ecological causes of speciation. In *Endless Forms: Species and Speciation,* ed. DJ Howard, SH Berlocher, pp. 114–29. Oxford, UK: Oxford Univ. Press

Seattle Times. 2002. Leery of grizzly tests,

legislator demands review of all endangered-species studies. *Seattle Times*, Jan. 10

Simberloff DS. 2000. Nonindigenous species: a global threat to biodiversity and stability. In *Nature and Human Society: The Quest for a Sustainable World*, ed. P Raven, T Williams, pp. 325–34. Washington, DC: Natl. Acad. Press

Smith C. 1979. *Salmon Fishers of the Columbia*. Corvallis: Oregon State Univ. Press

Smith CJ, Castle P. 1994. *Puget Sound chinook salmon* (Oncorhynchus tshawytscha) *escapement estimates and methods—1991. Washington Dep. Fish Wildl. Proj. Rep. Ser. No. 1*. Wash. Dep. Fish. Wildl., Olympia, WA

Smith CT, Nelson RJ, Wood CC, Koop BF. 2001. Glacial biogeography of North American coho salmon (*Oncorhynchus kisutch*). *Mol. Ecol.* 10:2775–85

Stokstad E. 2002. Fur flies over charges of misconduct. *Science* 295:250–51

Storfer A. 1996. Quantitative genetics: a promising approach for the assessment of genetic variation in endangered species. *Trends Ecol. Evol.* 11:343–48

Stouder DJ, Bisson PA, Naiman RJ, eds. 1996. *Pacific Salmon and Their Ecosystems: Status and Future Options*. New York: Chapman & Hall

Stouder DJ, Fresh KL, Feller RJ, eds. 1994. *Theory and Application in Fish Feeding Ecology*. Columbia: Univ. SC Press

Strassel KA. 2002. The missing lynx. *The Wall Street J.*, Jan. 24

Tallman RF, Healey MC. 1991. Phenotypic differentiation in seasonal ecotypes of chum salmon, *Oncorhynchus keta. Can. J. Fish. Aquat. Sci.* 48:661–71

Tallman RF, Healey MC. 1994. Homing, straying, and gene flow among seasonally separated populations of chum salmon. *Can. J. Fish. Aquat. Sci.* 51:577–88

Taylor EB. 1991. A review of local adaptation in Salmonidae, with particular reference to Atlantic and Pacific salmon. *Aquaculture* 98:185–207

Tear TH, Scott JM, Hayward PH, Griffith B. 1993. Status and prospects for success of the Endangered Species Act: A look at recovery plans. *Science* 262:976–97

Tear TH, Scott JM, Hayward PH, Griffith B. 1995. Recovery plans and the Endangered Species Act: Are criticisms supported by data? *Conserv. Biol.* 9:182–95

Teel DJ, Milner GB, Winans GA, Grant WS. 2000. Genetic population structure and origin of life history types in Chinook salmon in British Columbia, Canada. *Trans. Am. Fish. Soc.* 129:194–209

US v. Washington. The Boldt Decision. 384 F. Suppl. 312; 1974 U.S. *Dist. LEXIS* 12291

Utter F, Milner G, Ståhl G, Teel D. 1989. Genetic population structure of Chinook salmon, *Oncorhynchus tshawytscha*, in the Pacific Northwest. *Fish. Bull.* 87:239–64

Verhovek SH. 2002. "Saving" wild salmon's bucket-born cousins. *New York Times*, Feb. 4

Vitousek PM. 1994. Beyond global warming: ecology and global change. *Ecology* 75:1861–76

Waples RS. 1991. Pacific salmon (*Oncorhynchus* spp.) and the definition of "species" under the Endangered Species Act. *Mar. Fish. Rev.* 53:11–22

Waples RS. 1995. Evolutionarily significant units and the conservation of biological diversity under the Endangered Species Act. See Nielsen 1995, pp. 8–27

Waples RS. 1998. Separating the wheat from the chaff: patterns of genetic differentiation in high gene flow species. *J. Hered.* 89:438–50

Waples RS. 1999. Dispelling some myths about hatcheries. *Fisheries* 24:12–21

Waples RW, Ford MJ, Schmitt D. 2002. Empirical results of salmon supplementation: a preliminary assessment. In *Ecological and Genetic Implications of Aquaculture Activities*, ed. T Bert. The Netherlands: Kluwer Academic. In press

Waples RW, Gustafson RG, Weitkamp LA, Myers JM, Johnson OW, Busby PJ, et al. 2001. Characterizing diversity in salmon from the Pacific Northwest. *J. Fish Biol.* 59(Suppl.A):1–41

Wash. Dep. Fish Wildl. Point-No-Point Treaty

tribes (WDFW & PNPTT). 2000. *Summer Chum Salmon Conservation Initiative. An implementation plan to recover summer chum in the Hood Canal and Strait of Juan de Fuca region.* Olympia, WA: WDFW. 424 pp.

Wash. Dep. Fish Wildl. West. Wash. Treaty Indian Tribes (WDFW & WWTIT). 1993. *1992 Washington State Salmon and Steelhead Stock Inventory.* West. Wash. Treaty Indian Tribes. Olympia, WA: WDFW

Weber B. 2002. Wildlife society clarifies its role in lynx studies. *The Wall Street J.*, Feb. 12

Weitkamp LA, Wainwright TC, Bryant GJ, Milner GB, Teel DJ, et al. 1995. *Status Review of Coho salmon from Washington, Oregon, and California. NOAA Tech. Memo. NMFS-NWFSC-24.* Northwest Fish. Sci. Cent., Seattle, WA

Welch DW, Ward BR, Smith BD, Eveson JP. 2000. Temporal and spatial responses of British Columbia steelhead (*Oncorhynchus mykiss*) populations to ocean climate shifts. *Fish. Oceanogr.* 9:17–32

Williams JG, Smith SG, Muir WD. 2001. Survival estimates for downstream migrant yearling juvenile salmonids through the Snake and Columbia River hydropower system 1966–1980 and 1993–1999. *N. Am. J. Fish. Manag.* 21:310–17

Williams JG, Tuttle ME. 1992. The Columbia River: fish habitat restoration following hydroelectric dam construction. In *Restoring the Nation's Marine Environment*, ed. GW Thayer, pp. 405–22. College Park: Maryland Sea Grant

Willson MF, Gende SM, Marston BH. 1998. Fishes and the forest. *BioScience* 48:455–62

Willson MF, Halupka KC. 1995. Anadromous fish as keystone species in vertebrate communities. *Conserv. Biol.* 9:489–97

Wipfli MS, Hudson JP, Chaloner DT, Caouette JR. 1999. Influence of salmon spawner densities on stream productivity in southeast Alaska. *Can. J. Fish. Aquat. Sci.* 56(9):1600–11

Wondolleck JM, Yaffee SL. 2000. *Making Collaboration Work. Lessons from Natural Resource Management.* Washington, DC: Island

Wood CC. 1987. Predation of juvenile Pacific salmon by the common merganser (*Mergus merganser*) on Eastern Vancouver Island. I: Predation during seaward migration. *Can. J. Fish. Aquat. Sci.* 44:941–49

Wood CC. 1995. Life history variation and population structure in sockeye salmon. See Nielsen 1995, pp. 195–216

Wood CC, Riddell BE, Rutherford DT, Withler RE. 1994. Biochemical genetic survey of sockeye salmon (*Oncorhynchus nerka*) in Canada. *Can J. Fish. Aquat. Sci.* 51:114–31

Zabel RW, Williams JG. 2002. Selective mortality in chinook salmon: what is the role of human disturbance? *Ecol. Appl.* 21(1):173–83

Zinn JL, Johnson KA, Sander JE, Fryer JL. 1977. Susceptibility of salmonid species and hatchery stocks of chinook salmon (*Oncorhynchus tshawytscha*) to infections by *Ceratomyxa shasta. J. Fish. Res. Board Can.* 34:933–36

Annu. Rev. Ecol. Syst. 2002. 33:707–40
doi: 10.1146/annurev.ecolsys.33.010802.150500
Copyright © 2002 by Annual Reviews. All rights reserved

Estimating Divergence Times from Molecular Data on Phylogenetic and Population Genetic Timescales

Brian S. Arbogast
*Department of Biological Sciences, Humboldt State University, Arcata,
California 95521; email: bsa2@humboldt.edu*

Scott V. Edwards
*Department of Zoology, University of Washington, Seattle, Washington 98195;
email: sedwards@u.washington.edu*

John Wakeley
*Department of Organismic and Evolutionary Biology, Harvard University, Cambridge,
Massachusetts 02138; email: jwakeley@oeb.harvard.edu*

Peter Beerli
*Department of Genome Sciences, University of Washington, Seattle, Washington 98195;
email: beerli@gs.washington.edu*

Joseph B. Slowinski[1]
California Academy of Sciences, San Francisco, California

Key Words ancestral polymorphism, coalescence theory, maximum likelihood, molecular clock, sequence saturation

■ **Abstract** Molecular clocks have profoundly influenced modern views on the timing of important events in evolutionary history. We review recent advances in estimating divergence times from molecular data, emphasizing the continuum between processes at the phylogenetic and population genetic scales. On the phylogenetic scale, we address the complexities of DNA sequence evolution as they relate to estimating divergences, focusing on models of nucleotide substitution and problems associated with among-site and among-lineage rate variation. On the population genetic scale, we review advances in the incorporation of ancestral population processes into the estimation of divergence times between recently separated species. Throughout the review we emphasize new statistical methods and the importance of model testing during the process of divergence time estimation.

[1]In memoriam: Joseph Slowinksi passed away in September 2001 as the result of a snake bite he received while conducting fieldwork in Myanmar. He will be dearly missed by his friends and colleagues.

0066-4162/02/1215-0707$14.00

707

INTRODUCTION

The molecular clock hypothesis, first advanced in the 1960s (Zuckerkandl & Pauling 1965), remains one of the most influential concepts in modern evolutionary biology. This hypothesis proposes that genes and gene products evolve at rates that are roughly constant over time and across evolutionary lineages. The implications of this hypothesis are powerful; if genetic divergence accumulates in a relatively clocklike fashion, then time scales can be developed for important evolutionary events even in the absence of fossil evidence. This realization, along with dramatic technical advances in molecular and computational biology over the past two decades, has revolutionized the way researchers address temporal questions in evolutionary biology. Along with traditional (nonmolecular) methods, molecular genetic approaches are now a key part of the toolkit of researchers interested in reconstructing historical patterns of organismal diversification through space and time. Molecular clocks have profoundly influenced modern views on the timing of many important events in evolutionary history, including those related to human evolution and migration (Cann et al. 1987, Underhill et al. 2000, Ke et al. 2001), Pleistocene speciation (Bermingham et al. 1992, Klicka & Zink 1997), and historical radiations of major groups of plants and animals (Doolittle et al. 1996; Hedges et al. 1996, 2001; Wang et al. 1999). In turn, the ability to provide dates for diversification events permits estimates of absolute rates of adaptive radiation, ecological diversification, and a host of other exciting evolutionary topics (Givnish & Sytsma 1997, Schluter 2000).

Despite the impact of molecular clocks on evolutionary biology, there are a number of controversies surrounding their use (Swofford et al. 1996). One of the most fundamental debates concerns the degree to which rates across lineages, genes, and genomic regions are heterogeneous; such heterogeneity will almost always confound attempts to accurately estimate evolutionary dates of divergence. Early on, several "universal" molecular clocks (clocks that could be applied across a broad spectrum of taxa) were proposed. These included universal clocks for such clades as bacteria (Ochman & Wilson 1987) and for silent sites across the genome as a whole (Wilson et al. 1987). However, by far the most prominent of universal clocks has been the "mtDNA clock" (Brown et al. 1979, 1982), which holds that animal mtDNA evolves at a rate of $\sim2\%$ sequence divergence per million years. Throughout the 1980s the validity of this clock was widely accepted. However, as comparative molecular data have accumulated over the past two decades, it has become clear that there is much more variation in the rate of mtDNA evolution across taxonomic groups (Vawter & Brown 1986) than originally thought. Investigation of other parts of the genome (i.e., nuclear genes) has also revealed considerable variation among lineages in the rate of molecular evolution. As a result, the idea of universal molecular clocks that can be applied across a broad range of taxa has been replaced by the notion of taxonomically "local" clocks that are useful primarily within the bounds of particular genes and closely related taxa (Swofford et al. 1996, Yoder & Yang 2000). The rationale behind local molecular clocks is based on the premise that differences in population size, metabolic rate, generation time, and DNA repair efficiency are among the most likely sources of among-lineage rate heterogeneity (Martin & Palumbi 1993, Rand 1994), and because these parameters are likely to be similar in closely related species, such clades should experience similar rates of molecular evolution. Indeed, it has been proposed that the gradual divergence of these factors may be responsible for the gradual divergence of evolutionary rates among evolutionary lineages (Thorne et al. 1998). A related concept, which could be called a genomically local clock, is based on the idea that differences in number of meiotic replications and DNA repair efficiency in different regions of the genome can result in markedly different rates of molecular evolution for different chromosomes, gene families, or genomic subcompartments (Hurst & Ellegren 1998, Ellegren 2000).

The difference in rate of molecular evolution among lineages is only one of the potential problems faced by the evolutionary biologist interested in using molecular clocks to date divergence events. All molecular clocks must be calibrated using independent evidence, such as dates of speciation events inferred from the fossil record or dates estimated for particular biogeographic events. In each case, these dates are best estimates based on the available data and can be subject to different interpretations. Thus, the calibration point(s) used to establish a given molecular clock may be a source of considerable error (Hillis et al. 1996). This problem is compounded by the fact that often only a single calibration point is used. In addition, the very nature of DNA substitution, which is most often viewed as a Poisson process, makes it difficult to estimate dates of divergence with the degree of precision required to adequately address many temporal questions. For example, Hillis et al. (1996) showed that even under idealized conditions, the 95% confidence limits for dates estimated via a molecular clock are quite large, and for natural populations we would expect them to be much larger. Thus, inherent in the exercise of estimating dates of divergence from molecular data are a variety of potential pitfalls. Still, with the advent of sophisticated methods for handling complex models of nucleotide substitution, among-lineage rate heterogeneity, and population genetic processes, molecular clocks are likely to continue to be important tools in evolutionary biology.

A number of recent developments make this review timely. First, new and more complex models of nucleotide substitution now dominate phylogenetic analyses, and likelihood and Bayesian methods have emerged as powerful tools that significantly broaden our ability to estimate divergence times from molecular data. Second, we hope to re-emphasize the tight link between systematics and population genetics, two fields that traditionally are treated as separate (Felsenstein 1988). For example, although saturation is generally viewed as a phenomenon affecting only ancient divergences, the complex models of substitution now available show that it can compromise estimation even when genetic divergence between lineages is relatively small. Likewise, recent studies have shown that ancestral population processes, which are generally thought to affect estimates of divergence times only for very recently separated species, can impact estimates of divergence time even for species that diverged several million years ago, depending on the size and structure of the ancestral species. Thus, in many cases researchers will need to address both phylogenetic and population genetics issues when estimating dates of divergence. Our review is therefore structured such that it moves from the phylogenetic to the population genetic time scale. We first discuss the complexities of DNA sequence evolution as they relate to estimating ancient divergences and outline recent methods to test for a molecular clock. We then move on to divergence time estimation between recently separated species, emphasizing the incorporation of ancestral population processes, and conclude with a look at problems for the future. Throughout the review we emphasize new statistical methods and the importance of model testing during the process of divergence time estimation.

ANCIENT DIVERGENCES

The Problem of Sequence Saturation

One of the major challenges to estimating rates of molecular evolution and dates of divergence is obtaining reliable estimates of the actual number of substitutions that have occurred in each gene lineage since they diverged from a common ancestor. Because the actual number of substitutions includes both the observed number plus those now masked because of saturation (multiple substitutions at single sites), this requires the use of an appropriate model of nucleotide evolution. If no such model is used, then the estimated rate of molecular evolution is based solely on the observed number of substitutions. That is, the rate of molecular evolution is equated with the observed proportional amount of sequence divergence (d) between two sequences divided by the amount of time (t) since they diverged. This is problematic because d/t decreases over time by virtue of saturation at the sequence level. Whereas t can increase linearly to infinity, the number of observed substitutions plateaus as more and more substitutions become superimposed over previous ones. The curvilinear relationship between d and t has long been recognized (Brown et al. 1979, 1982). It has traditionally been viewed as a problem only when relatively ancient divergence events are the subject of investigation, but this is not necessarily the case. When variation in the rate of substitution among nucleotide sites is high, this can have a pronounced effect on estimates of substitution rates and dates of divergence, even when the divergences in question are relatively recent.

MODELS OF NUCLEOTIDE SUBSTITUTION   A wide variety of models have been developed to describe the process of DNA nucleotide substitution. These models differ in the number and types of parameters that are free to vary. Common model parameters include the number of substitution types or classes, the frequencies of the four nucleotide bases, and variation in the rate of substitution among nucleotide sites. The most general model of nucleotide substitution usually considered is known as the general time-reversible (GTR) model (Rodriguez et al. 1990). Most other models of nucleotide substitutions are simply special cases of the GTR model, wherein fewer parameters are free to vary (Swofford et al. 1996, Posada & Crandall 1998). For each given data set there thus exists a series of nested models that one can choose from when estimating branch lengths of phylogenetic trees or values of sequence divergence between pairs of taxa. Some of these models are likely to fit a given DNA data set well, whereas others will fit the data set poorly. The likelihood ratio test (LRT) is a widely accepted method for testing the goodness of fit of competing nested models to empirical DNA data sets. The test statistic for the LRT is $-2\log \Lambda$, where

$$\Lambda = \max [L_0(\textit{Null Model} \mid \textit{Data})] / \max [L_1(\textit{Alternative Model} \mid \textit{Data})],$$

L_0 is the likelihood under the null hypothesis (the simpler of the two models being compared), and L_1 is the likelihood under the alternative model (which is the more complex, or parameter-rich model). When the models being compared are nested (which is the case if they are being evaluated in relation to the same phylogenetic tree), the test statistic will be asymptotically distributed as a χ^2 with the degrees of freedom equal to the difference in the number of free parameters between the two models. The computer program MODELTEST (Posada & Crandall 1998), which works in conjunction with the computer program PAUP* (Swofford 1998), makes comparing the fit of competing models with LRTs relatively straightforward and easy. Although the use of LRTs is well represented in the phylogenetic literature (Posada & Crandall 2001), various other approaches to model selection are also available. For example, both the Akaike information criterion (Akaike 1974) (AIC) and the Bayesian information criterion (Schwartz 1978) (BIC) can be used to evaluate competing models of molecular evolution (Morozov et al. 2000). The AIC, which does not require that models be nested, penalizes an increase in the number of parameters if the addition of each new parameter does not increase the likelihood by at least one unit of log-likelihood. The BIC provides an approximate solution to the Bayes factor, and, like the AIC, can be used on either nested or nonnested models [For detailed comparisons of these three methods, see Morozov et al. (2000) and Posada & Crandall (2001).] The null distribution for the test statistic can also be generated via a parametric bootstrapping approach, such as Monte Carlo simulation (Goldman 1993, Huelsenbeck & Rannala 1997).

   Although the methods outlined above are designed to provide an objective criterion for choosing among competing models, several concerns have been raised regarding model selection. For example, Posada & Crandall (2001) found that in some cases the model selection process could be influenced by both the order in which models are tested and the complexity of the initial model in the sequence of LRTs. A second concern is the adequacy of existing of molecular evolution. Because sequence evolution is so complex, even the most complex models may do a poor job of capturing this process (Sanderson & Kim 2000). Just because a model provides a significantly better fit to the data than do other competing models, this does not mean that the model provides an accurate description of the substitution process underlying the studied sequences (it may simply be the best model of the relatively small subset of all possible models being evaluated).

Furthermore, although models more complex than the GTR model are probably always in operation, in some cases we may not have enough sequence data to reject simpler models (Sanderson & Kim 2000). However, Sullivan & Swofford (2001) argued that "perfect models are not necessarily a prerequisite for reliable statistical inference," and that though it is important to incorporate certain features of molecular evolution into models, these features need not be modeled perfectly in order for approaches such as maximum likelihood to be robust.

Although a large number of models of nucleotide evolution have been developed over the past several decades, traditionally these models have not included a parameter corresponding to variation in the rate of substitution among different nucleotide sites or "among-site rate variation." Recently, however, this parameter has become the subject of considerable interest in phylogenetic inference (Yang 1993, 1994; Gaut & Lewis 1995; Sullivan et al. 1996) and in the estimation of branch lengths, evolutionary rates, and divergence times (Arbogast & Slowinski 1998, Buckley et al. 2001). One way that among-site rate variation has been incorporated into existing models of nucleotide substitution is through the use of a gamma distribution (Uzzell & Corbin 1971, Yang 1994). In this context, the gamma distribution has a shape parameter, α, that is inversely proportional to the amount of among-site rate heterogeneity present in the data (the equal-rates condition, i.e., no among-site rate variation, is a special case of this gamma distribution wherein $\alpha = $ infinity). Empirical estimates of α suggest that among-site rate variation can be substantial, i.e., $\alpha < 1$ (Arbogast & Slowinski 1998, Baldwin & Sanderson 1998). In practical terms, a low value of α means that for a given set of sequence data, a relatively small proportion of the sites are experiencing the overwhelming majority of substitutions; see Swofford et al. (1996) for a graphical depiction of different values of α.

AMONG-SITE RATE VARIATION AND ESTIMATES OF DIVERGENCE TIMES   What are the implications of high-levels of among-site rate variation in DNA sequences with regard to estimating substitution rates and dating evolutionary events? Recent studies have shown that failure to address among-site rate variation may lead to substantial underestimates of branch lengths and the rate of substitution (Yang et al. 1994, Arbogast & Slowinski 1998, Slowinski & Arbogast 1999, Buckley et al. 2001). This underestimation is due to the fact that when most nucleotide substitutions are occurring at relatively few nucleotide sites, rather than being more evenly distributed across all sites, the number of unobserved (superimposed) substitutions will be underestimated. As a result, the true number of substitutions that have occurred since the divergence of two lineages will be underestimated. What is not widely appreciated is that this phenomenom can lead to large underestimates of branch lengths even when the observed amount of divergence between two sequences is small. For example, if two sequences differ at 10% of the nucleotide sites being compared, and if all sites have an equal probability of undergoing a subsequent substitution (i.e., if α really equals infinity), then the probability of the next substitution masking a previous substitution is relatively small ($\sim$0.1). If, however, substitution rates are highly skewed (i.e., if α is small) such that only 10% of the nucleotide sites are responsible for most of the substitutions, then the probability of the next substitution being superimposed over a previous substitution is close to 1! Thus, while it is true that saturation will become increasingly more common as time of divergence increases, this does not mean that saturation cannot be substantial even when time of divergence is small. As such, researchers should be especially concerned about among-site rate variation when attempting to estimate evolutionary dates, regardless of the time frame under consideration.

To illustrate the influence that among-site rate variation can have on estimates of rates of substitution and dates of divergence, it is useful to revisit the mtDNA clock. Originally based on the divergence of chimpanzees and humans, dated at ~5 million years (My) before present (B.P.), this clock proposes that mtDNA diverges at a constant rate of ~2% per My (equivalent to a substitution rate of 0.01 nucleotide substitutions per site per lineage per My) (Brown et al. 1979, 1982). However, the relatively simple models (i.e., Jukes & Cantor 1969) used to estimate sequence divergence between chimpanzees and humans in the derivation of the 2% per My mtDNA clock do not address the high levels of among-site rate variation typically found in mtDNA sequences. What affect does this have on estimating the rate of mtDNA evolution and for estimating dates of divergence? Arbogast & Slowinski (1998) investigated this question by using maximum likelihood to infer a phylogeny of the great apes based on complete mtDNA cytochrome *b* sequences and then used LRTs to evaluate which of six increasingly complex models of molecular evolution provided the best fit to the sequence data. The LRTs revealed two important trends: (*a*) Incorporating unequal base frequencies and increasing the number of categories into which substitution rates are partitioned significantly improved the fit of the models to the data, but only to a point; and (*b*) when a gamma distribution was added to a model to address among-site rate heterogeneity, the fit of the model was improved in every case. Overall, the gamma-HKY85 model (Hasegawa et al. 1985) provided the best fit to the cytochrome *b* data with the fewest parameters. For each gamma model, the estimated value of α was quite low, indicating a high level of among-site rate heterogeneity in the sequences. The poorer fits of the nongamma models suggest that such heterogeneity cannot be addressed simply by increasing the number of rate categories into which the data are partitioned. Based on the best-fit gamma-HKY85 model and the same calibration point used in the 2% per My mtDNA clock (5 My B.P.), the estimated rate of substitution for the cytochrome *b* gene was 0.0259 nucleotide substitutions per site per lineage per My (Figure 1, color insert). This estimate is ~2.6 times greater than the substitution rate proposed by the 2% per My mtDNA clock. A reanalysis of the rate of mtDNA substitution in birds produced a similar estimate (Arbogast & Slowinski 1998).

Because it underestimates the actual rate of substitution, use of the 2% per My mtDNA clock will, even in the ideal case where there is no variation in the rate of molecular evolution among lineages, produce reasonable estimates for dates of divergence only around the calibration point (Figure 1). As actual dates of divergence become more and more recent, use of this clock will produce increasingly large overestimates of actual dates; similarly, as actual dates of divergence become more and more ancient, use of this clock will lead to increasingly large underestimates of the actual dates. This bias toward the calibration point is illustrated in Figure 1. In all cases, the rate at which the observable number of substitutions accumulates will decrease over time owing to saturation (red curves). In the original calibration of the 2% per My mtDNA clock, saturation was considered to have little impact prior to the calibration point of 5 My B.P. (Figure 1A). In contrast, the gamma-HKY85 model (Figure 1B; green line) suggests that the true rate of substitution in the mtDNA sequences is much higher but also that saturation begins much sooner. Superimposing the line representing the 2% per My clock (Figure 1A; blue line) over the curve representing the manner in which observed sequence divergence is predicted to accumulate under the gamma-HKY85 model (Figure 1B; red line) reveals that use of undercorrected distances (or branch lengths) will lead to estimated dates of divergence that are biased toward the calibration point (Figure 1C). This type of phenomenon is expected to occur anytime that measures of genetic divergence are undercorrected for saturation.

The above example illustrates a major pitfall that can arise when estimating rates of molecular evolution and dates of divergence. Because traditional models of molecular evolution fail to capture the highly skewed manner in which substitution rates are often distributed among different nucleotide sites, even within particular rate classes, they may often provide poor fits to DNA sequence data. As a result, the effects of saturation will be underestimated, and use of such models is likely to lead to poor estimates of substitution parameters, evolutionary rates, and evolutionary dates. By extension, the use of inadequate models also can lead to spurious correlations between the rate of molecular evolution and life history parameters (Slowinski & Arbogast 1999). Addressing these problems requires that all of the genetic divergences or branch lengths involved in a given analysis (i.e., those upon which the rate of substitution, or "clock," is based and those for which dates are being estimated) be adequately corrected for the effects of saturation. Only then can one hope to remove the potentially confounding effects of saturation on estimates of rates of substitution and dates of divergence.

TESTING FOR A MOLECULAR CLOCK

Like saturation, among-lineage rate heterogeneity will tend to be more problematic when dealing with ancient as opposed to recent divergences. However, regardless of the age of the divergence, there is no a priori way to know whether all of the lineages under consideration have similar rates of molecular evolution. Therefore, some methodology is required to test whether this is the case. The maximum likelihood framework and LRT method described above is one approach that can be used to evaluate whether the sequences of the taxa under consideration are evolving according to a molecular clock (i.e., to test if there are significant rate differences among the lineages). This consists of conducting an LRT (under the model determined to provide the best-fit to the data) with and without a molecular clock enforced (Huelsenbeck & Rannala 1997). The degrees of freedom for the LRT of the molecular clock are equal to the number of taxa minus two. If there is no significant difference between the model with and without the molecular clock enforced, then a molecular clock cannot be rejected. As a result, it would be reasonable to conclude that rates of molecular evolution among the taxa in question do not differ significantly from one another. In some cases, it may also be useful to conduct relative rate tests (Figure 2a) to test for rate differences among lineages (see Li 1997). However, these tests are typically conducted in a pairwise fashion, and as such they would seem less desirable than the tree-based approach made possible by the use of LRTs. In the following section we discuss methods of estimating divergence times when a molecular clock is rejected (i.e., when the lineages of interest exhibit disparate rates of molecular evolution).

Estimating Divergence Times in the Presence of Among-Lineage Rate Heterogeneity

Attempts to estimate divergence times are obviously simpler when the taxa in question share a similar rate of molecular evolution. However, in the real world researchers may often be faced with rate variation among lineages. A number of promising methods have emerged in recent years to deal with this problem.

METHODS THAT REMOVE NONCLOCK-LIKE SUBSETS OF THE DATA   Several recent methods for estimating divergence times do so only after the nonclock-like subsets of the data are removed. These include the linearized tree method (Takezaki et al. 1995) and the quartet method (Cooper & Penny 1997). The linearized tree method, or "two cluster" method, first tests for deviations from a molecular clock of various lineages in a phylogenetic tree. It does so by calculating a statistic δ, which is the difference in average branch length of all lineages on one of two sides of a node in a tree compared with all lineages on the other side of that node. The variance on δ is used to estimate its statistical significance via a Z-score. This method has been used in a variety of contexts, including avian biogeography (Voelker 1999), molecular evolution (Edwards et al. 1997), and mammalian (Takezaki et al. 1995, Hedges et al. 1996) and metazoan (Wang et al. 1999) diversification to determine which lineages deviate from a molecular clock. Takezaki et al. (1995) then propose to eliminate those lineages that are evolving at rates significantly higher or lower than the average rate across the tree to thereby produce a "linearized" tree that only includes taxa with approximately equal rates of molecular evolution. Dates of divergence of these taxa can then be estimated using ultrametric methods of tree-building, such as UPGMA (Takezaki et al. 1995).

Another method that involves elimination of lineages not conforming to a local clock of the taxa in question is the quartet method introduced by Cooper & Penny (1997). This method first identifies pairs of taxa that have good fossil data with which to calibrate absolute rates of molecular evolution between the pair. These

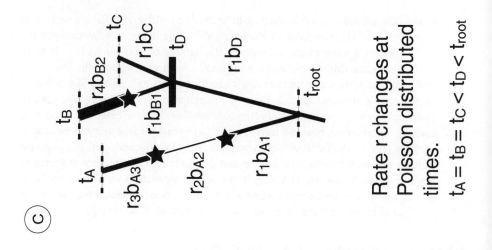

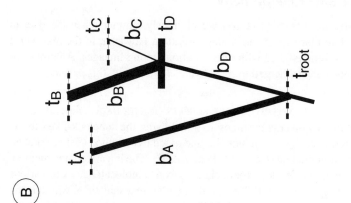

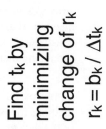

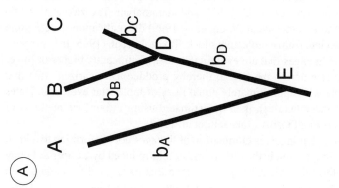

(A)

A   B   C

b_B b_C

b_A b_D D

E

$H_0: d_{AB} - d_{AC} = 0$

$d_{AB} = b_A + b_D + b_B$

$d_{AC} = b_A + b_D + b_C$

(B)

t_C t_B

b_C

b_B t_D

t_A

b_D

b_A

t_{root}

Find t_k by
minimizing
change of r_k

$r_k = b_k / \Delta t_k$

(C)

t_C t_B

$r_1 b_C$ $r_4 b_{B2}$

t_D

$r_1 b_{B1}$

$r_1 b_D$ $r_3 b_{A3}$ t_A

$r_2 b_{A2}$

$r_1 b_{A1}$

t_{root}

Rate r changes at
Poisson distributed
times.

$t_A = t_B = t_C < t_D < t_{root}$

pairs can in turn be assembled into quartets consisting of two pairs of taxa, each of which has a known fossil date of divergence. As long as the pairs are independent and do not subsume any common branches in the tree, their phylogeny is essentially known. The average rate between these pairs is then used to estimate the date of the common ancestor of these two pairs. The variance of the estimate can be reduced by combining information from different quartets that derive from the same common ancestral node (Steel et al. 1996). The method is conservative and has been used to argue that the divergence times of birds occurred earlier in the Cretaceous than fossil data would suggest (Cooper & Penny 1997). The quartet method has been extended to include a LRT of rate equality, which is used again to identify lineages that do not conform to a molecular clock (Rambaut & Bromham 1998). These lineages are then removed prior to estimating divergence times of splits encompassed by the remaining taxa. This method was shown to be reasonably robust to the model of sequence evolution assumed as well as length of sequence employed. Bromham et al. (1998) used this likelihood quartet method and metazoan quartets with known fossil calibrations to argue that the divergence of the protostome-deuterostome split as well as the vertebrate-echinoderm split occurred no earlier than 680 Mya, well before the base of the Cambrian. This date conformed closely to other recent studies (Ayala 1997). Because the different estimates of these two divergences were derived from nonindependent pairs of taxa, however, it was not possible to develop an estimate based on all the available data. In addition, both the linearized tree method and quartet methods suffer from throwing out taxa that could provide valuable information on the mode and pattern of rate heterogeneity in the clades under study.

METHODS THAT ESTIMATE MODEL PARAMETERS FOR ALL TAXA   Two methods developed by Sanderson (1997, 2002) overcome the problem of throwing out nonclock-like data. These methods include nonparametric rate smoothing (NPRS) and penalized likelihood. They are distinct from the previous two methods because, rather than throwing out nonclock-like data, they estimate local rates, i.e., for specific branches or clades (Figure 2). Such estimation is possible because the methods place a constraint (albeit a broad one) on the ways in which the rate of

←————————————————————————————————

Figure 2   Estimation of divergence times on phylogenetic trees. (*A*) Relative rate test; if the null hypothesis is rejected, a molecular clock cannot be assumed, and divergence times should not be estimated. (*B*) Nonparametric estimation of variable rates; each branch with branch length b_k has its own rate r_k that depends on rates on neighboring branches. The unknown divergence times t_k can be estimated even when branches have very different rates (t_{root} must be known). (*C*) Compound-Poisson method; rate changes are governed by two parameters: rate change, and rate change frequency per branch length. Dashed lines indicate divergence times with known dates and the thick lines indicate unknown, but estimable, divergence times.

molecular evolution can vary among lineages. In the case of the NPRS method, the constraint is the temporal autocorrelation of the rate of molecular evolution in related lineages throughout the tree (Figure 2*B*). The notion of autocorrelated rates of molecular evolution has previously been applied to the question of how well the neutral theory can accommodate the existence of variation in the rate of molecular evolution (Takahata 1987, Gillespie 1991). However, prior to the NPRS method such models had not been used to actually estimate absolute rates or dates of divergence. The NPRS method's basic goal is to estimate a local rate of evolution for each node in the tree (r_k), and then to minimize the difference in these rates across the tree. The optimal level of smoothing is determined by a numerical least-squares method in which drastic changes in rate along the tree are penalized. The method was shown to work well in simulations and in comparison to older methods of divergence time estimation. Richardson et al. (2001) used the NPRS method to document a very recent radiation for a species-rich genus characteristic of the Cape flora of South Africa. However, a second method also by Sanderson, the penalized likelihood method, outperformed NPRS in all cases tested. Like NPRS, penalized likelihood attempts to determine an optimal level of smoothing (autocorrelation) for a given data set on a tree. However, penalized likelihood finds the optimal value for the smoothing parameter using a "roughness," which increases as rate variation across the tree increases. Sanderson found that the estimated age of plant clades depended strongly on the smoothing parameter in some cases (e.g., Gnetales), but less so in others (e.g., Angiosperms).

Other recent methods for determining optimal but varying rates along a tree assign rates to parts of a tree according to prior distributions. Two recent methods assign rates according to lognormal (Thorne et al. 1998) and compound Poisson (Huelsenbeck et al. 2000) distributions (Figure 2*C*). These models are versatile in their ability to test for variation in a variety of parameters of molecular evolution (base composition, transition/transversion ratio) in addition to the rate of molecular evolution.

BAYESIAN METHODS   Recently, researchers have begun to use a Bayesian approach to infer phylogenies (Huelsenbeck et al. 2001). Under this approach, phylogenetic inference is based on the posterior probabilities of phylogenetic trees, which consist of the joint probabilities of the tree, branch lengths, and model of nucleotide evolution. The same models of nucleotide evolution used in maximum likelihood analyses can be used in Bayesian inference, and prior information, if available, can be incorporated into the analysis. Bayesian approaches have been used recently to test for a molecular clock (Suchard et al. 2001) and to estimate divergence times (under both strict and relaxed molecular clocks) (Thorne et al. 1998). Because Bayesian approaches are in their infancy with regard to phylogenetic inference, they require additional testing. However, they appear to be quite promising and are likely to play an increasingly important role in many facets of phylogenetics, including the estimation of divergence times (Huelsenbeck et al. 2001). One potential advantage of a Bayesian approach to phylogenetic inference, and by extension,

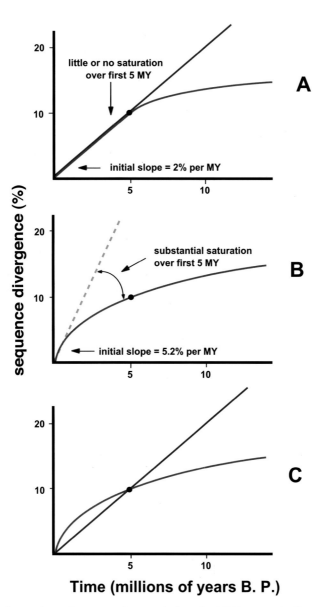

Figure 1 Comparison of patterns of accumulation of mtDNA sequence divergence in primates over time under the 2% per My mtDNA clock (*A*) (Brown et al. 1979, 1982) and the best-fit gamma-HKY85 model (*B*) (4). In all cases, the rate at which the observable (uncorrected) number of substitutions accumulates will decrease over time due to saturation (red curves). However, the two models predict the shape of this curve to be different. This comparison suggests that use of undercorrected distances (or branch lengths) will produce a phenomenon wherein estimated dates of divergence are systematically biased toward the calibration point (*C*), i.e., dates of divergence that are truly more recent than the calibration point will tend to be overestimated and those truly older than the calibration point will tend to be underestimated.

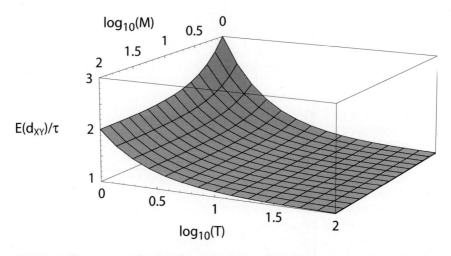

Figure 4 The expected fractional overestimation of the divergence, $\tau = 2ut$, when the uncorrected number of average pairwise differences between species, d_{XY}, is used as an estimator. It is assumed that the ancestral species is subdivided, and $M = 4Nm$. In terms of the parameters of the model, the value of $E(d_{XY})/\tau$ is given by $1 + (1+1/M)/T$, in which T is the time of divergence measured in units of $2N$ generations.

to the estimation of divergence times, is that it appears to make analysis of large data sets more tractable than other methods. For example, maximum likelihood approaches are powerful, but because they are computationally intense they are not well suited for large data sets. Rather than searching for an optimal tree, a Bayesian approach samples trees according to their posterior probabilities, and once such a sample has been created, common features among the trees can be assessed. The sample can then be used to construct a consensus tree with posterior probabilities assigned to each node. The result is similar to a maximum likelihood search with bootstrapping and includes the important parameter of branch length estimates, but the analysis itself is typically much faster. Huelsenbeck (Huelsenbeck & Ronquist 2001) recently developed the computer program MRBAYES for Bayesian inference of phylogenies. This program is versatile in that it allows for a variety of models of molecular evolution to be used and has several methods for incorporating among-site rate variation.

RECENT SPECIES AND POPULATION DIVERGENCES

Many workers adhere to the view that estimating divergence times between species thought to have diverged recently is problem-free compared with estimating divergence times between distantly related species. This mindset arises because sequence saturation, widely thought to be the main stumbling block to accurate estimation of divergence times, is often considered to be small for recent divergences. As discussed in the previous section, however, saturation likely impacts estimation of divergence times at all time scales to varying degrees. Yet, for recent divergences, problems associated with ancestral gene polymorphism are at least, if not more, daunting than those of sequence saturation in terms of estimating divergence times. Most researchers now appreciate that a variety of discordances between the gene tree (gene genealogy) and the species tree (organismal history) can occur through incomplete lineage sorting (Maddison 1997). What is less appreciated is that even in the face of complete genealogical concordance between the gene and species trees, an additional level of discordance still exists—that between the times of gene and population divergence. This distinction has been widely appreciated at least since the late 1970s (Gillespie & Langley 1979, Nei & Li 1979) and was promulgated during the restriction fragment length polymorphism (RFLP) era of phylogeography in the 1980s (Wilson et al. 1985, Avise et al. 1987); however, it has only recently become the focus of sophisticated population genetics models (Takahata et al. 1995, Takahata & Satta 1997, Nielsen 1998, Nielsen et al. 1998, Li et al. 1999).

The discrepancy between times of gene and population divergence arises because prior to species divergence, a degree of gene divergence has already accrued in the ancestral species (Figure 3). This gene divergence is simply the coalescent analogue of any sort of polymorphism at a locus in the ancestral species and has been known since the time of Wright to have an expectation of $2N$ generations when the ancestral species is a randomly mating population (Wright 1951). For

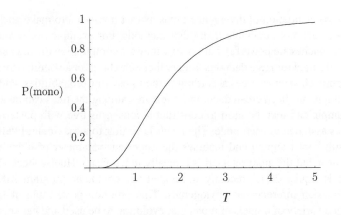

Figure 3 The probability of monophyly of a sample of size $n = 10$ as a function of the divergence time, T, measured in units of $2N$ generations. This is the probability that the sample has reached a single common ancestor, or coalesced, by time T in the past; see Equation 6.1 and 6.2 in Tavaré (Tavaré 1984).

very recently diverged species pairs, this ancestral gene divergence can comprise a substantial fraction of the total gene divergence observed between species, a fraction that will increase dramatically when the ancestral species is structured (Figures 3 & 4, see color insert) (Edwards & Beerli 2000, Wakeley 2001b). Bottlenecks during speciation do not erase the discrepancy because such bottlenecks will only affect coalescence times in the species undergoing founding, not the ancestral species from which they arose. Because the discrepancy between gene and population divergence becomes immeasurably small as the species divergence time increases (Figure 4), it will not impact ancient species divergences much; however, depending on the size of the ancestral population, it can have a substantial impact for the first several millions of years after divergence.

The issues that must be addressed in the estimation recent divergence times thus result from processes that act on two levels: the molecular level and the population or species level. The important molecular-level processes that affect DNA sequence or other genetic data are mutation and recombination. In contrast to the realistic and complex substitution models used in the estimation of ancient divergences, it is typical to use the infinite sites model for recent divergences. This model assumes that the mutation rate per nucleotide site is so small that the possibility of multiple mutations at single sites can be ignored. The infinite sites mutation model is routinely applied to DNA sequence data within species, probably without significant error. However, when rate variation among sites is extreme, some sites may have mutated more than once even between closely-related sequences. Even without such extreme rate variation, it is difficult to know how recently diverged a pair of species needs to be for the infinite sites model to hold. In any event, it is important to test the model, e.g., using the four-gamete test (Hudson & Kaplan 1985). A positive result for this test means that one or more of the assumptions of the infinite sites model have been violated (i.e., either some sites in the data have experienced multiple mutations or recombination has occurred between polymorphic sites in the sequences).

The population or species level processes that affect estimates of recent divergence times are modeled using the coalescent (Kingman 1982a, 1982c; Hudson 1983b; Tajima 1983). The coalescent is a stochastic model that describes the ancestral or genealogical process for a sample of gene copies. This genealogical approach to population genetics is well suited to data analysis because it generates testable predictions about variation in a sample and because it yields efficient simulation algorithms. Because the same coalescent process holds for a wide variety of different reproductive schemes, it is considered to be a very robust model. For example, it applies both under the commonly used Wright-Fisher model (Fisher 1930, Wright 1931), which assumes strictly nonoverlapping generations, and under the Moran model (Moran 1958), which assumes that generations overlap. For the most part, natural selection, changes in population size over time, and population subdivision preclude the application of the Kingman's coalescent. However, the range of possible kinds of species represented by the exchangeable models of Cannings (1974), to which the coalescent is applicable (Kingman 1982b), is impressive. Exchangeability means that the individuals or alleles in the population can be relabeled or permuted without affecting the predictions of the model (Kingman 1982b, Aldous 1985) and thus rules out most types of natural selection and population subdivision. The coalescent usually models a diploid species of effective size N_e in which case there are $2 N_e$ copies of each genetic locus. The fundamental result is that time to a common ancestor event in a sample of n lineages will be exponentially distributed with mean $4 N_e/n(n-1)$ generations. Thus, for a sample of size two, the time to a common ancestor will be $2 N_e$ generations on average. There are a number of reviews of the coalescent model available (Hudson 1990, Donnelly & Tavaré 1995, Nordborg 2001) that also describe some recent extensions of the model.

The problems associated with estimating divergence times between closely related species follow from the recognition that, in addition to describing the pattern of ancestry for intraspecific data, the coalescent applies to the lineages ancestral to the sample that existed at the time of speciation or divergence. That is, there is a stochastic genealogical component to divergence and this must be taken into account (Rosenberg & Feldman 2002). To do this we must know or assume something about the historical demography of the species and their ancestor. A simple general model of divergence between two species is the "isolation" model described in Wakeley & Hey (1997). In this model, a panmictic ancestral species of constant effective size splits into two descendant species at some time in the past. After the split, there is no gene flow between the two descendant species, and their effective sizes are constant up to the present. Thus, the coalescent applies to each of the three species (ancestor plus two descendents). This model has four parameters if each of the three species has its own effective size and the mutation rate remains constant over time. Wakeley & Hey (1997) used the four parameters—$\theta_1 = 4N_1u$,

$\theta_2 = 4N_2u$, $\theta_A = 4N_Au$, and $\tau = 2ut$—to describe the model, where N is the effective population size, t is the divergence time in generations, u is the neutral mutation rate at the sampled locus, and the subscripts refer to the three species. The isolation model has been used extensively (Li 1976, Gillespie & Langley 1979, Nei & Li 1979, Takahata & Nei 1985, Takahata 1986, Hudson et al. 1987, Sawyer & Hartl 1992) although typically with restrictive assumptions about the values of N_1, N_2, and N_A (e.g., $N_1 = N_2 = N_A$). This simple model is probably incorrect for many species but still provides a powerful starting point. Some of the likely deviations from it are discussed below.

The Issue of Monophyly

The concept of monophyly has been important in phylogenetic studies since Hennig (1966), and its use in intraspecific studies has helped clarify differences between work at these two fundamentally different levels (Tajima 1983, Pamilo & Nei 1988, Takahata 1989). In this context, samples from within a species are called monophyletic, with respect to a particular speciation event, if they share a most recent common ancestor (MRCA) among themselves before any coalescent events with samples from the other species (Figure 5). Reciprocal monophyly means that samples from both species are monophyletic. At present, seemingly basic questions, such as "what fraction of genetic loci in a typical recognized species is in fact monophyletic?" remain unanswered. The relevance of reciprocal monophyly to the estimation of divergence times is that different approaches to estimation are sometimes required depending on whether alleles are reciprocally monophyletic or not. Some loci appear predisposed to reciprocal monophyly and may even be directly associated with speciation (Lee et al. 1995, Metz & Palumbi 1996, Tsaur et al. 1998, Aguadé 1999, Wyckoff et al. 2000, Parsch et al. 2001). These fast-evolving loci code for proteins involved in male reproductive function, for example, gamete recognition (Metz & Palumbi 1996), and are subject to positive Darwinian selection (Lee et al. 1995, Metz & Palumbi 1996, Tsaur et al. 1998, Aguadé 1999, Wyckoff et al. 2000, Parsch et al. 2001). It is expected that loci such as these, whether they cause speciation (Ting et al. 2000) or simply undergo more fixation events than typical genes, will be reciprocally monophyletic even between closely related species. At the other extreme are loci such as *Mhc* that are subject to balancing selection at which alleles can be shared between surprisingly distantly related species (Figueroa et al. 1988, Lawlor et al. 1988, Mayer et al. 1988, McConnell et al. 1988, Takahata 1990, Takahata & Nei 1990, Edwards et al. 1997). Most loci fall between these two extremes, and their history can be described using the coalescent.

Under the coalescent model, the expected time to the MRCA of a sample of n gene copies is $4N(1 - 1/n)$ generations (Kingman 1982c, Hudson 1983b, Tajima 1983). Thus, if the divergence time between two species is much longer than $4N$ generations, samples of multiple gene copies from them will very likely be reciprocally monophyletic. For recently diverged species (less than $4N$ generations)

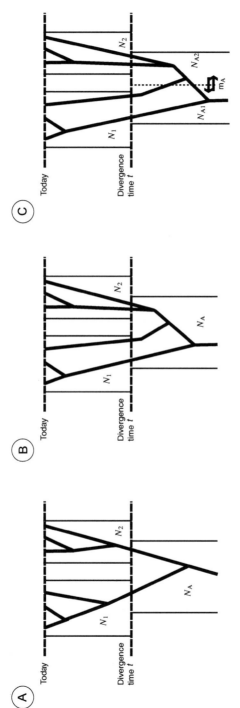
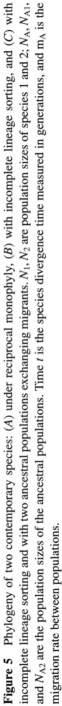

Figure 5 Phylogeny of two contemporary species: (*A*) under reciprocal monophyly, (*B*) with incomplete lineage sorting, and (*C*) with incomplete lineage sorting and with two ancestral populations exchanging migrants. N_1, N_2 are population sizes of species 1 and 2; N_A, N_{A1}, and N_{A2} are the population sizes of the ancestral populations. Time t is the species divergence time measured in generations, and m_A is the migration rate between populations.

the coalescent process for samples within each species will tend not to have reached a MRCA, and multiple ancestral lineages will trace back to the time of speciation. In this case, monophyly of samples becomes unlikely, and some samples will be more closely related to samples from the other species than they will be to other samples from their own species. Under the coalescent, it is possible to derive the probability, $g_{ij}(T)$, that a present-day sample of size i coalesces into j lineages by time $T = t/(2N)$, measured in units of $2N$ generations (Tavaré 1984). This, specifically $g_{nl}(T)$, is what is plotted in Figure 3. In general, the chance that a gene tree matches the species tree is greater when the times between speciation events are long relative to $4N$ generations. The question of whether a locus is reciprocally monophyletic between two species is only indirectly related to the problem of estimating divergence times in closely related species, which necessarily involves the joint estimation of the time and the effective population size of the common ancestor. There is always some chance that a locus is not monophyletic; this is difficult to assess a priori, and, ideally, estimators should take this into account.

Single Versus Multiple Loci

ADVANTAGES OF MULTILOCUS ESTIMATES OF DIVERGENCE TIME AMONG CLOSE RELATIVES   Phylogeography is still dominated by mtDNA, which, from a population genetic perspective, acts as a single locus because all mitochondrial genes are linked together and do not recombine (Avise 1991). [We regard this statement as true, but see recent discussion by Awadalla et al. (1999) and Eyre-Walker & Awadalla (2001)]. However, the number of phylogeographic analyses involving nuclear genes is increasing (Hare 2001). These nuclear gene analyses have a number of advantages for estimating divergence times among close relatives. However, they also pose some difficulties of analysis because the mutation rate, and hence resolving power of nuclear genes, is considerably less than mtDNA on a per locus basis. Additionally, recombination both within and between loci now becomes a real issue.

An appreciation of the advantages of nuclear gene dating comes from considering the multiple sources of variance in an estimate of recent divergence times. For a single locus under reciprocal monophyly there is variance in the estimate of d, the number of substitutions per site; this variance can be further reduced by sequencing a large number of sites for that locus. Although a formal analysis of variance components for divergence times has not yet been done, it is likely that this particular source of variance is small to moderate compared with the second major source of variance in estimation, the coalescent variance. The coalescent variance is the stochastic variability in gene divergence time that arises as a natural consequence of drift. Just as we can think of the fixation time of an allele being a stochastic process looking forward in time, we can think of the coalescent event that comprises the MRCA of alleles in two species as being stochastically variable looking backward in time. In both cases the variance is approximately equal to the square of the effective population size of the common ancestral species. Edwards

& Beerli (2000) showed that for a particular set of parameters this second source of variance was considerably larger than that arising from the small number of sites that can be sequenced for a given locus. This variance is reduced each time the variability from a new locus is incorporated into the estimate. Indeed, the variance associated with estimates of divergence time between recently diverged species can be minimized not by sequencing a large number of sites per locus, but by sequencing a large number of independently segregating loci. For a given two-species divergence, each locus is in essence an independent replicate of the coalescent process, an independent attempt to estimate the ancestral population size of the common ancestral species. The more loci that are brought to bear on this question, the more accurate the estimate of this ancestral population size will be, and hence the estimate of divergence time will also improve. It is known that maximizing the number of loci is the most efficient means of increasing accuracy of estimates of current population size from molecular data (Pluzhnikov & Donnelly 1996).

Nuclear genes are also much less variable than mitochondrial loci, and so it might be concluded that they are of less use in estimating population divergence times. However, the sheer numbers of loci that can be brought to bear on a question of divergence time will ultimately outweigh the high "resolving power" of the single mitochondrial locus. Only a multiplicity of nuclear genes can reduce the variance associated with coalescent processes. The variance associated with variable mutation rates at different loci is yet another source of variance, and providing empirical estimates of this among-locus variation in mutation rate will be important for eliminating this additional factor (Yang 1997). Rarely has this coalescent variance been actually calculated or visualized in an empirical study, hence the difficulty of appreciating its impact.

RECOMBINATION   Recombination is a double-edged sword with regard to estimating divergence times. On the one hand, free recombination between loci benefits many estimators of population divergence time because this ensures that different loci provide statistically independent estimates, conditioned only on the divergence time itself. However, recombination within nuclear loci can cause problems of analysis because in this case all the sites within a locus do not share the same coalescent history, making analysis as a single entity difficult. Hare (2001) reviews the recent empirical examples of nuclear gene phylogeography and points out that the effects of intralocus recombination can be removed by analyzing different data partitions separately, within each of which there is little or no recombination. Recombination will likely have a much greater impact in situations in which alleles have not achieved reciprocal monophyly between populations or species (Wakeley & Hey 1997); in these cases, variable loci still segregating in diverging populations will be molecular mosaics. Using HIV as an example, it has recently been shown that recombination drastically increases the likelihood of rejecting a molecular clock even when a clock applies (Schierup & Hein 2000). The enhanced levels of rate variation among gene lineages with recombination will no doubt affect estimates of population divergence time as well. However, when alleles have achieved reciprocal monophyly, recombinational processes in the common ancestral species, or in the two diverging extant populations, are unlikely to obscure the picture provided by the distribution of coalescent events. These coalescent events can still be used reliably to estimate population divergence time because recombination in extant populations does not affect the structure of these past events. Indeed, under reciprocal monophyly, sampling multiple alleles at a locus becomes a minor issue because the coalescent structure for those loci in the common ancestor is not affected by current processes.

Changes in Population Size over Time

Natural populations are thought to change frequently over time. Because population size is a critical parameter in estimates of divergence time, these fluctuations will cause problems and enhanced opportunities for estimating the timing of speciation and cladogenesis. Methods for estimating population size changes fall into two classes: those that analyze population size changes along a single lineage without cladogenesis (Sherry et al. 1994, Kuhner et al. 1998), and those that compare extant population sizes with those of ancestral populations (Takahata & Satta 1997, Wakeley & Hey 1997, Yang 1997). The former methods are of use in the divergence time problem because both pairwise and maximum likelihood methods provide estimates of the time of the population size change; such inferences have been used to gauge the timing of speciation in some studies (Zink 1997, Knowles et al. 1999). Single-lineage approaches have been employed frequently in studies of human evolution (Harpending et al. 1998); typical analyses imply a substantial population expansion in the human lineage over the last 100,000 years or so.

Methods built around a model of population isolation and divergence are somewhat more relevant to the divergence time problem because with single lineages, the estimate of divergence time provided may or may not be associated with a specific divergence event. A number of studies utilizing such methods have been published in recent years, particularly in primates (Takahata et al. 1995, Takahata & Satta 1997, Li et al. 1999, Chen & Li 2001). In the case of the human-chimpanzee divergence, which Takahata et al. (1995) estimated to have occurred ~4.5 MYA, the inclusion of an ancestral population size parameter did not change the divergence time estimate very much compared with traditional estimates when rate variation among lineages was taken into account (Arnason et al. 1996, 1998, 2000). The correction on the estimate of gene divergence time made by ancestral population size will usually be on the order of a few hundreds of thousands of generations, assuming that contemporary and ancestral N_e are of similar size in individuals. Those studies that have attempted to estimate ancestral population sizes at the time of speciation or population founding have inevitably achieved rather large estimates of ancestral population size, with the inference being that speciation has not been accompanied by a bottleneck. Such studies in "paleo-demography" promise to yield much useful insight into the historical demographic process. These estimates should be made cautiously, however, as there may be unknown biases in the estimate of ancestral population size; for example, it is unknown how difficult it is to estimate very small ancestral population sizes. Most such methods rely in some way on the variance in coalescence time among loci to estimate ancestral population size; multiple loci showing a large variation in coalescence time will imply large founding populations, whereas loci exhibiting a small range of coalescence times will suggest small ancestral populations. One source of bias may be hidden variation in substitution rate among loci. Yang (1997) showed that when among-locus variance in substitution rate is ignored, the result is an inflated estimate of ancestral population size and hence a more recent than actual estimate of divergence time. Small ancestral population sizes may be intrinsically difficult to estimate because the stochastic nature of nucleotide substitution can be large. In this case, variation in coalescence time among loci due to stochastic DNA substitutions is incorrectly attributed to large ancestral population size, again resulting in an inflated estimate. Further studies are needed to assess how readily ancestral bottlenecks at the time of speciation can be detected.

Population Subdivision and Migration

Subdivision within species, either descendant or ancestral or both, adds yet another level of complexity to isolation models. There are many ways in which this might be realized in a given species. Some species may be composed of isolated demes diverging without gene flow, and others may be subdivided into demes among which there is some pattern of migration. We consider only the latter situation here. Population subdivision with ongoing genetic exchange will produce distinctive patterns in samples of DNA or other genetic data, and these patterns have been identified in many species (Slatkin 1985, 1987). A number of different models have been proposed to explain the geographic structure of species. One of the earliest was Wright's (1931) island model, in which the migration rate is assumed to be the same for every pair of demes regardless of their geographic locations. More complicated and probably more realistic models, such as the one- and two-dimensional stepping-stone models (Kimura & Weiss 1964), the migration matrix model of Bodmer & Cavalli-Sforza (1968), and continuous habitat models (Wright 1943, Barton & Wilson 1995, Wilkins & Wakeley 2002), generate a further prediction that Wright dubbed "isolation by distance" (Wright 1943). The prediction is that genetic distance and geographic distance will be correlated. Despite great differences, all of these forms of ongoing migration will have two kinds of effects on genealogies: an effect on the patterns of relationship and an effect on the timescale of the ancestral or genealogical process.

The effect on the patterns of relationship will cause, for example, the probability of reciprocal monophyly to be greater when samples are taken only from a restricted geographic locality within each species. The effect on the timescale of the genealogical process can be understood by considering the effective size of the species. Subdivision increases the effective size of a species (Wright 1943, 1951). Under the island model, the effective size is the product of the total population size and a factor that depends inversely on the rate of migration (Hey 1991, Slatkin 1991, Nei & Takahata 1993). As the number of demes in the species becomes large, that factor rapidly approaches $1 + 1/(4Nm)$, where N is the deme size and m is the fraction of each deme that is replaced by migrants each generation. In the limit of a large number of demes, the effect of subdivision on genealogical topologies and on the timescale of the coalescent process can be separated in the model, and expressions for the effective size of the population can be obtained under less restrictive assumptions about the pattern of migration in the island model (Wakeley 1998, 2001a). The effect of restricted migration on the depth of genealogies, and thus on the magnitude of the problem of inferring divergence times (see Figure 4), embodied by the factor $1 + 1/(4Nm)$, is substantial even if the migration rate is large. For example, the excess divergence between two genes from a pair of species is twice that of the panmictic case when $Nm = 0.25$ in their (subdivided) ancestor. Methods of correcting for this, related to the use of net nucleotide differences to estimate divergence time (Nei & Li 1979), can be developed under some simple models of subdivision (Wakeley 2000).

A potentially more pressing problem than intraspecies subdivision for the accurate estimation of divergence times is incomplete isolation between species. For example, if an ancestral species gives rise to two descendents between which there is gene flow (Wakeley 1996b, Rosenberg & Feldman 2002), the concept of a divergence time becomes blurred. In addition, historical association (isolation without gene flow) and ongoing genetic exchange are very difficult to disentangle empirically (Templeton et al. 1995, Wakeley 1996a, Templeton 1998, Nielsen & Wakeley 2001). This is because simple summaries of genetic variation can be adjusted to fit either scenario (Wakeley 1996c) and because of the stochastic nature of migration. If migration is infrequent, a large amount of data will be required to rule out isolation without gene flow as a possibility. For species that currently do not or cannot exchange genes, incomplete isolation earlier in their history—for instance during speciation (Wakeley & Hey 1998)—will still cause problems in the estimation of divergence times. First, there is the issue of how to define the divergence time: Should it be the time the two species first began to diverge or the time when isolation between them became complete? The latter may be preferable, in which case the problem is a slightly more complicated version of the problem of a subdivided ancestral species. Second, there is the technical issue of how precisely to achieve such inferences in the face of more and more complex historical models.

Estimating Recent Divergence Times

A number of methods have been proposed to estimate the divergence time between pairs of closely related species. We treat the majority of these only briefly here and focus on just a few below. Edwards & Beerli (2000) and Rosenberg & Feldman (2002) provide comprehensive reviews. Our aim is to provide an overview of possible approaches and to outline the major conceptual issues confronting the field as it enters the genomic era.

The first methods of estimating genetic relatedness between populations used allele frequencies and measures such as F_{ST} (Cavalli-Sforza 1969, Nei 1972, Reynolds et al. 1983). Assuming the infinite alleles mutation model (Kimura & Crow 1964), Watterson studied the joint frequencies of alleles in a sample from the two species (Watterson 1985b) and developed a maximum likelihood estimator of divergence time (Watterson 1985a). Under the additional assumption that no mutation has occurred since divergence, Nielsen et al. (1998) present a maximum likelihood approach that does not require the assumption of equilibrium in the ancestral species. For sequence data for which a tree can be reliably inferred, Slatkin & Maddison (1989, 1990) give a method based on the inferred number of interspecies coalescent events. Methods for estimating divergence times also have been developed under the stepwise mutation model (Goldstein et al. 1995, Zhivotovsky 2001) that are appropriate for polymorphic, repeated sequences such as microsatellite loci. We consider a handful of other methods below, treating moment-based methods and maximum likelihood methods separately.

The goal in developing any method is to estimate the parameters of the isolation model(s) described above. Both moment-based methods and maximum likelihood methods have been proposed to do this, but they accomplish the task in very different ways. Moment methods seek parameter values that equate the observed and expected values of measures of DNA sequence polymorphism or divergence. The average numbers of pairwise differences within and between species are examples of such measures. The parameter values that give the best fit between observations and expectations are the point estimates of the parameters. In contrast, maximum likelihood methods compute the probability of the observed data under the model and seek the parameters that make this most likely. Maximum likelihood estimators can be developed either for summary measures, like average pairwise differences, or for the total data available, which we will consider to mean DNA sequences for which the haplotypes are known. A more detailed description of the conceptual differences between moment methods and maximum likelihood methods can be found in any introductory statistics text, e.g., Rice (1995). Here, an important distinction between the two is that it is possible to design moment methods in which the point estimates of parameters do not depend on the recombination rate in the sequences. This can be understood by considering the marginal coalescent process at a single nucleotide site (averaged over all possible histories at other sites), which is precisely the coalescent without recombination. Therefore, expected values of single-site measures of DNA sequence variation obtained under the assumption of no recombination apply regardless of the actual recombination rate. It is not possible to develop maximum likelihood methods that have this property, even ones based on single site measures of polymorphism because the likelihood always depends on the haplotype structure of the data. It is very important to note, however, that measures of statistical confidence or significance will depend on the recombination rate under both kinds of methods.

MOMENT METHODS   A naïve moment-based estimate of the divergence time between two species would be to count (or estimate) the number of differences between a pair of DNA sequences, one from each species, and then to equate this with the value $2ut$, where t is the species divergence time. If more than one sequence is sampled from each species, the average number of interspecific differences between pairs of sequences could be used. As already noted, the problem with this approach is that gene divergences predate species divergences (see Figure 4), sometimes by a large amount. Thus, this naïve estimate will be biased; it will be larger than the true divergence time. Nei & Li (1979) noted this and proposed to correct for the ancestral portion of the observed divergence using an estimate from the descendant species. The net nucleotide difference, $d = d_{XY} - (d_X + d_Y)/2$, subtracts the average of the intraspecific pairwise differences from the observed interspecific value. If all the species are of the same size ($N_1 = N_2 = N_A$), then the expected value of d is equal to $2ut$, and the method is unbiased. If the species are diverged enough that reciprocal monophyly of samples is guaranteed, the method will be unbiased as long as the size of the ancestral species is equal to the average of the sizes of the two descendant species: $N_A = (N_1 + N_2)/2$ (Hudson et al. 1987). Because average pairwise differences do not depend on the haplotype structure of the data, but only on the allele frequencies at variable sites—e.g., see Tajima (1989) and Fu (1995)—point estimates of divergence times made by this method do not depend on any assumptions about recombination. In contrast, confidence limits will depend on the rate of recombination. Takahata & Nei (1985) have studied the variance of net nucleotide differences assuming no recombination.

When just a single gene copy is available from each species, the above method could not be applied because there would not be any intraspecies sequence comparisons. However, for the case when such data are available for many loci, Takahata (1986) proposed to use the mean and variance of differences among loci to estimate the divergence time and the ancestral effective size, i.e., θ_A. As with d_{XY} above, the mean will not depend on the recombination rate. The variance, however, does, and Takahata used the variance expected among loci if there is no recombination within loci and free recombination between them. Thus, this is an example of a moment method in which the point estimates do depend on an assumption about recombination. This same framework was later used to implement a maximum likelihood method of inferring divergence times and ancestral population sizes (see below) (Takahata et al. 1995, Takahata & Satta 1997).

Another moment method in which the point estimates of parameters do not require any assumption about the rate of recombination is the segregating sites method of Wakeley & Hey (1997). In this method, every segregating site in a sample of multiple gene copies from both species is put into one of four mutually exclusive categories (shared, fixed, and exclusive in species 1 or in species 2). These counts are then set equal to the theoretical expectations that depend on θ_1, θ_2, θ_A, and $\tau = 2ut$. Approximate confidence limits can be generated using simulations if an estimate of the recombination rate at each locus is available. The method performs best when data come from many loci or there is a lot of intralocus recombination, in which case the segregating site counts will tend to be close to their expected values. It performs worst when data are from a single nonrecombining locus owing to the strong correlations of allele frequencies at different sites imposed by the single genealogy, which all sites share. An advantage of this method over the method of net pairwise differences and other methods (Hudson et al. 1987) is that it does not constrain the relationship between θ_1, θ_2, and θ_A; all three can be estimated in addition to the divergence time (Wang et al. 1997).

MAXIMUM LIKELIHOOD METHODS   Takahata et al. (1995) and Takahata & Satta (1997) turned the mean and variance approach for samples of single gene copies from two species at many loci (Takahata 1986) into an analytical maximum likelihood method. This method uses an expression for the probability generating function of the number of pairwise differences under the isolation model, which was obtained under the assumption of no intralocus recombination. Yang (1997) further extended the method to account for substitution rate variation among loci, and Edwards & Beerli (2000) allowed for finite sites, rather than infinite sites, mutation. Because of the sample size (one from each species), in this case it is relatively easy to envision the effect of recombination. Recombination will break up the sequences ancestral to the sample (Hudson 1983a) so that different segments of a single chromosome sampled today will be located on a number of different chromosomes at the time of divergence (Wiuf & Hein 1997). This decreases the correlation in ancestry among segments of a locus, which will make narrower (i.e., have a smaller variance) the distribution of the number of differences than if recombination is absent; e.g., see Figure 2 in Hudson (1990). Note that, technically, this is a violation of the assumptions of the maximum likelihood methods of estimating ancient divergence times but one that is not expected to compromise accuracy if the divergence time is long relative to the size of the ancestral species.

Two other methods that also assume no intralocus recombination, but which use full haplotype data in samples of multiple sequences from two closely related species, are the Markov Chain Monte Carlo (MCMC) methods of Nielsen (1998) and Nielsen & Wakeley (2001). Simulation-based methods like MCMC are rapidly becoming the methods of choice for computation of likelihoods and posterior distributions. Except for very small data sets, it is not possible to calculate the distributions of parameters for the full data analytically, which is the probability of the observed data under the model (including parameter values). One strategy for computing the likelihood is to condition on the genealogy of the sample, that is, to compute the likelihood given the genealogy and sum this over all possible genealogies weighted by their probabilities under the model. This is attractive because the likelihood of the data given the genealogy is easy to compute but is impossible either analytically or using simulations because the space of genealogies is too large. The only viable approach to the problem when it is stated in this manner is to try to focus the sampling of genealogies on those that contribute most to the likelihood. A few different methods of focusing on the important parts of the space of genealogies by "integrating" over them have been proposed (Griffiths & Tavaré 1994, Kuhner et al. 1995, Griffiths & Tavaré 1996), but this is still a very active field of research. Under the assumption of infinite sites mutation without recombination, Nielsen (1998) extended the single-population method of Griffiths & Tavaré (1994) to the two species isolation model where all population sizes are the same. Nielsen & Wakeley (2001) further extended the method to allow for different population sizes and for migration to occur between the two species.

PROBLEMS FOR THE FUTURE

Microevolutionary Clocks

We have entered an age in which the accumulation of large multilocus DNA sequences data sets will likely become the norm. At present, the number of loci that are typically brought to bear on a phylogeographic or divergence time estimation problem is still small, frequently less than five, although in *Drosophila* some recent studies have gathered more than ten loci in closely related species groups (Kliman et al. 2000, Machado et al. 2001). Though this is certainly an improvement over single locus studies, the number of loci that are typically required to estimate a particular recent divergence with confidence can be large (Edwards & Beerli 2000). It is only in humans, where the number of loci available for some purposes is now in the millions (The International SNP Map Working Group 2001) that such sampling is possible at present.

Although we may soon have enough data to make sound inferences under quite complicated historical (gene flow, changes in population size, etc.) and mutational (finite sites, recombination) models, the analytical framework for utilizing such data has yet to be fully developed. One major issue for the future is to provide efficient methods for combining information from nuclear and mitochondrial genomes, as well as the sex chromosomes. All of these chromosome sets have different population sizes and modes of inheritance and hence different population dynamics, which will affect estimates of divergence time. Under certain assumptions, such as equal sex ratios and similar patterns of migration between the sexes, it is likely that efficient methods for combining these diverse data sets can be developed; e.g., see Wang et al. (1997).

Both moment-based methods and full-data maximum likelihood methods are likely to play important roles in the future. For complicated histories, moment methods will be computationally feasible and will give unbiased estimates from large data sets in which intragenic recombination is present, probably long before maximum likelihood methods that can deal effectively with arbitrary recombination will be developed. Though nearly everyone would agree that maximum likelihood methods are preferable to moment methods, assuming both are feasible, a further point to consider is that almost nothing is known about how demographic history is manifest in the complicated divergence and linkage patterns of DNA sequence data. Thus, the development of sufficient statistics for DNA sequence data in the context of complicated historical models would be a particularly useful contribution to the field.

Macroevolutionary Clocks

The recent models of Sanderson (1997, 2002), Huelsenbeck et al. (2000), and others (Takezaki et al. 1995, Rambaut & Bromham 1998, Thorne et al. 1998) represent promising new advances in the measurement of absolute and relative rates of molecular evolution on phylogenetic trees. These models range from parametric to semiparametric and nonparametric, and as such they provide a much fuller account of all sources of error than in previous models. In addition, they provide robust statistical tests, several in the context of maximum likelihood, for asking whether rates of evolution are uniform or variable across a tree. As discussed above, the computational advantages of Bayesian approaches will likely lead to their increased use in studies aimed at estimating dates of divergence from molecular data. An important project for the future is the incorporation of uncertainty in fossil divergence times into molecular divergence time estimates, something that should be readily possible with some recent models, i.e., Huelsenbeck et al. (2000). Paleobiological methods for estimating confidence limits on times of clade origins are improving and should prove a useful complement to molecular evolutionary models.

CONCLUSIONS

The topics outlined in our review underscore the many complex, and often interrelated, issues involved in estimating times of evolutionary divergence from molecular data. In many ways the task of estimating dates of divergence has become increasingly more challenging as our knowledge and sophistication has increased; whereas estimating divergence times once seemed fairly straightforward and simple, we now have a much better appreciation of the variety of pitfalls that exist. Still, researchers should be encouraged by the many recent theoretical and computational advances that provide powerful new tools for detecting and avoiding these pitfalls. Future computational and theoretical advances, in both the phylogenetic and population genetic contexts, promise to greatly enhance our ability to accurately estimate dates of divergence from molecular data while also contributing importantly to our understanding of evolutionary processes on the molecular and organismal levels.

AKNOWLEDGMENTS

This work was supported in part by grants from the National Science Foundation (NSF) to J. Wakeley (DEB-9815367 and DEB-0133760) and S.V. Edwards

(DEB-9977039 and DEB-0129487). Peter Beerli was supported in part by NSF grant DEB-9815650 and National Institutes of Health (NIH) grants GM-51929 and HG-01989, all awarded to J. Felsenstein. We thank Yoko Satta for helpful discussions on the manuscript.

The *Annual Review of Ecology and Systematics* is online at
http://ecolsys.annualreviews.org

LITERATURE CITED

Aguadé M. 1999. Positive selection drives the evolution of the ACp29AB accessory gland protein in *Drosophila*. *Genetics* 152:543–51

Akaike H. 1974. A new look at the statistical model identification. *IEEE Trans. Autom. Contr.* 19:716–23

Aldous DJ. 1985. Exchangeability and related topics. In *Ecole d'Ete de Probabilites de Saint-Flour XII—1983*, ed. A Dold, B Eckmann, pp. 1–198. Berlin: Springer

Arbogast BS, Slowinski JB. 1998. Pleistocene speciation and the mitochondrial DNA clock. *Science* 282:1955a

Arnason U, Gullberg A, Burguete AS, Janke A. 2000. Molecular estimates of primate divergences and new hypotheses for primate dispersal and the origin of modern humans. *Hereditas* 133:217–28

Arnason U, Gullberg A, Janke A. 1998. Molecular timing of primate divergences as estimated by two nonprimate calibration points. *J. Mol. Evol.* 47:718–27

Arnason U, Gullberg A, Janke A, Xu X. 1996. Pattern and timing of evolutionary divergences among hominoids based on analyses of complete mtDNAs. *J. Mol. Evol.* 43:650–61

Avise JC. 1991. Ten unorthodox perspectives on evolution prompted by comparative population genetic findings on mitochondrial DNA. *Annu. Rev. Genet.* 25:45–69

Avise JC, Arnold J, Ball RM, Bermingham E, Lamb T, et al. 1987. Intraspecific phylogeography: the mitochondrial DNA bridge between population genetics and systematics. *Annu. Rev. Ecol. Syst.* 18:489–522

Awadalla P, Eyre-Walker A, Smith JM. 1999. Linkage disequilibrium and recombination in hominid mitochondrial DNA. *Science* 286:2524–25

Ayala FJ. 1997. Vagaries of the molecular clock. *Proc. Natl. Acad. Sci. USA* 94:7776–83

Baldwin B, Sanderson M. 1998. Age and rate of diversification of the Hawaiian silversword alliance. *Proc. Natl. Acad. Sci. USA* 95:9402–6

Barton NH, Wilson I. 1995. Genealogies and geography. *Philos. Trans. R. Soc. B* 349:49–59

Bermingham E, Rohwer S, Freeman S, Wood C. 1992. Vicariance biogeography in the Pleistocene and speciation in North American wood warblers: a test of Mengel's model. *Proc. Natl. Acad. Sci. USA* 89:6624–28

Bodmer W, Cavalli-Sforza L. 1968. A migration matrix model for the study of random genetic drift. *Genetics* 59:565–92

Bromham L, Rambaut A, Fortey R, Cooper A, Penny D. 1998. Testing the Cambrian explosion hypothesis by using a molecular dating technique. *Proc. Natl. Acad. Sci. USA* 95:12386–89

Brown W Jr, George M Jr, Wilson AC. 1979. Rapid evolution of animal mitochondrial DNA. *Proc. Natl. Acad. Sci. USA* 76:1967–71

Brown W, Prager EM, Wang A, Wilson AC. 1982. Mitochondrial DNA sequences of primates: tempo and mode of evolution. *J. Mol. Evol.* 18:225–39

Buckley T, Simon C, Chamb G. 2001. Exploring among-site rate variation models in a maximum likelihood framework using empirical data: effects of model assumptions on estimates of topology, branch lengths and bootstrap support. *Syst. Biol.* 50:67–86

Cann RL, Stoneking M, Wilson AC. 1987. Mitochondrial DNA and human evolution. *Nature* 325:31–36

Cannings C. 1974. The latent roots of certain Markov chains arising in genetics: a new approach. I. Haploid models. *Adv. Appl. Prob.* 6:260–90

Cavalli-Sforza LL. 1969. Human diversity. *Proc. 12th Int. Congr. Genet.* 2:405–16

Chen FC, Li WH. 2001. Genomic divergences between humans and other hominoids and the effective population size of the common ancestor of humans and chimpanzees. *Am. J. Hum. Genet.* 68:444–56

Cooper A, Penny D. 1997. Mass survival of birds across the Cretaceous-Tertiary boundary: molecular evidence. *Science* 275:1109–13

Donnelly P, Tavaré S. 1995. Coalescents and genealogical structure under neutrality. *Annu. Rev. Genet.* 29:401–21

Doolittle R, Feng D, Tsang S, Cho G, Little E. 1996. Determining divergence times of the major kingdoms of living organisms with a protein clock. *Science* 271:470–77

Edwards SV, Beerli P. 2000. Perspective: gene divergence, population divergence, and the variance in coalescence time in phylogeographic studies. *Evolution* 54:1839–54

Edwards SV, Chesnut K, Satta Y, Wakeland EK. 1997. Ancestral polymorphism of *Mhc* class II genes in mice: implications for balancing selection and the mammalian molecular clock. *Genetics* 146:655–68

Ellegren H. 2000. Microsatellite mutations in the germ line: implications for evolutionary inference. *Trends Genet.* 16:551–58

Eyre-Walker A, Awadalla P. 2001. Does human mtDNA recombine? *J. Mol. Evol.* 53:430–35

Felsenstein J. 1988. Phylogenies and quantitative characters. *Annu. Rev. Ecol. Syst.* 19:445–71

Figueroa F, Gunther E, Klein J. 1988. MHC polymorphism predating speciation. *Nature* 335:265–67

Fisher RA. 1930. *The Genetical Theory of Natural Selection.* Oxford, UK: Clarendon

Fu X-Y. 1995. Statistical properties of segregating sites. *Theoret. Pop. Biol.* 48:172–97

Gaut B, Lewis P. 1995. Success of maximum likelihood in the four-taxon case. *Mol. Biol. Evol.* 12:152–62

Gillespie JH. 1991. *The Causes of Molecular Evolution.* New York: Oxford Univ. Press. 352 pp.

Gillespie JH, Langley CH. 1979. Are evolutionary rates really variable? *J. Mol. Evol.* 13:27–34

Givnish T, Sytsma K, ed. 1997. *Molecular Evolution and Adaptive Radiation.* New York: Cambridge Univ. Press

Goldman N. 1993. Statistical tests of models of DNA substitution. *J. Mol. Evol.* 36:182–98

Goldstein DB, Ruiz Linares A, Cavalli-Sforza LL, Feldman MW. 1995. Genetic absolute dating based on microsatellites and the origin of modern humans. *Proc. Natl. Acad. Sci. USA* 92:6723–27

Griffiths RC, Tavaré S. 1994. Ancestral inference in population genetics. *Stat. Sci.* 9:307–19

Griffiths RC, Tavaré S. 1996. Monte Carlo inference methods in population genetics. *Math. Comput. Model.* 23:141–58

Hare M. 2001. Prospects for nuclear gene phylogeography. *Trends Ecol. Evol.* 16:700–6

Harpending HC, Batzer MA, Gurven M, Jorde LB, Rogers AR, Sherry ST. 1998. Genetic traces of ancient demography. *Proc. Natl. Acad. Sci. USA* 95:1961–67

Hasegawa M, Kishino H, Yano T. 1985. Dating of the human-ape splitting by a molecular clock of mitochondrial DNA. *J. Mol. Evol.* 21:160–74

Hedges SB, Chen H, Kumar S, Wang DY, Thompson AS, Watanabe H. 2001. A genomic timescale for the origin of eukaryotes. *BMC Evol. Biol.* 1:4

Hedges SB, Parker PH, Sibley CG, Kumar S. 1996. Continental breakup and the ordinal diversification of birds and mammals. *Nature* 381:226–29

Hennig W. 1966. *Phylogenetic Systematics.* Urbana: Univ. Ill. Press

Hey J. 1991. A multi-dimensional coalescent process applied to multi-allelic selection models and migration models. *Theoret. Pop. Biol.* 39:30–48

Hillis DM, Mable BK, Moritz C. 1996. Applications of molecular systematics: the state of the field and a look to the future. In *Molecular Systematics*, ed. D Hillis, C Moritz, BK Mable, pp. 515–43. Sunderland, MA: Sinauer

Hudson RR. 1983a. Properties of a neutral allele model with intragenic recombination. *Theoret. Pop. Biol.* 23:183–201

Hudson RR. 1983b. Testing the constant-rate neutral allele model with protein sequence data. *Evolution* 37:203–17

Hudson RR. 1990. Gene genealogies and the coalescent process. *Oxford Surv. Evol. Biol.* 7:1–44

Hudson RR, Kaplan NL. 1985. Statistical properties of the number of recombination events in the history of a sample of DNA sequences. *Genetics* 111:147–64

Hudson RR, Kreitman M, Aguadé M. 1987. A test of neutral molecular evolution based on nucleotide data. *Genetics* 116:153–59

Huelsenbeck JP, Larget B, Swofford D. 2000. A compound Poisson process for relaxing the molecular clock. *Genetics* 154:1879–92

Huelsenbeck JP, Rannala B. 1997. Phylogenetic methods come of age: testing hypotheses in an evolutionary context. *Science* 276:227–32

Huelsenbeck JP, Ronquist F. 2001. MRBAYES: Bayesian inference of phylogenetic trees. *Bioinformatics* 17:754–55

Huelsenbeck JP, Ronquist F, Nielsen R, Bollback P. 2001. Bayesian inference of phylogeny and its impact on evolutionary biology. *Science* 294:2310–14

Hurst L, Ellegren H. 1998. Sex biases in the mutation rate. *Trends Genet.* 14:446–52

Jukes TH, Cantor CR. 1969. Evolution of protein molecules. In *Mammalian Protein Metabolism*, ed. HN Munro, pp. 21–132. New York: Academic

Ke Y, Su B, Song X, Lu D, Chen L, et al. 2001. African origin of modern humans in East Asia: a tale of 12,000 Y chromosomes. *Science* 292:1151–53

Kimura M, Crow JF. 1964. The number of alleles that can be maintained in a finite population. *Genetics* 49:725–38

Kimura M, Weiss GH. 1964. The stepping stone model of population structure and the decrease of genetic correlation with distance. *Genetics* 49:561–76

Kingman JFC. 1982a. The coalescent. *Stochastic Process. Appl.* 13:235–48

Kingman JFC. 1982b. Exchangeability and the evolution of large populations. In *Exchangeability in Probability and Statistics*, ed. G Koch, F Spizzichino, pp. 97–112. Amsterdam: North-Holland

Kingman JFC. 1982c. On the genealogy of large populations. *J. Appl. Prob.* 19A:27–43

Klicka J, Zink RM. 1997. The importance of recent ice ages in speciation: a failed paradigm. *Science* 277:1666–69

Kliman RM, Andolfatto P, Coyne JA, Depaulis F, Kreitman M, et al. 2000. The population genetics of the origin and divergence of the *Drosophila simulans* complex species. *Genetics* 156:1913–31

Knowles LL, Futuyma DJ, Eanes WF, Rannala B. 1999. Insight into speciation from historical demography in the phytophagous beetle genus *Ophraella. Evolution* 53:1846–56

Kuhner MK, Yamato J, Felsenstein J. 1995. Estimating effective population size and mutation rate from sequence data using Metropolis-Hastings sampling. *Genetics* 140: 1421–30

Kuhner MK, Yamato J, Felsenstein J. 1998. Maximum likelihood estimation of population growth rates based on the coalescent. *Genetics* 149:429–34

Lawlor DA, Ward FE, Ennis PD, Jackson AP, Parham P. 1988. HLA-A and B polymorphisms predate the divergence of humans and chimpanzees. *Nature* 335:268–71

Lee Y, Ohta T, Vacquier VD. 1995. Positive selection is a general phenomenon in the evolution of abalone sperm lysin. *Mol. Biol. Evol.* 6:424–35

Li W-H. 1976. Distribution of nucleotide difference between two randomly chosen cistrons in a subdivided population: the finite island model. *Theoret. Pop. Biol.* 10:303–8

Li W-H. 1997. *Molecular Evolution.* Sunderland, MA: Sinauer. 432 pp.

Li YJ, Satta Y, Takahata N. 1999. Paleodemography of the *Drosophila melanogaster* subgroup: application of the maximum likelihood method. *Genes Genet. Syst.* 74:117–27

Machado CA, Kliman RM, Markert JA, Hey J. 2002. Inferring the history of speciation from multilocus DNA sequence data: the case of *Drosophila pseudoobscura* and close relatives. *Mol. Biol. Evol.* 19:472–88

Maddison WP. 1997. Gene trees in species trees. *Syst. Biol.* 46:523–36

Martin AP, Palumbi SR. 1993. Body size, metabolic rate, generation time and the molecular clock. *Proc. Natl. Acad. Sci. USA* 90:4087–91

Mayer WE, Jonker J, Klein D, Ivanyi P, van Seventer G, Klein J. 1988. Nucleotide sequences of chimpanzee MHC class I alleles: evidence for a trans-species mode of evolution. *EMBO J.* 7:2765–74

McConnell TJ, Talbot WS, McInode RA, Wakeland EK. 1988. The origin of MHC class II gene polymorphism within the genus *Mus. Nat.* 332:651–54

Metz EC, Palumbi SR. 1996. Positive selection and sequence rearrangements generate extensive polymorphism in the gamete recognition protein bindin. *Mol. Biol. Evol.* 13:397–406

Moran PAP. 1958. Random processes in genetics. *Proc. Camb. Phil. Soc.* 54:60–71

Morozov P, Sitnikova T, Churchhill G, Ayala F, Rzhetsky A. 2000. A new method for characterizing replacment rate variation in molecular sequences: application of the Fourier and Wavelet models to *Drosophila* and mammalian proteins. *Genetics* 154:381–95

Nei M. 1972. Genetic distance between populations. *Am. Nat.* 105:385–98

Nei M, Li W-H. 1979. Mathematical model for studying genetic variation in terms of restriction endonucleases. *Proc. Natl. Acad. Sci. USA* 76:5269–73

Nei M, Takahata N. 1993. Effective population size, genetic diversity, and coalescence time in subdivided populations. *J. Mol. Evol.* 37:240–44

Nielsen R. 1998. Maximum likelihood estimation of population divergence times and population phylogenies under the infinite sites model. *Theor. Pop. Biol.* 53:143–51

Nielsen R, Mountain JL, Huelsenbeck JP, Slatkin M. 1998. Maximum likelihood estimation of population divergence times and population phylogeny in models without mutation. *Evolution* 52:669–77

Nielsen R, Wakeley J. 2001. Distinguishing migration from isolation: a Markov Chain Monte Carlo approach. *Genetics* 158:885–96

Nordborg M. 2001. Coalescent theory. In *Handbook of Statistical Genetics*, ed. DJ Balding, MJ Bishop, C Cannings. Chichester, UK: Wiley

Ochman H, Wilson AC. 1987. Evolution in bacteria: evidence for a universal substitution rate in cellular genomes. *J. Mol. Evol.* 26:74–86

Pamilo P, Nei M. 1988. The relationships between gene trees and species trees. *Mol. Biol. Evol.* 5:568–83

Parsch J, Meiklejohn CD, Hartl DL. 2001. Patterns of DNA sequence variation suggest the recent action of positive selection in the *janus-ocnus* region of *Drosophila simulans. Genetics* 159:647–57

Pluzhnikov A, Donnelly P. 1996. Optimal sequencing strategies for surveying molecular genetic diversity. *Genetics* 144:1247–62

Posada D, Crandall KA. 1998. MODELTEST: testing the model of DNA substitution. *Bioinformatics* 14:817–18

Posada D, Crandall KA. 2001. Selecting the best-fit model of nucleotide substitution. *Syst. Biol.* 50:580–601

Rambaut A, Bromham L. 1998. Estimating divergence dates from molecular sequences. *Mol. Biol. Evol.* 15:442–48

Rand DM. 1994. Thermal habit, metabolic rate

and the evolution of mitochondrial DNA. *Trends Ecol. Evol.* 9:125–31

Reynolds J, Weir BS, Cockerham CC. 1983. Estimation of the co-ancestry coefficient—basis for a short term genetic distance. *Genetics* 105:767–79

Rice JA. 1995. *Mathematical Statistics and Data Analysis.* Belmont, CA: Duxbury

Richardson J, Weitz F, Fay M, Cronk Q, Linder H, et al. 2001. Rapid and recent origin of species richness in the Cape flora of South Africa. *Nature* 412:181–83

Rodriguez F, Oliver JL, Marin A, Medina J. 1990. The general stochastic model of nucleotide substitution. *J. Theor. Biol.* 142: 485–501

Rosenberg NA, Feldman MW. 2002. The relationship between coalescence times and population divergence times. In *Modern Developments in Theoretical Population Genetics*, ed. MW Slatkin, M Veuille. Oxford, UK: Oxford Univ. Press

Sanderson MJ. 1997. A nonparametric approach to estimating divergence times in the absence of rate constancy. *Mol. Biol. Evol.* 14:1218–31

Sanderson MJ. 2002. Estimating absolute rates of molecular evolution and divergence times: a penalized likelihood approach. *Mol. Biol. Evol.* 19:101–9

Sanderson MJ, Kim J. 2000. Parametric phylogenetics? *Syst. Biol.* 49:817–29

Sawyer SA, Hartl DL. 1992. Population genetics of polymorphism and divergence. *Genetics* 132:1161–76

Schierup MH, Hein J. 2000. Recombination and the molecular clock. *Mol. Biol. Evol.* 17:1578–79

Schluter D. 2000. *The Ecology of Adaptive Radiation.* Oxford, UK: Oxford Univ. Press. 296 pp.

Schwartz G. 1978. Estimating the dimensions of a model. *Ann. Stat.* 6:461–64

Sherry ST, Rogers AR, Harpending H, Soodyall H, Jenkins T, Stoneking M. 1994. Mismatch distributions of mtDNA reveal recent human population expansions. *Hum. Biol.* 66:761–75

Slatkin M. 1985. Gene flow in natural populations. *Annu. Rev. Ecol. Syst.* 16:393–430

Slatkin M. 1987. Gene flow and the geographic structure of natural populations. *Science* 236:787–92

Slatkin M. 1991. Inbreeding coefficients and coalescence times. *Genet. Res., Camb.* 58: 167–75

Slatkin M, Maddison WP. 1989. A cladistic measure of gene flow inferred from the phylogenies of alleles. *Genetics* 123:603–13

Slatkin M, Maddison WP. 1990. Detecting isolation by distance using phylogenies of genes. *Genetics* 126:249–60

Slowinski JB, Arbogast BS. 1999. Is there an inverse relationship between body size and the rate of molecular evolution. *Syst. Biol.* 48:396–99

Steel MA, Cooper AC, Penny D. 1996. Confidence intervals for the divergence time of two clades. *Syst. Biol.* 45:127–34

Suchard M, Weiss R, Sinsheimer J. 2001. Bayesian selection of continuous-time Markov chain evolutionary models. *Mol. Biol. Evol.* 18:1001–13

Sullivan J, Holsinger K, Simon C. 1996. The effect of topology on estimates of among-site rate variation. *J. Mol. Evol.* 42:308–12

Sullivan J, Swofford D. 2001. Should we use model-based methods for phylogenetic inference when we know that assumptions about among-site rate variation and nucleotide substitution pattern are violated? *Syst. Biol.* 50:723–29

Swofford D. 1998. *PAUP*: Phylogenetic Analysis Using Parsimony (*and Other Methods). Version 4.0.* Sunderland, MA: Sinauer

Swofford D, Olsen G, Waddell P, Hillis D. 1996. Phylogenetic inference. In *Molecular Systematics*, ed. D Hillis, C Mortiz, BK Mable, pp. 407–514. Sunderland, MA: Sinauer

Tajima F. 1983. Evolutionary relationship of DNA sequences in finite populations. *Genetics* 105:437–60

Tajima F. 1989. Statistical method for testing the neutral mutation hypothesis by DNA polymorphism. *Genetics* 123:585–95

Takahata N. 1986. An attempt to estimate the effective size of the ancestral species common to two extant species from which homologous genes are sequenced. *Genet. Res. Camb.* 48:187–90

Takahata N. 1987. On the overdispersed molecular clock. *Genetics* 116:169–79

Takahata N. 1989. Gene genealogy in three related populations: consistency probability between gene and population trees. *Genetics* 122:957–66

Takahata N. 1990. A simple genealogical structure of strongly balanced allelic lines and trans-species polymorphism. *Proc. Natl. Acad. Sci. USA* 87:2419–23

Takahata N, Nei M. 1985. Gene genealogy and variance of interpopulational nucleotide differences. *Genetics* 110:325–44

Takahata N, Nei M. 1990. Allelic genealogy under overdominant and frequency-dependent selection and polymorphism of major histocompatibility complex loci. *Genetics* 124:967–78

Takahata N, Satta Y. 1997. Evolution of the primate lineage leading to modern humans: phylogenetic and demographic inferences from DNA sequences. *Proc. Natl. Acad. Sci. USA* 94:4811–15

Takahata N, Satta Y, Klein J. 1995. Divergence time and population size in the lineage leading to modern humans. *Theor. Pop. Biol.* 48:198–221

Takezaki N, Rzhetsky A, Nei M. 1995. Phylogenetic test of the molecular clock and linearized trees. *Mol. Biol. Evol.* 12:823–33

Tavaré S. 1984. Lines-of-descent and genealogical processes, and their application in population genetic models. *Theoret. Pop. Biol.* 26:119–64

Templeton AR. 1998. Nested clade analysis of phylogeographic data: testing hypotheses about gene flow and population history. *Mol. Ecol.* 7:381–97

Templeton AR, Routman E, Phillips C. 1995. Separating population structure from population history: a cladistic analysis of the geographical distribution of mitochondrial DNA haplotypes in the tiger salamander. *Ambystoma tigrinum. Genetics* 140:767–82

The International SNP Map Working Group. 2001. A map of human genome sequence variation containing 1.42 million single nucleotide polymorphisms. *Nature* 409:928–33

Thorne J, Kishino H, Painter I. 1998. Estimating the rate of evolution of the rate of molecular evolution. *Mol. Biol. Evol.* 15:1647–57

Ting C-T, Tsaur S-C, Wu C-I. 2000. The phylogeny of closely related species as revealed by the genealogy of a speciation gene. *Odysseus. Proc. Natl. Acad. Sci. USA* 97: 5313–16

Tsaur S, Ting CT, Wu C-I. 1998. Positive selection driving the evolution of a gene of male reproduction, ACP26Aa of *Drosophila*: II. Divergence vs. polymorphism. *Mol. Biol. Evol.* 15:1040–46

Underhill P, Shen P, Lin A, Jin L, Passarino G, et al. 2000. Y chromosome sequence variation and the history of human populations. *Nat. Genet.* 26:358–61

Uzzell T, Corbin KW. 1971. Fitting discrete probability distributions to evolutionary events. *Science* 172:1089–96

Vawter L, Brown WM. 1986. Nuclear and mitochondrial DNA comparisons reveal extreme rate variation in the molecular clock. *Science* 234:194–96

Voelker G. 1999. Dispersal, vicariance, and clocks: historical biogeography and speciation in a cosmopolitan passerine genus (*Anthus*: motacillidae). *Evolution* 53:1536–52

Wakeley J. 1996a. Distinguishing migration from isolation using the variance of pairwise differences. *Theoret. Pop. Biol.* 49:369–86

Wakeley J. 1996b. Pairwise differences under a general model of population subdivision. *J. Genet.* 75:81–89

Wakeley J. 1996c. The variance of pairwise nucleotide differences in two populations with migration. *Theoret. Pop. Biol.* 49:39–57

Wakeley J. 1998. Segregating sites in Wright's island model. *Theoret. Pop. Biol.* 53:166–75

Wakeley J. 2000. The effects of population subdivision on the genetic divergence of populations and species. *Evolution* 54:1092–101

Wakeley J. 2001a. The coalescent in an island model of population subdivision with variation among demes. *Theoret. Pop. Biol.* 59:133–44

Wakeley J. 2001b. The effects of subdivision on the genetic divergence of populations and species. *Evolution* 54:1092–101

Wakeley J, Hey J. 1997. Estimating ancestral population parameters. *Genetics* 145:847–55

Wakeley J, Hey J. 1998. Testing speciation models with DNA sequence data. In *Molecular Approaches to Ecology and Evolution*, ed. B Schierwater, R DeSalle, pp. 157–75. Basel: Birkhauser-Verlag

Wang D, Kumar S, Hedges S. 1999. Divergence time estimates for the early history of animal phyla and the origin of plants, animals and fungi. *Proc. R. Soc. London. Ser. B* 266:163–71

Wang R-L, Wakeley J, Hey J. 1997. Gene flow and natural selection in the origin of *Drosophila pseudoobscura* and close relatives. *Genetics* 147:1091–106

Watterson GA. 1985a. Estimating species divergence times using multilocus data. In *Population Genetics and Molecular Evolution*, ed. T Ohta, K Aoki, pp. 163–83. Tokyo: Jpn. Sci. Soc.

Watterson GA. 1985b. The genetic divergence of two populations. *Theoret. Pop. Biol.* 27:298–317

Wilkins JF, Wakeley J. 2002. The coalescent in a continuous, finite, linear population. *Genetics* 161:873–88

Wilson AC, Cann RL, Carr SM, George M, Gyllensten UB, et al. 1985. Mitochondrial DNA and two perspectives on evolutionary genetics. *Biol. J. Linnaean Soc.* 26:375–400

Wilson AC, Ochman H, Prager EM. 1987. Molecular time scale for evolution. *Trends Genet.* 3:241–47

Wiuf C, Hein J. 1997. On the number of ancestors to a DNA sequence. *Genetics* 147:1459–68

Wright S. 1931. Evolution in Mendelian populations. *Genetics* 16:97–159

Wright S. 1943. Isolation by distance. *Genetics* 28:114–38

Wright S. 1951. The genetical structure of populations. *Ann. Eugenics* 15:323–54

Wyckoff GJ, Wang W, Wu C-I. 2000. Rapid evolution of male reproductive genes in the descent of man. *Nature* 403:304–9

Yang Z. 1993. Maximum likelihood estimation of phylogeny from DNA sequences when substitution rates differ over sites. *Mol. Biol. Evol.* 10:1396–401

Yang Z. 1994. Maximum likelihood phylogenetic estimation from DNA sequences with variable rates over sites: approximate methods. *J. Mol. Evol.* 39:306–14

Yang Z. 1997. On the estimation of ancestral population sizes of modern humans. *Genet. Res.* 69:111–16

Yang Z, Goldman N, Friday A. 1994. Comparison of models for nucleotide substitution used in maximum-likelihood phylogenetic estimation. *Mol. Biol. Evol.* 11:316–24

Yoder A, Yang Z. 2000. Estimation of primate speciation dates using local molecular clocks. *Mol. Biol. Evol.* 17:1081–90

Zhivotovsky LA. 2001. Estimating divergence time with the use of microsatellite genetic distances: impacts of population growth and gene flow. *Mol. Biol. Evol.* 18:700–9

Zink RM. 1997. Phylogeographic studies of North American birds. In *Avian Molecular Evolution and Systematics*, ed. DP Mindell, pp. 301–24. San Diego, CA: Academic

Zuckerkandl E, Pauling L. 1965. Evolutionary divergence and convergence in proteins. In *Evolving Genes and Proteins*, ed. V Bryson, HJ Vogel. New York: Academic

Annu. Rev. Ecol. Syst. 2002. 33:741–77
doi: 10.1146/annurev.ecolsys.33.010802.150520
Copyright © 2002 by Annual Reviews. All rights reserved
First published online as a Review in Advance on August 19, 2002

THE FATE OF CLADES IN A WORLD OF RECURRENT CLIMATIC CHANGE: Milankovitch Oscillations and Evolution

Roland Jansson and Mats Dynesius[1]

Landscape Ecology Group, Department of Ecology and Environmental Science, Umeå University, SE-901 87 Umeå, Sweden; email: roland@eg.umu.se, matsd@eg.umu.se

Key Words biodiversity gradients, geographical range-size, macroecology, phylogeography, speciation

■ **Abstract** Variations in Earth's orbit with periods of 10–100 thousand years (kyr) (Milankovitch oscillations) have led to recurrent and rapid climatic shifts throughout Earth's history. These cause changes in the geographical distributions of clades, which we term orbitally forced range dynamics (ORD). The magnitude of ORD varies geographically, e.g., with latitude. Climatic shifts cause extinction, splitting, and merging of gene pools and clades. They select among individuals and clades for traits enhancing the ability to survive in situ and to establish new populations. There is also nonadaptive sorting caused by the large geographical variation in ORD, as only gene pools that are in the right place when climate shifts survive. ORD lead to sorting at many levels of genealogic inclusiveness. Clades that have survived climatic shifts during at least one entire period of the longest significant Milankovitch oscillations (100 kyr), we name β-clades. The products of more recent cladogenesis are α-clades, which are always nested within a β-clade. We conclude that ORD may promote α-clade formation but curb rates of β-clade formation. In areas with little ORD, where gene pools persist without going extinct or merging, clade splits and divergence may accumulate leading to high rates of β-clade formation and β-anagenesis (evolutionary change persisting >100 kyr). High ORD should lead to low numbers of β-clades, β-clades with low levels of spatial genetic divergence, little geographical subdivision and large ranges, organisms with high vagility and low specialization, high proportions of β-clades formed by polyploidization, and little β-anagenesis. We predict global and interregional geographic patterns in these variables caused by differential ORD. Thus, ORD potentially explains a wide array of patterns, suggesting ORD as a fundamental factor in evolution. The vulnerability of biotas to many human activities should vary with the magnitude of ORD.

---

[1]The authors contributed equally to this work.

INTRODUCTION

Research in recent decades has led to a major expansion in the knowledge of long-term climatic variability and of how organisms respond to such changes. The discovery that Earth's climate varies with variations in its orbit (Berger 1988, Hays et al. 1976, Imbrie et al. 1993b) revolutionized modeling of past climates (Wright et al. 1993). Owing to increased sampling of fossils and new dating techniques, it is now possible to document evolutionary responses and changes in the distribution of organisms in relation to climatic events (Bennett 1997). The combination of methods and ideas from genetics, systematics, and biogeography in the field of phylogeography (Avise et al. 1987, Avise 2000) has also made it possible to infer past changes in geographical distributions and levels of gene flow from patterns of genealogical relatedness among clades (Cruzan & Templeton 2000). All this new information has led to the realization that recurrent climatic changes affect evolution and cause biogeographical and macroecological patterns (Bennett 1997, Dynesius & Jansson 2000, Hewitt 1996).

BACKGROUND

Orbital Forcing of Climates

Beyond seasonal changes, the amplitude of climatic changes increases towards longer timescales but has marked peaks on the timescale of 10–100 thousand years (kyr) (Figure 1A). These climatic variations are caused by periodical changes in Earth's orbit, called Milankovitch oscillations, resulting from gravitational interactions with other planets in the Solar System, primarily Jupiter because of its size and Venus because of its proximity to Earth (Berger 1988, Berger & Loutre 1991). The tilt of Earth's axis varies with a 41-kyr period, the eccentricity of the orbit varies with a 100-kyr period, and the annual timing of minimum Earth-Sun distance varies with 19 and 23 kyr periods (Berger 1988). These orbital variations modify the spatial and temporal distribution of solar radiation received on Earth (Berger 1988, Imbrie et al. 1989). Combined with earthbound feedbacks from, e.g., atmospheric CO_2, global ice volume, and surface albedo, the variations in insolation produce large, rapid, nonlinear climatic changes (Berger 1988; Imbrie et al. 1989, 1993b) and stable conditions generally lasting only a few thousand years at a time (Imbrie et al. 1989, Webb & Bartlein 1992, Wright et al. 1993) (Figure 1B). In fact, the climate during the present 10-kyr-long interglacial period has been more stable than during glacial periods, which make up more than 90% of the Quaternary (Kukla 2000). Superimposed on orbital variation, there are millennial-scale climatic variations, which have been more pronounced during glacial periods (Dansgaard et al. 1993). Milankovitch oscillations have been a feature of Earth's entire history, as documented in rhythmic sedimentation patterns having the same periodicity as orbital variations (Olsen 1986, Zachos et al. 1997), although the amplitude and the relative importance of variation in different orbital

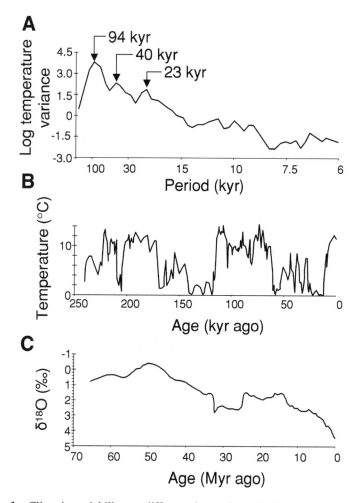

Figure 1 Climatic variability on different timescales. (*A*) The relative variance of climate is concentrated into peaks corresponding to variability in parameters of Earth's orbit. Data from the late Quaternary southern Indian Ocean. Adapted from Hays et al. (1976). (*B*) Sea-surface temperatures during the last 250 thousand years (kyr) at a site situated 50°N in the Atlantic, illustrating variability on Milankovitch timescales. Adapted from Imbrie et al. (1989). (*C*) Longer term variability in Earth's climate during the Cenozoic. Higher δ^{18}O-values indicate lower temperatures and larger global ice-volume. Myr = million years. Adapted from Zachos et al. (2001).

parameters have varied over time. For example, 100-kyr oscillations dominated during the past 800 kyr (Figure 1*A*), whereas 41-kyr oscillations dominated during the late Tertiary and the early Quaternary (Hooghiemstra & Sarmiento 1991, Imbrie et al. 1989, Webb & Bartlein 1992). Over millions of years, there are also long-term trends in Earth's climate caused largely by plate tectonics of oceanic and atmospheric flows (Figure 1*C*), determining the mean climate over many Milankovitch oscillations.

Orbitally Forced Range Dynamics

Paleoecological studies show that the dominant responses to recurrent climatic shifts of species and other clades recognizable in the fossil record, have been changes in the size and location of their geographical distributions (Coope 1979, Cronin et al. 1996, Davis 1976, Graham et al. 1996, Huntley & Birks 1983, Roy et al. 1996). Such range fluctuations have occurred globally, although the magnitude of change has varied tremendously among regions (Bennett 1997). In a previous paper (Dynesius & Jansson 2000) we named changes in geographical distributions in response to Milankovitch climate oscillations "orbitally forced species' range dynamics." However, the term applies to clades at any level of genealogical inclusiveness, from single gene pools to phyla, and therefore should be more generally rephrased "orbitally forced range dynamics" (ORD). ORD entail advances and retreats of range limits and also movements and extinctions of genes, genotypes, and subclades in nonmarginal parts of the geographical ranges caused by the moving environmental gradients.

The magnitude of climatic change varies geographically owing to variation in the amplitude of insolation change and to spatially varying effects of the earthbound feedbacks. For example, the 41- and 100-kyr oscillations cause larger temperature changes toward the poles (Imbrie et al. 1989, Wright et al. 1993), causing more ORD (Bennett 1997). For example, plant taxa in tropical rain forests persisted locally in large proportions of their ranges during the last glacial period (Flenley 1998). In contrast, tree taxa of the temperate deciduous forests of the Northern Hemisphere were often restricted to small refugia during the last glacial period and have since expanded up to many thousands of kilometers (Davis 1976, Huntley & Birks 1983). The latitudinal gradient in the 100-kyr oscillations is caused by earthbound feedbacks, whereas the 41-kyr orbital oscillations intrinsically lead to a latitudinal gradient in insolation change (Imbrie et al. 1989, 1993b; Wright et al. 1993). Therefore, we assume that a latitudinal gradient in ORD has been continuously present. Local to regional factors can buffer against regional climatic change. For example, tropical mountain cloud-forests have generally been considerably less climatically variable than the region to which they belong because local factors lead to persistent mist and cloud formation (Fjeldså 1995, Fjeldså et al. 1999). Oceanic islands are generally little affected by Milankovitch climate oscillations at least in tropical latitudes (Cronk 1997). Steep physical gradients reduce ORD because organisms only have to move short distances as temperature, moisture conditions, or water levels change (Brown 1995, Darlington 1943, Hewitt 1996, van der Hammen 1995). The prime examples are steep slopes, both on land and under water. Both local moderation of climate variability and steep gradients were probably important for the survival of tree species in the

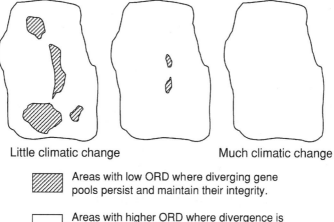

Little climatic change   Much climatic change

Areas with low ORD where diverging gene
pools persist and maintain their integrity.

Areas with higher ORD where divergence is
lost due to extinction or mixing of gene pools.

Figure 2   Schematic maps showing the total area occupied by a clade over an entire
period of the longest significant Milankovitch oscillations (about 100 kyr). The level
of orbitally forced range dynamics (ORD) of the constituent gene pools of the clade
varies within this area. With increasing magnitudes of regional climatic change, smaller
proportions of the total area maintain gene pools that both survive and remain separated
over the entire time period.

mountains of southern Europe during glacial periods (Bennett et al. 1991). All
else being equal, the lower the magnitude of climatic change, the larger the area
continually inhabited by a clade (hatched areas of Figure 2). Those of the clade's
gene pools that are in such areas will exhibit low levels of ORD.

ISSUES AND SCOPE

Concepts and Scale Dependence

A gene pool is the pool of genetic information carried by all individuals in a population. A gene pool and all its descendant gene pools constitute a clade (Williams
1992). Evolution is affected by sorting among genetic variants at any level of
genealogic inclusiveness (Williams 1992). The evolutionary effects of climate
change vary with genealogic inclusiveness and are thus spatially and temporally
scale dependent. Individual gene pools occupying small areas are more likely
to be strongly affected by climate shifts than more inclusive clades occupying
larger areas. Therefore, we use a hierarchical approach, considering how orbitally
forced climatic oscillations lead to sorting of genetic variation at different levels of
genealogical inclusiveness. Sorting can be by natural selection, e.g., clade selection

sensu Williams (1992), or occur by chance, e.g., stochastic extinctions of populations that happen to be in the wrong place when climate shifts. We do not restrict clade selection to selection on emergent group properties because of the difficulties in determining which properties are emergent and at what level selection actually occurs. Ultimately, of course, any property of a clade can be ascribed to characters of individuals. If any process operating at a lower level, such as selection of genes within gene pools, results in differences in fitness among gene pools, it has produced the kind of fitness variation required for a higher level of selection (Williams 1992).

Evolutionary effects of orbitally forced climatic change are strongly temporally scale dependent. Most cladogenetic events (splitting of gene pools) do not contribute to long-term evolution, because gene pools merge or go extinct after relatively short time periods. To make a conceptual distinction, we name clades that at any point in time have survived and remained genetically separate for more than 100 kyr β-clades (Figure 3). These clades have been sorted by all climatic challenges during at least one entire period of the longest significant Milankovitch

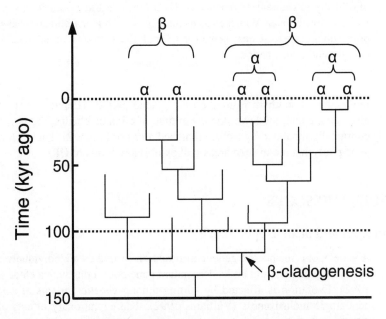

Figure 3 Conceptual distinction between recently produced clades and older ones that have been sorted during several climatic shifts. β indicates β-clades, i.e., clades older than 100 kyr at time zero. These clades have been sorted by all climatic challenges during at least one entire period of the longest significant Milankovitch oscillations. α indicates α-clades, i.e., clades produced by more recent cladogenesis. A single node, marked by an arrow, represents β-cladogenesis at time zero. All other nodes represent α-cladogenesis.

oscillations. A β-clade may or may not be reproductively isolated or morphologically distinguishable from its sister β-clade, and thus taxonomically recognized. The sister clades produced by more recent cladogenesis are α-clades, which are always nested within a β-clade (Figure 3). Of course, all β-clades started out as α-clades. In accordance with the naming of clades, we name all cladogenesis irrespective of the subsequent duration of the resulting clades α-cladogenesis. The subset of α-cladogenesis in which both sister clades survive and remain genetically separated at least 100 kyr becomes β-cladogenesis. Thus, a β-cladogenetic event must be at least 100-kyr old. The rate of α-cladogenesis equals the frequency of gene pool separation, whereas the rate of β-cladogenesis is also determined by later events. By analogy, we name any evolutionary change within a lineage over time α-anagenesis, and the change persisting more than 100 kyr β-anagenesis.

Orbitally Forced Range Dynamics as a Factor in Evolution

The primary role of recurrent orbitally forced climatic shifts is to put limits on how far evolutionary processes may go, rather than being a creative force in macro evolution. Orbitally forced climatic shifts can be seen as an overarching tier, interrupting evolutionary trends accumulating during the relatively short periods between climatic shifts, as gene pools mix or go extinct, or selective regimes change (Bennett 1990, 1997). Geographical variation in the magnitude of climatic shifts will cause variation in the degree to which evolutionary processes are interrupted. If the magnitude of ORD is low, cladogenesis and anagenesis, triggered by a wide range of processes, may continue, leading to an accumulation of β-clades and anagenetic change. In an influential paper, Hutchinson (1959) posed the question, "Why are there so many kinds of animals?" As pointed out by Blackburn & Gaston (1996), most models to explain large-scale, geographic patterns in numbers of taxa have, in the spirit of Hutchinson, considered high diversity, e.g., in the tropics, as the phenomenon. In contrast, we conclude that Milankovitch climatic oscillations curb rates of β-cladogenesis, leading to fewer taxa in a region than otherwise (Figure 4). Thus, Hutchinson's question might better be rephrased, "Why are there so few kinds of organisms in most parts of the world?" If the arguments presented here are correct, a world without orbitally forced climatic oscillations would have had biota with higher taxonomic diversity and more phenotypic disparity. This is in direct opposition to the idea that Milankovitch climate oscillations are primarily a factor raising diversity in the long term by fragmenting ranges, allowing divergence among isolated refugia owing to restricted gene flow (Clarke & Crame 1997; Haffer 1969, 1997; Valentine 1984; Vrba 1992). If this was shown to be generally true, our arguments would be undermined. However, molecular genetic data of rain forest faunas, for which the hypothesis was originally formulated (Haffer 1969), lend little support to the hypothesis that orbitally forced climatic changes favor speciation (Moritz et al. 2000). Instead, we argue that orbitally forced climatic shifts generally increase gene flow directly by

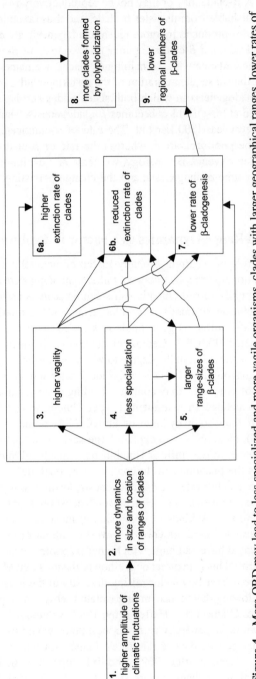

Figure 4 More ORD may lead to less specialized and more vagile organisms, clades with larger geographical ranges, lower rates of β-cladogenesis, more clades formed by polyploidization, and fewer β-clades.

reshuffling gene pools (Futuyma 1987), as well as indirectly by selecting for vagility and generalism. In areas where a clade is continuously present, gene pools may survive and maintain their genetic integrity, in allopatry or sympatry, over many climatic oscillations (García-Moreno & Fjeldså 2000). ORD reduces the extent of these areas and thus also curbs β-cladogenesis and β-anagenesis. For example, many European species have persisted and diverged genetically during both glacial and interglacial periods in the southern mountainous peninsulas, whereas northern expansions during interglacials were extinguished by each cold period (Bennett et al. 1991, Hewitt 1999). Good examples of radiations within stable areas are found in Lake Baikal, in the Cape flora, and among salamanders in Middle America (see below).

Beyond Milankovitch timescales, over millions of years, mean climates change (Figure 1C). These changes are potentially strong creative forces in macroevolution, in contrast to Milankovitch oscillations. Changes in mean climate are claimed to cause periods of increased speciation and species extinction, so-called turnover pulses (Vrba 1985, 1993). Turnover pulses may be produced in the following way: Changes in the long-term mean climate lead to a pulse in β-cladogenesis because regionally ephemeral habitats become persistent and because new habitats appear. Conversely, β-clade extinction rates should rise because persistent habitats may become regionally ephemeral or disappear. In addition, the change in mean climate adds to Milankovitch climate oscillations, increasing the challenges for organisms. On an even grander timescale, tropical climates have probably been continuously present, whereas arctic and boreal climates have only been present during some geological epochs. Therefore, it has been argued that tropical biotas have existed for longer periods, allowing more clades to accumulate (Fischer 1960).

Relation to Other Theories

Because Milankovitch climate oscillations are global in extent and have been working during Earth's entire history, their predicted consequences can be tested using many independent data sets. To do this, data on ORD with sufficient temporal resolution and spatial coverage are needed. As an alternative, levels of ORD might be inferred from climatic data. Because alternative explanations have been proposed to most of the patterns we predict, a catalogue of patterns confirming our individual predictions constitutes a weak test of our hypotheses. Stronger cases can be made if (*a*) several spatial patterns predicted by varying levels of ORD co-occur, (*b*) predictions that are unique to one hypothesis are tested, and (*c*) if hypotheses make contradictory predictions. We analyze the co-occurrence of patterns in three different contexts: latitudinal gradients, biota of the Cape Region of South Africa, and New World salamanders (see Co-Occurrence of Several Predicted Patterns, below).

To evaluate other hypotheses in relation to the role of ORD is beyond the scope of this paper, but some general comments are relevant. First, proper null models should be defined (e.g., Colwell & Lees 2000) to identify phenomena in need of explanations. Second, some hypotheses are only correlative models, lacking mechanisms to generate predictions beyond the empirical correlation they describe. Third, because the factors behind many hypotheses covary spatially, it is necessary to identify situations in which they make opposing predictions. For example, most hypotheses of diversity patterns have been developed to explain the latitudinal gradient in taxonomic diversity, and consequently the proposed factors covary with latitude and make similar predictions along the gradient. The mechanisms most discussed recently are gradients in productivity (Wright 1983) and in the available area (Rosenzweig 1995). We consider ORD, productivity, and area as potentially complementary explanations. The question lies in their relative importance in different situations and whether they taken together suffice to explain observed patterns. The productivity hypothesis and our ORD hypothesis make opposite predictions in situations with low levels of ORD (hereafter "low ORD") and low productivity. One such example is the extremely species-rich, low productive fynbos vegetation of the Cape region (Cowling & Pressey 2001), conforming to the predictions of ORD but not the productivity hypothesis. Furthermore, within the Cape Region, the climatically stable and low productive western parts support more species than the more productive, climatically variable areas to the east (Cowling & Lombard 2002).

Climate change affects the factors involved in the productivity and area hypotheses. Productivity as well as the extent of biological provinces and the range sizes of their taxa change with climate. This variation increases with the amplitude of climatic changes. For example, the temperate forest biome in Europe at the last glacial maximum about 21-kyr ago was only a small fraction of its early Holocene size about 12 kyr later.

We synthesize and review the evolutionary consequences of Milankovitch climate oscillations. First, we review relatively short-term processes at work within populations and gene pools during ORD. In the next section we explore higher-order consequences concerning clade sorting, rates of clade extinction, cladogenesis, and anagenesis. Then we deduce predictions about geographical patterns caused by spatial variation in ORD and review data to test them. We conclude by describing implications for conservation. The hypothesized consequences are summarized in Figures 4–6.

RESPONSES OF POPULATIONS TO ORBITALLY FORCED CLIMATIC CHANGE

ORD entail several different processes. Given that populations are locally adapted, changes in climate and associated biotic and abiotic factors should alter fitness optima of population phenotypes throughout the range of the clade (Davis & Shaw 2001). In response to this, local populations may either go extinct, adapt to the new conditions in situ, or individuals may disperse to and establish new populations in environments that have become suitable (Jackson & Overpeck 2000). Gene flow may increase as individuals from previously separated populations come into contact and interbreed (Futuyma 1987). The type of response depends on the magnitude and rate of the climatic change, but traits and locations of the organisms are also important.

Consequences of Advancing Range Limits

SELECTION FOR VAGILITY   When climate changes, range boundaries may advance. Marginal populations at range boundaries may become sources for new populations beyond the present range boundary. Phenotypes with high vagility (i.e., high dispersal ability and propensity) are more likely to found new populations and to initiate range advances. The descendents of the individuals arriving first preempt suitable habitats to the disadvantage of organisms arriving later (Hewitt 1999). This leads to selection for vagility, provided differences in vagility are heritable (Figure 4, *box 3*). Repeated founding events after long-distance dispersal would result in rapid directional selection for vagility (Cwynar & MacDonald 1987). Range advances are often achieved by long-distance dispersal. For example, the observed rapid advances in the beginning of the Holocene of north temperate tree taxa can only be explained this way (Cain et al. 2000, Clark et al. 1998, Skellam 1951). In such situations, selection for vagility should primarily affect traits associated with the tail of dispersal curves.

There are several examples in which advances of range limits are correlated with higher proportions of vagile phenotypes, although it has seldom been conclusively shown that the change is heritable. Lodgepole pine, *Pinus contorta*, has, into recent time, been advancing northwards from glacial refugia. Seed dispersability, measured as the size of the seed wing relative to seed mass, increases the more recently the population was founded ($r^2 = 0.77, P < 0.001, n = 14$) (Cwynar & MacDonald 1987). Two bush cricket species, *Conocephalus discolor* and *Metrioptera roeselii*, have expanded their distributions in Britain over the past decades (Thomas et al. 2001). In both species the frequency of the most long-winged, dispersive form was higher in recently founded populations, which is argued to at least partly have a genetic basis. Assuming that virtually all long-distance movements are achieved by the longer-winged morphologies, their higher frequencies in populations founded less than 10 years ago represent about 4-fold and 14-fold increases in long-distance dispersal for *C. discolor* and *M. roeselii*, respectively (Thomas et al. 2001). Individuals of the butterfly *Pararge aegeria* subsp. *tircis* from populations established after recent range advances were heavier and had relatively more flight muscles compared with older populations, which should make them able to fly longer distances (Hill et al. 1999). The individuals were raised in a common laboratory environment, excluding environmental (except maternal) effects on the phenotype.

Selection against vagility owing to local adaptations, dispersal-related mortality, and (at least among flying insects) tradeoffs with reproductive investment are ubiquitous. Given this, can we expect to see persistent trends in vagility among taxa from regions differing in ORD? The carabid beetle *Pterostichus melanarius* is dimorphic for wing length. In its native range in Europe the proportion of the long-winged form was low in old habitats (about 2%) but considerably higher (24%–45%) in new and unstable habitats (den Boer 1970). In its introduced North American range the proportion of the long-winged form increased from about 20% to 60% to 70% in successively more recently founded populations (Niemelä & Spence 1991). In one such recently established population, the proportions of long-winged individuals dropped from 61% to 43% over 5–6 years, suggesting that the long-term effect may be less conspicuous (Niemelä & Spence 1999). Cody & Overton (1996) reported that plants colonizing small islands along the coast of British Columbia evolved reduced seed dispersal distances within 10 years. In this case the fitness of dispersing seeds must be effectively zero because they are lost in the ocean. In continental settings selection against dispersal must be much less severe. Furthermore, less conspicuous range dynamics also occur at millennial and shorter timescales (Webb & Bartlein 1992), suggesting that traits enhancing vagility should be frequently favored in climatically variable areas.

SELECTION FOR LOW SPECIALIZATION   Organisms differ in the width of their environmental tolerance, habitat and resource specialization, and in how dependent they are on specific species (Futuyma & Moreno 1988). Less specialized, more tolerant individuals are more likely to find suitable habitats, establish in new areas, and survive while moving through heterogeneous environments. Thus, range advances should select for low specialization and wide tolerance (Figure 4, *box 4*). This conclusion is supported by a spatially explicit patch-occupancy model (Bowers & Harris 1994). In the model a broad-niched generalist species was able to track environmental changes better than a specialist species. In constant or slowly changing environments the specialist species was favored owing to better competitive abilities, but with moderate or rapid environmental change, the generalist species was able to limit the ability of the specialist to survive and/or track the shifting environment. An empirical example of selection for generalism is the northward advance of the butterfly *Aricia agestis* in Britain (Thomas et al. 2001). Northern populations formerly laid their eggs mostly on *Helianthemum chamaecistus*, but recent range expansions have been accomplished primarily by utilizing *Geranium* and *Erodium* growing in habitats that formerly were probably too cool. This has enabled the butterfly to bridge gaps greater than 14 km in the fragmented distribution of the original host plant. This would have been extremely unlikely without these stepping-stones. In a field choice experiment wild females from recently founded populations preferred to lay eggs on *Geranium molle*, the most widespread host plant used during the range advance, even though they naturally used other host plants at these localities. Butterflies retained their differences in host plant preference when reared in a common laboratory environment, strongly suggesting a genetic contribution to host plant choice (Thomas et al. 2001).

Improved environmental conditions at existing range boundaries are likely to initiate range advances without any evolutionary change. Once the advance is initiated, individuals that advance most rapidly are likely to be favored. Advancing range fronts become characterized by dispersive generalists or by specialists on habitats common in the colonized area (Thomas et al. 2001).

INTERBREEDING DURING RANGE ADVANCES   Individuals from previously separated populations of sexual organisms may come into contact and interbreed during range advances. This may result in hybrid speciation (e.g., polyploidization; see below) or various degrees of introgression between differentiated gene pools or clades. Introgression may be counteracted by the formation of hybrid zones (Hewitt 1975, 2001; Remington 1968). The postglacial northward advance of European white oaks, *Quercus robur* and *Q. petrea*, and silver fir, *Abies alba*, involved introgression. The white oaks spread from three refugial regions with distinctive chloroplast DNA types and mixed in central and northern Europe. The geographic pattern of chloroplast DNA types is independent of the taxon boundary, implying both intraspecific and interspecific introgression (Comes & Kadereit 1998). Silver fir appears to have survived in five refugial regions and advanced northwards from three of these. The entire French refugial gene pool appears to have merged with the north Italian gene pool. The north Italian and the Balkan refugial gene pools have mixed in two extensive areas in central Europe, but also retain their genetic integrity in the refugia and over large expansion areas (Konnert & Bergmann 1995).

Responses to Climatic Change in Other Situations

The same processes as those at work during advances of range limits are active in other parts of the range where climate changes. However, populations in other situations are more likely to go extinct without leaving offspring during climatic change. Organisms may fail to establish new populations, either because they cannot move fast enough, because they encounter barriers (Ashworth 1996, Brown 1995), or because the preferred habitat becomes rare or temporarily disappears regionally. Establishment in new areas with suitable habitat may also be impeded by competition from individuals of the same clade already present before the change. Survival in situ is hampered by competition with immigrants better adapted to the new environment at that site. Again, all this may lead to selection for high vagility and low specialization (Figure 4, *boxes 3 and 4*). Low specialization is selected for because it is less likely that the niche of such individuals will disappear completely in an area (Jackson & Overpeck 2000, McGlone 1996). Moreover, less specialized organisms do not need to disperse as rapidly to track their habitat. If peripheral populations go extinct, range limits retreat.

MACROEVOLUTIONARY CONSEQUENCES

The immediate responses of populations to climate change add up to a number of evolutionary consequences on longer timescales and/or for more inclusive clades, including clade sorting as well as long-term cladogenesis and anagenesis.

Clade Sorting

The probability of long-term survival varies among clades depending on characters of organisms, e.g., degree of specialization and vagility. Generalism and high vagility promote high population densities and wide distributions. All these factors are positively correlated with low extinction rates at the level of species or higher taxonomical rank (McKinney 1997). There is also nonadaptive sorting caused by the large geographic variation in ORD. Clades that are in the right place, i.e., in areas demanding the least distributional change, have higher probability of surviving climatic shifts and should thus have higher fitness over long timescales. Oceanic islands have high proportions of paleoendemics, which have disappeared from continents, probably because stable climates and poverty of competitors and exploiters have allowed island clades to persist longer (Cronk 1997). For most taxa of European forest trees, only clades that persisted during both glacial and interglacial times in southern and southeastern Europe have living European representatives (Bennett et al. 1991). The northern populations of most taxa went extinct in situ during the glacials. In the beginning of interglacials northern Europe was colonized from southeastern and southern refugia (Bennett et al. 1991).

The relationship between extinction rates and magnitude of ORD should differ between α- and β-clades. For α-clades, represented by terminal branches in a cladogram, the risk of going extinct should increase with higher ORD. However, β-clades, which have been sorted by several previous climatic shifts, include at least one gene pool with traits that enabled survival, and/or occupy an area requiring little movement (Figure 2). Such gene pools are more likely to survive similar climatic shifts in the future, making the β-clades to which they belong less extinction prone (Figure 4, box 6b). Thus, climatic shifts act as extinction filters (Balmford 1996), pruning phylogenetic trees from extinction-prone clades. The filtering will be stronger, reducing extinction rates of β-clades more, the higher the magnitude of ORD. Thus, for β-clades the positive relationship between their extinction rate and ORD should be less steep than for α-clades, or even negative, depending on the effectiveness of filtering at climatic shifts in reducing subsequent extinctions. Evidence from fossils of marine taxa (species and higher taxonomic ranks) indicate that tropical taxa (low ORD) are on average geologically more short-lived and have higher extinction rates than taxa at higher latitudes (high ORD) (Flessa & Jablonski 1996). This suggests a negative relationship between ORD and extinction rate of β-clades.

Another aspect of clade sorting by climatic changes is the effect on the size and composition of regional assemblages of clades. For example, Europe is poor in temperate-forest tree taxa compared with eastern Asia and eastern North America, because Europe has lost more taxa owing to late-Tertiary and Quaternary climatic oscillations (Latham & Ricklefs 1993). Clades including vagile, generalist individuals are selected for when groups of clades respond to climatic shifts. An empirical example is the 46 nonmigratory butterfly species that reach their northern limit in the United Kingdom. If temperature is limiting, these species should have benefited from climate warming during the past 30 years, but intensification of agriculture has led to 70% losses of seminatural habitats since 1940 (Warren et al. 2001). A higher proportion of the species that are mobile (45%) and/or habitat generalists (50%) expanded their ranges, compared with sedentary (8%) and habitat-specialist species (7%; habitat specificity and low mobility were highly correlated among species). Low mobility and/or high habitat specialization restrict species in fragmented habitats and slow down advances across patchy landscapes (Warren et al. 2001). Another example is the colonization of Scandinavia by alpine carabid beetles during the last deglaciation. Alpine environments generally select against wings in insects (Sømme 1989), but alpine specialist species of Scandinavia are instead better flyers than generalist species that occur in both alpine and lowland areas. Probably only the good flyers among the alpine specialists were able to colonize from glacial alpine refugia during the rapid postglacial warming, whereas flightless habitat generalists have been able to disperse slowly through nonalpine habitats (Nilsson et al. 1993). In the alpine flora of New Zealand the deglaciated regions have primarily been colonized by generalist species or species capable of long-distance dispersal (McGlone et al. 2001). The probability that a clade will colonize a new suitable area should generally decrease with distance. Grytnes et al. (1999) proposed this as an explanation for the strong latitudinal diversity gradient in Scandinavian plants, having their presumed glacial refugia mainly in southern Europe. The time since appearance of a new habitat may also effect how many species have colonized. In New Zealand the postglacial sea-level rise converted a number of freshwater lakes into marine fjords ~9–17 kyr ago. There is a strong positive correlation between fjord age and local species richness of marine epifaunal invertebrates of the subtidal zone (Smith 2001).

Cladogenesis

We argue that ORD may raise the rate of α-cladogenesis but curb β-cladogenesis (Figure 5). There is a wide range of mechanisms, abiotic as well as biotic, that could lead to separation of gene pools in sexual organisms. The separation occurs when gene flow is reduced or prevented by barriers to dispersal (allopatrically), distance (parapatrically), disruptive selection (sexual or ecological) within a locality (sympatrically), or abrupt speciation (e.g., polyploidization). If ORD is low, gene pools are more likely to survive and remain separated (Figure 5). Moreover, organisms may specialize and evolve reduced dispersal abilities without going extinct. This leads to decreasing gene flow, further favoring the integrity of gene pools and leading to radiation of β-clades.

The importance of isolate formation in controlling speciation rates has been emphasized, assuming, explicitly or implicitly, that speciation rates are limited by the rate of α-cladogenesis. However, gene pool separation (α-cladogenesis) leads to β-cladogenesis only if the gene pools survive and maintain their integrity. This fact has been accounted for in some speciation models, which argue that speciation is most likely at intermediate levels of disturbance. At low frequency

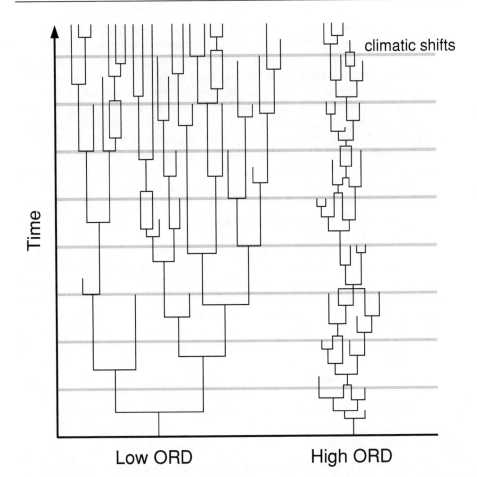

Figure 5 Cladograms depicting the evolution of lineages through several climatic shifts in regions with low and high ORD, respectively. Lineages may split (cladogenesis), merge (mixing of gene pools), or terminate (extinction). In situations with low ORD, α-clades persist longer, leading to deep-branched trees and accumulation of lineages. At high ORD, α-clades are more likely to go extinct or merge, keeping the number of β-clades low, even if α-cladogenesis may be more frequent. In this example, the probability of splitting and merging is more than twice as high whereas the persistence of lineages is less than half, at high compared to low ORD. Cladogenesis through hybrid speciation, e.g., polyploidization, is excluded from the cladogram.

or intensity of disturbance few isolates are formed, whereas at high levels isolates do not persist long enough to allow speciation (Allmon et al. 1998, Stanley 1986). We argue that the general relationship between rates of β-cladogenesis and ORD are monotonically negative under realistic conditions. In other words, even though ORD favor some factors leading to α-cladogenesis, the rise in rates of extinction of

α-clades (Bennett et al. 1991; Cronin 1985, 1987; Hewitt 1996) and of merging of α-clades through interbreeding (Coope 1994, Futuyma 1987, Glazier 1987) will be stronger. Moreover, these processes will also negatively affect cladogenesis not caused by ORD. Thus, the net effect of ORD will be to curb β-cladogenesis (Figure 5). We see two main reasons for this. First, there is much α-cladogenesis that is not caused by climatic change. Second, ORD indirectly disfavor all processes leading to α-cladogenesis (except polyploidization) because ORD select for characters enhancing gene flow (generalism and high vagility).

The magnitude of orbitally forced climatic change determines what proportion of a clade's gene pools remain unaffected by extinction and gene-pool merging. Such persistent gene pools will be primarily in areas continuously occupied by the clade (Figure 2). This concept of areas with persistent gene pools is more restrictive than that of glacial refugia (Bennett et al. 1991, Haffer 1997, Hewitt 1996) in being not only a place for clade survival, but a place where gene pools are not disrupted by climatic shifts. Regions with a large area with persistent gene pools should have high rates of β-cladogenesis and β-anagenesis.

POSITIVE EFFECTS OF ORD ON α-CLADOGENESIS   ORD should favor gene pool splitting in three principal ways: (*a*) by creating physically separated (allopatric) gene pools, (*b*) by favoring expansion into new areas where parapatric divergence can start, and (*c*) by causing a paucity of specialized organisms, leaving resources underexploited and thereby making ecological divergence in sympatry more frequent (Schluter 2000). Parapatric divergence occurred in the marine gastropod *Acanthinucella spirata* as it expanded northward along the Californian coast following the last glacial maximum. The recently established populations form a clade morphologically distinct from populations in refugial areas, fossil as well as recent (Hellberg et al. 2001). At high latitudes suddenly appearing freshwater habitats following glacial retreat have led to bursts of cladogenesis and phenotypic evolution in colonizing fishes by providing environments with few competitors (Schluter 2000). Pairs of such divergent, closely related fish clades usually divide resources in the same way, one being a pelagic zooplanktivore, and the other a benthic prey consumer. Postglacial ecological cladogenesis has been repeated in many freshwater bodies and for several fish taxa in northern areas. Such evolution of similar clade pairs repeated in several waterbodies is virtually nonexistent at lower latitudes (Schluter 1996).

GENERALISM AND VAGILITY REDUCE BOTH α- AND β-CLADOGENESIS   Indirectly, ORD reduces the probability of cladogenesis by selecting for generalism and vagility (Figure 4, *box 7*), thereby raising gene flow. For example, niche extension to a new host plant type by an herbivorous insect clade is more likely to lead to a new isolated gene pool, the narrower the host-plant use in the clade. If host plant specialization is low, gene flow between the individuals utilizing the new host plant and those using other host plants is more likely, preventing α-cladogenesis. Moreover, contact between two geographically separated but closely related clades is more likely to lead to gene flow between them if the organisms have broad niches. This leads to lower β-cladogenesis. Empirically, the frequency of β-cladogenesis is lower in generalist than specialist clades (Chown 1997, Kauffman 1978, Stanley 1979, Vrba 1989).

High vagility enhances gene flow, making cladogenesis less probable (Chown 1997, Mayr 1963). High frequency and long distance of dispersal decrease the probability of both parapatric (Gavrilets et al. 2000b) and sympatric (Kawata 2002) cladogenesis according to model simulations. Empirically, there is a general negative correlation between the geographical genetic structure among populations and the vagility of organisms (Bohonak 1999). In marine taxa, high fecundity and highly vagile larvae are associated with low speciation rates in the fossil record (Chown 1997, Jablonski 1986). The same pattern is found among land plants (Chown 1997). On small islands, only vagile taxa are unable to radiate (Terborgh 1992). In Hawaiian crickets the low vagility and high specialization compared with their continental relatives have led to a tremendous radiation resulting in a unique concentration of species (Otte 1989). In regions with a large area with persistent gene pools (Figure 2), organisms that for some reason (e.g., low vagility or patchy habitats) have low levels of intraclade gene flow may radiate extensively while others do not. Thus, the latter type of organisms, with little innate tendency for cladogenesis, should exhibit less spatial patterning in relation to ORD. For example, the latitudinal diversity gradient is much less pronounced or is absent in generally vagile groups such as protozoa, microalgae, meiofauna, and bryophytes (Churchill et al. 1995, Hillebrand & Azovsky 2001).

ORD AND REPRODUCTIVE BARRIERS   Reproductive barriers make clades resistant to merging with other clades, thus increasing clade persistence. Reproductive barriers may evolve gradually over many generations (allopatrically, parapatrically, or sympatrically) or form abruptly, e.g., by polyploidization (Mayr 1963). In this section, we discuss gradual formation of reproductive barriers; abrupt formation by polyploidization is treated in "Polyploidy," below. It has been argued that the short stable periods during Milankovitch oscillations (Figure 1B) are generally not long enough for reproductive barriers to evolve before isolation is lost (Coope 1995, Futuyma 1987, Glazier 1987). It is believed that evolving reproductive isolation takes tens of thousands to millions of years in most cases (Avise et al. 1998, Coyne & Orr 1997, Mayr 1963), but it can occur much faster if divergent selection leads to reinforcement of mating barriers (Hendry et al. 2000, Higgie et al. 2000). In a review of extensive data on genetic distances and degrees of reproductive isolation (derived from crossing experiments) among species pairs of *Drosophila* fruit flies, Coyne & Orr (1989, 1997) concluded that evolving reproductive isolation takes about 0.2 million years (Myr) for taxa that currently are sympatric and about 2.7 Myr for allopatric taxa. Using molecular data and phylogenetic relationships among and within about 300 vertebrate species, Avise and coworkers (Avise et al. 1998, Avise 2000) estimated the average speciation duration to be $\sim$1–3 Myr. McCune & Lovejoy (1998) concluded that the time for speciation ranged between

0.8 and 2.3 Myr for sister species of fish of putative allopatric origin and up to 0.5 Myr for fish species pairs assumed to have originated in the same lake. Eight pairs of marine clades that became separated by the Isthmus of Panama approximately 3 Myr ago have all evolved some degree of reproductive isolation, ranging from incomplete prezygotic to complete pre- and postzygotic isolation (Lessios 1998).

To conclude, although evolving reproductive barriers can be rapid if divergence affects mate choice, and seem to be rapid in radiations of freshwater fish in situations with much ecological opportunity, most estimates indicate that evolving reproductive barriers normally takes far longer than the stable parts of Milankovitch oscillations. Thus, the probability of evolving reproductive barriers should decrease with increasing magnitude of ORD, although the formation of hybrid zones may to some extent protect the genetic integrity of clades when partial reproductive barriers have evolved (Hewitt 1975, 2001; Remington 1968).

EXAMPLES OF ORD REDUCING β-CLADOGENESIS   There are many examples indicating a monotonically negative effect of ORD on β-cladogenesis. Even within the low-ORD tropics, β-cladogenesis is higher in areas with the lowest levels of ORD, such as mountains with steep gradients and locally stabilized climate. In tropical South American and African birds and in tropical African plants low climatic amplitude, inferred from the occurrence of paleoendemic species that are extinct elsewhere, is well correlated with the occurrence of recently formed species with restricted ranges (neoendemics) (Fjeldså 1994, 1995; Fjeldså & Lovett 1997). These hotspots of paleo- and neoendemic congruence are found in montane forests, e.g., in specific parts of the tropical Andes, whereas the lowland rainforests are dominated by old, widespread species. The endemism hotspots are in places with peaks in climatic stability, caused by local, orographic moderation of regional climate (Fjeldså 1995; Fjeldså et al. 1997, 1999). The correlation between current climatic stability and peaks in paleo- and neoendemic congruence indicates that local moderation of climatic extremes persisted through the shifting climates of the Quaternary, permitting gene pools to survive and maintain their integrity.

In Indo-Pacific shallow-water corals, the frequent sea-level changes during the Quaternary repeatedly laid the continental shelves dry, so that any particular area was available for coral growth for on average 3.2 kyr at a time. These frequent large-scale distributional changes probably prevented β-cladogenesis, and most of the coral species have little geographical genetic subdivision (Potts 1983). During the Great American Biotic Interchange, a number of North American mammalian taxa entering South America diversified extensively, whereas South American taxa entering North America exhibited very little β-cladogenesis (Marshall et al. 1982). Our interpretation of this is that clades entering a region with low ORD (South America) radiate, whereas clades entering a region with high ORD (North America) produce few β-clades. In addition, the initial cladogenesis in the south-heading clades had already taken place in tropical North America prior to the interchange

(Webb 1991), the part of the continent that had lowest ORD during the Pliocene and Quaternary. A parallel to this is found among pelagic foraminiferans in the early Tertiary, during which acquisition of algal symbionts is associated with radiations. However, one symbiotic clade, originating at high latitudes, remained a low-diversity taxon for almost two Myr. As the clade then expanded into low latitudes, it gave rise to several β-clades (Quillévéré et al. 2001). In some situations, however, ORD may favor β-cladogenesis by allowing clades to colonize otherwise inaccessible refugial areas. For example, flightless grasshoppers of the genus *Melanoplus* have speciated in the Rocky Mountains during the high-ORD Pleistocene (Knowles 2000). Cool climates during glacial periods made it possible for populations to colonize mountaintops inaccessible during interglacial periods. Populations then diverged among mountaintops (Knowles 2001). Merging of gene pools during subsequent climatic shifts may have been prevented by rapid evolution of reproductive barriers by sexual selection (Knowles 2000). However, it is unknown what the rate of β-cladogenesis would have been in this sedentary clade under low ORD.

Anagenesis

Anagenesis is phenotypic evolutionary change within a lineage over time, irrespective of whether changes evolve gradually or are punctuated. As long as there are stable differences in selection among gene pools and gene flow is not excessive, gene pools can develop any degree of difference in any number of characters. The geographic variation among gene pools in different parts of extant species' ranges (Endler 1986) shows that such differences are ubiquitous. The level of genetic differentiation among gene pools depends on the rate of divergence among gene pools, and for how long the gene pools survive and remain isolated.

All aspects of ORD affecting cladogenesis may also affect anagenesis. ORD may drive anagenetic change by recurrently providing new biotic and abiotic environments favoring α-anagenesis. Clades that expanded into new areas during the Holocene have adapted to local environmental conditions (Davis & Shaw 2001, Hellberg et al. 2001). High ORD may lead to habitats poor in specialized clades setting the stage for adaptive radiations into vacant niches, causing surges of α-anagenesis (Schluter 2000). The rate of differentiation can be very rapid under strong selection, e.g., following expansions into new areas (Bell 1988, Schluter 2000). However, ORD also curb anagenesis by favoring vagile and generalized organisms, thereby promoting gene flow and slowing down divergence. In addition, oscillating climates are likely to produce oscillating selection pressures, thereby counteracting β-anagenesis. ORD also curb β-anagenesis by making it more likely that diverging gene pools accumulating anagenetic change go extinct or merge with other gene pools before evolving reproductive barriers (see above). Analogous arguments have been used for differentiation in metapopulations on short timescales. Rapid turnover of populations means that local populations are ephemeral and are likely to go extinct before they have diverged much (Gavrilets et al. 2000a, Lande 1992, Slatkin 1977). However, when ranges expand by the successive founding of new populations, differentiation among populations in a metapopulation might be favored if the propagules come from only one source population (Harrison & Hastings 1996). In contrast, if the propagules for a new population come from many source populations, range expansion would lead to mixing and thus reduced differentiation among populations (Harrison & Hastings 1996).

NORMALIZING CLADE SORTING   ORD may curb β-anagenesis by imposing long-term normalizing sorting. The gene pools most likely to survive climatic shifts are either inhabitants of locally persistent as opposed to ephemeral habitats (Bell 1988, Williams 1992) or clades that survive by habitat tracking, thereby always experiencing environmental conditions to which they are sufficiently adapted (Eldredge 1989). The survivors are likely to be the ancestral forms because the habitats producing new forms periodically disappear from the region. Because diverging lineages have lower probability of survival, cladograms will look more like a raceme than a tree (Figure 6). On a broad scale, such as in the fossil record, this will look like long-term anagenetical stasis. Clades branch off during each stable period, but these diverging branches are aborted at subsequent climatic shifts, leaving primarily the ancestral forms. The more a clade has diverged from the ancestral form and acquired adaptations to the local environment, the more likely it will be disfavored by climatic change. Therefore, although sometimes there will be more α-anagenesis in regions with high ORD, β-anagenesis will be lower. If ORD is low, normalizing clade selection will be less strong, making new locally adapted gene pools almost as likely to survive as the ancestral form (Figure 6). The idea of more β-anagenesis in regions with low ORD parallels Sheldon's (1996) hypothesis of continuous phyletic evolution and much anagenetic change in stable environments, whereas stasis is hypothesized to prevail in unstable environments.

Marine three-spined sticklebacks, *Gasterosteus* spp., have from the Miocene onward repeatedly invaded freshwater drainages, where extensive phenotypic differentiation has taken place (Bell 1988). Divergent clades within lakes are generally maintained despite gene flow between them (Taylor 1999). The freshwater forms soon go extinct, as the lakes they inhabit disappear (Bell 1988). Most extant forms inhabit lakes formed since the last glacial period (Bell 1988). McCune (1987) reports similar selective extinctions of lacustrine Mesozoic Semionotid fishes and persistence of their fluvial relatives. In the African Lake Turkana basin several mollusc populations diverged morphologically during Pleistocene periods of isolation owing to water-level regressions, but these divergent forms disappeared and were replaced with molluscs of ancestral morphology during subsequent transgressions (Williamson 1981). Despite New Zealand's cold, highly seasonal climates during most of the Quaternary there are no tree species adapted to those conditions because those conditions disappeared during brief, highly oceanic and warm periods (McGlone et al. 2001). High-latitude insects show almost no morphological

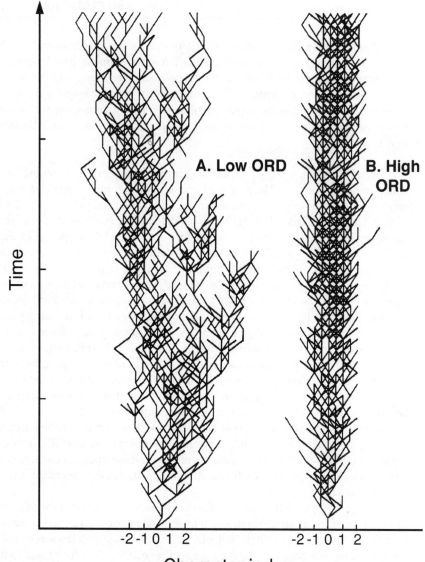

Figure 6 Normalizing clade sorting suppresses anagenetic change in lineages experiencing high ORD. The sorting becomes stabilizing as the gene pools most likely to survive inhabit persistent (as opposed to ephemeral) habitats, or are capable of moving sufficiently fast to track their preferred habitat. *A* (low ORD) shows a completely random phylogeny, whereas *B* (high ORD) shows random changes opposed by normalizing clade selection. At each time step, each lineage (gene pool) may change its character index. In *A*, the number of descendants is the same irrespective of character index, whereas in *B*, the number of descendant lineages is a decreasing function of character indices deviating from zero (simulating normalizing clade selection). Modified from Williams (1992). Used by permission from Oxford University Press, Inc.

change in the abundant Palearctic and Nearctic fossil record during the high-ORD Quaternary (Coope 1995) in spite of having short and often nonoverlapping generations, characters generally considered to promote evolutionary change.

EVOLUTIONARY INNOVATIONS   One aspect of anagenesis is the appearance of evolutionary innovations, i.e., new traits which in hindsight have proved to have significant long-term repercussions. More innovations and new higher taxa have appeared first in the tropics. In post-Paleozoic marine invertebrates first appearances of orders increase dramatically toward lower latitudes, suggesting higher rates of origination of new body plans close to the equator (Jablonski 1993). Angiosperm plants first appeared close to the equator and were absent from high latitudes well after the flora at low latitudes had diversified (Crane & Lidgard 1989). Thus, all the many derived traits defining this clade arose at low latitudes, as did the traits of the later-appearing huge angiosperm clade eudicotyledons (Crane & Lidgard 1989). Genera and families in late Cretaceous and Tertiary floras of North America consistently appeared later at high latitudes, as did taxa of North American dinoflagellates from the Cretaceous (Spicer et al. 1987). However, there are more β-clades to produce innovations in the diverse low-ORD areas. This means that, on the basis of the cited studies, we cannot say if ORD reduce β-anagenesis by limiting β-clade richness, by reducing the amount of anagenesis per β-clade, or both.

BIOGEOGRAPHICAL PATTERNS

Numbers of Clades and Spatial Genetic Divergence

Regions with high ORD should have fewer β-clades (Figure 4, *box 9*), owing to lower rates of β-cladogenesis. The sorting of clades during climatic shifts makes clades less extinction prone (Figure 4, *box 6b*). However, it is unlikely that high-ORD regions should accumulate more β-clades than a low-ORD region as a result of low extinction rates, because extinction rates are diversity dependent (Rosenzweig 1995). If numbers of clades increase, population densities and/or clade range sizes decrease, both factors that raise extinction rates. Rising local diversity also increases competition among organisms of different clades, leading to selection to become more specialized (Rosenzweig 1995) and making clades more extinction prone. Rising diversity will increase extinction rates in regions with low ORD as well, but numbers of clades will be sustained by high rates of β-cladogenesis. Thus, we predict that numbers of β-clades should be highest in regions with little orbitally forced climatic change and with steep environmental gradients. Numbers of species is a good proxy to test the prediction about β-clades, as there are plenty of data, but also because species are reasonably close to the tips of the phylogenetic tree but seldom so close as to merely represent α-cladogenesis. Higher taxa, representing branching events further away from the tips are less likely to have originated within the region and under similar geographical variation in ORD as during recent oscillations. There are several examples of extreme numbers of species in areas with low ORD, e.g., salamanders in Middle America, the Cape Floristic Region in South Africa, and the fauna of Lake Baikal (see below). Species diversity of South American birds peaks in the humid montane forests along the eastern slopes of the Andes (Rahbek & Graves 2001), where there is low ORD (Fjeldså 1995; Fjeldså et al. 1997, 1999). More generally, species numbers of mammals, birds, reptiles, amphibians, and vascular plants in different geopolitical units on the world's continents (differences in area controlled for) decrease with increasing amplitude of Milankovitch climate oscillations, measured as the estimated change in mean annual temperatures between the present and the last glacial maximum (R. Jansson, submitted).

There should be less spatial genetic divergence and less geographical genetic subdivisions within β-clades in regions with high ORD, all else being similar. The reasons are that higher proportions of the range should on average be recently colonized and that the β-clade has been subject to more extinction and merging of gene pools during climatic shifts. Such large-scale spatial genetic divergence is studied in the field of phylogeography (Avise 2000). The general pattern emerging from this field conforms to the prediction. Within β-clades, such as species, the genetic divergence among gene pools is deeper for clades that have experienced low ORD. In areas with high ORD, such as arctic and boreal regions, there is little genetic divergence among gene pools (Hewitt 2000, 2001), representing much α- but little β-anagenesis. In the climatically less variable southern Europe and southeastern United States, intraspecific genetic divergence is often high, and species are finely geographically subdivided genetically (Hewitt 2000, 2001). In southern Europe gene pools were able to survive and maintain their integrity in the mountains, where suitable habitats could be found continuously by small range changes (Hewitt 1996, 1999). For example, the Iberian endemic lizard, *Lacerta schreiberi*, has maintained separations between relatively closely situated inland and west coast gene pools since the Pliocene (Paulo et al. 2001). Divergence in mitochondrial DNA among phylogroups within fish species from Pleistocene glaciated areas in the Northern Hemisphere is much lower than within species from nonglaciated areas (Bernatchez & Wilson 1998, Billington & Hebert 1991). This pattern of less divergence in glaciated areas is even more evident within species with ranges spanning both glaciated and nonglaciated areas (Bernatchez & Wilson 1998). Gene pools within Neotropical bird species are genetically more deeply divided and more finely structured than temperate ones (Brawn et al. 1996). For example, in two Panamanian bird species genetic distances between two populations only 50 km apart exceed those found on a continental scale for many temperate species (Brawn et al. 1996).

LAKE BAIKAL   In freshwater systems that have persisted for long geological periods, such as Lake Baikal, nonvagile clades have evolved. Such clades typically form radiating species flocks, whereas vagile clades do not (Cohen & Johnston 1987). Despite the high-latitude position of Lake Baikal, the lake was not covered by glacial ice-sheets during the Quaternary (Kashiwaya et al. 2001), enabling clades to persist locally. The pelagic and deep benthic habitats harbor species flocks of both ancient, endemic clades and recently immigrated clades presently undergoing radiation, whereas widespread species primarily occur along the shorelines (Mashiko 2000, Sherbakov 1999, Sideleva 2000). One of the strongest radiations has been among amphipod crustaceans, and the lake contains 20% of the world's freshwater species of this group (Martens & Schön 1999). Spatial genetic divergence within species is also deep. The shallow-water, benthic crustacean *Eulimnogammarus cyaneus* has diverged for ∼60 kyr on either side of the lake outlet, having a maximum depth of only 4 m (Mashiko 2000). The high numbers of β-clades and narrow endemics in Lake Baikal contrast sharply with other large but recently glaciated or geologically young boreal lakes that have few and widespread β-clades.

Geographic Range Sizes

ORD should also increase geographic distributions of β-clades (Figure 4, *box 5*). First, clades with wide distributions are more likely to survive climatic shifts (Bennett 1997, Brown 1995). In the fossil record taxa with large range sizes are generally less extinction prone (McKinney 1997). Second, generalist organisms are likely to find habitats over large geographic areas. Third, clades consisting of vagile generalists attain larger ranges through effective colonization of areas with suitable habitats and by effectively maintaining marginal sink populations. In addition, high gene flow prevents cladogenesis, and thus subdivision of ranges. In many taxa, e.g., gastropods (Jablonski 1985, Stanley 1979), vascular plants (McGlone et al. 2001), and mosses (van Zanten & Pócs 1981), species with effective dispersal range more widely.

Regions with low ORD have higher proportions of range-restricted clades. The smaller the geographic range size of a clade, the more likely it is endemic to a specific area. Numbers of endemic species of mammals, birds, reptiles, amphibians, and vascular plants per geopolitical unit (differences in area controlled for) increase with decreasing amplitude of Milankovitch climate oscillations, measured as the estimated change in mean annual temperatures between the present and the last glacial maximum (R. Jansson, submitted). In sub-Saharan Africa range-restricted vascular plant species are concentrated into specific areas likely to have been buffered against Quaternary climatic oscillations (Linder 2001), leading to low ORD. Likewise, in the Australian flora all centers of range-restricted species are near-coastal, probably because variation in the extent of the central desert during the Quaternary limited the function of inland refugia for range-restricted species (Crisp et al. 2001).

Polyploidy

Abrupt formation of reproductive barriers mostly takes place as polyploidization, but also by diploid hybridization. Polyploidization accounts for about 2%–4% of the speciation events leading to extant species in flowering plants and about 7% in ferns (Otto & Whitton 2000). ORD favor the formation, establishment, and persistence of polyploid clades in at least four ways (Figure 4, *box 8*). First, large climatic oscillations lead to few clades being specialized to contemporary environments. This spurs ecological speciation by means not only of gradual divergence (Schluter 2000) but also polyploid establishment. Hybrid polyploids may rapidly provide reproductively isolated gene pools with new genetic combinations better adapted to new, suddenly appearing climates, habitats, or resources (Brochmann & Elven 1992; Stebbins 1971, 1984). Consequently, polyploids are more frequently found in human-made habitats or those most modified by recent climatic change (Hodgson 1987, Stebbins 1971). Second, during periods of large and rapid climatic change, a new hybrid clade can establish by expanding its distribution and thereby avoid competitive exclusion by its parental clades. Third, ORD make previously separated clades more likely to encounter each other and hybridize to form a new polyploid clade in a "secondary contact" (Stebbins 1971, 1984). Human introductions may play the same role. Several new allopolyploid plant species arose in the twentieth century, in each case involving an introduced taxon (Ramsey & Schemske 1998). Fourth, apomixis is generally associated with polyploidy, and facilitates dispersal and establishment, as one individual is sufficient for establishment and reproduction, and in plants no pollinator vector is needed to reproduce and spread in the colonized area. Consequently, apomictic representatives of clades more often range into previously glaciated areas (Bierzychudek 1985). For example, outbreeding diploids of the fern genus *Asplenium* in Europe are largely restricted to southern refugia, whereas their selfing polyploid descendants dominate the high-ORD parts of the continent (Vogel et al. 1999). Good colonizing ability of apomicts should lead to wide distributions, and there are many examples of temperate/arctic clades of plants in which diploids have small distributions and polyploids wide (Stebbins 1971). There are several well-supported examples of polyploid cladogenesis in high-ORD regions during the last great climatic shift in the early Holocene (Dufresne & Hebert 1997, Franzke & Hurka 2000, Nilsson & Jorde 1998).

Co-Occurrence of Several Predicted Patterns

We have deduced a number of evolutionary effects of ORD. From these effects we predict several geographic patterns caused by differential ORD. High ORD should lead to low numbers of β-clades; β-clades with low levels of intraclade genetic divergence, coarse geographical subdivision, and large ranges; organisms with high vagility and low specialization; high proportions of β-clades formed by polyploidization; and little β-cladogenesis and β-anagenesis. Thus, ORD parsimoniously explain a wide array of patterns, suggesting it is a fundamental factor in evolution. Below, we provide three good cases for the importance of ORD in which data concerning several of the predicted patterns are available.

LATITUDINAL GRADIENTS   The prime gradient in ORD is the latitudinal gradient (see "Orbitally Forced Range Dynamics," above). In a previous paper (Dynesius

& Jansson 2000) we explored the consequences of ORD along the latitudinal gradient. All nine patterns we predict along gradients in magnitude of ORD are found along the latitudinal gradient: (*a*) A latitudinal gradient in numbers of species has been recorded for most taxonomic groups (Rosenzweig 1995), within species. (*b*) Spatial genetic divergence increases and (*c*) geographical subdivision becomes finer towards lower latitudes (Bernatchez & Wilson 1998; Brawn et al. 1996; Hewitt 2000, 2001). (*d*) Vagility increases with latitude in a wide range of taxa. Diaspore size decreases, and vagility thus increases, with latitude in seed plants (Lord et al. 1997), lichens (Lawrey 1980), and polypore fungi (Ryvarden 1993). (*e*) Tropical organisms are often more specialized than high-latitude ones in, e.g., pollination systems (Johnson & Steiner 2000) and habitat requirements (Pagel et al. 1991). In high-latitude areas serpentine specialists are few and mostly represented by α-clades. The serpentine plants of the formerly glaciated Scandinavia are mainly ecotypes or subspecies of widespread, nonserpentine species (Rune 1953) and probably of postglacial origin. In contrast, in the tropics there are often many species or higher taxa specialized to serpentine soils (Whittaker 1954). (*f*) The geographical range sizes of taxa generally increase toward higher latitudes (Lawton 1994), a relationship called Rapoport's rule (Stevens 1989). (*g*) The proportion of polyploids increases with latitude in all taxa studied (Kuta & Przywara 1997, Rosenzweig 1995, Ward et al. 1994). However, in northwestern North America there is an inverted latitudinal gradient. The highest proportion of polyploid plant species is in formerly glaciated areas (higher ORD) to the south of the permanently ice-free northern Alaska (lower ORD), where diploids as well as range-restricted taxa are more frequent (Stebbins 1971, 1984). (*h*) Rates of β-cladogenesis increase toward lower latitudes, e.g., in New World mammals (Marshall et al. 1982), pelagic foraminiferans with algal symbionts (Quillévéré et al. 2001), and salamanders (García-Paris et al. 2000). (*i*) A latitudinal gradient with more β-anagenesis in the tropics has been recorded for a wide array of taxa (Crane & Lidgard 1989, Jablonski 1993, Spicer et al. 1987).

THE CAPE OF SOUTH AFRICA   In the Cape Floristic Region (CFR) of South Africa, the Quaternary climate has been comparatively stable, with low ORD (Meadows & Sugden 1993). Many clades have radiated here since the Miocene (Richardson et al. 2001). Most notable is the massive diversification of many vascular plant lineages, which has led to an extremely species-rich flora (Cowling & Pressey 2001). Bryophyte endemism is exceptionally high, with the ranges of more than 40% of the hepatic species confined to southern Africa (Schofield 1992). Most vascular plants in the CFR occur in the low-productive Mediterranean shrublands called fynbos, and have narrow habitat specializations, highly specific pollinators (Bond 1995, Johnson & Steiner 2000), and small geographical ranges (Cowling et al. 1998, Goldblatt 1997). Diversification has been concentrated in lineages with certain traits. Diversifying linages of woody plants are fire-killed (i.e., nonsprouting) shrubs with poorly dispersed seeds, small and weakly persistent seed banks, and insect-pollinated flowers (Cowling & Pressey 2001, Goldblatt 1997). These traits lead to restricted gene flow and rapid turnover of nonoverlapping generations, favoring genetic divergence among gene pools on a relatively small spatial scale. Specialist pollinators have driven β-cladogenesis in several lineages, especially in the large geophyte flora (Johnson 1996). The diversifying clades have disintegrated into many range-restricted species (Wisheu et al. 2000).

Polyploidy is considered to be infrequent in the Cape flora, although no comprehensive study has been done (Goldblatt 1978). Species-rich fynbos taxa such as *Erica*, *Aspalathus*, and Proteaceae have no recorded polyploid species (Goldblatt 1978). The CFR also has high proportions of range-restricted species of butterflies, fish, and amphibians, primarily in the fynbos (Rebelo 1992). On the Cape Peninsula local endemism of invertebrates is extremely high (Cowling et al. 1996, Picker & Samways 1996). The accumulation of nonvagile, specialized species with small ranges would not have been possible under high ORD. Thus, this is a good example of low ORD as a prerequisite for long-term accumulation of products of other evolutionary processes.

Plant diversification has been strongest within the western part of the fynbos biome (Cowling & Lombard 2002), which has had the least variable climatic conditions and lowest ORD (Cowling et al. 1999, Meadows & Sugden 1993). The western fynbos has more than twice as many species as the eastern part (Cowling & Lombard 2002, Cowling et al. 1992). Within the Cape flora the strong diversification and the associated characteristics of species are not confined to the fynbos but are equally pronounced in the adjacent succulent karoo biome, that also experienced low ORD during the Quaternary (Cowling et al. 1999). The succulent karoo is drier and less affected by fire than fynbos and has a different flora (Cowling et al. 1998). Thus, the strong diversification cannot be explained solely by ecological idiosyncrasies of the fynbos. To conclude, the Cape flora exhibits six character states that we predict for the β-clades of a region with low ORD: high regional richness of β-clades with small range sizes, organisms with high specialization, low vagility, high rates of β-cladogenesis, and few polyploids.

The only other area with Mediterranean climate with a density of plant species similar to the CFR is southwestern Australia. Here, shrublands and woodlands persisted through the last glacial maximum (Dodson 2001). This is in sharp contrast to the equal-sized Mediterranean region in southeastern Australia, which has only a tenth as many species. Here, the vegetation was much more disrupted during the last glacial period, and the current vegetation formations were assembled from refugial areas (Dodson 2001). Again, low ORD begets diversity.

NEOTROPICAL SALAMANDERS   Salamanders (order Caudata) initially diversified in northern temperate areas. Late in the history of the clade a tiny twig (the supergenus *Bolitoglossa* of the family Plethodontidae, lungless salamanders) entered and established in the mountains of tropical Middle America, an area with low ORD. This resulted in an extensive diversification, and the supergenus now contains more than 40% of the world's salamander species (García-Paris et al. 2000).

Strong differentiation is also evident within species over distances of a few kilometers, such as in *Bolitoglossa pesrubra* (García-Paris et al. 2000). García-Paris and coworkers (2000) state, ". . . there is long-term stability and geographic integrity of the distribution of haplotype clades and clusters of populations identified by protein variants; the phylogeographic units do not intermix or even overlap . . . . The combination of low vagility, high philopatry, . . . and nonmigratory life history . . . establishes the foundation for local differentiation and regional genetic fragmentation." Ecological specialization is higher, expressed by narrow altitudinal ranges of species, than in another salamander hotspot, the southern Appalachians (García-Paris et al. 2000). Once again we have the syndrome of low ORD, a remarkable radiation of clades, small range sizes, narrow specialization, and low vagility.

ASPECTS FOR CONSERVATION

The global warming of the past century has resulted in poleward and upward shifts of species distributions across a wide range of taxonomic groups and regions (Walther et al. 2002). Extant β-clades have experienced many climatic shifts, implying that they may have the potential to survive human-induced climatic changes as well. However, global temperatures have been warmer than today's for less than 5% of the past three Myr (Webb & Bartlein 1992). Future warming may extinguish habitats regionally, e.g., cloud forests on mountaintops (Pounds et al. 1999). Most species survived previous climatic shifts by tracking their habitats through space, which is becoming increasingly difficult because of human habitat transformation and fragmentation (Coope 1995). Furthermore, the insular habitats produced by human-made fragmentation select against vagility if the matrix is hostile. Thus, large human efforts will be needed to allow species to track their preferred environments in response to the projected rapid global warming.

If, as we argue, differences in ORD lead to differences in traits of organisms and clades among regions, conservation strategies may not be interchangeable. Regions with a history of low ORD are often hotspots of β-clades with small geographical ranges, harboring organisms with high specialization and low vagility. These traits render them vulnerable to both climatic change and habitat destruction, and therefore these regions should be of high priority for conservation. For example, of 25 areas identified globally as hotspots of endemism and threat of extinction (Myers et al. 2000), most have experienced low ORD, being situated in the tropics (17 areas), on oceanic islands (5), and/or in mountainous regions (10). Conversely, β-clades that have experienced high ORD should be more resilient to many human disturbances (Balmford 1996, Coope 1995, Mönkkönen & Welsh 1994). As humans make habitat tracking increasingly difficult, areas where global climatic change is buffered against and/or where steep environmental gradients reduce the need for migration, become more important for clade survival. Thus, characters related to low ORD, i.e., high climatic stability (Fjeldså et al. 1997) or steep environmental gradients, could be used to identify both regional and global priority areas for conservation.

ACKOWLEDGMENTS

We thank T.M. Cronin, O. Eriksson, J. Fjeldså, G.M. Hewitt, K. Hylander, M.S. McGlone, C. Nilsson, C. Petit, H.B. Shaffer, J. Silvertown, U. Zinko, and an anonymous reviewer for valuable comments on the manuscript. RJ was supported by grants from The Swedish Foundation for International Cooperation in Research and Higher Education (STINT) and The Swedish Research Council for Environment, Agricultural Sciences and Spatial Planning (FORMAS).

The *Annual Review of Ecology and Systematics* is online at
http://ecolsys.annualreviews.org

LITERATURE CITED

Allmon WD, Morris PJ, McKinney ML. 1998. An intermediate disturbance hypothesis of maximal speciation. In *Biodiversity Dynamics*, ed. ML McKinney, JA Drake, pp. 349–76. New York: Columbia Univ. Press

Ashworth AC. 1996. The response of arctic Carabidae (Coleoptera) to climate change based on the fossil record of the Quaternary Period. *Ann. Zool. Fenn.* 33:125–31

Avise JC. 2000. *Phylogeography: The History and Formation of Species*. London: Harvard Univ. Press

Avise JC, Arnold J, Ball RM, Bermingham E, Lamb T, et al. 1987. Intraspecific phylogeography: the mitochondrial DNA bridge between population genetics and systematics. *Annu. Rev. Ecol. Syst.* 18:489–522

Avise JC, Walker D, Johns GC. 1998. Speciation durations and Pleistocene effects on vertebrate phylogeography. *Proc. R. Soc. London Ser. B* 265:1707–12

Balmford A. 1996. Extinction filters and current resilience: the significance of past selection pressures for conservation biology. *Trends Ecol. Evol.* 11:193–96

Bell MA. 1988. Stickleback fishes: bridging the gap between population biology and paleobiology. *Trends Ecol. Evol.* 3:320–25

Bennett KD. 1990. Milankovitch cycles and their effects on species in ecological and evolutionary time. *Paleobiology* 16:11–21

Bennett KD. 1997. *Evolution and Ecology: The Pace of Life*. Cambridge: Cambridge Univ. Press

Bennett KD, Tzedakis PC, Willis KJ. 1991. Quaternary refugia of north European trees. *J. Biogeogr.* 18:103–15

Berger A. 1988. Milankovitch theory and climate. *Rev. Geophys.* 26:624–57

Berger A, Loutre MF. 1991. Insolation values for the climate of the last 10 million years. *Quat. Sci. Rev.* 10:297–317

Bernatchez L, Wilson CC. 1998. Comparative phylogeography of nearctic and palearctic fishes. *Mol. Ecol.* 7:431–52

Bierzychudek P. 1985. Patterns in plant parthogenesis. *Experientia* 41:1255–64

Billington N, Hebert PDN. 1991. Mitochondrial DNA diversity in fishes and its implications for introductions. *Can. J. Fish. Aquat. Sci.* 48:80–94

Blackburn TM, Gaston KJ. 1996. A sideways look at patterns in species richness, or why there are so few species outside the tropics. *Biodivers. Lett.* 3:44–53

Bohonak AJ. 1999. Dispersal, gene flow, and population structure. *Q. Rev. Biol.* 74:21–45

Bond WJ. 1995. Assessing the risk of plant extinction due to pollinator and disperser failure. See Lawton & May 1995, pp. 131–46

Bowers MA, Harris LC. 1994. A large-scale metapopulation model of interspecific competition and environmental change. *Ecol. Model.* 72:251–73

Brawn JD, Collins TM, Medina M,

Bermingham E. 1996. Associations between physical isolation and geographical variation within three species of Neotropical birds. *Mol. Ecol.* 5:33–46

Brochmann C, Elven R. 1992. Ecological and genetic consequences of polyploidy in arctic *Draba* (Brassicaceae). *Evol. Trends Plant.* 6:111–24

Brown JH. 1995. *Macroecology*. Chicago: Univ. Chicago Press

Cain ML, Milligan BG, Strand AE. 2000. Long-distance seed dispersal in plant populations. *Am. J. Bot.* 87:1217–27

Chown SL. 1997. Speciation and rarity: separating cause from consequence. In *The Biology of Rarity*, ed. WE Kunin, KJ Gaston, pp. 91–107. London: Chapman & Hall

Churchill SP, Balslev H, Forero E, Luteyn JL, eds. 1995. *Biodiversity and Conservation of Neotropical Montane Forests.* New York: NY Bot. Gard.

Churchill SP, Griffin D, Lewis M. 1995. Moss diversity of the tropical Andes. See Churchill et al. 1995, pp. 335–46

Clark JS, Fastie C, Hurtt G, Jackson ST, Johnson C, et al. 1998. Reid's paradox of rapid plant migration: dispersal theory and interpretation of paleoecological records. *Bioscience* 48:13–24

Clarke A, Crame JA. 1997. Diversity, latitude and time: patterns in the shallow sea. In *Marine Biodiversity. Patterns and Processes*, ed. RFG Ormond, JD Gage, MV Angel, pp. 122–47. Cambridge: Cambridge Univ. Press

Cody ML, Overton JM. 1996. Short-term evolution of reduced dispersal in island plant populations. *J. Ecol.* 84:53–61

Cohen AS, Johnston MR. 1987. Speciation in brooding and poorly dispersing lacustrine organisms. *Palaios* 2:426–35

Colwell RK, Lees DC. 2000. The mid-domain effect: geometric constraints on the geography of species richness. *Trends Ecol. Evol.* 15:70–76

Comes HP, Kadereit JW. 1998. The effect of Quaternary climatic changes on plant distribution and evolution. *Trends Plant Sci.* 3:432–38

Coope GR. 1979. Late Cenozoic fossil Coleoptera: evolution, biogeography, and ecology. *Annu. Rev. Ecol. Syst.* 10:247–67

Coope GR. 1994. The response of insect faunas to glacial-interglacial climatic fluctuations. *Philos. Trans. R. Soc. London Ser. B* 344:19–26

Coope GR. 1995. Insect faunas in ice age environments: why so little extinction? See Lawton & May 1995, pp. 55–74

Cowling RM, Cartwright CR, Parkington JE, Allsopp JC. 1999. Fossil wood charcoal assemblages from Elands Bay Cave, South Africa: implications for Late Quaternary vegetation and climates in the winter-rainfall fynbos biome. *J. Biogeogr.* 26:367–78

Cowling RM, Holmes PM, Rebelo AG. 1992. Plant diversity and endemism. In *The Ecology of Fynbos: Nutrients, Fire and Diversity*, ed. RM Cowling, pp. 62–112. Cape Town, S. Afr.: Oxford Univ. Press

Cowling RM, Lombard AT. 2002. Heterogeneity, speciation/extinction history and climate: explaining regional plant diversity patterns in the Cape Floristic Region. *Divers. Distrib.* 8:163–79

Cowling RM, Macdonald IAW, Simmons MT. 1996. The Cape Peninsula, South Africa: physiographical, biological and historical background to an extraordinary hot-spot of biodiversity. *Biodivers. Conserv.* 5:527–50

Cowling RM, Pressey RL. 2001. Rapid plant diversification: planning for an evolutionary future. *Proc. Natl. Acad. Sci. USA* 98:5452–57

Cowling RM, Rundel PW, Desmet PG, Esler KJ. 1998. Extraordinary high regional-scale plant diversity in southern African arid lands: subcontinental and global comparisons. *Divers. Distrib.* 4:27–36

Coyne JA, Orr HA. 1989. Patterns of speciation in *Drosophila*. *Evolution* 43:362–81

Coyne JA, Orr HA. 1997. "Patterns of speciation in *Drosophila*" revisited. *Evolution* 51:295–303

Crane PR, Lidgard S. 1989. Angiosperm diversification and paleolatitudinal gradients in Cretaceous floristic diversity. *Science* 246:675–78

Crisp MD, Laffan S, Linder HP, Monro A. 2001. Endemism in the Australian flora. *J. Biogeogr.* 28:183–98

Cronin TM. 1985. Speciation and stasis in marine Ostracoda: climatic modulation of evolution. *Science* 227:60–63

Cronin TM. 1987. Speciation and cyclic climatic change. In *Climate: History, Periodicity and Predictability*, ed. MR Rampino, JE Sanders, WS Newman, LL Königsson, pp. 333–42. New York: Van Nostrand-Reinholdt

Cronin TM, Raymo ME, Kyle KP. 1996. Pliocene (3.2–2.4 Ma) ostracode faunal cycles and deep ocean circulation, North Atlantic Ocean. *Geology* 24:695–98

Cronk QCB. 1997. Islands: stability, diversity, conservation. *Biodivers. Conserv.* 6:477–93

Cruzan MB, Templeton AR. 2000. Paleoecology and coalescence: phylogeographic analysis of hypotheses from the fossil record. *Trends Ecol. Evol.* 15:491–96

Cwynar LC, MacDonald GM. 1987. Geographical variation of lodgepole pine in relation to population history. *Am. Nat.* 129:463–69

Dansgaard W, Johnsen SJ, Clausen HB, Dahl-Jensen D, Gundestrup NS, et al. 1993. Evidence for general instability of past climate from a 250-kyr ice-core record. *Nature* 364:218–20

Darlington PJ. 1943. Carabidae of mountains and islands: data on the evolution of isolated faunas, and on atrophy of wings. *Ecol. Monogr.* 13:37–61

Davis MB. 1976. Pleistocene biogeography of temperate deciduous forests. *Geosci. Man* 13:13–26

Davis MB, Shaw RG. 2001. Range shifts and adaptive responses to Quaternary climate change. *Science* 292:673–79

den Boer PJ. 1970. On the significance of dispersal power for populations of carabid beetles (Coleoptera, Carabidae). *Oecologia* 4:1–28

Dodson JR. 2001. Holocene vegetation change in the mediterranean-type climate regions of Australia. *Holocene* 11:673–80

Dufresne F, Hebert PDN. 1997. Pleistocene glaciations and polyphyletic origins of polyploidy in an arctic cladoceran. *Proc. R. Soc. London Ser. B* 264:201–6

Dynesius M, Jansson R. 2000. Evolutionary consequences of changes in species' geographical distributions driven by Milankovitch climate oscillations. *Proc. Natl. Acad. Sci. USA* 97:9115–20

Eldredge N. 1989. *Macroevolutionary Dynamics.* New York: McGraw-Hill

Endler JA. 1986. *Natural Selection in the Wild.* Princeton: Princeton Univ. Press

Fischer AG. 1960. Latitudinal variation in organic diversity. *Evolution* 14:64–81

Fjeldså J. 1994. Geographical patterns for relict and young species of birds in Africa and South America and implications for conservation priorities. *Biodivers. Conserv.* 3:207–26

Fjeldså J. 1995. Geographical patterns of neoendemic and older relict species of Andean forest birds: the significance of ecologically stable areas. See Churchill et al. 1995, pp. 89–102

Fjeldså J, Ehrlich D, Lambin E, Prins E. 1997. Are biodiversity 'hotspots' correlated with current ecoclimatic stability? A pilot study using the NOAA-AVHRR remote sensing data. *Biodivers. Conserv.* 6:401–22

Fjeldså J, Lambin E, Mertens B. 1999. Correlation between endemism and local ecoclimatic stability documented by comparing Andean bird distributions and remotely sensed land surface data. *Ecography* 22:63–78

Fjeldså J, Lovett JC. 1997. Geographical patterns of old and young species in African forest biota: the significance of specific montane areas as evolutionary centres. *Biodivers. Conserv.* 6:325–46

Flenley JR. 1998. Tropical forests under the climates of the last 30,000 years. *Clim. Change* 39:177–97

Flessa KW, Jablonski D. 1996. The geography of evolutionary turnover: a global analysis of extant bivalves. In *Evolutionary Paleobiology*, ed. D Jablonski, DH Erwin, JH

Lipps, pp. 376–97. Chicago: Univ. Chicago Press

Franzke A, Hurka H. 2000. Molecular systematics and biogeography of the Cardamine pratensis complex (Brassicaceae). *Plant Syst. Evol.* 224:213–34

Futuyma DJ. 1987. On the role of species in anagenesis. *Am. Nat.* 130:465–73

Futuyma DJ, Moreno G. 1988. The evolution of ecological specialization. *Annu. Rev. Ecol. Syst.* 19:207–33

García-Moreno J, Fjeldså J. 2000. Chronology and mode of speciation in the Andean avifauna. *Bonn. Zool. Monogr.* 46:25–46

García-Paris M, Good DA, Parra-Olea G, Wake DB. 2000. Biodiversity of Costa Rican salamanders: implications of high levels of genetic differentiation and phylogeographic structure for species formation. *Proc. Natl. Acad. Sci. USA* 97:1640–47

Gavrilets S, Acton R, Gravner J. 2000a. Dynamics of speciation and diversification in a metapopulation. *Evolution* 54:1493–501

Gavrilets S, Li H, Vose MD. 2000b. Patterns of parapatric speciation. *Evolution* 54:1126–34

Glazier DS. 1987. Toward a predictive theory of speciation: the ecology of isolate selection. *J. Theor. Biol.* 126:323–34

Goldblatt P. 1978. An analysis of the flora of Southern Africa: its characteristics, relationships, and origins. *Ann. Miss. Bot. Gard.* 65:369–436

Goldblatt P. 1997. Floristic diversity in the Cape Flora of South Africa. *Biodivers. Conserv.* 6:359–77

Graham RW, Lundelius EL, Graham MA, Schroeder EK, Toomey RS, et al. 1996. Spatial response of mammals to late quaternary environmental fluctuations. *Science* 272:1601–6

Grytnes JA, Birks HJB, Peglar SM. 1999. Plant species richness in Fennoscandia: evaluating the relative importance of climate and history. *Nord. J. Bot.* 19:489–503

Haffer J. 1969. Speciation in Amazonian forest birds. *Science* 165:131–37

Haffer J. 1997. Alternative models of vertebrate speciation in Amazonia: an overview. *Biodivers. Conserv.* 6:451–76

Harrison S, Hastings A. 1996. Genetic and evolutionary consequences of metapopulation structure. *Trends Ecol. Evol.* 11:180–83

Hays JD, Imbrie J, Shackleton NJ. 1976. Variations in the Earth's orbit: pacemaker of the ice ages. *Science* 194:1121–32

Hellberg ME, Balch DP, Roy K. 2001. Climate-driven range expansion and morphological evolution in a marine gastropod. *Science* 292:1707–10

Hendry AP, Wenburg JK, Bentzen P, Volk EC, Quinn TP. 2000. Rapid evolution of reproductive isolation in the wild: evidence from introduced salmon. *Science* 290:516–18

Hewitt GM. 1975. A sex chromosome hybrid zone in the grasshopper *Podisma pedestris*. *Heredity* 35:375–87

Hewitt GM. 1996. Some genetic consequences of ice ages, and their role in divergence and speciation. *Biol. J. Linn. Soc.* 58:247–76

Hewitt GM. 1999. Post-glacial re-colonization of European biota. *Biol. J. Linn. Soc.* 68:87–112

Hewitt GM. 2000. The genetic legacy of the Quaternary ice ages. *Nature* 405:907–13

Hewitt GM. 2001. Speciation, hybrid zones and phylogeography: or seeing genes in space and time. *Mol. Ecol.* 10:537–49

Higgie M, Chenoweth S, Blows MW. 2000. Natural selection and the reinforcement of mate recognition. *Science* 290:519–21

Hill JK, Thomas CD, Blakeley DS. 1999. Evolution of flight morphology in a butterfly that has recently expanded its geographic range. *Oecologia* 121:165–70

Hillebrand H, Azovsky AI. 2001. Body size determines the strength of the latitudinal diversity gradient. *Ecography* 24:251–56

Hodgson JG. 1987. Why do so few plant species exploit productive habitats? An investigation into cytology, plant strategies and abundance within a local flora. *Funct. Ecol.* 1:243–50

Hooghiemstra H, Sarmiento G. 1991. Long continental pollen record from a tropical intermontane basin: late Pliocene and

Pleistocene history from a 540-meter core. *Episodes* 14:107–15

Howard DJ, Berlocher SH, eds. 1998. *Endless Forms: Species and Speciation*. New York: Oxford Univ. Press

Huntley B, Birks HJB. 1983. *An Atlas of Past and Present Pollen Maps for Europe: 0-13, 000 Years Ago*. Cambridge: Cambridge Univ. Press

Hutchinson GE. 1959. Homage to Santa Rosalia, or why are there so many kinds of animals? *Am. Nat.* 93:145–59

Imbrie J, Berger A, Boyle EA, Clemens SC, Duffy A, et al. 1993a. On the structure and origin of major glaciation cycles. 2. The 100,000-year cycle. *Paleoceanography* 8:699–735

Imbrie J, Berger A, Shackleton NJ. 1993b. Role of orbital forcing: a two-million-year perspective. In *Global Changes in the Perspective of the Past*, ed. JA Eddy, H Oeschger, pp. 263–77. Chichester, UK: Wiley

Imbrie J, McIntyre A, Mix A. 1989. Oceanic response to orbital forcing in the late Quaternary: observational and experimental strategies. In *Climate and Geo-Sciences*, ed. A Berger, A Schneider, JC Duplessy, pp. 121–64. Dordrecht, The Netherlands: Kluwer

Jablonski D. 1985. Marine regressions and mass extinctions: a test using the modern biota. In *Phanerozoic Diversity Patterns: Profiles in Macroevolution*, ed. JW Valentine, pp. 335–53. Princeton, NJ: Princeton Univ. Press

Jablonski D. 1986. Larval ecology and macroevolution in marine invertebrates. *Bull. Mar. Sci.* 39:565–87

Jablonski D. 1993. The tropics as a source of evolutionary novelty through geological time. *Nature* 364:142–44

Jackson ST, Overpeck JT. 2000. Responses of plant populations and communities to environmental changes of the late Quaternary. *Paleobiology* 26:194–220

Johnson SD. 1996. Pollination, adaptation and speciation models in the Cape flora of South Africa. *Taxon* 45:59–66

Johnson SD, Steiner KE. 2000. Generalization versus specialization in plant pollination systems. *Trends Ecol. Evol.* 15:140–43

Kashiwaya K, Ochiai S, Sakai H, Kawai T. 2001. Orbit-related long-term climate cycles revealed in a 12-Myr continental record from Lake Baikal. *Nature* 410:71–74

Kauffman EG. 1978. Evolutionary rates and patterns among Cretaceous Bivalvia. *Philos. Trans. R. Soc. London Ser. B* 284:277–304

Kawata M. 2002. Invasion of vacant niches and subsequent sympatric speciation. *Proc. R. Soc. London Ser. B* 269:55–63

Knowles LL. 2000. Tests of Pleistocene speciation in montane grasshoppers (genus *Melanoplus*) from the sky islands of western North America. *Evolution* 54:1337–48

Knowles LL. 2001. Did the Pleistocene glaciations promote divergence? Tests of explicit refugial models in montane grasshoppers. *Mol. Ecol.* 10:691–701

Konnert M, Bergmann F. 1995. The geographical distribution of genetic variation of silver fir (*Abies alba*, Pinaceae) in relation to its migration history. *Plant Syst. Evol.* 196:19–30

Kukla GJ. 2000. Paleoclimate: the last interglacial. *Science* 287:987–88

Kuta E, Przywara L. 1997. Polyploidy in mosses. *Acta Biol. Cracov. Bot.* 39:17–26

Lande R. 1992. Neutral theory of quantitative genetic variance in an island model with local extinction and colonization. *Evolution* 46:381–89

Latham RE, Ricklefs RE. 1993. Continental comparisons of temperate-zone tree species diversity. In *Species Diversity in Ecological Communities*, ed. RE Ricklefs, D Schluter, pp. 294–314. Chicago: Univ. Chicago Press

Lawrey JD. 1980. Sexual and asexual reproductive patterns in *Parmotrema* (Parmeliaceae) that correlate with latitude. *Bryologist* 83:344–50

Lawton JH. 1994. Population dynamic principles. *Philos. Trans. R. Soc. London Ser. B* 344:61–68

Lawton JH, May RM, eds. 1995. *Extinction Rates*. Oxford: Oxford Univ. Press

Lessios HA. 1998. The first stage of speciation as seen in organisms separated by the Isthmus of Panama. See Howard & Berlocher 1998, pp. 186–201

Linder HP. 2001. Plant diversity and endemism in sub-Saharan tropical Africa. *J. Biogeogr.* 28:169–82

Lord J, Egan J, Clifford T, Jurado E, Leishman M, et al. 1997. Larger seeds in tropical floras: consistent patterns independent of growth form and dispersal mode. *J. Biogeogr.* 24:205–11

Marshall LG, Webb SD, Sepkoski JJ, Raup DM. 1982. Mammalian evolution and the Great American Interchange. *Science* 215:1351–57

Martens K, Schön I. 1999. Crustacean biodiversity in ancient lakes: a review. *Crustaceana* 72:899–910

Mashiko K. 2000. Insights into the mechanism of speciation in gammarid crustaceans of Lake Baikal using a population-genetic approach. *Adv. Ecol. Res.* 31:221–35

Mayr E. 1963. *Animal Species and Evolution.* Cambridge, MA: Harvard Univ. Press

McCune AR. 1987. Lakes as laboratories for evolution: endemic fishes and environmental cyclicity. *Palaios* 2:446–54

McCune AR, Lovejoy NR. 1998. The relative rate of sympatric and allopatric speciation in fishes. See Howard & Berlocher 1998, pp. 172–85

McGlone MS. 1996. When history matters: scale, time, climate and tree diversity. *Global Ecol. Biogeogr.* 5:309–14

McGlone MS, Duncan RP, Heenan PB. 2001. Endemism, species selection and the origin and distribution of the vascular plant flora of New Zealand. *J. Biogeogr.* 28:199–216

McKinney ML. 1997. Extinction vulnerability and selectivity: combining ecological and paleontological views. *Annu. Rev. Ecol. Syst.* 28:495–516

Meadows ME, Sugden JM. 1993. The late Quaternary palaecology of a floristic kingdom: the southwestern Cape South Africa. *Palaeogeogr. Palaeoclimatol. Palaeoecol.* 101:271–81

Moritz C, Patton JL, Schneider CJ, Smith TB. 2000. Diversification of rainforest faunas: an integrated molecular approach. *Annu. Rev. Ecol. Syst.* 31:533–63

Myers N, Mittermeier RA, Mittermeier CG, da Fonseca GAB, Kent J. 2000. Biodiversity hotspots for conservation priorities. *Nature* 403:853–58

Mönkkönen M, Welsh DA. 1994. A biogeographical hypothesis of human-caused landscape changes on the forest bird communities of Europe and North America. *Ann. Zool. Fenn.* 31:61–70

Niemelä J, Spence JR. 1991. Distribution and abundance of an exotic ground beetle (Carabidae): a test of community impact. *Oikos* 62:351–59

Niemelä J, Spence JR. 1999. Dynamics of local expansion by an introduced species: *Pterostichus melanarius* Ill. (Coleoptera, Carabidae) in Alberta, Canada. *Diversity Distrib.* 5:121–27

Nilsson AN, Pettersson RB, Lemdahl G. 1993. Macroptery in altitudinal specialists versus brachyptery in generalists: a paradox of alpine Scandinavian carabid beetles (Coleoptera, Carabidae). *J. Biogeogr.* 20:227–34

Nilsson T, Jorde PE. 1998. Allozyme variation in allotetraploid *Saxifraga osloënsis* and its diploid progenitors. *Nord. J. Bot.* 18:425–30

Olsen PE. 1986. A 40-million-year lake record of early Mesozoic orbital climatic forcing. *Science* 234:842–48

Otte D. 1989. Speciation in Hawaiian crickets. In *Speciation and Its Consequences*, ed. D Otte, JA Endler, pp. 482–526. Sunderland, MA: Sinauer

Otto SP, Whitton J. 2000. Polyploid incidence and evolution. *Annu. Rev. Genet.* 34:401–37

Pagel MD, May RM, Collie AR. 1991. Ecological aspects of the geographical distribution and diversity of mammalian species. *Am. Nat.* 137:791–815

Paulo OS, Dias C, Bruford MW, Jordan WC, Nichols RA. 2001. The persistence of Pliocene populations through the Pleistocene climatic cycles: evidence from the phylogeography of an Iberian lizard. *Proc. R. Soc. London Ser. B* 268:1625–30

Picker MD, Samways MJ. 1996. Faunal diversity and endemicity of the Cape Peninsula, South Africa: a first assessment. *Biodivers. Conserv.* 5:591–606

Potts EC. 1983. Evolutionary disequilibrium among Indo-Pacific corals. *Bull. Mar. Sci.* 33:619–32

Pounds JA, Fogden MPL, Campbell JH. 1999. Biological response to climate change on a tropical mountain. *Nature* 398:611–15

Potts EC. 1983. Evolutionary disequilibrium among Indo-Pacific corals. *Bull. Mar. Sci.* 33:619–32

Quillévéré F, Norris RD, Moussa I, Berggren WA. 2001. Role of photosymbiosis and biogeography in the diversification of early Paleogene acarininids (planktonic foraminifera). *Paleobiology* 27:311–26

Rahbek C, Graves GR. 2001. Multiscale assessment of patterns of avian species richness. *Proc. Natl. Acad. Sci. USA* 98:4534–39

Ramsey J, Schemske DW. 1998. Pathways, mechanisms, and rates of polyploid formation in flowering plants. *Annu. Rev. Ecol. Syst.* 29:467–501

Rebelo AG. 1992. Red Data Book species in the Cape Floristic Region: threats, priorities and target species. *Trans. R. Soc. S. Afr.* 48:55–86

Remington CL. 1968. Suture-zones of hybrid interaction between recently joined biotas. *Evol. Biol.* 2:321–428

Richardson JE, Weitz FM, Fay MF, Cronk QCB, Linder HP, et al. 2001. Rapid and recent origin of species richness in the Cape flora of South Africa. *Nature* 412:181–83

Rosenzweig ML. 1995. *Species Diversity in Space and Time*. Cambridge: Cambridge Univ. Press

Roy K, Valentine JW, Jablonski D, Kidwell SM. 1996. Scales of climatic variability and time averaging in Pleistocene biotas: implications for ecology and evolution. *Trends Ecol. Evol.* 11:458–63

Rune O. 1953. Plant life on serpentine and related rocks in the north of Sweden. *Acta Phytogeogr. Suecica* 31:1–139

Ryvarden L. 1993. Tropical polypores. In *Aspects of Tropical Mycology*, ed. S Isaac, pp. 149–70. Cambridge: Cambridge Univ. Press

Schluter D. 1996. Ecological speciation in postglacial fishes. *Philos. Trans. R. Soc. London Ser. B* 351:807–14

Schluter D. 2000. *The Ecology of Adaptive Radiation*. Oxford: Oxford Univ. Press

Schofield WB. 1992. Bryophyte distribution patterns. In *Bryophytes and Lichens in a Changing Environment*, ed. JW Bates, AM Farmer, pp. 103–30. Oxford: Clarendon

Sheldon PR. 1996. Plus ça change: a model for stasis and evolution in different environments. *Palaeogeogr. Palaeoclimatol. Palaeoecol.* 127:209–27

Sherbakov DY. 1999. Molecular phylogenetic studies on the origin of biodiversity in Lake Baikal. *Trends Ecol. Evol.* 14:92–95

Sideleva VG. 2000. The ichthyofauna of Lake Baikal, with special reference to its zoogeographical relations. *Adv. Ecol. Res.* 31:81–96

Skellam JG. 1951. Random dispersal in theoretical populations. *Biometrika* 38:196–218

Slatkin M. 1977. Gene flow and genetic drift in a species subject to frequent local extinction. *Theor. Popul. Biol.* 12:253–62

Smith F. 2001. Historical regulation of local species richness across a geographic region. *Ecology* 82:792–801

Sømme L. 1989. Adaptations of terrestrial arthropods to the alpine environment. *Biol. Rev.* 64:367–407

Spicer RA, Wolfe JA, Nichols DJ. 1987. Alaskan Cretaceous-Tertiary floras and arctic origins. *Paleobiology* 13:73–83

Stanley SM. 1979. *Macroevolution*. San Francisco: Freeman

Stanley SM. 1986. Population size, extinction, and speciation: the fission effect in Neogene Bivalvia. *Paleobiology* 12:89–110

Stebbins GL. 1971. *Chromosomal Evolution in Higher Plants*. London: Arnold

Stebbins GL. 1984. Polyploidy and the distribution of the arctic-alpine flora: new evidence and a new approach. *Bot. Helv.* 94:1–14

Stevens GC. 1989. The latitudinal gradient in

geographical range: how so many species coexist in the tropics. *Am. Nat.* 133:240–56

Taylor EB. 1999. Species pairs of north temperate freshwater fishes: evolution, taxonomy, and conservation. *Rev. Fish. Biol. Fish.* 9:299–324

Terborgh J. 1992. *Diversity and the Tropical Rain Forest.* New York: Sci. Am. Libr.

Thomas CD, Bodsworth EJ, Wilson RJ, Simmons AD, Davies ZG, et al. 2001. Ecological and evolutionary processes at expanding range margins. *Nature* 411:577–81

Valentine JW. 1984. Neogene marine climate trends: implications for biogeography and evolution of the shallow-sea biota. *Geology* 12:647–59

van der Hammen T. 1995. Global change, biodiversity, and conservation of Neotropical montane forests. See Churchill et al. 1995, pp. 603–7

van Zanten BO, Pócs T. 1981. Distribution and dispersal of bryophytes. In *Advances in Bryology*, ed. W Schultze-Motel. Vaduz, Liechtenstein: Cramer

Vogel JC, Rumsey FJ, Schneller JJ, Barrett JA, Gibby M. 1999. Where are the glacial refugia in Europe? Evidence from pteridophytes. *Biol. J. Linn. Soc.* 66:23–37

Vrba ES. 1989. Ecology in relation to speciation rates: some case histories of Miocene-Recent mammal clades. *Evol. Ecol.* 1:283–300

Vrba ES. 1985. Ecology and environment: alternative causes of the temporal distribution of evolutionary events. *South Afr. J. Sci.* 81:229–36

Vrba ES. 1992. Mammals as a key to evolutionary theory. *J. Mammal.* 73:1–28

Vrba ES. 1993. Turnover-pulses, the red queen, and related topics. *Am. J. Sci.* 293A:418–52

Walther GR, Post E, Convey P, Menzel A, Parmesan C, et al. 2002. Ecological responses to recent climate change. *Nature* 416:389–95

Ward RD, Bickerton MA, Finston T, Hebert PDN. 1994. Geographical cline in breeding systems and ploidy levels in European populations of *Daphnia pulex. Heredity* 73:532–43

Warren MS, Hill JK, Thomas JA, Asher J, Fox R, et al. 2001. Rapid responses of British butterflies to opposing forces of climate and habitat change. *Nature* 414:65–69

Webb SD. 1991. Ecogeography and the Great American Interchange. *Paleobiology* 17:266–80

Webb T, Bartlein PJ. 1992. Global changes during the last 3 million years: climatic controls and biotic responses. *Annu. Rev. Ecol. Syst.* 23:141–73

Whittaker RH. 1954. The ecology of serpentine soils. IV. The vegetational response to serpentine soils. *Ecology* 35:275–88

Williams GC. 1992. *Natural Selection: Domains, Levels, and Challenges.* New York: Oxford Univ. Press

Williamson PG. 1981. Palaeontological documentation of speciation in Cenozoic molluscs from Turkana Basin. *Nature* 293:437–43

Wisheu IC, Rosenzweig ML, Olsvig-Whittaker L, Shmida A. 2000. What makes nutrient-poor mediterranean heathlands so rich in plant diversity? *Evol. Ecol. Res.* 2:935–55

Wright DH. 1983. Species-energy theory: an extension of species-area theory. *Oikos* 41:496–506

Wright HE Jr, Kutzbach JE, Webb T III, Ruddiman WF, Street-Perrott FA, Bartlein PJ. 1993. *Global Climates Since the Last Glacial Maximum.* Minneapolis: Univ. Minn. Press

Zachos J, Pagani M, Sloan L, Thomas E, Billups K. 2001. Trends, rhythms, and aberrations in global climate 65 Ma to present. *Science* 292:686–93

Zachos JC, Flower BP, Paul H. 1997. Orbitally paced climate oscillations across the Oligocene/Miocene boundary. *Nature* 388:567–70

SUBJECT INDEX

Cumulative Indexes

CONTRIBUTING AUTHORS, VOLUMES 29–33

CHAPTER TITLES, VOLUMES 29–33

Volume 29 (1998)

Volume 30 (1999)

Volume 32 (2001)